3rd International Conference on Tropical and Coastal Region Eco Development (ICTCRED 2017)

IOP Conference Series: Earth and Environmental Science
Volume 116

Yogyakarta, Indonesia
2 - 4 October 2017

Editor:

Munawar A. Riyadi

ISBN: 978-1-5108-5948-7
ISSN: 1755-1307

Printed from e-media with permission by:

Curran Associates, Inc.
57 Morehouse Lane
Red Hook, NY 12571

Some format issues inherent in the e-media version may also appear in this print version.

Printed by Curran Associates, Inc. (2018)

For permission requests, please contact the Institute of Physics
at the address below.

Institute of Physics
Dirac House, Temple Back
Bristol BS1 6BE UK

Phone: 44 1 17 929 7481
Fax: 44 1 17 920 0979

techtracking@iop.org

Additional copies of this publication are available from:

Curran Associates, Inc.
57 Morehouse Lane
Red Hook, NY 12571 USA
Phone: 845-758-0400
Fax: 845-758-2633
Email: curran@proceedings.com
Web: www.proceedings.com

Table of contents

Volume 116

3rd International Conference on Tropical and Coastal Region Eco Development 2017

2–4 October 2017, Yogyakarta, Indonesia

Accepted papers received: 31 January 2018
Published online: 8 March 2018

Preface

Papers

IOP Conf. Series: Earth and Environmental Science **116** (2018) 011001 doi:10.1088/1755-1315/116/1/011001

FOREWORD FROM THE CHAIR OF THE 3RD ICTCRED 2017

On behalf of the Organizing Committee, I would like to extend our warmest welcome to you at the International Conference on Tropical and Coastal Region Eco Development (ICTCRED) 2017. This annual conference is the third event after the second has been successfully conducted in 2016 at Bali. This conference is organized by Research and Community Services Institute (LPPM), Diponegoro University. The conference aims to provide a forum for researchers, academicians, professionals, and industries to expose and exchange innovative ideas, methods, and experience in the areas related to tropical life sciences and coastal development.This conference also provides forum for researchers and scientists to exchange ideas and their current achievements.

In this year, 215 abstracts from various universities and research centers from many countries have been received. However, after in-depth review, only 147 high quality papers are accepted for oral and poster presentation in this conference. In addition, we cordially invite seven highly respected researchers in various fields as keynote speakers in this conference, to share their knowledge and expertise. I am grateful of each one of them for setting aside their valuable time to participate in this conference.

Moreover, I would like to announce that the ICTCRED 2017 Committee has signed an agreement with the Institute of Physics (IOP) to publish the conference proceeding in their Scopus-indexed *IOP Conference Series: Earth and Environmental Sciences (EES)* after a series of review. We do hope that the collaboration with IOP will increase the visibility of this conference papers to international levels which also give benefits to authors and also their institutions.

Finally the success of this conference lies not only in the quality of papers but also on the dedicated team efforts of the organizing committee. We thanks to the keynote speakers for the participation in this conference. I would like to acknowledge Institute of Physics (IOP) for the collaboration in publishing the conference proceedings. Indeed, I would like to thank the Scientific Committee members for their effort in reviewing and evaluating the papers for maintaining the quality of the conference. Last but not least, all staffs of The Research and Community Services Institute, Diponegoro University, deserves our great appreciation for their unlimited supports.

To all delegates, I hope that the 3rd ICTCRED 2017 event will be memorable not only from the scientific perspective but in the joy of meeting with other scientists for mutual collaboration. I wish you enjoy the conference as well as the beautiful nature and great traditions of Yogyakarta.

Chair,
Organizing Committee of ICTCRED 2017

Dr. Agus Trianto

FOREWORD OF THE DIRECTOR

Dear distinguished speakers, delegates, ladies, and gentlemen

I am very pleased to welcome you all to this international conference, Tropical and Coastal Region Eco-development, which acts as a forum for those interested in tropical and coastal development issues. Diponegoro University commits to provide an opportunity for scientific society to always play important role in disseminating ideas and research results especially in the area of coastal and tropical development, which is the main research field of our university. Hence, this conference offers a platform for extensive sharing and exchange of knowledge for the development of coastal and tropical areas.

The topics presented in this conference cover marine biodiversity, sustainable marine utilization and development, climate change on coastal and marine ecosystems and coral reef ecosystems and coastal management. In the tropical field, this conference deals with relevant ideas and knowledge addressing vital life sciences, tropical health and nutrition, tropical diseases and tropical food and energy. In addition, this conference also covers socio-economic aspects such exploration of tropical rainforests, deforestation, rising immigration, etc. Thus, it is clear that the International Conference on Tropical and Coastal region eco-development is a unique blend of coastal and tropical that nicely fits the current interest among the community concerned with sustainable coastal and tropical ecosystems.

Finally, we would like to express our gratitude to our distinguished keynote speakers, Prof. Susilo Wibowo, Dr. Hadiyanto, Prof. Gerard Pals, Prof. Junichi Tanaka, Dr. Roel H. Bosma, Prof. Tao Liu, and Prof. Soottawat Benjakul, who had been traveling all the way to Yogyakarta. Certainly we will have an important benefit of preparing the next generation of Indonesian scientists with international exposure. We thank our participants to present their research papers, to share extensively and exchange of ideas thoughts and discussions so that this conference facilitates the formation of networks among participants. We thank all invited guests who have shown their interests in coastal and tropical region development field. Many thanks to the organizing and scientific committee of ICTCRED 2017 who have work very hard to run the conference.

I wish you all a productive and successful conference.

Yours sincerely
Director of Research and Community Services Institute
Diponegoro University

Prof. Heru Susanto

IOP Conf. Series: Earth and Environmental Science **116** (2018) 011001 doi:10.1088/1755-1315/116/1/011001

WELCOME ADDRESS OF THE RECTOR OF DIPONEGORO UNIVERSITY

It's a great pleasure and honour for our University to be the host of the 3rd International Conference on Tropical and Coastal Region Eco Development 2017 organized by Research and Community Service Institute, Diponegoro University. The special acknowledgement, I address to the distinguished speakers Prof. Susilo Wibowo and Dr. Hadiyanto from Diponegoro University-Indonesia, Prof. Gerard Pals from VU University Medical Center – Netherlands, Prof. Junichi Tanaka from University of the Ryukyus – Japan, Dr. Roel H. Bosma from Wageningen University – Netherlands, Prof. Tao Liu from Ocean University of China, and Prof. Soottawat Benjakul from Prince of Songkla University. Thank you for the valuable time to deliver knowledge and share scientific information at this conference. I believe that this opportunity will provide the valuable information for us and deliberate some new research ideas for participants of this conference. For all the participants, I would also like to welcome you at this conference.

The origin of the conference theme is reflected from the idea of our Center of Excellence (CoE) which was established in 2012 representing our priority as a research university. Since the declaration of Diponegoro Universityas a research university, the main theme of every research result will be enhanced to the level of international benchmarking.

Diponegoro University, consists of 13 faculties, has strong human resources and research background related to the coastal development and tropical life sciences. It is also supported by integrated laboratory of marine and fisheries, which are located at Teluk Awur Jepara.

Coastal development and tropical life sciences are two important issues in Indonesia and need to be actualized by the government. The enormous potencies of Indonesia, large resources of marine area and their potential value of natural marine products extract, provide an opportunity to contribute for health energy and food. Recent issue has been arisen that the food and energy also can be exploited from the sea. Indeed, the exploration and exploitation of marine products must be considered on the impact of the environmental devastation. These issues are interesting topics which are reflected by large number of abstracts submitted to this conference. These interesting issues need to be discussed in this conference by sharing research finding and ideas. I am gratefull to see that this conference has enormous responds from the participants either from domestic or from other countries such Japan, China, US, and France, as reported by organizing committee.

Number of publication indexed by reputable database has been set as an indicator for world university rank including Indonesia. Therefore, Diponegoro University also encourages all scientists and academic staffs to increase their publication records in these international reputation journals. Currently, Diponegoro Universityis in the 6th position among universities in Indonesia for the number of

IOP Conf. Series: Earth and Environmental Science **116** (2018) 011001 doi:10.1088/1755-1315/116/1/011001

publications in reputable International journals. I sincerely express appreciation to the organizing committee for their effort to realize this conference.

By the end of my short welcome address, I hope our foreign guests take advantage of their stay here to enjoy Yogyakarta and its wonderful places. It is a beautiful and historical city to visit with a wonderful and unique traditional art dance, stunning sunset, great sceneries and interesting shopping.

Once again, it is my great pleasure to welcome you all to the 3rd International Conference on Tropical and Coastal Region Eco Development 2017. I wish you a pleasant two fully scientific days of conferences and I hope you can get a fruitfull share with other scientists on current developed knowledge and perhaps seeking for potential collaboration of your interested field.

Thank you for your kind attention.

Prof. Yos Johan Utama
Rector

IOP Conf. Series: Earth and Environmental Science **116** (2018) 011001 doi:10.1088/1755-1315/116/1/011001

KEYNOTE SPEAKERS

Prof. Susilo Wibowo
Universitas Diponegoro – Indonesia
"Indonesia got obese; do we care? Genetics, epigenetic, and enviromental point of view"

Dr. Hadiyanto
Universitas Diponegoro – Indonesia
"Effects of sugar addition on the thermal degradation of phycocyanin from Spirulina sp."

Prof. Gerard Pals
VU University Medical Center – Netherlands
"Cancer and the environment"

Prof. Junichi Tanaka
University of the Ryukyus – Japan
"Exploration of coral reef organisms for bioactive molecules and related issues"

Dr. Roel H. Bosma
Wageningen University& Research – Netherlands
"Investing in climate change mitigation and adaptation on mangrove and aquaculture doubles benefits."

Prof. Tao Liu
Ocean University of China - China
"Research on complete mitochondrial genome of marine algae"

Prof. Soottawat Benjakul
Prince of Songkla University
"Valorization of fish processing by product"

Note from Editors

The 3rd International Conference on Tropical and Coastal Region Eco Development (ICTCRED) 2017 was held in Yogyakarta, Indonesia from 2nd to 4th October 2017. The conference was organized and fully supported by the Institute of Research and Community Services, Diponegoro University, Indonesia. Authors and participants from many countries made the conference truly international in scope. The scope of this conference varies from marine environment, marine products and their processing as well as oceanography technology, coastal environment management and policies. In addition, social science related to the coastal area development were also discussed.

This volume of *IOP Conference Series: Earth and Environmental Sciences* contains selected articles from those presented in the conference. After presentation, the revised papers were peer reviewed by fellow reviewers to ensure the quality of published materials. Finally, Editors decided to select and publish as many as 105 papers. It is hoped that the presented papers can offer more insight towards broad audience.

On behalf of Editors, we appreciate enormous work of all staffs and reviewers in the preparation of this volume. We would like to express our sincere thanks to all authors and presenters for their valuable contributions. We hope that the conference has been beneficial to all participants.

Guest Editors

Munawar A Riyadi, PhD
Diponegoro University, Indonesia
Chair of Scientific Committee - ICTCRED 2017

Scientific Committee/Editor

Munawar A Riyadi (Diponegoro University, Indonesia)
Tri Indah Winarni (Diponegoro University, Indonesia)
Diah Permata Wijayanti (Diponegoro University, Indonesia)
Makoto Tsuchiya (University of the Ryukyus, Japan)
Heru Susanto (Diponegoro University, Indonesia)
Jamari (Diponegoro University, Indonesia)
Hussein Gasem (Diponegoro University, Indonesia)
Ambariyanto (Diponegoro University, Indonesia)
Muhammad Zainuri (Diponegoro University, Indonesia)
Sultana MH Faraz (Diponegoro University, Indonesia)
Gerard Pals (Vrije Universiteit Amsterdam, Netherland)
Randi Hagerman (University of California Davis, United States)
Flora Tassone (University of California Davis, United States)
Herawati Sudoyo (Eijkman Institute, Indonesia)
Ocky Karna Rajasa (MRTHE, Indonesia)
Craig Starger (Colorado State University, United States)
CN Ravishankar (Central Institute of Fisheries Technology, India)
Agus Sabdono (Diponegoro University, Indonesia)
Maria Barbosa (Wageningen University, Netherland)
Yasuhiro Igarashi (Japan)
Irwandi Jaswir (IIUM, Malaysia) Michio Hidaka (University of the Ryukyus, Japan)

Organizing Committee

Agus Trianto	Chair
Diana Nur Afifah	Co-Chair of TLS
Ahmad N Al-Baarri	Co-Chair of CRED
Ima Wijayanti	Secretary
Muh. Arfan	Publication
Desrina	Program

This conference has been conducted by Institute of Research and Community Services, Diponegoro University, Indonesia 2-4 October 2017

Peer review statement

All papers published in this volume of *IOP Conference Series: Earth and Environmental Science* have been peer reviewed through processes administered by the proceedings Editors. Reviews were conducted by expert referees to the professional and scientific standards expected of a proceedings journal published by IOP Publishing.

Social Cognitive Predictors of Interest in Research Among Life Sciences Academics

Dian R. Sawitri [1]*, Harlina Nurtjahjanti[1], Anggun R. Prasetyo[1]

[1]*Faculty of Psychology, Diponegoro University, Jalan Prof Sudarto, S.H., Tembalang, Semarang 50275, Indonesia.*
**Dian. R. Sawitri, mobile: +628112798878, email: dian.r.sawitri@gmail.com*

Abstract. Research interest is the degree to which an individual is interested in conducting research-related activities. Nowadays, Indonesian higher education academics are expected to be research productive, especially those in life sciences. However, what predicts interest in research among life sciences academics is rarely known. We surveyed 240 life sciences academics (64.6% female, mean age = 31.91 years) from several higher degree institutions in Indonesia, using interest in research, research self-efficacy, and research outcome expectations questionnaires. We used social cognitive career theory which proposes that individual's interests are the results of the interaction between one's self-efficacy beliefs and outcome expectations overtime. Structural equation modelling demonstrated that research self-efficacy was directly and indirectly associated with interest in research via research outcome expectations. Understanding the social cognitive predictors of interest in research contributes to an understanding of the associations between research self-efficacy, outcome expectations, and interest in research. Recommendations for life sciences academics, faculties, and higher education institutions are discussed.

1. Introduction

Understanding the development of interest in research is an important foundation to building a model to explain research productivity[1]. Previous researcher investigated research motivation in business faculty members[2], other researchers found the relationships between achievement motivation, research environment, and interest [3-4]. More recently, researchers then developed a scale to measure research motivation[5]. In 2012, Deemer, Mahoney, and Ball used it to explain research engagement in academics. However, the participants were limited in science, technology, engineering, and mathematic (STEM) areas[6].

Social cognitive career theory explains the developmental processes of interests, choice, and performance outcomes in the career and education domains. From this theoretical perspective, interests are reflective of the interaction between individual's efficacy beliefs and outcome expectations over time[7]. Research self-efficacy is an estimation of individuals ability to successfully carry on research-related tasks, for example collecting data, analysing data. This belief is expected to affect the initiation and persistence of those research-related tasks[8]. Research outcome expectations are

beliefs about the outcomes of a course of research-related action[9]. Individuals act on the basis of their judgment about what they are able to do (i.e., self-efficacy), as well as on the beliefs about the expected consequences of their actions. Having high outcome expectations (i.e., a sense that individuals will be successful when their goals are attained) provides motivation during the goal striving process and influences how individuals progress in the career decision making process. Through learning experiences, outcome expectations may take various forms of behaviour, namely, social effects such as recognition and acknowledgment from others, physical effects such as financial benefit, and self-evaluation that is progressively shaped via individuals' learning experiences[10].

Previous researchers used social cognitive career theory (SCCT) as a theoretical lens for the first time to explain interest in research[1][11] and also research productivity in doctoral psychology students[12]. However, previous studies demonstrated that research self-efficacy served as a more consistent predictor of research productivity that interest in research[13]. A more recent study also showed that achievement goals were more significant than social cognitive and demographic variables in predicting interest in research[3], and research outcome expectations were found to mediate the association between mastery approach goal orientation and research interest[4]. While the relationship between research self-efficacy, outcome expectations, and interest in research in previous studies was robust, these variables have not been tested in academics, especially in life sciences academics. Therefore, we were interested in examining the relationships between research self-efficacy, research outcome expectations, and interest in research in life sciences academics. We hypothesized that research self-efficacy would be positively associated with interest in research directly and indirectly via research outcome expectations. See Figure 1.

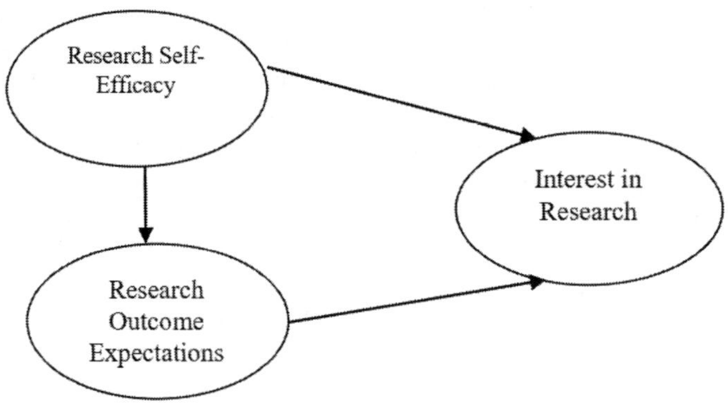

Figure 1. Hypothesized model

2. Method

Participants were 240 academics (64.6% female, mean age = 31.91 years) from several higher degree institutions in Indonesia. The data were collected using research self-efficacy, research outcome expectations, and interest in research scales. They came from medical, nursing, biology, animal sciences, and fisheries and marine sciences departments.

The 38-item Research Self-Efficacy Scale [8] was used to measure research confidence [8]. Participants were asked to assess the degree to which they feel confident in their ability to execute each task, such as "Discuss research ideas with peers" on a scale ranging from 1 (not confident) to 6 (totally confident). The 8-item Research Outcome Expectations Questionnaire-Revised [9] was used to measure research outcome expectations. Participants responded to items such as, "Involvement in research will enhance my job/career opportunities," on a 6-point scale (1 = strongly disagree and 6 = strongly agree). The 16-item Interest in Research Questionnaire [14] was used to assess interest in research. Participants were asked to rate the degree of interest they have in activities such as, "Reading a research journal article," on a 6-point scale (1 = very uninterested and 6 = very interested).

3. Results

Latent variable analysis (maximum likelihood estimation using AMOS V21) was used to examine the hypothesized structural model. We represented the latent variables using multi-item parcels to meet ideal ratios of sample size to parameters estimated (20:1) [15]. The measurement model, $\chi^2(24) = 63.09$, $p < .001$, $\chi^2/df = 2.63$, CFI = .98, RMSEA = .06 demonstrated good fit statistics, with factor loadings ranging from .40 to .85 ($p < .001$).

Table 1. Correlations, Means, Standard Deviations, and Cronbach's Alpha (N = 240)

Variables	1	2	3	*M*	*SD*	α
Research self-effcacy	-	.64	.75	196.12	17.63	.95
Research outcome expectations	.71	-	.62	40.62	4.40	.86
Interest in research	.78	.66	-	81.90	8.10	.93

Note. Correlations among scales reported above diagonal, correlations among variables reported below. ** $p < .01$, ***$p < .001$

The hypothesized structural model also demonstrated good fit statistics, $\chi2(24) = 63.08$, $p < .001$, $\chi2/df = 2.63$, CFI = .98, RMSEA = .08. All paths were significant: research self-efficacy → research outcome expectations (β = .71, $p < .001$), research outcome expectations → interest in research (β = .22 $p < .01$), research self-efficacy → interest in research (β = .63, $p < .001$). The model accounted for 50% of the variance in research outcome expectations, 63.4% in interest in research.

Using the AMOS bootstrapping procedure with 1,000 samples, we evaluated the standard errors and 95% bias corrected confidence intervals (CIs) for all direct and indirect estimates to test the mediation path. Mediation is established when the predictor is significantly related to the outcome variables, the mediator is significantly related to both the predictor and the outcome variables, and the 95% CIs of the indirect effect through the mediator do not include zero[16]. We tested a potential mediation path from research self-efficacy to interest in research (via research outcome expectations). When we tested the direct effect only, the path from research self-efficacy to interest in research was significant (β = .78, $p < .001$). Using 1,000 bootstrapped samples and 95% bias-corrected confidence interval to test the direct and indirect effects simultaneously. Research self-efficacy → research outcome expectations (β = .71, $p < .001$), research outcome expectations → interest in research (β = .22, $p < .01$) were all significant, showing all requirements for mediation were met. In the presence of research outcome expectations as the mediators, research self-efficacy remained significantly associated with interest in research (β = .63, $p < .001$), and as the indirect CI did not contain zero (CIs = .03 to .27), we concluded that the path from research self-efficacy to interest in research was partially mediated by research outcome expectations.

5. Conclusion

Based on the findings, it can be concluded that those who perceive that have higher levels of research self-efficacy are more likely to have higher outcome expectations in research activities and higher interest in research. The findings highlight the important contribution for research self-efficacy and outcome expectations in life science academics' interest in research. The results also highlighted the important role of the university in exposing life science academics to sources of research self-efficacy, such as role models, and facilitate positive research outcome expectations.

Refferences

[1] Bishop R M and Bieschke K J 1998. *Journal of Counseling Psychology.* **45** 182-188
[2] Chen Y, Gupta A, and Hoshover L. 2006. *Journal of Education for Business.* **81** 179-189
[3] Deemer E D, Martens M P, and Podchaski, E. J 2007. *Training and Education in Professional Psychology.* **1** 193-203
[4] Deemer E D, Martens M P, Haase, R F, and Jome L M 2009. *Training and Education in Professional Psychology.* **3** 250-260
[5] Deemer E D, Martens M P, and Buboltz, W C 2010. *Journal of Career Assessment.* **18** 292-309
[6] Deemer E D, Mahoney K T, and Ball J H. 2012. *Journal of Career Assessment.* **20** 182-195
[7] Lent R W, Brown S D, and Hackett G 1994. *Journal of Vocational Behavior.* **45** 79-122
[8] Greeley A T, Johnson E, Seem S, Braver M, Dias, L, Evans K, and Pricken P. 1989. *Research Self-Efficacy Scale. Unpublished scale,* The Pennsylvania State University, University Park.
[9] Bieschke K J 2000. *Journal of Career Assessment.* **8** 303-313
[10] Bandura A 1986. *Social foundations of thought and action: A social cognitive theory.* Englewood Cliffs, NJ: Prentice Hall.
[11] Mallinckrodt B, Gelso C J, and Royalty, G M 1990. *Professional Psychology: Research and Practice.* **21** 26-32
[12] Kahn J H and Scott N A 1997. *The Counseling Psychologist* **25** 38-67
[13] Kline R B 2011. *Principles and practice of structural equation modeling. (3rd ed.),* The Guildford Press, New York.
[14] Bishop R M and Bieschke K J 1994 *Interest in Research Questionnaire. Unpublished scale,* The Pennsylvania State University.
[15] Shrout P E and Bolger N 2002. *Psychological Methods* **7** 422-445

Library Development Strategy for The Community at Coastal Areas

Putut Suharso[1,2*], Bani Sudardi[1], Sahid Teguh Widodo[1], Sri Kusumo Habsari[1]

[1] Postgraduate Program in Cultural Studies, Sebelas Maret University, Jl. Ir. Sutami 36A, Surakarta 57126, Indonesia
[2] Departement of Library Science, Diponegoro University, Jl. Prof. H. Soedarto SH, Semarang 50275, Indonesia

*e-mail: putut.undip@gmail.com

Abstract. This study aims to reveal various phenomena of the existence of libraries in coastal areas, especially to pay a particular attention to the relation power between various interests of the library, especially the managers and the community as the users and to explore how both parties understand and imagine the idea of library through the developing the library as a center of learning and skill development. The design of the research is explorative qualitative. Applying case study approach, the location of this research is in a coastal area, in the floating library of Tambak Lorok coastal, Semarang. Data are collected through observation and in-depth interviews with several parties involved. To examine the validity of data, this study applies a triangulation technique of sources focusing on key informants living in the community. The result of this research shows that the existence of library in the coastal area is more than the common belief to the function of library. Although finding shows that the government's attitude to support the development of the library has been as expected through producing some regulations, the response of the community to the development of the library is far beyond the expectation. Some drawbacks found are in the handling of library management which is still lack of implementation from the initial planning and the absence of integrated coastal resources management resulted from low understanding and mastery of policy makers and implementers of the national library program. However, the community has developed a community-based management which is an approach that has been widely used in various regions as an effort to empower themselves through utilizing any existed facilities, in this case is the library. Community has successfully produced an initiative of developing integrated coastal resource management programs which are still hard to find in Indonesia, especially those involving library institutions.

Keywords: library; coastal communities; cultural studies

1. Introduction

Library has a function to educate the nation. It is stated in the mandate law no. 43 2007 on library, "that in order to improve the intelligence of the nation, it is necessary to develop a reading culture through the development and utilization of the library as a source of information in the form of papers, prints and or recordings". Library is as a means of educating the nation through the availability of potential intellectual work resources for the society use to develop their knowledge. Therefore, the library services should continue to expand to meet people's needs to access to information source. In addition, library could also provide a medium for individuals to use maximum facilities to improve their skills, productivity and personality.

Starting from the children to adults, ordinary citizens to the important people and educated in various levels of education can provide exuberant thinking as an attempt to solve problems as well as to improve the quality of life and life. Through continuous and serious efforts, libraries have an important role to play in helping the information-rich and up-to-date community, and are being developed with the support of all the resources around the coastal libraries.

The basis of the concept of community empowerment that the efforts undertaken should be directed directly at the root of the problem, namely improving the community's ability [1-2]. The part left in society should be enhanced by developing and dynamizing its potential, in other words, empowering it. Practically the effort that is the mobilization of resources to develop the potential of this people[1][3-4] will increase the productivity of the people so that both human resources and surrounding natural resources can be used maximally to increase productivity. Thus, the community and its environment are able to participate in generating and growing educational, social and cultural values. This study aims to reveal the various phenomena of the existence of libraries in coastal areas. Currently the existence of the library has become a place of learning that can be used by the general public to meet the information needs. Through the program of developing cultural reading and library development in coastal areas.

2. Research Methods

The research method used qualitative explorative research through case study approach in coastal area that is in floating library of Tambak Lorok coastal, Semarang. This research is a study designed in accordance with the scientific study of cultural paradigm that pays special attention to various interests in this case the existence of the library seen from the ideology and level of library development in coastal areas that have been established.

This research is to analyze the development strategy undertaken by library managers in the coastal areas in mobilizing the level of public participation utilizing public facilities that have been provided by the government through the Ministry of Public Works and Public Housing (PUPR). The existence of the Citizenship Meeting Hall which is integrated with this library is important in the scrutiny because it is a learning resource for the community that plays an important role in fostering a reading culture that can be obtained free of charge by the community. The research method used with the approach of cultural studies [5–8] through data collection from documents, observations, and in-depth interviews. Research sites in rural areas in Tambak Lorok include Tambakrejo and Tambakmulyo (from RW 12-15) Semarang City. Data were analyzed by qualitative descriptive method. The validity of the data was examined by triangulation techniques involving key informants residing in the research area [9].

3. Result and Discussion

3.1. Strategy of coastal library empowerment

The results of this study found the existence of libraries in coastal areas. Also it is found the role of government in taking the necessary attitudes and policies towards the library, and to determine the level of community resistance through community empowerment in managing the library. Currently the existence of the library is managed by some people with minimum tasks and services provided with limited library materials as well. The simple management system of library operation is the main attraction for researchers to look at the impact of such simple operation, especially in a coastal library. The existence of such library has been the responsibility, authority, and right of the local communities for their skills and literacy development. In this case, it needs to develop a sense of participation for all members of community to take a responsibility in maintaining the existence of library. People who care about and care for coastal libraries with those who are aware of the existence of libraries are not only important, but are indispensable for the community to gather.

From the findings obtained by researchers the lack of implementation of planning and management of coastal resources in an integrated way which leads to low understanding and mastery of policy makers and implementers of this library program. The existence of the library through community-based management is a community empowerment approach that has been widely used in various regions. Empowerment not only involves strengthening individual members of society [10], but also their institutions. Embedding modern cultural values, such as hard work, thrift, openness, and responsibility are central to this empowerment effort. Similarly, the renewal of social institutions and their integration into development activities and the role of society in them [11] [3][4][12]. However, the community empowerment program in integrated coastal resource management is still hard to find in Indonesia, especially those involving library institutions.

Figure 1. Location library in coastal Tambak Lorok

The stages of the community empowerment program is a cycle of change that seeks to reach a better level. In this case, the library Tambak Lorok (Figure 1) from the stage of community empowerment from the stage of assessment and implementation of the program runs as expected. However, in the evaluation stage as a process of supervision of citizens and officials of community empowerment programs have not run as expected. Developing libraries through planned activities is directed more towards facilitating structures and systems within the community that can ensure the development of a better community coastal community process. Currently, physical building facilities already exist, just waiting for the sustainability of the planned program from the beginning which must be supported by the community around Tambak Lorok.

3.2. Library from the perspective of cultural studies

A library institution is a readable-case object on how the representation of the power of the group in the form of library management, from the library management structure, to the activities undertaken by the surrounding community. Both the manager and his activities are ideological sites and battlegrounds among different actors of different backgrounds. This ideological battle ("interest") between these actors spawned discursive practice dynamics in libraries that directly or indirectly escape the vision of a professional and academic library.

In fact, the library as a site representing the various possible battles illustrated in the Foucault approach [13][14] can be seen in relation to the powers that operate in libraries in the time before Christ as in Erskine [15] entitled Culture and Power in Ptolemaic Egypt: The Museum and Library of Alexandria. This discursive practice dynamics as an active cultural process becomes an important discourse to be explored in the context of library management. The importance of the discourse can not be separated from the core presence of a library as the bearer of an ideal vision in the context of professional organizations. Unfortunately, libraries are generally understood by the public as practical and professional institutions and unimaginable to them about the possibility of a power discourse process that has implications in practical, technical, and clerical service processes [16-17]. It is the public attention that makes the discourse process effective without any control by them, either internally or externally. The consequences of the situation led the library into the place of the process of domination by the interests of certain powerful groups into the status quo which then resulted in the helplessness of libraries in carrying out their professional functions in the field of libraries and librarianship.

3.3. Improvement of community skills

Basic skills training is one of the empowerment programs aimed at improving the skills of the community as a stock of work and community welfare. Efforts to improve skills in the library to be able to independently exploit the potential of one area with knowledge and skills about entrepreneurship and handicraft making that can be used as a souvenir. The purpose of this program is to improve the skills, knowledge and awareness of the community, especially the group of women PKK to increase the value added of marine products so that it can increase family income. And promote souvenirs as a product of the local community [18-19].

Human resources is a very important capital in the development of coastal community. The linkage of coastal libraries to community empowerment is interesting. The goal of community empowerment is the community's independence in overcoming their problems through initiative and creativity to improve the quality of life [20-22]. Currently the community needs knowledge, skills and attitude to get out of the problems faced by the community [23-24].

Become a consideration for policy makers in development planners, when faced with the problem of increasing the capacity of human resources of coastal communities whose average formal education is limited. Even in some places still found those who are illiterate. Of course, it is necessary to choose the methods and educational media that suit the needs and conditions of society. Empowerment actors at the community level in this case libraries can develop training methods for adults to improve knowledge and skills of coastal communities. The key to adult education methods is the process of awareness of growing confidence, fostering a sense of need in the community to improve the quality of life with the support of existing facilities in their environment.

3.4. Library as a public space

In order to become a developed nation, all members of society must care, take responsibility, support each other, and synergize to improve reading interest by increasing knowledge about library. A low public interest reading is not merely the responsibility of the government, but the responsibility of all citizens in providing quality literature. Government through the Regional Library Agency, always encourage reading culture to the community, even encourage the parties of education observers to open or provide a community reading park around their residence. This is one of the government's efforts to increase interest in reading the community by protecting the community itself.

One of the functions of the library is the informational function that is in the provision of access to information freely, information democratization, and connectivity to global information. Information is at the core of the public sphere at an information or library [25-26]. With the ability of libraries to provide information for the development of independent opinions, the abstract library is a public space. But the library is physically a library building where people can gather to hold activities such as discussion, book review, seminars, and so on (Figure 1 and Figure 2). Before participating in these activities, participants of the discussion will provide themselves with the information provided in the library. When these activities are carried out in the library, there has been a widespread interaction.

Figure 2. Utilization of library by children after school.

Figure 3. Some parents use the library as a relaxing get together space.

In the perspective of public space [27-28], libraries must be democratic so that libraries are open to anyone who wants to use the information they have, whether poor, rich, officials, merchants, students, and others. Libraries should be able to receive and respond to input and criticism from users so that libraries can meet the needs of users/communities. Libraries must also be able to follow the development of information technology so that the user community can still make the library as a public space because it is always able to provide the needs of user information. To maintain the library public space, the library should always be introspection in order to be a place coveted by the community and do not forget to do promotions that invite people love to read.

4. Conclusion

Ideally, a library has no importance whatsoever in the content of the provider of information and knowledge. It does not discriminate against truth, objectivity, and value contained therein, but it is in its sole interest to make such a collection of information and knowledge available to the public. The library is a place with a system designed to maintain and facilitate information seekers. But if there is an unfairness in the implementation of the essence of library tasks as already manifested in the middle of coastal communities.

Acknowledgements

The first author would like to thank all the leaders of Diponegoro University who have given permission to join the doctoral program of cultural studies at postgraduate of Sebelas Maret University to explore library science through cultural studies. On this occasion the author also thanked the promotor team in dissertation writing.

References

[1] Adams R 2008 *Empowerment, Participation and Social Work* (New York: Palgrave Macmillan)
[2] Friedmann J 1992 *Empowerment: The Politics of Alternative Development* (Cambridge: Blackwell Publishing)
[3] Karsidi R 2001 Paradigma Baru Penyuluhan Pembangunan dalam Pemberdayaan Masyarakat *Mediat. (Jurnal Komunikasi)* **2**
[4] Wrihatnolo R R and Dwidjowijoto R N 2007 *Manajemen Pemberdayaan: Sebuah Pengantar dan Panduan untuk Pemberdayaan Masyarakat* (Jakarta: Elex Media Komputindo)
[5] Gray A 2003 *Research Practice for Cultural Studies* (London: SAGE Publications)
[6] Schwoch J and White M 2006 Introduction: The Questions of Method in Cultural Studies *Questions of Method in Cultural Studies* ed J Schwoch and M White (Victoria: Blackwell Publishing) pp 1–16
[7] Saukko P 2003 *Doing Research in Cultural Studies: An introduction to classical and new methodological approaches* (London: SAGE Publications)

[8] McRobbie A 2005 *The Uses of Cultural Studies: a textbook* (London: SAGE Publications)
[9] McCarthy A 2006 From the Ordinary to the Concrete: Cultural Studies and the Politics of Scale *Questions of Method in Cultural Studies* ed M White and J Schwoch (Victoria: Blackwell Publishing) pp 21–53
[10] Kay R and Alder J 2005 *Coastal Planning and Management* (London: Taylor & Francis Group)
[11] Cicin-Sain B and Knecht R W 1998 *Integrated Coastal and Ocean Management: Concepts And Practices* (Washington: Island Press)
[12] Retno S S, Rohmiyati Y and Husna J 2015 Pemberdayaan Masyarakat melalui Perpustakaan: Studi Kasus di Rumah Pintar "Sasana Ngudi Kawruh" Kelurahan Bandarharjo-Semarang *J. Ilmu Perpust.* **4**
[13] Moulaison H L, Dykas F and Budd J M 2014 Foucault, the author, and intellectual debt: Capturing the author-function through attributes, relationships, and events in knowledge organization systems *Knowl. Organ.* **41** 30–43
[14] Sturrock J 2004 Michel Foucault *Strukturalisme Post-strukturalisme: Dari Levi-Strauss ke Derrida* (Surabaya: Jawa Pos Press) p 129
[15] Erskine A 1995 Culture and Power in Ptolemaic Egypt: The Museum and Library of Alexandria *Greece Rome* **42** 38–48
[16] Laugu N 2013 *Representasi Kuasa Dalam Pengelolaan Perpustakaan (Studi Kasus Pada Perpustakaan Perguruan Tinggi Islam di Yogyakarta)* (S3 Kajian Budaya dan Media, Universitas Gadjah Mada)
[17] Buschman J E 2010 The Social as Fundamental and a Source of the Critical: Jürgen Habermas *Critical Theory for Library and Information Science: Exploring the Social from across the Disciplines* ed G J Leckie, L M Given and J E Buschman (Santa Barbara, CA.: Libraries Unlimited) p 161
[18] Jaeger P T, Gorham U, Bertot J C, Taylor N G, Larson E, Lincoln R, Lazar J and Wentz B 2014 Connecting government, libraries and communities: Information behavior theory and information intermediaries in the design of LibEGov.org *First Monday* **19**
[19] Taylor N G, Jaeger P T, McDermott A J, Kodama C M and Bertot J C 2012 Public Libraries in the New Economy: Twenty-First-Century Skills, the Internet, and Community Needs *Public Libr. Q.* **31** 191–219
[20] Scott R 2011 The Role of Public Libraries in Community Building *Public Libr. Q.* **30** 191–227
[21] Vårheim A 2007 Social capital and the multiethnic challenge: The role of the public library *Social Capital: An introduction* ed R. Sharma (Hyderabad: ICFAI University Press) pp 100–10
[22] Vårheim A 2014 Trust in Libraries and Trust in Most People; Social Capital Creation in the Public Library *Libr. Q. Information, Community, Policy* **84** 258–77
[23] Kenneally A, Sung H Y, Hepworth M, Scott R and McNicol S 2011 Modelling community engagement in public libraries *Public Libr. Q.* **30** 191–227
[24] Sung H Y and Hepworth M 2013 Modelling community engagement in public libraries *Malaysian J. Libr. Inf. Sci.* **18** 1–13
[25] Jochumsen H, Skot-Hansen D and Rasmussen C H 2015 Towards Culture 3.0 – performative space in the public library *Int. J. Cult. Policy*
[26] Jochumsen H, Skot-hansen D and Rasmussen C H 2010 The four spaces – A new model for the public library *New Libr. World* **113** 586–97
[27] Alstad C and Curry A 2003 Public Space, Public Discourse, and Public Libraries *LIBRES* **13**
[28] Connaway L S 2015 Meeting the Expectations of the Community: The Engagement-centered Library *The Library in the Life of the User: Engaging with People Where They Live and Learn* (Ohio: OCLC Research) p 201

IOP Conf. Series: Earth and Environmental Science **116** (2018) 012003 doi:10.1088/1755-1315/116/1/012003

The Effects of Treatments on Batu Banana Flour and Percentage of Wheat Substitution on The Resistant Starch, In Vitro Starch Digestibility Content and Palatability of Cookies Made with Banana (*Musa balbisiana* Colla) Flour

D Ratnasari [1], N Rustanti [1,2], F Arifan [3], DN Afifah [1,2*]

[1]Department of Nutrition Science, Faculty of Medicine, Diponegoro University, Jl. Prof Soedarto, SH, Tembalang, Semarang, Indonesia 50275
[2]CENURE (Center of Nutrition Research), Diponegoro University, Jl. Prof Soedarto, SH, Tembalang, Semarang, Indonesia 50275
[3]Department of Chemical Engineering, Faculty of Engineering, Diponegoro University, Jl. Prof Soedarto, SH, Tembalang, Semarang, Indonesia 50275

Email : d.nurafifah.dna@gmail.com

Abstract. Diabetes mellitus (DM) is the most common endocrine disease worldwide. Resistant starch is polysaccharide that is recommended for DM patient diets. One of the staple crops containing resistant starch is banana. It is the fourth most important staple crop in the world and critical for food security, best suited plant in warm, frost-free, and coastal climates area. Among banana varieties, Batu bananas (*Musa balbisiana* Colla) had the highest content of resistant starch (~39%), but its use as a food ingredient is limited. Inclusion of Batu banana flour into cookies manufacturing would both increase the economic value of Batu bananas and provide alternative snacks for DM patients. Here we sought to examine whether cookies made with modified Batu banana flour would be a suitable snack for DM patients. This study used a completely randomized design with two factors: substitution of Batu banana flour (25%, 50%,75%) for wheat-based flour and Batu banana flour treatment methods (no treatment, autoclaving-cooling, autoclaving-cooling-spontaneous fermentation). The resistant starch and *in vitro* starch digestibility levels were analyzed using two-way ANOVA and Tukey test, whereas the acceptance level was analyzed by Friedman and Wilcoxon tests. The content of resistant starch and *in vitro* starch digestibility of the different treatments ranged from 3.10 to 15.79% and 16.03 to 52.59%, respectively. Both factors differed significantly ($p<0.05$) with respect to Batu banana flour substitution, but not to processing method ($p>0.05$). Meanwhile, palatability in terms of color, aroma, texture, and flavor differed significantly among the different treatments and starch contents ($p<0.05$). Together these results show that Batu banana flour could be a promising ingredient for the production of snacks suitable for consumption by DM patients.
Keywords: Batu banana, cookies, resistant starch, *in vitro* starch digestibility

1. Introduction

Diabetes mellitus (DM) is the most common endocrine disease worldwide and represents a growing public health problem [1]. According to the International Diabetes Federation, 382 million adults worldwide (8.3 %) have DM, and 80% live in countries with low and middle income. If this trend continues, by 2035 592 million people, or nearly 10% of the adult population, will suffer from diabetes [2].

Type 2 diabetes mellitus is most common in adults. Type 2 diabetes mellitus patients can regulate and monitor their carbohydrate intake to control blood glucose levels. Moreover, type 2 diabetes patients are encouraged to consume complex carbohydrates or polysaccharides as these sugars require longer digestion times that in turn slow the glycemic response [3]. One type of polysaccharide recommended for type 2 diabetes patients is resistant starch, which undergoes unhydrolysis to D-glucose in the intestine within two hours of consumption before fermentation in the colon [4]. Resistant starch is present in rice, peas, potatoes, and bananas [5].

The banana is a tropical/subtropical plant best suited to warm, frost-free, coastal climates [6]. The production, consumption and trade of banana is in high volumes in the world. Banana is the fourth most important staple crop in the world and is critical for food security in many tropical countries. This makes banana to be the prime leading fruit crop in terms of volume and value in the world market [7-8].
Among banana varieties, Batu bananas (*Musa balbisiana* Colla) had the highest content of resistant starch (39.35%) [5][9]. Moreover, during processing the resistant starch content can be increased with autoclaving-cooling cycles and spontaneous fermentation. A previous study showed that spontaneous fermentation of sliced bananas combined with one cycle of autoclaving-coolingincreased the resistant starch content of banana flour more to more than 17% of dry weight [10].

Despite its potential benefits, the use of Batu bananas as a food or food ingredient is limited. Banana flour has benefits for inclusion in semi-finished products because it can be saved longer, is easy to incorporate, and can be cooked faster [11]. Cookies are a commonly consumed snack in Indonesia, as evidenced by a 2013 survey conducted by Basic Health Research (RISKESDAS) found that 13.4 % of Indonesians consume cookies more than once a day [12-13].
The objective of this study is to study resistant starch and *in vitro* starch digestibility contents as well as the palatability of cookies made with modified Batu banana flour.

2. Materials and methods

2.1. Materials
Cookies were made from flour produced from unripe Batu bananas obtained from the Pemalang district in Indonesia, wheat flour, stevia sugar, egg yolks, regular tub margarine, vanilla powder, single-acting baking powder, cocoa powder (windmolen), and table salt.

2.2. Production of cookies with banana flour
For production of banana flour cookies, 25 g egg yolks and 70 g margarine were mixed for 2-3 minutes before the other ingredients (3 g stevia sugar, 0.5 g vanilla powder, 0.3 g table salt, 0.2 g baking powder, 5 gr cocoa powder, wheat flour and batu banana flour formulation) were incorporated to produce a homogenous batter. The dough was applied to baking sheets in uniform amounts using a syringe before baking at 150 °C for 15-20 minutes.

2.3. Preparation of modified banana flour using spontaneous fermentation and autoclaving-cooling cycles
Methods used to increase resistant starch content in this study were autoclaving-cooling (AC) and autoclaving-cooling spontaneous fermentation (AF). Untreated banana flour (WT) was also examined. For all treatments, dried Batu bananas (microwave drying, 450 watt, 25 min) were pulverized and passed through a #60 mesh sieve (0.23 mm). Untreated Batu bananas were simply dried (450 watt, 25 min). Spontaneous fermentation was carried out at room temperature for 24 hours in distilled water. One cycle of autoclaving-cooling involved autoclaving at 121 °C for 15 minutes followed by cooling at 4 °C for 24 hours.

2.4. Chemical analysis
Resistant starch content and *in vitro* starch digestibility was determined according to methods described by Kim *et al.* (2003) and Anderson *et al.* (2002), respectively [14-15]. The water, ash, fat, protein, and carbohydrate content was determined by AOAC 1995 [16], AOAC 2006 [17], soxhlet method AOAC

1995, bradford method [18], and by difference, respectively. The materials used for chemical analysis were 0.05 M phosphate buffer (NaH_2PO_4 & $NaHPO_4$) pH 6, Endo α-amylase (termamyl), NaOH, protease enzymes (Pronase E), HCl, amyloglucosidase and α-amylase enzymes (Cargill), ethanol, Whatman filter paper 42, Na-phosphate buffer (0.02 M and 0.1 M pH 7), 3,5-dinitrosalisilat (DNS), Na-K tartaric acid, and distilled water.

2.5 Palatability analysis
Product palatability was assessed using a hedonic test with a four-point scale wherein, 1 = dislike very much, 2 = dislike, 3 = like, and 4 = like very much. Thirty panelists recruited by the Nutrition Science Department, Diponegoro University tested cookie palatability.

2.5. Statistical Analysis
Resistant starch content and *in vitro* starch digestibility data were analyzed using a two-way ANOVA followed by a Tukey test. Palatability data were analyzed using a Friedman test followed by a Wilcoxon test.

3. Results and discussion

Banana flour produced using the AF method had a lower ash content relative to the AC and WT methods. This decreased ash content likely resulted from the soaking and fermentation steps [10].

Table 1. Chemical composition of Batu banana flour

Composition	Autoclaving cooling + fermentation (AF)	Autoclaving cooling (AC)	Without treatment (WT)
Water (%)	5.19 ± 0.02*	4.93 ± 0,02	6.87 ± 0,01
Ash (%)	2.59 ± 0,01	5.46 ± 0,04	5.29 ± 0,03
Fat (%)	12.18 ± 0,03	10.00 ± 0,14	7.47 ± 0,03
Protein (%)	6.70 ± 0,14	6.93 ± 0,03	6.93 ± 0,04
Carbohydrate (%)	73.34 ± 0,21	72.67 ± 0,24	73.43 ± 0,11
Resistant starch (%)	36.74 ± 1,26	31.54 ± 1,90	25.94 ± 0,85

*Mean ± SD of two trials is shown

For all processing methods, the fat and protein content of the banana flour tended to be higher in this study compared to values obtained in an earlier [10]. This difference could be attributed to the different varieties of bananas that were used or differences in processing, particularly the use of microwave for drying. A microwave was used here for drying the fruit because microwaving shortened the drying time and yielded a final product that had better flavor and nutritional value compared to flours made using conventional methods [19]. Furthermore, less protein was lost with a shortened drying time [20].

3.1. Resistant starch content and in vitro starch digestibility of cookies made with Batu banana flour
There were significant differences in the mean content of resistant starch as a function of the percentage banana flour substituted ($p=0.000$), whereas the starch contents of flours produced using different methods were similar ($p=0.284$) (Table 2). Moreover, the interaction of Batu banana flour substitution and method also had no difference in resistant starch content ($p=0.987$). An advanced Tukey test showed significant differences in starch content for each amount of banana flour substituted ($p<0.05$).

The results for method of flour preparation were consistent with those of Nurhayati et al., who found that the amylose content in agung semeru banana flour made using three different methods (autoclaving-cooling spontaneous fermentation, autoclaving-cooling and no treatment) had some degree of significant difference, although the difference was not large. Thus, it is possible that banana flour treated by different methods and used for cookie production had similar amylose contents, which positively correlate with resistant starch content [21-22].

Meanwhile, higher amounts of resistant starch were seen with higher amounts of substituted banana flour, which was expected given that refined wheat-based flour has a low content of resistant starch (0.65% w/w) [5][23].

Table 2. Resistant starch content (%) of Batu banana flour cookies.

Batu Banana Flour Substitution	Resistant Starch Content – Flour Processing Method: Autoclaving cooling + fermentation (AF)	Autoclaving cooling (AC)	Without Treatment (WT)	P
25%	4.52 ± 1.16*	3.59 ± 0.51	3.10 ± 0.49	
50%	12.07 ± 1.59	10.57 ± 1.59	10.14 ± 0.67	0.000
75%	15.79 ± 2.81	15.48 ± 0.40	14.60 ± 2.46	
P		0.987		

*Mean ± SD of two trials is shown

A study by Musita [5] showed that the resistant starch content of Batu banana flour cookies was less than that of untreated Batu banana fruit. This result could be due to the release of amylose from granules during baking that could increase starch solubility, which is associated with decreased resistant starch content [24]. Another study showed that heat treatments such as baking could decrease the Insoluble Dietary Fiber (IDF) content of which resistant starch is a component of cereal and legumes [25].

Moreover, the addition of ingredients containing fats and proteins such as eggs and margarine can reduce resistant starch content, as was seen for potato starch treated with autoclaving and cooling followed by the addition of albumin before rehydration at 20 °C [26]. Furthermore, heating of starches to 100 °C in the presence of fats induced the formation of amylose-lipid complexes, which can reduce formation of resistant starch during recrystallization due to lipid binding to amylose [26].

Statistical analysis showed that there were differences in the mean in vitro starch digestibility among the different banana flour substitutions ($p=0,000$), whereas the methods did not have significant changes in starch content ($p=0.193$). The interaction of banana flour substitution and preparation method with in vitro starch digestibility also showed no significant differences ($p=0.973$). Among cookies prepared with different amounts of banana flour substituted for wheat flour, an advanced Tukey test showed differences between the concentrations ($p<0.05$) (Table 2).

Table 3. Analysis of in vitro starch digestibility (%) of cookies prepared with banana flour

Batu banana flour substitution	In vitro starch Digestibility – Flour processing method: Autoclaving cooling + fermentation (AF)	Autoclaving cooling (AC)	Without treatment (WT)	P
25%	50.18 ± 2.78*	51.92 ± 1.84	52.59 ± 1.73	
50%	33.28 ± 1.26	34.74 ± 2.45	35.92 ± 1.81	0.000
75%	16.03 ± 1.53	16.31 ± 1.29	17.31 ± 1.12	
P		0.193		

*Mean ± SD of two trials is shown

Starch digestibility is the ease with which a starch type is hydrolyzed into simpler units by disgestive enzymes [27]. The in vitro starch digestibility in cookies produced with Batu banana flour was similar among the various flour processing methods, but the amount of banana flour substituted during cookie production did have a significant effect on the in vitro starch digestibility (Table 3).

A low value for in vitro starch digestibility can be produced by autoclaving-cooling cycles, wherein starch molecules rearrange between amylose-amylose and amylose-amylopectin to yield a stronger bond between starch molecules that is more resistant to digestion [26]. Lower in vitro starch digestibility was associated with a higher content of resistant starch and vice versa (Tables 2 and 3).

In vitro starch digestibility can also be affected by interactions between starches and proteins that reduce the susceptibility of starch to enzymatic hydrolysis due to embedding of starch granules in protein matrices in the corneous endosperm of plants. Starch gelatinization arising during processing can form starch and protein complexes that reduce both starch digestibility and protein content [28].

3.2. Palatability of Batu banana flour cookies

Study participants noted significant differences in color (p = 0.001), aroma (p = 0.000), texture (p = 0.003), and flavor (p = 0.000) of cookies produced with varying amounts of Batu banana flour (Fig. 1; Wilcoxon test).

Based on a hedonic test of color, cookies with made with 25% Batu banana flour produced using the AC method had the highest hedonic scores, with a mean 3.00 (like), whereas cookies made with a substitution of 75% Batu banana flour made with the AC method had the lowest score (2.30, dislike), likely because these cookies had the darkest color. Flours produced by AC, AF, and WT methods had increasingly lighter color, suggesting that the brown color of the AC method arose from the Maillard reaction or other non-enzymatic browning reactions that occur when foods are cooked at high temperatures and sugar-reducing agents react with compounds bearing a NH_2 group (e.g., proteins, amino acids, peptides, and ammonium) [29-30].

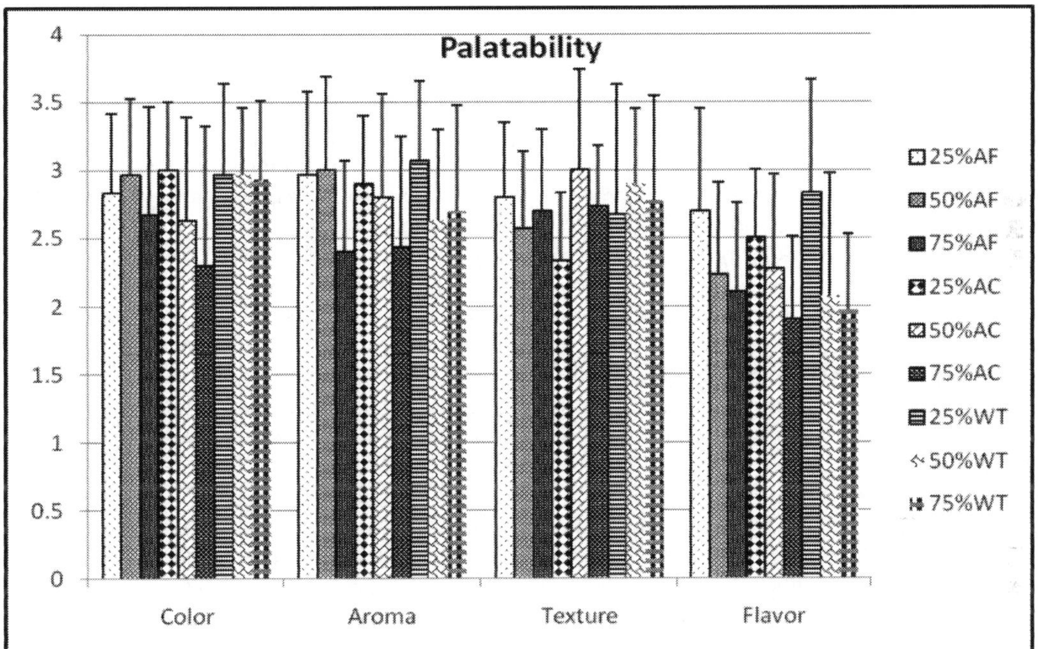

Figure 1. Palatability in terms of color, aroma, texture, and flavor of cookies made with different amounts of Batu banana flour processed using different methods. Flour processing methods are indicated by AF: autoclaving-cooling followed by spontaneous fermentation; AC: autoclaving-cooling only; WT: no treatment. Data represent mean ± SD.

Cookies produced with Batu banana flour processed by AF had lighter color than the AC method, likely because bananas that underwent spontaneous fermentation that involved soaking in distilled water for 24 hours can have delayed enzymatic browning reactions [30].

Meanwhile, scores for aroma were highest for cookies produced with 25% WT banana flour (mean=3.07, like), whereas cookies with 75% AF banana flour had the lowest preference scores (mean = 2.40, dislike). In terms of aroma, cookies made with banana flour with no treatment had higher scores than cookies with 75% banana flour processed either by the AC or AF method, which again may be due to Maillard or heat-induced reactions that result in loss of aromatic compounds [31], which, along with individual perception, form the basis for aroma [32].

Cookies made with substitution of 50% AC Batu banana flour were judged to have the best texture (mean =3.00, like), whereas cookies with substitution of 25% AC banana flour had the lowest scores (mean = 2.33, dislike) and were noted to have a less crispy texture. Indeed, cookies that had higher levels of banana flour had greater crispiness that may be due to gelatinization processes that occur during baking that break intramolecular hydrogen bonds in proteins and carbohydrates [33].

The presence of free hydroxyls will absorb water that swells starch granules. The more hydroxyl groups contributed by starch molecules in banana flour, the greater its capacity to to absorb water, such that when drying occurs, the water carrying capacity is reduced and the crispy texture increases [33]. For flavor, cookies with 25% WT Batu banana flour substitution had the best socres (mean = 2.83, like) and those with 75% substitution of AC banana flour had the lowest scores (mean = 1.9, dislike).

3.3. Consumption of resistant starch affects blood glucose levels
Some research suggest that the compsumtion of foods containing resistant starch has a positive effect on glycemic control. Research on mice, pigs, and humans indicate that resistant starch can lower fasting blood glucose concentrations, increase insulin secretion, and improve insulin sensitivity [27]. A study using mice with type 2 diabetes mellitus showed an increased pancreatic beta cell mass and insulin sensitivity after the mice consumed feed with 30% resistant starch content for 10 weeks. A study in humans showed improvements in glucose tolerance, blood glucose, insulin, and colonic fermentation condition in healthy or type 2 diabetes mellitus subjects after consuming food containing resistant starch [27].

Some evidence showed that the mechanism by which blood glucose levels are regulated might be associated with resistant starch fermentation in the colon and SCFA (Short Chain Fatty Acid) production. SCFAs are organic fatty acids with 1 to 6 carbon atoms with a straight and branched chain that are produced in the distal intestine by bacterial fermentation of fibrous food and allow resistant starches to escape the gastrointestinal tract and enter the colon. The most common forms of SCFA are formic, acetic, propionic, butyric, valeric, isovaleric, and caproic acid, but in the colon 90-95% of SCFA is acetic, propionic, and butyric acids [34].

One form of SCFA is butyrate that can induce L-cells to secrete an incretin as intestinal peptide YY (PYY, also known as peptide tyrosine tyrosine or glucagon-like-peptide-1/GLP-1 hormone) [27]. Incretin hormone secretion from endocrine cells of the small intestine occurs in response to consumption of carbohydrates, proteins, and fats. The incretin hormone has many functions, including appetite suppresion. The main form of incretin hormone is GLP-1, which controls post prandial glucose by increasing insulin secretion, stimulating insulin gene expression, inhibiting glucagon synthesis, and delaying gastric emptying. However, the mechanism by which butyric SFCA increases the secretion of GLP-1 is unclear [34],[35].

Several studies of patients with controlled type 2 diabetes mellitus showed increasing levels of GLP-1 after cosumption of 40 g resistant starch daily for 2 weeks. Thus, long-term supplementation with resistant starch may have the potential to increase blood glucose tolerance in patients with type 2 diabetes mellitus [36].

4. Conclusion
The levels of resistant starch and in vitro starch digestibility significantly differed according to amount of Batu banana flour included during production of banana flour cookies, but the flour processing method had no effect. All palatability parameters, including color, aroma, texture, and flavor, showed differences among the amount of banana flour incorporated. The optimal palatability was seen for cookies produced with substitution of 25% Batu banana flour produced using the AF method. These results show that Batu banana flour could be a useful ingredient to produce desirable snacks having high levels of resistant starch and in vitro digestibility that would allow type 2 diabetes patients to control their blood glucose levels.

5. Acknowledgements
The authors acknowledge Ministry of Research, Technology and Higher Education
of the Republic of Indonesia for support of this study.

6. References
[1] Marcia N K P S, Karen L and Sara L R 2010 *Nutrition Therapy and Pathophysiology* 2nd ed (Belmont: Wadsworth Cengage Learning) Part 4 Diseases of Endocrine System
[2] Federation ID 2013 *IDF Diabetes atlas* 6th ed International Diabetes Federation Chapter 2

[3] Evert A B, Boucher J L, Cypress M, Dunbar S A, Franz M J, Mayer-Davis E J, Neumiller J J, Nwankwo R and Verdi CL 2013 *Diabetes Care* **36** 3821 – 42

[4] Fuentes Z E, Riquelme N M J, Sánchez Z E and Pérez A J A 2010 Resistant starch as functional ingredient: a review *Food Research International*, **43** 931 - 942

[5] Musita N 2009 The study of resistant starch content and characteristic of resistant starch fom some banana types *Journal of Industrial and Agricultural Products* **14** 68 – 79.

[6] Deuter P, White N, Putland D, Mackenzie R and Muller J 2011 *Critical (temperature) Thresholds and Climate Change Impacts/adaptation in Horticulture* (Queensland : Department of Employment Economic Development and Innovation)

[7] Akcaoz H, Ozcatalbas O and Kizilay H 2009 *Journal of Food, Agriculture & Environment* **7** 283-294

[8] Alemu MM 2017 *Journal of Environmental Protection* **8** 319 -329

[9] Management Information System in Rural Development (BAPPENAS) 2000 *Banana (Musa spp)* Deputy State Minister of Research and Technology Utilization and Correctional Division of Science and Technology 1 – 13

[10] Nurhayati, Jenie B S L, Widowati S and Kusumaningrum H D 2014 *Journal of Agricultural* Technology **34** 146-149

[11] Histifarina D, Rahman A, Rahadian D and Sukmaya 2012 *Agrin Journal* **16** 125 – 128

[12] Claudia E J and Widjanarko S B 2016 *Journal of Food and Agro-Industry* **4** 391 - 399

[13] Setyowati W T and Nisa F C 2014 *Journal of Food and Agro-Industry* **2** 224 - 231

[14] Karimah, Ima 2011 *Glycemix index values of instant porridge prepared from cassava starch and cassava resistant starch* (Bogor : Institut Pertanian Bogor)

[15] Abdillah, F 2010 *Modification of tanduk banana (musa paradisiaca formatypica) flour through spontaneous fermentatition and autoclaving to increase resistant starch* (Bogor : Institut Pertanian Bogor)

[16] Association of Official Analytical Chemist 1995 Official methods of analysis of the AOAC 16th ed (Arington Washington DC)

[17] Association of Official Analytical Chemist 2006 The official methods of analysis of AOAC international 18th ed (Arington : The Association)

[18] Bradford M M 1976 *A rapid and sensitive method for the quantification of microgram quantities of protein utilizing the principle of protein dye binding* (Anal Biochem University of Georgia)

[19] Chang Y K, Steel C J and Clerici M T P S 2011 *Use of Microwave Radiation to Process Cereal-Based Products* Intech. Rijeka

[20] Boateng E D 2013 *International Food Research Journal* **20** 1569 – 73

[21] Wulan S N, Widyaningsih T D and Ekasari D 2007 *Journal of Agricultural Technology* **8** 80 – 87

[22] Mir J A, Srikaeo K and Garcia J 2013 *International Food Research Journal* **20** 1329 – 35

[23] Nigudkar M R 2014 *Current Research in Nutrition and Food Science* **2** 73 – 83

[24] Bavaneethan Y, Vasantharuba S, Balakumar S and Thayananthan K 2015 *Journal of Agricultural Science*, **11** 244 - 246

[25] Aziah N and Noor M, L H H 2012 *International Food Research Journal* **19** 1539 – 43.

[26] Faridah D N, Rahayu W P and Apriyadi M S 2013 *Journal of Agricultural Industry Technology* **23** 61-69

[27] Zhang L, Li H T, Shen L, Fang Q C, Qian L L and Jia W P 2015 *Biomedic Environment Science* **28** 291 – 297

[28] Siller A D C P 2006 *In Vitro Starch Digestibility and Estimated Glycemic Index of Sorghum Products* (Texas: Food Science and Technology Texas A&M University)

[29] Tamanna N and Mahmood N 2015 *International Journal of Food Science* 1 – 3

[30] Ayu DC and Yuwono SS 2014 *Journal of Food and Agro-Industry* **2** 110 - 120

[31] Boudhrioua N, Giampaoli P, Bonazzi C 2003 *Swiss Society of Food Science and Technology* **36** 633 – 642

[32] El Hadi M A, Zhang F J, Wu F F, Zhou C H, Tao J 2013 Molecules **18** 8200 – 99

[33] Silfia 2012 *Journal of Research and Development Industry* **2** 43 – 49

[34] Puddu A, Sanguineti R, Montecucco F and Viviani G L 2014 *Hindawi Publishing Corporation* 1 – 3

[35] Huether S E and McCance K L 2012 *Understanding Pathophysiology* 6th ed (Missouri: Elsevier)

[36] Bodinham C L, Smith L, Thomas E L, Bell J D, Swann J R, Costabile A, Russell J D, Umpleby A M and Robertson M D 2014 *Endocrine Connection* **3** 75 - 84

IOP Conf. Series: Earth and Environmental Science **116** (2018) 012004 doi:10.1088/1755-1315/116/1/012004

Stock Analysis of *Metapenaeus affinis* (H.Milne Edwards, 1837) on the North Coast of Central Java, Indonesia

Suradi Wijaya Saputra, Anhar Solichin, Wiwiet Teguh Taufani
Aquatic Resource Department, Faculty of Fisheries and Marine Science, Diponegoro University, Jl. Prof. H.Soedarto, SH, Tembalang, Semarang 50275, Indonesia.
Email:suradi@live.undip.ac.id

Abstract. This research aims at studying the length of first capture (L_c), length of sexual mature (L_m), and the stock analysis of *Metapenaeus affinis* in the North coast of Central Java, Indonesia. The field research activities were conducted from May 2016 to April 2017 using survey method and direct observations on the catch area of fishing units. The results showed that total length of the first capture for the male shrimp is 76,4 mm and 63 mm for females. The total length of first sexual mature is 116 mm. Y'/R result analysis on male shrimp shows 0,595 maximum exploitation rate (E_{max}), the rate of exploitation at $E_{0,1}$ is 0,521 and the rate of exploitation at $E_{0.5}$ is 0,352. While the female shrimp obtained is E_{max} = 0.637, $E_{0.1}$ = 0.562 and $E_{0.50}$ = 0,373. To increase the production and to preserve shrimp resources, it is ideal if $L_c = L_m$, i.e. at a total length of 116 mm. In conclusion, the increase in length of first captured will increase Y / R 'and B / R'

Key word :L_c, L_m, stock analysis, M.affinis, North Coast of Central Java

1. Introduction

The habitat of *Metapenaeus affinis* starts from subtidal edge to a depth of 60 m, prefer mud and sand-based mud substrates, while young shrimp found in intertidal areas [1- 2]. This species during pascalarva and juvenile stages live in the estuary, but spawn at the bottom of inshore and offshore [2].

The exploitation of *M.affinis* in the north coast waters of Central Java was done by trawls and seines nets, which have been banned by the Indonesian government based on Fisheries and Marine Ministerial Regulation No 2 KP / Permen-KP / 2015 about the prohibition of the use of trawls and seine nets in the fishery management territory of Indonesia's republic fisheries management, such as trawls, danish seine, etc.

Several studies have been conducted on *M. affinis*, such as their age and growth in Mumbai [3], abundance and seasonal migrations [4], biology, assessment, and management [5-6], population dynamics, and exploitation at Kuwait [7]. Previous research on *M.affinis* in the northern waters of Central Java mainly focuses on growth, mortality and exploitation rates [8]. The results show that mortality rates are high and have been over-exploited (E> 0,5). Therefore, further research is needed. This study aims to examine the optimum size that should be captured, the first size of mature sexual and the results per optimum recruit. Besides, for the exploitation of *M. affinis* shrimp can be sustainable and provide optimum benefits for fishermen.

2. Material and Methods

2.1. Sampling Method.

This study was conducted in the northern coastal waters of western Central Java, starting from the northern waters of Brebes Regency (in the west) to Kendal Regency (in the east). Sampling was taken 12 times (May 2016 until April 2017). Shrimp samples were obtained from catch result by using danish seine fishing gear, at 10 places of fish auction (TPI), i.e. Kaliwlingi and Kluwud (Brebes regency), Larangan and Suradadi (Tegal regency), Asemdoyong and Tanjungsari (Pemalang regency), Jambean (Pekalongan regency), East Roban (Batang regency) and Bandengan and Tawang (Kendal regency). Shrimp samples were taken using systematic sampling method. The number of sample fishing vessel units is adjusted to the number of boats landing on a daily basis. If in one day the number of landed vessels catching the *penaeid* shrimp is less than 5 (five) vessels, then it will be taken one as a sample. In addition, if there are 5-10 vessels, and then 2 (two) vessels samples are taken. Moreover, the vessel no 1 (one) and no 6 (six) are taken and so on every 5 (five) vessels, as presented by Potier and Sabdotomo [9]. When the shrimp catch is less than 3 (three) kilograms, then all the catch is taken as a sample. Meanwhile, if the shrimp catch is between 3-10 kg, then only 3 kg of the catch is taken as a sample. Moreover, if the catch is above 10 kg, then the sample is taken as much as 20% of the whole catch. The method has been used in a previous study **(8).**

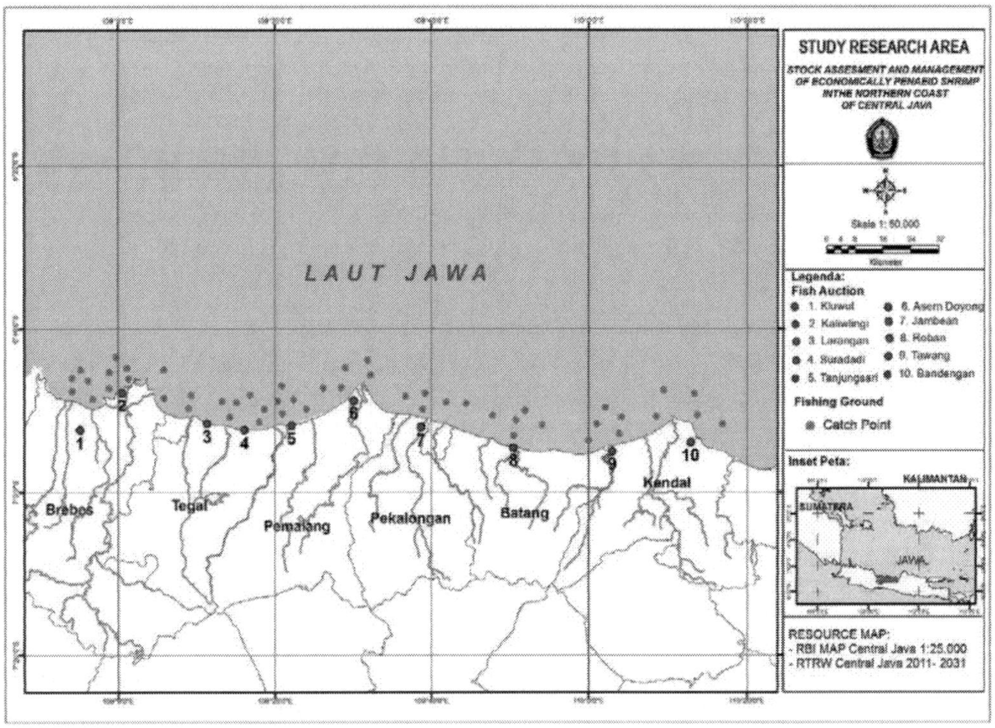

Figure 1. The sampling sites in the northern coastal waters of western Central Java, Indonesia

2.2. Data collection.

The recorded data included sex, carapace length (mm), total length (mm), body weight (gram), and gonad maturity level. Ttotal length and the carapace length were measured using a caliper with 0.5 mm accuracy, from the tip of the carapace (eye base) to the back of the rear carapace. The weight of shrimp and gonad was measured using an electric scale, with 0,05-gram precision. The sex of the shrimp was identified by observing the ventral part of the shrimp (in pleopods and periopods). The maturity level of gonads was identified morphologically by observing the color, shape and size of the gonads, with the help of loops based on [1][10-11] and King [12].

2.3. Data Analysis.

Standard logistic curve method from Spearman-Karber [13] is used to determine the length first capture using danish seine tool (L_c). The method to find out the length first mature gonad by plotting the cumulative frequency percentage of gonad shrimp with the length of the carapace, then calculated using the standard logistic model as follows [14]:

$$Y(\%) = \frac{100}{(1 + ae^{-bX})} \quad (1)$$

Where: Y (%) is the cumulative percentage, a is a constant, b is a slope, and X is the total length. Plotting Y and X will be obtained the logistic curve, where the cut point between the curve with 50% cumulative is the length when 50% of gonad ripe or also called the length of first matured (L_{m50}).

Stock analysis is counted by the result per recruit relative (Y'/R) and biomass per recruit relative (B'/R) from Beverton and Holt [15], as follows:

$$\gamma = \frac{Y}{(RW_\infty)} = E\sum_{n=0}^{3} \frac{Un.(1-c)n}{1+(\frac{nK}{M})(1-E)} \quad (2)$$

Where γ = yield per new addition without unit, E = exploitation rate, M = natural mortality rate, c = ratio of L_c / L_∞, K, L_∞ and t_0 is von Bertalanffy growth parameter, Y = result catch, R = abundance in age group L_c, and Un = is the sumasi coefficient, taken value 1, -3, 3, -1 for n = 0, 1, 2, 3.

To examine the effect of Lc change on results per new addition of reparameterization was made by Beverton and Holt (15) as follows:

$$Y' = \frac{y'}{R} = \frac{Y}{N(t_o)W_\infty} = (1-c)^{M/K}.\gamma \quad (3)$$

To facilitate the calculation, the equation can be written in another way to be:

$$(Y/R)' = E * U^{M/K} * (1 - \frac{3U}{1+m} + \frac{3U^2}{1+2m} + \frac{U^3}{1+3m}) \quad (4)$$

The relative yield (Y') is a function of the rate of exploitation (E), c=(L_c/L_∞) and M/K. The first two parameters, E and c can be controlled, while M/K is only the biological parameters needed in the analysis. This model can be used to determine the optimum combination of the number of fishing attempts (measured by fishing mortality rate, F) and size when first caught (L_c), which will result in maximum sustainable yield. The variables required in this analysis are L_∞ (mm), K (per year), Z (per year), M (per year), F (per year), E, L_c (mm) and t_o (year). The variables are obtained from previous.

3. Results

3.1. The length of first captured (L_{c50}).

The total length of first caught using a danish seine (L_{c50}) is 76,4 mm for male and 63 mm for female, weighing 3,.44 grams (male) and 1,8 grams (female) (Figure 2). On the total length of *M. affinis* shrimp, the shrimp is allegedly aged 102 days or 3,37 months (male), while the female is only 64 days or 2,13 months. Based on these data, it shows that female shrimp is more quickly caught, with smaller size and younger age.

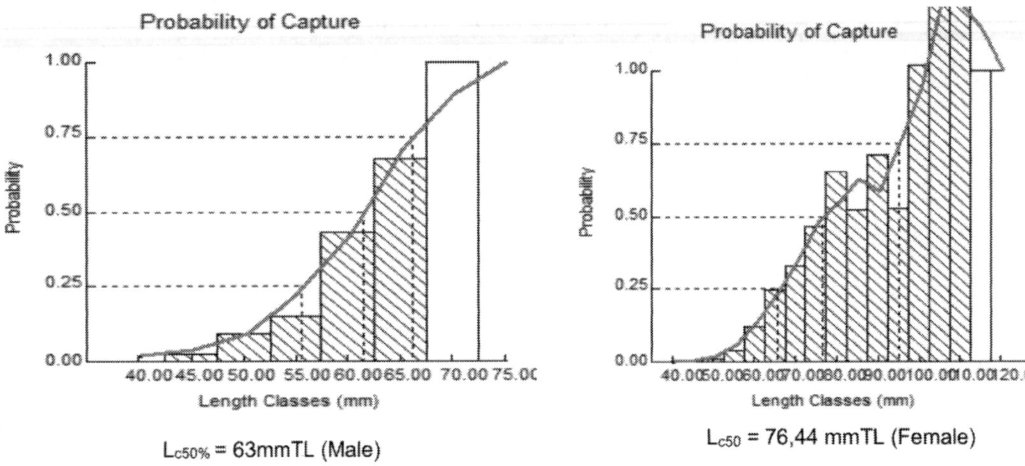

Figure 2. Total length of the first captured (L_{c50}) of *M. affinis* shrimp in the northern coastal waters of western Central Java

3.2. The total length of the first sexual mature (L_{m50}).

The calculation result shows that the total length of first sexual mature *M. affinis* female is 116 mm (Figure 3).

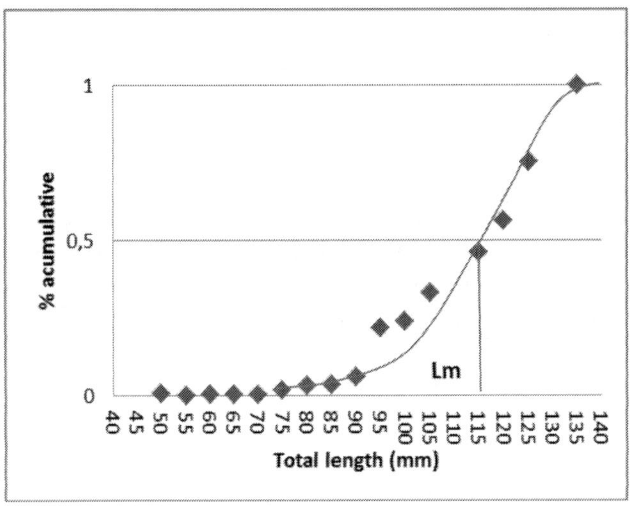

Figure 3. Total length of first sexual mature *M. affinis* (L_{m50}) in the Northern waters of western Central Java

3.3. Stock analysis based on the yield per recruit relative (Y/R)'.

The number of young *M. affinis* shrimp entering fishery on the northern coast of Central Java waters is not known, so the amount of new shrimp addition is assumed using relative yield, i.e. the yield per recruit relative (Y'/R) and biomass per recruit relative (B'/R). Y/R method can be used to determine an optimum combination between the capture effort and the size of first captured (L_c) that will result in maximum sustainable yield (MSY). The variables required in this analysis are presented in Table 1.

Table 1. Variables required for the calculation of (Y'/R) of *M. affinis*

Parameter	Average	
	Male	Female
L_∞ (mm)	168	178.5
K (per year)	1.8	1.91
Z (per year)	6.9	4.6
M (per year)	1.23	1.6
F (per year)	5.68	3.02
E	0.82	0.62
L_c (mm)	76.4	63
t_o (year)	-0.053	-0.049

The total length of the first capture (L_c) of *M. affinis* was 63 mm (female), the maximum exploitation rate (E_{max}) is 0,595, the exploitation rate at $E_{0,1}$ is 0,521 and the rate of exploitation at $E_{0,5}$ (exploitation rate which will result in a biomass reduction of 50%) is 0,352. While male shrimp's Lc = 76,4 mm, with E_{max} = 0,637, $E_{0,1}$ = 0,562 and $E_{0,50}$ = 0,373 (Figure 4).

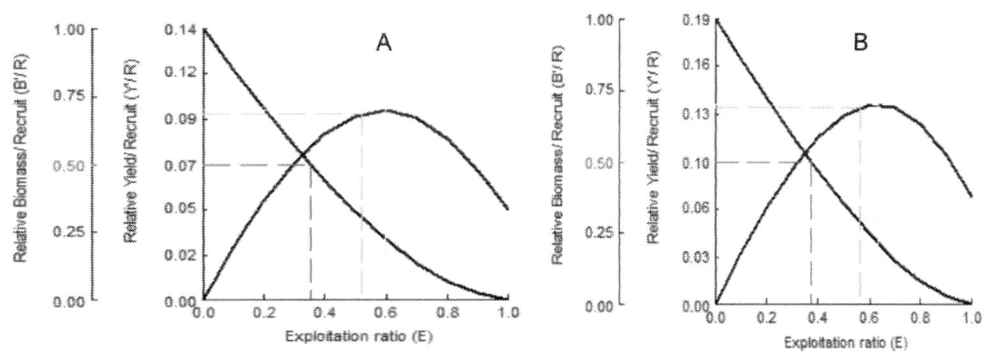

Figure 4. The value of relative yield per recruitment (Y/R') and biomass per relative recruitment (B/R') as a function of exploitation rate of *M. affinis* (A = female, B = male)

If the size of the length of first caught (L_{c50}) is increased in equal size to L_{m50} (the total length of the first mature = 116 mm), then E_{msy} = 0,8, $E_{0,1}$ = 0,73 and $E_{0.5}$ = 0,41. While male shrimp with L_{c50} is increased to 111 mm, then it obtains E_{msy} = 0,79, $E_{0.1}$ = 0,71 and $E_{0.5}$ = 0,41 (Figure 9). The increase in the average captured size will increase Y/R' and B/R'.

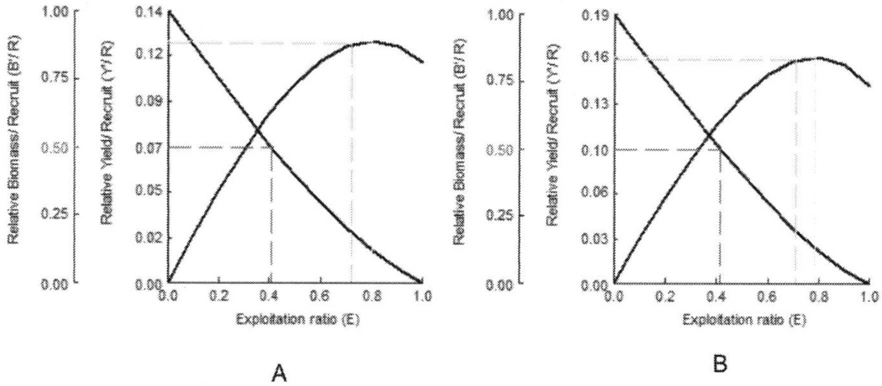

Figure 5. The Plots of yield per relative recruitment (Y/R') and biomass per relative recruitment (B/R') as the function of exploitation rate, with L_c value of female *M. affinis*, is 116 mm (A) and L_c of male *M. affinis* is 111 mm (B).

4. Discussion

The total length first capture of female *M. affinis* was 76,4 mm or at carapace length was 18,7 mm. While total length of male shrimp was 63 mm or at carapace length = 14,58 mm. Based on the length-weight relationship, it is known that at a total length of 74,6 mm and 63 mm, the shrimp weight is 3,26 grams (females) and 1,73 grams (male). So, the shrimp are very small. This indicates that the fishing gear used, i.e. danish seines, is not selective. The results showed that the first sexual matured *M. affinis* at 116 cm (total length) or 30,19 mm (carapace length). Gerami *et al.*, [16] in the Hormozgan Province shrimp fishing grounds obtained L_{m50} was 27,12 cm (carapace length), smaller than that obtained in the northern coast waters of Central Java.

Analysis result on the yield per recruit relative (Y'/R) shows maximum exploitation rate (E_{max}) of female *M. affinis* shrimp is 0,595, the exploitation rate of $E_{0,1}$ is 0,521 and the exploitation rate at $E_{0,5}$ is 0.352. While male shrimp's E_{max} = 0,637; $E_{0,1}$ = 0,562 and $E_{0,5}$ = 0,373. Saputra [17] got E_{msy} value on *M. elegans* shrimp in Segara Anakan waters of Central Java, Indonesia as equal to 0,54, $E_{0,1}$ = 0,47 and $E_{0,5}$ are 0,33. Nurul Amin *et al.* [18] described on *A. intermedius* shrimp in Malaysia obtained E_{max} = 0,65, $E_{0,1}$ = 0,5 and $E_{0,5}$ = 0,34. Saputra and Subiyanto [19] also reported on white shrimp (*P. merguiensis*) in Segara Anakan lagoon Cilacap, Central Java which obtained E_{max} = 0,681, $E_{0,1}$ = 0,552, and $E_{0,5}$ = 0,357. While Wahhab [20] at Yemeni Red Sea waters on male *P.semisulcatus* shrimp obtained greater E_{max} = 0,918; $E_{0,1}$ = 0,804 and $E_{0,5}$ = 0,391, female shrimp E_{max} = 0,855; $E_{0,1}$ = 0,753, and $E_{0,5}$ = 0,38. Compared to these studies, Y'/R of *M. affinis* shrimp in the northern coastal waters of Central Java is similar to that of *M. elegans* in the Segara Anakan lagoon of Central Java, but is smaller than *P.merguiensis* in the same waters and *P semisulcatus* in the waters of the Red Sea, Yemen as well as *A. intermedius* at Coastal Waters of Malacca, Malaysian Peninsula. Therefore, it is necessary to hold exploitation management in order to ensure the sustainability of resources and to maximize the production of *M. affinis* shrimp. Management is conducted by setting the size of the captured shrimp. The size of the shrimp which should be captured at L_c = 116 mm, with 12 grams individual weight. This length is equal to the size of first gonad mature ($L_{m50\%}$). In short, 50% of female shrimp at the length has ever spawned. If it can be conducted *M. affinis* shrimp consequently may have a chance to breed. In addition, the catch result will increase, as Y/R' increases to E_{max} = 0,705, $E_{0,.1}$ = 0,411, $E_{0,5}$ = 0,774.

Setting the mesh size to capture shrimp with certain size and length is generally less effective, because shrimp have many limbs. Therefore, the arrangement to obtain the size of shrimp at a given length is excellent if it is done by prohibiting non-selective fishing gears (such as trawl, danish seine and so on), and replacing them with a selective catching tool (such as trammel net). The banning of drag-pull fishing gear and replacing it with selective fishing tools such as trammel net to catch shrimp will considerably reduce the exploitation rate, increase efficient use of shrimp resources, and increase the fishermen's production and income.

In their previous study, obtained the length-weight relationship for *M. affinis* as $W = 0{,}000007TL^{3{,}02}$. Based on those equation then the total length of shrimp of 74,6 mm have weight of 3,26 grams [8]. Therefore one kilogram contains 307 individuals, and the price is IDR15.000/ kg. Saputra *et al* [8] also obtained the equation of length and age relationship as $Lt = 178{,}5(1\text{-}e^{-1{,}91}$ [t +0,049367]). Based on those equation, the total length of 74,6 mm shrimp is predicted to be 0,295 years old or 3,6 months. Based on the same equation, the total length of shrimp is 116 mm, weight was 12 grams (1 kg contains 83 individuals) of 0,62 years old (7,5 months).

Based on the previous research [8], the equation Nt = N (t + dt) * exp (-Z * (t + (t + dt)), then 307 individuals of shrimp 0.294 years old will be left 182 individuals at the age of 0,62 years (assuming Z = M, due to no catch), weighing 12 grams/indv., meaning shrimp production to 2,18 kg (182 * 12 grams). The shrimp price at that size is IDR 50,000/kg, so the fishermen's receipts to IDR 109,000 indicate that the initial fishermen income of only IDR 15.000 will increase to IDR. 109.000, - (increase 626%) in 4 months later.

5. Conclusion

Based on the result of this research, it can be concluded that total length of the first captured of male *M. affinis* is 63 mm with weight of 1,7 grams at age of 0,265 years old or 3,2 months, while the total length of female shrimps 74,6 mm with weight of 3,26 grams at age of 0,295 years old or 3,6 months. The total length of the first mature of female shrimp is 116 mm with weight of 12 grams at 0,62 years old or 7,5 months. Based on the values of L_c, Y/R' and B/R', the status of *M.affinis* shrimp fishery in the northern coastal waters of Central Java is categorized as overfishing or growth overfishing. To get sustainable yield of *M. affinis* as well as to get profitable business, the total length of the first captured must not be less than 116 mm with 12 grams weight.

Acknowledgement. This research was funded by a grant from PNBP Diponegoro University under Research Institutions and Community Service, scheme No. DIPA-278-21/UN7.5.1/PG/2017.

References

[1] Carpenter K E, Krupp F, Jones D A and Zajonz U 1997 FAO 293 p. 17 color plates.
[2] Chan T Y. Food and Agriculture Organization of the United Nation. Rome. 1998. Available at http://www.fao.org/3/a-w7192e/w7192e13.pdf (accessed on June 20, 2016)
[3] Leena K and Deshmukh V D 2009 *Indian J. Fish.* **56** (1) 1-5
[4] Salman S D, Ali M H and Al-Adhub A H Y 1990 *Hydrobiologia* **196** 79-90.
[5] Mathews C P 1986 *Annu.Res.Rep.Kuwait Inst.Sci.Res.* pp. 6-9.
[6] Mathews C P 1986 *Kuwait-Bull.-Mar.-Sci.* no. 10, pp. 3-36
[7] Mohammed HM. 1995 **18** (2) 38-41. http://www.worldfishcenter.org/Naga/na_2227. pdf
[8] Saputra SW, Solichin A, Taufani W T 2017 *Iranian Journal of Fisheries Sciences* (in prep.).
[9] Potier, M, Sadhotomo B 1995 *Biodynex. Aard/Orstom.* p. 195-214
[10] Lovett D L. 1981 *Occasionally Publication.*Faculty of Fisheries and Marine Science. Universitas Pertanian Malaysia.
[11] Croccos D J and Keer J D 1983 *J. Mar. Biol. Eco.* **69** (1): 37-59.

IOP Conf. Series: Earth and Environmental Science **116** (2018) 012004 doi:10.1088/1755-1315/116/1/012004

[12] King M A Division of Blackwell Science Ltd. London. 1995. Available on: http://as.wiley.com/WileyCDA/WileyTitle/productCd-140515831X.html.
[13] Udupa K S. Statistical method of estimating the size at first maturity in fishes 1986 *Fishbyte*, **4** (2) 8-10.
[14] Pauly D. 1984 *Fish-byte,* **2** (1) 17-19
[15] Beverton R J H and Holt S J 1966 *FAO Fish. tech. Pap,* **38** (Rev-1) 67p.
[16] Gerami, M H, Ghorbani R, Paighmabari S Y, and Momeni M. 2013 **1**(2) 48-54. Available access: https://www.researchgate.net/publication/236623234
[17] Saputra, SW 2010 *Journal of* Coastal *Development* **14** (1) : 18 – 25
[18] Nurul Amin SM, Arshad A, Bujang J S and Siraj S S. *Pertanika J. Trop. Agric. Sci.* **31** (1) 55 - 66.
[19] Saputra SW and Subiyanto 2007 Prosiding Seminar Nasional Perikanan dan Kelautan: Available on: http://eprints.undip.ac.id/48738/1/
[20] Abdul-Wahab M M 2014 Iranian *Journal of Fisheries Sciences* **13** (3) 585-596

IOP Conf. Series: Earth and Environmental Science **116** (2018) 012005 doi:10.1088/1755-1315/116/1/012005

Potential of L-fucose isolated from Brown Seaweeds as Promising Natural Emulsifier compare to Carboxymethyl Cellulose (CMC)

A N Al-Baarri[1,3], A M Legowo[1], Widayat[2,3], S B M Abduh[1], F P Lestari[1*], D Desnasari[1], I P M Santoso[1]

[1]Food Technology Department, Faculty of Animal and Agricultural Sciences, Diponegoro University, Semarang, Indonesia.
[2]Chemical Engineering Department, Faculty of Engineering, Diponegoro University, Semarang, Indonesia
[3]Laboratory of Food Technology, UPT Integrated Laboratory, Diponegoro University, Semarang, Indonesia

E-mail: albari@undip.ac.id

Abstract. L-fucose has been understood as sulfated polysaccharides and it could be extracted and fractionated from brown algae. These polysaccharides contains carbohydrate, sulfate, and protein that may be used as emulsifier. This research was aimed to study the emulsification properties of L-fucose through the determination of total dissolved solids (TDS), color CIE L*a*b* and stability of oil-in-water emulsion. As much as 0.5% of high concentrated L-fucose and 0.5% of carboxymethyl cellulose (CMC) were used as emulsifier in a 10% (v/v) oil-in-water (O/W) emulsion. The emulsifier was added to O/W emulsions and then heated at 72°C. Result of stability emulsion and TDS showed that L-fucose was comparable to the CMC but remarkable changed the color of O/W emulsion. Heating process significantly reduced the stability O/W emulsion when L-fucose was applied. As conclusion, L-fucose might be used as natural emulsifier in O/W emulsion but in the low heat treatment of food processing. This study may provide valuable information for utilizing natural emulsifier from abundant resources from nature.

Keywords: L-fucose, CMC, O/W emulsion, natural emulsifier, total dissolved solid.

1. Introduction

Fucoidans, the family of sulfated heteropolysaccharides, built up essentially of α-1,3-L-fucose. Its constituted more than 40% of the component monosaccharaides and classified as sugars that present abundantly in brown algae [14]. From the previous study, brown seaweeds possess great potential in biomedical issued due to their bioactive component which generally composed of mannose, glucoronic acid and galactose residues [6]. Recently reports have shown that these bioactive component use for multiple pharmacological activities such as anticancer, antitumor, antioxidant, anti-inflammatory, anticoagulant and immune system boosting properties [15]. Considering to the benefits that is mostly exposed the functional properties, it will be interesting if this component was exposed base on their physical properties.

Oil-in-water (O/W) emulsions system are basically containing two immiscible liquids, one dispersed as droplet (oil) throughout the other (water). These defined by the volume ratio and the composition of oil and aqueous phase and the nature or synthetic emulsifiers. Its usually used in food industry to increase the retention and stability of active compounds. Generally, emulsions are prepared with surface active agents or amphiphilic polymers which decrease interfacial tension when absorbing at the O/W interface [10]. L-fucose which extracted from brown seaweed cell wall are potential sourced of soluble and insoluble dietary fibres. These compounds exhibits potential as water holding capacity and demonstrate the ability to increase viscosity, gels forming and act as emulsifier [7]. As the current demand of consumers prefer to healthy and natural food product, L-fucose has led as great potential of natural food additive. L-fucose are potentially used as natural emulsifier compare to carboxymethyl cellulose (CMC). These polysaccharides contains carbohydrate, sulfate, fats, fibres and protein that might be reacted easily with oil and other aqueous substance [13].

CMC is formed from cellulose after heating with alkali, and the mixture will react with monochloroacetic acid leading to an alteration of the glycopyranose by esterification of the hydroxyl groups with methylcarboxyl groups [8]. Nowadays, CMC has widespread application due to its good properties in binding, thickening and stabilizing properties. These emulsifier is often chosen as active agent instead of other gum or other synthetic emulsifier such as Tween 80. In general, CMC has a linear structure composed of a backbone of β-1-4-glucopyranose that will holds insoluble microcrystalline cellulose in a network with other hydrocolloid component [2].

This study was examined the potential of using L-fucose as natural emulsifier in O/W emulsions. This was aimed to determine the emulsification properties of L-fucose through the determination of total dissolved solids (TDS), color CIE L*a*b* and stability of oil-in-water emulsion comparing with a synthetic (CMC) emulsifier to establish the influence of physical properties during heating and non-heating treatment.

2. Materials and Methods

2.1. Chemicals and reagents

L-fucose was extracted from brown seaweed and harvested from Jepara beach, Central Java, Indonesia. Palm oil "Bimoli" was purchased from modern store in Semarang, Indonesia. Acetone (99%) was ordered from Jedang, Co. Ltd., Indonesia. Aquadest and Sodium Chloride powder was obtained from UPT Integrated Laboratory, Diponegoro University, Semarang, Indonesia.

2.2. Extraction of Brown Seaweed

The isolation and separation of water-soluble polysaccharides were carried out using the method from previous researcher [14] with some modifications. The brown seaweeds were stored in laboratory and dried at 50°C under slow air circulation in oven, ground using a grinder and incubated with acetone to eliminate lipids and pigments. About 50 g of powdered algae was suspended with five volumes of 0.25 M NaCl and pH was adjusted to 8.0 with 1 M NaOH. Peptone enzyme was added as much as 1 g and let them for 24 h. After incubation for 24 h at 60°C, the mixture was filtered through Whatman No.1 and precipitated with increasing amounts of ice-cold acetone (600 ml) under gentle agitation. Precipitates formed were collected by centrifugation at 10,000 rpm for 10 min and dried under freeze condition.

2.3. Preparation of Oil-in-Water Emulsions
The sample was prepared in tube using 0.5% L-fucose and 0.5% carboxymethyl cellulose (CMC) into different formulas using a ratio of 1:10 (v/v) for emulsifiers and palm oil. Water-soluble (O/W) liquid emulsion was prepared with 5% palm oil and 0.5% natural emulsifier of L-fucose or a synthetic emulsifier of CMC. The mixture then was added with aquadest into the maximum volume of 15 ml. The emulsion was heated at 72°C and unheated. The final mixture then was shaken using vortex that was set to 1000rpm for 1 min. The sample mixture were labelled as fucan+water+oil and CMC+water+oil for the mixture with the addition of L-fucose and CMC, respectively. The mark "H" was used to determine the heating treatment.

2.4. TDS (Total Dissolved Solids) measurement
The total dissolved solids was examined with 8200 (Ezdo, Indonesia) instrument. Firstly, calibrations the instrument was done using standard solution 1413 μS/cm. The solution was stirred gently and let them until the display showed stable then press call for 3 sec. and flashing with 1413 μS/cm. The instrument was ready to use by dipping the cell into the sample solution and data were taken for 3 times then was calculated the average±SD.

2.5. Colour analysis measurement
The appearance of solution colour was measured using a Digital Color Meter that was done previously [12]. The instrument was calibrated on the CIE LAB color space system using a white tile (Dc: L=97.79, a=−0.11, b=2.69). The L* value represents lightness and a* and b* values represents redness and yellowness, respectively. Colour measurements of O/W emulsions using CMC and L-fucose were recorded base on heating treatment.

2.6. Stability of Oil-in-Water Emulsions.
Stability of oil-in-water emulsions was measured with micrometre screw. After centrifugation, the precipitation of emulsion sample was measured to obtain the thickeness of separated upper part of emulsion.

3. Result and Discussion

3.1. Total Dissolved Solids
Total dissolved solids analysis using 8200 (Ezdo, Indonesia) was measured 3 times and the average calculation used as the data. The result indicates that TDS in O/W emulsion with CMC without heating treatment was less than L-fucose (Fig. 1). Those might be influenced by the charactheristic of CMC to improve water solubility and it will maximumlly be happened by heating the cellulose with alkali [8]. This statement was in agreement with the study of other researcher [5] that the solids of matrix from CMC would dissolve in hot water resulting a good heat barrier even at low consentration. The best result of heating treatment was emulsion with CMC as emulsifier as shown in Figure 1 that CMC resulted TDS higher than L-fucose when heating was applied. This indicated that CMC was work well in hot water rather than L-fucose. It might be explained because hydrocolloids from brown seaweed were generally acted as physical functions in stabilizing emulsions, viscose behaviour, gelation, suspension and foams, and control of crystal growth and all characteristics would give adverse effects during heating [4].

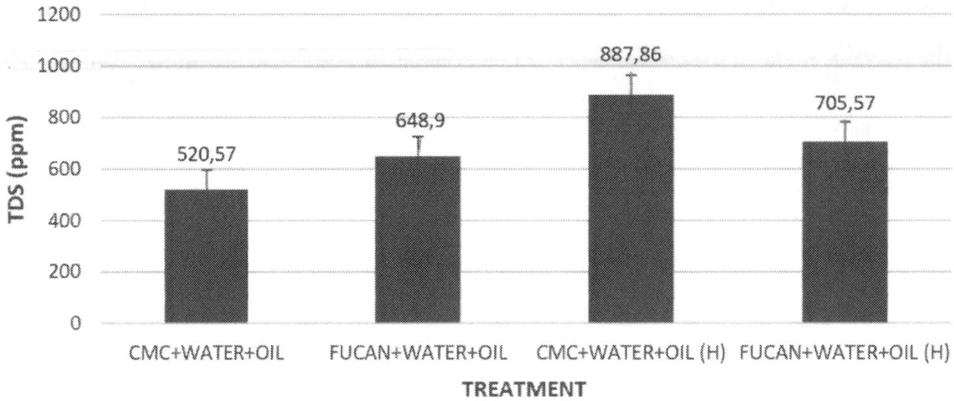

Figure 1. Total dissolved solution of O/W emulsions with 0.5% L-fucose or 0.5% of CMC that was prepared with heating and non-heating treatment. The label "H" at the sample means heated sample.

3.2. Colour Analysis

The Lab color model consists of three components (lightness or L*) starting from 0 (black) to 100 (white) and component a* represents the red-green level with the level (+60 red, -60 green) and component b* represents blue to yellow that are two chromatic components with a range of values from -120 to +120 [11]. The emulsion with heat treatment was applied at 72°C. The heated and non-heated sample were done to determine the effect of heating on the Lab color model. The results show remarkable changes in L-fucose color before and after heating. This resulted that by heating treatment, the color of L-fucose was more clear and closed to CMC (Fig. 2). Heat treatment would naturally degrade colour of L-fucose by destroying the glycosidic bonds. While some previous study reported that the extraction of brown seaweeds usually happened in low temperature to keep their bioactive component and the colour turned up into dark resulting the problem during the application of L-fucose. But after heating treatment, lightness of L-fucose was almost same value as CMC. This suggests that L-fucose was potentially used as additive in food as a natural emulsifier without changing colour. This is in line with statements other researcher [3] which suggest that the lightness was important factor in oil-in-water emulsions to show better results when high temperatures was applied, suggesting the possible application in pasteurization process.

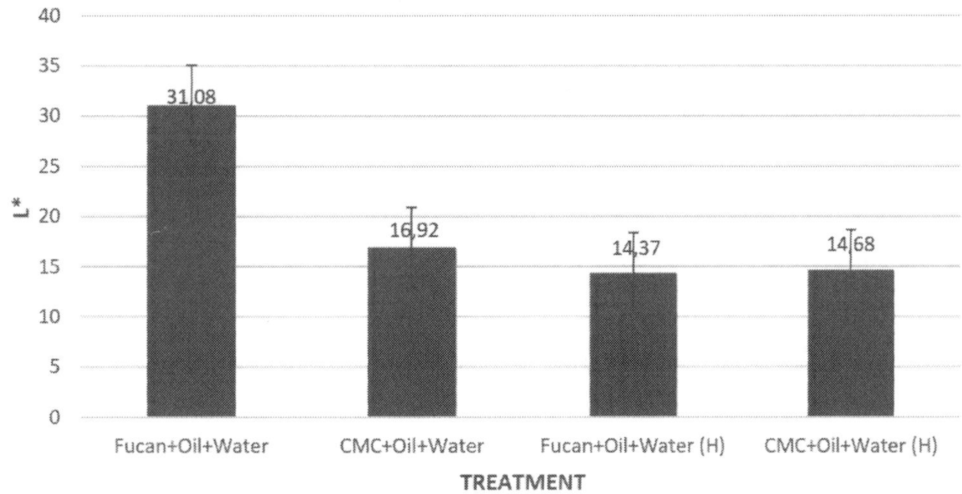

Figure 2. Lightness of O/W emulsions with 0.5% L-fucose or 0.5% of CMC prepared with heating and non-heating treatment. The label "H" at the sample means heated sample.

3.3. *Stability of Oil-in-Water Emulsions*

In the emulsion, the separation of emulsion may appear the as a time of storage. Figure 3 showed the separation in upper part of emulsion. In the sample with unheated treatment, CMC remained 0.2 mm while L-fucose remained 0.23 mm in upper part. In the heated treatment, CMC remained 0.089 mm while L-fucose remained 0.114 mm. The separation of emulsion may be explained by the natural appearance of oil in a water. The stability of emulsion with CMC and L-fucose were comparable indicating the potent use as emulsifier of L-fucose. The speed of stirring may also affect the stability of emulsion. The increase in stability of emulsion might be achieved with higher stirring speed since the stirring was able to decrease particle size leading to an increase in the interfacial surface. Hence, utilization of high stirring intensity could stabilize the system [9]. The increasing of temperature often accelerates emulsion breaking [1] resulting the high stability of emulsion.

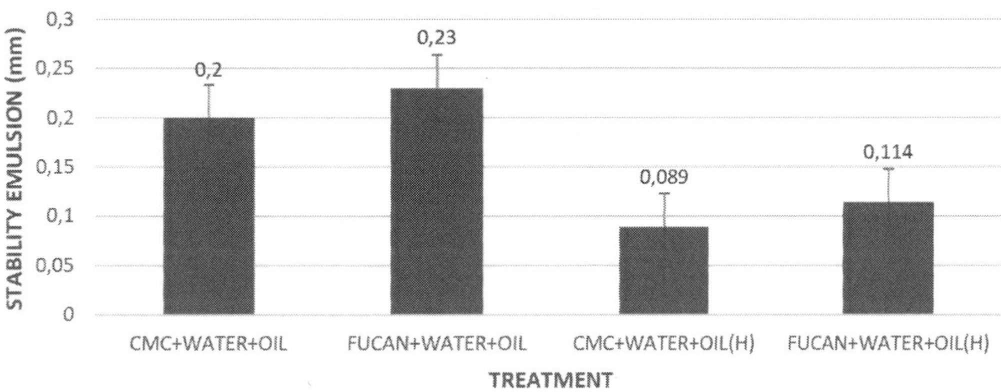

Figure 3. Stability of O/W emulsions with 0.5% L-fucose or 0.5% of CMC prepared with heating and non-heating treatment. The label "H" at the sample means heated sample.

4. Conclusion

The result of stability emulsion, TDS, and lightness showed that L-fucose was comparable to the CMC. Heating process might reduce the stability O/W emulsion with L-fucose. Thus, L-fucose might be used as natural emulsifier in O/W emulsion. This study may provide valuable information for utilizing natural emulsifier from abundant resources from nature.

Acknowledgment

The authors are thankful to Ministry of Research, Technology, and Higher Education of Republic Indonesia for providing the full support in this research.

References

[1] Bendjaballah M, JP Canselier and R Oumeddour 2010 *J. of Dispersion Science and Technology* **31**(9): 1260-1272
[2] Cai Z, J Wu, B Du and H Zhang 2017 *Journal of Food Hydrocolloids* 1-8
[3] Chanamai R and D J McClements 2000 *J. Colloid and Interface Science* **225** 214-218
[4] Chapman V 2012 *J. Springer Science and Business Media* **118** (3) 823-829.
[5] Chitprasert P, P Sudsai and A Rodklongtan 2012 *Journal of Carbohydrate Polymers* **90** 78-86
[6] Dore C M P G, M G C F Alves, L S E P Will, T G Costa, D A Sabry, L A R S Rego, C M Accardo, H A O Rocha, L G A Filgueira and E L Leite 2013 *J. Car. Pol.* **91** 467-475
[7] Elleuch M, D Bedigian, O Roiseux, S Besbes, C Blecker and H Attia 2011 *J. Food Chemistry* **124** (2) 411-421.
[8] Gibis M, V Schuh and J Weiss 2015 *Journal of Food Hydrocolloids* **45** 236-246
[9] Juntawong S, J Charoenteeraboon, G Chansiri, T Tuntarawongso, P Katewongsa and T Phaechamud 2010 *Research Journal of Pharmaceutical, Biological and Chemical Sciences* **1** (2) 360-365.
[10] Kaci M, E Arab-Tehrany, I Desjardins, S Banon-Desobry and S Desobry 2017 *Journal of Food Enginering* **194** 109-118
[11] Mendoza F, P Detmek and J M Aguilera 2007 *J. Food Res.* **40** 1146-1154
[12] Moroney N C, M N O'Grady, J V O'Doherty and J P Kerry 2012 *Journal of Meat Science* **92** 423-429 .
[13] Roohinejad S, M Koubaa, F J Barba, S Saljoughian, M Amid and R Greiner 2016 *Journal of Food Research International* 1-18
[14] Sellimi S, N Kadri, V Barragan-Montero, H Laouer, M Hajji and M Nasri 2014 *Int J. Biol Macromol.* **66** 281-288
[15] Senthilkumar K, P Mnivasagan, J Venkatesan and S Kim 2013 *J. Biological Macromolecules* **60** 366-374

Copigmentation Of Anthocyanin Extract of Purple Sweet Potatoes (Ipomea Batatas L.) Using Ferulic Acid And Tannic Acid

I Susanti, H Wijaya, F Hasanah, and S Heryani

Center for Agro-based Industry, Jln. Ir. H. Juanda No.11 Bogor 16122

email: irma.naura@gmail.com

Abstract. Copigmentation is one of the methods to improve the color stability and intensity of anthocyanin extract. This study aimed to do the copigmentation of the anthocyanin extract of purple sweet potato using ferulic acid and tannic acid. The anthocyanin extraction was conducted with distilled water at pH 7 and pH 2 while the copigmentation was conducted by varying the concentration of ferulic acid and tannic acid. The results showed that best anthocyanin extraction method of the purple sweet potato was using distilled water at pH 2. The yield of freeze dried anthocyanin on the extraction with distilled water at pH 2 was 1710 ppm, while the yield when using distilled water at pH 7 was 888 ppm. Ferulic acid and tannic acid can be used for the copigmentation of anthocyanin extract of purple sweet potato by observing the maximum wavelength shift (bathochromic effect, $\Delta\lambda max$) and increase of color intensity (hyperchromic effect, $\Delta Amax$). The bathochromic effect of ferulic acid began to occur at a concentration of 0.01M, while the hypochromic effect on tannic acid occurred at a concentration of 0.005M and remained up to a concentration of 0.02M. The best copigmentation concentration of ferulic acid was 0.015M, while tannic acid was 0.02M. The use of tannic acid 0.02M is recommended compared to ferulic acid 0.015M because with the same bathocromic effect ($\Delta\lambda maks$ = 3,9) results hyperchromic effect (ΔA = 0,258) higher by tannic acid.
Keywords : copigmentation, anthocyanin, purple sweet potato, ferulic acid, tannic acid

1. Introduction

Food colorant is a natural ingredient as well as a chemical that is added to the food in order to give a particular appearance or an attractive color. The use of colorant has been significantly commercialized since more than 25 years ago. The role of food coloring is quite important for the food industry because of its role to improve the appearance of food. Colorant that are widely used in the food industry so far are synthetic dyes. The use of synthetic colorant in food products is growing rapidly both in large industries and in the home industry, this is because the color produced by synthetic colorant more intens and stable. However, the use of excessive synthetic colorant is feared will cause health problems.

Colorants are widely used in food products to make the desired product color variations, to replace the color lost during processing and during storage, to provide color and appearance to products that are naturally colorless (jelly, cake, cake, beverage, etc.) provide identification of products consumed through color appearance, and improve the natural appearance of the product.

Natural colorant are widely used derived from plants such as suji leaves, cocoa fruit, turmeric, red peppers, carrots, caramel, brown sugar, grape wine, tomatoes, purple sweet potatoes and others. In Indonesia, purple sweet potatoes are common used as traditional food and snacks. the pigment in purple sweet potatoes known are anthocyanins which give red, purple and blue pigments common to plant species. Anthocyanins are amphoteric compounds, which have the ability to react with both acid and alkaline. In acid medium anthocyanins are red, and on alkali media turns into purple and blue.[1]

Anthocyanins may replace the use of synthetic colorant of rhodamine B, carmoisin, and amaranth as a red colorant in food products. Natural colorant preparations are needed so it can be an alternative for the food industry to improve the appearance and increase the color intensity of food products by minimizing the risk of health problems.

The natural anthocyanin color substance is susceptible to color degradation which results in the color becoming faded. In food matrix, Anthocyanins are highly unstable molecules. several factors that affect the stability of the color are pH, temperature, anthocyanin concentration, oxygen, light, enzymes, and other accompanying substances. Antocyanin degradation can be reduced by the presence of copigmentation. Copigment are defined as a colorless or colorless but very thin, generally yellowish, or colored molecule naturally present in plant cells around anthocyanins. Given the interaction with the copolymer compound, the intensity of the pigment color will increase (hyperchromic) or a maximum wavelength shift (batochromic) occurs [2][3][4][5]. The result obtained from copigmentation is colorantb expected that it can be use on a wider pH range. This study aimed to do the copigmentation of the anthocyanin extract of purple sweet potato using ferulic acid and tannic acid.

2. Materials and Methods

2.1. Materials

The research materials used were local purple sweet potatoes obtained from Bogor area, aquades, pH paper, tannic acid, fumeric acid, DMSO, amonium acetat, and citric acid.

2.2. Methods

2.2.1 anthocyanin extract. The sorted and washed sweet potatoes are cut into sizes about 3 cm x 0.5 cm x 1 cm. The sweet potato pieces are then blanched at 90 ° C for 10 minutes. After that it is then heated by adding 1: 2 (w/v) water at 70 °C for 1 hour. The anthocyanin extraction was conducted with distilled water at pH 7 and pH 2 , for pH 2 condition the water used has been added citric acid so that the pH of water becomes 2.The sweet potatoes is then mashed and filtered to obtain the extract. The extracts obtained are then dried with freeze drier and ready for copigmentation processed.

*2.2.2 Copigmentation.*Tannic acid and fumeric acid variations as copigment has concentration of 0.005 M; 0.010 M; 0.015 M; 0.02 M; 0.03 M; and 0.04 M. Tannic acid (MW 1701,2) or ferulic acid (MW 194,18) dissolve in 10% DMSO of 0,02 M amonium acetate solution. This means if the tannic acid copigment will be made 0.02 M as much 250 mL then it should be weighed as much 8,5059 g while ferulic acid 0,9709 g in 0,3854 g ammonium acetate (dissolved in 250 mL aquades) and 25 mL DMSO

3. Results and Discussion

Purple sweet potatoes is one type of sweet potato that is found in Indonesia in addition to the white, yellow, and red [6]. The plant is rich in anthocyanin content. Some research has been carried out about the anthocyanin from purple sweet potato used as the functional compound due to its obvious anticancerogenic, antioxidant, antimutation, and antineoplastic activities [7][8][9][10][11].The content of anthocyanin in sweet potatoes is also potential as a natural colorant, for food and beverages. Figure 1 showed the raw material that used in this reserearch.

Figure 1. Purple sweet potatoes

In fruits or vegetables, anthocyanin pigment is generally located on cells near the surface, it is expected that hot water can denature the cell membrane of the plant then dissolve the anthocyanin pigment out of the cell. Solvent extraction is based on the solubility of the component against other components in the mixture to extracted the pigment from a complex matrix [12]. The anthocyanin pigments are soluble in water because they are equally polar. Water has a high degree of polarity. This is indicated by the polarisability value of water molecules in an electric field or dielectric constant (Ɛ ') of 78.50 at 20 ° C. The value is greater than other solvents such as ethanol, methanol, hexane, and acetone. Of The two pH treatments performed there were differences in the results obtained from the sweet potato extract produced. The results showed in Table 1.

Table 1. The results of anthocyanin extract process

Sample	**Color appearance**	**Anthocyanin (ppm)**
Raw purple sweet potatoes	Dark purple	235,40
extraction with distilled water at pH 7	Browning	888
extraction with distilled water at pH 2	Bright red purple	1710,6

The results showed that best anthocyanin extraction method of the purple sweet potato was using distilled water at pH 2. It is seen from the appearance of the color and the resulting anthocyanin levels Chemically the anthocyanin is a derivative of a single aromatic structure, cyanidine, and all formed from cyanidine pigments by addition or subtraction of hydroxyl groups, methylation and glycosylation [13]. Anthocyanins are amphoteric compounds, which have the ability to react with both acid and alkaline. In red antosianin acid medium, and on basic media turns into purple and blue (1).

The color of the purple sweet potato antosianin is highly dependent on the pH of the solution. Figure 2. shows the color pattern of purple sweet potato extract from pH 2-12. The anthocyanin color of purple sweetpotato is influenced by the pH of the solution, ie, respectively red, purple, and blue under acidic, neutral and basic conditions [14].

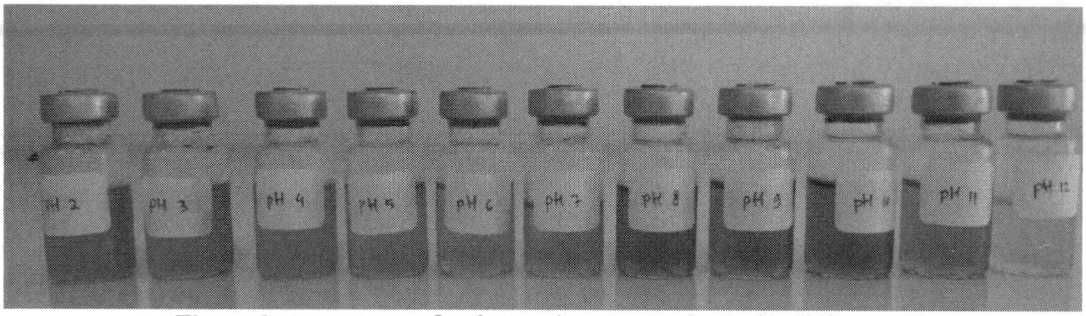

Figure 2. Appearance of anthocyanin extract color under different pH

The color of the anthocyanin extract tends to change as the pH 2-12 increases from red, purple, blue, green to yellow. Purple sweet potato extract at strong acid pH 2-3 is red, in weak acid pH 4-6 pink, pH 7 is blue, at pH of weak base 8-10 is green and pH 12 is yellow. Results of research conducted by Cevallos-Casals and Cisneros Zevallos [15], showed the color of red sweet potato anthocyanin extract at pH 0.9 to pH 4 red, at pH 5 to 7 green.

A study conducted by Nikijuluw 2013 on anthocyanin flowers of pearls at pH 1 and ph 2 is red, at pH 3 purple, at pH 4 samapai pH 9 is blue, at pH 10 to pH 12 is green and at pH 13 and pH 14 is colored yellow [16]. Differences in anthocyanin color in purple sweet potato, red sweet potato, flower of teleng caused by difference antosianin compilation compound. The main anthocyanin composer in purple sweet potato is a cyanidine compound. The color change of anthocyanin extract in purple sweet potato along with pH change is caused by a change of anthocyanin structure. The changes of anthocyanin structure are from flavilum cation (red) to pseudobasa carbinol (colorless), quinonoidal (blue), quinonide anion (blue) or kalkon (yellow).

In the process of copigmentation, it is necessary to optimize the use of copigments which will be used for the process of copigmentation. The copigment used for copigmentation is an organic acid. Copigment concentrations must be greater than those that will be digitized. In this study, the copolymers used are synthetic chemicals, which have higher purity than the crude extracts obtained from nature (eg tea leaves).Maximum wavelength shift (batokhromik effect) and increased color intensity (hyperchromic effect) are a cooling effect in the anthocyanin co-opment process. These two effects indicate the success of the copigmentation process and the increased intensity of color after added copigments. The copigmentation process in this study is to use tannic acid and ferulic acid which is a commercial organic acid that can function as copigment. The results of optimization of copigmentation with tannic acid and ferulic acid shown in Table 2.

Table 2. The results of optimization of copigmentation with tannic acid and ferulic acid

No	Copygment	Concentration (M)	λ max (nm)	Δ λ max	A	ΔA max
1	Blanko	0	527,9	0	0,382	0
2	Ferulic acid	0.005	527,9	0	0,398	0,016
3	Ferulic acid	0.01	531,0	3,9	0,430	0,048
4	Ferulic acid	0.015	531,0	3,9	0,489	0,107
5	Ferulic acid	0.02	532,0	4,9	0,441	0,059
6	Ferulic acid	0.03	531,0	3,9	0,437	0,055
7	Tannic acid	0.005	531,0	3,9	0,522	0,140
8	Tannic acid	0.01	531,0	3,9	0,578	0,196
9	Tannic acid	0.015	531,0	3,9	0,610	0,228
10	Tannic acid	0.02	531,0	3,9	0,640	0,258

11	Tannic acid	0.03	527,9	0	0,694	0,312
12	Tannic acid	0.04	527,9	0	0,737	0,355

Table 2 shows the result of anthocyanin purple sweet potato antopian coupling with ferulic acid and tannic acid. The concentration of ferulic acid used was 0.005M-0.03M and the tannic acid concentration used was 0.005M-0.04M. The reaction of copigmentation is influenced by pH, temperature, and concentration [17]. At low pH the antocyanine molecule is a red flavinium cation, while a higher pH will be a pale-colored carbinol pseudobase. Increased temperatures will lead to increasingly unstable co-opmentation. This happens because of partial damage to the hydrogen bond. The added copolymer concentration will affect the copigmentation process. The amount of copolymers added should be more than the anthocyanin. The concentration of pigment and copolymer ratio is expressed in molar.

The use of both copolymers observed the effect of maximum wavelength shift (batochromic effect, $\Delta\lambda max$) and increased color intensity (hyperchromik effect, $\Delta Amax$). The cytochromic effect on ferulic acid starts to occur at a concentration of 0.01M and the hypochromic effect on tanic acid occurs at a concentration of 0.005 and remains at a concentration of 0.02M. In the tannic acid copolymer, increased concentration results in an increase in hyperchromic effect while in ferulic acid there is a maximum hyperchromic effect at a concentration of 0.015 M. Based on these results, the best concentration of ferulic acid copolymer is 0.015M and the tannic acid copolymer is at a concentration of 0.02M. The use of tannic acid 0.02M is recommended compared to ferulic acid 0.015M because with the bathocromic effect ($\Delta\lambda maks$ = 3,9) results hyperchromic effect (ΔA = 0,258) higher by tannic acid. Figure 3 show the result of copigmentation.

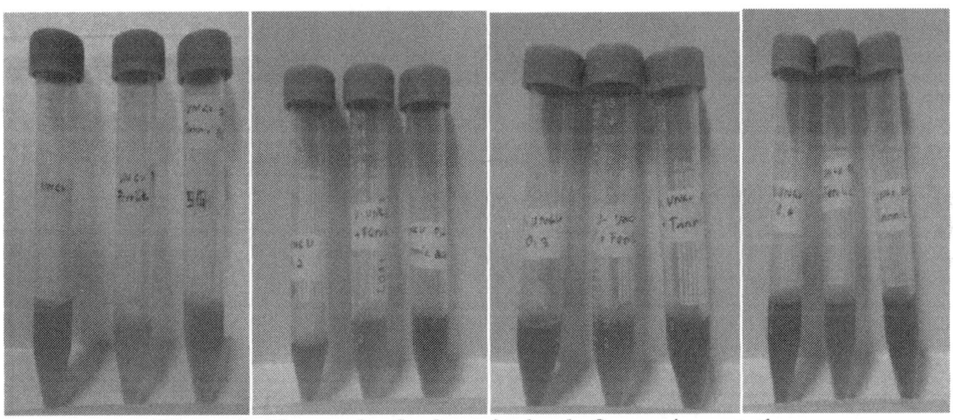

Figure 3. The result of colors obtained after copigmentation

The phenomenon shown when the process of copigmentation occurs is the effect of the batochromic ($\Delta\lambda max$) that is the absorption of the maximum wavelength (λmax) and in the visible distance there is an increase in bluing effect, such as anthocyanin color changing from red to bluish red or hyperchromatic effect ΔA) is the intensity of the anthocyanin color seen from the increased of absorbance value.

4. Conclusion

The use of tannic acid and ferulic acid which is added to the anthocyanin extract indicating the occurrence of the copigmentation process. The use of 0.02M tannic acid and 0.015M ferulic acid produce hyperchromic effect and bahtochromic effect. The use of tannic acid 0.02M is recommended compared to ferulic acid 0.015M because with the same bathocromic effect ($\Delta\lambda maks$ = 3,9) results hyperchromic effect (ΔA = 0,258) higher by tannic acid.

References

[1] Man, J. M. de 1997 Food Chemistry. Institut Teknologi Bandung. Bandung
[2] Asen, S, Stewart, R.N., and Norris, K.H 1972 Copigmentation of anthocyanins in plant tissues and its effect on color. In: Rein, M. 2005. Copigmentation Reactions and Color Stability of Berry Anthocyanins. Helsinky University. Helsinky.
[3] Bolivar AC, Luis 2004 *J Food Chem* **86** : 69-77
[4] Brouilard, R. and O. Dangel 1994 Food Chemistry **51**: 365-371.
[5] Hendry, G. A. F 1999 Plant pigments. in P. J. Lea and R. C. Leegood (eds.). Plant Biochemistry and Molecular Biology. John Wiley and Sons Ltd, New York. p 219-234.
[6] Lingga P 1995. Bertanam Umbi-Umbian. PT. Penebar Swadaya. Jakarta.
[7] Cevallos-Casals BA, Cisneros-Zevallos LA 2002 *Acta Horticulturae* **583**: 195-203.
[8] Z. Fang and Y. Ni, 2001 *Guangzhou Food Science and Technology*, vol. **17**, no. 3, pp. 60–62.
[9] Katsube N, Iwashita K, Tsushida T, Yamaki K, Kobori M 2003 *Journal of Agricultural and Food Chemistry* 51 : 68-75.
[10] Y. Liu, L. Zhu, Z. Han et al 2005 *Acta Acadimae Qingdao Umvesitatis*, vol. **41**, no. 1, pp. 46–51.
[11] X. Meng, X. Cu, F. Wang et al 2005 *Medical Journal of Qilu*, vol. **20**, no. 5, pp. 394–399
[12] Hendry, G. A. F. 1999 Plant pigments. in P. J. Lea and R. C. Leegood (eds.). Plant Biochemistry and Molecular Biology. John Wiley and Sons Ltd, New York. p 219-234.
[13] Harborne J B., Williams C.A. 2001 Nat. Prod. Rep. 18 : 310-333.
[14] Suda I, Oki T, Masuda M, Kobayashi M, Nishiba Y, Furuta S 2003 JARQ 37 167–173
[15] Cevallos-Casals BA, Cisneros-Zevallos LA 2004 *Food Chemistry* 86: 68-77
[16] Nikijuluw, Corazon and Andarwulan, Nuri 2013 Color Characteristic of Butterfly Pea (Clitoria ternatea L.) Anthocyanin Extracts and Brilliant Blue. Bogor Agricultural University
[17] Brouilard, R. and O. Dangel 1994 Food Chemistry 51: 365-371.

Growth Performance of Catfish (Clarias Gariepinus Burchell, 1822) Cultured in High Density on the Biofloc System

Fajar Basuki*, Tristiana Yuniarti, DickyHarwanto, TitikSusilowati

Aquaculture Department, Faculty of Fisheries and Marine Science, Diponegoro University, Jl. Prof. H. Soedharto SH., Tembalang 50275 Semarang Indonesia

Email: fajar.basuki@live.undip.ac.id

Abstract. Start your The fish farmers and researchers have made efforts to increase catfish production through high levels of dispersal and application of biofloc systems. This system, in addition to increase production, can also predict when the harvest is done, control the quality of water causing death so that the Survival Rate (SR) is high. The purpose of this study is to determine the growth performance and predict the harvest weight and catfish survival rate (SR) that are cultured with different high density on the biofloc system. The research method used is experimental with test fish in the form of catfish seeds weighing of 8 ± 0.2 gram/fish with each density as treatment: P1. Density of 1,100 fish/m3; P2. Density of 1,200 fish/m3 and P3. Density of 1,300 fish/m3. Pond: The pond used is made of 3m in diameter of tarpaulin, equipped with a 6mm diameter of iron frame. Total volume of water in each pond is 3 m3. Catfish seeds have been adapted and put into the pond at random with the length of cultivation period is 4 weeks. Growth performance analysis uses the SPSS regression statistic series 20. The research results showed that the growth performance for 4 weeks showed that P1 generates the regression equation Y1 = 6.46 +7.81 X with r2 = 0.971 means that if the harvest is done at week 12th then the weight of catfish has reached 100 grams; P2 generates the regression equation Y2 = 6.53 + 6.0 X den r2 = 0.969 means if the harvest is done at week 12th then the weight of new catfish has reached 78.53 grams and P3 generates the regression equation Y3 = 6.71 + 5.94 X with r2 = 0.976 means if harvest is done on week 12th then the weight of new catfish has reached 77.99 grams. Survival rate at P1 = 97.73%; P2 = 95.91% and P3 = 94.55% while the oxygenated water quality is more than 3.5 ppm; pH ranged from 6.7-7 and ammonia content at week 3 is P1 = 0.04 mg/l; P2 = 0.13 mg/l and P3 = 0.27 mg/l.

Keyword: Catfish Growth Performance ,density, biofloc

1. Introduction

As the population increases, there will be an increase in the number of food needs, for example the prediction of the population in Central Java by 2015 is as much as 16,750,900 people and in 2017 will increase to 16,988,100 people [1]. Fish as one of the foods that contain high protein can be a nutritious food supplier needed by the population, therefore fish production also needs to be improved. The production of catfish Clariasgariepinus (Burchell 1822) in Central Java is also continuously improved, the production of catfish in Central Java in 2012 reached 62,686 tons and in 2013 reached 75,236 tons, increased by 20.02% [2].

Many people cultivate catfish is caused by several things, including: biological reasons those are fast in growing, easy to cultivate, easy to consume artificial food, tend to resist disease; Social reasons i.e. good market prices and physical reasons that are resistant to environmental changes [3]. Cultivation of catfish with biofloc system in closed recirculation container is one techniques that can be done to increase catfish production, in accordance with opinion [4] which states that biofloc system allows cultivation of catfish with high density so that its production is increasing. Biofloc is a collection of various organisms including bacteria, fungi, protozoa or algae gathered in a floc. Cultivation of catfish with high density to produce more waste so that the need for stirring that allows waste can be suspended so that it can be decomposed aerobically into inorganic compounds by heterotrophic bacteria. The remaining organic material derived from feed or dirt will cause a decrease in water quality. Biofloc technology can add organic carbohydrates through an increase in C/N ratio and stimulate the growth of heterotrophic bacteria. Probiotics also serve as immunostimulants to decrease FCR, inhibit pathogenic bacteria and improve water quality [5].

The purpose of this study is to determine the growth performance and predict the harvest of catfish weight and survival rate (SR) that are cultured with different high density on the biofloc system.

2. Research materials and methods

2.1.Test Fish

Catfish seeds weighed of 8 ± 0.2 gram/fish or 8 ± 0.2 g/fish with each density as treatment: P1. Density of 1,100 e/m3 or 3,300 fish/pond; P2.Density of 1200 fish/m3 or 3,600 fish/pond and P3. Density of 1300 fish/m3 or 3,900 fish/pond.

2.2. Pond

The pond used is made of 3 m in diameter of tarpaulin, equipped with a 6mm in diameter of iron frame. Total volume of water in each pond is 3000 L. Catfish seeds have been adapted and put into the pond at random. Pond Preparation: Before the fish is stocked, water is inserted with height as high as 43 cm so that the volume is ± 3 m3, added by 600 gram/pond of salt or 200 gram/m3 and molasses of 300 cc/pond or 100 cc/m3. Pond water is replaced periodically if the water quality variable is above the allowed threshold. Pool Recirculation: For pond water recirculation use 3 water pumps on each pond, with each power of 33 watt.

2.3. Food Preparation

Fish food is in form of commercial floating pellets with protein content of 31-33%; 3-5% of fat; 4-6% of fiber and ash content of 10-13%, water content of 11-13%. Every 1 kg of food to be supplied is added by 5cc of EM 4 (effective micro organism 4) dissolved in 150cc of water. Every 1cc of EM 4 contains a minimum Lactobacillus casei 2.0x106 cells/ml and Sacharomycescerevisiae 3.5x105 cells/ml .

2.4. Experiment Design

The catfish seeds have been adapted and put into the pond randomly, according to the treatment of P1. Density of 1,100 e/m3 or 3,300 fish/pond; P2. Density of 1200 fish/m3 or 3,600 fish/pond and P3. Density of 1300 fish/m3 or 3,900 fish/pond. The collected data were analyzed regressionally with SPSS series 20.

2.5. Measuring of Fish Growth

Weight growth are measured every 7 (seven) days after fish is cultivated, the cultivation period is 21 days; as for the size of fish growth by using the calculation formula used [6]:

2.5.1. Individual Weight Growth/Absolute Weight (g/fish) (1)

2.5.2. Measuring of survival rate (SR) : $\frac{\text{Total fish end of research}}{\text{Initial total fish}} \text{ X } 100\%$ (2)

3. Result and discussion

3.1. Result.

3.1.1.Growth Performance

The observation result of the increasing of catfish weight cultured in biofloc system with different high density indicates that starting at the first week of observation, all treatments responded to an increase in absolute weight growth, indicating that the food given and the quality of water management had a positive impact on the pet fish, but the magnitude of response at treatment is diverse; in P1 showed an increase in absolute weight of 11.10333 ± 0.57 gr; In treatment P2 showed an absolute weight increase of 12 ± 0.4 gr and treatment P3 showed an increase of 10.70 ± 0.55 gr. In week IV, the best growth is still dominated by P1 that reaches 37,91667 g in weight, and growth of P2 reaches 32,41667 in weight higher than P3 that is 30.26 g. The weight growth data can be seen in Figure 1.

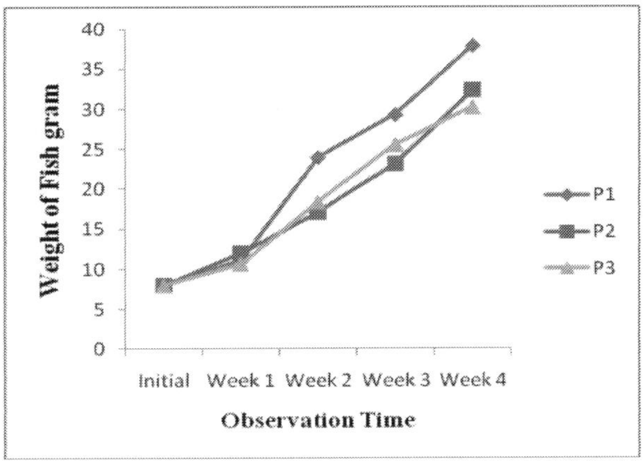

Figure. 1 The growth rate of catfish weights that are cultured with different high density

The growth performance analysis based on regression test showed that P1, P2, and P3 was YP1 = 6.46 +7.81 X (R2 = 0.971); YP2 = 6.53 + 6.0 X (R2 = 0.969); And YP3 = 6.71 + 5.94 X (R2 = 0.976), from the line equation then if catfish is cultured for 12 weeks or 3 months with treatment P1, it will be harvested with weight of 100 grams/fish, whereas if it is cultured with P2 will be harvested with weight of 78.53 grams/ fish and P3 will be harvested 77.9 grams/fish (especially for P2 and P3 weighing of 100 grams per fish takes 16 weeks or 4 months).

3.1.2. Survival Rate (SR)

After the pond is well prepared, the fish that has been adapted then put in the cultivation pond, then the cultivation stage is started. Initial cultivation of healthy fish condition, after being cultured on the first week, there are some fish that died in P1 as much as 2 (two) fish, in P2 and P3 as many as 5 (five) fish. On the 2nd week there were 3 (three) dead fish in P2, on P2 and P3 each of 10 (ten) and 15 (fifteen) fish. On week 3 the death increased more in P1 by 19 (Nineteen) fish, in P2 and P3 by 30 (thirty) and 40 (forty) fish. When entering the 4th week, water is replaced by 75%. The results showed there was a significant decrease in death rate, at week 4 the death decreased to P1 as much as 5 (five) fish, in P2 and P3 were 6 (six) fish.

After the fish cultured for 4 weeks, it has experienced a very real growth (Figure.1), although it also experienced death. While the total death on P1 as much as 25 (twenty five) fishes or its survival rate (SR) reached 97.73%, on P2 and P3 the number of deaths of 45 (forty five) fishes and 60 (sixty) fishes or survival rate (SR) reached 95.91% and 94.55% or as seen in Figure.2.

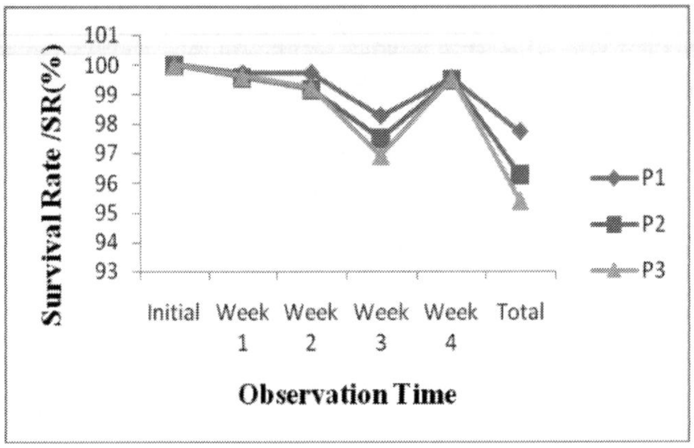

Figure. 2 Survival rate (SR) of catfish that are cultured with different high density

3.1.2. Water Quality

Water quality management in the cultivation of fish with high density with biofloc system becomes a very important factor, especially the factors that greatly affect the growth performance and survival rate (SR) those are the presence of dissolved oxygen (DO) and ammonia.

The results showed DO levels ranged from 3.1-4.2 mg/L from all treatments that have been applied and showed the fish can grow well (Figure 3).

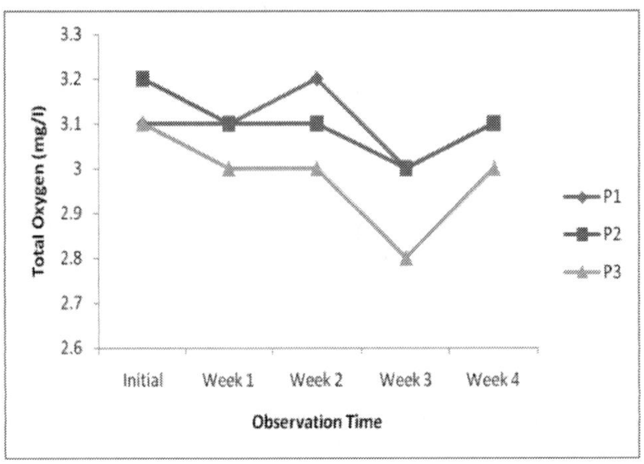

Figure 3. dynamics of oxygen in the water cultivation of catfish with different high density

While ammonia in all treatments showed an increase especially in the 3rd week, the ammonia content in P1 showed 0.13 mg / L, whereas in P2 showed 0.27 mg / L and in P3 showed 0.3 mg / L (Figure 4). The death increased by the increasing of ammonia content especially in 3rd week and the death decreased after water replacement, as well as ammonium content.

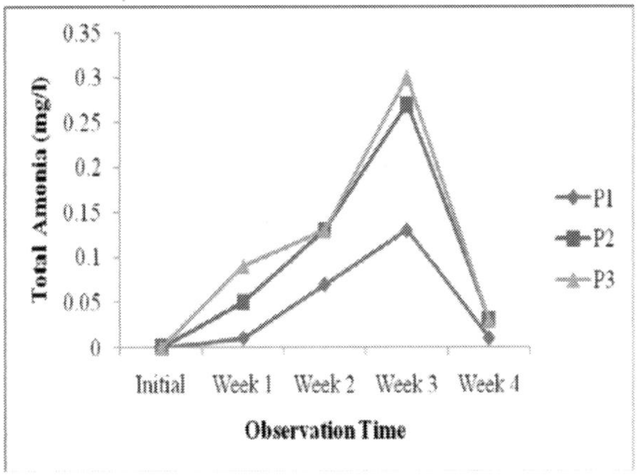

Figure 4. Dynamics of amonia in the water cultivation of catfish with different high density

3.2. Discussion

3.2.1. Growth Performance of Catfish

Result of observation indicated that, starting from the 1st week until the 4th week, there has been growth of absolute catfish weight which is cultured by using biofloc system with high density. All catfish grew well in accordance with opinion [4] that catfish can be cultured with very high density. In Fig. 1 can be seen that the more increasing of density, the more decreasing of fish growth. It is very clearly seen at week 4, the results of this treatment according to the opinion [7] which states that the more dense the cultivated fish, the more decrease in growth. The decrease in growth from low density to high density is assumed to be due to competition of narrower space or movement, oxygen competition and the increasing of ammonia accumulation.

Regression analysis showed that the growth performance was still increasing linearly, of the growth performance can be predicted that the weight of 100 gr per individual on P1 can be achieved at 12th week, so P1 would be harvested at the 12th week or 3rd month, while P2 and P3 would be harvested at the 16th week or 4th month. Production of catfish with biofloc is very productive, predicted in 1 m3 of water, can be harvested ± 100 kg fish or one pond can harvest 300 kg of fish. The catfish production with biofloc technique in high density is much more efficient than bamboo karamba as the opinion [8] states that when the catfish are kept in bamboo karamba, the production only reaches 75 fish/m3.

3.2.2. Survival Rate (SR)

After fish have been maintained for 4 weeks it has experienced very significant growth (Fig.1), it is suspected that water quality, feed, and cultivation management have supported the growth of fish, even though the environmental dynamics has influenced the survival rate (SR) of the fish cultured.

Base on fig. 2 the results showed that SR performance in the three treatments was very high, above 90%, compared to the resulted of the study [9] showed that the catfish culture of high density is 5 fish/liter with the initial weight of 0.82 ± 0.23 grams cultured in the cultivation tank for 21 days reached SR 56.66 ± 33.30. Based on the results of the study [10] Clarias gariepinus catfish weighing 46 gr, density 1 fish/10L water and maintained for 90 days then the survival rate (SR) ranged between 86.67 - 93.33%.

3.3.3. Water Quality

Water quality is critical in successful catfish farming, water quality observations have shown that DO levels range from 3.1-4.2 mg / L from all treatments that have been applied and these results have shown that fish can grow well (Fig3), this result is in accordance with [11] that in the standard water quality of catfish seed maintenance DO level of at least 3 mg / l, In accordance with opinion [12] states that in order for fish including catfish to be hdup well if the oxygen level is more than 3 mg / l. .

As the ammonia level during the maintenance of the fish continues to increase with time, by the third week of ammonia has exceeded the threshold has reached 0.13-0.3 mg/l (Fig 4), increase ammonia has caused death in fish [11], therefore do the replacement of water to decrease the ammonia, and seen fish start healthy again, water quality maintenance has ensured fish can grow well.

4. Conclusion and sugesstions

4.1. Conclusion

Based on the results and discussion then the cultured catfish in high density with biofloc system can be concluded as follows

4.1.1. Growth performance for 4 weeks of cultivation P1 generates the regression equation Y1 = 6.46 +7.81 X with r2 = 0.971 means that if the harvest is done at week 12th then the weight of catfish has reached 100 grams; P2 generates the regression equation of Y2 = 6.53 + 6.0 X with r2 = 0.969 means if the harvest is done at week 12th then the weight of catfish has new reached 78.53 grams and P3 generates the regression equation Y3 = 6.71 + 5.94 X with r2 = 0.976 means if the harvest is done on week 12th then the weight of catfish has new reached 77.99 grams.

4.1.2. The calculation of survival rate (SR) is very high, in P1 = 97.73%; P2 = 95.91% and P3 94.55%.

4.1.3. Water quality can be maintained in conditions which support the growth of catfish

4.2. *Suggestion*

Based on the results and discussion, the catfish that are cultured in high density with biofloc system suggests that to increase the production of catfish can be used biofloc system with a density of 1100 fish/m3.

References

[1] National Development Planning Agency 2013 *Indonesia Population Projection 2010 - 2035* National Development Planning Agency, Central Bureau of Statistics. Jakarta-Indonesia.

[2] Director General of Aquaculture 2013 *Data-Statistics-Series-Production-Fishing-Culture-Indonesia.*

[3] Jamabo N A and R I Keremah 2009 *Journal of Fisheries International* **4** (4): 55-57p

[4] Kareem O K and Olanrewaju A N 2015 Effect of Different Stocking Density on Nutrient Utilization Growth Performance and Survival of African Catfish (*Clariasgariepinus*, Burchell 1822) Fry in Recirculatory System *Journal of Fisheries and Aquatic Science* **10** (5): 400-404, 2015. ISSN 1816-4927 / DOI: 10.3923/jfas.2015.400.404 © 2015 Academic Journals Inc

[5] Yusuf, M.W, N.B. PriyoUtomo, M. YuhanadanWidanarni 2015 Growth Performance of Catfish (Clariasgariepinus) in Biofoc-Based Super Intensif Culture Added with *Bacillus sp. Journal Of Fisheries and Aquatic Science* **10** (6): 523-532

[6] Islam M Z Sarder Md R I and Akhand Md R I 2015 Growth performance of genetically male tilapia derived from YY male sex reversed male tilapia and mixed sex tilapia of *Oreochromisniloticus*in earthen pond aquaculture system in Bangladesh. ISSN: 2 347-5129 IJFAS 2015; 2(3): 186-191© 2015 IJFAS

[7] Akinwole A O A F Bankole A B Dauda and O E Saliu 2014 Growth and Survival of Clariasgariepinus (Burchell. 1822) fingerlings culture at different stocking dentities in Igbora, Oyo State, Nigeria. *Journal of Agriculture and Biodiversity Research* **3**(4) : 58-60

[8] Dasuki A Auta J and Oniye S J 2013 *Bajopas Volume 6 Number 1 June 2013* Bayero *Journal of Pure and Applied Sciences* **6**(1): 112 – 117 ISSN 2006 – 6996

[9] Nwipie G N Erondu E S and Zabbey N Influence of Stocking Density on Growth and Survival of Post Fry of the African Mud Catfish *Clarias gariepinus*. *Fish Aquac J* 2015 6:1
[10] Yildirim Y B and Turan F 2010 Effect of exogenous enzyme supplementation in diets on growth and feed utilization in Africa catfish *Clarias gariepinus Journal of Animal and Veterynary Advan* **9** (2) : 327 – 331
[11] Indonesian National Standard 2014 Dumbo Catfish Part 4: Seed Production SNI 6484.4:2014
[12] Boyd C E and Hanson T 2010 Dissolved-oxygen concentration in pond aquaculture. Global aquaculture advocate January/February 2010

Sex Diversity Approach of Spiny Lobster (*Panulirus* spp) to Marine Oil Spill Pollution in Southern Waters of Java

F E D Haryono[1*], Ambariyanto[2] and I Sulistyo[3]

[1] Marine Science Department, Fisheries and Marine Science Faculty, Jenderal Soedirman University, Address : FPIK Unsoed Jl. dr. Soeparno Karangwangkal Purwokerto 53123, Indonesia. Tepehone : +62 281 642360, Fax : +62 281 642360,
[2] Department Marine Science. Faculty Fisheries and Marine Science. Diponegoro University, Semarang. Address: FPIK-Undip Tembalang, Diponegoro University. Semarang – 50275, Indonesia, email : ambariyanto@undip.ac.id
[3] Department of Water Resources Management, Fisheries and Marine Science Faculty, Jenderal Soedirman University, Address : FPIK Unsoed Jl. dr. Soeparno Karangwangkal Purwokerto 53123, Indonesia Telephone : +62 281 642360, Fax : +62 281 642360, email : eastdee@gmail.com

Email: marine_2807@yahoo.com

Abstract. Coastal of southern Java waters is known as inhabit area of spiny lobster. Accumulation of hydrocarbon frequently occurs at the coastal waters as impact of oil pollution caused by oil leak from supplying ship of crude oil to Cilacap refinery. As shipping channel of oil, presence of oil spills is often detected at coastal areas of Cilacap. It can be indicated by range of sediment in the area which has risk levels in range of low to medium-low. It was, therefore, found that some locations suffered a greater impact on the ecological which giving high risk for marine organism life. Spiny lobster is one of many organism living at sea bed which threatened its life due to the presence of oil. Population of Spiny Lobster has to be protected because it has commercially valuable commodity for producing high nutrition. Considering the matters, it is therefore important to find a method for alleviating the problem. Investigation should be focused on biological aspect of spiny lobster in encountering extreme pollution at the coastal. For that purpose, a field research was conducted from January until July 2015. Using gillnet with 1 inch mesh size, the lobsters were randomly collected from southern Java districts waters. There were 1137 lobsters collected from six districts waters. Furthermore, the sample was morphologically identified and it was found that there were six species in the areas. In all area, *P. homarus* was found as dominant species, except in Gunung kidul district which was dominated by *P. penicillatus*. In term of sex diversity, there is statistically difference in number of female and male, each species no significant. Even though environment quality was very worse, there was found existence of ovigerous female in the research area as about 12% of the population. Those facts strongly indicated that the lobsters has a unique adaptation to survive in extremely low quality of environment due to marine oil spill.

Key words : Sex diversity, spiny lobster, oil spill, Southern Java waters

1. Introduction

Cilacap waters is known dense of marine ships transportation, at other side, the activity of fisheries fleet have lead more complex of maritime sector. Super marine tanker vessel supplying crude oil to Cilacap oil refinery, barges supplying coal to 3 electric power plants and cement factory. Those activity have impacted to water quality. Navigation ship channel to Tanjung Intan port have lead a lot of sinking of ships in Cilacap waters. The ships that have sunk were USS Langley on February 27, 1942, Niaga 34 loaded full of asphalt drums, Alisa XVII on November 1989 loading 11.000 drums of asphalt, MT King Fisher loaded full of crude oil on April 2000. The distribution of crude oil and bitumen spill shows toward to northwest when spring tide and to southeast at neap tide [1]. Banana VI and a ship were leak with 5000 tons of airplane oil on January 15, 2010, MT Alisa XVII 18.101 GT aground and leaked, including a lot of fishing boats. Four shipwreck of oil tanker had been accidents during period 2006-2008 [2]. The latter MT HHC were sunk loaded bulk of asphalt and MT Harmony Seven load 33,000 tons of diesel at October 17, 2015. Liquid and solid wastes extremely have polluted in Southern Java waters. Concentration of hydrocarbon in Cilacap sediment with risk levels ranged from low to medium-low and ecological risk for marine organism were obtained a greater impact at some locations [3]. The presence of oil spills has sensitive and very sensitive at coastal areas of Cilacap waters [4]. Concentrations heavy metal Cr, Ni, and Zn in sediments as result of Cilacap refinery site stations. The adverse effect of the conditions may cause more serious harmful effect to a wider range of organisms [5]. At other side, sewage disposal outlets produced a decrease in population density and changes in size spectra [6].

Variability of all organism that life on mainland, marine and aquatic ecosystem is known as diversity. Ecologically with their complexity of all sub part, include diversity of species, inter species and their ecosystem. The condition of diversity stated by [7] as diversity biology or biodiversity. Furthermore, the number and frequency of ecosystems, or genetic system are included the parts of degree of natural diversity [8]. Counting the number of species is done to obtain the diversity, although even more that can be expressed by a real number. Species composition, structure and function tropic ecology of a species is focus aspects of biodiversity [9]. Ecosystems under stress indicates diversity decline [10].

Fishery species under over-exploitation condition for some decade will occur diversity and long period of over exploitation impacted to diversity. The condition impacted to spiny lobster genus Panulirus [11]. Stocks of spiny lobster on some area may be close to collapse, the conditions as result of heavily overexploitation to the species [12]. Management of spiny lobster fisheries for international and regional cooperation is importance to manage to area that have high of degree of connectivity [13].

Phylum of crustaceans consists of 30.000 species, which crab and lobster are large species [14] and important crustacean fishery resource was spiny lobster [15]. The species have supported the global food stocks of fishery production currently around 260,000 tons per year [16]. Tropical spiny lobster consists several species, the species can survive without water for a certain periods. Lobster fisheries production was decrease for some decades and become an important issue in Indonesia now [17].

Biological aspect of spiny lobster, spawning circle starts to spawn that females migrate away from coastal waters to deeper waters, whish range number of egg between 159.000 – 1.925.000. After hatching the species living in the bottom waters for two years and moving to coral for mating and laying eggs [18]. The eggs fertilized externally by the male sperm [19]. Lobster age of 4-5 years old of coral lobster change of colour from red to pink before migrated to offshore [20]. Eggs of lobster found different at a large geographic scale [21]. The eggs hatch into transparent phyllosome larvae and metamorphic into puerulus. The distribution of the larvae follow ocean current for 6-12 months [post larvae]. The transparent plueruli settle on beach substrate as benthic organism and experience the dark pigmentation and moulting with 8-10 days into juvenile. Life a year as benthic organism in depth of 3-15 m, two years grew into adulthood and female migrate to deeper water to spawn [22]. Eggs of lobster hatch and the larvae grew to adult of 30 mm total length experiencing 22 times of moulting during a 256 days [23].

Male and female of spiny lobster exhibits dimorphism with clear differences and the key parameter for the management regime were size of sexual maturity. Lobster in Las Perlas found 297 females, 175 (59%) had eggs and ratio of male than female 1,29:1 [24]. [25] observed that in Iran waters obtained 260 of P. homarus which number of female more than male [female 143 and male

117]. Hereinafter referred at monsoon climate obtained 104 mature females and 42 were ovigerous and 36 were moulting. Female CL decline smaller than male and fishery activity impacted the difference of size between the sexes geographically [26]. Larvae with different length classes, sex, moulting condition and presence of eggs in female found at any significant difference between frequency of occurrence of each larvae [27].

Biological information such as size of sexual maturity, size distribution, relationship of CL, BW and sex ratio are important criteria to study the behaviour and characteristic of spiny lobster, significantly estimating size relation to the number taken by the fishery. Sexual maturity of lobster can attain at 6,02 cm of CL for male and 6.59 cm for female [28]. Most female of lobster during reproductive season choose to cohabit with large males and large male tended to cohabit with large female. Mate size of male and female associated freely with no preference [29]. Sexual mature of females can estimate the growth of lobster. The estimated sizes at 50%, physiological and functionally mature are [mean ± SE] CL : 56.46 ± 0.56 mm and CL : 66.63 ± 1.07 mm [30]. The oil spill pollution and high of fisheries activity have pressure extremely to coastal waters quality of the Southern of Java. This study aimed to analyze the influence of marine oil spill pollution to sex diversity of spiny lobster.

2. Method

Coastal waters of Southern Java is known as high of lobster catch area. The study area especially in Central Java Province and Special Region of Yogyakarta waters. Which Central Java waters consist four districts, namely Cilacap, Kebumen, Purworejo and Wonogiri. Special Region of Yogyakarta consist tree districts, namely Kulonprogo, Bantul and Gunungkidul.

The study were done from January to July 2015. Spiny lobsters were collected randomly encompassed in coastal waters of the districts. All sampling were performed six times at the location using 1 inch mesh size of gillnet. Captured lobsters were put in a cool box to be identified morphologically and sex in situ based on [31]. Total Length [TL] and Carapace Length [CL] measurement were measured using a caliper to the nearest 0.1 mm, based on [32] method. The wet weight was measured using electric balance to the nearest 0.1 g and data analyzed statistically ($\alpha_{0.05}$).

3. Result

Fisherman who chatch of spiny lobster as small scale fishery. Totally, 892 fisherman were capture of lobster at study area, using 0.5 ton of fiberglass canoe. The average of lobster net length were 1000 m or 10-20 piece of gillnets. Figure 1 shown the species distribution of spiny lobster was captured from districts. Spiny lobster from Gunungkidul waters were captured as highest percentage, which as much as 37.12%. Wonogiri district waters was not obtained of lobster, at the waters was not present of fisherman. Percentage of spiny lobster from Cilacap district were captured 24.195 %. Coastal morphology of Gunungkidul district was limestone cliffs and distance to Cilacap approximately 250 km. Spiny lobster catch activity by Gunungkidul fishermen was more productive than other district and the habitat support the growth of lobster. Gunung kidul and Cilacap were centre of spiny lobster capture activities, although the number of fishermen at Gunung kidul were at least compare with fishermen by other districts, lobster was captured at the district as the most numerous.

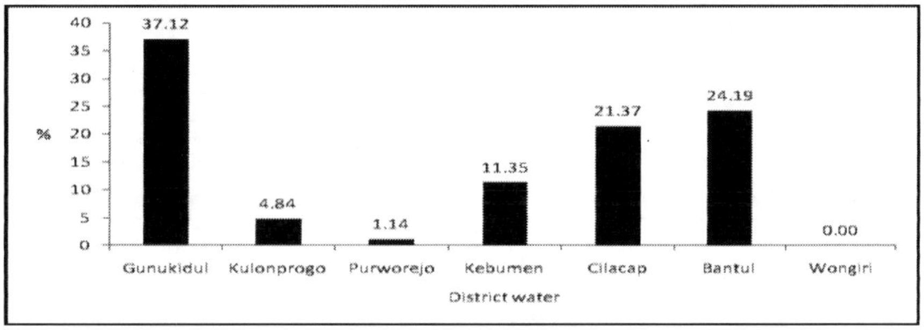

Figure 1. Percentage of *Panulirus* sp captured from districts

Morphologically identify of lobster were obtained six species, namely *P. homarus, P. ornatus, P. versicolor, P. penicillatus, P. longipes* and *P. polyphagus*. Diversity morphology of spiny lobster species at districts was dominated by *P. homarus* [64%]. *P penicillatus* species was captured 23%, *P. ornatus* 6%, *P versicolor* 4%, *P. longipes* 2% and *P. polyphagus* 1%. Seabed habitat of Gunungkidul and one fifths of Kebumen districts were rocky. Spiny lobster were captured from those distric dominated by *P. penicilatus*. Seabed substrate other those districts was sand or sand muddy. *P. homarus* [scallops spiny lobster] inhabit in sand seabed substrate. Figure 2 shown the percentage of dominant distribution of lobster, which each district was captured *P. homaruss*, except Gunungkidul district that dominated by *P. penicillatus*. The different of diversity of species between Gunungkidul and Cilacap districs than other districts were obtained that, at those districts more differ of species lobster. Each districts were obtained five species of spiny lobster, except *P. longipes* at Gunung Kidul and *P. penicillatus* at Cilacap.

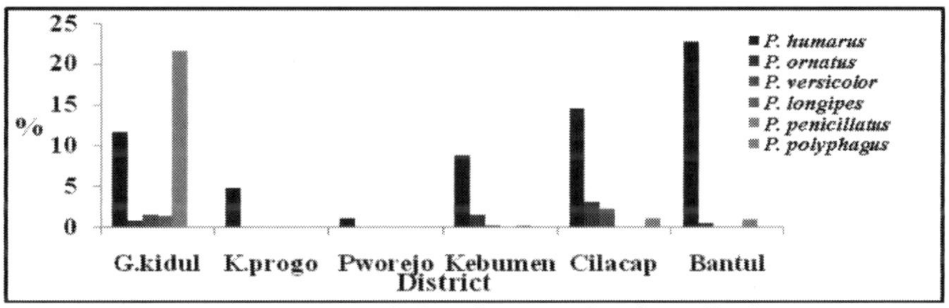

Figure 2. Percentage of *Panulirus* species from districts waters

Sex diversity of male in Figure 3 bellow shown that *P. homarus* as dominant species [64.01%] and secondly was *P. penicillatus* [22.06%]. Spiny lobster that inhabit in Gunungkidul district waters, which the district is known as most distant from Cilacap, which about 250 km. The Cilacap waters was center of extreme water quality due to result of frequent oil transport vessels that run aground and sunk.

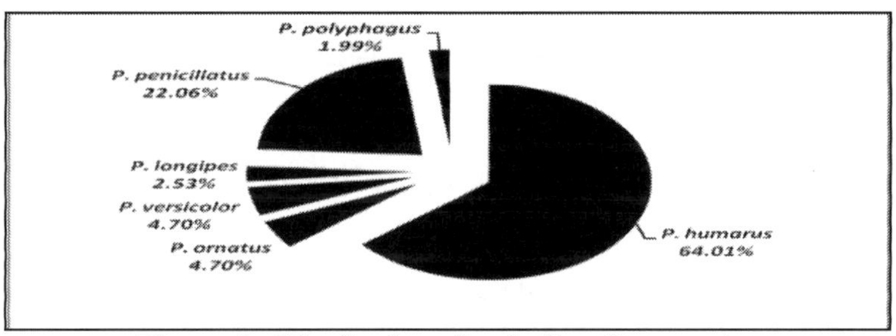

Figure 3. Distribution of male *Panulirus* species

Sex diversity of female spiny lobster species from southern Java waters [Figure 4] were shown that *P. homarus* species as the dominant [63.70%] , *P. penicillatus* species were obtained 23.97% and the fewest female was *P. polyphagus* [0.68%].

Gunungkidul, Bantul, Kulonprogo, Purworejo and Kebumen districts directly related to influence of Indonesian ocean waters. The condition differ with Cilacap district waters, which Cilacap district is protected from influence of Indonesian ocean by Nusakambangan island. Bantul and Kulonprogo districts as part of Yogyakarta province were found a lot of river, which springs at Merapi volcano. The seabed substrate at the districts waters was muddy sand and in 2014 the volcano was erupt. A lot of sand sediment as eruption material from the mountain moved into the coast.

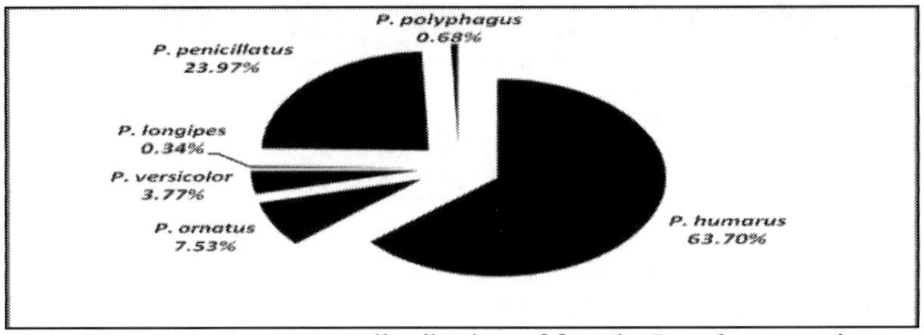

Figure 4. Percentage distribution of female *Panulirus* species.

The comparison of male to female of each district [Table 1] was acquired lobster from Gunungkidul and Cilacap districts more female than male. Male and female ratio for Gunungkidul spiny lobster were found 0.99 and 0.95 for Cilacap. Coastal waters quality of Gunungkidul was clearer than Cilacap and distance from sources of pollution of oil spill 250 km, thus supporting the existence the spiny lobster. Comparison between male and female lobsters in each district was found the number of male more than female at the two districts. Bantul district with ratio of males to females 1:03 and distance to Cilacap 200 km, and secondly Kulonprogo with ratio 1:4, distance to Cilacap 150 km. We assumed that the eruption of volcano recently and volcano material move into coastal waters have effect to the sex diversity of lobster.

Table 1. Diversity of males and females of *Panulirus* sp. in districts waters

District	Percentage		Ratio	
	Male	Female		
Cilacap	10.4	11.0	0.95	
Kebumen	4.4	6.9	0.64	
Purworejo	0.3	0.9	0.33	
Kulonprogo	2.8	2.0	1.40	
Bantul	12.3	11.9	1.03	
Gunungkidul	18.5	18.6	0.99	
Sum	48.7	51.3	0.949	$\alpha_{0.05}$: 0.741

Diversity of males and females for the entire district were obtained females [51.3 %] more than males [48.7 %], which the ratio of male compared to female 0.949, these value shown that 9 of male were obtained 10 of female lobsters. Statistically was showed significant differences [$\alpha_{0.05}$: 0.741]. Likewise, comparison of sex diversity of each lobster species which male compared to the female were obtained females more than male, except *P. Versicolor*. Sex diversity of lobster were shown no significant relationships [$\alpha_{0.05}$: 0.397].

Sex diversity condition of marine organism as a result of the pressure fishing activity expressed by [9] that diversity aspects, focus on species composition, community structure and ecological function of species was formed from integration of size diversity between richness of species and size of abundance size. Condition of females more than males allegedly were captured in long period can cause large scale change in stock other than increasing of exploitation.

Table 2 was shown the comparison between females and ovigerous female, that one-eigth of female have eggs, [12 %]. The comparison between female to male is very different from that observed by [24] in Iran waters was found by almost equal ratio, that is half of female have eggs, while for female lobster in Southern waters of Java was obtained only one-sixth from total of lobster females have eggs.

Table 2. Female and ovigerous female ratio of *Panulirus* sp.

Female [%]	Ovigerous Female [%]
88	12

Climatic conditions at the time of data collection in Southern Java waters was not different from the conditions of the season conducted by [24] research, but we found female have eggs differ of ratio. The shortest carapace length of spiny lobster is 46 mm and weigh 102.8 g, while the longest size was 92 mm with weight 1689 g.

4. Conclusion

Cilacap waters as center of oil spill accumulation was influent existence of spiny lobster. Sex diversity of spiny lobster in the Southern waters of Java was more female of lobster than male and the relationship was significant. Sex distribution ratio in six districts waters was found more male than female in Bantul and Kulonprogo districts. Eruption of Merapi vulcano have accumulate in bantul and Kulonprogo districts waters. Ratio of ovigerous female to non ovigerous female was found 0.12.

Lobster fishery production conditions was resulted that females more than males, and these conditions take place during the period of time, then the future will affect to availability of lobster in the waters and become extinct. Furthermore, conditions of districts waters is interconnected and open to exploitation by any fisherman or common property resources. Under the conditions, important to design the joint management to the fisheries resources as well as joint management of prevention and treatment of water pollution impacts. The aim of management is maintain the availability of fishery resources, particularly lobster stock, those step is importance to the long term lobster fishery.

5. Discussion

Oil spills was very often occurred and as a representative of chronic oil contaminated in Cilacap coastal waters [32], and managed based on ecosystem is importance to help the development of fisheries, especially species of spiny lobster [13]. Linked the condition of fish size and shrimp that mean annual production at Cilacap district for commercially important fish and shrimp were smaller. On the other side the health of water quality can be seen from the biological condition of female and male organism. Average length of species that 90% of the female length mature have egg indicate the water quality is health [33]. While diversity is a whole variability of living organisms from all resources, including the diversity of species, between species and ecosystems. Ecosystems under stress conditions indicating diversity is decrease [10].

Spiny lobster as a large-sized of decapod organism [35] influence of water quality is being limiting of life factor. The relation to oil spill pollution that crude oil settles on seabed and lobster live on seabed, which need ability to survive in order to retain its kind. The existence of lobster at this research is result of a form of adaptation to extreme environmental water quality and fishery pressure. As bottom organism, lobster must survive to adjust oil spill that settles on seabed. Lobster sample were obtained over 1100 species indicated that the species able to survive and proliferate. 51.3 % female from total number of lobster was obtained 12% as ovigerous female.

The grow and mature lobster species which subsequently produce eggs are strongly influenced by the conditions of water quality i.e. availability to feed, water quality, external fertilization. Furthermore, water quality conditions greatly is affect to growth, egg production and fertilization process that external system. Based on the high ratio of non-ovigerous females, we assumed the condition is caused by worse of water quality as direct and long term impact of frequent oil spill that have contaminate in Southern Java waters.

The coast line from Cilacap to Wonogiri districts approximately is 300 km and spiny lobster was captured by 826 fisherman that capture time depend upon to climate season. However, if the lobster were caught dominated by female for long period, the condition will affect to the stock. Problem solving to the condition that is appropriate the government to do more to manage lobster fishery by limiting of catch, including policies with sanctions appropriate as responsibilities to impact of pollution. Marine oil spill pollution was impacted to water quality and extreme impact to the level of tropic marine organisms, the existence of the fish resources are very urgent to be study with regard to the availability of food, especially the fish resources of high economic value. Study conditions fishery resources and water quality are very dynamic and interrelated. The assessment as the basis for further management and management that have been enacted should be evaluated, as recommended by [36] that economic impact of the policy liabilities seabed ship fleet that landed in Spain that operating in Iberian sea.

References

[1] Widhayanti, A., Ismanto, A. and Yulianto, B. 2015. Approach of Oil Spill Distribution with Hydrodynamics Model and Spill Analysis in Cilacap Waters. *J. Ose.* **4 (4)**:641 – 650

[2] Fitri, A. 2009. Economic Analysis Impact of Polution at Cilacap District waters. Thesis. Bogor Agricultural University. Bogor

[3] Syakti, A. D., Hidayati, N. V., Hilmi, E., Piram, A. and Doumenq, P. 2016. Source apportionment of sedimentary hydrocarbons in the Segara Anakan Nature Reserve, Indonesia. *Mar. Pol. Bulletin* **74 Vol. 1** p. 141-148

[4] Wibowo, M., Prijambodo, T. and Wibowo, M. T. 2012. The Mapping Environmental Sensitivity Index to The Oil Spill in Coastal Areas of Cilacap. Preceeding. The Second International Conference on Port, Coastal, and Offshor Engineering [2nd ICPCO], p. 15

[5] Syakti, A. D., Hidayati, N. H. 2015. Heavy metal concentrations in natural and human-impacted sediments of Segara Anakan Lagoon, Indonesia. *Env. Mon. and Ass.* **187 Vol. 1** p. 4079

[6] de-la-Ossa-Carretero, J. A., del-Pilar-Ruso, Y., Giménez-Casalduero, F., Sánchez-Lizaso, J. L. 2010. Sensitivity of tanaid *Apseudes latreillei* (Milne-Edwards) populations to sewage pollution. *Mar. Envi. Res.* **169**:309–317

[7] Dowle, J. and S. Rojas. 2003. Biodiversity . Delivering Resuls. United Nations Development Programme – Global Environment Facility Unit [UNDP-GEF]. New York.

[8] Cruz, R., Teixeira1, C. E. P., Menezes1, M. O. B., Santana, J. V. M., Neto, T. M., Juliana, G. C., Freitas, P. P. D., Silva, K. C. A. and Cintra, I. H. A. 2015. Large-scale oceanic circulation and larval recruitment of the spiny lobster *Panulirus argus* (Latreille, 1804). *Crustaceana* **88**(3) p. 298 – 323. doi: 10.1163/15685403-00003411

[8] Subade, R. F. 2005. Valuing Biodiversity Conservation in A World Heritage Site : Citizen's Non-Use Values for Tubbataha Reefs National marine Park, Philippines. Research Report. No. 2005-RR4. Co-published by the International Development Research Centre. Singapore. p. 6

[9] Gratwicke, B. 2000. Factors Affecting Fish Distribution in Coastal Habitats of the British Virgin Islands. Thesis of Doctor of Philosophy. Faculty of Biological Sciences, *Oxford University*

[10] Bracken, M. E. S., Bracken, B. E. and Bennett, L. R. 2007. Species Diversity and Foundation Species : Potential Indicators of Fisheries Yields and Marine Ecoystem Functioning. *Biodiversity*, Foundation Species and Marine Ecosystem Management. CalCOFI Rep. 48 Fishery Purposes. FAO. Rome. p. 972-1043

[11] Palero, F., Crandall, K. A. and Pascual, M. 2009. Phylogenetic relationship between spiny, slipper and coral lobster [Crustacea, Decapoda, Achelata]. Science Direct. *Mol. Phyl. and Ev.* **50** Vol.1 p. 152-162

[12] Andrade, H. A., 2015. Stock assessment of the red spiny lobster (*Panulirus argus*) caught in the tropical southwestern Atlantic. *Lat. Am. J. Aquat. Res.*, **43**(1): 201-214, 2015 . doi : 10.3856/vol43

[13] Truelove, N. K., Griffiths, S., Ley-Cooper, K., Azueta, J., Majil, I., Box, S. J., Behringer, D. C., Butler IV. J. and Preziosi, R. F. 2015. Genetic evidence from the spiny lobster fishery supports international cooperation among Central American marine protected areas. Springer. *Conservation Genetics* **16** Vo.2 p. 347-358

[15] de Carvalho Gaeta, J., de Brito Silva, M., de Godoy, T. And Cruz, R. 2015. Update on the obster species from Rocas Atoll marine Reserve, Brazil. Check List **11[4]**: 1705. doi: http://dx.doi.org/10.15560/11.4.1705

[16] de Lestang, S., Caputi, N. and Penn, J. W. 2015. A review of lobster fishery management: the Western Australian fishery for *Panulirus cygnus*, A case study in the development and implementation of input and output-based management systems. *ICES J. Mar. Sci.* **72**

[17] Haryono, F. E. D. and Ambaryanto, 2017. Approach Biodiversity For Spiny Lobster Fishery Management in Southern Waters of Java Using SWOT and AHP Analysis. *.J. Omni Aq.***13(1)**: 26-33

[18] Phillips, B. F., Cobb, J. S., and George, R. W. 1980. The Biologi and Managenet of Lobster. Volume I : Physiology and Behavior. Academic Press, Inc. New York. p. 5-10.

[19] Gongora, M. 2009. Assessment of Spiny lobster [*Panulirus argus*] of Belize based on fishery-dependent data. United Nation University Fisheries Training Program. Final Project 2009. Iceland.

[20] Melville-Smitha, R., Cheng, Y. W., and Thomson, A. W. 2003. Factors affecting colour change in 'white' western rock lobsters, *Panulirus cygnus*. Elsevier. *J. of Exp. Mar.Bio. and Eco.* **291** p. 111 – 129

[21] Tremblai, M. J., Silva, M. A., Reeves, A. R., Pezzack, D. S., and Denton, C. 2011. Sexual maturity in female Americanus Lobster-detectng major trends from the size of ovigerous female. Proceeding The 9th International Conference and Workshop on Lobster Biology and Management. Bergen. Norway. 19-24 June 2011.

[22] Smith, G. G., Kenway, M. and Hall, M. 2010. Starvation and recovery ability of phyllosoma of the Tropical Spiny Lobster *Panulirus ornatus* and *P. homarus* in captivity. *Mar. Biol. Ass. of India.* **52 V**ol.**2** p. 249-256

[23] Gusman, H, M., Cipriani, R., Vega, A. J., Lopes, M. and Mair, J. M. 2008. Population assessment of the Pacific Green spiny lobster *Panulirus asrgus* in Pacific Panama. *J. of Shellfish Res.* **27** Vol. 4 p. 907-917

[24] Mashaii, N., Rajabipour, F. and Shakouri, A. 2011. Feeding Habits of Scalloped Spiny Lobster (*Panulirus Homarus* (Linnaeus, 1758) (Decapoda : Palinuridae) from the South East Coast of Iran. *Turkish J. of Fish. and Aqua. Sci.* **11** p. 45-54

[25] Bertelsen, R. D. and Cox, C. 2001. Sanctuary roles in population and reproductive dynamic of Caribbean Spiny lobster. Spatial Processes and Management of Marine Populations. Alaska Sea Grant. Collage Program. AK-SG-01-02

[26] Ford, A. T. 2008. Can You Feminise a Crustacean ?. *Aq. Tox.* **88** Vol. 4 p. 316-321, doi.10.101

[27] Ikhwanuddin, M., Fatihah, S. N., Nurul, J. R., Zakaria, M. Z. and Abdol-Munafi, A. B. 2004. Biological features of mud spiny lobster, *Panulirus polyphagus* (Herbst, 1793) from Johor coastal water of Malaysia. *World Appl. Sci. J.* 31 Vol. 12 p. 2079-2086 (2004)

[28] Butler, M., Bertelsen, R. and MacDiarmid, A. 2015a. Mate choice in temperate and tropical spiny lobsters with contrasting reproductive system. *ICES J. of Mar. Sci.* **72** (Suppl. I) p.i101-i114

[29] Chang, Y. J., Sun, C. L., Chen, Y., Yeh, S.`Z. and Chiang, W.C. 2007. Reproductive biology of the spiny lobster *Panulirus Penicillatus* in the southeastern coastal waters off Taiwan. *Mar Biol* **151** p.553–564 DOI 10.1007/s00227-006-0488-9 (2007)

[30] Carpenter, K. E. and Niem, V. H. 1998. FAO Species Identification Guide for Fishery Purposes. The Living Marine Resources of The Western Central Pacific. Vol. 2. Cephalopods, Crustaceans, Holothurians and Sharks. FAO Species Identification Guide for Fishery Purposes. FAO. Rome. p. 972-1043

[31] Radhakrishnan, E. V., Thangaraj, R. and Vijayakumaran, M. 2015. Ontogenetic changes in morphometry of the spiny lobster, *Panulirus homarus homarus* [Linnaeus, 1758] from southern Indian coastriginal. *J. Mar. Biol. Ass.* India, **57** (1)

[32] Damayanti, A. 2016. Efficacy of Bioremediation Methods for Oil Contaminated Sandy Beach in Cilacap, Central Java, Indonesia. Disertation. Post Graduate School. Bogor Agricultural University. Bogor. p. 104

[33] Froese, R., Demirel, N. and Sampang, A. 2015. An overall indicator for the good enviromental status of marine water based on commercially exploited species. Elsivier. *Mar. Pol.* **51** p. 230-237

[34] Diniz, F. M., Ogawa, M., Cintra, I. H. A., Maclean, N. and Bentzen, P. 2010. Genetic Identification of Fishing Stocks: New Tools for Population Studies of the Spiny Lobster *Panulirus argus* (Latreille, 1804). *Bol. Téc. Cient. Cepnor* **10(1)** p. 95–111

[35] García, D., Prellezo, R., Sampedro, P., Da-Rocha, J. M., Castro, J., Cerviño, S., García-Cutrín, J. and Gutiérrez, M-J. 2017. Bioeconomic multistock reference points as a tool for overcoming the drawbacks of the landing obligation. *ICES J Mar Sci* **74** Vol. 2 p. 511-524. doi: https://doi.org/10.1093/icesjms/fsw030

IOP Conf. Series: Earth and Environmental Science **116** (2018) 012009 doi:10.1088/1755-1315/116/1/012009

Effect of The Phytase Enzyme Addition in The Artificial Feed on Digestibility of Feed, Feed Conversion Ratio and Growth of Gift Tilapia Saline Fish (*Oreochromis niloticus*) Nursery Stadia I

Diana Rachmawati[1], Istiyanto Samidjan, Tita Elfitasari
Aquaculture Department of Fisheries, Faculty of Fisheries and Marine Sciences, Diponegoro University, Jl. Prof. Soedarto, SH, Tembalang, Semarang, 50275 Indonesia

Email: dianarachmawati1964@gmail.com

Abstract. The purpose of this study was to determine the effect of adding the phytase enzyme in the artificial feed on digestibility of feed, feed conversion ratio and growth of gift tilapia saline fish (*Oreochromis niloticus*) nursery stadia I. The fish samples in this study used gift tilapia saline fish (*O. niloticus*) with an average weight of 0,62 ± 0,008 g/fish and the stocking density of 1 fish^1 L. Experimental method used in this study was completely randomized design with 4 treatments and 3 repetitions. The treatments were by adding phytase enzyme in artificial feed with the different level of doses those were A (0 FTU kg^1 feed), B (500 FTU kg^1 feed), C (1000 FTU kg^1 feed) and D (1500 FTU kg^1 feed). The results show that the addition of phytase enzyme was significantly ($P<0.01$) affected on apparent digestibility coefficient of protein (ADC_P), apparent digestibility coefficient of Phospor (ADC_F), feed conversion ratio (FCR), protein efficiency ratio (PER), and relative growth rate (RGR), on the other hand it insignificantly ($P>0.05$) affected on Survival Rate (SR) of gift tilapia saline fish. The optimum doses of phytase enzyme on RGR, FCR, PER, ADC_P and ADC_F of gift tilapia saline fish ranged from 1060 to 1100 FTU kg^{-1} feed.

Keywords : artificial feed, phytase enzyme, digestibility of feed, feed conversion ratio, gift tilapia saline fish (*Oreochromis niloticus*)

1. Introduction
Gift tilapia saline fish (*Oreochromis niloticus*) has several advantages, such as the fish can be grown in the used shrimp pond which has high salinity (0.5-30 ppt), easily cultivated, easily adapted, fast growth, able to eat any kind of feed (omnivores), and has high adaptability to various conditions. The intensive cultivation of the fish highly depends on the artificial (commercial) feed. Commercial feed currently still uses plant based ingredients such as soybean meal, as source of protein. Baruah *et al* [1] reported that the soybean meal has been commonly used in the artificial feed; however, it contains

anti-nutrient that can reduce the benefits of using plant based artificial feed. One of the anti-nutrient is Phytate acid [2]. Cao *et al* [3] reported that every kilogram of soybean meal contains Phytate acid 3.88 g. Phytate acid will bind minerals that have 2 or 3 valence (calcium, ion, zinc, and magnecium) to form complex that is difficult to absorb [4]. Besides bonding with minerals, Phytate acid also binds with protein and amino acid to cause decreasing in digestibility [5]. The digestibility of feed also depends on physical and chemical factors, types of feed, nutrient content, type and amount of digestive enzyme, and size and age of fish [6].

Moreover, an increase of plant based feed use creates another problem, that it produces phosphor pollutant into the water. Phosphor in the plant based feed cannot fully be utilized by the fish due to lack of phytase enzyme that decomposes Phytate acid [7][8][9][10]. Kumar *et al* [2] explained that phytate acid bound 80% phophor of the total phosphor available in the plant based feed. Phytate acid in the artificial feed is excreted along with feces into the environment, then it is degraded by microbe that produces phytase enzyme and the phosphor is released into the water. High phosphor concentration in the water will trigger eutrofication that hinders the cultivation of the fish [1]. Jagannathan and Nielsen [11] stated that phosphor is macro nutrient that is needed by animal including fish.

One of the solutions to solve the problem is by adding exogenous phytase enzyme [12] [13][14]. The addition of phytase enzyme into the artificial feed can increase nutrient absorption by the fish and regulate nutrient excretion (phosphor, nitrogen, and minerals) and hydrolyze phytate acid becoming inositol and phosphate acid [12]. Baruah *et al* [1] explained that the enzyme can hydrolyze phytate acid (mio-inositol hexahisphosphate) becoming mio-inositol mono, di, tetra and pentaphosphate and organic phosphate. Besides extracting phosphor from phytate acid, the enzyme can release other nutrients that are bound by phytate complex [5]. Moreover, Vielma *et al* [15] explained that the addition of phytase enzyme into the feed containing 50% soy meal can increase protein and phosphor digestibility of rainbow trout (*Oncorhynchus myskiss* Walbaum).

The objective of the study is to find the effects of the phytase enzyme addition into the artificial feed on feed digestibility, phosphor digestibility, feed conversion ratio, and growth of tilapia gifts saline.

2. Materials and Methods

The Gift tilapia saline fish (*Oreochromis niloticus*) with the average weight 0.62 ± 0.008 g were used in this study and obtained from Center for Brackish Water Aquaculture, Jepara, Central Java. The cultivation density for every treatment and repetition was one fish per liter [10]. The fish used in the study have been purposedly selected and healthy, no deformity, uniform in size and weight [6].

The feed used in this study was phytase enzyme supplemented feed in the form of pellet and with the protein content of 30%. The treatments were with the doses of A (0 FTU kg^{-1} feed), B (500 FTU kg^{-1} feed), C (1000 FTU kg^{-1} feed), and D (1500 FTU kg^{-1} feed). The feed contained fish meal as a source of animal protein; soybean meal as a source of plant based protein; corn meal, rice bran, wheat flour as sources of carbohydrate; fish oil and corn oil as sources of fat, vitamin mix and mineral mix as sources of vitamin and mineral; CMC as a binder; phytase enzyme as an unbinder of phytate acid; and Cr_2O_3 as much as 0.5-1%, these were used as indicators of feed digestibility [14].

Preparation on feed treatment done in this study consisted of proximate test for feed treatment [16], calculating feed treatment, and manufacturing feed treatment. The formulated feed ingredients and proximate analysis can be seen in Table 1. Phytase enzyme was Natuphos 5000G produced by PT. BASF Indonesia. Natuphos 5000G form was granule which contains active materials of *myo-inositol-hexakisphosphate* β-*phosphohydrolase* (EC 3.1.3.8) which was produced by *Aspergillus niger*. Natuphos 5000G contains of pythase enzyme 5.000 FTU/g. One unit of phytase activity (*Phytase Unit*/FTU) was defined as the amount of enzyme which release 1 micro molecule of non organic per minutes from 0,0051 mol/l of phytase acid on pH of 5,5 and 37°C [17]. To get 500 FTU of enzymatic activity needs 100 mg of phytase enzyme.

The procedure to prepare feeding experiment was first to dissolve an appropriate dose of phytase enzyme into warm water (45° C) and then mix with soybean meal evenly. The mixture was stored in the air sealed container for around 24 h [18]. Artificial feed made by mixing the least amount of the ingredients first and gradually adding and mixing the bigger amount of the ingredients except fat source (corn oil and fish oil) was added after all ingredients have been mixed. The evenly mixture of the artificial feed was formed into granules with the diameter of 1 mm to 2 mm. Then the artificial feed was dried in the oven with temperature of 40° C [14].

Table 1. Composition and Proximate Analysis in the Artificial Feed

Ingredients	Composition			
	A	B	C	D
Phytate enzyme (FTU)	0	0.1	0.2	0.3
Fish meal	25	25	25	25
Soybean meal	26	26	26	26
Corn meal	13	13	13	13
Rice bran	14	14	14	14
Wheat flour	13.5	13.4	13.3	13.2
Fish oil	2	2	2	2
Corn oil	2	2	2	2
Min.Vit	3	3	3	3
Cr_2O_3	0.5	0.5	0.5	0.5
CMC	1	1	1	1
Total (g)	100	100	100	100
Results of Proximate Analyses				
Protein (%) (*)	30.46	30.41	30.69	30.27
Fat (%) (*)	9.35	9.46	9.34	9.45
BETN (Extract without Nitrogen) (%) (*)	41.70	41.40	41.34	41.73
Energy (kkal)	286.62	286.55	286.42	286.81
Ratio E/P	9.41	9.42	9.33	9.48

Notes:
a.The values were calculated based Digestible Energy [19] for 1 g protein equals 3.5 kcal, 1 g fat equals 8.1 kcal, and 1 g carbohydrate equals 2.5 kcal.
b. According [20], the optimal E/P ratio for growth ranges from 8 kcal/g to 12kcal/g.
*Animal Nutrient Laboratory, Faculty of Husbandry and Agriculture, Diponegoro University (2017)

Containers used in this study were made of plastic with the volume of 25 l as many as 12 buckets. The buckets were first cleaned with calium permanganat to sterilize the containers from bacteria. After cleansing, the buckets were cleaned with water and then dried off. The buckets were equipped with sand filter and filled with water that has 20 ppt salinity. To provide enough oxygen during experiment was put blower. The fingerlings were raised in recalculated water in order to maintain water quality in optimum range. Finally the buckets were covered with plastic sheet to prevent fish from jumping out and to maintain temperature. The study was started by scaling the weight of the fish, and then the fish were put in the cultivating containers for 42 days. The fish were fed 3 times a day at satiation and fed in the morning at 06:00, in the noon at 12:00, and afternoon at 16:00. The fish were sampled and scaled the weight every week. The feces were siphoned one hour before feeding in order to keep the water media cleaned and viable to raise the fish.

Parameters include Relative Growth Rate (RGR), Protein Efficiency Ratio (PER), Feed Conversion Ratio (FCR) according to [21], Apparent Digestibility Coefficient of Protein (ADC_P) and Apparent Digestibility Coefficient of Phosphor (ADC_F) according to [22], and Survival Rate (SR) according to [14]. The chromic oxide levels in feeds and feces were analyzed using a modified colorimetric method [22]. The levels were measured with a spectrophotometer (540 nm) (Shimadzu UV-2102 PC, UV-visible Scanning Spectrophotometer) after perchloric acid oxidation and forming a colored complex with diphenylcarbazide (DPC). Samples were analyzed to determine phosphorous (P) concentrations by flame atomic absorption spectrophotometer on a Shimadzu AA6800 (Shimadzu, Japan). Variables of water quality that were tested were pH (Jenway 3510), DO (Jenway 970), temperature and Ammoniac (HANNA: HI. 8633). Aerator to recalculated the water was placed in every container.

$$RGR : \frac{\text{Final weight} - \text{Initial weight}}{\text{Initial weight xTime experiment}} \text{ x } 100\% \quad (1)$$

$$FCR : \frac{\text{The amount of feed consumed}}{\text{(Final weight + Total weight fish death)} - \text{Initial weight)}} \text{ x } 100\ \% \quad (2)$$

$$PER : \frac{\text{Final weight} - \text{Initial weight}}{\text{The amount of feed consumed x Protein content of feed}} \text{ x } 100\% \quad (3)$$

$$ADC_P : 100\% \left\{ \frac{Cr_2O_3 \text{ in the feed}}{\%\ Cr_2O_3 \text{ in the feces}} \text{ x } \frac{\%\text{ protein in the feces}}{\%\text{ protein in the feed}} \right\} \quad (4)$$

$$ADC_F : 100\% \left\{ \frac{Cr_2O_3 \text{ in the feed}}{\%\ Cr_2O_3 \text{ in the feces}} \text{ x } \frac{\%\text{ fosfor in the feces}}{\%\text{ fosfor in the feed}} \right\} \quad (5)$$

$$SR : \frac{\text{(Final count)}}{\text{Initial count}} \text{ x } 100\% \quad (6)$$

Analysis of variance (ANOVA) was used to analyze the data. Before analyzing, the data were first tested the normality, additively, and homogenity. The test was to make sure that the data were normal, homogen, and additive property. If the analysis of variance was significant ($p<0.05$) or highly significant ($p<0.01$), Duncan test was conducted to find out the difference of the treatmens, while water quality data were descriptively analyzed. To determine optimal dose of phytase enzyme, polynomial orthogonal test was conducted using SAS9 and Maple12 [23].

3. Results and Discussion

The results of study on Tilapia Gift Saline Fish (*O.niloticus*) for RGR, FCR, PER, ADC_P, ADC_F, and SR were shown in the Table 2.

Table 2. The Values of RGR, FCR, PER, ADC_P, ADC_F, and SR Tilapia Gift Saline Fish

Data	Treatment			
	A	B	C	D
RGR %/day)	9,46±0,40^{bc}	10,62±0,69^{b}	12,26±0,32^{a}	9,97±0,61^{bc}
FCR	2,65±0,10^{a}	2,36±0,24^{b}	1,75±0,07^{c}	2,30±0,22^{ab}
PER	0,90±0,04^{b}	1,19±0,23^{b}	1,95±0,05^{a}	1,24±0,25^{b}
ADC_P (%)	75.47±0.02^{c}	79.65±0.05^{bc}	83.93±0.05^{a}	77.65±0.04^{ab}
ADC_F (%)	71.57±0.03^{c}	74.64±0.04^{bc}	78.89±0.06^{a}	72.23±0.02^{ab}
SR (%)	80,00±5,00^{a}	88,33±2,89^{a}	88,67±2,89^{a}	83,±2,89^{a}

Note: The Values with the same superscripts in the column show that there was no difference

Table 2. shows that the addition of phytase enzyme in the artificial feed significantly ($p<0.01$) increased the growth of Gift Tilapia Saline. The results gave evidence that supplemented feed with the phytase enzyme could hydrolyze protein that was bind by phytate acid into amino acid. This amino acid was readily digested and could provide energy to grow [24][25]. According to Chung [12] the addition of phytase enzyme into the artificial feed can increase nutrient absorption by the fish and regulate nutrient excretion (phosphor, nitrogen, and minerals) and hydrolyze phytate acid becoming inositol and phosphate acid. The decomposition of phytate acid makes easier to methabolize protein and mineral compounds. The addition of the enzyme were also reported to increase the weight of *Marsupenaeus japonicus* [26], *Psetta maxima* L. [27], *Panaeus monodon* [6], and *Channos channos* [10].

Table 2 also shows that the addition of phytase enzyme in the artificial feed of 1000 FTU kg^{-1} feed significantly increased the growth of Gift Tilapia Saline compared to the addition of 500 and 1500 FTU kg^{-1} feed. The highest relative growth was obtained in the C treatment (12.26%). It can be concluded that the addition of 1000 FTU phytase on the every kg feed could reduce antinutrients or phytate acid on soybean meal. This finding was supported by Masumoto [6] that the dose of 1000 FTU kg^{-1} feed was the optimal dose to increase growth of *Penaeus monodon*. Yu FN and Wang [28] also reported that the addition of phytase 1000 FTU kg^{-1} feed could increase average weight of crucian carp *Carassius carassius* by 25 percent. Moreover Rachmawati *et al* [10] reported that the enzyme addition of 1000 FTU kg^{-1} feed can increase the weight of *Channos channos*. The same results were also found in carp [29], African catfish [30], striped bass [31], rainbow trout [15], Atlantic salmon [32], Korean Sebastes schlegeli [33].

The lowest growth was obtained in the A treatment as much as 9.46% per day. It was thought due to lack of inositol in the Gift Tillapia Saline, as reported by [14] that when the feed contained phytate acid, it could decrease fish' appetite, the growth, and cause anemia. The deficiency of inositol indicated that phytate acid has not been hydrolized into inositol and phosphate acid as reported by [12] [1]. They explained that the enzyme can hydrolize phytate acid (mio-inositol hexahisphosphate) becoming mio-inositol mono, di, tetra and pentaphosphate and organic phosphate. Moreover, phytate acid can obstruct the decomposition of complex mineral, therefore the availability of minerals, especially phosphore, cannot fullfil the need of Gift Tillapia Saline and in turn it reduce the growth of the fish. NRC [14] stated that phosphor is an important factor in the process of muscle constraction, bone development, and phosphate development that is needed to transform energy. Baruah *et al* [1] and Fox *et al* [18] reported that an increase on growth, raw protein coefficient, sulphur, phosphor total, and phytate phosphor have happened in rainbow trout which was given plant based feed with addition of phytase enzyme.

The results of orthogonal polynomial test on the relationship of phytase enzyme in the artificial feed and the relative growth (Figure 1) had cubical pattern with the equation, $Y = -5.8844x^3 + 9.7867x^2 - 1.1022x + 9.46, R^2=0.86$. The optimum dose of the phytase enzyme in the feed on the relative growth was 1060 FTU kg^{-1} feed with the maximum relative growth of 12.28% per day.

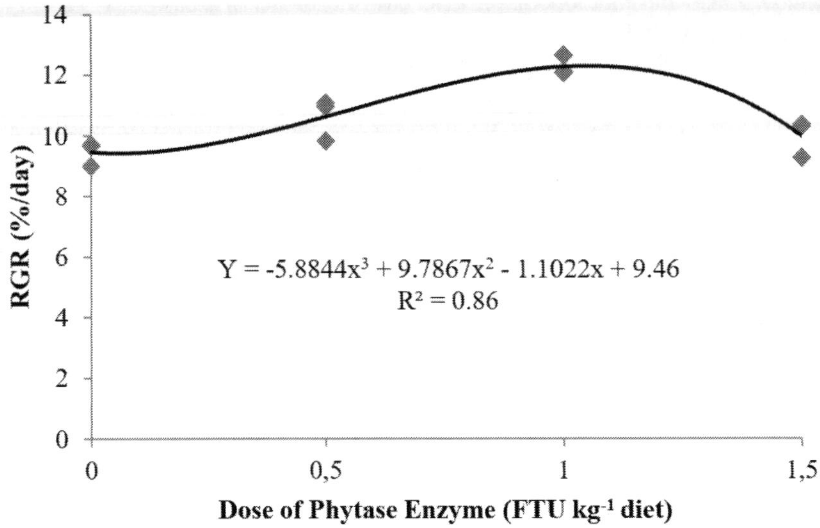

Figure 1. Graph of the Relationship between Phytase Enzyme Addition in the Artificial Feed and RGR of Gift Tillapia Saline (*O. niloticus*)

According to Stickney [34] feed conversion can be found by comparing between the amount of feed and the weight addition of the fish and the weight of dead fish during study. The less feed conversion the more efficient of feed utilization. Table 2 shows that artificial feed with the addition of phytase enzyme (B, C, and D treatments) has feed conversion ratio lower than the artificial feed without the addition of phytase enzyme (A treatment). It was suggested that phytase enzyme can catalyze phytate acid decomposition. Hydrolized reaction was able to unbind between phytate acid and protein and mineral complex that increased trypsinogen activity into trysin enzyme. The trysin enzyme can decompose protein into amino acids, therefore feed utilization efficiency became maximal and feed conversion ratio became low. The similar results of study have been done by [17] in the rainbow trout. Wang *et al* [35] also reported that artificial feed with phytase enzyme addition can improve feed conversion ratio in the rainbow trout. The same result was also found for *L. Rohita* [4].

The addition of phytase enzyme significantly ($P<0.01$) affected on the feed conversion ratio in the Gift Tilapia Saline *(O. niloticus)* (Tabel 2). The results shows that the highest feed conversion ration happened in the A treatment as much as 2.65, while the lowest was in the C treatment as much as 1.75. The C treatment (1000 FTU kg^{-1} feed) resulted in the lowest feed conversion among other treatments, B (500 FTU kg^{-1} feed), D (1500 FTU kg^{-1} feed), and A (0 FTU kg^{-1} feed). It was suggested that the addition of phytase enzyme could increase the efficiency in feed utilization and make feed conversion ratio low. Li and Robinson [36] studied on the addition of 250 units or more microbial phytase in the feed. The results show that there were higher feed consumption, higher weight gain, and lower feed conversion ratio than those without additional microbial phytase in the feed. Phytase enzyme in the feed had very important role since it could increase feed utilization [17].

The relationship between phyta.se enzyme and feed conversion ratio based on the orthogonal polynomial test, as shown in the Figure 2, was cubical. The equation was $Y = 1.9289x^3 - 3.4933x^2 + 0.6644x + 2.6533$, $R^2 = 0.84$. The optimum dose of the phytase enzyme in the feed on the feed conversion ratio was 1100 FTU kg^{-1} diet with the maximum value of feed conversion ratio 1.72.

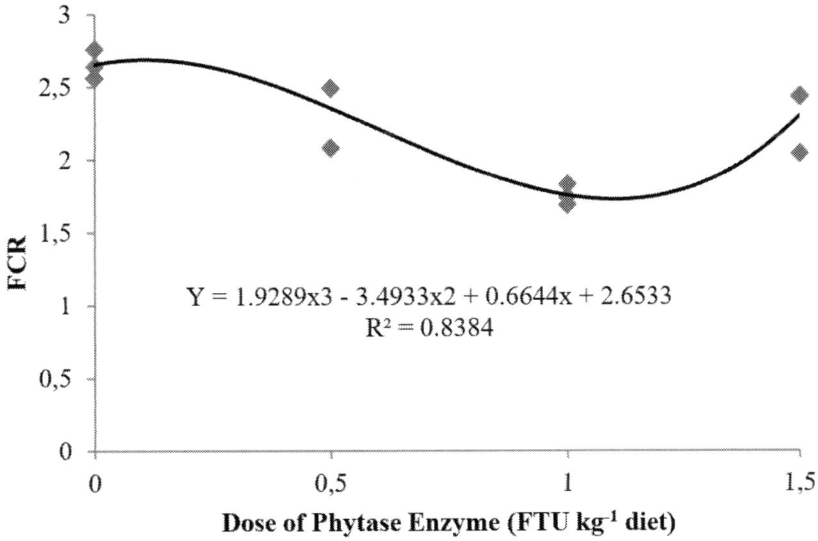

Figure 2. Graph of the Relationship between Phytase Enzyme Addition in the Artificial Feed and FCR of Gift Tillapia Saline (*O. niloticus*)

The addition of phytase enzyme in the feed significantly ($P<0.01$) affected Protein Efficiency Ratio on the Gift Tilapia Saline (*O. niloticus*) (Table 2). The highest protein efficiency ratio reached at C treatment with the value of 1.95, while the lowest was the A treatment with the value 0.90. The C treatment that gave the highest result suggested that the enzyme dose was very effective to reduce and break down the phytate acid, to dissociate between phytate acid and protein and minerals, therefore it affected digestibility enzymes to break down protein into amino acids. Li and Robinson [36] stated that the higher the protein conversion ratio indicates better efficient feed because the fish can utilize protein better. Maximum absorption of protein was due to phytase enzyme that break down phytate acid [37]. It also caused digestibility of phosphor to increase and anti-nutrient to decrease [39]. The breakdown of phytate acid provided favourable situation for the fish to absorb protein. The addition of phytase enzyme that could increase the utilization of protein has been reported by [39][40].

The results of orthogonal polynomial test (Figure 3) shows the relationship between phytase enzyme in the artificial feed and protein efficiency ratio. It had cubical relationship with the equation $Y = -2.6133x^3 + 4.88x^2 - 1.22x + 0.9033$, $R^2=0.88$. The optimum dose of the phytase enzyme in the artificial feed for the protein efficiency ratio was 1100 FTU kg^{-1} diet with the maximum value of the protein efficiency ratio of 1.98.

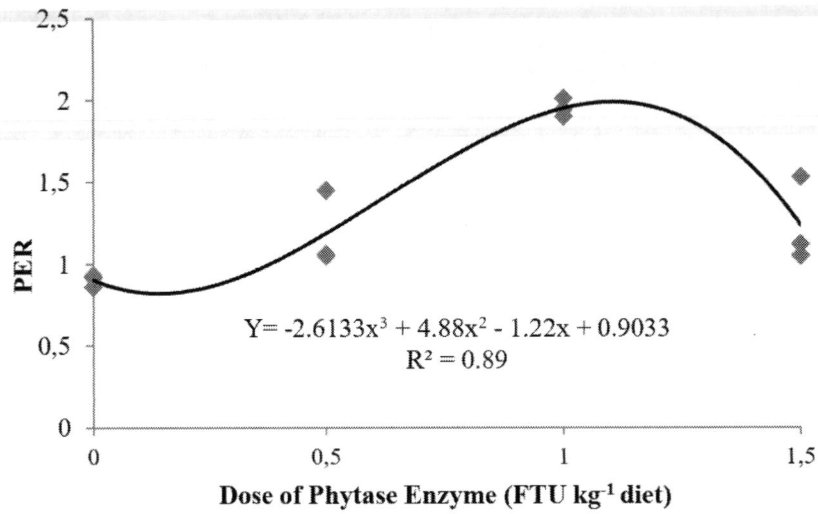

Figure 3. Graph of the Relationship between Phytase Enzyme Addition in the Artificial Feed and PER of Gift Tillapia Saline (*O. niloticus*)

The addition of phytase enzyme significantly ($P<0.01$) affected on the digestibility of protein and the digestibility of phosphor on the Gift Tilapia Saline (*O. niloticus*). Table 2 shows that the addition of phytase enzyme 500-1000 FTU kg^{-1} feed could increase the digestibility protein and the digestibility of phosphor. Storebakken *et al* [40] has already reported that the addition of phytase enzyme increased protein digestibility and protein retention. These results were also confirmed by Debnath *et al* [9] that the addition of phytase enzyme significantly increased protein utilization and digestibility on Atlantic salmon, otherwise they had low protein utilization and digestibility. Hunter [17] also found that the addition of phytase enzyme significantly increased protein digestibility from 84.5% to 87.7%. Similar results were found on carp [15], rainbow trout [40][41], Labeo rohita [42]. Kornegay and Qian [43] and Baruah *et al* [1] also reported that addition of phytase enzyme in the plant based feed increased protein digestibility due to breaking down of phytin-protein compound. The addition of phytase enzyme of 1000 FTU kg^{-1} feed (C treatment) was the optimal dose to disintegrate the anti-nutrient and increase feed digestibility as in the [17] findings. He found that phytase enzyme can break down anti nutrients in the feed, such as phytate acid, non-starch polysaccharide, and trypsin inhibitor. It could also increase feed digestibility.

The results of orthogonal polynomial test (Figure 4) shows the relationship between phytase enzyme in the artificial feed and ADC_P was cubical, with the equation $Y = -22.20x^3 + 36.64x^2 + 2.545x + 73.53, R^2 = 0.99$. The optimum dose of the phytase enzyme in the artificial feed for the protein efficiency ratio was 1060 FTU kg^{-1} feed with the maximum value of the protein efficiency ratio of 85.56%.

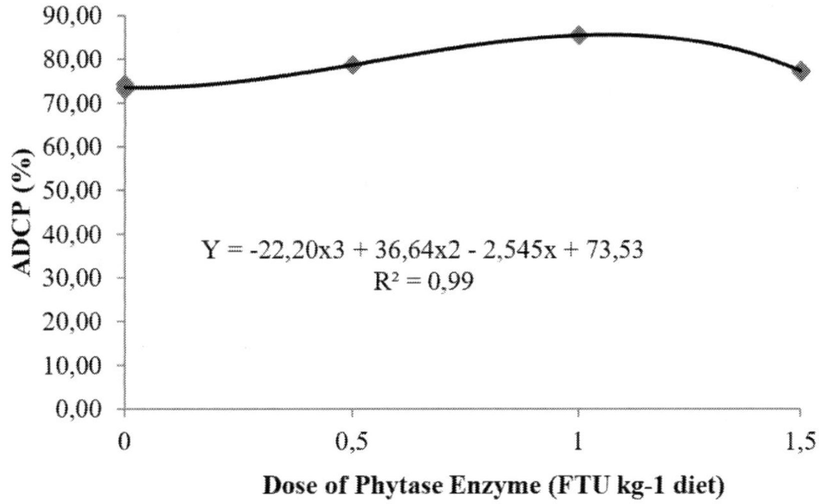

Figure 4. Graph of the Relationship between Phytase Enzyme Addition in the Artificial Feed and ADCp of Gift Tillapia Saline (*O. niloticus*)

The results of orthogonal polynomial test (Figure 5) shows the relationship between phytase enzyme in the artificial feed and ADC_F was cubical, with the equation $Y = -26.627x^3 + 49.753x^2 - 11.71x+72.213$, $R^2 = 0.99$. The optimum dose of the phytase enzyme in the artificial feed for the protein efficiency ratio was 1070 FTU kg^{-1} feed with the maximum value of the protein efficiency ratio of 81.80%.

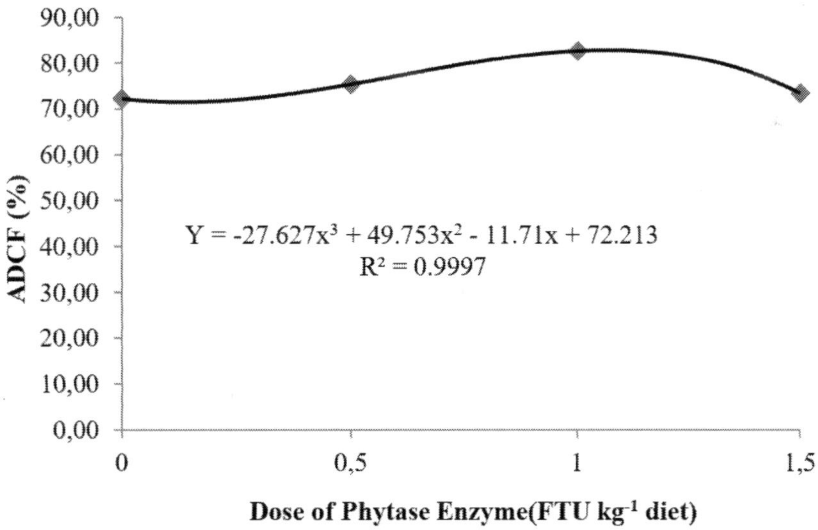

Figure 5. Graph of the Relationship between Phytase Enzyme Addition in the Artificial Feed and ADC_F of Gift Tillapia Saline (*O. niloticus*)

The addition of phytase enzyme in the artificial feed insignificantly affected on survival rate of Gift Tilapia Saline, as shown in the Table 2. The result was in line with the Li Robinson and Manning [44] finding that survival rate was insignificantly affected by addition of phytase enzyme in the feed. The survival rate was affected by internal factors such as gender, heredity, age, reproduction, disease resistance and external factors such as water quality, density, number and composition of amino acid in the feed [45]. Water of quality during the research still on condition that overpass to the cultivation of Gift Tillapia Saline. The measurement of water parameter during cultivation Gift Tilapia Saline can be seen in the Table 3.

Table 3. Parameters of Water Quality for Gift Tillapia Saline (*O. niloticus*) Cultivation

Treatment	Water Quality			
	Temperature (^{0}C)	pH	DO (mg/l)	NH_3 (%)
A	26 – 33	7.50 – 7.85	3.30 – 3.55	0.0072 – 0.0074
B	26 – 33	7.50 – 7.82	3.24 – 3.48	0.0072 – 0.0074
C	26 – 33	7.50 – 7.81	3.28 – 3.58	0.0072 – 0.0074
D	26 – 33	7.50 – 7.81	3.32 – 3.53	0.0072 – 0.0074
Feasibility	14-38*	6.5 – 8.5*	>2*	<0.1*

Note : * Rachmawati *et al.* (2017)

4. Conclusion

The addition of phytase enzyme in the artificial feed significantly increased on the growth rate, protein efficiency ratio, protein and phosphor digestibility, and decreased feed coversion ratio, otherwise it insignificantly affected on survival rate of the Gift Tillapia Saline (*O. niloticus*). The optimal doses of phytase enzyme in the artificial feed on RGR, FCR, PER, ADC_P and ADC_F of gift tilapia saline fish ranged from 1060 to 1100 FTU kg^{-1} feed.

Acknowledgments

Appreciation expressed to those who already helped in this study, especially to: 1) The head of BBPBAP (Centre for Freshwater and Brackish Aquaculture), Jepara, Central Java; 2) The head of Artificial Feed Laboratory of BBPAP; and 3) The head of Fishery Technology Laboratory, Department of Fisheries, Faculty of Fisheries and Marine Sciences, Diponegoro University who has done proximate analysis of feed preparation phytate acid, and feed experiments.

References

[1] Baruah K, Sahu N P, Pal A K, Debnath D and Mukherjee S C 2007 *J. Aquaculture Study* **38**(2):109-120
[2] Kumar V, Sinha A K, Makkar H P S, De Boeck G and Becker K 2011 *J. of Animal Physiology and Animal Nutrition*, **96**: 335–364
[3] Cao L, W Wang, C Yang, Y Yang, J Diana, Yakupitiyage A, Luo dan D Li 2007 *J. Enzym and Microbial Technology* **40**:497-507
[4] Baruah K, Sahu NP, Pal AK, Debnath D 2004 *NAGA World Fish Center Quart* : 27(3/4):15–9
[5] Ravindran V 2000 *Effect of Natuphos Phytase on the Bioavailability of Protein and Amino acids – a review* (Monogastric Research Center Institute of Food Nutrition and Human Health Massey University, Palmerston North Nezw Zealand) p : 1-10
[6] Rachmawati D and Istiyanto S 2016 *J. Teknologi*, **78**(4–2) : 39–43
[7] Masumoto T, Tamura B and Shimeno S 2001 *J.Fisheries Science*,**67**: 1075-1080
[8] Cheng Z J and Hardy R W *Aquaculture,* 218 : 501–514
[9] Debnath D, Pal A K, and Sahu N P 2005 *Aquacult Res*, **36**(2) :180–7
[10] Rachmawati D, Istiyanto S and Maizirwan M 2017. *Philippine J. of Science*, **146** (3):237-245
[11] Jagannathan K R and Nielsen P H 2013 *J. Cleaner Production*, **42** : 228-240
[12] Chung T K 2001 *Sustaining livestock production and environment* (Food and Agriculture Asia Pacific Development. Singapore) pp: 52-54
[13] Jobling M 2002 *Food Intake in Fish* ed Houlihan D, Boujard T, Jobling M (Oxford, UK :

Blackwell Science) pp : 25-48.
[14] National Research Council (NRC) 1993 *Nutrient Requirements of Fish* (Washington D. C.: National Academy Press)
[15] Vielma J, Ruohonen K, Gabaudan J and Vogel K 2002 *Aquaculture Research*, **35**(10): 955–964
[16] AOAC 1990 *Official Methods of Analysis* (Association of Official Analytical Chemists Fifteenth edition, Washington D C) 1298 pp
[17] Hunter B 2002 *Phytase Aplications in Aquaculture* (Roche Aquaculture Center Asia Pasific Bangkok) 425 p
[18] Fox J M, Addison L L, Anthony J S, D Allen D, Denis R M, Elizabeth C D, Tzachi M S 2006 *Phytase Supplementation In Aquaculture Feeds Improves Fish, Shrimp Growth Performance* (Global Aquaculture Alliance) 66 pp
[19] Wilson RP 1982 *Nutrition and Feeding of Channel Catfish* ed RR Stickney and RT Lovell (Southern Cooperative Series) 193-201 p
[20] De Silva SS 1987 *Finfish Nutrition Study in Asia* (Proceeding of The Second Asian Fish Nutrition Network Meeting Heinemann Singapore) 128 p
[21] Tacon A G 1995 *The Nutrition and Feeding of Farmed Fish and Shrimp-A Traning Mannual* (FAO of The United Nations Brazil) pp 106-109
[22] Fenucci J L 1981 *Studies on The Nutrition of Marine Shrimp of The Penaeus* Faculty of Department of Biology, University of Houston, Houston, Texas, USA (Ph D Thesis)
[23] Steel R G D, Torrie J H and Dickey DA 1996 *Principles and Procedures of Statistics* 3rd ed (New York: McGraw Hill International Book Company, Inc)
[24] Amoah Y T, Thorarensen H and O Sigurgeirrson 2011 *Effect of Feedary Protein Levels on Growth and Protein Utilization in Juvenile Arctic Char (Salvelinus alpinus)* (Fisheries Training Programme, United Nations University) 26 pp
[25] Haghbayan S and M S Mehrgan 2015 *The Effect of replacing fish meal in the feed with enzyme-treated soybean meal (HP310) on growth and body Composition of Rainbow Trout Fry. Journal of Molecules*, **20:**258-266
[26] Bulbul M, Md A Kader, M A Ambak, Md S Hossain, M Ishikawa dan S Koshio 2015 *Aquaculture Elsevier*, **438:**98-104
[27] Danwitz A, Von, C G J van Bussel, S F Klatt and C Schulz 2016 *J. Aquaculture Elsevier*, **450**:405-411
[28] Yu FN and Wang DZ 2000 *J. Fish SciChin,***7**(2):106–9
[29] Schaefer A, W M Koppe, K H Meyer-Burgdorff and K D Guenther 1995 *J. Water Science and Technology*, **31**:149–155
[30] Weerd V J H, Khalaf K H, Artsen EJ and Tijssen P A 1999 *J. Aquacult Nutr*, **5**(2):135–42
[31] Papatryphon E, Howell R A and Soares J H 1999 *J. World Aquacult Soc*, 30:161–73
[32] Sajjadi M and Carter C G 2004 *J. Aquacult Nutr,***10**(2):135–42
[33] Yoo GY, X Wang, S Choi, K Han, J C Kang and S C Bai 2005 *J. Aquacul.*, **243:** 315-322
[34] Stickney R R 1979 *Principles of Warm Water Aquaculture* (John Wiley and Sons Inc. New York) Pp 223-229
[35] Wang F, Y H Yang, ZZ Han, HW Dong, CH Yang and ZY Zou 2009 *J. Aqua. Int*, **17**: 143–157
[36] Li M H and Robinson EH 1997 *J. World Aquacult Soc,* **28**:402–6
[37] Tawwab M A 2012 *J.International Aquatic Study*, **4**(3):1-13
[38] Carter C G And Sajjadi M 2011 *J. Aquacult Int,* **19**:431-444
[39] Storebakken T, Shearer K D and Roem A J 1998 *J. Aquaculture,***161**(1):365–79
[40] Sugiura SH, Gabaudan J, Dong FM, Hardy RW 2001 *J. Aquacult Res,***32**:583–92
[41] Forster I, Higgs DA, Dosanjh BS, Rowshandeli M 1999 *J. Aquaculture,***179**(1):109–125
[42] Hussain S M, T Hameed, M Afzal, M S Mubarik, M Asrar, S Z H Shah, S Ahmad, M Z H Arzalan, D Riaz, N Tahir, F Amber, M M Shahzad and Tanwir A A K 2011 *International*

J. of Biosciences, 5 (12):173-181
[43] Kornegay E T and Qian H 1996 *Br J. Nutr*,**76**(4):563–78
[44] Robinson E H, Li M H and Manning B B 2002 *J. Appl Aquacult,*12:81–8
[45] Hepher B 1988 *Nutrition on Pond Fisheries* (Cambridge University Press Cambridge USA) p 388

Engineering Technology Of Fish Farming Floating Nets Cages On Polka Dot Grouper (Cromileptes Altivelis) Used Artificial Feed Enriched Phytase Enzyme

Istiyanto Samidjan, Diana Rachmawati

Aquaculture Departement, Faculty of Fisheries and Marine Science, Diponegoro University,Jl.Prof.H.Soedarto SH,Tembalang,Semarang,50275,Indonesia.

Email: istiyanto_samidjan @ yahoo.com

Abstract. One solution is to utilize engineering technology cultivation floating cage net polka dot grouper (ducker grouper), which is given artificial feed enriched with phytase enzymes. The objectives of this study was to examine the use of technology engineering floating net on ducker grouper on artificial feed that is enriched with different dose phytase enzymes to accelerate growth and survival. The research method used ducker grouper fish size 15,5 ± 0,5 cm in the net cages unit (1 m x 1 m x 1 m), 250 fish per cage, using 12 cages. Each net- cages was made of polyethylens netting, mesh size 12.5 mm. with complete randomized design (CRD) 4 treatment and 3 replication were feed Artificial enriched of phytase enzyme with the doses of A (0 FTU • kg–1 diet), B (200 FTU • kg–1 diet), C (500 FTU • kg–1 diet), and D (800 FTU • kg–1 diet) phytase enzyme. Feed was given 2 times a day in the morning and afternoon with 5% biomass per day. Data includes the growth of absolute weight polka dot grouper, FCR, and survival rate analyzed variety and Test Tukey.The result of the research showed that the difference of artificial feeding enriched phytase enzyme significantly (P <0,05) to growth, food conversion ratio (FCR), survival rete of polka dot grouper. The best treatment at C (500 mg / kg of feed) increase growth of absolute weight of 128.75 g, 1.75 (FCR), and a survival rate of 93.5%.

Keyword: cage, floating net, polka dot grouper, artificial feed, phytase enzyme.

1. Introduction

The Marine waters as maritime waters, especially in Jepara Karimun islands have not been utilized optimally. One solution was using of engineering technology cultivation floating net cage on Polka Dot Grouper fish or Humpback Grouper (*Cromileptes altivelis*) was given artificial feed enriched used phytase enzyme that can accelerate the production of fishery industry maritime.

The general condition of Karimunjawa Islands is included in Jepara Regency of Central Java Province. Located west of the city of Jepara with a distance of 45 nautical miles (83 km). Geographically this area is located at 5°40 '-5°57' L.S and 110°4 '-110°40' BT. Karimunjawa Islands has an area of 107,225 hectares of 100.105 Ha of ocean and 7,120 Ha of land spread over 27 Islands [1][2], Of 7,120 ha of land if 10% opened for aquaculture far from coral reefs then 712 Ha used for ponds, especially fish ponds 5% Tiger Grouper or Humpback Grouper (using 356 Ha pond), then the result per Ha 12,500 kg with size 250 g / price Rp.50.000,00 - so its value Rp.625.000.000,00 - with production cost Rp. 375,000,000.00, - still profit Rp. 250,000,000, - 3-month cycle of enlargement. It is a miracle that people on small islands whose livelihoods catch fish around the Karimunjawa Marine National Park can turn to Humpback Grouper culture with high economic value and eco-friendly cultivation technology.

Problems of fishmeal is also one of the most expensive and demanded ingredients and has become the main and most critical ingredient in aquafeed production. Because of the increasing cost and demand of fishmeal has encouraged feed manufacture to search for cheaper alternative protein sources using humpback grouper culture source as plant proteins. Fish nutritionists have tried to use less expensive plant protein sources to partially or totally replace fishmeal. Substitution with other ingredients, especially from plant origin, is likely to compromise nutrient balance and sometime fails to match the energy concentration achieved using fishmeal [1][3][4][5]. Plant protein sources contain high lipid and fiber and lack essential amino acids. Considering the increasing cost of fishmeal and doubt concerning its long-term availability, much research has been carried out to find alternative protein sources as fishmeal substitutes. [6] reported that for aquaculture to make a net contribution to human supplies, the present use of fishmeal in substitute with addition phytase enzyme for aquaculture diets of humpback grouper especially to cuture in floating net cages must be substantially reduced [2][4][5]. Alternative for addition phytase enzyme to increase growth fish culture in floating net cages reported by [7][8][9].

The best of Humpback Grouper fisheries product used fish culture of floating cage but although still experiencing problems in its cultivation of floating cage. The current condition of grouper cultivation with floating net cage system in Karimunjawa Jepara high mortality 70-80%[2][5]. One of the efforts to overcome the high mortality problem is by improving the technical cultivation and adding good quality feed with phytase enzyme, so the growth and survival of grouper is better [2][10][11][12][13]. The advantages of enzyme phytase in the addition of artificial feed given to fish because of phytate of 3.88 g • kg^{-1} diet chelates with mineral such as magnesium (Mg), manganese (Mn), iron (Fe), zinc (Zn), calcium (Ca) and protein which are very beneficial for growth of plant, animal, and human. These elements are also forming poorly soluble complexes which are difficult [14][15][16]. Phytase enzyme has characteristic properties of chelating characteristics. Phytic acid reacts with protein forming phytate-protein complexes and vitamins thus decreasing their bio-availabilty and digestibility [2][17][18][19] showed that protein and phytate compound were barely broken down by proteolysis enzyme, even phytate can hinder pepsin, amylopsin and amylase to work properly. Growth and high survival rate in addition to artificial feed with high protein content is also supported by the maintenance of groupers kept in floating net cages [1][2][4][5]. The purpose of the research is to examine the role of technology using of cultivation floating net cage of Humpback Grouper based on artificial feed used enriched of phytase enzyme with different dosage to accelerate growth and survival of Humpback Grouper fish and to support maritime fishery cultivation industry.

2. Material and methods

2.1 Experimental fish

Humpback Grouper (polka dot grouper) used for the present experiments were seed of Juvenile humpback grouper fish size 15,5 ± 0,5 cm in the net cages unit (1 m x 1 m x 1 m) from Center for Brackish Water Aquaculture, Jepara, Central Java.The examination of growth was done every week.The fingerling used in this study were healthy individuals and had uniform size as suggested modification by [3]. The present experiments were suspended from a floating wooden frame 9 m long by 5.4 m wide. Each net- cages was made of polyethylens netting, mesh size 12.5 mm. at Karimunjawa waters, Jepara. The research method used complete randomized design (CRD) 4 treatment and 3 replication were feed Artificial enriched of phytase enzyme addition.

2.2 Diet preparation

Ingredients and proximate composition of experimental diets are presented in Table 1. Trial feed formulation with addition of phytase enzyme at various doses are presented in table 2.

Table 1. Proximate analysis of compositiont of Feed Composers Materials

Ingredient (g per 100 g diet)	Analysis of materials					
	Moisture	Ash	Crude lipid	Coarse Fiber	Crude Protein	BETN
Fish meal	10,89	22,75	7,98	9,25	45,40	3,73
Soybean meal	11,06	5,65	9,23	5,46	38,71	29,88
Corn meal	13,71	1,77	2,03	0,01	9,38	73,09
Rice meal	12,43	9,25	10,97	18,94	13,62	34,79
Dextrin	10,60	0,20	0,59	0,00	0,10	88,51

The Source: Laboratory of Animal Nutrition and Feed Science, Faculty of Animal Husbandry, Diponegoro University.2017.

Treatment were feed Artificial enriched of phytase enzyme with dosage A (0 FTU • kg^{-1} diet), B (200 FTU • kg^{-1} diet), C (500 FTU • kg^{-1} diet), and D (800 FTU • kg^{-1} diet) based phytase enzyme dosage suggested modification by [3] was optimal dosage used phytase enzyme 500 FTU • kg^{-1} diet. Feed was given 2 times a day in the morning and afternoon with 5% biomass per day. The test feed was prepared with the formulation as follows (Table.2): Table 2. Feed Formulation Test in Research.

Table 2. Feed Formulation Test in Research

Ingredient (g per 100 g diet)	Treatment			
	A (0 FTU · kg–1 diet)	B (200 FTU kg–1 diet)	C (500 FTU kg–1 diet)	D (800 FTU kg–1 diet)
Phytase enzyme	0	0,02	0,05	0,08
Fish meal	34,4	34,5	34,5	34,5
Soybean meal	35	35	35	35
Corn meal	9	8,7	8,5	8,5
Rice meal	8,1	8,1	8,1	8,1
Dextrin	9,8	10	10,21	10,2
Fish oil	1,75	1,75	1,75	1,75
Corn oil	1,75	1,75	1,75	1,75
Vitamine& mineral	1,1	1,1	1,1	1,1
CMC	1,1	1,1	1,1	1,1
Total	102	102	102	102
Energi (kkal)	300,06	300,02	300,02	300,36
Ratio E/P	8,57	8,57	8,57	8,57

A= Experimental feed contained phytase enzyme 0 FTU kg–1 diet, B= 200 FTU kg–1 diet, C= 500 FTU kg–1 diet, and D= 800 FTU kg–1 diet.

2.3. Feeding Trial

The experiments were conducted at Karimunjawa Waters, Jepara, Central Java using Complete Random Design with four treatments and each treatment repeated three times. Phytase enzyme supplemented diet was used in the experiment with the doses of A (0 FTU • kg^{-1} diet), B (200 FTU • kg^{-1} diet), C (500 FTU • kg^{-1} diet), and D (800 FTU • kg^{-1} diet). The enzyme doses used in this study were the modified results from the study by [3][4][5]. The results suggested that the phytase enzyme as much as 500 to 750 FTU • kg^{-1} diet was the optimum level for the growth of humpback grouper with the average weight of 12.4±0.2 g • $fingerlings^{-1}$.

2.4. Culture Cages

The net cages unit (1 m x 1 m x 1 m) used for the present experiments were suspended from a floating wooden frame 9 m long by 5.4 m wide. Each net- cages was made of polyethylens netting, mesh size 12.5 mm. The capacity of the net-cage always submerged in water was calculated to be 3 m3, this capacity was used in calculation of initial stocking density of the fish. The general layout of the cages and their maintenance were as described ealier[1]. Sea cages were suspended from a floating raft in seawater approximately 10 m depth and 0.5 m daily tidal fluctuation. Before the treatment diets, 10 fish were sacrificed for carcass analysis to determine nutritional composition of the initial fish. During the experiment, fish were fed by hand as much related feed as they would consume in 30 min at 07.00 and 16.00. Feeding was carefully monitored to minimize any food wastage. The feeding trial lasted for 12 weeks. Fish were weighed individually at 2-weekly intervals. The individual fish in each cage were weighed using a top loading balance with a precision of 1 g. The experiment finished, four representative fish from each cage were randomly sampled and frozen for determination of the whole body composition according to standard methods[20].

2.5. Parameter of observation

The parameters observed include absolute growth (W), relative growth rate (RGR) and food conversion ratio (FCR), suggested by Tacon [51], ADCP and ADCF according to [28], and Survival Rate (SR) according to NRC [52]. The pH (Jenway 3510), DO (Jenway 970), temperature and Ammoniac (HANNA: HI. 8633, proximate analysis of the percentage of crude protein was determined by micro Kjeldahl analysis with distillation into 4% boric acid and titration with sulphuric acid using methyl red indicator for end point determination.Percentage lipid was determined gravimetrically following a chloroform: methanol extraction of the sample, and moisture was determined by drying at 105oC until constant weight.

The research method using complete randomized design were 4 treatment and 3 replication that are given artificial feed enriched of phytase enzime with different of dosage as follow as are the doses of A (0 FTU • kg^{-1} diet), B (200 FTU • kg^{-1} diet), C (500 FTU • kg^{-1} diet), and D (800 FTU • kg^{-1} diet).. Feed was given 2 times a day in the morning and afternoon with 5% biomass per day.

2.6. Data collection

2.6.1. Absolute Growth. The absolute growth formula is calculated by the formula Tacon [51] , with the following formula:

W = Weight of end test animal research – Weight of animal initial test of

$$RGR : \frac{\text{Final weight} - \text{Initial weight}}{\text{Initial weight xTime experiment}} \times 100\%$$

$$FCR : \frac{\text{The amount of feed consumed} \times 100\ \%}{(\text{Final weight} + \text{Total weight fish deat}) - \text{Initial weight})}$$

$$SR\ (Survival): \frac{(\text{Final count})}{\text{Initial count}} \times 100\%$$

2.7. Statistical Analysis

The experimental design used Completely Randomized Design (CRD) with four dosages of phytase enzyme content as a treatment and each treatment repeated three times. The effect of the treatments was tested using an analysis of variance (ANOVA)[49], which was then used to analyze the data. Before analyzing, the normality, additivity, and homogeneity of the data were first tested. If the analysis of variance was significant ($p<0.05$) or highly significant ($p<0.01$), Tukey test was conducted to find out the mean of the treatment[49]. To determine optimal dose of phytase enzyme, was conducted using Minitab 16 and Maple12. Water quality data were descriptively analyzed.

3. Results and discusson

The results showed that the difference of artificial feeding enriched with different enzyme phytase enzyme showed highly significant (P <0,01) to growth, food conversion ratio (FCR), survival of Humpback Grouper fish, and also to improve the aquaculture industry. The best treatment was at C (500 FTU kg^{-1} diet) with an absolute weight of 128.81 ± 0.85^{a} g, 1.74c (FCR), and a survival rate of 94.03 ± 1.06^{a} %. The quality of water during the maintenance of groupers in floating net cages is relatively good and feasible (Table.3).

Table.3. Absolute growth rate, Relative Growth Rate (RGR), Food Conversion Ratio, Survival of Humpback grouper fish kept in floating net cages in Kep.Karimunjawa waters, Jepara.

	Treatments			
	A (0 FTU · kg–1 diet)	B (200 FTU kg–1 diet)	C (500 FTU kg–1 diet)	D (800 FTU kg–1 diet)
1. Absolute weight growth (g)	123.35 ± 0.38^{b}	$127.24\pm1.00.37^{a}$	128.81 ± 0.85^{a}	126.31 ± 0.99^{a}
2. Relative growth rate (% day)	0.95 ± 0.01^{c}	1.48 ± 0.01^{b}	1.63 ± 9.01^{a}	1.45 ± 0.01^{b}
3. Feed conversion ratio (FCR)	4.11^{a}	3.41^{b}	1.74^{c}	3.05^{b}
4. Survival rate (%)	63.57 ± 0.50^{b}	73.78 ± 1.3^{b}	94.03 ± 1.06^{a}	86.19 ± 0.81^{a}

Information :

Treatmens are given artificial feed enriched of phytase enzime with the doses of A (0 FTU • kg^{-1} diet), B (200 FTU • kg^{-1} diet), C (500 FTU • kg^{-1} diet), and D (800 FTU • kg^{-1} diet).

The same supercrip marks showed no significantly (P> 0.05) difference (Tukey's Test) Based on the Tukey'S test indicating that the mean value of C-A, C-B treatment is very different, but not significantly different from C-D.

3.1. The absolute growth of the Humpback grouper

The results showed that the difference of artificial feeding enriched with different phytase enzyme significantly (P <0.01) on the growth of absolute weight of humpback grouper fish. The addition of phytase enzyme on the diet significantly ($p<0.01$) increased the relative growth rate of milkfish fingerlings as shown in the Table 2. The results gave evidence that supplemented diet with the phytase enzyme could hydrolyze protein that was bind by phytate into amino acid. This amino acid was readily digested and could provide energy to grow [3][4][5][41][42].In addition [3] result of research that the addition of phytase enzyme 500 FTU kg- diet could increase relative growth rate of humpback grouper initial growth weight 26.7 to 31.7 g and specific growth rate 1.4 % to 1.7%, [37]) reported crucian carp *Carassius carassius* by 25 percent better higth compare [2][42]was average weight lower was 9.95%. But compare with reported by Diana & Samidjan [42]. [42] showed (RGR) weight of milkfish 1.95 ± 0.04c %,was given of artificial feed and 2.95 ± 0.29^{b} was given of artificial feed addition of 500 FTU kg- diet phytase enzyme. The same results were also found in carp, African catfish, striped bass rainbow trout, Atlantic salmon Korean Sebastes schlegeli, nile tilapia, tiger prawn, rainbow trout, sea bass, snakehead (*Channa striata* & *Channa micropeltes*), grass carp [3][4][29][42][53][54][55][56], The results of orthogonal polynomial test on the relationship of phytase enzyme in the diet and the relative growth rate (Figure 1) had cubical pattern with the equation Y = - 19.36 X2 + 19.61 X + 123.0, R2=0.919. The optimum dose of the phytase enzyme in the diet on the relative growth rate was 506 FTU kg- diet with the maximum absolute growth of 128.015 g.

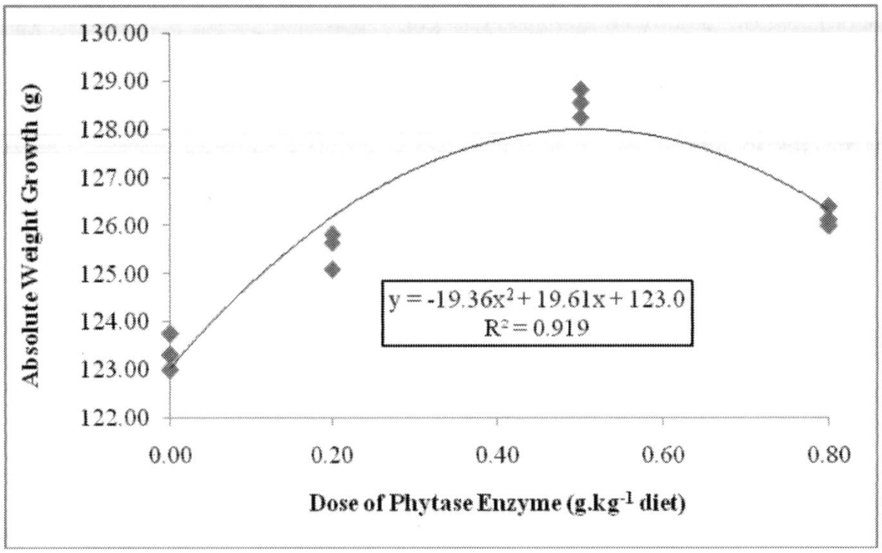

Figure 1. Graph of orthogonal polynomial absolute weight of Humpback grouper (C. altivelis). Information : 506 FTU kg- diet with the maximum absolute growth of 128.015 g.

Furthermore, based on the results of analysis of variance showed that the presence of artificial feeding enriched phytase enzyme different dosage highly significant (P <0.01) on the average growth rate of humpback grouper

It is assumed that there is a real influence due to grouper fish that can utilize artificial feed enriched fitase enzyme different dose independently and supported by good water quality condition, so as to increase tiger grouper fish growth. This is consistent with the opinion of Shapawi et al [4][5] reported that maintaining good water quality along with suitable and nutritious feeding can increase the growth and survival of humpback grouper. Based on the Tukey'S test indicating that the average value of C-A, C-B treatment is very different, but not significantly different from C-D. Based on Table 3, in the presence of different doses of enzyme phytase there is a very significant (P<0.0) effected on growth. It is found that the absolute growth very high were treatment C was 128.55 ± 0.29a g and the lowest A (123.35 ± 0.38[b]) .The growth of humpback grouper shows the high of this is the same opinion by Shapawi et al. [4][5], [27] showed final weight 26.7 g to 31.7 g lower with result experiment.

3.2. Relative Growth (RGR)

Relative growth rate shows that the highest at C (1.63±9.01%) (Table 3). The analysis of variance showed that the presence of the addition of phytase enzyme on the diet highly significant (p<0.01) increased the relative growth rate of humpback grouper as shown in the Table 3. The results gave evidence that supplemented diet with the phytase enzyme could hydrolyze protein that was bind by phytate into amino acid. This amino acid was readily digested and could provide energy to grow [2][3][4][5], [41][42]. The result showed elative growth rate of humpback grouper that the highest at C (1.63±0.01%) but compare with humpback grouper 1.4 %[4][5], Specific growth rate (SGR) of tiger grouper 0.96 to 1.06 %[3][4][5], SGR of tiger grouper 1.12 to 1.38 %[3][4], and milkfish lower RGR 2.95 ± 0.29b % [41][42]. The results of orthogonal polynomial test on the relationship of phytase enzyme in the diet and the relative growth rateof humpback grouper (Figure 2) had quadratical pattern with the equation , Y = -2.6857 X2+2.617 X +0.8495, R2=0.88, The optimum dose of the phytase enzyme in the diet on the relative growth rate was 487 FTU kg- diet with the maximum relative growth rate of 1.49 % per day.

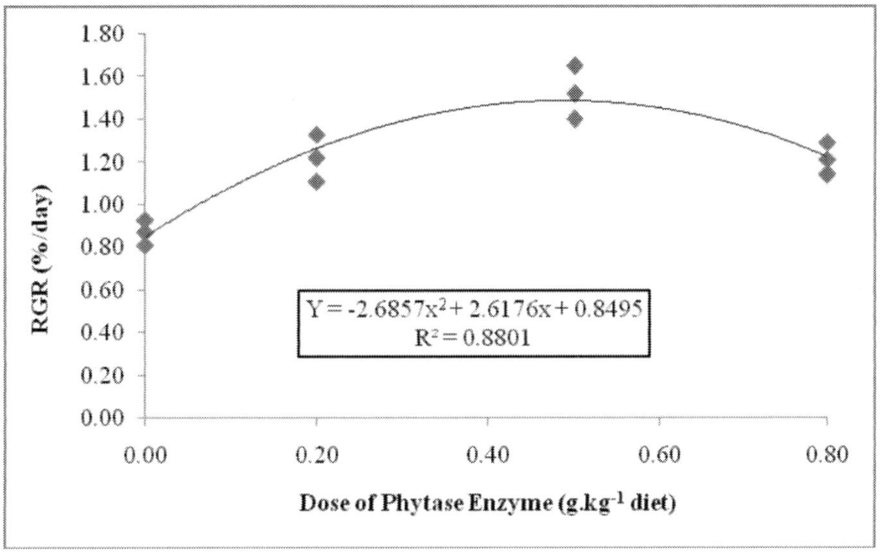

Figure 2. Graph of orthogonal polynomial RGR of Humphback grouper (*C. altivelis*).

The optimum dose of the phytase enzyme of RGR 487 FTU kg- diet with the maximum relative growth rate of 1.49 % per day.

[3] also reported of phytase enzyme 500 to 750 FTU kg- diet could increase average weight of humpback grouper final weight 26.7 g to 31.7 g and SGR 1.4% to 1.7 %. In addition by Yu & Wang (2000) that showed crucian carp Carassius carassius by 25 percent. The same results were also found in tiger grouper [2][3][4][5], African catfish [53][54], striped bass [55]. Based on the Tukey'S test indicating that the average value of C-A, C-B treatment was very different, but not significantly different from C-D. This shows that relatively good growth of C (1.63±9.01%) dose of phytase enzyme (500 FTU • kg–1 diet) compared with [3][4][5] that the daily growth rate of hump grouper SGR 1.4 to 1.7 %, during maintenance 60 days fed "pellet enrichment of phytase enzyme" with a protein content of final weight 26.7 to 31.7 g[3][4]. Furthermore ,[4] stated that the daily growth rate of humpback grouper and tiger grouper is different of SGR , because tiger grouper better growth rate and better feeding rate.The cultivation of humpback grouper must be used by feed in protein content of 41.20 to 41.55%, indicating that daily growth rate observations were still good [3][50].

The addition of phytase enzyme on the feed 500 FTU kg- diet significantly increased the growth of humback grouper fingerlings compared to the addition of 200 and 800 FTU phytase enzyme per kg- diet. It can be concluded that the addition of 800 FTU phytase enzyme on the every kg feed could reduce antinutrients or phytate on soybean meal. [3] also reported that the addition of phytase enzyme 750 FTU kg- diet could increase average weight of humpback grouper.

3.3. Food conversion Ratio

The lowest humpback grouper feed conversion showed that the feed was very efficient, this was found in treatment C(500 FTU kg–1 diet) of feed with FCR (1.74c) (Table 3).

The addition of phytase enzyme significantly ($P<0.01$) affected on the feed conversion ratio of humpback grouper. conversion ratio better. The same result was also found for absolute growth and relative growth rate. Based on the Tukey'S test indicating that the mean value of middle treatment of CA, CB is very different, but not significantly different from the CD. The low feed ratio C ratio with FCR (1.45 ± 0.3) is more efficient feed utilized by Humpback Grouper (C. altivelis) [3][4][50] was suggested that phytase enzyme can break down phytate, losing the binding between phytate acid and protein and minerals compound. It would positively increase activity to convert trypsinogen into trypsine enzyme which broke down protein into amino acids. In addition by Abrehouch, et al.[47], It can increase feed utilization and reduce feed conversion ratio. Wang et al. [6][35] also reported that the addition of phytase enzyme the soybean diet for rainbow trout also made feed conversion ratio better. The same result was also found for *L. rohita*[13][21]. C treatment (500 FTU kg⁻ diet) resulted in the lowest feed conversion ratio among other treatments, B (200 FTU kg⁻ diet), D (800 FTU kg- diet), and A (0 FTU kg⁻ diet). Therefore to improve the feed efficiency is required addition of phytase enzyme in diet utilization and make feed conversion ratio low. [3][4][51][52] reported using in addition of 250 units or more microbial phytase enzyme in the diet showed increase feed the consumption to increase growth but lower of food conversion ratio. The results show[15][16][21][22][23][24] [29][30] that there were higher feed consumption, higher weight gain, and lower feed conversion ratio than those without addition [21][22] [23] [24][25] [26][27] [28][31] [32][33][59]. The relationship between phytase enzyme in diet and the survival based on the orthogonal polynomial test, as shown in the Figure 3, was quadratical. The equation was Y= 11.24 X2- 11.022 X +4.5152 and R2 =0.6322. The optimum dose of the phytase enzyme in the diet on the convension ratio was 490 FTU kg- diet with the maximum of FCR 1.81 (Figure .3).

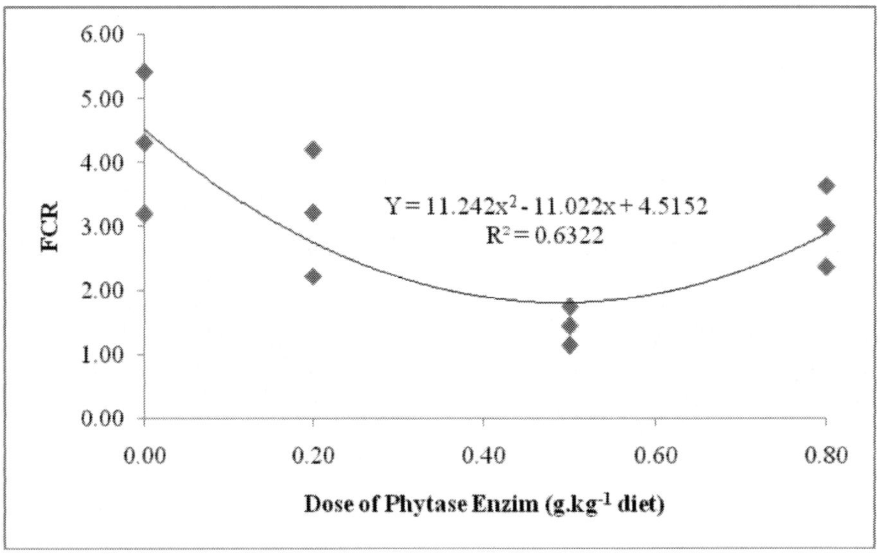

Figure 3. Graph of orthogonal polynomial FCR of Humphback grouper (*C. altivelis*).

3.4. Survival Rate of Humpback Groupers

The highest Humpback grouper life was found in treatment C (93.33%) (Table 3).The existence of feeding and good cultivation environment gave a highly significant effect on the survival of Humpback grouper fish (P <0.01), as shown in the Table 3. The result was in line with the [3][4][5][51] showed finding that survival rate was highly significant affected by addition of phytase enzyme in the diet. gender, heredity, age, reproduction, disease resistance and external factors such as water quality, density, number. The same result was also found [2][38][39][40][41][42], showed the survival rate was affected by internal factors such as using technology of floating net cages is very good and supporting of environment Karimunjawa waters in live of Humpback grouper [34][35][36][37] showed can increase the same of survival rate was highly significant affected by addition of phytase enzyme.

The relationship between phytase enzyme in diet and the survival based on the orthogonal polynomial test, as shown in the Figure 4, was quadratical. The equation was
Y= -81.03 X2+97.49 X +61.38 and R2 =0.917. The optimum dose of the phytase enzyme in the diet on the survival was 602 FTU kg- diet with the maximum value of survival 90.71% (Figure .4).

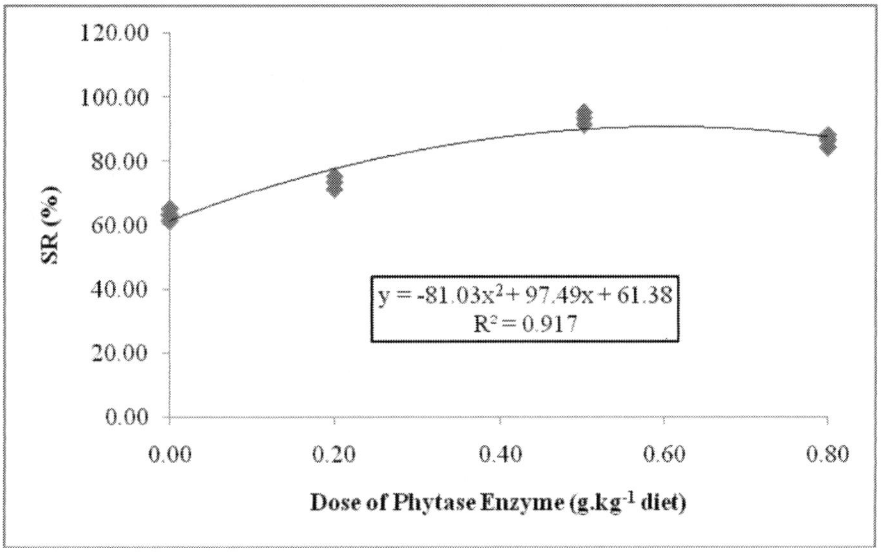

Figure 4. Graph of orthogonal polynomial survival of Humphback grouper (*C. altivelis*).
Information :
The same superkrip marks show no significant difference (Tukey's Test) P> 0.05
The optimum dose of the phytase enzyme in the diet on the survival was 602 FTU kg- diet with the maximum value of survival 90.71%

Shapawi et al [3][4][5] reported that high survival rate 90±10% to 96.67±5.77%, in addition by Moreover, [52] reported that the phytate content 0.5% in the diet could decrease the growth and feed efficiency for rainbow trout (*O. misskis*). [17][49][51][52] also suggested that 2.58% phytate in the diet could decrease growth, feed efficiency, protein efficiency, and also cause mortality and composition of amino acid in the diet [3][40][51]. The water quality during the research was suitable for cultivating the humpback grouper. The measurement of the observed parameters of water during cultivation of humpback grouper can be seen (Table.3). Result of treatment C showed better survival was 94.03±1.06a than B (200 FTU • kg–1 diet) was 73.78±1.3b %, D (800 FTU • kg–1 diet) 86.19±0.81a %and lowest was A (0 FTU • kg–1 diet) of survival 63.57±0.50b %

The existence of feeding and good cultivation environment gave a very real effect on the survival of humpback grouper fish showed highly significant (P <0.01). Based on the Tukey'S test indicating that the average value of C-A, C-B treatment was very different, but not significantly different from C-D.The result of our reserch different reported by Rachmawati & Samidjan [42] the addition of phytase enzyme in the diet insignificantly affected on survival rate of milkfish fingerlings. Moreover, [3][51][52] reported shows in line [4], finding that survival rate was insignificantly affected by addition of phytase enzyme in the diet. [3][4] the addition of phytase that survival rate was insignificantly affected. This is in accordance with the opinion of [3], with good feeding and cultivation environment gave a very real effect on the survival of Humpback grouper. Addition by Shapawi et al. [4][5] that by maintaining tiger grouper better SGR higher with humpback grouper whose maintenance media using biofilter combination of biology and chemistry as well as water intermediary of grouper fish maintenance can improve water quality and increase the life of Humpback Grouper [2][27] results showed that the survival of Humpback Grouper grouper fish kept in floating net cages in Karimunjawa waters of Jepara regency, Central Java province showed a relatively higher survival compared with those reported by Shapawi et al.[3][4], reported that the survival of tiger grouper fish kept in floating net cages ranged from 55 to 60%. In addition of opinion [34] reported to increase of survival rate to improve enviroment waters in floating cages, in the same reported by Shapawi et al [3][4] used addition of phytse enzyme 500 FTU increase survival rate of humpback grouper. In the same opinion by Chateau et al. [14] reported environment good support to increase survival rate fish culture in float net cages in the same to increase in floating cages of fish [45][49] [57].

3.5. *Water quality*

Water quality during the experiment of humpback grouper fish kept in floating cages showed its feasibility at the end of the study is still alive well (Table 4).

Table 4. The Water quality during the maintenance of humpback grouper in floating net cages

Parameter	unit	Range to (Research)	Reference	
Temperature	o C	27.5-30.5	26-33	Shapawi et al.(2007)
pH		7.2-8.1	7-8.2	Shapawi et al.(2007)
Salinity	ppt	25-30.5	25-32	Shapawi et al.(2007)
Dissolved Oxygen	mg/l	5.27-6.62	7-May	Shapawi et al.(2013[b,c])
NH_3	mg/l	0.011-0.017	0.02	Shapawi et al.(2007)
NO_2	mg/l	0.01-0.05	0-Jan	Shapawi et al.(2013[b,c])

Based on Table 3, it shows that the temperature range is 27-30 oC, pH 7.1-8.1, salinity 24.5-28.0 ppt, dissolved oxygen 5.25-6.50 mg / l, NH3 0.012-0.018 mg / l and nitrite 0.03-0.08 mg / l. It shows the proper water quality for tiger grouper cultivation[2][3][4][5][6] [32][33][41][42[43][44][45][46] added also that the role of aquatic environments and habitat of sediments elements of sand and gravel with 85% composition and calcareous algae from Halimeda sp can improve the survival of groupers and improve water quality for grouper life [2][12][13][20] [58].

3.6.*Aquaculture Industry Humphback grouper fishery with floating net cages system*

The development of tiger grouper fish breeding in Karimunjawa waters of Jepara with floating net cages, the floating nets cages ware very fast and can improve the welfare of the community around Karimunjawa, Jepara districts. The results shows that Grouper duck (*Cromileptis altivelis*) at the producer level costs about Rp 140,000, - to Rp 300.000, - kg^{-1} diet. Tiger grouper fish belongs to the type of euryhalyne, which has a relatively wide range of salinity tolerance, so that tiger grouper other than can be cultivated in the sea, can also be cultivated in ponds. While the price of grouper sunu (*Plectropomus leopardus* and *P. maculatus*) at the producer level cost around Rp 180.000, - to Rp 300.000, - . kg^{-1} diet Grouper sunu can be cultivated in floating net cages and easy on the sea. Muddy grouper '; This is in line with opinion. In adition by Shapawi et al. [3][4] there are several important types of coral reef fish in Indonesia, including *Serranidae, Lutjanidae, Kyphosidae, Lethrinidae, Acanthuridae, Mulidae, Siganidae, Labridae* and *Haemulidae* [3][4][10] [11][12].

Humpback grouper fish as one of the commodities studied in this study included the *Serranida*e family. The characteristics of *Serranidae* according to Shapawi et al [3][4][21] showed have characteristics such as solitary, usually hiding in caves or under corals, up to 2 meters long, weighing up to 200 kg, and belonging to carnivores, eating fish, shrimp and crustaceans. According to Suhartono & Samidjan [60] showed humpback grouper (*Cromileptes altive*lis) has the highest price, between Rp 350.000, - to Rp 400,000, - kg^{-1} diet at the producer level (fishermen and fish farmers).

As for the seeds of 4-5 cm, the price per tail reached Rp 5,000, - to Rp 7,000, -. Humpback Grouper (*C. altivelis*) at the producer level costs about Rp 140,000, - to Rp 300.000, - kg^{-1} diet. Tiger grouper fish belongs to the type of euryhalyne, which has a relatively wide range of salinity tolerance, so that tiger grouper other than can be cultivated in the sea, can also be cultivated in ponds. While the price of grouper sunu (*Plectropomus leopardus* and *P. maculatus*) at the producer level cost around Rp 180.000, - to Rp 300.000, - kg^{-1} diet. Grouper sunu can be cultivated in floating net cages and easy on the sea of Mud groupers. Business Feasibility Study of Coral Reef Fishery, Grouper and Seaweed Cultivation 2-3 (*Epinephelus suillus / E. tauvina* and *E. coioides*) cost around Rp 100.000, - to Rp 200.000, - per kg. Mud grouper fish can be cultivated in floating net cages, easy and ponds. While malabar grouper (*Epinephelus malabaricus*) is often also called as grouper. Added by Afero et al [10] examines the feasibility analysis of tiger grouper size 400 grams Indicator Value Conclusion NPV (Rp) 85,790,261 Deserves IRR (%) 52% Worth Payback Periods 1 Year 5 months 11 days Eligible BC Ratio (PV) 1.14 Eligible, but different Economic analysis of tiger grouper *E. fuscoguttatus* and Humpback Grouper *C. altivelis* commercial cage culture in Indonesia [10] showed analysis of the financial performance of humpback grouper that all production scales can be regarded as marginally viable with a projected cumulative cash flow for 5 years for small-scale (IDR 133,344,826.52), medium-scale (IDR 446,749,192.00) and large-scale (IDR 1,692,212,500.00) farms. Application of the 15% discount rate produced a NPV for small-scale (IDR 85,009,002.45), medium-scale (IDR 286,822,375.91) and large-scale (IDR 1,088,181,355.75) farms over a 5-year period. Further analysis of financial performance highlighted a benefit cost ratio over 2, an internal rate of return greater than 300% for all production scales and a payback period of\1 year.

4. Conclusions and recommendations.

4.1. *Conclusion*

The results showed that the difference of artificial feeding enriched with different phytase enzyme showed highly significant (P <0,01) to growth, food conversion ratio (FCR), survival of humpback grouper fish, and also to increase cultivation of fishery industry. The best treatment at C treatment (500 FTU kg- diet) were absolute growth (128.81±0.85^{a}g), relative growth rate (1.63±9.01^a %), Feed conversion ratio (FCR) (1.74c) and survival rate (94.03±1.06^a %).

The optimal doses of phytase enzyme on growth performance, Feed conversion ratio (FCR) and survival rate were in humpback grouwer range 490 FTU kg- diet to 602 FTU kg$^-$ diet.

4.2. *Recommendation*

It is necessary to have a follow-up on artificial feeding in addition phytase enzyme using of the optimal doses of phytase enzyme on growth performance, Feed conversion ratio (FCR) and survival rate were in humpback growth range 490 FTU kg- diet to 602 FTU kg- diet.

5. Acknowledgements

This work was supported by grants from Hibah programme Ristek 2012. The authors are indebted to Mr. Sutarmin chairman of POKDAN fish farmers in Karimunjawa who has lent his floating net cages to research and Dean of Fisheries and Marine Science Faculty Diponegoro University who has provided research infrastructure,

References

[1] Abrehouch A, A A Ali, K Chebbaki, H Akharbach, M Idaomar 2010 *J. Agriculture and Biology Journal of North America*. **1**(3): 175-184.
[2] BTN Karimunjawa 2002 *Statistics and Monographs of Karimunjawa Islands, Jepara Regency. Sea Park National Park, Karimunjawa Islands.*
[3] Shapawi R, W N Keong, S Mustofa 2007 *J. Aquaculture* **273** (118-126)
[4] Shapawi R, Ebi I, Yong A, Chong M, Chee L K, Sade A 2013 *J.Agricultural Sciences*

4(6A):19–24
[5] Shapawi R, I Ebi, A Yong 2013 *J. Agricultural Science* Vol **4**(7) 317-323
[6] Wang, M S, Yan Y. Wang Z R. Lun, and T B Yang 2013 *J. Aquaculture Research* 1-5
[7] Fox J M, Addison L L, Anthony J S, Allen D, Denis RM, Elizabeth CD, and Tzachi M S 2006 Phytase supplementation in aquaculture diets improves fish, shrimp growth performance.
[8] Gondwe M J S, S J Guildford, R E Hecky 2011 *J.of Great Lake Reaserch 37,* 93-101.
[9] Garcia F, O J Sabbag, JM Kimpara, D M Romera, N S Sousa, E M Onaka, L P Ramos 2017 *J. Aquaculture* 479, 838-844.
[10] Afero F, Miao S, Arlenie A Perez A A 2010 *J. Aquacult Int* **18**:725–739.
[11] Baliao D D, M A D Santos, N M Francos, N R S Jamon 2000 Grouper culture in floating net cages (Aquaculture Extension Manual No.29)
[12] Barcellos L, J G, L C Kreutz, R M Quevedo, I Fioreze, L Cericato, A B Sos, M Fagundes, J Conrad, R K Baldissera, A Bruschi and F Ritter 2004 *Journal Aquaculture* 232, 383-394.
[13] Baruah K, Sahu N P, Pal A K, Debnath D, Mukherjee S C 2007 *J.Aquaculture Research* 38(2):109–120.
[14] Chateau, P A,Y C A Huang.,C A Chen,Y Chang.,2015. *J.Ecological Modelling* 299,140-146.
[15] Chen C P, Hsien, H L and Chang K H 1980 *Some aspect of the sex chance and reproductive biology of the Grouper, Epinephelus diacanttus* (*Cuvier valenciencis*).
[16] Chor W K, L S Lim, M Chong, K C Lu, A Sade, R Shapawi 2015 *Malaysian J. of Science.* **34** (1): 58-68.
[17] Liu B, Rafig A, Tzeng Y M, and Rob A. 1998 *Enzyme Microbial Technology.j.* 22:415–424.
[18] Li M H, Robinson E H. 1997. J World Aquacult Soc **28**:402–406.
[19] [NRC] National Research Council 1993. *Nutrient requirements of fish.* [Washington DC: National Academy Press]. 124p.
[20] AOAC[Association of Official Analytical Chemists] 1990 *Official Methods of Analysis.* [15th ed. Washington DC: Association of Official Analytical Chemists)]. 1298p.
[21] Baruah K, Sahu NP, Pal AK, Debnath D 2004. J.*World Fish Center Quarterly* **27**:15–19.
[22] Chua T E and Teng S K 1974 *in* Chua and Teng, 1982. Effects of food ration on growth condition factor, food conversion efficiency, and net yield of estuary Grouper, *Epinephelus salmoides*. [Maxwell, culture in Floating net cages].
[23] Chua T E and Teng S K 1978 *Reared in floating net cages aguaculture,J.* **16**:219-232.
[24] Chua T E and Wong S K 1978. *Aquaculture,J.* **16**:219.
[25] Coleman F C, C C Koenig, K M Scanlon, S Heppell, S Heppell and M W Miller 2010 The Open Fish Science J, 2010, 3, 1-15.
[26] Huang Y, X Huang, Y Yan, J Cai, Z Ouyang, H Cui, P Wang and Q Qin 2011 *BMC Genomic J.* **12**(556), 1-12.
[27] Hussain N A and M Higuchi. 1980. *Aquaculture J.* 19. 339-350.
[28] Fenucci J L 1981 *Studies on the nutrition of marine shrimp of the Penaeus*. [Ph.D. Dissertation]. Houston, Texas, USA: Faculty of Department of Biology, University of Houston. 185p.
[29] Rachmawati D and I Samidjan 2017 *Philippine J. of Science* **146** (3):237-245.
[30] Forster I, Higgs D A, Dosanjh B S, and Rowshandeli M 1999. *Aquaculture J. 179(1):109–125.*
[31] Randall and B Tuvia. 1983. *Bull of Marine Sciences* J.:33(2):373-426.
[32] Ransangan J, B O Manin, A Abdullah, Z Roli., E F Sharudin 2011 *J. Aquaculture* 315, 327-334
[33] Sedgwick, R W 1979. *J. Aquaculture,* **16** : 7 - 30.
[34] Turja R, N Hoher, P Snoeijs, J Barsiene, L Butrimaviciene, T Kuznetsova, S V, Kholodkevich., M H Devier, H Budzinski, K K Lehtonen 2 014 *J. Aquacultre* 473-474, 398-409.
[35] Wang X, Choi S, Han K, Kang J C, and Bai S C 2005 *J. Aquaculture* 243: 315–322.
[36] Watanabe T, C Kitajima,S Fujita 1983. *Nutritional Values of Live Organisme Usea in Japan for Mass Propagational Fish. A Review Aquaculture 34 Elsever Science.* Publisher B-V

Amsterdam, Netherlandns.
[37] Yu F N, Wang D Z 2000. *J Fish Sci Chin* 7(2):106–109.
[38] Hassan MS, Soltan MA, H M Agouz, A M Badr 2013 *Egyptian J. of Aquatic Research 39(3):205–213.*
[39] Hassan S, Altaff K, and Satyanarayana T 2009 *Pakistan Journal of Nutrition J.* 8(4):341–344.
[40] Hepher 1988. *Nutrition on pond fisheries.* [Cambridge, USA: Cambridge University Press]. 388p.
[41] Rachmawati, D, I Samidjan, M, and Mel. 2016 *J. Teknologi* 78(2–4):39–43.
[42] Rachmawati D, I Samidjan, and M Mel 2017 *Philippine J. of Science.* 146 (3):237-245.
[43] Paquotte P L, Chim, J L M, Martin, E Lemos, M Stren, G Tosta. 1998 *J. Aquaculture 168*, 151-166.
[44] Pescod M B 1973 *Investigation Of Rational Effluent And Stream Standart for Tropical* [Countries. San Fransisco].
[45] Degefu, F S, Mengistu, M Schagerl 2011 *Aquaculture J.* 316, 129-135.
[46] Demetrio J A, L C Gomes, J D Latini 2012 *J. Aquaculture* 330-333, 172-178.
[47] Gatlin III D M, D M Barrows, F T Brown, P Dabrowski, P Gaylord, T G Hardy, R W Herman, E Hu, G Krogdad, S Nelson. 2007. *Aquaculture Research J.* 38:551-579.
[48] Gopakumar G 2009. *Cultured in floating net cages J.*
[49] Stell R G D, Torrie J H and Dickey D A 1996. *Principles and procedures of statistics,* [3rd ed. New York: McGraw Hill International Book Company, Inc]. 672p.
[50] Suloma.A and H.Y. Ogata 2011 *J Aquac Res Development 2011, 2:2*
[51] Taco 1987 *Nutrition and Farmed Fish and Shrimp.* [Training Manual]. Brazil: The Essential Nutrients Food anf Agricultural Organization of the United Nations].
[52] [NRC] National Research Council 1993. *Nutrient requirements of fish.* [Washington DC: National Academy Press]. 124p
[53] Papatryphon E, Howell R A and Soares J H 1999 *J World Aquacult Soc* 30:161–173.
[54] Sajjadi M, Carter C G. 2004 *J Aquacult Nutr 10(2):135–142.*
[55] Sajidan A, Farouk A, Greiner R, Jungblut P, Mueller E C, Borriss R. 2004 *Applied Microbiology and Biotechnology J.* **65**(1):110–118.
[56] Schaefer A, Koppe W M, Meyer-Burgdorff K H, Guenther D. 1995 *Water Science and Technology J.*; 31:149–155.
[57] Chua T E and S K Ten 1979 *Marine Biology J* **54**, 363-372.
[58] Madin, J, V C. Chong., B Basri 2009 *J. Aquaculture* 83, 19-29.
[59] Korsoen O J, T Dempster, P G Fjelldal, F Oppedal T S Kristianse 200 *Aquaculture J.* 296, 373-381.
[60] Suhartono and I Samidjan 2014 J.*Applied Mathematical Sciences*, Vol. **8**, 2014, no. 179, 8921 – 8928.

The Diversity of Vibrios Associated with Vibriosis in Pacific White Shrimp (*Litopenaeus vannamei)* from Extensive Shrimp Pond in Kendal District, Indonesia

Sarjito[1]*, Alfabetian Harjuno Condro Haditomo[1], Desrina[1], Ali Djunaedi[2] and Slamet Budi Prayitno[1]

[1]Aquaculture Department, Faculty of Fisheries and Marine Science Faculty, Diponegoro University,
Jl.Prof. H. Soedarto SH., Tembalang, Semarang 50275, Indonesia
[2]Department of Marine Science, Faculty of Fisheries and Marine Science Faculty, Diponegoro University,
Jl.Prof. H. Soedarto SH., Tembalang, Semarang 50275, Indonesia

Email: sarjito_msdp@yahoo.com

Abstract. Vibriosis out breaks frequently occur in extensive shrimps farming. The study were commenced to find out the clinical signs of white shrimp that was infected by the Vibrio and to identify the bacterial associated with vibriosis in the pacific white shrimp, *Litopenaeus vannamei*. Bacterial isolates were gained from hepatopancreas and telson of moribund shrimps that were collected from extensive shrimp ponds of Kendal District, Indonesia and cultured on *Thiosulfate Citrate Bile Salts Sucrose Agar* (TCBSA). Isolates were clustered and identified using repetitive sequence-based polymerase chain reaction (rep-PCR). Three representative isolates (SJV 03, SJV 05 and SJV 19) were amplified with PCR using primers for 16S rRNA, and sequence for further identification. The clinical signs of shrimps affected by vibrio were pale hepatopancreas, weak of telson, dark and reddish coloration of smouth, patches of red colour in part of the body on the carapace, periopods, pleuopods, and telson. A total of 19 isolates were obtained and belong to three groups of genus *Vibrios*. Result of the 16S DNA sequence analysis, the vibrio found in this study related to vibriosis in white shrimps from extensive shrimp ponds of Kendal were closely related to *Vibrio harveyi* (SJV 03); *V. parahaemolyticus* (SJV 05) and *V. alginolyticus* (SJV 19).

Keywords: White Shrimp, Vibriosis, Rep PCR, 16S r RNA

1. Introduction

Pacific white shrimp (*Litopenaeus vannamei*) is an important aquaculture species that has high economic value for export commodities from Kendal. This shrimp farms in this area usually use intensive and extensive technology to increase its production. However, inadequate intensive shrimp management or low quality management of extensively pond will cause disease problems, such as bacteria [1] and mass mortality [2]. Outbreaks of bacterial diseases, such as vibriosis, are still a problem in shrimp farming [3-4]. Vibrio infection in pacific white shrimp (*L. vannamei*) is characterized by pale or reddish carapace, pale hepatopancreas, root and redish of telson; red and root antenna [1]. Although this is a normal part of the bacterial flora in the environment of estuary and marine waters [4-5], some species of vibrio are considered opportunistic pathogens against finfish, shellfish and shrimp [5]. Therefore, the genus Vibrio can cause production failure due to vibriosis in shrimp culture [3-4][6], resulting on death at the stage of zoe, larvae and adult [7].

Various studies have been conducted to identify vibrio associated with vibriosis in the aquaculture industry [2]. Some vibriosis agents in shrimp have been identified, such as: *V. harveyi* [8-9]; *V. alginolyticus* [2]; *V. parahaemolyticus* [2][7]; *V. fluvialis, V. damsela and V. vulnificus* [7]. Previous studies using biochemical methods to identify vibrio diversity in this area. Hence, the classification of bacteria with molecular approaches should be applied to develop early detection of disease caused by viruses, bacteria, including vibriosis in shrimp [9].

2. Materials and Methods

2.1. Sample of Shrimp

A total 31 moribound pacific white shrimp *(L. vannamei)*, suspected having vibriosis infection, were collected from extensive brackish water pond in Kendal district located on Wakak River, Kendati, Indonesia. The identification of pacific white shrimp with vibriosis is based on the clinical symptoms of references suggested by Lighter [6]. This research applies exploratory method with purposive random sampling. The isolation of bacteria in the sample was done at the Aquaculture Laboratory of Fisheries and Marine Science Faculty of Diponegoro University.

2.2. Isolation of Bacteria.

The vibrio bacteria were isolated from hepatopancreas and telson from pacific white shrimp (*L. vannamei*) with streak plate methods on Thiosulfate Citrate Bile Salts Sucrose Agar (TCBSA) and it incubated at room temperature for 24-48 hours [10]. The colonies were randomly drawn and purified by making streak plates based on their morphological features [11].

2.3. Repetitive-PCR

Rep-PCR was done based on the method previously described by [10] which has been modified by [13]. BOX AIR (5'-CTACggCAAggCgACgCTgACg-3') was applied in the rep–PCR. The REP 1R-I and REP 2-I primers contain the nuclutide inosine (I) at ambigious potitions in the REP consensus. PCR reaction contained of 1 μL DNA template (diluted 100x), 1 μL primer, 7.5 μL Megamix Royal and sterile water up to a total volume of 05 μL.

Amplification was performed by thermal rotor model of Gene Amp PCR 9700 with the following temperature conditions as follows: initial denaturation at 95°C for 5 minutes; 30 cycles of denaturation (92° C for 1 minute), annealing (50°C for 1.5 minutes), extension (68°C for 8 minutes) and final extension at 68°C for 10 minutes. Five microliter aliquot PCR products were carried out using electrophoresis on 1% bromade gel ethers using 1X TBE buffer . The result was visualized in UV trans-illuminator and the image was captured by the gel documentation system (Biorad) [9].

2.4. Grouping of Isolates

The grouping of isolates was performed based on the method of making matrixes from the position of bands on the gel analyzed using the Free Tree program. This program uses UPGMA method for tree creation. Resampling was done by bootstrapping, 1000 replications [10]

2.5. PCR Amplification and Sequencing of 16s rRNA Gene Fragments

PCR amplification was initiated based on the method of Radjasa *et al.* [10][12] which has been modified by Sarjito *et al.* [13]. Two primers, GM3F (5'AGAGTTTGATCMTGGC-3') and GM4R (5'-TACCTTGTTACGACTT-3') were used to amplify the nearly complete 16S rRNA gene [12]. The genomic DNA of the causative agent of strain vibriosis for PCR analysis was obtained from the cell material taken from the agar plate, which was suspended in sterile water (Sigma, Germany) and subjected to five freezing cycles at -80°C and melting at 95°C. PCR amplification of 16S partial bacterial rRNA gene [14], PCR product purification and subsequent sequencing analysis were performed in accordance with [10][12-13]. The sequence of DNA strains determined then compared to the homology to the BLAST database. To determine the relationship of the genus vibrios, the phylogenetic stage begins in accordance with the software Mega 1.

3. Results and Discussion

3.1. Result

3.1.1. Characteristic of the Bacterial Isolates

The clinical signs of shrimp affected by vibrios were pale hepatopancreas, weak of telson, dark and reddish colorozation of mouth, patches of red colour in part of the body, periopods, pleuopods and telson (Figure 1).

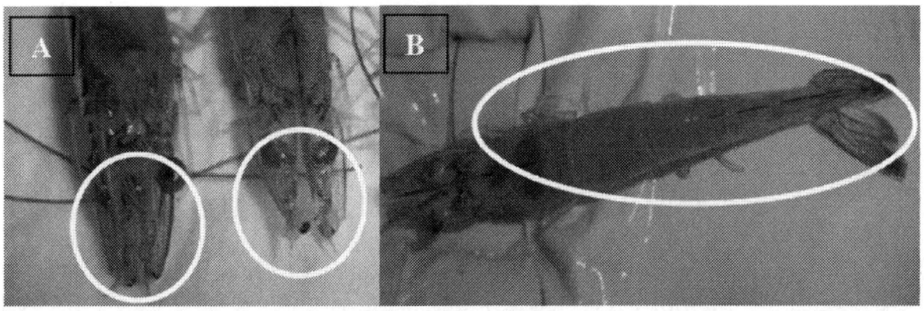

Figure 1. Clinical sign of pacific white shrimps, *L. vannamaei,* infected by vibriosis: (a.) Dark and reddish colorization of mouth; (b.) Reddish the carapace, periopods, pleuopods, mouth and telson.

Table 1. Characteristic of Isolates vibrios Associated with Vibriosis on Pacific White Shrimps, *L. vannaei,* from shrimp pond in Kendal District, Indonesia.

No.	Isolate code	Media	Source	Colony		
				Colour	Form	Characteristic
1	SJV01	TCBS	Hepatopancreas	Yellow	Rounded	Convex
2	SJV02	TCBS	Hepatopankreas	Yellow	Rounded	Convex
3	SJV03	TCBS	Hepatopankreas	Yellow	Rounded	Flat
4	SJV04	TCBS	Hepatopankreas	Yellow	Oval	Convex
5	SJV05	TCBS	Hepatopankreas	Yellow	Oval	Convex
6	SJV03	TCBS	Hepatopankreas	Yellow	Rounded	Convex
7	SJV03	TCBS	Hepatopankreas	Yellow	Rounded	Convex
8	SJV08	TCBS	Hepatopancreas	Yellow	Rounded	Convex
9	SJV09	TCBS	Hepatopancreas	Green	Rounded	Convex
10	SJV10	TCBS	Hepatopancreas	Yellow	Rounded	Convex
11	SJV11	TCBS	Hepatopancreas	Yellow	Rounded	Convex
12	SJV12	TCBS	Hepatopankreas	Yellow	Rounded	Convex
13	SJV14	TCBS	Hepatopankreas	White	Irregular	Convex
14	SJV05	TCBS	Hepatopancreas	White	Irregular	Rough
05	SJW17	TCBS	Hepatopankreas	Yellow	Rounded	Concave
16	SJV16	TCBS	Hepatopancreas	Yellow	Rounded	Rough
17	SJV17	TCBS	Hepatopancreas	Yellow	Oval	Convex
18	SJV18	TCBS	Hepatopancreas	Yellow	Rounded	Flat
19	SJV19	TCBS	Hepatopankreas	Yellow	Irregular	Rough

Bacterial isolation resulted in a total of 19 isolates were obtained from hepatopancreas, gill, haemolimph and carapaceous wounds from affected pacific white shrimp (*L. vannamei*) (Table 1)

3.1.2. Repetitive-PCR Analysis
Based on PCR-repetitive results and dendogram construction, there were three groups of Vibrios Associated with Vibriosis in pacific white shrimp, *L. vannamei*, from extensive shrimp pond in Kendal, Indonesia (Figure 2.)

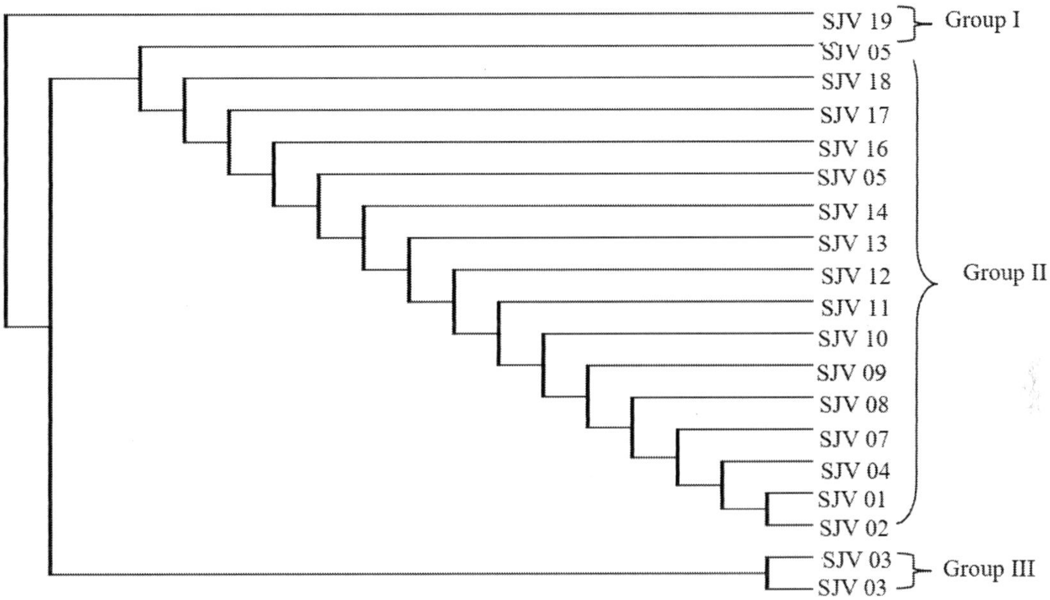

Figure 2. Dendogram of 19 Isolates of Vibrios associated in Pacific White Shrimps, *L. Vannamaei*, from Extensive brackish water Shrimp Ponds of Kendal, Indonesia

Three isolates (SJV 03, SJV 05 and SJV19) of 19 isolates (SJV01 - SJV 19) were selected for molecular identification. Based on molecular characterization, it was shown that all isolates belong to the Vibrio genus (Table 2)

Table 2. Molecular Characterization of Three Representative Of Genus Vibrio

No.	Isolates	Close Relative	Homology (%)	Acc. Number
1.	SJV 03	*Vibrio harveyi*	99	JF264473.1
2.	SJV 05	*Vibrio. parahaemolyticus*	98	HQ694830.1
3.	SJV 19	*V. alginolyticus*	99	EU055503.1

Vibrios associated with vibriosis in shrimp from extensive brackish water ponds in Kendal show a close relation with *Vibrio harveyi* (SJV 03); *V. parahaemolyticus* (SJV 05) and *V. alginolyticus* (SJV 19) with homology ranging from 98-99%. This was based on DNA sequence analysis of 16S. The complete phylogenetic of vibrio is shown in Figure. 3

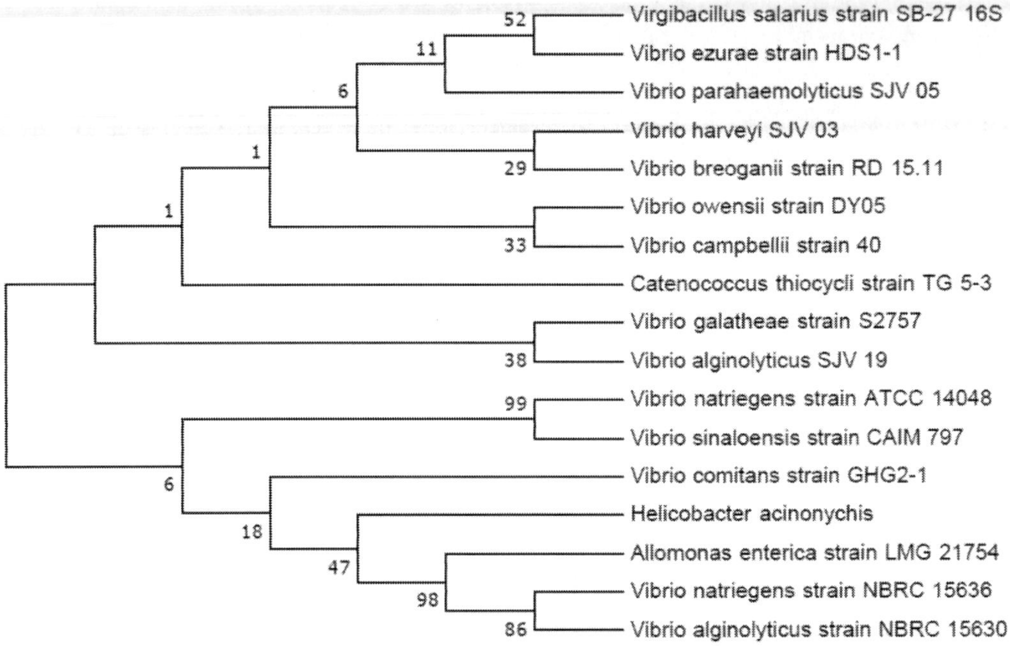

Figure 3. Phylogenetic of The Vibrios Related to Vibriosis in Pacific White Shrimps from Extensive Brackish Water Ponds of Kendal

3.2. Discussion

The genus vibrios are commonly found in marine and estuarine environments [4]. However, some species have demonstrated clinical significance for aquatic animals and are recognized as potential pathogens. Vibrio has been detected related to vibriosis in shrimp [9][15]. Therefore, the genus vibrio has the potential to cause vibriosis in shrimp [16]. These vibrio strains are found in aquatic ecosystems, such as water or sediment [17] or waters [18]. Most *Vibrio* spp. commonly reported pathogens for aquatic animals, but [2] reported that *Vibrio* sp. 33 is an avirulent strain.

The pacific white shrimp (*L. vannamei*) infected by the vibrio were identified by clinical signs as follows: pale hepatopancreas, weakened telson, darkness and redness colorization the mouth, and red spots on the carapace on the body, periopods, pleuopods, and telson. Similar clinical signs have been reported by Sarjito *et al.* [1].

The study also revealed that there were three Vibrio strains that infected pacific white shrimp in extensive brackish water pond, Kendal, i.e. *Vibrio harveyi* (SJV 03); *V. parahaemolyticus* (SJV 05) and *V. alginolyticus* (SJV 19). These results indicate that the vibrios diversity associated with vibriosis in pacific white shrimp, *L. vannamei*, was higher than the diversity reported on tiger shrimp, *Penaeus monodon,* [1]. However, this diversity was lower than the diversity found in grouper in Karimunjawa [13], pacific white shrimp, *L. vannamei*, in India [5].

V. harveyi is naturally found in marine and estuarine environments [16]. These bacteria are also known as pathogenic bacteria associated with black tiger shrimp, *P. monodon,* and the Pacific white shrimp *L. vannamei*, [20], Atlantic mediterranian invasive algae *Caulerpa cylindracea* [20], black tiger shrimp, *P. monodon,* from brackish water pond and coastal the region of India [16], marine invertebrate and vertebrate [3]. *V. harveyi* is also the main causative agent in luminous vibriosis [15] and vibriosis in shrimp ponds [22]. Furthermore, Thongkao [8] and Ramesh *et al.* [21] stated that *V. harveyi* is known as a bacterial pathogen in shrimp around the world that causes mortality up to 100% in the culture system.

V. parahaemolyticus has been reported as an important causative agent of disease in shrimps and resulted on mortality, and economical losses in this industry [22]. *V. parahaemolyticus* was also detected in pacific white shrimp, *L. vannamei* [23]. According to Stalin and Srinivasan [4], *V. parahaemolyticus* is a pathogenic opportunistic causing symptomatic infection of shrimp. This can be attributed to outbreak disease between pacific white shrimp, *L. vannamei* cultures in India [16][23]. The vibrio was also known as the causative agent of vibriosis in pacific white shrimp *L. vannamei* [24]. The bacteria was also identified the dominant vibrio species associated with the mass mortality of juvenile Chinese shrimp, *Fenneropenaeus chinensis* [25].

Meanwhile, *V. alginolyticus* was reported as normal bacteria of estuarine and marine environment [27]. *V. alginolyticus* was reported as a pathogenic bacterium in a giant tiger shrimp growing pond, *P. monodon*, in India. Furthermore, Bunpa *et al.*[26] also stated that this bacterium was important bacterial and pathogenic bacteria associated with infections caused by seafood, fish diseases, shrimp diseases, sediment and seawater.

4. Conclusion

Clinical signs of shrimp infected by the genus vibrio associated with vibriosis from Kendal district, Indonesia were pale of hepatopacreas weaken of telson, dark and reddish colorization mouth, patches of red colour in part of the body on the carapace, periopods, pleuopods and telson. The study found that vibrios associated with vibriosis in shrimp from Kendal's extensive brackish water were closely to *Vibrio harveyi* (SJV 03); V.*parahaemolyticus* (SJV 05) and *V. alginolyticus* (SJV 19)

5. Acknowledgement

The authors would like to thank the Dean of the Faculty of Fisheries and Marine Sciences, Diponegoro University for his permission to begin this research. Thanks also to our students: Desta VP, Dyah, and Kristina Rhesi who have helped this research. We are also grateful to the Head of Diponegoro University Integrated Laboratory, Aquaculture Laboratory Faculty of Fisheries and Marine Sciences, Diponegoro University which provides an opportunity to conduct this research.

6. References

[1] Sarjito, Evrina N, Prayitno, S B Radjasa OK 2013 *J. Coast Dev* **2** (1) 14-21
[2] Liu N, Zhang S, Zhang W, Li C 2017 *Aquaculture* **40** 68 – 73
[3] Austin B, Zhang XH 2003 *J Appl Microbiol* **43** 119
[4] Stalin N and Srinivasan P 2016 *J. Microbial Pathogenesis* **97** 110-118
[5] Austin B and DA Austin 2007 *Bacterial Fish Pathogens.* Fourth ed. Ellis Horword limited (Chichester: England) p 383
[6] Ligthner DVA 1996 *Hand book of Shrimp Pathology and Diagnostic Procedures for Disesases of Cuture Penaeid Shrimp in Asia.* (Baton Rounge, LA USA) p 350
[7] Chythanya R, Karunasagar I, Karunasagar I 2002 *Aquaculture* **208** 1–10
[8] Thongkao K, 2005 *Proc. 7th World Conference on Educational Sciences (Athens, Greece)* (Procedia - Social and Behavioral Sciences) pp 1627 – 1633
[9] Setiawan WA, Widyastuti U, Yuhana M 2005 *Journal of Biosciences HAYATI* (22) **2** 60-66
[10] Radjasa OK, T Martens, HP Grossart, T Brinkoff, A Sabdono and M Simon 2007 *J. Biol Sci*, **7** 239-246
[11] Brock TD and MT Madigan 1991 *Biology of Microorganisms.* 6th ed. Prentice Hall, (New Jersey)
[12] Radjasa OK, A Sabdono, Junaidi and E Zocchi 2007 *Int. J. Pharmacol* **3** 275-279
[13] Sarjito, OK Radjasa, A Sabdono, SB Prayitno and S Hutabarat 2009 *J Current Research in Bacteriology* **2** (1) 14-21.
[14] Muyzer G, Teske A, Wirsen C O and Junnasch H W 1995 Phylogenetic Relationship of *Thiomicrospira* Species and Their Identification in Deep-sea Hydrothermal Vent Samples by Denaturing Gradient Gel Electrophoresis of 16S rDNA Fragments *Arch Microbiol.* **f164** 165
[15] Robertson PAW, Calderon J, Carrera L, Stark JR, Zherdmant M, Austin B 1998 *J Dis Aquat Organ* **32** 051e5.

[16] Stalin N and P Srinivasan 2017 *J.Vetmic* **207** 83-96.
[17] Hai NV 2005 *J. Appl. Microbiol* **119** (4) : 917–935
[18] Thompson FL, CC Thompson, B Hoste, K Vandemeulebroecke, Guillan, and J Swings 2003 *Int. J. Syst. Evol. Microbiol* **53** 1495 – 1501.
[19] Runggrassamee W, Klanchui, A Mainbukaew S, Karoonuthaisiri 2016 *J. of Invertebrate Pathology* **132** 12 – 19
[20] Rizzo L, Fraschetti, S Alifano, Predici MS, Stabili L 2016 *J. Exp. Marine Biology and Ecology* **475** 129-136.
[21] Ramesh K, Natarajan M, Sridhar H, Umamaheswari S 2014 *Glob J. Bio-Sci Biotechnol* **3**(1) 109–114
[22] Lomelí-Ortega C O, Díaz S F M 2014 *J. Aquaculture* **434** 208
[23] Raja R, A Sridhar, R Balachandran, C Palanisammi, A Ramesh, S Nagarajan K 2017 *Journal Fish and Shellfish Immunology* 67: 268-381
[24] Kumar B K, Deekshit V K, Raj J R M, Rai P, Shivanagowda B M, Karunasagar I, Karunasagar, I 2014 *Aquaculture* **433** 247
[25] Zhang X, Song X, Huang J 2016 *Chinese Journal of Oceanology and Limnology* **34** 1278
[26] Bunpa S, N Sermwittayawong and V Vuddhakul 2016 *Proc. Procedia Chemistry* **18** 12 – 17

IOP Conf. Series: Earth and Environmental Science **116** (2018) 012012 doi:10.1088/1755-1315/116/1/012012

Factors Affecting Husband Participation in Antenatal Care Attendance and Delivery

R Rumaseuw[1], S M Berliana[2], N Nursalam[3], F Efendi[3], R Pradanie [3], P D Rachmawati[3], and G E Aurizki[3]

[1]Badan Pusat Statistik, Supiori District of Papua Province, Indonesia
[2]Sekolah Tinggi Ilmu Statistik – Badan Pusat Statistik, Jakarta, Indonesia
[3]Faculty of Nursing, Universitas Airlangga, Surabaya, Indonesia

Email: praba-d-r@fkp.unair.ac.id

Abstract. The government has implemented several programs to prevent and reduce a mother's mortality rate by enhancing active role of the family. The most responsible family member on maintaining the pregnancy and delivery process is the husband. The husband must be active to take care of his wife. Active participation of the husband in accompanying his wife during pregnancy and the delivery process is one of the substantial factors, which helps the husband to take decisions related to the health of his wife. This study aimed to identify variables and its trends, which significantly affect a husband's participation in accompanying his wife during pregnancy and the delivery process. The data used in this study was from an Indonesian Demographic Health Survey 2012. The study used binary logistic regression as the analysis method. The result showed as many as 8,237 husbands accompanied their wife in antenatal care and the delivery process. The significant variables affecting the husband participation are the age of the wife, the education of wife, the education of the husband, the occupational status of the wife and the husband, the number of children, pregnancy status, and residency region. The possibility for a husband to accompany his wife is larger in several factors, such as the wife being between the ages of 21 – 35 years old, a husband who minimally graduated from junior high school, a working husband, as well as a wife, and the number of children less than and equal to two and the expected pregnancy. The government should consider those factors to create policy related women's health and integrate the factors into various sectors.

Key words: Husband participation, Indonesian DHS, Women's Health

1. Introduction

Indonesia is the country with the fourth largest population in the world after China, India and the United States, and the third in Asia. The total population of Indonesia from the Census in 2010 is 237,641,326 people and continues to increase each year. It is estimated that the population until 2035 will reach 305,652,400 people. This large population makes Indonesia have many kinds of problems related to population, one of which is high a mortality rate.

The Maternal Mortality Rate (MMR) is an important indicator of the health status of a region. MMR describes the number of women who died per 100,000 live births. The cause of deaths are related to pregnancy or treatment disorder (excluding accidents or incidental cases) during pregnancy, childbirth and in the puerperium (42 days after delivery) without taking into account the length of pregnancy [1]. The World Health Organization (WHO) 2014 estimates that 800 women die every day due to pregnancy complications and the birth process. Approximately 99% of all maternal deaths occur in developing countries. Approximately 80% of maternal deaths result from increased complications during pregnancy, labor and after delivery. Based on the Millennium Development Goals (MDGs), the Ministry of Health Republic Indonesia, targeted to reduce maternal mortality (MMR) to 3/4 in the period 1990 - 2015 at 102 per 100,000 births [2].

The Indonesian Demographic Health Survey (SDKI) report recorded that MMR in Indonesia decreased from 1991 to 2007, to 228 deaths per 100,000 live births, and again increased in 2012, to 359 per 100,000 live births. Although the mortality rate in 2012 is decreasing and is slightly smaller than what it was in 1991, the potential to achieve the 5th MDG target to reduce maternal mortality rate is off track, meaning that it takes hard work and earnestly to achieve it [2]. Many of the things that resulted in the death of the mother in Indonesia is high, the Ministry of Health Republic of Indonesia divide the cause of death of this mother into causes ie directly and indirectly. The direct maternal death is caused by bleeding, hypertension/eclampsia, infection, serotinous and abortion. Meanwhile, the indirect causes were 3 delays and 4T (too young to give birth, too old to give birth, too close to birth spacing and too frequent childbirth). The three delay phases are, firstly, late recognizing the alarm, and are late in making a decision to seek help individually, or either from family or both. The second one is late for referring to a health care facility. Factors affecting this phase too are the duration of transportation, road conditions and transportation costs. The third late is late for a lack of service because of limited facilities and human resources [3-5].

The government has made various efforts to prevent and reduce maternal mortality. Some government efforts include Making Pregnancy Safer (MPS). One of the MPS strategies of encouraging the empowerment of women and families, and the expected output of the strategy is to establish family involvement in promoting maternal health and enhance the family's active role in pregnancy and childbirth. In addition, the government also launched the Birth Planning and Complication Prevention Program (P4K), where the general objective of P4K is to increase coverage and the quality of health services for pregnant women and new infants through enhancing the active role of families and communities in planning a safe delivery and preparing for complications, and maternal obstacles for giving birth to healthy babies [5].

These programs show the importance of family participation in pregnancy and childbirth. The most responsible family member in maintaining pregnancy and childbirth is the husband, where a husband should be able to be SIAGA (*Siap*/Ready, *Antar*/Delivery, and *Jaga*/Guard), a husband who is always ready when his wife needs it, be more alert and act/or anticipate if you see a sign. He should also be aware of the dangers of pregnancy, always take the wife wherever the wife wants to go, especially when examining the womb by the doctor or midwife. The husband should accompany the wife during the process of examination of pregnancy and childbirth, and maintain the health and safety of the wife and baby. Being a standby husband is clear proof that the husband fully supports the wife's pregnancy [6] [7].

Active participation of husbands during pregnancy can help the success of a wife through pregnancy and childbirth. According to [8] in his book entitled "Pregnant Mother Care", the husband's support is important during pregnancy because the wife is sometimes faced with the situation of fear and solitude, so that the husband is expected to always motivate and accompany his wife. In addition, the support provided by the husband during the pregnant can also reduce anxiety and restore confidence in the expectant mother in the process of pregnancy. This study aims to find out how the general description of husbands' participation in accompanying wives during pregnancy and delivery, and also knows what variables significantly affect the participation of husbands, and knowing how the trends of significant variables affect the participation of husbands in accompanying the wife during the examination of pregnancy and childbirth in Indonesia.

2. Methods

This study uses secondary data, the Indonesia Demographic and Health Survey (SDKI) 2012. The data is used to analyze the participation of the spouse/husband in accompanying the wife during pregnancy and childbirth examination. Some of the topics of SDKI-WUS 2012 used in this research concern background respondents, the birth history, pregnancy care, husband background and occupation of respondent and husband.

Implementation of SDKI 12 was conducted from May to August 2012 in 33 provinces. The population in this study were all women of reproductive age between the ages of 15 – 49 years, who had children up to 5 years old in the survey. Furthermore, the sample in this study were women who had given birth and who had the last child in the last 5 years period, that is 2007 to 2012 and married/live together with the husband/spouse. The number of samples used in this study was as many as 8,237 respondents.

The data collection in the SDKI 2012 was conducted through direct interviews using four questionnaires, namely household questionnaires, female fertile age questionnaires, marriage man questionnaires, and teenage male questionnaires who had never been married. This study used nine independent variables, namely wife age, husband /spouse, wife education, husband/spouse education, wife's work status, husband's working status, the number of children, pregnancy status, and residence area. Besides, this research uses three variables bound by the participation of husbands in accompanying the wife during the examination of pregnancy, childbirth, and the combination of both. The method of analysis used in this research is descriptive analysis and inferencing analysis. In the inferential analysis, the analysis used was Binary Logistic Regression Analysis.

3. Results

The general description of a husband's participation in pregnancy and delivery in Indonesia can be explained through the results of descriptive analysis. Based on the result of the data processing of SDKI-WUS 2012, most of the husbands accompanying their wives during pregnancy examination is equal to 81.4%, meanwhile, the number of husbands accompanying their wives at the time of delivery is greater that 86.5%, and the rest do not accompany their wives, and, and if seen percentage of husband who accompanied his wife at the time of checking pregnancy & childbirth, the percentage decreased that is equal to 73.2%.

Regarding the 95% confidence level, the results among variables were varied. For the wives' education, the wife who has a minimum of junior high school education has a greater tendency to be accompanied during the pregnancy examination 1,632 times and accompanied during the examination of pregnancy and labor 1,411 times, compared to wives who have education under junior high school level. The wife who has a husband with a minimum of junior high school education has a greater tendency to be accompanied during pregnancy examination 1,822 times, during labor 1,218 times, and during pregnancy and labor 1,651 times, compared to the wife who has a husband with education in under graduate junior high school.

Furthermore, the working wife had a greater tendency to be accompanied during the pregnancy examination 1,303 times, during labor 1,277 times, and, at the time of pregnancy and labor 1,273 times, compared to the unemployed wife. The wife who had a husband that worked had a greater tendency to be accompanied during the pregnancy examination 2,062 times and during the examination of pregnancy and labor 1,799 times, compared to the wife whose husband did not work.
When compared to the number of children, the most respondents accompanied by their husbands in the examination of pregnancy, childbirth, or both, are respondents who have children less than equal to 2, although, at the time of birth, not too different participation of husbands who accompany his wife. The respondents whose pregnancies were desirable had a greater tendency to be accompanied during pregnancy checks 1,429 times, during labor 1,246 times, and during pregnancy and labor 1.385 times, compared to wives who did not want pregnancy at the time of pregnancy. Based on the 95% confidence level, it was concluded that wives with an amount of two, or less than two, had a greater tendency to be accompanied during pregnancy checks 1,641 times and, at the time of pregnancy and labor 1,463 times, compared to wives with more than 2 children.

IOP Conf. Series: Earth and Environmental Science **116** (2018) 012012 doi:10.1088/1755-1315/116/1/012012

Table 1. Estimation of the Parameters and the Ratio of the Husband's Inner Parent's Tendency - Pregnancy and Delivery Checks

Husbands' Participation	Demographic (%)	Pregnancy (%)		Delivery (%)		Pregnancy & Delivery (%)	
		B	OR	B	OR	B	OR
Participating		81.40		86.50		73.20	
Not-Participating		18.60		13.50		26.80	
Total		100		100		100	
Participation Percentage in the Demographic Data							
Age of Wife							
<21	4.95	80.39		86.30		72.10	
		0.008	1.008	-	-	-	-
21 – 35	72.72	83.67		86.50		75.10	
		0.175*	1.191	-	-	-	-
>35	22.33	74.44		86.30		67.40	
		-	1.00	-	-	-	-
Age of Husband							
<26	9.03	83.10		87.20		74.50	
		-	-	-	-	-	-
26 – 35	50.87	84.20		86.40		75.30	
		-	-	-	-	-	-
>35	40.10	77.60		86.40		70.40	
		-	-	-	-	-	-
Education of Wife							
<Junior High School	21.12	69.30		85.40		62.50	
		-	1.00	-	-	-	1.00
≥Junior High School	78.88	84.70		86.80		76.10	
		0.490**	1.632	-	-	0.344**	1.411
Education of Husband							
<Junior High School	22.00	69.10		84.40		61.70	
		-	1.00	-	1.00	-	1.00
≥Junior High School	78.00	84.90		87.10		76.50	
		0.600**	1.822	0.198**	1.218	0.502**	1.651
Occupation Status of Wife							
Working	47.91	79.50		85.10		70.80	
		0.264**	1.303	0.245**	1.277	0.242**	1.273
Not Working	52.09	83.60		88.00		75.90	
		-	1.00	-	1.00	-	1.00

Table 1. Continued

Husbands' Participation	Demographic (%)	Pregnancy (%)		Delivery (%)		Pregnancy & Delivery (%)	
		B	OR	B	OR	B	OR
Occupation Status of Husband							
Working	98.4	69.90		85.70		61.70	
		0.724**	2.062	-	-	0.587**	1.799
Not Working	1.6	81.60		86.50		73.40	
		-	1.00	-	-	-	1.00
Number of Children							
≤2	70.86	85.00		86.90		76.50	
		0.495**	1.641	-	-	0.381**	1.463
>2	29.14	72.70		85.50		65.30	
		-	1.00	-	-	-	1.00
Pregnancy Status							
Planned	84.64	73.20		84.10		64.70	
		0.357**	1.429	0.220**	1.246	0.326	1.385
Not Planned	15.36	82.90		86.90		74.80	
		-	1.00	-	1.00	-	1.00
Residential Status							
Rural	36.70	78.40		86.60		70.90	
		-	1.00	-	-	-	-
Urban	63.30	83.20		86.40		74.60	
		-0.157*	1.170	-	-	-	-
Constant		-0.992		1.409		-0.856	

The respondents whose residence was in the urban area is mostly accompanied by the husband in the pregnancy examination and the examination of pregnancy and childbirth, while, at the time of more labor accompanied by the husband is the respondent/wife residing in rural areas, but the difference is not too far away. Based on the 95% confidence level, it was concluded that wives living in urban areas had a greater tendency to be accompanied during pregnancy checks 1,170 times, compared to wives living in rural areas.

The variables that significantly influence the participation of husbands in accompanying wives during pregnancy and labor can be researched using binary logistic regression. By using this method, we can know the influence between a wife's age, the husband's age, the wife's education, husband's education, wife's working status, husband's work status, number of children, pregnancy status and residence area to a husband's participation in accompanying his wife during pregnancy and delivery.

The results show that with a 95% confidence level, independent variables that significantly influence a husband's participation in accompanying his wife during pregnancy examination are the wife's age, wife's education, husband's education, wife's working status, husband's working status, number of children, and pregnancy status. Independent variables that significantly influence the participation of husbands in accompanying wives during childbirth are a husband's education, the wife's working status and pregnancy status.

4. Discussion

The variables that significantly influence the participation of a husband in accompanying his wife during pregnancy examination are the wife's age, wife's education, husband's education, wife's working status, husband's work status, number of children and pregnancy status. Some of these independent variables also have a significant effect on a husbands' participation in pregnancy and delivery, but only the wife's age variable does not have a significant effect on the participation of husbands during pregnancy and childbirth.

Pregnancy is a time when a mother undergoes major changes, both psychologically and physiologically, these changes in pregnant women can cause anxiety and stress during pregnancy [9]. Stress in pregnant women is a very easy condition to occur [10] [11] and research says that the age of respondents who are in the early adult age range, most of whom will experience moderate stress during pregnancy [9].

Most of the respondents in this study had an age in the range of 21 - 35 years; this is the age of early adulthood. Children, less than 2 children, this shows little experience in the delivery process. Support and role as husband SIAGA is very important during the process of pregnancy [12].

A mother's mortality rate increases according to the 2012 IDHS, as an effort to prevent Indonesia to mobilize the SIAGA husband program. SIAGA means the husband is ready to take and keep his wife who is pregnant and facing the process of childbirth, and the role of husband as husband SIAGA as this is a form of a husband's participation or support to the wife during pregnancy [13]. One of the factors that significantly influence the participation of a husband in accompanying his wife during pregnancy examination is the wife's age. The wife's age in this research mostly ranged in the age of a young adult. The task of early adulthood development is the age at which one begins living with a partner, starting a new family and performing roles and functions within the family. According to MOH, the role of a husband in accompanying his wife during pregnancy is always ready to provide time to take the faithful wife while including checking her pregnancy [14]. Characteristics of wives in this study are mostly located in the age of 21 - 35 years, where, at that age, is in the stage of development of the family and at the stage of married couples and families with children aged under five [15]. Newly-married couples also have the characteristics to plan to have children, and this is in accordance with the results of research that most respondents who have husbands participate in the examination of pregnancy and have a planned pregnancy status. This wife's age factor does not have a significant effect on the husband's participation during pregnancy and labor due to the age of 21 - 35 years being a safe age for a wife to conceive and give birth [16]; therefore, in this study, the participation of husbands focused only on the examination of pregnancy only.

The number of children also significantly influenced the participation of husbands in the examination of pregnancy and childbirth, and the results of this study are that most of the new couples have less than two children. Having less than two children indicates the experience of pregnancy and childbirth is still less, so that the husband participates in the examination of pregnancy with the aim to learn together about the condition of his pregnant wife, and this can hopefully play an active role in keeping the wife safe during pregnancy. Experience effect on one's behavior [17]. This is in accordance with the results of the Marina research that the husband who did not provide support SIAGA occurs in multigravida, meaning that the husband has a lot to learn to have a smooth pregnancy experience with his wife.

Another factor that significantly affects the participation of husbands in the examination of pregnancy and childbirth is the occupation of the wife and the husband. Characteristics of respondents in this study indicate that the number of percentage of wives and husbands who work and participate in the examination of pregnancy is greater than the number of husbands and wives who do not work. This suggests that the economic conditions affect the role of the husband in participating in the wife's pregnancy examination. The higher the economic conditions affect the husband's support during pregnancy [18]. Family income will affect the behavior of a person in using health services as the lower the family income, the lower the family activity on the health of pregnant women. This is due to the high cost of daily living that they prioritize, compared with the maternal health of pregnant women.

The results showed that from the participation of husbands during pregnancy, labor and childbirth, the largest percentage is the husband accompanying the wife during childbirth. The role of the husband is very important in the process of the pregnancy of the wife, both during pregnancy and during childbirth. A husband's support will affect the success of labor [19]. Good interaction between husband and wife during childbirth will affect the wife's emotional state during childbirth, thus decreasing anxiety and improving the success of labor and early breastfeeding initiation [21-[21].

A husband's participation in the delivery process is influenced by the tendency of the wife to get emotional support during childbirth. The process of labor is a condition that requires a good physical and mental readiness for a mother. The results showed that 86% of pregnant women experience stress in the 3rd trimester; one of the causes of stress in pregnant women is the mother's fear of labor. A husband's support is very important in the process of childbirth, and research indicates that the wife with her husband accompanying her in childbirth will experience fluency during labor, compared to the wife who was not accompanied by the husband during childbirth [11]. A husband's support during labor will reduce anxiety during postpartum wives [12].

Educational factors are also one of the factors that influence the participation of husbands in accompanying their wives during childbirth. This is in line with the statement [8], that factors affecting the role of the assistant in delivery are socioeconomic, cultural, environmental, experiential, age, and educational factors. Education will affect the need for maternal assistance during childbirth [22]. The educational background of respondents in this study mostly have more education than junior high school levels. Education will greatly affect a person's knowledge and coping. A person with a higher education may also have good health literacy [13]; thereby, impacting on improving a good knowledge of the importance of the husband's role in accompanying during labor.

The wife's working status also becomes one of the factors influencing husband participation during childbirth. Within this research, wives who work have a tendency to be accompanied during labor. One of the factors that influence the accompaniment of the husband during childbirth is socioeconomic [16]. The higher the socioeconomic person, the better the husband in accompanying the wife during the labor process. Wives who work in a family can help the family economy, so tend to have a better economic level, compared with wives who do not work.

In this study, most of the respondents had a planned pregnancy status. Planning a pregnancy itself is also a form of co-operation and effective communication between the husband and wife; thus, the husband also has the responsibility to participate in maintaining the health of the wife and the fetus that she contains during pregnancy until the birth process. One of the responsibilities of the husband is accompanying the wife during the examination of pregnancy until the process of childbirth. In addition, mothers with experience who were accompanied during the previous labor had a tendency to be recharged during labor. The psychological state of the mother will affect the process of childbirth so that it takes the support of a husband, at least in the form of touch and words of praise that can make her comfortable, give reinforcement, and improve family harmony [23].

5. Conclusion

The variables that significantly influence the participation of a husband in accompanying his wife during pregnancy examination are the wife's age, wife's education, husband's education, wife's working status, husband's working status, number of children and pregnancy status. Independent variables that significantly influence the participation of husbands in accompanying wives during childbirth are the husband's education, wife's working status and pregnancy status. Independent variables that significantly influence a husband's participation in accompanying his wife during pregnancy and delivery are his wife's education, husband's education, wife's work status, husband's employment status, number of children and pregnancy status.

6. References

[1] Ministry of Health RI 2012 *Maternal Mortality Rate.* (Jakarta: Ministry of Health RI)
[2] Ministry of Health RI 2014 *Maternal Health Situation.* (Jakarta: Ministry of Health RI)
[3] Kurniati A, Chen CM, Efendi F and Berliana SM 2017 *Health Care Women Int* 1–17,
[4] Efendi F, Chen CM, Kurniati A and Berliana SM 2016 *Women Health* **57** 614–629,
[5] Ministry of Health 2010 *National Strategy for Making Pregnancy Safer (MPS) in Indonesia 2001-201.(*Jakarta: Team of Directorate General of Community)
[6] Mullany BC, Becker S and Hindin M 2006 *Health Educ. Res* **22** 166–176
[7] Setyobudihono S, Istiqomah E and Adiningsih S 2016 *Procedia - Soc. Behav. Sci* **222** 160–168
[8] Kusmiyati *The Caring Of Pregnant Women (Perawatan Ibu Hamil)* (Yogyakarta: Fitramaya)
[9] Armini NKA, Wahyuni ED and Rachmawati PD 2017 *J. Ners* **3** 72–76
[10] DiPietro JA 2012 *J. Adolesc. Heal,* **51** 3–8
[11] M. Michael W.O'Hara;Katherine L.WisnerMD 2014 *Best Pract. Res. Clin. Obstet. Gynaecol.*, **28** 3–12
[12] Liamputtong P and Naksook C 2003 *Midwifery* **19** 27–36
[13] Rakhmawati M and Indawati R 2014 *Biometrika dan Kependudukan*
[14] Ministry of Health of the Republic of Indonesia 2001 *Guidelines for the implementation of Neonatal Improvement Efforts.* (Jakarta: Ministry of Health of the Republic of Indonesia)
[15] Friedman MM 2010 *Family Nursing Teaching Textbook: Research, Theory, and Practice.* (Jakarta: EGC)
[16] Bobak 2005 *Maternity Nursing* (Jakarta: EGC)
[17] Azwar 2009 *Human Attitude: Theory and Measurement* (Jakarta: Pustaka pelajar).
[18] Yanuasti 2001 *Husband's Social Support for ANC Services.*
[19] Sapkota S, Kobayashi T and Takase M 2013 *Midwifery* **29** 11 1264–1271
[20] Retno S, Nursalam NN, Santoso BS and Rachmat RR 2016 *J. NERS* **11** 224
[21] Story WT, Burgard SA, Lori JR, Taleb F, Ashraf Ali N and Emdadul Hoque D 2012 *BMC Pregnancy and Childbirth* **12**
[22] Suto M, Takehara K, Yamane Y and Ota E 2017 *J. Affect. Disord.* **210** 115–121
[23] Umiati T, Rakhmawati M, Simangunsong W, Lukitowati B, Fauzi E, Setiawan E 2013 *Indones. J. cancer*

Nitrate and Phosphate Contents on Sediments Related to The Density Levels of Mangrove *Rhizophora* Sp. in Mangrove Park Waters of Pekalongan, Central Java

E Supriyantini, A Santoso* and N Soenardjo

Department of Marine Science, Faculty of Fisheries and Marine Science, Diponegoro University, Jl. Prof. H. Soedarto SH, Tembalang, Semarang 50275 Indonesia

Email: adisantoso1959@gmail.com

Abstract. Mangrove Park waters area of Pekalongan City, Central Java, used to be an aquaculture field, now turning the function into a restoration-based mangrove area, and now it has undergone rehabilitation. The conditions may affect the distribution of nitrate and phosphate content. The objective of this study was to determine the content of nitrates and phosphates in sediments related to the density levels of mangrove *Rhizophora* sp. The method used in this research was a descriptive method, and sampling was done by purposive sampling method. Water and sediment sampling were conducted at three stations respectively, representing: no mangrove area but used as a residential and tourist area (station 1); less dense mangrove (station 2); and, the previously aquaculture field - or medium dense mangrove (station 3). The results showed that the content of nitrate and phosphate in the whole sediment showed a low fertility rate. Average nitrate content for station 1, station 2 and station 3 were 0.86 mg/100 g, 0.94 mg/100 g and 0.81 mg/100 g, respectively. The average phosphate content at each station were 1.14 mg/100 g, 0.04 mg/100 g and 0.05 mg/100 g, respectively. Except to the station 1 that was no vegetation anymore, the mangrove density levels at two other stations at study sites were relatively low to medium; at station 2 was 0.8 ind/10 m^2 and at station 3 was 1.2 ind/10 m^2. The role of nitrate and phosphate were for mangrove growth at the site.

Keywords: Nitrate, Phosphate, sediment, water, mangrove density

1. Introduction

Mangrove is an ecosystem with high organic productivity, and high biota diversity [1-2]. Mangroves usually grow in tropical and subtropical regions, growing only on coastal protected from wave action [3]. The role of mangroves other than as abrasion holder also has an ecological function as a provider of nutrients for aquatic biota [4].

Mangrove can bind sediment dissolved from the river [5]. The mangrove efficiency of suspended bonding of water column depends on various factors such as particle size, salinity, the tidal, and wide area of intertidal zone [6]. The value of binding efficiency can be very high at 15-44% [7]; 30-60% [8]; even up to 80% [9]. The presence of mangroves causes a lack of water movement to give a significant effect on the sedimentation of fine sedimentary particles that tend to congregate at the bottom [10]. Due to typical and dense root systems, it can cause soluble particles in water to settle around their roots, thus forming a collection of sedimentary layers [11]. Sediments that exist around this mangrove vegetation are then mixed with the litter [5]. Nutrient elements of organic matter will be deposited in the sediments and will be distributed by environmental factors [5]. This condition makes the mangrove forest as a contributor of nutrients to other ecosystems in the vicinity [5]. Nitrate in the waters comes from the breakdown of organic and inorganic nitrogen in soils derived from the decomposition of organic materials with the help of microbes [12]. Phosphate in waters naturally comes from weathering of rocks and also decomposition of organic matter [13].

Mangrove density is influenced by the growth factor of the mangrove. According to Hossain and Nuruddin [14], mangroves typically grow in areas that have large river mouths, and their streams are rich in mud and sand. Mangroves derive nutrients from inorganic mineral ions and organic matter as well as from nutrient recycling internally through detrital based food web [5][15]. Inorganic nutrients such as N and P are essential nutrients that are needed by the organism because it can not be replaced by other elements. Nitrate (NO_3) and phosphate (PO_4) are the major nutrients that determine the stability of vegetation growth such as mangroves [2]. Inorganic nutrient sources are rain, surface flow, sediment, seawater, and degraded organic materials [2]. Nitrate (NO_3) is the main form of nitrogen in natural waters and is a major nutrient for plant growth [16]. NO_3-N is easily dissolved in water and stable [17]. This compound is produced from the perfect oxidation process of nitrogen compounds in the waters. Nitrification which is the process of oxidizing ammonia to nitrite and nitrate with the help of microorganisms is an important process in the nitrogen cycle [5].

Nitrification is a process consisting of two reactions namely the oxidation of ammonia to nitrite and subsequently oxidation of nitrite to nitrate [5]. The decomposed organic matter is the source of ammonia, which is the beginning of nitrate formation through the breakdown of organic and inorganic nitrogen present in soil and water with the help of microbes and fungi [5]. According to Reef et al [2], in mangrove ecosystems, nitrogen fixation is found to occur in sediments even though only a few centimeters at the top of the sediment layer [18], nitrogen fixation in sediments with upper mangrove vegetation is higher than the plantless sediments present above it, due to differences in detritus content present in the soil.

According to Kolliopoulos et al [19], the phosphorus element is not found in free form as an element but in the form of dissolved organic compounds (orthophosphates and polyphosphates) and particulate organic compounds. Phosphorus forms complexes with iron and calcium ions in aerobic conditions, soluble and deposited in sediments. The phosphorus present in seawater comes from the decomposition of dead organisms [20]. Orthophosphates are a form of phosphorus that can be directly exploited by aquatic plants, while polyphosphates must be reduced to orthophosphate before use. Phosphorus in the form of phosphate is a necessary micronutrient in small amounts but is essential for aquatic organisms [19]. Phosphate deficiency can also inhibit plant growth. Monovalent orthophosphate anions (H_2PO_4) and divalent (HPO_4) are absorbed by plants from 50% of total soil solution at almost neutral pH (pH 6-7). At pH 4-6, H_2PO_4 is absorbed about 100 percent, and at pH 8, H_2PO_4 is absorbed by 20 percent and HPO_4 is 80 percent of total P [21].

In sediments, the main source of phosphorus is from terrestrial sediments that are eroded and agricultural fertilizers carried by streams [22]. The source of phosphorus (P) from anthropogenic activity can increase the level of soil erosion from soil to sea by about 300% [23]. Other fractions of soluble phosphate which are partially in the form of colloids are derived from the excretion of the organism and from the resulting autolysis of dead organisms [24]. The existence of various forms of phosphate in the ocean is controlled by biological and physical processes [25][24]. In sediments, these phosphate minerals are absorbed by hydrolyzed sediments, especially clays [26]. Increased orthophosphate is proportional to the increase in sediment concentration. Suspended materials may also carry the absorbed phosphate therein [26].

Pekalongan city is one of the urban areas in Central Java Province which has a tourism object based on learning place and mangrove restoration. Mangrove Park Pekalongan City was established in 2013 and is under the authority of DPPK (Dinas Pertanian, Peternakan, dan Kelautan / Department of Agriculture, Husbandry, and Ocean) of Pekalongan city. Land area used for mangrove restoration reached 5.7 Ha with more than 50% planted with various types of mangroves, and dominated by mangrove species of *Rhizophora* sp. The condition of the waters in Mangrove Park, Pekalongan is generally calm not affected by the currents or sea waves.

Mangrove park Pekalongan city, formerly a pond that has long not functioned. At the location is now built mangrove tourist attractions entitled restoration and used as a place of learning mangrove.

This is because the existing mangrove conditions in the region of Pekalongan increasingly apprehensive.

Research on the content of nitrates and phosphates in mangrove sediments has been widely practiced. But specifically in the waters of Mangrove Park, Pekalongan has never done research on nitrate and phosphate. Logically, the older the mangrove will be the more nitrogen content. This is because the root system is getting stronger so the plants can precipitate the mud that contains many nutrients. Also explained by Alongi [27], that the N content along with the increase in height and diameter of trees on *Rhizophora* sp mangrove species, but unlike the phosphate content in sediments have no linear relationship to tree age. The value of phosphate content is probably more influenced by input other sources derived from industry and fertilizer. While the total value of total organic carbon content is more influenced by the source of input from the river and the location around the mangrove plant.

Based on these considerations, the research on the content of nitrate (NO_3) and phosphate (PO_4) in the sediment to the level of mangrove density of *Rhizophora* sp in Mangrove Park waters, Pekalongan is necessary. The aim was to find out the content of nitrate (NO_3) and phosphate (PO_4) in the sediment in the waters of mangrove park, Pekalongan and its relation to mangrove density. Sediments play an important role in the provision of nutrients and stability of plant growth, because the sediment is the storage of chemical compounds accumulated from compounds in the water column. Described by Chaubey et al [28], observation of boundary nutrient content can help identifying the relationship between nutrient availability and plant growth. The results of this study were expected to provide information about the presence of nitrate and phosphate in the mangrove sediments in the waters of Mangrove Park, Pekalongan City to become a reference in the management of the mangrove forest at the future.

2. Methodology

This research was conducted in Mangrove Park waters, Pekalongan City in January 2016. The research material used was water and sediment samples, taken from three stations (station 1 = settlement & tourism, no mangrove, station 2 = less mangrove density, station 3 = previously aquaculture field - or moderate mangroves density). Distance between stations one to another is 87-94 meters. Determination of the station based on purposive sampling, and each station was done repetition sampling 3 times (Figure 1). The coordinates of each station was determined using the Global Positioning System (GPS) tool. Water samples were analyzed for Total Organic Matter (TOM) and Biological Oxygen Demand (BOD_5), while sediment samples were analyzed for nitrate (NO_3) and phosphate (PO_4) conducted at BPIK ((Balai Pengujian dan Informasi Konstruksi), Semarang. Measurement of parameters of the waters was done to support the research results. All data obtained were analyzed descriptively and the results were interpreted through histogram tables and graphs and the results of those analyzes were compared with the quality standard.

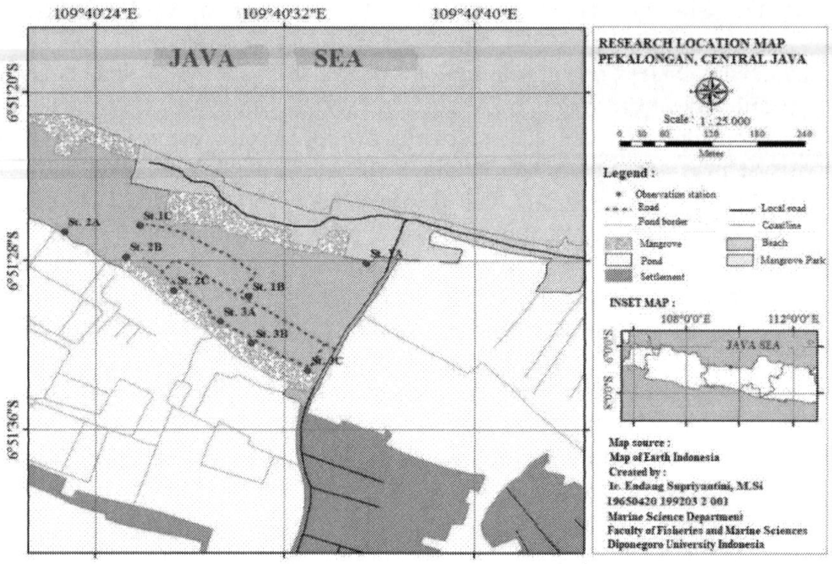

Figure 1. Study Site at the Pekalongan Mangrove Park Central Java

Water sampling was done at each substation. Samples of 600 ml of water were taken directly from the surface of the waters by using a polyethylene bottle [29], then TOM and BOD5 were tested [30].

Sediment sampling was conducted at each substation of 500 g taken at a depth of ± 20 cm using sediment grab. Sediment samples were analyzed for phosphate, nitrate, and sediment grain analysis. Phosphate analysis used Ascorbic Acid method and read the absorbance using spectrophotometer at wavelength of 880 nm [31]. Analysis of nitrate content by brucin method using spectrophotometer at 410 nm wavelength [31]. Sediment grain measurement analysis was performed to find out the substrate type with the method of sieving and hydrometer analysis and the result was read according to triangle of USDA texture triangle [32].

The measurement of environmental parameters was carried out directly during in situ sampling which includes temperature, salinity, pH, dissolved oxygen (DO) measurements.

Calculation of mangrove density at each station was done to determine the condition of mangrove vegetation. Calculation method of mangrove was by using plot sampling method. Samples were taken randomly from selected sampling point by using transect method (Figure 2.) which was expected to represent the characteristics of each station.

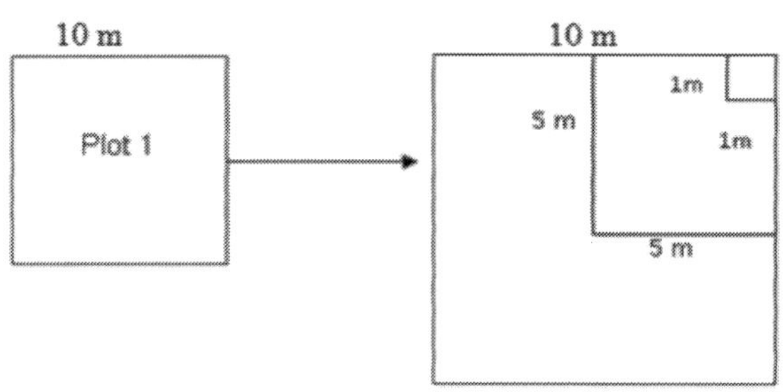

Figure 2. Plotting Method: 10 m x 10 m for Tree Category; 5 m x 5 m for Sapling Category; and, 1m x 1m for Seedling Category

Calculation of the amount of mangrove tree stands was done at an altitude of ± 1.3 m and has a diameter at Breast Height or dbh ≥ 10 cm in subplots of 10 m x 10 m; Sapling counting (seedling) with stem diameter 2≤ dbh <10 cm and height> 1 m; And Seedling sample calculation (seedlings) were performed on mangrove vegetation with height <1 meter in subplot measuring 1 m x 1 m [33]. According to Pellegrini et al [33], the density is the number of individuals per unit area. The calculated density value has an ind /m^2 unit, using the formula:

$$\text{Density} = \frac{\text{Number of individuals of a species in all plots (ind)}}{\text{Plot area (m}^2\text{)}} \qquad (1)$$

3. Results and Discussion

3.1. *Water Quality Measurement, Phosphate & Nitrate in Sediment and Substrate Type*

In situ field, water quality measurements and the results of phosphate nitrate analysis on sediment and substrate type ex situ (laboratory) were presented in Table 1.

Table 1. Water Quality Measurement, Phosphate and Nitrate Analysis in Sediments and Substrate Type at Study Site ($\bar{x}$ ± SD, n=3)

Station	Water Quality					Sediment Analysis				Substrate Type
	pH Air	Salinity (ppt)	Temperature (^{o} C)	DO (mg/L)	BOD_5 (mg/L)	TOM (mg/L)	NO_3 (mg/100 g)	PO_4 (mg/100g)	N:P Ratio	
I	7.30-8.70	24.00-26.00	30.00-31.00	8.30-8.70	11.04-12.60	24.76-25.40	0.86±0.65	1.14±0.09	0.75:1	Silty Sand
II	7.50-7.90	24.00-26.00	29.80-30.30	5.90-7.30	16.50-17.00	29.15-29.69	0.94±0.19	0.04±0.02	23.5:1	Sandy Silt
III	6.30-6.87	26.00-27.00	29.80-31.00	4.20-7.90	16.30-16.46	26.99-27.50	0.81±0.32	0.05±0.001	16.2:1	Silty Sand
Standard Quality	7-8.5*)	0-34*)	28-32*)	>5*)	20*)	0.01-30**)				

Sources : *) [34]
**) [35].

Based on the results of the measurement of water quality at the site, it could still support the life and growth of the mangrove optimally because it was still in the range of existing quality standards according to the [34].

3.2. *Nitrates (NO_3) in Sediments*

Based on the result of measurement of nitrate content (NO_3) in sediment, the result showed that varied 0.81-0.94 mg /100 g. The highest nitrate content was in station II of 0.94 mg/100 g; Station I of 0.86 mg/100 g and the lowest at station III is 0.81 mg/100 g. (Table 1 & Figure 3).

Decomposition of mangrove litter is related to bacterial activity that is influenced by chemical physics parameters such as temperature, salinity, and pH. The salinity value affects the number of littering, the higher the salinity, the more litter produced [36]. However, salinity is negatively correlated with decomposition rate of mangrove litter, the higher the salinity, the slower decomposition rate of mangroves, and the lower the decomposition rate, the faster. The measured salinity at the time of observation in the three station locations had a low value that was between 24-27 ppt when compared with the quality standard according to Decree of the State Minister of Environment No. 51 of 2004 that was 0-34 ppt.

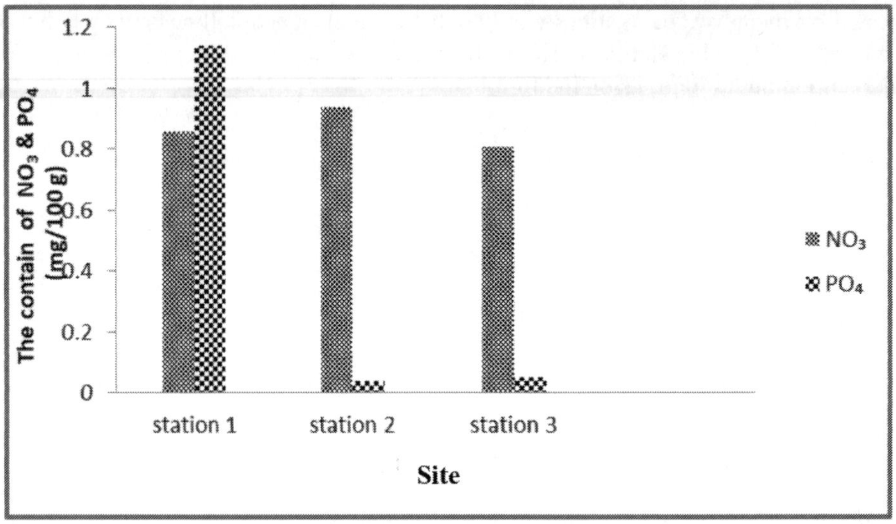

Figure 3. NO_3 and PO_4 Contents in Sediments at the Study Site at the Pekalongan Mangrove Park

The content of nitrates in sediments is also influenced by the rate of nitrification. According to Reef et al [2], the most significant environmental factors in controlling nitrification rates are pH, temperature, substrate, and oxygen (DO) availability.

According to Hossain and Nuruddin [15], pH is a determinant of the absence of nutrients absorbed by the soil and affect the development of microorganisms. Low pH causes the nitrification process to run slowly so that the resulting nitrate is low. In contrast, high pH (base) is the optimum pH for nitrification process. It is proven that the content of NO_3 in sediment at station II has a higher value than other stations, presumably because the pH at station II approaching the value of 8 is around 7.50- 7.90.

At high pH (base) ammonia the degradation of organic matter can be present in non-ionized form (NH_3) which is toxic, otherwise low pH ammonia present in ionized form (NH_4^+) [2]. By dissolved oxygen the ammonia compound (NH_3 and NH_4^+) will undergo oxidation process to nitrate (NO_3). This oxidation process is assisted by nitrification bacteria nitrosomonas and nitrobacter bacteria [2]. Nitrification reaction according to APHA [35] as follows:

$$2NH_4^+ + 3O_2 \longrightarrow 2NO_2^- \text{ (nitrit)} + 4H^+ + 2H_2O \text{ (Nitrosomonas bacteria)} \quad (2)$$

$$2NO_2^- + O_2 \longrightarrow 2NO_3^- \text{ (Nitrobacter bacteria)} \quad (3)$$

The nitrite compound (NO_2^-) produced is a compound between the oxidation of the ammonia compound, which then continues to produce nitrate (NO_3) by the aid of nitrobacter bacteria.

The high content of nitrate in sediment at station II was also due to the type of substrate in the location dominated by silt which was 60%, while in station I and III sand was dominated by 53% and 58% respectively (Figure 4). In accordance with the results of research [16], the higher the percentage of silt or Clay in the sediment, the higher the nitrate content.

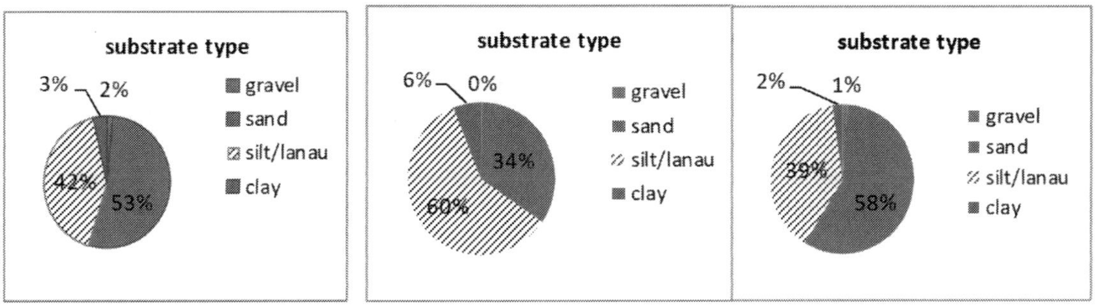

Station I Station II Station III

Figure 4. The Substrate Type Composition at Each Study Site

According to Zhu et al [37], the smaller sediment size, the greater the availability of nutrients N and P in the substrate, as well as the larger the size of the sediment the availability of N and P elements is lesser [38]. This is because the fine-textured soils have a larger surface area so that the ability to hold water and provide high nutrients, while sand-textured soils have a smaller surface area making it difficult to hold water and nutrients [39]. Silt and clay particles have a higher ability to trap nutrients because they have finer particles than sand [40].

The content of nitrates in sediments is also influenced by the content of organic matter in the waters. According to Reef et al [2], the availability of nutrients in mangrove production is controlled by various biotic and abiotic factors such as tidal inundation, elevation in the tidal framework, soil type, redox status and soil microbial activity, plant species, litter production and decomposition [2]. The total organic matter content (TOM) in station II has a higher value than other stations that is equal to 29.15-29.69 mg/L. The amount of TOM content in these waters is suspected to come from the mangrove litter, waste disposal from human activities, considering that in the area other than close to the residential area is also the center of the area based tourist attractions and mangrove restoration. The decomposition of organic matter will increase the number of soil organisms and result in increased phosphate and organic nitrate in the network of microorganisms and can improve the environment to be more suitable for plant growth.

Based on the results of nitrate content analysis on sediments in the study sites belong to the classification of less fertility, because the value ranged from 0.81 to 0.94 mg/100 g. According to Effendi [41], nitrate content < 0.227 mg/L (< 2.27 mg/100 g) classification of less fertility; 0.227-1.129 mg/L (2.27-11.29 mg/100 g) classification of moderate fertility; and 1.130-11.250 mg/L (11.30-112.50 mg/100 g) classification of high fertility.

In Mangrove Park waters, Pekalongan is still possible to do mangrove growth effort, because almost every location of TOM content station ranges from 24.76-29.69 mg/L. According to Afu, [35] waters condition with organic matter content of 0.01-30 mg/L is still possible to do coastal conservation effort (support growth of mangrove).

3.3. Phosphate (PO_4) in Sediments

The result of measurement of phosphate content (PO_4) at the site, obtained an average value of 1.14 mg/100 g (station I); 0.04 mg /100 g (station II); and 0.05 mg/100 g (station III) (Table 1 & Figure 3).

At the station 1, it showed that the high content of phosphate was derived from the remnants of agricultural waste (fertilization), considering the Mangrove Park in Pekalongan city was once a farming area. Phosphate comes from terrestrial deposits that are eroded and from domestic waste input (detergent) deposited into the sediments. According to Childers et al [22], the main source of phosphorus in sediments is from terrestrial sediments that are eroded and agricultural fertilizers carried by streams, as well as from domestic waste [42]. Anthropogenic activity can also increase the level of soil erosion from the ground to the sea by about 300% [23]. According to Zhuang et al [43], suspended materials may also carry the phosphate absorbed therein. Reported by Vicente [44], organic matter in sediments also contributes to phosphorus retention by sediments.

The content of phosphates in sediments is influenced by the type of substrate. According to Strauch et al [15], sandy soils tend to have higher PO_4^{-3} alkalinity than litter dominated soil. This statement is by the results of research in Mangrove Park that at the station I has the highest phosphate value compared to stations II and III, suspected because the type of substrate is dominated by sand (sand) by 53%.

The high content of phosphate at station I is suspected due to the absence of mangrove vegetation grown on the site so that most of the phosphate is not utilized and settles in sediment. This condition is supported by the condition of waters in Mangrove Park which tend to calm and not much affected by the tides so that the content of phosphates in the sediment tends to be high.

Based on the measurement of phosphate content in sediment, the station I belong to very good fertility, while station II and III belong to low fertility. According to Effendi [41], the phosphate content of 0.00-0.20 mg /100 g classified as low fertility, 0.21-0.50 mg /100 g classified as sufficient fertility, 0.51-1 mg /100 g classified as good fertility , and phosphate values> 1 mg/100 g belong to excellent fertility.

3.4. Relationship of N:P with Mangrove Rhizophora sp Density

Based on the calculation of mangrove density, for station, I was not found mangrove plant, while station II has a density of 0.8 individuals/10 m^2 and station III 1.2 individuals /10 m^2 (Table 2)

Table 2. Mangrove *Rhizophora* sp. Density at Study Site ($\bar{x}$ ± SD, n=3)

Plant Category	Station I (ind)	Station II (ind)	Station III (ind)
Seedling	-	6 ± 1,73	9 ± 2,64
Sapling	-	1 ± 0,57	1 ± 0,58
Tree	-	1 ± 0,57	2 ± 1,00
Total	-	8	12
Density (ind/10 m^2)	-	0,8	1,2

Note: - : no mangrove

Based on the calculation of mangrove density level, in Mangrove Park Pekalongan the site was moderate density. According to Giesen [45], the medium category of mangrove density if having ≥1000 - <1500 trees/hectare.

In Mangrove Park location, Pekalongan is dominated by Rhizophora sp. This is by the type of substrate in the location that is dominated by sand and silt substrate. According to Windusari et al [46], mangrove vegetation grows on substrate variation (sand, clay, and silt) where the condition of organic materials also varies according to substrate composition. Mangrove vegetation is more fertile at low salinity; high salinity mangroves consume more energy to maintain water balance and ion concentration than for primary production and growth [47]. This is proven on the salinity content at the research sites in Mangrove Park has a lower value that ranges from 24-27 ppt compared to the standard quality of 0-34 ppt.

The growth and structure of mangroves, in addition, to correlating with soil chemical physics characteristics also correlates with the ratio of nitrate to phosphate, soil moisture content, sedimentation rate, and soil quality [14].

The ratio value of N: P in sediment at station I is N:P = 0.75: 1; Station II ratio N:P = 23.5: 1; and station III ratio N:P = 16.2: 1. According to Sofawi et al [10], the N: P ratio on the substrate above 30 is considered to be evidence of P-limitation and N: P ratio less than 25-30 is considered to indicate N limitations. Based on this, the location of the study was classified as the criterion of N limitation, since in the third station has an N:P ratio of 0.75 to 23.5. The low value of the ratio is thought to be due to nutrients re-utilized by mangroves for growth. Mangrove requires elements of N and P for growth and primary production. Nitrogen (NO_3-N) was required mangroves for protein formation, improved vegetative growth of plants and photosynthesis [16]. While phosphates are nutrients that are needed by all organisms for energy synthesis (ATP, NADPH), nucleic acids, protein and amino acid formation, and other important compounds [48].

Giesen et al [45], said that environmental factors such as tides, substrate textures also affect mangrove growth. The mangrove litter that is submerged in river water or tidal will experience faster decomposition, as well as nitrogen-rich organic materials will be broken-down quickly [16]. This statement is evidenced by the observations at station II that have high TOM will result in high nitrate decomposition of organic material, and silt particles (silt) have a higher ability to trap nutrients because it has finer particles than sand. Therefore, the ratio value of N:P generated at station II is higher than other stations.

The availability of nutrients is related to the allocation of root biomass. According to McKee [49], when nutrient availability (N:P ratio) is high enough, mangrove seeds will invest more in above ground biomass (by maximizing carbon acquisition) than at root, while when nutrient availability is low, mangrove seeds divert the source to improve root biomass. This is evident at the site at station 3 has a ratio of N:P is lower than station 2, it is suspected that the nutrient is used for the growth of root biomass so that the organic material produced is low.

4. Conclusion

The conclusion that can be drawn from this research is nitrate content (NO_3) and phosphate (PO_4) on sediment in Mangrove Park waters, Pekalongan is still classified into less fertility category, which ranged from 0.81-0.94 mg /100 g (nitrate) and 0.04-1.14 mg /100 g (phosphate). Mangrove density at Mangrove Park is classified into medium density, which is about 0.8 ind /10 m^2 - 1.2 ind/10 m^2. The role of nitrate and phosphate for mangrove growth is to help regenerate the growth of mangrove it self. The high content of phosphate at station I was thought to be derived from the remnants of agricultural waste (fertilization) considering the Mangrove Park in Pekalongan city was once a farming area and from domestic waste input (detergent) deposited into the sediments.

5. Acknowledgments

The author conveyed his gratitude to DPPK (Agriculture, Livestock and Marine Service) and PIM (Mangrove Information Center) Pekalongan City which has given opportunity to do research in PIM region

6. References

[1] Alongi DM *et al* 2005b *Estuar. Coast.* Shelf Sci. **63** 605
[2] Reef R, Feller IC and Lovelock CE 2010 *Tree Physiol* **30** 1148
[3] Ceesay A, Dibi NH, Njie E, Wolff M and Kone T 2017 *Environment and Ecology Research* **5** 145
[4] IUCN 2010 National Diagostric Report for The Gambia on Climate Change Vulnerability International Union for Conservation of Nature (IUCN) Banjul The Gambia p 132
[5] Kristensen E, Bouillon S, Dittmar T and Marchand C 2008 *Aquatic Botani* **89** 201
[6] Wolanski E 1995 *Hydrobiologia* **295** 31
[7] Victor S, Golbuu Y, Wolanski E and Richmond RH 2004 *Wetl. Ecol. Manag.* **12** 277
[8] Kitheka JU, Ongwenyi GS and Mayuti KM 2002 *Ambio* **31** 580
[9] Furukawa K, Wolanski E and Mueller H 1997 *Estuar. Coast. Shelf Sci.* **44** 301
[10] Sofawi AB, Rozainah MZ, Normaniza O and Roslan H 2017 *Marine Biology Research* ISSN: 1745-1000 :1
[11] McKee KL 2001 *J. Ecol.* **89** 876
[12] Kristensen E and Suraswadi P 2002 *Hydrobiologia.* **474** 197
[13] Gadd GM 2010 *Microbiology* **156** 609
[14] Hossain MD and Nuruddin 2016 *Journal of Environmental Science and Technology* **9** 198
[15] Strauch AM, Cohen S and Ellmore GS 2012 *J. Wet. Eco.* **6** 16

[16] Hogart P 2007 *The Biology of Mangroves and Seagrasses* (New York: Oxford University Press)
[17] Hollister CC, Bisogni JJ and Lehmann 2013 *Journal of Environmental Quality* **42** 137
[18] Corredor J and Morell J 1994 *Estuar. Coasts.* **17** 295
[19] Kolliopoulos AV, Dimitrios K, Kampouris and Banks CE 2015 *Anal. Chem.* **87**(8) 4269
[20] Sharma SB, Sayyed RZ, Trivedi MH and Gobi TA 2013 *Springerplus* **2** 587
[21] Black CA 1988 *Soil-plant relationships* (New York: John Wiley & Sons)
[22] Childers DL, Corman J, Edwards M and Elser JJ 2011 *Bio. Science* **61** 117
[23] Bennett EM, Carpenter SR and Caraco 2001 *Bioscience* **51** 227
[24] Filippelli GM 2011 *Chemosphere* **84** 759
[25] Wu FC, Qing HR and Wan GJ 2001 *Water Res.* **35** 1334
[26] Zhu Y, Wu F, Feng W, Liu S and Giesy JP 2016 *Physicochem. Eng. Aspects* **495** 46
[27] Alongi DM 2011 *Journal of Experimental Marine Biology and Ecology* **397** 85
[28] Chaubey I, Sahoo D, Haggard BE, Matlock MD and Costello TA 2007 *Transactions of the ASABE* **50**(1) 35
[29] Hutagalung HP, Setiapermana D and Riyono SH 1997 *Methods of Sea Water, Sediment, and Biota Analysis* (Jakarta: LIPI) **182** pp 59-77
[30] APHA (American Public Health Association) 2005 *APHA 5210-B-2005 : 5-Days BOD Test* (Washington American: Public Health Association)
[31] APHA (American Public Health Association) 1989 *Standard Method for Examination of Water and Waste Water 14 th Edition APHA-AWWA-WPFC* (Washington DC: Port Press)
[32] Ashman MR and Puri G 2002 *Essential Soil Science: A clear and concise introduction to soil science* (Oxford: Blackwell Science)
[33] Pellegrini JAC, Soares MLG, Chaves FO, Estrada GCD and Cavalcanti VF 2009 *Journal of Coastal Research, SI 56 (Proceeding of the 10 th International Coastal Symposium)* Lisbon Portugal:443
[34] Decree of State Minister for The Environment 2004 No: 51/MNKLH/1/2004 *about the guidelines for stipulation of seawater quality standards* (Jakarta: Ministry of Environment) pp 6-7
[35] Afu LOA 2005 *Influence of organic waste to water quality in Kendari bay South east Sulawesi* (Bogor: Institut Pertanian Bogor)
[36] Yaping C and Ye Y 2014 *Journal List.* **9**
[37] Zhu Y *et al.* 2012 *Environ Earth Sci.* **68**(4) 1041
[38] Tang XQ, Wu M, Dai XC and Chai PH 2014 *Ecol. Eng.* **64** 276
[39] Zakaria RM, Aslezaeim N and Sofawi AB 2017 *Pak J Bot.* **49** 33
[40] Nguyen HYT, Cao DM and Schmitt 2013 *Journal of Coastal Conservation* **17**(1) 93
[41] Effendi H 2003 *Study of Water Quality for Resource Management and Water Environment* (Yogyakarta: Kanisius)
[42] Islam MS 2005 *Marine Pollution Bulletin* **50** 48
[43] Zhuang W *et al.* 2014 *Mar. Pollut. Bull.* **83** 343
[44] Vicente MAF *et al.* 2016 *Phosphorus Fractionation Distribution in Guapimirim Estuary* (SE Brazil)
[45] Giesen W *et al.* 2007 *Mangrove Guidebook For Southeast Asia. FAO and Wetlands International* (Bangkok)
[46] Windusari Y, Sarno, Saleh E and Hanum L 2014 *Journal of Biological Researches* **19** 82
[47] Kathiresan K and Bingham BL 2001 *Adv. Mar. Biol.* **40** 81
[48] Khlyntseva SV *et al.* 2009 *J. Anal. Chem.* **64** 657
[49] McKee KL 1995 *Am. J. Bot.* **82** 299

Analysis of the Survival of Children Under Five in Indonesia and Associated Factors

Annisa Nur Islami Warrohmah [1], Sarni Maniar Berliana [2], Nursalam Nursalam[3], Ferry Efendi[3*], Joni Haryanto[3], Eka Misbahatul M Has[3], Elida Ulfiana[3], Sylvia Dwi Wahyuni[3]

[1]Statistics Indonesia (BPS), Kota Metro, Lampung
[2]Institute of Statistics- Statistics Indonesia (BPS), Jakarta
[3]Faculty of Nursing, Universitas Airlangga, Surabaya

Email: ferry-e@fkp.unair.ac.id

Abstract. The under-five mortality rate (U5MR) remains a challenge for developing nations, including Indonesia. This study aims to assess the key factors associated with mortality of Indonesian infants using survival analysis. Data taken from 14,727 live-born infants (2007–2012) was examined from the nationally representative Indonesian Demographic Health Survey. The Weibull hazard model was performed to analyse the socioeconomic status and related determinants of infant mortality. The findings indicated that mother factors (education, working status, autonomy, economic status, maternal age at birth, birth interval, type of births, complications, history of previous mortality, breastfeeding, antenatal care and place of delivery); infant factors (birth size); residence; and environmental conditions were associated with the childhood mortality. Rural or urban residence was an important determining factor of infant mortality. For example, considering the factor of a mother's education, rural educated mothers had a significant association with the survival of their infants. In contrast, there was no significant association between urban educated mothers and their infants' mortality. The results showed obvious contextual differences which determine the childhood mortality. Socio-demographic and economic factors remain critical in determining the death of infants. This study provides evidence for designing targeted interventions, as well as suggesting specific needs based on the population's place of residence, in the issue of U5MR. Further interventions should also consider other identified variables while developing programmes to address infant's needs.

Keywords: survival analysis, infant's mortality, Indonesian DHS

1. Introduction

Under-five Mortality (U5MR) remains a huge challenge in Indonesia despite concerted efforts by all sectors [1]. Although U5MR in Indonesia showed a significant decreased in recent years, achieving the Sustainable Development Goals (SDGs) target requires greater attention on this issue [2-3]. In addition, progress has been very slow, particularly among ASEAN countries [3].
Based on the 2007 Indonesia Demographic and Health Survey, the decline in Indonesian U5MR from 1991 to 1997 is quite sharp, but then sloped until 2012 [4]. Innovative initiatives have been deployed to address maternal and child health problems such as the *Suami Siaga* program [5], *Desa Siaga* [6] and other initiatives among stakeholders. Looking at the provincial data, there were differences in the achievement of U5MR between provinces across Indonesia [7]. As stated in the SDGs target, Indonesia should reduce Infant Mortality Rate (IMR) and U5MR by 25 deaths per 1000 live births

by2030 [3,8]. To achieve this target, it is imperative to examine factors associated with U5MR, particularly in Indonesia.
Previous studies have shown that, among Ethiopian mothers, U5MR is associated with mothers' education level, sex and mothers age at first birth [9]. Similar findings from Bangladesh have shown that size of child and mother's education level as having potential determinants in U5MR [10]. Other factors, such as inadequate housing and moderate housing material, were among predictors of U5MR in Nigeria [11]. In Indonesia's context, extensive research has been done to explore this issue [12, 13]. However, study with big sample size and utilising rigorous study methods has not been done to address this problem. In this paper, we focus on the factors associated with U5MR, and also examine the extent to which factors relate to geographical condition (urban and rural group).

2. Method

The data from the 2012 Indonesia Demographic and Health Survey (IDHS) as nationally representative data are utilised in this study. IDHS is a nationally representative household survey that collects data on a wide range of population, health and nutrition indicators [14], and has been conducted approximately every five years since 1987–1992 with the aim of improving the health of Indonesia's population [14]. Two-stage stratified cluster sampling was employed to determine the sample size. The data used in the present study were derived from the 2012 surveys. Survival analysis was used for analysis to scrutinise the event of death or progression as equal to the time from diagnosis to death [15].

Survival information was obtained from 14,727 singleton live-born infants of the most recent birth of 2012 IDHS. Our analysis was restricted to ever-married women aged 15–49 years. The dependent variable in this study was childhood mortality. The independent variables or risk factors for childhood mortality were mother factors (education, working status, autonomy, economic status, maternal age at birth, birth interval, type of birth, complications, history of previous mortality, size at births, breastfeeding, antenatal care and place of delivery); infant factors (birth size); residence; and environmental conditions. These identified factors were taken from the Mosley and Chen framework of factors determining child survival in developing countries [16]. Descriptive and inferential statistics were utilised to identify factors associated with U5MR.

3. Results and discussion

Of the 14,727 singleton live-born infants, there were 300 under-five child deaths across provinces of Indonesia. Table 1 presents the characteristics of respondents of children under the age of five for each covariate considered in the analysis. It shows that, among the educational level of mothers, 46.5% had attended middle schools. More than half the mothers had a job or were economically empowered (55%) and had a high level of autonomy (71.5%). The data show that the majority of Indonesian mothers went to hospital for delivery of their babies (89.7%). The table also summarises the variables for all the other covariates considered in this study.

In the survival analysis with Weibull distribution, the results showed the likelihood test value of $526.98>X^2_{(19;0.1)}$ with a p-value less than 0.1, which means that the model with covariate is better in describing the data of survival of Indonesian U5MR. Based on the hazard model, all variables significantly affected the survival of childhood mortality (Table 2).

A child from a low-educated mother has a risk of 1,602 times to experience under-five mortality compared with a toddler from a high educated mother. The role of education as the key to reducing child mortality is well known in general in literature worldwide [17-19].
Indonesian mothers who were working having a risk of 2.22 times to experience U5MR compared with children from mothers who are not working. This aligns with a study by Basu who stated that occupation might influence a mother's time in paying attention to the baby [20].

Table 1. Demographic characteristics of mother and child (n=14,727)

Variable	Category	n	%
Mother's education	Low	1878	12.8
	Middle	6839	46.5
	High	6006	40.8
Working status	No	6632	45.0
	Yes	8095	55.0
Autonomy	Low	818	5.6
	Medium	3384	23.0
	High	10525	71.5
Economic status	Low	7202	48.9
	Middle	2718	18.5
	High	4807	32.6
Age	< 20 years	1376	9.3
	20-34	10936	74.3
	> 35 years	2415	16.4
Birth interval	<24	1147	7.8
	>24 & single	13580	92.2
Type of birth	Single	14634	99.4
	Twins	93	0.6
Complications	No	12868	87.4
	Yes	1859	12.6
History of previous mortality	No	13341	90.6
	Yes	1386	9.4
Residence	Urban	6766	45.9
	Rural	7961	54.1
Size at births	Average	7791	52.9
	Risk	6936	47.1
Environmental conditions	Not poor	8114	55.1
	Poor	6609	44.9
Breastfeeding	Yes	14031	95.3
	No	696	4.7
Antenatal care	At least 4	9947	67.5
	Less than 4	4780	32.5
Place of delivery	Hospital	13215	89.7
	Out of hospital	1512	10.3

Table 2. Estimation of Weibull regression model parameters of childhood survival (n=14,727)

Variables	Coefficient	Hazard ratio	P>z
Education			
Low	0,471	1,602	0,014**
Middle	0,265	1,303	0,078*
Occupation	0,774	2,169	0,000**
Autonomy			
Low	0,584	1,793	0,004**
Medium	0,474	1,607	0,000**
Economic status			
Low	0,352	1,421	0,068*
Medium	0,241	1,273	0,185
Age			
<20 years	0,451	1,569	0,015**
>=35 years	0,285	1,330	0,044**
Birth interval	0,631	1,880	0,000**
Type of birth	0,807	2,240	0,027**
Complications	0,363	1,438	0,018**
History of previous mortality	0,495	1,641	0,002**
Size at birth	0,212	1,236	0,072*
Residence	0,248	1,281	0,072*
Environmental conditions	0,310	1,363	0,045**
Breastfeeding	2,689	1,471	0,000**
Antenatal care visit	0,490	1,632	0,000**
Place of delivery	0,347	1,415	0,093*
_cons	-6,765		
/ln_p		-1,282	
P		0,278	
1/p		3,603	

Children from mothers of low autonomy category had a risk of 1.79 times to experience mortality compared to children from mothers who had high autonomy category. This supports the research of Williamson and Boehmer using data from 96 countries, including Indonesia, which found strong evidence showing autonomy is a significant variable and an important factor of survival of under-five children [21].

Children from low economic status households have a risk of 1.42 times to experience under-five mortality compared with households from high economic status. In a similar vein, studies in 55 developing countries using household asset indexes to measure wealth found that child mortality was significantly higher in low-status households than in high-status households [22].

This study shows that children born to a mother at the age of less than 20 years have a risk of 1.57 times to experience under-five mortality compared to under-five children born to ideal age (20-34 years old). This view is supported by Bhalotra and Soest (2008) who write that women who give birth too young or too old have a higher risk of child mortality [23].

Children born with birth interval less than two years have a risk to experience under-five mortality of 1.88 times compared to infants born with birth interval of 24 months or more. Children born as twins have a risk of 2.24 times to experience under-five mortality compared to children born who are not twins. Interestingly, children who have experience of complications while in the pregnancy period have a risk of 1.44 times to experience under-five mortality compared to children who have no experience of complications. Children born with a history of sibling death has a risk of 1.64 times for experience under-five mortality compared to children born without a previous sibling's death. The findings also similar direction with bigger size at birth, reside in rural, worse environment, never had breastfeeding and less than four antenatal care visits having higher risk to experience under-five mortality. These results match those observed in earlier studies [20-25].

Findings have shown that children born in other than hospitals have a risk of 1,415 times to experience under-five mortality compared to infants who were born in hospital. Study conducted in Pakistan showed that the place of delivery significantly associated with childhood survival [24]. Similarly, Buwembo stated that the risk of mortality is 1.9 times in under-fives who were born in places other than the hospital compared to children who were born in hospital [25].

These results indicate that more specific interventions may be those that target individual level rather than group or communities level. In the same vein, another study of Indonesia revealed that improving access for women itself may be benefit to enhance maternal health and diminish mortality and morbidity [26]. The interventions should be embedded in social intervention conducted by the Government of Indonesia.

4. Conclusion and suggestions

The U5MR of Indonesian children indicates a decline, but still slow progress despite the new indicators determined by SDGs. Socio-demographic and economic factors remain critical in determining the health of infants. Government programmes must address issues in education, empowerment, age and women's health issues. The results also suggest that government initiatives should direct towards the economic status of population. Specific targets may be better rather than community targets in order to accelerate the interventions. Poor and marginal mothers, particularly from coastal regions, need greater attention to improve the outcome.

References

[1] Hodge A, Firth S, Marthias T 2014 PLOS ONE **9**(7) e103597
[2] Efendi F, Chen CM, Kurniati A, Berliana SM 2017 Women Health **57**(5) 614-629
[3] Victora CG et al 2016 Lancet **387**(10032) 2049-59
[4] Statistics Indonesia - Badan Pusat Statistik - BPS and Macro International, *Indonesia Demographic and Health Survey 2007* 2008 Calverton, Maryland USA: BPS and Macro International
[5] Kurniati A, Chen CM, Efendi F, Elizabeth Ku LJ, Berliana SM 2017 Health Policy Plan **32**(8) 1-9
[6] MOH 2010 *Pedoman Umum Pengembangan Desa Dan Kelurahan Siaga Aktif (Guidelines for the Development of Active Alert Villages)* Jakarta: Pusat Promosi Kesehatan Kementerian Kesehatan
[7] BPS 2017 *Angka Kematian Bayi menurut Provinsi 1971, 1980, 1990, 1994, 1997, 2000, 2002, 2007, 2010, 2012 dan Kematian Dibawah Usia Lima Tahun menurut Provinsi 1971, 1980, 1990, 1994, 1997, 199 dan 2007, 2012.* [cited 2017 August 1]; Available from: https://www.bps.go.id/linkTabelStatis/view/id/1270.
[8] World Health Organization 2015 *Health in 2015: from MDGs, Millennium Development Goals to SDGs, Sustainable Development Goals* World Health Organization.
[9] Getachew Y, Bekele S 2016 Journal of Biosensors & Bioelectronics **7**(3) 1-10
[10] Mohammad KA, Fatima-Tuz-Zahura M, Bari W 2017 BMC Int Health Hum Rights **17**(1) 3

[11] Adebowale SA, Morakinyo O, Ana G 2017 BMC Pediatr **17**(1):30-40
[12] Mahmudah U, Cahyati HW, Wahyuningsih AS 2011 Jurnal Kesehatan Masyarakat **7**(1):20-30
[13] Afifah TS, Djaja, Irianto J, 2003 Buletin Penelitian Kesehatan **31**(6): 50-70
[14] Statistics Indonesia, Badan Pusat Statistik, BPS, et al., *Indonesia Demographic and Health Survey 2012* 2013 Jakarta: BPS, BKKBN, Kemenkes, and ICF International
[15] Kleinbaum DG, Klein M 2012 *Evaluating the proportional hazards assumption*, in *Survival analysis* 2012 Springer 161-200
[16] Mosley WH, Chen LC 1984 Population and development review **10**(0):25-45
[17] Bicego GT, Boerma JT 1993 Social science & medicine **36**(9):1207-1227
[18] Marindo R, Hill K *Trends and differentials in child mortality: Zimbabwe 1970-1994* 1997 Demographic Health Survey
[19] Rajna P, Mishra AK, Krishnamoorthy S 1998 Asia-Pacific Population Journal **13**(2):27-38.
[20] Basu AM, Basu K 1991 Health Transition Review **1**(1):83-103
[21] Williamson JB, Boehmer U 1997 Social Science & Medicine **45**(2):305-317
[22] Hertzman C et al., 2010 BMJ 340(2):468-487
[23] Bhalotra S, Soest AV 2008 Journal of Econometrics **143**(2): 274-290
[24] Zahid GM 1996 The Pakistan Development Review **35**(4):719-731
[25] Buwembo P *Factors associated with under-5 mortality in South Africa: trends 1997-2002.* 2010

[26] Kurniati A, Chen CM, Efendi F, Berliana SM 2017 *Health Care Women Int.* **0**(00):1–17

Formulation of Emergency Food in Biscuit-Form Made From Proso Millet Flour (*Panicum miliaceum*) and Snakehead Fish (*Channa striata*) –Tempeh Flour Koya

R B K Anandito*, S R Kurniawan, E Nurhartadi, Siswanti
Department of Food Science and Technology, Faculty of Agriculture, Universitas Sebelas Maret
Email: anandito_ito@yahoo.com

Abstract. Indonesia is a natural disaster-prone country. Food availability appears as the post-disaster main problem. Emergency food is an effort to overcome it. An ideal emergency food is expected to have a complete nutritional composition. Fulfilling macronutrient needs, proso millet flour and snakehead fish-tempeh flour *koya* were being used in this research. The objective of this research was to determine the formulation of emergency food in biscuit form made from proso millet flour and snakehead fish-tempeh flour *koya* based on its sensory, chemical and physical properties. This research was using a completely randomized design with one factor affecting. The variations of proso millet flour and snakehead fish-tempeh flour *koya* composition were 100:0 ; 80:20 ; 60:40 ; and 40:60. All treatments were carried out sensory evaluation using hedonic test, proximate analysis, calories analysis using bomb calorimetry methods. and physical analysis using texture profile analyzer. The compensatory model was used to determine the best formula based on all parameters. The result indicated that the best biscuit formula had compositions of proso millet flour 80% and snakehead fish-tempeh flour *koya* 20%. It had a brownish color, not-strong-fishy aroma, a strong-enough bitter aftertaste and crunchy texture. Chemical analysis showed that its moisture, ash, protein, fat, carbohydrate and total calories content were 1,146%; 4,827%; 12,536%; 26,325%; 56,312% and 5,033 kcal/gram, respectively. The physical analysis shows that its hardness and fracturability values were 8,037 N and 2,403 N, respectively.
Keywords : emergency food, biscuit, proso millet, snakehead fish, tempeh flour, *koya*

1. Introduction

Natural disaster is an incident that cannot be avoided or prevented. In Indonesia, it is increasing each year in amount [1]. These lead to many new post-disaster problems. One of them is food availability problem. Food becomes a scarce stuff at that time, whereas the victims need adequate nutrition. Resolving these nutrition need, Emergency Food Product (EFP) is intended to provide a compact, self-contained, high-energy, nutrient-dense emergency food for refugees and victims of disasters for a short duration at the initial stages of an emergency. EFP should contain 10-15% of protein, 35-45% of fat, and 40-50% of carbohydrates. Furthermore, the EFP must be safe, palatable, easy to dispense, easy to use, and nutritionally complete [2].

There are many alternative forms of EFP, one of them is biscuit [3]. Biscuit is a food made from wheat flour with the addition of other foodstuffs, with the process of heating and molding [4]. Biscuit form is used because it is easy to be fortified or substituted, high palatability and good shelf-life. Past studies showed that many designed EFP (e.g. food bars) didn't meet emergency food standards, especially on its calories and macronutrient composition. Fulfilling macronutrient needs, proso millet flour and snakehead fish-tempeh flour koya were being used in this research.

Proso Millet is the fourth economic ingredient after rice, wheat, and corn. Proso millet is a group of cereals used as a bird feed. Chemical analysis of proso millet flour showed that water, ash, fat, protein, carbohydrate and crude fiber content were 9.19%; 1.80%; 2.58%; 11.29%; 74.52%; And 2.01% [5].

Koya is a tasty food supplement in the form of powder [6]. This research used snakehead fish-tempeh flour koya with the percentage of 60% snakehead fish and 40% tempeh flour which has been studied in previous research. Koya has a protein content of 42.201% [7]. The use of koya is expected to increase the content of macro nutrient biscuits made in order to meet emergency food standards.

Koya is made from snakehead fish as main ingredient and tempeh flour as an additional ingredient. The nutritional value contained in snakehead fish is high enough, that is with protein equal to 42% and 1.7 % of fat [8]. Snakehead fish is one of good albumin source. Albumin is a soluble protein that is soluble in water and plays an important role in the formation of new cell tissues, boosts the recovery of damaged cell tissues and maintains fluid balance in the blood vessels [9].

Tempeh flour is added to the koya to disguise a distinctive fishy scent. In addition to disguising the fishy aroma, which is not desirable, the use of tempeh flour increases protein content in koya. Tempeh has a content of protein, fat, carbohydrate, water, and ash in a row is 46%; 24.7%; 19.3%; 7.7%; And 2.3% [10].

2. Materials and methods

2.1. Preparation of Koya Flour

Koya flour was made using snakehead fish obtained from Cengklik Lake, Boyolali with a length of 16-23 cm and tempeh obtained from Manang Village, Sukoharjo with incubation period for 2 days. First, snakehead fish was filleted and steamed and tempeh was cut and dried at a temperature of 70°C for 6-7 hours. Then, steamed fish meat was cooked with spices and coconut milk until dry. Tempeh flour was added afterward, then roasted for 5 minutes.

2.2. Preparation of Proso Millet Flour

Proso millet flour was made using millet obtained from Depok Market, Surakarta city. Millet was sorted from dirt then its epidermis was removed from the endosperm by crushing. Afterward, it was sorted again from the epidermis using winnowing method. Then, the endosperm was diminished in size and sieved with a particle size of 80 mesh [5].

2.3. Preparation of Biscuit Samples

Preparation of biscuit samples began with the calculation of total calories predictions. The formula was made using the previous study about biscuit (**Table 1**) [11]. Afterward, all ingredients in each formula were mixed until homogeneous. Then, the dough was flattened with a thickness of 3 mm and molded in a circle shape. Then, it was baked with a temperature of 200°C for 10 minutes.

Table 1 Biscuit's Formula

Ingredients	**Formula**			
	Control	**F1**	**F2**	**F3**
Proso Millet Flour (gram)	50.0	40.0	30.0	20.0
Snakehead Fish-Tempeh Flour Koya (gram)	0.0	10.0	20.0	30.0
Sugar (gram)	6.0	6.0	6.0	6.0
Salt (gram)	1.5	1.5	1.5	1.5
Full Cream Milk (gram)	8.0	8.0	8.0	8.0
Baking Soda(gram)	0.5	0.5	0.5	0.5
Vanilla (gram)	0.5	0.5	0.5	0.5
Shortening (gram)	20.0	20.0	20.0	20.0
Egg yolk (gram)	7.0	7.0	7.0	7.0
Egg white (gram)	8.0	8.0	8.0	8.0
Water (ml)	5.0	5.0	5.0	5.0
Total	105.0	105.0	105.0	105.0

2.4. Determination of Sensory Properties

The sensory properties were determined using the hedonic testing method. Color, aroma, taste, texture and overall of all biscuit formula was observed by 40 semi-trained panelists [12].

2.5. Determination of Chemical Properties

The chemical properties observed using standard analysis protocol from AOAC [13] in koya flour, proso millet flour and all biscuit formulas are moisture (935.29), ash (942.05), protein (total) (984.13), fat (Crude) (991.36), carbohydrate (986.25) and total calories content (only for all formulas) [14].

2.6. Determination of Physical Properties

The physical properties were determined using Texture Profile Analyzer (TPA) [15]. Observed parameters are hardness and fracturability.

2.7. Statistical Analysis Methods

All collected were analyzed using one-way analysis of variance (ANOVA) using SPSS Statistics 24. Difference between means was determined using Duncan's least significant difference. Then additive weighting technique of compensatory model used to determine the best formula [16].

3. Result and discussion

3.1. Chemical properties of ingredients

Main ingredients in this study are proso millet flour and snakehead fish-tempeh flour koya. These need to be observed to make a proper formula as an emergency food. Chemical properties of proso millet flour and snakehead fish-tempeh flour koya are showed in Table 2.

Table 2 Chemical Properties of Proso Millet Flour and Snakehead Fish-Tempeh Flour Koya

Chemical Component	Proso Millet Flour	Snakehead Fish-Tempeh Flour Koya
Moisture (%wb)	11.459±0.033	4.266±0.042
Ash (%db)	1.651±0.100	4.897±0.186
Protein (%db)	11.715±0.076	48.186±0.106
Fat (%db)	5.754±0.326	18.429±0.396
Carbohydrate (%db)	80.880±0.392	28.488±0.511

Notes: wb : wet basis
db : dry basis

As seen in Table 2, proso millet, which used in this study, has similarly protein content to the wheat flour which has 11.6 grams of protein in 100 grams wheat flour [17]. All chemical component of proso millet flour used in this study had meet standard requirements such as moisture content (maximum 14,5%) and protein content (minimum 7%) [18].

Koya used in this study has a high protein content of 48,186%. Koya was suggested in previous studies to be able to establish emergency food products that can meet the needs of protein by 10-15 percent [7]. All chemical component of Snakehead Fish-Tempeh Flour Koya used in this study had meet standard requirements such as moisture content (maximum 7%), ash content (maximum. 7%), fat content (maximum 30%) and protein content (minimum 15%). But, it has contrast in moisture content and protein content. The difference in moisture content is caused by cooking time. The previous study cooked koya for approximately 45 minutes, but in this study, it needs 90 minutes of cooking. Another difference, protein content, is caused by the length of snakehead fish used in this study. Some research reported that snakehead fish with 16-23 cm of length has higher protein content than others [19].

3.2. Sensory properties

The result of hedonic (Table 3) test shows that the use of koya causes a decrease in the level of liking on biscuit's color. It is caused by non-enzymatic browning. When reducing sugar (e.g. glucose) meets amino acid (e.g. lysine), Maillard reaction was done and led the raising of a brown pigment called melanoidin [20]. Furthermore, caramelization occurs during the process of biscuits making and bring a charred color up. Caramelization occurs when sucrose is heated at 200°C and forming caramelan.

Table 3 Sensory Properties of Biscuit

Formula	Color	Aroma	Flavor	Texture	Overall
Control	$4.075^{c}\pm0.888$	$3.825^{b}\pm0.844$	$3.100^{b}\pm0.982$	$3.925^{b}\pm0.730$	$3.500^{b}\pm0.906$
F1	$2.650^{a}\pm0.975$	$3.125^{a}\pm0.911$	$2.575^{a}\pm0.712$	$4.075^{b}\pm0.859$	$2.750^{a}\pm0.630$
F2	$3.075^{b}\pm0.917$	$3.075^{a}\pm0.971$	$2.725^{ab}\pm0.933$	$3.325^{a}\pm0.694$	$2.875^{a}\pm0.790$
F3	$2.625^{a}\pm0.839$	$2.800^{a}\pm0.939$	$2.475^{a}\pm0.847$	$3.225^{a}\pm0.768$	$2.625^{a}\pm0.774$

Notes : Values within the same row followed by the same letter are not statistically different ($p < 0.05$)
1= very dislike, 2= dislike, 3= Neutral, 4= like, 5= very like.
Control : 100% proso millet flour and 0% snakehead fish-tempeh flour koya
F1 : 80% proso millet flour and 20% snakehead fish-tempeh flour koya
F2 : 60% proso millet flour and 40% snakehead fish-tempeh flour koya
F3 : 40% proso millet flour and 60% snakehead fish-tempeh flour koya

Biscuits with greater use of koya have worse aromas. Showed in Table 3, greater koya used in the formula made the biscuit more disliked by the panelist. This is caused by poly carbonyls and mono carbonyls compounds that comes up during Maillard reaction and formed burnt aromas (pungent, empyreumatical) [20].

Biscuits with greater use of koya have worse flavor. in Table 3, greater koya used in the formula made the biscuit more disliked by the panelist. This is caused by caramelan pigment which is soluble in water or ethanol and has a bitter taste. This pigment comes up during caramelization process [20].

Biscuits with greater use of koya have worse texture. Showed in Table 3, greater koya used in the formula made the biscuit more disliked by the panelist. It is caused by waxy mouth feels contributed by koya's fats [21]. As seen in Table 4, fat content in biscuit increases with the addition of koya. These results bear a resemblance to the results of the physical analysis. The overall parameter is an evaluation of all parameters (e.g. Color, Aroma, Flavor, and Texture). As seen in Table 3, palatability of the biscuit decreases with the addition of koya.

3.3. Chemical properties of biscuits

In Table 4, the Moisture content of biscuit increases with the addition of koya. Koya donates sucrose and polar amino acids. Sucrose and polar amino acid have the ability to bind with water [22]. The moisture content of all formula has met the standard at a maximum of 5% [4].

Ash contains minerals. In Table 4, Ash content of biscuits increases with the addition of koya. As known in Table 2, koya donates greater minerals than proso millet flour. This makes biscuits with the greater addition of koya has a greater amount of minerals within. These results haven't met the standard at a maximum of 1.6% [23].

Protein content increases with the addition of koya as seen in Table 4. It is caused by donated koya has a greater amount of protein than proso millet flour. As seen in Table 2, the protein content of koya is 48.186% whereas flour only has a protein content of 11,715%. These results have met protein requirement as a biscuit and as an emergency food product. The minimum protein content of a product in the form of the biscuit is 5% [4]. As an EFP, the product should contain 10-15% of the protein [2].

Table 4 Chemical Properties of Biscuit

Formula	Mositure (%wb)	Ash (%db)	Protein (%db)	Fat (%db)	Carbohydrate (%db)	Total Calories (kcal/gram)
Control	2.556^{b}±0.139	3.941^{a}±0.085	9.909^{a}±0.292	25.828^{a}±0.150	60.322^{a}±0.498	5.364^{b}±0.004
F1	1.146^{a}±0.032	4.827^{b}±0.017	12.536^{b}±0.058	26.325^{a}±0.678	56.312^{b}±0.610	5.033^{a}±0.017
F2	2.677^{b}±0.012	5.083^{c}±0.068	15.999^{c}±0.105	29.773^{b}±0.276	49.145^{c}±0.312	5.390^{b}±0.019
F3	3.041^{c}±0.755	5.448^{d}±0.027	18.049^{d}±0.222	31.131^{c}±0.447	45.372^{d}±0.501	5.579^{c}±0.067

Notes : Values within the same row followed by the same letter are not statistically different ($p < 0.05$)
Control : 100% proso millet flour and 0% snakehead fish-tempeh flour koya
F1 : 80% proso millet flour and 20% snakehead fish-tempeh flour koya
F2 : 60% proso millet flour and 40% snakehead fish-tempeh flour koya
F3 : 40% proso millet flour and 60% snakehead fish-tempeh flour koya

Fat content increases with the addition of koya as seen in Table 4. It is caused by donated koya has a greater amount of fat than proso millet flour. As seen in Table 2, the fat content of koya is 18.429% whereas flour only has a protein content of 5.754%. These results have met fat requirement as a biscuit but not as an emergency food product. The minimum fat content of a product in the form of the biscuit is 9.5% [4]. As an EFP, the product should contain 35-45% of fat [2].

In Table 4, the carbohydrate content of biscuit decreases with the addition of koya. It's happened because koya doesn't has sufficient amount of carbohydrate replacing carbohydrate donated by proso millet flour. These results have met carbohydrate requirement as an emergency food product. As an EFP, the product should contain 40-50% of carbohydrate [2].

The results of total calories analyses showed in Table 4. It's showed that with a greater addition of koya make the biscuit has greater total calories. These results have met total calories requirement as a biscuit and as an emergency food product. The minimum total calories of a product in the form of the biscuit is 4.000 kcal/gram [23]. As an EFP, the product must have minimum 4.660 kcal/gram [2].

3.4. Physical Properties

Hardness is the perceived force required to break the sample into several pieces during the first bite by the molars. Hard as a texture characteristic describes a product which displays substantial resistance to deformation or breaking [15]. The results of total physical analyses showed in **Table 4**. It's showed that with a greater addition of koya make the biscuit has a harder texture. The result relates to the moisture content of each formula. Higher moisture content, harder the texture of the biscuits. A product at moisture content lower than 10% has a brittle state. In this state, the cracks formation is so frequent that this process becomes to be the important source of strength decreases with decreasing moisture content [24]

Table 5 Physical Properties of Biscuit

Formula	Hardness (N)	Fracturability (N)
Control	12.878^{b}±0.336	6.906^{ab}±0.230
F1	8.037^{a}±1.030	2.403^{a} ±0.305
F2	20.380^{c}±2.372	9.859^{b} ±1.491
F3	29.329^{d}±0.427	22.247^{c} ±3.014

Notes : Values within the same row followed by the same letter are not statistically different ($p< 0.05$)
1= very dislike, 2= dislike, 3= Neutral, 4= like, 5= very like.
Control : 100% proso millet flour and 0% snakehead fish-tempeh flour koya
F1 : 80% proso millet flour and 20% snakehead fish-tempeh flour koya
F2 : 60% proso millet flour and 40% snakehead fish-tempeh flour koya
F3 : 40% proso millet flour and 60% snakehead fish-tempeh flour koya

Fracturability (originally called brittleness) was defined as the force of the initial break in the curve on the first bite [15]. The results of total physical analyses showed in Table 4. It's showed that with a greater addition of koya make the biscuit has more firm texture. The result relates to the moisture content of each formula. Higher moisture content, harder the texture of the biscuits. A product at moisture content lower than 10% has a brittle state. In this state, the cracks formation is so frequent that this process becomes to be the important source of strength decreases with decreasing moisture content [24]

3.5. Determination of the best formula

Compensatory model was used in this study to determine the best formula. The determination was done based on its sensory, chemical and physical parameters. The method was started with non-dimensional scaling and followed by the additive weighting technique [16]. The highest total score showed the best formula. As seen in Table 6, F1 formula is the best formula followed by F2 and F3 formula.

Table 6 Determination of the best formula using Compensatory model

Component	Relative Rank	Normalized Weight	F1		F2		F3	
			DV	Score	DV	Score	DV	Score
Sensory								
Color	1	0.077	0.056	0.004	1.000	0.077	0.000	0.000
Aroma	1	0.077	1.000	0.077	0.846	0.065	0.000	0.000
Flavor	1	0.077	0.400	0.031	1.000	0.077	0.000	0.000
Texture	1	0.077	1.000	0.077	0.118	0.009	0.000	0.000
Overall	1	0.077	0.500	0.038	1.000	0.077	0.000	0.000
Chemical								
Moisture content	1	0.077	1.000	0.077	0.192	0.015	0.000	0.000
Ash content	1	0.077	1.000	0.077	0.588	0.045	0.000	0.000
Protein content	1	0.077	1.000	0.000	0.283	0.048	0.000	0.000
Fat content	1	0.077	0.000	0.077	0.628	0.022	1.000	0.077
Carbohydrate content	1	0.077	1.000	0.077	0.345	0.027	0.000	0.000
Total calories	1	0.077	0.000	0.000	0.654	0.050	1.000	0.077
Physical								
Hardness	1	0.077	1.000	0.077	0,420	0.032	0.000	0.000
Fracturability	1	0.077	1.000	0.077	0,624	0.048	0.000	0.000
Total	**13**		**8.956**	**0.689**	**7,498**	**0.592**	**2.000**	**0.154**

Control : 100% proso millet flour and 0% snakehead fish-tempeh flour koya
F1 : 80% proso millet flour and 20% snakehead fish-tempeh flour koya
F2 : 60% proso millet flour and 40% snakehead fish-tempeh flour koya
F3 : 40% proso millet flour and 60% snakehead fish-tempeh flour koya

Conclusion

The best biscuit formula contains of 80% proso millet flour and 20% snakehead fish-tempeh flour koya. It had a brownish color, not-strong-fishy aroma, a strong-enough bitter aftertaste and crunchy texture. Chemical analysis showed that its moisture, ash, protein, fat, carbohydrate and total calories content respectively were 1.146%; 4.827%; 12.536%; 26.325%; 56.312% and 5.033 kcal/gram. The physical analysis showed that its hardness and fracturability values respectively were 8.037 N and 2.403 N.

References

[1] BNBP 2014 *Info Bencana: Informasi Kebencanaan Bulanan Teraktual.* Pusdatinmas Badan Nasional Penanggulangan Bencana

[2] Zoumas B L, Amstrong L E, Backstrand J R, Chenoweth W L, Chinachoti P, Klein B P, Lane H W, Marsh K S, and Tolvanen M 2002 *High-Energy Nutrient-Dense Emergency Relief Food Product* (Washington D.C: National Academy Press)

[3] VanLaanen P 2010 *Emergency Food and Water Supplies* (Texas A&M : Agrilife Extension E11 S99)

[4] Badan Standarisasi Nasional 2011 *SNI 2973:2011 Biskuit*. Jakarta

[5] Anandito R B K, Siswanti, Edhi N dan Rini H 2016 Formulasi Pangan Darurat Berbentuk *Food bars* Berbasis Tepung Millet Putih (*Panicum milliaceum l.*) dan Tepung Kacang Merah (*Phaseolus vulgaris l.*). *Jurnal Agritech.* 1(26):23-29.

[6] Handayani T H dan Marwanti 2011 *Pengolahan Makanan Indonesia* (Yogyakarta:UNY Press)

[7] Ashary H S 2016 Formulasi dan Pendugaan Umur Simpan Koya Ikan dengan Bahan Dasar Ikan Gabus (Channa striata) dan Tepung Tempe sebagai Pelengkap Makanan Sumber Protein. Program Studi Ilmu dan Teknologi Pangan UNS, Surakarta

[8] Utomo D, Wahyuni R dan Wiyono R 2013 Pemanfaatan Ikan Gabus Menjadi Bakso dalam Rangka Perbaikan Gizi Masyarakat dan Upaya Meningkatkan Nilai Ekonomisnya. *Jurnal Teknologi Pangan.* 1(1): 38-55

[9] Sudarmadji S, dan Suhardi 2007Analisis Bahan Makanan dan Pertanian (Yogyakarta: Liberty Yogyakarta)

[10] Bastian F E, Ishak A B, Tawali dan Bilang M 2013 Daya Terima dan Kandungan Zat Gizi Formula Tepung Tempe dengan Penambahan Semi Refined Carragenan (SRC) dan Bubuk Kakao. *Jurnal Aplikasi Teknologi Pangan.* 2(1):5-8

[11] Sarofa U, Yulistiani R dan Mardiyah 2012 Pemanfaatan Tepung Buah Lindur (*Bruguiera gymnorrhiza* dalam Pembuatan Crackers dengan Penambahan Gluten. *Jurnal Rekapangan* 6(1): 13-18

[12] ASTM 2012 *Sensory Testing Methods: Second Edition* (United States of America: American Society for Testing and Materials International)

[13] AOAC 2002 *Official Methods of Analysis of AOAC International 17th* (Washington DC: Association of Official Analytical Chemists)

[14] Mulyaningsih Y dan Rosida J 2002 Membandingkan Hasil Analisis Energi Total Menggunakan Bom Kalorimeter dengan Hasil Analisi Proksimat. *Temu Teknis Fungsional Non Peneliti.* Bogor

[15] Bourne M C 2002 *Food Texture and Viscosity: Concept and Measurement 2nd edition* (New York, USA: Elsevier Science & Technology Books by Academic Press)

[16] Sullivan W G, Wicks E M and Koelling C P 2015 *Engineering Economy 16th Edition* (London: Pearson Education Ltd)

[17] Suarni 2012 Potensi Sorgum sebagai Bahan Pangan Fungsional. *IPTEK Tanaman Pangan* 7(1):58-66

[18] Badan Standarisasi Nasional 2002 *SNI 01-3751-2002 Tepung Terigu Sebagai Bahan Makanan.* Jakarta

[19] Gam L H, Leow C Y and Baie S 2006 Proteomic Analysis of Snakehead Fish (*Channa striata*) Muscle Tissue. *Malaysian Journal of Biochemistry and Molecular Biology.* 14:25-32

[20] deMan J M 1999 *Principle of Food Chemistry Third Edition* (Gaithersburg, Maryland: Aspen Publisher)

[21] Young S. P. C. a. L. S. 2006 *Baked Product: Science, Technology and Practice* (United Kingdom: Blackwell Publishing)

[22] Triyono A 2010 Mempelajari Pengaruh Penambahan Beberapa Asam pada Proses Isolasi Protein Terhadap Tepung Protein Isolat Kacang Hijau (*Phaseolus radiates L.*). *Seminar Rekayasa Kimia Dan Proses*. Semarang

[23] Badan Standarisasi Nasional 1992 *SNI 01-2973-1992 Mutu dan Cara Uji Biskuit*. Jakarta

[24] Blahovec J 2007 Role of Water Content in Food and Product Texture. *International Agrophysics.* 21:209-215

IOP Conf. Series: Earth and Environmental Science **116** (2018) 012016 doi:10.1088/1755-1315/116/1/012016

The Relation of Environmental Quality and Fishery Sector in Indonesia

Shanty Oktavilia[1], Reikha Habibah Yusfi[2], Firmansyah[3], FX Sugiyanto[3]

[1]Faculty of Economics, Universitas Negeri Semarang and Doctoral Program in Economics, Diponegoro University
[2]Master of Economics Development, Diponegoro University
[3]Department of Economics, Diponegoro University

oktavilia@mail.unnes.ac.id

Abstract. The condition of fishery sector is currently stagnating, even tending to decline, which is indicated by the decrease of production in some areas in Indonesia. Environmental degradation in marine waters is due to global climate change and uncontrolled fish exploitation impact on the decline of marine fisheries production. While in aquaculture, the environmental quality is also indicated to influence the production. Nevertheless, the increase of production of both marine and terrestrial fisheries has an effect on the quality of the environment. This study aims to analyze the interrelationship between the influence of environmental quality on the production of fishery sub-sector and the influence of fishery subsector production on environmental quality. This research employs environmental quality data and output of fishery of 34 provinces in Indonesia during 2011-2015. The study finds that output of fishery sector affects the environmental quality, which proves the Environment Kuznets Curve in the fishery sector in Indonesia. Since a certain threshold is achieved, the increase in revenue followed by the increase in environmental quality. The study also finds that the environmental quality has a positive effect on the production of fishery. Implication of the study is the increase of income of fishery households can be encouraged the ability of the community to protect the environment and increases the willingness of households to sacrifice other goods to environmental protection.

1. Introduction

Fish and fish products are the essential commodities required by humans in fulfilling their needs on food meterials. FAO predicts that in 2010, fish and seafood commodities may contribute 17 percent of the world's protein supplies. Fish products are essential commodities for the developing countries because they contribute 75 percent of their population's protein consumptions [1]. Fish Commodities have been captured and cultivated for thousands of years and remained becoming the essential food sources. Fish products contain micronutrients and omega-3 fatty acids which are essential for the mothers' and children's health throughout the world.

Several studies [2], [3], suggest that although statistically the number of global fish catches increase, the fish-catch growth actually has annually decreased since the 1990s. In Indonesia, the production of capture fisheries in some provinces shows a declining growth, although there are still many provinces showing a positive growth in the production of capture fisheries. The provinces showing a declining growth include West Sumatra, Riau, Bangka Belitung Islands, Special Region of

Yogyakarta, Bali, West Nusa Tenggara, West Kalimantan, East Kalimantan, North Sulawesi, Central Sulawesi, Southeast Sulawesi and Papua.

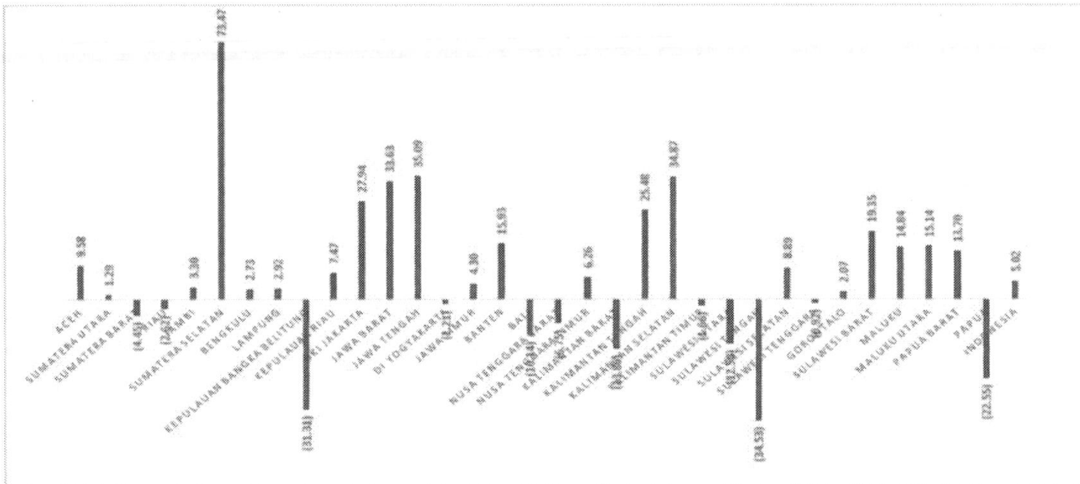

Source: [4], processed

Figure 1. Average Growth Rate in 2011-2015 on the production of capture fisheries based on Each Province (Percent)

The problem on the declining production of capture fisheries are due to several factors, inlcuding climate changes, offer fishing, and fish catches without paying attention to the environmental impacts. Climate changes are long-term impacts due to the humans' behaviours which pay no attention on the environment. Environmental pressures appear due to the increasing human activities. The increasing greenhouse concentrations and atmospheric ozone-depleting gases, climate changes, some species extinctions, energy source depletion and the declining number of natural resources which do not pay attention to their life cycles [5], [6]. Some ecological scientists and economists warn that there will be an environmental degradation due to the economic activities exceeding the Earth's ecological capacity to regenerate [7]. In the fishery sector, the relationship of environmental quality dimension with sustainable economic and social development becomes real because fishery sector is the driving force of human activities, processes and patterns since fishery sector is the nutrition sources, employment providers, entertainment centers, wealth sources, which also provides multiplier influences to the other economic sectors [8].

Non-environmentally friendly fish catches without paying any attention to the fish cycles are considered causing problems on the environmental quality. The declining environmental quality, especially the quality of marine ecosystems may eventually cause problems on fish availability in the future. This problem is also a universal occurence. In some studies, the impact of the declining capture fishery stocks due to the environmental degradation mainly occurs at the Atlantic waters of the Middle East and the Northeast Pacific, the Black Sea and Mediterranean [1]. The environmental degradation phenomena may also occur in Indonesia's marine waters. This research aims at analyzing the influence of capture fishery production and fishery sub-sector on the environmental quality and analyzing the environmental impacts on the production of capture fisheries in Indonesia.

2. Method

This research is conducted to analyze the influence of capture fishery production on the environmental quality in Indonesia and vice versa, to analyze the influence of environmental quality on the capture fishery production. The data used are the secondary data of 34 provinces in Indonesia in the period of 2009-2016. Then, the analytical model used is the econometric regression of data panel model intended to figure out the most efficient estimation result due to the increasing number of observations which implications may improve the degree of freedom. The approaches used for data panel are both fixed effect model (FEM) and random effect model (REM) [9].

By entering the variables analyzed in this study, there are two empirical models with data panel regression equations as follows:
The first model analyzes the influence of capture fishery production (FISH) and revenues proxied with the regional domestic gross (GDRBF) fishery subsector on the environmental quality index (Z). This model may also prove whether the Environmental Kuznets Curve (EKC) hypothesis occurs on the fishery subsector in Indonesia.

$$Z_{it} = \beta_0 + \beta_1 LFISH_{it} + \beta_2 LGDRBF_{it} + \beta_3 (LGDRBFit)^2 + u_{it} \quad (1)$$

The second model analyzes the influence of environmental quality (Z) and the capture fishery production (FISH) in Indonesia on the regional domestic gross (GDRBF) fishery subsector.

$$LGDRBF_{it} = \alpha_0 + \alpha_1 Z_{it} + \alpha_2 LFISH_{it} + e_{it} \quad (2)$$

In both models, (L) shows the natural logarithm, (u) and (e) are the error terms, (i) shows the observations based on provinces and (t) shows the observations based on the time series.

3. Result and Discussion

The analytical results both using FEM and REM model show that the first finds that individual capture fishery production variable significantly influences the environmental quality variable. However, the Hausman test (chi-square statistics = 14.1238) shows that REM model may be rejected that FEM model is used. Subsequently, the interpretation of results may only be analyzed based on FEM results.

Table 1. Data Estimation Results of the Panel Model: Environmental Quality (Z) as the Dependent Variable

	FEM		REM	
Variable	Coefficient	t-Statistic	Coefficient	t-Statistic
C	38.00837	1.950840*	31.75860	1.648604**
LOG(FISH)	2.780590	2.861419*	2.052783	2.159718*
LOG(GDBRF)	1.713498	0.719993	1.492188	0.632936
$LOG(GDBRF)^2$	-0.162617	-1.720482**	-0.059349	-0.664534

Notes: *) significant at 5%; **) significant at 10%

The estimation result presented in table 1 shows the statistic value of t-test showing the fishery production variable has a significant and positive effect on environmental quality (t-statistic 2.8614). This means that the increase of capture fishery by 1 percent will increase the environmental quality index by 2.7 units. It means that the regeneration cycle of fishery commodity in Indonesia is still very good making fishing able to encaurage the the quality of ecosystem. Meanwhile the regional income variable of fishery sub-sector shows a significant positive relationship on GDRBF and significant negative relationship on (GDRBF) 2.

Such type of relationship suggests that the empirical model does not represent environmental Kuznet curve phenomenon on fishery subsector in Indonesia. It is proveny by the regional income of fishery subsector having positive relationship for environment. Even though it is found to be insignificant, the characteristics of fishery sub-sector in Indonesia tends to have positive effect on environment showed by the coefficient value $\beta_1 > 0$ and $\beta_2 < 0$. This environmental Kuznet curve of fishery subsector for this empirical model does not represent phenomenon of inverted U-curve because the variable used for proxy the environment is the environmental quality index instead of the negative variables such as pollution, emission, or other degrading indicators.

Table 2. Estimation Result of Panel Data Model:
Dependent Variable of Regional Income for Fishery Subsector (GDBRF)

	FEM		REM	
Variable	Coefficient	t-Statistic	Coefficient	t-Statistic
C	11.13801	8.418026*	11.15457	0.0000*
Z	-0.027825	-3.216833*	-0.027630	0.0017*
LOG(FISH)	0.279326	2.570663*	0.276862	0.0118*

Notes: *) significant at 1%

The estimation result using the second model aims to see the effect of environmental qulity and fishery production on the regional income of fishery subsector in Indonesia. It shows that the both independent variables have significant effect on degree of error of 1 percent with negative direction on the environmental quality variable and have positive on capture fishery production variable. The result of analysis either using FEM or REM Model shows that the environmental quality and capture fishery production individually has significant effect on regional income of fishery subsector. However, the Hausman test (chi-square statistic 9.3609) shows that the REM model is rejected to use over the FEM model. Then, the second model will be analyzed and intrepreted using FEM model.

The estimation result using FEM panel data shows that the average intercept value for the whole model is 11.13 (significant at alpha 1 percent). It means tat the FEM panel data are able to explain the different behaviour of regional income per capital on 34 provinces which are analyzed in the empirical model of dependent variable. Then, it is found that the environmental quality variable has negative and significant effect on regional income of fishery sector with t-statistics 8.4180. It means that a decrease of one unit for environmental quality index will improve the regional income of fishery subsector by 0.027 percent. This confirmes that in developing countries like Indonesia, the increase of fishery production is influenced by bad behaviour on the environment. While the fishery production variable has positive effect on regional income of fishery subsector. An increase of one percent fishery production will improve regional income of fishery sub sector by 0.27 percent.

4. Conclusion

In this study, the ecological problem related to fishery production can be summarized in two points. First, how the effect of the capture fishery on environmental quality is and how the phenomenon of EKC of fishery subsector occurs in Indonesia. Second, the pressure on fish resources especially in coastal areas due to its quality of fishery resources. This pressure has caused the decline of fishery stock. The decline occurs likely because of the environmentally unfriendly human behaviour such as the use of fishing gear disrpting the ecosystem. Based on both models applied in this research, they confirm the result of the research in developing countries related to environmental quality and fishery subsector. The result of empirical model in Indonesia shows that fishery production variable positively affects the quality of environment. Then, the second model shows that the decrease of environmental quality index improves the fishery production in Indonesia. This research has some data limitations so that it does not represent all provinces. Besides, the research uses aggregate variables making it unable to give more detail relationship behaviour. Then, the next research of EKC model can be expanded and have more complex environmental indicators to see more detail other representing environmental components. This research can also show the relation of environmental quality with fishery commodity and other marine products per commodity.

References

[1] FAO 2001 *The State of World Fisheries and Aquaculture 2000* (FAO, Rome)
[2] Pauly, D, Christensen, V, Guenette, S, Pitcher, TJ, Sumaila, R, Walters, CJ, Watson, R and Zeller, D 2002 Towards sustainability in world fisheries *Nature* **418**, 689-695

[3] Sainsbury, K and Sumaila, UR 2001 Incorporating ecosystem objectives into management of sustainable marine fisheries, including 'best practice' reference points and use of marine protected areas *Paper to Reykjavik Conference on Responsible Fisheries in the Marine Ecosystem*, Iceland 1-4 October 2001 (FAO, Rome)

[4] Badan Pusat Statistik https://www.bps.go.id/linkTabelStatis/view/id/1706 (BPS, Indonesia)

[5] Tacon, A. G. J., M. Metian, G. M. Turchini, and S. S. de Silva 2010 Responsible Aquaculture and Trophic Level Implications to Global Fish Supply *Reviews in Fisheries Science* **18**(1) 94–105

[6] Garcia, SM, Staples, DJ and Chesson, J 2000 The FAO guidelines for the development and use of indicators for sustainable development of marine capture fisheries and an Australian example of their application *Ocean and Coastal Management* No 43 pp537- 556

[7] Esty, Daniel C., and Michael E. Porter. 2005. "National Environmental Performance: An Empirical Analysis of Policy Results and Determinants", Environmental and Development Economics **10**(4) 391–434.

[8] Deere, C 2000 *Net gains: linking fisheries management, international trade and sustainable development* (IUCN,Washington D.C.)

[9] Gujarati D N and Porter D C 2009 *Basic Econometrics* 5th ed (New York: McGraw-Hill)

Oceanographic Factors in Fishing Ground Location of Anchovy at Teluk Cenderawasih National Park, West Papua : Are These Factors Have an Effect of Whale Sharks Appearance Frequencies?

Evi Nurul Ihsan[1,2*], Siti Yasmina Enita[2,3], Kunarso[3] and Anindya Wirasatriya[3,2]

[1]WWF Indonesia Site Teluk Cenderawasih National Park –Jln Huntap Iriati 2 Lorong 1 sebelah Kiri, Teluk Wondama 98362 Papua barat, Indonesia.

[2]Marine Diving Club, Marine Science Diponegoro University, Prof. H. Soedarto, S.H. – Tembalang Semarang 50275, Indonesia.

[3] Department of Oceanography, Diponegoro University, Prof. H. Soedarto, S.H. – Tembalang Semarang 50275, Indonesia.

Email : evinurulihsan@gmail.com

Abstract. The appearance frequencies of whale sharks in Teluk Cenderawasih National Park (TCNP) is unique because they appear throughout the year and are not related to any seasons. WWF Indonesia's monitoring results from 2011 untill now showed the appearance of whale sharks is closely related to the presence of fisherman liftnet and anchovies. This study focuses on oceanographic factors in fishing ground areas such as chlorophyll-a and satellite imagery data including chlorophyll-a, Sea surface Temperature (SST) over the last four years resulting in AQUA MODIS level 2 recording with 1km resolution in 2013-2016. The catch of anchovies and chlorophyll-a values showed a fairly good correlation during the west monsoon, 0.305, in addition the relationship between the anchovies with SST has a good correlation value starting in the transition period 1 towards the east monsoon. Based on these data, oceanographic factor conditions in TCNP provide a good opportunity for anchovies that live in warm waters to grow and reproduce. This also influences the appearance of whale sharks closely related to the presence of the anchovies where a positive correlation value is found between the catch of anchovies and whale sharks during the west monsoon and east monsoon in sequence 0,912 and 0.819. Chlorophyll-a and SST have a significant effect on fish catch of anchovies and the appearance frequencies of whale sharks even it were not in all seasons.

Keywords : Whale Shark, Anchovy, Chlorophyll-a, Sea Surface Temperature, Aqua MODIS.

1. Introduction

Teluk Cenderawasih National Park (TCNP) is the largest marine national park in Indonesia located in Nabire Regency of Papua Province (30.98%) and in Teluk Wondama Regency of West Papua Province (69.02%). One of TCNP's diversity is the presence of a unique species of shark family that is the largest whale shark in the world. Whale sharks love small meals, even though they are large. The type of zooplankton which is the food of the whale shark is copepods, larvae of crab, molluscs, crustaceans, coral eggs, and fish eggs [1] In recent years, whale sharks have been reported to often roam near TCNP's coasts, allegedly as a response to the booming of *bagan* (stationary fishing vessels with lift nets) along the coast. This has resulted in TCNP gaining popularity as a tourist destination, the results of which has been both economically rewarding and potentially damaging to the whale shark population itself. The increasing frequency of whale sharks in Kwatisore waters followed by average fish catches of anchovies. Anchovies is one of the most abundant fish catches at Kwatisore area, with the percentage of 26% in 2014 and increasing in the next year to 32.1% by 2015. Based on the case, allegedly appearance of whale sharks in TCNP is closely related to the fish catch of anchovies as their main food. The distribution and abundance of biological resources in the ocean is inseparable from the conditions and variations of oceanographic parameters, so it is indispensable for the purpose of sustainable resource use management. This shows the oceanographic parameters will affect the presence of fish and the establishment of potential fishing areas. There are three environmental parameters that become the focus and core attention in this study, those are the results of fish catches of anchovies, concentration of chlorophyll-a and sea surface temperature (SST) as these two parameters are very important role in the presence of fish in the ocean [2]. This study aims to determine the oceanographic factors in the form of distribution of chlorophyll-a, SST and nutrient concentration on the appearance of whale sharks in the Kwatisore waters, Nabire regency.

2. Methods

Research was carried out in Cenderawasih Bay National Park, in the waters off the coast of Nabire Regency, Papua Province, Indonesia (approximately 2°53'04.5"S 134°50'15.4"E to 3°13'47.2"S 134°57'43.5"E), in August 2016.

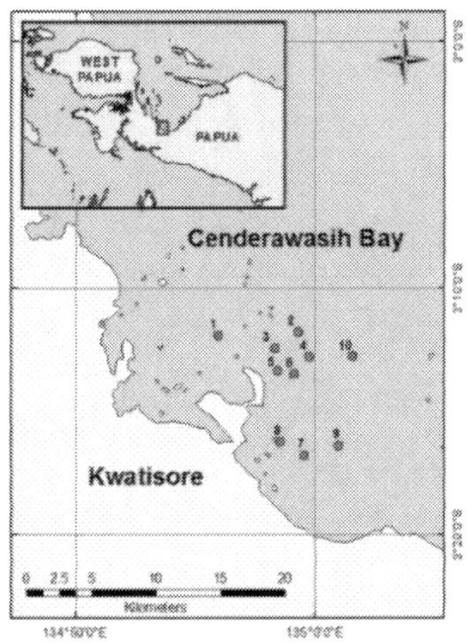

Figure 1. Approximate placement of sampling stations.

This research uses quantitative method. According Sugiyono [3] quantitative method is a research method that uses data in the form of numbers of data collection and the results of images, tables, graphics, or other display and systematic. The SST and chlorophyll-a data in this study used daily level 2 aqua MODIS (Moderate Resolution Imaging Spectroradiometer) satellite imagery with 1 km image resolution in the form of .NET Common Data File (NetCdF) format during 4-year period from January 1st, 2013 Until December 31st, 2016 downloaded from the website https://oceancolor.gsfc.nasa.gov/cgi/l3. Data on physical variables such as chlorophyll-a data and nutrient content at 10 stations were determined taking into account the results of the chlorophyll-a distribution of Aqua MODIS imagery and the largest number of appearance of whale shark location data over the past three years [4]. The measurement of chlorophyll-a and nutrient in situ using APHA method (America Public Health Association). The results of the nitrate and phosphate solution were analyzed using a UV-Vis spectrophotometer in a chemical oceanography laboratory with 885 nm wavelength for phosphate and 543 nm for nitrate and searched for correlation value (r). The results of chlorophyll-a concentration analysis were obtained by using the formula:

$$\text{Chlorophyll-a (Ca)} = 11{,}85.(E_{664}\text{-}E_{750}) - 1{,}54.(E_{647}\text{-}E_{750}) - 0{,}08.(E_{630}\text{-}E_{750}) \quad (1)$$

$$\text{Mg Klorofil/m}^3 = \frac{C\,x\,v}{V\,x\,d} = \frac{1\,x\,10\,ml}{3000\,ml\,x\,1} = \frac{mg}{m3} \quad (2)$$

Information :
E = Absorbance value
C = Ca, Chlorophyll-a Concentration
V = Number of acetone used (ml)
V = Number of samples used (ml)
d = Width of cuvette (cm)

3. Results

The Effect of Oceanographic Factors on the Emergence of Whale Sharks in Kwatisore Waters

Based on the results of data processing of the emergence of whale shark with the data of fish catches in the Kwatisore waters in 2013-2016 which has been overlaid into a map of distribution shows that the emergence of whale sharks and the catch of anchovies were centered around Tanjung Paus area (fig 2). The area were bordered is the area to be analyzed. Figure 3 shows no similar fluctuation trend between the appearance of whale sharks and fish catches. Based on the correlation value (r) between the occurrences of whale sharks with the fish catch of anchovies in Table 1 shows a value of 0.090 which means it belongs to very low correlation classification.

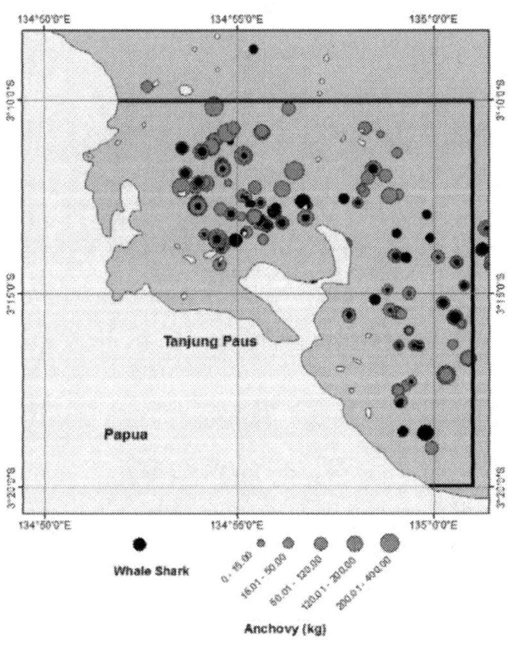

Figure 2. Distribution of anchovies and whale shark appearance during 2013-2016. Red dot shows the location of whale shark appearance, white dot shows the anchovies catch.

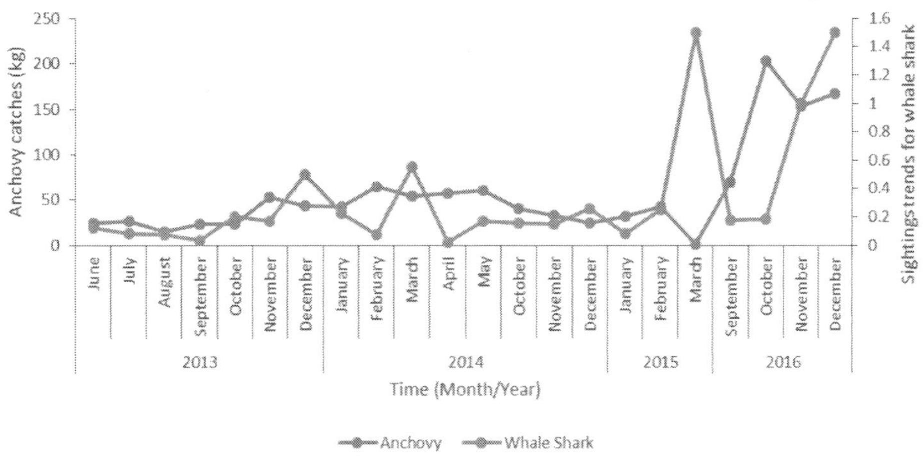

Figure 3. Fish catch and whale sharks appear temporally

Figure 4 shows the frequency of occurrence the whale sharks and chlorophyll a showing the highest occurrence being in the range of chlorophyll-a concentration of 0.38-0.48 mg / m^3 with a frequency ranging from 0.4-0.8, with a Pearson Correlation of 0.535 (Table 1). In the temporal data shown in Fig. 5, the peak of whale sharks appearance occurs frequently in December and March, followed by high chlorophyll-a concentration on these month during the 4-year period (2013-2016).

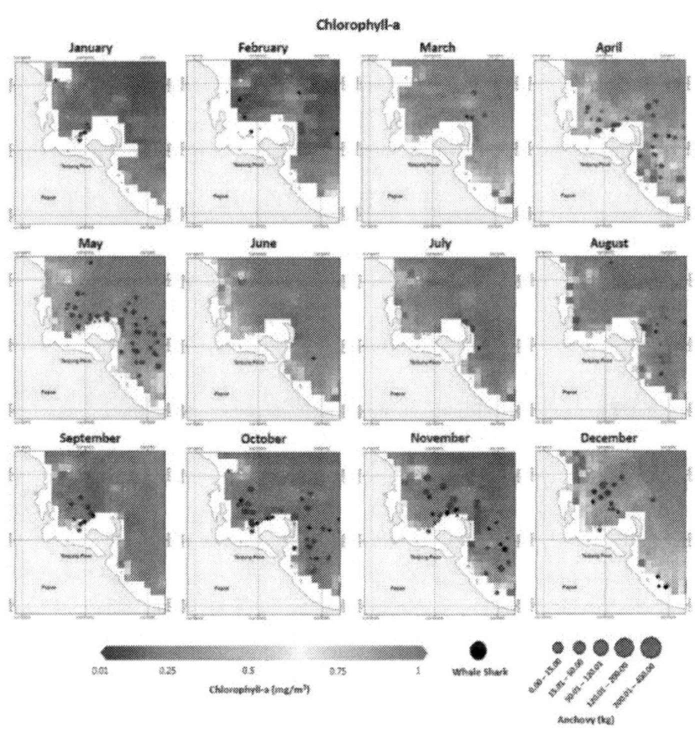

Figure 4. Overlaid map of chlorophyll-a distribution, the appearance of whale sharks and the fish catch at Kwatisore waters.

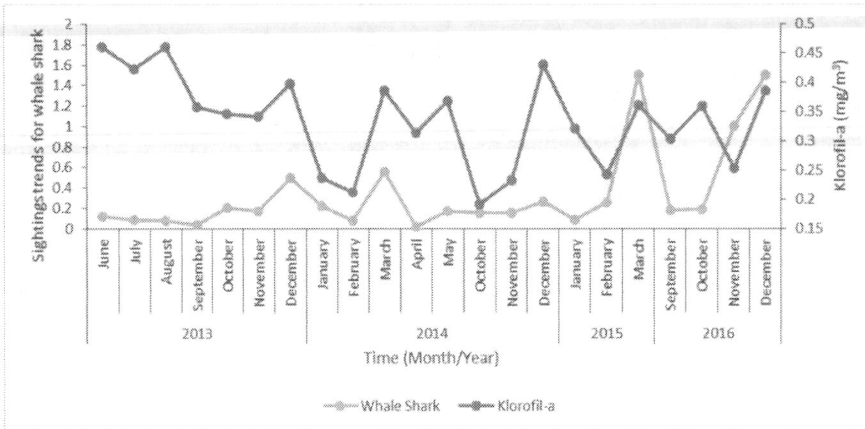

Figure 5. The appearance of whale sharks and the concentration of chlorophyll-a temporally.

Figure 5 shows a good relationship between the appearance of whale sharks with a concentration of chlorophyll-a in the waters of Kwatisore. The appearance of whale sharks to the surface usually coincides with the hordes of small fish on the water surface [5]. The presence of small fish on the surface indicates that the plankton concentration which is also the food is abundant enough and then the energy produced will be utilized by tropical top level organisms in this example is the anchovies [6]. Whale shark feed in Mexico consists of ~ 85% Copepods. In Australia, whale shark feed consists of krill shrimp (*Pseudeuphausia latifrons*), Copepods, and small fish hordes [7]. In Tanzania, the composition is over 50 percent of the sergestid shrimp species (*Lucifer hanseni*) [8]. Copepods include the main constituents of various types of zooplankton on the coast and the high seas and the presence of Copepods is closely related to food availability (phytoplankton) [9]. Based on the above reference it can be said that the association of the appearance of whale sharks and chlorophyll-a concentration in this research indicates the occurrence of food chain process in normal Kwatisore waters with intermediate tropic level suspected of krill shrimp, Copepods and or sergestid shrimp.

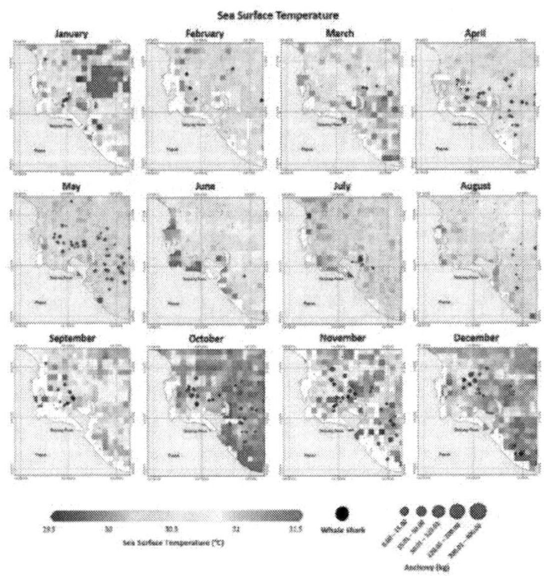

Figure 6. Map of SST distribution, the appearance of whale sharks and the catch of anchovies in the Kwatisore waters.

Figure 6 shows the highest climatological frequency of whale sharks in the SST range of 30.5 to 31.3 °C. Pearson correlation results between the appearance of whale sharks with SST showed a value of 0.53 (Table 1). Figure 7 shows that the rise in whale shark frequency is not always accompanied by an increase in SST, but the peak of whale sharks increases at an SST above 30.5 °C. The range is between of tropical areas that are the habitat of whale sharks. The data indicate a link between the appearance of whale sharks and SST in the Kwatisore waters.

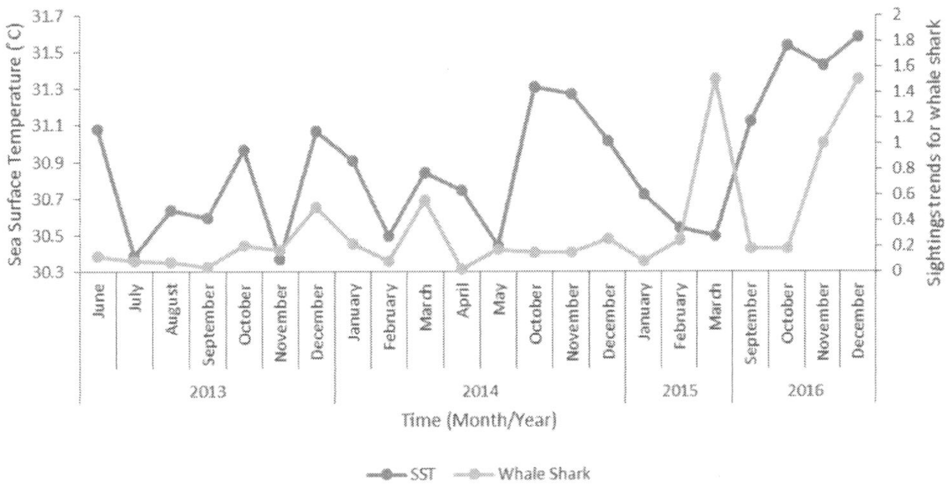

Figure 7. The appearance of temporary whale sharks and SST concentrations.

Table 1. Bivariate Correlation between monthly climatological variables in Kwatisore waters (N = 12)

N = 12	Chlorophyll-a	SST	Whale Shark	Anchovies
Chlorophyll-a	1			
SST	0,201	1		
Whale Shark	0,548	0,543	1	
Anchovies	0,227	-0,105	0,090	1

Table 2. Results of field data from 10 stations in Kwatisore waters

Station	x	y	Chlorophyll-a	Phosphate	Nitrate
1	134.9328	-3.1987	4.057711	0.006572	0.003584
2	134.9882	-3.19646	0.335064	0.003348	0.002492
3	134.9723	-3.20727	0.417528	0.004309	0.002744
4	134.9962	-3.21341	0.865244	0.003255	0.001554
5	134.9737	-3.22265	1.859378	0.006014	0.001946
6	134.9856	-3.22469	1.212439	0.003441	0.002002
7	134.9922	-3.27972	0.739933	0.004371	0.00154
8	134.9758	-3.2705	2.913667	0.004402	0.001806
9	135.0161	-3.27283	0.680173	0.003038	0.001274
10	135.0258	-3.2129	1.93666	0.003224	0.001176

Table 3. Bivariate correlation between stations (N = 10)

N = 10	Chlorophyll-a	Nitrate	Phospate
Chlorophyll-a	1		
Nitrate	0,402	1	
Phospate	0,683	0,644	1

4. Discussion

There are many whale sharks found at sea surface temperatures above 29°C. Particularly for unfamiliar or non-clumped whale sharks found around 90% of the Indian Ocean are found in waters with sea-surface temperatures ranging from 25°C to 35°C [10]. Whale sharks can still tolerate temperatures up to 10°C during whale sharks to dive in and also found whale sharks in Fundy Bay on the east coast of North America with a sea surface temperature of 44°C [10]. So it can be said that the appearance of whale sharks in the waters of Kwatisore more in October until December due to SST in waters Kwatisore is in accordance with the habitat of whale sharks basically, it is also supported by the high concentration of chlorophyll-a in those months.

The assumption that the puri fish is a whale shark food in Kwatisore is allegedly due to a change in feeding habits on the whale shark at Kwatisore. According to field observations and interviews with the chart fishermen it is known that throwing the fish from the catch to the sea can provoke the whale shark to rise to the surface. The existence of the anchovies on the surface becomes a sign for whale sharks to feed, where the presence of small fish indicates the high concentration of zooplankton in waters which is a food for the anchovies and also the whale shark. While the anchovy fishing takes place throughout the year in Kwatisore, it is thought to be the factor of changing the habit of feeding on the whale sharks in Kwatisore.

There is the possibility that the anchovies became one of the main food of whale sharks in TNTC area. However, it is also possible that the appearance of whale sharks in locations where abundance of anchovies is due to both species has similar preference types of diet [11]. The high biomass of the anchovies caused by the abundance of primary producers, in this case phytoplankton which is also an abundant zooplankton food. Thus the catch of the anchovies on the chart is thought to be strongly influenced by the water fertility factor. The abundance of the anchovies is highly dependent on the abundance of food [12].

Based on analysis of Fig. 4, it appears that chlorophyll-a in coastal waters is higher than offshore. This is indicated by the range of chlorophyll-a in coastal waters being between 0.5-1 mg / m^3. The range of chlorophyll-a in offshore waters is between 0.01-0.5 mg / m^3. The distribution of chlorophyll is higher concentration in coastal and coastal waters, and low in offshore waters. The high distribution of chlorophyll-a concentration in coastal waters is due to the supply of nutrients in large quantities through run-offs from the mainland, whereas the low concentration of chlorophyll-a in offshore waters in the absence of direct supply of nutrients from the land [13].

Table 2 shows the results of data from 10 stations, 4 points located near the coast i.e: point 1, 3, 5 and 8, while the other 6 points are more toward the high seas i.e: point 2, 4, 6, 7, 9 and 10. Phosphate concentration At 4 points are in the range of 0.0043-0.0065 ppm where the range is higher than the 6 points located in the high seas with a range of 0.0032-0.0043 ppm. Nitrate concentration at 4 points is in the range of 0.0018-0.0035 ppm where the range is higher than the 6 points located in the high seas with a range of 0.0011-0.0024 ppm. This suggests that nutrient concentrations are affected by the innate nutrients of the land. The concentration of chlorophyll-a and the concentration of nitrate and phosphate (Table 3) shows the correlation (r) values of 0.402 and 0.683, respectively, the correlation rate is moderate to strong.

This indicates a similar fluctuation trend, where as the phosphate content increases the concentration of chlorophyll-a also increases. The high correlation value indicates that phosphate is the main indicator of Kwatisore water fertility. High nutrient concentrations supported by sufficient sunlight intensity will increase primary productivity, as evidenced by indicators of elevated chlorophyll-a levels. Chlorophyll-a is often used as a measure of phytoplankton abundance and primary marine productivity. Nitrate and Phosphate are the nutrients that serve as the material for photosynthesis. The nutrient supply is thought to come from the mainland through the river [14].

In addition to nutrient factors, other factor that is likely to cause high concentrations of chlorophyll-a is the light factor. In certain seasons sunlight in the waters of Kwatisore has a high intensity and good water brightness, it is used by phytoplankton to produce their food through the process of photosynthesis. Light is one of the factors that determine the distribution of chlorophyll-a at sea. On the high seas, on a mixed surface layer is available enough sunlight to process photosynthesis. In addition to high-chlorophyll a-concentrations in coastal areas, in offshore waters are also found areas with high chlorophyll-a concentrations, although in general they have low chlorophyll-a concentrations in the absence of nutrient supplies derived from mainland. The high concentration of a-chlorophyll in offshore waters due to the high concentration of nutrients produced through the physical process of water mass, for example upwelling process [15].

5. Conclusions

Based on the results obtained from the research, it can be concluded that the frequency of appearance of whale sharks has no significant correlation with the abundance of fish catches of anchovies with a correlation of 0.09. The occurrence of whale sharks has the same fluctuation trend with the concentration of chlorophyll-a and SST in Kwatisore waters with the correlation of 0.548 and 0.543 respectively. The appearance of whale sharks and chlorophyll-a concentrations increased in December and March, with the range of chlorophyll-a concentrations of 0.38-0.48 mg / m^3 with SPL above 30.5 °C. The range is a range of tropical areas that are the habitat of whale sharks.

Acknowledgments

The authors would like to thank Oceanography Departement Diponegoro University for helping make this study comes true, Teluk Cenderawasih National Park Autorithy and for all the people at Kali Lemon Resort and Kwatisore Village for helping at the field.

References

[1] Tania C 2015 *Pemantauan dan Studi Hiu Paus di Taman Nasional Teluk Cenderawasih: Laporan Pemantauan Tahun 2014–2015* (Taman Nasional Teluk Cenderawasih/WWF Papua)
[2] Gaol JL and Sadhotomo B 2007 *Journal of Indonesian Fisheries Research* **3** (1): 201-211.
[3] Sugiyono. 2012. *Qualitative and Quantitative Research Methods R & D.* (Bandung: Alfabeta)
[4] Sudjana. 2005. 6th *Statistical Method.* (Bandung: Tarsito)
[5] Noviyanti, S.N. 2015. *Characteristics of the Whale Shark Habitat Rhincodon typus Smith, 1828 (Elasmobranchii: Rhincodontidae) in Probolinggo Coastal District, East Java.* Bogor Agricultural University: Bogor.
[6] Nelson JD and Eckert SA 2007. *Fisheries Research* **84** (1): 47-64.
[7] Taylor JG, 2007 *Fisheries Research* **84** (1): 65-70.
[8] Rohner CA, Armstrong AJ, Pierce SJ, Prebble, CE, Cagua EF, Cochran JE, Berumen ML and Richardson AJ 2015 *Journal of Plankton Research* **37** (2): 1-11.
[9] Folt CL and Burns CW 1999 *Trends in Ecology & Evolution* **14** (8): 300-305.

IOP Conf. Series: Earth and Environmental Science **116** (2018) 012017 doi:10.1088/1755-1315/116/1/012017

[10] Stacey N, Karam J, Dwyer D, Speed C and Meekan M. 2008. *Assessing Traditional Ecological Knowledge of Whale Sharks (Rhincodon typus) in Eastern Indonesia: a Pilot Study with Fishing Communities in Nusa Tenggara Timur*. (Canberra: DEWHA)
[11] Marliana SN 2016. *Ecological Studies of Natural Shark Feed (Rhincodon typus) in the Context of Fishing Activities in Cenderawasih Bay National Park*. Wasior: WWF-Indonesia.
[12] Dinisia A and Adiwilaga EM 2016 *Journal of Marine Fisheries* **6** (2): 143-154
[13] Rashid A 2009 *Journal of Science & Technology* **9** (2): 125-132.
[14] Rousseaux, C.S. and Gregg, W.W., 2012 *Journal of Geophysical Research* **117** (C10006): 1-10.
[15] Ridha U, Muskananfola M R, and Hartoko A 2013 *Diponegoro Journal Of Maquares* **2** (4): 53-60.

Probiotic Candidates from Fish Pond Water in Central Java Indonesia

Alfabetian Harjuno Condro Haditomo*, Desrina, Sarjito, and S. Budi Prayitno

Department of Aquaculture, Fisheries and Marine Science Faculty, Diponegoro University, Tembalang, Semarang, Indonesia

*Corresponding author: condrohaditomo@gmail.com

Abstract. *Aeromonas hydrophilla* is a major bacterial pathogen of intensive fresh water fish culture in Indonesia. An alternative method to control the pathogen is using probiotics. Probiotics is usually consist of live microorganisms which when administered in adequate amounts confer a health benefits on host. The aim of this research was to determine the probiotic candidates against *A. hydrophilla* which identified based on the 16S rDNA gene sequences. This research was started with field survey to obtained the probiotic candidate and continue with laboratory experiment. Probiotic candidates were isolated from fish pond water located in Boyolali, and Banjarnegara Regency, Central Java, Indonesia. A total of 133 isolates bacteria were isolated and cultured on to TSA, TSB and GSP medium. Out of 133 isolates only 30 isolates showed inhibition to *A.hydrophilla* activity. Three promising isolates were identified with PCR using primer for 16S rDNA. Based on 16S rDNA sequence analysis, all three isolates were belong to Bacillus genus. Isolate CKlA21, CKlA28, and CBA14 respectively were closely related to Bacillus sp. 13843 (GenBank accession no. JN874760.1 -- 100% homology), Bacillus subtilis strain H13 (GenBank accession no.KT907045.1 -- 99% homology), and Bacillus sp. strain 22-4 (GenBank accession no. KX816417.1 -- 97% homology).

Keywords: Probiotic candidates, Fish pond water, *Aeromonas hydrophila*, 16S rDNA.

1. Introduction

The Aquaculture has considerably grown during the last few decades as well as becoming an economically significant business [1]. Aquaculture is aimed as a solution to meet the growing demand for fish and shellfish. To meet the ongoing globalization of food shortage, improving aquaculture practices by new technological innovations is a challenging assignment for scientists and biologists [2]. There are some important aquaculture commodity in Indonesia such as walking catfish (*Clarias gariepinus*), tilapia (*Oreochromis niloticus*) and gouramy (*Osphronemus gourami*), etc. Increasing demand for domestic markets encourages the fish farmers to increase their production. With the escalation intensification of aquaculture production, diseases and declining of environmental conditions are major problems and may caused massive economic losses [3].

The negative impacts of intensification of aquaculture were bacterial disease outbreaks such as *Aeromonas hydrophila* which is still became a major problem in fresh water aquaculture. *A.*

hydrophila is a ubiquitous aquatic bacterium which has been implicated as a pathogen of fish [4-5] and another aquatic organisms [6-7]. Antibiotics was frequently used for prevention and treating the diseases and to keep the fish live. However this causes adverse effects on the environment and bacteria that become increasingly resistant to antibiotics administered to ailing organisms.

The need for fish protein and environmentally friendly fish cultivation requires an alternative substance to replace antibiotics and chemicals in order to support sustainable quality production of aquaculture [8-10]. An alternative solution that can be used to overcome this problem is by using probiotics. It usually contained live microorganisms that can be administered in adequate amounts in order to confer a health benefits on host. Various studies on the use of probiotics in the field of aquaculture demonstrate the ability to control potential pathogens to developed [5]. The growth performance and various immunologic parameters may be enhanced by the use of probiotics, prebiotics, immunostimulant as additional feed routine in aquatic organisms [11-12].The types of bacteria and mechanisms by which probiotics exert their effects are still being studied, and there are still many open research opportunities on this subject.

Water as a fish culture medium is the site of various bacteria. With the bacterial screening process is expected to obtain potential isolates to increase the productivity of aquaculture in various ways. On the other hand probiotics from the same species and/or its natural environment could be best approach for better efficacy in host [13-14]. Several researches had been performed to find out potential bacteria on freshwater fish based on the 16S rDNA gene sequences. There were limited reports being documented so far describing the application of polymerase chain reaction on the diversity of potential probiotic bacteria on fresh water aquaculture derived from fish pond of central fish production in Central Java, Indonesia.

2. Materials and Methods

2.1. Isolation of probiotics candidates bacteria from fish pond water.

Bacteria isolates were collected from fresh water fish aquaculture ponds of central production area in Boyolali Regency (from *C. gariepinus* pond), Klaten Regency (from *O. niloticus* pond), and Banjarnegara Regency (from *O. gouramy* pond). The best production of the aquaculture pond in the last harvest period became one of the requirements for sampling on selected fish ponds. Bacteria were isolated from pond water by spread plate method and cultured in to TSB (Mereck). Then, 0.1 ml of appropriate dilutions was spread on Tryptic Soybean Agar (TSA-Merck) . After 24 to 72 h of incubation, the bacterial colonies were purified several times on TSA agar to get pure culture and isolated colonies. Preliminary identification of the isolates was carried out using microscopic observation.

2.2. Gram staining

Morphological features of the colonies were picked and purified by a single colony to the plate. Gram stain test have done based on Gram stain protocol Animal Health Diagnostic Center, Cornell University(15).

2.3. *Antimicrobial activity test*

Probiotic candidates were co-cultured with *A. hydrophilla* in various concentrations to evaluate their antagonistic activities using disc method. The bacterial that shows the antagonistic activity against pathogens can be classified as potential isolates. Based on the method developed by David Stout, the division of antibacterial activity can be done based on the size of the inhibitory diameter formed [16]. Based on the method, an antibacterial agent producing an inhibitory area of <5 mm is categorized as a weakly antibacterial substance; 5-10 mm are categorized as medium-strength antibacterial; 10-20 mm categorized as strong antibacterial; And > 20 mm are categorized as very powerful antibacterial. Morphological features of the colonies were picked and purified by a single colony to the plate.

2.4. Determination of pathogenicity

The potential isolates obtained were tested for their pathogenicity against 3 species of freshwater fish (*C. gariepenus*, *T.nilotica* and *O. gouramy*) using healthy fish that have been acclimatized for 14 days in a well-labelled glass aquarium. Fish were injected either intramuscularly (IM) with pre-prepared 0.1 ml of bacterial suspension contains 10^8 cfu ml^{-1}. Control groups were injected either IM with 0.1 ml of sterile PBS. The each group (n = 10) of fish were monitored for 14 days with daily replacement of water. The parameters measured were survival rate and apparent clinical symptoms.

2.5. Molecular identification of the most potent isolate

The potential probiotic isolate was identified based on 16S rDNA sequencing. Genome DNA purification kit (Promega, Madison, USA) was used to extract the DNA for the amplification of 16S rRNA genes by the polymerase chain reaction (PCR) [24]. PCR products were resolved by electrophoresis in agarose gels and visualized by ethidium bromide staining and purified. The 16S rDNA sequence alignment was done using BLAST with the non-redundant 16S rRNA sequence database of NCBI GenBank (www.ncbi.nlm.nih.gov). Features of the colonies were picked and purified by a single colony to the plate and used for cross check.

3. Results and Discussions

3.1. Candidates Probiotic againts A. hydrophila.

A total of 133 bacterial isolates were obtained as probiotic candidates derived from cultivation of catfish, tilapia and gourami cultivation pond water. In the end there are 79 bacterial isolates left which consisted of 18 Isolates from Boyolali, 31 isolates from Klaten and 30 isolates from Banjarnegara. Further, antimicrobial activity test results obtained fourteen [14] isolates probiotic candidate which showed inhibiting zone for the growth of *A. hydrophila* bacteria with a range of 10,33-19,67 mm. Antimicrobial activity test againts A hydrophila results are given in Table 1.

Table 1. Antimicrobial activity test againts *A hydrophila* (mm).

No	Isolate Code	1	2	3
1	CB A_{14}	19	18	16
2	CB A_{16}	13	10	8
3	CKl A_{14}	11	12	11
4	CKl A_{18}	13	12	13
5	CKl A_{20}	17	14	14
6	CKl A_{21}	21	23	15
7	CKl A_{22}	15	15	12
8	CKl A_{23}	10	10	12
9	CKl A_{27}	13	15	15
10	CKl A_{28}	20	20	13
11	CBj A_{14}	16	11	10
12	CBj A_{18}	11	13	16
13	CBj A_{19}	15	13	9
14	CBj A_{21}	9	13	15

The fourteen selected isolates were included in the strong category with a zone of> 10 mm inhibition [1]. To obtain more detailed information from selected bacteria carried out the morphological characteristics of bacterial colonies Characteristics of Probiotics candidate isolates bacteria are given in Table 2.

Table 2. Characteristics of Probiotics candidate isolates bacteria.

No	Isolates	Antimicrobial activity (mm)	Colony Characteristic	Gram stain and Shape
1	CB A_{14}	17,67	White yellowish, entire, L shaped, wet, raised	Positive, short rod
2	CB A_{16}	10,33	White pale, round, wavy, and spreading, dry, convex	Positive, long rod
3	CKl A_{14}	11,33	White milk, round, wavy, wet, convex	Positive, long rod
4	CKl A_{18}	12,67	White pale, filiform, wavy, dry, convex	Positive, long rod
5	CKl A_{20}	15	White pale, filiform, irregular, dry, raised	Positive, long rod
6	CKl A_{21}	19,67	White pale, irregular, wavy, dry, raised	Positive, short rod
7	CKl A_{22}	14	White pale, filiform, irregular, dry, raised	Positive, long rod
8	CKl A_{23}	10,67	White pale, round, entire, wet, convex	Positive, long rod
9	CKl A_{27}	14,33	White pale, rhizoid, wavy, dry, convex	Positive, long rod
10	CKl A_{28}	17,67	White pale, filiform, irregular, dry, raised	Positive, short rod
11	CBj A_{14}	12,33	White pale, round, wavy, wet, convex	Positive, short rod
12	CBj A_{18}	13,33	White pale, round, wavy, dry, convex	Positive, short rod
13	CBj A_{19}	12,33	White pale, round, entire wet convex	Positive, short rod
14	CBj A_{21}	12,33	White pale, irregular, dry, convex	Positive, short rod

Gram staining test results showed that all isolates were Gram positive (blue color) with long rod shape, which means its enter into *Bacillus* spp genus. Some probiotics, especially of the genus Bacillus have been widely used in disease control in aquatic animals. Similar results were also reported [17], that the genus Bacillus bacteria is a Gram-positive bacteria and is included in probiotic bacterial candidates.

There is no fish mortality that caused by the all isolates as a probiotics candidate. This result is better and also in accordance with the opinion that fish mortality that caused by probiotic candidates less than 33% are considered capable of controlling pathogenic bacterial infections [18]. It has succeeded in isolating the potent probiotics of *Bacillus* spp. which is used to improve the immune system of cyprinids (*Labeo rohita*) against the attack of *A. hydrophila* pathogens. The success rate obtained was very significant with the mortality range of only 20-40% compared to controls that achieved 80% mortality. The summarized sequence analysis of CBA14, CKIA21 and CKIA28 isolate are presented in Table 3.

Table 3. Genetic Diversity of Potential Probiotics Candidates Based on Gen Blast

No.	Isolate Code	Result	Homology (%)	Access Number
1	CBA14	*Bacillus sp.* strain 22-4	97	KX816417.1
2	CKIA21	*Bacillus sp.* 13843	100	JN874760.1
3	CKIA28	*Bacillus subtilis* strain H13	99	KT907045.1

Results showed that CKIA28 isolate had a similarity level of 99% with *Bacillus subtilis* H13 strain. *Bacillus subtilis* H13 strain is reported on the research of cow dung as a cheap substrate for polyhydroxy butyrate production by soil bacteria. Genus *Bacillus* spp. commonly found from various sources. Bacterial genera *Bacillus* sp isolated from the healthy intestine of cyprinids (*Labeo rohita*) [2-3]. *Bacillus* spp genus is also commonly found in giant prawns [4]; fish cyprinids[5], *Bacillus* spp. reported on indigestion of tilapia while another reported that *Bacillus* sp. in research related to natural resources of agriculture; even *Bacillus* spp. also found in the Mulberry plant endophytes.

Probiotic therapy offers a suitable alternative for controlling pathogens thereby overcoming the adverse consequences of antibiotics and chemotherapeutic agents. In fish culture, probiotics either in diet or bioencapsulation help in achieving natural resistance and high survivability of larvae and post larvae of fishes [6-8].

4. Conclusions

In the present study, 133 isolates bacteria were isolated from different sources and only fourteen isolates were selected on the basis of their high antimicrobial activity. Three bacteria that have potential as a probiotic candidate able to inhibit *A. hydrophilla* with the strong category were aligned with the 16S r-DNA sequences from the Gen Bank database (website) to identify the studied microorganism. The Results isolate CBA14 97% homology with *Bacillus* sp. strain 22-4, CKIA21

5. Acknowledgement

The authors would like to thank the Dean of the Faculty of Fisheries and Marine Sciences, Diponegoro University for his permission to begin this research that Funded by Others APBN DPA SUKPA Faculty of Fisheries and Marine Sciences University of Diponegoro Fiscal Year 2017. Thanks also to Pak Marsudi, Hendra, Hanifah, Yosef, Tiara, Falah, Vava, Putri, Femy, Ilham, Fina, Thesa, Sofa, Sarah, and Mada who have helped this research. We are also grateful to the Head of Diponegoro University Integrated Laboratory, Aquaculture Laboratory Faculty of Fisheries and Marine Sciences, Diponegoro University which provides an opportunity to conduct this research.

6. References

[1] FAO *The state of world fisheries and aquaculture* 2016;160 (State World Fish Aquac)
[2] Ibrahem MD *J Adv Res* 2013; **6**(6):765–91
[3] World Bank *Fish to 2030: Prospects for Fisheries and Aquaculture* 2013; **3** (83177):102 (Agric Environ Serv Discuss Pap)
[4] Burton GA, Lanza GR *Water, Air, and Soil Pollution* **34** (1987) 199-206
[5] Haditomo AHC, Lusiastuti AM, Widanarni *Jurnal Saintek Perikanan* 2016;**11**(2):111–4
[6] Rajesh KM, Shankar KM, Mohan C V, Mridula RM *Dis Asian Aquac* **VI** 2008;245–58.
[7] Shao J, Liu J, Xiang L *Aquaculture* 2004;229(1–4):11–23.

IOP Conf. Series: Earth and Environmental Science **116** (2018) 012018 doi:10.1088/1755-1315/116/1/012018

[8] Amara AA, Shibl A *Saudi Pharmaceutical Journal* 2013 **23**:107–14
[9] Ibrahem MD *J Adv Res* 2013;**6**(6):765–91.
[10] Ghosh AK, Bir J, Azad MAK, Hasanuzzaman AFM, Islam MS, Huq KA *J Aquac Reports* 2016;**4**:112–7
[11] Mona MH, Rizk E-ST, Salama WM, Younis ML *J Basic Appl Zool* 2015;**69**:17–25
[12] Aly SM, Abd-El-Rahman AM, John G, Mohamed MF *Aquaculture* 2008;**277**(1–2):1–6
[13] Verschuere L, Rombaut G, Sorgeloos P, Verstraete W *Microbiol Mol Biol Rev* 2000; **64** (4): 655–71
[14] Gatesoupe FJ *Aquaculture* 1999 **180**;147–65
[15] Smith AC, Hussey MA *Gram Stain Protocol* (ASM Lab Protoc) 14852
[16] Davis WW, Stout TR *Appl Microbiol* 1971;**22**(4):666–70
[17] Nayak SK *Fish Shellfish Immunol* 2010;**29**(1):2–14
[18] Ramesh D, Vinothkanna A, Rai AK, Vignesh VS *Fish Shellfish Immunol* 2015;**45**(2):268–76
[19] Dharmaraj R, Rajendren V *Int J Sci Res Publ* 2014;**4**(7):1–6
[20] Rengpipat S, Phianphak W, Piyatiratitivorakul S, Menasveta P *Aquaculture* 1998; **167** (3–4): 301–13
[21] Abareethan M, Amsath A, *International Journal of Pure and Applied Zoology* 2015; **3**(2):148–53
[22] Reid G *Int Dairy Journal* 2015; **49**:16–22
[23] Pandiyan P, Balaraman D, Thirunavukkarasu R, George EGJ, Subaramaniyan K, Manikkam S, *Drug Invent Today* 2013; **5**(1):55–9
[24] Jiin-Ju Guo, Kuan-Fu Liu, Shin-Hong Cheng, Chin-I Chang, Jiunn-Jyi Lay,Yueh-O Hsu1, Jan-YenYang & Tzyy-Ing Chen *Aquaculture Research* 2009; **40**:609–618

IOP Conf. Series: Earth and Environmental Science **116** (2018) 012019 doi:10.1088/1755-1315/116/1/012019

Flavor Enhancer From Catfish (*Clarias batrachus*) Bekasam Powder and Angiotensin-I-Converting Enzyme (ACE) Inhibitory Activity in Various Dishes

Yanesti N Lestari[1,5], Retno Murwani[1,2,3] and Tri W Agustini[4]

[1]Master Program in Nutrition Program, Faculty of Medicine, Diponegoro University, Semarang, Indonesia
[2]Faculty of Animal Science and Agriculture, Diponegoro University, Semarang, Indonesia
[3]Natural Product Laboratory, Centre of Research and Services, Diponegoro University, Semarang, Indonesia
[4]Faculty of Fisheries and Marine Science, Diponegoro University, Semarang, Indonesia
[5]Bachelor Program in Nutrition, Faculty of Health, Nahdlatul Ulama University, Surabaya, Indonesia

*Corresponding author Email: yanestinur.al@unusa.ac.id,

Abstract. Flavor enhancer is characterized by high glutamic acid content and it can be obtained from fermented food such as Bekasam. Fermented food had inhibitory effect on Angiotensin-I-Converting Enzyme (ACE) activity which is advantageous for hypertension. However, such activity was not known to sustain in food system. The aim of this research was to study addition of flavour enhancer from Catfish Bekasam Powder (CBP) in various food systems and to determine the ACE inhibitory (ACEI) activity in the food system. Four food system consisted of carrot, champignon, and chicken meat dishes were boiled in water and added with CBP or MSG. Each food system was added with graded level of CBP (0%; 0.5%; 0.8%; 1.1%; and 1,4%) and for control monosodium glutamate (MSG) was used. ACEI activity in each food system and organoleptic test using multiple comparison differentiation on 15 semi-trained panellists were determined. The results showed that there were fluctuation of ACEI activity in the carrot, champignon, and chicken meat dishes ($p=0.017$; 0.043; and 0.032). The MSG containing dishes showed the lowest ACEI activity. Addition of graded level of CBP on carrot, champignon, and chicken meat dishes were directly proportional to glutamic acid content but inversely proportional to ACEI activity ($p<0.05$). The addition of commercial MSG on all dishes increased glutamic acid content but reduced ACE-inhibitory activity significantly ($p<0.05$). Comparing CBP to MSG addition in champignon dish revealed that increasing level of CBP increased the flavour preference of the panellists. On the contrary the higher the addition CBP in noodle and chicken meat dishes the worse were the flavour score ($p<0.05$). It can be concluded that the addition of CBP as flavour enhancer on various dishes can deliver better flavour and ACE-inhibitory activity than the addition of commercial MSG.

Key words: umami, taste, natural, fish, diet

1. Introduction
Hypertension is still recently becomes a major problem in the world. According to WHO, hypertension is the number one cause of death in the world. Data 2010 in the United States shows that 28.6% of adults aged 18 years and over are suffering hypertension [1]. Based on Riskesdas (Basic Health Research) in Indonesia of 2013, 25.8% of adults aged 18 and over are suffering hypertension [2]. Hypertension should be prevented and treated both by medication and dietary modification. Low salt diet program is one form of dietary modification for patients with hypertension. Adding flavor enhancer to the dish may become one effort to reduce the use of salt but still can provide taste sensation to the diet program of people with hypertension. The flavor enhancer enhances the attractiveness at low flavor dishes [3]. Monosodium Glutamate (MSG) is the most favorite flavor enhancer and common used as a flavoring which is a combination of sodium salt and glutamic acid [4]. MSG strengthen salt sensation taste of the dish. The addition of 0.38% MSG at 0.4% to the salt may provide the same saltiness with 0.8% NaCl [5].

Bekasam (soaking salt fish) is a product of preserved freshwater fish using high salinity spontaneous fermentation. Research shows that the *bekasam* has an inhibitory activity of Angiotensin-I-Converting Enzyme (ACE) which causes hypertension. Fermentation process of milkfish *bekasam* on 6th day shows 51.77% ACE inhibition [6]. Catfish (*Clarias batrachus*) is one freshwater fish that can be used as a flavor enhancer. Catfish becomes favorite because of its savory taste, reasonable price, and high nutrient (protein, essential fatty acids, lysine, leucine, and phosphorus) [7,8]. *Bekasam* has specific flavor that can increase appetite, but due to its uninteresting color and shape so that people is not interested. Therefore, it is necessary to reform the food processing technology of the products to be more interesting and safe to consume for people who suffers hypertension. This study aims to analyze ACE inhibitory activity and organoleptic properties of catfish *bekasam* powder (CBP) flavor enhancer of several food products.

2. Materials and Methods

2.1. Preparation of Catfish Bekasam Powder (CBP)
The local catfish (*Clarias batrachus*) weighing + 200-225 g/head was bought from the local market in Semarang. Its viscera and head were discarded then the flesh was split into two parts. Afterward, they were washed and added 10% salt (w/w) and let them for 24 hours in a sealed container (37°C). Next procedure was adding *samu* (roasted sticky rice) by the ratio of fish: sticky rice 1: 1 then ripened for 6 days. *Bekasam* that was ready-made then steamed for 40 minutes and drained. The lees was dried with a drying cabinet (45°C for 6 hours) then finely ground. CBP were stored at room temperature inside closed containers which treated with silica gel prior to analysis.

2.2. Sample Preparation of Organoleptic Test
The sample consisted of 4 types of dishes, namely noodles (carbohydrate source), carrot (vegetable source), *Champignon* (vegetable protein source), and chicken meat (source of animal protein). Raw ingredients were boiled until cooked for 15 minutes then drained. About 50 grams of dishes were weighed and then boiled again in 250 mL boiling water and added salt and complementary spices (garlic powder, onion, bombay onion, white pepper, ginger, galangal and coriander) as the main flavors which each added as much 0.25% of weight of dish. Decoction of dishes was then added by CBP to the treatment group and added MSG for the standard sample. Boiling continued for + 2 minutes.

2.3. Properties Tests of Organoleptic Flavor

The organoleptic test used multiple comparisons. This test assessed the presence or absence and enormity of the difference intensity of the sample organoleptic properties if compared with the standard sample [9]. The organoleptic properties which studied were the flavors. The flavor are the complexity of combination of sensations of aroma, taste and involving the somatosensoric senses [10]. The assessment was conducted by 15 semi-trained panelists. Six samples were presented together in a tray which consist of one standard sample containing 0.38% MSG, one sample of processed dishes without CBP, and four processed dishes samples added with CBP (dose 0.5 %; 0.8%; 1.1%; and 1.4%). Each sample was consecutively coded using 3 (three) digit numbers, namely 189; 492; 278; 760; 251; And the standard samples are coded R. Organoleptic properties used Likert scale [11]. The enormity and intensity differences of flavor were scored by ranging from 1-5 (1 = much better than R and 5 = is much worst than R for the difference assessment of flavor, while 1 = there is no difference and 5 = very much for the difference intensity assessment).

2.4. *Sample Extraction*

The samples were firstly blended and finely diluted with ratio of 1:3 (volume between sample and deionized water). Samples were then homogenized for ±10 minutes and centrifuged at 20.000 x g for 10 minutes. The supernatant at the top of the tube was piped and filtered. Fresh extracts sample were directly used for analysis.

2.5. *Determination of Glutamic Acid Level*

Glutamic acid levels were analyzed by spectrophotometric method at 565 nm wavelength. 20 μL sample extracts were mixed with 60 μL buffers, 1 μL enzyme, 5 μL (Nicotinamide adenine dinucleotide) NAD, and 14 μL [3-(4,5-Dimethylthiazol-2-yl)-2,5-Diphenyltetrazolium Bromide] MTT and homogenized. The solution was then incubated for 30 minutes at room temperature. The analysis was conducted for three times repetition. Standard curves used standard glutamate (containing 100 mM glutamate / mL) [12].

2.6. *Determination of Angiotensin-I-Converting Enzyme (ACE) Inhibition*

ACE inhibitory activity was conducted by Cheung & Chusman (1971) modified spectrophotometric method. 65 μL sample extracts were mixed with 75 μL Hip-His-Leu (HHL) substrate 5 mM in HEPES buffer (pH 8.3 containing 1 M NaCl) and 25 μL of ACE (3.3 mU) enzyme.The solution was incubated at 37°C for 1 hour and then added 125 μL HCl 1 M to stop the reaction. Hypuric acid was extracted and added by 750 μL of ethyl acetate and then solution was homogenized. 200 μL of the solution was piped and evaporated (waterbath 65°C for 30 minutes) until ethyl acetate dried. The formed residual were dissolved by 1 mL of demineralization water. Absorbance was measured on 228 nm wavelength. The analysis was conducted 3 times repetition. Inhibitory activity by the formula: [13]

$$\%\ \text{ACE Inhibition} = (B / A) / (B-C) \times 100\% \quad (1)$$

Information:
A = absorbance with enzymes and ACE enzyme inhibitory compounds
B = absorbance without enzymes and without ACE enzyme inhibitory compounds
C = absorbance without enzymes but with enzyme inhibitors

2.7. *Data Analysis*

Data were analyzed using SPSS 16.0 for Windows. Data of glutamic acid levels and ACE inhibitory activity were analized by variance test using the One-Way Anova by 95% reliable level. It was continued by Duncan Multiple Range Test (DMRT) test. Further analysis by comparing the addition of CBP with standard value (addition 0.38% MSG of processed) to each dishes group using One-Sample T-Test [14]. Properties of organoleptic data were analyzed by Kruskal-Wallis test and continued by Wilcoxon Test (Wilcoxon Sign rank test) with 95% reliable level. The analysis was continued by comparing the treatment of addition CBP doses with standard values to each foodstuff group using the Wilcoxon test (*Wilcoxon sign rank test*) [15]. The analysis was performed only for the addition of CBP doses (0%, 0.5%, 0.8% , 1.1%, and 1.4%) and conducted to each foodstuff group.

3. Result and Discussion

3.1. Glutamic Acid Content of Various Dishes

The results showed that CBP contains fewer glutamic acids (6.22 + 0.008) than fresh *bekasam* (11.88 + 0.025). Reduction of glutamic acid levels as a result of processing and drying of fresh *bekasam* for a long time. Glutamic acid is formed as a result of microorganism's proteolytic activity during fermentation which breaks the protein complex molecules into smaller components of peptides and amino acids. Miroorganism's proteolytic growth during fermentation was affected by salt concentration of substrate. The increased of salt concentration on substrate, microorganism's proteolitic growth would be declined [16,17].

Table 1. Glutamic Acid Content of Various Dishes

Treatment	Glutamic Acid Level (mg/g)	Difference (Δ)	P-Value
Noodle			
Dose 0%	3.65 ± 0.011^{a}	0.28 ± 0.007	0.0001*(#)
Dose 0.5%	3.93 ± 0.004^{b}	0.13 ± 0.012	
Dose 0.8%	4.06 ± 0.009^{c}	0.10 ± 0.014	
Dose 1.1%	4.16 ± 0.006^{d}	0.07 ± 0.002	
Dose 1.4%	4.23 ± 0.004^{e}		
Standard (0.38% MSG)	4.43 ± 0.003		0.0001*(^)
Carrot			
Dose 0%	2.71 ± 0.003^{a}	0.14 ± 0.006	0.0001*(#)
Dose 0.5%	2.85 ± 0.007^{b}	0.11 ± 0.008	
Dose 0.8%	2.96 ± 0.006^{c}	0.09 ± 0.100	
Dose 1.1%	3.05 ± 0.005^{d}	0.13 ± 0.003	
Dose 1.4%	3.18 ± 0.006^{e}		
Standard (0.38% MSG)	3.32 ± 0.003		0.0001*(^)
Champignon			
Dose 0%	5.03 ± 0.005^{a}	0.60 ± 0.003	0.0001*(#)
Dose 0.5%	5.63 ± 0.003^{b}	0.59 ± 0.003	
Dose 0.8%	6.21 ± 0.005^{c}	0.23 ± 0.008	
Dose 1.1%	6.44 ± 0.004^{d}	0.33 ± 0.030	
Dose 1.4%	6.77 ± 0.032^{e}		
Standard (0.38% MSG)	7.32 ± 0.006		0.0001*(^)
Chicken Meat			
Dose 0%	9.12 ± 0.009^{a}	0.61 ± 0.012	0.0001*(#)
Dose 0.5%	9.73 ± 0.006^{b}	0.67 ± 0.017	
Dose 0.8%	10.40 ± 0.024^{c}	0.51 ± 0.020	
Dose 1.1%	10.91 ± 0.004^{d}	0.45 ± 0.019	
Dose 1.4%	11.35 ± 0.015^{e}		
Standard (0.38% MSG)	11.84 ± 0.007		0.0001*(^)

* Significant difference average at level of 0.05
(#) *One-Way Anova* (comparison among treatment)

Treatment	Glutamic Acid Level (mg/g)	Difference (Δ)	P-Value

(^) *One Sample T-Test* (comparison between treatment and standard)
Letter notation applies to the average of each foodstuff group
Notation differences show significant average differences
Δ = level of the initial addition dose of glutamic acid – subsequent addition dose of glutamic acid (for each foodstuff group)

The results stated that the addition of CBP on various dishes did not show any synergistic effect. This was indicated by the difference (Δ) of glutamic acid levels that get lower even though the dose of CBP was further enhanced. This is related to the change of crystalline form of glutamic acid molecule due to the saturated glutamic acid concentration in high temperature boiling water, so the result of the analysis of glutamate levels showed low.

Table 2. ACE Inhibitory Activity of Various Dishes

Treatment	ACE Inhibitory Acctivity (%)	P-Value
Noodle		
Dose 0%	10.42 ± 0.46	0.287 (#)
Dose 0.5%	9.92 ± 0.51	
Dose 0.8%	9.87 ± 0.64	
Dose 1.1%	10.02 ± 0.42	
Dose 1.4%	9.42 ± 0.50	
Standard (0.38% MSG)	9.56 ± 0.56	0.017*(^)
Carrot		
Dose 0%	10.49 ± 0.72[b]	0.017*(#)
Dose 0.5%	10.49 ± 0.62[b]	
Dose 0.8%	8.80 ± 0.52[a]	
Dose 1.1%	9.83 ± 0.58[ab]	
Dose 1.4%	9.09 ± 0.54[ab]	
Standard (0.38% MSG)	6.41 ± 0.17	0.0001*(^)
Champignon		
Dose 0%	9.02 ± 0.63[ab]	0.043*(#)
Dose 0.5%	8.59 ± 0.63[a]	
Dose 0.8%	9.83 ± 0.58[bc]	
Dose 1.1%	10.27 ± 0.64[c]	
Dose 1.4%	9.93 ± 0.67[bc]	
Standard (0.38% MSG)	5.93 ± 0.77	0.0001*(^)
Chicken Meat		
Dose 0%	7.73 ± 0.41[b]	0.032*(#)
Dose 0.5%	7.86 ± 0.47[b]	
Dose 0.8%	6.83 ± 0.50[a]	
Dose 1.1%	6.76 ± 0.38[a]	
Dose 1.4%	7.73 ± 0.41[b]	
Standard (0.38% MSG)	6.33 ± 0.41	0.0001*(^)

* Significant difference average at level of 0.05
(#) *One-Way Anova* (comparison among treatment)
(^) *One Sample T-Test* (comparison between treatment and standard)
Letter notation applies to the average of each foodstuff group

Treatment	ACE Inhibitory Acctivity (%)	P-Value

Notation differences show significant average differences

The process of drying the catfish *bekasam* was assumed may alter the molecular structure of L-glutamic acid into α-glutamic acid crystals. The α-glutamic acid crystal would not undergo transformation into β-glutamic acid form as long as it was still in powder form. When the CBP was added to boiling liquid during the processing of the food, the α-glutamic acid crystals dissolve in the boiling liquid and if all the crystalline α-glutamic acid was dissolved it would induce form change to β-glutamic acid. A solution with a high glutamic acid concentration at 45°C would produce β-glutamic acid crystals, whereas a solution with a low glutamic acid concentration would only produce α-glutamic acid crystals in both high and low temperature conditions [18].

As or the initial levels of glutamic acid in foodstuff is known to contribute to glutamic acid levels in the final product. The higher the initial level of glutamic acid in the food stuff, will result in higher glutamic acid level. It is also known that the processed food added 0.38% MSG always shows the highest levels of glutamic acid (Table 1). This is because of the high level of glutamic acid contained in MSG, equal to 733.29 mg/g so as to produce high glutamate levels also on the dishes added MSG [19].

Table 3. The Assessment of Panelists Against Comparative Level of Dish Flavor Against Standard Sample

Treatment	Flavor Comparison against R (n=17)						Median Score*)	P-Value
	Better		Equal		Worse			
	n	%	N	%	N	%		
Noodle								
Dose 0%	1	5,9	9	52,9	7	41,2	4	
Dose 0.5%	2	11,8	1	5,9	14	82,4	3	0.007*(#)
Dose 0.8%	0	0	2	11,8	15	88,2	3	
Dose 1.1%	2	11,8	2	11,8	13	76,5	3	
Dose 1.4%	2	11,8	3	17,6	12	70,6	3	
Standard (0.38% MSG)								0.0001*(^)
Carrot								
Dose 0%	6	35,3	3	17,6	8	47,1	2	
Dose 0.5%	4	23,5	2	11,8	11	64,7	3	0.054(#)
Dose 0.8%	4	23,5	4	23,5	9	52,9	4	
Dose 1.1%	4	23,5	4	23,5	9	52,9	4	
Dose 1.4%	1	5,9	0	0	16	94,1	3	
Standard (0.38% MSG)								0.0001*(^)
Champignon								
Dose 0%	3	17,6	7	41,2	7	41,2	2	
Dose 0.5%	2	11,8	2	11,8	13	76,5	2	0.530(#)
Dose 0.8%	2	11,8	4	23,5	11	64,7	3	
Dose 1.1%	4	23,5	3	17,6	10	58,8	2	
Dose 1.4%	3	17,6	1	5,9	13	76,5	3	
Standard (0.38% MSG)								0.0001*(^)
Chicken Meat								0.008*(#)
Dose 0%	6	35,3	4	23,5	7	41,2	4	

Treatment	Flavor Comparison against R (n=17)						Median Score*)	P-Value
	Better		Equal		Worse			
	n	%	N	%	N	%		
Dose 0.5%	4	23,5	5	29,4	8	47,1	2	
Dose 0.8%	2	11,8	3	17,6	12	70,6	3	
Dose 1.1%	4	23,5	0	0	13	76,5	3	
Dose 1.4%	1	5,9	2	11,8	14	82,4	3	
Standard (0.38% MSG)								0.0001*(^)

* Significant difference average at level o 0.05
(#) *Kruskal-Wallis Test* (median comparison among treatments)
(^) *Wilcoxon Signed Rank Test* (comparison between median of treatment and standard)
*) Median Score : 1=so much better; 2=better; 3=just as good; 4=worse; 5=very worse

3.2. *Angiotensin-I-Converting Enzyme (ACE) Inhibitory Activity*

The results of the analysis indicated that the CBP had ACE inhibitory activity of 34.025%. This is in accordance with previous study which stated that milkfish *bekasam* had ACE inhibitory activity of 35.67 + 6.78% on 5th day of fermentation and 51.77 + 8.80% on 6th day of fermentation [6].

Table 4. The Assessment of Panelists of Intensity of The Different Dishes Flavor

Treatment	Diffference of Flavor Intensity to R (n=17)						Median Score*)	P-Value
	Almost Nothing		Moderat/ Medium		Many			
	N	%	N	%	n	%		
Noodle								
Dose 0%	5	33,3	6	40,0	4	26,7	3	
Dose 0.5%	9	60,0	1	6,7	5	33,3	2	0.366(#)
Dose 0.8%	9	60,0	1	33,3	5	6,7	2	
Dose 1.1%	6	40,0	4	26,7	5	33,3	3	
Dose 1.4%	5	33,3	6	40,0	4	26,7	3	
Standard (0.38% MSG)								0.0001*(^)
Carrot								
Dose 0%	7	46,7	5	33,3	3	20,0	3	
Dose 0.5%	9	60,0	2	13,3	4	26,7	2	0.808(#)
Dose 0.8%	6	40,0	5	33,3	4	26,7	3	
Dose 1.1%	7	46,7	5	33,3	3	20,0	3	
Dose 1.4%	4	26,7	6	40,0	5	33,3	3	
Standard (0.38% MSG)								0.0001*(^)
Champignon								
Dose 0%	13	86,7	2	13,3	0	0	2	
Dose 0.5%	9	60,0	3	20,0	3	20,0	2	0.0001*(#)
Dose 0.8%	3	20,0	8	53,3	4	26,7	3	
Dose 1.1%	2	13,3	3	20,0	10	66,7	4	
Dose 1.4%	4	26,7	4	33,3	6	40,0	3	
Standard (0.38% MSG)								0.0001*(^)
Chicken Meat								
Dose 0%	9	60,0	0	0	6	40,0	2	
Dose 0.5%	8	53,3	2	13,3	5	33,3	2	0.895(#)
Dose 0.8%	7	46,7	4	26,7	4	26,7	3	
Dose 1.1%	8	53,3	6	40,0	1	6,7	2	
Dose 1.4%	6	40,0	6	40,0	3	20,0	3	
Standard (0.38% MSG)								0.0001*(^)

Treatment	Diffference of Flavor Intensity to R (n=17)						Median Score*)	P-Value
	Almost Nothing		Moderat/ Medium		Many			
	N	%	N	%	n	%		

* Significant difference average at level o 0.05
(#) *Kruskal-Wallis Test* (median comparison among treatments)
(^) *Wilcoxon Signed Rank Test* (comparison between median of treatment and standard)
*) Median Score : 1=no differene; 2=little bit; 3=moderat/medium; 4=many; 5=very much

The results also stated that ACE inhibitory activity decreased to 12.799% after being processed into CBP. This was associated with the process of making CBP in a long time and high temperature. High temperatures would cause enzyme inactivation. Most enzymes exhibited optimum activity at 30-40°C and started denatured at 45°C [20]. Besides catfish *bekasam* and CBP, ACE inhibitory activity was also studied on various dishes and compared to MSG (Table 2).

Differences of ACE inhibitory activity that fluctuate at the carrots, champignon, and chicken meat dishes were associated with the difference content of specific ACE inhibitory peptides in each foodstuff. Peptides which had optimum ACE inhibitory activity, at the chain base had to have a series of amino carboxyl acids which was composed of three amino acids (Phe-Ala-Pro). The side chains of these three amino acid sequences residue were assumed to interact specifically with the active part of the ACE enzyme (S1, S'1 and S'2). The active part of the enzyme was a carboxyl group attached to a Zn mineral. The presence of a series of three Phe-Ala-Pro amino acid carboxyls in the foodstuff component could increase the ACE inhibition potential by 1000-fold [21].

The low ACE inhibitory activity on dish which was added 0.38% MSG (standard) associated with the active side of the ACE enzyme crystalline structure. Glutamate is the constituent molecule of one of the active sides of the crystal structure of the ACE enzyme, so that MSG is able to provide large amounts of glutamate donors for the activation of the ACE enzyme. The active Zn side of the ACE enzyme structure binds to two molecules of histidine residue and one glutamate molecule [22,23]. The higher the glutamic acid content in a dish, the more ACE enzyme will be activated. Therefore, ACE inhibitory activity will get lower along with increasing amount of activated ACE enzyme.

3.3. Flavor Organoleptic Properties of Various Dishes

The result of panelist assessment showed that the higher the dose of addition of CBP on the noodle and chicken meat dishes, the worse flavor was produced. The higher addition of glutamic acid in a dish, would not make the flavor and taste better, but worse. The addition of excessive dose of glutamic acid to dishes, would not make the taste of dish tastier but would worsen the flavor. The dosage of glutamic acid addition which was capable to improve the taste and flavor umami of dish was only 0.1-0.8% by weight of total ingredients [24].

The addition of CBP on dishes might produce flavor that tended to equal even better than standard (Table 3). The results of the varying panelist assessments were attributed to panelist perceptions that vary on the sensory nature of the dish flavor. This variation was due to the differences of panelist flavor thresholds, because each individual had different threshold sensitivity to a particular flavor. When the food was digested, the taste component of the foodstuff contacted with the receptors on taste cell through the taste bud hole so that it caused a taste sensation. While the nervous system in the oral cavity would act as a signal receiver and could identify the taste sensation of different chemical components. The sensitivity of the tongue as a sense of taste in recognizing a flavor depend on the threshold, when the tongue just able to taste of a flavor with the minimum concentration at the flavor component [25]. Sweet and salty flavors also had varying optimum concentrations of each-individual cases. Some individuals might like the sensation of the flavor but the other hand some individuals might also disliked it, depending on the sensitivity belong to each individual [24].

The threshold values and perceptions of a food's taste were often subjective and different for each individual. Moreover, the results of a panelist assessment of the sensory properties of dishes flavor might also be influenced by the presence of different flavor components of a foodstuff. This resulted an interaction and rose the different perceptions of taste. Such interactions could be either mutually enhancing or eliminating flavors, flavor modifications, multiplying and flavor adjustments [25].

The result showed that the higher the dosage addition of CBP on *Champignon* dishes, the greater the intensity of the flavor difference. When it was compared with the standard, the intensity of the *Champignon* dishes flavor added by the CBP has a moderate / medium difference. It also meant that the *Champignon* dishes added with CBP have a better flavor than the standard (Table 4). Dishes which were added by CBP had savory flavor that was not much different than the standard. The savory flavor on dishes which were added by CBP, was produced from free glutamic acid contained in dishes, CBP and complementary spices. Savory taste or better known as umami arose because of the natural glutamic acid content in dishes and spices. Although CBP had a mixture of sour, salty, and slightly bitter taste, but when it was combined with herbs and ingredients, the sour and bitter tastes would disappear, and the sweet and salty taste would become sharper. Free glutamic acid not only increases the sweetness and salt perception but also minimizes sour and bitter taste. The addition of glutamic acids in the dishes could also increase the food's palatability through the umami flavor, although in fact umami was not palatable. Glutamic acid had no effect on aroma but can increase overall flavor intensity [26,27].

Panelist perceptions of food flavors may also be influenced by the presence of glutamic acid components in dishes. Basically the food already contains glutamic acid in certain levels and if the dish was added by glutamic acid, certainly it would change the basic flavor of the foodstuff. The higher the addition of glutamic acid, the dish taste would not be better but worse [25].

This study had limitations such as catfish samples used were obtained from the market so it was not known how if the catfish from cultivation (feed type, feeding method, and environment). Moreover, this study had not been supported by research data and physicochemical powder properties standard. This study also had not been able to describe in detail about the flavor because it only used multiple comparison test.

4. Conclusion

Carrots, *Champignon*, and chicken meat dishes showed varying ACE inhibitory activity, while MSG added dishes showed the lowest ACE inhibitory activity. On the one hand, the addition of CBP on the processed of carrots, champignon, and chicken meat was directly proportional to glutamic acid levels but inversely proportional to ACE inhibitory activity. On the other hand, the addition of MSG to all dishes could increase glutamic acid levels but could significantly reduce ACE inhibitory activity. Based on the results of organoleptic tests, the *Champignon* dish which was added by CBP resulted in better flavor preference compared to *Champignon* dish which was added by MSG. Conversely, the higher the dose of addition of CBP to the processed wet noodles and chicken meat, resulting worse flavor.

The researcher suggests that the addition of CBP as much as 0.5% has been able to provide better cooking flavor and show ACE inhibition activity which is higher than MSG. The researcher also suggests the need of further research on in-vivo ACE inhibitory activity, identification of appropriate use of processing technology to optimize the catfish *bekasam* typical flavor as a flavor enhancer of food, identification of ACE-inhibiting peptides and the descriptive of organoleptic properties testing.

References

[1] Kartika U 2014 *Hipertensi bukan sekadar tekanan darah tinggi* Publish at March 7 2014. Available from http://health.kompas.com [Accessed at April 13 2014]

[2] Badan Penelitian dan Pengembangan Kesehatan Kementerian Kesehatan RI 2013 *Riset Kesehatan Dasar (Riskesdas)*

[3] Loliger J 2000 *Function and importance of glutamate for savory foods. J. of Nut.* **130** 915S–920S

[4] Rangan C and Barceloux D G 2009 *Food additives and sensitivities Disease-a-Month* **55**(5)

292–311
[5] International Glutamate Information Service 2014 *Umami dan keasinan* Available from http://www.glutamate.org/id/media/Umami_dan_ keasinan.php [Accessed at May 12 2014]
[6] Wikandari P R 2011 *Potensi bakteri asam laktat indigenous sebagai penghasil angiotensin-i-converting enzyme inhibitor pada fermentasi bekasam* Available from http://etd.ugm.ac.id/index.php?mod=penelitian_detail&sub=PenelitianDetail&act=view&typ=html&buku_id=53183&obyek_id=4 [Accessed at April 13 2014]
[7] Direktorat Jenderal Perikanan 2014 *Budidaya kementerian kelautan dan perikanan: perikanan budidaya indonesia* January 7 2014 Available from http://www.djpb.kkp.go.id/berita.php?id=955 [Accesed at March 25 2014]
[8] Pusat Penyuluhan Kelautan dan Perikanan 2011 *Pengolahan ikan lele* Available from http://www.sharepdf.com/4698693212fa4ab98563_f38fcdca53fe/ikanlele.pdf [Accessed at May 11 2014]
[9] Soekarto T S 1985 *Penilaian organoleptik* (Jakarta: Bhatara Karya Aksara)
[10] Monasterio P R 2014 *Advance in food science and nutrition Volume 2 Chapter 9 : transformations of food flavor due to industrially processing of elaboration* (USA : John Wiley & Sons)
[11] Heymann H and Lawless H T 1998 1998 *Sensory evaluation of food principles and practices* (USA: Chapman & Hall)
[12] Bio Assay System 2009 *EnzyChrom TM glutamate assay kit (EGLT-100): quantitative colorimetric determination of glutamate at 565 nm* from http: // www.gentaurpdf.com/pdf/EnzyChrom%20Glutamate%20Assay%20Kit.pdf [Accessed at February 20 2013].
[13] Wikandari P R Suparmo Marsono Y and Rahayu E S 2011 *BIOTA* **16**(1) 145 – 152
[14] Yitnosumarto S 1993 *Percobaan perancangan, analisis dan interpretasinya* (Jakarta: PT Gramedia Pustaka Utama)
[15] Uyanto S S 2009 *Pedoman analisis data dengan spss edisi 3* (Yogyakarta: Graha Ilmu)
[16] Molliet A 2008 *Molecular dynamic stimulations of glutamic acid in solution thesis for the degree of doctor philosophy, school of engineering and physical science heriot-watt university* 3-13
[17] Murwani R Supriyadi Subagio Trianto A Ambariyanto 2015 *Isolation and identification of thermophilic and mesophylic proteolytic bacteria from shrimp paste "terasi" (Conference Paper) AIP Conference Proceedings* 1699
[18] Pavia M and Carmen 2001 A*spects of msg intolerance thesis from school of food science and technology university of new south wales* 32
[19] Andarwulan N Nuraida L Madanijah S Lioe N H and Zulaikhah 2011 Food *and Nut. Sci.* **2** 764-69
[20] Fennema R 1985 *Food chemistry 2nd edition* 402-406 (New York : Marcel Dekker, Inc)
[21] Chusman D W and Ondetti M A 1999 *Design of angiotensin converting enzyme inhibitors Nature Med.* **5** (10) 1110-112
[22] Corradi H R Sylva L U S and Aloysius T N 2006 *Crystal structure of the n domain of human somatic angiotensin-i-converting enzyme provides a structural basis for domain-specific inhibitor design J. Mol. Biol.* **357** 964-74
[23] Zeng X and Huang G 2012 *Food flavor In Wang D H Lin J Kan et al food science and technology : food chemistry* (USA : Nova Science Publishers, Inc)
[24] Jinap S Ilya-Nur A R Tang S C Hajeb P Shahrim K and Khairunnisak M 2010 *Sensory attributes of dishes containing shrimp paste with different concentrations of glutamate and 5'-nucleotides Appetite* **55** 238-244
[25] Shi H and Ho C T 1994 *chapter 4 : the flavour of poultry meat. in flavor of meat and meat product* edited by Fereiddoon S 52-70 (United States : Springer)
[26] Yamaguchi S and Ninomiya K 2000 *Umami and food palatability. J. Nutr* **130** 921S-26S.

[27] Jinap S and Hajeb P 2010 . *Appetite* **55** 1-10

The Use of Different Diets for Feeding Rate and Growth of Shortfin Eel (*Anguilla bicolor bicolor*)

N Taufiq-Spj[1,3], S Sunaryo[1], A Wirasatria[2,3] I Pratikto[1], D H Ismunarti[2] and M I Syaputra[1].

[1] Departement of Marine Science, Faculty of Fisheries & Marine Science, Diponegoro University, Jl. Prof. H. Soedarto, SH, Tembalang, Semarang 50275, Indonesia.
[2] Departement of Oceanography, Faculty of Fisheries & Marine Science, Diponegoro University, Jl. Prof. H. Soedarto, SH, Tembalang, Semarang 50275, Indonesia.
[3] Center for Coastal Disaster Mitigation and Rehabilitation Studies, Center of an Excellence Research University, Diponegoro Univ. Semarang, Cent. Java, Indonesia.

E-mail: taufiqspj_1999@yahoo.com

Abstract. Growth and feed rate accompanied with proximate, texture and condition of the organism. The objective of this study was to obtain the effect of different diets to the feed rate and growth of shortfin eel *Anguilla bicolor bicolor.* The study used three replications and 3 treatments of different sources of feed diet during 84 days culture. The fish were fed with proximation of 2.2% wet basis d^{-1} (ratio of dry feed and water = 4 : 3). Initial density of approx. 15 kg seed stock of eel fingerling size (early elver) per m^3 water in the recirculating water system with ranging of means±SD of temperatures, pH and DO were 27.14±0.11 to 27.66±0.05 °C, 7.44±0.09 to 7.66±0.05, and 3.54±0.19 to 4.66±0.09 ppm respectively. The result shows that feeding rate has relation to lipid content of feed but not significantly different among treatments ($\alpha > 0.05$), the highest *FR* shows by F_2 followed by F_3 and F_1, where: F_1= 2.04±0.07% d^{-1} and F_2= 2.12±0.05% d^{-1} and F_3= 2.10±0.06% d^{-1}). Absolute growth of individuals and biomass were related to the protein content; the highest was F_1 followed by F_2 and F_3. Growth margin F_1, F_2 and F_3 of individual were 25 g (46%), 22 g (44%), and 15 g (27%) respectively, whilst biomass margin were 3,855 g (26%), 3,629 g (24%), and 2,834 g (18%) respectively during 70 days culture. Marginal mean of Individual-specific growth rate (SGRi) shows no significantly different ($\alpha > 0.05$) between F_1 (0.53±0.05% d^{-1}) and F_2 (0.52±0.01% d^{-1}), but both of F_1 and F_2 have significantly different ($\alpha < 0.05$) to F_3 (0.35±0.07% d^{-1}). Biomass-specific growth rate (SGRb) also shows no significantly different ($\alpha > 0.05$) between F_1 (0.33±0.09% d^{-1}) and F_2 (0.31±0.02% d^{-1}), but both of F_1 and F_2 have significantly different ($\alpha < 0.05$) to F_3 (0.24±0.08% d^{-1}).

Keywords: Eel, *Anguilla bicolor bicolor*, Feeding Rate, Specific, Marginal growth.

1. Introduction

Anguilla sp. is mainly as carnivorous species [5] [12] [15] [16] with shorter gut line compare to either omni and herbivorous species. As nocturnal species, *Anguilla* preys mainly pelagic swimmer or drifting organism and some of the benthic organism. Atlantic menhaden *Brevoortia tyrannus* had been used for baiting American eel *A. rostrata* at Hudson River [15]. Early glass eel of *A. bicolor bicolor* and *A marmorata* pray crustacean and zooplankton in Segara Anakan (southern seawater of Cilacap) Central Java [16]. As predators, eels become increasingly piscivorous by increasing body length. [12] conclude that most research on the diet of eels has been in nutrient-rich basic habitats and the eels which less than 300 mm in length fed mostly on Salmonid ova and parr, *Trichoptera* sp. larvae, *Simulium* sp. larvae, Asellus sp., Nymphs *Ephemeroptera, Gammarus pulex,* and *Diptera larvae.*While eels more than 300 mm in length fed on *Gammarus, Asellus,* crayfish, and fish of Sculpin and Stickleback. Due to this feeding habit, the sources of feed are important in racing this species.

Study on feed intake mainly related to feeding efficiency and economic sustainability should take into consideration. The relationship between feed intake and growth in finfishes has been conducted extensively using bioenergetic approaches [6] [7] [1]. However, Heinsbroek et al [8] stated that as bioenergetic models are focussed on the energy they do not take into account the differences in attributes and functions of the different energy-containing nutrients as well as the differences in transformation efficiencies, including a deposition. Due to these energy-containing nutrients it has been indicated that the non-protein macronutrients, lipids, and carbohydrates, are to some extent interchangeable in their effect on feed intake and growth and can be grouped as nonprotein energy (NPE), provided that the level of carbohydrates does not exceed the limit of the fish species investigated [14].

The shortage of eel in Asia (Japan), Europe, and America made Indonesian government enforcing regulation for eel seed export [17]. Nevertheless, eel aquaculture in Indonesia nowadays is just starting to develop compared to other species, i.e., prawn (*Penaeus monodon*), shrimp (*P. merguensis*), milkfish (*Chanos canos* F), or Nile (*Oreochromis niloticus*). Hence, food industry of those species has already been found in the right formula with reasonable price. However pasted feed for eel has still being imported or produced with relatively high price and low supply assurance. By using the pelleted feed of shrimp, hopefully, can minimize the production cost with consistent assurance in supply. The previous study using pasted feed of 2.2 % in dry basis [17] gave the best absolute growth slightly more than 100 % during 100 days culture. Trial on wet basis needs to be conducted to get another figure of feed ratio and growth. For this reasons, the study on different diets in wet basis will hopefully get the figure of feeding rate and weight gain of eel. Hereafter, this study aimed to evaluate different diets for feeding rate and growth performance.

2. Methodology

Fingerling elver (approx. 50 gr) of eel *A. bicolor bicolor* were obtained from seed collector farm in Purwokerto (Central Java). This study used a biomass of fish seeds (approx. 15 kg) which was distributed to 1 m^3 of 9 tanks with basic recirculating system conducted by [17]. The fish were maintained at 26.86±0.58°C water temperatures, 7.39±0.18 pH, 6.48±0.11 ppm of dissolved oxygen and normal photoperiod in the semi-enclosed farm (HDPE roof and aluminum fence). Two kind of commercial pasted feeds F_1 (Sakae), and F_3 (UNS) and commercial shrimp feed (Feng Li) F_2 (Table 1) was given by 2.2 % d^{-1} in wet basis and give twice a day. The best-pasted feed performance using water weighed 75 % of dry food given. Hence the actual feed powder (dry) given was 1.26 % and mixed with water of 0.94 % (with ratio feed water of 4 : 3). CMC, cellulose gum (5% of food weight) was added for F_2 as a stabilizer and food binder. Uneaten food was collected and weighted after two hours feeding. Feed given assembled with biomass of eel weight and weight sampling of eel were taken in every fortnight. Feeding Rate (*FR*) were observed daily during 84 days and calculated by using the formula, i.e., amount of food given (*fg*) per day minus uneaten food (*uf*) divided by fish biomass (*b*) in every tank (*Form.1.*). To reduce mistaken calculation, the feed was put on floating net, and weighted if any remainder and the residue collected by flushing and netting the water.

$$FR = \frac{fg - uf}{b} \times 100\% \qquad (1)$$

Table 1. Feeds proximate analysis given to fish

Feeds	Water (%)	Ash (%)	Protein (%N)	Lipid (%)
F_1	10,51	13,18	39,67	12,17
F_2	10,18	8,65	34,11	4,84
F_3	9,25	10,98	27,71	12,04

Specific growth rate (*SGR*) of individual and biomass of fish per tank in group of treatments (F_1, F_2, and F_3) was calculated during 70 days observation by the difference of normal logarithmic of weight at "*t*" time (W_t) and initial weight (W_0) divided by time (*t*) in percent d^{-1} (*Form.2.*) [17]. Statistical analysis uses a completely random design and analysis co-variance with three treatments and three replications. The estimated marginal mean was used for feeding rate (*FR*). Individual and biomass-specific growth were analyzed by using generalized linear model repeated measurement ANOVA and post hoc test using multiple comparisons based on observed means.

$$SGR = \frac{Ln(Wt) - Ln(W0)}{t} \times 100\% \quad (2)$$

Water was a pump with a maximum capacity of approx. 5,000 l h^{-1} by using Ø ¾ inc inlet hose, recirculated through biological filter. Medium aeration was administered in every tank. For growing nitro bacteria (*Nitrosomonas* and *Nitrobacter*) and reduce metabolite excess, filter bucket filled with solid phase media of coral, zeolite stone, and charcoal. To prevent accumulated uneaten food and feces, bioplastic filter (dacron) were placed on the surface of the solid phase. The dirt produced were separated between uneaten food and feces. Medial aeration was also administered in every tank during the culture.

3. Result and Discussion

As a carnivorous species, i.e. *Anguilla bicolor bicolor* grow related to the protein content in feed given, the growth apparently increases by increasing protein. Feed rate fluctuates every fortnight accompanied by a condition of an organism, the texture of feed but seems to not correlate with the degree of the lipid content in the feed. As nocturnal species, eel tends to be active in the night time, but during feeding time especially in the morning, the fish tend to be active to prey while the food is moving onto floating net due to dynamic schooling of fish.

3.1. Feeding Rate

Feed rate observation conducted every day which assembled with fish biomass [17] weighted fortnightly. By using one way ANOVA shows that the feeding rate seems to be stable or no significantly different during 84 days observation ($\alpha > 0.05$). The *FR* value in between 1.98 to 2.16 % (Table 2), fluctuation condition apparently depend on the condition of the organism. The wet basis of feed treated mainly to compare the previous study which used dry basis with *FR* between 1.30 – 2.07% [17]. Hence present study has a higher value of *FR* and more efficient in term percent of feed swollen by fish compare to the previous study of dry basis, and will probably give economic advantages. Due to a smaller quantity of feed given with smaller uneaten and debris of feed, will automatically give a better water quality.

Table 2. Mean of Feeding Rate (% d^{-1}) of eel during 84 days observation

Treat-ments	Days of observations							
	01-14	15-28	29-42	43-56	57-70	71-84	Mean	±SD
F_1	1.98	2.05	2.09	1.99	1.99	2.16	2.04	0.07
F_2	2.05	2.08	2.12	2.15	2.16	2.15	2.12	0.05
F_3	1.99	2.10	2.15	2.13	2.08	2.16	2.10	0.06

Nevertheless, treatment F_2 had a highest mean value with lowest deviation (2.12±0.05% d^{-1}) compare to F_1 (2.04±0.07% d^{-1}) and F_3 (2.10±0.06% d^{-1}) (Table 2). In fact, this F_2 of commercial shrimp feed which was added with cellulose gum (5%) have slightly more compact than original pasted feed with a similar water content of 75%. The influencing factors for feeding rate are food and feeding habit [17], once type of food introduced, the fish will take the feed to go along with starving conditions. Even rancidity rate was not measured, but the trend of these *FR* apparently related to lipid content where the feed rate increased by decreasing lipid content (Table 1). In correlation of this rancidity condition, all of the treatment replications of treatment F_1 have lower DO than others (Table 3).

Observation on temperatures, salinities, hydrogen ion concentrations (pH) and dissolved oxygen (DO) conducted in every second day remain stable (Table 3). Hydrogen ion concentration (pH) was between 7.44 and 7.66. This pH will be related to Alkalinity, and the pH will affect to NH_3 and NO_2^- concentration. This corresponds with stating *Nitrosomonas sp* which can grow better in pH between 7 to 8 [9], this bacteria will be responsible for converting NH_4 to NO_3.

The highest *FR* in this study was 2.16% with a natural temperature of 27.66±0.05°C, while [11] by using water heater of 33°C have the best *FR* for *A. bicolor pacifica* was 2.41±0.04% day^{-1} and *A. marmorata* (2.23±0.10% day^{-1}). Meanwhile, by using temperature of 28°C [11] just can attain of *FR* only 1.82±0.01% day^{-1} for *A. marmorata* and 2.13±0.03% day^{-1} for *A. bicolor pacifica*. Natural environment conditions of this study seem to have a better *FR* compare to what which [11] have conducted. Hereafter, in regulating feeding of aquatic organisms, natural environment, and adjacent conditions may as an important role to provoke a feeding rate.

Table 3. Temperatures, salinity, pH and dissolved oxygen measured every second day in every replication of treatments.

Treatment	t (°C)	S (‰)	pH	DO (ppm)
$F_{1.1}$	27.14±0.11	1.22±0.04	7.44±0.09	3.78±0.11
$F_{1.2}$	27.24±0.05	1.28±0.08	7.44±0.05	3.54±0.19
$F_{1.3}$	27.26±0.05	1.36±0.05	7.50±0.00	3.64±0.15
$F_{2.1}$	27.50±0.07	1.22±0.04	7.44±0.05	4.28±0.11
$F_{2.2}$	27.54±0.05	1.28±0.04	7.56±0.05	4.42±0.18
$F_{2.3}$	27.66±0.05	1.30±0.00	7.52±0.05	4.36±0.09
$F_{3.1}$	27.56±0.09	1.16±0.09	7.62±0.08	4.58±0.11
$F_{3.2}$	27.66±0.05	1.20±0.10	7.63±0.04	4.56±0.22
$F_{3.3}$	27.64±0.05	1.18±0.04	7.66±0.05	4.66±0.09

3.2. Growth performance

3.2.1. Specific growth rate

Specific growth rate gives a figure or performance in certain technique, and methods had been used in giving percent daily weight attainment during the culture of organisms. Base on ANOVA analysis of individual fish growth repeated measurement during five times observation the grown shows significantly different among treatment ($\alpha < 0.05$), in continues test, indicated that F_1 similar to (=) F_2, but F_1 and F_2 different to ($\neq$) F_3. Table 2. shows that individual specific growth fluctuates fortnightly, the lowest *SGRi* occurred in the local pasted feed (UNS) of $F_{3.1.}$ during second measurement (0.05% d^{-1}), mean while the highest *SGRi* on Korean formula pasted feed (Sakae) of $F_{1.3.}$ during third measurement (0.80% d^{-1}). Base on standard deviation repeated measurement of *SGRi* during 70 days culture, F_1 shows lowest followed by F_2 and F_3 (Table 2). This conditions indicate that treatment F_1 gave a stable impact of growth in every replication of tanks compared to F_2 and F_3, meanwhile, F_2 more stable than F_3.

When the feeding rate is over maintenance requirement since the energy consumed equals to the energy required by aquatic organisms to maintain them self without mobilizing endogenous energy reserves. Hence the growth rate will rapidly increase [3]. Furthermore, the growth will increases at a decreasing rate to a maximum point of feeding rate. For this conditions, this study shows that fish *A. bicolor bicolor* have a fluctuation of growth, but remain increase above zero line (Figure 1 and 2) even the *FR* remain stable (Table 2).

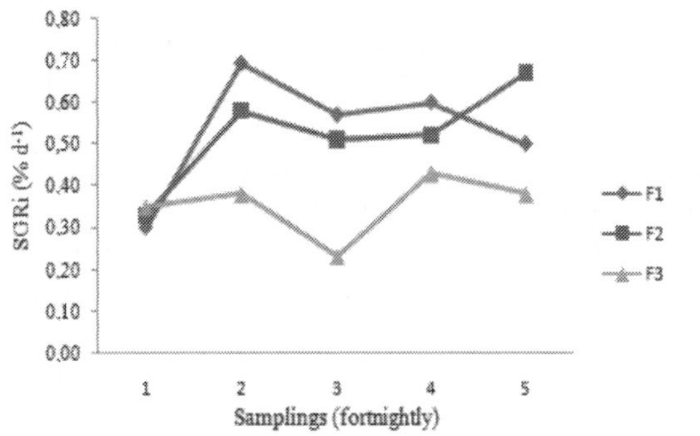

Figure 1. Individual-specific growth rate (%) of fish (*A. bicolor bicolor*) measured in every fortnight during 70 days culture fed with different feed.

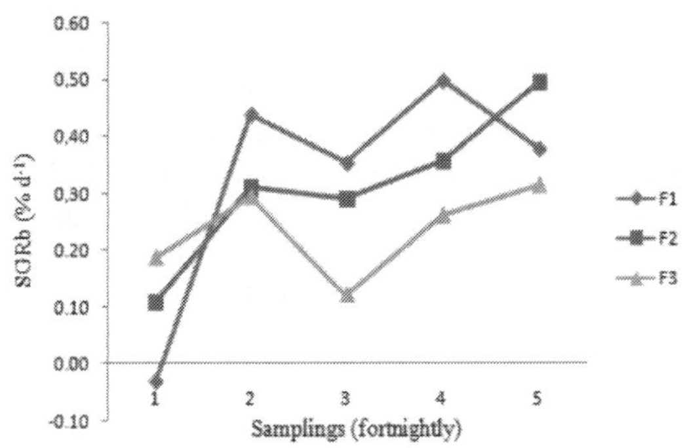

Figure 2. Biomass-specific growth rate (%) of fish (*A. bicolor bicolor*) measured in every fortnight during 70 days culture fed with different feed.

Figure 1 and 2 shows that the treatment 1 (F_1) gave the best growth either in individual and biomass of fish followed by F_2 and F_3 base on time series of samplings. The growth pattern of both individual and biomass seems to be similar. During previous study growth performance seems related to the feeding rate, the highest feeding rate gave the best growth, while, the lowest feeding rate followed by low growth of *A. bicolor bicolor* [17].

Nevertheless, unknown cause made quite high death rate of fish in one of the replications of F_1 ($F_{1.1}$, 7.4%) compare to other replications during the first period of sampling (1st-fortnight culture). Hence, total biomass gone drop under zero line after two weeks innoculation (Figure 2). Adaptation processes may affect in certain individual of fish, [17] stated that adaptation of captive fish requires high energy. They also stated that protein is sparing effect of lipid if the energy in the feed for an organism is inadequate to be used for environmental adaptation. Hence protein energy will be used for pertaining life, not for cell development. In this case, even though feed in F_1 treatment have significantly higher protein content (Table 1), but during first-week innoculation, in replication 1 of

F_1, some of the fish bundle may face longer handling during transportation. Moreover, during second samplings of treatments, first treatment of F_2 also have high (6.9%) death rate. Hence, this death rate has no correlation with food given. Holmgren et al [10] have experience in an initial decrease in weight of *A. anguilla* , this decrease can be regarded as a common phenomenon for the eels to complete their metamorphosis and to adapt to their new environment. Tesch [18] has also described an initial reduction in length and weight of European glass eels in aquaria, either fed or unfed, while pigmentation developed.

Possibility causes of unhealthy seed in a certain bundle during transportation were temperatures, feeding before transportation, extruded undigested feces, and finally low dissolved oxygen. This temperature fluctuation during transportation will lead to unstable metabolic rate hence feed given before transported will digest then protruded as undigested feces. Hereafter, The organism will take more energy and oxygen consumption as the water temperature increases [2]. Furthermore [11] stated that as tropical and subtropical species *A. marmorata* and *A. bicolor pacifica* could tolerate higher water temperatures, also [4] mention that some of the Indonesian shortfin eels can tolerate to 41°C temperature. As a proof [11] tried the experiment on *A. bicolour pacifica* and *A. marmorata* exposed at different temperatures, i.e., 8, 13, 18, 23, 28, and 33°C, both species most died within 10 days at both 8 and 13°C and both of them can survive at ≥18°C.

The graphs either individual and biomass-specific growth rate have a similar pattern (Figure 1 and 2), after second-fortnight samplings (28 d) the growth increase steeply either F_1 and F_2 treatments. Third samplings show a decrease in the percent of growth to all of the treatments. During this fortnight of culture (days of 29th to 42nd) there were some obstructions of noises and tremble waves from building construction. Even though the obstacles still happen during next fourth samplings, the growth of fish can still increase may due to the ability of environmental adaptation. Correlation of feed rate in this case of obstruction seems to be slightly increase to go along with increasing feces during 29 to 42 days of culture. However, the next samplings (4th, 43 – 56 d) the *FR* slightly decrease assembled with continues obstruction of noises and tremble wave, but the number of feces was lower. Hence F_1 and F_3 individual *SGR* remain increased. Moreover the 5th samplings (57 – 70 d) the *FR* remain similar to 4th samplings and both *SGR* of F_1 (pasted feed of Sakae) and F3 (pasted feed of UNS) tend to be decreased. On the other hand, *FR* of F_2 remains consistently increased (slightly) during obstructions (days 29th to 70th), hereafter, both individual and biomass growth also consistently increased after first decreased in the 3rd samplings (Table 2, Figure 1 and 2).

From this study seems to have no correlation between growth performance and feeding rate (Figure 1 and 2, Table 2), but apparently related to protein available in the feed (Table 1). The *FR* seems to have a correlation with the lipid content in the food. Taufiq-Spj et al [17] stated that protein is sparing effect of lipid, if the energy in the feed for an organism is inadequate to be used for environmental adaptation, protein energy will be used for adaptation, not for growth. Hereafter, the growth itself as one of captive fish adaptations and this process require a high energy. Moreover, even the figure of feeding rate remains stable and tend to be increased at the end of sampling in F_1 and F_3 treatments (Table 2), but specific growth rate figures show a decrease for F_1 and F_3 but increase for F_2 (Figure 1 and 2).

3.2.2. Growth margin

Absolute growth of individual eel *A. bicolor bicolor* was bigger than biomass absolute growth. This calculation of the margins from the initial individual weight of fish and at the end of the measurement. The calculation shows that F_1 gave highest individual absolute growth (25 g, 46%) followed F_2 (22 g, 44%) then F_3 (15 g, 27%) during 70 days of culture. Biomass absolute growth gave values of F_1, F_2, F_3 were 3,855 g (26%), 3,629 g (24%), and 2,834 g (18%) respectively during 70 days culture (Calculated from Table 4). No different between F_1 and F_2 ($\alpha > 0.05$) either in individual and biomass, but F_1 and F_2 have different to F_3 ($\alpha < 0.05$) also in individual and biomass. F_2 of shrimp feed gave more consistent in weight among tanks replication where the deviation was smaller than F_1 and F_3.

Table 4. Initial and t time weight of Individual (I_0, I_t), Biomass (B_0, B_t), Specific Growth Rate (%) of Individual (SGR_i) and Biomass (SGR_b) of *A. bicolor bicolor* fed with different diets during 70 days culture.

Treatments	Observed mean of growth margin					
	I_0 (kg)	I_t (kg)	% SGR_i	B_0 (kg)	B_t (kg)	% SGR_b
$F_{1.1.}$	0,056	0,078	0,48	15,025	17,481	0,22
$F_{1.2.}$	0,054	0,080	0,55	15,007	19,463	0,37
$F_{1.3.}$	0,053	0,079	0,57	14,988	19,641	0,39
25Mean	0,054	0,079	0,53	15,007	18,862	0,33
±*SD*	0,001	0,001	0,05	0,018	1,199	0,09
$F_{2.1.}$	0,049	0,072	0,53	14,843	18,490	0,31
$F_{2.2.}$	0,050	0,071	0,52	14,910	18,751	0,33
$F_{2.3.}$	0,051	0,074	0,52	14,833	18,233	0,29
22Mean	0,050	0,072	0,52	14,862	18,491	0,31
±*SD*	0,001	0,001	0,01	0,042	0,259	0,02
$F_{3.1.}$	0,055	0,068	0,30	16,595	18,908	0,19
$F_{3.2.}$	0,055	0,069	0,32	15,947	18,256	0,19
$F_{3.3.}$	0,054	0,073	0,43	14,915	18,796	0,33
15Mean	0,055	0,070	0,35	15,819	18,653	0,24
±*SD*	0,001	0,003	0,07	0,847	0,349	0,08

Total margin of individual-specific growth (*SGRi*) during 70 days culture F_1 shows the best mean value of 0.53±0.05% followed by F_2 (0.52±0.01%) and F_3 (0.35±0.07%) (Table 3). Moreover, the margin in biomass specific growth (SGR_b) tend to be smaller than *SGRi*. The trend of the growth pattern of SGR_b (total margin of biomass) was similar to *SGRi* either in statistical analysis. Table 3 shows that percent of SGR_b on F_1 show the best growth (0.33±0.09%) followed by F_2 (0.31±0.02%) and F_3 (0.24±0.08%).

The previous study by using the bigger size of *A, bicolor bicolor* (±70 g tail^{-1}) [17] shows the absolute biomass growth 52.94±6.17% with an initial density of ±20 kg m^{-3} and the *FR* of 1.89±0.06% d^{-1} during 100 days culture. While, present study using a smaller size of eel (±50 g tail^{-1}) absolute growth attained 24.42% (calculated from Table 4) with initial density of ±15 kg m^{-3} and the *FR* of 2.12±0.05% d^{-1} during 70 days culture. If the T_2 in a previous study [17] equal to F_2 in the present study in term of water volume and protein content of the feed, hence during 100 days culture, F2 should attain approximately 34.88% absolute growth (calculated from Table 4). For this reasons, bigger sizes of eel will grow faster than smaller. According to [3] stated that the FCR decreases as the feeding rate increases towards to the maximum rate, and the level of *FR, FCR* as well as growth rate varies according to especially the size of cultured organisms, species, and water temperature.

4. Conclusion

From 3 different feed treatments, the feeding rate of *A. bicolor bicolor* do not correlate with protein content but apparently, follows the degree of lipid in the foodstuff. Higher lipid content gave slightly lower of feeding rate, on the other hand, a higher protein on feed gave higher growth performance.

The initial weight of eel seeds gave a figure of survival rate followed by biomass growth; smaller seed inoculated will gave smaller survival rate and smaller biomass growth. The calculation of efficiency value of this study represents by biomass density per liter of water volume, have a value of 3.8 g l^{-1} during 70 days culture.

Acknowledgements
This work was financially supported by Diponegoro University, Funding source of PNBP DIPA UNDIP fiscal year of 2017. Authors thank every single person who helps and involve for this study especially V Devinta, S Nabilla.

References

[1] Bureau D P, Hua K and Cho C Y, 2006 *J Aquac. Res.* **37**, 1090–1098.
[2] Clark A and Johnston N M 1999 *J. of Animal Ecology* **68** 893-905.
[3] De-Silva S S and Anderson T A 1995 Fish Nutrition, 1st Ed.Chapman and Hall pp 32-35.
[4] Fahmi M R 2010 Phenotypic Plastisity Kunci Sukses Adaptasi Ikan Migrasi: Studi Kasus Ikan Sidat *(Anguilla sp.) In*: Fahmi MR (ed) *Prosiding Forum Inovasi Teknologi Akuakultur*; Balai Riset Budidaya Ikan Hias, Desember 2010. Depok
[5] Haryono, 2008. Sidat, belut bertelinga: potensi dan aspek budidayanya. Pusat Penelitian Biologi-LIPI. Fauna Indonesia 8 (1) 22-26.
[6] Heinsbroek L T N, 1987. Effects of body weight, feeding level and temperature on energy metabolism and growth in fish. In: Verstegen M W A, Henken A M (eds.), Energy Metabolism in Farm Animals. Martinus Nijhoff Publishers, Dordrecht, pp. 478–500.
[7] Heinsbroek L T N, Tijssen P A T, Flach R B, de-Jong G D C 1993. Energy and nitrogen balance studies in fish. In: Kaushik, SJ, Luquet, P (eds.), Fish Nutrition in Practice. INRA Editions, Versailles, pp. 375–390.
[8] Heinsbroek L T N, P L A Van-Hooff, W Swinkels, M W T Tanck, J W Schrama and J A J Verreth 2007 *Aquaculture* **267**: 175–187
[9] Hill D R and Webster T 2008 High pH Inhibit Nitrifying Bacteria *Des Moines Water Works* (www.dmww.com) *American Water Works Association*. Opflow July 2008
[10] Holmgren K, Wickstrô M H and Fredg A K 1991 Difficulties when estimating growth of eel in a mesoscale experiment *European Inland Fishery Advisory Commission (EIFAC) Working Party on Eel*, Dublin, Ireland, 20-25 May 1991, 1-19.
[11] Luo M, Guan R, Li Z and Jin H 2013 *Aquaculture* **400**-401: 61-64.
[12] May D and S Marshall 2008. The 'West Sutherland Elver Survey' project, Communities Project for Highland Biodiversity Ref: CPHB12
[13] Pawson M, B Knights, M Aprahamian, R Rosell, T Williams and M Beveridge 2005 UK – Report on Eel Stocks and Fisheries in the United Kingdom and Northern Ireland, in 2004. Report ICES WGEEL 99-117pp.
[14] Ruohonen K, Simpson S J, Raubenheimer D 2007 *Aquaculture* **267**, 147–156.
[15] Secor D H, J E Baker, W E Morrison and J C Steinbacher 2002 Ecology and Contamination of the Hudson River American Eel, DH Secor and JE Baker, Co-P.I.s. *UMCES Tech. Series No. TS-367-02-CBL.*
[16] Setijanto, Yuwono E, Sukardi P and Sulistyo I, 2003 Study on feeding behavior of eels and the larvae occurrence in Segara Anakan. Proc. of Tropical Eel Fishery Resources, BPPT, Ministry of Research and Technology, Indonesia.
[17] Taufiq-Spj N, S Sunaryo, A Wirasatria, D N Sugianto 2017 *IOP Conf. Series*: Earth and Environmental Science **55** 012015 doi:10.1088/1755-1315/55/1/012015
[18] Tesch E W 1977 The Eel: Biology and Management of Anguillid Eels. (*Chapman and Hall*), London.

Impact of Monsoon to Aquatic Productivity and Fish Landing at Pesawaran Regency Waters

Kunarso[1,3]*, Muhammad Zainuri[1,3], Raden Ario[2], Bayu Munandar[1], Harmon Prayogi[1]

[1] Department of Oceanography, Faculty of Fisheries and Marine Science, Diponegoro University, Jl. Prof. H. Soedarto,SH, Tembalang, Semarang 50275, Indonesia
[2] Department of Marine Science, Faculty of Fisheries and Marine Science, Diponegoro University, Jl. Prof. H. Soedarto,SH, Tembalang, Semarang 50275, Indonesia
[3] Center for Coastal Rehabilitation and Disaster Mitigation Studies (CoREM), Diponegoro University, Center of Excellence Science and Technology (PUI), Indonesia

*Corresponding author E-mail: kunarsojpr@yahoo.com

Abstract. Monsoon variability influences the productivity processes in the ocean and has different responses in each waters. Furthermore, variability of marine productivity affects to the fisheries resources fluctuation. This research has conducted using descriptive method to investigate the consequences of monsoon variability to aquatic productivity, sea surface temperature (SST), fish catches, and fish season periods at Pesawaran Regency waters, Lampung. Variability of aquatic productivity was determined based on chlorophyll-a indicator from MODIS satellite images. Monsoon variability was governed based on wind parameters and fish catches from fish landing data of Pesawaran fish market. The result showed that monsoon variability had affected to aquatic productivity, SST, and fish catches at Pesawaran Regency waters. Maximum wind speed and lowest SST occurred twice in a year, December to March and August to October, which the peaks were on January (2.55 m/s of wind speed and 29.66°C of SST) and September (2.44 m/s of wind speed and 29.06°C of SST). Also, Maximum aquatic productivity happened on January to March and July to September, which it was arisen simultaneously with maximum wind speed and the peaks was 0.74 mg/m^3 and 0.78 mg/m^3, on February and August respectively. The data showed that fish catches decreased along with strong wind speed and low SST. However, when weak wind speed and high SST occurred, fish catches increased. The correlation between Catch per Unit Effort (CPUE) with SST, wind speed, and chlorophyll-a was at value 0.76, -0.67, and -0.70, respectively. The high rate fish catches in Pesawaran emerged on March-May and September-December.

Keywords: Monsoon variability, aquatic productivity, Pesawaran Regency, sea surface temperature, fisheries resources

1. Introduction

Pesawaran Regency waters, located contiguity to Sunda Strait, has considerable marine fisheries resources. The area of this waters is supposed around 689 km^2 or 68.900 Ha with 96 km of coastline and 50 m of mean water depth. The waters condition has fish catches prospectively, but it is not employed yet. Until now, fish demand consumed by tourists is imported from outside Pesawaran Regency.

The problem has been occurred in fish-catching industries to increase the catches because there are difficulties to determine the lowest fish catches season and to find potential fishing ground [1]. Moreover, it is caused by Unpredicted annual climate variability conditions [2]. Due to the limitation of understanding about these problems, fish catches activity still has some drawbacks, such as ineffectiveness, inefficiency, also time and fuel consuming.

Annual climate variability caused by monsoon variability which the effect to aquatic productivity has been investigated by some researchers including [3] [4] [5] and [6]. Correlating to biological conditions of the ocean, monsoon variability, particularly, can trigger the change of marine productivity parameter including the change of chlorophyll-a indicator [7] [8] [9].

The different responses of each waters to monsoon variability become the problem of this research. In Pesawaran Regency, there were not researchers studying the responses of this waters to monsoon variability yet. Therefore, this research was focused to investigate Pesawaran waters responses to monsoon variability. The research area at Pesawaran waters and its round are described in Figure 1.

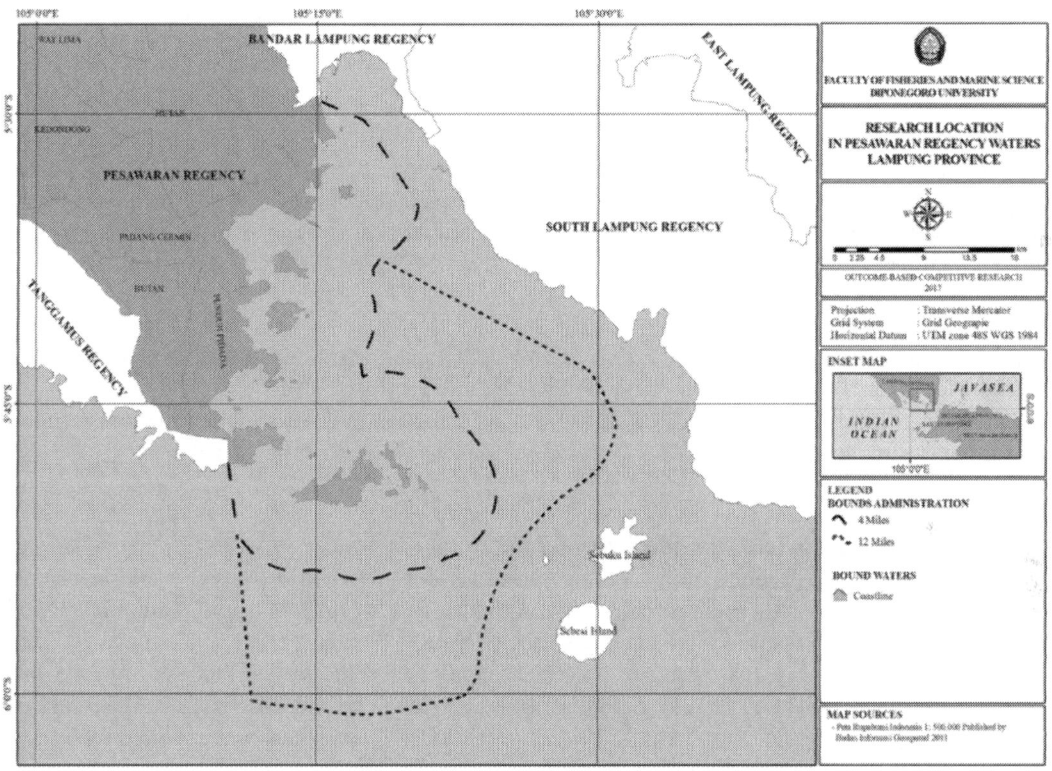

Figure 1. The research location at Pesawaran waters and its round. The waters bound 4 miles and 12 miles are marked by broken line and dotted line.

2. Materials and methods

The primary materials in this research were satellite-based remote sensing data [10] [11] which were MODIS data, spatial distribution of chlorophyll-a and monthly sea surface temperature (SST), from Aqua and Terra satellite, and monsoon variability including wind parameters, wind speed and direction, also fisheries data. Because monsoon variability was considered per month, the data used in the research were monthly data.

This research employed descriptive method. Firstly, it was done by considering monthly climatologist mean of monsoon variability in a year; then it was correlated to variability of productivity using chlorophyll-a indicator at Pesawaran waters. Secondly, SST data were described to know what phenomena was occurring at the time of research. Finally, all parameters were analysed to understand the influences to the lowest and the highest catches based on monthly fish landing data. To achieving the clarity in this research, problem solving methods were divided into three steps.

2.1. Primary and secondary data

Primary and seconday data were gathered from websites, national governments, and field survey including:

1. Monsoon data, wind speed and direction, collected from European Centre for Medium-Range Weather Forecasts (ECMWF), downloaded from http://www.ecmwf.int, have 1/12 degrees (9.17 km) spatial resolution. These data were analysed using SeaDAS 7.0 and mapping software.
2. Monthly SST and chlorophyll-a data were collected from MODIS (Moderate Resolution Imaging Spectroradiometer) image that was downloaded from: http://www.oceancolor.gfsc.nasa.gov in NetCDF (*Network Common Data Form*) format. The data, 4 km spatial resolution, were analyzed in SeaDAS 7.0.
3. Time series fish landing data from 2014 to 2016 were gathered from Pesawaran Regency fish market that were collected by Marine and Fisheries Resources Control Unit, Lampung. From these data, we obtained CPUE (Catch Per Unit Effort) data.

2.2. MODIS satellite images, chlorophyll-a and SST data

Chlorophyll-a and SST data from MODIS satellite image were level-3 data in NetCDF format which were resulted from Aqua and Terra satellite. These data, spatial distribution of chlorophyll-a and SST at Pesawaran waters, were extracted to get text file (.txt) using SeaDAS 7.0, and the mean of monthly climatologist from 2001 to 2016 were calculated and analyzed the correlation to other data.

2.3. Wind speed and direction

Wind data was collected from ECMWF in NetCDF format, and SeaDAS 7.0 was used to extract the data in research area on text file which were wind vector direction, u (west-east) and v (north-south). Then, from these data, we got total wind speed and direction that were plotted in wind rose to get the dominant direction.

3. Results and discussions

3.1. Monsoon influences to aquatic productivity and SST

The variability of aquatic productivity at Pesawaran areas correlated to monsoon, wind speed and wind direction. There was two peaks of wind speed that occurs in different period. In first period, maximum wind speed was occurred from December to March. In second period, it was occurred from August to October (Figure 2). Increasing in wind speed influenced the productivity of waters which can be seen from strengthening of chlorophyll-a and SST in west and east season. The peak of increasing chlorophyll-a concentration and lowest level of SST was happened collectively with maximum wind speed in January, February, August, or September (Figure 2 and Figure 3).

Maximum wind speed at Pesawaran waters was occurred twice which were in January about 2.55 m/s and September about 2.44 m/s, but minimum wind speed was occurred in April and November about 1.4 m/s and 1.7 m/s respectively (Figure 2). Also, maximum chlorophyll-a concentration was occurred twice which were in February about 0.74 mg/m^3 and in August about 0.78 mg/m^3, but the minimum was occurred in December and May about 0.44 mg/m^3 and 0.46 mg/m^3 respectively (Figure 4).

According to the explanation before, primary productivity based on chlorophyll-a concentration at Pesawaran waters in east season is higher than in west season. It is expected because of two factors. Firstly, it is caused by mixing processes that increase along with wind speed becoming greater, and it was also stated by [12]. This is in accordance with the research [13] which explained that winds and surface chlorophyll-a had high correlation in shallow mixed layer. Secondly, it is correlated with waters turbidity which is higher in east season than west season, so light intensity goes through the waters more frequently [14]. Thirdly, there is an upwelling phenomenon from the southeast Pesawaran waters centred in the west coast of South Lampung Regency (Figure 6) explained that limiting factor bounding primary productivity was the light intensity that must be sufficed. In east season light intensity through Pesawaran waters is higher than in west season. Consequently, it caused photosynthesis processes more optimum in east season, so chlorophyll-a intensity in that season higher than another season. The correlation between wind speed and chlorophyll-a showed value of 0.48, and it explained that monsoon variability intercorrelated to chlorophyll-a variability in moderate category.

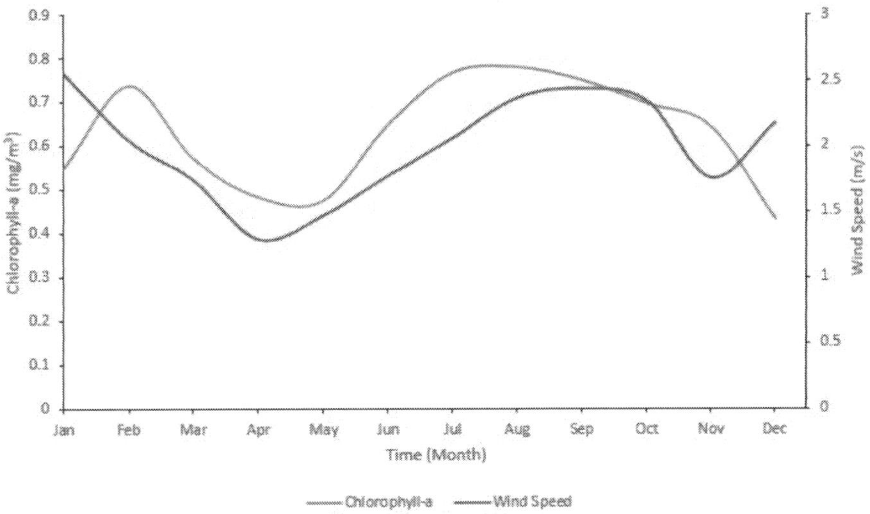

Figure 2. The average variability of wind speed in m/s and chlorophyll-a in mg/m^3 at Pesawaran waters from 2014 to 2016.

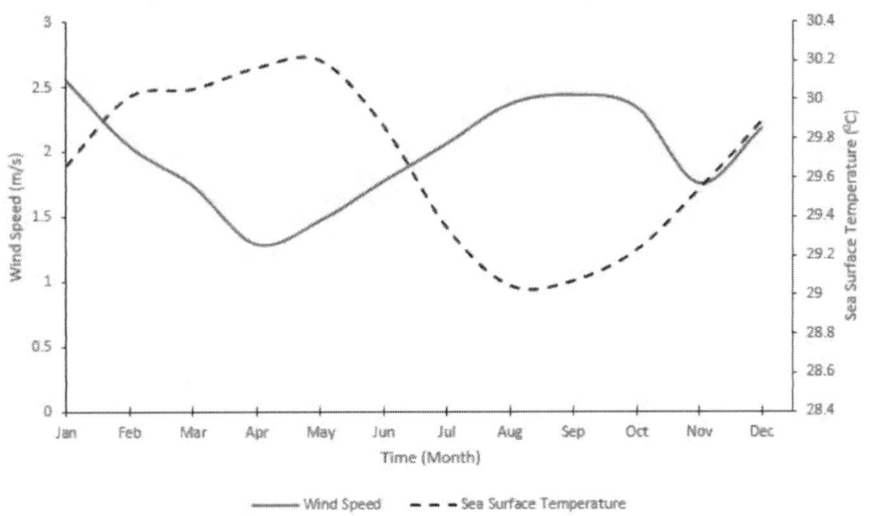

Figure 3. The average variability of wind speed in m/s and SST in ^{0}C at Pesawaran waters between 2014 and 2016.

In spite of increasing the aquatic productivity, monsoon also influencing SST [15] [16] [17], in this research, the monsoon decreased SST. Figure 3 shows that SST and wind speed have opposite correlations. In other word, maximum peak of wind speed is equal to minimum peak of SST. The lowest SST was happened in west season around January and in east season around September. Cold SST in west and east seasons are considered due to three factors. Firstly, it is the mixing processes that occur due to the increase of wind speed, so there is mixing process between cool water mass in the lower layer and hot water in the upper layer and it also described by [12] and [18]. Secondly, strong wind speed makes the water surface disturbed [19]. Unsteady sea surface reduces the process of penetrating sunlight into the waters, which consequently reduce water temperature. On the contrary, increased temperatures occur when the water level is calm as the penetration of sunlight into the water column increases [20]. Finally, increasing in wind speed is expected by improving the advection process (heat-to-air release) and evaporation processes that also releasing heat from the water column and this is also argued by [21]. The correlation between wind speed and SST shows a value of -0.74, so it proof that wind variability strongly influences the variability of SST in the opposite relationship.

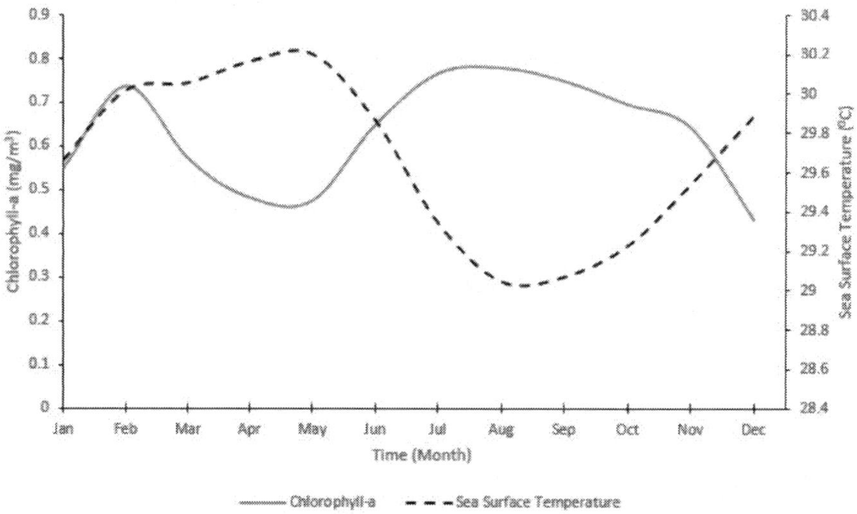

Figure 4. The average variability of chlorophyll-a and SST at Pesawaran waters between 2014 and 2016.

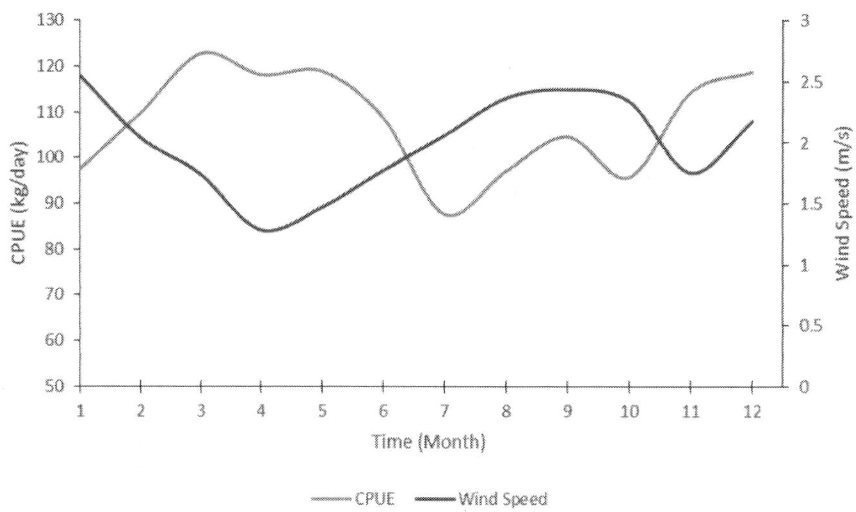

Figure 5. The average variability of wind speed from 2014 to 2016 and CPUE from 2014 to 2015 at Pesawaran waters.

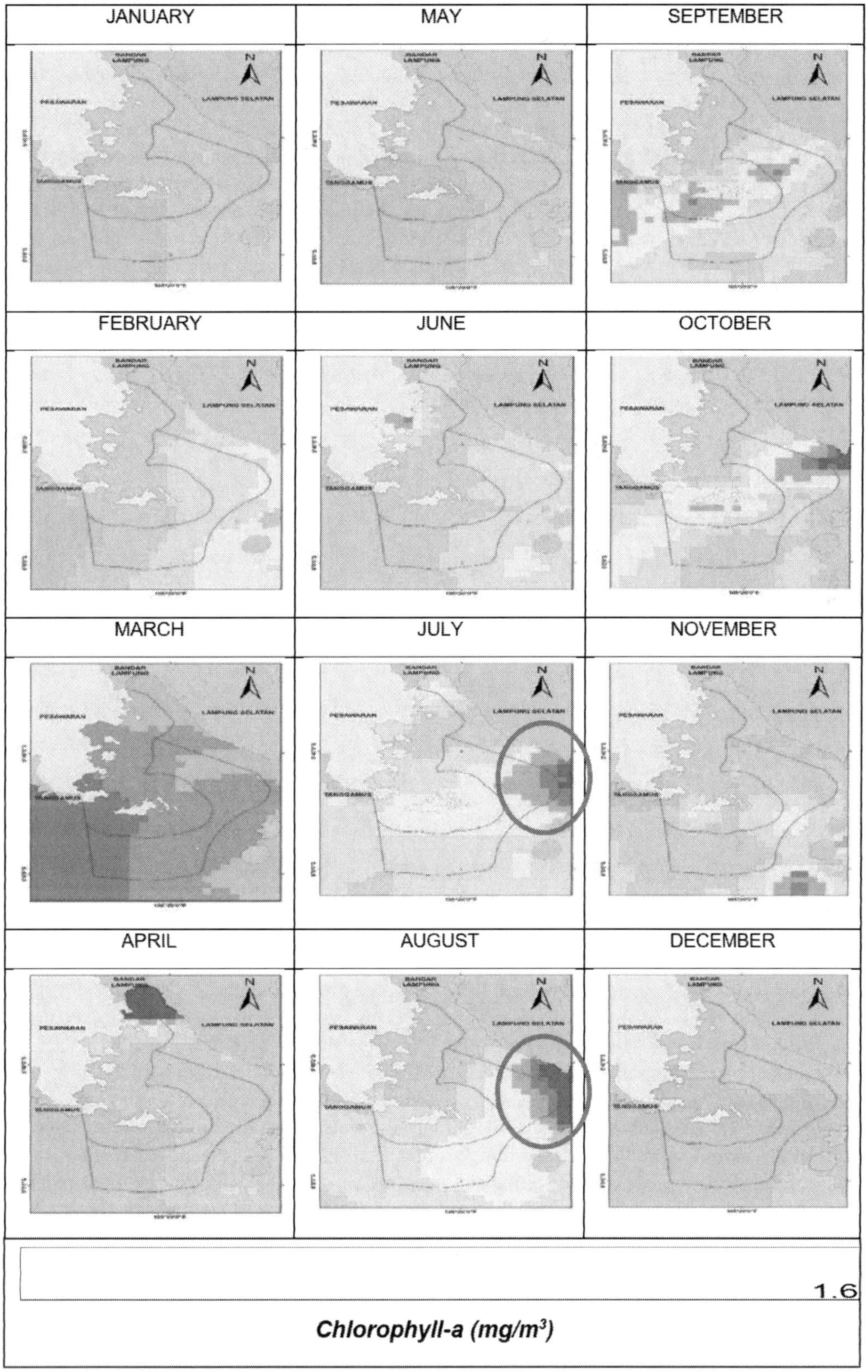

Figure 6. Monthly chlorophyll-a distribution (mg/m^3) at Pesawaran Regency Waters from 2002 to 2016, circled red is considered to be an upwelling centre.

3.2. Monsoon influences to fisheries productivity

The influence of monsoon variability is not only on the increase of aquatic productivity, but also on fishery productivity. Strong wind variability associated with SST variability further affects the productivity of fisheries. The wind velocity and chlorophyll-a concentration show the inverse relationship with the CPUE (Figure 5 and Figure 7), whereas the SST with CPUE shows a proportional relationship (Figure 8 and Figure 9). Figure 9 show that positive coherence between SST and CPUE (indicator the direction of the arrow to the right) very strong at the coherence value 0.8-1.0 in the 9th to 12th months. In general, there is lag times about 2 months between the peak of temperature and the peak of CPUE. The linkage can be seen when high wind speeds and low temperatures then the fishing results decreased. Meanwhile, when the wind speed is low and high temperatures then the fish catch increases. The correlation between temperature, wind, chlorophyll-a and CPUE can be categorized as strong including the correlation between wind and CPUE shows the value of -0.67. The correlation between SST and CPUE shows a value of 0.76, while the correlation between chlorophyll-a and CPUE shows a value of -0.70. The association of SST with CPUE is of course limited to the limits of fish tolerance that are subjected to capture.

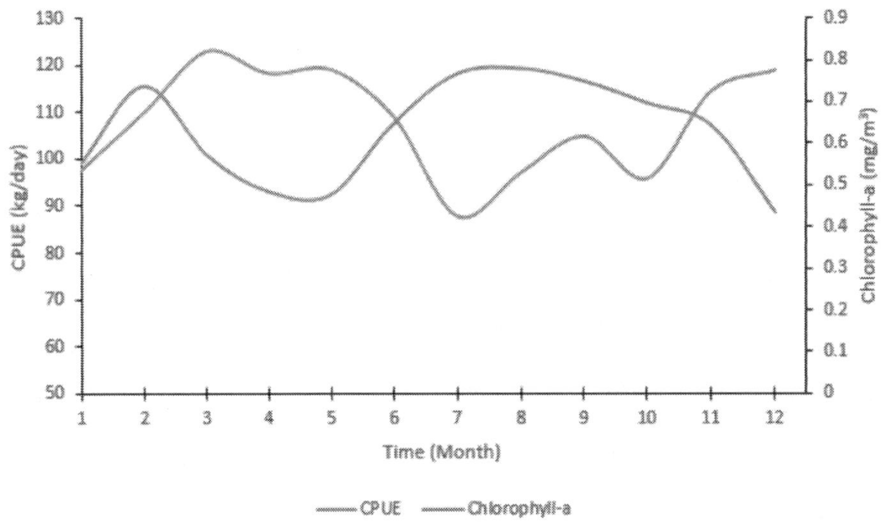

Figure 7. The average variability of chlorophyll-a from 2014 to 2016 and CPUE from 2014 to 2015 at Pesawaran waters.

Reduced wind velocity and increased SST associated with increased CPUE, in [22] SST had significant correlation to albacore CPUE, allegedly related to the optimization of the catching process by purse seine fishing gear and character of the fish target that favored warm waters. When the wind speed is weak in the waters in a calm condition, many fishermen get the process of catching with purse seine fishing gear. Purse seine fishing gear is assisted by light to attract the attention of the fish. When the condition of calm waters is expected to light into the waters become optimum that will attract fish, so that results more fish catches. Water temperatures that tend to be warmer when the wind speed is weak is presumably still within the optimum temperature tolerance. The dominant of fish caught in Pesawaran waters is mainly tuna, skipjack, shrimp, needle fishes and swordfish. These kind of fish have optimum temperature ranges from 29-30 ^{0}C which is in accordance with the Pesawaran water. The optimum temperature range for fish, especially small pelagic ranges from 29-30 ^{0}C [20].

IOP Conf. Series: Earth and Environmental Science **116** (2018) 012021 doi:10.1088/1755-1315/116/1/012021

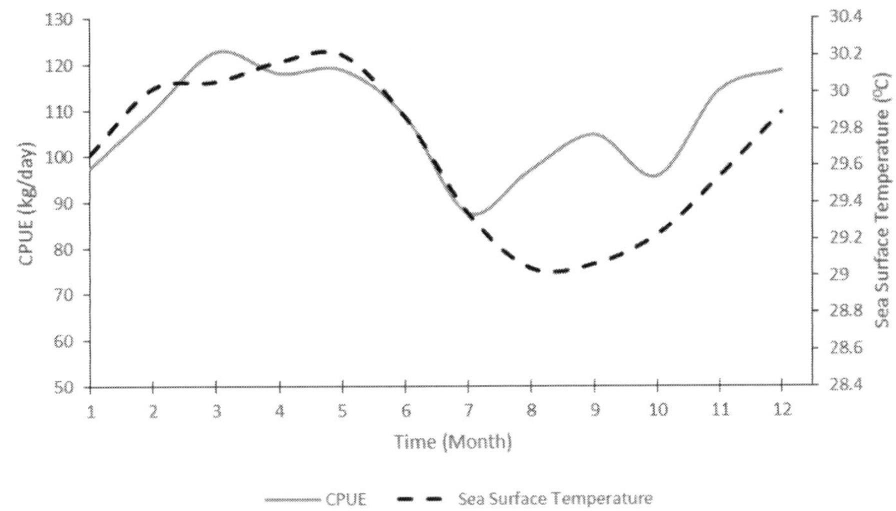

Figure 8. Average SST variations from 2014 to 2016 and CPUE from 2014 to 2015 at Pesawaran waters.

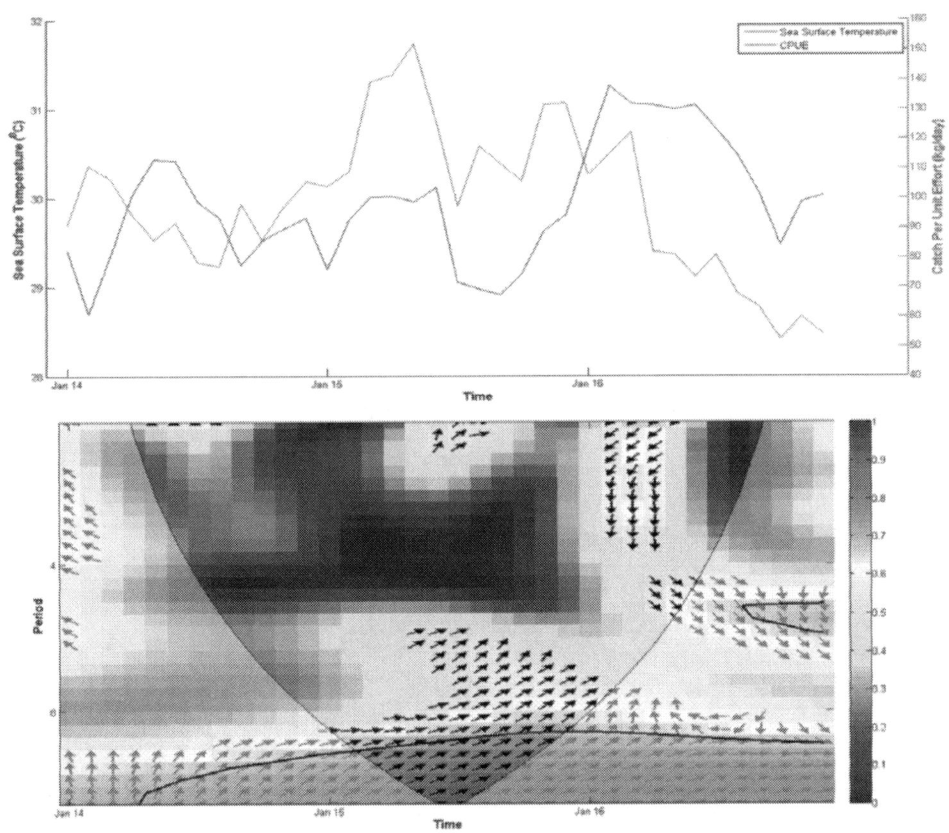

Figure 9. Coherence between SST and CPUE, show that coherence very strong with value of 0.8-1.0 in the 9 - 12 months, and there were lag times about 2 months between the temperature peak and the CPUE peak

Fishing season in Pesawaran waters generally occurs in March-May and October-December. The peak of the catches occurred in May and December (Figure 5). The highest catches occur to coincide with the maximum SST and minimum wind speed (Figure 5 and Figure 7). Increasing aquatic productivity in west season (the highest season) either in January or February and in east season occurred either in July or August are followed by increasing of fisheries productivity in 1 – 4 months after the highest peak of chlorophyll-a (Figure 6). This finding was unexpected and suggest that fish season at Pesawaran waters generally occur in 1 – 4 months after the highest peak of chlorophyll-a both in transitional season I (March-May) and transitional season II-West (September-December). As a result, the highest fish season occur in March and December. The fish season found is almost the same as that found by [21] in the Java sea.

4. Conclusions

Monsoon variability, that is wind speed and direction, effect on productivity, SST and fish catches at Pesawaran Lampung waters. Strong winds and cold SST occur twice a year, from December to March and from August to October, the peak is in January (wind speed 2.55 m/s, SST 29.66 ^{0}C) and September (wind speed 2.44 m/s, SST 29.06 ^{0}C). High productivity in Pesawaran waters also occurs two times a year with almost the same time with strong wind speeds, ranging from January to March and July to September, peak around February (0.74 mg/m^3) and August (0.78 mg/m^3).

The variability of fish catches based on landing data at Pesawaran waters appears to be more closely related to wind and SST variability than with aquatic productivity, when strong winds and low catchments of fish catches decreased, and when winds were weak and high catchments of fish catches increased. The measured temperature range is presumed to be within the optimum tolerance range of catch fish in Pesawaran waters. Correlation (r) between CPUE fish with SST, wind, and chlorophyll-a were at value 0.76, -0.67, and -0.70, respectively. Fishing season in Pesawaran waters generally occurs in March-May and September-December and the peak of fishing season generally occurs in March and December coincides with the hottest SST and the weakest wind speed.

References

[1] Kunarso 2014 Pengaruh monsun, El Niño-southern Oscillation dan Indian Ocean Dipole terhadap waktu dan daerah penangkapan ikan tuna di Samudra India bagian timur. Disertation, Program Studi Sains Kebumian, Fakultas Ilmu dan Teknologi Kebumian, ITB, Bandung
[2] Brander K M 2007 *Proceedings of the National Academy of Sciences of the United States of America* **104**(50) 19709-14
[3] Pramanik D S and S S Mohanty 2016 *International Journal of Fisheries and Aquatic Studies* **4**(2) 331-6
[4] Basit A, M R Putri and W M Tatipatta 2012 *Mar. Res. Indonesia* **37**(1) 47-56
[5] Currie J C, Lengaigne M, Vialard J, Kaplan D M, Aumont O, Naqvi S W A and Maury 2013 *Biogeosciences* **10** 6677-6698
[6] Amri K, Suman A, Irianto H E, Wudianto W 2015 *Indonesian Fisheries Research Journal* **21**(2) 75-90
[7] Susanto D and J Marra 2005 *Journal Oceanography* **18**(4) 124-7
[8] McPhaden M J 2012 *Geophys. Res. Lett.* **39** L09706
[9] Syamsuddin M L, Saitoh S I, Hirawake T, Bachri S, Harto A B 2013 *Fishery Bulletin* **111**(2) 175-188
[10] Tseng C T, Su N J, Sun C L, Punt A E, Yeh S Z, Liu D C and Su W C 2013 *ICES Journal of Marine Science* **70**(5) 991-999
[11] Syah A F, Saitoh S I, Alabia I D and Hirawake T 2017 *IOP Conf. Series: Earth and Enviromental Science* **54** 012074
[12] Wyrtki K A. 1961. Naga Report Volume 2: Physical oceanography of the Southeast Asean Waters. Handbook. The University of California, California, 195 p
[13] Kahru M, S T Gille, R Murtugudde, P G Strutton, M Manzano-Sarabia, H Wang and B G Mitchell 2010 *Journal of Geophysical Research* **115** C12040

[14] Thurman H V. 1991. Introductory oceanography, sixth edition. Macmillan Publishing Company. New York, 526p
[15] Wündsch M, Biagioni S, Behling H, Reinwarth B, Franz S, Bierbaß P, Daut G, Mäusbacher R and Haberzetti T 2014 *The Holocene* **24**(12) 1743-1756
[16] Fan L, Shin S I, Liu Z and Liu Q 2016 *Climate Dynamics* **47**(7-8) 2501-2514
[17] Pradhan M, Yadav R M, Dandi A R, Srivastava A, Phani M K and Rao S A 2016 *International Journal of Climatology* **37**(5) 2280-2294
[18] Nurdin S, M A Mustapha and T Lihan 2013 *AIP Conference Proceedings* **1571** 466-72
[19] Pelikán P and Marková J 2013 *Acta Universitatis Agriculturae et Silviculturae Mendelianae Brunensis* **LXI (203)** 1823-1828
[20] Rasyid J A 2010 *Torani* **20**(1) 1-7
[21] Sachoemar S I, T Yanagi and R S Aliah. 2012 *Mar. Res. Indonesia* **37**(2) 75-87
[22] Lan K W, Evans K and Lee M A 2013 *Climatic Change* **119**(1) 63-77

IOP Conf. Series: Earth and Environmental Science **116** (2018) 012022 doi:10.1088/1755-1315/116/1/012022

Business Profile of Boat Lift Net and Stationary Lift Net Fishing Gear in Morodemak Waters Central Java

Trisnani D Hapsari*, Bogi B Jayanto, Aristi D P Fitri and I Triarso

Department of Capture Fisheries, Faculty of Fisheries and Marine Science, Diponegoro University, Jl. Prof. H. Soedarto, SH, Tembalang, Semarang 50275 Indonesia

*Corresponding author Email : trisnanihapsari@live.undip.ac.id

Abstract. Lift net is one of the fishing gears that is used widely in the Morodemak coastal fishing port (PPP) for catching pelagic fish. The yield of fish captured by these fishing gear has high economic value, such as fish belt (*Trichiurus* sp), squids (*Loligo* sp) and anchovies (*Stelophorus* sp). The aims of this research were to determine the technical aspects of boat lift net and stationary lift net fishing gear in Morodemak Waters Demak Regency; to find out the financial aspect of those fishing gears and to analyze the financial feasibility by counting PP, NPV, IRR, and B/C ratio criteria. This research used case study method with descriptive analysis. The sampling method was purposive sampling with 22 fishermen as respondents. The result of the research showed that the average of boat lift net acceptance was Rp 388,580,000. The financial analysis of fisheries boat lift net with the result of NPV Rp 836,149,272, PP 2.44 years, IRR value 54%, and B/C ratio 1.73. The average of stationary lift net acceptance was Rp 27,750,000. The financial analysis lift net with the result of NPV Rp 37,937,601; PP 1.96 years, IRR value 86%, and B/C ratio 1.32. This research had a positive NPV value, B/C ratio >1, and IRR > discount rate (12 %). This study concluded that the fishery business of boat lift net and stationary lift net in Morodemak coastal fishing port (PPP) was worth running.

1. Introduction

Lift net is one kind of fishing gear, which is quite widely used in Indonesia. The large number of lift net usage cannot be separated from regional development, technological ease, low investment level, and one day fishing method of catching. In addition to these technical matters, the high use of the lift net also includes the importance of the catching unit of the lift net for the pelagic fish.

According to the decision of the Minister of Marine and Fisheries of Republic of Indonesia No. 06/MEN/2010 lift net is fishing tool group made of rectangular-shaped nets fitted frame bamboo or other materials as order, operated in a way to be immersed in the waters while setting and lifted to the surface while hauling which is equipped with or without a light-collecting fish, to catch pelagic fish (SNI 7277.9:2008). The fishing operation was limited by lunar cycle [1]. The *FAO* stated that lift nets are active a few meters below the surface; when light attraction is used, the active depth depends on the attraction power of the light [2]. The light functioned as attractant to various types of plankton-eating organisms [3].

The boat lift nets are fishing gear operated by means of descending into the water column and being lifted back after many fish on it, in operation using a boat to move to locations that are predicted many fish. This boat lift net consists of two boats on the front of the back linked with two bamboo stems so that the square as a place to hang the lift net's nets. In addition there is still a boat lift net that uses only one boat only.

Stationary lift net is a lift net mounted by placing the body frame into the water, so the position of the lift net can be planted only once and can't moving during the fishing season. Fishing operation of the stationary lift net is done at night. Most use light coming from petromax lamp, although some use electric lights.

Boat lift net and stationary lift net are fishing equipment that are mainly caught pelagic fish and quite plenty observed in Morodemak waters. Based on survey results, there are 180 units of fishing

gear lift net, which consist of 69 units of boat lift net and 111 units of stationary lift net. It is necessary to observe the income from each fishing business, given the investments incurred for the two different types of fishing gear, despite having the same fishing target. The result of financial aspect analysis will results in feasibility of catching business in the future so that it can be cultivated in a sustainable manner with good planning.
The objectives of this study were to evaluate the technical tools of boat lift net and stationary lift net in Morodemak Waters; to analyze the amount of capital, income, and profits from boat lift net and stationary lift net; and to analyze the financial feasibility of fishing effort by boat lift net and stationary lift net.

2. Materials and Methods

The research method used in this research is case study method with descriptive analysis. The subjects studied in this research are technical aspect and financial aspect from fishing effort by using fishing gear of boat lift net and stationary lift net in Morodemak waters of Demak Regency. Sampling method used in this research is purposive sampling. The population used for this sampling is the fisheries business unit of the boat lift net and stationary lift net in Morodemak waters of Demak Regency.

The number of respondents used in this study can be calculated using the cross sectional formula. The population size of boat lift net fisheries business unit in Morodemak waters of Demak Regency is 69 units. Fishing gear of boat lift net can be taken sample was 11 respondents. The population size from fisheries business unit of stationary lift net in Morodemak waters of Demak Regency is 111 units. The fishing gear can be sampled by the formula was 11 samples of respondants.

Analysis of the feasibility of lift net fisheries business in Morodemak waters of Demak Regency, used some of the financial approach that includes : NPV (Net Present Value), B / C Ratio (Benefit Cost Ratio), IRR (Internal Rate of Return), and return period analysis Capital (Payback Period).

1. NPV (Net Present Value)

NPV is the difference between the Present Value of the investment and the present value of net cash receipts (cash flow operations and cash flow terminals) in the future. To calculate the present value it is necessary to determine the relevant interest rate. NPV analysis can be known by the formula [4] :

$$NPV = \sum_{t=1}^{n} \frac{CFt}{(1+i)^t} - Co$$

CFt	:	cash flow per year in period t
Co	:	initial investment in year 0
i	:	interest rate (discount factor)
t	:	year to-
n	:	number of years

If :
NPV > 1; the business is feasible
NPV = 0; the business can be feasible
NPV < 1; the business is not feasible.

2. B/C Ratio (Benefit Cost Ratio)

$$B/C \quad Ratio \quad = \frac{total \quad revenew}{total \quad cost}$$

The criteria used are :
B / C Ratio > 1; the business generates profits so feasible to continue.
B / C Ratio = 1; the business is not profit and no loss (break even)
B / C Ratio < 1; the business suffered losses so it is not feasible to continue [5].

3. IRR (Internal Rate of Return) [6]

$$IRR = i_1 + \times (i_2 - i_1)]$$

i_1	:	The 1st interest rate
i_2	:	The 2nd interest rate
NPV1	:	NPV at the 1st interest rate
NPV2	:	NPV at the 2nd interest rate

4. Analysis of Capital Return Periode (Payback Periode)

Payback period in capital budgeting refers to the period of time required to recoup the funds expended in an investment, or to reach the break-even point [7]. It is used to know the duration of investment capital rotation used in doing business or in other words to know the time that can be used to recoup the investment expenditure by using the profit as comparison.

3. Results and Discussion

3.1. Technical Aspects of Boat Lift Net

The construction of a boat lift net in Morodemak waters of Demak Regency consists of wooden frames, nets, motorized boats as a means of transportation at sea, as well as tools to ease the operation of this boat lift net that is the spade, lamp and roller that serves to lower or lift the net. Boats used in this boat lift net only use one boat only. The position of the boat is in the midst of the construction of the lift net framework. The boat is 9 GT, the length of the boat L is 23 m, its keel size is 15 m, the width (B) of the boat is 3 m, and the inner size (D) of the boat is 2.5 m.

The frame material on the boat lift net is made of bamboo. The length of deck in the boat lift net (P X L) is 22.5 m x 22.5 m. The amount of material on the boat lift net construction is 200 bamboos. In the center of boat lift net there is a house with size (P x L x T) is 6 m x 2.2 m x 2.4 m which serves as a resting place, where electric generators, fuel and other fishing equipment.

The net material on the boat lift net is made from polyethylene. The size of the net (P X L) is 45 m x 45 m. The net is square with mesh size of 0.6 - 1 cm. The net is strung one by one to form a large square. The length of the towing rope is 60 m for one part, with a diameter of 1.8 cm made from polyethylene. Its rotator made from wood with a length of 1 cm. The nets vary in depth from 18 – 30 meters.

The type of lamp used by the boat lift net as fish attractor is a mercury lamp with a 400 watt and the brand is Philips. The number of lamps used ranged from 40 to 70 pieces. In addition to the lights, the boat lift net uses a rip and roller to pick up the catch.

The fishermen of boat lift net usually leave in the afternoon at around 5 p.m. to go to the fishing ground or the location of the boat lift net and to reach the location of the boat lift net about 2.5 hours or about 6 miles. Appointment of nets in one operation of fishing gear usually fisherman do every 3 hours once and finish at 4 a.m. High average catches before midnight might be generally caused by the domination of positive phototaxis fish caught that came to lighting area because the fish likes approaching the light, not for searching food [8].

Figure 1. Boat lift-net in Morodemak waters

3.2. Technical Aspects of Stationary Lift Net

The construction of the stationary lift net in Morodemak waters of Demak Regency consists of nets, building, spade, roller and lamps. Frame of stationary lift net made from bamboo with the size is 9 m x 9 m. The height of the stationary lift net from the sea surface is about 3.5 meters. Then the house is located on the floor of the stationary lift net that serves as a place for catching activities and to observe the movement of fish gathered around the lights, inside the house there are fisherman's supplies. The house at the stationary lift net is about 1 x 2 m.

Other construction is net's stationary lift net made from polyethylene (PE). Nets has a mesh size of 0.5 - 1 cm and the length of the nets is 8.5 m x 8.5 m. The depth of the nets is about 4.8 meters, while in Muncar waters Banyuwangi (East Java) and Karangantu waters Banten (West Java), lift-net operated at 11m – 13m depth [9,10] and in Muncar waters (Pang pang strait) to obtain optimal catch results then suggested water depth is ≥ 17 m [11].

Stationary lift net using lights as a tool to stimulate or attract the attention of fish to gather in the light of the lamp. The type of lamp used is petromax lamp. The number of lamps used ranged from 2 to 3 pieces. In addition to the lights, the stationary lift net using a spade and roller or pulling tool used is made from strappy bamboo. There are two ropes that are used that is a small rope to install the lights and a large rope to towing nets with a diameter of 1.24 cm.

The fishermen of stationary lift net usually leave in the afternoon at around 5 p.m. to go to fishing ground or location of the stationary lift net, the fisherman use fishing boat to reach the location of the stationary lift net, it takes about 30 minutes to 1 hour or about 1.5 miles.

Figure 2. Stationary lift-net in Morodemak waters

Catch Composition of Lift Net

Catch composition of lift net in Morodemak waters was dominated by small-pelagic fishes such as greenback mullet (*Mugil dussumieri*), white shrimp *(Litopenaeus vannamei),* indian squid (*Loligo duvauceli*), mud crab (*Scylla serrata*), anchovy (*Stolephorus spp.),* goldstripe sardinella (*Sardinella gibbosa*), short-bodied mackerel (*Rastrelliger negletus*) and yellowstripe scad (*Selaroides leptolepis*). From the catch composition, it is known in the catchment boat lift-net that is dominated by anchovy, where the fish does not exist on the catch of stationary lift-net. The type of anchovy (*Stolephorus sp.)* obtained by boat lift-net fishing gears is the type of *Stolephorus indicus* and *Stolephorus commersonnii* where many obtain at a depth of 25-30 meters. The appropriate mesh size for selective catching of relatively adult anchovy was 16 mm mesh size [12].

IOP Conf. Series: Earth and Environmental Science **116** (2018) 012022 doi:10.1088/1755-1315/116/1/012022

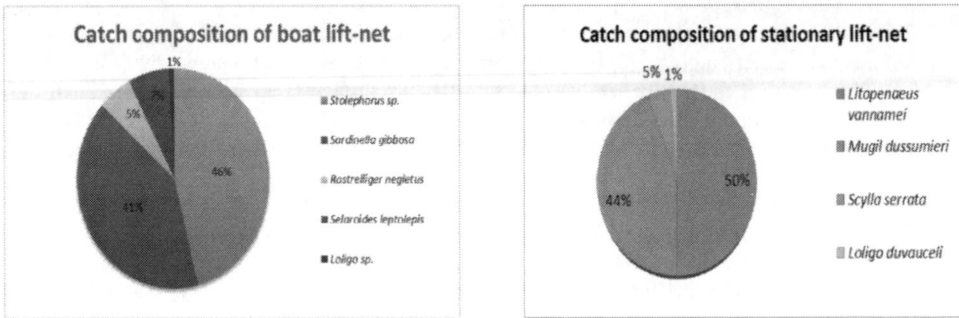

Figure 3. Catch Composition of Lift Net

3.3. *Capital Investment of Lift Net*

Capital needed in a fishing business is the total amount of capital invested in the form of lift net, nets, lamps, boats and other fishing equipment. Based on the research results obtained data on the average capital size of fishing business using lift net in Morodemak waters of Demak Regency can be seen in table 1.

Table 1. Investment Capital Average of Fishing Business Using Lift Net

Number	Description	Value (IDR)	
		Boat Lift Net	Stationary Lift Net
1.	Minimal	234,500,000	6,560,000
2.	Maximal	264,500,000	7,180,000
	Average	251,027,273	6,781,818

Based on Table 1 it is known that the boat lift net requires more capital than the stationary lift net because the capital required of each fishing gear is also different. Capital required in the boat lift net includes boat, main engine, lift net, lights, nets, and generator, while the capital on the stationary lift net includes lift net, nets, and lights.

Based on the interview, the price of boat propulsion machine varies according to the brand, the year of purchase, and the condition of the purchase. The price of the machine ranges from 17,0400,000 IDR until 18,600,000 IDR. Generator's price on the boat lift net ranges from 18,000,000 IDR until 22,000,000 IDR. Boat's price ranges from 50,000,000 IDR until 75,000,000 IDR. While the net's price used in the boat lift net ranges from 10,000,000 IDR until 14,500,000 IDR. The net's price used in the stationary lift net ranges from 1,000,000 IDR until 1,300000 IDR.

Based on the data obtained, it is known that the capital or purchase price of each fisheries business unit of lift net in Morodemak waters there is a difference. The price difference is caused by several factors, including the year of purchase. Years of purchasing of boats, nets, and machines of each business unit are not the same. A business unit with a longer year of purchase has a cheaper price than a starting business unit. It is also due to the purchase conditions of both boats and engines. Boats and machines that buy in a new circumstance must have a much higher price than ships or machines that buy in used conditions.

3.4. *Production Cost of Lift Net Fishing Business*

Production cost consist of fixed cost or the amount of cost incurred in a fixed amount for one year and variable costs is the amounts that changed each time do the fishing operation. Costs that are not dependent on production activities are fixed costs. It will always be calculated as long as the business is still conducted so that no matter how much the amount of the production does not give any influence toward the fixed cost [13]. Because the cost is unavoidable as long as the business is still running. Fixed costs consist of depreciation cost, maintenance cost, licensing cost, and sea alms cost.

Table 2. Average of the Fixed Cost of Fishing Business Using Lift Net

		Value (Rp/year)	
Number	Description	Boat Lift Net	Stationary Lift Net
1.	Depreciation Cost	53,932,424	5,252,364
2.	Maintenance Cost	39,327,273	12,375,364
3.	Licensing Cost	2,200,000	-
4.	Sea Alms Cost	100,000	-
	Total	95,559,697	17,627,728

Variable costs include retribution cost, operational cost, and labor cost. Operational costs of fishing business using lift net in Morodemak waters include costs incurred for the purchase of diesel fuel, and supplies (food, beverages, ice cubes). Based on the results of interviews with fisherman in Morodemak waters the amount of operational costs are different on each fishing gear. Fuel requirements of each fisherman is different depending on the distance to fishing ground.

Table 3. Average of the Variable Cost of Fishing Business Using Lift Net

		Value (Rp/year)	
Number	Description	Boat Lift Net	Stationary Lift Net
1.	Operational Cost	151,763,545	8,654,545
2.	Labor Cost	20,736,634	-
3.	Auction Cost	12,644,127	-
	Total	185,144,306	8,654,545

Overall fisherman of boat lift net in Morodemak waters are average spend diesel cost per year 88,125,000 IDR, where every fishing trip requires 120 liters of diesel. The annual cost of supplies is 61,860,000 IDR, while the cost of ice cubes is 1,316,182 IDR. Fisherman of stationary lift net are average spend diesel cost 1,348,500 IDR each year and 2 liters each trip. Diesel cost each year is 4,021,136 IDR and 4 liters each trip. The supplies cost each year is 2,840,000, while the cost of ice cubes is 393,727 IDR.

Labor cost represents the expense that fisherman must pay to the crew as a consideration in carrying out catching operation at sea to get the catch with a certain amount. The average cost of the fisherman of boat lift-net operated in Morodemak waters is Rp 20,736,634 per year. Labor costs obtained by the cost of revenue minus operational costs in a year later divided equally on each crew.

The number of crew involved in the fishing operation is around 7 - 9 people. In the stationary lift net in Morodemak Waters there is no division of labor costs because only 1 fisherman on the stationary lift net. The total cost is the overall cost of a business unit. Total cost is derived from the sum of fixed costs and non-fixed costs. The total cost per year of boat lift net average 278,419,003 IDR each year or 2,032,225 IDR each operation, while for the stationary lift net 21,029,909 IDR each year or 148,098 IDR each operation.

Table 4. Average of the Total Cost of Fishing Business Using Lift Net

Number	Description	Boat Lift Net	Stationary Lift Net
		(Rp/year)	(Rp/year)
1.	Minimal	271.018.375	19.054.500
2.	Maximal	286.681.238	22.779.000
	Average	280.698.003	21.029.909

3.5. Revenue and Profit of Lift Net Fishing Business

Revenue on fishing business of the lift net in Morodemak waters is obtained from the sale of fish catching. The average revenue of fishing business of the boat lift net in the Morodemak waters each year is 388,580,000 IDR. Where the results are obtained by using the average operation each year, that is 137 operations each year, consisting of 40 operations when harvest season, 75 operations when regular season, and 20 operations when famine season. Revenue of fishing business of the stationary lift net each year is 27,750,000 IDR. The result is obtained by assuming the average operation of fisherman each year is 142 operations each year, consist of 50 operations when harvest season, 72 operations when regular season, and 20 operations when famine season.

Table 5. Average of the Revenue Each Year of Fishing Business Using Lift Net

Number	Description	Value (Rp/year)	
		Boat Lift Net	Stationary Lift Net
1.	Revenue in Harvest Season	142.054.545	14.306.364
2.	Revenue in Regular Season	233.065.455	11.567.273
3.	Revenue in Famine Season	13.460.000	1.876.364
	Average of the Revenue	388.580.000	27.750.000

The profits are derived after receipt of the sale of the production of the catch is reduced by the total cost. Assuming if the result is high then the acceptance will be high and so will the higher profit. The boat owner's profit is 737,773 IDR each operation, while ABK gets 149,825 IDR each person (fisherman of boat lift net consists of 7-9 people), then for the stationary lift net the profit each operation of 47,776 IDR, which on the stationary lift net not use more than fisherman because it can operated alone.

3.6. Financial Analysis

The financial aspect is a key aspect of a feasibility study. It is said that, because even if other aspects are considered feasible, if the financial aspect studies give unfeasible results, then the proposed project will be rejected because it will not provide economic benefits.

Table 6. Financial Detail of Fishing Business Using Lift Net with Criteria (NPV,B/C, PP, and IRR)

Number	Description	Value (Rp)	
		Boat Lift Net	Stationary Lift Net
1.	NPV	836,149,272 IDR	37,973,601 IDR
2.	IRR	54 %	86 %
3.	B/C Ratio	1.73	1.32
4.	PP	2.44	1.96

Based on the results of the research, financial analysis on the boat lift net and stationary lift net in Morodemak waters has a positive NPV value, B / C Ratio> 1, and IRR> discount rate (discount rate used in this research is 12% using local bank's interest rate). The rate of return on the two fishing gear is categorized as fast because the PP value is less than 3 years. Based on the investment criteria used, the fishing business with the boat lift net and stationary lift net is feasible and efficient to do.

4. Conclusion and Recommendation

4.1. Conclusion

Based on the results, it can be taken some conclusions, among others:

1. The method of operation of the boat lift net and stationary lift net are ranging from setting, immersing, and lifting. Fishermen usually leave in the afternoon at around 5 p.m., with a distance of about 6 miles for the boat life net and 1.5 miles for the stationary lift net. Upon arrival at the fishing ground and the day before night, the lights are turned on and immediately lower the net. Lifting of nets in one operation of boat lift net is usually done by fisherman once every 3 hours, whereas for stationary lift net usually lifting is done after 30 minutes of immersion.
2. Investment cost of the boat lift net was Rp 251.027.273 and and stationary lift net was Rp 6.781.818. Revenue from the boat lift net was Rp 388.580.000 each year and stationary lift net was Rp 27.750.000 each year. The profit from the boat lift net was Rp 110,160,997 each year and the stationary lift net was Rp 6,720,091 each year.
3. Based on the financial analysis, it is known that the fishing business of boat lift net and stationary lift net in Morodemak Waters is feasible and efficient to proceed. However, it was more profitable the stationary lift net than the boat lift net because a higher IRR and lower payback period.

4.2. Recommendation

Based on the results obtained from the research that has been done, fisherman of stationary lift net step on the long term need to enlarge the scale of business considering the large B/C ratio slightly larger than one. Furthermore, there needs to be research about the efficiency of the use of fishing gear for the boat lift net and stationary lift net so that it can optimize the use of production factors in order to improve the fisherman's welfare of catch fishery of the lift net. In Banten Bay [14], application LED lamps in fixed lift net had significant effect to reduce fuel consumption with mean saving energy up to 35%, and in Sulawesi waters, light from White LED lamp used in bagan lift-net could penetrate to deeper water and caught more white anchovy (*Stolephorus indicus*) than mercury lamps [15].

5. Acknowledgment

On this occasion the authors would like to thank the Faculty of Fisheries and Marine Sciences of Diponegoro University who have contributed to this research funding. Thanks go to M.Ulin Nuha for the assistance in data collection.

References

[1] Prasetyo, A.P., Nugroho, D., Sadiyah, L., Purwoko, R.M., Faizah, R. and A Setyawan. 2014 *Indonesian Fisheries Research Journal* **20**(2):99-106
[2] FAO Fisheries and Aquaculture Department [online]. Updated 13 September 2001. [Cited 11 August 2017]. http://www.fao.org/fishery/geartype/105/en
[3] Puspito G, Thenu I M, Julian D, Tallo I 2015 *AACL Bioflux* **8**(2):159-167
[4] Umar, H 2005 Business Feasibility Study, 3rd edition. PT Gramedia Pustaka Utama Jakarta
[5] Raharjo F 2007 Technical Economics: Decision Making Analysis. Yogyakarta: ANDI publisher
[6] Dahlan M N 2011 Capture Fisheries Development in Belitung District: An Economic Trade-Off Local Based Analysis. [*Dissertation*]. IPB Graduate Schools. Bogor
[7] Farris, PW, NT Bendle, PE Pfeifer, and JR David 2010 *Marketing Metrics: The Definitive Guide to Measuring Marketing Performance.* Upper Saddle River, New Jersey: Pearson Education, Inc. ISBN 0-13-705829-2

[8] Arimoto T, Glass CW and Zhang X 2011 Fish Vision and Its Role in Fish Capture. In : Pingguo He (Ed.), Behaviour of Marine Fish : Capture Processes & Conservation Challenges (pp. 25–43)
[9] Aliyubi FK, Boesono H, and Setiyanto I 2015 *Journal of Fisheries Resources Utilization Management and Technology*, **4**(2):93 – 101
[10] Rahmawati E, R Irnawati, and A Rahmawati 2017 *Jurnal Perikanan dan Kelautan*, **7**(1):40 – 49.
[11] Ekawaty, R 2015 *Journal of Marine and Aquatic Sciences* **1** : 7–13
[12] Alba E B, Chiuco M B and M C Rubia. 2016. *International Journal of Fisheries and Aquatic Studies* **4**(2): 265-273
[13] Hapsari, T D and ADP Fitri 2016 *Aquatic Procedia* 7 : 254 – 264

[14] Susanto A, R Irnawati, Mustahal and MA Syabana 2017 *Turkish Journal of Fisheries and Aquatic Sciences* **17**(2):283-291
[15] Sulaiman M, MS Baskoro, A A Taurusman, SH Wisudo and R Yusfiandayani 2015 *International Journal of Sciences: Basic and Applied Research (IJSBAR)* **20**(1):228-239

IOP Conf. Series: Earth and Environmental Science **116** (2018) 012023 doi:10.1088/1755-1315/116/1/012023

Impact of Fishery Policy on Fishery Manufacture Output, Economy and Welfare in Indonesia

Firmansyah[1*], Shanty Oktavilia[12], F.X. Sugiyanto[1], Ibnu N Hamzah[1]

1) Department of Economics, Diponegoro University, Jl. Prof Sudharto SH, Tembalang, Semarang, Jawa Tengah Province, Indonesia
2) Faculty of Economics, Universitas Negeri Semarang, Jl. Sekaran, Gunung Pati, Semarang, Jawa Tengah Province, Indonesia

*Corresponding author Email : firmansyah@live.undip.ac.id

Abstract. The fisheries sector and fish manufacturing industry are the bright prospect sectors of Indonesia, due to its huge potency, which has not been worked out optimally. In facts, these sectors can generate a large amount of foreign exchange. The Government has paid significant attention to the development of these sectors. This study simulates the impact of fishery policies on the production of fish manufacturing industry, national economic and welfare in Indonesia. By employing the Input-Output Analysis approach, impacts of various government policy scenarios are developed, covering fisheries technical policy, as well as infrastructure development policies in the fisheries sector. This study indicates that the policies in the fisheries sector increase the output of fishery, the production of fish manufacturing industry, the sectoral and national outputs, as well as the level of national income.

1. Introduction.

The fishery sector, especially the capture fisheries in marine waters of Indonesia is a very potential and strategic sector to become the mainstay of economy. It supported by the fact that Indonesia is an archipelagic country with 17,504 islands [1] and has the coastline of 99,000 km with the marine fisheries area is 5.8 million km^2 – with maximum potential sustainable yield of fish resources in marine waters at 6.5 million tons per year, with an allowable catch of 5.2 million tons/year (80% of maximum sustainable yield). In addition, the sector provides employment for about 2,667,440 people and 2,186,900 of whom are fishermen who capture fisheries in marine waters [3].

Based on the Food and Agricultural Organization [4], the total capture of production in marine waters was 81.5 million tons in 2014, and Indonesia ranked the second largest in the world after China. The Indonesia's production of fisheries capture in marine waters was 6,204,669 tons in 2015 [1]. According to the Indonesia Central Bureau of Statistics or BPS [1], it was known that the contribution of the fishery sector as a whole - including aquaculture - reached 65957.5 trillion rupiah in 2014, was 7 percent of the growth rates, and compared to the Gross Domestic Product (GDP), the contribution of this sector was 2.27 percent. Table 1 displays the development of Indonesian fishery sector along 2000-2014. The capture fishery sector alone exported 95.4 thousand tons or 268.1 million US $ in 2016 [5]. The value of potency of the fishery sector is greater when includes the calculation of inter-linkages with other sectors. The all of related sectors also absorb workers and exports.

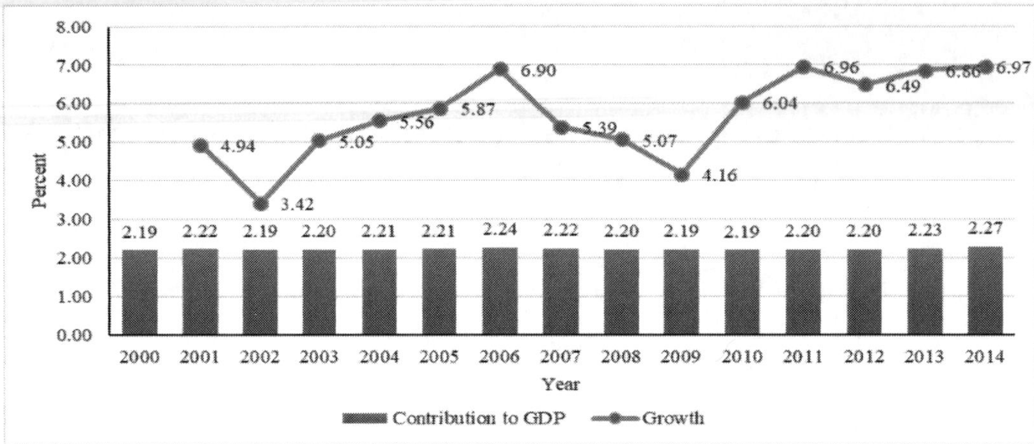

Source: [6], processed

Figure 1. Growth and contribution of Fishery sector to the GDP at Constant Prices (2000-2014)

According to the Development Planning Agency (Bappenas) study [2], it identified several problems of the Indonesian fishery sector, which cause the sector performance are not optimal and sustainable yet. Some issues which are related to marine capture fisheries are: 1) the rampant of illegal fishing activities; 2) the overfishing in some coastal waters of Indonesia; 3) the using of destructive fishing gear; and 4) the weaknesses of the control system of fish resource utilization.

Due to the various problems of fisheries, the Government of Indonesia issued some policies, and this study examines the impact of the policies on the national and sectoral economy and welfare. The scope of policies which be analyzed in this study is limited to the marine capture fishery policies, based on the 2015-2019 Strategic Plan of the Directorate General of Capture Fisheries of Ministry of Marine and Fisheries Affairs [3]. These policies are in the scope of marine capture fisheries, namely the investment of fishery business and the construction of fishery ports.

To capture the comprehensive impact of government policy on the fisheries industry on economy and welfare, the Input-Output (I-O) analysis is employed. The I-O analysis is a powerful in measuring the impact on the economy through inter-sectoral linkages, and is commonly used in many countries. Several Input-Output models of fisheries industry and the impact toward economy which are developed in several countries, among others the Commercial Fishing Model of Northeast U.S. [7], UK Sea Fishing [8], models of fisheries Input-Output in U.S. [9], Pacific Coast Fisheries [10], Wildlife and Fisheries Recreation in Mississippi, U.S. [11], and a hybrid model for fish harvester [12].

2. Method

2.1 Data

The data employed in this study are the input-output table with the based price transaction of 185 sectors of the 2010 Input-Output (I-O) table of Indonesia, the latest I-O Table published by BPS [13]. Sector list can be seen in the 2010 I-O Table of Indonesia [13] .

2.2 The development of shock scenario

The policy scenarios used in this study are obtained from the 2015-2019 Strategic Plan [3], which covers the amount of bussiness investment in the marine capture fisheries sector and the fishery port construction fund throughout 2015-2019. The forming of the shock for the policies simulation is explained with the information in Table 1.

Table 1. Shock Scenarios

Description	Total
The investment value of capture fishery business 2015-2019 (trillion rupiah)	127
Number of fishing ports that meet operational standards 2015-2019	262
Average cost of 1 fishing port construction	90
Value of fishing port construction (trillion rupiah)	23.580

The data of the business investment value of marine capture fishery and the amount of port construction along 2015-2019 is obtained from 2015-2019 Strategic Plan of Directorate General of Capture Fisheries Ministry of Marine and Fishery [3]. The cost of constructing a fishing port is obtained from the Fishery Port Information Center, Ministry of Marine Affairs and Fisheries [14]. The magnitude of investment value and cost of post construction are the value of policies shock on selected sectors in Table I-O, namely Sector 33 (Fish) and Sector 152 (Road, Bridge and Port).

2.3 Impact of fishery sector policy on national and sectoral economy: Input-Output model approach

Input-Output Analysis is a general equilibrium analysis, based on data in the form of transaction matrices of goods and services, with emphasis on the sectoral production side. The transaction matrices include several blocks that are grouped into 4 quadrants ie intermediate transaction quadrant, final demand quadrant, primary input quadrant, and the primary input to final demand quadrant [15]. The inter-block linkage scheme in the Input-Output model is explained briefly through the following Figure 2:

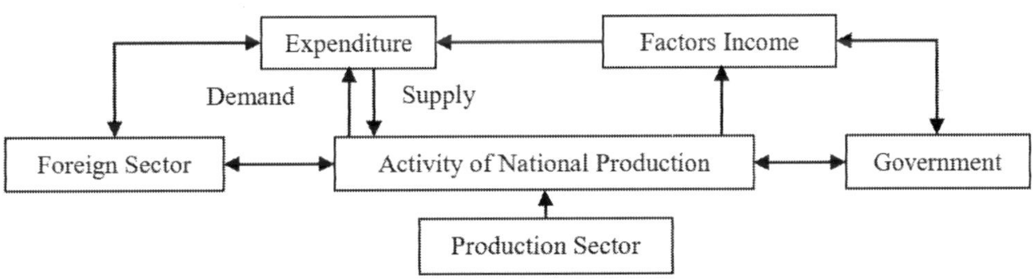

Figure 2. Scheme of Input-Output Model

Briefly, it can be explained that the production sector absorbs input from the household sector, namely primary inputs: capital, labor, and land. The production sector produces consumer goods for household needs, and intermediate goods for other sectors. The production sector also exports. Households receives income from producers as remiuneration of primary input, and transfer from the government. Government obtaines tax from household. The production activity box at the national level in Figure 2 displays the production side of GDP, the factor's revenue box shows the income side of GDP and the expenditure side exhibits the expenditure side of GDP.

The impact of the changes in investment of fishery business and the construction of fishery port to sectoral output changes, is calculated by the following formula:

$$\Delta X = (I-A)^{-1}\Delta Y \quad (1)$$

where ΔY represents a vector of changes in the final demand (including investment) and ΔX vector of output change. A is the technological coefficient matrix and $(I-A)^{-1}$ is Leontief inverse matrix [15].

The impact of the policy shocks on sectoral household income is calculated by the following formula:

$$\Delta H = H_R (I\text{-}A)^{-1} \Delta Y \quad \text{.........(2)}$$

where ΔH represents the income changes vector, H_R represents income coefficient of sectoral household income [15].

3. Results and Discussion

Tables 2 and 3 present the simulation results of 185 sector classification of 2010 Indonesia Input-Output, which covers the sectoral and national impacts of output and household income. The results which are presented in the two tables cover the ten sectors which hace the greatest impact, and special observations on two sectors of the fish processing industry namely the sector of Dried Fish and Salted Fish (code 55) and the Fish Processing and Preservation (code 56).

Table 2. Impact of the largest sectoral output and Fish Processing Industry

Rank	Code	Sector	Output Change (Million Rupiah)
1	33	Fish	128,557,932.00
2	152	Road, Bridge, and Port	23,612,197.58
3	95	Oil and Gas Refinery Products	7,215,167.72
4	156	Trade of the other than Cars and Motorcycles	5,895,496.74
5	72	Animal Food Processed	5,378,009.81
6	30	Agriculture, Forestry and Fisheries Services	3,572,377.33
7	38	Crude Oil	2,941,945.73
8	48	Various Mining Goods	2,204,996.54
9	94	Other Goods from Non-Metal Material	1,975,938.98
10	39	Natural Gas and Geothermal	1,827,347.95
97	56	Fish Processing and Preservation	59,291.87
150	55	Dried Fish and Salted Fish	8,193.88
		National Output Change	213,099,172.00

From Table 2, it can be seen that the two sectors which experiences the increase of investment, have the highest impact of output compared to other sectors. The increase of investment in fishery business and fishery port directly and indirectly also lead to the increase of other sector outputs, namely the Sector 95 (which the product is fuel) and Sector 156 (which one of its activities is motor boat trade), as the two other largest impacted sectors. These sectors have highly interlinkage to the Sector 33 and 152. The investment increase will promote the increase of fishing activities and push the use of fuel and trade in ships and other fishery equipment. Sectors that utilize the output of Sector 33 and Sector 152 as its inputs for their production processes (such as Sector 56 and 55), are had relatively small output compared to other sectors as listed in Table 1. It implies that the backward link (input path) of Sector 33 and Sector 152 are higher than the forward link (output path). In the national level, the investment of fihery bussiness by 127 trillion rupiah and the the fishery port construction by 23.58 trillion rupiah will increase the national output by 213.10 trillion rupiah.

Tabel 3. Impact of the largest sectoral income and Fish Processing Industry

Rank	Code	Sector	Income Change (Million Rupiah)
1	33	Fish	23,093,052.85
2	152	Road, Bridge, and Port	3,516,128.16
3	156	Trade of the other than Cars and Motorcycles	1,114,137.77
4	95	Oil and Gas Refinery Products	874,893.99
5	30	Agriculture, Forestry and Fisheries Services	755,651.49
6	48	Various Mining Goods	648,386.33
7	72	Animal Food Processed	423,294.77
8	94	Other Goods from Non-Metal Material	337,765.44
9	158	Land Transportation Services beside Rail Transport	234,520.56
10	170	Financial and Banking Services	223,006.06
125	56	Fish Processing and Preservation	3,156.28
161	55	Dried Fish and Salted Fish	591.70
		National Income Change	35,185,261.61

Table 3 displays the largest increase in sectoral household income due to increase of the investment in Sector 33 and Sector 152. Other sectors that have the largest impact in income, almost in line with the impact of output, ie Sector 156, Sector 95, and so on. In addition to having the high linkage to the Sector 33 and Sector 152, these sectors also have higher household income coefficients than other sectors, which indicate that they have more labor-intensive than other sectors. In the national level, the income will increase by 35.2 trillion rupiah due to the increase in the investment.

4. Conclusion

The direct and indirect impacts of fisheries policy through the investment increase in fisheries bussiness and fishery port construction, are enjoyed highest by sectors which has high interlinkage with the fishery and port construction sectors. The increase of fisheries sector activity by investment increasing in the sector will attract inputs such as fuel-producing and trade of boats and fishery equipment, and promote the other sectors which use fishery output as its inputs, such as the fish preservation sector. From the the impact rankings, the sectors which preserve inputs for the fisheries and port sectors have the greatest impact in comparison with user-output sectors. It is reasonable because the output of the Fish sector are largely the final product that is consumed by people and is also exported.

The implication of the simulation results are the needed of a systemic policy that also supports the sectors which preserve input for fishery sector – which strengthens the domestic sector – as the investment policy increase the output, consumption and export of fisheries, as well as benefit the sector that preserve the inputs for fisheries sector.

To produce a more comprehensive study, and with some limitations in this study, there may be some development in the future. First, the development of shock scenarios by employing econometric model to obtain a more precise estimate of the value of shocks. Second, the use of global injection rates throughout 2015-2019 which combines the annual impact, ignores the emerging annual impacts for the next year, so in the future, a dynamic I-O model can be developed which can adapt the annual policy scenarios.

References

[1] Badan Pusat Statistik 2017 *Statistical Yearbook of Indonesia 2017* (Jakarta: BPS)
[2] Badan Perencanaan dan Pembangunan Nasional 2014 *Kajian Strategi Pengelolaan Perikanan Berkelanjutan* (Jakarta: Direktorat Kelautan dan Perikanan Bappenas) https://www.bappenas.go.id/files/7614/4401/4206/Strategi_Pengelolaan_Perikanan_Berkelanjuta

IOP Conf. Series: Earth and Environmental Science **116** (2018) 012023 doi:10.1088/1755-1315/116/1/012023

n.pdf

[3] Direktorat Jenderal Perikanan Tangkap Kementerian Kelautan dan Perikanan 2015 *Rencana Strategis Tahun 2015-2019* (Jakarta: Kementerian Kelautan dan Perikanan)

[4] Food and Agricultural Organization 2016 *The State of World Fisheries and Aquaculture 2016: Contributing to Food Security and Nutrition for All* (Rome: Italy)

[5] Badan Pusat Statistik 2017 *Analisis Komoditi Ekspor 2010-2016: Sektor Pertanian, Industri, dan Pertambangan* (Jakarta: BPS)

[6] Badan Pusat Statistik 2017 *PDB Atas Dasar Harga Konstan 2000 Menurut Lapangan Usaha (Miliar Rupiah)* 2000-2014 https://www.bps.go.id/linkTabelStatis/view/id/1200

[7] Steinback S R and Thunberg E M Northeast 2006 Region Commercial Fishing Input-Output Model *NOAA Technical Memorandum NMFS-NE-188* https://www.nefsc.noaa.gov/publications/tm/tm188/tm188.pdf

[8] Seafish 2006 The economic impacts of the UK sea fishing and fish processing sectors: *An input-output analysis* (Edinburgh: Sea Fish Industry Authority) http://www.seafish.org/media/Publications/2006_I-O_Key_Features_Final_090108.pdf

[9] Seung C K and Waters E C 2006 A Review of Regional Economic Models for Fisheries Management in the U.S. *Marine Resource Economics* 21101–124

[10] Leonard J 2013 Input-Output Model for Pacific Coast Fisheries, 2013 Revisions and Extensions http://www.pcouncil.org/wp-content/uploads/IOPAC_SSC_Econ_Review_April_NWC.pdf

[11] Henderson J E, Grado S C, Munn I A, Jones W D 2010. Economic Impacts of Wildlife- and Fisheries Associated Recreation on the Mississippi Economy: An Input-Output Analysis. *Research Bulletin FO429*, (Mississippi: Forest and Wildlife Research Center)

[12] Failler P, Pan H, Thorpe A, and Tokrisna R 2014 On Macroeconomic Impact of Fishing Effort Regulation: Measuring Bottom-Up Fish Harvesters' Economy-Wide Contribution *Natural Resources* 5 269-281 http://dx.doi.org/10.4236/nr.2014.57025

[13] Badan Pusat Statistik 2015 *Tabel Input Output Indonesia 2010* (Jakarta: BPS)

[14] Fisheries Port Information Centre 2015 *Susi Janji Bangun Pelabuhan Untia* http://pipp.djpt.kkp.go.id/detail_berita/1662

[15] Firmansyah 2006 *Operasi Matrix dan Analisis Input-Output (I-O) untuk Ekonomi: Aplikasi Praktis dengan Microsoft Excel dan MATLAB* (Semarang: BP UNDIP)

IOP Conf. Series: Earth and Environmental Science **116** (2018) 012024 doi:10.1088/1755-1315/116/1/012024

Cytotoxicity and Phytochemical Profiling of *Sargassum Sp.* Extract As Anti-Mdr Bacteria

Wilis A. Setyati[1], Rini Pramesti*[1], Muhamad Zainuddin[2], Maya Puspita[1] and Person P Renta[3]

[1]Marine Science Department, Fisheries and Marine Science Faculty, Diponegoro University, Semarang

Jl. Prof. H. Soedarto, S.H. Tembalang Semarang, 50275 Indonesia

[2]Aquaculture Departement, University of Islam Nahdlatul Ulama, Jepara

Jl. Taman Siswa, Pekeng, Tahunan, Jepara, Jawa Tengah 59427, Indonesia

[3]Marine Science Departement, Bengkulu University

*Corresponding author Email : rinipramesti63@gmail.com

Abstract. *Sargassum* sp. contains bioactive compounds having the potential as an antibacterial agent. *Sargassum sp.* was collected from five different locations, i.e., Teluk Awur, Panjang Island, Bandengan, Ujung Piring, and Bondo. There were several different species of *Sargassum sp.* identified from each sampling locations. The collected seaweeds were washed, naturally dried, and ground to the powder-sized dry material. Dry seaweed was extracted gradually using n-hexane, ethyl acetate, and methanol (1:3 w/v). The antibacterial analysis was conducted based on Agar Diffusion method using Zobell as media. Resistance analysis was performed to evaluate the resistance of pathogen against commercial antibiotics, namely, chloramphenicol, ampicillin, erythromycin, amoxycillin, and tetracycline. Each *Sargassum sp.* extract was tested against three candidates of MDR bacteria, i.e., *Staphylococcus aureus*, *Escherichia coli,* and *S. epidermidis*. Results showed that *S. aureus* was resistant towards four out of five commercial antibiotics. *E. coli* and *S. epidermidis* were not susceptible to two and three out of five commercial antibiotics, respectively. N-hexane, ethyl acetate and methanol yielded 0.1-0.3%, 0.3-0.7 % and 0.8-4.7% of dry extract. Ethyl acetate extract of Sargassum from Teluk Awur performed the best antibacterial activity and contained an alkaloid, flavonoid, and phenolic compounds. Toxicity analysis showed that this ethyl acetate extract had LC_{50} at 463 ppm and categorized as chronic toxicity.

1. Introduction

Applications of drugs particularly synthetic antibiotics have been commonly used to control the disease. The excessive use of antibiotics becomes the single most important factor leading to a resistance increase of bacteria towards antibiotics. This is due to the adaptability of disease to the toxicity of antibiotic and bacterial resistance [1]. Bioactive compounds from marine resources offer a promising solution to overcome the resistance problem. Also, it also becomes an alternative replacement of synthetic drugs to treat infection disease. World Health Organization (WHO) has recommended the use of natural products in maintaining the well being of the community, in preventing and resolving chronic and degenerative disease [2]. *Sargassum* sp. is one of the marine resources that can be proposed as alternative natural products for disease control. *Sargassum sp.* has been reported to show interesting and promising pharmacological properties serving as anticancer, anti-inflammatory, antibacterial, and antivirus [3], antioxidant [4], antifouling [5]. And antifungal [6]. *S. fulvellum* and *S. thunbergii* showed antipyretic, analgesic, and anti-inflammatory activities in mice [7]. *S. echinocarpum, S. duplicatUM* and *S. polycystum* performed potential antioxidant [8]. *S. aquifolium, S. ilicifolium* and *S. polycystum* [9]. showed antibacterial activity against *Bacillus subtilis, Pseudomonas aeruginosa* and *Escherichia coli* [9]. *S. pallidum* can serve as antioxidant and anti-hemolysis agents exhibiting potential for further exploration as functional food or complementary medicine [10]. *S. tenerrimum* from India is reported to have antibacterial activity against *Klebsiella pneumonia, Escherichia coli, Staphylococcus aureus, Proteus sp., Streptococcus sp., Pseudomonas aeruginosa, Vibrio parahaemolyticus, Salmonella sp., Shewanella sp., Vibrio fluvialis, Vibrio splendidus, Vibrio cholera, Shigella flaxneri, Staphylococcus epidermidis, Aeromonas liquefaciens* and *Bacillus subtilus.* Furthermore, *Sargassum sp.* also have antifungal activity against *Aspergillus niger, A. flavus, A stetreus, Candida albicans* and *Penicillum* sp. [11]. This study aims to evaluate the inhibition ability of *Sargassum sp.* extracts against pathogenic bacteria categorized as multidrug resistant bacteria. Using conservation and biotechnological approach, the study is expected to resolve the resistance of pathogenic bacteria towards synthetic drugs and to prevent more infections.

2. Materials And Methods

The samples *Sargassum sp.*were collected from five different locations, i.e. Teluk Awur (6° 7' 36.48" S, 110° 24' 0" E), Panjang Island (06°34"30' LS, 110° 37"44' BT), Bandengan (°07'36"LS 110°24'00"BT / 6,1268°LS 110,4°BT), Ujung Piring (6° 30' 40.91" S, 110° 40' 06.42" E) and Bondo (6° 28' 26.79" S, 110° 42' 33.13" E) at Jepara, Central Java.

The solvents used : n-hexane, ethyl acetate, and methanol as solvents to extract the bioactive compounds in *Sargassum sp.* Antibacterial activity against *S. aureus, E. coli,* and *S. epidermis* was evaluated through the Agar Diffusion method by using Zobell as media. The media used consisted of peptone, yeast, and agar [12].

2.1 Samples preparation

The collected seaweeds were washed with tap water to remove the epiphytes and the remaining debris. A whole thallus of seaweed was taken for taxonomic identification based on morphology study. Before natural drying under the shade, seaweeds were cut to pieces (± 5 cm). Further, the dried seaweeds were ground with a multi use blender to obtain powder material.

2.2 Extraction of Sargassum sp.

Dry seaweed of 25 gr was extracted gradually by using 100 mL n-hexane, ethyl acetate, and methanol, respectively. The extraction was performed for 24 h at room temperature in obscurity. The extracted seaweed was filtered by Whatman Filter Paper to separate the solute from the residues. The remaining residues of dry seaweed material were re-extracted using the same process and solvents. Filtrates obtained from the extraction process were pooled. N-hexane, ethyl acetate, and methanol filtrates were evaporated to dryness with vacuum rotary evaporator at 40 °C in 500 mmHg. n-hexane, ethyl acetate and methanol extracts of *Sargassum sp.* were dried using N_2 before storage in -20 °C [13]. Extraction yield was determined with following formula :

$$Ce = \left(\frac{W_2}{W_1}\right) \times 100\%$$

Where :
Ce = Extraction yield (%)
W_2 = Weight of extract (gr)
W_1 = Weight of initial sample (gr)

2.3 Resistance analysis of Multidrug-Resistant (MDR) bacteria

Staphylococcus aureus, *Escherichia coli,* and *S. epidermidis* were analyzed against five commercial antibiotics. The antibiotics were chloramphenicol, ampicillin, erythromycin, amoxicillin, and tetracycline.

2.4 Antibacterial activity analysis

Bacterial culture in liquid media was centrifuged at 5000 rpm for 10 min and was washed with PBS.

0.1 mL of bacterial solution was read with a spectrophotometer at 600 nm to obtain optical density between 0.6-0.8 [14].
MDR bacterial solution of 0.1 mL previously cultured in liquid media was spread evenly on to solid media in the petri dish using spread technique [14]. The inoculated bacterial solution was left for 30 min to diffuse on the media. Extract and chloramphenicol served as a positive control was prepared in distilled water at 2500 ppm. The sterile paper disc was impregnated with 20 μL of antibiotics solution at 50 μg/disc [13]. And 20 μL of solvents without extracts as a negative control. Paper discs were placed on the solid media in Petri dish previously inoculated with MDR bacteria. All Petri dishes were incubated at 37 °C for 24-48 h.

2.5 Artemia salina Brine Shrimp Lethality Test (BSLT)
BSLT was performed to evaluate the toxic effect and cytotoxicity of chemical compounds against the larvae of shrimp *Artemia salina.* The test was based on [15], [16], [17]. *A. salina* was immerged in fresh water for 15-30 min to clean the attached feces. 0.25 gr of eggs were put in a box filled with 500 mL seawater at 25-30 °C with pH at 7-8. Irradiation from TL 40 watt light was placed above the box to light it during the hatching for 48 h. After 48 h, the eggs hatched to nauplii instar III/IV and were ready to use for BSLT test [18].

Ten larvae of *A. salina* were introduced into a test tube filled with extract showing the best antibacterial activity. Concentrations of extracts applied for the BLST were: 1000, 500, 100, 50 and 10 ppm with ethanol 2% [15]. Negative control used was methanol concentrated at 6, 4, 2 and 1%. The observation was made at 1, 3, 6, 12, 18, 24 and 36 h after the introduction of larvae into extract by using a loop. Determination of time interval was according to the lethal concentration of a compound whether it was acute or chronic lethal. LC_{50} was calculated from the linear regression equation based on the observation made. All tests were performed in triplicate.

2.6 Phytochemical profiling of Sargassum sp. extracts
This analysis was meant to examine the presence of certain chemical compounds in the tested samples, in this case, *Sargassum sp.* By determining a presence of certain compounds in plant matrix, a correlation can be made between the compounds and the biological activities performed. Thus it facilitates the phytopharmacological procedures [12]. Phytochemical profiling of *Sargassum sp.* extracts was carried out to evaluate the presence of alkaloid, flavonoid, saponin, steroid, and triterpenoid. *Sargassum sp.* dry material of 0.05 gr was dissolved in 10 drops of sulfuric acid 2 N, and three drops of this solution was introduced to Marquis, Dragendorff, Meyer and Wagner reagent. These reagents, previously prepared in separated test tubes, were designated to test the alkaloid content in a sample. For flavonoid, 0.05 g of dry seaweed material was added with 0.1 mg magnesium and 0.4 mL amyl alcohol, a mixture of HCl 37% and ethanol 95% in the same volume. Further, four mL of alcohol was added to the solution. To detect the presence of saponin, 0.05 g of dry seaweed material was dissolved in hot water. This test was also called as foam test due to the foam established on the surface of the solution. For the steroid/triterpenoid test, 0.05 g of dry seaweed material was dissolved in 2 mL of chloroform in test tube added with Lieberman Burchard reagent.

2.7 Statistical analysis
Data from resistance and antibacterial analysis were statistically analyzed using SPSS 16. Test of normality by Shapiro-Walk and homogeneity by Levene test with the significant level at 0.05 were performed. Multivariable analysis of variance (MANOVA) was used to analyze the interaction between variables, *i.e.,* sampling locations, *Sargassum sp.*, type of solvents and bacteria. Post-hoc test of Tuckey was applied to acknowledge the significant difference between tested variables (p<0.05).

3. Results And Discussion

Use of different solvents in extracting the bioactive compound from *Sargassum sp.* yielded dried extract differently. N-hexane resulted the lowest yield of extraction with 0.23 ± 0.05% of dry extract compared to ethyl acetate and methanol with 0.56 ± 0.12% and 2.18 ± 1.23% of dry extract, respectively. It indicates that the bioactive contained in *Sargassum sp.* tend to dissolve in polar solvents than in semi or no-polar one, implying that the extracted compounds are polar. Basic principle of extraction is the appropriate selection of solvents used to extract the solute. Non-polar solvents dissolve non-polar solutes while polar solvents dissolve polar solutes [19]. Bactericidal activity of *Sargassum sp.* extracts showed a diverse activity based on different locations. It might be suggested that there is a geographical variation in biological activity of *Sargassum sp.* Inhibition zone of extracts from *Sargassum sp.* collected at Bondo, Ujung Piring, and Bandengan were significantly different one to each other ($p<0.05$). Also, these same samples were also significantly different to those from Panjang Island and Teluk Awur ($p<0.05$). It is possible that these two stations, Panjang Island and Teluk Awur, have similarity in their water qualities. Furthermore, extracts of *Sargassum sp.* from Panjang Island and Teluk Awur showed the most active against the MDR bacteria than the other extracts. It was based on the biggest inhibition zone shown by the extracts from these two locations. It leads to an assumption that the environmental conditions in Panjang Island and Teluk Awur are favorable for the production of bioactive compounds.

Table 1 shows the statistical analysis of the antibacterial activity of *Sargassum sp.* based on different stations, species, and solvents. As can be seen from Table 1, antibacterial activity was best performed by *S. crassifolium* extracted with ethyl acetate. Moreover, samples from Panjang Island showed the best activity.

Table 1. Analysis of variance from each variable analyzed

Variance	*df*	*F*	*p*	*Tukey p <0.05*
Station	4	493.4	0	Bondo^{a}< Ujung piring^{b}< Bandengan^{c}< Teluk awur^{d} < Pulau panjang^{d}
Species	6	99.65	0	Bin^{a}< Echi^{b}< Plag^{b}< Cine^{b}< Poly^{c}< Dupli^{cd} < Cras^{d}
Solvent	2	4602	0	n-hexana^{a}< Metanol ^{b}< Etil asetat^{c}

Df: degree of freedom; *F*: F value; *p*: probability; Bin: *S. binder*; Cine: *S. cinereum*; Cras: *S crassifolium;* Dupli: *S. duplicatum;* Echi: *S. echinocarpum;*Plag: *S .plagyophyllum;* Poly: *S. polycystum;* Sa: *S. aureus.* The order of variables was based on the smallest of the biggest average value (<); different letters indicate significant difference ($p>$ 0.05).

Factors influencing the difference in inhibition zone from the antibacterial assay are the sensitivity of organisms, the media of culture, the condition of incubation, and the agar diffusion rate. Meanwhile, the agar diffusion rate depends on the concentration of tested samples, the composition of media, temperature and period of incubation [20]. Aside from these factors, it can also be suggested that different type of bioactive compounds extracted from *Sargassum sp.* play an important role in their bactericidal effects towards the MDR bacteria.

Type of solvents also contributes to the antibacterial effect of *Sargassum sp.* extract. Ethyl acetate and methanol extracts exhibited the highest inhibition zone; meanwhile, the n-hexane extracts had the smallest inhibition zone. Such difference might be due to different polarity of bioactive compounds from *Sargassum sp.* As non-polar solvents, n-hexane attracts non-polar compounds. As for semi-polar to polar compounds, they prefer to dissolve in ethyl acetate and methanol. Ethyl acetate is an aromatic compound characterized by its semipolar with $CH_3CH_2OC(O)CH_3$ as its chemical structure. This compound can dissolve analytes tending to be semipolar and polar [21]. It indicates that the bioactive compounds extracted from *Sargassum sp.* are semipolar.

From 42 extracts tested, one best extract with a wide spectrum of antibacterial effect and the biggest inhibition zone was chosen. It was *S. duplicatum* from Teluk Awur extracted with ethyl acetate. This extract was further tested for its toxic effect using BSLT and results are presented in Table 2.

Table 2. BSLT represented by the value of LC_{50} of *S. duplicatum* in different time interval

ta; *S. duplicatum*; ea							
Time (hours)	y = a + b*x	R^2	R	Correlation (%)	LC_{50} Concentration (ppm)	LC_{50} Criteria	
1	-	-	-	-	-	-	-
3	y = 0.017 x - 1.051	0.989	0.9945	99.45	3003	Not Toxic	-
6	y = 0.025 x + 2.130	0.904	0.9508	95.08	1915	Not Toxic	-
12	y = 0.034 x + 5.311	0.891	0.9439	94.39	1314	Not Toxic	-
18	y = 0.036 x + 13.27	0.906	0.9518	95.18	1020	Not Toxic	-
24	y = 0.082 x + 12.04	0.946	0.9726	97.26	463	Toxic	Chronic
30	y = 0.076 x + 28.47	0.930	0.9644	96.44	283	Toxic	Chronic
36	y = 0.143 x + 30.42	0.915	0.9566	95.66	137	Toxic	Chronic

Toxicity of *S. duplicatum* extracted in ethyl acetate was evaluated towards the larvae of *A. salina* through the BSLT method. It seems that the toxicity of this extract started 24 h after introducing the larvae of *A. salina* to ethyl acetate extract of *S. duplicatum*. Value of LC_{50} after 24 h was 463 ppm implying that this concentration could cause lethal effect towards 50% of the larvae population.

Table 2 displays that ethyl acetate extract *S. duplicatum* is considered as active or toxic because of its ability to cause 50% of mortality to the tested organisms with less than 1000 ppm [23]. Based on the category of toxicity from a compound (15). It generates toxic effect during the time interval 18 h. BSLT method is widely used as a method to evaluate a toxicity of natural product as mentioned by several studies [22]; [23]. The advantages of this method are the simplicity. The high reproductivity (24 h). cheap and can be used to determine the intensity of cytotoxicity.

Table 3. Phytochemical profiling of *S. duplicatum* extracted with ethyl acetate

Phytochemical profile	*S. duplicatum*	Standard (color)
Alkaloid		
Dragendorf	+	Red to orange
Meyer	+	Murky white
Flavonoid	+	Layer of amyl alcohol is red/yellow/green

Steroid	-	Green/blue
Phenolic	+	Green or greenish blue
Triterpenoid	-	Violet or purple
Saponin	-	Foaming on top of the layer
Tanin	-	Greenish brown or dark blue (bluish black)

Table 3 shows that the ethyl acetate extract of *S. duplicatum* contained alkaloid. Flavonoid and phenolic compounds. These compounds are assumed to be responsible for the antibacterial effect of *S. duplicatum* against *S.aureus*. *E. coli* and *S. epidermidis*. Inhibition of bacterial growth in the presence of antibacterial agents is provoked by the combination of these agents with bioactive compounds. These bioactive compounds might be from the secondary metabolites such as alkaloids. Peptides. Terpenes. Pigments and sterols [24]. Mechanisms of antibacterial from a compound can take place in several ways, which are : 1) the destruction of bacteria cell wall; 2) the change of permeability in a membrane cell and 3) prevention of protein and nucleic acid synthesis. Furthermore, concentration of extracts and bacteria species of bacteria the availability of organic material, temperature, and pH have been suggested as factors affecting the antibacterial effect of a compound [25].

Alkaloid serves as antibacterial agents and provokes the death of bacteria cell due to the presence of alkali groups containing nitrogen in its structure [26]. It leads to a change in amino acid structure and the genetic balance experiences a lysis. Moreover, flavonoid could obstruct the growth of bacteria by destroying the permeability of bacteria cell wall. Flavonoid can release transduction energy to the cytoplasmic membrane of bacteria and to inhibit the motility of bacteria. Hydroxyl group in flavonoid contributes to the alteration of the organic component and transport of nutrition generating a toxic effect towards the bacteria. Phenolic compounds also have a hydroxyl group in their chemical structure. The mechanism to eliminate the bacteria is mainly based on the difference of polarity between lipid and hydroxyl group. If the bacteria contain more lipids, then higher concentration is needed to rupture the bacteria cell. The mechanism of phenolic activity as an antimicrobial is the ability of phenolic to cause shrinkage. also known as astringent [27]. Phenols can form a complex with a microbial enzyme to reach the membrane cell of bacteria through the cell wall. Bacteria cell wall consists of different polysaccharides and proteins; thus. it allows part of phenols to penetrate the cell wall.

4. Conclusion

The present study reveals that *Sargassum* sp. extract can inhibit the growth MDR bacteria. It implies that *Sargassum* sp. contains bioactive compounds having the property as an antibacterial agent. of this extract. Results obtained unlocks the potential of *Sargassum* to replace commercial antibiotic against the MDR bacteria. Nevertheless. it is important to further consider the type of solvent used to extract the target compounds that are active against the MDR bacteria. Apparently. certain solvents might have a negative impact towards the living cell. as presented in this study. Eventhough ethyl acetate extract of *Sargassum* sp. showed to be the most effective against MDR bacteria, this extract was also toxic. As a consequence. it is crucial to perform further research related to the application of a more eco-friendly method to extract the target compound from *Sargassum* sp.

5. Acknowledgment

Authors would like to express gratitude to DRPM-SIMLITABMAS RISTEK DIKTI- Jakarta period 2016/2017 for the financial supports for this study.

6. References

1. Cappucino N.S J. 2012. (Rockland Community College). State University of New York.
2. Sari LORK. 2012. *Pharm Sci Res*. **3**(1).
3. Liu L, Heinrich M, Myers S, Dworjanyn SA. 2012. *J Ethnopharmacol.*;**142**(3):591 - 619
4. Bhaigyabati T.2011. *Int J Pharm Res Dev* ;**3**(974):25 - 30
5. Plouguerne E. Ioannou E. Georgantea P. Vagias C. Roussis V. Hellio C. 2010. *Mar Biotechnol*. **12**(1): 52–61.
6. Khaled N. Hiba M. Asma C. 2012.*Adv Environmen Biol*. **6**(1):8 - 42
7. Kang JYY, Khan MNANA, Park NHH, Cho JYY, Lee MCC and Fujii H. 2008. *J Ethnopharmacol*. **116**(1):90 - 187
8. Widowati I, Susanto AB, Puspita M, Stiger-pouvreau V, Bourgougnon N. 2014. *Int J Mar Aquat Res Conserv Co-exist*. **1**(1):63-67
9. Puspita M, Deniel M, Widowati I, Radjasa OK, Douzenel P, Bedoux G. *IOP Conf.Ser : Earth Environ. Sci*. **55** 012507
10. Li C, Li X, You L, Fu X, Liu RH. 2017. *Carbohydr Polym*. ;**155**: 70 - 261.
11. Manivannan K, Karthikai Devi G, Anantharaman P, Balasubramanian T. *Asian Pac J Trop Biomed*. 2011;**1**(2): 20 - 114.
12. Farnsworth NR. 1966. *J Pharm Sci*. **55**(3): 76 - 225.
13. Kanjana K, Radtanatip T, Asuvapongpatana S, Withyachumnarnkul B. Wongprasert K .2011. *Fish Shellfish Immunol*. 30(1): 89–96.
14. Lalitha MK. 2004. *Manual on Antimicrobial Susceptibility Testing* (Under the Auspices of Indian Association of Medical Microbiologist). Departement of Microbiology, Christian Medical College, Vellore, Tamil Nadu p. 43
15. Meyer BN, Ferrigni NR, Putnam JE, Jacobsen LB, Nichols DE , McLaughlin JL. 1982. *Planta Med*. ;**45**(5):4 – 31
16. Mclaughlin JL, Rogers LL, Anderson JE. 1998. *Drug Inf J*. ;**32**(2): 13–24.
17. Carballo JL, Hernández-Inda ZL, Pérez P, García-Grávalos MD. 2002. *BMC Biotechnol. ;2(1):17.*
18. Suparno. 2012. *Kajian Pertumbuhan dan Bioakvitas Antibakteri Spons Laut Petrosia nigricans yang Ditransplantasikan pada Lingkungan Perairan yang Berbeda*. (Institut Pertanian Bogor).
19. Palma M. Barbero G,. Pineiro Z, Liazid A, Barroso CG, Rostagno MA. 2013. *Extraction of Natural Products: Principles and Fundamental Aspects*. In: Rostagno MA. Prado JM. editors. (Natural Products Extraction: Principles and Applications. Cambridge. UK: The Royal Society of Chemistry). p. 58–86.
20. Schlegel HG. 1994. *Mikrobiologi Umum*. Keenam. Baskoro R. Wattimena J. (Gadjah Mada University Press). Yogyakarta.
21. Snyder LR, Kirkland JJ, Glajch JL. 2012. *Practical HPLC Method Development*. (John Wiley & Sons).
22. Fajarningsih ND, Januar HI, Nursid M, Wikanta T. 2006. *J Pascapanen dan Bioteknol Kelaut dan Perikan*. **1**(1):35–42.
23. Annie Selva Sonia G, Lipton AP, Paulraj R. 2009. *J Mar Biol Assoc India*. **51**(1): 5 - 122.
24. Barbosa M, Valentão P, Andrade P. 2014. *Mar Drugs*. **12**(9):4934–4972.
25. Pelczar MJ and Chan ECS. *Dasar-Dasar Mikrobiologi*. 1986. Jilid ke-1. Penerjemah: HadioetomoRS, Imas. T., Tjitrosomo. SS. and Angka. SL. (Universitas Indonesia Press). Jakarta.
26. Farida R, Dewa M, Titis N, Endrawati T. 2010. *J Kedokteran dan Kesehatan Indonesia*. ;**1**(7):10–25.

27. Fitrial Y, Astawan M, Soekarto SS, Wiryawan KG, Wresdiyati T, and Khairina R.2008. *J. Teknol dan Ind Pangan.* ;**19**(2):158–64.

Mapping of HABs Contaminated In Green Shells (Perna viridis) in Semarang Bay

Churun A'in[1*], Suryanti Suryanti[1], Haeruddin Haeruddin[1]

[1] Aquatic Resources Department, Faculty of Fisheries and Marine Sciences, Diponegoro University Jl. Prof Soedarto, SH, Tembalang, Semarang, Indonesia 50275

*Corresponding author Email : ainchurun@yahoo.com

Abstract. The existence of *Harmful Algae Blooms* (HABs) can adversely affect the water like a mass death of fish and oxygen depletion. Some types of HABs can be contaminated with seafood and contain biotoxins that are detrimental to the health of humans who consume them. Green mussels *(Perna viridis)* has the properties of *filter feeders* so vulnerable to contamination HABs. This research was conducted to produce spatially thematic maps contaminated HABs in *P. viridis* so providing information about risk prediction *P. viridis* when consumed by humans. Sampling was done *purposively* in three (3) stations that represent the Bay Semarang namely western boundary waters (Kendal), middle (Semarang) and the eastern boundary (Demak). Sampling done two (2) times, namely East season (June -July) and the second transitional season (September) 2016. Analysis of HABs done either in water or body tissues *of P. viridis* through the analysis of *food habit.* The results shows that *P. viridis* genus positive contaminated HABs phytoplankton Tricodesmium and Ceratium. Spatial distribution and abundance of Tricodesmium genus Fitoplankton is wider and taller than the HABs Phytoplankton genus Ceratium. Group HABs are found in the tissues *of P. viridis* no potential as biotoxin that does not cause adverse health risks.

Keywords: Green Shells, HABs, Semarang's Bay

1. Introduction

Semarang's Bay is a water that stretched from Delta Bodri Kendal to Delta Wulan Demak. It's under increasing pressure as long as development of settlement areas, agriculture, aquaculture, ports and industries along the Gulf coast Semarang [1] The used of land in the coastal area impacts on the water quality conditions in Semarang Bay and potentially be pollution [2]. The mechanism of distributing the impact from upstream toward downstream areas Semarang Bay can through the mechanism of precipitation, infiltration and runoff . There are five main watersheds that lead to Semarang Bay and within the DAS are many large and small rivers that flow into Semarang Bay [3], so many pollutants from various sources will accumulate in Semarang Bay and can affect aquatic organisms such as Green Mussels. In 2015 Semarang Bay's water showed mesotrofik until to eutrofik status that have nutrient enrichment, so potentially be make a pollution [4].

Green Mussels as a filter feeder organisms have a high vulnerability to absorb harmful contaminants, but has a privilege that adaptability or the ability to survive in a high ecological pressures without interference, metabolically Green Mussels are not disrupted by the presence of contaminants. *P. viridis* exposed with harmful contaminants, are not able to be detected by naked eyes without laboratory analysis and further analysis [5]. The threat of contaminants it will endanger the *trophic level* organisms or humans who consume *P. viridis*. Information on the effects of pollution on biota Green mussels still very low so we need studies that examine this particular risk prediction *(Risk Assessment)* of *P. viridis*. There are two (2) contaminants that could potentially get into the body tissues *of P. viridis,* namely inorganic contaminants such as heavy metals and organic contaminants such as phytoplankton. Phytoplankton is an autotrophic organism that plays an important role in the food chain in waters, phytoplankton is a good source of food for Green Shells. However, the increase in the population of some species of phytoplankton due to nutrient enrichment waste from domestic activities, agriculture and fisheries could lead to a population explosion of phytoplankton containing toxins or better known as *Harmful Algae Blooms* (HABs) can have a negative impact. An increase in the population of toxic phytoplankton in the waters will produce toxins that can be transferred through the food chain *(food web)* or the environment. The content of toxins contained in these phytoplankton can affect the growth and even kill the organisms that consumed [6].

This study aims to determine risk prediction information based on the analysis of HABs on Green Shells spatially in Semarang Bay. This research is useful to provide information to the public about information on Green Shell Risk Prediction so as to give consideration and choice to the community to consume Green Shell caught in Semarang Bay safely.

2. Materials And Methods

2.1. Sampling Methods

The research location is in the Semarang Bay, sampling was done *purposively* in three (3) representing the Semarang Bay stations are: Station I, the waters of the western boundary of Semarang Bay (Kendal); Station II, is the central waters of Semarang Bay (Semarang); And Station III, is the eastern boundary of Semarang Bay (Demak). Sampling time is 2 (two) times in East (June-July) and transitional period II (September) in 2016. Data taken are sea water sample data: plankton to identify and study HABs concentration and water quality data Supporters. Sampling was done by *purposive P. viridis* with consideration for the uniform samples (obtained both in terms of size or age) and have same of absorption, metabolism and response to HABs. At each point of observation in station, sample's 30 shells *P. viridis* taken with a total weight of shell @ ± 5-10 grams, in order to obtain the minimum weight for the analysis of *P. viridis* HABs on the network without a shell 25 grams.

2.2. *Biotoxin analysis*

Biotoxin analysis of HABs was detected through the analysis of the Food Habit *P.viridis*. The shells are prepared in advance by blending and then diluted with aquadest before being analyzed. Phytoplankton identification in the network based on the book identification *P.viridis* Yamaji essay, M, Sachlan by using a checklist.

2.3. *Spatial analysis*

Spatial analysis using ARGIS software with IDW interpolation method. All parameters that have been interpolated overlay done then reclass method, weighting or scoring quantitatively on each indicator both Biotoxin and Bioakumulasi. The end result of these processes is the risk prediction maps obtained spatially *P. viridis*.

3. Results And Discussion

3.1. Location Description

Semarang Bay is one of the coastal areas susceptible to environmental degradation such as coastal erosion, sedimentation and rob floods. Industries located in the vicinity of Semarang Bay like Terboyo Industrial Area, LIK Genuk, PT. Indonesia Power, Port of Tanjung Emas Semarang, Kendal Port, PT

Pokphan Sayung, PT KLI Kendal, and several other industries provide threats of pollution in Semarang Bay, so the study of potential contaminants both organic and non-organic that could harm biota and the environment It is important to do it as a preventive measure as well as management in coastal areas.

3.2. Phytoplankton Analysis of HABs in Waters

Identification of HABs in the waters is done for comparison and supporter of the data within the network HABs *P. viridis.* Based on the list of HABs phytoplankton groups (Wiadnyana, 1996) and GEOHAB (2001), there are 2 (two) genera including HABs in Semarang Gulf waters, namely Tricodesmium *(Tricodesmium erythreaum)* and Ceratium *(Ceratium fusus)* (Figure 1). Although the percentage of phytoplankton HABs abundance is low (<15%) compared with Non-HABs Phytoplankton found in waters, the presence of HABs Phytoplankton still needs to be wary because accroding Nontji [7], HABs produce toxins in the body which can then be diverted to shellfish or fish through the food chain *(food chain).* The presence of toxic in the body of the shell may not cause death to the shell, but when eaten by humans it can cause health problems and even death.

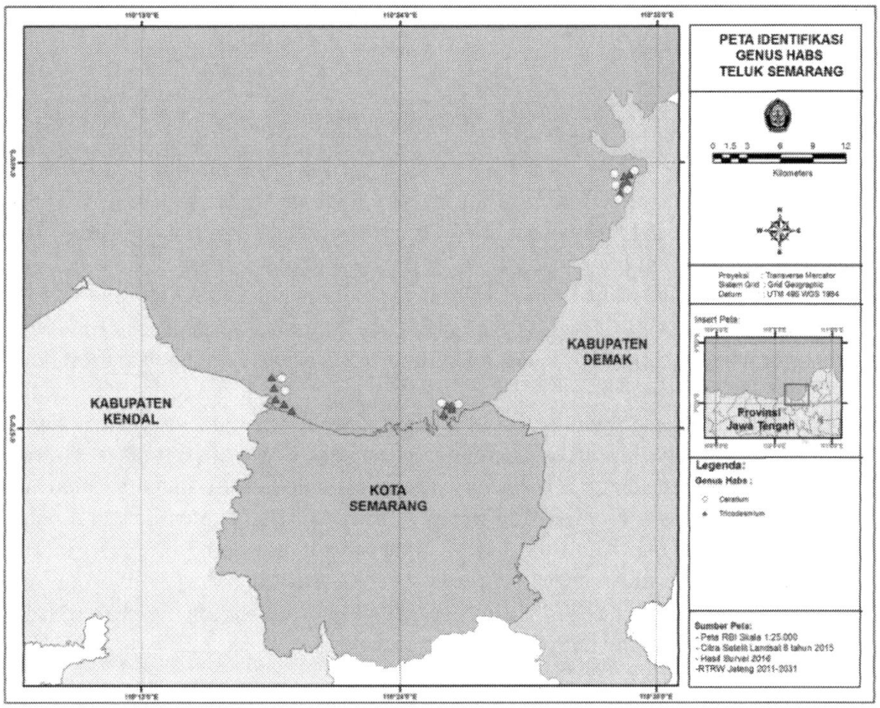

Figure 1. Map of Identification of Genus HABs in Teluk Semarang Waters

3.3. Phytoplankton Analysis of Contaminated HABs in the viridical Perna Network

Analysis of HABs contaminated phytoplankton in Green Shell tissue was analyzed by looking at the habit of eating the biota The. While the large percentage of Phytoplankton HABs approach *of Prefoderance Index (%)* as presented in Table 1, 2 and 3.

Table 1. Plankton on Green Shells at Station 1. Demak

No	Genus	*Index of Prefoderance (%)*	
		St 1 (East Season)	St 2 (Transitional Season II)
1	Coscinodiscus	58.24	30.07
2	Thalasiotrix	14.29	2.20
3	Synedra	12.31	-
4	*Tricodesmium erythreaum* *	5.41	14.95
5	Guinardia	2.64	3.34
6	Thalasionema	0.26	-
7	Nitzschia	3.74	11.74
8	Bacteriastrum	0.92	2.66
9	Pyrocitis	0.79	-
10	Rhizosolenia	1.32	24.79
11	Minidiscus	0.04	4.89
12	Pleurosigma	0.04	2.63
13	Tintinnopsis	-	0.16
14	Protoperidium	-	1.82
15	Leptocylindrus	-	0.45
16	*Ceratium fusus* *	-	1.07

Description (-): unidentified, (*) Phytoplankton HABs

Analysis of the food habits of *P. viridis* at the station I (the territorial waters of Demak) has found 12 genera of phytoplankton in East season, while in the transition season II 13 genera. Plankton found on Green Shell's gastric at Station 1. Demak is relatively small compared to other stations, it is presumably because the size found at Demak station is smaller than other stations, so the absorption capacity or filter feeder capability is not as big as the shells found. Differences in the size of Shell between stations are caused by slightly different planting times (charts) on each station. Phytoplankton types of HABs found in East season just *Tricodesmium erythreaum* while in the transition season, HABs found Phytoplankton is *Tricodesmium erythreaum* and *Ceratium fusus*. The present of *Ceratium fusus* same as research in a coastal region of Sagami Bay, Japan that seasonal abundance of the dominant dinoflagellate, *Ceratium fusus*, was investigated from January 2000 to December 2003. In Sagami Bay, *C. fusus* increased significantly from April to September, and decreased from November to February, though it was found at all times through out the observation period. Rapid growth was observed over a salinity range of 24 to 30 that *C. fusus* has the ability to grow under wide ranges of water temperatures (14–28°C), salinities (20–34), and photon irradiance (50–800 μmol $m^{-2}s^{-1}$); it is also able to grow at low nutrient concentrations [8].

Table 2. Plankton on Green Shells at Station 2. Semarang

No	Genus	*Index of Prefoderance (%)*	
		St 1 (East Season)	St 2 (Transitional Season II)
1	Thalasiotrix	20.43	-
2	*Tricodesmium erythreaum* *	17.15	1.24
3	Coscinodiscus	29.78	60.02
4	Miniduscus	2.52	4.17
5	Guinardia	2.35	1.79
6	Triceratium	0.07	-
7	Rhizosolenia	1.74	12.28
8	Synedra	17.08	-
9	Navicula	0.18	-
10	Thalasiosira	0.72	-

Table 2

No	Genus	*Index of Prefoderance (%)*	
		St 1 (East Season)	St 2 (Transitional Season II)
11	Thalasionema	2.89	-
12	Pyrocytis	1.14	-
13	Bacteriastrum	0.11	-
14	Skeletonema	0.80	-
15	Nitzschia	3.66	6.34
16	Pleurosigma	0.33	4.68
17	Leptocylindrus	0.25	3.89
18	Isthmia	0.07	-
19	Chaetoceros	0.15	0.63
20	Biddulphia	0.07	-
21	Favella	0.22	-
22	Surirella	0.02	-
23	Detonule	0.02	-
24	Microstella	0.01	-
25	Stephanopyxis	-	1.63
26	Protoperidinium	-	0.67
27	Prorocentrum	-	2.66

Description (-): unidentified, (*) Phytoplankton HABs

Analysis of the food habits of *P. viridis* in station II (Semarang territorial waters) has found 24 genera of phytolankton on East season, while in the transition season II 12 genera. HABs Phytoplankton are found in both East and Transition II season is kind *Tricodesmium erythreaum.*

Table 3. Plankton on Green Shell's at Station 3. Kendal

No	Genus	*Index of Prefoderance (%)*	
		St 1 (East Season)	St 2 (Transitional Season II)
1	Thalasiotrix	15.31	5.18
2	*Tricodesmium erythreaum* *	4.61	8.34
3	Coscinodiscus	44,78	30.60
4	Miniduscus	6.03	0.97
5	Guinardia	0.39	-
6	Triceratium	0.25	-
7	Rhizosolenia	2.90	29.47
8	Synedra	12.56	-
9	Thalasiosira	0.53	-
10	Thalasionema	1.74	-
11	Pyrocytis	0.06	-
12	Bacteriastrum	0.17	-
13	Skeletonema	0.10	-
14	Nitzschia	7.97	1.40
15	Guinardia	0.79	-
16	Pleurosigma	1.31	3.84
17	Leptocylindrus	0.21	1.97
18	Skeletonema	0.26	0.08
19	Chaetoceros	0.26	9.74
20	Biddulphia	0.04	-

Table 3.

No	Genus	*Index of Prefoderance (%)*	
		St 1 (East Season)	St 2 (Transitional Season II)
21	Detonule	0.08	-
22	Stephanopysis	0.43	3.81
23	Dictyoca	0.13	-
24	*Ceratium fusus* *	1.43	2.06
25	Thalasiotrix	15.31	-
26	Tintinnopsis	-	0.40
27	Asterionella	-	0.51
28	Melosira	-	0.08
29	Protoperidinium	-	1.02

Description (-): unidentified,, (*) Phytoplankton HABs

Analysis of the food habits of *P. viridis* in station III (Kendal territorial waters) showed on East season found 25 genera were found, while in the transition season 16 genera. Species HABs Phytoplankton *Tricodesmium erythreaum* and *Ceratium fusus* found on east season and transition season. Comparison of Phytoplankton type composition in each station is presented in Figure 2.

Based on the analysis of food habits, phytoplankton HABs Tricodesmium contaminants in water was found also in the stomach tissue *of P. viridis,* while the genus Ceratium only found in *P. viridis* in Kendal waters and Demak in the transition season (Figure 2). Factors thought to cause these differences and at least Ceratium found on *P. viridis* are:

- The low abundance of Ceratium in the waters
- Ceratium eaten not only by *P. viridis* but other consumer level I like the fish of planktonic
- Suitability Ceratium size with openings and uptake *P. viridis,* if Ceratium size larger than *P. viridis,* it will most likely not be absorbed and filtered out in the network *P. viridis.*

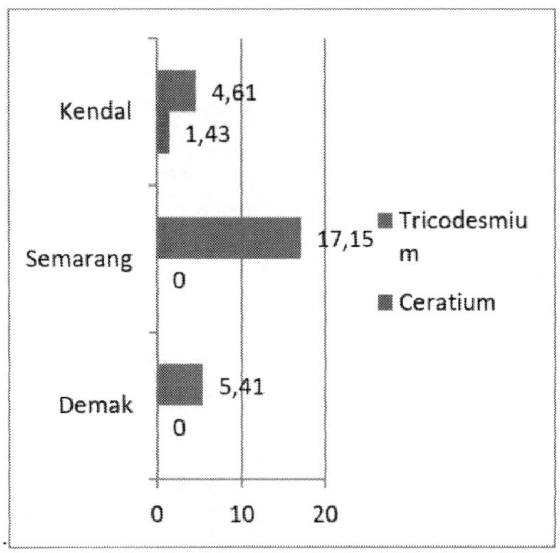

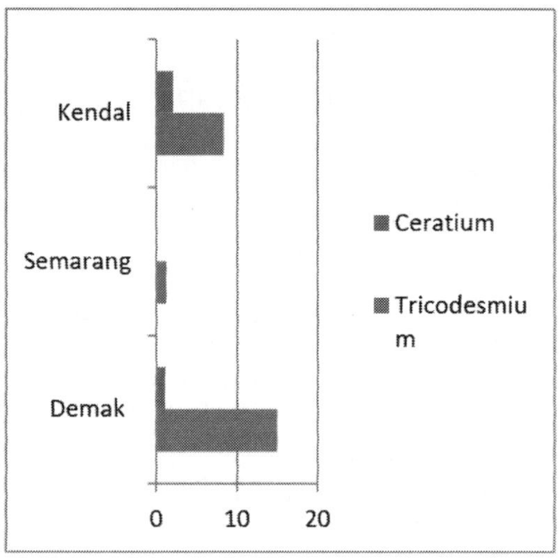

Figure 2. Composition of Phytoplankton Contaminated HABs on the Network *P. viridis* East season (A) and the Transition season (B)

HABs dominant genus of phytoplankton found in both aquatic and *P. viridis* is Tricodesmium. Which belongs to the Cyanophyceae class. Two (2) types of Tricodesmium genus that cause HABs are

Tricodesmium thiebautii and *Tricodesmium erythareum*. Sediadi *in* Anggita *et al.,* [9], this type of shaped colonies Tricodesmium important role in the biogeochemical processes in marine waters. Tricodesmium type distribution. Found in subtropical marine waters to tropical marine waters. *Tricodesmium* sp. in high densities in these waters need special attention, because of the impact of *blue-green algae bloom* can cause oxygen depletion. Another characteristic *Tricodesmium* spp. to watch out for is when overflow *(blooms)* in tropical waters, which can cause a lack of oxygen resulting in the decay process that eventually can lead to the death of marine life, such as fish [10].

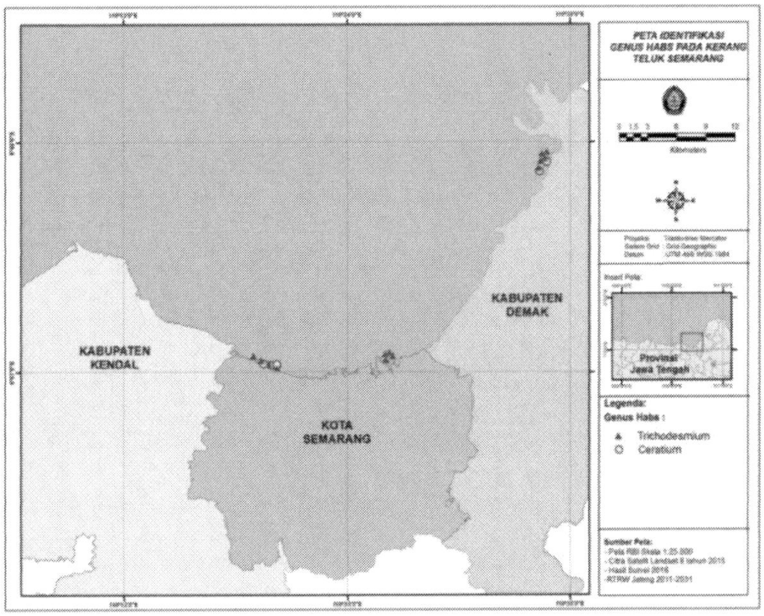

Figure 3. Mapping the identification of contaminated HABS on *P.viridis* in Semarang Bay

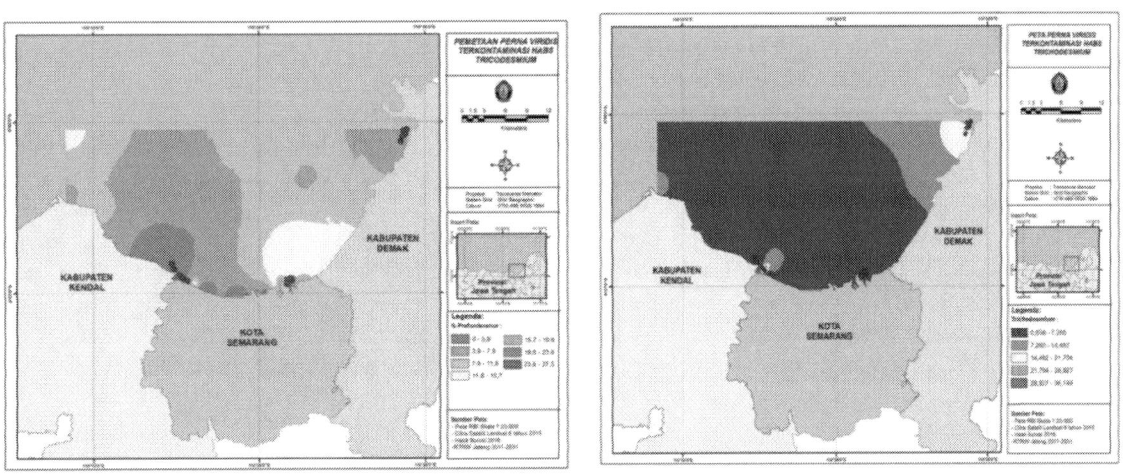

Figure 4. Mapping *P. viridis* Contaminated Tricodesmium HABs in East season (A) and Transition season (B)

Ceratium sp. is the second highest HABs phytoplankton genera were found either on the water or on *P. viridis* in the Semarang Bay. *Ceratium* s p.population explosion Can lead to mass death of marine biota due to decreased oxygen levels. According to Baek *et al*[8], a population explosion of *Ceratium* sp. can cause anoxia and sea water hypoxia, in addition there are species of this genus that can injure the gills of fish, but the mechanism can not be known. *Ceratium Furca* able to compete with other phytoplankton species, especially in the availability of nutrients, sunlight, and other environmental factors [11]. *Ceratium Furca* often found in abundance and dominate other species. Therefore, *C. Furca* more often *blooming,* resulting in the mass death of marine organisms due to oxygen depletion can occur in waters and affect other resources.

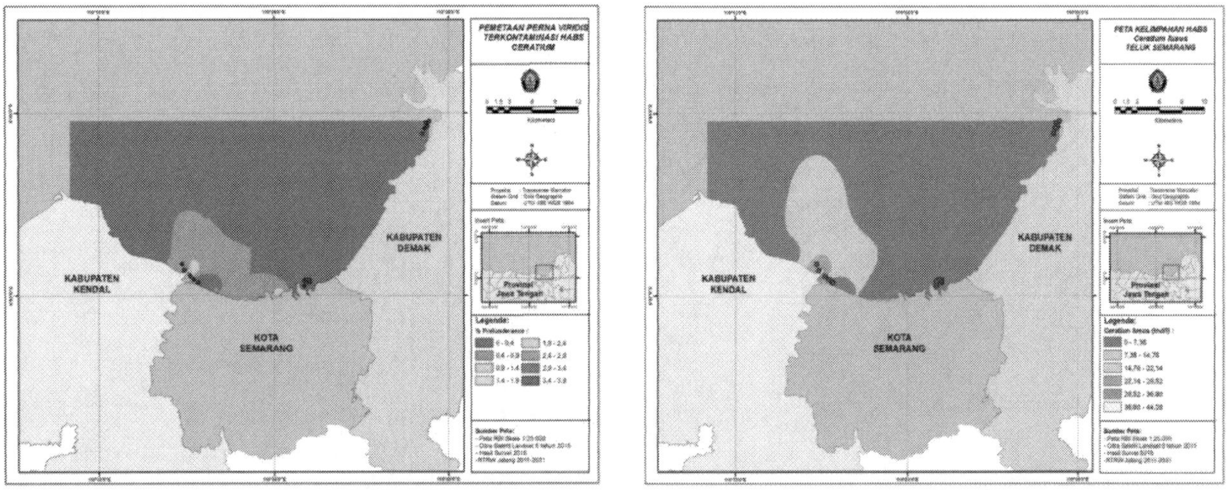

Figure 5. Mapping *P. viridis* Contaminated Ceratium HABs in East season (A) and the Transition season (B)

There are five (5) factors that cause the *blooming of* phytoplankton and harmful phytoplankton trigger HABs [12], namely: eutrophication, or nutrient enrichment; The existence of the clan *Pyrodinium* cyst at the base of the waters back into phytoplankton HABs when lifted to the surface layer through two mechanisms, namely: The mechanism through rising water masses *(upwelling)* and mechanism under the influence of tectonic earthquakes; biological, which means that the lack of predators as predators of the HABs-causing species; Large-scale hydrometeorological changes; And Heavy rains and the entry of fresh water into the sea in large numbers.

Biota cause HABs produce toxins in the body which can then be transferred to shellfish or fish through the food chain. The presence of toxic in the body of the shell may not cause death to the shell, but when eaten by humans it can cause health problems and even death [7]. Toxins produced is a stable molecule, which can not be decomposed by high-temperature cooking methods and food processing [13]. The process of accumulation of toxins takes place in the food chain, where the last consumer is the largest toxic accumulation. Results of research have shown that contaminants Tricodesmium and Ceratium positive contaminated *on P. viridis*. However, the genus has no characteristics as a biotoxin, its adverse effects the environment such as oxygen depletion.

4. Conclusion

Based on the results of the research, the conclusions obtained are as follows:

1. Green mussels *(Perna viridis)* positive contaminated HABs phytoplankton. They are from genus Ceratium and Tricodesmium, that does not have the characteristics as biotoxin
2. Spatial distribution and abundance of Tricodesmium genus Fitoplankton is wider and taller than the HABs Phytoplankton genus Ceratium.

5. Acknowledgment
Authors acknowledge the Diponegoro University for funding this research through Development and Application Research (Riset Pengembangan dan Penerapan) Scheme for fiscal year of 2016.

6. References

[1] Adji K 2008 *Evaluation of contamination of pathogenic bacteria in fresh fish in Semarang Bay waters* Graduate Program Master of Coastal Resources Management
[2] Marfai, MA and King L 2008 J.Indonesia Environmental Geology **55**(7) 1507–1518
[3] BPDAS Pemali Jratun 2009 *Watershed Management Information System* Central Java BPDAS
[4] Ain C, Jayanto B B, Latifah N 2015 J. Saintek Perikanan **11** 11-19
[5] Wu H and Wang W X 2011 J. Environmental Toxicology and Chemistry **30** (4) 806–812
[6] Takahashi E, Qiming Y, Geoff E, Des W C, James M 2007 J. Marine Environmental Research **64** (4) 429–442
[7] Nontji A 2006 *No Life on Earth Without Plankton Presence* Lembaga Ilmu Pengetahuan Indonesia (LIPI) Jakarta
[8] Baek SH, Shimode S and Kikuchi T 2007 J. Oceanography **63** 35-45
[9]Anggita WT, Haeruddin, Ain C 2016 J. Saintek Perikanan **12** 52-59
[10] Garno YS 2008 *Water Quality and Phytoplankton Dynamics In Hope Island Waters* J. Hidrosfir Indonesia **3** 87-94.
[11] GEOHAB 2001 *Global Ecology and Oceanography of Harmful Algal Bloom Science Plan* SCOR and IOC Paris
[12] Widnyana NN 1996 J.Oceanography and Limnology in Indonesia **29** 15-28
[13] Anderson D M, J M Burkholder, W P Cochlan, PM Gilbert, C J Gobler, CA Heil, RM Kudela, L Parsons, JE Jack Rensel, DW Townsend, V L Trainer, G A Vargo 2008 *Harmful Algall Blooms And Eutrophication: Examining Linkages From Selected Coastal Region of The United Stated, Harmful Algae* p.39-53

Nutritional Composition Changes During *Tempeh Gembus* Processing

Ruth Nazaretha Sandessy Damanik[1], Dwi Yanti Winda Pratiwi[1], Nurmasari Widyastuti[1], Ninik Rustanti[1], Gemala Anjani[1], Diana Nur Afifah[1,2]
[1] Department of Nutrition Science, Faculty of Medicine, Diponegoro University, Jl. Prof. H. Soedarto, SH, Tembalang, Semarang 50275, Indonesia
[2] Center of Nutrition Research (CENURE), Diponegoro University, Jl. Prof. H. Soedarto, SH, Tembalang, Semarang 50275, Indonesia

d.nurafifah.dna@fk.undip.ac.id

Abstract. This paper was aimed to analyze nutritional composition during *tempeh gembus* processing. Processing of *tempeh gembus* may cause an increase or decrease in its nutritional value such as amino acids, fatty acids, vitamins, and minerals. Regardless, *tempeh gembus* may still be potentially beneficial to help fulfill nutrient needs. This was a descriptive study with three samples: soybeans, tofu residue, and *tempeh gembus*. Soybeans used were varieties of Grobogan, Central Java. Minerals identified in all three samples include sodium, calcium, potassium, phospor, and iron. Soybeans, tofu residue, and *tempeh gembus* each contain 34.12%, 5.40%, and 4.80% of amino acids. Saturated fatty acids were the lowest in soybeans (12.01%), marginally surpassed by tofu residue (12.41%) and *tempeh gembus* (12.55%). Total levels of monounsaturated fatty acids were also found in a similar pattern (34.1%, 36.5%, 36.7% respectively). In contrast, soybeans had significantly more polyunsaturated fatty acids than other samples, containing up to 43.6% in comparison with 38.29% in tofu residue and 30.18% in *tempeh gembus*. Overall, the nutritional profile of soybeans was generally better than tofu residue and *tempeh gembus*. However, fermentation process may play a role in enchancing the fatty acid profile.
Keywords: nutritional profile, soybeans, tofu residue, *tempeh gembus*

1. Introduction

Food not only be acceptable and have a high nutritional value, but also have functional properties. Therefore, the function of food was varied increasingly, in addition to supplying nutritional needs with delicious flavor, but also served to maintain our health. The development of functional food requires raw materials that can fulfill these aims [1].

Functional foods are defined as foods that contain bioactive ingredients thought to enhance health, not only fulfill of basic energy needs, macronutrients (proteins, carbohydrates, fats) and micronutrients (vitamins and minerals) but also provide nutritional and physiological benefits for our body [2]. Functional food ingredients are associated with the prevention and treatment of several leading causes of death, such as cancer, diabetes, hypertension, and heart disease [3]. Soybeans had many beneficial health effects when we consumed. Soybeans are an excellent source of digestible protein. The content of vitamins and minerals in soybeans is high, and saturated fatty acids in soybeans is low and contains high unsaturated fatty acids such as linoleic and linolenic acids [4].

One of the nutrients which contained in functional food is protein (amino acids). Amino acids are nutrients that have an important function in the human body, such as leucine, isoleucine, and valine that can prevent protein catabolism caused by decreased levels of glycogen in muscle. Soybeans are one of the foods that contain high protein [5]. In addition, the most commonly used bioactive compounds of the oil or fat groups in functional food formulations are Omega-3 fatty acids, which are classified as polyunsaturated fatty acids (PUFAs) [6]. PUFA plays an important role in fat transport and metabolism, immune function, maintaining fumigation and integrity of cell membranes. PUFA has been showed to reduce levels of VLDL and LDL in the blood because the liver will not convert it into VLDL [7]. Vitamins are organic compounds essential for a proper functioning of the human body. They required in small amounts and are obtained from a correct diet [8]. A balanced and varied diet will probably supply the amounts of vitamins and minerals necessary for a proper functioning of the body [9].

In Indonesia, soybeans can be processed into several processed food products, one of them is tofu. Processing of tofu will produce solid waste and liquid waste. Solid waste is known as tofu residue. However, the nutrients value of tofu residue is lower than soybeans, since the tofu residue is the soybeans residue filtration. The tofu residue in wet conditions per 100 grams contains 11.07% of carbohydrate, 4.71% of protein, 1.94% of fat and 0.08% of ash [10]. The tofu waste can still be further processed through fermentation process and produce new processed product that is *tempeh gembus*. *Tempeh gembus* is a fermented food made from tofu residue with the help of the same microorganisms in the process of making soybeans of tempeh, namely *Rhizopus oligosporus* [11]. Fermentation process will affect the nutritional value of food. The nutrient content of a material will change during fermentation. In addition, fermentation causes the product to have a higher value, such as organic acids, single cell proteins, biopolymers and antibiotics [12].

Based on these of *tempeh gembus* was containing nutrient components such as amino acids, fatty acids, vitamins, and minerals that are beneficial of health. However, the processing of food in particular fermentation causes changes in the value of these nutrients. Until now there has been no research on the change in fatty acid profile in the process of *tempeh gembus* from soybeans, tofu residue, to become *tempeh gembus*. Therefore, researchers interested to analyze the profile of amino acids, fatty acids, vitamins, and minerals in soybeans, tofu residue, and *tempeh gembus*.

2. Methods

2.1 *Materials*

Sample was using soybeans, tofu residue, and *tempeh gembus*. Soybeans was using Grobogan varieties were obtained from soybean farmers in Grobogan Village, while the tofu residue and *tempeh gembus* were obtained from tempeh producers in Wulung Village, Blora.

2.2 Processing of tempeh gembus

Tempeh gembus were processed traditionally, including the process of extortion, steaming, cooling, the addition of tempeh inoculum containing *Rhizopus oligosporus* spores, packaging with hollow plastic bags, and incubation. The result is a solid, grayish-white substrate, since the entire surface is covered in mushroom mycelium and has a distinctive odor.

2.3 Chemical analysis

Analysis of amino acids protein constituents with HPLC (High Performance Liquid Chromatography) Shimadzu CBM 20A made by North America, as well as fatty acids with Gas Liquid Chromatography Shimadzu GC 2010 made by North America.

2.4 Data analysis

This research is descriptive research to know the profile of amino acids, fatty acids, vitamins and minerals in soybeans, tofu residue, and *tempeh gembus*. Samples will be analyzed by duplo to know the profile of amino acid, fatty acid, vitamins and minerals of soybeans, tofu residue, and *tempeh gembus*.

3. Result

3.1 Proximate composition

In this present study the energy, moisture, ash content, fat, protein, and carbohydrate of soybeans, tofu residue, and *tempeh gembus* were investigated. Soybeans contain a high content of energy, moisture, ash content, fat, protein, and carbohydrate than tofu residue and *tempeh gembus*. Table 1 and figure 1 show the proximate composition of soybeans, tofu residue, and *tempeh gembus*.

Table 1. Proximate composition of soybeans, tofu residue, and *tempeh gembus* (%w/w)

Component	Mean ± SD			
	Soybeans	Tofu residue	*Tempeh gembus*	Unit
Energy	438,98 ± 0,42	107,32 ± 0,58	109,08 ± 0,26	kkal/100 g
Moisture	10,69 ± 0,01	77,22 ± 0,35	75,75 ± 0,01	%
Ash content	4,48 ± 0,02	0,8 ± 0,01	0,78 ± 0,03	%
Fat	19,94 ± 0,05	3,88 ± 0,07	3,04 ± 0,01	%
Protein	34,95 ± 0,04	3,7 ± 0,02	6,7 ± 0,04	%
Carbohydrate	29,94 ± 0,02	14,4 ± 0,00	13,73 ± 0,07	%

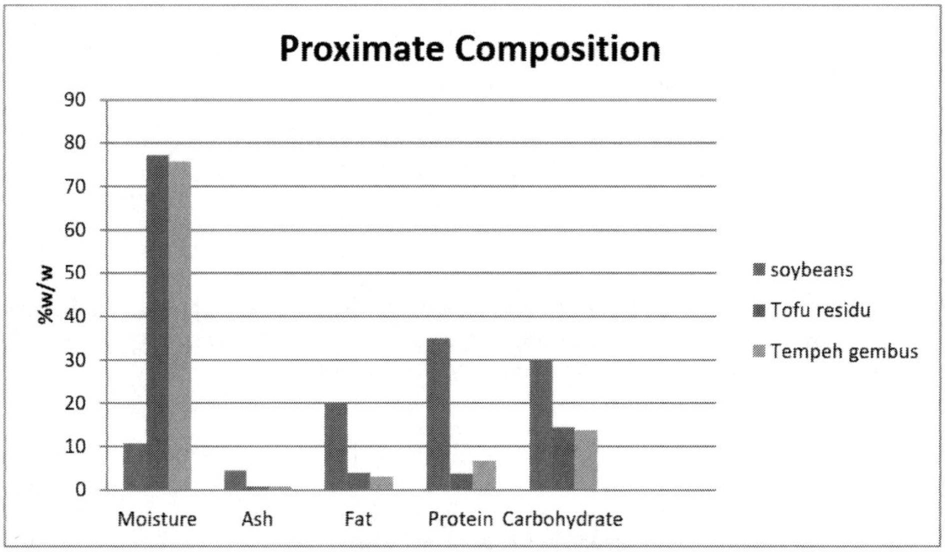

Figure 1: Proximate composition of soybeans, tofu residue, and *tempeh gembus*

3.2 Amino Acids

The present investigation shows totally fifteen amino acids present in soybeans, tofu residue, and *tempeh gembus*. The leucine levels are the highest level of essential amino acids of soybeans, tofu residue, and *tempeh gembus* whereas in non-essential amino acids glutamic acids are the highest level. The amino acids are shown in Table 2 and Figure 2

Table 2. Amino acids component of soybeans, tofu residue, and *tempeh gembus* (%w/w)

Amino Acids	Mean ± SD		
	Soybeans	Tofu residue	*Tempeh gembus*
Essential amino acids			
Histidine	0,82 ± 0,01	0,14 ± 0,01	0,12 ± 0,01
Isoleucine	1,93 ± 0,00	0,33 ± 0,02	0,29 ± 0,02
Leucine	2,92 ± 0,00	0,52 ± 0,02	0,44 ± 0,02
Lysine	1,63 ± 0,01	0,24 ± 0,01	0,24 ± 0,04
Methionine	0,42 ± 0,01	0,07 ± 0,01	0,05 ± 0,00
Phenylalanine	2,15 ± 0,02	0,36 ± 0,02	0,31 ± 0,02
Threonine	1,16 ± 0,22	0,22 ± 0,01	0,21 ± 0,01
Valine	1,94 ± 0,02	0,36 ± 0,02	0,31 ± 0,02
Non essential amino acids			
Tyrosine	1,21 ± 0,14	0,19 ± 0,01	0,20 ± 0,02
Alanine	1,74 ± 0,04	0,13 ± 0,01	0,31 ± 0,02
Arginine	2,48 ± 0,12	0,34 ± 0,02	0,28 ± 0,00
Aspartic acid	4,50 ± 0,07	0,67 ± 0,04	0,59 ± 0,04
Glutamic acids	7,51 ± 0,22	1,05 ± 0,08	0,88 ± 0,07
Glycine	1,70 ± 0,13	0,29 ± 0,00	0,26 ± 0,02
Serine	1,83 ± 0,00	0,28 ± 0,00	0,28 ± 0,02

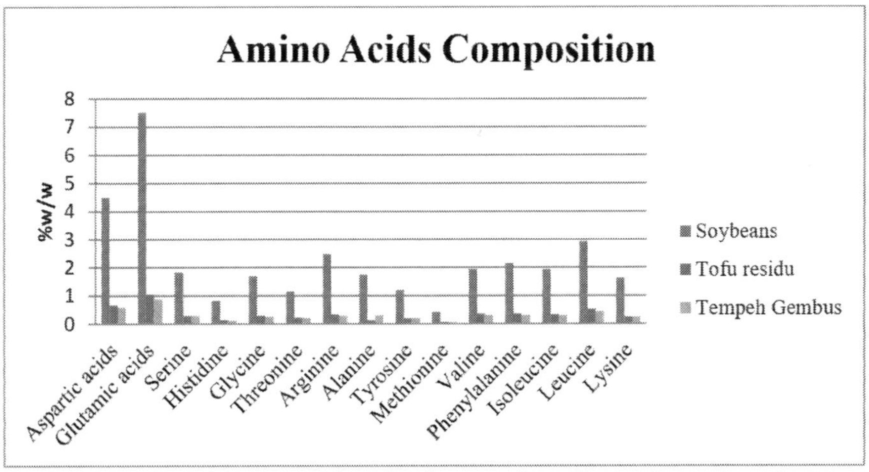

Figure 2: Amino acids composition of soybeans, tofu residue, and *tempeh gembus*

3.3 Fatty acids

In this present study, there are seventeen fatty acids were observed. Linoleic acids are the highest levels of fatty acids in soybeans, tofu residue, and *tempeh gembus*. In *tempeh gembus*, there is a fatty acid that is not contained in soybeans namely lauric acids, and one amino acids loss in *tempeh gembus* is heneikosanoat acids. The fatty acids are shown in Table 3

Table 3. Fatty acids composition of soybeans, tofu residue, and *tempeh gembus* (w/w)

Fatty acids	**Mean** ± SD		
Saturated Fatty Acid (SFA)	Soybeans	Tofu residue	*Tempeh gembus*
Lauric acid (C12:0)	-	-	0,04 ± 0,01
Miristic acid (C14:0)	0,07 ± 0,00	0,05 ± 0,01	0,08 ± 0,00
Palmitic acid (C16:0)	9,42 ± 0,13	9,62 ± 0,55	9,21 ± 0,39
Heptadekanoat acid (C17:0)	0,05 ± 0,01	0,06 ± 0,00	0,06 ± 0,01
Stearic acid (C18:0)	1,64 ± 0,08	1,85 ± 0,12	2,12 ± 0,07
Arakhidat acid (C20:0)	0,21 ± 0,02	0,23 ± 0,01	0,23 ± 0,01
Heneikosanoat Acid (C21:0)	0,02 ± 0,02	0,01 ± 0,01	
Behenic acid (C22:0)	0,40 ± 0,02	0,38 ± 0,02	0,41 ± 0,01
Trikosanoat acid (C23:0)	0,04 ± 0,00	0,04 ± 0,01	0,05 ± 0,00
Lignoserat acid (C24:0)	0,14 ± 0,01	0,15 ± 0,01	0,33 ± 0,00
Total	**12,01 ± 0,25**	**12,41 ± 0,7**	**12,55 ± 0,45**
Monounsaturated Fatty Acid (MUFA)			
Palmitoleic acid (C16:1)	0,08 ± 0,01	0,09 ± 0,00	0,1 ± 0,01
Oleic acid (C18:1n9c)	33,99 ± 0,76	36,36 ± 1,52	36,53 ± 1,42
Elaidic acid (C18:1n9t)	0,03 ± 0,00	0,04 ± 0,01	0,07 ± 0,01
Total	**34,1 ± 0,77**	**36,5 ± 1,53**	**36,7 ± 1,39**
Polyunsaturated Fatty Acid (PUFA)			
Linoleic acid (C18:2n6c)	38,01 ± 0,44	33,79 ± 0,72	26,83 ± 1,49
γ-linoleic acid (C18:3n6)	-	0,05 ± 0,01	0,82 ± 0,01
Linolenic acid (C18:3n3)	5,53 ± 0,02	4,38 ± 0,01	2,46 ± 0,11
Cis-11.14 Eikosedienoat acid (C20:2)	0,05 ± 0,01	0,06 ± 0,00	0,06 ± 0,01
Total	**43,6 ± 0,42**	87,21 ± 2,95	**30,18 ±1,58**
Fatty acid total	**89,72 ± 0,59**	**87,21 ± 2,95**	**79,44 ±3,44**

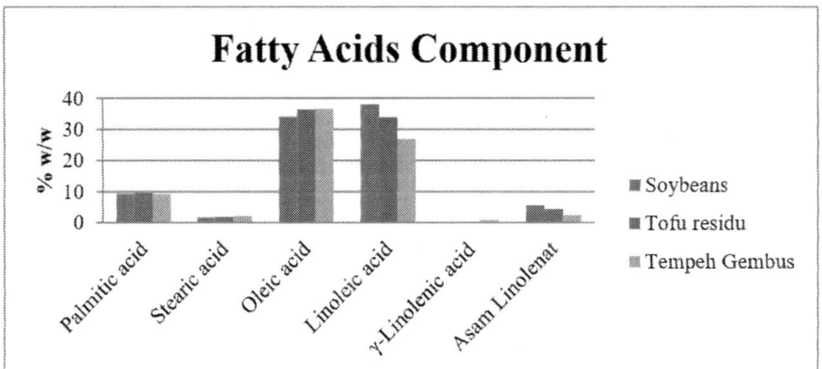

Figure 3: Fatty acids composition of soybeans, tofu residue, and *tempeh gembus*

3.4 Vitamins and Minerals

In this present study sodium, calcium, potassium, phosphorus, iron, vitamin B1, vitamin B12, vitamin A, vitamin C were investigated. However, mostly vitamins were not detected. Table 4 shows minerals and vitamins composition of soybeans, tofu residue, and *tempeh gembus*.

Table 4. Vitamins and minerals composition of soybeans, tofu residue, and *tempeh gembus* (%w/w)

	Mean ± SD			
Parameter	Soybeans	Tofu residue	*Tempeh gembus*	Unit
Mineral				
Sodium	19,2 ± 0,19	10,89 ± 0,23	11,69 ± 0,76	mg/100 g
Calsium	197,84 ± 0,26	223,07 ± 1,55	232,09 ± 2,46	mg/100 g
Potassium	914,46 ± 4,63	28,94 ± 0,03	28,73 ± 0,09	mg/100 g
Phospor	4367,8 ± 36,37	815,7 ± 3,17	751,42 ± 1,21	Ppm
Iron	32,39 ± 0,39	4,52 ± 0,79	20,04 ± 0,04	Ppm
Vitamin				
Vitamin B1	Not detected	Not detected	Not detected	mg/100 g
Vitamin B12	Not detected	Not detected	Not detected	Ppb
Vitamin A	Not detected	165,31 ± 0,42	Not detected	mcg/100 g
Vitamin C	Not detected	Not detected	Not detected	mg/100 g

4. Discussion

4.1 Proximate, vitamin, and mineral

The processing method of making *tempeh gembus* such as soaking, steaming, and fermentation caused decreasing or increasing proximate, mineral, and vitamin contents in soybeans, tofu residue, and *tempeh gembus*. Time and temperature can affect the value of water content (moisture) of food. The longer processing time and higher temperature used will result in more water released. The moisture content of soybeans decreased during fermentation. The decrease in soybean moisture was due to the activity of mold metabolism during fermentation releasing heat that made the fermentation temperature increased. Increasing temperature was followed by evaporation of the water content in soybeans. Thus, the next fermentation results decreasing of soil moisture.

The ash content on tofu residue decreased compared with soybeans. The decrease of ash content is caused as a result of the loss of solids during soaking and cooking. The decrease of ash content indicates that the minerals contained in soybeans are found in the epidermis. In addition, the carbohydrate in both tofu residue and *tempeh gembus* decreased. Decreasing occured due to enzymatic activity of mold during fermentation. Kapang digested carbohydrates resulting in a drastic decrease in hexoses and slow hydrolysis of stakiosa.

Soaking process of soybeans into tofu residue caused changing of fat content. At first, soybeans contained 19.94 % and decreased to 3.88%. Soaking can alter the fat content in soybeans due to temperature changes during soaking. In addition, fat changes occur during the fermentation process. Fermentation caused fat content decreased to 3.04% in *tempeh gembus*. During the fermentation the mold will synthesize the lipase enzyme that will hydrolizethe triacylglycerol into free fatty acids. Furthermore, fat is used by mold to be source of energy to grow so fat content decreases.

The process of washing, soaking, dehulling, and cooking affects the loss of protein soybean. Soybeans contained 34.9% protein, after processing the protein of tofu residue decreases to 3.7%. However, during fermentation process, the proteint content increased to 6.7 %. The changes that occur during fermentation were caused by enzyme produced by the mold. Generally, mold produced alpha amylase, betha amylase, phosphorilase, isoamylase, maltase, and amyloglukosidase enzyme. The increase in protein content is thought to be caused by rhyzopus oligosporus synthesizing urease enzymes used to break down urea into ammonia acid and CO_2.

There was a decrease in mineral content from soybeans to tofu residue. The sodium content in *tempeh gembus* was 11.69 mg/100 mg. The decrease of sodium level in *tempeh gembus* was 7.51 mg/100 mg while *tempeh gembus* increases 1.07 mg/100 mg. Levels of potassium in soybeans was 914 mg/100 mg, in tofu residue decreased 885,5 mg/100 mg and in *tempeh gembus* was 885.73 mg/100 mg. Level of phosphor in soybean was 4367.8 ppm, level tofu residue decreased 3552.1 ppm, and in *tempeh, gembus* decreased 3616,38 ppm. Level iron in soybeans was 32,39 ppm, in tofu residue decreased 27,87 ppm, and in *tempeh, gembus* decreased 12,35 ppm. However, iron content increased 15.52 ppm in *tempeh gembus*. Calcium level increase 25.23 mg/100mg in tofu and 34.25 mg/100mg in *tempeh gembus*.

The process of decreasing mineral content is caused by a heating process that resulted in food cell wall broken and occurs discharging of component composing material. Mineral will be damaged in most processing because the mineral is sensitive to pH, oxygen, light, and heat. The heating process occurred in soybeans producing tofu residue. There was a decrease in sodium, potassium, and iron content caused by heating process. Loss of sodium during processing can not be retained because sodium melts at 97,5°C. Level of phosphor on tofu residue decrease caused by high temperature leading to evaporation of water molecules. The iron content in tofu residues decreased caused by heating process.

4.2 Amino Acids

Amino acids analyzed in soybeans, tofu, and *tempeh gembus* was nine essential amino acids and six nonessential amino acids. The essential amino acids are arginine, phenylalanine, histidine, isoleucine, leucine, lysine, methionine, threonine, and valine. The nonessential amino acids are alanine, aspartic acid, glutamic acid, glycine, serine, and tyrosine. The glutamic acid was higher than other amino acids. It can be caused by glutamine deamination into glutamic acid [12].

The total content of amino acids decreased 28.70% from soybean to tofu residue, 0.60% from tofu residue to *tempeh gembus*, and 29.32% from soybeans to *tempeh gembus*. Decreased amino acids can be caused by high temperature made during boiling and steaming that cause protein denaturation due to amino acids reactive to heat. In addition, soaking can decrease amino acids level because amico acids are generally powder-shaped and soluble in water but not soluble in nonpolar organic solvent [13].

Table 5. The need for amino acids

Amino acids	Requirements	
	mg/hari	mg/kgW per day
Histidine		10
Isoleucine	700	10
Leucine	1100	14
Lysine	800	20-30
Methionine	1100	13
Phenylalanine	1100	14
Threonine	500	7
Valine	800	10
Tyrosine	1100	14

Source: WHO (1985)

Fermentation consists of three strages namely growth stage (0-30 hours), transition stage (30-50 hours), and advanced fermentation stage (50-90 hours). Mold grwoth in tempeh reachs optimal 48 hours fermentation. However, protein decrease at 72 hours and subsequent [14]. Likewise in fermentation in *tempeh gembus*, there is a tendency to increase and decrease protein content according to fermentation time. The texture of tofu residue also affects the amout and component of nutrition in *tempeh gembus*. After fermentation, *Rhyzopus oligosporus* secretes proteinase enzyme affecting quality substrate [17].

Total amino acids in *tempeh gembus* is less than soybean and tofu residue. However, the amino acid profile has not changed. The highest essential amino acid in *tempeh gembus* is leucine. The leucine requirement according to WHO is 1100 mg [15] or 14 mg/kgW (table 5) [16]. Leucine in *tempeh gembus* was 0,44 mg/100mg. This amount can meet 40% requirement of leucine requirement. Although amino acids in *tempeh gembus* have decreased significantly but amino acids can still be utilized to complemen the fulfllment of protein needs. Research conducted on Sparque Dawley mice showed that there was a greater weight gain when given *tempeh gembus* that not given *tempeh gembus* [16].

4.3 Fatty Acids

The result of the analysis of fatty acids in soybean, tofu residue, and *tempeh gembus* consist of satureated fatty acids (SFA), monounsaturated fatty acids (MUFA), and polyunsaturated fatty acids (PUFA). In this study, the level of unsaturated fatty acids was higher (soybean: 77.69%, tofu residue 74.77% and *tempeh gembus* 68.87%) while compared with saturated fatty acids (soybean: 11.99%, tofu residue 12.39% and *tempeh gembus* 12.53%). The dominant saturated fatty acids are palmitic acid and sterat 9.21 ± 0.39% and 2.12 ± 0.07% respectively. The same study war conducted on *tempeh gembus* that dominant saturated fatty acids were lauric acid (3.46%), palmitic acid (8.23%), and stearic acids (3.42%) [18].

Soybeans is source of linoleic, oleic, and linolenic acids. In this study linoleic acid content (38.01 ± 0.44 %), linolenic acid (15.53 ± 0.02 %), and oleic acid (33.99 ± 0.76 %). This is similar to research on soybeans that linoleic acid content (25-64%), linoleic acid (1-12%), oleic acid (11-60%) while in *tempeh gembus*, linoleic acid (26.83 ± 1.49 %), linolenic acid (2.46 ± 0.11 %), and oleic acid (36.53 ± 1.42 %). The results are consistent with other studies conducted on *tempeh gembus* that linoleic acid, linolecin acid, and oleic acid is higher than other fatty acids in *tempeh gembus*. In this study, oleic and linoleic acids level in *tempeh gembus* have higher levels compared with previous studies, in other studies, *tempeh gembus* has linoleic and linolenic acid 21.51% and 1.82 respectively [18]. It is because of different types, varieties of raw materials, and microorganisms used for fermentation.

The *tempeh gembus* processing such as soaking until fermentation stage causes changes composition of fatty acids in soybeans, tofu residue, and *tempeh gembus*. Processing from soybeans to tofu residue through soaking and heating process decreased linolei and linolenic acids. This study is in line with research conducted on rice bran. The heating process can cause the conversion of fatty acid component into volatile compound such aldehydes, ketones, acids, and hydrocarbons. These compound will evaporate when heat-treated that makes the fatty acid decrease [19].

The fermentation process in *tempeh gembus* caused a decrease linoleic acid and linolenic acid. This is similar to the research on fermentation of tiger nut in Nigeria that linoleic acid and linolenic acid decreased, wherein linoleic acid with 9.00 ± 0.01% to 7.74 ± 0.00%. The oxidation process of linoleic and linolenic acids are influenced by temperature and light intensity. The higher temperature makes formation of fatty acids may be decreased [20]. However, the processing of soybeans into *tempeh gembus* causes oleic acid levels to increase, where oleic acid from 33.99 ± 0.76% to 36.53% ± 42%. This is similar to Nigerian study of the process of fermentation of nuts flour which increased oleic acid from 18.59% to 20.15% [21].

A study conducted on tempeh fermentation occurs an increase essential fatty acids level that occcur at 24, 72, 69 hours of fermentation time. Soybean changes to be tempeh after incubation for 12 to 24 hours of fermentation time. Rhyzopus oryzar secretes lipase enzynes that can breakdown fat into fatty acids [22]. While the decline occured in 48 hours of fermentation time due to microbial activity, where microbes that grow in *tempeh gembus* increased slightly and a decrease free fatty acids [23]. A study sugegests that fatty acids have increased again in the 24 hours. This is because, *Rhyzopus oligosporus* has lipolytic hydrolize and hydrolize one third of fatty acid for 72 hours of fermentation [24.] After 96 hours of fermentation the levels of essential fatty acids increased. In the phase of decay or further fermentation occur an increase of number bacteria, the amount of free fatty acids and the growth, the growth of the mold decresed [23].

The fermentation process cause an increase γ-linolenic acid. It was caused during syntjhesis of fatty acids, *Rhyzopus sp*, only produce γ-linolenat instead of α-linolenat. The data show that there is γ-linolenic acid in tofu (0,05 ± 0,01) and *tempeh gembus* (0,82 ± 0,01). A research showed that fermentation using *Rhizopus s.p* can produce γ-linolenic acid. Another study on fermentation in glutinous rice showed that γ-linolenic acid in fermented rice, which in terms of length of processing, γ-linolenic increased to 1,4 ± 1,1% [25].

During fermentation of soybean products occured degradation process of fat by mold, which can eventually be liberated into fatty acids. The main components of fatty acid from triglyceride in tempeh are unsaturated fatty acids which are dominated by linoleic acid, linoleic acid, and oleic acid. Although, the essential fatty acids in *tempeh gembus* average decreased, but in *tempeh gembus* still contain linoleic and linolenic fatty acids which if consumed will get any benefits, among others, pay a role in the maintenance of cell membranes, regulation of cholesterol metabolism, loweing blood pressure, inhibit hepatic lipogenesis, and lipid transport. It is also a precusor of group eikosnoid compounds similar to hormones, namely prostaglandins, prostacyclines, thromboxanes, and leukotrienes. These compounds regulate blood pressure, heart rate, immune function, nervous system stimulation, muscle contraction and wound healing [26].

The bioactive compound of the oil or fat group most widely used in functional food formulation is the Omega-3 fatty acid, which is classified as polyunsaturated fatty acid (PUFA). The replacement of a diet high in saturated fatty acids from foods by consuming unsaturated fatty acids such as linoleic and linolenic acids is recommended in an attempt to prevent coronary heart disease [27]. WHO recommends a sufficient intake of PUFAs ranging from 6 to 10% of daily intake [25]. PUFA can lower levels of VLDL and LDL in the blood because the liver will not convert it to triglyceride VLDL. The liver transports unsaturated fatty acids to tissues for oxidation without leaving lipoprotein remnant in the form of LDL. Meanwhile, ketone compounds will be removed through the lungs (respiration), urine or sweat, and some can be converted into fatty acids or glucose to be used as an energy source [29].

5. Conclusion

Levels of nutritional composition in soybeans is higher than tofu residue and *tempeh gembus*. However, there are some increase of some fatty acids caused by fermentation process.

6. Acknowledgements

The authors would like to acknowledge Diponegoro University.

References

[1] Badan Pengawasan Obat dan Makanan 2001 *Kajian proses standarisasi produk pangan fungsional di badan pengawas obat dan makanan* Lokakarya Penyusunan Standar Pangan Fungsional. Badan Pengawasan Obat dan Makanan Jakarta

[2] Jenie BSL 2003 *Pangan fungsional penyusun flora usus bagi kesehatan dan kebugaran* Institut Pertanian Bogor Bogor

[3] Bruulsema TW 2000 vol **84**

[4] Setiawan JP, Hardian *Pengaruh pemberian tablet asam amino terhadap kelelahan otot* (Skripsi) Universitas Diponegoro:1-20

[5] Almatsier S 2001 *Prinsip dasar ilmu gizi* Gramedia Pustaka Utama Jakarta

[6] Badan Pengawasan Obat dan Makanan 2001 *Kajian proses standarisasi produk pangan fungsional di badan pengawas obat dan makanan* Lokakarya Penyusunan Standar Pangan Fungsional Badan Pengawasan Obat dan Makanan Jakarta

[7] Higdon J 2003 *An Evidence-Based Approach to Dietary Phytochemicals. Health Benefits and Intake Recommendations* Thieme Medical Publishers New York

[8] Lieberman S Brunning N 2003 *Real Vitamin and Mineral Book* Penguin Books New York

[9] Sutriswati RE 2012 *Teknologi proses pembuatan tahu* Kanisius Yogyakarta

[10] Ervina, Suseno, Thomas I 2000 *Potensi tersembunyi flake tempeh gembu.* seminar nasional industri pangan

[11] Lesatri S 2001 *Pengaruh kadar ampas tahu yang difermentasi terhadap efisiensi pakan dan pertumbuhan ikan mas (Cyprinus carpio)* [Skripsi] Fakultas Perikanan dan Ilmu Kelautan, Institut Pertanian Bogor, Bogor

[12] Murray RK, Granner DK, Mayes PA, Rodwell VW 2003 *Biokimia Harper* Penerbit Buku Kedokteran EGC Jakarta (ID)

[13] Sitompul S 2004 *Analisis asam amino dalam tepung ikan dan bungkil kedelai Buletin Teknik Pertanian* **9**(1): 33-37

[14] Sulistyowati E, Arianingrum R, Salirawati D 2004 *Studi pengaruh lama fermentasi kedelai terhadap aktivitas tripsin* (Skripsi) Universitas Negeri Yogyakarta Yogyakarta

[15] Rose WC 1957 *Nutr Abstr Rev* **27**(3):631–647

[16] Meredith CN et al 1986 *Am J Cin Nutr* **43**:787–794

[17] Wangh L, Vespaj B, Hesseltince W 1974 *J App. Microbiol* **27**; 906-911

[18] Sulchan M, Endang NW 2007 *Nilai gizi dan komposisi asam amino tempeh tempeh gembus serta pengaruhnya terhadap pertumbuhan tikus* Majalah Kedokteran Indonesia **57**(3):80-85

[19] Gumilar GG, Zackiyah, Dwiyanti G, HM Siti 2009 *J Pengajaran MIPA* **14**(2); 143-150

[20] Agbaje RB, Oyetayo VO, Ojokoh AO 2015 *Afr J of Biochem Res* **9**(7): 89-94

[21] Ijarotimi OS, Keshinro OO 2012 *Acta Sci Pol Techno Aliment* **11**(2): 151-165

[22] Sudaryatiningsih C, Supyani 2009 *Nusantara Bioscience* **3**: 110-116

[23] Handoyo T 2006 *Int J of Food Propories*. **9**: 347 - 355

[24] Deliani 2008. *Pengaruh lama fermentasi terhadap kadar protein, lemak, dan asam fitat pada pembuatan tempeh* [Tesis] Universitas Sumatera Utara Medan

[25] Gui-you L, Sheng Y, Chuan-chao D 2004 *Int J Food Microbiol* **21**:299-304

[26] Mayes PA 2003 *Biosintesis asam lemak* In: Murray RK Granner DK Mayes PA Rodwell VW editors Biokimia Jakarta

[27] Iskandar, Y 2004 *Penentuan kadar asam linoleat pada tempeh secara kromatografi gas* [Skripsi] Universitas Padjajaran Bandung

[28] WHO 2013 *Diet, nutrition and the prevention of chronic disease* Report of the joint WHO/FAO Expert Consultation

[29] El Sohaimy SA 2012 *World Appl Sci J* **20** (5): 691 – 708

Characterizations of milkfish (*Chanos chanos*) meatballs as effect of nanoencapsulation liquid smoke addition

Fronthea Swastawati[1*], Ambaryanto[2], Bambang Cahyono[3], Ima Wijayanti[1], Diana Chilmawati[4]

[1] Departement of Fish Product Technology, Faculty of Fisheries and Marine Science, Diponegoro University Jl. Prof. Soedarto, SH, Tembalang, Semarang 50275, Indonesia
[2] Departement of Marine Science, Faculty of Fisheries and Marine Science, Diponegoro University Jl. Prof. Soedarto, SH, Tembalang, Semarang 50275, Indonesia
[3] Departement of Chemist, Faculty of Sain and Mathematics, Diponegoro University Jl. Prof. Soedarto, SH, Tembalang, Semarang 50275, Indonesia [3)]
[4] Departement of Aquaculture, Faculty of Fisheries and Marine Science, Diponegoro University Jl. Prof. Soedarto, SH, Tembalang, Semarang 50275, Indonesia [4)]

Email: fronthea_thp@yahoo.co.id , Telp: +628562681613*

Abstract. Milkfish meatballs are one of the value-added product that is potential to be developed because this product is favored by consumers. The aims of this study were mainly to evaluate the effect of nano encapsulation liquid smoke addition to the properties of milkfish meatballs. The different concentration (0%, 1%, 3% and 5%) of nano encapsulation liquid smoke was treated into milkfish meatballs with three replications. Test parameters used include texture (hardness, deformation and gel strength); proximate (moisture, protein, fat, ash and carbohydrates content); sensory (organoleptic and hedonic). Parametric data were analysis using ANOVA and non-parametric parameters using Kruskal Wallis test. Analysis of variant shows that nanoencapsulation gave significant effect on the hardness, deformation, and gel strength; protein, lipid, ash and water content($P<0,05$), but gave no significant effect on carbohydrate ($P >0.05$). The best texture characteristic was obtained by milkfish meatball added with 1% of nanoencapsulation liquid smoke indicated by the moisture water; protein; fat; ash and carbohydrate content: 66,35%; 12,69%; 6,75%; 2,42% and 5,93% respectively. The best organoleptic and hedonic value of milkfish meatballs was achieved by a panelist at the sample with the addition of 1% liquid smoke, but this sample found gave no significance different to control in terms of sensory valle.

Key words: milkfish meatball, liquid smoke, nanoencapsulation

1. Introduction

Fish is one of the most important food sources because it contains the essential amino acids and essential fatty acids the body needs [1-4]. Nevertheless, some societies are less fond of fish in their original form to increase the joy and level of public consumption. One form of processed fish that is popular with meatballs. Meatballs are one of the fish pasta products known to originate from the Chinese plains and are now widely distributed in Asian countries. The meatball industry has grown in

Asian countries since 1980 [5]. Fish meatball is popular enough for the community for this product is still being developed. All types of fish, in general, can be used as raw material for meatballs, but not from white fleshy fish so that the resulting gel can be optimal. Nonwhite fleshy fish can be used as raw material for meatballs, but it is necessary to increase the strength of the gel.

Milkfish is one of the cultivated fish whose production continues to increase from year to year. The high production makes milkfish not only consumed fresh but also processed to provide added value. Milkfish can be processed into various products such as soft-boned milkfish, meatballs, meatballs and others. Diversification of processed milkfish continues to be developed to provide a variety of products so that milkfish consumption can be improved and also provide added value. Meatball fish that exist today has the original taste of fish as raw materials. High competition, high technology investment is required to obtain high-quality products. Fish meat that now has the original taste, the addition of a different flavor in the meatballs will provide a variety of products preferred by consumers. The addition of different flavors to fish meatballs is expected to provide a variety of products and attract consumers to try. One of the natural flavors that can be used is liquid smoke.

Liquid smoke is a liquid form of wood burning through pyrolysis process. Liquid smoke is used to provide flavor, color, texture and in some cases can increase the shelf life of the product. Liquid smoke is obtained from thermal degradation reactions of cellulose, hemicellulose, and lignin [6](Ramakrishnan and Moeller, 2002). The smoke flavor had several advantages over traditional fumigation techniques such as ease of application, speed, product uniformity, consistent end product characteristics, in-app cleanliness and decreased polycyclic aromatic hydrocarbon content[7]. In relation to consumer preferences, indicating that consumers do not like the same type of product, some people prefer strong smells and flavors, others prefer flavors of certain smoke [8]. Research on liquid smoke has been conducted among others as an antibacterial agent [9-10]; preservatives in foodstuffs [8][11] [12] [13][14]. Currently, liquid smoke has been developed in powder form. The powder of liquid smoke is obtained from the process of encapsulation of liquid smoke with an encapsulation material. Research on liquid smoke encapsulation has been done by Saloko et al. [15]; Saloko et al. [16] and Petzold et al. [17] showing the use of chitosan and maltodextrin and ca-alginate as encapsulation. Liquid smoke in the form of powder is easier in handling and application. A study of the application of nanopedic liquid smoke has not been widely practiced. The purpose of this study was to determine the effect of various types and concentrations of nano-encapsulate liquid smoke on the texture, sensory and proximate fishballs meat-fish.

2. Materials and Methods

2.1. Materials

The material used in the production of meatballs is milkfish with size 250 ± 25 g. Milkfish is obtained from cultivation in Central Java which is sold in the fish market. Liquid smoke used is nanostsulated liquid smoke. Addition of meatballs used includes wheat flour, bread, pepper, and onion. Materials for proximate analysis, among others, H2SO4, HCl, NaOH, H2SO4, HCl, NaOH, HBrO3, Kjeldahl Tablets, N-hexane. The tools used in the production of meatballs include filet knife, food processor, steamer. The tool for texture analysis is TATX Texture type analyzer. Tools for proximate analysis include Oven; Kjeldahl Apparatus, Soxhlet Apparatus, digital scales, furnace.

2.2. Preparation of Fishballs

Milk with size 250 ± 25 g in fillet then ground into the pulverized meat. Spices consisting of pepper, onion, and salt are crushed. Furthermore, the dough is made by mixing the flesh of creamed fish, seasonings, flour and liquid smoke with different concentrations (1%, 3%, and 5%). The type of liquid smoke used is panoptic liquid smoke. The dough is then boiled twice with different temperature. The temperature was 50 ° C for 20 minutes, and the temperature was 80 ° C for 30 minutes and cooled. The parameters performed on milkfish meatballs, i.e.

texture (Hardness, deformation and Gel strength); Sensory (organoleptic and hedonic); proximate levels (water, protein, fat, ash and carbohydrate content).

2.3. Procedure Analysis

2.3.1. Textural Analysis

The texture analysis consists of hardness, deformation and gel strength. The texture analysis of milkfish meatballs was done using a texture analyzer (Model TA Plus, LLOYD Instrument, UK). Three milk fish meatball samples of each treatment (height 2.5 cm, width 2 cm) prepared and tested. Hardness and deformation were measured using texture analyzer textures equipped with ball probes (5 mm in diameter, depression rate of 60 mm per minute). The gel strength is calculated by the following equation:

gel strength (g.cm) = Hardness (gf) x Deformation (cm)

2.3.2. Proximate Analysis

Proximate analysis consists of moisture, protein, fat, ash and carbohydrates based on AOAC [18]

2.3.3. Sensory Analysis

The sensory analysis consists of organoleptic and hedonic tests. Panelists used as many as 30 people. Organoleptic test according to Indonesian National Standardization Agency No: SNI-01-6683-2002 and hedonic based on Indonesian National Standardization Agency No: SNI 01-2346-2006

2.3.4. Statistical Analysis

All treatments were performed in three replications. Analysis of parametric data using ANOVA. If the actual treatment of further test is done by Tukey test (Steel & Torrie, 1980). Nonparametric data were analyzed by Kruskal Wallis Test and continued with Mann Whitney test. The analysis was performed using SPSS (SPSS 17.0 for Windows, SPSS Inc., Chicago, IL, USA

3. Results and Discussion

3.1. Texture

The texture of fish balls with additional nano encapsulation is presented in Table 1. Statistical analysis shows the concentration of liquid smoke encapsulation has significant effect on all characteristic parameters of meatball texture (Hardness, deformation and gel strength) ($P < 0.05$)

Table 1. Characteristics of Meatball Texture with Additional Nanoencapsulated Liquid Smoke

Parameter	Concentration			
	0%	1%	3%	5%
Hardness (gf)	1560.61±47.23^{a}	1727.09±31,69^{a}	1253.73±160.37^{b}	1132.81±147.74^{b}
Deformation(cm)	1.97±0.14^{c}	2.33±0.04^{ab}	2.43±0.05^{a}	2.071±0.15^{bc}
Gel strength (g.cm)	3072.25±202.71^{b}	4017.81±60.06^{a}	3040.24±371.02^{b}	2333.96±149.57^{c}

Data with different superscripts on the same line show a significantly different

Statistical analysis shows the addition of nano-encapsulated liquid smoke significantly affect the hardness of fish balls. The concentration of 1% liquid smoke nano-encapsulation improves hardness 11% compared to control. At concentrations of 3% and 5%, addition of liquid smoke powder decreases the hardness of meatballs.

The addition of nano-encapsulated liquid smoke had a significant effect on the deformation of milkfish meatballs (P <0.05). The highest deformation of meatballs with the addition of liquid smoke powder 3% but no significant effect on the meatballs with the addition of 1%. The control treatment showed lower values of treatment of 1% and 3% but not significantly different from the 5% treatment.

Variance analysis showed that the addition of nanoencapsulation liquid smoke had a significant effect on the gel strength of fish balls (P <0.05). The highest gel strength is achieved by fish meatballs with 1% liquid smoke powder while meatballs in liquid smoke powder 3% and 5% lower than control. This shows that the concentration of 1% can increase the strength of milkfish meatballs gel up to 30% compared to control.

Increasing the texture characteristics of hardness, deformation and gel strength in milkfish meatballs only at 1% concentration whereas a higher concentration indicates a decrease in texture. Improved texture of fish meatballs may be due to the addition of liquid smoke with chitosan encapsulation. Liquid smoke with a high phenol content can interact with proteins forming complex crosslinks that enhance the texture. The ability of phenolic compounds to interact with proteins (especially proline-rich proteins including casein) has been described Luck et al. [19] (1994), and both hydrophobic interactions and hydrogen bonds appear to be the main driving force of the reaction between liquid smoke in this case phenol with proteins [20]. In excess of liquid smoke, this can reduce the texture because the high phenol content has lower solubility, so difficult to interact with the protein so that the impact on texture decline [21].

3.2. *Proximate Content*

The proximate data of milkfish meatballs with different liquid smoke is presented in Table 2. The type and concentration of liquid smoke had a significant effect on water content, protein, fat, and ash, but no significant effect on carbohydrate of milkfish meatballs.

Table 2. Proximate Content of Milkfish meatball with Different Concentration of nanoencapsulation of Liquid Smoke

Parameter	Concentration			
	0%	1%	3%	5%
Moisture	70.02±0.66^{a}	66.35±0.62^{c}	68.11±0.66^{b}	65.96±0.03^{c}
Protein	10.93±0.30^{c}	12.69±0.05^{a}	12.09±0.21^{b}	11.81±0.06^{b}
Lipid	6.13±0.02^{b}	6.75±0.13^{a}	6.88±0.06^{a}	6.85±0.07^{a}
Ash	1.99±0.05^{c}	2.42±0.02^{b}	2.57±0.03^{a}	1.85±0.07^{d}
Carbohydrate	5.84±0.39^{a}	5.93±0.10^{a}	5.15±0.58^{a}	6.62±1.78^{a}

different letter (a,b, c,) in the same line indicate significances different (P<0.05)

The concentration of nano-capsulated of liquid smoke had a significant effect on the water content of milkfish meatballs (P <0.05). The control water content (without liquid smoke powder) showed the highest water content, while the lowest was at 5% concentration but not significantly different with the treatment of 1% concentration. The affluence of treatment especially the control treatment showed higher moisture value. The lowest moisture content was achieved in milkfish meatballs with the highest concentration (5%). The addition of liquid smoke powder impact on the composition of solids in fish balls.

Analysis of variance showed that nano-capsulated concentration of liquid smoke had a significant effect on the protein content of milkfish meatballs (P <0.05). The highest protein content was achieved in milkfish meatballs treatment of 1% nanoencapsulation of liquid smoke, while the lowest protein content in non-smoked liquid nano-fish meatballs. All treatments showed that protein content had met Indonesian national standards of at least 10%.

The fat content of milkfish meatballs was influenced by the concentration of nanoencapsulation liquid smoke (P <0.05). The nano-encapsulated liquid smoke treatment showed the highest fat content, but all treatments with the addition of liquid smoke showed no significant difference in the fat content.

Addition of liquid smoke nanoencapsulation liquid smoke significantly affects the ash content of fish balls (P <0.05). The highest ash content was found in fish meatballs with the addition of 2% nano-encapsulated liquid smoke and the lowest in milkfish meatballs with nano-encapsulated 5% liquid smoke.

The addition of liquid smoke has no significant effect on the carbohydrate content of fish balls (P> 0.05). Carbohydrate content ranged from 5.15% -6.62%. Carbohydrates in fish meatballs is a contribution of adding tapioca starch to fish meatballs

3.3. Sensory

3.3.1. Organoleptic

The organoleptic value of milkfish meatballs is described in Table 3. The results of the organoleptic analysis showed that the concentration of liquid smoke had a significant effect on the organoleptic value of fish balls (P <0.05). Overall, fish meatballs meet Indonesian national standards with a value of> 7 except on fish meatballs with the addition of nano-encapsulated 5% liquid smoke.

Table 3. Organoleptic of Milkfish Bakso with Different type and Concentration of Liquid Smoke

	Specification				
Concentration	Appearance	Smell	Taste	Texture	Total
0%	8.00±0.23	7.44±0.1	8.27±0.24	8.38±0.17	8.02±0.10
1%	8.02±0.14	7.42±0.10	8.33±0.33	8.47±0.18	8.06±0.13
3%	7.87±0.20	5.98±0.10	7.11±0.84	7.82±0.34	7.19±0.14
5%	6.84±0.50	5.76±0.04	4.84±0.37	6.82±0.37	6.07±0.15

Overall, the organoleptic value of the largest fish balls in the addition of 1% but not significantly different from the control, while the lowest was found in the meatballs with the addition of nano-encapsulated 5% liquid smoke.

The addition of liquid smoke significantly affects the value of the appearance of milkfish meatballs. The appearance of fish meatballs at concentrations of 0% and 1% were not significantly different. The higher concentration of liquid smoke indicates the value of the appearance decreases. This is due to the addition of liquid smoke nanoencapsulation causes the appearance of fish meatballs tend to be dark so that the appearance of less clean so that the value becomes down.

The organoleptic value of the odor shows the same trend that is with increasing concentration indicating the organoleptic value of the odor decreases. The 1% concentration was not significantly different from the control; it showed 1% increase indicated the fish's distinctive odor in meatballs was still detected by panelists while at higher concentrations the fish meatball smell was reduced and the dominant smell of liquid smoke.

The addition of liquid smoke significantly affects the taste of fish meatballs. In taste specifications also showed 1% concentration was not significantly different with control. The taste of fishfish meat-fish with the addition of liquid nanoencapsulation smoke up to 3% still by Indonesian national standards, but the concentration of 5% is considered unfit for consumption by panelists.

The addition of liquid smoke significantly affects the value of the texture specification on fish meatballs. The highest texture value is found in fish meatballs with 1% liquid smoke addition and decreases with the liquid smoke enhancer.

3.3.2. Hedonic

Hedonic test method based on BSN [21] (2006) according to SNI 01-2346-2006 on Guidelines for sensory testing on fishery products. Hedonic test consists of 4 specifications, the appearance, smell, taste and texture with a score of 9 scala i.e. 1 = strongly dislike; 2 = dislikes; 3 = rather dislike; 4 = ordinary; 5 = rather like; 6 = likes; 7 = very like; 8 is very fond of; 9 = very, very fond.

The hedonic value of milkfish meatballs with different concentrations of nano-capsulated liquid smoke is presented in Table 4. The results of the hedonic data analysis show the addition of different nano-encapsulated liquid smoke to give different effects on the hedonic level. Overall, milkfish meatballs with 0% and 1% liquid smoke additions are favored by panelists (> 7), while concentrations of 3% and 5% tend to be favored by panelists (> 6). 1% The addition of nano-encapsulated liquid smoke provides a unique flavor without leaving a taste of the fish, so the panelist gives an average rating above 7 which means very fond.

Table 4. Hedonic Scale of Milkfish Bakso with Different Concentration of Nano-encapsulation Liquid Smoke

Concentration	Specification					TOTAL
	Appearance	Colour	Smell	Taste	texture	
0%	7.48±0.30	7.41±0.19	7.27±0.35	7.50±0.24	7.34±0.05	7.40±0.19
1%	7.43±0.28	7.46±0.27	7.08±0.21	7.16±0.10	7.73±0.14	7.37±0.13
3%	7.49±0.18	7.09±0.08	5.91±0.30	5.87±0.32	7.39±0.29	6.75±0.07
5%	7.14±0.29	7.11±0.14	4.36±0.59	4.31±0.26	7.41±0.33	6.07±0.20

The appearance of milkfish meatballs with the addition of liquid smoke filtration 1% higher than the concentrations of 3% and 5%, but overall is very favored by the panelists. The highest value of the meatballs achievement was achieved with the liquid smoke of nano-encapsulated 3% but not significantly different with concentrations of 0% and 1%.

The color of the milkfish meatballs was rated highly favorably by panelists with values above 7 in all treatments with the highest value in fish meatballs with 1% nano-encapsulated liquid smoke, but not significantly different from the treatment at 0% concentration. The lowest color value is fishballs meatballs with the addition of nanoencapsulation 5% liquid smoke. Liquid smoke gives a distinctive color of smoke in fish balls. The phenol component plays an important role in providing a distinctive color to the meatballs. The lowest value in the color of fish balls with nano-capsulated 5% liquid smoke may be due to the dark color of the meatball so that the panelists give the lowest value but are still in the category favored because the value is above 7.

Analysis of hedonistic value data showed that the difference of nano-strucan liquid smoke concentration significantly affected the smell and taste of milkfish meatballs. The panelist's assessment of the specification baud a flavor implies the same tendency. At the treatment of 0% and 1% baud, a flavor of milkfish meatballs is favored by the panelists. At concentrations of 3%, the panelists tend to be rather fond of whereas at 5% concentration the usual rate panelists. This indicates the increasing concentration of liquid smoke baud fishballs tastes increasingly less favored by the panelists. The odor and smell of smoke that is too sharp on the fish meatballs with higher concentrations are less favorable to the panelists.

Different concentrations of liquid smoke have a significant effect on hedonic texture value of milkfish meatballs, but overall the texture of meatballs is well-liked. Concentrations of 1% nanostructured liquid smoke had the highest texture hedonic value compared to concentrations of 3% and 5%, as a whole the addition of liquid smoke had a preferred texture over the control.

4. Conclusion

The addition of liquid nano-encapsulated liquid with different concentrations significantly affected the value of hardness, deformation, gel strength, moisture content, protein, fat, ash, organoleptic and hedonis, but no significant effect on the carbohydrate content of milkfish meatballs. The best treatment was achieved in fish balls with liquid smoke nano-encapsulation concentration of 1%

5. Acknowledgment

Further thanks to the Directorate of Research and Community Service, the Ministry of Research. Technology and Higher Education, which has funded this research through the Research Incentive Scheme SINas 2016 DIPA budget was funded by the Ministry of Research. Technology and Higher Education in 2016

References

[1] Goodman-Lowe GD, Carpenter JR, Atkinson S, Ako H. 1999 *Comparative Biochemistry and Physiology Part* **A 123**: 137–146
[2] Osibona AO, Kusemiju K, and GR Akande 2009 *African Journal of Food Agricultural Nutrition and Development* **9** (1): 608-621
[3] Mohanty B, et al. 2014 Journal of Amino Acids 2014: 1-7
[4] Ibrahim HMI 2015 *Int.J.Curr.Microbiol.App.Sci* **4**(4): 917-933
[5] Boran M and Köse S. *2007 Turkish Journal of Fisheries and Aquatic Sciences* **7**: 65-70
[6] Ramakrishnan S and Moeller P. 2002 *Fuel Chemistry Division Preprint* 47(1): 366.
[7] Simon R, de la Calle B, Palme S, Meier D, Anklam E 2005 *J. Sep Sci.* **28**:871–882.
[8] Swastawati F, Eko Susanto, Bambang Cahyono, and Wahyu Aji Trilaksono 2012 *International Journal of Bioscience, Biochemistry, and Bioinformatics,* **2** (3): 212-216
[9] Zuraida, I., Sukarno, and Budijanto, S. 2011 *International Food Research Journal* **18**: 405-410
[10] Swastawati F, Herry Boesono S, and Dian Wijayanto 2014 *International Conference on Food Security and Nutrition IPCBEE vol.67 IACSIT Press, Singapore*
[11] Gedhela S 2005 Application Of Liquid Smoke Alone and in Combination With Pre-And Post-Package Pasteurization Against *Listeria Monocytogenes* on Ready-To-Eat Meats. Oklahoma State University
[12] Yusnaini, Soeparno, E. Suryanto, and R. Armunanto 2012 *J.Indonesian Trop.Anim.Agric*. **37**(1): 27-33
[13] Purba R, Sugeng Heri Suseno, Ayu Fitri Izaki, Syahrizal Muttaqin 2014 *International Journal of Applied Science and Technology* **4** (2):212-217
[14] Krisen, S. S., Setiaji, B., Trisunaryanti, W. and Pranowo, H. D. 2014 *Asian Journal of Science and Technology* **5**(9):573-576
[15] Saloko S, Darmadji P, Setiaji B, and Pranoto Y. 2012 *J. Teknol. Dan Industri Pangan* **23** (2): 173-179
[16] Saloko S, Darmadji P, Setiaji B, Pranoto Y and Anal A.K. 2013 *International Food Research Journal* **20** (3): 1269-1276
[17] Petzold G, María Pia Gianelli; Graciela Bugueño; Raymond Celan; Constanza Pavez; Patricio Orellana 2014 *J Food Sci Technol* **51**(1):183–190
[18] [AOAC] Association of Official Analytical Chemist 2005 *Official Methods of Analysis of The Association of Official Analytical Chemist*. Arlington, Virginia USA: AOAC Inc.
[20] Harbourne N, Jean Christophe Jacquier, Dolores O'Riordan 2011 *International Dairy Journal* **21**: 185-191
[21] Balange, A. dan Benjakul, S. 2009 *Food Science dan Technology* **42**: 1059–1064.

Clinical Outcome And Arginine Serum of Acute Ischemic Stroke Patients Supplemented by Snakehead Fish Extract

Dwi Pudjonarko[1,2], **Retnaningsih**[1], **Zainal Abidin**[1]
1. Department of Neurology Faculty of Medicine Diponegoro University / Kariadi Central Hospital. Semarang – Indonesia.
2. Corresponding Author, E-mail: dr.onang@fk.undip.ac.id

Abstract:
Background: Levels of arginine associated with clinical outcome in acute ischemic stroke (AIS). Arginine is a protein needed to synthesis nitric oxide (NO), a potential vasodilator and antioxidant. Snakehead fish is a source of protein which has antioxidant activity. Snakehead fish contains mineral, vitamin, and amino acids. One of the amino acids that were found quite high in snakehead fish extract is arginine. The aim of this study was done to determine the effect of snakehead fish extracts (SFE) on serum arginin levels and clinical outcome of AIS patients.
Methods: It was double-blind randomized pretest-posttest control group design, , with. AIS patients were divided into two groups i.e. snakehead fish extracts (SFE) and control. SFE group were administered 15 grams SFE for 7 days . Arginine serum levels and clinical outcome (measured by National Institute of Health Stroke Scale= NIHSS) were measured before and after treatment, other related factors were also analyzed in Logistic regression.
Results: A total of 42 subjects who were performed random allocation as SFE or control group. There was no differences in subject characteristics between the two groups. There was a differences Δ arginine serum levels between SFE and control (33.6±19.95 μmol/L; 0.3±2.51 μmol/L; $p<0.001$). Change in NIHSS score in SFE improved significantly compared to the control group (4.14 ± 2.03; 2.52 ± 1.81;$p=0.009$). Logistic regression analysis showed only female gender factor that affected on improvement of NIHSS (OR=7; p=0,01).
Conclusion: There is Clinical outcome improvement and enhancement of arginine serum levels in AIS patient with snakehead fish extract supplementation.

Keywords: snakehead fish extract, arginine serum, acute ischemic stroke, NIHSS

1. Introduction

Stroke if not handled properly can have a negative impact on quality of life, social life, psychology, and patient's economy. Therapy for ischemic stroke in general is to overcome the emergency problem that occurs, treatment of underlying disease such as hypertension, handling complications and special therapies such as prevention of reperfusion and protecting brain tissue.

Previous studies by Cherubinni showed that antioxidant levels of patients with acute ischemic stroke were associated with the output of ischemic stroke. Low levels of antioxidants associated with the incidence of death or severe neurological dysfunction.[1] Based on this, antioxidant treatment are put into consideration as an additional therapy in ischemic stroke. The benefits of antioxidants in ischemic stroke are through the mechanism of inhibiting the production of free radical, elimination of free radicals and increasing free radical degradation.[2]

Study done by Jurcau showed that administration of antioxidants are improving neurological functionalities, assessed by the *National Institutes of Health Stroke Scale* (NIHSS)

score.[3] Metaanalysis result by Schurks M, et al showed vitamin E supplementation may reduce the risk of ischemic stroke but, on the other hand, may increase the risk of hemorrhagic stroke, which concluded that administration of Vitamin E should be done carefully.[4]

Another alternative in the administration of antioxidants in ischemic stroke is to use Nitric Oxide (NO). NO is an antioxidant that can effectively break the chain reaction of lipid oxidation caused by free radicals. NO will react with the peroxyl radical ($OH^{\bullet}$).[5] Clinical trials by Jara using a combination of L-Arginine (NO precursor), Lamotrigine (inhibitor of glutamate and aspartate release) and Tianeptine (inhibitor of plasma serotonin) in patients with ischemic stroke may decrease the NIHSS score compared with the control group.[6]

Snakehead fish extract is a source of nutrient that has potential therapeutic effect. Snakehead fish extract rich with amino acids, vitamins and minerals as antioxidant. One of amino acids in snakehead fish extract is Arginine (10.74±0.15/ 100 grams snakehead fish protein).[7] Arginine used to NO Synthesis that has potential effecs. NO has several benefits in ischemic stroke therapy. NO can potentially used as vasodilator, can inhibits platelet aggregation and adhesion, inhibits leucocyte adhesion and chemotaxis, and blocks *N-Methyl-D-aspartate* (NMDA)receptors. One of NO's source is NO donor and L-arginine. Administration of NO donor and L-arginine was also reported to reduce the extent of infarction in ischemic stroke.[8] On the other hand, there is currently insufficient evidence to recommend the use of NO donors, L-arginine or NOS-I in acute stroke, and only transdermal glyceryl trinitrate (GTN), an NO donor has been assessed. In people with acute stroke, GTN reduces blood pressure, increases heart rate and headache, but does not alter clinical outcome.[9]

Objective

This study was aimed to determine the effect of snakehead fish extracts (SFE) on serum arginin levels and clinical outcome of acute ischaemic stroke (AIS) patients.

2. Methods

This study was a *double-blind randomized pretest-posttest control group design*, performed on 42 patients with acute ischemic stroke. The subjects were differentiated into treatment groups with snakehead fish extract (SFE) therapy and control group that received placebo within 48 hours after the onset of stroke for 1 week. The snakehead fish extract dosage was 15 grams per day, divided to 5 grams every 8 hours. Variable of snakehead fish extract were considered independent variable, Serum arginine sample as intermediate variable was analyzed with *Microplate reader*, Neurological clinical outcome as dependent variable was measured with *National Institutes of Health Stroke Scale* (NIHSS) score. All data were measured on day 1 and day 8. This study has received approval from the Commission of Ethics Faculty of Medicine UNDIP / RSUP dr. Kariadi with the number 944/EC/FK-RSDK/IX/2016. Data were analyzed using statistics software. The p value was considered significant if $p<0.05$

3. Results

A total of 42 subjects were included and then classified randomly to SFE/treatment group or control group. The characteristic feature of the subjects by age showed no significant mean age difference between the two groups ($p=0.5$). Similarly, other characteristics include sex ($p = 0.06$), hypertension ($p = 0.3$), diabetes mellitus ($p = 0.2$) and smoking ($p = 0.6$). (Table 1)

Serum arginine levels before treatment were significantly different ($p=0.002$), whereas in the snakehead fish extract group, the majority of subjects had low serum arginine levels/11 people (52.4%) and subjects with normal serum arginine levels/10 people (47.6%). In control group, serum arginine level before treatment was extreemly low in all subjects.

Serum arginine level after treatment was significantly different ($p <0.001$), seen in snakehead fish extract group, most subjects had normal serum arginine levels (90.5%) and the rest of it had low serum arginine levels (9.5%). In the control group, all of subject's serum arginine levels after treatment remain low (100%)

Table 1. Characteristics of study subject in groups

Characteristics	Group		p
	Treatment (n=21)	**Control (n=21)**	
Gender			
- Male	7 (33.3%)	13 (61.9%)	0.06*
- Female	14 (66.7%)	8 (38.1%)	
Age (years old)	58.3 ± 9.39	56.7 ±6.17	0.5¶
Age Category(years old)			
- < 55	6 (28.6%)	8 (38.1%)	0.4¶
- 55-65	10 (47.6%)	11 (52.4%)	
- > 65	5 (23.8%)	2 (9.5%)	
BMI	24.3±4.04	21.4 ±2.40	0.01¶
BMI Index			
- ≥ 25	10 (47.6%)	2 (9.5%)	0.06*
- < 25	11 (52.4%)	19 (90.5%)	
Hypertension			
- Present	20 (95.2%)	17 (81.0%)	0.3§
- Not present	1 (4.8%)	4 (19.0%)	
Diabetes Mellitus			
- Present	9 (42.9%)	5 (23.8%)	0.2*
- Not present	12 (57.1%)	16 (76.2%)	
Smoking Habit			
- Present	3 (14.3%)	1 (4.8%)	0.6§
- Not present	18 (85.7%)	20 (95.2%)	

¶Independent t-test; *χ^2test; § Fisher-Exact test
Value in table are : mean (value±Standard Deviation); n(%) percentage are calculated in the column

There was no correlation between change in serum arginine level with change in NIHSS score, with value in snakehead fish extract group ($r = -0.03$ and $p = 0.90$) and value in the control group ($r = - 0.12$, $p = 0.60$).(Fig.1) But the change of serum Arginine in the snakehead fish extract group are higher than the control group (33.6±19.95 µmol/L; 0.3±2.51 µmol/L; p<0.001) (Table 2), while the change in NIHSS score in snakehead fish extract group improved significantly compared to the control group (4.14 ± 2.03; 2.52 ± 1.81:p=0.009). (Table.3)

Table 2. Serum arginine level before and after treatment in SFE and control group

Serum Arginine Level	Group		t	p
	SFE (n=21) (µmol/L)	**Control** (n=21) (µmol/L)		
Before	41.2 ±22.20	23.7±=8.78	3.344	0.002*
After	74.8 ±21.36	24.0±=8.10	10.178	< 0.001*
Δ	33.6±19.95	0.3±=2.51	7.594	<0.001*

Table 3. NIHSS before and after treatment in SFE and control group

NIHSS	Group SFE (n=21)	Group Control (n=21)	t	p
Before	8.67± 4.39	7.57 ± 4.73	0.777	0.442
After	4.59± 3.23	5.05± 4.63	-0.425	0.673
Δ	4.14± 2.03	2.52 ± 1.86	-2.729	0.009*

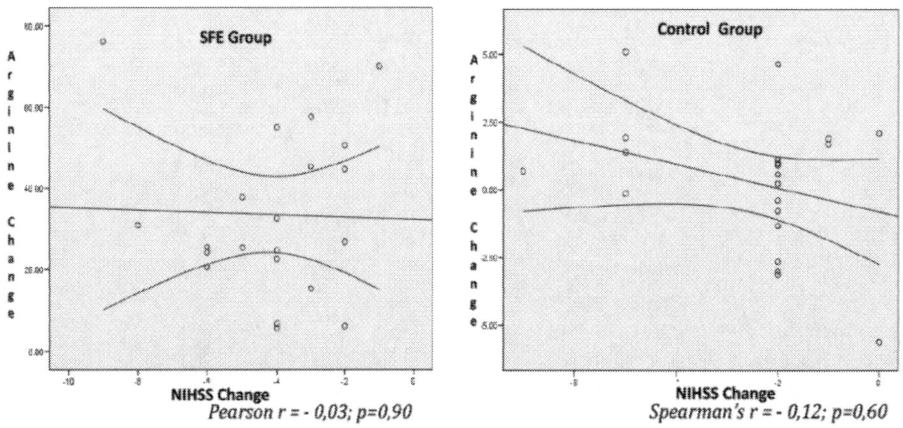

Figure 1. Correlation between change in arginine serum level with change in NIHSS score in both group

4. Discussion

The role of Arginine in ischemic stroke is as NO donor. NO synthesized by eNOS is an important endogen mediator in the regulation of cerebral blood flow and protection of brain blood vessel. NO synthesis by eNOS is through the oxidation of L-arginine to L-sitrulin. NO formed by endothelial blood vessels diffuses to adjacent cells such as capillaries, platelets and leukocytes, activates guanilate cyclase enzymes, and increases intracellular cGMP levels. NO from eNOS will cause neuroprotective effects such as vasodilation, anti thrombotic, anti-inflammatory and anti-proliferative effects.[10]

This study found that the change in NIHSS score in snakehead fish extract group improved significantly compared to the control group (*4.14 ± 2.03: 2.52 ± 1.81; p-0.009)*) and it is consistent with the previous study. The administration of NO donors such as L-arginine is reported to provide benefits in reducing the extent of infarction in ischemic stroke..[11] L-arginine may improve endothelial function through several mechanisms, namely increased intracellular transport, competitive antagonist Asymetric Dimethyl Arginine (ADMA), antioxidant effect, stimulates Histamine mast cells release, inhibition of norepinephrine activity, increased insulin secretion and change in intracellular pH. Previous study has shown that arginine supplementation is a major source of the availability of intracellular L-arginine required for the manufacture of eNOS.[12] Arginine also plays a role in inhibition of ADMA formation. ADMA is a competitive inhibitor of eNO synthase. Increased levels of ADMA inhibit NO synthesis and therefore impair vascular endothelial function.[13]

NO synthesized by eNOS is an important endogen mediator in the regulation of cerebral blood flow and brain blood vessel protection. NO synthesis by eNOS is through the oxidation of L-arginine to L-sitrulin. NO formed by endothelial blood vessels diffuses to adjacent cells such as capillaries, platelets and leukocytes, activates guanilate cyclase enzymes, and increases intracellular cGMP levels. NO from eNOS have neuroprotective effects such as vasodilation, anti thrombotic, anti-inflammatory and anti-proliferative effects.[10]

The content of arginine in snakehead fish extract has an important role in wound healing process. Arginine stimulates the release of various chemical compounds for wound healing, such as growth hormone and insulin. Arginine is also described to cause a positive nitrogen balance. Arginine is a substrate of NOS and arginase that plays a role in the wound healing process. Arginine is metabolized in the wound through arginase. During the formation of connective tissue, arginine is used to produce hydroxy-proline which is important in the formation of collagen. The high content of arginine in snakehead fish extract [14-15] can potentially be utilized as NO donor for additive therapy of ischemic stroke.

Snakehead fish extract in addition to having a high nutritional content also has medical benefits, especially because the content of albumin in snakehead. Snakehead fish extract is used for wound healing, reduce postoperative pain, as antioxidant and various other benefits for the nervous system and cardiovascular, as carrier of metabolism substances like free fatty acids and bilirubin, provides osmotic pressure in capillaries, the albumin can maintain blood plasma fluid balance for the purpose of prevention of edema, and the formation of new cell tissues [15-16].

Mean serum arginine level in SFE group before treatment was 41.2± 22.20 μmol/L, higher compared to control group 23.7± 8.78 μmol/L, statistic result showed that the difference was significant ($p=0.002$). Serum arginine level in SFE group after treatment is 74.8 ±21.36 μmol/L, higher compared to control group, but the statistic result showed significant difference in arginine levels between treatment group after treatment and control group *($p < 0.001$)*. Table. 2 also showed a delta increase in serum arginine levels in SFE group was significantly greater than in control group (33.6 ± 19.95: 0.3 ± 2.51; $p<0.001$).

According to normal serum arginine levels in the laboratory is 41-114 μmol/L, all subjects in control group still categorized to low serum arginine level (100%). The statistic result showed that the serum arginine level differences after treatment in hospital between the SFE group and the control group have a significant value *($p<0.001$)*. Previous study by Worthmann on 67 patients showed findings of arginine deficiency characterized by increased Asymmetric Dimethyl Arginin (ADMA) and its analog counterpart, Symmetric Dimethyl Arginin (SDMA) in stroke patients.. Both an increase of ADMA known as a marker of cardiovascular risk and atherosclerosis and an increase of SDMA plasma levels after acute ischemic stroke predict poor functional outcome.[17] Similar results were found in this study of 42 patients with acute ischemic stroke in which 32 of them had low serum arginine levels (<41 μmol/L).

The statistic result showed that serum arginine level differences in the hospital admission between the SFE group and the control group have a significant value *($p=0.002$)*. The results of the study were also supported by characteristic data of research subjects by Jara who reported on the neuroprotective effects of arginine on ischemic stroke and cerebral trauma. Of the 49 research subjects, 44 of whom had low serum arginine level (<41 μmol/L), 5 other subjects had normal serum arginine level. This study also showed that the serum arginine levels at admission to hospital between treatment group and control group was significant *($p<0.001$)*.[6]

In this study, logistic regression analysis showed only female gender factor that affected on improvement of NIHSS (OR=7; p=0,01) (Table.4), it means that female got more benefit than male. This study also found that the correlation of change in serum arginine with change in NIHSS score of SFE and control group was not significant, it is possible because there is a content or mechanism of snakehead fish extract outside arginine mechanism which more influential as variable between the effect of snakehead fish extract on improvement of stroke clinical output such as albumin, zinc, glutamate and BCAA (Brain Chain Amino Acid) consisted of leucine, isoleucine and valine, and daily dosage of small and peroral arginine may also be the cause of the study.

5. Conclusion

Administration of snakehead fish extract may increase serum arginine levels in AIS patients. Snakehead fish extract administration can improve the clinical outcome of AIS patients. Snakehead fish extract administration may be considered to support the recovery of stroke patients.

Acknowledgement

We would like to thanks to all research team as well as to all patients who participate in this study.

References

[1] Cherubini A, Polidori M C, Bregnocchi M, Pezzuto S, Cecchetti R, Ingegni T, Di Iorio A, Senin U and Mecocci P 2000 Antioxidant profile and early outcome in stroke patients *Stroke* **31** 2295–300

[2] Shirley R, Ord E and Work L 2014 Oxidative Stress and the Use of Antioxidants in Stroke *Antioxidants* **3** 472–501

[3] Jurcau A 2007 The role of antioxidant treatment in acute ischemic stroke: a clinical study *Rom J Neurol* **6** 181–8

[4] Schurks M, Glynn R J, Rist P M, Tzourio C and Kurth T 2010 Effects of vitamin E on stroke subtypes: meta-analysis of randomised controlled trials *Bmj* **341** c5702–c5702

[5] Hummel S G, Fischer A J, Martin S M, Schafer F Q and Buettner G R 2006 Nitric Oxide as a Cellular Antioxidant: A little goes a long way *Free Radic Biol Med* **40** 501–6

[6] Jara H, García F, Torres F and Lechin F 2011 An Effective Neuroprotective Treatment in Ischemic Stroke and Cerebral Trauma with Low Doses of *Sch. Dr. Stud. (European Union) J.* 121–33

[7] Tan B H and Azhar M E 2014 Short Communication: Physicochemical properties and composition of Snakehead fish (*Channa striatus*) whole fillet powder prepared with pre-filleting freezing treatments *Int. Food Res. J.* **21** 1219–24

[8] Willmot M, Gray L, Gibson C, Murphy S and Bath P M W 2005 A systematic review of nitric oxide donors and L-arginine in experimental stroke; effects on infarct size and cerebral blood flow *Nitric Oxide - Biol. Chem.* **12** 141–9

[9] Bath P M, Krishnan K and Jason P Appleton 2017 Nitric oxide donors (nitrates), L-arginine, or nitric oxide synthase inhibitors for acute ischaemic stroke *Cochrane Database Syst Rev* **4**

[10] Sawada N and Liao J K 2009 Targeting eNOS and beyond: Emerging heterogeneity of the role of endothelial Rho proteins in stroke protection *Expert Rev Neurother* **9** 1171–86

[11] Davis M and Barer D 1999 Neuroprotection in acute ischaemic stroke. II: Clinical potential *Vasc. Med.* **4** 149–63

[12] Loscalzo J 2004 Arginine Metabolism: Enzymology, Nutrition, and Clinical Significance *J. Nutr* **134**2798S–2800S

[13] Böger R H 2004 Asymmetric dimethylarginine, an endogenous inhibitor of nitric oxide synthase, explains the "L-arginine paradox" and acts as a novel cardiovascular risk factor *J. Nutr.* **134**2842S–2847S

[14] Mohd Shafri M A and Abdul Manan M J 2012 Therapeutic potential of the haruan (*Channa striatus*): From food to medicinal uses *Malays. J. Nutr.* **18** 125–36

[15] SH B and KA S 2000 The wound healing properties of *Channa striatus*-cetrimide cream-- tensile strength measurement *J Ethnopharmacol* **71** 93–100

[16] Mustafa A, Widodo M A and Kristianto Y 2012 Albumin And Zinc Content Of Snakehead Fish (*Channa striata*) Extract And Its Role In Health *IEESE Int. J. Sci. Technol.* **1** 1–8

[17] Worthmann H, Chen S, Martens-Lobenhoffer J, Li N, Deb M, Tryc A B, Goldbecker A, Dong Q, Kielstein J T, Bode-Böger S M and Weissenborn K 2011 High plasma dimethylarginine levels are associated with adverse clinical outcome after stroke. *J. Atheroscler. Thromb.* **18** 753–61

Table 4. Results of Logistic Regression Analysis of factors influencing the change in NIHSS score of patients with acute ischemic stroke

Variables	Bivariate Analysis						Multivariate Analysis			
	Change in NIHSS Score		p	OR	IK 95%		p	Exp(B)	IK 95%	
	Improvement	No Improvement			Min	Max			Min	Max
Change in serum arginine level										
- Increased	17 (94.4%)	17 (70.8%)	Fisher's = 0.11§	7.00	0.78	63.21	0.186	4.79	0.47	48.81
- Not increased	7 (29.2%)	1 (5.6%)								
Age (years old)										
- > 65	5 (71.4%)	2 (28.6%)	Fisher's = 0.12§	4.23	0.72	25.02	0.301	2.80	0.40	19.74
- ≤ 65	13 (37.1%)	22 (62.9%)								
Gender										
- Female	8 (36.4%)	14 (63.6%)	χ^2= 0.004*	7.00	1.73	28.34	0.01	7.00	1.73	28.34
- Male	16 (80.0%)	4 (20.0%)								
Diabetes										
- Present	6 (42.9%)	8 (57.1%)	χ^2= 1.0*	1.00	0.27	3.66				
- Not present	12 (42.9%)	16 (57.1%)								
Hypertension										
- Present	16 (43.2%)	21 (56.8%)	Fisher's = 1.0§	1.14	0.17	7.67				
- Not present	2 (40.0%)	3 (60.0%)								
Smoking Habit										
- Present	2 (50.0%)	2 (50.0%)	Fisher's = 1.0§	1.38	0.18	10.82				
- Not Present	16 (42.1%)	22 (57.9%)								
BMI Category										
- ≥ 25	6 (50.0%)	6 (50.0%)	χ^2= 0.5*	1.50	0.39	5.77				
- < 25	12 (40.0%)	18 (60.0%)								

*χ^2test
§Fisher-Exact test

Implementation of Water Safety Plans (WSPs): A Case Study in the Coastal Area in Semarang City, Indonesia

Budiyono[1*], P Ginandjar[2], L D Saraswati[3], D R Pangestuti[4], Martini[5], S P Jati[6]
[1,2,3,4,5,6] Faculty of Public Health, Diponegoro University, Jl. Prof. H. Soedarto, SH, Tembalang, Semarang 50275, Indonesia

E-mail:kenang92@yahoo.com

Abstract. An area of 508.28 hectares in North Semarang is flooded by tidal inundation, including Bandarharjo village, which could affect water quality in the area. People in Bandarharjo use safe water from deep groundwater, without disinfection process. More than 90% of water samples in the Bandaharjo village had poor bacteriological quality. The aimed of the research was to describe the implementation of Water Safety Plans (WSPs) program in Bandarharjo village. This was a descriptive study with steps for implementations adopted the guidelines and tools of the World Health Organization. The steps consist of introducing WSPs program, team building, training the team, examination of water safety before risk assessment, risk assessment, minor repair I, examination of water safety risk, minor repair II (after monitoring). Data were analyzed using descriptive methods. WSPs program has been introduced and formed WSPs team, and the training of the team has been conducted. The team was able to conduct risks assessment, planned the activities, examined water quality, conduct minor repair and monitoring at the source, distribution, and households connection. The WSPs program could be implemented in the coastal area in Semarang, however regularly supervision and some adjustment are needed.

Keywords: implementation, WSPs program, risk assessment, tidal inundation, coastal area.

1. Introduction

In 2011, tidal inundation (the local term is rob) in the urban area of Semarang affected 1,538.8 hectares. North Semarang is the largest rob-affected area as many as 508.28 hectares [1]. North Semarang consists of 9 villages [2]. Bandarharjo and Tanjung Mas are affected by tidal inundation (rob) that reached 20-60 cm in height. Compare to other villages, Bandarharjo and Tanjung Mas experienced the deepest subsidence, which reached 4 cm/year [3].

Rob influences the quality of clean water in the areas [4]. A study in 2007 showed more than half (59 %) of 39 respondents in Bandarharjo village used unsafe water in term of bacteriological quality [5]. Other study revealed 88.9 % out of 27 sample water from wells in Bandarharjo had high MPN, a measurement of total coliform [6]. By 2010, people in Bandarharjo started to use deep groundwater (artesian wells). The wells belong to some owners with many customers. Counted 45% out of 20 water samples of deep groundwater were not met the requirement in terms bacteriological quality. Microbial contamination of major urban has the potential to cause outbreaks of waterborne disease [7].

World Health Organization (WHO) has developed guidelines and tools to promote water safety plans (WSPs) to improve the drinking water quality through risk identification, communication, and management approach using HACCP (hazard identification and critical control points), a tool previous developed and used for food quality control [8]. The WSPs is an instrument in water quality management in floods and natural disasters areas. In this study, tidal inundation (rob) affected area of Semarang was chosen for WSPs implementation.

2. Methodology

2.1. Study design

The location was in RW II Bandarharjo, Semarang is chosen based on initial bacteriological water quality of deep groundwater. Research activities used the framework of WSPs guidelines [8], steps included: introduction of WSPs to community was conducted by presentation and discussion method; the formation of WSPs team building by workshop method; training the team by presentation, exercise, simulation, and field visit method; conduct risk assessment of water safety by WSPs team using tools of WHO; conduct minor repair I after risk assessment; monitoring of water safety risk; minor repair II after monitoring of WSPs field trial in rob area [9]. This work used descriptive study and the data are analyzed by describing both qualitatively and quantitatively [10],[11].

2.2. Introducing WSPs program and team building

The introduction of WSPs was facilitated by Faculty of Public Health, Diponegoro University and WHO representatives of Indonesia which were motivated by the health problem and quality of drinking water in Bandarharjo. Health problems are based on the incidence of waterborne diseases i.e., diarrhea, dysentery, shigellosis, typhoid, and skin problem [12]. These diseases are getting an increase during 2011-2013 as well as lower in clean water coverage. The problem also was supported by a previous study [5],[6] and preliminary study. These data are intended to make problem awareness of the stakeholders. The meeting was held and attended by stakeholders in Bandarharjo, there are village officer, women group, youth group, community empowerment board at the village level, health officer, local government water supply officer, deep groundwater owners, village security group, and the customers. Facilitator and an expert presented the background of a health problem-related water supply in the area as well as discussion. The most important in the meeting was understand the health problem by the stakeholders and made the commitment to them for overcoming the problem. Each participant or group brainstorms and declares the willingness to implement the WSPs program.

A team of WSPs was developed after had commitment among the stakeholders to implement WSPs program. The team structure of WSPs consisted of the head of WSPs team (coordinator), secretary, technical unit, monitoring, and evaluation unit, and others were the member. The team of WSPs was legalized under the decree had signed by the head of the Bandarahrjo village office.

2.3. Risk assessment

A semi-quantitative risk matrix had been used to assess the risk. The matrix consists of 5 columns and 5 rows. The columns contain the degree of severity (consequences), whereas the rows contain the appearance or frequency of risk. The description of how to rate the assessment point is as follow a. The frequency of occurrence: once a day (5), once a week (4), once in a month (3), once in a year (2), once in 5 years (1); b. Degrees of severity: no effect (1), a small impact of compliance (2), a medium impact of aesthetic (3), impact on regulation (4), impact (disaster) on public health (5); c. Risk value and degrees of risk: low risk (< 6), medium risk (6-9), high risk (10-15), very high risk (>15) [8],[10], [13]. An instrument for observation and risk assessment consists of items to the risk of contamination of the drinking water supply system. The instrument is in the form of checklist to observe and assess the risk of water supply system: at a source, process, distribution and consumer levels. This instrument is used by the team for assessing the risk.

2.4. Minor repair
Minor repairs are made based on the results of risk assessment and monitoring. This activity was carried out by the WSPs team along with the owners of artesian wells. The provision of materials and tools for minor repairs is done by sharing resources between owners and the WSPs team. Minor repairs included at the source, distribution, or house connection. For example repair of reservoirs, distribution pipes, house connection pipes, and water meters.

2.5. Monitoring of risk
Monitoring was conducted by the WSPs team consisting of health workers, artesian wells owners, security groups, and technical units. Monitoring activities are carried out in accordance with the agreed schedule and used an instrument/checklist. The monitoring team will perform the survey of water supply system ranging from water sources, distribution systems, and households connections that had minor repair and unconfirmed improvements. Monitoring officers observed and recorded if there were damage, leaking pipes, loose pipes, broken pipe, or submerged water meters. The results of the follow-up monitoring activities were discussed with the WSPs team for planning for further improvements [14],[15].

3. Results and Discussion

3.1. Results
In 2011, tidal inundation (the local term is rob) in the urban area of Semarang affected 1,538.8 hectares. North Semarang is the largest rob-affected area as many as 508.28 hectares [1]. North Semarang consists of 9 villages [2]. Bandarharjo and Tanjung Mas are affected by tidal inundation (rob) that reached 20-60 cm in height. Compare to other villages, Bandarharjo and Tanjung Mas experienced the deepest subsidence, which reached 4 cm/year [3].

3.1.1. Introducing of WSPs
Thirty persons participated in this activity, consisted of owners of the deep groundwater, customers/users, the Bandarharjo village officer, staff of local government of drinking water supply system, youth community organization, women organization, technical water sanitation of the village, health officers from primary health care (PHC), and others. First session discussion revealed: (a) WHO and Faculty of Public Health (FPH) Diponegoro University intended to apply Water Safety Plan (WSPs) in rob area of Bandarharjo due to unsafe water in term of bacteriological quality. The purpose was to secure water in order to ensure the water quality; (b) The quality of water in Bandarharjo is deficient as rob and flood contaminate the water; (c) Most residents obtain water from deep ground wells; (d) There are 10 owners of deep ground wells in the area, each serves 50-100 customers; e. Customers of deep groundwater are less informed about water quality they use and will be satisfied if there is a study to determine the quality of their water; (f) Customers want to know how to deal with high pH and acid water, also the management of healthy drinking water; (g) Before implementation of WSPs in Bandarharjo, there would be an examination of the biological quality of 40 water samples; (h) The WSPs needed full cooperation from the owners of deep groundwater; (i) All participants agreed to monitor the quality of deep groundwater.

Second session discussion revealed: (a) local government of drinking water supply system suggested all deep ground wells to be examined. The result then might be used to formulate priorities in WSPs implementation; (b) Purpose of examination was to get an idea of the quality of water from deep groundwater wells. Test results will be used as the basis for the implementation of WSPs. The activity is expected to securing clean water that is adversely affected by rob; (c) This activity aims to undertake efforts to secure the provision of drinking water, through the establishment of WSPs team work together with security systems that have been prepared; (d) Water providers should have a commitment to support WSPs implementation; (e) WSPs team will be formed to monitor the quality of drinking water in

Bandarharjo; (f) The team should approach the deep ground well owners; (g) Training for WSPs team will be held after Idul Fitri.

3.1.2. Preliminary water survey

Preliminary water survey was conducted in order to: (a) describe and identify water supply system of deep groundwater wells in Bandarharjo village; (b) assess the bacteriological parameter (total coliform) for standard requirement; and (c) compare bacteriological quality (total coliform) with the standard of Permenkes No 416/1990. The main parameters of drinking water are total coliform, turbidity, salinity, and pH.

Water supply system in Bandarharjo starts from water provider, upper reservoir, distribution and then used by customers. Most of the water supplies in Bandarharjo village are deep groundwater wells. Water was pumped up from 85 meters down in the ground into higher water reservoir (e.g., 6-20 m). Water reservoir will provide sufficient pressure of water distribution to customers. Water was distributed to customers by using PVC pipe with diameters 4-6 ". The result of observation found that pipe line distribution was frequently drowned by rob. This system is a difference to a local government drinking water supply system.

3.1.3. WSPs team building and training

WSPs team was formed on the first day of training. The team consisted of coordinator, secretary, and field technician, and the members. As consultant was the PHC of Bandarharjo, a local government drinking water supplies system, and deep groundwater owner.

On day 2, Management simulation (exercise) was given on day 1. For effective training WSPs team was divided into three small groups for risk assessment. Group 1 focused on water sources and discussed a condition of the catchment, piping, water tank (upper and bottom reservoir), present of disinfection, surround the area, contamination possibility (Table 1). Group 2 focused on water distribution and discussed a condition of piping, pipe joining, a branch of distribution, contamination possibility (Table 2). While group 3 focused on customers and discussed the condition of water, households connection, water meter, piping, valve, hose, water tank (Table 3).

For the exercise and discussion in the training, a team conducted field visit for measurement of the risk assessment using instrument [8][13]. Risk/hazard occurrence, for example, are pipe leakage, broken pipe, the loose connection of pipe, submerged water meter, broken water meter, no treatment (i.e., disinfection, filtration), lack of cleaning or draining upper a reservoir, no tight cover of the reservoir, sediment in inner pipe.

Table 1. Result of the group 1 for working discussion component system at source level

Problem	Hazard	Occurrence	Severity	Rank
Flood tide	Contamination (salt/smell)	5	2	10
Leakage	Contamination/germ/smell	3	2	6
Underground pipes	Leak	1	2	2
Precipitation	Soiled water, mossed water	4	3	12

Rank: 1. Precipitation/Sediment, 2. Flood tide/rob, 3. Leakage

Table 2. Result of the group 2 for working discussion component system at distribution level

Problem	Hazard	Occurrence	Severity	Rank
Leaking pipe	Contamination of sewage	2	1	2
Clogged pipe	Containing moss	2	1	2
Pipe connections	Loose, cracked	3	1	3

Rank: 1. Pipe connections, 2. Clogged pipe, 3. Leaking pipe

Table 3. Result of the group 3 for working discussion component system at customer level

Problem	Hazard	Occurrence	Severity	Rank
Has a salty taste	Diseases, germs	5	2	10
Soiled water (dirty)	Itching, diarrhea	5	5	25
Smelly water	Respiratory problem	1	2	2
Broken pipe	Clogged water	5	3	15
Blackout	Water does not flow	3	3	9
Water faucet has a moss plant	Germs	4	3	12
Sediment in the pot after boiling water	Kidney stones	4	4	16
Reservoir rarely drained	Mosquito larvae appears	4	5	20

Rank: 1. Soiled water (dirty), 2. Reservoir rarely drained, 3. Sediment in the pot after boiling water

On day 3, module 8-11 (adapted from WHO training module) was delivered [16]. The activity consisted of formulating the plan of action and explaining a plan of action after training. Plan of action which comes from participants was dissemination/socialization water safety plan to the community and WSPs team legality (decree) from the head of Bandarharjo village. Explanation of plan of action included measuring a parameter and doing sanitation survey in Bandarharjo village and explaining the description of WSPs implementation after training. The plan of action was made by the WSPs team and lead by the coordinator of WSPs and the result in Table 4. In addition, Table 5 for handling the emergency response.

Table 4. Result of the working discussion for plan of action due to result in table 1, 2, and 3

Hazard	How (what to do?)	Person in charge	When	Cost (rupiahs)
Sediment	Cleaned/ drained	Owner of deep groundwater well	1 year, 1 times	200,000
Pipe connections	Pipe replacement	Owner of deep groundwater well	Incidental	100,000
Soiled water (dirty)	Complain to owner, place a filter	Owner of deep groundwater well	Incidental	25,000

Table 5. Result of the working discussion for handle emergency condition at the source, distribution, and customer

At the source	How to handle?
- Condition of source fine - Condition of distribution fine - Trouble came from customer: soiled water (dirty), trouble on water meter	Check behind of water meter, if the water ticket comes out and we can see the result then we can place it normally
At the distribution	**How to handle?**
- Leakage on pipe connection - Unstable machine - Blackout - Flood tide - Leaking pipe crashed by truck - Dirty reservoir - Broken pipe - Submerged sewer pipe	- Replace - Repairmen - Call state power company - Cleaned - Replace/check how severe the broken pipe - Be elevated
At the customer	**How to handle?**

- Trouble in water meter	- Open the water meter and clean it
- Soiled water (dirty)	- Check water source. If it is fine then it might be the reservoir and distribution pipe had a clogged - Cleaned through high pressure flushing

3.1.4. Examination of water and risk assessment of water safety by WSPs team
In term of pH and salinity, all water sample qualify the standard, yet two of them exceed turbidity standard (> 25 mg/liter). Eighty percent of water sample at customer level did not meet specified standard according to Decree of Ministry Health of Republic of Indonesia number 416/1990, which is maximum 10/100 ml water sample. Water quality does not require BOD parameter. The BOD examination was performed to estimate a possibility of pipe leakage, loose connections, or water contamination by sewer or tidal inundation, also possibility sedimentation in inner pipe. In addition, from the results of our spot check by cutting distribution pipeline, we found in the inner wall of the pipe has a lot of stick impurity deposition.

Field risk assessment in previous study [17] ranged from sourcing, processing, distribution, households connection (costumer). A very high degree of risk at source system (risk value = 25), caused by the presence of holes (space) between lid and reservoir. In term of turbidity, a reservoir has a high degree of risk (risk value = 15) due to the existence of dust, moss, and spore contaminants. At processing system, we found a very high degree of risk (risk value = 25) due to no physical or chemical treatment. At distribution system, we obtained a high degree of risk (risk value = 15). It because of sediment in inner pipe and loose of connection. At costumer/household system, we observed connection pipes, water-meters, and faucets in the houses, and it obtained a very high risk (risk value = 25).

WSPs team gave priority to the handling improvement of drinking water supply systems, based on categories of very high and high degree of risk. The priorities include: at the source, the reservoir tub cover less dense and dirty water reservoir; in processing, the lack of primary disinfection treatment (based on initial measurements which given total coliform exceeded the standards); on the customer system, the priority is to water meter which submerged by rob, sewer water, or soil undercurrent.

3.1.5. Corrective action plan and minor repair I: after risk assessment
Hazard event at source system: (1) Little space between tub and cover allows contamination from dirt or dust; (2) Dirty tub. Corrective action would be made a concrete of tub cover by WSPs team and owner. Draining and manufacture of concrete tubs cover by owner. Hazard event at processing system: (1) Bacterial contamination (coliform total); (2) Performing disinfection should be done socialization. Disinfection uses chlorine by a technician from the officer of local water drinking water supply system Semarang. Before disinfection so do socialization by health officers on advantages and disadvantages of disinfection. Hazard event in distribution pipe system: sediment in inner pipe. Hygienic reasons for maintaining the internal cleanliness of pipework. The method used to clean the dirt is flushing. This is the simplest of the pipe-cleaning techniques. Hazard event at customer system: Water meter was submerged by trench/gutter and tidal inundation. The submerged pipe was elevated by giving additional pipe so that it will not re-submerge by tidal inundation by WSPs team and customer.

The budget plan had been prepared and cistern cap/cover foundry design which conducted by WSPs team and expert designer. Concrete thickness approximately 10 cm in total (two cisterns) which approximately 12 m^2. Concrete framed with steel with a concrete slab 2.5-3 meters long and 50 cm wide which were arranged in rows. Further, on it cast with concrete. Two manholes 60x60 cm were built to simplify for monitoring the water, cleaning, and drainage the cisterns. In addition, to easing climbing up to the cistern, iron stairs with a height of 7 meters were made.

Implementation of the foundry was conducted on November 26, 2013, which was supposed to do in the first week of November 2013. This setback/tardiness due to bad weather (rain and rob) which means it didn't support for a foundry. Implementation of foundry involves WSPs team along with competent personnel to perform casting/foundry.

Repairs also carried out on the exhaust pipe or pipes to drain along with the faucet, so that can be opened and closed when done the draining. As it is known that the cistern tank has not drainage since it was made (almost ten years). From observations when drained it clearly be seen dirty brown sewage on the ground floor, tub/cistern and walls. Dewatering was done by the owner of the wells and executed a week after casting/foundry. It appears that the inner wall of tub/cistern was very dirty, dark brown colored, so as the water in the tub. Dewatering was conducted on November 30, 2013.

For improvement in the system of water processing, before the disinfection socialization to as many as 25 heads of households (consumers) who subscribe to the deep ground well belongs to the owner (Mrs. Warsi) were carried out. Socialization is intended to inform that deep ground well water were possibly contaminated, therefore it is necessary to perform disinfection treatment in order to make people understand that water is contaminated with bacteria would be harmful to health. Socialization was carried out by sanitarians from Primary Health Care and WSPs team. Information of benefit and disadvantage of the water processed by disinfection (chlorination) were given in socialization.

Depletion/dewatering on the distribution pipe was done by cutting the tip of the end of distribution pipe. There were 3 distribution pipes. After the cutting, it is very noticeable that the water coming out of the distribution pipes was very dirty. To ensure that all the impurities in the water in the pipe have been carried out, the pipes were stamped at the end of the pipe so that all the dirt out of the pipe wall. Before it, first, we opened the faucet distribution from above tub/cistern, so that the flow coming out of the distribution pipeline is quite strong, so it will be able to flush the dirt that was in the pipelines. The time required for cleaning/flushing pipeline distribution approximately 30 minutes. After 30 minutes streamed clearly, the water that comes out of the distribution pipes has been very clear. In addition to performing flushing, also carried off the connecting distribution pipe those loosen. The activity was conducted on December 7, 2013.

Improvements to the customer were elevated and replacement water meter submerged pipe sewer, water and soil were not done yet, because after the double-checked there were still in good condition (data/ information less valid). To do so, there were re-assessment and observation (December 2, 2013) to the whole household connections (approximately 73 customers), to look at the risks that are likely for more. In addition, it is intended to recalculate the need for tools and materials used to repair and time estimate. The problem is when there is no customer list of an owner for certainty. The observation and reassessment were done by WSPs team Bandarharjo village.

From observation and re-identification observed from 70 customers on Monday, December 2, 2013, there were 9 water meters submerged land and water from sewerage. It also made improvements in the household connection. Activities of household plumbing repair connection and water meter elevation performed on Saturday, December 7, 2013. Distribution pipe flushing is done after the entire water meter level and household connections have been repaired.

3.1.6. Monitoring and minor repair II

On December 27, 2013, WSP team observed and monitored 20 customers. There were 2 leaky connection pipes, 1 submerged pipe that connected to a water meter, 1 submerged water meter, 4 leaky faucets, 1 leaky indoor pipe, and 1 submerged indoor pipe. As for indoor reservoir, which is generally in the form of a bucket, there were 4 reservoirs in turbid condition.

Next monitoring was held on December 28, 2013. WSP team found showed 3 leaky/seeped faucets, 1 broken faucet, 3 rushed of a water meter, 1 leaky pipe, and 2 turbid reservoirs. Further water safety monitoring was held on December 30, 2013. Almost all equipment was in good condition. WSP team found showed only 1 broken faucet. Monitoring on January 2-4, 2014 found one turbid reservoir. The rest were all in good condition (water meter, pipes, and faucets).

Other than at customer level, monitoring was also performed at a water source and distribution level, both by observation. WSPs team found a cover of a reservoir, a concrete cover, was in good condition, clean and no crack observed. Pipes and faucets for leaching were also in good condition and properly function, as well as a pipe from a reservoir for distribution. Pipes of distribution, which have been cleaned or flushed with water, remains in good condition and tight. Besides, an owner has cleaned floor at the water point that previously mossy and slippery. The result of observation showed water in the reservoir was very clear (bottom of the reservoir could be seen). The physical quality of water obtained from the reservoir also clear, tasteless, and odorless.

Before conducting minor repairs coordination have been carried out between WSPs team and owner of deep ground wells. This activity was done by identifying distribution pipeline that was leak/rupture. Repair leaky pipes and faucets by replacing faucets/cellophane tape or glue that has been separated. Elevate a water meter level by replacing the pipe and cellophane tape and replaced the broken one with a new water level meter. Similarly, the distribution pipe and connection pipe that loose were replaced and gluey. Pipes were submerged with litter/soil replaced with new pipes. This work was done by the WSPs team and owner of deep ground wells.

Observation result showed that during rainy season and flood, the casing is not cemented submerged by floods. This potential for seepage on the side of the casing pipe and can be infiltrated in the existing water in the pipe. Therefore, to prevent contamination of the casting is done by providing cement, sand, and bricks. On monitoring and observation also obtained the result that many distribution pipes are submerged soil, water sewer, and trash. Therefore, it needs efforts to prevent damage and contamination as well as the monitoring of the distribution pipe by replaced and elevated distribution pipes that submerged soil, sewers and garbage. Distribution pipes that replaced had a 3 " as much as 60 rods and distribution pipes with size ¾ " as much as 4 rods, including pipe connection 'T' and gluing using cellophane tape and glue.

Searching of leaking distribution pipes requires quite a long time (2 weeks); this is because most of the distribution pipes submerged in the soil and submerged sewer, and garbage. The searching of leaking pipe had been done for 3 times, involving owners, WSPs team and from Faculty of Public Health Diponegoro University. During observation, it was found cover water reservoir (manhole) was rusty and corroded. Based on that, need efforts to do the painting so that it has a protection from rust.

3.2. Discussion

3.2.1. Introducing of WSPs program.

It is not easy to adopt new programs or innovations (i.e., WSPs) by a community. This is due to various educational background, culture, experience and socio-economic conditions [18],[19]. People will be willing to accept new programs or innovations if they feel they need or get benefit [20]. However, it needs a good socialization and it is one of the diffusion mechanisms [21].

Before introducing the WSPs program to the community, the community needs to be shown or informed about the health problems they face. The several considerations to decide the priority in health problem based on magnitude, severity or important or seriousness, and benefit [22]. The introducing of health problems to the community is very important because some members of the community may not aware of any health problems in the region. Some inhabitants did not realize that quality of drinking water is important, it was revealed in the discussion. To convince the community of the importance of water quality requirements, we present the results of research related to water quality in the region to the representative community. To stimulate a collective sense of belonging to drinking water among community members as they confront the significant problem about contaminated water and its negative impacts on the customers [23]. Based on Ministry of Health Republic of Indonesia regulation No. 416/MEN-KES/SK/IX/1990, a maximum number of total coliform allowed is 10/100 ml sample (piping water) and 50/100 ml sample (non-piping water). Seven out of 20 wells in Bandarhajo had a poor hygienic condition (>50/100 ml) in term of bacteriological quality. This scientific perspective introduced important to the drinking water quality and triggers community to protect the drinking water [24].

The introducing of WSPs program was also to communicate the value of drinking water (i.e., reduce future water borne disease) to increase people's awareness of the importance of this work [25]. Therefore this program is based on the quality of drinking water.

The first step to a pre-trigger community is selecting community who concern with the problem. The representative community would be respected people with skills who provide services like village midwives, community leaders, religious leaders, teachers, youth, and women [23]. Coordination should be done before the training and aimed to keep a commitment. High committed to the team in the program is a key success of WSPs. Manager's performance on the job and the team members' technical background and commitment are most critical for project success [26].

3.2.2. WSPs team building and training

The WSPs team included a representative the communities group, government sector, and private sector and at many levels of knowledge and experience within the small community. At the initial stage to identify the people who have experience in community activities were selected by village officer that well-known the people in the area. Members of the WSPs team have varying work, education and social backgrounds, such as sanitarian, youth group, deep ground or artesian wells owners, women's groups, urban community empowerment agencies, village officials, community protection officer, health cadres, the local government of drinking water supply system, and community leaders. The characteristics of WSPs team are levels of education ranged from elementary school to higher education, gender (16 males and 8 females), and age (20-60 years old) which were distributed in each of RWs in Bandarharjo village. It is expected that with various backgrounds and levels the community will strengthen the work of a team. High-performance partnerships employ practice designed to ensure they get the right people and resources (i.e., resources, skills, and expertise) on board and it will reduce the financing burden [27],[28]. The WSPs implementation is also taken into account the flexibility and adapted to local needs [16]. An initial investment of time and resources can be characterized as a mutual selection process [29]. Assess people's capacities for collaborative work is important due to their technical abilities and scientific [30]. Although they may have a level of knowledge and experience in water supply system, by delivering the training module of WSPs, they had the same knowledge and perspective.

The WSPs team was legalized by decree of village office to increase commitment, clear their role, involvement, and responsibilities such as coordination of all activities, water quality monitoring, pipe checking (i.e., at the source, distribution, and household connection), monitoring of supply system of drinking water. The team could be in a large or small team, but the principles of the team are shared goals, clear roles, mutual trust, effective communication, and measurable processes and outcomes [31]. Building the team is the efforts to increase the likelihood that effective WSPs programs will be sustained and it may be developed with individuals, groups, teams, organizations, inter-organizational coalitions, or communities [32].

Moreover, the WSPs team should have the strong leadership and community participation (i.e., deep groundwater owner) to ensure program sustainability. The collaboration between community, the member of the team and team leader of WSPs will reinforce the teamwork. Shortell et al [33], collaboration tend to develop a committed core leadership that helps to create and consistently of objectives, tasks, projects, and programs. Community member enables to take control, maintenance, and sustainability of the water facilities and eliminate barriers and effective participation in water supply system [34]. The members of the team were fully involved in the planning, implementation, operation, and maintenance of water supply facilities in the area [35].

Module 1-6 (adapted from WHO training module) was delivered on day 1 training [36]. In the training, the team members were divided into a large and small group to ease delivered the materials. In the large group generally explained the material of module and discussion, while in the small group for exercise, field visit, and working in a team. It was expected more effective to deliver the training module and increased relationships. According to Steinert, there are 12 tips in which small-group teaching can become more effective and more enjoyable. In the small group, teaching is to promote understanding, critical thinking and problem solving to enhance communication skills and to foster self-directed learning [37]. Effective results of working group discussion, exercise, field visit, and simulation can be seen in Table 1,2,3,4, and 5. This mixed method makes the participants not bored, fun, and productive [38], because of little explanation from the facilitator and or an expert, then the participants work with the tools already provided. In addition, this way makes it easier for participants to understand the WSPs materials, to conduct simulations, field trips, or working group.

The participants able to identified the problem, hazard, and made priorities the problem at the source, distribution, and customer of water supply system. The working group discussions were guided by facilitators from Faculty of Public Health Diponegoro University and WHO representatives as an expert to ensure the effectiveness of training. Participants were also able to make simulation the planning and budgeting at small scale to tackle the problems at routine and emergency conditions. The role of the leader of a small group is important to directing and clarifying the discussion [39],[40].

3.2.3. Water quality

The local government drinking water supply system consists of intake of raw water, physical and chemical treatment, distribution and household connection (customers). Complete water treatment included screening, coagulation, sedimentation, filtration, disinfection, and distribution [41]. However, the minimum requirement for artesian wells (deep groundwater) is disinfection. One way to disinfect water is using chlorine. A very effective disinfection using chlorine kills bacteria, viruses, and protozoa such as Giardia and Cryptosporidium [42].

3.2.4. Risk assessment and corrective action plans

At costumers (households) level, we observed connection pipes, water meters, and faucets in the houses, and it obtained a very high risk (risk value = 25) [17]. Concrete tubs cover will protect the water in a reservoir from dirt or feces from the wild animal. The fecal coliform source includes wild and domestic animals feces, which can be spread by the wind [43].

Problems that often occur in the distribution system, among others, the presence of microbial growth and biofilm, cross connection, backflow, rusty and aged, contamination during service and nitrification [44]. Deposits material in the inner of distribution system was removed by flushing because this method is the cheapest and simplest methods for removing of loose deposits [45]. Even this method also has disadvantages of the quite large amount of water used and unsuitability for large diameter pipes to achieve the desired the flushing velocity. The flushing involves the discharge of water from pipes, generally through hydrants and washouts, to generate velocities in the pipe capable of removing accumulated material and biofilms inside the pipe and attached to its walls. Although there are no reports of health effects directly attributed to deposits in pipes, they do provide conditions for the proliferation of microorganisms and animals [46]. Flushing can be obtained periodically to minimize deposits in the distribution system, including microbial [47].

The water meter assembly must be fully supported with a minimum ground clearance of 150 mm, and should not be greater than 300 mm from the finished ground level to the base of the water meter assembly. On a case by case basis consideration will be given to varying the height of the water meter up to a maximum of 1.5 m. Water meters must not be located within garages, roof cavities, ceiling spaces or inside pits [48],[49].

Monitoring in accordance to identified the drinking water supply system which not work properly and so be fixed and prevent contamination. Treated drinking water quality monitoring would be able to inform of potential contamination of drinking water [50].

4. Conclusion

The WSPs programs could be implemented by the WSPs team in the coastal area of Semarang city. The team able to identified risks of the drinking water supply system and the actions were taken to make improvements to the water supply system. The minor repairs were done by sharing resources. However, there is still need to provide assistance to the WSPs team and educate community for the sustainability of the WSPs program.

5. Acknowledgment

Authors thank the owners of artesian wells, customers, the WSPs team, the village officers of Bandarharjo, the local government of water supply agency, primary health care of Bandarharjo. This work was sponsored by World Health Organization (grant number: SE INO1206445).

References

[1] Ramadhany A S, Nugroho A D S and Subardjo P 2012 *J. Mar. Res.* **1** pp 174–80 (in Bahasa)
[2] Suprianto H 2011 Monograph of Subdistrict North Semarang July-December 2011 (in Bahasa), Semarang
[3] Adhy D S 2007 *J. Pondasi.* **13** pp 14–26 (in Bahasa)
[4] Waskito 2008 *Maj. Ilm. Pawiyatan.* **17** pp 77–92 (in Bahasa)
[5] Budiyono 2007 *Media. Kesehat. Masy. Indones.* **64** pp 5–9 (in Bahasa)
[6] Rahajoe T 2008 *Media Kesehat. Masy. Indones.* **1** pp 19–25 (in Bahasa)
[7] World Health Organization 2006 *Guidelines for drinking-water quality First Addendum to Third Edition Volume 1 Recommendations* (Electronic Resource) (Geneva: WHO Press) p 35
[8] Bartram J, Corrales L, Davison A, Deere D, Drury D and Gordon B 2009 *Water safety plans manual: Step by step risk management for drinking water suppliers* (Geneva: WHO Press) pp 8–65
[9] World Health Organization 2008 *Guidelines for drinking-water quality: incorporating 1st and 2nd addenda, Vol.1, Recommendations.– 3rd ed.* (Geneva : WHO Press) p 282
[10] Githenya M S and Ngugi K *2014 Journal of Business Management* **1** (11) pp 230–253
[11] Samuel G, Mbabazize M and Shukla J 2016 *European Journal of Business and Social Sciences* **5** (1) pp129–145
[12] Primary Health Care of Bandarharjo 2013 *Health Profile of Bandarharjo Semarang*
[13] Deere D, Stevens M, Davison A, Helm G and Dufour A 2001 *Management Strategies in Fewtrell L, Bartram J (eds) Water Quality Guidelines Standard Health Risk Management: Water-related Infectious Disease* (Geneva: IWA Publishing) pp 257–88
[14] Bartram J, Brocklehurst C, Fisher M B, Luyendijk R, Hossain R, Wardlaw T and Gordon B 2014 *Int. J. Environ. Res. Public Health* **11** 8137-8165; doi:10.3390/ijerph110808137
[15] Wescoat Jr. J L, Fletcher S and Novellino M 2016 *Water Policy* **18** 1015–1032 doi: 10.2166/wp.2016.158
[16] World Health Organization 2012 *Water Safety Planning for Small Community Water Supplies Step-by-step risk management guidance for drinking-water supplies in small communities* (Geneva: WHO Press) pp 8–37
[17] Budiyono, Ginandjar, P, Saraswati L D, Pangestuti D R, Jati S P and Rahfiludin Z 2015 *Procedia Environ. Sci.* **23** pp 93–8 doi: 10.1016/j.proenv.2015.01.014
[18] Wejnert B 2002 *Annu. Rev. Sociol.* **28** pp 297–326 doi: 10.1146/annurev.soc.28.110601.141051
[19] Cicek H, Cevger Y and Tandogan M 2008 *Ankara. Üniv. Vet. Fak. Derg.* **55** pp 183–187
[20] Montgomery M A, Bartram J and Elimelech M 2009 *Environmental Engineering Science* **26** (5) pp 1017-1023 doi: 10.1089=ees.2008.0388
[21] Bennett S, Dalglish S L, Juma P A and Rodrıguez D C 2015 *Health Policy and Planning* **30** pp ii26–ii35 doi: 10.1093/heapol/czv071
[22] World Health Organization 2012 *WHO recommendations: optimizing health worker roles to improve access to key maternal and newborn health interventions through task shifting* (Geneva: WHO Press) pp 15–6
[23] Kar K and Chambers R 2008 *Handbook on Community-Led Total Sanitation vol 44* (London: Institute of Development Studies) pp 13–41
[24] Isunju J B, Schwartz K, Schouten M A, Johnson W P and van Dijk M P 2015 *Public Health* **XXX** pp 1-9 doi:10.1016/j.puhe.2011.03.008

[25] Baum R, Amjad U, Luh J and Bartram J 2015 *International Journal of Hygiene and Environmental Health* **218** pp 677–685 http://dx.doi.org/10.1016/j.ijheh.2014.12.004
[26] Belassi W and Tukel O I 1996 *Int. J. Proj. Manag.* **14** pp 141–51
[27] Corrigan M B, Hambene J, Hudnut III W, Levitt R L, Stainback J, Ward R and Witenstein N 2005 *Ten Principles for Successful Public/Private Partnerships* (Washington D.C.: ULI–the Urban Land Institute) pp 2-7
[28] Thadani K B 2014 *Procedia - Social and Behavioral Sciences* **157** pp 307–316 doi: 10.1016/j.sbspro.2014.11.033
[29] Weiner B J and Alexander J 1998 *Health Care Manage. Rev.* **23** pp 39–55
[30] Bennett L M and Gadlin H 2012 *J. Investig. Med.* **60** (5) pp 768–775 doi:10.231/JIM.0b013e318250871d
[31] Mitchell P, Wynia M, Golden R, McNellis B, Okun S, Webb C E, Rohrbach V and Von Kohorn I 2012 *Core principles & values of effective team-based health care. Discussion Paper* (Washington, DC: Institute of Medicine) pp 6–20
[32] NSW Health Department 2001 *A Framework for Building Capacity to Improve Health* NSW Health Department p 3-7
[33] Shortell S M, Zukoski A P, Alexander J, Bazzoli G J, Conrad D, Hasnain-Wynia R, Sofaer S, Chan B Y, Casey E and Marcolin F S 2002 *J. Heal. Polit. Policy Law.* **27** pp 49–91
[34] Mahama V A and Badu-Nyarko S K 2014 *African Educational Research Journal* **2** (2) 54–60 ISSN 2354-2160
[35] Chukwuma O M 2016 *American Journal of Water Resources* **4** (3) pp 58–67 doi:10.12691/ajwr-4-3-2
[36] World Health Organization 2012 *Participant workbook: Water Safety Plans – Training package* (Geneva: IWA Publishing) pp 8–26
[37] Steinert Y 1996 *Med. Teach.***18** pp 203–7
[38] Lateef F 2010 *J. Emerg. Trauma Shock.* **3** (4) pp 348–352 doi:10.4103/0974-2700.70743
[39] Frederick P. Morgeson F P, DeRue D S and Karam E P 2010 *Journal of Management* **36** (1) pp 5–39
[40] Algahtani A 2014 *Journal of Management Policies and Practices* **2** (3) pp 71–82 DOI: 10.15640/jmpp.v2n3a4
[41] Department of Housing and Infrastructure 2004 *Guidelines of construction and building: Plan for installation of water treatment system with mobile method* (in Bahasa), Jakarta
[42] World Health Organization 2006 *Guidelines for drinking-water quality, a first addendum to the third edition, volume 1. Third* (Geneva: WHO Press) pp 1–7
[43] Cho K H, Pachepsky Y A, Kim J H, Guber A K, Shelton D R and Rowland R 2010 *J. Hydrol.* **391** pp 322–32
[44] United State Environmental Protection Agency 2001 *Potential contamination due to cross-connections and backflow and the associated health risks* (Washington DC: EPA)
[45] Rubulis J, Verbeek J, Vreeburg J, Gruškevica K and Juhna T May 22–23, 2008 *The 7th International Conference, Faculty of Environmental Engineering Vilnius Gediminas Technical University Saulėtekio* ave 11, LT-10223 Vilnius, Lithuania pp 695–702
[46] Vitanage D, Pamminger F and Ourtsanis T 2004 *Maintenance and survey of distribution systems. In: Ainsworth R, editor. Safe piped water: Management Microbial Water Quality in Piped Distribution System* (London: WHO and IWA Publishing) pp 69–86
[47] Douterelo I, Husband S, Loza V and Boxall J 2016 *Appl. Environ. Microbiol.* **82** (14) pp 4155–4168 doi:10.1128/AEM.00109-16.
[48] City West Water, South East Water, Varra Valley Water 2013 *Water metering and servicing guideline. vol. 3. 3rd ed.* (Victoria: City Water) pp 6–11
[49] City West Water, South East Water, Yarra Valley Water 2015 *Water metering and servicing guideline Version Six* (Victoria: City Water) pp 7−10

[50] Rizak S and Hrudey S 2007 *Strategic water quality monitoring for drinking water safety*. (Salisbury SA: CRC for Water Quality and Treatment) pp 27−44

Effect of Different Coating Materials on The Characteristics Of Chlorophyll Microcapsules from *Caulerpa racemosa*

R A Kurniasih[1], E N Dewi[1], and L Purnamayati[1]

[1]Department of Fish Product Technology, Faculty of Fisheries and Marine Science, Diponegoro University, Jl. Prof. Soedarto, SH, Tembalang, Semarang 50275, Indonesia

Email: retno_ayuspi@yahoo.com

Abstract: The sea grape (*Caulerpa racemosa*) has a chlorophyll pigment that can be extracted using a non-polar solvent. Chlorophyll as a natural dye has unstable characteristics of temperature, pH, and light. Microencapsulation by the freeze-drying method can be used to protect chlorophyll from degradation caused by external influences where the type of coating material can affect the characteristics of the chlorophyll microcapsules. The objective of this study was to determine the characteristics of chlorophyll microcapsules with various types of coating material. Chlorophyll was microencapsulated using maltodextrin (CM), maltodextrin-alginate (CMA), and maltodextrin-fish gelatin (CMG). Chlorophyll encapsulated with maltodextrin-alginate resulting in the highest yield. The results of FTIR analysis indicated the presence of following functional groups in chlorophyll microcapsules viz., inter- and intra-molecular bonded alcohol OH, C = N stretching imine/oxime or C = O stretching conjugated ketone or alkenes, OH phenol, and CN stretching amine. CM had a particle size between 9,061 – 469.9 nm, CMA between 9,707 – 363.5 nm, and CMG between 11.49 – 433.2 nm. Based on the observation of morphology by using SEM, it showed that the all of the chlorophyll microcapsules were in the form of flake shape and porous. CM and CMA looked more fragile than CMG it can be seen from the cracks in some parts of CM and CMA. Therefore, CMG release time was longer than CM and CMA.

Keywords: Characteristics, Microcapsules Chlorophyll, Maltodextrin, Alginate, Gelatin

1. Introduction

Indonesia is one of *Caulerpa rasemorsa* (sea grape) producer, especially in Jepara District, Central Java Province. By the locals, this kind of seaweed is consumed as a vegetable. *C. racemosa* categorized as Chlorophyta, a green seaweed that has a natural pigment called chlorophyll. According to Kokilam and Vasuki [1], *Caulerpa taxifolia* also categorized as *Chlorophyta* which taken from Pondicherry waters and Mandapam waters, India had Chlorophyll content around 0,076 ± 0,002 mg/g dry weight.

On the food industry, the request of green color is quite high. Chlorophyll is a green pigment which can be used as a colorant and natural antioxidant. However, the natural chlorophyll is not suitable as the food colorant as its easily degraded by enzymatic reaction or other factors such as acid, oxygen, light, and heat. Chlorophyll can be degraded and produced chlorophyll derivative such as pheophytin, pheophorbide, pyropheophytin, and pyropheophorbide [2]. Therefore, the chlorophyll microencapsulation is conducted to facilitate its handling and improve stability against oxidation and solubility.

Microencapsulation is one of the methods when active compound as core being coated in tiny particles. Microencapsulation able to create a physical barrier between core and coating materials, so able to protect the sensitive compound and control the time release of that compound. This method also aimed to repair the pigment dispersibility in water and improve the shelf-life of active compound [3, 4, 5].

Pigment microencapsulation can be conducted using the freeze-drying method, which is also known as lyophilization. Freeze-drying is one of drying method which freezes the material first followed by sublimation. Although need long period of time and expensive compared to the spray drying method, the freeze-drying method is often used to dry the materials which sensitive with high temperature [6].

Besides of the drying process on microencapsulation technology, the selection of coating material may change the physical, chemical, and biological properties of the microcapsule. Bio-polymer which often used as coating materials includes maltodextrin, alginate, and gelatin. The selection of maltodextrin as the coating materials due to its low viscosity which can form the film layer [7]. The alginate encapsulant is used to decrease the capsule permeability by improving the immunoprotective properties of microcapsule [8]. Fish skin gelatin has high solubility, white, no smell and taste, can form bio-film, so can be used as encapsulant [9]. According to the description above, the further study about the using of the various encapsulant to encapsulate the chlorophyll *C. racemosa* should be conducted. The aim of this study was to discover the effect of maltodextrin, maltodextrin-alginate, and maltodextrin-fish as the coating material against the characteristic of *C. racemosa* chlorophyll which encapsulated

2. Methods

2.1. Chlorophyll Extraction

The chlorophyll was obtained from the maceration of dried *Caulerpa racemosa* by using acetone 90% with ratio 1:4 (w/v) for 2 days at room temperature. Next, the filtrate was evaporated using rotary evaporator at 20°C.

2.2. Microencapsulation of Chlorophyll

The chlorophyll extract plus coating material solution 1:4 (v/v) and Tween 80 were homogenized using homogenizer (WiseTis HG-15D, Germany) for 2 minutes at 10.000 rpm. The encapsulant solution was made by dissolving the coating materials into the distilled water until its concentration becomes 10% (w/v). The coating materials used consist of maltodextrin, maltodextrin-alginate 9:1 (w/w), and maltodextrin-fish gelatin 9:1 (w/w). Then freeze-drying was performed (Heto Powerdry LL 1500, Japan) at temperature -24°C. The yield assay, Fourier Transform Infrared Spectroscopy (FTIR), morphology observation, particle size distribution, and chlorophyll time release.

2.3. Yield

The yield was determined according to the percentage of total mass of chlorophyll microcapsule produced and the total mass of solid before drying [10].

2.4. FTIR

The microcapsule of chlorophyll was determined by FTIR spectrometer (Shimadzu FTIR 8400, Japan) using the spectral resolution of 1 cm^{-1} and the spectral profile was collected in the range between 4 000 cm^{-1} and 400 cm^{-1} [11].

2.5. Morphology Observation

The morphology of chlorophyll microcapsule was determined using Dewi et al. method with modification [12], chlorophyll microcapsule was coated by gold then observed using (SEM) (Jeol JSM 6510LA, Japan) at 15 Kv and 5000x.

2.6. Particle Size Distribution

The particle size distribution of chlorophyll microcapsule was determined by dispersion using a laser scattering particle size distribution analyzer (Malven Zetasizer Nano ZS ver. 6.20, UK) [13].

2.7. Time Release

Time release of chlorophyll was determined by modifying Belscak-Cvitanovic et al. method [14], chlorophyll microcapsule 1% (w/v) on distilled water was stirred continuously at speed 400 rpm. Next, 1 ml of solution was taken every 1 minute to analyze its chlorophyll content, the sample then spectrophotometrically measured at 646 and 663 nm by spectrophotometer UV-Vis (Shimadzu, Japan) [15]. The chlorophyll content was measured according to the equations (1), (2), and (3).

chlorophyll a (mg/L): 12,25 A663 nm – 2,79 A646 nm (1)

chlorophyll b (mg/L): 21,5 A646 nm – 5,1 A663 nm (2)

Total chlorophyll (mg/L): 20,2 A646 nm + 8,02 A663 nm (3)

2.8. Statistical Analysis

All experiments were conducted in triplicate and statistical analysis was carried out by using SPSS version 17.0 (International Business Machines Corporation, USA). The data were analyzed with one-way analysis of variance (ANOVA) and for mean separation using Duncan range test.

3. Results and Discussion

3.1. Yield

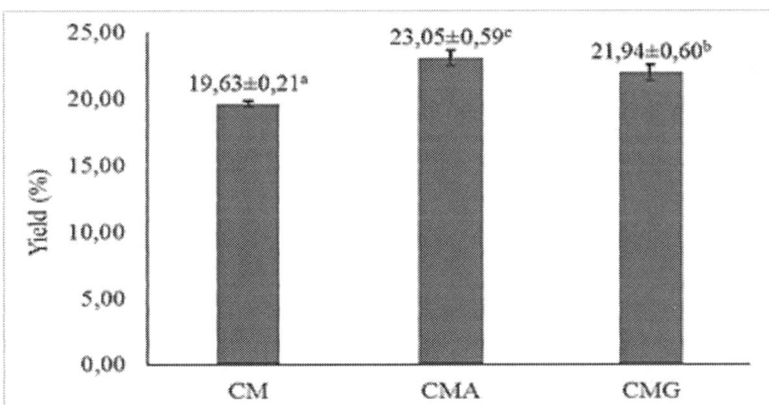

Figure 1. Yield of microcapsule chlorophyll with maltodextrin (CM), maltodextrin-alginate (CMA) and maltodextrin-fish gelatin (CMG) as coating materials

Note: The data was the average of triplication ± standard deviation.

Different superscript on the same column indicates significantly different ($p<0,05$)

The using of different coating materials with the same concentration able to affect the yield of chlorophyll which encapsulated by freeze-drying method (Figure 1). The chlorophyll which encapsulated using maltodextrin-alginate produced the highest rendement, while the lowest yield was produced from chlorophyll which encapsulated using maltodextrin. The different results showed from the previous study, the phycocyanin which encapsulated using different coating materials, such as maltodextrin, maltodextrin-alginate, and maltodextrin-carrageenan did not affect the yield of the microcapsule. Freeze-drying was used as the microencapsulation method [16]. The result of yield was affected by the ratio of encapsulated materials and coating materials used. The yield of microcapsule also causing the thickness of the layer which wrapping the core [17].

3.2. FTIR

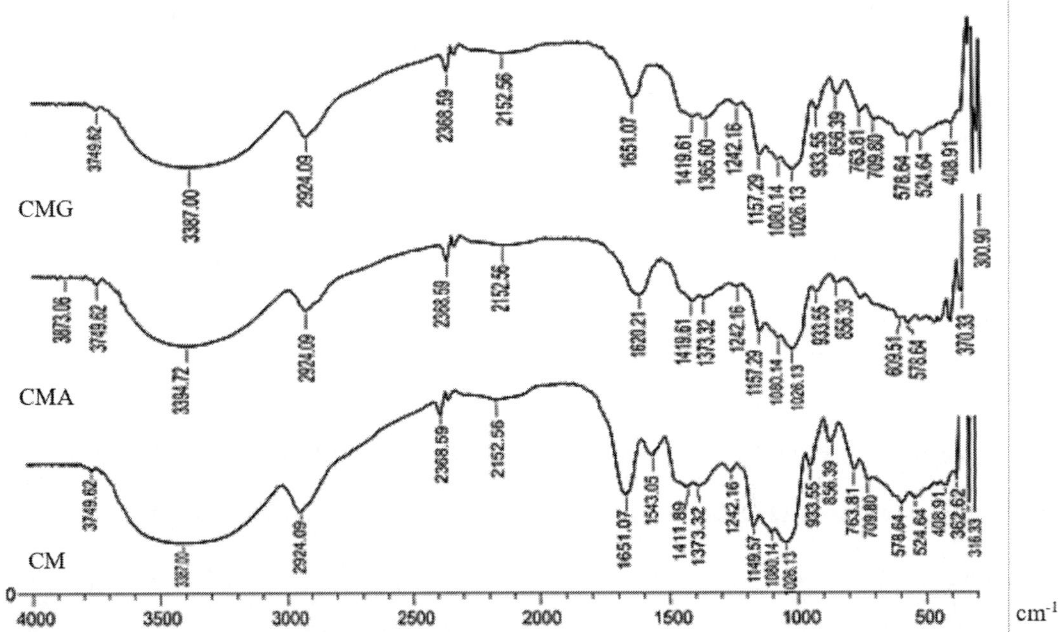

Figure 2. The yield of microcapsule chlorophyll with maltodextrin (CM), maltodextrin-alginate (CMA) and maltodextrin-fish gelatin (CMG) as coating materials.

Table 1. The functional group on chlorophyll microcapsule.

Functional Groups	CM	CMA	CMG
Unknown	3749.62	3749.62	3749.62
Intermolecular bonded alcohol O-H stretching	3387	3394.72	3387
Intramolecular bonded alcohol O-H Stretching	2924.09	2924.09	2924.09
C=N stretching imine/oxime or C=O stretching conjugated ketone or alkenes	1651.07	1620.21	1651.07
O-H banding phenol	1365.6	1373.32	1373.32
C-N stretching amine	1026.13	1026.13	1026.13
	1080.14	1080.14	1080.14
	1157.29	1157.29	1149.57
	1242.16	1242.16	1242.16

FTIR spectrophotometer used to identify the presence of chlorophyll within the microcapsule. Figure 2 indicated the same wave number showed by the third sample, either the chlorophyll microcapsule with maltodextrin, maltodextrin-alginate, or maltodextrin-gelatin as coating materials. The same wave number indicated the same functional group on maltodextrin and chlorophyll, the same wave number showed on Table 1. The results of FTIR analysis showed the presence of following functional groups in

IOP Conf. Series: Earth and Environmental Science **116** (2018) 012030 doi:10.1088/1755-1315/116/1/012030

chlorophyll microcapsule viz., inter- and intra-molecular bonded alcohol OH, C = N stretching imine/oxime or C = O stretching conjugated ketone or alkenes, OH phenol, and CN stretching amine. The same result also showed by Kumar et al. [18], which extracted the chlorophyll from the various kind of green vegetables (*Hibiscus cannabinus* defatted, *H. sabdariffa* defatted, *Basella alba* defatted, *B. rubra* defatted, *Rumex vesicarius* defatted, *H. cannabinus* native, *H. sabdariffa* native, *B. alba* native, *B. rubra* native, and *R. vesicarius* native).

3.3. SEM

According to the morphology observation using SEM, the chlorophyll microcapsule produced was a flake and porous, either the chlorophyll encapsulated by maltodextrin, maltodextrin-alginate, or maltodextrin-fish gelatin (Figure 3). On the freeze-drying method, the porous form was produced as the result of ice crystal formation during freezing and the sublimation process during drying. After drying, the formed brittle matrix can be ground into smaller pieces [19-20]. The same study also showed by Karthik and Anandharamakrishnan [21], a docosahexaenoic acid which encapsulated by freeze-drying was in the form of flake, irregular, and porous. Fish oil which encapsulated by freeze-drying also had the similar shape with our studies [22].

Figure 3. SEM photograph of microcapsules chlorophyll with a.) maltodextrin, b.) maltodextrin-alginate and c.) maltodextrin-fish gelatin as coating materials

3.4. Particle Size Distribution

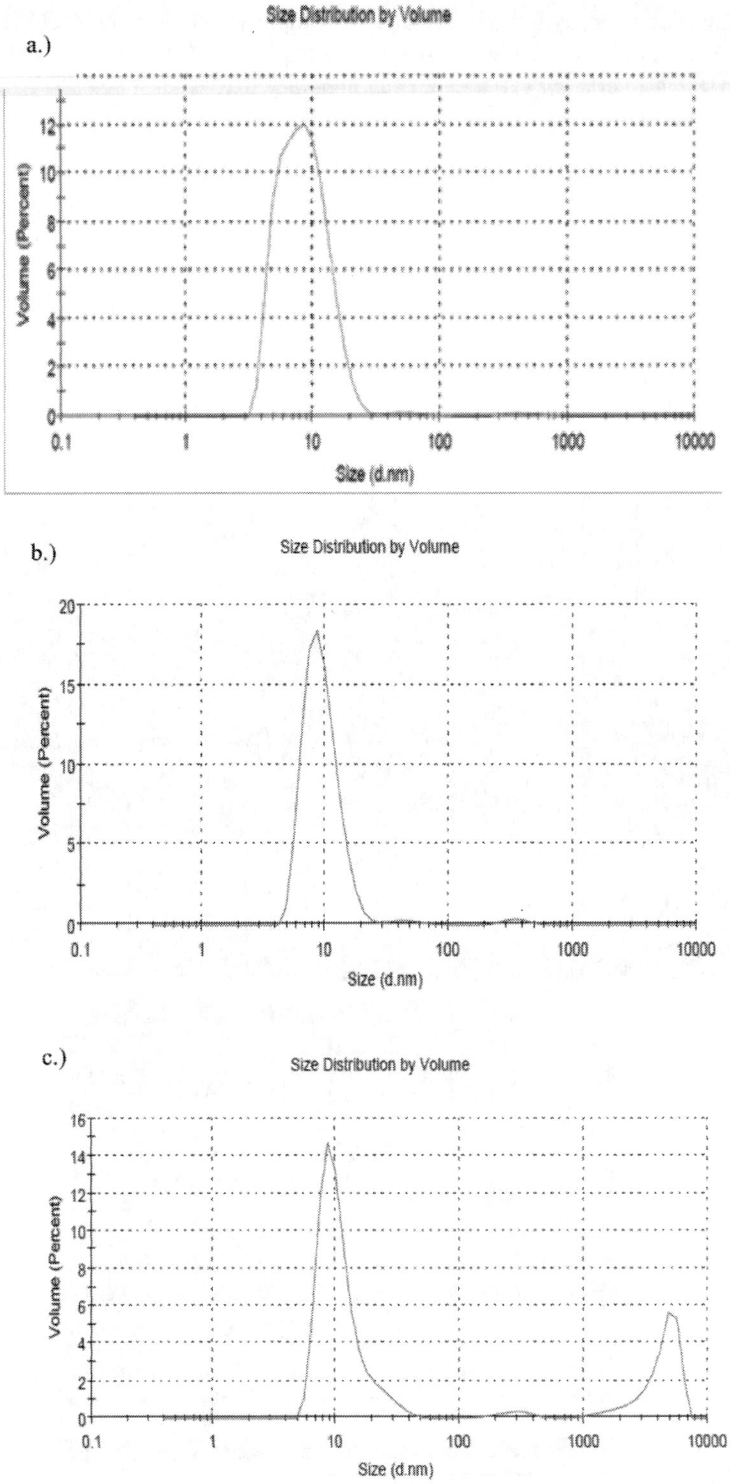

Figure 4. Particle size distribution of microcapsules chlorophyll with a.) maltodextrin, b.) maltodextrin-alginate and c.) maltodextrin-fish gelatin as coating materials

The chlorophyll microcapsule either encapsulated using maltodextrin, maltodextrin-alginate, or maltodextrin-fish gelatin had various size (Figure 4). According to the result, the particle size of chlorophyll which encapsulated using maltodextrin had the particle size distribution as follows: 98,3% sized 9,061 nm, 1,0% sized 58,30 nm, and 0,8% sized 469,9 nm. According to the result, the particle size of chlorophyll which encapsulated using maltodextrin-alginate had the particle size distribution as follows: 96,9% sized 9,707 nm, 1,6 % sized 47,65 nm, and 1,5% sized 363,5 nm. The chlorophyll microcapsule with maltodextrin-fish gelatin as coating material had size 11,49 nm as of 73,5%, 282,7 nm as of 2,0%, and 4332 nm as of 24,5%. The variety of particle size caused by product from freeze-drying process which in the form of aggregations then fined using mortar [23].

The using of maltodextrin-alginate and maltodextrin-fish gelatin as the coating material able to produce chlorophyll microcapsule in the bigger particle size compared to chlorophyll microcapsule with maltodextrin as the coating material. The size of microcapsule particle caused by the using of more than one coating material, it can form a tissue between the coating material and led the enhancement of layer thickness which wrapped the core [24-25]. The size of the particle also often attributed to the number of active compounds which able to be encapsulated. The bigger the size, the more of active compound which can be trapped [26].

3.5. Time Release

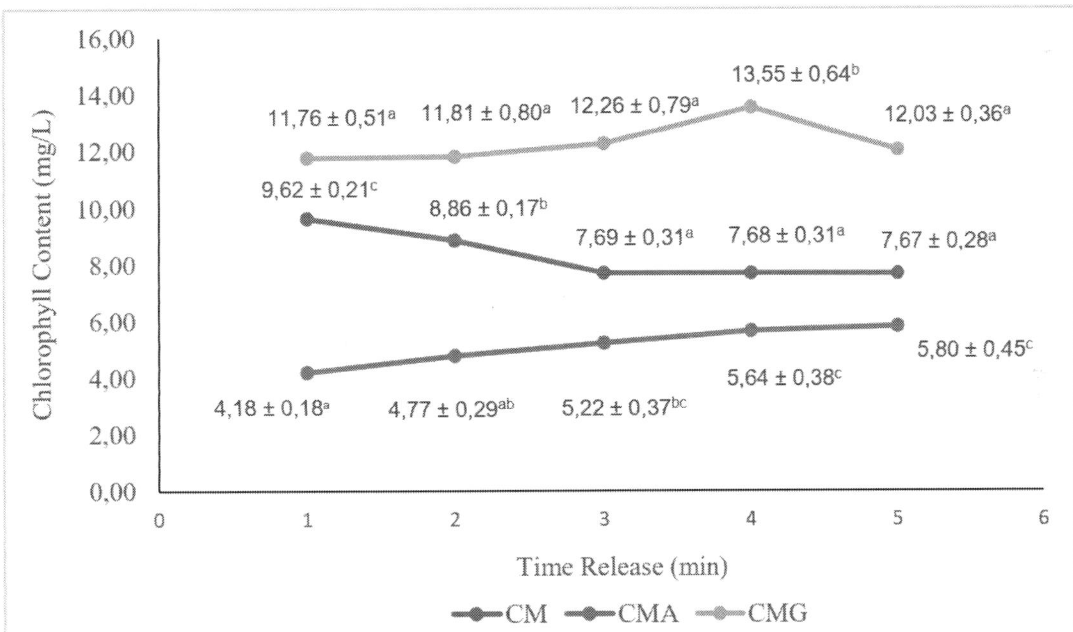

Figure 5. Time release of microcapsule chlorophyll with maltodextrin (CM), maltodextrin-alginate (CMA) and maltodextrin-fish gelatin (CMG) as coating materials

The result showed that chlorophyll which encapsulated with maltodextrin had been released within 1 minute (Figure 5). This due to maltodextrin was water soluble. maltodextrin had high ability to absorb water. This allegedly due to maltodextrin had a high swelling capacity. The ability to absorb water was affected by the involvement of hydroxyl in order to form hydrogen chain and covalent between starch chain [27]. Therefore, chlorophyll was easily released from microcapsule.

The using of maltodextrin-alginate and maltodextrin-fish gelatin can extend the chlorophyll time release. The active compound can release from microcapsule in the controlled way and certain condition.

The release of active compound can be affected by the type of coating material used [28]. The using of two kinds of coating materials can protect the active compound, so it cannot easily release. This due to the presence of the cross-linking between two coating materials used. The result of Belscak-Cvitanovic et al. [14] research showed that the release of raspberry extract polyphenol from microcapsule using alginate and chitosan as the coating material was slower than using alginate as the coating material. This due to the cross-linking between alginate and chitosan.

The maltodextrin contains polyanion while gelatin contains polycation. Polyanion only able to protect bio-active compound on the first layer and polycation able to protect the second layer after maltodextrin, it can protect the active compound well. The two kinds of coating materials, polyanion and polycation will form spiral tissue to protect Bio-active content [13]. Moreover, chlorophyll which encapsulated using maltodextrin-fish gelatin had the longest time release.

4. Conclusion

The using of maltodextrin-fish gelatin as the coating material can produce chlorophyll microcapsule with bigger yield compared to chlorophyll microcapsule with maltodextrin, but still lower compared to chlorophyll microcapsule with maltodextrin-fish gelatin as the coating material. The chlorophyll which encapsulated using maltodextrin-fish gelatin had the bigger particle size and longest chlorophyll time release compared to chlorophyll microcapsule with maltodextrin and maltodextrin-alginate as the coating material. According to FTIR analysis, indicated the presence of chlorophyll in each sample. Each chlorophyll microcapsule was in the form of flake and porous.

5. Acknowledgement

This research was funded by grants of Fisheries and Marine Science Faculty, Diponegoro University, 2017.

References

[1] Kokilam G and Vasuki S 2014 Biochemical and phytochemical analysis on *Ulva fasciata* and *Caulerpa taxifolia*. *International Journal of Pharmacy and Pharmaceutical Science Research*. **4**(1) 7–11.

[2] Ozkan G and Bilek S E 2014 Microencapsulation of natural food colorants. *International Journal of Nutrition and Food Sciences*. **3**(3) 145–156.

[3] Dubey R, Shami T, Bhasker Rao K 2009 Microencapsulation technology and application. *Defence Science Journal*. **59** 82–95.

[4] Dolly P, Anishaparvin A, Joseph G S, and Anandharamakrishnan C 2011 Microencapsulation of *Lactobacillus plantarum* (mtcc 5422) by spray-freeze-drying method and evaluation of survival in simulated gastrointestinal conditions. *Journal of Microencapsulation*. **28**(6) 568–574.

[5] Pillai D S, Prabhasankar P, Jena B S, and Anandharamakrishnan C 2012 Microencapsulation of *Garcinia cowa* fruit extract and effect of its use on pasta process and quality. *International Journal of Food Properties*. **15** 590–604.

[6] Rautela S, Badola A, Verma R, Baluni S, and Kothiyal P 2015 A review: the lyophilization process for the development of unstable drug. *European Journal of Biomedical and Pharmaceutical Sciences*. **2**(4) 1137–1151.

[7] Balasubramani P, Palaniswamy P T, Visvanathan R, Thirupati V, Subbayaran A, and Maran J P 2014 Microencapsulation of garlic oleoresin using maltodextrin as wall material by spray drying technology. *International Journal of Biological Microcapsule*. **72** 210–217.

[8] Al-Rammah T Y 2014 Alginate microencapsulation of stem cells as alternative source to the limited supply of donor tissue. *Biomedical Research*. **25**(2) 276–280.

[9] Wandrey C, Bartkowiak A, and Harding S E 2010 Materials for encapsulation *In*: Zuidam N J and Nedovic V A (Eds.) Encapsulation technologies for active food ingredients and food processing. Spinger, New York.

[10] Nunes I L and Mercadante A Z 2007 Encapsulation of lycopene using spray drying and molecular inclusion processes. *Brazilian Archives of Biology and Technology an International Journal*. **50**(5) 893–900.

[11] Krishnaiah D, Sarbatly R, and Nithayanandam R 2012 Microencapsulation of *Morinda citrifolia* L. extract by spray-drying. *Chemical Engineering Research and Design*. **90** 622–632.

[12] Dewi EN, Purnamayati L, Kurniasih RA. 2016. Antioxidant activities of phycocyanin microcapsules using maltodextrin and carrageenan as coating materials. Journal Technology 78(4–2): 45–50.

[13] Saloko S, Darmaji P, Setiaji B, and Pranoto Y 2014 Antioxidative and antimicrobial activities of liquid smoke Nano capsules using chitosan and maltodextrin and its application on tuna fish preservation. Journal of Food Bioscience. 7: 71–79.

[14] Belscak-Cvitanovic A, Stojanovic R, Manojlovic V, and Komes D 2011 Encapsulation of polyphenolic antioxidants from medical plant extracts in alginate-chitosan system enhanced with ascorbic acid by electrostatic extrusion. *Food Research International*. **44** 1094–1101.

[15] Sumanta N, Haque C I, Nishika J, and Suprakash R 2014 Spectrophotometric analysis of chlorophylls and carotenoids from commonly grown fern species by using various extracting solvents. *Research Journal of Chemical Sciences* **4**(9) 63–69.

[16] Dewi E N, Purnamayati L, and Kurniasih R A 2017 Physical characteristics of phycocyanin from *Spirulina* microcapsules using different coating materials with freeze-drying method. *IOP Conf. Series: Earth and Environmental Science* **55** 1–7.

[17] Quispe-Condori S, Saldana M D A, and Temelli F 2011 Microencapsulation of flax oil with zein using spray and freeze-drying. *Food Science and Technology*. **44** 1880–1887.

[18] Kumar S S, Manoj P, and Giridhar P 2015 Fourier transform infrared spectroscopy (FTIR) analysis, chlorophyll content and antioxidant properties of native and defatted foliage of green leafy vegetables. *J. Food. Sci. Technol*. **52**(12) 8131–8139.

[19] Anandharamakrishnan C, Rielly C D, and Stapley a G F 2010 Spray-freeze-drying of whey proteins at sub-atmospheric pressures. *Dairy Science and Technology*. **90**(2–3) 321–334.

[20] Samakradhamrongthai R, Thakeow P, Kopermsub P, and Utama-ang N 2015 Encapsulation of *Michelia alba* D.C. extract using spray drying and freeze-drying and application on thai dessert from rice flour. *International Journal of Food Engineering*. **1**(2) 77–85.

[21] Karthik P and Anandharamakrishnan C 2013 Microencapsulation of docosahexaenoic acid by spray-freeze-drying method and comparison of its stability with spray-drying and freeze-drying methods. *Food Bioprocess Technology*. **6**(10) 2780–2790.

[22] Chen Q, Zhong F, Wen J, McGillivray D, and Quek S Y 2013 Properties and stability of spray-dried and freeze-dried microcapsules co-encapsulated with fish oil, phytosterol esters, and limonene. *Drying Technology* **31** 707–716.

[23] Gurak P D, Cabral L M C, and Rocha-Leao M H 2013 Production of grape juice powder obtained by freeze-drying after concentration by reverse osmosis. *Brazilian Archives Biology and Technology* **56**(6) 1011–1017.

[24] Carvalho A G S, Silva V M, and Hubinger M D 2014 Microencapsulation by spray drying of emulsified green coffee oil with two-layered membranes. *Food Research International*. **61** 236–245.

[25] Najafi M N, Kadkhodaee R, and Mortazavi S A 2011 Effect of drying process and wall material on the properties of encapsulated cardamom oil. *Food Biophysics*. **6** 68–76.

[26] Briones A V and Sato T 2010 Encapsulation of glucose oxidase (GOD) in polyelectrolyte complexes of chitosan-carrageenan. *Reactive & Functional Polymers*. **70** 19–27.

IOP Conf. Series: Earth and Environmental Science **116** (2018) 012030 doi:10.1088/1755-1315/116/1/012030

[27] Ahmed M, Akter M S, Lee J C, and Eun J B 2010 Encapsulation by spray drying of bioactive components, physicochemical, and morphological properties from purple sweet potato. *International Journal of Food Science and Technology*. **43** 1307–1312.
[28] Shekhar K, Madhu M, Pradeep B, and Banji D 2010 A review on microencapsulation. *International Journal of Pharmaceutical Sciences and Research*. **5**(2) 5862.

Effect of Melanin Free Ink Extracted From Squid (*Loligo* sp.) on Proximate and Sensory Characteristics of Soft-Bone Milkfish (*Chanos chanos*) During Storage

Tri Winarni Agustini[1*], Hadiyanto[2],Ima Wijayanti[1], Ulfah Amalia[1], Soottawat Benjakul[3]

[1]Department of Fish Products Technology, Faculty of Fisheries and Marine Science, Diponegoro University, Jalan Prof. Soedarto, SH Tembalang Semarang, 50275, Indonesia
[2]Department of Chemical Engineering, Faculty of Engineering, Diponegoro University, Jalan Prof. Soedarto, SH Tembalang Semarang, 50275, Indonesia
[3]Department of Food Technology, Faculty of Agro-Industry, Prince of Songkla University, Hat Yai, Songkhla 90112 Thailand

*Corresponding author Email: tagustini@yahoo.com

Abstract. Antioxidant could be extracted and isolated from squid inks. Squid ink in the form of melanin free ink (MFI) could be act as an electron donor which can stabilize free radicals in lipid oxidation. This study was carried out to assess the antioxidant activity of squid inks converted into MFI in different dilution and to optimize the extraction conditions for the application of MFI as an antioxidative agent on fish product. Three different types of MFI extracts i.e : pure squid ink, squid ink with 5 times dilution and squid ink with 10 times dilutions by using cooled ionized water (4°C). The ink was then centrifuged at 18.000 x g for 30 minutes at cooled centrifuge (4°C) followed by DPPH analysis. The results showed that the IC50 of MFI extracts were 2.84 ppm; 1.11 ppm and 0.34 ppm, respectively ($p < 0.05$). The results indicated that squid ink with 10 times dilution in extraction of MFI had the highest value in free radical inhibitory. Although the IC50 of three different dilutions are equally low, and are considered as very strong antioxidative agent, however, it showed that the MFI extracted from squid ink had the ability to prevent free radical

Keywords : *antioxidant, squid ink,* melanin *free ink, extract, lipid oxidation*

1. Introduction

Squid (*Loligo* sp.) production in Indonesia increase 30.40% starting from 2000-2010, within year 2009-2010 significant increase of 80.8% was occurred. Squid production in 2010 reached 10,860 tons [1]. Processing of squid has results in some wastes including viscera and ink which is potential as raw material for bioactive compound resources which can be utilized as antioxidant, antibacterial and so on. Squid ink can be used as antioxidant in the form of squid ink polysaccharides [2] and melanin free ink [3]. Application of squid ink on some fisheries products in Indonesia as preserving agent is still lack in information. Therefore, it is necessary to conduct research on the use of squid ink and extract the melanin free ink to preserve the fish product.

Soft bone milkfish (*Chanos chanos*) is one of specific fish product produced and becoming typical food in Semarang city. Milkfish is one of fish species that is commonly cultured and developed in Central Java and its potential production reaches 6,975 Ha. Milkfish is one of prime commodity in Central Java and its production increase by year. In 2012 the production increase by 26% and this increase due to increasing demand of soft bone and boneless-milkfish [4]. This composition has brought about the fish for susceptible oxidation process. Therefore, it is necessary to involve antioxidant to preserve fish product based on milkfish during processing and storage. The use of synthetic antioxidant has been considered as non-safe because it is potential for cancer. Natural antioxidant has been developed and squid ink in the form of melanin free ink can be of potential antioxidant for milkfish and the like species.

Squid ink can be used as antioxidant agent as it contains L-dopa and dopamine which have hydroxyl group and capable to be oxygen donor. Squid ink in the form of melanin free ink could be electron donor which can stabilize free radicals. In addition this melanin-free ink capable to act as metal chelating agent so that lipid oxidation can be retarded especially in initiation phase [3].

Study on utilization of squid ink as antioxidant for fish product has been conducted by Vate and Benjakul [3]. Ink extract as melanin free ink has been applied into surimi of short-bodied mackerel. The results showed that concentration of melanin free ink of 100 and 200 mg/Kg (ppm) have significant in reducing lipid oxidation during 15 days of cold storage. Application of melanin free ink into different fish product has not been carried out. This study was aimed to observe the effect of melanin free ink in reducing lipid oxidation process of milkfish based product.

2. Materials and methods

2.1. *Materials*
Materials used in this study was squid ink extracted from squid (*Loligo* sp.). Squid was obtained from fish market in Semarang. Fish products intended to be observed include: soft-bone milkfish (year 1) and fish ball milkfish (year 2). The equipment used including for production of milkfish-based products (soft-bond milkfish and fish ball milkfish), extraction tools for MFI, cold centrifuge, apparatus for quality assessment of the products.

2.2. *Methods*
Study was started by extraction of squid ink into MFI (melanin free ink). Extract MFI is objected for antioxidant activity test. Year 1 focus on potential assessment of MFI as antioxidant agent and applying on soft-bond milkfish product. MFI was applied by spraying on the surface of product as edible coating. Concentration of MFI were 0.5%; 1% and 1.5% and then followed by storage at different temperatures of 0°C, 10°C and 30°C and left for several days with interval assessment of 0, 5, 10, and 15 days. Parameters of the product evaluated include fat content, protein content and moisture content, PV, and sensory.

2.2.1. Preparation of MFI
Preparation of MFI accorded to Vate and Benjakul [3]. Squid was separated from its ink by cutting ink truck from the body and collect them inside the bowl. This squid ink was diluted 10 times by using cooled ionized water (4°C). The ink was then centrifuged at 18.000 x g for 30 minutes at cooled centrifuge (4°C).

2.2.2. Analysis of DPPH
DPPH test was carried out by using Bloish method (1958) in Vate and Benjakul [3]. Sample of approximately 1.5 ml was added with 0.15 mM DPPH in 95% ethanol solution. Sample

and DPPH were mixed by using vortex and left at room temperature and dark condition for 30 minutes. Absorbance was detected by using spectrophotometer at 517 nm. Blank sample was also prepared using the same procedure

2.2.3. Proximate analysis
Proximate analysis was conducted for protein, lipid and moisture by using AOAC methods [5].

2.2.4. Analysis of Data
This study was designed for completely randomized design with Factorial of 4x4. First factor was MFI concentration (0%; 0.5%; 1% and 1.5%). Second factor was storage time (0, 5, 10 and 15 days). Parametric data include: lipid, protein, water content, PV) were analysed by ANOVA [6], DPPH data was analyzed descriptively and compare with the references.

3. Results and discussion

3.1. DPPH Analysis of MFI
The results for DPPH of the MFI was summarized in Table 1 and it was confirmed that MFI contain bioactive compounds such as flavonoids and phenol. As stated by Vate and Benjakul [3] and Hakim [8], MFI contain approximately 0.004% and 0.021% for flavonoid and phenol, respectively. Flavonoids can act as antioxidant which can prevent lipid oxidation on food matrix. The flavonoid as antioxidant can take up free radicals by release hydrogen atom from hydroxyl group because flavonoid has very reactive hydroxyl. According to Formagio et al [8], most of flavonoids has antioxidant activity because the present of hydroxyl phenolic compounds in their molecule structure.

Table 1. The antioxidant activity of squid inks

Antioxidant activity	A (pure squid ink)	B (squid ink with 5 x dilution)	C (squid ink with 10 x dilution)
IC_{50} (%)	31.07	15.39	9.46

Explanation for IC_{50} (the antioxidant activity):
< 50 : very strong
50-100 : strong
101-150 : medium
>150 : weak

Following DPPH analysis, the results showed that the IC_{50} of MFI extracts were 31.07%, 15.39% and 9.46%, respectively ($p < 0.05$). These results were comparable with Nicy [9] in which maximum DPPH radical scavenging activities of MFI of cuttlefish were 8.83% (*Sepia pharaonis*); 20.12% (*Sepia prabahari*) and 11.81% (*Sepia ramani*). The results indicated that squid ink with 10 times dilution in extraction of MFI had the highest value in free radical inhibitory. The IC_{50} of three different dilutions were low, meaning that all MFI have very strong antioxidant activity. It's showed that the MFI extracted from squid ink had the ability to prevent free radical. MFI of the splendid squid (*Loligo formosana*) showed antioxidative activities and was effective in scavenging DPPH. It was observed that DPPH scavenging activities of such squid was 179.6±2.1 protein [3]. Furthermore Vate and Benjakul [3] stated that melanin free ink of the splendid squids (*Loligo formosana*) was able to prevent lipid oxidation in surimi gels during refrigerated storage. Vate et al [10] informed that MFI could prevent lipid oxidation in surimi gels during refrigerated storage. Sudhakar and Nazeer [11] reported that cuttlefish peptide could be used as natural antioxidant in preventing oxidation reactions in food processing. Susilo *et al.*[12], state that inhibitor concentration (IC_{50}) was effective concentration in sample which could inhibit 50% of DPPH absorbance. The result shows that the dilution of squid inks increased the antioxidant activity of MFI to a significant extent. Luo and Liu [2] stated further that marine bioactive squids ink polysaccharides not only have strong scavenging activity on DPPH and hydroxyl radicals as well as total reducing power and are called antioxidant ability, but also protect DNA from oxidative damage induced by free radicals originated from combined action of UV and H_2O_2.

Derby [13] reported that cephalopod ink has anti-oxidant activity in both the melanin and melanin-free fraction of ink. Liu et al [14] showed squid ink has antioxidant function in growing broiler chickens; Fahmi and Soliman [15] informed that *Sephia officinalis* ink has invitro antioxidant activities; Zahara and Rabita [16] declared squid ink powder has antioxidant activities.

3.2. Proximate content

The different concentrations of MFI showed significantly different effect on proximate of water, protein, fat and ash content during storage at different temperatures (10°C, 20°C and 30°C.

3.2.1. Water content

The different concentrations of MFI and storage temperature showed significant different effect on water content of soft bone milkfish during storage. As shown at Table 2, there is no effect of MFI to water content of soft bone milkfish during storage at 10oC, whereas for temperature storage of 20oC and 30oC the effect were varied and for control showed that the water content decrease in the 9th days of storage at temperature of 30°C. The decrease in water content at temperature 30°C higher than 20°C, with water content was 44.53%. This is due to higher storage temperatures resulting in higher water evaporation during storage. Soft bone milkfish with the addition of 0.75% and 1.5% MFI concentration showed that the water content changes especially for storage temperature of 20oC and 30°C. Temperature storage of 10°C performed more stable water content on the samples, indicating that the suitable storage for soft bone milkfish is 10°C. Interestingly, during storage at 20°C, it was found that mold was growth on the samples causing some defect on the texture.

Table 2. Water Content of Soft bone Milkfish with Different Concentration of MFI during Storage in 10°C, 20°C, 30°C

Storage Temperature (°C)	Storage Time (day)	Concentration of MFI		
		0%	0.75%	1.50%
10	0	46.67±2.47^{b}	50.74±0.99^{b}	47.4±2.28^{b}
	3	51.83±0.78^{a}	49.52±1.49^{a}	51.54±1.85^{a}
	6	50.85±0.88^{a}	51.97±3.25^{a}	50.32±2.21^{a}
	9	51.21±0.92^{a}	50.75±0.77^{a}	51.69±0.37^{a}
20	0	54.82±0.77^{a}	50.97±0.81^{c}	52.89±1.39^{abc}
	3	53.38±0.74^{abc}	51.86±0.30^{abc}	54.80±1.72^{a}
	6	51.87±1.37^{bc}	52.70±0.59^{abc}	47.71±1.03^{d}
	9	51.50±0.02^{bc}	52.79±0.12^{abc}	54.05±0.63^{ab}
30	0	52.22±0.69^{b}	51.87±0.95^{b}	55.82±2.63^{ab}
	3	54.09±1.08^{ab}	55.27±0.85^{ab}	55.89±1.41^{ab}
	6	51.36±2.16^{bc}	46.81±2.21^{d}	46.96±1.86^{cd}
	9	44.53±0.02^{d}	56.73±0.57^{a}	51.95±1.61^{b}

At 20°C and 30°C showed the addition of 0.75% MFI concentration can increase the water content, it is due to the MFI that were added can form a layer of edible coating that can inhibit the release of water so that the evaporation process in soft bone milkfish did not occur. The increase levels of water content was due to the possibility of soft bone milkfish hydrolysis during the storage time and stuck in the flesh of fish because of the MFI coating

3.2.2. Protein content

Protein content of the samples were analyzed during storage as shown on Table 3. By applying MFI as edible coating on the soft bone milkfish, it was expected that product deterioration can be retarded to some extend due to the role of MFI as antioxidant. Based on the data resulted, samples added MFI at all storage temperatures (10°C, 20°C and 30°C) showed the same result in which protein content decreased at the end of the storage.

Table 3. Protein Content of Soft bone Milkfish with Different Concentration of MFI during Storage at 10°C, 20°C, 30°C

Storage Temperature (°C)	Storage Time (day)	Concentration of MFI		
		0%	0.75%	1.50%
10	0	29.27±0.27^{cde}	29.64±0.57^{cd}	32.85±0.73^{ab}
	3	29.48±0.258^{cde}	34.55±1.19^{a}	29.87±1.33^{cd}
	6	28.14±0.11^{def}	28.65±0.80^{cde}	27.16±0.64^{ef}
	9	30.78±0.02^{bc}	28.24±1.28^{def}	26.28±0.64^{f}
20	0	28.60±0.33^{ab}	29.64±0.79^{ab}	28.58±0.88^{abc}
	3	28.07±0.61^{bcd}	26.20±0.35^{cd}	23.21±1.5^{f}
	6	28.85±1.35^{ab}	25.03±1.08^{ef}	29.59±0.6^{ab}
	9	30.63±0.37^{a}	25.87±0.39^{de}	22.72±0.15^{f}
30	0	30.57±1.10^{ab}	28.95±0.70^{abc}	25.82±0.13^{ef}
	3	28.39±0.31^{bcd}	28.19±0.22^{bcde}	30.94±0.55^{a}
	6	21.06±0.21^{g}	25.55±1.25^{f}	26.12±1.36^{def}
	9	30.36±0.40^{ab}	25.27±0.59^{f}	27.28±1.35^{def}

At temperature storage of 10°C soft bone milkfish with the addition of 0%, 0.75% and 1.5% MFI concentration showed that the protein content tend to decrease compared to the initial protein content. The decrease of soft bone milkfish's protein content at 20°C and 30°C higher than 10°C, soft bone milkfish with the addition of 0.75% MFI concentration showed that the protein content decrease 12.7% compared to the initial content. At 20°C soft bone milkfish with the addition of 1.5% MFI concentration showed that the protein content decrease 20.5% and at 30°C soft bone milkfish with the addition of 1.5% MFI concentration showed that the protein content decrease 11.8%.

3.2.3. Fat content

The use of MFI on soft bone milkfish was intended for preserving the product during storage. MFI which can act as antioxidant was expected for retarding lipid oxidation on the product so that it can prolong the shelf life. In the control treatment at 10°C showed that the fat content decrease 32.8% compared to the initial content. At 20°C showed that the fat content decrease 14.5% at the end of the storage and at 30°C showed that the fat content decrease 17.5% at the end of the storage.

At the end of the storage, soft bone milkfish with the addition of 0.75% and 1.5% MFI concentration showed that the fat content higher than control and the fat content with the addition of 1.5% MFI higher than 0.75% MFI. This phenomenon was considered due to the possibility of edible coating of these MFI as antioxidants which can reduce the occurrence of fat degradation during storage. At 10 °C at the end of the storage, MFI 1.5% was the best treatment to maintain the fat content, it is shown that the highest fat content was at 10°C. It was due to the possibility of MFI acts as an antioxidant and supported by low storage temperature so that damage or fat oxidation can be inhibited to some extent. At 20°C and 30°C the fat content of soft bone milkfish control and with the addition of MFI was lower than 10°C, but the addition of 1.5% of MFI in each storage temperature showed at the end of storage the fat content is higher than the control treatment and MFI 0.75%.

Table 4. Fat Content of Soft bond Milkfish with Different Concentration of MFI during Storage in 10°C, 20°C, 30°C

Storage Temperature (°C)	Storage Time (day)	Concentration of MFI		
		0%	0.75%	1.50%
10	0	13.40±0.39[a]	6.93±0.57[de]	7.34±0.98[cde]
	3	8.06±0.21[bcd]	8.88±0.28[bc]	7.46±0.97[bcde]
	6	5.56±0.49[de]	8.60±0.42[bcd]	14.81±0.95[a]
	9	9.00±0.36[b]	9.37±0.01[b]	13.50±0.37[a]
20	0	8.73±0.29[e]	11.11±0.75[cd]	9.55±0.95[de]
	3	10.93±0.48[cd]	16.39±0.47[a]	10.83±0.69[cd]
	6	13.19±0.06[b]	12.25±0.92[bc]	15.48±0.41[a]
	9	9.55±0.05[de]	11.834±0.30[bc]	12.16±0.62[bc]
30	0	8.11±0.87[fg]	12.10±0.75[cd]	9.55±0.32[ef]
	3	11.07±0.47[de]	5.84±1.03[h]	7.53±0.32[gh]
	6	17.06±0.76[a]	14.22±0.30[b]	13.18±0.56[bc]
	9	9.15±0.06[efg]	9.12±0.32[efg]	10.68±0.26[de]

3.2.4. Ash content

The ash content of soft bone milkfish with the addition of MFI were higher than control treatment at all temperatures at the beginning of storage. The addition of MFI gave the additional of mineral so that the ash content increased. The ash content changed during storage. In all treatments showed a decrease in ash content during the storage time.

Table 5. Ash Content of Soft bond Milkfish with Different Concentration of MFI during Storage in 10°C, 20°C, 30°C

Storage Temperature (°C)	Storage Time (day)	Concentration of MFI		
		0%	0.75%	1.50%
10	0	4.45±0.16[ef]	9.15±0.68[a]	5.76±0.40[cde]
	3	1.96±0.31[g]	1.29±0.21[g]	3.88±0.74[f]
	6	6.43±0.59[bc]	3.57±0.38[f]	4.17±0.24[f]
	9	8.17±0.53[ab]	5.91±0.57[cd]	4.94±0.39[def]
20	0	3.95±0.03[bc]	5.331±0.58[ab]	7.02±0.76[a]
	3	3.60±0.15[c]	3.701±0.81[bc]	4.21±0.43[bc]
	6	4.48±0.11[bc]	4.589±0.52[bc]	3.66±0.50[bc]
	9	3.41±0.26[c]	6.713±1.21[a]	3.33±0.41[c]
30	0	3.23±0.37[cde]	3.59±0.70[bcd]	3.63±0.19[bc]
	3	2.59±0.62[de]	4.48±0.512[ab]	1.97±0.42[e]
	6	3.62±0.29[bcd]	5.58±0.672[a]	4.71±0.76[ab]
	9	2.86±0.68[de]	3.597±0.317[bcd]	4.75±0.17[ab]

3.3. Sensory analysis

At 10°C MFI concentration gave significant effect on sensory value. At the beginning of the storage, sensory value of control treatment were higher than that of MFI treatments, but during the storage, control treatment showed decreased on sensory value. At the end of the storage, MFI 1.5% has the highest sensory value compared to that of 0% and 0.75% MFI. At 10°C all treatments are acceptable. This occurrence profound that using MFI as edible coating on soft bone milkfish can significantly retard deterioration of the product.

Table 6. Organoleptic value Soft bone Milkfish with Different Concentration of MFI during Storage at 10°C, 20°C, 30°C

Storage Temperature (°C)	Storage Time (day)	Concentration of MFI		
		0%	0.75%	1.50%
10	0	8.24±0.544	7.594±0.928	7.587±0.651
	3	8.027±0.620	7.713±0.589	7.247±0.893
	6	7.827±0.486	7.573±0.598	7.46±0.538
	9	7.727±0.595	7.447±0.692	7.813±0.911
20	0	8.213±0.378	6.53±0.788	6.86±0.916
	3	6.3±0.787	5.28±0.823	6.047±0.664
	6	4.467±1.091	4.033±1.269	3.373±1.034
	9	4.707±0.486	4.407±0.801	4.327±0.653
30	0	7.92±0.511	8.06±0.479	7.753±0.532
	3	8.12±0.629	7.72±0.57	7.367±0.675
	6	7.447±0.545	7.347±0.55	7.153±0.500
	9	4.98±0.539	4.707±0.74	4.707±0.529

At 20°C organoleptic value on all treatments are not acceptable in the 3rd day. At 20°C the control and MFI treatment was overgrown with molds in the 3rd day. At 20°C MFI cannot able to inhibit the growth of mold, it was due to the possibility of 20°C is the optimum temperature for molds growth.

At 30°C MFI concentration did not give significantly different effect in organoleptic value but the storage time gave significantly different effect. At 30°C organoleptic value of soft bone milkfish was better than 20°C. Organoleptic value on all treatments decreased and still suitable for consumption in the 6th day, but not acceptable in the 9th day.

4. Conclusion

Result of this tentative study concluded that the squid ink potentially used as an antioxidant agent after processing into Melanin-Free Ink.

5. Acknowledgement

Further thanks to Diponegoro University, which has funded this research through the International Scientific Publication (RPI) 2016 budget.

References

[1] Ministry of Marine and Fisheries Affairs 2011 Marine and Fisheries in Figures (Ministry of Marine and Fisheries Affairs, Indonesia)
[2] Luo P and Liu H 2013 *African J. of Pharmacy and Pharmacology* **7**(21) : 1382-1388.
[3] Vate NK and Benjakul S 2013 *International Aquatic Research* **5**(9): 1-12
[4] Department of Marine and Fisheries of Central Java 2013 Profile of Aquaculture Field 2013. http://diskanlut-jateng.go.id.
[5] [AOAC] Association of Official Analytical Chemist 2005 *Official Methods of Analysis of The Association of Official Analytical Chemist* (Arlington, Virginia USA : AOAC Inc)

IOP Conf. Series: Earth and Environmental Science **116** (2018) 012031 doi:10.1088/1755-1315/116/1/012031

[6] Steel dan Torrie. 1991. *Principle and Procedure of Statistic* (PT Gramedia Pustaka Utama, Jakarta)
[7] Hakim 2017 *Effect of Melanin Free Ink as Antioxidant on Milkfish Fishbal during Refrigerated Storage* Research Report of Bachelor Degree Faculty of Fisheries and Marine Science, Diponegoro University Indonesia
[8] Formagio ASN, Carla Roberta Ferreira Volobuff, Matheus Santiago, Claudia Andrea Lima Cardoso, Maria do Carmo Vieira and Zefa Valdevina Pereira 2014 *Antioxidants* **3,** 745-757
[9] Nicy Brita 2016 *Antioxidant And Antibacterial Properties Of Cuttlefish Ink Collected From Selected Cuttlefish Landed At Thoothukudi Coast* Thesis of Master Degree Fisheries College And Research Institute Tamil Nadu Fisheries Unversity Thoothukudi
[10] Vate NK, Benjakul S, Agustini TW 2015 *J. Sci. Food Agric.* **95**(11): 2201-7
[11] Sudhakar and Nazeer 2015 *Lebensmittel-Wissenschaft und-Technologi*e **64**(2):593-601
[12] Susilo J, Istianatus S, Syamsul R 2013 Uji Aktivitas Antioksidan Ekstrak Etanol Daun Poslen (*Talinum triangulare* (*jacq.*) *Wild*) Dengan Metode DPPH (*2,2 – Diphenyl – 1-picrylhydrazyl*) Farmasi STIKES Ngudi Waluyo Jawa Tengah
[13] Derby CD 2014 *Marine Drugs* **12**: 2700-2730.
[14] Liu H, Ping Luo1, Shaohong Chen and Jianghua Shang 2011 *Asian-Australian Journal Animal Science* **24** (12) : 1752 – 1756
[15] Fahmy SR, Soliman AM 2013 *Afr. J. Pharm. Pharmacol. 7*, 1512–1522.
[16] Zaharah F MY and Rabeta 2018 *Food Research* 2-7

Nutrition Quality and Microbiology of Goat Milk Kefir Fortified with Vitamin B_{12} and Vitamin D_3 during Storage

EP Dianti [1], G Anjani[1,2], DNAfifah[1,2], N Rustanti[1,2], B Panunggal[1,2*]

[1]Department of Nutrition Science, Faculty of Medicine, Diponegoro University, Jl. Prof. H. Soedarto,SH, Tembalang, Semarang 50275, Indonesia
[2] Center of Nutrition Research (CENURE), Diponegoro University, Diponegoro University, Jl. Prof. H. Soedarto,SH, Tembalang, Semarang 50275, Indonesia

Corresponding author, +62 81285376785, panunggalbinar@gmail.com

Abstract Goat milk kefir fortified vitamin B12 and vitamin D3 can be an alternative to improve insulin resistance. Fortification of vitamin B12 and vitamin D3 could affect the balance of complex ecosystems of bacteria and yeasts in goat milk kefir. We analyzed the nutritional quality and microbiology of goat milk kefir during storage. This experiment was done with completely randomized design consisted of five treatments, ie storage day 0, 3rd, 6th, 9th, and 15th. Vitamin B12, vitamin D3, protein were analyzed using spectrophotometric, fat using babcock method, total lactic acid bacteria using total plate count, pH with pH meter, and viscosity using ostwald method. During 15 days in refrigerated storage, there was no significant difference ($p>0,05$) of vitamin B12 concentration, protein, fat, and viscosity of kefir. Storage time of the final product caused the decrease of vitamin D3 concentration between day 9 and 15 ($p=0,038$), pH ($p=0,023$), and total lactic acid of kefir ($p=0,0001$). Total lactic acid bacteria increased until 9 days of storage (6161 ± 1.296x107cfu / ml), but decreased to 24% on day 15 (150 ±7.78x107cfu /ml). pH kefir tends to fluctuate during storage with the highest pH value of 4,38 on day-3, and lowest 4.13 on day-9.

Keywords : goat milk, kefir, storage, fortification, vitamin B12 and vitamin D3

1. INTRODUCTION

Indonesia is a developing country that has a tropical climate, is now facing new health problem. The epidemiological transition has resulted in a double burden of infectious diseases and chronic non comunicable diseases [1]. Nowadays, diabetes has become a major public health problem in Indonesia and needs national strategies to screen, prevent and treat the disease [2].
Insulin resistance is the stage of diabetes mellitus. This is a condition of cells or tissues which could not use insulin effectively to regulate the transport of glucose into target cells or tissues resulting in impaired blood glucose absorption[1]. Riskesdas 2013 reported 36,6% of Indonesian population aged >15 years old experienced fasting blood glucose state disrupted, this condition lead to diabetes mellitus type 2.

Based on study, patient with insulin resistance tend to have low level of vitamin D3 and vitamin B12 in the plasma [3-5]. Vitamin D supports pancreatic beta cell activity. Beta cells possess vitamin D receptors. Animal studies show that missing vitamin D receptors result in poor insulin secretion and vitamin D supplementation improves insulin secretion[3]. Vitamin B12 deficiency is well documented in adults with inadequate intake, gut malabsorption or pernicious anaemia. Malabsorption of B12 is also associated with metformin therapy, an insulin sensitizer used for the treatment of type 2 diabetes and increasingly in obese, insulin resistant, adolescents.[4-5]

Food fortification is an effort to prevent vitamin B12 and D3 deficiency and improve insulin resistence through serving food which rich of vitamin B12 and vitamin D3. The success fortification depends on the choice of fortified media. Fortification vehicle should be easy to obtain, affordable prices, and contain high nutritional value[6].

Goat milk kefir is a good vehicle for fortifying vitamin B12 and vitamin D3 because it has high nutritional value [7]. It has been proven to have many health benefits, such as maintaining the balance of intestinal microflora, strengthening the immune system, lowering blood sugar and cholesterol[8]. Kefir is a type of dairy fermented from kefir grain containing probiotic bacteria complexes, such as lactobacilli, lactococi, leuconostoc, and acetobacteria and yeast[7]. The taste of the natural kefir beverage has been described as a yeasty taste, sparkling and prickling have been used to describe the mouth-feel of the kefir beverage caused by the liberation of CO2. The distinctive taste of kefir is a result of several flavour compounds which are produced during fermentation. In this study, kefir is made of goat milk. Goat milk has the advantage of rich minerals of calcium, phosphorus, magnesium, vitamin A, vitamin C and medium chain trygliserides (MCT) rather than cow's milk. However, goat milk contained only slight vitamin B12 and D3[9].

Fortification of vitamin B12 and D3 affect the quality of nutrition and microbiology because goat milk kefir contained complex of probiotic and yeast bacteria that are active and sensitive to environment changes. Previous study reported that there was an increase in CO2 and ethanol due to Saccaromyces cerevisiae activity during storage[10]. In addition, the activity of lactic acid bacteria and [11-12]. In this study, the nutritional quality of goat milk kefir vitamin B12 and vitamin D3, which includes vitamin B12, vitamin D3, carbohydrates, proteins, fats, and total bacteria during storage of day 0, 3, 6, 9, and 15 were analyzed.

2. MATERIALS AND METHODS

2.1 Preparation of Goat Milk Kefir Fortified with Vitamin B12 and Vitamin D3

The main ingredients of goat milk kefir were kefir grain, goat milk, vitamin D3 and vitamin B12. Goat's milk and kefir grain were obtained from Omah Kefir Ungaran. Vitamin B12 was obtained from the nutritional support product "Blackmores Vitamin B12", while vitamin D3 is obtained from the product of "Vitamin D3 1000 IU Healthy Care Australia".

The first step to make kefir kefir was pasteurized of goat milk up to 70^{0}C. Furthermore, goat milk cooled until 25^{0}C, then inoculated with kefir grain (5% of the total milk). Fermentation duration was 24 hours and it stirred every 6 hours once. Fortification of vitamin B12 20mcg / 100mL and vitamin D3 42 IU / 100mL at 12 hours fermentation[13-14]. After fermented 24 hours, then kefir filtered to separate kefir and kefir grain. Furthermore, kefir was divided into 5 groups of storage treatment ie kefir day 0, 3, 6, 9, and 15 for tested vitamin B12, vitamin D3, protein, fat, total lactic acid bacteria, pH, and viscosity.

2.2 Nutrition quality assay

Vitamin was analyzed by spectrophotometry. The wavelength of vitamin B12 was 530 nm and 264 nm for vitamin D3. Determination of protein concentration using Bradford method then analyzed by spectrophotometry at wavelength 595 nm. Preparation of bradford reagent was done by dissolving 10 mg of coomassie brilliant blue G-250 in 5 mL of 95 % ethanol. Then, 10 mL of 85% phosphoric acid was added. The solution was diluted with distilled water to a volume of 100 mL. The samples (each 0.1 g) were added bradford reagent, then it was incubated for an hour. The solution was homogenized using a vortex an it was incubated at room temperature for an hour. Fat was analyzed using the babcock method, total lactic acid bacteria with total plate count method using MRSA, viscosity, and pH using pH meter

2.3 *Analysis Data*
All data were processed using statistical software. The data were tested for normality using the Saphiro-Wilk test, then the data were tested with anova oneway test, Kruskall Wallis test, and Mann Whitney test based on normality test results.

3. RESULTS
The nutrition quality of goat milk kefir fortified with vitamin B12 and D3 during storage is presented in Table 1.

Table 1. Nutrition quality of goat milk kefir fortified with vitamin B12 and D3 during storage

Time storage	Mean Vitamin D3 (IU)** **p=0,038***	Median Vitamin B12** (µg) **p=0,261**	Mean Protein (%)* **p=0,190**	Mean Fat (%)* **p=0,259**
Day-0	57,63 ± 9,590 [a]	678,12	2,320 ± 0,268	7,0443 ± 0.43935
Day-3	44,83 ± 1,1946 [ab]	609,06	1,880 ± 0,1149	8,0028 ± 2.3191
Day-6	41,16 ± 4,3787 [b]	560,62	1,595 ± 0,3517	8.0151 ± 0.6965
Day-9	45,58 ± 6,0774[ab]	836,25	1,615 ± 0,5567	7.3100 ± 1.1399
Day-15	27,33 ± 6,3024 [c]	867,81	1,449 ± 0,3911	7,9505 ± 0.8298

*Tested with oneway anova
** Tested with Kruskall wallis

Data in table 1 showed that Vitamin B12, protein, and fat remained fairly stable over the period of the study. Storage time of the final product caused the decrease of vitamin D3 concentration. There was significant difference at day-0 to 3rd day (p=0,05) and 9th to 15th day (p=0,05).

Table 2. Viscosity, pH, and total lactic acid bacteria of goat milk kefir fortified with vitamin B12 and D3 during storage

Time storage	**Mean Viscosity (cm^2/s)* p=0,038**	**Mean pH* p=0,023**	**Total lactic acid bacteria *(cfu/mL) p=0,000**
Day-0	0,0429	4,28 ± 0,0461 [a]	5,41 ± 4,419[ad]
Day-3	0.0500	4,38 ± 0,05774 [b]	53,38 ± 6,301[ad]
Day-6	0.0535	4,30 ± 0,0288 [ac]	6982 ± 8,732[b]
Day-9	0.0470	4,13 ± 0,0346 [d]	6161 ± 1,296[bc]
Day-15	0.0446	4,26 ± 0,0416 [ac]	150 ± 7,78[d]

*Tested with oneway anova
** Tested with Kruskall wallis

Results on viscosity, pH, and total lactic acid bacteria of kefir during storage are presented in table 2. There was significant difference on pH and total lactic acid bacteria. The highest of total lactic acid bacteria was on 6th day, and the lowest on day-0. It had been reported that total lactic acid bacteria got reduced between 9th day and 15th day. pH kefir tend to fluctuate during storage, there was significant different on day-0 to 3rd, 3rd day to 6th day, and 9th day to 15th day. Meanwhile, viscosity remained stable until the end of storage time observation.

4. DISCUSSION
4.1 *Vitamin B12*
Kefir fortified vitamin B12 resulted in 40 times higher in vitamin B12. The high concentrate of Vitamin B12 produced by biosynthesis of bacterial fermentation [15-16]. Propionibacterium shermani B369 and Propionibacterium freudenreichi spp were bacteria that produce vitamin B12 during fermentation. The role of vitamin B12 is as an essential cofactor in fatty acids, amino acids, carbohydrates, and nucleic acids[17].

There was no difference of vitamin B12 in goat milk kefir fortified with vitamin B12 and vitamin D3 during storage. The stability of vitamin B12 was affected by enhancement *propionic acid bacteria* which produced vitamin B12. Increasement of *propionic acid bacteria* followed also by increasement of etanol, acetic acid, butiric acid, glucose, dan galactose[18].

4.2 Vitamin D3

The result of vitamin B12 was different from vitamin D3. There was difference concentration of vitamin D3 during storage. Vitamin D3 slightly decreased between day 9 to 15. The highest concentration vitamin D3 on day-0 (57,63 ± 9,590 IU) and the lowest on day-15 (27,33±6,3024 IU). Fortified goat milk kefir had higher concentration of vitamin D3 than non-fortified kefir. This results was different from the result of fortification vitamin D3 on yogurt, whereas vitamin D3 was stable until 4 weeks storage[19]. Vitamin D3 was stable in yogurt due to the bacterial in yogurt was not as complex as that in kefir consisted of bacteria and yeast that had abilty to proteolisis and lipolisis.

Degradation of protein might affected by the stability of concentration of vitamin D3 because vitamin D3 was binding to β -laktoglobulin A dan β –kasein (peptide). β-laktoglobulin A and β-kasein were rich in whey and casein. [20]. β-laktoglobulin A and β-kasein was binding vitamin D3 strongly. Interaction of vitamin D3 and protein was dependable each other. Vitamin D3 could be binding to β-lactoglobulin and β-kasein even in the gastric environment[21]. However, the proteolisis activity of complex bacterial and yeast cause degradation β-laktoglobulin A and β-kasein being arginin and glutamin. It affected on instability of vitamin D3[11][33].

In this study, total of lactic acid bacteria decreased during storage, indicated there was increasement of yeast oxidative activity which produced ethanol, Vitamin D3 was sensitive to oxidation. Other factors that affected vitamin D3 instability were light and lipolisis. Vitamin D3 was sensitive on light and vitamin D3 dissolve in lipid, whereas lipolisis might be impaired the stability of vitamin D3[23].

4.3 Protein

Dissolved protein or often called protein digestibility was the ability of a protein to be hydrolyzed into amino acids by digestive enzymes. Protein digestibility is one of the factors that determine protein quality because it shows the availability of amino acids biologically[24]. In this study, the dissolved protein concentration of goat milk kefir vitamin B12 and vitamin D3 fortified stable for 15 days storage. The results of this study were in accordance with studies on kefir without fortification which mentions no difference in protein concentration during storage[12].

During fermentation, protein hydrolysis had occurred by the activity of proteolytic enzymes that break down proteins into smaller protein fractions or peptides. Lactoglobulin and lactalbumin and other components contained in whey will be converted into several peptides. Lactoglobulins which are the main components of whey are converted into β-lactoglobulin-A, β-lactoglobulin-B and angiotensin converting enzyme (ACE). Breakdown of protein into smaller fractions during fermentation caused the value of protein kefir stable during storage because it is degraded during fermentation[24-25].

4.4 Fat

Fat of goat milk kefir did not change significantly during storage. The fat concentration remained stable, although there was an increase in lactic acid bacteria up to the 9th day. Previous study reported the growth of lactic acid bacteria will cause lipolysis of fat tobe free fatty acids[26]. The results of this study differ from other studies which suggest that long storage may lead to fat loss and viscosity[11]. Previous study reported fat concentration decreased by 7.9% on day 14, occurs due to the growth of bacteria and yeasts that are lipolisis[26,27]. The type and quality of milk were factors that affects the stability of fat so that the results obtained different from previous study[26].

4.5 Lactic acid bacteria

Kefir and kefir grain contain a wide variety of lactic acid bacteria that had been identified through physiological and chemical tests. There were *Lactobacillus acidophilus, Lb. Brevis, Lb. Paracasei subsp. Paracasei, Lb. Delbrueckii subsp, Lb. Helveticus, Lb. Kefiri, Lb. Kefiranofaciens, Lb. Plantarum, Leuconostoc mesenteroides subsp., Lactococcus lactis subsp, Streptococcus* thermophilus [28]. The highest total lactic acid bacteria was obtained on day 6 (6.96 x 1010cfu / mL) and the lowest on day 0 (5.31x107cfu / mL). Increased lactic acid bacteria made drastically affect the physical appearance of kefir, namely the separation of the mixture (clear yellow and milk white) which showed that there was degradation of macro nutrients by lactic acid bacteria[8]. At day 0, the amount of lactic acid bacteria just a little because lactic acid bacteria was still in the adaptation phase[11]. Lactic acid bacteria populations began to increase on day 3, lactic acid bacteria were able to adapt and utilized the energy source of lactose, yeast would also take the opportunity to hydrolyze lactose, resulting in CO2 and alcohol. The OH-compound of the alcohol reacted with the H^+ from lactic acid[8][12]. Increased population of lactic acid bacteria continued to day 6 and 9, but there was a decrease in lactic acid bacteria on day 15 due to the activity of yeast metabolism that degraded lactose[10]. This result is in line with other studies which suggest that lactic acid bacteria decreased on day 7 to day 14 because of the increased in yeast[27].

4.6 Acidity (pH)
pH of goat milk fortified with vitamin B12 and vitamin D3 differently during storage. However, the pH of goat's milk fortified with vitamin B12 and vitamin D3 was still considered good and met the pH standard of kefir which ranged from 4.2 to 4.6[16]. The stability of the pH value was influenced by lactic acid bacteria. The more population of lactic acid bacteria, the pH value would be decreased. [27] Differences in pH values occured from the first 3 days. There was an increased in pH of kefir from 4.28 on day 0 to 4.38 on day 3. This occured due to the activity of yeasts which changed lactose into alcohol. Based on the previous the study, yeast population in kefir increased in the first 7 days[21]. Furthermore, the pH of kefir was significantly different on days 6 and 9, whereas pH got decreased. The lactic acid was formed too much, causing the pH kefir getting down. On the 15th day, pH got increased again (4.13 to 4.26). The declined in lactic acid bacteria population and increased yeast activity that changed lactose to alcohol affected the stability of pH. This result was not different from other studies, which stated pH decreased during the first 7 days, then pH constant until day 14[27-28].

4.7 Viscosity
The viscosity of goat milk kefir fortified with vitamin B12 and vitamin D3 did not differ during storage ($p> 0.05$). The viscosity of kefir ranges from 0.0393-0.0544 cm^2 / s. The results of this study differ from previous studies, whereas there was a decreased in viscosity during 14 days storage[30]. Viscosity was affected by lactic acid bacteria, total solids, fats, and proteins which was contained in kefir[31]. Viscosity of kefir increased during a 24-hour fermentation process due to the exopolysaccharide or kefiran. Kefiran is the main polysaccharide produced by the microorganisms found in kefir and its biosynthesis is generally ascribed to Lactobaccilus kefir or Lactobacillus kefiranofaciens. Lactic acid bacteria are responsible for the production of extracellular polysaccharides which contribute to a significant degree to the texture of the fermented dairy products in which they are present. Variety of lactic acid bacteria namely *Lactobacilius, Streptococcus, Lactococcus, and Leuconostoc. S. tomophilus* which interacted with milk proteins which caused increasing viscosity. Furthermore, the reduction of the viscosity can be attributed to a degradation of the protein structure caused by the proteolytic action of microorganisms[17][34].

5. Conclusion
Vitamin B12, protein, fat, and viscosity of goat milk kefir vitamin B12 and D3 fortified stable for 15 days storage. The pH value of kefir followed the standard range of pH kefir, ie 4.2-4.6. This shows that kefir is still worth consuming. However, there was a decrease in vitamin D3 and total lactic acid bacteria on the 15th day. Although total lactic acid bacteria decreased, but still meet the standard of *Codex Alimentarius*, which is 10^7cfu / ml.

6. Acknowledgement
The author would like to thank Faculty of Medicine fiscal year 2016 for funding this research.

References

[1] Folli F, Corradi D, Fanti P, Davalli A, Paez A, Giaccari A, and et al 2011 Curr Diabetes Rev **7** 313–24
[2] Badan Penelitian dan Pengembangan Kementerian Kesehatan RI 2013 Riset Kesehatan Dasar. p xvi
[3] Gandhe MB, Jain K, Gandhe SM 2013 J Clin Diagn Res **7** 2438
[4] Ho M, Halim JH, Gow ML, El-Haddad N, Marzulli T, and Baur LA 2006 Arch Pediatr Adolesc Med **160** 933–6
[5] Allen L, Benoist B, Dary O, Hurrell R, editor 2006 WHO, Food and Agricultural Organization of the United Nation 126–30.
[6] Farnworth ER 2005 Food Science and Technology Bulletin: Functional Foods **2** 1–17
[7] Urdaneta E, Barrenetxe J, Aranguren P, Irigoyen A, Marzo F,and Ibáñez FC 2007 Nutr Res **27** 653–8
[8] Zenebe T, Ahmed N, Kabeta T, Kebede G, Medicine V, Box PO 2014 Nutr Res **3** 30–9
[9] Grønnevik H, Falstad M, and Narvhus JA 2011 Int Dairy J **21** 601–6
[10] Vieira CP, Álvares TS, Gomes LS, Torres AG, Paschoalin VMF, and Conte CA 2015 PLoS One 2015 **10** 1–18
[11] Mal R and Radiati LE 2013 Universitas Brawijaya
[12] Sadeghian S, Fallahi F, Salarifar M, Davoodi G, Mahmoodian M, Fallah N, and et al 2006 BMC Cardiovasc **6** 38
[13] Grossmann RE, Tangpricha V 2011 Mol Nutr Food Res 2011 **54** 1055–61
[14] Stanton C, Ross RP, Fitzgerald GF, dan Sinderen DV 2005 Curr Opin Biotechnol **16** 198–203
[15] Fang H, Kang J, Zhang D 2017 Microb Cell Fact **16** 15
[16] Bruinenberg PG, Smid EJ 2006 European Patent Office
[17] Leite AMO, Leite DCA, Aguila EMD, Alvares TS, Peixoto RS, Miguel MAL, et al 2011 Dairy Sci **96** 4149-59
[18] Lehmann U, Hirche F, Stangl GI, Hinz K, Westphal S, Dierkes J 2013 J Clin Endocrinol Metab **98** 4339-45
[19] Leskauskaite D, Jasutiene I, Malinauskyte E, Kersiene M, and Matusevicius P 2015 Int J Dairy Technol 177–83
[20] Forrest SA, Yada RY, Rousseau D 2005 J Agric Food Chem **53** 8003–9
[21] Bulgari O, Caroli AM, Chessa S, Rizzi R, and Gigliotti C 2013 Journal Molecules **18** 10122-31
[22] Kazmi SA, Vieth R, and Rousseau D 2007 International Dairy Journal 753–9
[23] Kitts DD and Weiler KA 2003 Current Pharmaceutical Design 1309-23
[24] Belkaaloul K, Chekroun A, Abdessalam A, Saidi AD, anf Growth KO 2010 Afr J Biotechnol **9** 1463-69
[25] Antonioua KD , Topalidoua S, Tsavaliaa G, and Dimitreli G Department of Food Technology ATEI of Thessaloniki
[26] Irigoyen A, Arana I, Castiella M, Torre P, and Ibanez FC 2005 J Food Chem 613-20
[27] Simova E, Beshkova D, Angelov A, Hristozova TS, Frengova G, and Spasov G 2002 J Ind Microbiol Biotechnol **28** 1–6
[28] Viljoen BC 2001 International J Food Microbiol **69** 37-44
[29] Magra TI, Antoniou KD, and Psomas EI 2012 J Texture Stud 1-9
[30] Yoo SH, Seong KS, and Yoon S 2013 Korean J Food Sci An **33** 744-51
[31] Codex Alimentarius Milk and Milk Products 2011 FAO dan WHO p 56-58
[32] Semo E, Kesselman E, Danino D, and Livney YD 2007 J Food hydrocoll **21** 936-42
[33] Beshkova DM and Simova ED 2002 J Food Microbiology **19** 537-44

Total Lactic Acid Bacteria (LAB), Antioxidant Activity, and Acceptance of Synbiotic Yoghurt with Binahong Leaf Extract (*Anredera cordifolia* (Ten.) Steenis)

R P Lestari[1,*], C Nissa[1,2], D N Afifah[1,2], G Anjani[1,2] and N Rustanti[1,2,*]

[1] Departement of Nutrition Science, Faculty of Medicine, Diponegoro University, Jl. Prof. H. Soedarto,SH, Tembalang, Semarang 50275, Indonesia
[2] Center of Nutrition Research (CENURE), Diponegoro University, Jl. Prof. H. Soedarto,SH, Tembalang, Semarang 50275, Indonesia

* Corresponding author, E-mail: ninik.rustanti@gmail.com

Abstract Alternative treatment for metabolic syndrome can be done by providing a diet consist of functional foods or beverages. Synbiotic yoghurt containing binahong leaf extract which high in antioxidant, total LAB and fiber can be selected to reduce the risk of metabolic syndrome. The effect of binahong leaf extract in synbiotic yoghurt against total LAB, antioxidant activity, and acceptance were analyzed. The experiment was done with complete randomized design with addition of binahong leaf extract 0% (control); 0.12%; 0.25%; 0.5% in synbiotic yoghurt. Analysis of total LAB using Total Plate Count test, antioxidant activity using DPPH, and acceptance were analyzed by hedonic test. The addition of binahong leaf extract in various doses in synbiotic yoghurt decreased total LAB without significant effect (p=0,145). There was no effect of addition binahong leaf extract on antioxidant activity (p=0,297). The addition of binahong leaf extract had an effect on color, but not on aroma, texture and taste. The best result was yoghurt synbiotic with addition of 0,12% binahong leaf extract. Conclusion of the research was the addition of binahong leaf extract to synbiotic yogurt did not significantly affect total LAB, antioxidant activity, aroma, texture and taste; but had a significant effect on color.

Keywords: total lactic acid bacteria, antioxidant activity, binahong leaf, yoghurt, synbiotic.

1. Introduction

The prevalence of metabolic syndrome in some developed and developing countries has increased, by 7.9 - 43.6% in men and 7 - 56.7% in women [1]. National Health and Nutrition Examination Survey (NHANES) 2003-2012 The prevalence of metabolic syndrome in the United States averaged 33%, 35.6% for women, men 30.3% [2]. The prevalence of metabolic syndrome in Jakarta was 24.8%, respectively in women by 30.4%, male by 25.4% [3]. The metabolic syndrome defined by the National Cholesterol Education Program Adult Treatment Panel III (NCEP / ATP III), if a person has 3 or more of the following data; Waist circumference greater than 102 cm in men or greater than 88 cm in women, serum triglyceride levels of 150 mg / dL or greater, high-density lipoprotein (HDL) cholesterol levels less than 40 mg / dL in men or less than 50 mg / dL in women, systolic / diastolic blood pressure 130 / 85mmHg, or fasting plasma glucose levels of 100 mg / dL or more [4].

Risk factors which play a role in the development of metabolic syndrome included overweight, lack of physical activity and diet. Diet contributed to the reduction of some metabolic syndrome markers, one of the alternatives of dietary feeding or functional beverage [5,6]. Research development of food for people with metabolic syndrome is very necessary, especially in terms of the use of foodstuffs one of which can lose weight, anti hypertension, anti hyperlipidemia, and anti-hyperglycemia.

Yoghurt is a functional drink made from cow's milk fermented using probiotic bacteria *Lactobacillus bulgaricus* and *Streptococcus thermophilus*[7,8]. Probiotics produce organic acids such as glucoronic acid, propionic acid, folic acid, and lactic acid that potentially lower non-HDL cholesterol levels. Cholesterol reduction by *Lactobacillus* strain anaerobically can reach 27-38% [9]. Probiotics also have a hypoglycemic effect that can decrease the inflammatory status in patients with DM type 2 [10,11]. Probiotics provide health effects on the human body that is at a minimum of 10^7 CFU / ml before consumed [12]. The number and activity of BAL can be increased by the addition of prebiotics. The combination of probiotics and prebiotics is called sinbiotik. In yogurt, probiotics and prebiotics work synergist [13]. One of the prebiotics is inulin. The addition of inulin is intended to increase the survival and activity of probiotics during the shelf life. As a water-soluble food fiber, inulin can affect homeostasis and decrease fat stores, triglycerides and cholesterol [10].

The addition of high antioxidant herbs is also necessary in preventing the progression of the metabolic syndrome due to the high oxidative stress due to increased ROS. Natural antioxidants in foods containing flavonoid components can inhibit enzymes that produce ROS and reduce oxidized ROS. The phytochemical process of binahong leaf *(Anredera cordifolia* (Ten.) Steenis) contains flavonoids, saponins, steroids / triterpenoids and coumarin. Saponin (bousingosida A1) leaf binahong inhibits cholesterol formation by inhibiting hydroxy enzyme methylglutaril co-acetate reductase at cell phosphorylation stage. The content of saponin fresh binahong leaf that is equal to 28.14 mg / g. Provision of leaf extract of binahong lowers blood cholesterol levels of male white mice induced foods high in fat and propylthiouracil significantly at doses of 500 mg / kgBB with the duration of administration 21 days [14].

High-fiber ingredients are also needed in the metabolic syndrome diet. Carrageenan are a natural polysaccharide type of Rhodophyceae seaweed. A study of 344 mice showed feeding containing 15% kappa/lambda-carrageenan from *Gravicela radula* produced a significant cholesterol-lowering effect [15]. The addition of carrageenan into the yogurt as stabilizer, emulsifier, gel-forming thickener is intended to improve the texture. In addition to carrageenan, stevia is added as a sweetener to increase acceptance.

The development of functional beverage products form of synbiotic yoghurt with binahong leaf extract which high antioxidant, fiber, total lactic acid bacteria can be an alternative treatment for metabolic syndrome. The purpose of this study to determine the effect of addition of binahong leaf extract to total lactic acid bacteria, antioxidants activity and acceptance.

2. Materials dan Methods

That was Food Production research with true experimental research with one factor's complete randomized design, that was addition of binahong leaf extract in yoghurt of synbiotic herb by 0%, 0,12%, 0,25% , and 0.5%. Each treatment performed 3 replications for total analysis of lactic acid bacteria and antioxidant activity, while for the acceptance was not any repetition. Determination of the formulation of synbiotic yoghurt yoghurt for inulin addition based on previous research that was functional drink product of srikaya yoghurt. The addition of stevia and carrageenan was determined by preliminary test. The process of making binahong leaf extract was done based on previous research using maseration method with 95% ethanol solvent for 5 days.

The process of making synbiotic yoghurt with the addition of binahong leaf extract starts from pasteurized fresh cow's milk at 85 ° C for 15 minutes then cooled to 40 ° C. Pasteurized milk added 4% inulin as prebiotics, 0.2% stevia, and 10% starter *Streptococcus thermophilus* and *Lactobacillus bulgaricus*. Incubated at incubator at 37 ° C for 24 hours. Incubation process added 6% carrageenan and extract of binahong leaf, then be homogenization.

Fresh milk obtained from the Faculty of Animal and Agriculture Diponegoro University, inulin obtained from PT DPO Indonesia, karagenan obtained from Suket Segoro SMEs Semarang, and bacteria *Streptococcus thermophilus* and *Lactobacillus bulgaricus* obtained from PAU Gajah Mada University Yogyakarta.

This research was done by analyzing total lactic acid bacteria using Total Plate Count (TPC) method, antioxidant activity was analyzed using DPPH method and acceptance was done by hedonic method on 30 panelists of Nutrition Science Study Program Diponegoro University.

The data obtained is processed using statistical software. Before analyzing all the data was tested normality using *Shapiro-Wilk* test because the sample size <30. The effect of binahong leaf extract treatment on total lactic acid bacteria and antioxidant activity was analyzed by One Way Anova. The recording analysis is carried out by Friedman test and continued with Wilcoxon test if there was a significant effect.

3. Results

3.1. Total Lactic Acid Bacteria

That was no significant effect of addition binahong leaf extract to total lactic acid bacteria yoghurt sinbiotik (p = 0,145), although there decreased total lactic acid bacteria (Table 1). Total lactic acid bacteria in synbiotic yogurt with addition of binahong leaf extract still fulfillled probiotic criteria according to Indonesian National Standard that is more than 10^7 CFU / ml.

Table 1. Results of Total Lactic Acid Bacteria Analysis

Dose	Total Lactic Acid Bacteria (10^{13} CFU/ml)[1]
0%	90,37±65,02
0,12%	86,96±53,13
0,25%	42,60±22,56
0,5%	8,93±5,22
	p=0,145

[1]= One Way Anova Test

3.2. Antioxidant Activity

The addition of various dose binahong leaf extracts had no significant effect on antioxidant activity of synbiotic yoghurt (p = 0,297). The highest antioxidant activity in synbiotic yoghurt was by adding 0.5% binahong leaf extract with antioxidant content of 8.66% (Table 2).

Table 2. Results of Antioxidant Activity Analysis

Dose	Antioxidant Activity (%)[1]
0%	5,63±4,08
0,12%	4,33±1,52
0,25%	5,33±3,05
0,5%	8,66±0,57
	p=0,297

[1]= One Way Anova Test

3.3. Acceptance
The acceptance of synbiotic yoghurt with the addition of binahong leaf extract includes the aspect of the assessment of flvor, color, texture and taste (Table 3). The addition of binahong leaf extract significantly influenced the color acceptance (p=0.001), but not on aroma, texture and taste.

Table 3. Results of Acceptance Analysis on Yoghurt Sinbiotik with Addition Binahong Leaf Extract

Dose	Aroma		Color		Texture		Taste	
	Mean	Result	Mean	Result	Mean	Result	Mean	Result
0%	2,93±0,82	Like	3,30±0,59[a]	Like	2,46±0,68	Like	2,40±0,62	Dislike
0,12%	2,90±0,60	Like	3,13±0,51[a]	Like	2,86±0,73	Like	2,53±0,57	Like
0,25%	2,96±0,66	Like	2,63±0,55[a]	Like	2,60±0,67	Like	2,50±0,86	Like
0,5%	2,60±0,93	Like	1,76±0,67[b]	Dislike	2,53±0,89	Like	2,50±1,16	Like
	p=0,346		p=0,001		p=0,062		p=0,877	

3.3.1. Aroma
The aroma of synbiotic yoghurt with the addition of binahong leaf extract was a typical fermentation sour aroma. The addition of various dose binahong leaf extracts had no significant effect on the f;avor of acceptance (p = 0.346). The more addition of the binahong leaf does not decrease level of aroma preference.

3.3.2. Color
More addition of binahong leaf extract on synbiotic yoghurt decreases the level of color preference. Sinbiotik yogurt will be more brown if the amount of extract is added more and more. The addition of binahong leaf extract of 0% and 0.12% has the highest level of favorite, with 0% binahong extracts of 3.30 and 0.12% binahong extract of 3.13. The product color is not preferred in the addition of 0.5% extract.

3.3.3. Texture
Yoghurt sinbiotik's texture with addition of binahong leaf extract that was homogeneous. There was no significant influence between the addition of binahong leaf extract with texture acceptance level (p = 0.062). The higher levels of leaf extract of binahong added to the yoghurt sinbiotik did not decrease level of texture preferences.

3.3.4. Taste
The addition of leaf extract of binahong in various doses did not show a significant effect on the acceptance of the taste of synbiotic yoghurt (p = 0.877). The taste of synbiotic yoghurt with the addition of binahong leaf extract is a typical yoghurt sour taste. Control of synbiotic yoghurt (0%) is least favored compared with the other three dose.

4. Discussion

4.1. Total Lactic Acid Bacteria
The results showed that the addition of binahong leaf extract did not affect the total lactic acid bacteria although decrease total lactic acid bacteria when added amount of extract of binahong leaf. The decrease in the number of lactic acid bacteria can be caused by the presence of secondary metabolite compounds contained in binahong leaf extracts such as flavonoids, steroid saponins, and terpenoids. These secondary metabolite compounds proved to act as antibacterials, so added more binahong leaf extracts will lead to a decrease in the total number of lactic acid bacteria. The content of saponins in fresh leaf of binahong is 28.14 mg / g. The result of antibacterial activity of binahong leaf extract to gram positive bacteria showed that flavonoid of binahong leaf extract able to inhibit the growth of *Staphylococcus aureus* [16].

IOP Conf. Series: Earth and Environmental Science **116** (2018) 012033 doi:10.1088/1755-1315/116/1/012033

4.2. Antioxidant Activity

The addition of various dose binahong leaf extracts did not significantly influence antioxidant activity (p = 0.297). Compared with controls, the addition of 0.12% binahong extract decreased the value of antioxidant activity, but slowly increased when added 0.25% and 0.5% binahong leaf extract. In the preliminary study known the average of antioxidant activity of pure binahong leaf extract that is equal to 30%. Yoghurt sinbiotik with addition of leaf extract of binahong have antioxidant content from binahong leaf extract. Binahong leaf contain flavonoids, saponins, steroids / triterpenoids and coumarins. Flavonoids are unstable to changes in the effects of oxidation, light, and chemical changes, so that when the oxidized structure changes and its function as an active ingredient decreases and even disappears and its solubility is low [17]. Decrease in antioxidants due to antagonistic interactions of antioxidant extracts of binahong leaf with components of synbiotic yoghurt.

4.3. Acceptance

4.3.1. Aroma

The addition of binahong leaf extract did not decrease the level of preference to aroma. This is due to the similarity of the distinctive aroma of yoghurt is a typical fermented acid [18]. There was fluctuating aroma acceptance, but did not reduce the scale of the level of aroma acceptance. The preferred level of aroma was the addition of binahong leaf extract with a content of 0.25%. The typical fermentation aroma was derived from the activity of *Lactobacillus bulgaricus bacteria* that play an important role in producing aroma of yoghurt [18]. The aroma of yoghurt sinbiotik with the addition of leaf extract of binahong that was aroma typical of fermentasi. Aroma of leaf extract of binahong when mixed into yoghurt sinbiotik didn't give significant influence (p = 0,346).

4.3.2. Color

Statistical analysis on color acceptance had significant effect with giving of binahong leaf extract (p = 0,001). The highest color acceptance is on synbiotic yogurt with the addition of 0% binahong leaf extract. The results showed a decrease in color acceptance on the addition of the higher binahong leaf extract. This is due to the color of brown leaf extract binahong fresh maceration. Binahong leaf extract will be green when using the method of extraction by means of sokletasi and dry maceration, while the fresh maceration of leaf extract binahong will be brown. Soaking in 95% ethanol solution causes a change of green leaf color to brown due to browning reaction in maseration process [19,17].

4.3.3. Texture

The statistic analisis showed that the addition of binahong leaf extract did not significantly affect the texture acceptance (p = 0.062). This is because the addition of carrageenan in each treatment equal to 6%. The addition of carrageenan 0.6% based on preliminary test results that have been done previously.

4.3.4. Taste

The addition of binahong leaf extract did not significantly influence the taste acceptance (p = 0,877). Synbiotic yoghurt is given stevia sugar of the same amount at each treatment, which is 0.2%, but it does not affect the taste of synbiotic yoghurt. The taste of synbiotic yoghurt with the addition of binahong leaf extract is a typical yoghurt sour taste. Fermented milk using *Lactobacillus bulgaricus* bacteria and *Streptococcus thermophilus* produce lactic acid. Both types of bacteria will decompose lactose (milk sugar) to lactic acid and various components of aroma and flavor. *Lactobacillus bulgaricus* is more concerned with the formation of aroma, while *Streptococcus thermophilus* is more concerned with the formation of yoghurt flavors [18].

5. Best Products

Based on total lactic acid bacteria, antioxidant activity and acceptance the best product that is synbiotic yoghurt with addition of binahong leaf extract equal to 0,12%, that have total lactic acid bacteria equal to 86,966x10^{13} CFU / ml fulfill criteria minimum probiotic amount based on Indonesian National Standard more Of 10^7 CFU / ml, 4.33% antioxidant activity and acceptance of taste, color, texture and aroma favored by the panelists rather than the addition of other binahong leaf extracts.

6. Conclusion

The addition of binahong leaf extract to synbiotic yoghurt had an effect on the color acceptance, but it did not significantly affect the aroma, texture and taste; Antioxidant activity and total lactic acid bacteria. The addition of binahong leaf extract in various doses of synbiotic yoghurt decreased total lactic acid bacteria, but still qualified probiotics based on Indonesian National Standard that is more than 10^7 CFU / ml. No effect of addition of binahong leaf extract on antioxidant activity. At reception gives effect to color, but not on aroma, texture and taste. The best product is synbiotic yoghurt with addition of binahong leaf extract with amount of 0.12%.

7. Acknowledgements

This study was financially supported by Riset Pengembangan dan Penerapan PNBP UNDIP Anggaran 2016.

References

[1] Cameron A, Shaw JE and Zimmet PZ 2004 Endocrinol Metab Clin North Am **33** 351-75
[2] Aguilar M, Bhuket T, Torres S, Liu B and Wong RJ 2015 JAMA **313** 1973-4
[3] Soewondo P, Purnamasari D, Oemardi M, Waspadji S and Soegondo S 2010 Acta Med Indones **42** 199-203
[4] Grundy S, Becker D and Clark L 2002 J NIH Res **3**
[5] Kaur J 2014 Hindawi Publishing Corporation Cardiology Research and Practice **20** 21
[6] Nels M, Lacey K, Rorth S and Sucher K 2011 Nutrition Therapy and Pathophysiology. 2nd Ed. Cosio Y, Editor. USA: Cengange Learning
[7] AY Tamime, RK Robinson. 2010 Tamime And Robinson's Science And Technology. 3rd Ed. Woodhead Publishing Limited; 302.
[8] Presetyo H. 2010 Surakarta:Universitas Sebelas Maret
[9] Orviyanti and Gerin. 2012 Semarang:Universitas Diponegoro
[10] Ooi L-G, Liong M-T. 2010 International Journals of Molecular Sciences **11** 2499-522
[11] Mohamadshasi M, Veissi M, Haidari F, Shahbazian H 2014 Bioimpact **4** 83–88.
[12] FAO/WHO 2002 Guidelines For Evaluation On Probiotics In Food. Report of joint FAO/WHO Working Group on drafting Guidelines for the evaluation of probiotics in food
[13] Roberfroid MB 2000 American Journal of Clinical Nutririon **71** 1682–7
[14] Fauziah F, Arifin H and Elisma A 2014 Proseed
[15] Prajapati V, Maheriya P, Jani G and Solanki H 2014 Elsevier Ltd 97–112
[16] Makalunsenge F, K Salimi Y, Duengo S 2014 [Skripsi]. Gorontalo: Universitas Negeri Gorontalo
[17] Selawa W, Revolta M, Runtuwene J, and Citraningtyas G 2016 PHARMACON.Manado: UNSRAT 1–7.
[18] Widodo W 2002 Bioteknologi Fermentasi Susu.Pusat Pengembangan Bioteknologi Universitas Muhammadiyah Malang. Malang
[19] Murni Astuti S 2016 Malaysia: Universits Malaysia Pahang

The Effectiveness of Heterotrophic Bacteria Isolated from Dumai Marine Waters of Riau, Used as Antibacterial against Pathogens in Fish Culture

F Feliatra[1*], Nursyirwani[1], A Tanjung[2], DS Adithiya[1], M Susanna[1], I Lukystyowati[3]

[1]Laboratory of Marine Microbiology, Faculty of Fisheries and Marine Sciences, University of Riau, Pekanbaru 28286 Riau Province, Indonesia
[2]Laboratory of Marine Biology, Faculty of Fisheries and Marine Sciences, University of Riau, Pekanbaru 28286 Riau Province, Indonesia
[3]Laboratory of Parasite and Diseases of Fish, Faculty of Fisheries and Marine Sciences, University of Riau, Pekanbaru 28286 Riau Province, Indonesia

Corresponding author: Email :feliatra@yahoo.com

Abstract. Heterotrophic bacteria have an important role as decomposer of organic compounds (mineralization) derived from industrial waste, decomposition of unconsumed feed, faecal, excretion of fish, and have the ability to inhibit the growth of pathogenic bacteria. We investigated the role of heterotrophic bacteria used as antibacterial against pathogens in fish culture.This research was conducted from January until March 2017. The phylogenitic of the isolated bacterial was determined by 16S rDNA sequences analysis. Antagonism test showed that the bacteria had the ability to inhibit the growth of pathogenic bacteria (*Vibrio alginolyticus, Aeromonas hydrophila* and *Pseudomonas* sp.) Three isolates (Dm5, Dm6 and Dm4) indicated high inhibition zones which were classified into strong category with the average from 10.5 to 11.8 mm toward *V. alginolitycus*. Other isolates were classified into medium and weak category. Based on DNA analysis of heterotrophic bacteria isolated from marine waters of industrial area and low salinity of estuarine waters twelve strains of bacteria were identified, and all had highest level of homology to *Bacillus sp.*,one isolates has similarity to *Enterobacter cloacae*,other isolates to *Clostridium cetobutylicum*. Most of isolated bacteria obtained from the waters of industrial area due to it received much of nutrients that very influenced the growth of bacteria.
Keywords: heterotrophic bacteria, antagonism, pathogenic bacteria, 16S rDNA sequence,marine waters.

1. Introduction

Dumai marine waters is one of the Sumatera's marine waters which is very crowded with navigation, industrial activity as well as residents' bustle around its coast. The activity is seemed to be the main contributors of both organic and inorganic materials, causing pollution to the river mouth. Numerous physical processes, industrial activity and anthropogenic and marine transportation [1] influence the concentration of organic and inorganic chemical compound affecting the distribution and activity of bacteria. Basically, marine microorganisms are as various as the ones exist in land. Organisms that can be classified intomarine microbes are protista, cyanobacteria, bacteria, mold and viruses. These microorganisms have a critical role in the process occurring in sea water column [2]. In fact, bacteria are the organisms that are mostly huge in number and dissemination compared to other living things. The existence of bacteria population indicates sanitation, showing that the water has been polluted by waste. Generally, bacteria are facultative and heterotrophic, which allow them to live completely anaerobic condition.

Heterotrophic bacteria and other microorganisms are organisms having a crucial role in biogeochemical cycles within aquatic ecosystem for their ability to decompose and remineralize organic materials to simple inorganic material, which contains nutrition for phytoplankton, perifiton, and other aquatic microflora. Thus, bacterial activity in marine waters is very important [3]. The existence of heterotrophic bacteria is extremely affected by the entry of organic materials from the river mouth to Dumai marine waters.

This research aims to isolate and characterize heterotrophic bacteria using 16S rDNA sequences, and analyze their ability as an antibacterial agent toward pathogenic bacteria in mouth of River Mesjid and Dumai marine waters of Riau Province.

2. Materials and Methods

The research was conducted from January until March 2017. The sample was collected from four stations with different salinity, those were the mouth of River Mesjid, and marine waters of industrial area in Dumai of Riau Province. Heterotrophic bacteria was isolated in The Microbiology and Marine Biotechnology Laboratory, Marine Science Department, Faculty of Fisheries and Marine Sciences. The DNA identification was conducted in Molecular Laboratory of Biology Department, Faculty of Mathematics and Natural Sciences, Riau University, while the DNA purification and sequencing processes were performed by the Malaysian *First Base* sent by *PT. Genetika Science Indonesia*, West Jakarta.

The bacteria was isolated on medium Nutrient agar (NA, Merck, Cat.No. 1.05450.0500). For DNA isolation cultures of heterotrophic bacteria were grown in 4 mL of liquid Luria-Bertani medium (LB) at 37°C for 24 hours [4]. The cultures were then removed to1.5 mL microtube and were centrifuged at a speed of 13,000 rpm (revolutions per minute) at 4°C for 2 minutes. The supernatant was discarded,the pellet was dried and was dissolved in 40 µL of Tris-EDTA (TE) buffer solution at pH 8, then 20 µL of lysozyme was added to the solution. The solution was inverted 3-5 times, and incubated for 10 minutes at 37°C. After incubation, 50 µL of 10% sodium dodecyl sulfate (SDS) was added, the mixture was gently rotated (20 rpm) for 5-10 minutes in a rotamix rotator. After that, 550 V of Phenol: Chloroform: Isoamyl alcohol (25:24:1) was added to the mixture, and the mixture was further rotated gently (20 rpm) for 10 minutes. The mixture was then centrifuged at 7.000 rpm for 10 minutes. The upper aqueous phase was transferred to a new microtube, and an equal volume of Chloroform:Isoamyl alcohol (24:1) was added to the tube. The mixture was gently rotated (20 rpm) in rotamix rotator for 10 minutes. The mixture was then centrifuged at 7.000 rpm for 10 minutes. The upper aqueous phase was transferred to a new microtube and 3 M sodium acetate was added (40% volume) to the aqueous phase. Two times volume of 99% cold alcohol was then added to this solution to precipitate the DNA, and the solution was chilled at -20°C for 30 minutes. The precipitated DNA was centrifuged at 13.000 rpm and 4°C for 10 minutes, and the supernatant was discarded. The DNA pellet was washed in 1 mL 70% cold ethanol and centrifuged for another 10 minutes at 13.000 rpm and 4°C. The supernatant discarded, resulting DNA pellet was then air dried. The pellet was dissolved in 40 µLTE-buffer (pH 8) and 2.5 µL of RNAse solution were added and homogenized using vortex mixer and incubated at 37°C for 1 hour. Finally, the DNA of probiotic candidate bacteria was ready to be used, or stored at -20°C.

2.1. Polymerase Chain Reaction (PCR).

IOP Conf. Series: Earth and Environmental Science **116** (2018) 012034 doi:10.1088/1755-1315/116/1/012034

The PCR reaction was performed by using 8 µL DNA ofeach probiotic candidate bacteria as template, and adding 23 µL of H2O, 4 µL of 10x Taq polyrmerase, 0.5 µL of 12 mM dNTP mix, 2 µL for each of universal primers consisting of forward primer 24F (5'AGAGTTTGATCCTGGCT-3') and reverse primer 1541R (5'AAGGAGGTGATCCAGCCGCA-3') and 0.5 µL Taq polymerase. PCR conditions were: warming up for 2 minutes at 94°C, followed by 30 cycles of denaturation for 1 minute at 94°C, annealing for 40 seconds at 50°C, and strand extension for 1 minute at 72°C.

2.2. Purification of DNA PCR fragments

After PCR reaction, the DNA obtained was separated in a 1% Agarose gel by electrophoresis. PCR DNA bands were purified from the Agarose gels using the Geneaid® Gel/PCR DNA *Fragments Extraction Kit*following instructions by the manufacturer.

2.3. Sequencing,BLAST and phylogenetic analysis.

Purified PCR products were sequenced with ABI 3130 XL Genetic Analyzer Applied Biosystem. Sequencing was done at the First Base Malaysia, from PT. Genetika Science Indonesia, West Jakarta. BLAST analysis was done using the tools and data of Gen Bank accessed at http://www.nebi.nih.gov/blast. DNA sequences were aligned using CLUSTAL W 2.1. Phylogenetic trees were generated using the tools provided by CLUSTAL W, and trees were visualized using MEGA version 0.5 program. The data obtained from the isolation and identification of heterotrophic bacteria, and the activity against pathogenic bacteria (*Vibrio algynolyticus, Aeromonas hydrophila* and *Pseudomonas* sp.) were presented in tables and figures.

3. Result and Discussion

3.1. Water Quality Measurement

Water sample was collected from four stations, and each station sampling was determined at three sites of which the distance between them was 50 m (Figure 1). At the same time, water quality parameters were measured. Data of water quality parameters were presented in Table 1.

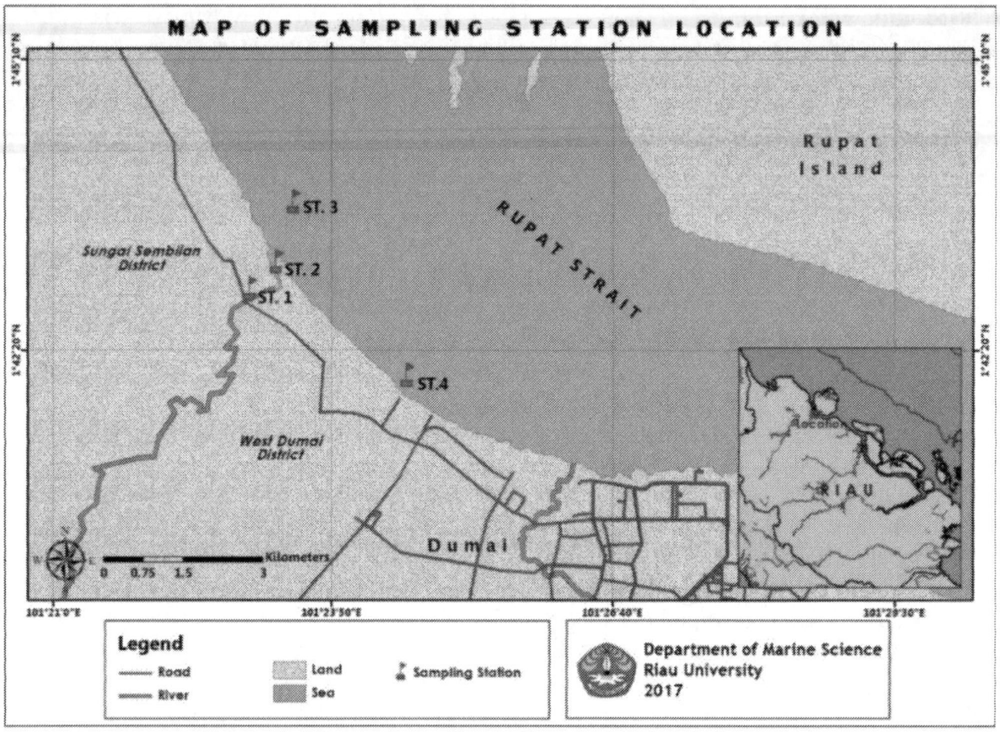

Figure 1. The stations of sampling sites in Dumai waters

Table 1. Water quality condition in each sampling station of Dumai waters.

Station	Coordinate	Time (WIB)	pH	Salinity (ppt)	Transparancy (cm)	Temperature (°C)	Current velocity (m/s)
1	N 01° 42'09.717'' E 101°22'38.745''	10.16	6,5	10	49.5	27	0.13
2	N 01°42' 13.154'' E 101°22' 44.457''	10.00	6,5	20	52.5	28	0.19
3	N 01°43' 38.500'' E 101°23' 06.906''	11.40	6,5	25	93.5	29	0.18
4	N 01°41' 57.706'' E 101°24' 30.942''	13.43	7	30	60	30	0.34

Condition of water quality measurement was as follows:samples were collected from 08:00 am until 14:00 pm; pH value ranged from 6.5 to 7.0;water salinities at station 1, 2, 3 and 4 were 10‰, 20‰, 25‰ and 30‰, respectively; water temperatures were 27-30°C, and the current velocities range was 0.13-0.34 m/s (Table 1). This area is located in the east part region of Sumatera island which is directly opposite to Rupat island.

Dumai city is located at the east coast of Sumatera island. This area integratedly develops its coastal area as a productive region for growing and cultivating its primary export commodity fish such as grouper, white snapper, crabs and pomfret through the restoration function of mangrove area. Dumai also has a harbour that can be functioned as a transit area for traveling to neighboring countries, such as Singapore and Malaysia. Besides, Dumai marine waters is considered as a place of industrial activity for several companies such as Pertamina UP II Dumai, PT. Patra Dock, PT. Chevron Pasifik Indonesia, PT. Sarana Sawitindo, and palm oil refinery Bukit Kapur Raksa and PT. Splar Cooperation.

3.2. Bacteria isolation

After being cultured and purified repeatedly, 14 different bacteria isolates wereselected based on morphological characters from each the bacterial colony, and the ability to produce antibacteria towards the pathogenic bacteria. The isolates were DM1, DM2, DM3, DM4, DM5, DM6, DM7, DM8, DM9, DM10, DM11, DM12, DM13 and DM14.The morphological characteristic of those isolates were presented in Table 2.

Table 2. Morphological characteristics of 14 bacterial isolates from Dumai waters

Isolate Name	Plate No	Diameter (cm)	Color	Colony Form	Border	Surface
DM1	1.3 $(-5)_1$	0.4	Yellowish white	Round	Waved smooth	Even
DM2	2.2 $(-5)_1$	0.5	White	L-shape	Curved	relief
DM3	1.1 $(-5)_2$	1.5	White	Irregular	Smooth	Even
DM4	2.2 $(-6)_2$	1.4	White	L-complex form	Irregular	relief
DM5	2.2 $(-4)_1$	1.8	Yellowish white	Circular with relief edge	Wavy	Bumpy
DM6	2.1 $(-5)_3$	1.2	Yellowish white	Rhizoid L-complex	Branched	Drop-like
DM7	2.2 $(-4)_1$	1.8	Yellowish white	Having Rhizoid Thread	Wavy	Bumpy
DM8	20 $(-6)_2$	1.9	White	Wrinkled	Curved	Even
DM9	20 $(-5)_2$	1.4	White	Irregular	Wavy	Bumpy
DM10	20 $(-4)_3$	0.7	White	Irregular	Wavy	Even
DM11	20 $(-5)_3$	1.0	White	Round with reef edge	Wavy	Even
DM12	30 $(-5)_2$	1.2	Red	Round	Smooth	Even
DM13	30 $(-6)_3$	0,9cm	White	Round	Irregular	Even
DM14	30 $(4)_3$	1,3cm	White	Round	Smooth	Even

The observation result indicated that the colony size varies between 0.4-1.9 cm which varied from diverges to white, yellowish white, and even red (DM12). The colony shape varied from round, irregular, and even coral-like edge. As for the border, there were also various like waved smooth and irregular. The colony surfaces were diverse to even, relief, and bumpy. This indicated that the colonies were indeed distinctive from each isolate to others (Table 2).

3.2. Antagonism of bacteria isolates

Results of antagonism test conducted in triplicate were presented in Table 3. Each bacterial isolate had an ability to inhibit the growth of pathogenic bacteria, indicated by the appearance of clear zone around the paper disc. There were three isolates (DM4, DM5 and DM6) demonstrated strongest inhibition power of 10.5-11.8 mm clear zone toward *V. algynolyticus* (Table 3). Other three bacteria isolates indicated the biggest clear zone toward pathogen *A. hydrophila* were DM14, DM8 and DM 5 with 5.8-7.8 mm clear zone (Table 3). Regarding to pathogen *Pseudomonas* sp., three isolates (DM1, DM 11 and DM 13), indicated clear zones of 6.5-7.6 mm (Table 3).

According to Pratama [5] when the diameter of inhibition zone is more than 20 mm, the response classification of bacterial growth inhibition will be intense. In addition,when the diameter of inhibition zone is in the range of 10-20 mm, the response classification of bacteria growth inhibition is still categorized intense. Yet, when the diameter of inhibition zone is merely around 5-10 mm, the response of bacteria growth inhibitionis medium, and if it is less than 5 mm, the response will be weak.To conclude, this research found there were only threebacterial isolates that werepowerful toward pathogen *V.algynoliticus*, and the rests showed medium and weak response.

Antimicrobial test of bacteria isolated from Dumai waters indicated that one potential isolate (DM11) in inhibiting pathogenic bacteria *V. alginolyticus* (9.6mm), *A. hydrophila* (5.6mm) and *Pseudomonas* sp (6.8mm) at medium levelin comparison to other 13 isolates. Meanwhile, the isolate

DM5 performed the largest clear zone toward *V. alginolyticus* (11.8mm). Then, isolate DM14 showed the largest clear zone against *A. hydrophila* (7.8 mm), and finally, isolate DM1 had the largest clear zone toward *Pseudomonas* sp.(7.5 mm) as shown in Table 3.

The heterotrophic bacteria inhibited pathogenic bacteria (*V. alginolyticus, A. hydrophila* and *Pseudomonas* sp.) by the production of antibiotic products, bacteriocin, or certain organic acid. This finding was in agreement to Verschuere *et al.* [6] who noted that microbial population can actually release the chemical materials having bactericidal or bacteriostatic capacity, which can prevent the growth of other bacteriaby several factors, such as antibiotic production, bacteriocin, siderophore, lysozim, protease, and hydrogen peroxide. In addition, it can also influence pH of media by generating certain organic acid. The antibacterial agent such aslactic acid and bacteriocin contained by probiotic bacteria were able to inhibit the development of pathogenic bacteria [7]. This occursdue to the antibacterial agent could decrease or lower the pH, thus the pathogenic bacteria are difficult to survive [8].

Beeneva *et al.* [9] performed a research on heterotrophic bacteria in several marine waters, and discovered 68.9% isolates from temperate zone and 56.76% of Vietnamese strain showed antimicrobial activity.The strains showing the greatest activity were of the tropical origin.

Table 3. Test result of bacteria isolate antagonism towards pathogenic bacteria

Isolat	Inhibition Zone Diameter (mm)														
	V. algynolyticus					*A. hydrophila*					*Pseudomonas sp.*				
	(+)	U_1	U_2	U_3	R	(+)	U_1	U_2	U_3	R	(+)	U_1	U_2	U_3	R
DM1	4	2	2	1	1,6	1.5	1	3	4	2.6	8	1	9	12.5	7.5
DM2	1.5	8	9.5	8	8.5	3	4.5	3.5	3	3.6	6.5	1	3	1.5	1.8
DM3	8	6.5	8	4.5	6.3	2	2	2	1.5	1.8	5.5	5	5	6.5	5.5
DM4	5.5	3.5	15	13	10.5	5	2	3	3	2.6	6	3	2.5	3	2.8
DM5	2	10	12.5	13	11.8	6	6	6	5.5	5.8	7	3	3	6.5	4.1
DM6	5.5	4	13	17	11.3	6	2	3.5	3.5	3	3	5	5.5	5.5	5.3
DM7	3	6	6	0.5	4.1	5	0	2	3	2.5	6	1	11	7	6.3
DM8	5	5	5.5	5	5.1	4	5	6.5	8	6.5	8	4	3	4	3.6
DM9	12	11	12	9.5	10.8	3.5	7	3	3	4.3	5.5	4	5	4.5	4.5
DM10	4.5	5	5	4	4.6	3	2.5	4.5	5	4	10.5	3	7	3	4.3
DM11	14	12	11	6	9.6	6	2	6	9	5.6	7.5	4	7.5	9	6.8
DM12	5	2.5	2.5	3	2.6	5	2.5	4	4.5	3.6	10.5	3.5	4	2.5	3.3
DM13	2	0	4	5	3	3	2.5	4	5.5	4	2	6	9.5	4	6.5
DM14	2	3.5	7	2.5	4.3	13	10	6.5	7	7.8	1	0	1	1	0.6

The result above have been multiplied by 6 mm paper disc
Information : (+) : positive control; (0) U1: 1st recurrent; U2 : 2nd recurrent, U3: 3rd recurrent,; R: average

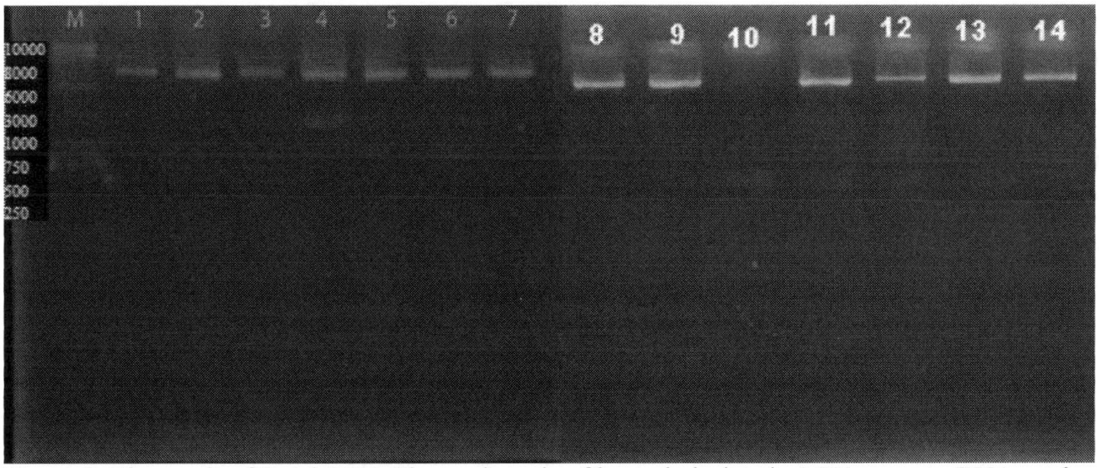

Figure 2. The result of Total DNA Electrophoresis of bacteria isolate in 1.2% agarose. DNA marker was 1kb ladder.

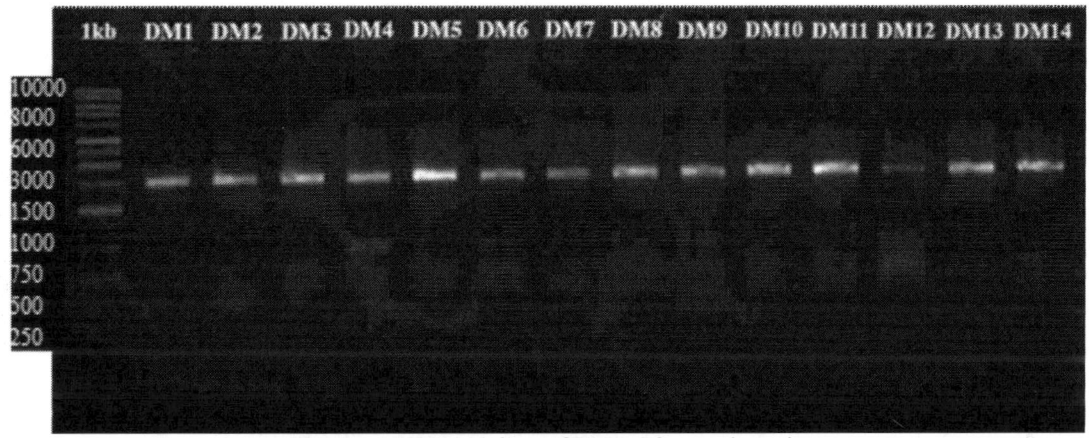

Figure 3. PCR Product of DNA Electrophoresis

From the electrophoresis shown in Figure 3, the size of PCR was going on for 1500 bp. This great size was as the expected result from the 16s rRNA genes. Sabdono stated that bacteria isolate amplification possessing single band revealed that primer used was specific one to beapplied to amplify the 16S rRNA of bacteria. The 16S rRNA amplification had become a standard to study phylogenetic and diversity of marine microorganism.

From the sequence analysis of bacteria isolate that had been *Blasted* in Table 4, it was found that there were several bacteria having higher level compared to other kind of bacteria. There was a possibility of close relationship of each isolate such as the DM1 isolate which had 97% homology rate to strain ASK16 *Bacillus cereus,*DM2 indicated 96% homology rate to strain A C *Bacillus toyonensis*, OPR1150Xg of DM3 had 94% homology rate to strain F1-8*B.pseudomycoides*. Furthermore, the association was as well occured to DM4 which had 97% homology rate to strain SBFW51 *Bacillus cereus*, DM5 possessed 97% homology rate to strain OPP5 3-2 *Bacillus cereus*, DM6 had 96% homology rate towards strain SBFW5S *Bacillus cereus*, and DM7 which had 97% homology rate towards strain B4 *Bacillus cereus*.

Moreover, the homology of DM8 isolate was similar to strain ACOPR1ISOXg *Bacillus toyonensis* with 95% homology rate and had 1541bp base length. It implies that the homology rate was only similar at the genus level and was different at species level. Based on a research *B. toyonensis* forming spores, was appropriate to be used as an active element of

TOYOCERIN additive. The notion is well supported by Jimenez *et al.,* [10] on his research declaring that the TOYOCERIN has shown inhibition toward the growth of pathogenic bacteria.

As for DM9 isolate, it has homology similar to strain Y37 *B. subtilis*, with 96% homology rate and 1609bp base length. Implicitly, its homology rate is identical merely until the genus level but not at the species level. As noted by Pant *et al.,* [11], *B. subtilis* looks like a trunk and has the ability to compose endospore. This bacterium has also play a role in health sector to yield bacitracin antibiotic, and in industrial sector to produce bacterialamylase or other materials that could modify extract, glue papers, and to release glue for textile. The *B. subtilis* bacteria can be found in either river or marine sediment [12].

Then, the DM10 isolate was known for its similar trait to strain Enc-3 *Enterobacter cloacae* with 99% homology rate and 1469 bp. This indicated that the homology rate was identified until the species level. According to Lina *et al.* [13] *Enterobacter cloacae* lived in diverse environment, both in land and water (soil and food waste). This strain lives as a microflora commensal within human and animal intestines, and play a fundamental role as a pathogen for plants and insects.

Next for the DM11, its homology was the same to strain S512 *Clostridium acetobutylicum*, having 96% homology rate and base length as 1542 bp. Unlike DM10, the homology level was similar merely until the genus level. *Clostridium acetobutylicum* was a commercial bacterium, belongs to *Clostridium* genus. This bacterium was a trunk-like Gram positive and mostly found in mainland, thought it could also be found in various circumstances. According to Logan and Devos [14], *Clostridium acetobutylicum* could be found in land, river sediment, wells, intestines, cattle excrement, dogs and human feces.

Bacillus cereus exists in DM12 and DM13 isolates. The homology of DM13 isolate was similar to strain 4PLGES *B. cereus* with 99% homology and 756 bp base length. In other words, the homology rate was similar until the species level. Meanwhile, F isolate homology was similar to strain ML267 *B. cereus* with 96% homology rate and 1517 bp length. Similar to DM11, the homology level was similar onli until genus level.

The primary habitat of *B. cereus* is the surroundings and the canal of digestion. Water and soil have a significant function for this bacterium in polluting food. This bacterium can also adhere to shoes, clothes, skin or through air and dust [15].

According to Khetan [16], the *B. cereus* bacterium is one of pathogenic agents possessing great potential to be utilized as a biological control. The bacterium has particular host sharmless to pests' natural enemies and other non-target organisms, which are easily disjointed by the environment, and its pathogenicity can be improved using genetic engineering technique. The *B. cereus* bacterium is a probiotic bacterium having the ability to hinder the pathogenic bacteria growth as *Vibrio* sp and *Aeromonas* sp [17].

The DM14 isolate has homological resemblance to strain C17 *B. thuringiensis* with 97% homology rate and 1546 bp base length. Its homological rate is identical until the species level. *B. thuringiensis* bacterium is a heterotrophic bacterium living in established and fine environment. As stated by Lantang and Runtuboi (2012), *B. thuringiensis* exists in natural realm and could also be discovered in various habitats, like soil, water, and mud.

B. thuringiensis could produce 2 types of toxin: crystal toxin (Crystal, Cry) and cytolytic toxin (cytolytic, Cyt). Cyttoxin could fortify the Cry toxin so that it is considerably used to develop the effectiveness in controlling the insects. Several studies have been conducted to isolate and purify the bacteriocin from *Bacillus* sp generated by *B. cereus*, and toxin yielded by *B. thuringiensis* [19].

From the end results, it can be understood that the fourteen heterotrophic isolate bacteria gotten from Dumai marine waters are genus from *Bacillus* bacterium. The bacteria of *Bacillus, Bifidobacteria, Pseudomonas, Lactobacillus* and *Micrococcus* genus have been proved useful and could dwell both inside and outside a normal organism, and are also believed to be beneficial probiotic bacteria [20]. The bacteria were also assumed to have ability in inhibiting the pathogenic bacteria growth, which was known based on the antagonism test result,showing the existence of clear zone around the disk. This is also indicated in a research conducted by Suwardi [21]. The bacteria resulting antimicrobia are *Bacillus* sp., A1 and A2. To find out the characteristics of antimicrobial producer bacteria, morphology, physiology, and analysis of protein were carried out using SDS-PAGE. The morphology observation exposes that the colony has a circular shape with medium size and convex surface. The color of the colony is white and that its edge is even. Further, from the Gram staining and BA1 and BA2 bacteria, the morphological traits are obtained. The bacteria appear as trunk bacteria cell and possess Gram positive traits (purple). As in the spora staining, BA1 and BA2 are identified to have oval shape positioned in the middle of the bacterium. Russell *et al,* [22] conducted a study on inland waters using 16s RNA technique and found that 3 bacteria are able to respire nitrate and oxygen. The cultures represent members of abundance phyla, as determined by amplicon sequencing of environment DNA extract, and allow for further studies into geothermal factors impacting life in the deep sub surface.

Table 4. The channel accessed from Gen Bank database used in determining the phylogenetic relationship towards bacteria isolate

Isolate	Species	Strain	Access Code	Homology
DM1	*Bacilluscereus*	ASK16	KY750685	97%
DM2	*B. toyonensis*	ACOPR1150Xg	KY750686	96%
DM3	*B. pseudomycoides*	FI-8	KY750687	94%
DM4	*B. cereus*	SBFW51	KY750688	96%
DM5	*B. cereus*	OPP5 3-2	KY750689	97%
DM6	*B.cereus*	SBFW5S	KY750690	96%
DM7	*B. cereus*	B4	KY750691	97%
DM8	*B. toyonensis*	ACOPR1ISOXg	KX129781	95%
DM9	*B. subtilis*	Y37	KF641801	96%
DM10	*Enterobacter cloacae*	Enc-3	KU747082	99%
DM11	*Clostridium cetobutylicum*	S512	JQ086380	96%
DM12	*B. cereus*	4PLGES	LC215052	99%
DM13	*B. cereus*	ML267	KC692161	96%
DM14	*B. thuringiensis*	C17	LC146717	97%

From the BLAST analysis,a *Phylogenetic tree* is created. The *tree* functions to connect node and is considered as a taxonomy unit, like species or genetrees' root is the oldest part of all analyzed organisms. The *Allignment* sequence of sample from Gen Bank database is undertaken by using Mega 0.6 and Clustal W programs (Figure 2).

Majority of the 14 observed isolates had particular relationship to *Bacillus* which is different from one isolate to another. For instance, DM1 and DM2 isolates have closer association. In contrast, isolates DM12,DM5, DM14, DM10 and DM11, had closest relationship to *Bacillus* sp. However, all isolates had close relationship to *Bacillus* sp. (Figure 4).

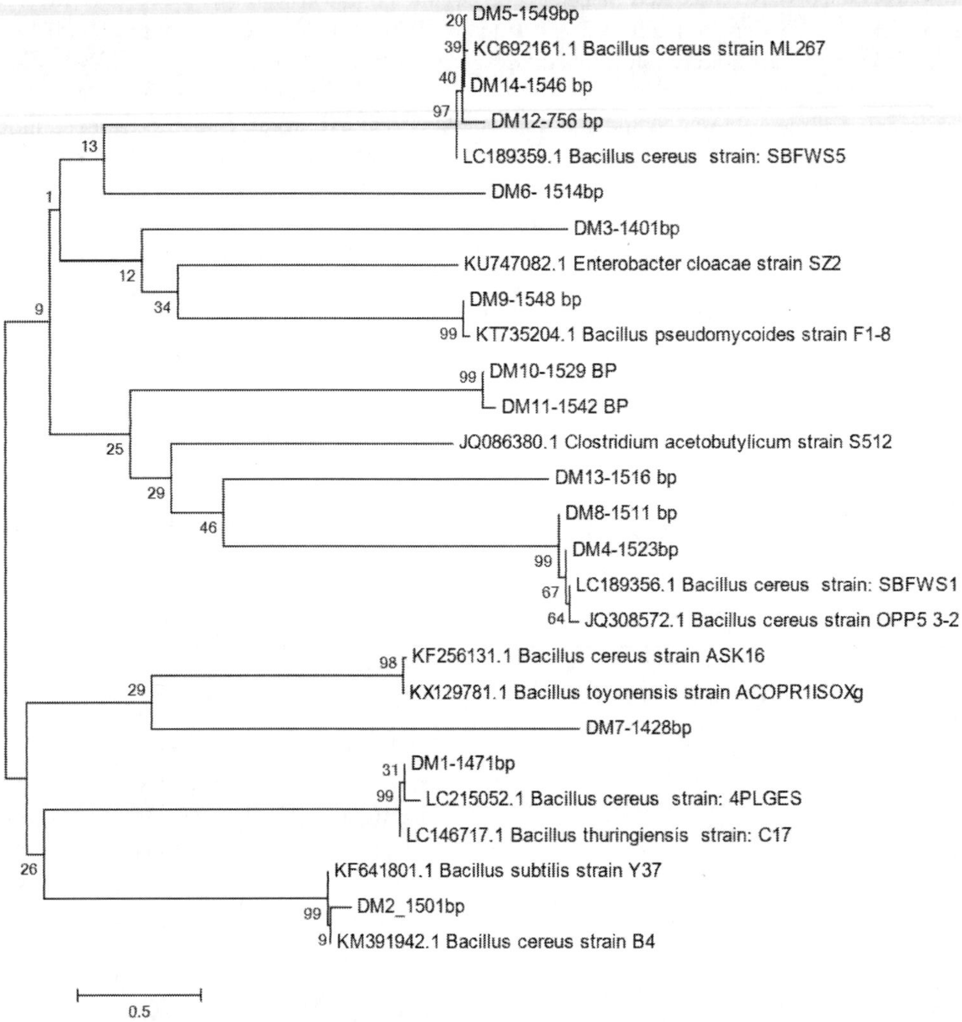

Figure 4. *Phylogenetic tree* constructed with Mega 5 program

4. Conclusion

From the research results, it can be identified that the four heterotrophic bacteria isolates indeed have relationship to *Bacillus* sp. All isolates had homology more than 90%. In addition, the fourteen heterotrophic bacteria have an ability to inhibit the growth of pathogenic bacteria for fish. In the antagonism test, the inhibition zones showed by isolates DM5, DM6, and DM4 were highest towards *V.alginolyticus*. Meanwhile, other isolates had medium and weak inhibition zone. Isolate DM11 was to be the best isolate in inhibiting the growth of the three pathogenic bacteria, *V. alginolyticus* (9.6mm), *A. hydrophila* (5.6mm) and *Pseudomonas* sp. (6.8mm), in comparison to other 13 isolates.

5. Acknowledgement

The authors wish to thank the Ministry of Research Technology and Higher Education, for funding this research in accordance with the Competence Grant No. 486/UN.19.5.1.3/PP/2017.

6. References

[1] Santos, L. M. P., Santos, A. L., Coelho, F. J. R. et al 2013 heterotrophic activities of neustonic and planktonic bacterial communities in an estuarine environment (Ria de Aveiro) *J. Plankton Res*. **36** (1): 230–242

[2] Feliatra 2010 marine microbiology. Faculty of Fisheries and Marine Sciences of Universitas Riau. 220p Pekanbaru.

[3] Kunarso, D. H and T.I. Agustin 2012 a study on heterotrophic bacteria in Halmahera sea waters. *Marine Science*. Vol. **17** (2). 63-73

[4] Feliatra, F. L .Iesje, Y. Dessy, R. Haqqy,M. Deasy ,H. Wahid, T.N. Titania, R.F. Andi, Y. Rofiza. 2016. phylogenetic analysis to compare populations of acid tolerant bacteria isolated from the gastrointestinal tract of two different prawn species *Macrobrachium rosenbergii* and *Penaeus monodon*. *AACL Bioflux*. Volume **9** (2) 360-368

[5] Pratama, M. R 2005 the effect of siwak wood powder extract (*Salvadora persica*) on the growth of *Streptococcus mutans* and *Staphylococcus aereus* bacteria with different diffusion methods.*Thesis*. IPB. Bogor

[6] Verscheure, L., Rombaut, G., Sorgeloos, B., dan Vestraete, W 2000 probiotik bacteria as biological control agents in aquaculture. *Microbial and mol. Biol. rev*. **64**: 655-671.

[7] Fauziah, P. N., J. Nurhajati, and Chrysanti 2012 the inhibition of adhesion of various *Klebsiella pneumoniae* strains by *Lactobacillus bulgaricus* in soyghurt in vitro on Hep-2 cell lines with various treatment processes *Thesis* Bandung: Universitas Padjadjaran. Pages. **66-75**

[8] Tambekar, D. H. and S. A. Bhutada2010an evaluation of probiotic potential of *Lactobacillus* sp. from milk of domestic animals and commercial available probiotic preparations in prevention of enteric bacterial infections. *Recent Research Science and Technology*, **2**(10): 82-88

[9] Beeneva I A, Kukhlevsky A D, KharchenkoV 2013 antimicrobial activity of heterotrophic bacterial strains of marine origin. *Jundishapur J Microbiol*. **6** (2):166-175.DOI:105812/jjm.5039

[10] Jimenez, G. M., H. Urdiain, A. Cifuentes, Lopez, A. R. Blanch, J. Tamames, P. Kampfer, A. B. Kolsto, D. Ramon, J. F. Martinez, F. M. Codonerg, and R. Rossello-Mora 2013 description of *Bacillus toyonensis* sp. nov., a novel species of the *Bacillus cereus* group, and pairwise genome comparisons of the species of the group by means of ANI calculations *Syst. Appl. Microbiol*., vol. **36**, pp. **383–391**

[11] Pant, G., A. Prakash, J. V. P. Pavania, S. Bera, G.V.N.S Deviram, A. Kumar, M. Panchpuriand R. G. Prasuna 2015 production, optimization and partial purification of protease from *Bacillus subtilis*. *Journal of Taibah University forScience*, **9**, 50–55. Department of Chemistry Faculty of Mathematics and Natural Sciences. Universitas Hasanuddin, Makasar

[12] Tariq, A. L., S. Sudhadan A. L. Reyaz 2016 Isolation and screening of bacillus peciesfronsedients and aplication in bioremediation. *Journal of current Microbiology and Applied Science*. Vol **5** (6) 916-924

[13] Lina M. M., G. Florianaand S. Stefania 2012 *Enterobacter cloacae* complex: clinical impact and emerging antibiotic resistance *Future Microbiol* **7** (7):887- 902.

[14] Logan, N. A., P. Devos 2009 bergey's manual of systematic bacteriology, Volume 3: The Firmicutes, Springer, 21-127

[15] Fatmasari 2015 sensitivity test of antibiotics of chloramphenicol, ciprofloxacin, erythromycin and clindamycin against *Bacillus cereus* isolated from beef meat in traditional and modern market of makassar city. *Thesis*. Makasar: Universitas Hasanuddin.

[16] Khetan S. K 2001 microbial pest control. Marcell Dekker, Inc. New York, 300 pp.

[17] Feliatra, Y. Fitria, and Nursyirwani 2012 antagonists probiotic bacteria isolated from bowel and stomach grouper ducks (*Cromileptes altivelis*) against pathogenic bacteria *Journal of Fisheries and Marine Affairs*.**17**,1: 16-25.

[18] Lantang, D. and D. Y. Runtuboi 2012 characterization of *Bacillus thuringiensis* bacteria from protected forests of Uncen Jayapura Campus, and the detection of its toxicity to *Anopheles* Mosquito larvae. *Journal*. Papua: Universitas Cenderawasih

[19] Paik H. D., S. S. Bae, S. H. Park, J. G. Pan 1997 identification and partial characterization of tochicin, a bacterion produced by *Bacillus thuringiensis* subsp. *tochigiensis*. *Journal of Industrial Microbiology and Biotechnology* **19**:294-298

[20] Feliatra.,Y. Dessy., L .Iesje., T.N. Titania and H. Wahid 2015 the potential of the isolated probiotics bacterial from giant prawns' digestive tract (*Macrobrachium rosenbergii*, De Man) With 16s rDNA sequencing technique. *International Journal of Oceans and Oceanography*. Vol **9** (1) 1-10.

[21] Suwardi, C.N.E., Kusuma, H. Nurhayati 2012 isolation and characterization of *Bacillus* sp. the producer of antimicrobial from chicken digestive tract (*Gallus domesticus*). Proceeding of SNSMAIP III-2012 **306**

[22] Russell JA, Leon-Zayas R, Wrighton K and Biddle JF 2016 deep subsurface lefe from North Pond: enrichement, isolation, characterization and genomes of heterotrophic bacteria. Front. *Microbiol* **7** :678.doi:10.3389/fmicb.2016.00678

Preliminary Study On Gonad Maturity Stages of the Sea Cucumber *Paracaudina australis* from Kenjeran Water, Surabaya, Indonesia

Widianingsih Widianingsih[1], Muhammad Zaenuri[2], Sutrisno Anggoro[3], Hermin Pancasakti Kusumaningrum[4], Retno Hartati[5]

[1]*Doctoral Programme of Coastal Resources Management, Faculty of Fisheries and Marine Science, Diponegoro University, Jln. Prof Soedarto, SH, Kampus UNDIP Tembalang Semarang 50275 Indonesia* [2]*Departement of Oceanography, Faculty of Fisheries and Marine Sciences, Diponegoro University*
[3]*Departement of Aquatic Resources, Faculty of Fisheries and Marine Sciences, Diponegoro University* [4]*Departement of Biology, Faculty of Sciences and Mathemathic, Diponegoro University*
[5]*Departement of Marine Sciences, Faculty of Fisheries and Marine Sciences, Diponegoro University*

e-mail: widia2506@yahoo.com; widia2506@gmail.com

Abstract. The holothurian *Paracaudina australis* is belong to family Caudinidae, ordo Molpadida and class Holothuroidea. This species is among the most common holothurian widely distributed in the tropical water. The purpose of this reseach is to do preliminary study on maturity stages of sea cucumber *Paracaudina australis* from Kenjeran Water, Surabaya, Indonesia. This research was conducted on April 2016. Samples were collected randomly on the Kenjeran Water, Surabaya. The result showed that there are five stages of gonad maturity. At the stage of maturity 1, the gonad was not clearly distinguished, there were unbranched small tubule. At the stage of maturity 2, there were small branched of tubules. At this stage, gonad can be differentiated between male and female. At the stage of maturity 3, tubule can been branched not only for male but also female. At the stage of maturity 4, the gonad was good mature, there were clearly branched tubule. At the stage of matury 5, there were generally had empty tubule except for a few relict unreleased spermatozoa. At female gonad, there were shrunken tubule and relict oocytes were presented in the lumen of the tubule.

Keyword: *Sea cucumber, Paracaudina australis, Stage of maturity*

1. Introduction

The demand for sea cucumber consumption is increasing with the high demand of the world and it makes the stock population of sea cucumbers in the tropical and surrounding areas decreases. This necessarily requires the cultivation of sea cucumbers, restoration of stock population of sea cucumbers in nature by enhancing the cultivation of sea cucumber [1] and [2]. Artificial breeding techniques in sea cucumbers have been developed for some species. Juvenile production techniques in the *Holothuria scabra* species have well developed in several tropical and subtropical countries such as Indonesia, China, Japan and India [3]. However, several species such as *Holothuria scabra* [4] and

* Corresponding author. Tel.: +62 813 2558 1564.
E-mail address: widia2506@gmail.com.

IOP Conf. Series: Earth and Environmental Science **116** (2018) 012035 doi:10.1088/1755-1315/116/1/012035

Paracaudina australis poorly known about reproduction organ system which including stage of gonad maturity and maturity index of gonad [5].

The high demand for cracker materials and snacks from *P. australis* from Kenjeran waters is increasing and this will result in a decreasing in *P. australis* population stock in the Madura Strait water. There is highly demand for sea cucumber commodities in Indonesia because some types of sea cucumbers have good nutritional value such as fatty acids, amino acids, fats, proteins, phosphorus, calcium, sodium and minerals [6] dan [7]. Sea cucumbers of this species (*Phyllophorus* sp and *P. australis*) also have potential as immunostimulant material against bacterial infections *Escherichia coli* [8].

Teripang *P. australis* is one of the many exploited sea cucumbers in the Eastern Coastal Waters of Surabaya, especially the waters of Kenjeran, Madura Strait. Based on the study of sea cucumbers population were found in Madura Strait waters, there were 7 species namely *Paracaudina australis, Phyllophorus* sp, *Colochirus quadrangularis, Holothuria sanctori, H. forskali, H. turriscelsa* and *Holothuria* sp. [5]. Sediments where the living sea cucumber population has the characteristics of sand mud and clay type of sediment [9].

Based on the observation in the field, sea cucumbers *P. australis* have been captured in large quantity. The over exploitation of sea cucumber could be reduced stock population in waters around Kenjeran, Madura Strait. So it is necessary to increase the stock of population in nature. This is certainly not separated from some reproductive studies on sea cucumbers *P. australis* such as maturity level of gonad and gonad maturity index and stock population of sea cucumbers in these waters [5].

The histology study of gonad development for the determination of stages of gonad maturity of sea cucumber *P. australis* has been done for three months in February, March and April, 2012 [5]. This observation will certainly support the study of the determination of pattern and reproduction cycle of sea cucumber *P. australis*. A macro observational study on male and female gonads was performed to supplement data on histology studies of gonad development of *P. australis*. So hopefully study about pattern and cycle of reproduction of *P. australis* in Kenjeran Waters, Madura Strait East Coast Surabaya can support data of population of sea cucumber *P. australis*. The recorded data were reproduction study on the type of sea cucumber *H. scabra* [4], sea cucumber (*H. vagabunda*) [10], fission reproduction of Stichopudidae from Karimun Jawa Island [11].

2. Materials and Methods

The material used in this research is the *P. australis* sea cucumber collected from the waters of Kenjeran Madura Strait (Surabaya East Waters) with geographical location between 07 ° 14'12.3 " to 07 ° 13'52.4" south latitude and between 112 ° 48'58.2 "to 112 ° 49'28.8" east longitude. The sampling of 50 sea cucumbers was conducted on the second week of April 2016. Seacucumber *P. australis* was directly taken from a depth of 3 meters with the help of divers by using hands. The observation of the stage of gonad maturity refers to [1], [4], [5] and [7], which consists of stage 1 (gametogenesis), stage 2 (maturation), stage 3 (spawning), stage 4 (early post spawing) and stage 5 (late post spawning). Whereas according to Perezrul et al.,[12], the stage of gonad maturity consist of stage 1 (immature stage), stage 2 (maturing stage), stage 3 (ripe stage) and stage 4 (spent stage). In this study stage of gonad maturity consists of stage 1 (immature stage), stage 2 (gametogenesis), stage 3 (maturation), stage 4 (spawning) and stage 5 (post spawning).

3. Results and Discussion

3.1. Morphology Gonad P. australis

According to this research, there are clearly differences morphological in male and female gonad. In Females, the gonads look more yellow to orange for stages of gonad maturity 1, 2 and 3, whereas in the stage of gonad maturity 4 and 5, the gonad looks more pink (Figure 1 and 2). Tubular branching in female stage of gonad maturity 1 has not been clearly visible, whereas tubules have short branching for stage of maturity gonad 2. Furthermore, for female, tubules have been seen filled in stage of gonad maturity 3 and the number of tubular branching were more nomerous. Similarly, in female of stage of gonad maturity 4 have seen a full tubule. Whereas in female for stage of maturity gonad 5 has already seen partially empty tubules (Figure 2). According this research, in the immature stage, tubules had short branched, thick, yellow-white in colour. Furthermore maturing stage had characteristic tubules long, thick and branched ovary ovary pale red, and for the ripe stage had characteristic tubules thick, branched and red colour. Although different species in gonad maturity level, but for maturity level, ripe stage and spawning stage for *Holothuria spinifera* [2] *Isostichopus fuscus* [11], *Holothuria atra* [12], *Holothuria fuscocinerea* [13] *Bohadschia vitiensis* [14] is similar with *Paracaudina australis*.

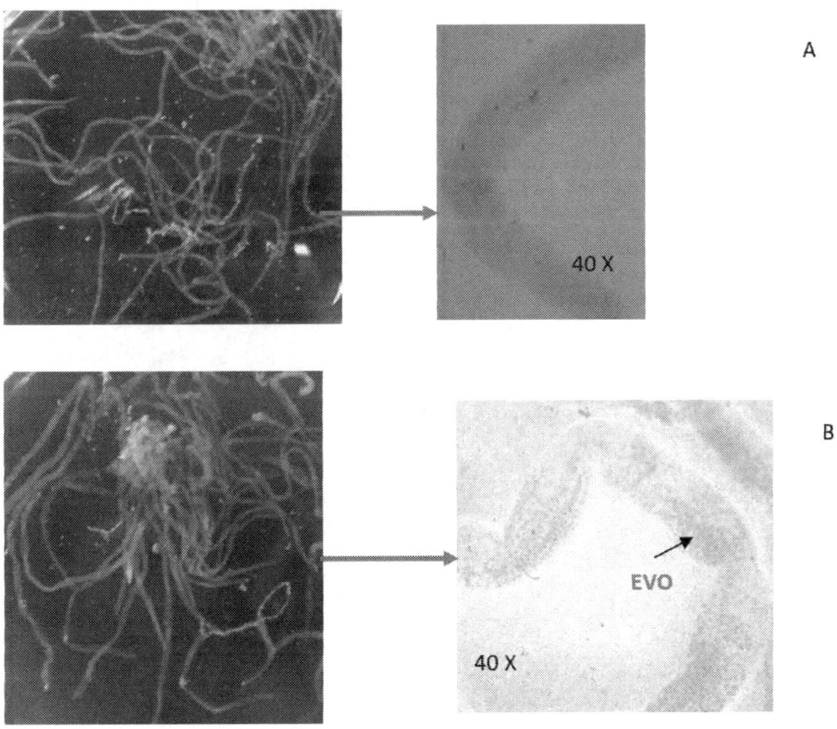

Figure 1. Macroscopic and microscopic of Stage of gonad maturity at female sea cucumber *P. australis*.(A) SGM 1; (B) SGM 2; EVO (Early Vitellogenic Oocytes)

IOP Conf. Series: Earth and Environmental Science **116** (2018) 012035 doi:10.1088/1755-1315/116/1/012035

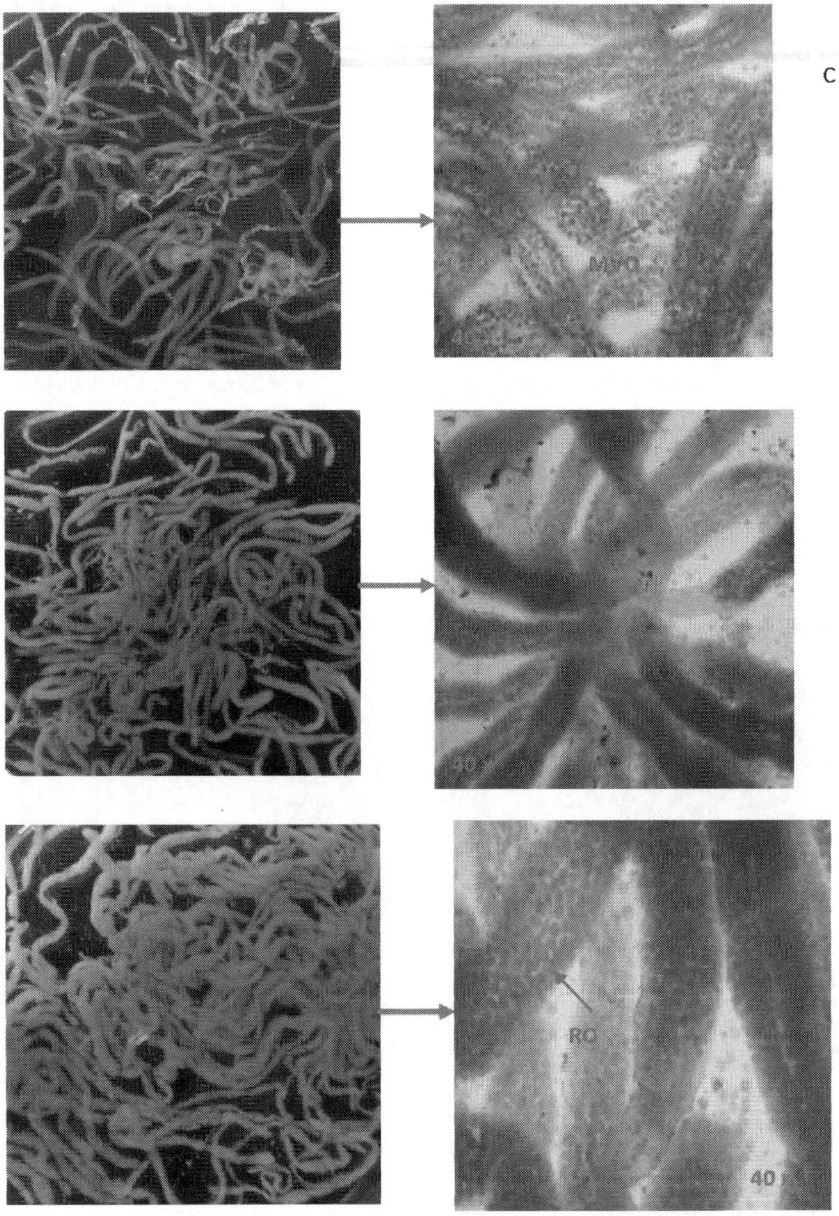

C

D

E

Figure 2. Macroscopic and microscopic at stage of gonad maturity (SGM) for female *P. australis*. (C) SGM 3; (D) SGM ; (E) SGM 5; MVO (mid Vitellogenic Oocytes); RO (Relict Oocytes)

Based on morphological observations on male individuals, the colour of gonads tends to be pale yellow at stage of gonad maturity 1, 2 and 3, whereas in stage of gonad maturity 4 and 5 were white ivory (Figure 3 and 4). At the maturity level of gonad 1 in male individuals, tubular branching has not been seen clearly and tubules are still small and unfilled with spermatozoa. Furthermore, at the stage of TKG 2 has seen a branching in the tubule and after entering TKG 3 tubules have started to fill with spermatozoa. Spermatozoa removal in tubule is getting fuller on TKG 4 has a lot of branching. While at the 5th stage where it has entered the spawning stage (spawning) has seen the tubules that have started empty. (Figure 3.). It is also in accordance with the opinion of [12] , that at the final stage of gonad maturity, the tubules in the *Holothuria atra* have been partially empty while the 2nd and 3rd stages are the maturity stage.

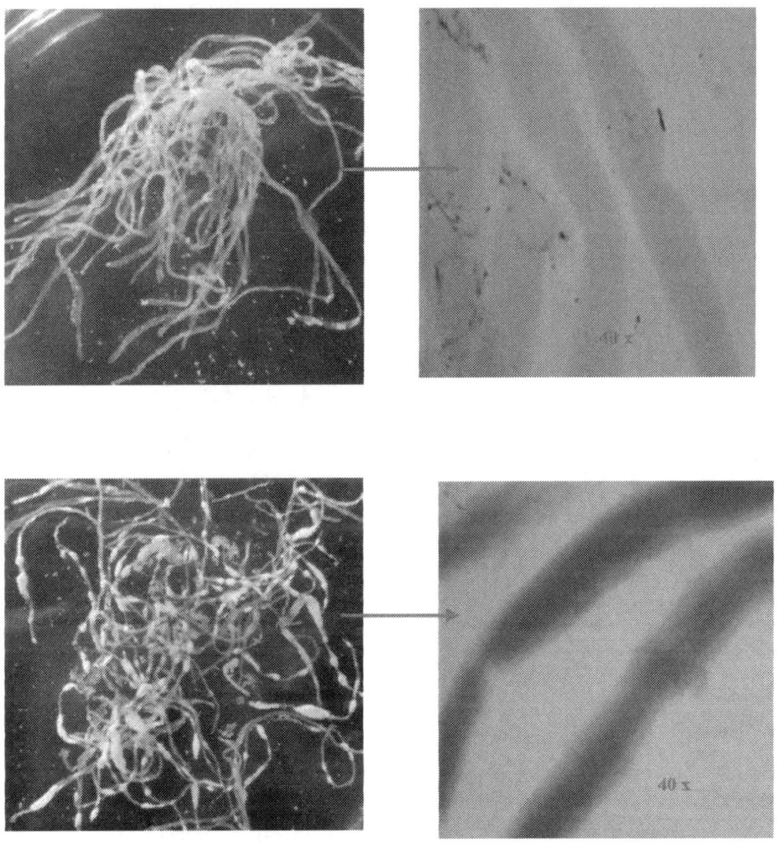

A

B

Figure 3. Macroscopis and microscopis Stage of Gonad Maturity (SGM) Tingkat on male of *P. australis*. (A) SGM 1; B) SGM 2.

In the observation of individual female sea cucumbers *P. australis* for stages 1 and 2 had a slightly different colour (orange colour) while in stages 3, 4 and 5 the colour becomes pink (Figure 3 and 4). It also appears in the *Holothuria atra*, where the colour of the tubules at stage of gonad maturity 2 and 3 were pink [2] [12]. The morphology structure of maturity level of gonad sea cucumber *P. australis* almost equal to the condition of sea cucumber *Holothuria vagabunda* [7], *Holothuria scabra* [4], *H. fuscocinerea* [14].

IOP Conf. Series: Earth and Environmental Science **116** (2018) 012035 doi:10.1088/1755-1315/116/1/012035

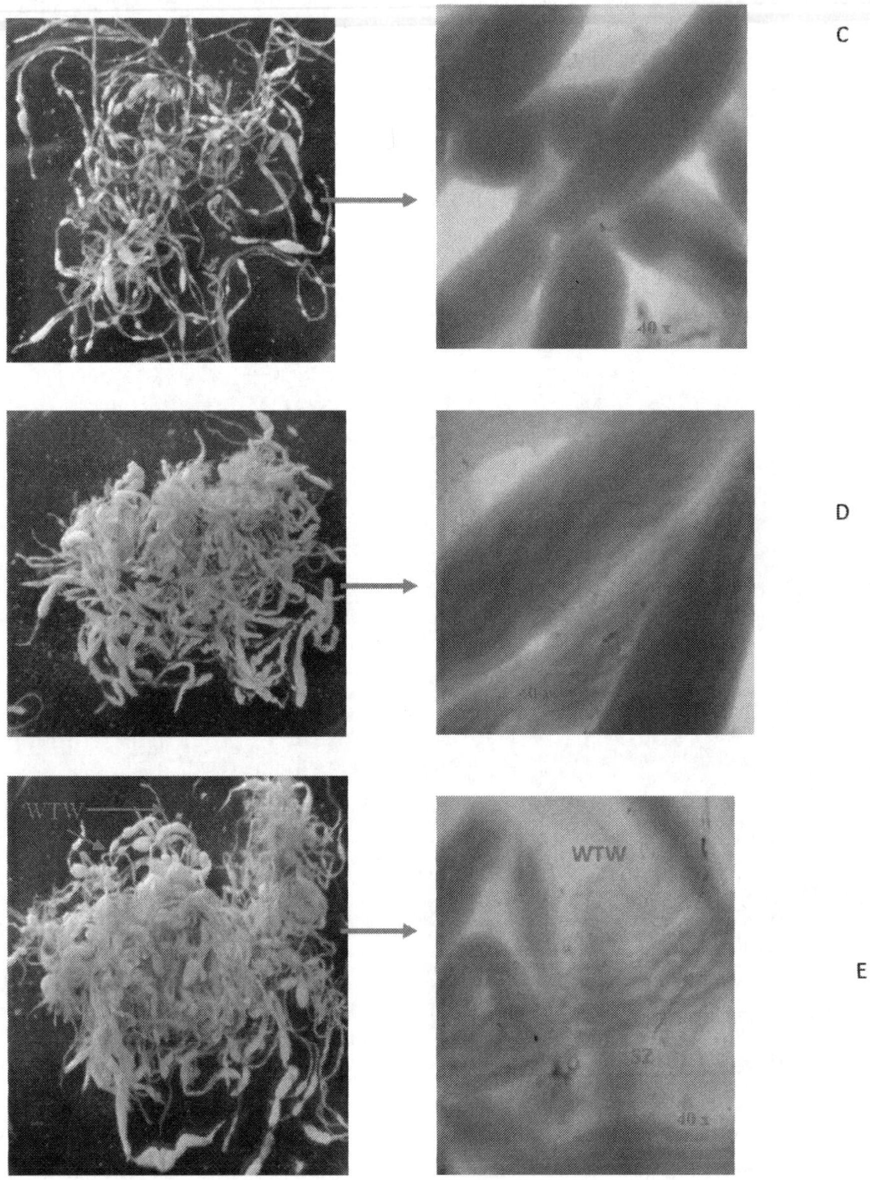

Figure 4. Macroscopis and microscopis Stage of Gonad Maturity (SGM) Tingkat on male of *P. australis;* C) SGM 3, D) SGM 4 ; E) SGM 5; WTW (wrinkled tubule wall); SZ (Spermatozoa)

In both male and female organism, the tubule at the first stage of gonad maturity are not yet clearly visible branching forms. While in the second stage has been formed branching followed by the filling of tubules with spermatozoa in males and oocytes in females. The maturation stage is usually called the third stage. While in the stage of gonad maturity 4 in tubules both in males and females already appear fully filled. Then in the fifth stage has begun to see the partially empty tubules where both the female and male individuals have spawning in natural.

According to the observation on the maturity level of gonads both in males and females, sea cucumbers *P. australis* is differenced between characteristic every stage gonad maturity not only for female but also for male. The characteristic gonad of *P. australis* is not similar with the other species of such as *Isostichopus fuscus* [11], *Holothuria atra* [12], *Holothuria fuscocinerea* [13] *Bohadschia vitiensis* [14], *Cucumaria frondosa* [15], *H. sanctori* [16] .

Conclusion

At low stage of gonad maturity in females showed that tubules with short branches in light yellow. While the higher stage of gonad showed that tubules have been long branches with pink to red in females. At low stage of gonad maturity levels for the males showed that tubules have been short branches in light yellow. While the higher stage of gonad maturity showed that indicate tubules with long branches with ivory colour up to white.

Acknowledgments

This research was supported by Doctor Grant Program with contract 182-03 / UN7.5.1 / PG / 2016. The funds come from the Directorate of Research and Community Service. The author also thank to all of people who help me during this research.

References

[1] Conand, C. 1993. *Marine Biologi.* **116**: 439-450.

[2] Asha, P.S. & P. Muthiah. 2008. *Aquacult Int.* **16**: 231-242.

[3] James, D.B. 2001. *Naga* **24 (1-2)**: 4-8.

[4] Purwati, P. 2006. *Indonesian Journal of Marine Science.* **30**: 47-55.

[5] Winarni, D., M. Affandi., E.D. Masithah., H. Damayanti and M. Nisa. 2012. *Proceeding of IFS. Seed production & Aquaculture system*: 34-40.

[6] Rustam. 2006. *Budidaya Teripang. Coremap Tahap II. Kabupaten Selayar. Yayasan Mattirotasi.*

[7] Widianingsih W, M. Zaenuri, S.Anggoro, H.P.S. Kusumaningrum. 2016. 2nd International Symposium on Aquatic Products Processing and Health (ISAPPROSH 2015). *Aquatic Procedia* 7: 271-276.

[8] Pradipta R. 2011. Imunitas alami mencit (*Mus musculus*) setelah pemberian ekstrak tiga jenis teripang lokal Pantai Timur Surabaya berdasarkan aktivitas dan kapasitas fagositosis. **Laporan Penelitian.** Universitas Airlanggga.

[9] Yusron, E. 2003. *Oseanologi – LIPI. Jakarta*: 129 – 133.

[10] Hartati, R dan H. Yanti. 2006. *Indonesian Journal of Marine Sciences* **11 (3)**: 126 – 132.

[11] Perezrul, H., H. Reyes Bonilla, F. Garcia Dominggues, C.E. Cintra Buenrostro. 1999. . *Journal of Marine Biology*. **135**: 521- 532.

[12] Abdel-Razek, F.A., S.H. Abdel-Rahman, N.A. El-Shimy, H.A. Omar. 2005. *Egyptian Journal of Aquatic Research.* **31 (2)**: 383-402.

[13] Villalobos, F.B., O.H. Avila –Poveda., IS. Gutierrez-Mendez. 2013. *Sexuality and Early Development in Aquatic Organisms.* **1** : 13-24.

[14] Omar, H.A., F.A. Abdel Razek., S.H. Abdel Rahman ., N.A. El Shimy. 2013. *Egyptian Journal of Aquatic Research.* **39**: 115-123.

[15] Hamel, JF and A. Mercier. 1996. *SPC Beche-de-mer Information Bulletin* **8:** 22-33.

[16] Navaro, P.G., S.G. Sanz, F. Tuya. 2012. *Scientia Marina.* **76(4):** 741-752

EXPLORING THE POSITION OF OLD SEMARANG SEA PORT: Based on Javanese City Pattern

R. Siti Rukayah [1]*, Endang Sri Susilo [2], Muhammad Abdullah [3] , & Siddhi Saputro [4]

[1] *Department of Architecture, Engineering Faculty, Diponegoro University Jl.Prof.H. Soedarto, SH, Tembalang, Semarang 50275, Indonesia*
[2] *Department of Marine Science, Faculty of Fisheries and Marine Science, Diponegoro University, Jl.Prof.H. Soedarto, SH, Tembalang, Semarang 50275, Indonesia*
[3] *Culture Science Faculty Universitas Diponegoro Jl.Prof.H. Soedarto, SH, Tembalang, Semarang 50275, Indonesia*
[4] *Department of Oceanography, Faculty of Fisheries and Marine science, Diponegoro University, Jl.Prof.H. Soedarto, SH, Tembalang, Semarang 50275, Indonesia*

[1] tututrsiti@yahoo.com

Abstract. According to the world traveller that have visited the old cities on the Java Island in the 15th century, the cities have a seaport. The sea port became a place for transit and stay until grown as a coastal city, Unfortunately, some of these cities have disappeared due to the process of secretions. Semarang in the 14th century also had a sea port as a landing area in Kampung Melayu. Recently the village no longer located in the coastal but in the middle of the city. How to prove that the area was a coastal area? By tracing some architectural remains, toponimy and comparing with the Javanese urban design pattern revealed the old phenomena in the past. Architectural sketches will make a clearly information about the old coastal line of Semarang. It is estimated that the seaport directly connected to the *alun-alun* / traditional square as central government. The city founders in Java puts the center of government using the same city spatial structure as waterfront city. The architecture remaining as a landmark of coastal city should be conserved.

Keywords: old, Semarang, seaport, *alun-alun*, local wisdom

1. Introductions

Hartono stated that the understanding of urban history is very important for urban planners and designers, since most urban planning techniques depend on extrapolations from past [1]. Roosmalen stated that the planning of the oldest city in Java is the palace of Majapahit. In the Majapahit city planning, referring from its ruins, the pattern of the city is composed of a port and the center of government [2-3] Furthermore, the next urban planning in Java refered to the unwritten concept of Majapahit (1293 to 1500) [4]. The planning of the city center that has the composition of *Alun-alun*, palace, mosque [5-7]. The structure could be attributed by the dominance of the North-South axis [8-9]. The concept of urban planning from the era of Majapahit kingdom continued in the era of Islamic sultanate such as Demak, Cirebon, and Banten [10]. The north-south axis concept still preserved in the Islamic era due to the location of the city on the north coast of Java [10-12]. By the way, the north axis was the city layout to face to the sea. So, the concept of the cities was a waterfront of the city.

From that explanation above it appears that in the 1950s the condition of *alun-alun* Semarang was full of market buildings and traders [13]. In the Dutch colonial era, *alun-alun* as old square cut off by the groote post weg road (1809-1811) (Pratiwo.P.Nas, 2002). In Semarang city the remaining oh the *alun-alun* after cut off by the post weg had a to kite-shape [14]. In the 1930s the East side of *alun-alun* was covered by the new modern market building (Johar market). In the era of the 1970s the remaining *alun-alun* in west side turned into Jaik market and hotel Metro and Kanjengan office on the south side of the square turned into a shopping center [5]. The identity of traditional city and surrounding were finally lost their identity in 1970s. However, long time before the area changed due to development, the area located facing through to the sea. Now the area is no longer on the coast of Semarang. In the 14th century the location of the port of Semarang located in the area of Sleko [15-16]. The port that connects the city of Semarang with ancient Mataram kingdom. [18]. Similarly, according to geologists, the coastline of Semarang was on the north side of Semarang's central government area around Sleko area in Kampung Melayu [19]. Geologically, some cities along the north coast of Java including Semarang city undergoing sedimentation process. Due to sea sedimentation process, using satellite imagery Semarang city progressed to the north as far as 8 km since 14th century [15-17]

Based on that development process (naturally or human disaster), the identity of Semarang city was lost as a traditional city that was waterfront. This paper is aimed to reveal the position of old Semarang as waterfront city. Where was the position of coast line and seaport on the old Semarang city planning in the past?

2. Research Method

The research using qualitative method is strengthened by using the historical approach. The process of collecting old data was obtained from KITLV, newspapers and history books. Analysis of information in old data, old maps, old photographs and map in the recent condition will be re-drawn using overlapping system until easy to do interpretation. The city pattern in traditional cities in Java will use as references. The conclusion of the research is based on the city development stage (layering morphology) of the city center that has been described.

To explore the situation in the past, some of the old data of Semarang city in 1695, 1719, 1875, 1880, 1909, 1917, 1945, and the 17th century obtained from KITLV will be the basis to reveal the pattern of the waterfront city at that time. Through an old image images and old maps the researchers will do 'seeing by believing'[18]. According to Pole, 2014 pictures and maps speak more about the past than a narrative. This approach is also used to strengthen that narrative history in Indonesia is very limited [19].

The location of this research is the old city center of Semarang, especially in the old Malay kampong until the old formerly capital city Semarang in Kanjengan. That area expressed by marine historians and marine geologists as the coastline of Semarang in the past. The model of traditional city in Java prototype which has composition consisting of the city port - the central square government.

3. North Axis of the Sultanate Cities in Java

In the 13th to 15th centuries, there was a substantial increase in commerce in Southeast Asia in general and Java in particular. As a result of the progress in the field of shipping, many foreign traders come to the cities on the north coast of Java. They came from India (Jambudwipa), Cambodia, China, Vietnam (Yawana), Campa, South India, Bengali and Siam [21].

The cities on the north coast of Java became a bustling port city. Geographically the city of Java can be divided into two parts. That is the coastal city and inland city. Both coastal and inland cities at the beginning of its development had the same structure. In the later historical development that the coastal city has a different structure from the inland city. This is because the coastal city will more interact with foreigners from across as a result of progress in the field of shipping. So the inhabitants of coastal city are more hiterogin when compared with the inhabitants of inland city [21]

The sultanate cities on the coast such as Cirebon, Banten have the concept of city arrangement that can still be traced. The structure of the old city had a type of cosmic city, a hierarchical-concentric city, and includes a planned settlement pattern, although not geometrically. In urban history, cities in the early days of their growth were usually well planned, ie, as a human effort to organize the environment in which to live for the better . The old city map of Banten and Cirebon as coastal city which can be obtained from the atlas of mutual Heritage shows the pattern of cosmic city with the composition of the square, mosque, palace and north axis toward sea area.

The concept of the sultanate city is actually regardless of cosmic thought. Nevertheless, it is suspected that the cosmic concept which originally was the view of traditional societies in Southeast Asia in general and Java-Bali based on Hinduism was continued by the founders of the imperial/ sultanate city. In Hindu terms it has two distinct views. The first is the holy-religious view and the second is 'profane-political'. Based on this view the royal capital was built. Thus, the grand design will include decisions on where religious areas as sacred place and where the political landscape as profane will be placed. The sacred space must be separated from the profane space. In the sacred space (microcosm) humans can regularly organize their activities based on cosmology and they need cosmography to make decisions. The effort to present this philosophy in their physical form is manifested in area planning and building planning following physical elements: Center (Axis Mundi & Imago Mundi), Hierarchy, Axis and Orientation.

In the core area that is considered as a sacred area lay an alun-alun / the traditional square. Alun-alun is a venue for important state, community and religious events. Religious events take place in the alun-alun with the mosque as its controller. Meanwhile, the government event took place in the alun-alun with the palace or dalem as the controlling activities. This square is very important because it is the outermost part of the palace or *dalem* . The square is a transitional area between the world outside the city and the world in the palace [9]. So the position of alun-alun and corridor as imaginary northern axis connecting to sea and palace/ *dalem*.

It can be concluded that the application of the North South axis in traditional cities with the Hindu concept is constantly on the urban governance of the Islamic Sultanate or the Islamic city center. The traditional urban design in the era of Dutch colonialism still maintains the Javanese traditional city concept wich contains of local wisdom in designing the city center of government.

4. *Boom* area on North side of Alun-alun

There is an interesting city pattern and structure in the imperial coastal city and imperial inland city. Some cities in coastal and inland Java have ports / *booms* that are directly connected to the *alun-alun* as the center of the city. Literally the meaning of the word *boom* in Dutch equals to seaport in English or *Pelabuhan* in Indonesian.

The city of Tuban on the north coast of Java, the physical form of *boom* is the land that jutted the sea. Geographically located just a few hundred meters north of the *alun-alun* Tuban. The boom serves as a dock where the ship docked. The boom was made as a result of the silting of the coast of Tuban since the 16th century, so large ships had to be anchored in the middle of the sea. With the *boom* is expected to the large/ medium size of the ship can dock in the *boom*. [1].

The Solo City, located on the banks of the great river Bengawan Solo in the interior of Java also has a kind of port called a bandar. The city of Solo was originally formed by the coolie community (Javanese: *soroh* until its leader was called *ki-soloh* or *ki-solo* or *ki sala*) located at *bandar* Nusupan.

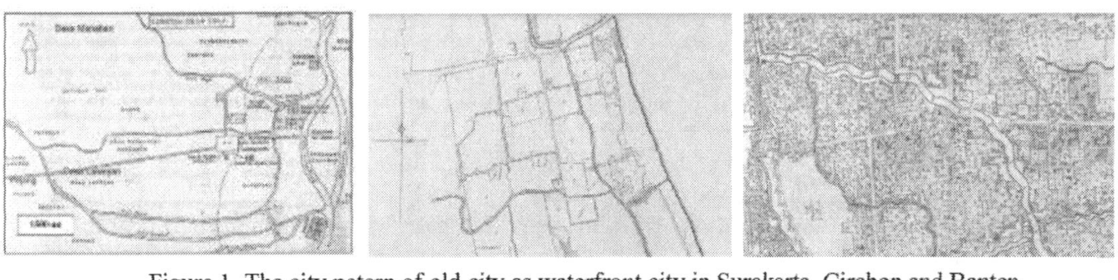

Figure 1. The city patern of old city as waterfront city in Surakarta, Cirebon and Banten.
Resources: atlas mutual heritage

Similarly, with the city of Banten. As the city of sultanate and township became the center of the highest market network for mid-market centers outside the city and palace areas, as well as local markets in rural areas. The location of the city and the market is on the north side of the square. It is also in accordance with the life of the central cities of the kingdom and port cities of Islamic maritime empires in Indonesia. The central cities of the kingdom and port cities such as Samudra Pasai, Aceh, Malacca, Demak, Banten, Gresik, Jaratan, Jepara, Surabaya, Ternate, Banda, Gowa-Makasar, Banjarmasin, Palembang and so on, many visited by big traders from various foreign countries as well as from royal areas in Indonesia.

It can be concluded that as a city of sultanate or a waterfront city based on waterfront and harbor city has a boom, a kind of dock of land that is useful as a landing site. It is estimated that the founders of the city set up the concept and lay out the city with the center of the city leads to the port area. Market position, shop and commercial area become the boundary of public activity to the alun-alun and the central government or palace as the core space of the city.

Both theories above about traditional city layout patterns in Java and boom positions that directly relate to the center of government can be a new knowledge and method to track the form of a traditional city that has lost its identity due to natural or human disasters.

5. The Layur Corridor in the Old Map

The existence of the Layur corridor has appeared on maps sourced at the atlas of mutual heritage in 1719 (see Figure 2 and 3). Layur Corridor is a road in Kampung Melayu Semarang forming by two rows

Figure 2. Old Semarang Map in 1787 AD. In this photo seen the structure of the old Semarang city, Layur Corridor perpendicular to the port and the square of Semarang.

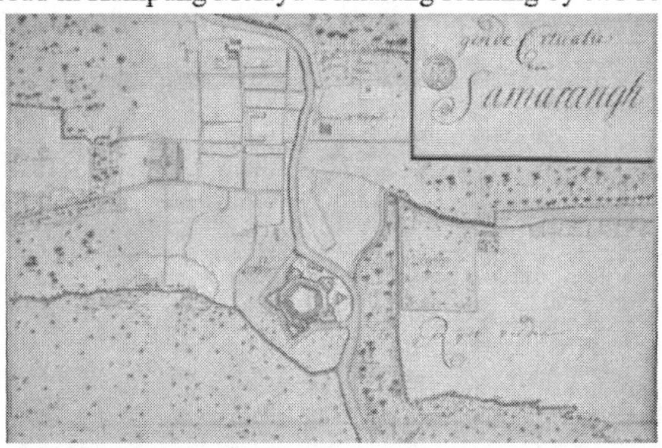

Figure 3. Old Semarang Map in 1719 AD, the blue-marked region (the end of the corridor) is a river encounter. It is supposed to be a landing place from the boats in the sea.

of China and Arabs shop houses. The row of merchant's house in that corridor served as a trade corridor. Layur Corridor location is strategic area because connecting the seaport – the center of government

Recently. Semarang city increasingly develops. Unfortunately, the attention of the local government in conservation effort does not reach this area. The area has a history as the main gate to entered Semarang as one of Asia's largest trading centers. In this village area was once the entrance to the city of Semarang. Based on the remaining of the toponym, the name of the kampong *Darat* means the place for landing from the boats. The remaining toponym *boom* lama has a meaning as seaport [20-21]. It can be concluded that from the two remaining of the toponym, the village once was the initial port of Semarang city before it was transferred to the port which is now called Tanjung Mas. In 1875 AD, the Dutch East Indies government opened a new channel as a new river line from Kampung Melayu to the seaport [20].

IOP Conf. Series: Earth and Environmental Science **116** (2018) 012036 doi:10.1088/1755-1315/116/1/012036

6. Seaport - North Axis – Capital City

The existence of Layur corridors at this time looks shabby. Many buildings are not functioning, broken and collapsed. While the straightening the

houses of Arab merchants are not working. Problems that arise today due to several factors and aspects in relation to the development of the city of Semarang. The existence of trade corridors such as the corridor Layur in Semarang has begun to fade, because it has been replaced with another commercial area such us in Johar market area, jalan Pemuda and Simpang Lima.

One of the architectural buildings indicates that this area formerly was a seaport area in the past is *Menara masjid Layur* mosque. This mosque has a tower that served as light tower at the time. The mosque has a historical value so the local government rewarded the Layur mosque as a cultural heritage building since 1992 AD.

The pictures bellow is juxtaposed of the present maps and old map in 1719 AD.

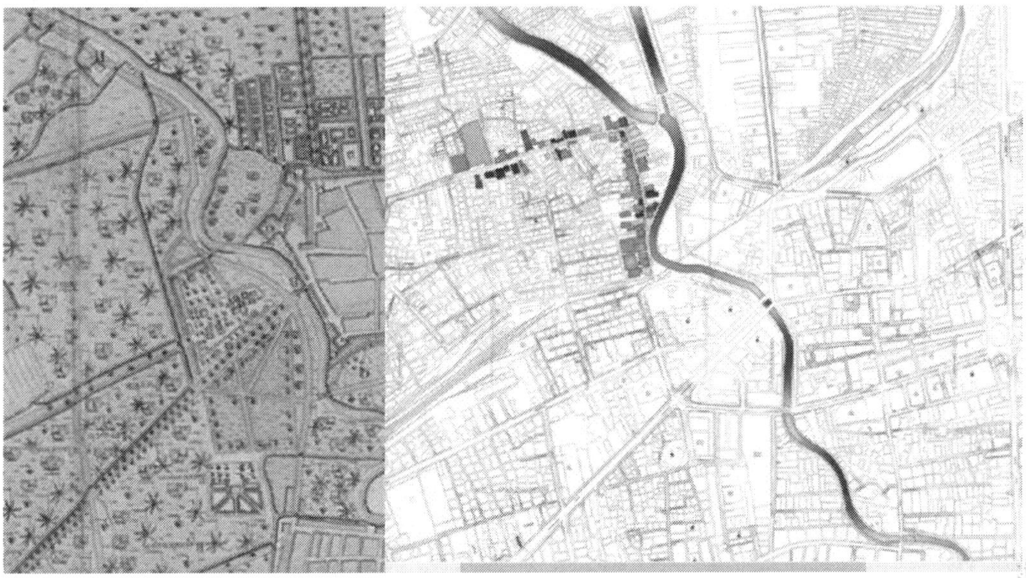

Figure 4. The Layur corridor in 1719 and recent condition

The Layur road has a function as an imaginary north axis connected seaport in *boom lama* with the *alun-alun* in Kanjengan area. This concept is similar to the Islamic sultanates along north coast Java and the sultanates that built based on waterfront city concept in Surakarta.

The urban patern of old Semarang designed by Ki Ageng Pandanaran at the beginning of the 15th century seems look similar to the concept of urban planning of the Demak Sultanate Therefore the old Semarang city had arranged based on the concept of the Demak waterfront city.

The development of the northern axis of Layur corridor has increasingly become the direction of the orientation of buildings and villages since the construction of Groote Post weg (the great mail road) in 1809-1811. The road position is on the south side Layur Corridor and splitting *alun-alun* Kanjengan. Since the construction of this road, Semarang river is no longer as the orientation and the direction all of building in Layur Corridor. The progressive sedimentation also causes the river to be no longer navigable. The condition of the land grew forward to the north and left the traces of Malay village as a coastal area.

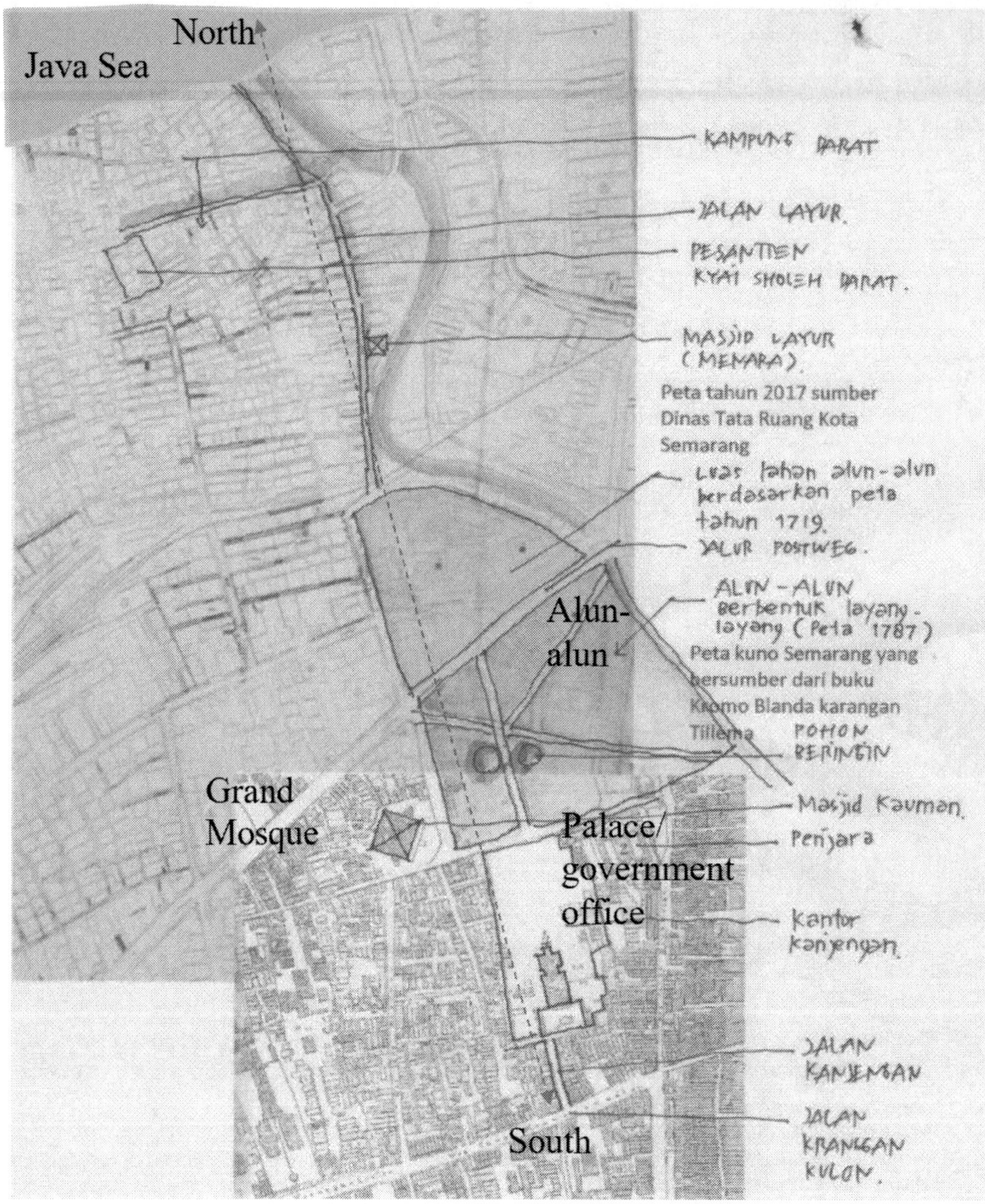

Figure 5. A simple sketch of the positions of ports and central government

Based on the compilation of the old map of Tillema's in chromo Blanda book, and the recent map is found that Layur Corridor directly connect s to the city center government. The city centre of Kanjengan consists of the the square/ *alun-alun*, the grant mosque and Kanjengan office. (Figure 5)

7. Conclusions

Based on the analysis of urban structures in Java where the city pattern is port- corridors- city government the position of Malay kampong was the coastal area of Semarang in the past. Based on field survey, matching with ancient Semarang maps, and historical references, the Layur corridor as northern imaginary axes to the centre of government has been lost.

The researcher suggests for the next research to be aimed to make a whole mapping of the old coastal Semarang city. For the city government, researchers suggest that the government has a serious concern for the conservation effort in old coastal area.

Acknowledgement
Thanks to the Architecture Department and Penelitian Terapan Unggulan Perguruan Tinggi that has provided funding for this research activity in 2017.

References

[1] Hartono S 2005 Alun-Alun Dan Revitalisasi Identitas Kota Tuban. *Dimensi Teknik Arsitektur.* **33** 131–141
[2] Noorduyn J 1978 Majapahit In The Fifteenth Century. *kitlv-journals.nl.* **134** 207-274
[3] Kementerian Pekerjaan Umum dan Perumahan Rakyat 2015 Mengembalikan Identitas Kota Pusaka. *Buletin Ciptakarya.*
[4] Rukayah, Siti R, Dhanang R, and Endang S 2016 Morphology of Traditional City Center in Semarang. Towards Adaptive re- use in urban heritage. *Environ Proc J.* **1**109–118
[5] Rukayah, Siti R 2010 *Simbiosis Di Ruang Terbuka Kota Simpang Lima Di Kota Semarang*. Phd Disertation, not published.
[6] Rukayah, Siti R, Malik A 2012 Between Colonial, Moslem, and Post-Independence Era, Which Layer of Urban Patterns should be Conserved?. *Procedia - Social Behavior Science* **68** 775–789
[7] Rukayah, Siti, R and Hartuti S 2013 Islamic City Structure In Sekayu Semarang (The Viewpoint of Urban Design). *Journal of Architecture, Planning and Construction* . **3** 1
[8] Gomperts A, Haag A, and Carey P 2010 Rediscovering the royal capital of Majapahit.*Kitlv.nl.* **2** 12–23
[9] Wessing R 1992 A Tiger In The Heart : The Javanese Rampok Macan.*Kitlv.nl.* **148** 287–308
[10] Rukayah, Siti R 2005 *Satu Lagi Potensi Wisata Sejarah, Temuan Typologi Konsep Alun-alun di Jawa, Kreasi Muslim Cina Anak Buah Cheng Ho*. Pemenang ke 3 Lomba Penulisan Tingkat Nasional Cheng Ho dan Kaitannnya dengan Indonesia, dalam rangka peringatan 600 tahun pendaratan Cheng Ho di Indonesia
[11] Permana RCE 2004 Kajian Arkeologi Mengenai Keraton Surosowan Banten Lama , Banten. *Makara, Sos Humaniora*, **8** 112–119
[12] Gomperts A, *et al* 2008 Stutterheim's enigma The mystery of his mapping of the Majapahit kraton at Trowulan. 1941.*Kitlv.nl.* **4** 411-430
[13] Rukayah, Siti R, Bharoto 2012 Bazaar in Urban Open Space as Contain and Container Case study: Alun-alun Lama and Simpang Lima Semarang, Central Java, Indonesia. *Procedia - Social Behavior Science*, **50** 741–755
[14] Rukayah, Siti R, Respati PD, and Endang, S 2016 Morphology of Traditional City Center in Semarang: Towards Adaptive re- use in urban heritage. Environ *Procedia Social Behaviour.* **1** 109–118
[15] Maziyah S 2012 Kondisi Jawa Tengah Pada Abad VIII Sampai Abad XV M. *Humanika* **15** (9)
[16] Taufik M 2004 Teknik Pendeteksian Garis Pantai Pada Delta Dengan Pendekatan Pengolahan Citra Satelit Resolusi Tinggi. *Pertemuan Ilmiah Tahun I Tek Geod - ITS,* 133–136
[17] Sardiyatmo 2004 Kajian Perubahan Garis Pantai Semarang d engan Foto Udara dengan Pankromatik Hitam Putih. *Ilmu Kelautan* **9** 160–168
[18] Pole CJ 2004 Seeing Is Believing ? *Studies in Qualitative Methodology* **7**

[19] Kuntowijoyo 1994 *Metodologi Sejarah.* Tiara Wacana Yogya
[20] Tio J 2004 *Kota Semarang dalam Kenangan*. Semarang
[21] Joe LT 2004 *Riwayat Semarang*. Hasta Wahana

Total lactic acid bacteria, antioxidant activity, and acceptance of synbiotic yoghurt with red ginger extract (*Zingiberofficinale var. rubrum*)

B A Larasati[1,*], B Panunggal[1], D N Afifah[1], G Anjani[1]and N Rustanti[1]**

[1]Department of Nutrition Science, Medical Faculty of Diponegoro University Semarang, Indonesia
* Presenting author; E-mail: amelarasati@hotmail.com; Tel: 0818-0745-0101
**Corresponding author; E-mail: ninik.rustanti@gmail.com; Tel: 0818-0427-2770

AbstractAntioxidant related to oxidative stress can caused the metabolic disorders. A functional food that high in antioxidant can be use as the alternative prevention. The addition of red ginger extract in yoghurt could form a functional food, that high in antioxidant, synbiotic and fiber. The influence of red ginger extract on yoghurt synbiotic against lactic acid bacteria, antioxidant activity and acceptance were analyzed. This was an experimental research with one factor complete randomized design, specifically the addition of red ginger extract 0%; 0,1%; 0,3% and 0,5% into synbiotic yoghurt. Total plate count method used to analyze the lactic acid bacteria, *1-1-diphenyl-2-picrylhydrazyl* (DPPH) method for antioxidant activity, and acceptance analyzed with hedonic test. The higher the dose of extract added to synbiotic yoghurt, the antioxidant activity got significantly increased (ρ=0,0001), while the lactic acid bacteria got insignificantly decreased (ρ=0,085). The addition of 0,5% red ginger extract obtained the antioxidant activity of 71% and 4,86 x 10^{13} CFU/ml on lactic acid bacteria, which the requirement for probiotic on National Standard of Indonesia is >10^7 CFU/ml. The addition of extract had a significant effect on acceptance (ρ=0,0001) in flavor, color, and texture, but not aroma (ρ=0,266). The optimal product in this research was the yoghurt synbiotic with addition of 0,1% red ginger extract. To summarize, the addition of red ginger extract in synbiotic yoghurt had significant effect on antioxidant activity, flavor, color, and texture, but no significant effect on lactic acid bacteria and aroma.
Keywords: yoghurt, synbiotic, red ginger, lactic acid bacteria, antioxidant activity.

1. Introduction

The National Cholesterol Education Program's Adult Treatment Panel III (NCEP-ATP III), identified metabolic syndrome as a condition in which there are 3 or more of the risk factors such as central obesity, elevated triglycerides, decreased HDL cholesterol, elevated blood pressure, and fasting glucose.[1] Research about the prevalences of metabolic syndrome, in the elderly obtained rates of 18,2% in elderly women and 6,6% in elderly men.[2] Meanwhile, in studies with adolescent subjects, obtained the prevalence by 34% of 50 obese adolescents, aged 10-19 years old.[3]Studies on adolescents with overweight and obesity, obtained rates of 23%.[4]

Metabolic syndrome, can be treated with a regulation of diet, physical activity, and weight.[1] The administration of functional food, in form of synbiotic yoghurt is one of the example of diet regulation. Synbiotic is a combination of probiotic and prebiotic.[5] In correlation with metabolic syndrome, probiotic in form of lactic acid bacteria, can help lower the serum cholesterol levels because of the Bile

IOP Conf. Series: Earth and Environmental Science **116** (2018) 012037 doi:10.1088/1755-1315/116/1/012037

Salt Hydrolase (BSH) enzyme, which deconjugates bile salts that synthesized from cholesterol. In addition, the consumption of probiotic could help lower the blood pressure.[7]

Inulin, which is a prebiotic, known to elevate the growth and viability of lactic acid bacteria.[8] The consumption of inulin has effect on glycemic status and lipid profile, by lowering HbA1c level, fasting blood sugar, total cholesterol, triglycerides, LDL cholesterol, LDL/HDL cholesterol ratio, triglyceride/HDL cholesterol rasio and increasing HDL cholesterol.[9]

Oxidative stress has an important role in the incidence of metabolic syndrome, because the increasing of reactive oxygen species (ROS).[10] Diet high in antioxidants can help to prevent ROS. Red ginger (*Zingiber officinale var. Rubrum*) is high in antioxidants, with phenol content (95,34 mg/100 g) and flavonoid (53,67 mg/100 g), that significantly higher ($p < 0,05$) than white ginger (61,89 mg/100 g and 34,55 mg/100 g).[11] In addition, there was study about the administration of red ginger, which stated that by given 3,2 ml/kg BB of red ginger for 21 days, can lower the cholesterol levels in women with dyslipidemia as much as 8,64%.[12]

Beside the addition of red ginger extract, the diversification of functional food in form of synbiotic yoghurt, can also be done by adding carrageenan. Carrageenan is a soluble fiber, that extracted from red seaweed or *Rhodophyceae*, which can be used for thickening, gelling, and emulsifying.[13] Carragenan also can reduce the level of total cholesterol and LDL, and lower the biomarker for chronic inflammation, which are leukocytes, fibrinogen, and c-reactive protein (CRP).[14] There was a study stated that administration of carrageenan in diet can improve the blood plasma lipid profile in mouse, by lowering the total cholesterol, LDL, and triglycerides, while also elevating the HDL cholesterol.[15] The innovation of functional food in form of synbiotic yoghurt with the addition of red ginger extract (*Zingiber officinale var. rubrum*) and carrageenan, which is the combination of synbiotic, high in antioxidant, and fiber, is expected to be an alternative product that could treat metabolic syndrome. The total lactic acid bacteria, antioxidant activity, and acceptance in synbiotic yoghurt with the addition of red ginger extract will be analyzed.

2. Methods

This was a food production research, that took place in *Laboratorium Terpadu*, Diponegoro University, from August until September 2016. Materials needed to make synbiotic yoghurt with the addition of red ginger extract are fresh cow's milk from Faculty of Animal Husbandry and Agriculture, Diponegoro University, starter *Lactobacillus bulgaricus* and *Streptococcus thermophillus*from PAU Gadjah Mada University, Yogyakarta, red ginger from *Pasar Bulu*, Semarang, inulin from PT DPO Indonesia, carrageenan from UKM Suket Segoro, and stevia from market in Semarang.

This was an experimental research with one complete randomized design, which is the addition of red ginger extract with 3 treatment levels; 0,1%; 0,3% and 0,5% and one control group without the addition of red ginger extract. Repitition did 3 times for calculating total lactic acid bacteria and antioxidant activity, and no repitition for acceptance.

A prelimenary research was conducted to determine the pecentage of stevia and carrageenan added to synbiotic yoghurt, whereas the addition of inuline was based on the research of functional beverage product jelly drink yoghurt srikaya. Extraction of red ginger conducted on prelimenary research, with maceration method for 36 hours using 95% ethanol as much as 125 mL. The 95% ethanol need to be replaced with another 125 ml of 95% ethanol every 12 hours during the maceration process.

The process of making synbiotic yoghurt with the addition of red ginger extract, begins with the pasterurization process of cow's milk up to ± 70°C and keep at that temperature for 15 seconds, then cooled down to 40°C. After that, added 4% inulin, 0,2% stevia, and 10% starter*Lactobacillus bulgaricus* and *Streptococcus thermophillus* to the cow's milk, then incubated for approximately 24 hours using an incubator at 37°C. Last, added 6% carrageenan and red ginger extract to the yoghurt.

In this study, the test used for determine the total lactic acid bacteria was Total Plate Count (TPC) method, DPPH method for antioxidant activity, and hedonic test for acceptance, on 30 panelist semi-trained that were students from Nutrition Sciences, Diponegoro University. The normality test used was Shapiro-Wilk, because the sample size was <30. The total lactic acid bacteria used One way Anova test. Antioxidant activity used One Way Anova, followed by Duncan test. Acceptance used Friedman, followed by Wilcoxon test.

3. Results

3.1 Total Lactic Acid Bacteria (LAB)

The addition of red ginger extract in synbiotic yoghurt, had no significant effect on total lactic acid bacteria (ρ = 0,085). However, as the dose of red ginger extract increased, there were some decreased on total lactic acid bacteria in synbiotic yoghurt all tables, figures and illustration should be written in the main text.

Table 1. Total Lactic Acid Bacteria (LAB) (should be refers in the main text)

Dose	Total LAB (10^{13}CFU/ml)[a]
0%	90,37±65,02
0,1%	81,65±23,68
0,3%	27,19±41,35
0,5%	4,86±6,02
	ρ = 0,085

[a] = One Way Anova Test

3.2 Antioxidant Activity

The higher the dose of red ginger extract added to synbiotic yoghurt, the antioxidant activity also got higher. The addition of red ginger extract had significant effect on antioxidant activity with ρ = 0,0001. The highest antioxidant activity was the synbiotic yoghurt with the 0,5% addition of red ginger extract.

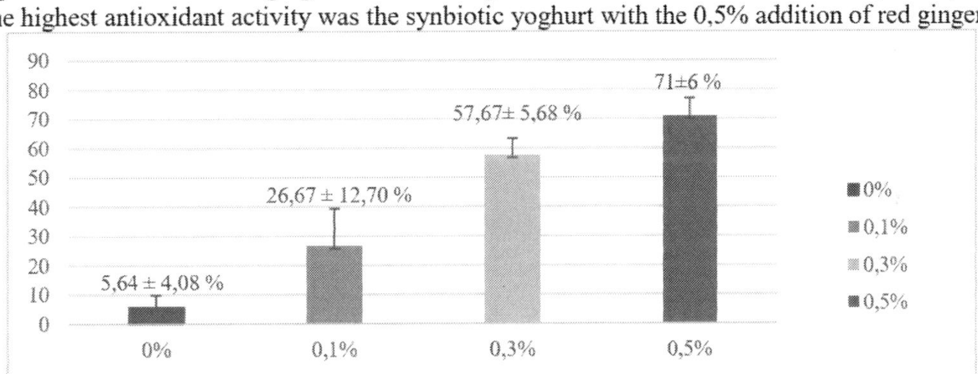

[1] = One Way Anova Test

Graphic 1. Antioxidant Activity should be mentioned in the text

3.3 Acceptance

Acceptance was done with hedonic test, in terms of flavor, color, aroma, and texture of synbiotic yoghurt with the addition of red ginger extract.

Table 2. Acceptance on Synbiotic Yoghurt with the Addition of Red Ginger Extract

Dose	Flavor[1]		Color[1]		Aroma[1]		Texture[1]	
	Mean	Exp	Mean	Ket	Mean	Exp	Mean	Exp
0%	2,80±0,66[a]	Like	3,10±0,61[a]	Like	3,07±0,74	Like	2,80±0,71[c]	Like
0,1%	2,77±0,77[b]	Like	3,03±0,49[b]	Like	2,80±0,71	Like	3,07±0,69[a]	Like
0,3%	2,30±0,75[c]	Dislike	2,37±0,62[d]	Dislike	2,77±0,77	Like	2,93±0,58[b]	Like
0,5%	1,80±0,66[d]	Like	2,80±0,66[c]	Like	2,80±0,76	Like	2,30±0,79[d]	Like
	ρ = 0,0001		ρ = 0,0001		ρ = 0,266		ρ = 0,0001	

Explanation: Numbers followed by different superscript letter (a,b,c,d) showed the real different.

[1]= Friedman Test

*3.3.1 Flavor.*Synbiotic yoghurt with the addition of red giger extract had significant effect on acceptance in term of flavor (ρ = 0,0001).As the dose get higher, there were some decreases on panelists' acceptance in flavor, whereas the synbiotic yoghurt with doses of red ginger extract as much as 0,3% and 0,5% were not preferred.
3.3.2 Color. The addition of red ginger extract on synbiotic yoghurt had significant effect on acceptance in color (ρ = 0,0001). Panelists liked all the color of synbiotic yoghurt with the addition of red ginger extract, except the 0,3% one.
3.3.3 Aroma. In term of aroma, there were no significant effect on synbiotic yoghurt with the addition of red ginger extract (ρ = 0,266). Panelists liked the aroma of synbiotic yoghurt with the addition of red ginger extract in every dose.
3.3.4 Texture. The most liked yoghurt in term of texture were the synbiotic yoghurts with the addition of red ginger extract of 0,1% and 0,5%. The addition of red ginger extract had significant effect on texture of synbiotic yoghurt (ρ = 0,0001). Panelists liked all the texture of synbiotic yoghurt except the one with o,5% red ginger extract addition.

4. Discussion

4.1 Total Lactic Acid Bacteria

The addition of red ginger extract in synbiotic yoghurt, had no effect on total lactic acid bacteria. However, there was a decrease on total lactic acid bacteria from synbiotic yoghurt, as the dose of red extract get higher. Despite the degression, the amount of lactic acid bacteria in synbiotic yoghurt with the addition of red ginger extract, still fulfilled the amount of probiotic needed acording to *Standar Nasional Indonesia*(SNI), in order to have benefits on health (>10^7 CFU/mL)

The degression of total lactic acid bacteria, might happened because of the antibacterial activity, possessed by ginger. The antibacterial activity, derived from the chemical components in ginger, which are *sesquiterpenoid* with *zingiberene* as its main component. There are also *β-sesquiphellandrene, bisabolene*, and *farnesene*, which are still part of *sesquiterpenoid*.[17]

The decrease in total lactic acid bacteria, as the dose of red ginger extract get higher, is compatible with another study about the addition of ginger juices in various levels into yoghurt, which proved that as the dose of ginger juice added to yoghurt got higher, the growth of S. thermophilus and L. bulgaricus would be more inhibited.[18] In other fermented product with addition of ginger extract, there was also a decrease on total lactic acid bacteria.[19]

The strength level of antibacterial activity in ginger, also influenced by the treatment process on ginger. In this study, the extraction process of ginger used maceration method with 95% ethanol as the solvent. Extraction with ethanol as the solvent, will provide a stronger antibacterial activity, compared to the extraction with water as the solvent.[17] Before the maceration process, red gingers used in this study were sun-dried. There were no significant difference between using fresh ginger or dried ginger, in related to total lactic acid bacteria. The significant difference or decrease in total lactic acid bacteria, more related to the dose of ginger extract added to product.[20]

4.2 Antioxidant Activity

The addition of red ginger extract had a significant effect on antioxidant activity in synbiotic yoghurt (ρ = 0,0001). The higher the dose of red ginger extract, the higher the antioxidant activity would be. The increasing of antioxidant activity could happen because ginger has some main components, which are 6-*gingerol*, 6-*shogaol*, and 10-*gingerol*, that have strong antioxidant activities.[21]

In the prelimenary study, it is known that the red ginger extract had a very high antioxidant activity, which was 92,5%. It was compatible with another study about the high antioxidant activity in ginger, which was 90,1%, whereas the ginger is dried and also extracted with ethanol as the solvent.[22],[23] Eventhough the value of antioxidant activity in red ginger extract was pretty high, when it added to the synbiotic yoghurt, the antioxidant activity value got decreased compared to the raw ginger extract.

Each ingredients in synbiotic yogurt has antioxidant activity. Lactic acid bacteria got the antioxidant from proteolitic activity.[24],[25] Carrageenan, which extracted from red seaweeds (*Rhodophyta*), also got natural antioxidant.[26] Inulin got the antioxidant from its bioactive component, synthesisized by oligosaccharides [27] and stevia also got antioxidant value.[28] The differences between bioactive components, lead to differences in antioxidant activity. Those different bioactive component, either can cause synergetic or antagonistic effects. In this study, combination of bioactive components in synbiotic yoghurt with the addition of red ginger extract, caused antagonistic effect, indicated by the decrease in antioxidant activity.[29]

The higher the dose of red ginger extract added to synbiotic yoghurt, the antioxidant activity got increased, but compared to the antioxidant activity in raw ginger extract, it got decreased. The result was compatible with one study about jelly drink product with the addition of ginger essence. There was an increased in antioxidant activity of jelly drink, as the concetration of ginger essence got higher, but compared to the antioxidant activity value in raw ginger essence, it got decrease.[30]

4.3 Acceptance

4.3.1 Flavor. As the dose of red ginger extract got increased, the acceptance level in term of flavor got decreased. Panelists described the flavor of synbiotic yoghurt with the addition of red ginger extract as spicy and strong. The flavor of synbiotic yoghurt became bitter with the addition of 0,3% and 0,5% red ginger extract, that might caused the decrease in acceptance of flavor.

Synbiotic yoghurt has a sour taste, that happened because fermentation process of lactose that becomes lactic acid, by starter *Lactobacillus bulgaricus* and *Streptococcus thermophillus*, that causes thesour taste in yoghurt.[31] The spicy and strong flavor in synbiotic yoghurt with the addition of red ginger extract, came from the components named *gingerol* and *shogaol* in ginger.[32] *Gingerol* is the main component in ginger that caused the spicy flavor, because of the [6]-*gingerol* (1-{4'-*hydroxy*-3'-*methoxyphenyl*}-5-*hydroxy*-3-*decanone*), the liquid oil that contained in *gingerol*.[33]

The red gingers that used in this study were sun-dried. The drying process can bring out the spicy flavor, that comes from *shogaol*, which is the nonvolatile phenylpropanoid-derived component from *gingerol*. Beside those two components, there also a component named zingerol which produced by *gingerol*, that brings out the slightly spice taste.[33]

For that reasons, the higher the dose of red ginger extract added to synbiotic yoghurt, the flavor of synbiotic yoghurt would become more spicy and bitter, that might lead to decrease in acceptance level of flavor.

4.3.2 Color. The addition of red ginger extracthad significant effect on the color of synbiotic yoghurt. The color of synbiotic yoghurt with the addition of red ginger extract was pale yellow, and the more of extract added to synbiotic yoghurt, the color would become more yellow-ish. The yellow collor came from the pigment in ginger.[18]

There are 3 main components of pigment that bring out the yellow in ginger, which are *6-dehydrogingerdione*, *curcumin*, and *demethoxycurcumin*, with 6-*dehydrogingerdione*as the main pigment, that becomes the base of pale yellow color in ginger. 6-*dehydrogingerdione*oxidized from 6-*gingerol*, which is the main component in ginger. The more of 6-*gingerol* contained in an extract, the more of 6-*dehydrogingerdione*will be, so the color will become more popped.[34]

In term of color, the product that got dislike from panelists was the synbiotic yoghurt with the the addition of 0,3% red ginger extract. That migt happened because the red ginger extract, in form of liquid gel, not mixed throughtully into synbiotic yoghurt, so the color didn't perfectly showed.

4.3.3 Aroma. The addition of red ginger extract had no correlation with the acceptance level of aroma in synbiotic yoghurt ($\rho = 0{,}266$). Panelists like the synbiotic yoghurt added with red ginger extract, in all kind of doses. There were no significant differences of aroma in every dose added to synbiotic yoghurt. Red ginger has its own signature aroma that comes from volatile oil, which contained a lot of *sesquiterpenoids*, with *α-zingibereneas* its main component.[35]

4.3.4 Texture. The acceptance level of texture, influenced by the addition of red ginger extract into synbiotic yoghurt ($\rho = 0{,}0001$). In term of texture, the only product that got dislike from panelists was the synbiotic yoghurt with 0,5% addition of red ginger extract. It might happened because the higher the dose of extract added to synbiotic yoghurt, the water content would increase that might affect the consistency of synbiotic yoghurt. This result was compatible with another study of organoleptic in yoghurt with the addition of ginger juice, whereas the panelists' acceptance of texture got decreased along with the increased of the dose of ginger juice added into synbiotic yoghurt.[18]

5. Conclusion (please add more fact in conclusion part)

The addition of red ginger extract in synbiotic yoghurt, had no effect on lactic acid bacteria and aroma, but had effects on antioxidant activity, and acceptance in flavor, color, and texture. The optimal product from this research was the synbiotic yoghurt with the addition of 0,1% red ginger extract.

6. Acknowledgement

The writer would like to thank*Riset Pengembangan dan Penerapan PNBP*Diponegoro University for the financial support and for all who might involved in this research.

References

[1] National Cholesterol Education Program Expert Panel. Third Report of the National Cholesterol Education Program (NCEP) Expert Panel on Detection, Evaluation, and Treatment of High Blood Cholesterol in Adults (Adult Treatment Panel III) final report. Circulation 2002; 106: 3143–421.

[2] Kamso S. Body mass index, total cholesterol, and ratio total to HDL cholesterol were determinants of metabolic syndrome in the Indonesian elderly. Medical Journal of Indonesia. 2007; 16(3): 195–200.

[3] Pulungan AB, Puspitadewi A, Sekartini R. Prevalence of insulin resistance in obese adolescents. Paediatrica Indonesiana. 2013; 53(3): 167–172.

[4] Malonda AA, Tangklilisan HA. Comparison of metabolic syndrome criteria in obese and overweight children. Paediatrica Indonesiana. 2010; 50(5): 295–299.

[5] Ooi L-G, Liong M-T. Cholesterol-lowering effects of probiotics and prebiotics: a review of in vivo and in vitro findings. International Journal of Molecular Sciences. 2010; 11: 2499–522.

[6] Kumar M, Nagpal R, Kumar R, Hemalatha R, Verma V, Kumar A et al. Cholesterol-lowering probiotics as potential biotherapeutics for metabolic diseases. Experimental Diabetes Research. 2012; 902917: 1-14

[7] Khalesi S, Sun J, Buys N, Jayasinghe R. Effect of probiotics on blood pressure: a systematic review and meta-analysis of randomized, controlled trials. Hypertension. 2014; 64: 897–903.

[8] Oliveira RPDS, Perego P, Oliveira MN De, Converti A. Effect of inulin as prebiotic and synbiotic interactions between probiotics to improve fermented milk firmness. Journal of Food Engineering. 2011; 107: 36–40.

[9] Dehghan P, Pourghassem Gargari B, Asgharijafarabadi M. Effects of high performance inulin supplementation on glycemic status and lipid profile in women with type 2 diabetes: a randomized, placebo-controlled clinical trial. Health Promotion Perspectives. 2013; 3(1): 55–63.

[10] Bonomini F, Rodella LF, Rezzani R. Metabolic Syndrome, Aging and Involvement of Oxidative Stress. Aging and Disease.2015; 6(2): 109–120.

[11] Oboh G, Akinyemi AJ, Ademiluyi AO. Antioxidant and inhibitory effect of red ginger (*Zingiber officinale var. Rubra*) and white ginger (*Zingiber officinale Roscoe*) on Fe(2+) induced lipid peroxidation in rat brain in vitro. Experimental and Toxicologic Pathology. 2012; 64: 31–6.

[12] Sari RP, Rahayuningsih HM. Pengaruh Pemberian Jahe Merah (Zingiber officinale var. rubrum) Terhadap Kadar Kolesterol Total Wanita Dislipidemia [Tesis]. Semarang: Universitas Diponegoro; 2014.

[13] Necas J, Bartosikova L. Carrageenan : a review. Veterinarni Medicina. 2013; 2013(4): 187–205.

[14] Sokolova EV, Bogdanovich LN, Ivanova TB, Byankina AO, Kryzhanovskiy SP, Yermak IM.

Effect of carrageenan food supplement on patients with cardiovascular disease results in normalization of lipid profile and moderate modulation of immunity system markers. PharmaNutrition. 2014; 2: 33–37.

[15] Subroto T. Efek anti hiperkolesterolemik karagenan rumput laut dalam diet terhadap plasma lipid tikus putih [Tesis]. Bandung: Universitas Padjadjaran; 2011.

[16] Suryani L. Optimasi metode ekstraksi fenol dari rimpang jahe emprit (*Zingiber officinale var. Rubrum*). Jurnal Agrisains. 2012; 3: 58–65.

[17] Malu SP, Obochi GO, Tawo EN, Nyong BE. Antibacterial Activity and Medicinal Properties of Ginger (*Zingiber officinale*). Global Journal of Pure and Applied Sciences. 2009; 15(3): 365–368.

[18] Yang GH, Guan JJ, Wang JS, Yin HC, Qiao FD, Jia F. Physicochemical and sensory characterization of ginger-juice yogurt during fermentation. Food Science Biotechnology. 2012; 21: 1541–1548.

[19] Okwute L., Olafiaji B. The effects of ginger (*Zingiber officinale*) on the microbial load of a nigerian traditionally fermented maize paste (ogi). American Journal of Research Communication. 2013; 1: 84–98.

[20] Abd El-Khalek AB, El-Sayed HS, Ibrahim GA, El-Shafei K, El-Din HMF, Sharaf OM *et al.* Phenolic compounds, microbial content and sensory evaluation of synbiotic labneh conaining ginger and probiotic. International Journal of ChemTech Research. 2016; 9(2): 238–247.

[21] Maizura M, Aminah A, Wan Aida WM. Total phenolic content and antioxidant activity of kesum (*Polygonum minus*), ginger (*Zingiber officinale*) and turmeric (*Curcuma longa*) extract. International Food Research Journal. 2011; 18: 526–531.

[22] Stoilova I, Krastanov A, Stoyanova A, Denev P, Gargova S. Antioxidant activity of a ginger extract (*Zingiber officinale*). Food Chemistry. 2007; 102: 764–770.

[23] Li Y, Hong Y, Han Y, Wang Y, Xia L. Chemical characterization and antioxidant activities comparison in fresh, dried, stir-frying and carbonized ginger. Journal of Chromatography B. 2016; 1011: 223–232.

[24] Sah BNP, Vasiljevic T, McKechnie S, Donkor ON. Effect of probiotics on antioxidant and antimutagenic activities of crude peptide extract from yogurt. Food Chemistry. 2014; 156: 264–270.

[25] Şanlıdere Aloğlu H, Öner Z. Determination of antioxidant activity of bioactive peptide fractions obtained from yogurt. Journal of Dairy Science 2011; 94: 5305–5314.

[26] Sun Y, Yang B, Wu Y, Liu Y, Gu X, Zhang H *et al.* Structural characterization and antioxidant activities of κ-carrageenan oligosaccharides degraded by different methods. Food Chemistry. 2015; 178: 311–318.

[27] Stoyanova S, Geuns J, Hideg É, Van Den Ende W. The food additives inulin and stevioside counteract oxidative stress. International Journal of Food Sciences and Nutrition. 2011; 62(3): 207–214.

[28] Geuns JMC, Hajihashemi S, Claes A. Radical-scavenging by steviol glycosides and crude Stevia extracts. In: Stevia: Months beyond authorization. Proceedings of the 6th EUSTAS Stevia symposium. Euprint: Belgium.2012. 157–180.

[29] Gupta RK, Chawla P, Tripathi M, Shukla AK, Pandey A. Synergistic antioxidant activity of tea with ginger, black pepper and tulsi. International Journal of Pharmacy and Pharmaceutical Sciences. 2014; 6(5): 477–479.

[30] Febriyanti S, Yunianta Y. Pengaruh konsentrasi karagenan dan rasio sari jahe emprit (*Zingiber officinale var. Rubrum*) terhadap sifat fisik, kimia, dan organoleptik jelly drink jahe. Jurnal Pangan dan Agroindustri. 2014; 3: 542–550.

[31] Jannah AM, Legowo AM, Pramono YB, Al-baarri AN, Abduh SBM. Total Bakteri Asam Laktat, pH, Keasaman, Citarasa dan Kesukaan Yogurt Drink dengan Penambahan Ekstrak Buah Belimbing.Jurnal Aplikasi Teknologi Pangan 2014; 3:7-11.

[32] Yeh H yu, Chuang C hung, Chen H chun, Wan C jen, Chen T liang, Lin L yun. Bioactive

components analysis of two various gingers (*Zingiber officinale Roscoe*) and antioxidant effect of ginger extracts. LWT - Food Science and Technology. 2014; 55: 329–334.

[33] Srinivasan K. Ginger rhizomes (*Zingiber officinale*): A spice with multiple health beneficial potentials. PharmaNutrition. 2017; 5: 18–28.

[34] Iijima Y, Joh A. Pigment Composition Responsible for the Pale Yellow Color of Ginger (*Zingiber officinale*) Rhizomes. Food Science and Technology Research. 2014; 20: 971–978.

[35] Ali BH, Blunden G, Tanira MO, Nemmar A. Some phytochemical, pharmacological and toxicological properties of ginger (*Zingiber officinale Roscoe*): A review of recent research. Food and Chemical Toxicology. 2008; 46: 409–420.

Edco-tourism; A Coastal Management Program to Improve Social Economics

Arsi Rakhmanissazly[1], Anggun Intan Permatasari [1,2], Ely Chandra Peranginangin[1]

[1] Department of Legal & Relations, PT PERTAMINA EP Asset 1 Pangkalan Susu Field, Langkat, North Sumatera, Indonesia 20858
[2] Department of Social Development and Welfare, Faculty of Social and Political Sciences, Gadjah Mada University, Yogyakarta, Indonesia 55281

Email: arsi.rakhmanissazly@pertamina.com,

Abstract. Coastal erosion is a natural process even in pristine environments. Mangrove is tropical coastal vegetation that may prevent and controls the soil erosion. As the longest coastline country in the world, Indonesia has great ecological potential of mangrove. Sadly, according to Food and Agriculture Organization (FAO) since 2007 Indonesia has lost 40% of mangrove forest because of deforestation for the last three decades. Langkat regency placed as the most destructive mangrove forests in North Sumatera about 25% of total area hectares due to severe damage. Green Belt Lubuk Kertang is a Corporate Social Responsibility (CSR) program held by PT PERTAMINA EP Asset 1 Pangkalan Susu Field (PEP) aims to overcome deforestation problem by implementing the concept of edco-tourism mangrove. Edco-tourism is a concept that combines educational and environmental as the basis of local tourism management. By implementing this concept PEP may create multiplier effect not only by saving environmental issues but also socio - economical for local communities. This study shows the impacts of the program for some aspects. By using Shannon – Wienner indicator shows the biodiversity index has increased since the program began three years ago. Not only biodiversity number but also group's income of local community increase obtained from the retributions. Furthermore, this program is social investment for PEP to engage company – community relations during operational by gaining license to operate. PEP attempt to implement integrated coastal zone management to balance environmental, technical, socio – economic and recreational objectives, into an eco – coastal sustainable system.
Keywords: *corporate social responsibility, coastal management, edco-tourism, environment, socio-economic, sustainable development.*

1. Introduction

As the biggest archipelago's country in the world, Indonesia has more than 90.000 km coastlines with enormous undiscovered of coastal potential. About 14 millions of Indonesia's people or at least 7,5% from total rely their livelihoods in coastal areas [1]. According to Sorensen and McCreary in Institutional Arrangement for Managing Coastal Resources and Environments, coastal zone is the borders where two main environments, the land and the water, elaborated. This definition is in line with UU No 27/2007 about Coastal and Small Islands Management. The supply of nutrients from the mainland through river and surface water makes the coastal area as a complex natural ecosystem. One of the ecosystems is mangrove that performs as production, protection and preservation area.

About 3 million hectares of mangrove forest grow along Indonesia's coastline. This is 23 percent of all mangrove ecosystems in the world [2]. Ecologically, Mangroves perform several functions such as inundation control, protection from erosion, storm, floods, and wave damage, recreation/ tourism, *and generate tangible goods (such as fishes, shellfishes and other forest products).* They possess distinct ecological attributes at multiple levels of organization such as site or stand, ecosystem and landscape levels [3]. Mangroves create double impact for the environment, as below the water as breeding and nursing grounds for fish and marine species, as above the water the mangrove trees and canopy serves as a habitat for birds, insects, mammals and reptiles also create economic value by community surround.

Unfortunately, over the past three decades, *according to FAO (Food and Agriculture Organization)* Indonesia has lost 40 percents of its mangroves. *caused by the deforestation, conversion of mangroves area into ponds, settlements, industries and plantations.* This means Indonesia has the fastest rate of mangrove destruction in the world [4]. Mangrove deforestations in Indonesia result in a loss of 190 million metric tonnes of CO_2 annually. This accounts for 20 percents of land – use emissions in Indonesia [5] where emissions are estimated to be 700 million metric tonnes of CO2-eq [6]

This problem also happens in Lubuk Kertang, Langkat Regency, North Sumatera. In Early 2000s, there are about 80 hectares of mangroves converted into palm oil which made the fishermen loss their livelihoods. About only 25 percents of mangrove forest in Langkat categorized in good condition and the other 75% are damaged and rehabilitation phase. [7] It made Langkat as the most destructive mangroves forest in North Sumatera. As a company which has main working area in Langkat, PT PERTAMINA EP Asset 1 Pangkalan Susu Field (PERTAMINA EP) consequently designed some strategic programs to answer the problems of the society by utilizing some local potentials people barely realized. Green Belt Lubuk Kertang is one of Corporate Social Responsibility program held by PERTAMINA EP which aim to overcome deforestation issues by implementing Edco-tourism Mangrove concept. This concept is not only to promote mangroves tourism; but also to make positive and direct contribution between visitors and the local mangrove guardian community (named 'MEKAR')

The combination of triple bottom of CSR (people – planet – profit) mixed into a strategic coastal management program. PERTAMINA EP realizes that being a part of the community may become social investment for the company in the future. PERTAMINA EP sees the problem of land converting will harm the community, coastal ecosystem, natural environment and also indirectly may become as obstacles for the operation of PERTAMINA EP. By implementing this program, PERTAMINA EP creates multiplier effects which improve community in economical and environment benefits. Intangible benefits of mangrove ecosystems include social and cultural functions. Mangrove forests support religious and spiritual values, as well as recreational and aesthetic values for edco-toursim [8].

2. Coastal Management Program: Edco-tourism Green Belt Lubuk Kertang

Coastal management is a coordinated strategy to undertake coastal problem and emerging issues such land erosion, public welfare, nature resources and also socio – cultural to minimize the negative impacts and assist coastal society to preserve the environment and create sustainable living based on coastal zone. On the other hand, CSR has been defined in many different ways. At bottom, it refers to companies integrating "social and environmental concerns in their business operations and in their interaction with their stakeholders on a voluntary basis (…) not only fulfilling legal expectations but also going beyond compliance [9]. Green Belt Lubuk Kertang is PERTAMINA EP's social strategy addressing deforestation issues. PERTAMINA EP elaborates the concept of coastal management with CSR itself. Developing the coastal management system in this program by cooperate not only with the community but also government agencies. (see Fig 2.1)

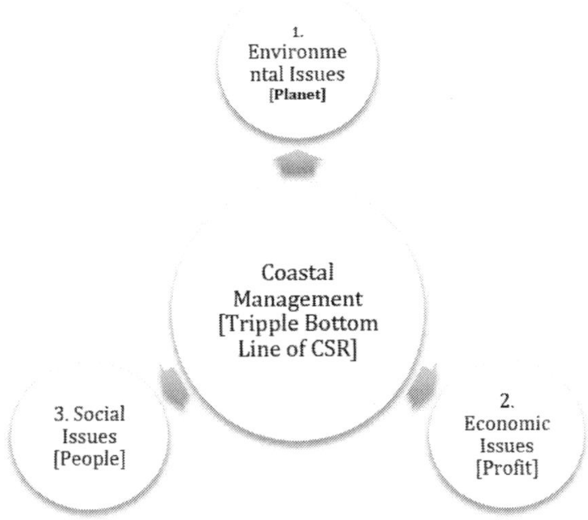

Figure 1. Correlation between Coastal Management and Triple Bottom Line of CSR

There are similarities between the basic concept of CSR and coastal management following elements (Fig 1.1). Through CSR program, PERTAMINA EP takes the initiatives to implement community development strategy based on local wisdom and its potential to achieve the well–being of local communities. Basically, there are three main points that PERTAMINA EP concerns: (i) nature conservation, (ii) ecological education and (iii) economic improvement. Corporate social responsibility (CSR) program can become a source of tremendous social progress, as the business applies its considerable resources, expertise and insights to activities that benefit society [10].

2.1 What is Edco-tourism?

The Edco-tourism (education in eco tourism) concept basically based on potential tourism resources. According to The International Ecotourism Society (TIES:2000), "Ecotourism is not only a tourist, but also a responsible natural tourism trip by conserving the environment and improving the welfare of local communities." PERTAMINA EP acquaint the interactive learning conservation for visitors in participating planting mangrove seeds and cultivation of natural ponds. Not only educate but also preserve mangrove vegetation the multiplier effects from the implementation of this program.

IOP Conf. Series: Earth and Environmental Science **116** (2018) 012038 doi:10.1088/1755-1315/116/1/012038

3. Methodology

Social mapping defined as society description systemic processes with conduct data collection and society's information including profiling and social problem in the specific area. According to Netting, Kettner and McMurtry (1993) social mapping can also defined as social profiling. Social mapping is one of many approaches in community development as Twelvetrees [11] said that social mapping is the process of assisting ordinary people to improve their own communities by undertaking collective actions [12]. We're using social mapping research as our based document before creating strategic plan.

Figure 2. Social Mapping Formed

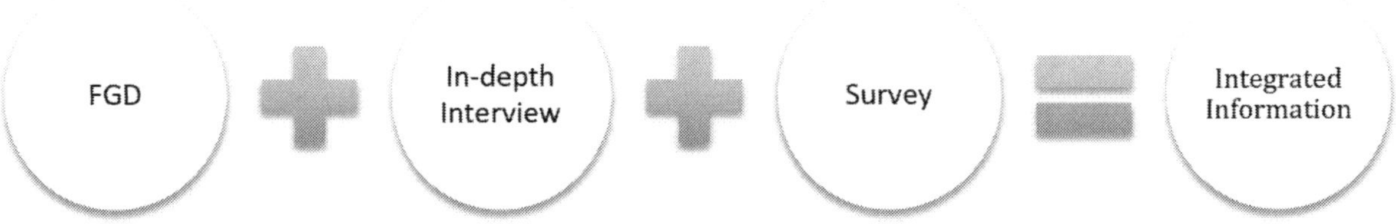

As the result of social mapping and needs assessment PERTAMINA EP will get the recommendation program which we use as groundwork of our CSR program in our main working area.

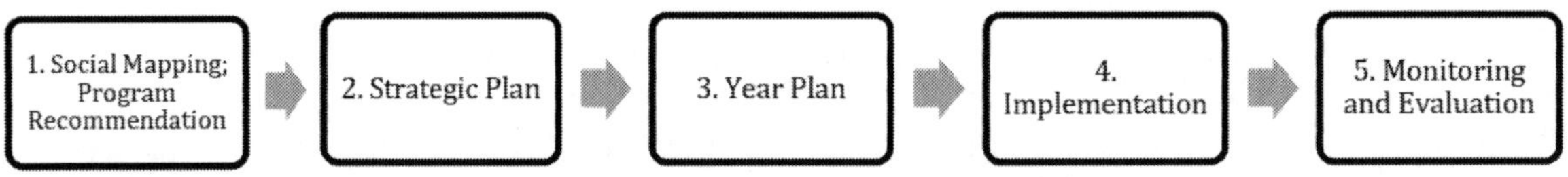

Figure 3. Process of CSR Program

3.1 Method of Effects Measurements

The implementation of Edco-tourism program creates multiplier effects that involve three main aspects.

3.1.1 Environment Effects

Some theories and approaches are used to measure environmental effects. We divided environment category into three main aspects; mangrove, local fauna and the whole conservation area. This research also used as our biodiversity report that describe how PERTAMINA EP practice to protecting environment issues during operational.

3.1.2 Economy Effects

Using quantitative approach by collecting visitors data each month to see total visitors and the revenue that MEKAR get from managing the program. The visitors and revenue graphic used to monitor and evaluate the program to develop the next development strategy.

3.1.3 Social Effects

Public Satisfaction Index (IKM) is data and information derived from costumer satisfaction, in this case community, using two research approaches. Community survey for the quantitative measurements and in-depth interview for the qualitative. This method considering community opinion as database to compare the expectations and needs about the existing CSR program.

To measure these aspects, we used different method and indicators as classified in **Table 1** as seen below:

Table 1. Classification of Effects Measurements Method

No	Effects Category	Indicator	Method	Output
1	Environment	Mangrove vegetation inventory	Systemic Cluster Analisis	Shannon Wienner Index
		Reptile and Aves Inventory	Count Conservation Analysis	Shannon Wienner Index
		Insects Inventory	Systemic Cluster Analisis	Shannon Wienner Index
		Life below water Inventory	Systemik Trap Identification	Shannon Wienner Index
		Conservation Area	GPS tematic overlay map	Biodiveristy index
2	Economy	Visitors	Amount of visitors	Visitors graphic
		Income	Financial report	Financial graphic
3	Sosial	Program congeniality	Public Satisfaction Index (IKM)	Quality of services

4. Results

At first, Green Belt Lubuk Kertang program is expected to help restore critical land due to deforestation. Programs that initially to save the environment only now develop and creating multiplier effects for the economic and social conditions of the community. This may give the community chance to improve their socio economic by preserving the nature and their coastal local potential for living. PERTAMINA EP also trying to improve the relationship between stakeholders by creating social stability and gain license to operate from the community surround.

4.1.1 Environment Effects

At the beginning of the program, PERTAMINA EP commits to plant mangroves gradually. PERTAMINA EP started to plant mangroves in 2014-today, we have rescued the critical land through mangrove plantation, with the success ratio about 90% based on Mangrove Forest Management Center (BPHM) research. The areas for the program of Green Belt Lubuk Kertang are Lubuk Kertang as the critical area about 8 hectares and 19.5 hectares in Paluh Tabuan Timur. [Details of indicators and results of Shannon Wienner Index for every indicator can see in these tables below:

Tabel 2. Mangrove Vegetation Inventory

No			**Vegetation Analysis**	**Buffer Level**	**Seeding Level**
	Location	Zone Critera	Shannon Wienner Index (H')		
1	Paluh Tabuhan Timur (19 Ha)	Main Zone	0,13	0	0,63
		Buffer Zone	0,08	0,4	0,56
		Used Zone	0,2	0,9	1,03
2	Lubuk Kertang (8 Ha)	Main Zone	1,9	1,24	0,8
		Buffer Zone	-	-	-
		Used Zone	-	-	-

From **Table 2** The highest Shannon Wienner Index (H') only remains in buffer level of Lubuk Kertang Main Zone 1,24 categorized as medium diversity. That shows that other zone has low diversity caused of unstable ecosystem, high ecology pressure, with low productivity. It makes sense, because to restore heavily damaged ecosystem requires a long period of time.

Table 3. Insects Inventory

No	Location	Number of Ordo	Number of Famili	Shannon Wienner Index (H')
1	Paluh Tabuhan Timur (19 Ha)	7	16	2,48
2	Lubuk Kertang (8 Ha)	8	26	3,02

On the other hands, from **Table 4.2** The Shannon Wienner Index (H') for insects inventory shows that Paluh Tabuhan Timur and Lubuk Kertang has a good environmental hygiene. Because the 2,48 and 3,02 categorized as high diversity and the diversity of insects is a positive bioindicator for potential ecosystem.

Table 4. Reptile and Aves Inventory

No	Location	Observation Time	Number of Species	Clasification Status	Shannon Wienner Index (H')
1	Paluh Tabuhan Timur	Day	7	Normal	1,40
			2	Endanger	
		Night	8	Normal	1,48
			1	Endanger	
	Total Species		**18**		
2	Lubuk Kertang	Day	6	Normal	1,58
			3	Endanger	
		Night	5	Species	0,61
	Total Species		**14**		

Table 4, shows that some species are classified in endanger status and by managing a good coastal management program and conserve their habitat may help these species to evolve. The endanger animals are Otter (*Enhydra lutris*), Black Stork (*Ciconia episcopus*), Long-tail Monkey (*Macaca fascicularis*), Falcon (*Haliastur indus*), and Gull (*Tringa hypoleucos*).

Table 5. Life below water inventory

No	Location	Number of Species	Shannon Wienner Index (H')
1	Paluh Tabuhan Timur	12	0,62
2	Lubuk Kertang	24	0,74

Table 5shows that low categorization of Shannon Wienner Index in both places. The damaged caused by deforestation in mangroves land also give high impact for the life below water.

Table 6. Mangrove Conservation Area

No	Years of observation	Biodiversity Index	
		Paluh Tabuhan Timur	Lubuk Kertang
1	2012	3,173	3,173
2	2013	3,173	3,173
3	2014	3,607	3,607
4	2015	3,902	3,902
5	2016	3,949	3,980

There is an increasing number for mangrove biodiversity in conservation area in **Table 6.** Though is still in small number but the environment keep improving and evolving. In 2016, we observed that the mangrove biodiversity has increased from 2013 to 2016. In Paluh Tabuhan Timur, biodiversity index increased from 3.173 to 3.95. It also happens in Lubuk Kertang, it increased from 3.173 to 3.98 in 2016. Another impatcs by implementing this program is reduction of carbon (CO_2) emission from mangroves zone with total 16, 119 ton / Ha eq.

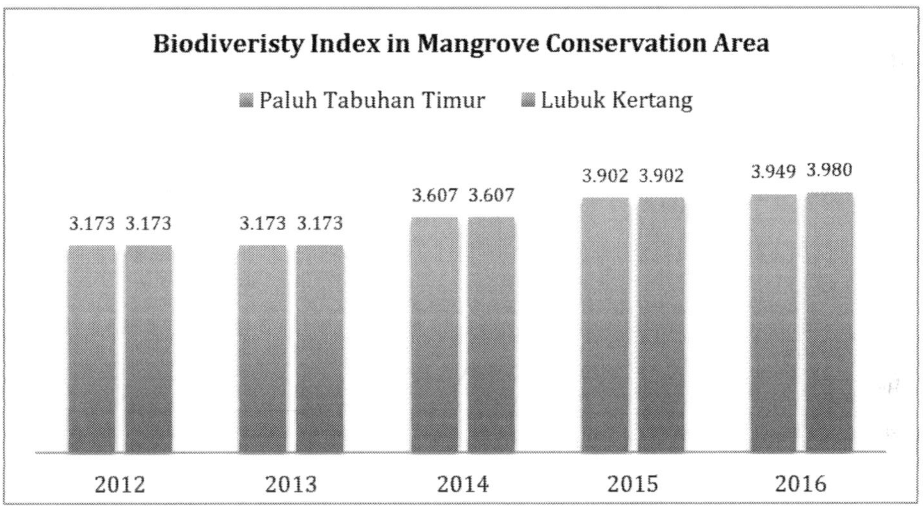

Figure 4. Biodiversity Index in Mangrove Conservation Area

4.1.2 Economic Effects

Since the implementation of development Edco-tourism program in the beginning of 2017 to April not less than 1000 visitors came to the site. From the retribution, at least IDR 15,000,000/month become the earning of MEKAR's group. This condition also contributes to increase the income of every member from IDR 300,000 to IDR 1,000,000/month. The graphics below are the recordable data from Edco-tourism management. (see **Fig 4.1.2** and **Fig 4.1.3**)

Figure 5. Visitors Graphic

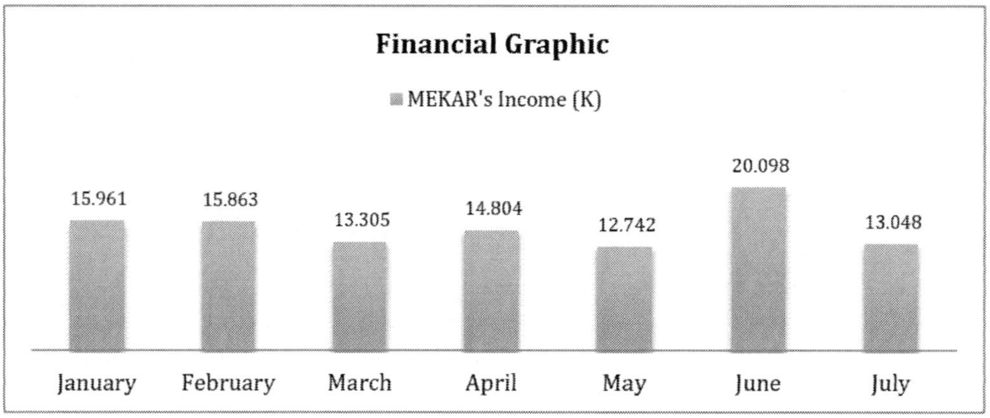

Figure 6. Visitors Graphic

4.1.3 Social Effects

Not only services but also program evaluation are needed to create a sustainable development of coastal management. Using Public Satisfaction Index (IKM) PERTAMINA EP try to elaborate with public expectation about the company.

Table 7. Public Satisfaction Index of Green Belt Lubuk Kertang Program

No.	Criteria	Aspects Value	Category level
1	Company's commitment	3,4	Good
2	Society contribution	4,00	Excellent
3	Program relevance	3,4	Good
4	Program sustainability	2,35	Good
5	Plan and implementation program	3,5	Excellent
6	Company's response	3,5	Excellent
7	Clarity and fairness in program implementation	3,4	Excellent
8	Officer resposibility	3,75	Excellent
9	Officer capacity	2,4	Good

10	Officer acceptance by society	4,00	Excellent
11	Budget	2,9	Good
12	Company's relationship with stakeholders	2,39	Poor
13	Scope of program	3,3	Excellent
14	Program benefit	2,8	Excellent
	Total	**44,8**	
Respondence			20
Index value			**3,18**
Conversion index value			**79,52**
Quality level			**B**
Category level			**Good**

From **Table 7** above, the category level of the program 'Good' means that there are so many room of improvement. Edco-tourism is a tourism concept that may combine between education and nature (eco) friendly tourism. This concept is not only to promote mangroves tourism; but also to make positive and direct contribution between visitors and the local community. Other social effect is the increasing number of MEKAR's member. Starting with only 2 members in 2012 now MEKAR has grown to 45 members. Through social media optimization, in 2017, Lubuk Kertang Mangroves Conservation Area is one of the tourist destinations in Langkat and becomes research centre for studies from domestic and international institutions. This mangroves area has also been published on National Geographic Indonesia in September 2016. The program "Green Belt Lubuk Kertang" was awarded Indonesia Green Awards in 2015, 2016 and 2017 for conservation biodiversity categories.

5. Conclusion

The focus of this paper will be on how PERTAMINA EP encourages the community's local potential to confirming their business sustainability around our main working areas of operation. Though, this program has not giving a massive impact yet but it has shown that commitment of company – community relationship is mutual. The relationship develops into social establishment and license to operate for the company. Value creating through CSR program may help not only nature preservation but also reduce negative reputation from the community. It is about making the company has humanism point of view by take social responsibility seriously into the business process. [14]
As for the community and mangroves area, the results are:

1. CO_2 emission reduction from mangroves zone with total 16,119 ton/Ha eq.
2. Evolving the diversity around mangroves conservation area.
3. Help to prevent the conversion of 27 ha of mangrove forests into oil palm plantations and illegal logging (deforestation).
4. Reducing 15% of Methane from Green House Effect for mangroves planting and seeding.
5. Increasing number of community empowerment in tourism activities.
6. Income of beneficiary members increasing for almost 70% since managing mangroves edco-tourism program.
7. Help to prevent Langkat's Coast from abrasion for Mangrove Cultivation Program.
8. Enlarging social engagement and coastal education for the community.

6. Acknowledgement

This study and strategy was supported and funded by PT PERTAMINA EP Asset 1 Pangkalan Susu Field. By implementing the program based on this study, PERTAMINA EP has awarded for GREEN PROPER since 2014 held by Ministry of Environment and Forestry Indonesia.

REFERENCES

[1] Data Source: Ministry of Fisher and Marine

[2] Giri et al., 2011. Global Ecology and Biogeography: Status and Distribution of Mangrove Forests of The World Using Earth Observation Satellite Data. p. 157. Retrieved from: https://forestsnews.cifor.org/31178/indonesian-mangroves-special-fact-file-a-global-treasure-under-threat?fnl=en

[3] Schaeffer-Novelli et al., 2005. Climate Changers in Mangrove Forests and Salt Marshes. Brazilian Journal of Oceanography.

[4] Campbell,A., & Brown,B. (2015). Indonesia's vast mangroves are a treasure worth saving. *The Conversation*. Retrieved from http://theconversation.com/indonesias-vast-mangroves-are-a-treasure-worth-saving-39367

[5]_Murdiyarso, D., Purbopuspito, J., Kauffman, J. B., Warren, M., Sasmito, S., Donato, D., . . . Kurnianto, S. (2015). The potential of Indonesian mangrove forests for global climate change mitigation. *Nature Climate Change. Vol.5, DOI: 10.1038/NCLIMATE2734*

[6] Ministry of Environment Republic of Indonesia. (2010). *Indonesia second national communication under the United Nations Framework Convention on Climate Change.* Jakarta.

[7] PT Pertamina EP Asset 1 Pangkalan Susu Field. (2015). Social Mapping Document: Desa Lubuk Kertang. Langkat.

[8] UNEP. (2014). *Importance of Mangrves to People: A Call to Action:* United Nations Environment Programme World Conservation Monitoring Centre, Cambridge. Retrieved from: https://forestsnews.cifor.org/31178/indonesian-mangroves-special-fact-file-a-global-treasure-under-threat?fnl=en

[9] European Commissin: Promoting a European Framework for Corporate Social Responsibility, Green Paper, July 2001, p.8 in Luetkenhorst, Willfried, Corporate Social Responsibility and the Development Agenda, Intereconomics, June 2004, p. 158.

[10] Michael E. Porter and Mark R. Kramer. December 2006. *Strategy & Society – The Link Between Competitive Advantage and Corporate Social Responsibility.*

[11] Twelvetrees, A. (1991). *Community Work* (2nd edn). London: Macmillan.

[12] Suharto, Edi. Metode and Teknik Pemetaan Sosial. Retrieved from: http://www.policy.hu/suharto/modul_a/makindo_18.htm

[13] Monitoring report of biodiversity index.2016. PT Pertamina EP Asset 1 Pangkalan Susu Field.

[14] Strategy Plan of CSR Document. 2016. PT Pertamina EP Asset 1 Pangkalan Susu Field.

Exploration of Sea Cucumbers *Stichopus hermanii* from Karimunjawa Islands as Production of Marine Biological Resources

Delianis Pringgenies[1], Siti Rudiyanti[2] and Ervia Yudiati[1]

[1] Dept of Marine Sciences, Faculty of Fisheries and Marine Scinces Diponegoro University-Semarang, Indonesia
[2] Dept of Fisheries, Faculty of Fisheries and Marine Scinces Diponegoro University-Semarang, Indonesia

Email: pringgenies@undip.ac.id; pringgenies@yahoo.com

Abstract. This research aim was to study the potential of *Stichopus hermanii* to determine the amino acid, chondroitin, and glucosamine contents, to discover its antibacterial and anti-cancer agent. The samples were rinsed prior to separation, with only the corpus being used in the study. Sea cucumber extract was then processed using HPLC to trace contents of amino acid, chondroitin, and glucosamine contents. The samples were then put into test against several strains of pathogenic bacteria by means of diffusion for any biological activity. The anti-cancer test was performed by *human ovarian cancer cell line* (KOC7C) method. The study showed that the extract of *Stichopus hermanii* has the potency to inhibit the growth of active ovarian cancer cells. The qualitative test of the sea cucumber extract showed that it is capable of suppressing the growth of several strains of pathogenic bacteria identified as *Staphylococcus aureus*, *Escherichia coli*, *Vibrio voinivica*, and *Pseudomonas* sp. HPLC results showed that the extract contained amino acid (mg/100g), the highest being Collagen (11200), followed by Glycine (3760), Glutamic Acid (3700), Aspartic Acid (2540), Alanine (2140), Proline (2050), Arginine (2050), Tyrosine (1430), Threonine (1270), Leucine (1170), Valine (1050), Serine (971), Isoleucine (816), Phenylalanine (713), Lysine (639), Methionine (383), Cystine (263) and Histidine (208). The extract also contained Chondroitin Sulfate (4200) and Glucosamine Hydrochloride (<5.00). In conclusion, the study found that the extract of *Stichopus hermanii*

has potential as anti-cancer, particularly against ovarian cancer, with the highest content being Collagen within amino acids, as well as chondroitin and glucosamine.

Key words: *Stichopus hermanii*, amino acids, chondroitin, glucosamine, anti bacteria, anti cancer

1. Introduction

Stichopus hermanii is a species of sea cucumber classified as marine invertebrate from the Echinodermata phylum. The morphology of this species is of elongated, cylindrical and soft body. Most sea cucumbers are nocturnal in nature/active during the night time and are inactive or in hiding during the daytime [1]. This caused fishermen to seek out sea cucumber catch during the night, as hauls during the day have been proven to yield less results. *Stichopus hermanii*, locally known in Indonesia as *teripang emas* (viz: golden sea cucumber), is classified as a deposit feeder species and a sedentary marine life whose natural habitat are hard surfaces or rocks, often feeding off prey on the seabed with its tentacles[1]

Stichopus hermanii mainly feed on subsists on small organisms, detritus (decomposed organic materials), diatoms, protozoans, nematodes, filamentous algae, copepods, ostracods and seaweeds. Also included in the diet of this species are radiolarians, foraminifera, sand/coral debris particles, and shells of various marine animals[2]

In several countries in Asia and Russia, sea cucumbers are exploited for traditional medicines to treat wound and a number of diseases. In Malaysia, coastal and fisherman communities use lotion or other salves made of sea cucumber as topical medication for wound treatment, known locally as *gamat*[3]. Information on the medical potency of extracts from various species of sea cucumber shows efficacy in their use in antimicrobial, antifungal, anti-tumor, anti-oxidant, anti-proliferation, wound recovery, immuno-modulator, anticoagulant, and antithrombotic purposes. Several researches on the sea cucumber extract also show its potency in controlling excessive cholesterol levels and blood fat contents, in addition to being effective antivirus, anti-malaria and anti-arthritic agents[4-6] [3].

Based on the field survey performed in the waters of Karimunjawa islands on the sea cucumber from the Stichopus family, one characteristic of the species could be readily observed, namely its body integrity. During the collection of the samples, the sea cucumbers were observed to be solid, rigid frame. However, when they were taken out of the water, the body of the sea cucumbers would quickly soften and its body walls would drip. When the taken samples were returned into the coast, their bodies would stop disintegrating and they showed rapid body structure regeneration rate. Therefore, sea cucumbers of the Stichopus family in the survey shows rapid body cellular regeneration rate, tissue restoration, and possible anti-body or immune system formation as well as enzyme and hormonal balancing. Based on the observation, this study aims to discover the contents of secondary metabolites in sea cucumber of the *Stichopus hermanii* species which acts as potential antimicrobial and anti-cancer as well as to discover the contents of amino acids, chondroitin and glucosamine.

2. Methodology

2.1. Sample Collection

Sea cucumber samples were collected from the waters of Karimunjawa islands, Indonesia. The samples were stored in a cool box to prevent degradation and were transported to Laboratory of Diponegoro University, Semarang. The samples were prepared by intestinal organ removal, ensuring that only the meat remains, which were then diced into 1 cm size. The diced samples were dried under indirect sunlight, by covering the container with black fabric during the day. During the night, sample drying process was carried out by exposure to 5-watt light bulbs. The samples were dried for 5 consecutive days.

2.2 Sample Extraction

Sample extraction was carried out using the methodology in Montano and Glorioso (1994). Each prepared sample was cut into 3-10 cm pieces. The extraction process was accomplished by dissolving 200 grams of prepared sample in 1000 mL of methanol soluent. The filtration result was extracted from the soluent on a rotary evaporator.

2.3 Phytochemical Screening

Phytochemical screening was performed on the sea cucumber extract to determine the contents of secondary metabolites. The Saponin screening is performed by shaking 10 mL of filtrate in reaction tubes for 5 seconds, after which observation for changes in pattern. Formation of bubbles after the addition of diluted HCl shows positive reaction in saponin screening. Tannin screening was performed by adding 0.5 grams of samples into 5 drops of $FeCl_3$ 1 %. Dark blue or greenish black coloration signifies positive result in tannin screening. Flavonoid screening is performed by adding 1 mL of extract into 1 gram of sample and 190 mL of concentrated HCl. Red coloration identifies flavonoid content. Alkaloid screening is performed by shaking 1 mL of sample with 10 drops of H_2SO_4, after which the acidic layer is separated into a different reaction tube. The sulfuric acid layer is tested on a test plate with Dragendrof catalyst. Positive result is signified by reddish orange coloration. Terpenoid and Steroid Screening is performed by adding drops of ascetic anhydride or one drop of sulfuric acid H_2SO_4. Red coloration confirms terpenoid content, whereas blue coloration signifies steroid content.

2.4 Analysis of The Type and Contents of Amino Acids in Sea Cucumber Samples Using High Performance Liquid Chromatography

The samples in this amino acid analysis were the extraction of sea cucumber from the *Stichopus hermanii* species. Prior to High Performance Liquid Chromatography (HPLC), the samples were prepared by protein levels analysis by Kjeldahl method (LC 20AD, Shimadzu, Kyoto, Japan). This was followed by acid and base hydrolysis process on the sample. The results of hydrolysis process would then be ready for analysis using standard-procedure HPLC. This analysis essentially involves converting protein into amino acids in order to be detected by the chromatograph. The detector detects the amino acids, after which the results are saved by the recorder. The computer will then display the recorded data in retention time and area graphs with peaks corresponding with the characteristic of each amino acid.

2.5 Antimicrobial Screening

The antimicrobial screening of sea cucumber extract was performed against multi-drug resistant test bacteria from the species *Staphylococcus aureus*, *Escherisia coli*, *Vibrio anguila*, *V. voinivica*, *Bacillus subtilis*, and *Pseudomonas sp* provided from Kariyadi Hospital Semarang, Indonesia. Each paper disc was prepared by adding 10 μL drop of methanol. Observation was then made on the paper disc _showing signs of antimicrobial activity, upon which said paper disc would be marked positive[7].

2.6 Anti-cancer Cell Screening

Cancer cells were incultured at 3000 cells per 96-well microplate and pre-incubated. Cancer cell: human ovarian cancer cell line (KOC7C)_obtain from Laboratory of Natural Medicinal Chemistry, Kobe Pharmaceutical Ubiversity Japan. After 24 hours, they were treated without or with test samples and then incubated. After 72 hours, WST-8 reagent was added to each and the cultures were incubated further for 1 hour. The relative viable cell number was determined by measuring the absorbance at 450 nm (Ref. 630 nm). The IC50 value, which reduce the viable cell number by 50 %, was determined.

Cells were incubated with SAB E extracts for two hours before being collected and washed three times with PBS prior to addition of 0.9 ml staining solutions and mixing using vortex. Cells suspension was incubated for 10 minutes in ice prior to addition of 0.1 ml 1.5 M NaCl. Samples were analyzed by flow cytometry. This assay was carried out in duplicate. Staining solution for 10-11 samples consisted of 4.5 ml Propidium iodide stock solution (55.6 mg/500 ml DDW), 1 ml Na-citrate stock solution (0.588/50 ml DDW), 0.45 ml RNAse (10 mg ml-1), 0.1 ml 10% Triton X-100, 3.95 ml DDW.

3. Results and Discussion

3.1 Results

3.1.1 Phytochemical Screening

The screening results showed positive on the saponin, tannin, flavonoid, terpenoid and steroid in the samples. However, no alkaloid content was detected.

3.1.2 Antimicrobial Screening

The sea cucumber extract showed positive activity against pathogenic bacteria from the species *Escherisia coli, Pseudomonas* sp., *V.voinivica* and *Staphylococcus aureus* as seen in Table 1 below.

Table 1. Qualitative test result of the antimicrobial activities of sea cucumber extract against test bacteria.

Sea Cucumber Species	Test Bacteria Species					
	S.aureus	*E.coli*	*Pseudomonas sp*	*V.voinivica*	*V.anguila*	*B.substilis*
S.hermanii	+	+	+	+	-	-

3.1.3 Anti-cancer screening

Preparation of Sample for Growth Inhibitory Test against Ovarian Cancer Cells (1)

Anti-cancer test used is the ovarian cancer cell K0c7c. Test results the effect of the concentration of sea cucumber extract, unfiltered and filtered e on viability shows that Ic50 between sea cucumber extract, unfiltered (Ic50 : 26.6 μg/mL) and filtered (Ic50 : 24.38 μg/mL) showed that almost the same results as seen in the picture below (Figure. 1)

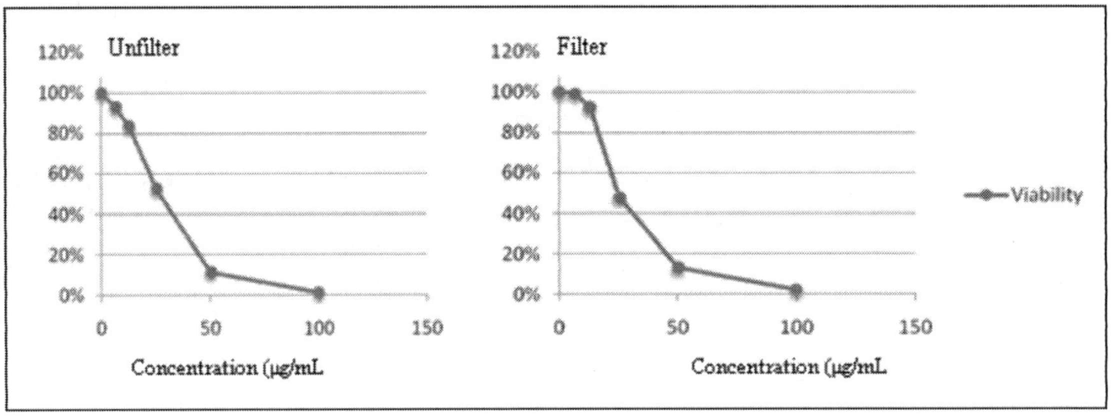

Figure 1. Test the effect of the concentration of sea cucumber extract, unfiltered and filtered e on Viability

Test cell cancer cell (K0c7c line) to extract the results showed that the extract of sea cucumber sea cucumber *Stichopus hermanii* against ovarian cancer cell active as shown in Figure 2.

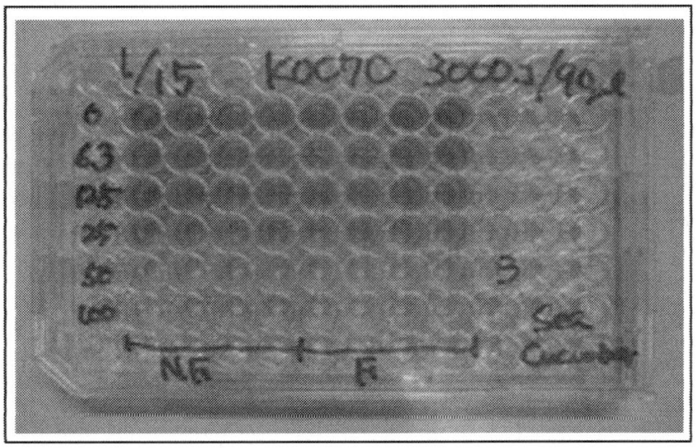

Figure 2. Test results with ovarian cancer cell method (K0c7c line) against the sea cucumber extract of *Stichopus hermanii*

3.1.4 Amino Acid Contents

HPLC analysis of the sea cucumber extracts revealed, from the highest to the lowest concentration, the following amino acids; Collagen (11200), Glycine (37600, Glutamic Acid (3700), Aspartic Acid (2540), Alanine (2140), Proline (2050), Arginine (2050), Arginine (2050), Tyrosine (Tyrosine), Threonine (1270), Leucine (1170), Valine (1050), Serine (971), Isoleucine (816), Phenylalanine (713), Lysine (639), Methionine (383), Cystine (263) and Histidine (208). The analysis also found Chondroitin Sulfate (4.200) and Glucosamine Hydrochloride (<5.00). Figure 3 showed the findings above in graphic.

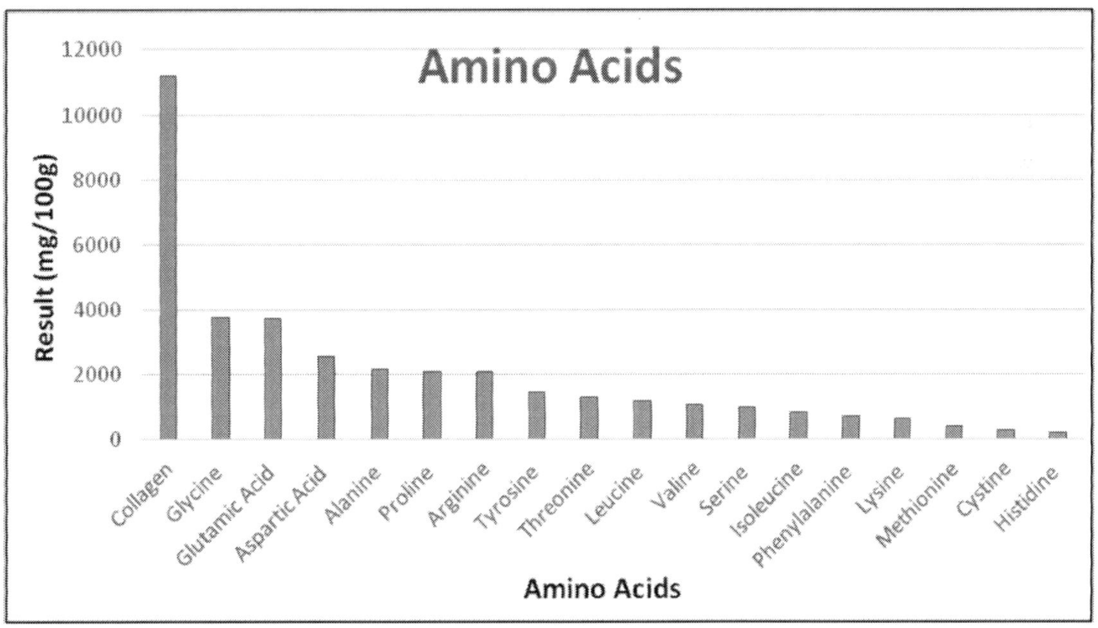

Figure 3. Amino acid contents of *Stichopus hermanii* extract

3.2 Discussion

The results of antimicrobial activity test of *Stichopus hermanii* extract showed positive activity in inhibiting the growth of test bacteria *Escherisia coli, Pseudomonas* sp., *V. voinivica* and *Staphylococcus aureus*. The saponin content of the extract was thought to be the main component in this antimicrobial properties. Phytochemical screening of *Stichopus hermanii* extract confirmed that saponin is one of the secondary metabolite contents of the extract, along with tannin, flavonoids, terpenoids and steroids. On the other hand, no alkaloid content was found in the extract. Saponin is a chemical compound with bioactive properties[8]. The first saponin to be found in sea cucumber is categorized as *holothurin,* signifying a wide spectrum in physiological activity [9]. The saponin was found on the skin, in the flesh and in the gut *(tubula cuverian*) of Bohadschia sp[10]. Saponin is used by sea cucumbers in high concentration as a self-defense mechanism against predators. Saponin can be used as an antimicrobial compound. The pure sea cucumber extract tend to contain *holotoxin*, of which effect was found to be similar to that of antimycin in 6,25-25 μg/mL dose[11].

The antimicrobial sensitivity test on the sample of *Bohadschia mamorata* and *B. argus* against *Staphylococcus aureus, Escherichia coli, Vibrio anguila, V. voinivica, Bacillus subtilis, Pseudomonas* sp test bacteria resulted in positive activities[8]. Sample of the Stichopus variegatus sea cucumber was found to inhibit the growth of *Escherichia coli, Pseudomonas* sp., *V. voinivica* pathogenic bacteria [8]. Antimicrobial test of *Stichopus chloronatus* and *Stichopus hermanii* found that both sea cucumber sample were tested positive against *Staphylococcus aureus, Escherichia coli, Vibrio anguila, V. voinivica* pathogenic bacteria.

Cancer cell line test (K0c7c line) of *Stichopus hermanii* extract showed positive activity against the growth of ovarian cancer cells. Saponin is thought to be the key compound in the cancer-inhibiting properties of the sea cucumber extract. Saponin is categorized as a glycoside, consisting of sapogenin (*steroid*; C27) or *triterpenoid* (C30) group, hexose or pentose group, or uronic acid[12]. These compounds have found application in medical industry as a base in steriodal medicine sythesis. The saponin also has detergent component which causes membrane disintegration, instigating hemolysis of erythrocyt[13]. Saponin has bitter taste, cause bubble in water, toxic for cold-blooded animals and shows anti-exudative and anti-inflammatory properties. Known saponin compounds other than holothurin includes holotoxin, stichoposide and thelothurin, all of which are isolated from sea cucumbers[9]. Another derivative compound with bacteria grwoth-inhibiting properties is lectin. Mitogenic of nature, this class of protein can develop qiuickly and be used as an antimicrobial agent. Lectin agglutinates lymphoid cells. Due to its potency in killing would-be predators, sea cucumber toxins are have now been developed as anti-inflammatory and anti-cancer natural antisceptic[14]. Coombs [14] described lectin as a class of proteins which bind glycoproteins on cell surface, inducing agglutination. Lectin showed potency in inhibiting the growth of muscle and lungs cancer cell[11]. This cellular agglutinating properties of lectin in sea cucumbers is thought to explain the fact that sea cucumbers can recover their natural physiology after being taken out from sea water for some time.

Analysis of amino acids contents of sea cucumber extract by HPLC discovered 18 kinds of amino acids. These amino acids, from the highest content to the lowest content (mg/100g), were Collagen (11200), Glycine (37600, Glutamic Acid (3700), Aspartic Acid (2540), Alanine (2140), Proline (2050), Arginine (2050), Arginine (2050), Tyrosine (Tyrosine), Threonine (1270), Leucine (1170), Valine (1050), Serine (971), Isoleucine (816), Phenylalanine (713), Lysine (639), Methionine (383), Cystine (263) and Histidine (208)

Amino acids are the building blocks of proteins. Protein is vital for the body in that it helps in maintaining tissues, in forming other amino acids, in synthesizing various hormones, in neurotransmitter production for the brain and other neural functions, in maintaining body immunity system, in maintaining liquid balance and as a source of energy. A gram of protein can be converted into 4 calories worth of energy[15]. Of all 20 amino acids which form the protein, of them cannot be synthesized by the human body and can only be obtained from nutritional intake. These amino acids are isoleucine, leucine, lysine, methionine, phenylalanine, threonine, trypthophane dan valine, otherwise known as the essential amino acids. Other than the essential amino acids, there are the conditionally essential amino acids, which include histidine and arginine. Isoleucine plays a vital role in erythrocyte formation. Leucin prevents muscle degradation. Methionine acts as a precursor of cysteine and creatine, lowers blood cholesterol level, helps liver in detoxification, and regenerates tissue in liver and kidneys. Threonine acts as a detoxifier and helps prevent the accumulation of fat in liver. Tryptophan is an essential material in producing serotonin (a hormone involved in relaxation) and stimulate the release of growth hormone. Valine plays a role in transporting other amino acids, such as tryptophan, phenylalanine, and tyrosine to the brain. Arginine is an important ingredient in creatine formation, and stimulates the production of growth hormones. Histidine is vital in the formation of erythrocyte and leucocyte. Lysine assists in the process of collagen formation as well as in the regeneration of other connective tissues in the body (cartilage and joints). Content analysis of sea cucumber extract showed that chondroitin was the highest (4.200) after collagen (11200) and glycine (3760). As chondroitin is vital for bone formation, it is very useful for those suffering from or vulnerable to osteoporosis, such as mature women, and during formative years of children and teenagers.

4. Conclusion

Based on the findings of this study, it can be concluded that,

1. Sea cucumber has saponin, tannin, flavonoid, terpenoid and steroid content. However, alkaloid content was not found in sea cucumber.
2. Stichopus hermanni sea cucumber extract showed antimicrobial activity in inhibiting the growth of pathogenic bacteria from the species *Escherisia coli*, *Pseudomonas* sp., *V.voinivica* and *Staphylococcus aureus*.
3. The highest content found in sea cucumber (mg/100gr) was collagen (11200), chondroitin (4.200), Glycine (3760), and Glutamic acid (3700).

Acknowledgements

The author would like to thank PT Kalbe Farma tbk for their support in the sample analysis. The author also extends her deepest gratitude to Dr. Nishiyama Yumi and Dr. Shitan Nobukazu of Laboratory of Natural Medicinal Chemistry, Kobe Pharmaceutical University, Japan for their cooperation in the research.

References

[1] Barnes, R. D. 1991. Invertebrata Zooology.ed Sounders College Publishing USA.P : 966 - 975.
[2] Martoyo, J. 1994. Systematic Animals: Invertebrates and Vertebrata For Universities. Second Edition. Publisher Sinar Wijaya . Surabaya: 322 – 333 hlm
[3] Farouk A.E., Ghouse F.A.H., Ridzwan B.H , 2007 *American Journal of Biochemistry and Biotechnology* **3** (2): 60
[4] Abraham TJ, Nagarajan J, Shanmugam SA, 2002 *Indian Journal of Marine Sciences* **31** (2) : 161-164
[5] Althunibat OY, Hashim RB, Taher M, Daud JM, Ikeda MA, Zali BI 2009 *European Journal of Scientific Research* **37** (3): 376-387
[6] Avilov et al. 2007. Process for isolating sea cucumber saponin frondoside A and immunomodulatory methods of use. United States Patent no. : US 7,163, 702 B1. Available at http://www.freepatentsonline.com/7163702.pdf, accessed November 2010.
[7] Mojica E R, E RJ Layson, MCA Rodil and CC Deocaris 2007 Marine Invertebrates As Source of Potential Antitumor and Antibacterial Agents. http:// www.pcamrd.dost.gov.ph/zone2/papers/8th/mojica1.pdf (25 November 2007).

[8] Pringgenies 2013 Antibacterial Activity of Sea Cucumbers Harvested From Karimunjawa. *Squalen, Bulletin of Marine and Fisheries Postharvest and Biotechnology.* **8** (2): 87-94
[9] Hashimoto, Yoshiro 1979 *Marine Toxins and Other Bioactive Marine Metabolites* (Tokyo: Japan Scientific Societies Press) 268-270 pp
[10] Jayasree, V., R. Sen Gupta and P. V. Bhavanarayana. 1991. Bioactive Compound from Marine Organisms, a Toxin from *Holothuria leucospilota.* An Indo-United States Symposium. A.A. Balkema. Rotterdam. 111-120.
[11] Martoyo, Joko, Nugroho Aji dan Tjahjo Winanto. 2006. *Budidaya Teripang*. (Jakarta: Penebar Swadaya) p 75
[12] Winarno, F.G. 1982. *Kimia Pangan dan Gizi. Gramedia Pustaka* Utama. Jakarta. 96 hlm.
[13] Coombs, J. 2007. *Dictionary of Biotechnology*. Oxford. Amsterdam. 330 hlm.
[14] Ayu. 2007. Teripang, Potensi di Dasar Samudra Indonesia. http:/www.suara merdeka/ap_ 2007 (10 Agustus 2007). 2 hlm.
[15] Schlenker, E.D. & Long Roth, S. (2011). *William's Essentials Of Nutrition And Diet Therapy.* (St. Louis, Missouri.: Elsevier)

Microbiological Characteristic and Nutrition Quality of Goat Milk Kefir Based on Vitamin D_3 Fortification Time

F Fauziyyah[1*], B Panunggal[1,2], D N Afifah[1,2], N Rustanti[1,2], G Anjani[1,2,]**

[1]Department of Nutrition Science, Faculty of Medicine, Diponegoro University, Jl. Prof. H. Soedarto,SH, Tembalang, Semarang 50275, Indonesia
[1]Center of Nutrition Research (CENURE), Diponegoro University, Jl. Prof. H. Soedarto,SH, Tembalang, Semarang 50275, Indonesia
*Presenting Author
**Corresponding author, +62 81285376785, gemalaanjani@yahoo.com

Abstract. Goat milk kefir fortified with vitamin D_3 is expected to benefit individual with insulin resistance. Different vitamin D_3 fortification time allegedly effect microbiological characteristic and nutrition quality of goat milk kefir due to its microbial growth curve, thus this study aimed to analyze those parameters. This study was an experimental research. This study contains five treatments (vitamin D_3 fortification at 0, 6, 12, 18, or 24 hours of fermentation) and a group of control. Total lactic acid bacteria, vitamin D_3, protein level, fat contain, crude fiber, viscosity, and pH was analyzed by Total Plate Count, spectrophotometry, Bradford method, Babcock method, gravimetric analysis, Ostwald method, and pH meter respectively. Time of vitamin D_3 fortification significantly effect vitamin D_3 content (p=0,021), fat content (p=0,001), crude fiber (p=0,0001), viscosity (p=0,010), and total lactic acid bacteria (p=0,048). The highest vitamin D_3content was found on the group fortified at 6 hours of fermentation. All treatment groups has lower fat content and crude fiber content than control group. Total LAB in all group meet the Codex standard ($\geq 10^7$ CFU/ml). Control group and fortification group at 24 hours of fermentation have higher viscosity than other groups. There was no significant difference found in goat milk kefir protein level (p=0,262) and pH (p=0,056) despite the difference of fortification time. Vitamin D_3 fortification time effect vitamin D_3 content, fat content, crude fiber, viscosity, and total lactic acid bacteria of goat milk kefir, but did not effect protein content and pH of goat milk kefir.

Keywords : goat milk kefir, fortification time, vitamin D_3, lactic acid bacteria, nutrition quality.

1. Introduction

Indonesia as developing country with a tropical climate, is now facing new health problems which is shifting from infectious disease to degenerative disease and malignancy[1]. Diabetes melitus is a kind of degenerative disease occuring in Indonesia. A study stated that the prevalence of diabetes in urban Indonesia was 5,7%, consisting of 1,5% diagnosed diabetes mellitus, 4,2% undiagnosed diabetes mellitus, and 10,2% encounter impaired glucose tolerance[2].*RisetKesehatanDasar* 2013 stated that there was 6,9% of population (age of $\geq$ 15 years) in Indonesia whose diabetes mellitus, which 90% of the case classified as diabetes mellitus type 2[3].Individual with obesity and insulin resistance will partially develop diabetes mellitus type 2[4]. Insulin resistance is a condition where the lack of cell insulin receptor led to cell inability to uptake glucose[5]. Vitamin D deficiency apparently play role in pathogenesis of Diabetes Mellitus type 2. Vitamin D help stimulates the expression of the insulin receptor in peripheral tissue to increase glucose uptake[6]. In addition, a good status of vitamin D may suppress chronic inflamation, thus insulin sensitivity will increased[7][8]. Human body could synthesize vitamin D with the help of the sun UV irradiation of 7-dehydrocholesterol on skin[9]. Unfortunately, vitamin D synthesis in human body limited to clothes they are wearing and depends on other factors like the intensity of ultraviolet, race, and age[10]. Sunny country in tropical and subtropical region was not an exception to suffer from vitamin D deficiency[11]. Vitamin D deficiency in Southeast Asia varied from 6-70%[12]. Indonesia, a tropical country with plentiful sunshine, also suffered a high number of vitamin D deficiency. There are only 5% children with adequate 25(OH)D in Indonesia[13]. Vitamin D deficiency itself could ensue all population groups due to its various risk factors[14]. Therefore, diabetes melitus type 2 and vitamin D deficiency are both important concerns in Indonesia.

Goat milk, a kind of milk which can be found in Indonesia, has multiple benefit. Goat milk contains high level of MCT and short chain fatty acids, which can easily digested and used for metabolism. In addition, goat milk contains higher protein, vitamin A, thiamin, riboflavin, niacin, panthotenat, calcium, and phospor compared to cow milk. But, both goat milk and cow milk are deficient to vitamin B6, vitamin C, and vitamin D[15].

Kefir, a kind of fermented milk, can be made by inoculating kefir grains into milk. Kefir grains contain microorganisms such as *Lactobacillus* spp., *Lactococcus* spp., *Streptococcus* spp., *Enterococci* spp., *Leuconostoc* spp, *Acetobacter* spp., *Bacillus* spp., *Kluyveromyces marxianus, Saccharomyces* sp., *Torulaspora delbrueckii, Brettanomyces anomalus,* and *Issatchenkia occidentalis*[16][17]. Kefir grains itself is a potential probiotic and antioxidant, which may benefit individual who have insulin resistance[18]. Probiotic drinks could lower fasting glucose and HbA1C in patient with diabetes mellitus type 2[19]. Furthermore, kefiran inside kefir could activate PI 3-kinase, thus help insulin signaling[20].

Kefir contains lactic acid bacteria and yeast. The minimum amount of lactic acid bacteria and yeast in kefir is 10^7 CFU/ml and 10^4 CFU/g, respectively[21][22][23].Chemical composition in milk could contribute to yeast growth. Yeast existence can help provide nutrition that are important for lactic acid bacteria growth, such as vitamin and amino acid[24]. Yeast has proteolitic and lypolitic activity[25][26]. Lactic acid bacteria also has proteolytic[27] and lypolitic activity[28]. Thus, activity of yeast and lactic acid bacteria could affect kefir composition and characteristics. Kefiran, an exopolisaccharide inside kefir, is suitable as viscofying, stabilizing, gelling, and emulsifying factor[29]. Bacteria activity that produce kefiran will later effect the viscosity of the product[30]. Acidity (pH) also affected by lactic acid bacteria and yeast contain[31]. Acidity (pH) of kefir ranged from 4,2 to 4,6[32].

Fortification aimed to increase nutrition value of food and to offer health benefits for consuments[10]. Milk and its products usually be used as media for vitamin D_3 fortification[10][33].Vitamin D_2 dan vitamin D_3 are commonly used as food fortificant[10][34].Vitamin D_2 retention in fortified milk is 76,96%[35], whereas vitamin D_3 retention in milk products such as cheese, yoghurt, and ice cream have higher retention (95-97%, 96,6-97,8%, and 99,8-99,3%, respectively)[33].

Microorganisms in kefir have different growth curve. At 0 hour of fermentation, those microorganisms just started to began the fermentation. At 6 hours of fermentation, lactic acid bactria, acetic acid bacteria, and yeast start to increasing. At 12 hours of fermentation, lactic acid bacteria and acetic acid bacteria increased significantly. Yeast keep growing untill 12 hours of fermentation, meanwhile for the rest fermentation it shows no significance growth. At 18 hours of fermentation, lactic acid bacteria population reach its peak, meanwhile acetic acid bacteria shows a little increase. At 24 hours of fermentation, acetic acid bacteria reach its optimum number[36]. There was no previous research that study about the effect of different time of vitamin D_3 fortification on microbiological characteristic and nutrition quality of goat milk kefir. Different vitamin D_3 fortification time allegedly effect microbiological characteristic and nutrition quality of goat milk kefir. Goat milk kefir fortified with vitamin D_3 produced in this research is expected to have good microbiological characteristic and nutrition quality which could benefit individual with insulin resistance.

2. Method

The making of goat milk kefir, vitamin D_3 fortification, microbiological (lactic acid bacteria) analysis, nutrition quality (vitamin D_3, protein, fat, and crude fiber) analysis, viscosity analysis, and acidity (pH) measurement was conducted in Laboratorium Terpadu Universitas Diponegoro. Preliminary study was conducted to analyze microbiological characteristic and nutrition quality of goat milk used for this research. Goat milk for this research was obtained from Oemah Kefir. Kefir grains used from this research was first obtained from Oemah Kefir, and then reared by the researcher. Vitamin D_3 was sourced from *Healthy Care*Australia.

This research was experimental, completely-randomized designed. This research were performed in five group of different fortification time, namely group of fortification on 0, 6, 12, 18, and 24 hours fermentation and a group of control. Group of control was goat milk kefir without any treatment. Vitamin D_3 was added 42 IU/100 ml kefir[37]. Vitamin D_3 used for this research was in form of emulsified oil. All of treatments was replicated three times. All control group and treatment groups was fermented for 24 hours[38].

The making of goat milk kefir was begun with milk pasteurization on 72°C for 15 second, then was cooled untill 25°C[38]. Milk was divided into 5 groups of treatment and a group of control, then was inoculated wih 5% of kefir grains. All of samples were fermented for 24 hours. During fermentation, vitamin D_3fortification was held based on each group (0, 6, 12, 18, and 24 hours of fermentation). All samples were homogenized every 6 hours, and then filtered after the fermentation fnished.

Vitamin D_3 was analyzed by sphectrophotometry (Shimadzu). Samples was dissolved in a solution mixture of kloroform:methanol = 1:9. Afterwards, the mixture was read by sphectophotometer at 264 nm[39]. Fat content was analyzed by Babcock method, which samples were dissolved with hexane. Crude fiber was analyzed by gravimetric analysis. Viscosity was analyzed by Ostwald method. Acidity (pH) was measured withpH meter (LaMotte). Total lactic acid bacteria was analyzed with*Total Plate Count* (TPC) method, which the bacteria was inoculated on M.R.S Agar (Oxoid).

Data obtained from this study were analyzed by statistic software. Data distribution were examined by *Saphiro Wilk*. Effect of vitamin D_3 fortification time on vitamin D_3 content, protein content, fat content, crude fiber, and total lactic acid bacteria was examined by ANOVA *one way*. Meanwhile, effect of vitamin D_3 fortification time on viscosity and pH was examined by *Kruskall-Wallis*.

3. Results

Table 1. shows the results of preliminary study on goat milk used in this research. This data was used for better understanding of the characteristic of goat milk used by this study. Table 2. Shows nutrition quality of goat milk kefir based on various fortification time and nutrition quality of the control group.

Table 1. Goat Milk Characteristics

Sample	Vitamin D_3 (IU)	Vitamin B_{12} (µg)	Protein (%)	Fat (%)	Crude Fiber (%)	Total LAB (10^7)	Viscosity (cm/s^2)	pH
A1	69,7	735,3	1,4	14,8	3,2	4,2	0,0134	6,5
A2	72,5	805,0	2,4	14,6	3,4	3,4	0,0134	6,6
A3	70,4	918,7	2,6	14,6	2,4	3,9	0,0139	6,6
Mean	70,9$\pm$1,4	819,7 $\pm$ 92,6	2,2 $\pm$ 0,6	4,7 $\pm$ 0,1	3 $\pm$ 0,53	3,83 $\pm$ 0,43	0,0136$\pm$ 0,0002	6,6 $\pm$ 0,008

There was significant different found on vitamin D_3 content in goat milk kefir due to its fortification time (p=0,021). The highest concentration of vitamin $D_3$was 34,65 ± 5,63 IU, which was found on group of fortification at 6 hours of fermentation (**Table 2.**).

Protein content does not show significant different due to its fortification time (p=0,262). Fat content show significant different (p=0,001) due to its fortification time. After *post hoc* test, it was found that group of control had significantly higher fat content than all treatment groups (**Table 2.**).

Table 2. Nutrition Quality of Goat Milk Kefir Based on Fortification Time

Fortification Time	Mean Vitamin D_3 Concentration		Mean Protein Content		Mean Fat Content		Mean Crude Fiber Content	
	IU	*p value*	%	*p value*	%	*p value*	%	*p value*
Control	22,87 ± 0,57[b]	0,021*	0,62 ± 0,07	0,262*	8,47 ± 0,39[a]	0,001*	23,27 ± 1,504[a]	0,000*
0 hour of fermentation	28,19 ± 5,34[ab]		0,93 ± 0,29		5,93 ± 0,73[b]		3,93 ± 1,83[b]	
6 hours of fermentation	34,65 ± 5,63[a]		0,63 ± 0,18		6,23 ± 0,59[b]		2,77 ± 0,50[b]	
12 hours of fermentation	26,55 ± 1,47[ab]		0,82 ± 0,09		6,67 ± 0,54[b]		3,90 ± 1,31[b]	
18 hours of fermentation	23,54 ± 3,29[b]		0,78 ± 0,14		6,44 ± 0,52[b]		3,80 ± 0,95[b]	
24 hours of fermentation	25,59 ± 2,58[ab]		0,81 ± 0,13		5,92 ± 0,38[b]		6,47 ± 2,44[b]	

* Tested with *one way* ANOVA

There was significant effect of vitamin D_3 fortification time on crude fiber content (p=0,000). Control group has significantly higher crude fiber content than all treatment groups. Group of fortification at 24 hour of fermentation has slightly higher crude fiber content than the other treatment groups, but the difference was not significant (**Table 2.**).

Table 3. Total Lactic Acid Bacteria, Viscosity, and Acidity (pH) of Goat Milk Kefir Based on Fortification Time

Fortification Time	Mean Total LAB		Median Viscosity		Median pH	
	x 10^9 CFU/mL	*p value*	cm^2/s	*p value*		*p value*
Control	13,4±6,54[b]	0,048*	0,1384[b]	0,010**	4,70	0,056**
0 hour of fermentation	70,0±27,07[a]		0,0563[d]		4,45	
6 hour of fermentation	27,0±21,23[ab]		0,0532[d]		4,55	
12 hour of fermentation	17,9 ± 15,47[ab]		0,0710[c]		4,55	
18 hour of fermentation	37,1 ± 29,85[ab]		0,0576[d]		4,55	
24 hour of fermentation	17,5 ± 13,81[ab]		0,1652[a]		4,45	

* Tested with *one way* ANOVA
*Tested with *Kruskall Wallis*

Table 3.shows results of total lactic acid bacteria (LAB), viscosity, and pH of goat milk kefir based on fortification time. Viscosity and pH was measured to make sure kefir made in this study met the characteristic of kefir. All of kefir samples made in this study had met the Codex standards for lactic acid bacteria ($\geq 10^7$ CFU/ml). There was significant difference of total LAB due to vitamin D_3 fortification time (p=0,048). The highest number of total LAB was found on group of fortification at 0 hour of fermentation. Based on the*post hoc* test, significant different was found between control group with group of fortification at 0 hour of fermentation, meanwhile the other differences between group was not significant.

Fortification time of vitamin D_3 also effect the viscosity (p=0,010). Control group and group of fortification at 24 hours of fermentation have higher viscosity than the other groups. Acidity of goat milk kefir does not differ by the vitamin D_3 fortification time(p=0,056), although there was tendency that all treatment groups had lower pH than control group.

4. Disscussion

4.1 Vitamin D₃

The vitamin D_3 content in milk used for this study was 70,9±1,4 IU, meanwhile vitamin D_3 content in control kefir group was 22,87 ± 0,57 IU.Vitamin D_3 is a fat soluble vitamin, but there was lypolitic activity during kefir fermentation[24][30]. This study performed vitamin D_3 fortification on various time during goat milk kefir fermentation in order to increase vitamin D_3 content in goat milk kefir final product. It was obtained that group ofvitamin D_3 fortification at 6 hours of fermentation had the highest vitamin D_3 content (34,65 IU) in final product of goat milk kefir compared to the other group fortification time. This found was different from the other study. Preceeding study was found that vitamin D_3 retention in yoghurt could reach 97,8%[33]. The result could be different due to vitamin D sensitivity of light, heat, and oxidation[33][40]. Acidity could also effect vitamin D_3 stability[33][41].

As stated before, the highest vitamin D_3 content in this study was found on the group fortification at 6 hour of fermentation. This result could be caused by the presence of β-laktoglobulin A (β-LG A) and β-kasein (β-CN), proteins that could stabilized vitamin D_3. In dairy products, if vitamin D_3 was not protected by its fat matrix, it will be stabilized by β-laktoglobulin A (β-LG A) and β-kasein (β-CN)[42][43].The presence of β-LG A dan β-CN could effect stability and availability of vitamin D_3 in the products[42]. During kefir fermentation, there was proteolysis activity[24][30]. In fermented dairy products, vitamin D_3 is bound tightly to β-LG A[43]. During fermentation, β-LG is was not easily hydrolized by lactic acid bacteria, meanwhile β-CNwas hydrolyzed significantly. After 6 hours of fermentation, β-CN was hydrolized about 35%, and the hydrolization was increasing untill incubation finished[25]. As more of β-CN hydrolyzed after 6 hours of fermentation, vitamin D_3 could not be stabilized by β-CN.

The increasing of lypolysis and proteolysis activies allegedly contribute to the loss of protein and fat that stabilized vitamin D_3. Lypolysis and proteolysis that happened during kefir fermentation affected by lactic acid bacteria and yeast[24][30]. Fortification at 12 and 18 hours of fermentation concurres with the growth peak of Lactic acid bacteria and yeast. Total lactic acid bacteria in kefir fermentation reached its peak at 12 hours of fermentation (*Lactococcus* spp.) and 18 hours of fermentation (*Lactobacillus* spp.), meanwhile yeast reached its peak at 12 hours of fermentation[36]. Addition of vitamin would effects the growth of kefir grains[44], thereby fortification at a certain time would be used by microorganisms for growth. At 12, 18, and 24 hours of fermentation, lypolitic and proteolytic activities continue, thereby it would effect the presence of vitamin D_3 in goat milk kefir.

Yeast in kefir could produce vitamin D_2 from ergosterol with the help of UV light[45]. Vitamin D_3 was analyzed by sphectrophotometer at 264 nm[39]. Four metabolic form of vitamin D, namely vitamin D_3, vitamin D_2, 25-hidroxyvitamin D_3 dan 25-hidroxyvitamin D_2, could be analyzed with the same wavelength (254 nm)[46]. The small difference of the wavelength could cause shifting peak, thus vitamin D_2 could unintentionally read by the sphectrophotometer.

4.2 Protein

Lactic acid bacteria (BAL) inside kefir could degradate milk protein into peptide and amino acids during fermentation. Free amino acids and small peptides used by lactic acid bacteria as nutritional source for growth[27]. Yeast inside kefir also has proteolytic activities[26].

Casein was build from some fractions, namely α-, β-, dan κ-kasein, meanwhile whey was formed by fractions of α-laktalbumin dan β-laktoglobulin[25]. The presence of β-LG A dan β-CN effect the stability and availability of vitamin D_3 in the product[42]. Fortification of vitamin D_3 at various time resulting that the highest vitamin D_3 content was found if vitamin D_3 fortified at 6 hours of fermentation. A research stated that beta-casein had been hydrolyzed about 35% at 6 hours of fermentation, and then keep increasing. At 24 hours of fermentation, it already hydrolyzed about 85%. Meanwhile, β-laktoglobulin dont significantly decreased during fermentation[25]. The more protein hydrolyzed may cause vitamin D_3 could not be stabilized by protein. Based on results in this study, it shows that on group fortification at 6 hours of fermentation vitamin D_3 was the highest, but it also shows that protein contents seems pretty lower than the other treatment group although the difference was not significant. This can be explained because the binding of vitamin D_3 with β-laktoglobulin A or β-kasein was stronger when affinity within protein and other protein is lower. Low molecular affinity may cause the presence of monomer in the solvent[42]. Meanwhile, the method being used, Bradford, has characteristic which could not detect the monomer in the solvent[47].

Protein content in this study was not significantly different in all groups. Protein content affected by the kind of the milk that being used in the experiment. The protein content could be affected by the feeding method and the breeding of the goat[48]. Beside, kefir grains and pH of fermentation may contribute to the protein content of the kefir[38]. In this study, the same amount of kefir grain, the same source of milk, and thesame pH fermentation was used for every groups, thereby it explains the unsignificant different on the protein content of all goat milk kefir samples.

4.3 Fat

The fat content in all treatment groups were lower than the control group. During fermentation, fat was degraded by microorganisms in kefir grains[49]. Vitamin D_3 fortification was performed during fermentation (at 0, 6, 12, 18, or 24 hours of fermentation), it means that vitamin D_3 was added during lactic acid bacteria and yeast growth. The addition of vitamin could effect the growth of kefir grains[44]. Lactic acid bacteria has intracellular and extracellular lipase that cause fat degradation to fatty acids and glycerols[30].

As stated before, this study shown that fat content of treatment group was lower than control group. That results was concured with the results of total lactic acid bacteria in this study. The number of total lactic acid bacteria on all treatment group was higher than the control group. The increase of total lactic acid bacteria may cause higer lipase produced by them, thus there were more fat being degraded and show a lower fat content as a result[21]. Group fortification at 0 hour of fermentation had the highest number of total lactic acid bacteria, thereby this group also show a low fat content. Yeast in kefir also performed lypolytic activity[26].*Saccharomyces cereviseae*has esterase which can hydrolyze short and medium chain fatty acids[50].

Fat content in this study (5,93-8,47%) was higher than other study (5,55%) who used the same concentration of kefir grains and used the same pH[38]. This difference may caused by the kind of milk that being used[48]. The fat content of control group wan higher than the treatment groups. This fat content may also effect the viscosity of kefir[51].

4.4 Fiber

Exopolysaccharide (EPS) in kefir, called kefiran, was produced by lactic acid bacteria namely *Lactobacillus, Streptococcus, Lactococcus,* dan *Leuconostoc*[49]. Kefir grains has α-galaktosidase activity[22]. Lactose in kefir was hydrolyzed, then galactose produced by this process was being used to build kefiran polymer[49]. In this study, all treatment groups has lower crude fiber content than the control group. Kefiran production may affected by culture condition, such as temperature, agitation rate, carbon cource, vitamin, and mineral provided in the culture. The addition of vitamin may effect the growth of kefir grains[44]. thus the addition of vitamin at various fortification time may be used by various bacteria or yeast that was being growth at the time. *Lactococcus* spp. reached its growth peak at 12 hours of fermentation, *Lactobacillus* spp. reached its growth peak at 18 hours of fermentation, meanwhile yeast reached its growth peak at 12 hours of fermentation[36].

Yeast contain ergosterol, which are majority located on plasma membrane. Ergosterol played role on yeast mating, through feromon signaling and help plasma membrane fusion[52]. A previous study learned the effect of temperature, pH, and the addition of yeast extract to the growth of EPS-producing bacteria and the production of EPS itself. From that study, it was obtained that higher yeast extract added may increase the production of EPS by bacterias. However, the higher number of yeast could also increase lactose consumption by yeast, thus there were lower lactose available for biosynthesis of EPS by bacteria[53].Kefiran content could contribute to the viscosity of kefir[30].

4.5 Total Lactic Acid Bacteria

There were homofermentative lactic acid bacteria (such as *Lactobacillus* spp., *Lactococcus* spp., *Streptococcus thermophilus*) and heterofermentative lactic acid bacteria (such as *L. kefiri*, *L. parakefiri*, *L. fermentum*,*L. brevis*)[17]. Based on Codex standard of fermented dairy products, kefir has to contain 7 log CFU mL^{-1} at minimum. All kefir samples made in this study already met the minimum Codex standard of total lactic acid bacteria. Total lactic acid bacteria in group of fortification at 0 hour of fermentation was higher than the other treatment groups and the control group. Nutrition available in the environment would be used by lactic acid bacteria for support their growth[30].Total lactic acid bacteria on group fortification at 0 hour of fermentation has the highest number. It could be caused because the addition of vitamin D_3 was being used by lactic acid bacteria to growth. A previous study showed that at 0 hour of fermentation, the amount of lactic acid bacteria and acetic acid bacteria was about 6 log unit, it means the amount was higher than yeast[36]. Thus, lactic acid bacteria in group of fortification at 0 hour of fermentation performed high lypolytic activitiy. On group of fortification at 6, 12, 18, and 24 hours the addition of vitamin may be used by yeast and acetic acid bacteria. It cause the number of lactic acid bacteria on these groups was lower than the group fortification at 0 hour of fermentation and the control group. Microorganisms in kefir perfomed symbiosis, so it keep the stability of the products[24][54].A previous study showed that at 6 hours of fermentation, the amount of yeast and *Lactobacillus* spp. increased significantly. At 12 hours of fermentation, lactic acid bacteria and acetic acid bacteria show a significant increase, meanwhile yeast reached its growth peak at this hour and remained stable for the rest of fermentation. At 18 hours of fermentation, *Lactobacillus* spp. reached its growth peak. At 24 hours of fermentation, acetic acid bacteria reached its growth peak[36].

Lactic acid bacteria has symbiosis interaction with yeast. *Lactococcus*hydrolyze lactose, then produce lactic acid, thus create a good environment for yeast growth. Bacterial growth would increase, then release organic acids that would be fermented by yeast. Besides that, yeast provide nutrition for the growth of bacteria such as amino acids, vitamins, and other component. Both of the microorganisms may also competing each other in order to get nutrition for supporting their growth, and could also produce metabolite that suppress or stimulate each other's growth. This symbiosis interactions will create the stability of the products[24][54].

4.6 Viscosity

Lactic acid bacteria inside kefir could produce exopolysaccharide (EPS) during fermentation. EPS that was generated would effect rheology of the fermented products[30]. EPS that produced during kefir fermentation is called kefiran[22]. In this study, kefir that fortified at 0, 6, 12, and 18 hours of fermentation has lower viscosity compared to control group and group hat was fortified at 24 hours of fementation. Fortification at 0, 6,12, or 18 hours of fermentation mean vitamin was added when lactic acid bacteria on growing phase[36]. The decreased of the viscosity could be affected by EPS production of lactic acid bacteria. The amount of EPS that was produced could be affected by temperature, pH and the amount of yeast that was inoculated. The more yeast inoculated may increase the production of EPS by lactic acid bacteria, but may also decrease the lactose available for EPS production[53]. Fortification at 24 hours of fermentation means the growth of lactic acid bacteria reach its optimum[36]. Moreover, the number of fat content may also contribute to the viscosity, whereas in this study the fat content in treatment group is lower than control group[51].

4.7 Acidity (pH)

Acidity (pH) of kefir in this study ranged from 4,45 to 4,7. Kefir has characteristic which has pH 4,2-4,6[32]. Kefir on other study has pH 4,85 after fermentation[36]. Lactic acid bacteria inside kefir could degradade lactose and yield lactic acid during fermentation. The presence of lactic acid could lower milk pH[31]. A study found that the presence of yeast could reduce the pH lower than single lactic acid bacteria culture[26]. Yeast use milk nutrition, such as protein, fat, lactose, and citrate, then release amino acid, vitamin, and other components to support the growth of lactic acid bacteria[24][26]. Yeast could metabolize lactic acid[26]. The higher kefir grains inoculated, the lower pH would be[49]. In this study, the amounts of kefir grains inoculated was the same (5%), thereby it explain why the results of pH obtained in all groups were unsignificantly different.

5. Conclusion

Vitamin D_3 fortification time on goat milk kefir effects vitamin D_3 content, fat content, and viscosity of goat milk kefir after fermentation. The highest vitamin D_3 content was found on the group forticication after 6 hour of fermentation. In general, fat and crude fiber content on all fortification groups were lower than control group. Meanwhile, viscosity of group fortification at 0, 6, 12, and 18 of fermentation were lower than control group and group of fortification at 24 hour of fermentation. Acidity (pH) of kefir in this study ranged from 4,45 to 4,7. All kefir samples in this study had met Codex Standard of Total Lactic Acid Bacteria in kefir ($\geq 10^7$ CFU/ml). Vitamin D_3fortification performed at 6 hour of fermentation was recommended for future research. Encapsulation of vitamin D_3 should be considered to increase the availability of vitamin D_3.

6. Acknowledgement

The author would like to thank kefir team and supervisor to supporting this research. We are also grateful to Faculty of Medicinefiscal year 2016 for funding this research.

References

[1] Litbang Kementerian Kesehatan RI 2013*Riset kesehatan dasar* (Jakarta: Kementerian Keseharan RI)p.247
[2] Mihardja L, Delima, Siswoyo H, Ghani L, Soegondo S 2009*Acta Med Indones-Indones J Intern Med***41** 169–74
[3] Kementrian Kesehatan RI 2014 *Situasi dan Analisis Diabetes*(Jakarta: Kementerian Keseharan RI) p.1-2
[4] Gallagher EJ, LeRoith D, Karnieli E 2010*Mt Sinai J Med***77** 511–23
[5] Nelms M, Sucher KP, Lacey K, Roth SL 2011*Nutrition therapy and pathophysiology*(Wadsworth: Cengage Learning) p. 199-200
[6] Alissa EM, Alnahdi WA, Alama N, Ferns GA 2015*J Clin Transl Endocrinol***2**42–7
[7] Von Hurst PR, Stonehouse W, Coad J 2010*Br J Nutr***103** 549
[8] Kampmann U, Mosekilde L, Juhl C, Moller N, Christensen B, Rejnmark L, Wamberg L, Orskov L 2014*J Metabolism***63** 1115–24
[9] Bender DA 2008*Introduction to nutrition and metabolism*(Chicester: Wiley-Blackwell) p. 335-6
[10] World Health Organization and Food and Agriculture Organization of the United Nations 2006*Guidelines on food fortification with micronutrients*(France) p.22, 24, 81-4, 130-1.
[11] Oren Y, Shapira Y, Agmon-Levin N, Kivity S, Zafrir Y, Altman A, Lerner A, Shoenfeld Y 2010 *Isr Med Assoc J***12** 751–6
[12] Nimitphong H, Holick MF 2013*Dermatoendocrinol***5**34–7
[13] Poh BK, Rojroongwasinkul N, Nguyen BKL, Sandjaja, Ruzita AT, Yamborisut U, Hong TH, Ernawati F, Deurenberg P, Parikh P 2016*Asia Pac J Clin Nutr***25** 538–48.
[14] Hidayat R, Setiati S, Soewondo P 2010 *Acta Med Indones***42** 123–9
[15] Park YW, Juarez MJ, C MR, Haenlein GFW 2007*Small Rumin Res***68**88–113
[16] Leite AMO, Miguel MAL, Peixoto RS, Paschoalin VMF, Mayo B 2015*J Dairy Sci***98**3622–32
[17] Nielsen B, Gurakan GC, Unlu G. 2014 *Probiotics & Antimicro Prot***6** 123–35
[18] Liu JR, Lin YY, Chen MJ, Chen LJ, Lin CW2005 *J Anim Sci***18** 567-73
[19] Ostadrahimi A, Taghizadeh A, Mobasseri M, Farrin N, Payahoo L, Beyramalipoor Gheshlaghi

Z, Vahedjabbari M2015 *Iran J Public Health***44** 228–37
[20] Teruya K, Yamashita M, Tominaga R, Nagira T, Shim SY, Katakura Y, Tokumaru S, Tokumaru K, Barnes D, Shiharata S2003 *J Cytotechnology***40** 107–16
[21] Martharini D, Indratiningsih I 2017*Agritech***37** 22–9
[22] Farnworth ER 2003*Food Science and TechnologyBulletin: Functional Foods* **2** 1–17
[23] World Health Organization and Food and Agriculture Organization of the United Nations 2011 *Codex Standard for Fermented Milks 243-2003*(Rome: Codex Aliment) p.7-8.
[24] Viljoen BC 2001 *Int J Food Microbiol***69** 37–44
[25] Ferreira IMPLVO, Pinho O, Monteiro D, Faria S, Cruz S, Perreira A, Roque AC, Tavares P 2010 *J Dairy Sci***93** 27–31
[26] Álvarez-Martín P, Flórez AB, Hernández-Barranco A, Mayo B2008 *Food Control***19** 62–70
[27] Hafeez Z, Cakir-kiefer C, Roux E, Perrin C, Miclo L, Dary-mourot A 2014*J Foodres***63** 71–80
[28] Chen M, Liu J, Lin C, Yeh Y 2005*J Anim Sci***18**711-5
[29] Jenab A, Roghanian R, Emtiazi G 2015*J Med Bacteriol***4** 55–66
[30] Hayek SA, Ibrahim SA 2013 *Food Nutr Sci***4** 73–87
[31] Haryadi, Nurliana, Sugito 2013*J Medika Veterinaria***7** 1–4
[32] Ayustanigwarno F 2013*Ilmu dan Teknologi Pangan*(Semarang: Logo Q-A Undip) p. 383
[33] Arif S, Vieth R 2007*International Dairy Journal***17** 753–9
[34] Cashman KD 2015*J Steroid Biochem Mol Biol***148**19–26
[35] Kaushik R, Sachdeva B, Arora S, Kapila S, Wadhwa BK 2014*J Food Chem***147** 307–11
[36] Leite AMO, Miguel MAL, Peixoto RS, Ruas-Madiedo P, Paschoalin VMF, Mayo B, Delgado S 2013*J Dairy Sci***96**4149–59
[37] Patterson KY, Phillips KM, Horst RL, Byrdwell WC, Exler J, Harnly JM, Holden JM, Lemar LR, Pehrsson PR, Wolf WR 2008*Beltsville Human Nutrition Research Center*
[38] Setyawardani T, Rahardjo AHD, Sulistyowati M, Wasito S 2014 *Animal Production***16** 48–54
[39] Kumar ASHOK, Kumar RG 2011*Int J Pharmagenes***2**43–8
[40] Hasanvand E, Fathi M, Bassiri A, Javanmard M 2015*Food Bioprod Process***96** 264–77
[41] Tangpricha V, Koutkia P, Rieke SM, Chen TC, Perez AA, Holick MF 2003*Am J Clin Nutr***77** 1478–83
[42] Forrest SA, Yada RY, Rousseau D 2005*J Agric Food Chem***53** 8003–9
[43] Bulgari O, Caroli AM, Chessa S, Rizzi R, Gigliotti C 2013*Molecules***18** 10122–31
[44] Zajšek K, Goršek A, Kolar M 2013 *Food Chem***139** 970–7
[45] Hohman EE, Martin BR, Lachcik PJ, Gordon DT, Fleet JC, Weaver CM 2011 *J Agric Food Chem***59**2341–6
[46] Jones G. 1978 *Clin Chem***24** 287–98
[47] Purwanto MGM 2014*J Ilm Sains Teknol***7** 64–71
[48] Satir G, Guzel-seydim ZB 2016*J Small Rum Res***134** 1–7
[49] Irigoyen A, Arana I, Castiella M, Torre P, Ibanez FC2005 *J Food Chemistry***90** 613–20
[50] Bintsis T, Vafopoulou-Mastrojiannaki A, Litopoulou-Tzanetaki E, Robinson RK 2003*J Appl Microbiol***95** 68–77
[51] Antoniou KD, Topalidou S, Tsavalia G, Dimitreli G. Effect of Starter Culture, Milk Fat and Storage Time on the Rheological Behaviour of Kefir.
[52] Jin H, McCaffery JM, Grote E 2008*J Cell Biol***180** 813–26
[53] Gorret N, Maubois JL, Engasser JM, Ghoul M 2001*J Appl Microbiol***90** 788–96
[54] Leite AMO, Miguel MAL, Peixoto RS, Rosado AS, Silva JT, Paschoalin VMF 2013*Brazilian Journal of Microbiology***44** 341–9

Effect of Protein-Based Edible Coating from Red Snapper (*Lutjanus* sp.) Surimi on Cooked Shrimp

I Rostini, B Ibrahim and W Trilaksani

[1] Department of Fishery Products Processing, Faculty of Fisheries and Marine Science, Padjadjaran University, Jl. Raya Bandung Sumedang Km. 21 Jatinangor 45363, Indonesia
[2] Department of of Aquatic Processing Technology, Faculty of Fisheries and Marine Science, Bogor Agricultural University, Jl. Agatis Babakan Dramaga Bogor 16680, Indonesia

E-mail: iis.rostini@unpad.ac.id

Abstract. Surimi can be used as a raw material for making protein based edible coating to protect cooked shrimp color. The purpose of this study was to determine consumers preference level on cooked shrimp which coated by surimi edible coating from red snapper and to know the microscopic visualization of edible coating layer on cooked shrimp. The treatments for surimi edible coating were without and added by sappan wood (*Caesalpinia sappan Linn*) extract. Application of surimi edible coating on cooked shrimp was comprised methods (1) boiled then coated and (2) coated then boiled. Edible coating made from surimi with various concentrations which were 2, 6, 10 and 14% of distillated water. The analysis were done using hedonic test and microscopic observation with microscope photographs. Effect of surimi edible coating on cooked shrimp based on the hedonic and colour test results showed that the 14% surimi concentration, added by sappan wood (*Caesalpinia sappan Linn*) extract on edible coating was the most preferable by panellist and giving the highest shrimp colour. The edible coating surimi application on cooked shrimp which gave the best result was processed by boiling followed by coating.

Keywords: edible coating, cooked shrimp, surimi, shrimp colour, shrimp coating

1. Introduction

Cooked shrimp is a value added product with high protein content, specific taste, ready to eat, have an interested colour for consumers, special flavour, and easy to serve. Cooked shrimp meat colour is red or orange. The redness and yellowness of shell were obviously higher than those of muscle, resulting from the higher content of carotenoids in shell than those in muscle [1]. Astaxanthin is the most effective pigment on shrimp colour [2]. Shrimps contain the red carotenoid astaxanthin, which is combined with proteins in the carapace, forming a green carotenoid–protein complex [3]. Protein denaturation caused red astaxanthin was released during the process of shrimp cooking, so that the shrimp colour change to be red [4]. The problems in the cooked shrimp during storage are discoloration, protein denaturation, and changes in texture [5].

Cooked shrimp must be protected from quality deterioration during storage. The effort to protect shrimp from deterioration during storage, in general is glazing [6]. According to [7], to expand the storage life and maintain product quality can be used edible coating. Edible coating also can be used to inhibit the microbial growth on the surface of fresh processed product [8]. Surimi in food industry can be used as material to make edible packaging or more famous as a protein based edible film and edible coating [9]. Edible film and coating are potential as packaging materials since it has potential to increase the food quality, food safety, and shelf life of the product.

The colour of cooked shrimp was one of the main sensory attribute that affect the quality and acceptance of the product. The ability of surimi edible coating in maintaining cooked shrimp colour can be applied with sappan wood (*Caesalpinia sappan Linn*) as natural dyes. Sappan wood has antioxidant activity [10] and antimicrobial activity [11], so that can be produced with good colour and high quality product. Edible coating and antimicrobial agent can be combined during the film making process to increase the safety and shelf life of ready-to-eat food [8]. The aim of the study were to investigate consumers accepted level on cooked shrimp which coated by surimi edible coating and to know the thickness of edible coating layer on cooked shrimp.

2. Materials and Methods

2.1. Materials

Meat of red snapper fillet waste was obtained from Muara Angke, North Jakarta. Peeled undevined_vannamei shrimp (*Litopenaeus vannamei*) with 60-70 size (in 1 kg exist 60-70 PUD shrimp), sappan wood (*Caesalpinia sappan Linn.*), water, salt, ice, cryoprotectant (sugar), filter paper Whatman No.1, cling film, and Styrofoam.

2.2. Methods

The research consisted of several steps, preparation of surimi from meat of red snapper fillet waste, sappan wood (*Caesalpinia sappan Linn*) extraction, preparation of edible coating in various surimi concentration (2, 6, 10, and 14%), and application surimi edible coating to cooked shrimp.

2.2.1. Preparation of surimi from meat of red snapper (Lutjanus sp.) fillet waste [12]. Fish meat that has been separated from the rest of the fillet was milled using a meat grinder in order to produce a smooth and homogeneous meat without bones, thorns, and filth. After that, fish meat was washed twice using cold water (15±1) °C and 0.3% saline solution (w/w). Immersion in cold water (ratio water : the meat is 3 : 1) performed for 10 minutes to clean the dirt still attached to the meat and mashed to dissolve the sarcoplasmic proteins. Fish meat then dewatering by using calico cloth to remove the water. Second submersion with saline solution 0.3% (w/w) (the ratio of the volume saline solution : meat is 3 : 1) for 10 minutes, then filtered again using calico cloth, made extortion. Cryoprotectant as much as 2% (w/w) was added and mixed using a food processor until homogeneous. The addition of cryoprotectant was done to prevent protein denaturation in surimi. Surimi was packaged with polyethylene plastic and stored in a freezer at -15 °C, then used as raw material for edible coating.

2.2.2. Sappan wood (Caesalpinia sappan Linn) extraction. Dry sappan wood milled to reduce the size by using the Hammer Mill screened with 40 mesh sieve. Sappan wood shaving used for the next stage of the extraction stage. Wooden sappan pigment extraction is done using [13] method with water as solvent. Material (100 g) extracted with 1 L of water and repeated 3 times for 30 minutes at 80 °C. after that was filtered with a vacuum filter using Whatman No.1 paper and the filtrate pH was measured. Extract was concentrated by vacuum evaporator at 40 °C to remove residual solvent in order to obtain the extract in the form of dry powder. Sappan wood extract powder was mixed into edible coating as a natural dye for cooked shrimp as much as 2.5 mg/ml. Sappan wood extract on that concentration had the highest antioxidant activity [14], so hopefully defend cooked shrimp color during storage.

2.2.3. Preparation of the surimi edible coating. Surimi based edible coating from meat of red snapper fillet waste method is a modification from [9], found that a stable edible film has been formed from

Alaska Pollack with a concentration of 2%. According to [15], showed that elected edible film from by trash fish surimi was edible with added by 10% surimi. Based on these results, the range used in this study the addition of surimi for edible coating are 2, 6, 10, and 14% (w/v) of distilled water. Frozen surimi thawed for 20 minutes. After the thawed surimi was stirred in distillated water to 150 ml for 30 min, 55 °C temperature, pH the coating solution was adjusted to 11 with 1 M NaOH. The surimi solution dispersed thoroughly using 150 mesh nylon, the filtrate that resulted was surimi edible coating. The treatments for surimi edible coating were without and added by sappan wood extract.

2.2.4. Preparation of cooked shrimp. Shrimp used in the study was Vannamei shrimp (*Litopenaeus vannamei*) PUD (peeled undevine) form with size of 60-70 (in 1 kg contained as much as 60-70 PUD shrimp). Fresh shrimp peeled and then washed with cold water. Shrimp simmered in boiling water for 5 minutes. Determination of long boiling based on [16]. After cooked, drained for the next shrimp dipped in edible coating and packed.

2.2.5. Application of edible coating to cooked shrimp. Surimi based edible coating that has formed, then applied as a coating of cooked shrimp with dipping method (30 minutes). Determination of long immersion was based on [17], state that immersion cooked shrimp for 30 min in a solution of edible coating can maintain the shelf life of cooked shrimp. The steps of surimi edible coating application consist of five types:
K = without edible coating
A = coated, boiled- without sappan wood extract
B = coated, boiled- added by sappan wood extract
C = boiled, coated– without sappan wood extract
D = boiled, coated- added by sappan wood extract.
Cooked shrimp that has been coated with edible coating, thickness of the coating was observed using an electron microscope with a magnification of 10 times. Surface cooked shrimp was also observed using an electron microscope to determine the brightness and color of the surface.

2.2.6. Organoleptic test.
Organoleptic test was conducted to determine the level of preference of panellist to appearance, colour, aroma, and taste of cooked shrimp coated edible coating. organoleptic assessment done by the hedonic test [18], that is somewhat trained panellist used as many as 30 people. Material presented at random with a given code numbers, then panellists were asked to pass judgment on any of the criteria hedonic scale. Observations presented in numbers of 1-7, in the following order: 1 (strongly dislike), 2 (do not like), 3 (somewhat like), 4 (neutral), 5 (somewhat like), 6 (likes), and 7 (strongly like).

2.2.7. Method of slice
Edible coating thickness measurements on cooked shrimp, done by making preparation with glass objects to do a photo shoot under the microscope [19]. The method of slice sliced by hand. Slice method is a method of making preparation by making an incision with a certain thickness, so it can be examined under a microscope. Method of slice by hand is a follows: a piece of tissue is held between the thumb pointer, then the network is cut crosswise with a sharp knife several times in rapid, parallel and as close as possible to the surface of tissue to be cut, in order to obtain slices as thin as possible. Furthermore, these thin slices that can be observed under the microscope.

3. Result and Discussion

3.1. Organoleptic assessment with hedonic test
Hedonic test performed in this study aimed to determine the level of consumer acceptance of cooked shrimp coated edible coating. Edible coatings composed of two types, there are without and added by sappan wood extract as much as 2.5 mg/ml. Characteristics tested include appearance, colour, aroma

and flavour. Hedonic test result cooked shrimp are coated surimi edible coating presented in Figure 1. And the hedonic test result for cooked shrimp that coated surimi edible coating with added by sappan wood extract presented in Figure 2.

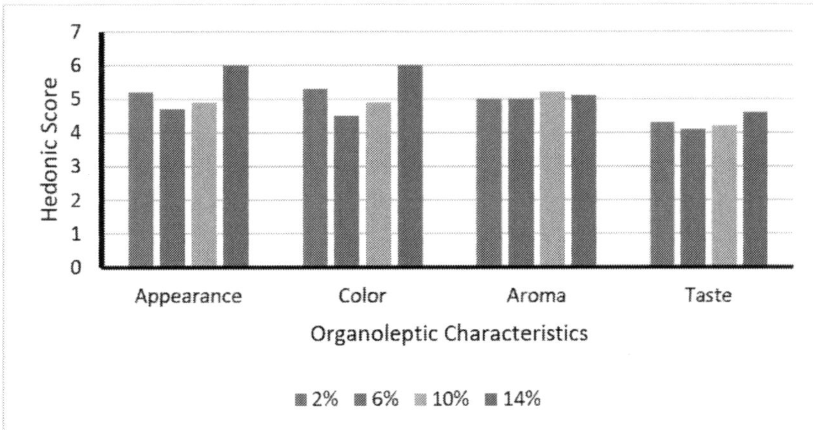

Figure 1. Hedonic test result on cooked shrimp coated surimi edible coating without sappan wood extract.

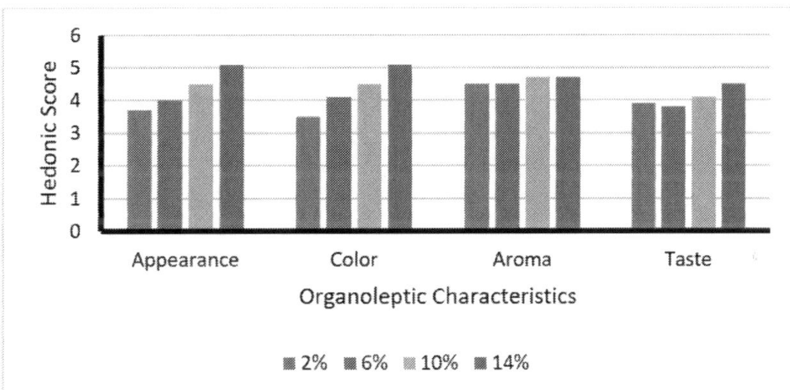

Figure 2. Hedonic test result on cooked shrimp coated surimi edible coating added by sappan wood extract 2.5 mg/ml.

The range average value of cooked shrimp appearance coated by edible coating is 4.7 to 6.03. The lowest value was given cooked shrimp surimi edible coating by 6%, while the highest values were found in cooked shrimp coated with edible coating concentration of 14% surimi. The range average value of cooked shrimp appearance which coated by surimi edible coating with sappan wood extract addition is 3.7 to 5.1. The highest value at a concentration of 14%. Based on Kruskal-Wallis test, the result that surimi concentration of surimi edible coating did not give significant effect on the appearance of cooked shrimp, surimi edible coating with added sappan wood extract providing significant effect on the appearance of cooked shrimp. The average value of the appearance of cooked shrimp are coated surimi edible coating and cooked shrimp are coated surimi edible coating with addition of sappan wood extract has the highest value at a concentration of 14% surimi. This happens because the concentration of 14% surimi to form edible coating well. Edible coatings can cover the surface with a perfectly when applied, so as to make the surface of cooked shrimp looks clear, transparent, shiny and bright. According to [20], the use of edible coatings can reduce the rate of damage during the process, improve the texture and appearance of the product.

Colour is a very important sensory attributes and should always be considered, because it has a direct influence on the quality of a product [21]. Good pigmentation and homogenized in a food is to the quality characteristics that determine consumer acceptance. Products with attractive colours will be more accepted by consumers, although with a more expensive price [22]. The average value of the range of colour coated cooked shrimp surimi edible coating is 4.5 to 6. The lowest value was given cooked shrimp surimi edible coating 6%, while the highest values were found in cooked shrimp coated with edible coating concentration of 14% surimi. The range average value of cooked shrimp colour that coated edible coating surimi given the sappan wood extract is 3.5 to 5.03. Lowest value at a concentration of 2% surimi and surimi highest value at a concentration of 14%. Based on the analysis using the Kruskal-Wallis test, the result that the surimi concentration of the surimi edible coating and surimi edible coating with sappan wood extract giving significant effect to the colour of cooked shrimp. The average value of cooked shrimp colour that coated by surimi edible coating and cooked shrimp with edible coating with sappan wood extract addition has the highest value at a concentration of 14% surimi. Concentration of 14% in the surimi edible coating was applied to improve the colour of cooked shrimp, cooked shrimp have brighter colours and shiny, and also liked by the panellists. Edible coatings are effective in reducing the decline in sensory quality products that include colour, smell, and firmness [23].

The average value of cooked shrimp aroma that coated by edible coating ranged from 5.06 to 5.13. The lowest value was given cooked shrimp surimi edible coating of 2%, while the highest values were found in cooked shrimp coated with edible coating concentration of 14% surimi. The average value of cooked shrimp aroma that coated by surimi edible coating with sappan wood extract ranged from 4.50 to 4.73. The lowest value at a concentration of 6% surimi and the highest value at a surimi concentration of 14%. Based on the analysis using the Kruskal-Wallis test, resulted that the surimi concentration of the surimi edible coating and surimi edible coating with sappan wood extract giving non significant effect on the aroma of cooked shrimp. This happens because the surimi edible coating has a neutral flavour, so when applied to the aroma of cooked shrimp poses no scent in defiance of cooked shrimp.

Taste is an important factor on which to base decisions taken by the customer to the receipt of a product. If a product has a bad taste, then the product will not be accepted by consumers, although the colour and smell good [24]. Average yield assessment panellist to taste cooked shrimp are coated surimi edible coating ranged 4.16 to 4.66. The lowest value was given cooked shrimp surimi edible coating 6%, while the highest values were found in cooked shrimp coated with edible coating concentration of 14% surimi. Averaging the results of the taste panel assessment of cooked shrimp are coated surimi edible coating with sappan wood extract ranged from 3.83 to 4.56. Lowest value at a concentration of 2% surimi and surimi highest value at a concentration of 14%. Based on the analysis using the Kruskal-Wallis test results obtained surimi concentration in the edible coating and surimi edible coating with sappan wood extract giving significant effect to the taste of cooked shrimp.

Based on the results of the hedonic test, the application of surimi edible coating on cooked shrimp showed that the concentration of the edible coating was applied to the cooked shrimp, which is most preferred by the panellists was 14%. The concentrations of both most preferred edible coating without or added by sappan wood extract on edible coating.

3.2. Visualization of surimi edible coating application on cooked shrimp

Coating method used in this study is the method of immersion (dipping). This method is a method of coating applications, where the product to be coated is dipped in surimi edible coating is used as a coating on the packaging or cooked shrimp. The entire surface of the cooked shrimp will be covered by edible coating after the dipping process. Edible coating thickness can affect the appearance of the packaged product. Thickness is formed due to the expansion or development of the denatured protein molecules in surimi red snapper fillet waste so open reactive groups that exist in the polypeptide chain. Bonds between protein reactive groups that hold all the liquid, it will be formed gel [24].

Based on the results of a microscopic photograph (Figure 3), reflecting the differences among all treatments. Visually, the difference in the step of application an edible coating of cooked shrimp has different effects on edible coating thickness. Controls which were not given edible coating surface appears shrimp with a clear surface lines. Shrimp fed edible coating before cooking, it does not look clear edible coating layer on the surface of cooked shrimp once through the cooking process. This happens because during the process of cooking shrimp, edible coating suffered due to denaturation temperature is as high as 100 ° C and the physical pressure of the boiling water movement, so that the edible coating that covers the shrimp become detached as a result of unstable layer (Figure 3, A and B). Gel structure on edible coating contained surimi also be destroyed at temperatures above 50 °C. When heating the gel increased to above 50 °C, the gel structure will be destroyed, the enzyme will break down the structure of the three-dimensional gel network that has been formed so that surimi gel will be brittle and lose its elasticity [12]. Edible coating thickness results visualization under a microscope is presented in Figure 3.

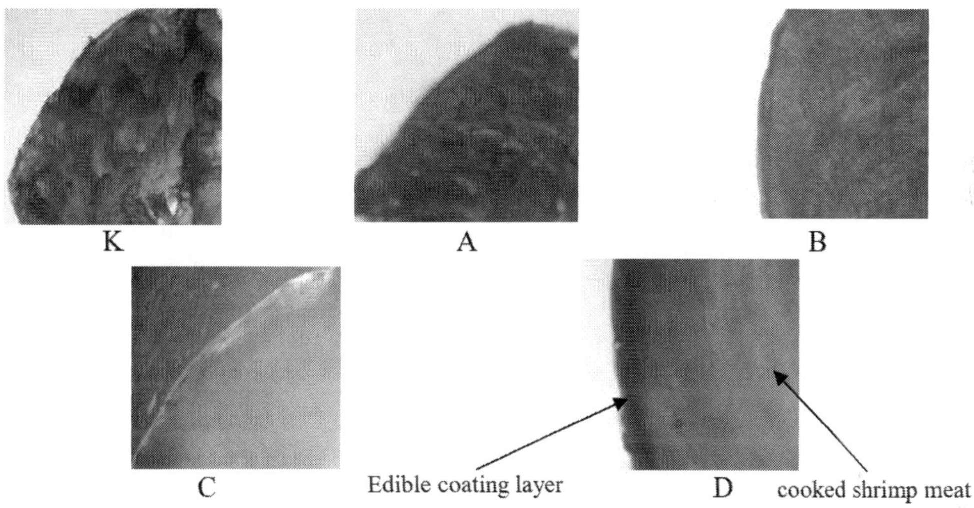

Figure 3. Cross section of cooked shrimp on a variety of surimi edible coating treatments (magnification 10 times). K = without edible coating; A= coated, boiled-without sappan woos extract; B=coated, boiled-added sappan wood extract; C = boiled, coated-without sappan wood extract; D = boiled, coated-added sappan wood extract.

The application step of edible coating to shrimp after cooking, edible coating layer on the surface of the shrimp can be seen clearly (Figure 3, C and D). This happens because the entire surface is covered shrimp stick coating evenly. Surimi edible coating has formed gel strength is stable so that when applied to the cooked shrimp capable shrimp paste and cover the surface. Fish myofibril proteins have the ability to form a three-dimensional gel network is stable [25]. Edible coatings are also applied to the cooked shrimp after untreated can cause damage to the stability of edible coating. Edible coating added by sappan wood extract layer shows the results of bright red on the surface of cooked shrimp. Edible coatings without sappan wood extract shows the results of a transparent coating, bright and shiny. Giving edible coating after the cooking process the meat showed was more soft texture and compact. The use of edible coating to reduce the rate of damage during the process, improve the texture and appearance of the product [20].

Cooked shrimp coated surfaces by edible coating has a bright and shiny appearance. Cooked shrimp are coated edible coating with the addition of sappan wood extract has a red colour. The red colour on cooked shrimp sensory attributes that affect the quality and acceptance of food products. Microscopic observations of the surface of cooked shrimp is presented in Figure 4.

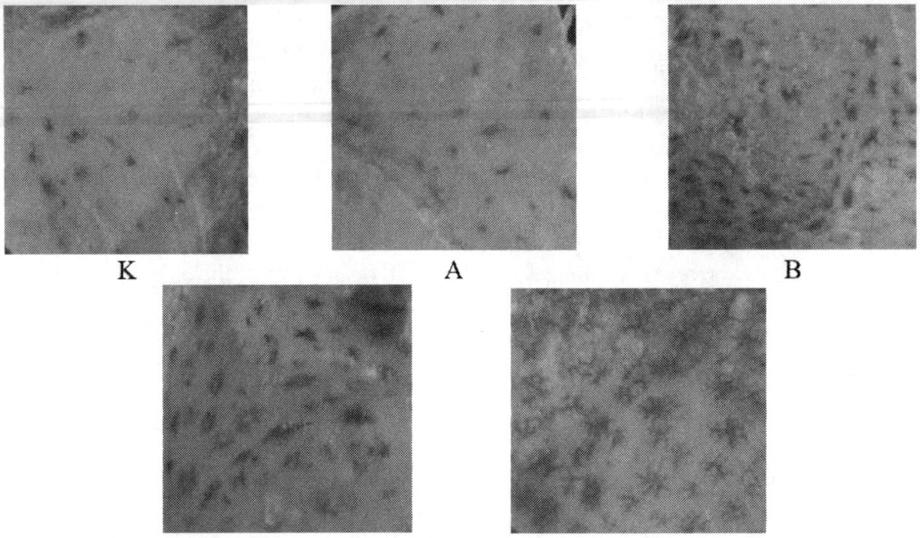

Figure 4. Cooked shrimp surface with a variety of surimi edible coating, treatment was observed by microscopy (magnification 10 times).

Microscopic observations on surface cooked shrimp without coated by edible coating possess a faint appearance compared to all treatments. Cooked shrimp which were edible coating all surfaces have a transparent appearance, bright and shiny. Edible coatings are added by sappan wood extract gives red colour to the cooked shrimp.

4. Conclusion

Effect of surimi edible coating on cooked shrimp based on the hedonic test results showed that the 14% surimi concentration, added by sappan wood extract on edible coating was the most preferable by panellist and giving the highest shrimp colour. The edible coating surimi application on cooked shrimp which gave the best result was processed by boiling followed by coating.

Reference

[1] Bono G, Gai F, Peiretti PG, Badalucco C, Brugiapaglia A, Siragusa G, Palmegiano GB 2012 *Food Chem.* **130**:104-110.
[2] Arredondo-Figueroa JL, Pedroza-Islas R, Poce-Palafox JT, Vernon-Carter EJ 2003 *Revista Mexicana de Ingeniería Química, Iztapalapa 2*: 101-108.
[3] Belitz HD, Grosch W, Schieberle P 2004 Food chemistry. Berlin: Springer-Verlag.
[4] Martinez-Alvarez O, López-Caballero ME, Gómez-Guillén MC, Montero P 2009 *LWT-Food Sci Tech.* **42**: 1335-1344.
[5] Erdogdu F, Balaban MO, Otwell WS, Garrido L 2004 *J. Food Eng.* **64** : 297-300.
[6] Bottino NR, Lilly ML, Finne G 1979 *J. Food Sci.* **44**: 1778-1779.
[7] Kilincceker O, Dodan IS, dan Kucukoner E 2009 *Food Sci Tech.* **42** : 868-873.
[8] Cagri A, Zeynep U, dan Elliot T R 2004 *J. Food Protect.* **67** : 833-848.
[9] Shiku Y, Hamaguchi PY, Benjakul S, Visessanguan W, Tanaka M 2004 *J. Food Chem.* **86** : 493-499.
[10] Lim DK, Choi U, Shin DH 1997 *Korean J. Food Sci. Technol.* **28**: 77−82.
[11] Winarti C, Sembiring, BS 1998 Warta Tumbuhan Obat Indonesia **4(3)**: 17−18.

[12] Suzuki T 1981 *Fish and Krill Protein, Processing Technology*. (London: Applied Science Publ.Ltd).
[13] Ye Min, Xie W, Lei F, Meng Z, Zhao Y, Su H, Du L 2006 *J. Int Immunopharmacol* **6**: 426-432.
[14] Weningtyas H 2009 Efek pencampuran pigmen kayu secang *(Caesalpinia sappan L.)* dengan beberapa sumber antosianin terhadap kualitas warna merah dan sifat antioksidannya [skripsi]. (Bogor: Fakultas Teknologi Pertanian, Institut Pertanian Bogor).
[15] Neviana Y 2007 *Edible film* berbahan dasar protein surimi ikan rucah [skripsi] (Bogor: Fakultas Perikanan dan Ilmu Kelautan, Institut Pertanian Bogor).
[16] Julikartika EP 2003 Karakterisasi *edible coating* dari alginat hasil ekstraksi rumput laut *Sargassum* sp. untuk pelapis udang [tesis] (Bogor: Sekolah Pascasarjana, Institut Pertanian Bogor).
[17] Riyanto B 2006 Pengembangan pelapis *edible* dari *isinglass* dan aplikasinya untuk mempertahankan mutu udang masak [tesis] (Bogor: Sekolah Pascasarjana, Institut Pertanian Bogor).
[18] Soekarto ST 1985 *Penilaian Organoleptik*. (Jakarta: Bina Aksara).
[19] Suntoro SH. 1983 *Metode Pewarnaan* (Jakarta: Penerbit Bhratara Karya Aksara).
[20] Krochta JM 1992 Control of Mass Transfer in Food with Edible Coating and Film. Dalam: Singh RP (Ed) *Advance Food Engineering* (Boca Raton: CRC Press).
[21] Niamnuy C, Devahastin S, Soponronnarit S, Raghavan GSV 2008 *J. Food Eng.* **87**: 591–600.
[22] Delgado F, Paredes VO, and Lopez 2003 *Natural Colorant for Food and Nutraceutical Uses* (Boca Raton: CRC Press LLC).
[23] Mastromatteo M, Danza A, Conte A, Muratore G, Matteo Nobile MAD 2010 *Int J. Food Microbiol.* **144**: 250–256
[24] Winarno FG 2008 *Kimia Pangan dan Gizi* Jakarta: PT. Gramedia Pustaka Utama.
[25] Yoon WB, Gunasekaran S, Park JW 2004 *Applied Rheology* : 133-139.

The Environmentally Sound Aquaculture Strategies Based on Bioaccumulation of Heavy Metal of Lead (Pb) on Seaweed of *Gracilaria verrucosa* on Aquaculture Areas of MuararejaVillage, Tegal City

Nurjanah[1*], Ambariyanto[2], Supriharyono[2], BambangYulianto[2]

[1]Faculty ofFisheries and Marine Science, Pancasakti University, Tegal, Indonesia
[2]Faculty of Fisheries and Marine Science, Diponegoro University Jl. Prof.Soedharto, SH, Tembalang, Semarang.
*nur_janah1963@yahoo.com

Abstract. Community activities such as industry, trade, animal husbandry and agriculture and ssettlements resulting in heavy metals of lead (Pb) can be accumulated in water, sediment and seaweed *Gracillaria verrucosa*. It can contaminate ponds and affect aquaculture activities in Tegal. Seaweed *Gracilaria verrucosa* is afisheries commodity that has economical value and cultivated in the area of aquaculture MuararejaTegal. It can serve as fitoremedian that will help reduce the impact of heavy metal pollution due to its ability to accumulate pollutants. The objective of this study was to analyze bioaccumulation of heavy metals of lead (Pb) and its relationship with water quality management in order to develop seaweed cultivation of *Gracillaria verrucosa* in ponds in the area of aquaculture MuararejaTegal. The method used in this study is a survey, analysis of heavy metals of lead (Pb) in pond water, sediment and seaweed using *Atomic Absorption Spectrophotometer* (AAS) and the data were analyzed by descriptive quantitative. Bioconcentration of lead (Pb) during the dry season in pond water, sediment and seaweed *Gracillaria verrucosa* was measured from 0.003 to 0.025 ppm,5.543 to 23.699 ppm and 0.209 to 0.326 ppm respectively. While in the rainy season bioconcentration of lead (Pb) are from 0.003 to 0.015 ppm, sediment from 6.377 to 9.858 ppm and 0.209 to 0.326 ppm respectively. Bioconcentration of Pb in dry season was higher than in the rainy season and the biggest bioconcentration was found in the sediment pond waters. Pb bioaccumulation low and still below the quality standards of the Ministry of Environment decision 51 of 2004 so that the product is safe for consumption.

Keywords: Pond management strategies, environmental, heavy metals Pb, seaweed *Gracilaria verrucosa*

1. Introduction

Population activities such as agriculture, animal husbandry, fisheries, industryand settlements cause various pollutions including heavy metals such as lead (Pb). This pollutant will contaminate the aquatic environment including the pond, mainly in the Muarareja village. The village has the widest pondin Tegal.

Heavy metals Pb can be derived from nature or impact of population activities. Generally, heavy metals enter into the aquatic environment through atmospheric deposition, erosion of geological matrix or due to anthropogenic activities caused by industrial effluents, domestic sewage and mining wastes [1].

Qari and Shidiqy [2] said that heavy metals are toxic and not degradable, but can be absorbed through the water and the food chain and can accumulate inseaweed.

Pollutants that enter the aquatic environment will undergo a process of accumulation, either physical, chemical or biological. The accumulation of physical and chemical pollutants will precipitate in the bottom waters as sediment of pond, while biological pollutant will be absorbed by organisms.
Wardhana [3] states that the process of moving a toxic substance from the water to the organism is called bioconcentration; while the movement of toxic compounds through the food chain to an organism is called biomagnification. Toxic materials such as Pb if absorbed will accumulate in water, sediment and seaweed. It will be having an effect on the sustainability of aquaculture. The content of Pb high in the water can also be a negative impact on the organisms in it as said by Palar [4] that the Pb concentration reached 188 mg/l, can kill fish.

Pb containe in the water will cause the process of bioaccumulation in the body of biota in the waters, such as fish. The content of Pb in the body would interfere with the activity of enzymes, such as aminolevulinic acid dehidrase (-ALAD), Hem synthetase, and another enzyme involved in the systemhemotopoietik.

Gracilaria verrucosa, one of a class of marine algae Rhodophyta (red algae) [5-6] does not have different structure between the roots, stems and leaves, and overall plant organ is known as thallus. Thallus has the ability to produce gelatin widely used in the food industry, one of which is used as a thickener or food stabilizer. Common characteristics of *Gracilaria verrucosa* is body-type branched form clumps with irregular branching, dichotomous, thallus form lamellar or cylindrical, cartillagenous, generally tapered tip thallus, smooth surface thallus or nodule thallus, with diameters ranging from 0.5 - 4.0 mm, a length of 30 cm or more and a greenish brown color [7].

Lead is more commonly known as the black lead, in scientific language is called plumbum, and the metal is symbolized by Pb. These metals included in the categoryofmetal IV- A on the periodic table of chemical elements, having atomic number (NA) 82 and has atomic weight (BA) 207.2 [4]. Pb in water comes from dust containing Pb that is caused by the burning of gasoline containing Pb tetra ethyl, erosion and industrial waste.

The objective of this study was to analyze the bioaccumulation of heavy metals Pb and its relationship with water quality and environmentally sound aquaculture management strategiesfor the development of environmentally sound aquaculture in seaweed ponds of *Gracilaria verrucosa* in the area Muarareja Tegal.

2. Material and Methods

This research was conducted in the seaweed (*Gracilaria verrucosa*) pond in Muarareja, Tegal. The method used in this study is a survey. Data were analyzed by quantitative descriptive variables in the study. GIS methods were used to determine the location of the research.

Data collected included primary data and secondary data. Primary data were obtained from the survey results/observations in the field and interviews. Secondary data were obtained through references from several institutions associated with the research. The variables in this study, namely heavy metals (Pb), temperature, salinity, pH, DO, ammonia, nitrate, phosphate, organic matter, and the substrate. Water quality analysis was conducted in the Integrated Laboratory of the Faculty of Fisheries and Marine Sciences, University of PancasaktiTegal. Analysis of heavy metals of lead (Pb) from the pond water, sediment and seaweed was carried out in the Laboratory of the University of Jenderal Soedirman Purwokerto using *Atomic Absorption Spectrophotometer* (AAS). Sampling technique is purposive/judgmental sampling or based on certain considerations.

Determining the location of sampling based on certain considerations, such as ease of reaching the point of sampling, as well as time and cost efficiency are based on the interpretation of the initial study site. Sampling is limited to the sample unit in accordance with certain criteria (characteristics of the soil) that is determined by objective research. Sampling for water and soil was conducted in eight seaweed ponds and two estuary of Kemiri River in aquaculture Muarareja.

3. Results and Discussion

Topography Tegal has a sloping beach, weak currents and waves is ramps. These topography are suitable for aquaculture of seaweed *Gracilaria verrucosa*, fish and shrimp vanname, both monoculture and polyculture systems.

3.1 Water Quality Parameter

Tegal's coast is low-lying, sandy clay substrate, tropical climate, and the average temperature is 27,9°C. Results of water quality measurements during the study are presented in Figure 1. In general, water quality parameters are still within the limits that can be tolerated by seaweed of *Gracilaria verrucosa*. The quality of aquaculture is aquaculture of seaweed *Gracilaria verrucosa*, fish and shrimp vanname, both monoculture and polyculture systems. Tegal's coast is low-lying, sandy clay substrate, tropical climate, and the average temperature is 27,9°C. The quality of aquaculture is strongly influenced by environmental conditions. If there is contamination it will greatly affect the cultivation.

According to Widigdo [8], other factors affecting the productivity of the pond is soil fertility. With good water quality conditions and fertile land, natural food is expected to grow well. Besides the fertility of the soil, the content of substances that are toxic are the the factors that affect the quality of the production.

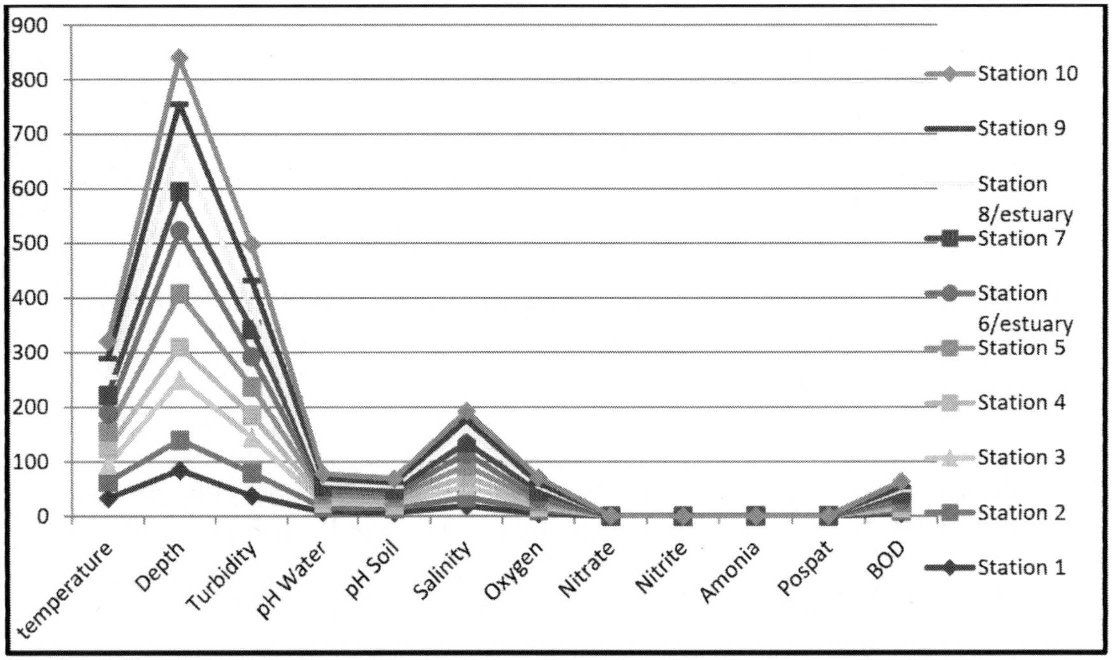

Figure 1. Measurement Water Quality of Pond in Rainy Season

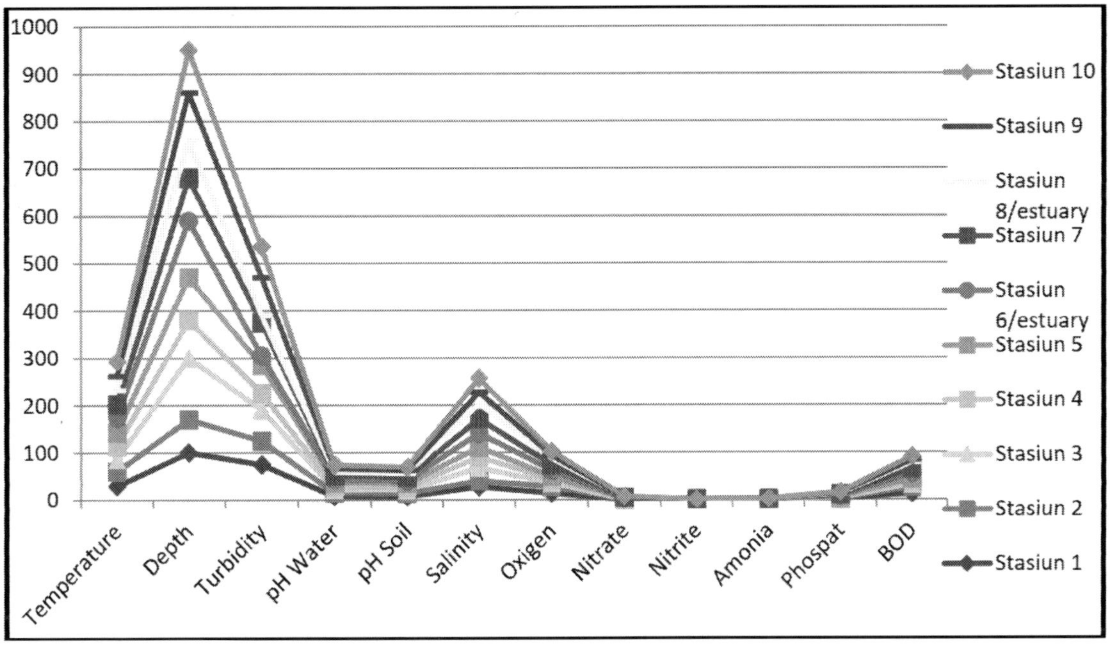

Figure 2. Measurement Water Quality of Pond in Dry Season

The parameters of water quality during the research were still within acceptable limits for the life and growth of seaweed *Gracilaria verrucosa*.

3.2 Heavy Metals Lead (Pb) in The Water

Darmono [9] also explains that the industrial waste containing Pb, such as the chemical industry, printing, metal, and paint will be wasted in the waters; it will increase the metal content of lead in the waters. The content of heavy metals of lead (Pb) or bioconcentration in pond water during the study is illustrated in Figure 3.

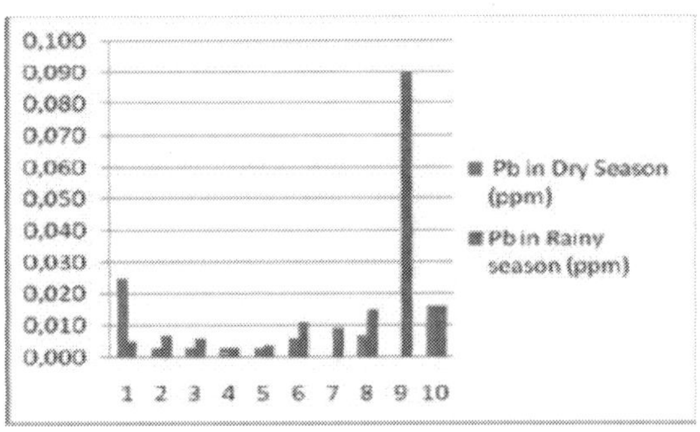

Figure 3. Content of Heavy Metals Lead (Pb) in the Water of Pond Research

The content of Pb in the water of pond was 0.003 to 0.016 ppm during the rainy season and from 0.003 to 0.025 ppm during the dry season. The highest content of Pb was in the pond near the grave because the pond is close to a water channel that carries various waste products from industrial waste, settelments waste, agriculture and fisheries. The content of Pb in pond water during the dry season is higher than in the rainy season. This is caused at the rainy season and there is a process of dilution, so the concentration of Pb is relatively lower. According to Palar [4], concentration of Pb in waters is influenced by the seasons. In the rainy season, the metal concentrations in water would be smaller than the dry season because during the rainy season the dilution process occurs, while in the dry season, the metal will be concentrated [9].

Argawala [10] states that the tolerance limits of aquatic organisms are at pH 5- 9; if there are pollutants that disrupt the aquatic buffer system, it will cause serious disruptions to aquatic organisms. The content of Pb in the water pond of research is still low, and the parameters of water quality during the study were still within acceptable limits for the life and growth of seaweed *Gracilaria verrucosa*. Normal Pb levels on the earth's surface <12.5 ppm [11-12]. Therefore Pb level in these waters including the normal category.

3.3 Heavy Metals Lead (Pb) in Seaweed Gracilaria verrucosa

Heavy Metals Lead (Pb) in Seaweed *Gracilaria verrucosa* in dry season and rainy season can be in Figure 4. Seaweed, *Gracilaria verrucosa*, is known of having an ability to absorb Pb in the water [13].

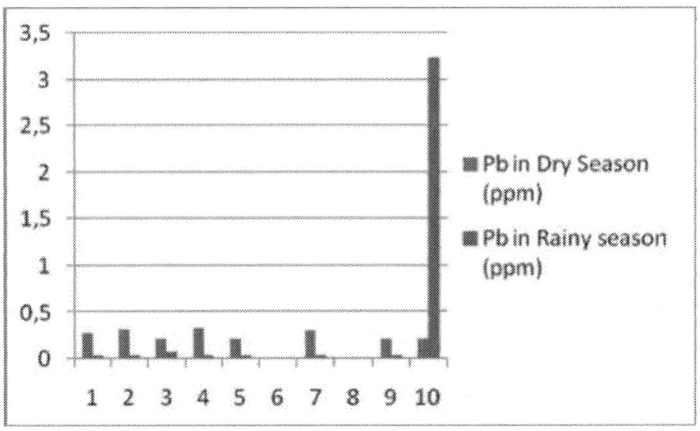

Figure 4. Content of Heavy Metals Lead (Pb) on Seaweed *Gracilaria verrucosa*

The content of Pb in seaweed *Gracilaria verrucosa* in ponds is 6.469 to 9.858 ppm in the rainy season and from 5.543 to 23.699 ppm in the dry season. The highest content of Pb was in settlement ponds because the waste from the settlement contains Pb. For station 6 (estuary of Kemiri II river) and station 8 (mouth of Kemiri River I) there is no Pb containing in seaweed. Pb bioconcentration in seaweed *Gracilaria verrucosa* in the dry season was higher than in the rainy season. Sorensen [14] stated that the higher the content of heavy metals in the water, the higher concentration of heavy metals accumulated in the body of the organism.

Darmono[15] also stated that the accumulation of heavy metals in the body of the organism depends on the concentration of heavy metals in the water. Bioaccumulation is the net result of which the absorption of contaminants by organisms beyond the ability of the organism to eject (depuration) from the body. Bioaccumulation can occur through respiration, digestion or direct contact with contaminated water or sediment [16].

3.4 *Heavy Metals Lead (Pb) in Sediment*

Haevy metals Pb can accumulate in sedimen because of the organic material produce by humus. Heavy Metals Lead (Pb) in Sediment can see in Figure 5.

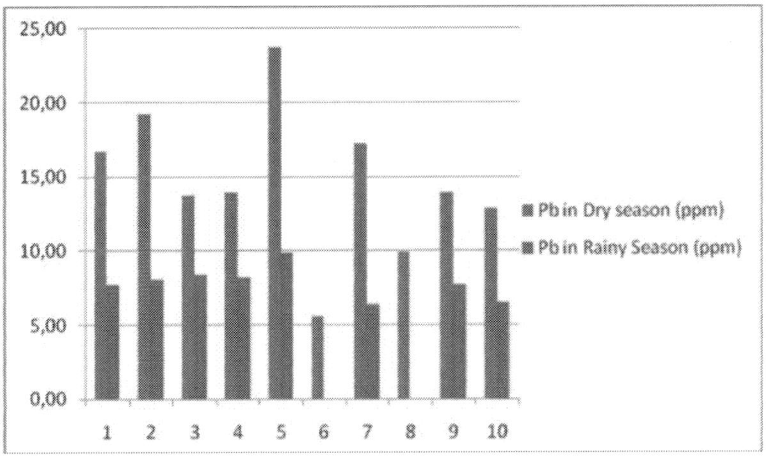

Figure 5. Bioconcentration of Heavy Metals Lead (Pb) in the Sediments of Ponds Research

The content of Pb in the pond sediment is 6.469 to 9.858 ppm in the rainy season and from 5.543 to 23.699 ppm in the dry season. The content of Pb was highest in the pond near the settlement. The content of Pb in the sediment in the dry season was higher than in the rainy season because of the increase in lead concentration. Heavy metals have easy binding properties of organic materials; settle to the bottom of waters, and blends with sediment, so the concentration of heavy metals in sediment was higher than in water [17]. In general the metal content of Pb decreases with increasingly thick layers of sediment [18]. Pb in water comes from dust containing Pb as a result of the combustion of gasoline containing tetra ethyl lead, erosion and industrial waste.

Table 1. The Range of Pb's Content during Research

Sample	Pb's content (ppm)	The Ministry of Environment decision 51/2004(ppm)
Pond's Water	0–0,025	0,008
Pond's Sediment	5,543–23,6990	>30
Seaweed *Gracilaria verrucosa*	0,209–0,326	2

Darmono [9] also explains that the industrial waste containing Pb, such as the chemical industry, printing industry, metal industry, and the paint will increase the concentration of Pb in the water. *Gracilaria verrucosa* is one class of marine algae Rhodophyta (red algae) which can function as a biofilter because it has the ability to absorb heavy metals. One way to reduce the pollution of Pb is using seaweed *Gracilaria verrucosa* as fitoremedian and can be used as bioindicators of water pollution, it is reinforced by research from [19-23].

The highest bioconcentration of Pb was in sediments in the pond water of seaweed (*Gracilaria verrucosa*) in Muarareja Tegal. Sediment of ponds is an accumulation medium of various organic materials as shown by results of research conducted by Ikbal *et al.*[24] and Ishak *et al.*[25]. Thus it can be said to have occurred bioaccumulation of Pb, where Pb in the pond water is lower than in seaweed (*Gracilaria verrucosa*) and the highest is in the sediment and it is a appropriate Suryono [26]. The Canadian Council of Ministers for the Environment (CCME) [27] and Ministry of Environment [28] set the Threshold Limit Threshold Pb value in sediments for biota protection respectively of 35 ppm and 36.8 ppm.

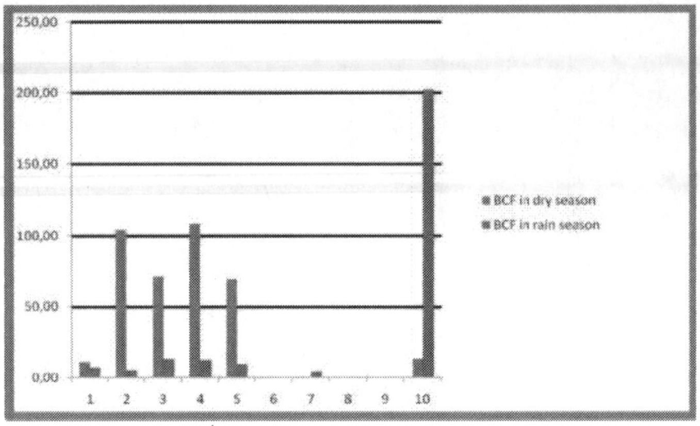

Figure 6. Bioaccumulation of Heavy Metals Lead (Pb) in Dry Season and Rainy Season

This is consistent with research by Ata *et a,* [29] and Tamilselvan *et al* [22] that *Gracilaria verrucosa* is able to absorb heavy metals Pb effectively; and Chakraborty *et al* [30] stated *Gracilaria verrucosa* can absorb heavy metals. Among the heavy metals, Pb, Cd, and Hg are highly toxic heavy metals than other heavy metals. The heavy metal pollutants are very dangerous if they contaminate the waters, because they are toxic, carcinogenic, bioaccumulative and biomagnification.

According to Matos *et al* [31], the integrated culture system of fish and seaweed can improve the feasibility of land for cultivation, because it can reduce the risk of damage to water quality and nutrients are released into the environment, and one of the good biological filter used in aquaculture is seaweed *Gracilaria sp.* Sohrab *et al* [32] research about The Environmental Monitoring of Heavy Metals in Seaweed and Associated Sediment.

The integration is an aspect which is essential in a system of resources management of coastal and marine. This not only ensures the synchronization between the internal policies and action programs, the projects and programs, but also ensures linkages between planning and implementation. Based on the type, integration can be divided into three types of integration, namely (1) the integration of systems, (2) integration of function and (3) policy integration [33]. According to Dahuri *et al* [34], integration of planning and integrated coastal zone management includes four (4) aspects, namely: (1) integration of the region/ecological; (2) integration of the sector; (3) the integration of science disciplines; and (4) the integration ofstakeholders.

Pollution in water can occur either because of industrial waste and domestic waste are discharged into waters without being processed first, or processed but the pollutant levels are still above the Standard Quality specified. The Pb's content from research results when compared to the quality standard of the Ministry of Environment Decree No. 51 [35] was still below the quality standard so that the seaweed is still safe for consumption.

4. Conclusion

Based on the results of the study, itcan be summarized as follows:

1. Pb bioconcentration during the dry season in pond water, sediment and seaweed (*Gracilaria verrucosa*) are respectively 0.025, 0.003 ppm, 5.543 to 23.699 ppm and0.209 to 0.326 ppm. While in the rainy season bioconcentration of lead (Pb) is respectively 0.003-0.015 ppm, sediment from 6.377 to 9.858 ppm and 0.209 to 0.326. Pb bioconcentration is greater during the dry season than during the rainy season.
2. Bioaccumulation of heavy metals is still relatively low, and the content of Pb is still below the quality standard of the Minister of the Environment Decree No.51 of 2004 so that the product is safe for consumption.
3. Water quality parameters during the study arestill within the limits that can be tolerated by seaweed (*Gracilaria verrucosa*).

References:

[1] Sudharsan SP, Seedevi, R Ramasamy, N Subhaprada, S Vairamy, and Shanmugamd 2012 *J.Chem.Pham.Res* **4** (9) 4240-4244.
[2] Qari R and SA Shidiqy 2010 *Indian Journal Marine Sciences* **39** (1) 27-42
[3] Wardhana WA 2004 *Impact of Environmental Pollution.* (Yogyakarta: Publisher Andi,) p 459
[4] Palar H 1994 *Contamination and Toxicology of Heavy Metals* (Jakarta: PT Rineka)
[5] Barsanti L, and Gualtieri P 2014 *Algae: Anatomy, Biochemistry, And Biotechnology*. (CRC Press)
[6] Pereira L and Neto J M (Eds.) 2014 *Marine Algae: Biodiversity, Taxonomy, Enviromental Assesment and Biotechnology* (CRC Press) p 398
[7] Anggadiredja JT, A Zatnika, H Purwoto and S Istini 2006 *Seaweed* (Jakarta: Publisher Swadaya) p169
[8] Widigdo B. 2000. Necessary Prediction of Ecobiological Criteria for Determining Natural Potential of Coastal Area for Shrimp Culture. Proceedings *of Training for Integrated Coastal Area Management Trainers.* Bogor 21 - 26 February 2000. PKSPL IPB, Bogor. 5p.
[9] Darmono 1995 *Metals in the Biological Systems of Organism. First Edition.* VII : **8**, 21-23, 37 (Jakarta :University of Indonesia).
[10] Argawala SP 2006 *Environmental Studies* (New Delhi Chennai Mumbai Kolkata.: Narosa Publishing House PVT. LTD.)
[11] Harikumar PS and TS Jisha 2010 *International Journal of Enginering Science and Technology* **2** (5): 840-850
[12] Mohiuddin MK, HM Zakir, K Otomo, S Sharmin., N Shikazono 2010 *International Journal Environmental Science Technology*, **7** (1): 17-28.
[13] Handhani AR, Ambariyanto and E.Supriyatini 2017 *AACL Bioflux* **10** (4) : 703-708.
[14] Sorensen. 1998. *Impact of Beach Pollution on Health.* (Bogor: Director General of Protection and Nature Conservation)
[15] Darmono 2001 *Environment and Pollution.*(Jakarta: UI Press) p144.
[16] Ambariyanto 2011 *Water Pollution Biomonitoring.* (Semarang: Publisher Diponegoro University. Semarang) p120
[17] Hutagalung HP 1991 *Marine Pollution by Heavy MetalStatus of Marine Pollution in Indonesia and The Solutions.* (Jakarta: *PO3-LIPI*)
[18] Siaka IM 2008 *Chemistry Jounal* **2** (2) : 61-70.
[19] Yulianto B, R. Ario, and A Triono 2006 *Journal of Marine Science* **II** (2) : 72-78.
[20] Huo Y, H Wu, ZC Hai, S Xu, F Han, L Dong, and P He 2012 *China.Aquaculture* **326329** 99-105.
[21] Schettino T, R Caricato, A. Calisi, M E Giordano and MG Lionetto 2012 *Open Environmental Science* **6** (12*)*:20-27.
[22] Tamilselvan N, J Hemachandran, T Thirumalai,ChV Sharma, K Kannabiran, and E.David 2013 *Journal of Coastal Life Medicine* **1** (2):102-107.
[23] Ihsan YN, A Aprodita, I Rustikawati, T Dewi, K Pribadi 2015 *Marine Journal* **8** (1): 10 – 17.
[24] Ikbal A and Noor Maming 2013 *Marina Chimica Acta* **14** (2) :1411-2132.
[25] Ishak J, B Amin and Thamrin 2014 *Periodical Fishing* **42** (2): 18 -27.
[26] Suryono, Ch A 2016 *Tropical Marine Journal* **19** (1): 37-42.
[27] Canadian Council of Ministers for the Environment (CCME)2002. Canadian sediment qualityguidelines for the protection of aquatic life summary table. Winnipeg, MB. 7 p.
[28] Ministry of Environment 2010 Decree of State Minister of Environment, No.03 dated January 18, 2010 concerning Quality Standard of Wastewater for Industrial Estate. Ministry of Environment. Jakarta, 7 p.
[29] Ata A, OO Nalcaci and B Ovez 2012 *Algal Research* **I** :194-204.
[30] Chakraborty S, T Bhattacharya, G Singh and JP Maity 2014 *Ecotoxicology and Enviromental Safety* **100** : 61- 68.
[31] Matos JS Costa, A Rodrigues, R Pereira and IS Pinto 2006 *Aquaculture* **252**:31-42.
[32] Sohrab A, Dadolahi, A. Nikvarz 2011 *World Journal of Fish and Marine Sciences* **3**(6): 576-

589.
[33] Supriharyono 2004 Role of Biological Indicators in the Evaluation of Level for Waters Pollution. Papers Submitted in The Event: *Bin.Tek.Bappedal Central Java Province*. 25 s.d. 27 September 2004. Semarang.
[34] Dahuri R, J Rais, SP Ginting and J Sitepu 2001 *The Integrated Management Resources Coastal and Marine*. Revised Edition (Jakarta: *Publisher PT.Pradnya Paramita*)
[35] Ministry of Environment. 2004. Decree of State Minister of Population and Environment, No. 51 of 2004, dated April 8, 2004 on the Quality Standards of Sea Water. (Jakarta: Ministry of Environment) p11

Determination Hypoiodous Acid (HIO) By Peroxidase System Using Peroxidase Enzyme

A N Al-Baarri[1, 3], A M Legowo[1], Widayat[2, 3], S B M Abduh[1], M Hadipernata[4], Wisnubroto[4], D K Ardianti[1*], M N Susanto[1], M Yusuf[1], E K Demasta[1]

[1]Food Technology Department, Faculty of Animal and Agricultural Sciences, Diponegoro University, Semarang, Indonesia.
[2]Chemical Engineering Department, Faculty of Engineering, Diponegoro University, Semarang, Indonesia
[3]Laboratory of Food Technology, UPT Integrated Laboratory, Diponegoro University, Semarang, Indonesia
[4]Center of Agricultural Post-Harvest Research and Development, Bogor, Indonesia

E-mail: albari@undip.ac.id

Abstract. It has been understood that peroxidase enzyme including peroxidase serves as catalyzer to enzymatic reaction among hydrogen peroxide and halides, therefore this research was done for generating hypoiodous acid (HIO) from peroxidase system using peroxidase enzyme. Hydrogen peroxide, potassium iodide, and peroxidase enzyme were used to produce HIO. Determination the amount of formed HIO was done using 2,2'-azino-bis(3-ethylbenzothiazoline-6-sulphonic acid) or ABTS as substrate through the colorimetric measurement of hydrogen peroxide residue during reaction process using at 412 nm. The result indicated that residual hydrogen peroxide showed the minimum concentration after 60 minutes reaction time. Because the reaction started at the beginning time of mixing, hydrogen peroxide was unable to be eliminated totally to produce HIO. The reaction of peroxidase system was able to determine the beginning of mixing process but the reaction process could not eliminate the initial concentration of hydrogen peroxide indicating the maximum amount of production of HIO could be determined. In conclusion, the less of H_2O_2, higher HIO obtained and peroxidase enzymes can accelerate the formation of HIO.

Keywords: Peroxidase system, hypoiodous acid, residual hydrogen peroxide, iodate, ABTS.

1. Introduction

Enzymes are known as good catalyzer. One of important enzyme in food is peroxidase. Peroxidases are widely distributed in nature and can be extracted from most plant cells, some animal tissues and fungus [1]. Plant peroxidase is an enzyme that catalyzes the one electron oxidation of various organic and inorganic substrates in the presence of hydrogen peroxide [2]. Plant peroxidases have been isolated and purified from several plant sources, such as *Leucaena leucocephala, Viscum angulatum, Vigna mungo, Solanum melongena, Beta vulgaris, Roystonea regia, Tamarix, Jatropha curcas* [3]. Peroxidase enzyme is one of the enzymes that can catalyse reaction between substrate and oxygen molecule. Peroxidases are much less effective as catalysts of oxygen-transfer reactions than cytochromes P450 [4]. Iodine is a compound of anion halide or (X^-), IO^{-3}, so application peroxidase can also catalyse the combination of hydrogen peroxide (H_2O_2) and potassium iodate (KIO_3) to produce Hypoiodite (OI^-) or Hypoiodous Acid (HIO).

Peroxidase enzyme activity test was generally performed using UV-VIS spectrophotometer with 412 nm wavelength and using 2,2'-azino-bis(3-ethylbenzothiazoline-6-sulphonic acid) or ABTS as substrate [5]. ABTS was used as an indicator because it was easily transformed into a radical form. ABTS oxidized with H_2O_2, then would form a radical compound of metastable cations and blue or green chromophore [6]. This research was done to analyse concentration of hypoiodous from the reaction mixture of H_2O_2 and KIO_3. Since hypoiodous has antibacterial properties, the minimum concentration of substrates H_2O_2 and KIO_3 was very important in order to generate maximum HIO through peroxidase system.

2. Materials and Methods

2.1. Chemicals and enzyme

H_2O_2 (0.15 mM), KI (0.15 mM), 2,2'-azino-bis [3-ethylbenzothiazoline-6-sulphonic acid] (ABTS) and horseradish peroxidase were obtained from Roche (Germany). Aquadest was obtained from UPT Integrated Laboratory, Diponegoro University, Semarang, Indonesia.

2.2. Preparation of HIO

There were two methods for determination of HIO. A 450 µl of 0.15 mM H_2O_2 was mixed with 450 µl of 0.15 mM KI and 100 µl of enzyme peroxidase. The other one was 450 µl of 0.15 mM H_2O_2 was mixed with 450 µl of 0.15 mM KI and 100 µl of aquadest. Solution was reacted, then measured using ABTS method every 10 minutes for an hour. This method was adopted from previous researcher [5]. The mixture of H_2O_2 and KI was determined as HIO solution.

2.3. Residue H_2O_2 with ABTS method

A 450 µl of HIO solution was added to a reaction mixture containing 450 µl of 0.55 mM ABTS in Phosphate Buffer (0.1 M, pH 7) and 100 µl of horseradish peroxidase (4.5 U). ABTS as substrate through the measurement of hydrogen peroxide residue during reaction process using spectrophotometer at wavelength of 412 nm. This method was derived from previous researcher [7].

3. Result and Discussion

3.1. Residue of H_2O_2

Measurement of H_2O_2 residues is enabled to determine the acceleration of HIO formation using peroxide enzyme catalyst or not using a catalyst as shown at Figure 1 and 2. The residue of H_2O_2 was measured by the absorbance value of the HIO solution using ABTS as substrate. The addition was intended for produces a green colour of the reaction between the rest of H_2O_2 and ABTS. The thicker the green colour the higher of absorbance. The decrease of absorbance value could be interpreted as the acceleration of HIO acid formation. Based on Fig. 1 and 2, it can be seen that the reaction between H_2O_2 and KI without enzyme run slower than those of enzyme addition. The velocity reaction of both compounds could be seen at the interval of spectrophotometric absorbance. This is because enzymes accelerate the reaction process of the formation of new compounds and the new compound from that reaction was HIO. This is in accordance with the opinion from other researcher [8] which states that peroxidase enzyme as catalysts of the oxidation of iodide into HIO. The more enzymes were given, the faster the reaction changes may occur. The speed of reaction may also increase with increasing concentration enzyme.

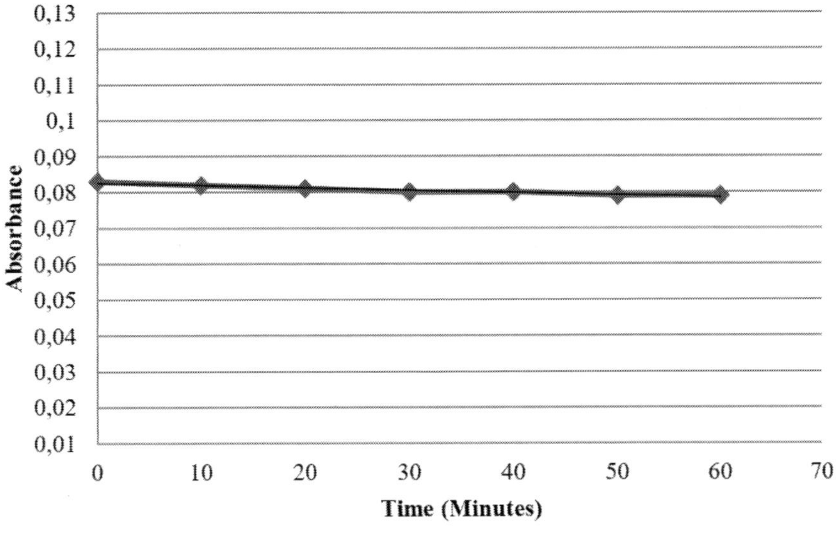

Figure 1. Residue H_2O_2 without Enzyme Peroxidase

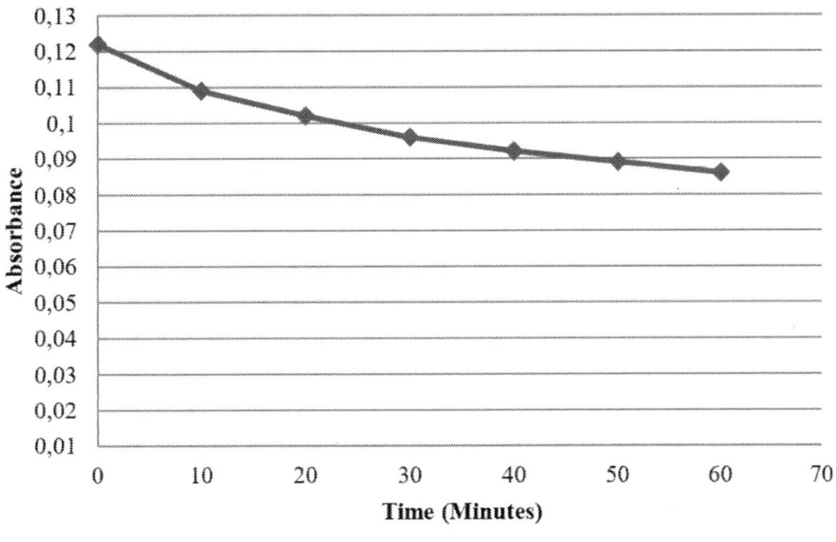

Figure 2. Residue H_2O_2 with Enzyme Peroxidase

The concentration of enzymes was expressed in units of units (U) and this research used 4.5 U. This amount of concentration was determined as eligible concentration to analyse the rate of reaction. As time of reaction, the enzyme might stop working resulting the decrease in reaction speed. The initial reaction speed generated large value then followed by smaller concentration of substrate's reduction. This can be seen at Table 1 where at the absorbance of H_2O_2 residue with enzyme was 0.092 to 0.089 in ten minutes starting from 40 to 50, then the reaction started to run slowly. Based on Table 1, the higher decrease in residual value of H_2O_2 indicated the more HIO was formed, meanwhile the absorbance of residue H_2O_2 without enzyme showed that the high residual value of H_2O_2 and it indicated that the HIO formed was still less in concentration. This is in accordance with the opinion from other researcher [5] which states that as the decline in the value of hydrogen peroxide concentration and the halide ions used in the peroxidase system, the higher concentration in final product of peroxidase systems.

3.2. Accelerated Reaction of Residual H_2O_2

Table 1. Accelerated Reaction Results of Residue H_2O_2 With and Without Enzyme Peroxidase

Time (Minutes)	Enzyme (Absorbance)	Acceleration of reaction	Non Enzyme (Absorbance)	Acceleration of reaction
0	0.122	0.013	0.083	0.001
10	0.109	0.007	0.082	0.001
20	0.102	0.006	0.081	0.001
30	0.096	0.004	0.08	0
40	0.092	0.003	0.08	0.001
50	0.089	0.003	0.079	0
60	0.086		0.079	

One of the factor that might affected the formation of HIO compound was the composition of the substrate. This was due to the peroxidase enzyme's ability to catalyse the oxidation reaction of KI with H_2O_2. According to other researcher [9], lactoperoxidase would be actively optimized when a state of sufficient ionic halide or pseudohalide. Meanwhile, the presence of an abundant substrate was also not providing the high concentration of HIO. According to previous researcher [10], the redox reactions existed in the peroxidase system could only occur if the reduction capability of the enzyme was weaker than the reduction capability of the substrate used. In addition to these factors, several things might also affected redox reactions in peroxidase systems such as anion size, anion access and anion ability to bind [11]. Besides these factors, the acidity level could also affect the reaction between H_2O_2 and KIO. According to other researcher [12], the acidity might hinder the action of hydrogen peroxide to oxidise iodine.

4. Conclusion

The peroxidase enzyme might accelerate the reaction process of peroxidase system for generating HIO and resulting minimum concentration of residual substrate. The maximum amount of production of HIO could be determined through the minimum rate of reaction that was reached at minute 40th of reaction time.

References

[1] Balasubramanian, M and R Boopathy 2013 *Molecular catalysis B: Enzymatic* **90**: 33-42
[2] Bozzo GG, K G Raghothama, and WC Plaxton 2004 *Biochem. J.* **377**: 1369-1379
[3] Pandey VP and UN Dwivedi 2011 *Journal of Molecular Catalysis B: Enzymatic*. **68** (2): 168-173
[4] Veitch NC 2004 *Phytochemistry* **65**: 249–259
[5] Touch V, S Hayakawa, S Yamada and S. Kaneko 2004 *J. of Food Microbiology* **93**: 175–183.
[6] Glucin I and A Dastan 2007 *Enzyme Inhibition and Medicinal Chemistry* **22**(6): 685–695
[7] Al-Baarri AN, M Hayashi, M Ogawa and S Hayakawa 2011 *Journal of Food Protection* **74**: 134–139
[8] Kupper FC, N Schweigert, E Ar Gall, JM Legendre, H Vilter and B Kloareg 1998 *Planta.* **207**: 163-171. Doi
[9] Seifu E, E M Buys and E F Donkin 2005 *Trends in Food Science and Technology* **16** (4) : 137-154
[10] Bafort F, O Parisi, JP Perraudin and MH Jijakli 2014 *Enzyme Research* 1–13.
[11] Davies MJ, CL Hawkins, DI Pattison and MD Rees 2008 *Antioxidants & Redox Signaling* **10** (7): 1199–1234
[12] Schmitz G 2001 *Phys. Chem.* **3**: 4741-4746

Antimicrobial activity of *tempeh gembus* hydrolyzate

A Noviana[1], F F Dieny[1,2], N Rustanti[1,2], G Anjani[1,2], D N Afifah[1,2]

[1]Department of Nutrition Science, Faculty of Medicine, Diponegoro University, Jl. Prof. H. Soedarto,SH, Tembalang, Semarang 50275, Indonesia
[2] Center of Nutrition Research (CENURE), Diponegoro University, Diponegoro University, Jl. Prof. H. Soedarto,SH, Tembalang, Semarang 50275, Indonesia

d.nurafifah.dna@gmail.com

Abstract. Tropical disease can be prevented by consumming fermented foods that have antimicrobial activity. One of them is *tempeh gembus* that has short shelf life. It can be overcome by processing it into hydrolyzate. This study aimed to determine antimicrobial activity of *tempeh gembus* hydrolyzate. *Tempeh gembus* was made of local soybean from Grobogan. They were added 5,000 ppm, 8,000 ppm, and 10,000 ppm of bromelain enzyme (TGH BE). Antimicrobial effects of TGH BE were tested against *Staphylococcus aureus, Escherichia coli, Bacillus subtilis,* and *Steptococcus mutans*. Antimicrobial test was carried out using *Kirby-Bauer Disc Diffussion* method. Soluble protein test used *Bradford* method. The largest inhibition zone against *S. aureus* and *S. mutans* were shown by TGH BE 8,000 ppm, 0.89±0.53 mm and 2.40±0.72 mm. The largest inhibition zone of *B. subtilis,* 7.33±2,25 mm, was shown by TGH BE 5,000 ppm. There wasn't antimicrobial effect of TGH BE against *E. coli.* There weren't significant differences of soluble protein (P=0.293) and the inhibition zones againt *S. aureus* (P= 0.967), *E. coli* (P= 1.000), *B. subtilis* (P= 0.645), *S. mutans* (P=0.817) of all treatments. There were antimicrobial activities of TGH BE against *S. aureus, B. subtilis,* and *S. mutans.*

Keywords : Antimicrobial activity, Hydrolyzate, *Tempeh Gembus,* Bromelain enzyme.

1. Introduction

Tropical disease is a common disease in tropics and subtropics. One of the causes of this disease is a bacterial infection. Some pathogenic microbes are *Staphylococcus aureus, Escherichia coli, Bacillus subtilis,* and *Streptococcus mutans. S. aureus* can lead to septicemia, endocarditis, cerebral abscess, puerperal sepsis, and pneumonia [1]. *E. coli* is a normal microflora in the gut that can cause disease. This microbe becomes pathogenic when its number in the digestive tract increase or it's in outside of the intestinal. Some diseases caused by *E. coli* are urinary tract infections, diarrhea, sepsis, and meningitis. *B. subtilis* can induce the insident of meningitis, diarrhea, endocarditis, eye infections, acute gastroenteritis, and immune dysfunction diseases [2]. Whereas, *S. mutans* can cause dental caries [3].

Foods that have antimicrobial function can be consumed to prevent infectious disease. *Tempeh gembus* is one of traditional foods in Indonesia. It is a fermented foods made of solid waste-tofu by *Rhizopus oligosporus* [4]. Eventhough *tempeh gembus* is made of by waste-tofu, it contains several nutrition contents, such as essential fatty acids (linoleic acid (21.51%), linolenic acid (1.81%), and oleic unsaturated fatty acid (16.72%)), protein, carbohydrate, fiber, calcium, iron [5]. The fermentation process of soybean generates functional properties changes owing to hydrolysis process, e.g protein hydrolysis into amino acids and peptides by proteolytic enzymes which can act as antimicrobial compounds [6]. Additionally, *Rhizopus* used in this fermentation process can produce antibiotics against pathogenic microbes as well [7].

Tempeh gembus which generally can not be stored for a long time, has a large size and the limited availability in some areas, as well as a distinctive flavor that sometimes is not liked by some people, can be tackled by processing it into flour or hydrolyzate [8]. Therefore, in this research, *tempeh gembus* was processed into hydrolyzate by adding bromelain enzyme. Besides, adding bromelain can hydrolyze the protein or polypeptide into simple compounds, such as peptide which may have bioactive function [5].

Tempeh extract has been shown to inhibit *B. subtilis* and *S. aureus* [9]. Study of antimicrobial activity of tempeh has done by Kuligoswki et al [10] against *B. subtilis, Bifidobacterium, E. coli, Lactobacillus acidophilus,* and *Lactobacillus paracasei* in the simulation of human intestinal condition. Tempeh extract has been also shown to inhibit *Enterotoxigenic Escherichia coli* (ETEC) adhesion to Caco-2 cell in the intestinal epithelium [11]. Hence, The objective of the present work is to study the antimicrobial activity of *tempeh gembus* hydrolyzate.

2. Materials and Method

2.1. Preparation of tempeh gembus

Tempeh gembus was made of soybean from Grobogan. Soybean was sorted and soaked for 24 h. Furthermore, the epidermis was stripped and washed. Soybean which is already clean was milled using hot water, with a ratio of soybean : water (1:8). Subsquently, the soy bean was boiled for 15 min and be filtered using filter cloth to separate the filtrate and the pulp. The obtained pulp was steamed for 30 min and be cooled. The fermentation process was carried out by adding 1% of yeast, then it was wrapped using polypropylene plastic and allowed to stand for 24 h.

2.2. Preparation of tempeh gembus hydrolyzate (TGH BE)

Tempeh gembus hydrolyzate was made by hydrolyzing *tempeh gembus* using bromelain enzyme. Firstly, *tempeh gembus* was weighed 300 g, and be steamed for 10 min. The steamed *tempeh gembus* was divided into three parts, each weighing 100 g. Afterwards, they were blended and added water to each 100 mL. Bromelain enzyme (Bromelyn, Yogyakarta,Indonesia) was added to each concentration of 5,000 ppm, 8,000 ppm, and 10,000 ppm using Phosphat Buffer Saline as the solvent. The mixture was incubated at 55 °C in water bath incubator for 1.5 h. *Tempeh gembus* hydrolyzate was heated by adding dextrin and NaCl (each 0.5 g/sample). Furthermore, *tempeh gembus* hydrolyzate added bromelain enzyme (TGH BE) condensed materials were dried using freezedryer [8].

2.3. Preparation of microbes

Stock culture was made by scratching *S. aureus, B. subtilis, E. coli,* and *S. mutans* on Nutrient Agar slant, After that, they were incubated in incubator at a temperature of 37 °C for 24 h. The colonies of microbes taken from the stock cultures using loopful and be suspended in a sterile tube containing 10 mL of NaCl solution.

2.4. Antibacterial assay using disk diffusion method

The sterilized Mueller Hinton Agar (MHA) was poured in a petri dish. After the media solified, microbial suspension was leveled at media surface using a sterile swab. Media containing microbes was allowed to stand about 10 min and put the paper discs dipped to TGH BE 5,000 ppm, 8,000 ppm, and 10,000 ppm, that have been diluted using sterilized distilled water with a concentration of 30%, for the comparison there was positive control using amoxycilin 2%, and negative control using sterilized distilled water. Subsquently, they were incubated in an incubator at a temperature 37 °C for 24 h [12].

2.5. Soluble protein assay using Bradford method

Preparation of bradford reagent was done by dissolving 10 mg of coomassie brilliant blue G-250 in 5 mL of 95% ethanol. Then, 10 mL of 85% phosphoric acid was added. The solution was diluted using distilled water to a volume of 100 mL. The samples (each 0.1 g) were added bradford reagent, then it was incubated for an hour. The solution was homogenized using a vortex and it was incubated at room temperature for an hour. Afterwards, each sample was measured using a spectrophotometer (Shimadzu UV-1280,Kyoto, Japan) at wavelength of 595 nm to measure the absorbance [13].

2.6. Statistical analysis

The results were presented as mean ± standard deviation. Antimicrobial activity (inhibition zone) of TGH BE was expressed in mm and soluble protein was expressed in percentage. The data were analyzed using non-parametric statistical computer program. The univariate analysis was conducted by calculating a mean of TGH BE inhibition zone. The bivariate analysis was conducted using one way ANOVA (Analysis of Variance) statistical test to find out the difference of inhibition zone and soluble protein [14].

3. Results

3.1. Antimicrobial activity

All three TGH BE were tested on selected Gram-positive and Gram-negative bacteria which is presented in Table 1. The largest diameter of inhibition zone against *S. aureus* and *S. mutans* was indicated by TGH BE 8,000 ppm that is respectively 0.89±0.53 mm and 2.40±0.72 mm. Whereas, the smallest diameter of inhibition zone by TGH BE 5,000 ppm against *S. aureus* that is 0.82±0.53 mm, and TGH BE against *S. mutans* that is 1.93±0.76 mm. While, TGH BE 5,000 ppm showed the largest diameter of inhibition zone against *B. subtilis* (7.33±2.25 mm) and TGH BE 10,000 ppm showed the smallest diameter of inhibition (5.37±4.09 mm).

Table 1. Mean of Inhibition Zone Diameter of TGH BE 5,000 ppm, 8,000 ppm, 10,000 ppm, Negative Control, and Amoxycilin

Treatment	Mean of Inhibition Zone (mm) ± SD[a]			
	S. aureus	*E. coli*	*B. subtilis*	*S. mutans*
TGH BE 5,000 ppm	0.82 ± 0.53	0.00 ± 0.00	7.33 ± 2.25	2.10 ± 1.40
TGH BE 8,000 ppm	0.89 ± 0.53	0.00 ± 0.00	5.60 ± 0.46	2.40 ± 0.72
TGH BE 10,000 ppm	0.88 ± 0.33	0.00 ± 0.00	5.37 ± 4.09	1.93 ± 0.76
Kontrol (-)	0.00 ± 0.00	0.00 ± 0.00	0.00 ± 0.00	0.00 ± 0.00
Amoxycilin 2%	18.33 ± 0.23	13.00 ± 1.00	17.47 ± 0.80	7.53 ± 0.92

[a]SD = Standard Deviation; TGH = *Tempeh Gembus* Hydrolyzate; BE = Bromelain Enzyme.

Table 2. showed that the largest diameter of inbition zone of TGH BE 5,000 ppm, 8,000 ppm, and 10,000 ppm were shown to against *B. subtilis,*that is 7.33±2.25 mm, 5.60±0.46 mm, and 5.37±4.09 mm, respectively. However, all samples of TGH BE did not indicate any inhibiton zone against *E. coli.*They were served in Figure 1. There was not significant difference in the inhibition zone against *S. aureus* (P= 0.967), *E. coli* (P= 1.000), *B. subtilis* (P= 0.645), *S. mutans* (P=0.817) of all treatments.

Table 2. Mean of Inhibition Zone Diameter of TGH BE 5,000 ppm, 8,000 ppm, and 10,000 ppm.

Microbes	Mean of Inhibition Zone (mm) ± SD[a)]			P value
	TGH BE 5.000 ppm	TGH BE 8.000 ppm	TGH BE 10.000 ppm	
S. aureus	0.82 ± 0.53	0.89 ± 0.53	0.88 ± 0.33	0.976
E. coli	0.00 ± 0.00	0.00 ± 0.00	0.00 ± 0.00	1.000
B. subtilis	7.33 ± 2.25	5.60 ± 0.46	5.37 ± 4.09	0.645
S. mutans	2.10 ± 1.40	2.40 ± 0.72	1.93 ± 0.76	0.817

[a)]SD = Standard Deviation; TGH = *Tempeh Gembus* Hydrolyzate; BE = Bromelain Enzyme.

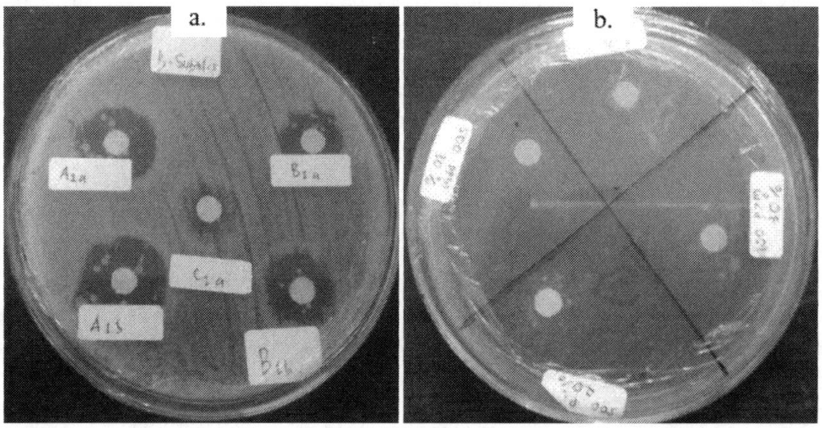

Figure 1. The largest inhibition zones were shown by TGH BE against a. *Bacillus subtilis;* and there wasn't inhibition zone of TGH BE against b. *Escherichia coli*

3.2. Soluble Protein

Table 3. presented the soluble protein of TGH BE. The highest content of soluble protein contained in TGH BE 5,000 ppm. While, TGH BE 10,000 ppm contained the lowest soluble protein. There was not significant difference in soluble protein content of TGH BE 5,000 ppm, 8,000 ppm, and 10,000 ppm (P= 0.293).

Table 3. Mean of Soluble Protein Percentage of TGH BE 5,000 ppm, 8,000 ppm, and 10,000 ppm.[a]

Treatment	% Mean of Soluble Protein	P value
TGH BE 5,000 ppm	0.73 ± 0.08	
TGH BE 8,000 ppm	0.65 ± 0.12	0.293
TGH BE 10,000 ppm	0.60 ± 0.03	

[a]TGH = *Tempeh Gembus* Hydrolyzate; BE = Bromelain Enzyme.

4. Discussion

This study showed that TGH BE 5,000 ppm, 8,000 ppm, and 10,000 ppm had antimicrobial activity against *S. aureus, B. subtilis,* and *S. mutans* which include Gram-positive bacteria. Similarly Bintari et al [15] has reported that antibacterial agents such as glycoproteins were produced during fermentation by *Rhizopus*, so that the growth of Gram-positive bacteria, *Micrococcus luteus,* which is one of the contaminants in tempeh was inhibited.

Bioactive compounds that have antibacterial activity are formed during soybean fermentation [16]. Saraswaty et al [16] revealed that glycoprotein compounds, protein groups (amines) which have been isolated from media that has inoculated *R.oligosporus,* and flavonoids (isoflavones) are several antimicrobial compounds contained in tempeh. According to Jawetz et al [17], the microbe growth inhibition or the death of microbes due to antimicrobial agents can be through four ways, that is cell wall synthesis inhibition, cell membrane dysfunction, protein synthesis inhibition, nucleic acid synthesis inhibition.

Glycoprotein is one of the antimicrobial compound that can bind to iron in bacteria. Most microorganisms require iron for growth and glycoproteins have the potential to inhibit the growth of bacteria, and even kill them by depriving them of iron. Iron is an essential nutrient for growth and metabolism in virtually all microorganisms and an important cofactor in many metabolic and enzymatic processes. Iron is a component of cytochromes (electron bearer) on the respiratory system [18].

Antimicrobial peptide compounds are capable to regulate the cell target to modify the outside structure of the cell in order to be more sensitive to antimicrobial agents, resistance prevention, non-competitive working to other antimicrobial agents. The working mechanism of antimicrobial peptides are causing the membrane dysfunction (permeabilization, depolarization, and other membrane malfunctioning), the biopolymer extracellular synthesis inhibition, and intracellular function inhibition (protein, RNA, and DNA synthesis inhibition) [19]. Whereas, flavonoids have the ability to bind to extracellular proteins and integral proteins that join to the bacterial cell wall, and disrupt cell wall permeability [20]. Flavonoids are phenolic compounds having glycoside bounds. Phenolic compounds will interact with the bacterial cell membrane proteins through adsorption process by bounding to the hydrophilic part of the cell membrane. Afterwards, Phenolic compounds will enter into the cell membrane and induce cell protein precipitation. It interferes cell membrane permeability, therefore cell membrane can go through lysis [20].

Based on the mechanism of antimicrobial action contained in soybean tempeh that could be also contained in TGH BE, TGH BE have antimicrobial function through four ways, namely cell wall synthesis, cell membrane dysfunction, protein synthesis inhibition, and nucleic acid synthesis inhibition.

According to the David Stout method, antimicrobial activity is very strong if the diameter of inhibition zone is more than 20 mm, antimicrobial activity is strong if the diameter of inhibiton zone is 10-20 mm, antimicrobial activity is medium if the diameter of inhibiton zone is 5-10 mm, and antimicrobial activity is weak if the diameter of inhibition zone is less than 5 mm [21]. Hence, Antimicrobial activities of TGH BE 5,000 ppm, 8,000 ppm, and 10,000 ppm against *S. aureus* and *S. mutans* were included in weak antimicrobial as the diameters were less than 5 mm. While, they were included in medium antimicrobial against *B. subtilis* because the diameters were in the range of 5-10 mm. However, TGH BE of various concentrations didn't have antimicrobial effect against *E. coli* because there wasn't any inhibition zone.

The largest inhibition zone of TGH BE of various concentrations were against *B. subtilis*. *B. subtilis* included in Gram-positive bacteria is more easily inhibited than *E. coli* which is included in Gram-negative bacteria. The cell stucture of Gram-positive microbe is relatively simpler and it makes antimicrobial substances easier to enter the cell and find a target to work [22]. This cell wall is thick, about 25-30 μm, only has one layer with the largest components which consists of peptidoglycan, and the low content of lipid (1-4%) [23]. There are several factors that can affect microbial inhibition, these are the antimicrobial agent concentration, the number of microorganisms, the species of microorganisms, the temperature, and the presence of other organic microorganism spesies [23]. According those factors, *B. subtilis* was known to be one spesies of microbes that was more sensitive to antimicrobial compound contained in TGH BE than *S. aureus* and *S. mutans* which were also Gram-positive microbes.

Tempeh gembus hydrolyzate with all treatments did not show inhibition zone against *E. coli*. This result was similar to study by Kiers et al [24] using disc method, recorded there was antibacterial activity in soybean tempeh against *Bacillus stearothermophilus* and *B. subtilis*, but soybean extract did not show antimicrobial activity against *E. coli.* This is because *E. coli* is included in Gram-negative bacteria that has a better resistance to antimicrobial compounds. The Gram-negative microbes have a selection system against foreign substances that is in the lipopolysaccharide layer. The structure of Gram-negative microbe cell wall is relatively more complex, it has three layers namely the outer layer in the form of lipoproteins, the middle layer in the form of lipopolysaccharide and peptidoglycan layer [17]. The outer membrane contains protein molecules called porin that is hydrophilic. Porin contained in the outer membrane can cause antimicrobial components more difficult to get into the bacterial cell. Porin leads to the passive diffusion of hydrophilic compounds which have low molecular weight such as glucose and amino acids, while the high molecular weight molecules such as antimicrobial molecules are difficult to penetrate to the membrane. The differences in the structure and components of the cell wall causes *E. coli* as Gram-negative microbe more resistant because one of the mechanisms of antimicrobial work is damaging of the cell wall structure in the plasma membrane, therefore, the plasma membrane ability of bacterial cell as osmosis barrier is decrease and disturbing of the number of biosynthesis process required by the membrane [25].

The study by Kiers et al [24] showed that the percentage of piglets excreting *Enterotoxigenic Escherichia coli* (ETEC) K88 was lower in the group fed tempeh or cooked soy compared to a group fed roasted soy. This was due to several factors such as lactic acid bacteria contained in cooked soybeans and *Rhizopus* fermented soybeans. *Rhizopus microsporus* LU 573 fermented soybean could inhibit adhesion of ETEC K88 *in vitro* and it was similarly to the result of .the ETEC K88 colony decrease observed in piglets fed *Rhizopus* fermented soybean [25]. In addition, the mean incidence of diarrhea was lower in piglets fed *Rhizopus* fermented soybean [26].

The diameters of inhibition zone against *B. subtilis* showed by TGH BE 8,000 ppm and 10,000 ppm were lower than TGH BE 5,000 ppm. Whereas, TGH BE 5,000 ppm and 10.000 ppm showed the lower inhibition zones than TGH BE 8,000 ppm. One of the factors that influence this antimicrobial activity is antimicrobial peptide content within each sample. The process of thermal can deactivate antimicrobial agents. This result was similar to a study by Nowak and Steinkraus [27] which showed that peas tempeh cooked for 10 min did not show antimicrobial properties against *Clostridium perfringens*. Roubus-van den Hil et al [7] also observed the antimicrobial properties of peptides isolated from tempeh were reduce after heating to 60^{0}C. The heating over 60^{0}C caused the lost of the antimicrobial activity approximately 20-60% in the 24-h fermented tempe extract. Antibacterial agents isolated by Roubus-van den Hil et al [7] were susceptible to protease, their research was using pronase E or proteinase K. Thus, these enzymes can degrade the antimicrobial peptides found in soybean tempeh. The high temperature of heating and the increasing of enzyme concentration (bromelain enzyme) were factors that might contribute to the decline of antimicrobial activity of the TGH BE in this study.

High soluble protein content does not mean that the antimicrobial activity which is demonstrated also high. The highest diameters of inhibition zone against *S. aureus* and *S. mutans* in this study were shown by TGH BE 8,000 ppm, eventhough the results showed that the highest protein soluble content was in TGH BE 5,000 ppm, instead. It was because the high soluble protein content of substance does not mean it contains high antimicrobial peptides as well. Research by Kusumaningtyas et al [28] about antimicrobial and antioxidant activity in goat milk hydrolyzate showed that peptides that have high antimicrobial activities were peptides having a molecular weight less than 3 kDa. In this study, the test to determine antimicrobial peptide content of TGH BE did not conducted.

5. Conclusion

Tempeh gembus hydrolyzate added 5,000 ppm, 8,000 ppm, and 10,000 ppm of bromelain enzyme showed their antimicrobial activities against *S. aureus, B. subtilis,* and *S. mutans*. However, there was not antimicrobial activity against *E. coli.* There was not difference in soluble protein and the diameter of inhibition zones against *S. aureus, E. coli, B. subtilis,* and *S. mutans* of all three treatments.

6. Acknowledgement

The authors would like to acknowledge Diponegoro University.

References

[1] Aulia A, Thihana, and Mirhanuddin 2007 *Bioscientiae* **4** 37-42
[2] Rahmaningsih S, Willis S, and Mulyana A 2012 *Ekologia* **12** 1-5
[3] Alfath C R, Yulina, and Sunnati 2013 *J dent Indones* **20** 5-8
[4] Afifah DN, Sulchan M, Syah D, Yanti, Suhartono MT, and Kim JH 2014 *Prev Nutr Food Sci* **19** 213-9
[5] Kurniasari R, Sulchan M, Afifah DN, Anjani G, Rustanti N 2017 *Rom J Diabetes Nutr Metab Dis* **24** 203-11
[6] Pablo D R, Gabriela T P, Maria C A, and Alberto E L 2010 *Food Bioproc Tech* **3** 395-405
[7] Roubus-van den Hil P J, Dalmas E, Nout M J R, and Abee T 2010 *J Appl Microbiol* **109** 137-45
[8] Subagio A., Hartanti S, Windrati W S, Unus, Fauzi M, and Herry B 2002 *Jurnal Teknologi dan Industri Pangan* **13** 204-10
[9] Mambang D E P, Rosidah, and Suryanto D 2014 *Jurnal Teknologi dan Industri Pangan* **25** 115-8
[10] Kuligowski M, Kuligowska I J, and Nowak J 2013 *Pol J Microbiol* **62** 189-94
[11] Roubus-van den Hil P J, Nout M J R, Beumer R R, Meulen J, and Zwietering 2009 *J Appl Microbiol* **106** 1013-21
[12] Bauer A W, Kirby W M M, Sherris J C, and Turck M 1966 *Am J Clin Pathol* **36** 493-6
[13] Bradford M 1976 *Anal Biochem* **72** 248–54
[14] Steel R G D and Torrie J H 1993 *Principle and procedures of statistics: a biometric approach* (Jakarta: Gramedia Pustaka Utama) p 633-648
[15] Bintari S H, Anisa D P, Veronika E J, and Rivana C R 2008 *Biosaintifika* **1** 1-8
[16] Saraswaty V, Zainal A, and Dewi R 2002 *Proc. of the Seminar on Chemical Research Challenges* 67-74
[17] Jawetz, Melnick, and Adelberg's 2001 *Medical microbiology* (Surabaya: Salemba Medika)
[18] Jawetz, Melnick, and Adenberg's 2005 *Medical microbiology* (Jakarta: Salemba Medika)
[19] Yeaman M R and Yount N Y 2003 *Pharmacol Rev* **55** 27–55
[20] Murphy M C 1999 *Clin Microbiol Rev* **12** 564-82
[21] David and Stout 1971 *J Microbiol* **22**
[22] Zuhud E A M, Pudji R W, Hanny W C, and Puspita S P 2001 *Jurnal Teknologi dan Industri Pangan* **7** 6-12
[23] Pelczar M J Jr and Chan E S C 1988 *Elements of microbiology* (New York-McGraw-Hill Companies Inc) p117,145-8
[24] Kiers J L, Nout M J R, Rombouts F M, Nabuurs M J A, and Van der Meulen J 2002 *Lett Appl Microbiol* **35** 311–5
[25] Brogden K A 2005 *Nat Rev Microbiol* **3** 238-50
[26] Kiers J L, Nout M J R, Rombouts F M, Nabuurs M J A, and Van der Meulen J 2003 *J Appl Microbiol* **95** 545–52
[27] Nowak J and Steinkraus K H 1988 *Nutr Rep Int* **38** 1163–72
[28] Kusumaningtyas E, Widiastuti R, Kusumaningrum H D, and Suhartono M T 2016 *Glob Vet* **16** 105-10

The Abudance Of Makrozoobenthos On Different Break Water In Semarang And Demak Coastal Area

A Kristiningsih[1], D N Sugianto[2,3], Munasik[4], R Pribadi[4], J Suprijanto[4]

1. Faculty of Fisheries and Marine Science, Diponegoro University. Jl. Prof. Soedarto, SH. UNDIP, Tembalang, Semarang, 50275, Indonesia
2. Departement of Oceanography, Fisheries and Marine Science Faculty, Diponegoro University. Jl. Prof. Soedarto, SH. UNDIP, Tembalang, Semarang, 50275, Indonesia.
3. Center for Coastal Disaster Mitigation and Rehabilitation Studies, Diponegoro University. Jl. Prof. Soedarto, SH. UNDIP, Tembalang, Semarang, 50275, Indonesia.
4. Departement of Marine Science, Fisheries and Marine Science Faculty, Diponegoro University. Jl. Prof. Soedarto, SH. UNDIP, Tembalang, Semarang, 50275, Indonesia.

Email : kristiningsihari@gmail.com

Abstract. The coast of Semarang and Demak has suffered some damage to its coastal areas. This damage is caused by natural factors and also human activities. There are number of mitigation methods such as hard, soft and hybrid that available for mitigation erosion. In Semarang and Demak coastal area using hard and hybrid option as their mitigation erotion. Breakwater is one of the way beach structure that often used as mitigation erosion di coastal area. Breakwater will cause sediment deposits that will become the living place of various organisms such as makrozoobenthos. The aim of this research is compare the abudance of makrozoobenthos in different type breakwater in Semarang and Demak coastal area.This research held on December 2016 – January 2017 in five different location with different breakwater type. Hard structure in Mangkang (West Semarang), Morosari (Demak district) and Tambak Lorok (North Semarang) and the hybrid engineering in Morosari 2 (Demak district) and Timbulsloko (Demak district). The method used in this study is descriptive comparative. Makrozoobenthos has been found in each station and the highest indeks is in hybrid engineering location. Polychaeta is a genus that dominates at every location because muddy sand is its main habitat.

Keywords: mitigation, breakwater, makrozoobenthos, Semarang, Demak

1. Introduction

The coastal region is a potential area for various activities, since the coastal area is a large zone for settlements, it has an important role in the wealth of many countries [1]. Because the area is very potential for various activities, then this region is also very vulnerable to the environmental changes [2]. The changing of coastal area has influenced by several factors. It has regularly affected by natural mechanism and has rapidly received human activities [3]. As in some coastal areas in Semarang and demak has been damaged in its coastal areas.The coastal damage of Semarang has also been investigated by FoE (Friend of the Earth) Japan (2009) [4] stating that since 1998 many residents in the coastal area of Tugu sub-district lost the ponds due to the abrasion by the sea water into the river with a high intensity so that the access road becomes drowned. Furthermore [5] in the period of 1991 - 2011 there was a change in the characteristics of the beach that is the length of the coastline increased 9.70 km (103.6%) and the area of abrasion in 2011 = 22.7% or 1,198.8 ha. If this is continue, the citizens will lose farmland and aquaculture areas. In Timbulsloko area (Morosari, Demak district), coastal change mainly because of the land conversion [6]. Furhermore [8] land conversion is not the only cause damage of Timbulsloko coat, but there are other triggering factors, namely the extension of the breakwater of Tanjung Emas harbor, land subsidence and sea-level rise Mangkang beach (West Semarang) is one of the beach that used for livelihood by the fishermen, and Morosari beach (Demak district) also became a tourist attraction for the people of Demak and surrounding areas.

To overcome various environmental damage can be through several ways as mitigation methods. Such as build some permanent structure or soft structure as breakwater in some coastal area. Mangkang (West Semarang), Morosari (Demak district) and Tambak Lorok (North Semarang) coastal area has breakwater to reduce the wave. The breakwater structure made from different material with different type. The type was hard structure in Mangkang and Tambak Lorok, and hybrid enggineering in Morosari. The hard structure is made from the concrete block and stone block and the second type is hybrid enggineering, this type using Paralon and bamboo as its main material. Hybrid engineering is a new methods thats being developed. The basic principle of the development of the breakwater with hybrid engineering type is to mimic the function of mangrove in protecting coastal, which dampen wave energy and trap sediment using environmentally friendly materials[6]. The impact of breakwater development is the formation of sediments on every breakwater. Sediment is a habitat for some living things such us makrozoobenthos. The aim of this research is to see the difference of macrozoobenthos abundance in the breakwater structure that has different material arrangement. Macrozoobenthos is used as an indicator, because macrozoobenthos is a living organism (sesile) and has varying adaptability to several environmental conditions [10]. Some macrozoobentos are also used to indicate whether a water is being polluted or not like Oligochaeta that has been used as an indicator of polluted waters [6]. [8] also explain that the diversity of macrozoobenthos can increase along the age of mangrove species in the rehabilitation programs. Thus, the purpose of this research is to see what is the difference of macrozoobenthos between hard structure and the hybrid enggineering breakwater type.

2. Methodology

This research was conducted in December 2016 - January 2017 in Mangkang (West Semarang), Morosari (Demak) and Tambak Lorok (North Semarang) coastal area. The research material includes sediment substrate, organic matter content, water quality and macrozoobenthos in Mangkang, Morosari and Tambaklorok coastal area. This research method is descriptive comparative method.

2.1 *Study area*

The location of the research is in five stations: Mangkang beach that breakwater made from concrete buis, Morosari beach first station that breakwater made from block stone, Morosari beach second station with hybrid enggineering type from paralon as the main material, Timbulsloko stasiun with hybrid enggineering from bamboo and the last one is Tambak Lorok station with break water from concrete buis. Each station is taken with 5 sampling points. Map of the study shown in Figure 1.

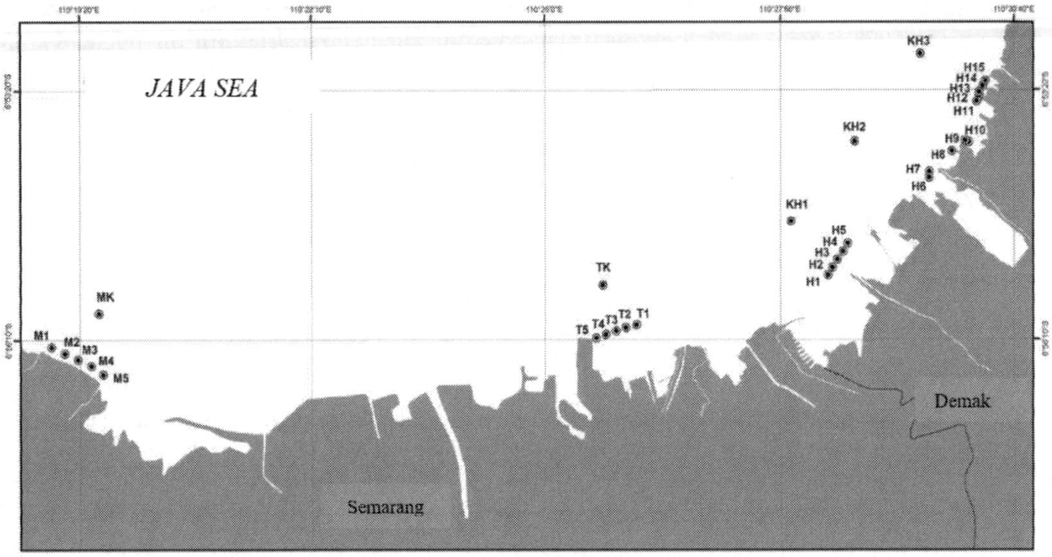

Figure 1. Area of Study

2.2 *Sediments sample*

Samples of mud containing macrozoobenthos were taken using Van Veen Grab tool, mud samples separated ± 25 grams for organic material analysis, 250 grams for grain size analysis and the rest sieved using bentos filter made from wire. The macrozoobenthos sample was then preserved by using 10% formalin which had been mixed with Rose Bengale solution. After reaching the labolatorium the sample was washed and sorted and then samples in the form of worms preserved with a 70% alcohol solution. Furthermore the sample is identified up to the genus level. Data analysis is done by determining the value of diversity index (H'), uniformity index (E) and dominance index (C). Measurements of water quality parameters include temperature, salinity, dissolved oxygen and pH.

3. Result and Discussion

Mitigation of coastal erosion is a difficults task since a number of parameters and processes are involved in the same [7]. Some parameters like biological, chemical and physic parameter will involved during the process. Most of conventional protection methods are build breakwater. Breakwater can build by three type which is soft structure, hard structure and hybrid structure. Hard structure usualy expensive because its using material permanent such us concerete and rubble mound, masive in size and sometimes not eco-friendly[7]. The soft structure is using green belts, dune rehabilitation, dune vegetation. This type is not expensive, not massive in size and mostly eco-friendly. The last type is hybrid that is combination between the hard and soft structure type. Its also calles as permeable dams [8]. This type is hard structure with soft engginering. The hybrid type using bamboo, wood or paralons as their materials.

Breakwater on each coastal area may vary, depend on the amount of the damage [9]. Hydroocenography and substratum are the major aspects need to be identified accurately prior to design the permeable breakwater. Factors that need to be identified include: water depth, bed slope, type of soil and its substrate, salinity, height and direction of wave, tidal, velocity and direction of currents, and sediment source.

The breaking wave which has been existed in surf zone will be transporting sediment along coastal area. With the breakwater among the coastal area it will caused some effect. One of the effect is sedimentasion. The accumulated sediments are home to a variety of organism, such as makrozoobenthos. From the Grain size analyze for all station indicated that the substrat dominated by the silt and clay which is the main habitat of makrozoobenthos. This research goal is to know the different breakwater with hard type and hybrid type based the abudance of makrozoobenthos.

Water quality indeks each station shown in Figure 2. Based on the data we can see that the lowest Disolved Oxygen (DO) is on Morosari 2 and Timbulsloko that have breakwater with hybrid engineering system with bamboo, probably it is because of high turbidity. Temperature has no significant difference in every station, salinity range below salinity standar its because of heavy rain before sampling, organic matter range is high in every station.

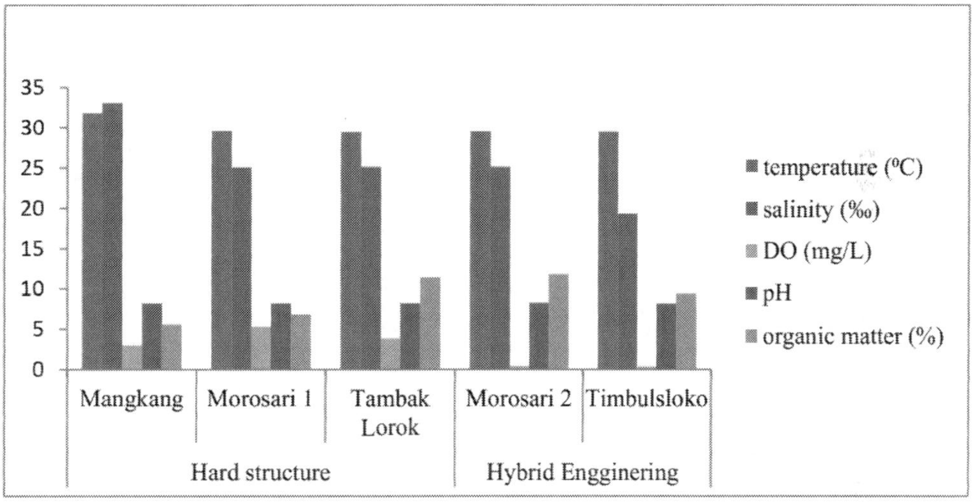

Figure 2. Water quality and organic matter each station

Based on research results has been identified that Mangkang station has 25 genus with 20 genus from Polychaeta, 2 genus from Bivalve, 2 genus Bivalve and 1 genus from Ophioroidea. Morosari 1 station has 18 genus 12 genus from Polychaeta, 3 genus from Bivalve, 1 genus Gastropoda, 1 genus from Ophioroidea dan 1 genus from Crustacea. Morosari 2 station has 30 genus 23 genus from Polychaeta, 3 genus from Bivalve, 1 genus Gastropoda, 1 genus from Ophioroidea dan 2 genus from Crustacea. Timbulsloko station has 20 genus with 16 genus from Polychaeta and 4 Bivalve. Tambak Lorok has 21 station with 17 genus from Polychaeta, 3 Bivalve and 1 Ophiouroidea. Detail of genus in each station showed by the Table 1 until Table 5. Polychaeta class has dominated in every station because in this area their habitat is muddy beaches.

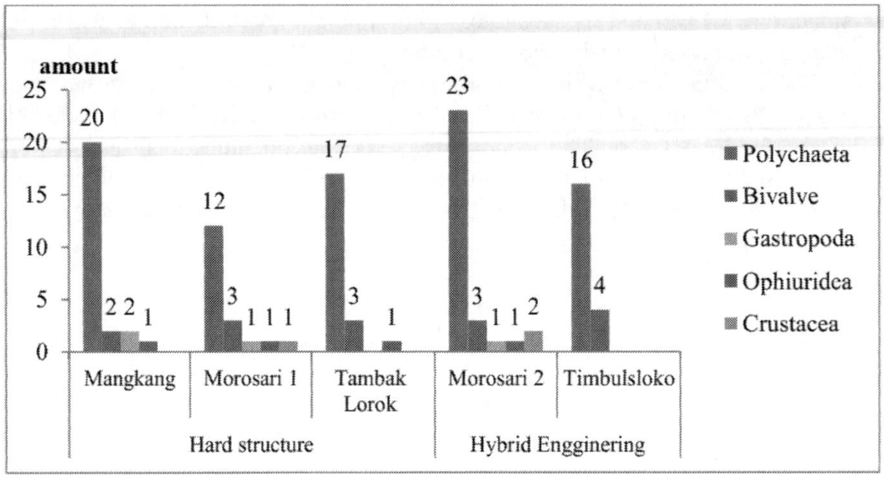

Figure 3. Genus amount in every station

Table 1. Genus on Mangkang station

Genus	Sampling location				
	M1	M2	M3	M4	M5
Polychaeta					
Aricidae	-	1	-	-	-
Capitella	-	6	10	6	2
Ceratocephale	-	-	1	-	1
Chrimia	-	-	1	-	-
Chone	-	1	-	-	-
Glycera	-	2	-	1	-
Glycinde	-	-	1		1
Goniada	-	3	1	3	-
Heteromastus	3		5	1	
Lumbrineris	1	2	-	-	3
Nereis	-	2	1	-	1
Notomastus	1	-	2	-	-
Perineris	-	-	-	1	-
Platynereis	-	-	-	-	1
Poechilochaetus	-	2	3	1	3
Prionospio	-	-	1	-	-
Pygospio	-	-	-	1	-
Sipuncula	-	-	-	1	-
Spionida	1	-	-	-	-
Sternapsis	-	1	1	1	-
Bivalve					
Asaphis	-	-	-	-	1
Tellina	1	-	-	1	-
Gastropoda					
Cerithium	-	-	1	1	-
Littorina	-	-	-	1	-
Ophiuridea					
Ophicuma	-	-	56	8	-
Total individu	7	20	84	27	13
Total amount		151 individu			

IOP Conf. Series: Earth and Environmental Science **116** (2018) 012045 doi:10.1088/1755-1315/116/1/012045

Polychaeta has dos dominated in Mangkang station with genus Capitella the most amount in Polychaeta class. In this station genus Ophicuma from Ophiuridea genus has been found 64 organism. Maybe its their breeding season, so the amount is abundance. Detail genus on Morosari 1 station shown by table 2. The number of organism in Morosari station less than Mangkang station. Maybe because on the location, behind breakwater at Mangkang station has mangrove vegetation. Its also called green belts, mangrove also produce nutrien to water.

Table 2. Genus on Morosari 1 station

Genus	Sampling location				
	H1	H2	H3	H4	H5
Polychaeta					
Capitella	1	-	-	2	1
Cossura	-	-	-	2	-
Echiura	-	-	-	1	-
Glycera	-	-	-	1	-
Goniada	-	-	1	-	2
Heteromastus	1	-	-	1	-
Lumbrineris	-	-	-	1	-
Platynereis	-	-	-	1	-
Poechilochaetus	1	-	-	2	-
Polynoidea	-	-	-	-	-
Prionospio	-	-	-	1	-
Spionida	-	-	1	-	-
Bivalve					
Anadara	-	2	-	-	-
Asaphis	-	1	1	-	-
Tellina	-	-	1	-	3
Gastropoda					
Cerithidea	-	1	-	-	-
Ophiuridea					
Ophiureidea	2	-	-	-	-
Crustacea					
Amphipoda	-	1	-	-	-
Total individu	5	5	4	12	6
Total amount	32 individu				

The amount of Tambak Lorok station is more than Morosari 1 station. Genus Anadara from Bivalve class at Tambak Lorok has dominated this station. Because this station near fishing pond. Polychaeta still dominated this station. Polycaheta almos dominated in every station because it is theirm main habitats. Breakwater with hard type with permanent material Polychaeta class has been dominated in each station even its not the biggest amount. In Mangkang station the biggest amount is Ophicuma from Ophiuroidea class and Tambak Lorok station is Anadara from Bivalve class. Genus from Tambak Lorok station shown on Table 3.

Table 3. Genus on Tambak Lorok station

Genus	Sampling location				
	T1	T2	T3	T4	T5
Polychaeta					
Aglaophamus	-	-	-	-	1
Capitella	1	3	1	-	1
Ceratocephale	2	3	3	1	5
Chaeptopterus	-	-	-	-	1
Cossura		1	-	-	-
Glycera	-	-	-	-	1
Goniada	5	1	1	-	1
Heteromastus	1	-	-	-	2
Hipponoa	-	1	-	-	1
Mediomastus	-	1	-	-	-
Neanthes	-	-	-	-	1
Nereis		1	-	1	-
Notomastus	2	-	-	1	-
Platynereis	-	1	-	-	-
Poechilochaetus	-	1	-	-	-
Polynoida	-	-	-	-	5
Prionospio		-	-	-	1
Bivalve					
Anadara	-	38	5	10	18
Asaphis	-	1	-	-	-
Perna viridis		4	-	-	-
Ophiuridea					
Ophicuma	1	1	-	2	4
Crustacea					
Amphipoda	-	-	-	-	3
Jumlah individu	12	57	10	15	36
Total amount	130 individu				

Morosari 2 station and Timbulsloko station is breakwater with hybrid type. Morosari 2 using paralons as their main structure and Timbulsloko using bamboo as their main struture. Detail genus on Morosari 2 station shown at Table 4. Polychaeta are dominated this station, Sternapsis genus is dominated this biggest amount in this station. At Morosari 2 station surounded by mangrove vegetation. Mangrove also has important effect to coastal environment, besides it role as nursery and spawning ground its also can be sediments trap.

Table 4. Genus on Morosari 2 station

Genus	Sampling location				
	H6	H7	H8	H9	H10
Polychaeta					
Anaspio	1	-	-	-	-
Aricidae	-	-	-	-	1
Capitella	4	7	-	-	1
Ceratocephale	3	1	-	3	-
Chaetozone	-	1		1	-
Cossura	1	5	-	-	-
Drilonesis	1	-	-	-	-
Glycera	1	2	-	-	1
Glycinde	-	-	-	1	-
Goniada	5	1	-	-	-
Heteromastus	-	2	5	9	-
Mediomastus	2	2	-	-	-
Nassarius	-	1	-	-	-
Nereis	-	-	-	-	2
Neanthes	-	-	-	-	1
Notomastus	1	2		3	-
Paraonella	-	1	-	-	-
Perineris	1	-	3	-	-
Poechilochaetus	-	3	-	1	1
Polynoidea	1	3	-	2	-
Spionida	1	-	-	-	-
Sternapsis	7	3	2	21	1
Terebelida	-	2	-	-	-
Bivalve					
Macoma	-	-	-	-	1
Solen	-	-	-	-	1
Tellina	1	-	-	-	1
Gastropoda					
Cerithidea	-	-	-	1	-
Ophiuridea					
Ophicuma	-	-	-	-	1
Crustacea					
Amphipoda	-	1	-	-	-
Uca	-	-	-	-	1
Total individu	30	37	10	43	13
Total amount	133 individu				

Timbulsloko station is breakwater with hybrid type with bamboo as their main materials. Detail genus shown on Table 5. Polychaeta has dominated this station with Sternapsis has the biggest amount. This station also surounding by mangrove vegetation but not as much as Morosari 2.

Table 5. Genus on Timbulsloko station

Genus	Titik Sampling				
	H11	H12	H13	H14	H15
Polychaeta					
Anaspio	-	1	-	-	-
Capitella	2	2	6	-	2
Ceratocephale	-	7	1	-	1
Glycera	-	-	4	-	1
Glycinde	2	-	-	-	1
Goniada	-	-	-	-	2
Heteromastus	-	-	4	-	-
Lumbrineris	-	-	4	-	-
Mediomastus	3	-	-	-	3
Nereis	-	-	5	1	-
Notomastus	1	1	-	-	3
Platynereis	-	2	3	3	-
Poechilochaetus	1	1	6	1	-
Polynoidea	1	-	1	-	-
Prionospio	-	-	2	-	-
Sternapsis	7	8	-	-	-
Bivalve					
Anadara	-	1	-	-	-
Asaphis	-	-	-	-	1
Macoma	1	-	-	-	-
Tellina	-	2	-	-	-
Total individu	18	25	36	5	14
Total amount			98 individu		

Every breakwater has been found various organism with different class and genus. The index shown at Figure 4.

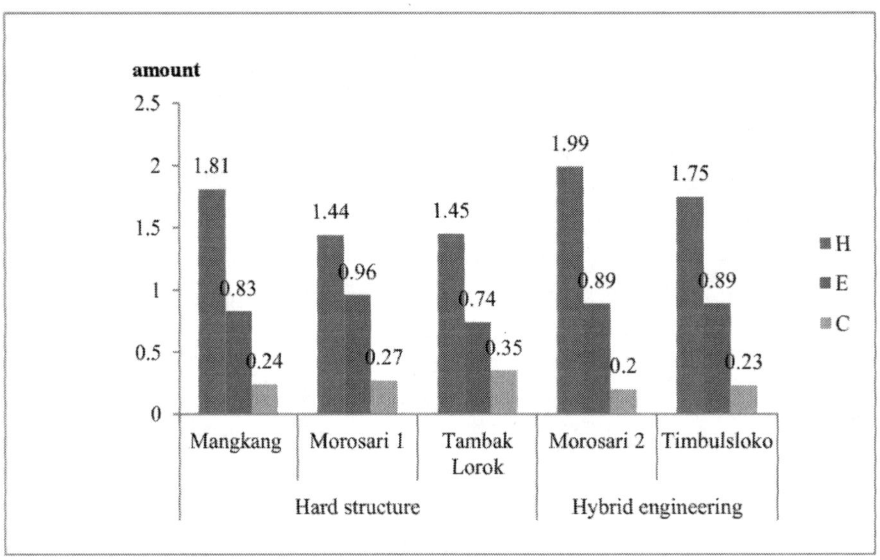

Figure 4. Diversity Index (H'), Equatibility index (E) and Dominance Index (C)

Breakwater with hybrid enggineering type has higher indeks comparing with the hard structure type based on Figure 3. Its because of mangrove mud beaches as a result of the hybrid enggineering type breaking waves naturally provide a variety of ecosystem needs such as coastal protection, fishery resources, marine life spans and improved water quality[4].

Each breakwater has their own advantages and disdvantages, for example the hard type its expensive, not eco-friendly and more durable than soft type. The soft cheaper than hard, but eco-frienly but not durable. One the option is using hybrid type that using permanent structure with eco-friendly materials. The breakwater that has been choosed to manage the damage can be different each location. Its depends on the charateristic each coastal area, and if it allows combine it with green belts so the proctection become double.

4. Conclusion

Based on the result by the research, that in all types of coastal retaining structures there are macrozoobenthos populations with varying degrees of diversity. Thus, the breakwater structure will form a new population of organisms. The abundance of macrozoobenthos is one of the supporting data in the determination of the right breakwater structure. Required other supporting data to be able to determine the right building, such as substrate type, current pattern, sedimentation, tidal vegetation type etc.

5. Acknowledgement

We would thanks to to Marine Science and Oceanography Departemens, Diponegoro Univercity for helping this research and also to Beasiswa Unggulan for the Master Degree Scholarship in Diponegoro Univercity.

References

[1] Zanuttigh, B 2011 *Coastal flood protection: What perspective in a changing Climate?* The THESEUS Approach: Environmental Science & Policy, v.14, p 845 – 863.

[2] Parman, S 2010 *Deteksi Perubahan Garis Pantai Melalui Citra Penginderaan Jauh Di Pantai Utara Semarang Demak.* Jurnal Geografi **7** (1)

[3] Purnomo H, Denny NS, Sutrisno A., Anindya W and Sugeng W 2016. *Waves Induce Sediment Transport at Coastal Region of Timbulsloko Demak.* IOP Conf. Series: Earth and Environmental Science **55** 012048

[4] Retno H, Rudhi P., Retno WA, Renny . and Itsna Y 2016 *Kajian Pengamanan dan perlindungan Pantai di wilayah pesisir kecamatan Tugu dan Genuk, kota Semarang* Jurnal Kelautan Tropis **19** (2): 95-100

[5] Prayogi, Johanes H and Jusup S 2016 *Kajian Perubahan Tata Guna Lahan Kawasan Pesisir Kecamatan Sayung Kabupaten Demak.* Thesis. Magister Ilmu Kelautan Universitas Diponegoro

[6] Suripin, Denny NS and Muhammad H 2017 *Mangrove Restoration With Environment Friendly Permeable Breakwater* Asian Jr. of Microbiol. Biotech. Env. Sc. **19** (1) : 2017 : 102-107

[7] Arkal VH 2010 *Coastal erosion and mitigation methods- Global state of art* Indian Journal of Geo-Marine Sciences **39** (4), December 2010, pp. 521-530

[8] Ahmad GDZ, Alfi S, Heriyoso S 2014 *Analisa Pengaruh pasang purnama (Spring) dan Perbani (Neap) terhadap laju sedimentasi di perairan Timbulsloko, Demak* Jurnal Oseanografi **3** (4): 540-548

[9] Zaky, A.R., Chrisna AS, Rudhi P 2012 *Kajian Kondisi Lahan Mangrove di Desa Bedono, Kecamatan Sayung, Kabupaten Demak dan Kelurahan Mangunharjo, Kecamatan Tugu, Kota Semarang* Journal of Marine Research. **1** (2) : 88-97.

[10] Pratiwi, N., Krisanti, N., I. Maryanto I, R. Ubaidillah, W.A. Noerdjito 2004 *Panduan Pengukuran Kualitas Air Sungai* Bogor: Institut Pertanian Bogor, Bogor

Wastewater Treatment from Batik Industries Using TiO2 Nanoparticles

Fahmi Arifan[1]*, FS Nugraheni[1], Hafiz Rama Devara[1], Niken Elsa Lianandya[1],

[1]Chemical Engineering, Vocation School, Diponegoro University, Jl Prof.Sudarto Semarang

Email: fahmiarifan80@gmail.com

Abstract. Batik is cultural patterned fabric, and the this industries produce wastewater that can pollute the aquatic environment. Besides dyes, batik wastewater also contains synthetic compounds that are hard degraded, such as heavy metals, suspended solids, or organic compounds. In this study, photocatalitic membrane TiO2 coated plastic sheets have been used to degrade batik wastewater under solar exposure. A total of 8 pieces of catalyst sheets are added on 1000 ml of the waste, and managed to degrade 50.41% of the initial concentration during 5-days irradiation. In this study, several parameters of the water quality such as chemical oxygen demand (COD), biological oxygen demand (BOD), and total suspensed solids (TSS) of the wastewater were observed to be decreasing during photodegradation process. The catalyst sheet also is stable to be used repeatedly in wastewater treatment.

Keywords: Batik wastewater; COD, BOD, TiO2 nanoparticles,TSS.

1. Introduction

Batik has been recognized by UNESCO as the world cultural heritage from Indonesia. Batik and textile industries are producers of wastewater originating from the dyeing process. Besides dyes, batik wastewater also contains synthetic ingredients which are recalcitrant to be degraded. Wastewater generated from batik or textile industries is generally a non-biodegradable organic compound, which can cause the environmental pollution, especially of the aquatic environment.

Some of the most commonly used in textile industry are remazol black, red and golden yellow. In the coloring process, these compounds can only be used about 5% while the remaining 95% will be discarded as a liquid waste. Because they are stable enough, their presence in nature are very difficult to be degraded, and for high concentrations these compounds are harmful to the environment because they can raise the COD and BOD levels of water.

Advanced oxidation processes (AOPs) with TiO2 as a photocatalyst has been widely used in wastewater treatment. This method has several benefits include: the process is simple, environmentally friendly, can operate in the environmental conditions, as well as low costs. In addition, the TiO2 catalyst is photostable, non-toxic, low cost, and stable in the water for a various of experimental conditions [1, 2, 3].

The implementation of TiO2 in wastewater treatment can be classified in two ways: slurry suspension (TiO2 nanoparticles directly dispersed in the wastewater) and immobilized system (TiO2 nanoparticles being fixed to a carrier)[4]. Slurry suspension, although it is simple in technique and high surface area to adsorption and reaction, its application is limited by the low-efficient utilization of light coming and difficult to separate nanoparticles after reaction [5]. In this case, it is necessary to immobilize

TiO2 nanoparticles on a support to overcome the deficiency of the slurry system for the practical application. In addition, the immobilized TiO2 can be used repeatedly and it can be regenerated.

Some simple methods have been developed to coat TiO2 nanoparticles on a transparent substrate. As performed by Isnaini *et al.* who coated the surface of the plastic fiber with TiO2 particles using the vibration method. TiO2-coated fibers were then applied to treat organic pollutants in wastewater [6]. Meanwhile, Aliah *et al.* using the thermal milling method, have managed to coat the surface of plastic granule with TiO2 nanoparticles [7]. Sutisna *et al.* using a combined method of the electrostatic and thermal treatment, have succeeded in superimposing TiO2 on the surface of the plastic sheet [8].

In this study, TiO2-coated plastic sheets were applied in photodegradation of batik wastewater. Parameters of the water quality such as color, COD, BOD, and TSS have been evaluated to see the performance of catalyst sheets.

2. Material and Methods

2.1. Reagent and materials

A technical grade of TiO2 (Bratachem, Indonesia) was used as a photocatalyst (density: 4.32 g cm-3, average particle size: 166 nm, BET surface area: 7.005 m2 g-1, total pore volume: 9.69x10-2 cc g-1, average pore diameter: 55.31 nm, crystalline structure: anatase). Compared with pure TiO2, this material is much less expensive and it has a fairly wide absorption band [9]. TiO2 particles were coated on plastic sheets of polypropylene (purchased from the local supplier) with a thickness of 0.08 mm and a melting point of 130-170 °C. As a treated solution, we used batik wastewater obtained from a batik industry in Jember, East Java, Indonesia.

2.2. Catalyst preparation

Before TiO2 catalyst is applied, first it was immobilized on surfaces of plastic sheets using a simple and cheap method. This coating method is a combination of electrostatic and thermal treatment [8]. The coating process was initiated by generating electrical charges on the plastic sheet by rubbing it with a wool cloth. The friction between surfaces of the plastic and wool cloth causes electrons to move from fabric to plastic. TiO2 particles were then sown on the surface of charged plastic and because of the electrostatic induction, they adhered on the plastic surface. To bind strongly TiO2 on the plastic, heating under pressure was done by using a modified lamination machine. Based on the amount of TiO2 in spite of the plastic surface during the washing process after coating, we identified that the optimum coating quality are obtained when it was heated at temperatures of 110 ° C for 4 minutes per sheet [8]. A SEM (scanning electron microscope) image of the sample proved that the TiO2 nanoparticles have been attached properly on the surface of the plastic.

2.3. Photocatalysis experiments

A total of 1000 ml batik wastewater was placed in a container with dimension 13x19x9 cm3. Degradation of batik wastewater was performed under different experimental conditions: (i) control solution without catalyst under solar irradiation (photolysis reaction), and (ii) solution with catalyst sheets under solar irradiation (photocatalytic reaction).

All experiments were carried out in Bandung city, Indonesia with geographical coordinates latitude = -6.9o and longitude =107.6o on August - September 2015. Different locations will have different solar intensity. Experiments conducted during 8 hours (from 08:00 o'clock to 16:00 o'clock) on each day. The intensity of the sun light during experiment was measured using a TM 206 Solar Power Meter (TENMARS, Taiwan). During photocatalytic process, the solar illumination changed with time with the average intensity of about 622 W/m2. Photodegradation of wastewater has been analyzed by measuring absorbance spectrum, COD, BOD, and TSS.

2.4. Analytic Methods,

Brunauer-Emmett-Teller (BET) specific surface area analyzer (Nova Quantachrome 1200e) was used to determine the specific surface area and pore size distribution of TiO2 nanoparticles. A SEM (JSM-651OLA, JEOL, Japan) was used to analyze morphological feature of the TiO2 coated plastic sheet. Decolorization of dye compounds have been analyzed by absorbance measurement during photodegradation process using a UV-Vis spectrometer (Mikropack Brands, NanoCalc 2000, Florida, USA). The measurement of COD, BOD, and TSS are based on the Standard Methods for the Examination of Water and Wastewater 22nd edition 2012 (SMEWW).

3. Results and Discussion

3.1. Decolorization of batik wastewater by photolysis and photocatalysis reactions

Photodegradation test of batik wastewater was done by placing 1000 ml wastewater in a container. In this experiment, effect of number of catalyst sheets on degradation efficiency has been investigated by varying the number of catalyst from 5-10 pieces. Each sheet has a cross-sectional area of 12x18 cm2. Batik wastewater without catalyst has been used as a control experiment. During the experiment, the container was covered by acrylic transparent (thickness of 2 mm) in order to avoid the effect of evaporation on the photodegradation reaction. Acrylic was selected for its transparency properties to pass solar light spectra in the wavelength range from UV-B until visible light. Photons in this spectra range are needed to stimulate the photocatalytic reaction. The experimental setup and transmission spectra of acrylic can be seen in Fig. 1.

Decolorization process of dye compounds contained in batik wastewater was investigated by measuring the absorbance of the solution. According to Lambert-Beer law, the concentration of a solution is directly proportional to the amplitude of the absorption spectra. Therefore, a decrease in absorption spectra peaks of a solution during degradation process indicates a decrease in the concentration of the dye in the wastewater. UV-Vis absorption spectra of batik wastewater without and with a catalyst can be seen in Fig. 2.

The decomposition of dye in batik wastewater during photodegradation process for different variations of the number of catalyst is shown in Fig. 3. It can be seen that the addition of catalyst sheets from 5 to 8 pieces have increased the photodegradation rate of batik wastewater. Nonetheless, this does not apply to the addition of the next catalyst sheets. The use of 9 to 10 pieces of catalyst sheets can no longer increase the photodegradation ability of the system. This phenomenon can be explained as follows: an increase in the amount of the catalyst will provide an increase in active sites of the catalyst that contact with contaminants in the wastewater and role in the photocatalytic reaction. On the other hand, the increase in the number of catalyst sheets will reduce the intensity of sunlight that penetrates and activate the catalyst. The number of catalyst sheets that can be penetrated by sunlight depends on the transparency of the catalyst sheet and the concentration of batik wastewater. In this case, the optimum number of catalyst sheets is 8 pieces.

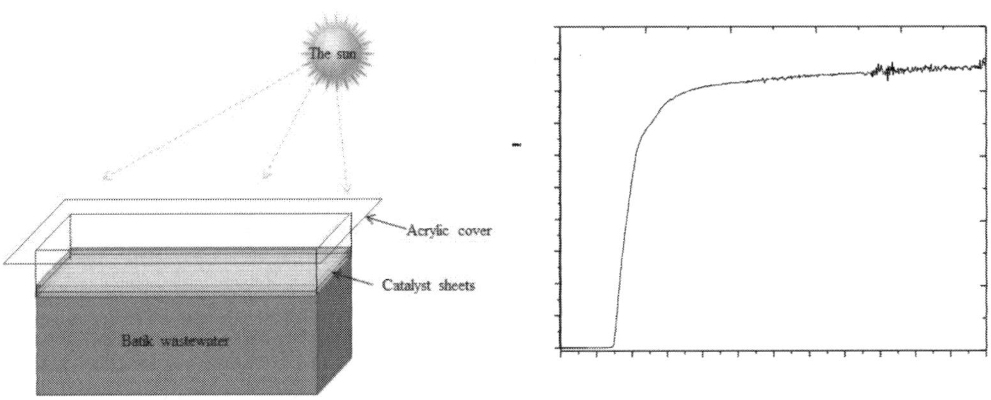

(a) (b)

Figure 1. Experiment setup, (b) transmission spectra of acryl

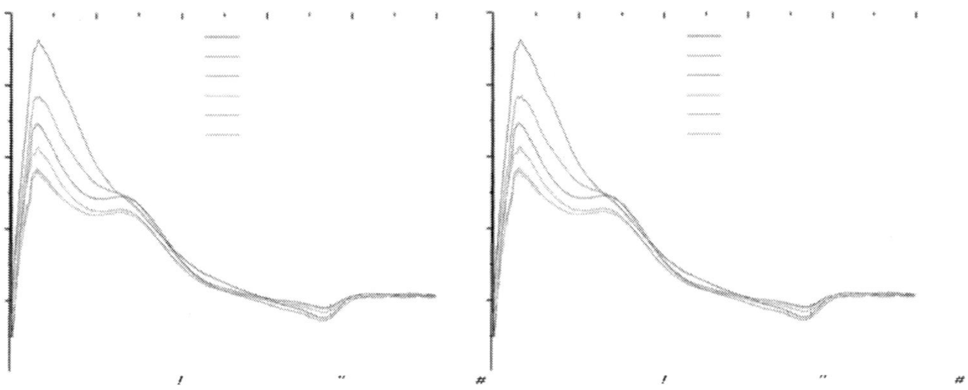

Figure 2. Absorption spectra of batik wastewater during degradation process, (a) no catalyst, (b) using 8 catalyst sheets.

The decomposition of dye in batik wastewater during photodegradation process for different variations of the number of catalyst is shown in Fig. 3. It can be seen that the addition of catalyst sheets from 5 to 8 pieces have increased the photodegradation rate of batik wastewater. Nonetheless, this does not apply to the addition of the next catalyst sheets. The use of 9 to 10 pieces of catalyst sheets can no longer increase the photodegradation ability of the system. This phenomenon can be explained as follows: an increase in the amount of the catalyst will provide an increase in active sites of the catalyst that contact with contaminants in the wastewater and role in the photocatalytic reaction. On the other hand, the increase in the number of catalyst sheets will reduce the intensity of sunlight that penetrates and activate the catalyst. The number of catalyst sheets that can be penetrated by sunlight depends on the transparency of the catalyst sheet and the concentration of batik wastewater. In this case, the optimum number of catalyst sheets is 8 pieces.

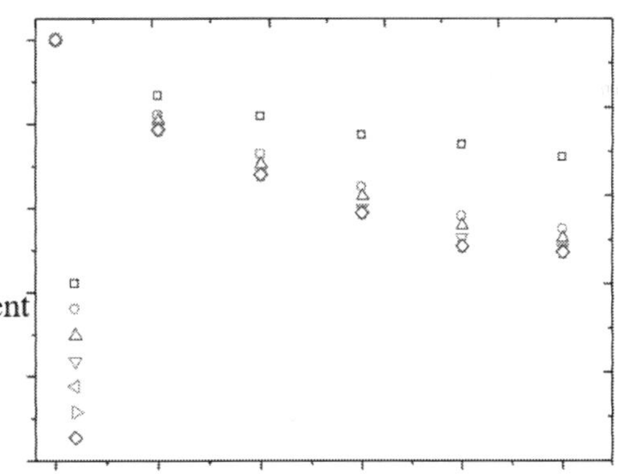

control experiment
photodegradation with 5 catalyst sheets
photodegradation with 6 catalyst sheets
photodegradation with 7 catalyst sheets
photodegradation with 8 catalyst sheets
photodegradation with 9 catalyst sheets
photodegradation with 10 catalyst sheets

Figure. 3. Decomposition of batik wastewater during 5 days of irradiation for different variations of the number of catalyst sheets.

In the same figure is also shown that, very low degradation rate is observed in the control experiment that without catalyst. Photolysis mechanism was able to degrade the dye in batik wastewater at a rate that is much smaller than the photocatalytic mechanism. One other interesting thing is that not all the dye in batik wastewater can be decomposed. From the experimental data by using 8 pieces of the catalyst, it is observed that the mechanism of photodegradation as if suspended after the 4th day of irradiation. On the 5th day of irradiation, the dye concentration is almost unchanged compared to the 4th day. We suspect that the remaining dye after the 4th day of irradiation is inorganic compounds that can't be decomposed by the photocatalytic reaction.

3.2. COD, BOD, and TSS reductions of the treated batik wastewater

In addition to dye, water quality is also determined by parameters of COD, BOD, and TSS [1, 10]. Fig. 4 shows the level of parameters for batik wastewater before and after treatment. It is seen that after 5 days under solar exposure, the photocatalyst has been able to reduce the levels of parameters of water quality. COD, BOD, and TSS levels were decreased by 97, 23 and 71 points, respectively. A decrease in the level of COD, BOD, and TSS indicates an increase in the water quality.

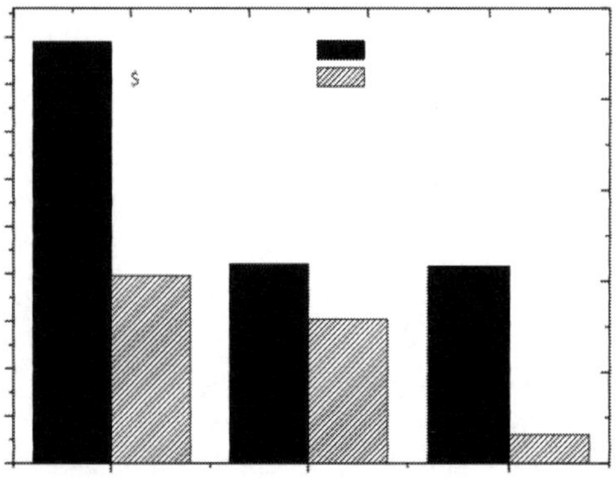

Figure 4. Reduction of parameters of water quality in the treated batik wastewater.

3.3. Repetitive operation performance

One of the benefits of the immobilized TiO2 is that it can be used repeatedly. To evaluate the stability of the catalyst sheet to decompose batik wastewater, photodegradation experiment of 1000 ml batik wastewater was repeated up to 4 cycles. In each cycle of experiments we used 8 layers of TiO2 coated plastic sheets. After each operation, catalyst sheets were washed using distilled water. The figure shows that over 49% of dye is already decomposed after 5 days of photocatalysis in each operation. The four repetitive operations gave a standard deviation of 0.57%, indicating that the stability of the catalyst sheets is almost constant. **Fig. 5** shows SEM images of the catalyst sheet before and after four times used. Based on the figure, it is clear that almost no differences in the morphology of the catalyst sheet before and after four times used. This fact shows that the TiO2 nanoparticles are stably attached on the surface of the plastic. The stability of the catalyst sheets causes its performance to degrade organic compound are also very stable when it is used repeatedly.

Figure 5. SEM images of the catalyst sheet, before (a) and after (b) four times used in the photodegradation of batik wastewater.

4. Conclusion

TiO2-coated plastic sheets were used to degrade wastewater produced in the dyeing process of batik. A total of 1000 ml batik wastewater was placed in a container and a number of catalyst sheets were the added to the solution. Optimization of the catalyst number was carried out by varying the number of catalyst sheets of 5 to 10 sheets. As a result, eight pieces of the catalyst are the optimum number of catalyst sheets with its ability to degrade batik wastewater upto 50.41% for 5 days irradiation. The decline in COD, BOD, and TSS were also observed during photodegradation which decreased by 97, 23 and 71 points, respectively. In this study also found that the catalyst sheet is stable to be used repeatedly in wastewater treatment.

References

[1] T. Zhang, X. Wang, X. Zhang, Review article: Recent progress in TiO2-mediated solar photocatalysis for industrial wastewater treatment, International Journal of Photoenergy 2014 (2014) 1-12.

[2] M.N. Chong, B. Jin, C.W.K. Chow, C. Saint, Recent developments in photocatalytic water treatment technology: A review, Water Research 44 (2010) 2997-3027.

[3] T. Zhang, X. Yan, D.D. Sun, Hierarchically multifunctional K-OMS-2/TiO2/Fe3O4 heterojunctions for the photocatalytic oxidation of humic acid under solar light irradiation, Journal of Hazardous Materials 243 (2012) 302-310.

[4] Y. Boyjoo, M. Ang, V. Pareek, Review: Some aspects of photocatalytic reactor modelling using computational fluid dynamics, Chemical Engineering Science 101 (2013) 764-784.

[5] P. Du, J.T. Carneiro, J.A. Moulijn, G. Mul, A novel photocatalytic monolith reactor for multiphase heterogeneous photocatalysis, Appl. Catal. A 334 (2008) 119-128.

[6] V.A. Isnaini, O. Arutanti, E. Sustini, H. Aliah, Khairurrijal, M. Abdullah, A novel system for producing photocatalytic titanium dioxide coated fibers for decomposing organic pollutans in water, Environ. Prog. Sustainable Energy 32 (2013) 42-51.

[7] H. Aliah, M.P. Aji, Masturi, E. Sustini, M. Budiman, M. Abdullah, The TiO2 nanoparticles-coated polypropylene copolymer as photocatalysts on methylene blue photodegradation under solar exposure, Am. J. Environ. Sci. 8 (2012) 280-290.

[8] Sutisna, M. Rokhmat, E. Wibowo, R. Murniati, Khairurrijal, M. Abdullah, Application of immobilized titanium dioxide as reusable photocatalyst on photocatalytic degradation of methylene blue, Advanced Material Researches 1112 (2015) 149-153.

[9] M. Abdullah, I. Nurmawati, H. Subianto, Khairurrijal, H. Mahfudz, Very wide band absorption of sunlight spectra using titanium dioxide particles with distributed band gap, Jurnal Nanosains and Nanoteknologi 3 (2010) 10-14.

[10] W.-J. Lau, A.F. Ismail, Polymeric nanofiltration membranes for textile dye wastewater treatment: preparation, performance evaluation, transport modelling, and fouling control - A review, Desalination 245 (2009) 321–348.

The Application of Microencapsulated Phycocyanin as a Blue Natural Colorant to the Quality of Jelly Candy

E N Dewi[1], R A Kurniasih[1] and L Purnamayati[1]

[1]Faculty of Fisheries and Marine Science, Diponegoro University, Jl. Prof. Soedarto, SH, Semarang – 50275, Indonesia

Email: nurdewisatsmoko@yahoo.com

Abstract. Phycocyanin is a blue color pigment which can be extracted from *Spirulina* sp. makes it potential to use as an alternative natural dye in the food product. The aim of this research was to determine the application of microencapsulated phycocyanin processed using spray dried method to the jelly candy. As a natural blue colorant, phycocyanin was expected to be safe for the consumer. The jelly candy was evaluated on the characteristics of its moisture, ash, Aw, pH, color appearance, and phycocyanin spectra with FTIR. The phycocyanin was microencapsulated using maltodextrin and Na-alginate as the coating materials (maltodextrin and Na-alginate in ratio 9:1.0 w/w). The spray drying process was operated with an inlet temperature of 80°C. The various concentrations of microencapsulated phycocyanin were added to the jelly candy such as 0%, 1%, 3%, 5% and jelly candy with brilliant blue used as comparison, each called PC, PS, PT, PL, and PB. The results showed that the various concentrations of phycocyanin added on the jelly product had significantly different on moisture content, Aw, and blue color. The FTIR spectra indicated that phycocyanin still persisted on the jelly candy. PL was the best jelly candy with the bluest color under PB.

Keywords: blue colorant, jelly candy, microencapsulated, natural dye, phycocyanin

1. Introduction

Candy is one of food which favored by various age groups. Jelly candy is made by gelling agent and sweetener which forming a certain texture [1]. Jelly candy usually added food colorant to attract the consumer attention, where 85,7% of candy contain synthetic dyes [2]. One of the synthetic dyes which often added to candy is Brilliant Blue FCF [3]. Brilliant Blue is not resistant to oxidation [4], consuming in the large quantities able to cause carcinogenic. Phycocyanin is one of natural blue colorant that can be used and contained in Spirulina [5].

Phycocyanin is more resistant to oxidation because of its function as antioxidant, anti-cancer, anti-inflammatory, and antiviral [6]. The phycocyanin weakness is its blue color is not resistant to pH and temperature [7] which causes the color to fade. Therefore, microencapsulation to phycocyanin is performed to change the form from liquid to powder [8] and preserve the color during storage.

Phycocyanin microcapsule addition to the jelly candy formulation is expected to affect its quality, besides bright blue color also has properties as an antioxidant. The aim of this study was to determine the effect of phycocyanin microcapsule addition to the quality of jelly candy.

2. Methods

2.1. Materials
The ingredients for the making of jelly candy includes sugar, glucose, and seaweed purchased from local market in Semarang. Phycocyanin was extracted from *Spirulina* sp. powder (PT. Neoalga, Sukoharjo, Indonesia) [9] with modification Coating materials: maltodextrin DE 10 (CV. Multi Kimia Raya, Semarang, Indonesia) and Na-alginate (PT. Selalu Lancar Maju Karya, Jakarta, Indonesia). Brilliant blue FCF synthetic dyes (CV. Indrasari, Semarang, Central Java).

2.2. Phycocyanin Microencapsulation
Phycocyanin microencapsulation [10] with modification was performed by homogenization using homogenizer (Ultraturrax T50 Basic Ika Werke, Germany) on phycocyanin extract with coating materials maltodextrin : Na-alginate with ratio 9%: 1% (w/v) toward the phycocyanin extract at speed 4500 rpm for 2 minutes. The homogeneous sample then processed into microencapsulation by the spray dryer (PlantLab, England) with an inlet temperature of 90°C. The microcapsule obtained then stored at dark glass bottle coated with aluminum foil.

2.3. The Making of Jelly candy
First, weighing seaweeds and sugar each for 200 grams. Seaweeds then blended with 500 ml of warm water (±40°C) until fine, while sugar dissolved into 600 ml of water. The fine seaweed was poured into sugar solution and stirred until homogeneous. The phycocyanin microcapsule which had been diluted with 100 ml of water was added in a different concentration of 0%(PC), 1% (PS), 3% (PT), and 5% (PL). After being well-mixed then formed using a pan. As the comparison, synthetic dyes Brilliant Blue FCF 1% (PB) was used.

2.4. Ash Content
The ash content was measured according to SNI 01-2891-1992. Several samples were ashes by the furnace at 550°C until becoming an ash, then cooled for 15-30 minutes at desiccator then weighed. Ash content is a difference between sample before and after ashes divided by initial sample weight multiplied by 100% [11].

2.5. Moisture content
The moisture content was measured according to SNI 01-2891-1992. The sample was heated using an oven at temperature 105°C for 24 hours. Ash content is a difference between sample before and after ashes divided by initial sample weight multiplied by 100% [11].

2.6. Aw and pH
The Aw measured by using Aw meter (rotronic HYGROPALM), while pH measured by using pH meter (pH meter TPX-90i Chemical Laboratories Co., Ltd.)

2.7. Color
The color was measured by Chroma-meter (CR-200 Minolta) where L (Lightness) indicated brightness, a indicated green-red color, and b indicated blue-yellow color [12].

2.8. Fourier Transform Infrared Spectroscopy (FTIR)
The molecule structure of phycocyanin on jelly candy was measured by FTIR using spectra IR 4000-400 cm^{-1} at room temperature [13].

2.9. Statistical Analysis

This study using Completely Randomized Design with 1 factor, phycocyanin microcapsule concentration. The data obtained from triplication of an assay and analyzed using SPSS 17. The advanced test was performed using Tukey analysis.

3. Results and Discussion

Table 1. Ash content, moisture content, Aw, and pH of jelly candy

No	Phycocyanin concentration	Ash Content (%)	Moisture content (%)	Aw	pH
1	PC	1,757 ± 0,015^{c}	54,527 ± 0,374^{a}	0,863 ± 0,002^{a}	6,400 ± 0,200^{a}
2	PS	1,290 ± 0,225^{b}	54,560 ± 0,503^{a}	0,869 ± 0,005^{ab}	6,600 ± 0,265^{a}
3	PT	0,950 ± 0,010^{a}	58,683 ± 0,270^{b}	0,875 ± 0,002^{bc}	6,567 ± 0,208^{a}
4	PL	0,897 ± 0,109^{a}	61,693 ± 0,905^{c}	0,879 ± 0,001^{c}	7,100 ± 0,100^{b}
5	PB	1,003 ± 0,055^{ab}	62,203 ± 0,261^{c}	0,895 ± 0,004^{d}	7,467 ± 0,153^{c}

Note: PC: jelly candy with the addition of 0% phycocyanin microcapsule
PS: jelly candy with the addition of 1% phycocyanin microcapsule
PT: jelly candy with the addition of 3% phycocyanin microcapsule
PL: jelly candy with the addition of 5% phycocyanin microcapsule
PB: jelly candy with the addition of 1% brilliant blue
The data was the average of triplication ± standard deviation.
Different superscript on the same column indicates significantly different at level α 0,05

3.1. Ash Content

The jelly candy with the addition of phycocyanin reduces the ash content. This caused by the increasing of phycocyanin concentration, as well increasing the moisture content on jelly candy so the ash content decreased. Meanwhile, compared to jelly candy with the addition of 1% brilliant blue was not significantly different to jelly candy with phycocyanin microcapsule. The ash content ranged between 0,897%-1,757%. This result was higher compared to Buntaran *et. al.*, [11] which adding tomatoes extract on candy and produced ash content around 0,6-0,8.

3.2. moisture content

jelly candy is an intermediate moisture food, rich in sugar and other compounds which are hygroscopic and hard to dry [14]. Moisture content plays an important role to determine the quality of jelly candy, especially to form the texture. Based on the table above, the higher of phycocyanin microcapsule would increase the moisture content on jelly candy. If compared to jelly candy with the addition of brilliant blue, jelly candy with 5% of phycocyanin microcapsule was significantly different on its moisture content. The moisture content of jelly candy in this study ranged between 54,527%-61,693%. The higher level of moisture on jelly candy due to water was not able to evaporate perfectly as the temperature applied was not too high and phycocyanin will be damaged at high temperature [6]. On the other hand, the existence of sugar compound will cause the color to change if heated at high temperature. This result was higher compared to Delgado and Banon [14] which producing the jelly candy with the moisture content around 21%. Whereas Muzzaffar *et. al.*, [16] add pumpkin to candy resulting in the moisture content around 20,1%. According to SNI 3547.2-2008 about soft cotton candy, the maximum moisture level is 20%.

3.3. Aw

Activity water is an important factor related to food decay and affects toward shelf-life, while fungi start to grow at Aw 0,7 (Utomo, 2014). The Aw value associated with the moisture content. The addition of phycocyanin microcapsule on jelly candy affects the Aw value. The Aw value increased along with the increasing of phycocyanin microcapsule, ranged around 0,863-0,879. This result was higher compared to Charoen *et. al.*, [18] where *Psidium guajava Linn.* leaves extract was added into jelly candy so produced Aw value around 0,75-0,79.

3.4. pH

The pH of jelly candy with phycocyanin microcapsule addition increased along with the increasing of phycocyanin microcapsule percentage. However, pH produced was neutral between 6-7. The pH value of this study was higher or around neutral because it aimed to maintain the phycocyanin as a natural colorant. The phycocyanin was stable at pH 5,5-6 and able to preserve effectively at pH 7 [7].

Table 2. Jelly candy Color

No	Phycocyanin concentration	L	a	B
1	PC	42,943 ± 0,919^{c}	0,033 ± 0,194^{b}	0,923 ± 0,471^{a}
2	PS	38,550 ± 0,654^{b}	-0,290 ± 0,087^{ab}	-1,303 ± 0,202^{b}
3	PT	40,743 ± 1,253^{bc}	0,623 ± 0,080^{c}	-4,707 ± 0,499^{c}
4	PL	41,200 ± 0,850^{bc}	1,063 ± 0,275^{c}	-7,297 ± 0,869^{d}
5	PB	31,050 ± 2,453^{a}	-0,423 ± 0,100^{a}	-9,660 ± 0,867^{e}

Note: PC: jelly candy with the addition of 0% phycocyanin microcapsule
PS: jelly candy with the addition of 1% phycocyanin microcapsule
PT: jelly candy with the addition of 3% phycocyanin microcapsule
PL: jelly candy with the addition of 5% phycocyanin microcapsule
PB: jelly candy with the addition of 1% brilliant blue
The data was the average of triplication ± standard deviation.
Different superscript on the same column indicates significantly different at level α 0,05

3.5. Color

Based on Table 2, the higher of phycocyanin microcapsule percentage added to jelly candy, the color was getting blue, it showed by the increasing level of Lightness (L) and blue (-b). This indicated that phycocyanin able to persist during jelly candy processing. However, at the addition of phycocyanin microcapsule 5%, the intensity of blue color was under jelly candy with 1% of brilliant blue FCF addition. The different result showed by Charoen *et. al.*, [18] where the addition of *Psidium guajava* leaves extract on jelly candy declining the value of L and b.

3.6. Fourier Transform Infrared Spectroscopy (FTIR)

jelly candy observed its phycocyanin spectra by FTIR. The result can be seen in Figure 1. The constituent component of jelly candy by FTIR spectrophotometrically measured at 400-4000 cm^{-1}. According to FTIR spectra, it can be seen that the five samples of jelly candy had the similar wavelength, it means that the constituent compound of jelly candy was same, seaweed and sugar (sucrose). Seaweeds constituent compound visible at 845-930 cm^{-1} of wavelength [19]. Whereas, the result of this study was around 848,68-910,40 cm^{-1}, while at 848,68 cm^{-1} wavelengths was the peak for Esther sulfate which was a functional group of carrageenan. While the peak for sucrose was detected at 995,27 cm^{-1} wavelengths. This result was similar with Adina *et. al.*, [21] where sucrose was detected at 995 cm^{-1} wavelengths. Phycocyanin was detected at 1651,07 cm^{-1} wavelengths and not contained on PC and PB. This result was similar with Gang *et.al.*, [22], that the C-Phycocyanin's absorption peaks was 1650 cm^{-1}. But the different result showed by Suzery *et. al.*, [23] that phycocyanin spectra appeared at 1550-1600 cm^{-1} wavelength.

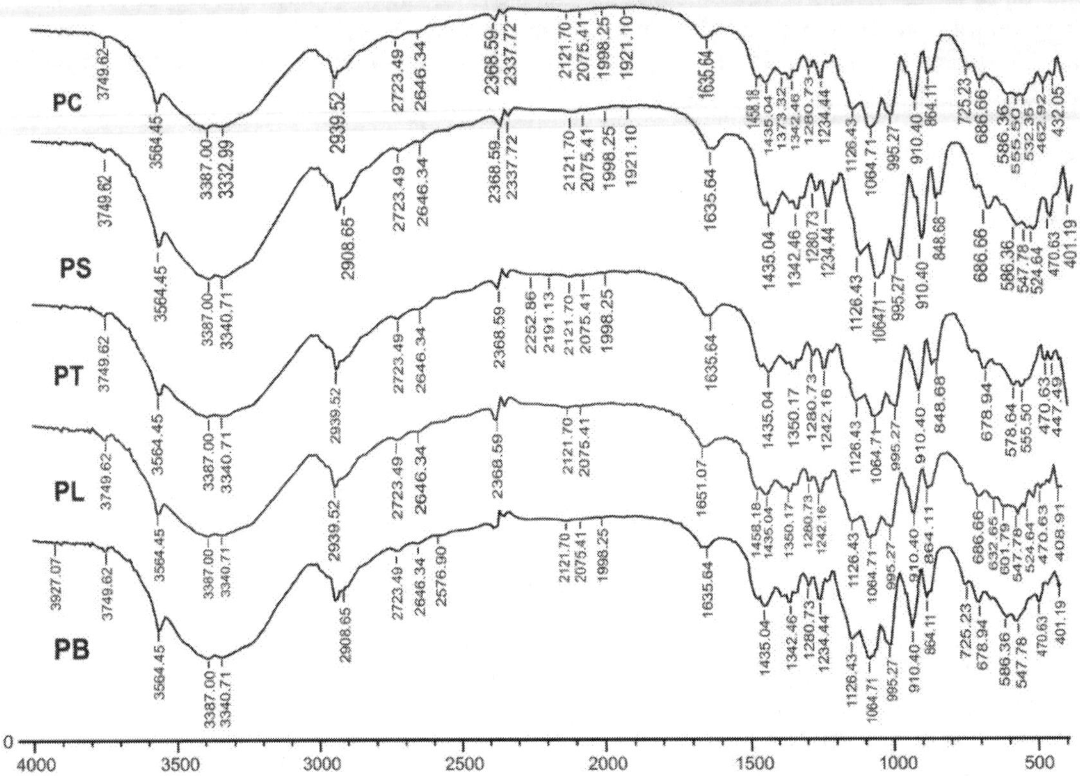

Figure 1. FTIR jelly candy

Note: PC: jelly candy with the addition of 0% phycocyanin microcapsule
PS: jelly candy with the addition of 1% phycocyanin microcapsule
PT: jelly candy with the addition of 3% phycocyanin microcapsule
PL: jelly candy with the addition of 5% phycocyanin microcapsule
PB: jelly candy with the addition of 1% brilliant blue

4. Conclusion

The application of phycocyanin microcapsule on jelly candy has not yet produced the moisture content which complies with Indonesian National Standard. However, the addition of 5% phycocyanin microcapsule able to produce bright blue color under brilliant blue dyes. This showed by measurement using chroma-meter and FTIR, that phycocyanin still persists during the processing of jelly candy.

5. Acknowledgement

This research funded by Diponegoro University through research scheme of Riset Pengembangan dan Penerapan (RPP) year 2017.

References

[1] Habilla C, Sim S Y, Azizah N and Cheng L H 2011 The properties of jelly candy made of acid-thinned starch supplemented with konjac glucomannan or psyllium husk powder. *International Food Research Journal.* **18** 213-220.

[2] Zahra N, Alim-un-Nisa, Kalim I, Fatima S, Khan H, Akhlaq F, Butt I F and Hina S 2016 Identification of synthetic food dyes in various candies. *Park. J. Biochem. Mol. Biol.* **49**(1) 09-17.

[3] Saleem N, Umar Z N and Khan S I 2013 Survey on the use of synthetic food colors in food samples procured from different educational institutes of Karachi city. *The Journal of Tropical Life Science*. **3**(1) 1-7.
[4] Allam K V and Kumar G P 2011 Colorant-The cosmetics for the pharmaceutical dosage forms. Review. *International Journal of Pharmacy and Pharmaceutical Sciences*. **3**(3) 13-21.
[5] Moraes C C, Sala L, Cerveira G P and Kalil S J 2011 C-Phycocyanin extraction from Spirulina platensis wet biomass. *Brazillian Journal of Chemical Engineering*. **28**(01) 45-49.
[6] Jerley A A and Prabu D M 2015 Purification, characterization and antioxidant properties of C-Phycocyanin from Spirulina platensis. *SIRJ-APBBP*. **2**(1) 7-15.
[7] Chaiklahan R, Chirasuwan N and Bunnag B 2012 Stability of phycocyanin extracted from Spirulina sp.: influence of temperature, pH, and preservatives. *Process Biochemical*. **47** 659–664.
[8] Chanana A, Kataria M K, Sharma M and Bilandi A 2013 Microencapsulation: advancements in applications. *International Research Journal of Pharmacy*. **4**(2) 1-5.
[9] Chaiklahan R, Chirasuwan N, Loha V, Tia S and Bunnag B 2011 Separation and purification of phycocyanin from *Spirulina* sp. using a membrane process. *Bioresource Technology*. **102** 7159–7164.
[10] Dewi E N, Purnamayati L and Kurniasih R A 2016 Antioxidant activities of phycocyanin microcapsules using maltodextrin and carrageenan as coating materials. *Journal of Technology*. **78**(4-2) 45-50.
[11] Buntaran W, Astirin O P and Mahajoeno E 2010 Effect of various sugar solution concentration on characteristics of dried candy tomato (*Lycopersicum esculentum*). *Bioscience*. **2**(2) 55-61.
[12] Dewi E N, Purnamayati L and Kurniasih R A 2017 Physical characteristics of phycocyanin from Spirulina microcapsules using different coating materials with freeze drying method. *IOP Conf. Series: Earth and Environmental Science*. **55** 1-7.
[13] Venkatesan S, Pugazhendy K, Sangeetha D, Vasantharaja C, Prabakaran S and Meenambal M 2012 Fourier Transform Infrared (FT-IR) Spectoroscopic analysis of *Spirulina*. *International Journal of Pharmaceutical & Biological Archives*. **3**(4) 969-972.
[14] Delgado P and Banon S 2015 Determining the minimum drying time of gummy confections based on their mechanical properties. *Cyta-Journal of Food*. **13**(3) 329-335.
[15] Ergun R, Lietha R and Hartel R W 2010 Moisture and shelf life in sugar confections. *Critical Reviews in Food Science and Nutrition*. **50** 162-192.
[16] Muzzaffar S, Baba W N, Nazir N, Masoodi F A, Bhat M and Bazaz R 2016 Effect of storage on physicochemical, microbial and antioxidant properties of pumpkin (*Cucurbita moschata*) candy. *Cogent Food and Agriculture*. **2** 1-13.
[17] Utomo B S B, Darmawan M, Hakim A R and Ardi D T 2014 Physicochemical properties and sensory evaluation of jelly candy made from different ratio of κ-carrageenan and konjac. Squalene Bulletin of Marine & Fisheries Postharvest & Technology. **9**(1) 25-34.
[18] Charoen R, Savedboworn W, Phuditcharnchnakun S and Khuntaweetap T 2015 Development of antioxidant gummy jelly candy supplemented with *Psidium guajava* leaf extract. *KMUTNB Intl J Appl Sci Tecnol*. **8**(2) 145-151.
[19] Pereira L, Gheda S F and Ribeiro-Claro P J A 2013 Analysis by vibrational spectroscopy of seaweed polysaccharides with potential use in food, pharmaceutical, and cosmetic industries. International Journal of Carbohydrate Chemistry. 1-7.
[20] Dewi E N, Darmanto Y S and Ambariyanto 2012 Characterization and quality of semi-refined carrageenan (SCR) products from different coastal waters based on fourier transform infrared technique. *Journal of Coastal Development*. **16**(1) 25-31.
[21] Adina C, Florinela F, Abdelmoumen T and Carmen S 2010 Application of FTIR spectroscopy for A rapid determination of some hydrolytic enzymes activity on sea buckthorn substrate. Romanian Biotechnological Letters. **15**(6) 5738-5744.

[22] Gang Y, Zheng L, Fei L, Chen L, Fu-xin D and Nai-ju Y 1999 Isolation and characterization of C-phycocyanin from Spirulina platensis. Chemical Research in Chinese Universities. **15**(1) 35-38.

[23] Suzery M, Hadiyanto, Majid D, Setyawan D and Sutanto H 2017 Improvement of stability and antioxidant activities by using phycocyanin – chitosan encapsulation technique. IOP Conf. Series: Earth and Environmental Science. **55** 1-7.

IOP Conf. Series: Earth and Environmental Science **116** (2018) 012048 doi:10.1088/1755-1315/116/1/012048

Mapping of trophic states based on nutrients concentration and phytoplankton abundance in Jatibarang Reservoir

Siti Rudiyanti, Sutrisno Anggoro, Arif Rahman

Department of Aquatic Resource, Faculty of Fisheries and Marine Science, Diponegoro University, Jl. Prof. H. Soedarto, SH, Tembalang, Semarang 50275, Indonesia

E-mail: st_rudiyanti@yahoo.com

Abstract. Jatibarang Reservoir is one of the Indonesian Reservoirs, which used for human activities such as tourism and agriculture. These activities will provide input of organic matter and nutrients into the water. These materials will impact water quality and eutrophication process. Eutrophication is the water enrichment by nutrients, especially nitrogen and phosphorus which can promote the growth of phytoplankton. Some indicators of eutrophication are increasing nutrients, trophic states, and change of phytoplankton composition. The relationship between water quality and phytoplankton community can be used as an indicator of trophic states in Jatibarang Reservoir. The aim of this study was to analyze the effect of nutrients concentration and phytoplankton abundance to the trophic states and mapping trophic states based on nutrients concentration and phytoplankton in Jatibarang Reservoir. This study was conducted in June and July 2017 at 9 stations around Jatibarang Reservoir. The results showed that average concentration of nitrate, phosphate, and chlorophyll-a in Jatibarang Reservoir was 0.69 mg/L, 0.27 mg/L, and 1.66 mg/m^3, respectively. The phytoplankton abundance ranged 16-62,200 cells/L, consists of 21 genera of four classes, i.e. *Chlorophyceae, Cyanophyceae, Bacillariophyceae*, and *Dinophyceae*. *Cyanophyceae* was a dominant phytoplankton group based on the composition of abundance (>80%). High nutrient concentrations and phytoplankton dominated by *Anabaena* (*Cyanophyceae*) which indicated that the waters in Jatibarang Reservoir were eutrophic.

Keywords: Jatibarang Reservoir, nutrients, phytoplankton, trophic states, eutrophic

1. Introduction

Jatibarang Reservoir is a reservoir located in Gunungpati District, Semarang City. Jatibarang Reservoir has a water volume capacity of 20,400,000 m^3. Jatibarang Reservoir catchment area is 54 km^2 and maximum water level ± 155.30 m. Jatibarang Reservoir has an area of about 46.56 ha which functions as flood control in Semarang City, power plant, tourism area, fishing activities and irrigation. Jatibarang Reservoir that is used to various activities needs to be reviewed the water quality as a method to manage resources in the reservoir.

One common problem that often occurs in the reservoir is eutrophication. Eutrophication is a worldwide problem that occurs in both freshwater and marine ecosystems. Eutrophication is an increasing process of trophic states from oligotrophic, mesotrophic to eutrophic. Some of the eutrophication caused by domestic waste pollution [1] and watershed management issues [2]. Some of the eutrophication effects are causing changes in water quality, such as decreased dissolved oxygen and water transparency, organic matter enhancement, and phytoplankton blooms.

This study is important because the trophic state is one of the basic determination of pollution load capacity. The pollution load capacity is a consideration for the determination of spatial planning of reservoir catchment area and allowance activity that can affect the water quality. The trophic state can also use as a basis for water management for the reservoir to function properly. Water management is essential because water is an important natural resource for life and useful for maintaining ecological balance.

The aim of this study was to analyze the effect of nutrients concentration and phytoplankton abundance to the trophic states and mapping trophic states based on nutrients concentration and phytoplankton in Jatibarang Reservoir.

2. Materials and methods

2.1. *Materials*

Materials and tools used during the study were materials and tools for water sampling, water quality analysis, and phytoplankton identification in the laboratory. The tools used for water sampling are Boat, Van Dorn water sampler, GPS, Secchi disk, plankton net and sample bottles. The tools and materials used for phytoplankton identification were microscopes and phytoplankton identification books.

2.2. *Sampling methods*

This study was conducted two times during the period from June to July 2017 consisted of water and phytoplankton sampling around Jatibarang Reservoir. The sampling sites were 9 stations located in the waters around Jatibarang Reservoir (Fig. 1). The stations were chosen based on different waters and land use conditions around Jatibarang Reservoir.

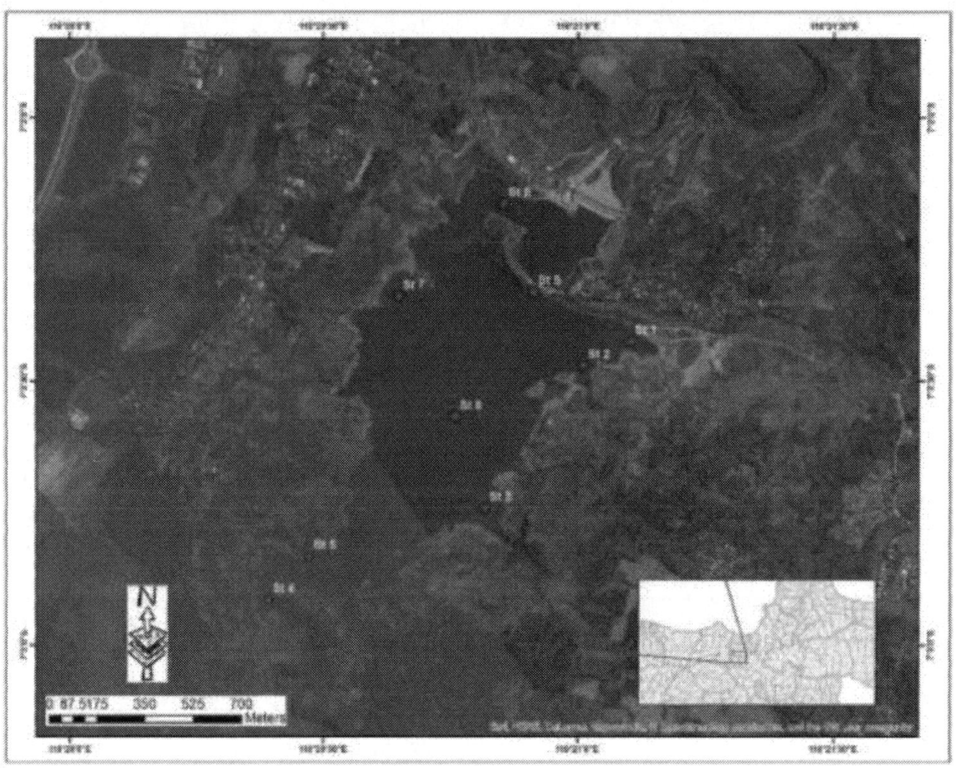

Figure 1. The sampling location at Jatibarang Reservoir

Samples were collected at a depth about 1 m below the surface at each location. Samples were carried out using Van Dorn water sampler at depth of 1 m from surface. Water and phytoplankton sample then inserted into the bottle. Water samples preserved with cooling in the cool box. Phytoplankton preserved immediately with 1% of acidified Lugol iodine solution for counting under a microscope in the laboratory.

2.3. *Analytical methods*

The water quality in Jatibarang Reservoir, including the water temperature, pH, dissolved oxygen, were measured in situ with Water Quality Checker. Transparency was measured with a Secchi disk. A sample of approximately 1.5 L of water was taken from each site to measure the chlorophyll-a and nutrient concentrations. Quantitative samples of the phytoplankton were collected each sampling site and preserved immediately with 1% of acidified Lugol iodine solution for counting under a microscope in the laboratory.

Nitrate and phosphate were analyzed with spectrophotometry method according to the standard method. In the filtered water samples, nitrate and phosphate with the ascorbic acid method. A spectrophotometer was used to conduct the measurements automatically. The phytoplankton samples were counted directly in counting chamber using a compound microscope at an objective magnification 10x. Phytoplankton species were identified according to plankton identification books [3][4].

2.4. *Data analysis*

Data analysis used in this study were phytoplankton community structure analysis and trophic states. The phytoplankton community structure analysis was determined using the diversity index, similarity, and dominance index according to Krebs [5]. The trophic state's analysis was determined based on the Nygaard index [6].

3. Result and Discussion

3.1. *Variations of the environmental factors*

The physical and chemical conditions in Jatibarang Reservoir are shown in Fig. 2. Fig. 2 shows the time course of water temperature, pH, transparency, and dissolved oxygen concentration, respectively. The physical-chemical conditions of the waters varied considerably during the study. The results showed that the water temperature varied from 27.60 to 29.05 °C with the highest value 29.05 °C in Station 3. The pH value of water varied from 7.14 to 8.03 with average 7.53. Water transparency varied from 0.44 to 0.77 m with average 0.62 m.

Dissolved oxygen (DO) varied from 6.92 to 7.94 mg/L with average 7.51 mg/L. The highest DO concentration reaching 7.94 mg/L in Station 2 and 7.85 mg/L in Station 9 and the lowest concentration (6.92 mg/L) was recorded in Station 3 when the highest temperature value occurred. A high DO concentration in water may cause by sampling time of day during when the oxygen produced by photosynthesis is higher than that consumed by respiration [7].

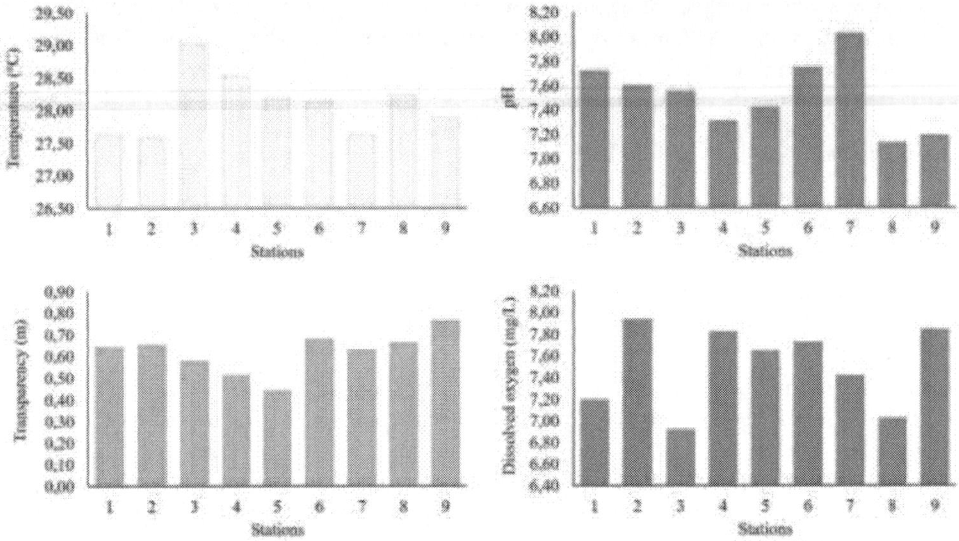

Figure 2. The physical and chemical conditions in Jatibarang Reservoir

Chlorophyll-a concentrations ranged between 1.29-2.22 mg/m^3 with an average of 1.66 mg/m^3. Nitrate concentrations ranged from 0.56 to 0.82 mg/L with an average of 0.69 mg/L. Nitrate concentration> 0.2 mg/L indicates that the waters around Jatibarang Reservoir belong to the eutrophic waters [8]. Nitrate is the main form of nitrogen in the water, stable and very soluble in water. Nitrate concentrations between 1-2 mg/L indicate the presence of fertilizer pollution from agricultural activities [9].

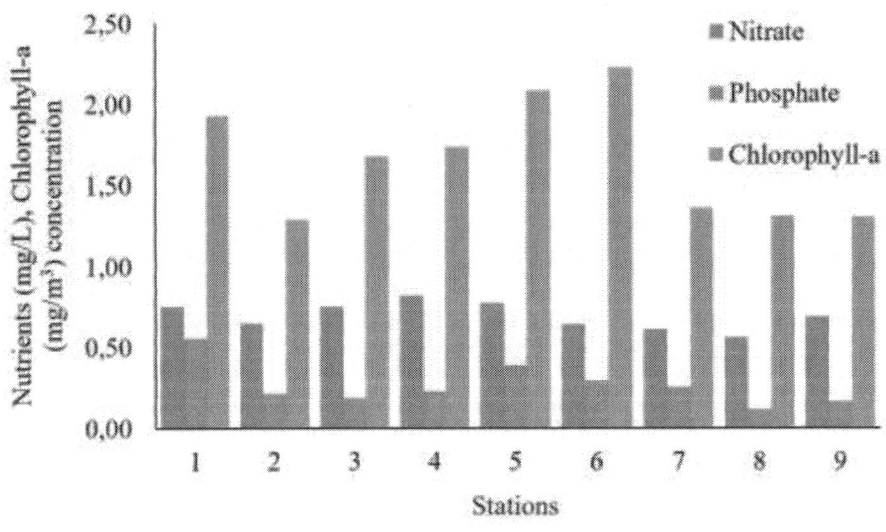

Figure 3. Nutrients and chlorophyll-a concentrations

Phosphate concentrations ranged between 0.11-0.56 mg/L with an average of 0.27 mg/L. Phosphate concentration in the waters around Jatibarang Reservoir showed a high or eutrophic condition (>0.20 mg/L) [10]. Agricultural land is one source of nutrients that enter the waters. One such agricultural activity as fertilizer will contribute to the increase of nutrient concentration in the waters and can lead to eutrophication of the waters. The use of fertilizers in agricultural activities around the lake can increase the concentration of nitrogen (ammonia and nitrate) in the waters [11].

3.2. Variations of the phytoplankton community

A total of 21 phytoplankton genera were identified. Among these genera, 2 genera in *Cyanophyceae*, 6 genera in *Chlorophyceae*, 11 genera in *Bacillariophyceae*, and 2 genera in *Dinophyceae*, respectively, were determined from the reservoir samples. The phytoplankton abundance varied between 16-62,200 cells/L. *Cyanophyceae* was greatest in 9 stations. As a phylum with the maximal genera, *Cyanophyceae* algae were determined in all samples. The composition of phytoplankton abundance is presented in Fig. 4.

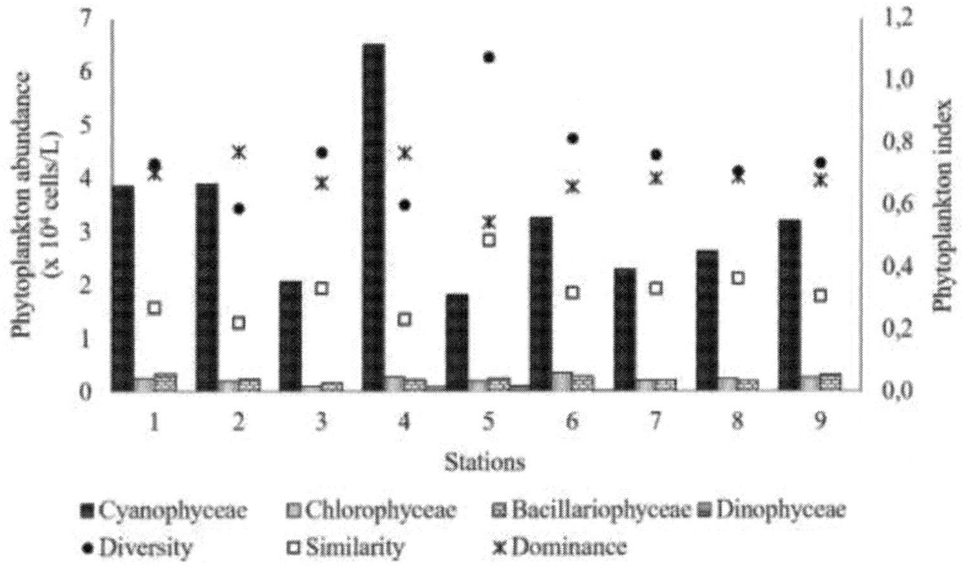

Figure 4. Composition of phytoplankton abundance in Jatibarang Reservoir

Cyanophyceae was absolutely the dominant genera of the community in Jatibarang Reservoir despite having a small number of species. *Cyanophyceae* represented ±85% of the total phytoplankton abundance, *Chlorophyceae* and *Bacillariophyceae* represented ±7%, and *Dinophyceae* represented ±1% of the total phytoplankton abundance. Cyanophyceae dominance was caused by the influence of eutrophic water conditions with high concentrations of nutrients.

Cyanophyceae was common in waters with high phosphate concentrations [12]. The presence of *Cyanophyceae* was influenced by phosphate concentrations where high *Cyanophyceae* abundance was present in waters with high phosphate concentrations [13][14]. These findings suggest that Jatibarang Reservoir has become entirely eutrophic waters.

Anabaena dominated the phytoplankton community. *Anabaena* was found at all sites with a total abundance of 285,800 cells/L. *Anabaena* is one of phytoplankton genera that became the founder of eutrophic waters. Some of the most phytoplankton abundant in eutrophic waters are *Anabaena*, *Microcystis*, *Chroococcus*, and filamentous species such as *Aphanizomenon* [14][15][16]. *Anabaena* and *Microcystis* are the toxic phytoplankton and causes of problems associated with hypoxia and changes in the structure of biological communities [17][18].

Dominant phytoplankton in this study was same with another study in Lake Toba. The same of condition dominant phytoplankton can cause by same in water quality conditions, especially the nutrients concentrations. In this study, nutrient concentrations, especially nitrate and phosphate, have high concentrations same within Lake Toba [19]. The others genera of phytoplankton were *Synedra* and *Staurastrum* with a total abundance of 20,097 cells/L and 15,615 cells/L, respectively. The others genera, i.e. *Cymbella, Rhizosolenia,* and *Pleurosigma*, had low abundance in all sites.

The presence of phytoplankton is influenced by several factors, such as nutrient concentration, light conditions, temperature, pH, and predation by zooplankton and planktivorous fish [14][20]. The development of phytoplankton community in every water is dynamic. A type of phytoplankton may be more dominant than other types at relatively short intervals. The dominant phytoplankton type at any given time can be a rare species at a later time and may be replaced by another more dominant species[3].

The diversity index during this study of 0.59-1.07 indicates that low phytoplankton diversity and phytoplankton communities are unstable. The value of the diversity index is similar to that of Lake Sentani of 0.8-2.3 [21]. The similarity and dominance index during this study varied from 0.22 to 0.49 and 0.54 to 0.77, respectively. The similarity index indicating that the phytoplankton composition type was low to evenly distributed and showed the dominance of certain phytoplankton genera.

Some stations with low diversity indexes show a higher index of dominance than the similarity index (Fig. 4). Both the index of diversity and similarity in this study were low. The low of the diversity index and similarity indicates that there has been a decrease in water quality. The low index value of phytoplankton diversity can be caused by poor water quality conditions, so only tolerant types of phytoplankton can live in those waters[22].

3.3. Trophic states

The trophic states of water were determined using the Nygaard index[6]. The Nygaard index calculation was based on the composition of the number of phytoplankton species found during the study. The Nygaard index results in this study varied from 3 to 9 indicating that the trophic states around Jatibarang Reservoir were eutrophic waters (Fig. 5).

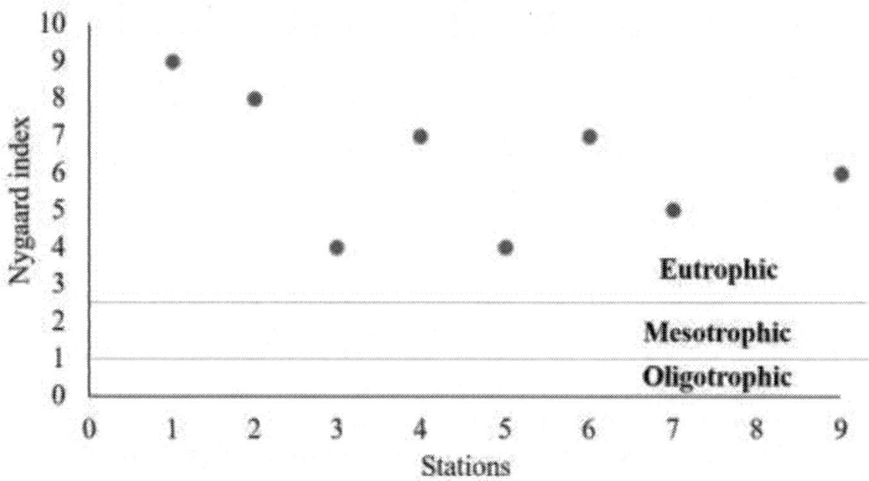

Figure 5 Trophic states in Jatibarang Reservoir

Based on the Nygaard index, nitrate and phosphate concentrations, trophic states in the waters Jatibarang Reservoir has shown eutrophic levels (Fig. 6). These results indicate that Jatibarang Reservoir has eutrophication. Eutrophication is a natural process that can occur in a wide variety of waters (especially lakes and reservoirs) and can rise very rapidly with human activities [23].

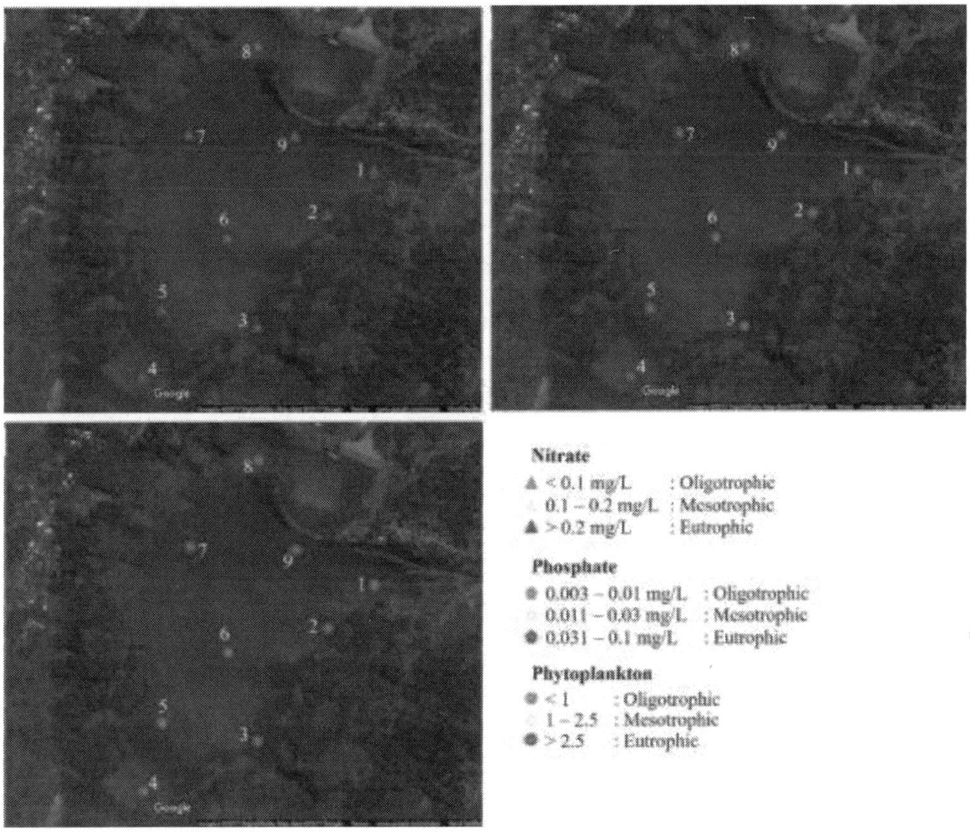

Figure 6 Trophic states distributions in Reservoir Jatibarang

The eutrophic condition is a description of the human activities impact on waters [24]. Some of the things that can affect nutrient concentration in waters are nutrient input from water catchment and lake sediment [25], fertilizer from agricultural, phosphate from detergent, soil erosion, industrial waste [26], and nutrient cycles [27].

4. Conclusion

Jatibarang Reservoir has a high nitrate and phosphate concentrations. Phytoplankton community was dominated by *Cyanophyceae*, especially *Anabaena*. High nutrient concentrations and phytoplankton dominated by *Anabaena* (*Cyanophyceae*) which indicated that the waters in Jatibarang Reservoir were eutrophic.

5. Acknowledgement

Authors would like to thank the Faculty of Fisheries and Marine Sciences, Diponegoro University for funding this research.

References

[1] Lehmusluoto P. 2000. Lake Toba, The first sound science initiative to abate change in the lake environment. *Research and Monitoring for Basin Management Decisions.* 1-12.

[2] Saragih B, Sunito S. 2001. *Lakes and Reservoirs: Research and Management.* **6**: 247-251.

[3] Davis CC. 1955 *The Marine and Freshwater Plankton*. Michigan State University Press. Michigan.

[4] Mizuno T. 1979. *Illustrations of the Freshwater Plankton of Japan*. Hoikusha Publishing Co Ltd. Osaka.

[5] Krebs CJ. 1989. *Ecological Methodology*. Harper Collins Publishers. Inc. New York.

[6] Rawson DS. 1956. *J. Fish Res*. **1**(1): 18-25.

[7] Xue PF, Chen CS, Qi JH, Beardsley RC, Tian RC, Zhao LZ, Lin HC. 2014. *J. Mar. Syst.* **131**: 102–119.

[8] Goldman CR, Horne AJ. 1983. *Limnology*. United States of America: McGraw-Hill Book Company. xvi+464 p.

[9] Weiner ER. 2008. *Applications of Environmental Aquatic Chemistry*. US: CRC Press.

[10] Wu FF, Wang X. 2012. *Procedia Environmental Sciences*. **13**: 1030-1036.

[11] Kattner E, Schwarz D, Maier G. 2000. *Limnologica*. **30**: 261-270.

[12] Sulastri. 2011. *Limnotek*. **18**(1): 1-14.

[13] Lu J, Wu H, Chen M. 2011. *Limnologica*. **41**: 48-56.

[14] Jiang YJ, He W, Liu WX, Qin N, Ouyang HL, Wang QM, Kong XZ, He QS, Yang C, Yang B, Xu FL. 2014. *Ecological Indicators*. **40**: 58-67.

[15] Abrantes N, Antunes SC, Pereira MJ, Goncalves F. 2006. *Acta Oecologica*. **29**: 54-64.

[16] Elliott JA, May L. 2008. *Freshwater Biology*. **53**: 32-41.

[17] Carmichael WW. 2001. *Human and Ecological Risk Assessment*. **7**(5): 1393-1407.

[18] Chen W, Song LR, Peng L, Wan N, Zhang XM, Gan NQ. 2008. *Water Research*. **42**: 763-773.

[19] Rahman A, Pratiwi NTM, Hariyadi S. 2016. *Jurnal Ilmu Pertanian Indonesia (JIPI)*. **21**(2): 120-127.

[20] Lau SSS & Lane SN. 2002. *The Science of the Total Environment*. **288**: 167-181.

[21] Astuti LP, Satria H. 2009. *Limnotek*. **16**(2): 88-98.

[22] Soedibjo BS. 2006. *Oseanologi dan Limnologi di Indonesia*. **40**: 65-78.

[23] Rohlich GA. 1969. *Eutrophication: Causes, Consequences, Correctives. National Academy of Sciences*. Washington DC (US). 681 p.

[24] Kagalou I, Papastergiadou E, Leonardos I. 2008. *Journal of Environmental Management.* **87**: 497-506.

[25] Levine SN, Schindler DW. 1992. *Limnology and Oceanography.* **37**(5): 917-935.

[26] Alvarez-Vazquez LJ, Fernandez FJ, Martinez A. 2014. *Journal of The Franklin Institute*. **351**: 4142-4182.

[27] Schindler DW, Bayley SE, Schlesinger WH. 1993. *Global Biogeochemical Cycles*. **7**: 717-734.

Characterization of Lactic Acid Bacteria (LAB) isolated from Indonesian shrimp paste (terasi)

U Amalia[1*], Sumardianto[1], and T W Agustini[1]

[1]Department of Fish Product Technology, Faculty of Fisheries and Marine Science, Diponegoro University, Jl. Prof. H. Soedarto, SH, Tembalang, Semarang 50275, Indonesia

Email : ulfah_amalia@yahoo.com

Abstract : Shrimp paste was one of fermented products, popular as a taste enhancer in many dishes. The processing of shrimp paste was natural fermentation, depends on shrimp it self and the presence of salt. The salt inhibits the growth of undesirable microorganism and allows the salt-tolerant lactic acid bacteria (LAB) to ferment the protein source to lactic acids. The objectives of this study were to characterize LAB isolated from Indonesian shrimp paste or "Terasi" with different times of fermentation (30, 60 and 90 days). Vitech analysis showed that there were four strains of the microorganism referred to as lactic acid bacteria (named: LABS1, LABS2, LABS3 and LABS4) with 95% sequence similarity. On the basis of biochemical, four isolates represented *Lactobacillus,* which the name *Lactobacillus plantarum* is proposed. *L.plantarum* was play role in resulting secondary metabolites, which gave umami flavor in shrimp paste.

Keywords: Fermentation, Lactic acid bacteria, Salt, Shrimp paste

1. Introduction

Shrimp paste was one of food fermentation, popular as a taste enhancer in many dishes in Indonesia, made by naturally fermentation between shrimp and salt. This product is made from salted small shrimps through a complicated process of grounding, drying, and solidifying [1]. A fermented shrimp paste (terasi), produced in Central Java, Indonesia, contains many compounds of macro and micro-nutrient which permits the growth of various micro-organisms, one of them is bacteria. The presence of bacteria in shrimp paste product could produce the secondary metabolites to give a specific taste and aroma of shrimp paste, which named *umami* taste [2]. However, there were less information about the secondary metabolites resulted from fermentation process, especially related to the micoorganism fermenter.

The intensity of umami taste in food is usually produced throughout synergistic interactions of free glutamate and 5'-ribonucleotides [3]. In shrimp paste, glutamate came through fermentation process, degradation of protein source of shrimp. Numerous studies on the source of umami taste on shrimp paste were previously performed from various scientific [4]. In previous studies, Kaewklom et al. [5] found *Bacillus amyloquefaciens* from shrimp paste *kapi,* traditional ingredient in Thailand. Tanasupawat and Visessanguan [6] stated that the bacteria isolated from fermented food are widely known to be safe for human consumption for their safety to human health. Similarly, Sumardianto et al. [7] NaCl and time of fermentation affected glutamate content of shrimp paste. The aim of this study was to characterize the physiological and biochemical of lactic acid bacteria isolated from shrimp paste (terasi).

2. Methods

2.1. Shrimp paste samples
Samples of 9 types of *terasi* used in the study were obtained from previous research that conducted 8 months before. The samples were *terasi* with time of fermentation for 30 days, 60 days and 90 days, and all conducted for 3 times repeated.

2.2. Isolated preparation [8]
The lactic acid bacteria characterized in this study were isolated from shrimp paste samples. The striking out-plate technique was used on agar plates of MRS (de Man Rogosa Sharpe Code: CM0361) medium (Oxoid); incubation was performed at 37 ^{0}C for 7 days. Liquid cultures were cultivated in microtube containing MRS CM0359 broth medium (Oxoid) and incubated on rotary shaker.

2.3. Physiological characteristics [8]
The physiological by determination of Gram was done by Gram Stain Method : Fix air-dried films of food sample in moderate heat. Stain films 1 min with crystal violet-ammonium oxalate solution. Wash briefly in tap water and drain. Apply Gram's iodine for 1 min. Wash in tap water and drain. Decolorize with 95% ethanol until blue color is no longer released (about 30 s). Alternatively, flood slides with ethanol, pour off immediately, and reflood with ethanol for 10 g. Wash briefly with water, drain, and apply Hucker's counterstain (safranin solution) for 10-30 g. Wash briefly with water, drain, blot or air-dry, and examine.

2.4. Biochemical characteristics of LAB [9]
Biochemical characteristics of LAB were done by VITECH tools and then characterized using Bergey's Manual method.

2.5. Statistical Analysis

Those data were analyzed using descriptive method.

3. Results and Discussion

3.1. Physiological characteristics of strains

(a) (b)

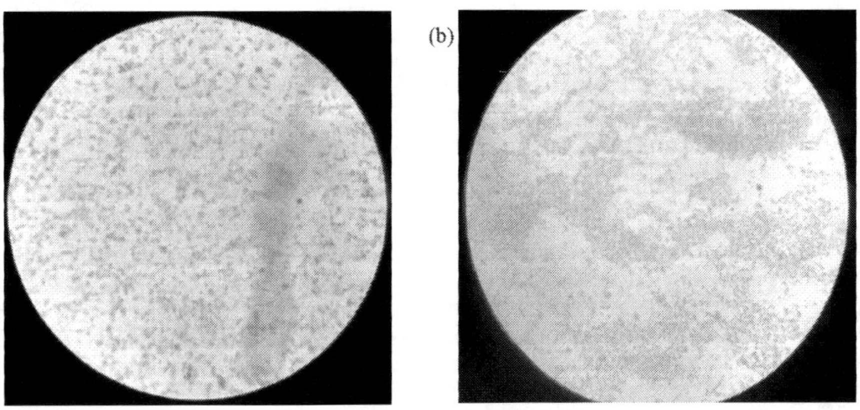

(c) (d)

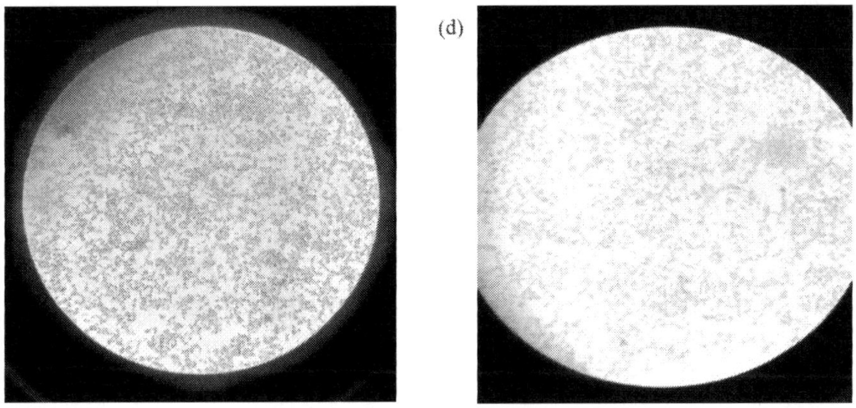

Figure 1. Gram staining of 4 strain isolated from shrimp paste, (a) LABS1; (b) LABS2 and (d)LABS4; (c) LABS3

Based on Figure 1, can seen that 3 of 4 cells strain (LABS1, LABS2 and LABS4) are Gram-positive, slender - rod - shaped, stricly aerobic, and red-pigmented, while the other one (LABS3) is dominated Gram-negative. Gram-positive cells found in shrimp paste samples with 60 days fermentation, while in shrimp paste samples with 30 days fermentation tends to Gram-negative with purple-pigment dominated. Those result indicated that different time of fermentation treated in the processing of shrimp paste influenced the presence of *lactic acid bacteria*. 30 days fermentation of shrimp paste allegedly did not optimum enough for microbes to produce secondary metabolites as well as fermentation time for 90 days, where secondary metabolite production tends to decrease. This decline is expected due to the availability of nutrients medium or shrimp paste samples are increasingly discharged by the 90th day so that no single isolates were found in shrimp paste samples with fermentation length for 90 days, and of course it would affected the glutamate levels of shrimp paste as a source of umami taste. The result are comparable to those obtained by previous study [7]. From 9 samples isolated, there were 4 presented of bacterial colonies and be expected as *Lactobacilli*. In fermented product, *Lactobacilli* had complex nutrient and requirement as indicator of many assays, such as the amount of amino acids in food products [10,11,12] as seen in Table 1.

Table 1. Microorganism used for assay of amino acids

Amino Acid	Microorganism
Arginine	*Streptococcus faecalis*
Aspartic acid	*Leuconostoc mesenteroides*
Cysteine	*Leuconostoc mesenteroides*
Glutamic acid	*Lactobacillus plantarum*

Glysine	*Leuconostoc mesenteroides*
Histidine	*Leuconostoc mesenteroides*
Isoleucine	*Lactobacillus plantarum*
Leucine	*Lactobacillus plantarum*
Lysine	*Leuconostoc mesenteroides*
Methionine	*Leuconostoc mesenteroides*
Phenylalanine	*Leuconostoc mesenteroides*
Proline	*Leuconostoc mesenteroides*
Serine	*Leuconostoc mesenteroides*
Threonine	*Streptococcus faecalis*
Tryptophan	*Lactobacillus plantarum*
Tyrosine	*Leuconostoc mesenteroides*
Valine	*Lactobacillus plantarum*

Banwart (1989)

It can be seen in Fig. 1 that the growth cells of strain be expected gave impact in cell change, whereas although *Lactobacillus* sp considered to be Gram-positive, as the culture ages, the cell may become Gram-negative [10].

3.2. Biochemical characteristics of LAB

Table 2. Characteristics of strain isolated from shrimp paste (*Shrimp paste samples, A3: 30 days fermentation for 1st repeated; B1: 60 days fermentation for 1st repeated; B3: 60 days fermentation for 3rd repeated); +: positive; -: negative

Characteristic	Isolated strains			
	LABS1	LABS2	LABS3	LABS4
Sample origin	*B1	*B1	*A3	*B3
Cell size (μm)	±0.1-0.3 x 0.8-2.5	±0.1-0.3 x 0.8-2.5	±0.1-0.3 x 0.8-2.5	±0.1-0.3 x 0.8-2.5
Colony color	white	white	white	white
pH range for growth	5-9	5-9	5-9	5-9
Optimum pH for growth	7.0	7.0	7.0	7.0
NaCl range for growth (%)	10-15	10-20	10-20	5-15
Temperature range for growth (°C)	25-38	25-40	34-37	25-40
Optimum temperature for growth (°C)	37	37	37	37

Acid produce from :				
D-fructose	+	+	-	+
D-galactose	+	+	-	+
D-glucose	+	+	-	+
Glycerol	+	+	+	+
D-mannitol	+	+	-	+
D-mannose	+	+	-	+
D-ribose	+	+	-	+
Sorbitol	+	+	-	+
Sucrose	+	+	-	+
Inulin	-	-	-	-
Lactose	-	-	-	-
Catalase	+	+	+	+
Oxidase	+	+	+	+
Urease activity	+	+	+	+
Nitrate reduction	+	+	+	+
Hydrolyses arginine and gelatin	+	+	+	+
Starch	-	-	+	-
Xanthine	-	-	-	-

Three strains tested positive. One strain tested negative

The G + C content of the type strain is 46 mol%

This study isolated 4 strains of *Lactobacilli,* Gram-positive rod vary from long to short coccobacilli, strictly aerobic, approximately 0.1 - 1.2 μm in diameter. Flagella were not observed. Colonies were clear, circular and white in colour. They grew in MRS medium CM 0359 and CM 0361 with 5-20% NaCl, at pH 5-9 and at 25-40 ^{0}C, but no growth was observed at 10 or 50 ^{0}C. The phenotypic and biochemical characteristics of the 4 strains are detailed in the species description below in Tables 2. Representative strain LABS1, LABS2 and LABS4 contained diaminopimelic acid in the cell-wall peptidoglycan. Positive in tests for catalase, oxidase, and urease activity and nitrate reduction. Hydrolyses arginine and gelatin, but not starch and xanthine. Produces acid from D-fructose, D-galactose, D-glucose, glycerol, D-mannitol, D-mannose, D-ribose, sorbitol and sucose, but not from inulin and lactose. Phosphatidylglycerol, diphosphatidylglycerol and two unidentified glycolipids are predominant in the polar lipid profile. The DNA G+C content of the type strain is 41.6 mol%.

And as shown in Table 1 and the biochemical characteristics, 4 strains isolated from shrimp paste (*terasi*) samples represent a species within the genus *Lactobacillus,* for which we propose the name *Lactobacillus plantarum.*

4. Conclusion

This study conclude that strain *Lactobacillus plantarum* was dominant isolated from fermented shrimp paste (*terasi)* produced in Central Java . The secondary metabolites of *L.plantarum* was glutamic amino acid, which play roled as umami taste in shrimp paste.

5. Acknowledgement

This study was funded by Non Tax Revenue research grant from Diponegoro University 2017.

6. References

[1] Kobayashi, T., Kondo, A., Terahara, T., Agustini, T.W., Kajiwara, M., Imada, C., Ibrahim, R, Ooka, Y., Kamei, K., Nakamura, H. 2016. Production of bacteriocin by *Virgibacillus salexigens* isolated from ''terasi'': a traditionally fermented shrimp paste in Indonesia. *World J Microbiol Biotechnol* **32**: 47. DOI 10.1007/s11274-015-1991-2

[2] Ohshima, T. and Giri, A. 2014. Traditional Fish Fermentation Technology and Recent Development. *A revision of the previous article by J H Al-Jedah, M Z Ali , Elsevier Ltd* **2**: 753-759.

[3] Jinap, S., Ilya-Nur, A.R., Tang, S.C., Hajeb, P., Shahrim, K., Khairunnisak, M. 2010. Sensory attributes of dishes containing shrimp paste with different concentrations of glutamate and 5'-nucleotides. *Appetite Journal* **55:** 238-244.

[4] Kijima, K. and H. Suzuki. 2006. Improving the Umami Taste of Soy Sauce by the Addition of Bacterial-glutamytranspeptides as a Glutaminase to the Fermentation Mixture. *Journal of Enzyme and Microbial Technology*, **41**(2):80-84.

[5] Kaewklom, S., Lumlert, S., Kraikul, W., Aunpad, R. 2013. Control of *Listeria monocytogenes* on sliced bologna sausage using a novel bacteriocin, amysin, produced by *Bacillus amyloliquefaciens* isolated from Thai shrimp paste (Kapi). *Food Control* **32**:552–557.

[6] Tanasupawat, S., Visessanguan, W. 2014. Fish fermentation. In: Boziaris IS (ed) Seafood processing: technology, quality and safety. Wiley, UK, pp 177–207.

[7] Sumardianto, Agustini, T W., Amalia, U. 2017. Investigation of the Quantity Effect of NaCl Salt on Generating Glutamic Acid Contents During Shrimp Paste Production by Fermentation Method. Adv. Sci. Lett. **23**, 6513–6515.

[8] Bacteriological Analytical Manual. 2015. R32 Gram Stain. [https://www.fda.gov/Food/FoodScienceResearch/LaboratoryMethods/ucm062229.html]

[9] Moraes, P., M., Perin, L.,M., Junior, A., S., Nero, L., A. 2013. Comparison of phenotypic and molecular test to identify lactic acid bacteria. *Brazilian Journal of Microbiology* (44): 109-112.

[10] Banwart, G., J. 1989. Basic Food Microbiology 2nd Edition. Van Nostrand Reinhold. New York

[11] Jockers, D.2013. Learn about the Importance of Good Bacteria, Part II: Lactobacillus Plantarum. Natural News.

[12] Fredriksen, L., Kleiveland, C.R., Hut, L.T.O., Nygaard, C.S., Eijsink, V.G.H. Mathiesen, G. 2012. Surface display of N-terminally anchored invasin by *Lactobacillus plantarum* activates NF-κB in monocytes. *Appl. Environ. Microbiol.* **78**: 5864-5871

Effect of Time Lenght Fermentation to *Katsuobushi* Oxidation Rate As Fish Flavor Based

U Amalia[1*], L Rianingsih[1], and I Wijayanti[1]

[1]Department of Fish Product Technology, Faculty of Fisheries and Marine Science, Diponegoro University, Jl. Prof. H. Soedarto, SH, Tembalang, Semarang 50275, Indonesia

Email : ulfahamalia0@gmail.com

Abstract. *Katsuobushi* or dried smoked skipjack had a distinctive flavor and widely used in traditional Japanese cuisine. This study aimed to evaluate the oxidation rate of Katsuobushi with different lenght fermentation. The processing treatment of the product were the differences of fish boiling time (30 min and 60 min) and the lenght of fermentation: 1 week, 2 weeks and 3 weeks. The glutamic acid content, the oxidation rate (thiobarbituric acid and peroxide value) and Total Plate Count of katsuobushi were analyzed statistically using analysis of varians. Significant differences were found among 3 weeks of fermentation compare to 1 weeks fermentation ($P < 0.05$). The conclusion of this study was katsuobushi with 60 min boiling and 3 weeks fermentation was potential to be developed become basic ingredients for the fish flavor.

Keywords: Fermentation, Fish flavor, Katsuobushi, Oxidation

1. Introduction

Skipjack is one of marine fishes consumed and produced in Indonesia, including in Central Java. The statistical data of Marine and Fisheries Bureau, Central Java province in last 5 years (2011-2016) noted the skipjack on the first position as the marine fish living in Central Java, Indonesian waters around Central Java, which potentially to be exploited and further processed. The potential value of skipjack is about 29,000 tons per year with the 114 % utility [1]. *Smoked fish* was one of traditional product as skipjack as a raw material, processed with conventional method using coconut shell, coconut husks, corn cobs, and firewood as fuel. Smoked fish products could be stored within 2-3 days at room temperature. The high amount of water content caused microbial growth rapidly and fish spoilage [2]. Therefore, its necessarily needed to create skipjack-based products which had a long-term storage. In Japan, there was a product made from dried smoked fish, known as *Katsuobushi,* with a variety of processing, among others, boiling, drying in the sun, smoking, and fermentation [2]. The specific taste and aroma of katsuobushi was to be suitabled for enhanced *umami* taste of many dishes. Fermentation processing on katsuobushi made the presence of glutamic amino acid (umami taste producer) as resulted from fish protein degradation. The range of free glutamate in other savoury condiments ranges widely, but is comparable. It must be noted that the above condiments are added to foods during food preparation in a relatively small amount and they interact with the original food matrix, including the present glutamate and other free amino acids [3]. Numerous studies on the source of umami flavor were previously performed from various scientific [4]. In previous studies, the concentration of free glutamate found in 11 *belacan* samples ranged from 601 to 4207 mg/100 g [3], which is comparable to values reported previously [5, 6]. Research related to the processing of fish flavor have been reported, including by Giyatmi et. al. [7] stated that the fermentation conducted for 3 weeks was the optimum time to produce the most preferred of katsuobushi. The differences of smoking method affect the characteristic of smoked fish flavor. This study aimed to to evaluate the oxidation rate of Katsuobushi with different lenght fermentation

2. Methods

2.1. Preparation of skipjack samples

Raw materials used in this study were fresh skipjack weight in the range of 200-300 gram for each fish and obtained from TPI Tambak Lorok Semarang, Indonesia. All samples were cleaned and stored at < 5 ^{0}C or chilled at -18 ^{0}C until use.

2.2. Experimental method and apparatusses

The chemicals reagents and microbiological reagents tests were distilled water, potassium chromate, $AgNO_3$, nutrient agar and others. The tools used in this study were cutting boards, pots, stoves, spinner, table drainer and digital scales. The equipments used for the analysis of the quality of katsuobushi were oven (Memmert), kjeltec system (Kjeltec 2300 Analyzer Unit; Foss Tecator AB), furnace (Memmert), soxhlet apparatus (Soxtec Avanti 2050 Auto System; Foss Tecator AB, Hoganas, Sweden), spectrophotometers (Prestige-21) and HPLC (Shimadzu RF-138). The procedure to make katsuobushi followed method proposed by Giyatmi et al., [7] with slight modifications in the boiling and fermentation process. The production of katsuobushi is in Fig. 1. After became katsuobushi, all samples were then wrapped in plastic and placed in osed container plastic for 3 weeks at temperature room (30-32 ^{0}C).

2.3. Analysis of katsuobushi

The quality analysis of katsuobushi include the total number of bacteria [8], The content of glutamic acid [9], Thiobarbituric Acid Reactive Substance (TBARS) [10], and Peroxide Value (PV) [11]. And also the content of Free Fatty Acids (FFA) [11]: 2 mL 0.5N NaOH/methanol was added to 20 mg of fat, which was later saponified for 10 minutes at 105 ^{0}C. It was examined after applying 2 mL boron trifluoride/methanol, and methylated. Then, 2-3 mL hexane (HPLC grade) and 2 mL saturated NaCl solution were added. The supernatant of the mixture used the separated funnel was analyzed by gas chromatography (Hewlett Packard 6890 series; Palo Alto, CA,USA). The column was set up with an HP-FFAP capillary column (25 m x 0.32 mm internal diameter, 0.5 μm film thickness); initial oven temperature of 130 ^{0}C (1 minute), increased at 2.5 ^{0}C/min to a final temperature of 230 ^{0}C (10 minutes); injector temperature 230 ^{0}C, detector temperature 250 ^{0}C; helium carrier gas with a spilt ratio of 20 : 1, and flow rate of 1 mL/min.

IOP Conf. Series: Earth and Environmental Science **116** (2018) 012050 doi:10.1088/1755-1315/116/1/012050

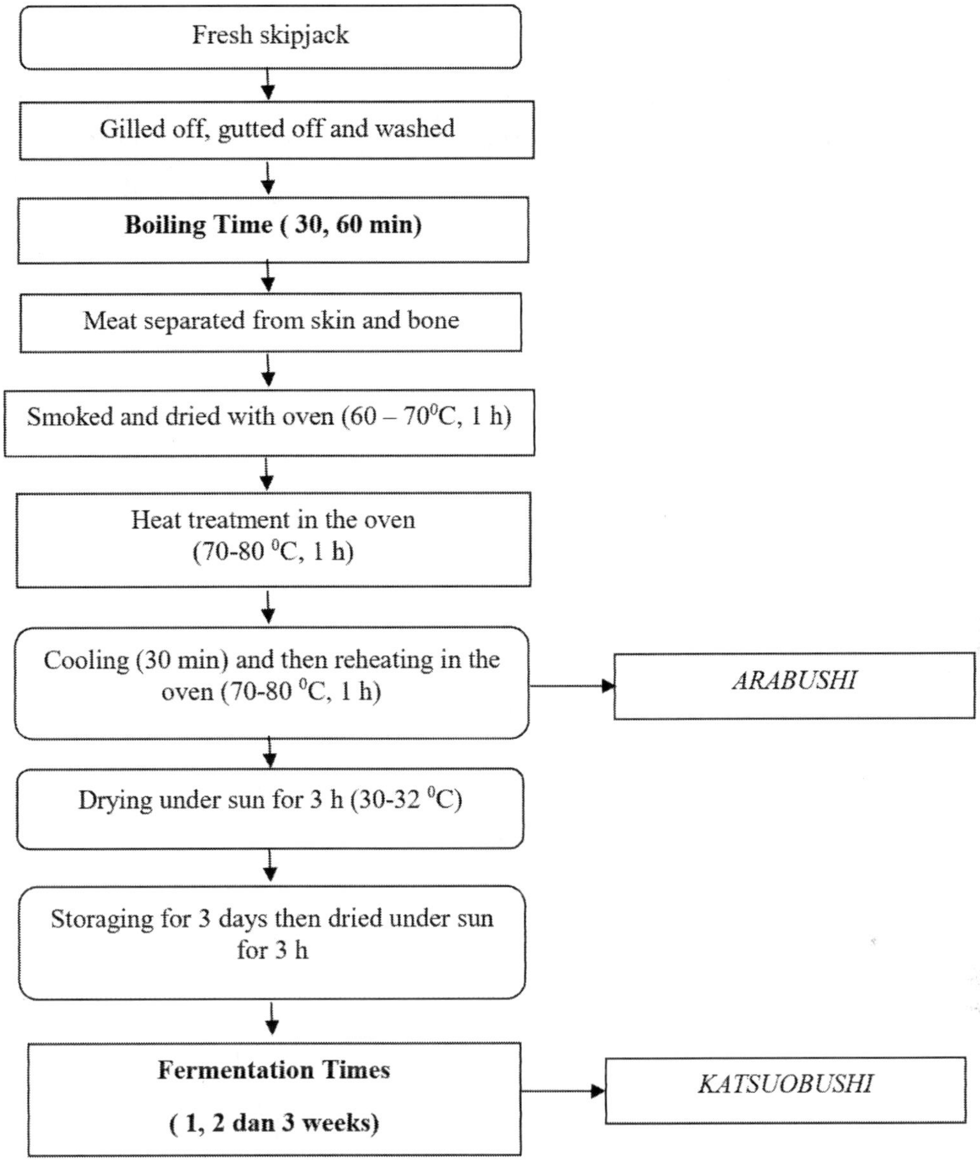

Figure 1. The flow chart of katsuobushi production

2.4. Data analysis

The contents of glutamic acid, TBA, PV, and the number of bacteria on katsuobushi were analyzed using analysis of varians

3. Results and Discussion

Katsuobushi Oxidation Rate

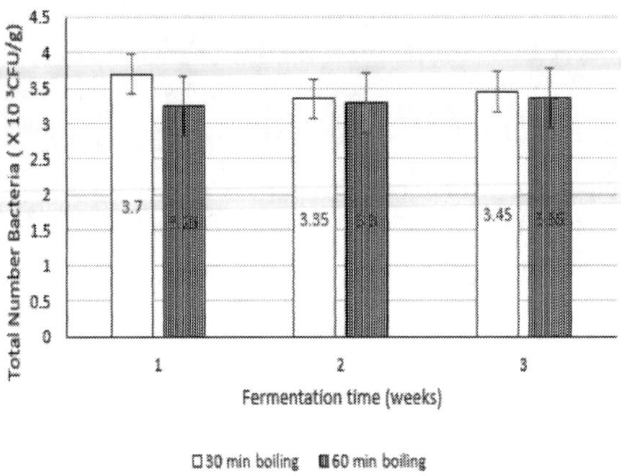

Figure 2. The total number of bacteria (cfu/g) on katsuobushi

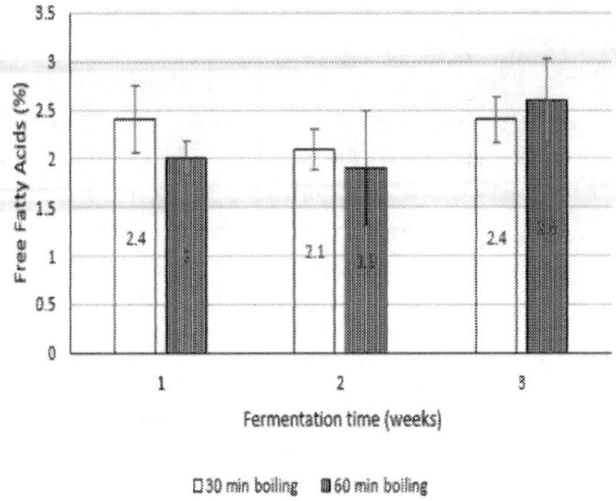

Figure 3. The Free Fatty Acid of katsuobushi

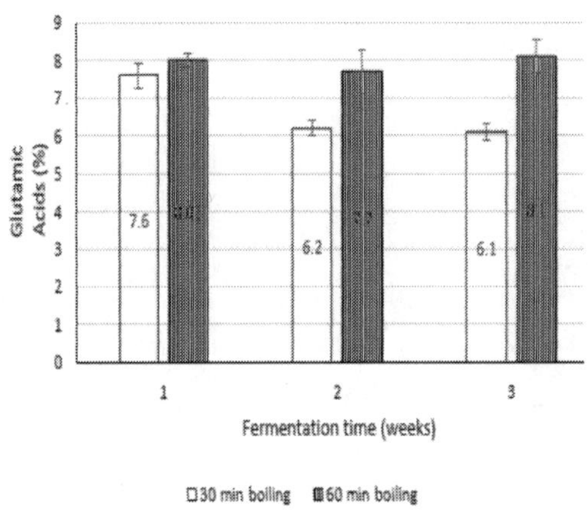

Figure 4. Glutamic acids content (%) of katsuobushi

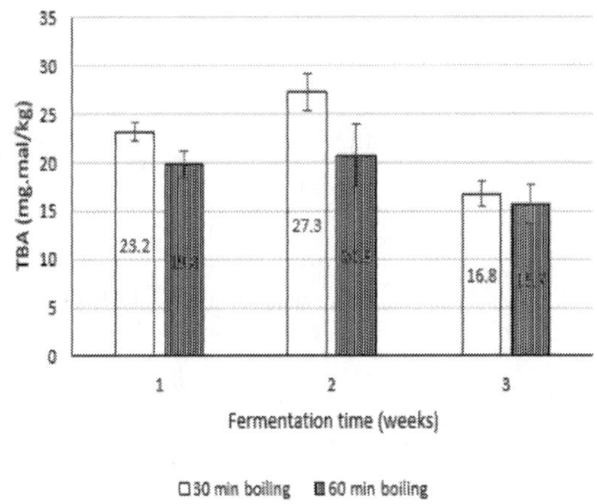

Figure 5. TBARS (mg.mal/kg) of katsuobushi

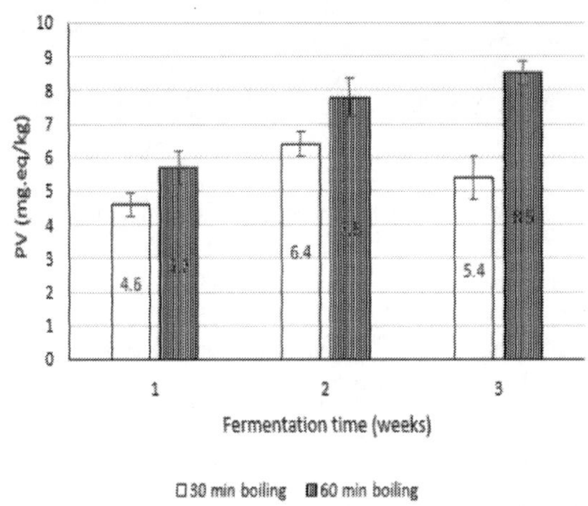

Figure 6. Peroxide value (mg.eq/kg) contained in katsuobushi

3.1. Total plate count (TPC) contents of katsuobushi

Fig. 2 shown that TPC contents of katsuobushi boiled for 60 minutes was lower than katsuobushi boiled for 60 minutes, during storage at room temperature (28 ± 1^0C). The total microorganism growth of katsuobushi with 30 min boiling, after 1 week fermentation was 3.7 x 10^5 CFU/g. After 3 weeks, there was decreased in microbial growth, giving 3.45 x 10^5 CFU/g. This would indicate that nutrition contained at katsuobushi was decreased, and there was competition between microorganism contained in katsuobushi. Meanwhile, katsuobushi with 60 min boiling shown the normal trend compared with katsuobushi with 30 min boiling, whereas trend of TPC tend to increase until 3 weeks fermentation. This trend indicated that the boiling process combined with the liquid smoke from coconut shells were abled to suppress microbial growth. Zuraida et al. [12] reported that liquid smoke contained bioactive compounds such as phenols, carbonyls and organic acids. Therefore, the coconut shell liquid smoke has the potential in increasing the shelf life of proteinaceous food products [12, 13].

3.2. Free fatty acids of katsuobushi

Based on Fig. 3, we showed that the values of free fatty acids (FFA) of katsuobushi for both treatment were fluctuative after 1 until 3 weeks of fermentation. It could be done because skipjack as one of marine fish contained more quantity of unsaturated fatty acid compared with saturated fatty acids. The component of long-chain unsaturated fatty acids, such as linoleic and linolenic acids, directly deriving from skipjack as a raw material of katsuobushi and also from microbial metabolism. This case was importance because their oxidative degradation may lead to the formation of a characteristic aging flavor [14].

Katsuobushi consist of many compounds (volatil during fermentation, and non-volatile) that affect katsuobushi flavor, many of these aroma compound are synthesized by bacteria, other derive directly from the raw materials. The combination of raw material condition, presence of bacteria, and degree of aeration would influenced medium-chain fatty acids such as hexanoic, octanoic and decanoic acid [15]. And then, as the composition of these acids increases, it will produce off flavors.

3.3. Glutamic acid contens of katsuobushi

Based on Fig. 4, we can seen that the differences of boiling time gave effect on katsuobushi glutamic acid content ($p < 0.05$). This is presumably because the protein was degraded into several amino acids, especially the generation of umami taste, glutamic amino acid caused by boiling process on the production of katsuobushi. This result is in line with research conducted by Mc Gee [16] and Daniel et.al. [17], explored katsuobushi as products that can be used as an umami flavor associated manufacturing process.

3.4. Thiobarbituric acid reactive substances (TBARS) of katsuobushi

The malondialdehyde (MDA) contents in katsuobushi samples are shown in Fig.5.TBARS values represent the content of secondary lipid oxidation products, mainly aldehydes and carbonyls of hydrocarbons, that contribute to off-flavors and flavors in meat [18]. As the period fermentation, the TBARS values are increased significantly ($P < 0.05$) from 1 week toward to 2 weeks fermentation, but then decrease after 3 weeks fermentation. This case done in both boiling treatment of katsuobushi. According to Ozer and Sariçoban [19], even though MDA is a secondary product of lipid oxidation, this does not unavoidably mean that TBARS value continues to increase throughout the storage. Also, this trend was probably caused by losses in the oxidation products formed, particularly the volatile counterparts. MDA and other short-chain carbon products of lipid oxidation are not stable and decompose to organic alcohols and acids, which are not detected by the TBARS test [20]. The decrease in the

TBARS values after 3 weeks fermentation is thought to be the result of MDA reactions with proteins [21].

3.5. Peroxide values of katsuobushi

The concentration of peroxide katsuobushi determined by hydroperoxides. Differences in heat process can be caused by differences in fish texture as a based raw for prepared katsuobushi. In this study, katsuobushi with 60 min boiling more easy to oxidized. As well as done with the katsuobushi with 30 min boiling that reached a peak at 2 weeks fermentation and decrease after 3 weeks fermentation. This case in line with previous study done by Lee et al. [22], who reported that peroxides value of cooked pork patties increases and thereafter decrease with the time of storage. Juntachote et al. [23] and Teets and Were [24] stated that peroxides are very reactive and may actually decrease during the storage of lipid-containing foods . This result is in line with the research conducted by Yusnaini et.al. [25] and Hwang et al. [26] which stated that heat temperatures caused high levels of oxidation. In addition, boiling and fermentation times affected the content of katsuobushi peroxide value ($p < 0.05$) (Fig. 6). Skipjack as red fleshed fish species had a high heme pigment especially myoglobin. Heme pigment is a source of Fe which is an indicator of heating and oxidation catalysts. Myoglobin will react with peroxide of polyunsaturated fatty acids fish. The average of katsuobushi PV with 60 min of boiling time was higher than katsuobushi PV boiled in 30 min. Fe^{2+} concentration increased dramatically during boiling so that resulted in at oxidation in fish. The increase of peroxide value supports the view that the formation of new hydroperoxides is occurring at a faster rate than the degradation of hydroperoxides into secondary oxidation products during the first part of refrigerated storage. As the lipid oxidation progresses, the decomposition of hydroperoxides into secondary products increases at a higher rate as compared to the formation of new hydroperoxides, resulting in decreased peroxide value [24].

4. Conclusion

In conclusion, this study showed that the longer time of boiling fish raw material and time of fermentation so the higher contents of glutamic acid, PV and TBA would rised. The free fatty acids was also rised, thus katsuobushi more easy to oxidized. So, for further research, it is necessary to conduct a study to make katsuobushi as the basic ingredient of making fish flavor more protected so as not easily oxidized.

Acknowledgment

The authors would like to express gratefully thanks to Diponegoro University for financial support funding year of 2017.

References

[1] Bureau of Marine and Fisheries Central Java Province. 2016. http://diskanlut-jateng.go.id. (accessed on January, 2 2017)

[2] Mitou, M., Shigemori,Y., Aoshima, H., and Yokoyama, S. 2008. Effect of dried bonito (katsuobushi) and some of its components on $GABA_A$ receptors. *Food Chemistry Journal*, Vol. **108**, Issue 3: 840-846.

[3] Jinap, S., Ilya-Nur, AR., Tang, SC., Hajeb, P., Shahrim, K., Khairunnisak, M. 2010. Sensory attributes of dishes containing shrimp paste with different concentrations of glutamate and 5'-nucleotides. *Appetite* (55): 238-244.

IOP Conf. Series: Earth and Environmental Science **116** (2018) 012050 doi:10.1088/1755-1315/116/1/012050

[4] Kijima, K. and H. Suzuki. 2006. Improving the Umami Taste of Soy Sauce by the Addition of Bacterial-glutamytranspeptides as a Glutaminase to the Fermentation Mixture. *Journal of Enzyme and Microbial Technology*, **41**(2):80-84.

[5] Khairunnisak, M., Azizaha, A. H., Jinap, S., & Nurul Izzah, A. 2009. Monitoring of free glutamic acid in Malaysian processed foods, dishes and condiments. Food Additives and Contaminants, **26**(4), 419–426.

[6] Yoshida, Y. 1998. Umami taste and traditional seasonings. Food Review International, **14**, 213–246.

[7] Giyatmi, Jamal, B., C. Hanny Wijaya, Srikandi, F. 2000. Pengaruh Jenis Kapang dan Lama Fermentasi terhadap Mutu Ikan Kayu (*Katsuobushi)* Cakalang. B*uletin Teknologi dan Industri Pangan*, Vol. XI, No. 2:11 pp.

[8]Bacteriological Analytical Manual. 2011. *http://www.cfsan.fda.gov/~ebam/bam.html* (accessed on January, 2 2017).

[9] AOAC . 2007. Official methods of analysis of AOAC (18th ed.)Washington, DC: Association of Official Analytical Chemists.

[10] Vate, NK and Benjakul, S. 2013. Antioxidative Activity of Melanin Free Ink from Splendid Squid (*Loligo formosona*).*International Aquatic Research*, **5**(9): 1-12.

[11] Bravi, E., Sensidoni, M., Floridi, S., & Perretti, G. 2009. Fatty acids composition in beer lipids and determination of corn adjuncts MBAA Technical Quarterly, **46**(4). http://dx.doi.org/10.1094/TQ-46-3-0916-01

[12] Zuraida, I., Sukarno, I., & Budijanto, S. 2011. Antibacterial activity of coconut shell liquid smoke (CS-LS) and its application on fish ball preservation. *International Food Research Journal*, **18**, 405–410.

[13] Saloko, S., Darmadji, P., Setiaji, B., Pranoto, Y. 2014. Antioxidative and antimicrobial activities of liquid smoke nanocapsules using chitosan and maltodextrin and its application on tuna fish preservation. *Food Bioscience* (7): 71-79.

[14] Bravi, E., Marconi, O., Sileoni, V., Perretti, G. 2017. Determination of free fatty acids in beer. *Food Chemistry* (215): 341-346

[15] Horak, T., Culik, J., Cejka, P., Jurkova, M., Kellner, V., Dvorak, J. 2009. Analysis of free fatty acids in beer: Comparison of solid-phase extraction, solid-phase microextraction, and stir bar sorptive extraction. Journal of Agricultural and Food Chemistry, **27**, 11081–11085.

[16] McGee, H. 2004. On Food and Cooking: The Science and Lore of the Kitchen. Scribner, New York, USA.Scribner, New York, USA.

[17] Daniel F., Daniel B., David C. 2011. Defining microbial terroir: The use of native fungi for the study of traditional fermentative processes. *International Journal of Gastronomy and Food Science.* Vol. **1**. 64-69.

[18] Juntachote, T., Berghofer, E., Siebenhandl, S., & Bauer, F. 2006. The antioxidative properties of holy basil and galangal in cooked ground pork. Meat Science, **72**(3), 446–456.

[19] Ozer, O., & Sariçoban, C. (2010). The effects of butylated hydroxyanisole, ascorbic acid, and a-tocopherol on some quality characteristics of mechanically deboned chicken patty during freeze storage. Czech Journal of Food Sciences, **28**(2), 150–160.

[20] Fernandez, J., Perez-Alvarez, J. A., & Fernandez-Lopez, J. A. (1997). Thiobarbituric acid test for monitoring lipid oxidation in meat. Food Chemistry, **59**(3), 345–353.

[21] Maqsood, S., & Benjakul, S. (2010). Comparative studies of four different phenolic compounds on in vitro antioxidative activity and the preventive effect on lipid oxidation of fish oil emulsion and fish mince. Food Chemistry, 119(1), 123–132.

[22] Lee, M. A., Choi, J. H., Choi, Y. S., Kim, H. Y., Kim, H. W., Hwang, K. E., Chung, H. K., & Kim, C. J. 2011. Effects of kimchi ethanolic extracts on oxidative stability of refrigerated cooked pork. Meat Science, **89**(4), 405–411.

[23] Juntachote, T., Berghofer, E., Siebenhandl, S., & Bauer, F. 2007. The effect of dried galangal powder and its ethanolic extracts on oxidative stability in cooked ground pork. LWT — Food Science and Technology, **40**(2), 324–330.

[24] Teets, A. S., & Were, L. M. 2008. Inhibition of lipid oxidation in refrigerated and frozen salted raw minced chicken breasts with electron beam irradiated almond skin powder. Meat Science, **80** (4), 1326–1332.

[25]Yusnaini, Soeparno, Edi S, and Ria A. 2015. The Effect of Heating Process using Electric and Gas Ovens on Chemical and Physical Properties of Cooked Smoked-Meat International Symposium on Food and Agro-biodiversity (ISFA2014). *Procedia Food Science* **3** : 19 – 26.

[26] Hwang, K., E., Choi., Y., S., Cho., S., M., Kim, H., W., Choi, J., H., Lee, M., A., and Kim, C., J. 2013. Antioxidant action of *ganghwayakssuk (*Artemisia princeps *Pamp.)* in combination with ascorbic acid to increase the shelf life in raw and deep fried chicken nuggets. *Meat Science* (95): 593-602.

IOP Conf. Series: Earth and Environmental Science **116** (2018) 012051 doi:10.1088/1755-1315/116/1/012051

Community Participation Of Coastal Area On Management Of National Park, Karimunjawa Island

Bambang A Wibowo[1]*, Aryo B Aditomo [2], Kukuh E Prihantoko[1]

[1]Department of Fisheries Resource Utilization, Faculty of Fisheries and Marine Science, Diponegoro University
Jl. Prof. Soedharto, SH Tembalang, Semarang, Central Java
[2]Student at Coastal Resources Management Study Program, Faculty of Fisheries and Marine Science, Diponegoro University
Jl. Prof. Soedharto, SH Tembalang, Semarang, Central Java

*Corresponding author Email : argobambang@gmail.com

Abstract. Karimunjawa island located in Jepara Regacy, Central Java has potential marine and fishing resources. Since 1998, this area has been selected as conservation for its natural resources. National park of Karimunjawa is managed by Balai Taman Nasional Karimunjawa (Karimunjawa National Park Beuroue). Some activities involved community have been done in order to get effective management. Community participation is an important component for success in coastal area management. The level of community/people awareness anual on natural resource conservation can increate sustainable resource. However, it is necesssary to provide tools in resource utilization for the community, so that their economic life can be secured. This study observe the level of community participation in the effort of Karimunjawa National Park management. Descriptive method and purposive random sampling were used to carry out the study parameters observed in this study include community participation related to level of knowladge and obedience on the rule of area zonation, an its impact to community. The result show that community knowledge was quite high (40%) with obedience (56%) on the rule of area zonation. Impact area zonation rule was less significant to community. The level of community participation to Karimunjawa National Park management was performed will low to medium level.
Keywords: Community participation, coastal, Karimunjawa

1. Introduction

Biodiversity can be maintained by conservation for ensuring ecosystem balance [1]. Based on The Decree No 5- 1990 [2] biodiversity conservation can be defined as natural resource management that reflecting its utilization wisely for guaranty stock sustainability with concerned on maintaining and increasing biodiversity quality and its value. Conservation is done to protect natural resource within the area as well as ensuring community access to sustainable use of natural resource [3].

Zonation is an effort made for managing the conservation area. Based on Ministerial Regulation of Forestry Affairs No P.56/Menhut-II/2006 [4] National Park Zonation consists of 7 (seven) zones: core zone, wilderness zone, utilization zone, traditional zone, rehabilitation zone, religious zone, and special zone. This zonation is a guideline for creating an effective and optimal National Park management system in accordance to its function [5]. Implementation of conservation often influences community acces to natural resources utilization which are their source of livelihood. Utilization of natural resources and sustainable develompment require an appropriate concept to overcome arising problems. There is no community inavertently inhibits their environmental survival, however with the ongoing environmental problems caused by negative impact of human activities, it is a sign that sustainability is still in doubt [6]. Sustainability is not a final achievement, but a target that must continually negotiated while the community learns to recognize the indications of unsustainability [7].

In 1986, Karimunjawa Island has been selected as marine conservation area because of its strategic values. In 1999, Karimunjawa Conservation Area was changed into Karimunjawa National Park based on The Decree of Ministry of Forestry and Estate Crops No.78/kpts-II/1999 [8], with area of 111.625 ha. In 2001, some area of Karimunjawa National Park was selected as Nature Reserve Area. The status change was conducted to accommodate the community existence in the area of Karimunjawa Island.. The implementation was through zonation system, therefore community around the area can utilize the natural resources in order to supply their needs [9].

Marine protected area is an area with the functions of protection and maintenance of biodiversity, natural resource and local culture, managed under valid laws or regulations [10].Establishment of marine protected area is an instrument related to ecological and institutional aspects simultaneously. However, establishment of Karimunjawa National Park as conservation area caused the community to adapt in supplying their needs, particularly due to its zonation system [11].

Since 1998, Karimunjawa National Park Agency as administrator area has been conducted many programs in order to improve community participation in maintaining natural resources within Karimunjawa National Park. Those programs were community empowerment through economic improvement, institutional strengthening, alternative livelihood finding, also community capacity improvement. According to Supardan *et al.* (2006) [12], community participation through programs conducted by administrator area could place the community as an object instead of a subject. Therefore, further study about the effects of conservation area through Zonation of Karimunjawa National Park on fishermen community and biodiversity is needed. Moreover, the fishermen only fish around the shallow coral area, which are very close in breaking the National Park Zonation system [13] .

The presence of the MPA expected to be increse the fish stock in the area from the spill over of the MPA[14]. Deep-water fishing activities are impacting deep-water habitats worldwide, yet the extent of the damage caused is mainly unseen and not quantified [15]. Habitat relations are complex and bi-directional. Echinoid community structure is affected by the habitat structural and environmental conditions [16] Understanding changes in trophic group interactions following the implementation of marine protected areas (MPAs) is critical in understanding their success, or otherwise[17].

This research aimed to find the community participation rate in maintaining Karimunjawa National Park involving community knowledge and obedience in zonation system within the area as well as its impact on the surrounding community. The research was conducted in March-April 2012 in Pelabuhan Perikanan Pantai Karimunjawa (Fisheries Port of Karimunjawa), Jepara, Central Java. Sponges within the 300 km2 Trænadypetcoral MPA (Marine Protected Area) were mapped using photo- graphic techniques[18]. Nearly all coral reefs bordering nations have experienced net losses in reef biodiversity, goods and services, even without considering the ever-developing global change impacts [19].

2. Materials and Methods

The method used in this research was descriptive, a case study with advantage in supporting substantial later studies. A case study can give hypothesis for later researches. From educative perspective, case study can be used as an illustrative example whether in problem formulation, statistical use in analyzing the way of general problem formulation and conclusion [20]. Data were obtained through observation, interview, and literature review. Samples were taken through Purposive Random Sampling method. Purposive Random Sampling is a method that the sampling is not based on random or regional, but based on a certain purpose [20]. Number of samples were calculated with Slovin formula in [21] and the value of n (number of samples) was 97.76~100. The following requirements were used in determining the sample:

1. Community with main livelihood as fisherman,
2. Fishermen who live or stay in Karimunjawa Village,
3. Active fishermen, fishermen who are still active in fishing activities at sea,
4. Fishermen who conduct fishing activities in the waters of Karimunjawa National Park

3. Results And Discussion

3.1. Fishing Gear and Fishing Vehicle

Fishing gears used by the fisherman around Karimunjawa National Park were dominated by passive fishing gears which categorized as environment friendly. As seen on Figure 1, the most used fishing gear was Squid line (48.5%). Besides Squid line, other fishing gears were Spool Hand line (16.2%), Spear gun (14.9%), Trolling line (11.5%), Trap (6%), Boat lift net (2.6%), and Gill net (0.4%). Those fishing gears were used considering the operation location was near the coral waters. Therefore, the fishermen chose to use passive and environment friendly fishing gears. Moreover, the fish targets were dominated by coral fish. The fishermen of Karimunjawa used engine-driven fishing vehicles in their operation activities, such us Motor Boat and Outboard Motor Boat. The number of Motor Boats reached 250 units and Outboard Motor Boards were 140 units.

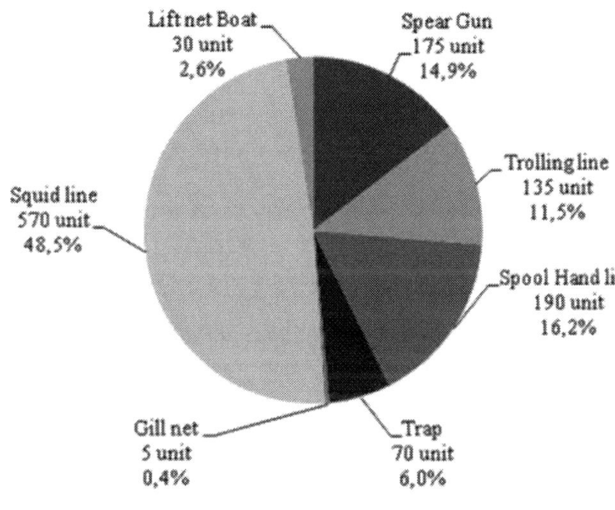

Figure 1. Fishing gears used by the fisherman around Karimunjawa National Park

3.2. Fishermen's Knowledge and Obedience on Zonation System

The participation rate on zonation system can be determined from the level of knowledge and obedience (Figure 2). Knowledge level of Karimunjawa Fishermen on zonation system was categorized high (41%), only 9% was categorized very high. Intermediate category was still found around 27%, and 23 % was categorized low, while very low category was not found. It showed that there were still many fishermen who didn't have enough knowledge on zonation system. The obedience level of fisherman community on zonation system was dominated high (56%) and only 14% was categorized very high. Very low category was still found, around 8% fishermen community did not have the obedience on zonation system.

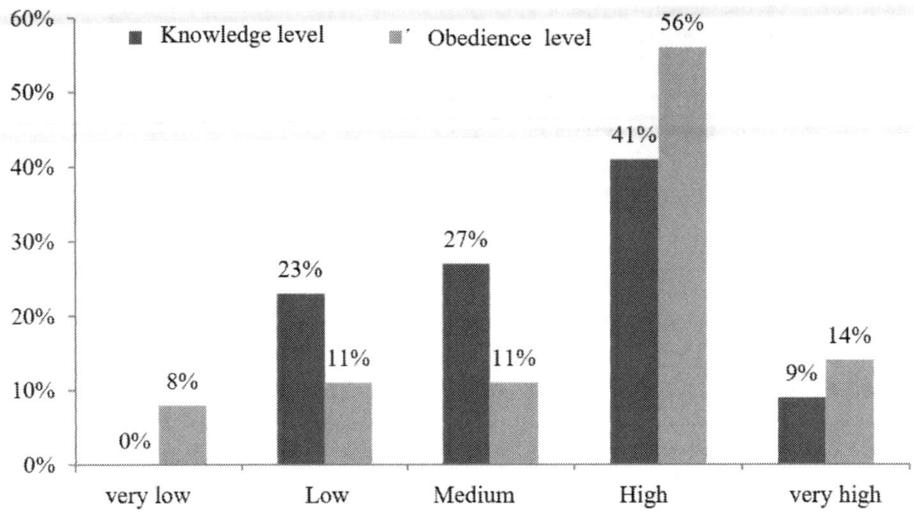

Figure 2. Fishermen's Knowledge and Obedience on Zonation System

3.3. Fishermen's Perception on Zonation System

Fishermen community participation on zonation system was an obligation considering the zonation will also impact on the activities of fishermen community. During the forming of zonation system, Karimunjawa National Park Administrator has been involving the fishermen community. However, information about fishermen community perception on zonation system was still needed. Interview had been conducted to gather information related to the zonation system thoroughly and also the confirmed protection zone (Figure 3 and Figure 4). Protection zone was important due to its limitation access of utilization by fishermen community.

From respondent interview showed that 51% fishermen community perception on Karimunjawa National Park Zonation system was quite unfair. However there was still 28% of fisherman community who considered the zonation system was fair. The high level of fishermen community who considered the zonation system was unfair indicated that there were still some fishermen community did not fully accept the zonation regulation yet. This unfair perception of the zonation system was due to the fishing ground area was belong to the zonation area, therefore the fisherman had to move to another fishing ground.

Protection zone is a part of zonation which was regulated on zonation system of Karimunjawa National Park. Protection zone is a water area, also serves as species protection area, habitat or ecosystem that supports the function of the core zone. These are several water areas included in protection zone of Karimunjawa: Gosong Tengah, Cemara Kecil, P. Sintok, P. Geleang, P. Burung, P. Katang, Gosong Selikur, dan Tanjung Gelam. The land areas were including low altitude tropical rain forest and mangrove forest. Based on the survey, 32 % of respondent perception agreed that the protection zone were used for fishing activity. While 21% stated strongly agree, 18% less agreed, 23% disagreed, and 6% strongly disagreed. Perception of protection zone for fishing activity represented the fishermen level of understanding. The high level of fisherman perception that agreed the using of protection zone for fishing activity had releveance to the aspect of justice which dominated by less fair category. Since the fishing ground was included in the protection zone, there was less fair perception on zonation system. The high level of agreement on using the protection zone for fishing activity indicated the fishermen had really high expectation to open access the fishing activity. On the other hand, expectation to open access the protection zone also indicated the fishermen limitation in understanding the protection zone.

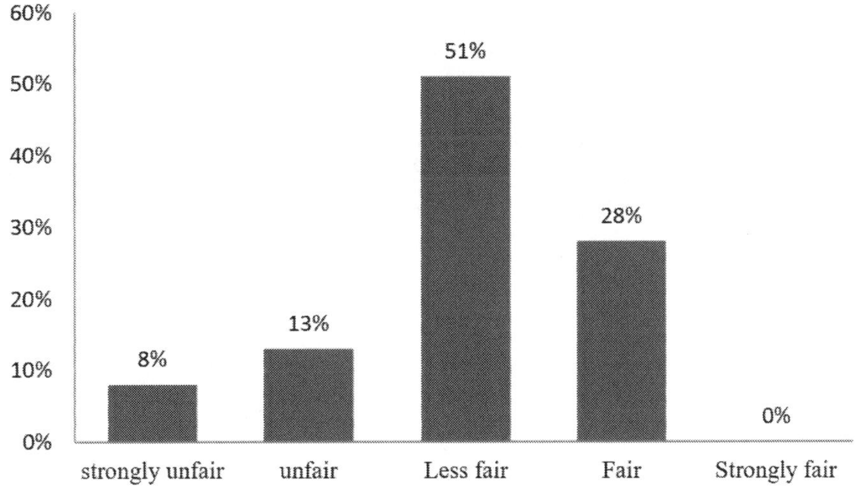

Figure 3. Fishermen Perception on Zonation Acceptance

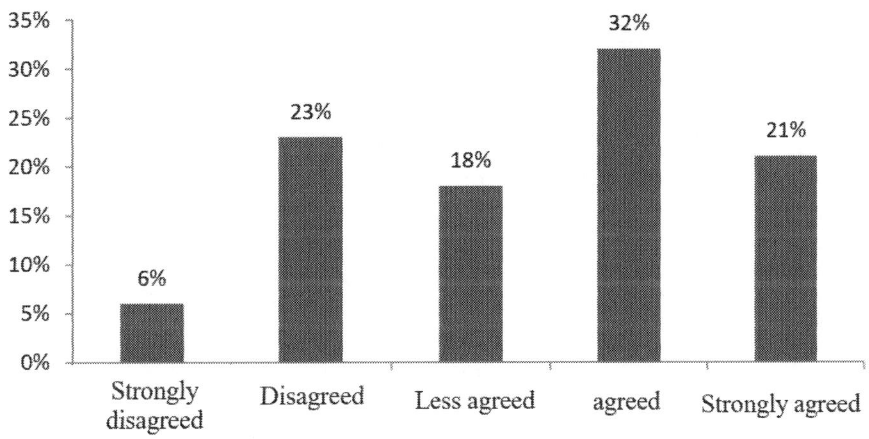

Figure 4. Fishermen Perception on Fishing Activity in Protection Zone

3.4. Effect of Zonation System on Fishermen Economy

Fishermen of Karimunjawa had fishing activities mostly on dry season between May-September. During the dry season, the wave condition was relatively small compared to the rainy season when the wave was relatively high. Moreover, due to the fishing vehicles and fishing gears were relatively small, fishermen tent to stay home during the high wave season. The survey on fishermen catches and incomes was conducted to know the effects of zonation system on fishermen economy. The catch and income of fishermen were some economic impacts on zonation system. Fig 5, showed the effect of fishermen catch and income on zonation system in Karimunjawa in percentage.

From the survey, it showed that 34% respondents stated that their catches were not impacted by zonation system. While the zonation system also known to have no effect on fishermen income (33%). Only small percentage of fishermen stated that zonation system affected their catch and income. Fishermen catch and income can be used in indicating the economic impacts on zonation system from fishermen perception. The relatively high percentage which stated the zonation system gave less impact on the fishermen's catch and income, showed that there was still no economic impact that affect the fishermen community directly.

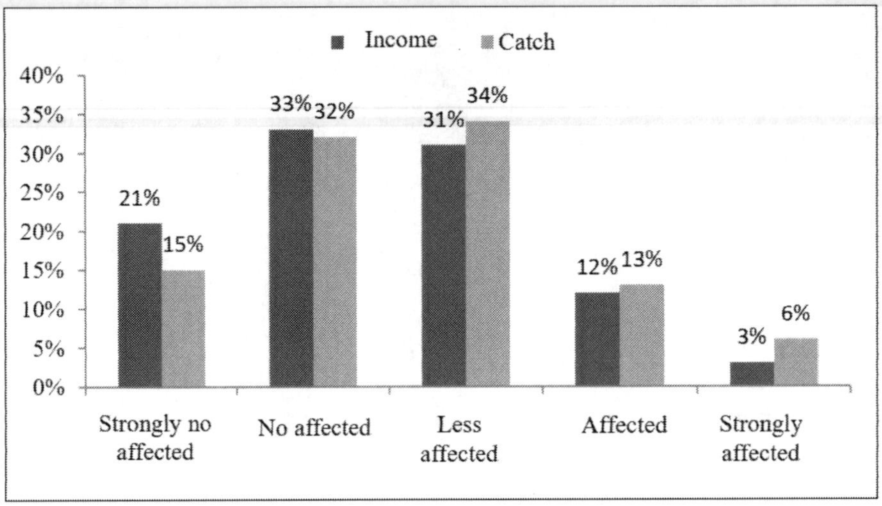

Figure 5. Effect of Zonation System on Fishermen Catch and Income

3.5. Fishermen Participation on Karimunjawa National Park Management
The perception of Karimunjawa fishermen on some aspects including knowledge level and obedience, zonation system, and fishermen economy indicated the participation rate of fishermen community on zonation system of Karimunjawa National Park. Zonation is an implementation on managing the conservation area, where the area is divided into certain zones according to their function and based on regulations related to activities that are allowed and not allowed.

Basically, fishermen community already had the knowledge and understanding on Karimunjawa National Park management. It was showed from the perception survey on fishermen level of knowledge and obedience which was dominated high. Thus showed that the information related to the management of the area had already known by the fishermen community. However, there were other perceptions from the fishermen related to the zonation system. Fishermen community still held the lack of justice in deciding the zonation area by the inclusion of fishing ground in protection zone area. With relatively high on knowledge and obedience level on area management, socialization and communication related to area management needed to be given continually therefore the injustice perception could be minimized, considered the zonation forming had already involved the community.

4. Conclusion
Fishermen on Karimunjawa National Park were categorized as small scale fishermen (artisanal fisheries). Fishing gears were categorized environment friendly, with relatively small fishing vehicles. The participation rate of Karimunjawa fishermen on Karimunjawa National Park management was relatively high. The injustice perception on zonation system could be minimalized through continued socialization and communication on fishermen community. Fishermen community should have space on area management, therefore whenever the change is needed, it should be done based on the admitted regulations, thus the administrator, community, and other related stakeholders can reach the expected goals together.

REFERENCE
[1] Dahuri R 2002 Fakultas Perikanan dan Ilmu Kelautan. IPB Bogor

[2] Menteri Kehutanan Republik Indonesia. 1999. Keputusan Menteri Kehutanan No. 78/Kpts-II/1999 Tentang Penetapan Kawasan Taman Nasional Karimunjawa Sebagai Kawasan Pelestarian Alam. Departemen Kehutanan, Jakarta

[4] Monintja D. dan Yusfiandayani R 2001 56-65

[5] Maksum MA 2005 [Tesis] Sekolah Pascasarjana, Institut Pertanian Bogor, Bogor

[6] Sulaksono, A dan A Setiawan 2004

[7] Mitchell B, Setiawan B, Rahmi D H 2007

[8] Menteri Kehutanan Republik Indonesia. 2001. Keputusan Direktur Jenderal Perlindungan Hutan dan Konservasi Alam No. 79/IV/Set-3/2005 Tentang Revisi Zonasi/Mintakat Taman Nasional Karimunjawa. Departemen Kehutanan, Jakarta

[9] Irnawati R 2008 [Tesis] Sekolah Pascasarjana Institut Pertanian Bogor

[10] Surochiem 2001 **8** (1) 50-56

[11] Faiza R. 2004. [Disertasi]. Sekolah Pascasarjana Institut Pertanian Bogor, Bogor.

[12] Supardan A, Haluan J, Manuwoto, Soemokaryo S. 2006. Buletin PSP (**15**):35-49.

[13] Nurmalasari Yessy. 2010. [Tesis]. Sekolah Pascasarjana, Institut Pertanian Bogor, Bogor.

[14] Edgar, GJ, GR, Russ and RC, Babcook. Oxford University, 2007

[15] Bo, Marzia, S Bava, S, Canese, M. Angiolio, RC, Vietti and G, Bavestrello, 2013

[16] Bronstein, O and Y, Loya, 2014, Journal of Experimental Marine Biology and Ecology

[17] Gilby Ben L and Tim Stevens, 2014, Global Ecology and Conservations

[18] Kutti, T, RJ, Bannister, and JH Fosca, 2013, Continental Shelf Research

[19] Rinkevich B 2015 Journal of Enviromental Management

[20] Sugiyono 2007 CV. Alfabeta, Bandung.

[21] Panayotou 1985 Ottawa-Canada. IDRC. 283

Study of AUTO-LION (Automatic Lighting *Rumpon*) on Fisheries of Stationary Lift Net in Semarang, Central Java

S Chairunnisa[1*], N Setiawan[1], Irkham[2], K Ekawati[1], A Anwar[2], A DP Fitri[1]

[1]Faculty of Fisheries and Marine Science, Diponegoro University, Semarang, Indonesia
[2]Department of Electrical Engineering, Faculty of Engineering,Diponegoro University, Semarang Central Java

E-mail: shasa.chairunnisa@gmail.com

Abstract. Fish Aggregation Device (FAD) is a fishing tool that serves to collect fish at a place to facilitate fishermen in the process of fishing. The use of light is also proven to help the process of fishing at night. AUTO-LION (Automatic Lighting *Rumpon*) is a FADs innovation equipped with fish-eating sound and solar-powered lights that can be activated automatically when it is dark or nighttime.The purpose of this study was to determine the effect of AUTO-LION use on fishermen catch. The research method used is experimental fishing.The research was conducted on May 2017 on the stationary lift net in Semarang Waters. The results showed the catch as much as 10.55 kg without the use of AUTO-LION, 15.05 kg on the use of FADs, 19.08 kg on the use of FADs with sound, 27.04 kg on the use of FADs with light, and 40.01 kg on the use of AUTO-LION. Based on these results it can be seen that the use of AUTO-LION can increase the catch of fishermen, especially when the light is activated.
Keywords: Fish Aggregation Device, lift net, *rumpon*

1. Introduction

Semarang City is located on the north coast of Central Java, at position 06005'07 "LS - 110035'28" East Longitude, with total area reaches 37,366,838 ha or 373,7 km^2. Seeing the existence of fishermen who live and fish in coastal Semarang, at this time they are looking fish far to the north to Karimunjawa islands, because the potential of fisheries in the coast and sea of Semarang City is much reduced. The results obtained are also not balanced with expenditures for fuel oil and for other operations.

Charts are one of a kind of rectangular fishing gear and are generally operated in the waters near the coast. The principle of the chart capture tool is to concentrate and collect fish in the center of the chart, then catch it. Acording to Sukandar dan Fuad (2015)[1].the step chart is a passive catch tool, where the tool is a building in the middle of the sea with a net under it, the net will be lifted after many fish gathered in the middle of the chart. Based on the way the operation of the chart is grouped in lift net, but because it uses light to collect the fish it is also called light fishing (Subani and Barus, 1989 in Fuad *et al.*, 2016) [2]. Chart type can be divided into 3 that is step tancap, floating chart and boat chart. The timepiece is fixed while the floating and boat charts can move.

FADs is one of the collecting fish tool that serves as a shelter for fish from predators. According Suwarsih (2012) [3], FADs is one of the fishing aids to increase the catch where it has a plant-like construction that placed in the sea waters and serves as a shelter, to eat, spawn, and gather fish. According to Regulation of the Minister of Marine and Fisheries No. 26 / PERMEN-KP / 2014 [4], FADs is a fish collecting tool that uses various forms and types of atractors from solid objects, it's serves to atracting fish to gather, which is utilized to improve the efficiency and effectiveness of fishing operations. FADs can be located in any place depending on the direction of current or wind, or a combination of both (Mandagi *et al.*, 2015 [5]; Nurani *et al.*, 2014[6]; Simbolon *et al.*, 2013[7]. Traditional FADs is generally made of coconut leaves that are hooked on a main rope. FADs can be placed on the surface, middle or bottom of the water. According to Yusfiandayani *et al.* (2012) [8], the traditional FADs atractor is generally made of coconut leaf or wrecked rickshaw.

The use of solar cells as a solar light tower is a technology that has just been developed as a renewable energy. The use of solar cells can save fossil fuels that until now become a decreased resource. The working principle of this tool is the absorption of sunlight by solar panels during the day. Then the energywill be stored in the battery to be used for the fishing operations of the sound and the light attractor. According to Yandri (2012) [9], solar energy is the energy gained by converting solar thermal energy (sun) through certain equipment into other form resources. According to Widayana (2012) [10], solar energy is one of the energies currently being developed by the Indonesian government because as a tropical country, Indonesia has considerable potential solar energy. Solar energy is extraordinary because it is not pollutant, cannot run out, can be trusted and does not buy. There are many ways to use the energy from the sun. The term "solar power" means converting sunlight directly into heat or electrical energy for our utility [11].

The use of *rumpon* has been widely applied by fishermen before, but there is no the use of *rumpon* which is equipped with the fish eating sound attractor and automatic flame lights. AUTO-LION is a *rumpon* innovation, which is a tool that works to help in catchingthe fish by collecting the fish in a place. When the fishes are concentrated in a point, the fishing operation can be more easily done and the catches will increase. The sound attractorwhich used is the recording of the sound of the fish while eating. Some fishes produce a sound as an acoustic cue when doing activities, such as communicating with other individuals, searching for a partner, prey detection, stress, and eating (Fitri *et al.*, 2010 [12]; Enger *et al.*, 1993[13]. So when the sound is played in the water, then the fish will come and gather around the *rumpon* because of the sound trick(Myrberg, 2001[14]; Sabet *et al.*, 2016[15], while the light attractor which used is white LED lights (Amorim *et al.*, 2015[16]; Taufiq *et al.*, 2015[17]; Yoshizawa and Sadao, 2008[18]. The light attractor is made waterproof so that the lamp is not easily damaged when it is inserted into the water. White color is used because this color is indicated as the best color to collect the fish. This is confirmed by Urbasa *et al.* (2015)[19], fish is more interested in white and green color than blue and red color. The most preferred color is white. According to Sukandar and Fuad (2015) [1], the light function in catching the fish is to collect the fish up to a certain catchable area, then the catch is done by means of net and fishing line.

The purpose of this study is to examine the effect of AUTO-LION usage on the catch of tancap charts in Semarang Waters.

2. Research Methods

The research method which used was experimental method with several treatments i.e. without AUTO-LION; the use of *rumpon*; the combination of *rumpon* and sound attractor; and the combination of *rumpon* and light attractor. The first step was to solve and process the absorption of solar light with solar panels at 09.00 - 15.00. Solar panels and batteries were placed on the top of the chart and allowed to stand for 6 hours. The second stage was the test of fishing without aid at 09.00-11.00. The third stage was an AUTO-LION raiding test at 12:00 to 14:00. At this stage, the *rumpon* was placed in the center of the tread. *Rumpon* was inserted into the water after the chart net was insertedinto the waterin order to not collide and damage the net. The fourth stage was the test with the combination of *rumpon* and sound attractor at 15.00 - 17.00. At this stage the work step was similar as before, but the the sound attractor turned on. And then the test of the combination of *rumpon* with light attractors at 19.00 - 21.00. And the last treatment is with AUTO-LION at 22.00 – 24.00. This test was performed during the night for a positive phototactic fish could approach the light attractor when the situation around it was dark. Treatment for 2 hours is based on the time of fishing habits when conducting fishing operations with the tancap chart.

This research was conducted on May 2017 at TambakLorok, Semarang with the coordinate position of 6 ° 56 '19 "LS and 110 ° 25' 57" BT. The analysis which used is the *IBM SPSS Statistics 20 License Authorization Wizard*. The hypothesis which used is the estimation of the influence of fish catch by using various capture treatments.

Ho : There is no influence between the catch fish and the capture treatment.
Ha : There is an influence between the catch fish and the capture treatment.

The decision making was done by doing a comparison to the sign value = 0.05. If the sign value <0,05 then Ho is rejected whereas the sign value> 0.05 then Ho is accepted. The observation was made on the stationary lift net 8x8m with a depth of 4m. The tools which used were AUTO-LION, stopwatch and fish finder as a tool to observe the existence of the fish. AUTO-LION is composed of triangular wood-shaped skeletons with a size of 1x0.5x0.5m, coconut leaves, 6 white LED lights 10 watts(Taufiq *et al.*, 2015) [17], speaker and MP3 player 800 Hz with the sound of recording fish, 4 pieces of concrete weights, Polyethylene ropes, 1 piece solar cell, solar controller cell, and 10 accu (Purnomo, 2010)[20]. The construction of AUTO-LION and the position of AUTO-LION on the chart can be seen in Figure 1

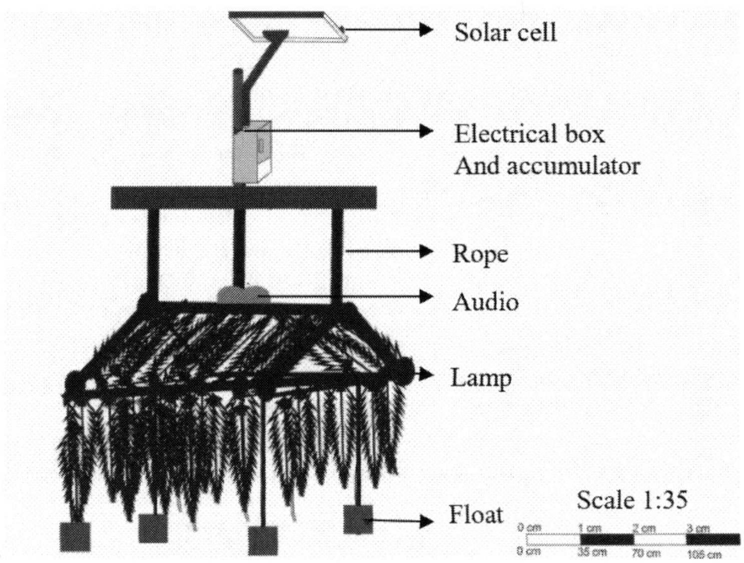

Figure 1. AUTO-LION Construction

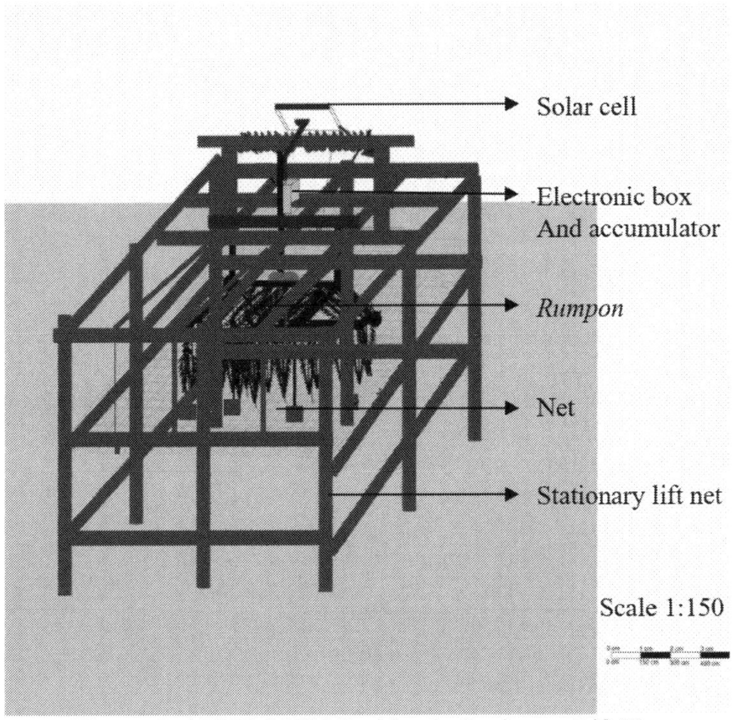

Figure 2. AUTO-LION Position in Stationary Lift Net

3. Results and Discussion

The results showed that the comparison of each treatment was in the following graph:

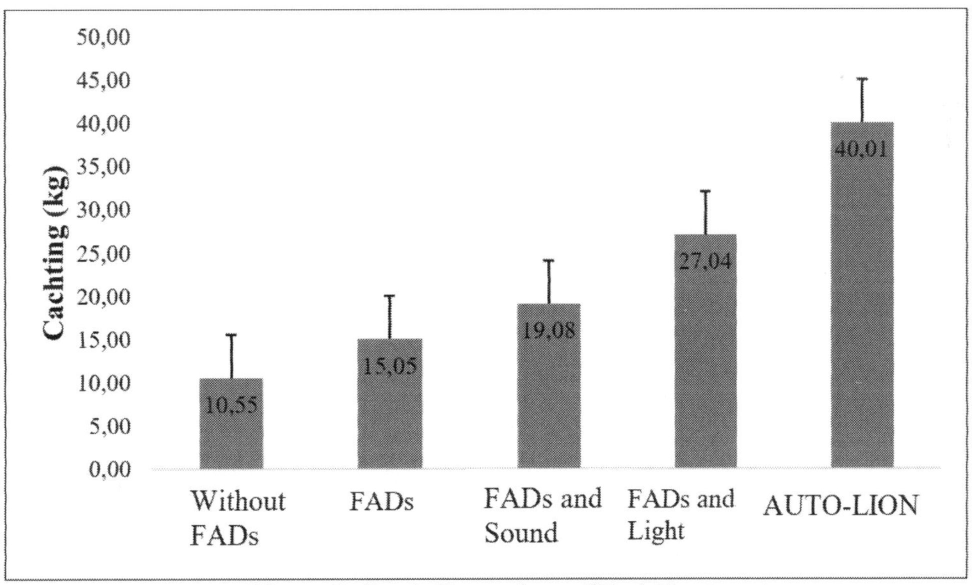

Figure 3. Graph Comparison of Each Treatment

Based on the observations on the fishing toolwhich wastested showed the significant results. The catch in each treatment had a difference that was when the fisherman used the fishing chart tool without using *rumpon* had a catch of 10,55 kg. The fishing tool that was added by the *rumpon* tool had a catch of 15,05 kg. The *rumpon* tool then was assigned a sound attractor and produced a catch of 19,08 kg, in contrast with when it given a light attractor of 27,04 kg. A *rumpon* tool that was given light and sound attractor at the same time produced the highest catch of 40,01 kg.Calculation result (sign 0,05) so that there is influence between catch with the treatment.

FADs added with light and sound attectors (AUTO-LION) are a combination of attractors who have 3 more power to collect fish. The resulting catch is subject to considerably higher changes than capture without using FADs. This fish aggregating device is suitable for as a catching tool performed on passive and static fishing equipment (Haruna, 2010[21]; Sudirman and Mallawa, 2004[22]. Fish aggregating device using one of the light and lower-light attractors in collecting fish compared to the two when combined. One strength of the attractor will affect the attractiveness of the fish. Light attractor will produce more catches than sound attractor because of the nature of the propagation light attractors can enter the vine into the water faster than the sound attractor so the fish will be more sensitive to light attractor. Fish will approach a catchy sound and make fish feel secure because of its acoustactic properties (Enger et al., 1993[13]; Fitri et al., 2010[12]). The indication is that the fish has a positive phototoxis of light, and the fish sound attractor will indicate its acoustax [14].

Fish has an organ that can be used to select the object that attracts it. One is the eye organ, the fish will be attracted by the things that stand out. Fish associated with the use of his eyes is a fish that has phototaxis properties. Phototaksis is the nature of fish that can be attracted to or away when there is a light. Fish attracted to lamu because it has positive phototaxis properties, if the fish away from the light, it means the fish has negative phototaxis properties. Nicol (1963) in Gustaman et al. (2014) [23] stated that the majority of the sea fish eye is very high in sensitivity to light. Light is the most important thing on the net fishery lift. It is attractant of plankton feeders around the light source [24].

At this time has grown a lot of FADs that uses lights but constrained that the lamp requires electrical energy. Usually the fishermen use electrical energy that stored in battery. The use of solar cells is a new stage that began to ogled the makers of fish collection equipment, which does not require home electricity but uses solar energy stored in solar cells. Solar cells will be used to turn on lights automatically. So the use of solar energy supports energy efficient capture. Solar energy becomes one source of power plants other than water, steam, wind, biogas, coal, and petroleum.

Fish that have positive phototoxis properties are fish that are usually in adult size. One of them is anchovy, according to Sudirman and Natsir (2011) in Fuad *et al.* (2016)[2], anchovies are thought to have very high positive phototaxic properties, while other species in addition to the nature of phototaxis also because of the impulse to find food. Visible axis is serve to identify fish habits in food or other objects. The axis of vision is obtained after the value of the con cell density of each part of the retina of the eye is known by drawing a straight line from the retina that has the highest concentration of conic cell to the center of the eyepiece. Fish have the ability to adapt to the incoming light in the retina of their eyes. Such capabilities include the ability to distinguish colors or wavelengths received.

The catches caught at the time of the study were fish cured (Plotusus canius), kwiper (Scatophagus argus), mullet (Mugil sp.), Sriding (Ambasius marianus), sumedar, shrimp (penaeus monodon), anchovy (stolephorus sp.). The main catch in this study is the anchovy which is the target fish in the fishing gear chart. Anchovy will approach the light source and survive to linger in the lighting area. In addition to being attracted by light, according to Baskoro (2007) in Gustaman et al. (2014) [23] occurrence of anchovy is also caused by the presence of food that usually gathered under the lights of petromaks ie plankton, shrimp and smaller fish.

Fish have hearing power with the organ of hearing that is owned, usually the fish is possessed linea lateralis, or line in the middle of the body. Fish that possess the lateral linea will respond to the sound source. Examples of fish that have lineal lateralis are gulamah, mullets, milkfish, srinding, and others.

4. Conclusion

The conclusion that can be obtained from the study of AUTO-LION study in the tap fishing gear is the highest catch obtained after the installation of AUTO-LION that is equal to 38.83% of the catch without using the aid of fish collecting equipment. The fish-raiser apparatus added by the voice enhancer increased the catch by 18.45%, while the added with the light attachment increased the catch by 26.21% and with AUTO-LION was more effective. AUTO-LION needs to be re-developed to become a fish-gatherer tool that meets the expected capabilities.

References

[1] Sukandar dan Fuad 2015 *J. Innovation and Applied Tech.* **2** pp 101-105
[2] Fuad, Sukandar, A Jauhari 2016 *J. Kelautan*. **9**(2016)1 pp 7-11
[3] Suwarsih 2012 *Prospektus* pp 181-191
[4] Regulation of the Minister of Marine Affairs and Fisheries of the Republic of Indonesia Number 26 Year 2014 about *Rumpon*
[5] Mandagi, Channia, Patrice N I Kalangi, Alfret Luasunaung 2016 *J. Ilmu dan Teknologi Perikanan Tangkap*. 2(2016)3 pp 105-108
[6] Nurani, T W, S H Wisudo, P I Wahyuningrum, dan R E Arhatin 2014 *J. Ilmu Pertanian Indonesia* **19**(2014)1 pp 57-65
[7] Simbolon, D, B Jeujanan, E S Wiyono 2013 Effectiveness of *Rumpon* Utilization in Fishing Operation in Southeast Maluku Waters
[8] Yusfiandani, R, I Jaya, dan A Baswantara 2014 *J. Fisheries and Marine Technology*. **5**(2014)1 pp 75-82
[9] Yandri, V R 2012 *J. ILMU FISIKA (JIF)*, **1** pp 14-19
[10] Widayana, G 2012 *J. PTK, UNDIKSHA*, **9** pp 37-46
[11] Asy'ari, H, A. Rozaq., dan F. S. Putra 2010 *J. Emitor*. **14**(2010)1 pp 33-40
[12] Fitri, A D P, Asriyanto, H Sutanto, and Widatini 2010 *J. Coastal Development*. **13** (2010) pp 205-214
[13] Enger, P. S., H. E. Karlsen., F. R. Knudsen and O. Sand 1993 *ICES mar. Sci. Symp.*, **196** pp 108-112
[14] Myrberg, A. A. 2001. Enviromental Biology of Fishe. **60** pp 31-45
[15] Sabet. S S, K Wesdrop, J Campbell, P Snelderwaard, and H Slabbekoom 2016 Science Direct **116** pp 1-11
[16] Amorim, M C P R O, Vasconcelos, and P J Fonseca 2015 ISPA. Instituto Universitário, Rua Jardim Do Tabaco 34, 1149-041 Lisbon, Portugal
[17] Taufiq, W Mawardi, M S Baskoro, dan Zulkarnain 2015 *J. Teknologi Perikanan dan Kelautan*. **6**(1) pp 51-67
[18] Yoshizawa, K and S Nogami 2008 Science Direct. 85(2008)1 pp 128-130
[19] Urbasa, F, F E Kaparang dan H J Kumajas 2015 *J. Science and Technology of Capture Fisheries* **2** pp 39-43
[20] Pernomo, W 2010 Electrical engineering major, FTI, Universitas Gunadarma. Margonda Raya.
[21] Haruna 2010 *J. Amanisal PSP FPIK Unpatti-Ambon* **1** pp 22 – 29
[22] Sudirman dan Achmar Mallawa 2004 Fishing Technique. Rineka Cipta, Jakarta.
[23] Gustaman, G, Fauziyah, dan Isnaini 2014 *J. Maspari*, **4**(2004)1 pp 92-102
[24] Puspito, G, S Ahmad, and M Sururi 2017 Egyptian *J. Aquatic Research*. **43** pp 155-160

Biodiversity of Cryptofauna (Decapods) and Their Correlation with Dead Coral *Pocillopora* sp. Volume at Bunaken Island, North Sulawesi

Muhammad Danie Al Malik[1*], Nenik Kholilah[1], Eka Maya Kurniasih[1,3], Andrianus Sembiring[2,3], Ni Putu Dian Pertiwi[2,3], Ambariyanto Ambariyanto[1], Munasik Munasik[1], Christopher Meyer[4]

[1]Department of Marine Science , Faculty of Fishery and Marine Science, Diponegoro University, Jl. Prof. H. Soedarto, S.H., Tembalang Semarang, 50275 Indonesia
[2]The Indonesia Biodiversity Research Center, Jl. Raya Sesetan Gg. Markisa No. 6 Denpasar, Bali, 80223 Indonesia
[3]Yayasan Biodiversitas Indonesia (BIONESIA) Foundation, Jl. Sulatri Gg XII No. 4 Denpasar, Bali, Indonesia
[4]National Museum of Natural History, Smithsonian Institution, 10th and Constitution Ave, NW Washington, DC 20560-0163

Email: daniealmalik2@gmail.com

Abstract. Decapod is known as cryptofauna which is also important component of coral reef biodiversity. Dead corals are one of the area which usually used by decapods to live. This research aims to observe the diversity of cryptofauna (decapods) and the correlation between the number of decapods with the volume of dead corals. Ten dead corals, *Pocillopora* sp., were collected at 5 m depth at Bunaken Island. These dead corals were measured their volume and all decapods found were counted and identified up to family level. The richness and abundance were analyzed using ACE (Abundance-Based Coverage Estimates) and Chao 1. The results show that there were in total 474 decapods from 13 families found within all ten dead corals. Xanthidae was showed as the most abundance family among all, with 161 individual. Diversity index of decapods was found at medium category with value of 2.01. Rarefaction curve based on richness and abundance showed an estimation of 13 families. The result also indicated that the asymptote stage was reached on the 10th dead coral samples. The correlation between decapod with the volume of dead coral were showed significant positive correlation ($r = 0.673$, $p<0.05$). This result provides benefits to basic knowledge about diversity of decapod which one of cryptofauna as component fauna have a habitat on coral reef ecosystem.

Keywords: Decapod, Dead Coral, Biodiversity, Bunaken Island

IOP Conf. Series: Earth and Environmental Science **116** (2018) 012053 doi:10.1088/1755-1315/116/1/012053

1. Introduction

Coral reef ecosystems are known as habitat of various organisms with its high biodiversity. One important component of coral reef biodiversity is cryptofauna [1] which include decapods and lives dwelling on the crack and crevices of dead or living coral. Cryptofauna is a small-sized fauna which difficult to distinguish and identify, including several fishes and invertebrates. These organisms usually use the cavity at substrate either temporarily or permanently. Several decapods can create self-cavity on coral reefs while the others are the opportunist occupier of existing space [2]. Decapod is also known as a key character on coral reef ecosystem, which will act as a suppressor for the growth of herbivore animals [3], and also indirectly give contribution on maintaining habitat [4]. Decapod can even be found on dead corals such as *Pocillopora* sp. [5].

The island of Bunaken is a part of Bunaken National Park [6, 7] North Sulawesi is one of the important marine parks in Indonesia. This park is part of the Coral Triangle which is known with its high biodiversity within coral reef ecosystem [8, 9, 10]. Understanding biodiversity and ecosystem health are very crucial especially in establishing proper management of an ecosystem [11, 12, 13] as well as avoiding degradation and endangering organisms [14]. However, little is known on the diversity of cryptofauna especially decapods in dead corals of *Pocillopora* sp., which can be used to estimate the hidden biodiversity at Bunaken island.

Study on life organism on dead coral of *Pocillopora* sp. will also give an overview about their diversity that inhabit the entire coral reefs around it [15]. The aims of this study were to investigate the diversity of cryptofauna decapods, and the correlation between the number and volume of dead coral (*Pocillopora* sp.) collected at Bunaken Island.

2. Material and Methods

Ten dead corals of *Pocillopora* sp. were collected on the reef of Bunaken Island at a depth of 5 m. See Figure 1. Each dead corals was put into plastic bag and brought to the laboratory for further treatments. Sample collection was based on the standard protocol from Plaisance *et al.* [5].

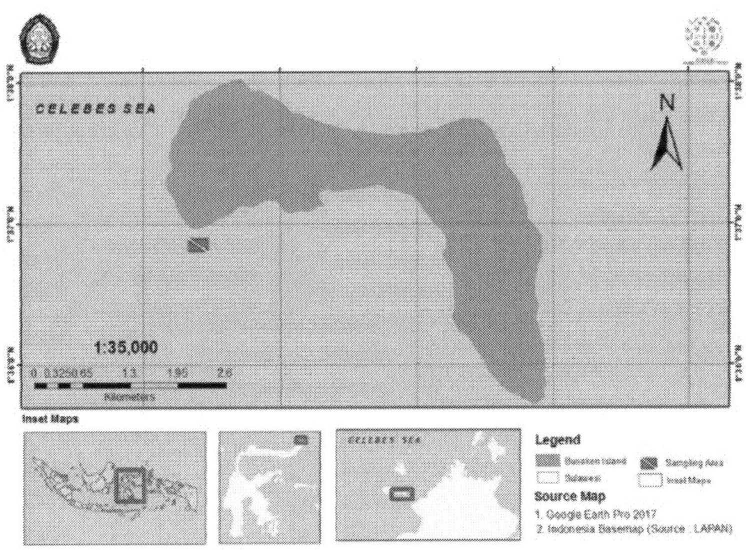

Figure 1. Sampling location at Bunaken Island, North Sulawesi, Indonesia

Measurement of the volume was done by putting each dead coral into bucket filled with water, and measures its volume using water displacement method [5]. While decapod samples were collected first by shaking dead corals in the bucket full of water and collect all organisms found. Each dead corals then were broken into small pieces using hammer and all organisms found were collected. These organisms then were sorted using morphological characters and identified up to family level based on Poore [16].

After being classified, decapods were put into a clove oil, photographed and being documented both the number and family names.

Diversity Indices of decapod were calculated using Shannon-Weaver formula [17], and diversity estimation were calculated using Chao 1 and ACE (Abundance-Based Coverage Estimates) analysis using software Estimates V.9.1.0 [18]. The correlation between the diversity of decapod and the volume of dead coral were calculated using software SPSS with bivariate Spearman analysis and 95% (*p<0.05*) confident interval.

3. Result and Discussion

Total number of decapods found within ten dead corals of *Pocillopora* sp. collected at Bunaken Island were 474 indiviual from 13 families (Figure. 2).

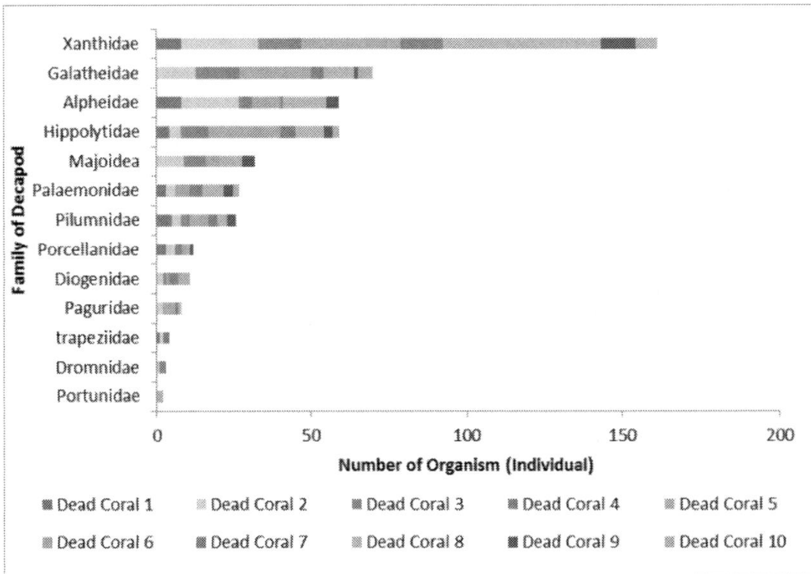

Figure 2. Family of decapod abundance on dead coral of *Pocillopora* sp. (Individual).

The most common and abundance family found was Xanthidae with 161 individual (Figure 2). This family was found at all ten dead corals of *Pocillopora* sp. on this location which have characteristic as branching coral. Xanthidae is the organisms that commonly found at dead coral with branching lifeform [19]. Both of these families are grazer of algae or detritus feeder [16]. When a coral died, it will be overgrown by sessile organisms (crustose algae and moss, bryozoan, sponge, and foraminifera) which will be the source of food for decapod [20].

Index of decapod diversity showed the value of 2.01 which indicated medium diversity [21]. The same research about decapod diversity index have been done by Pertiwi *et al.* [15] at Menjangan Besar Island, Karimunjawa and Pemuteran, Bali, the result of their study showed the value 1.9 at Menjangan Besar Island, and 2.35 at Pemuteran, Bali which indicated low diversity. These results suggest that diversity from reef crustacean found in Bunaken Island is more diverse than Menjangan Besar Island, Karimunjawa. These different values of diversity may be caused by coral conditions, geographic influences, and the diversity of other organism between areas [22]. Other possible causes are the availability of food and environmental influences, because both are very influential on the existence of an organism [23-24].

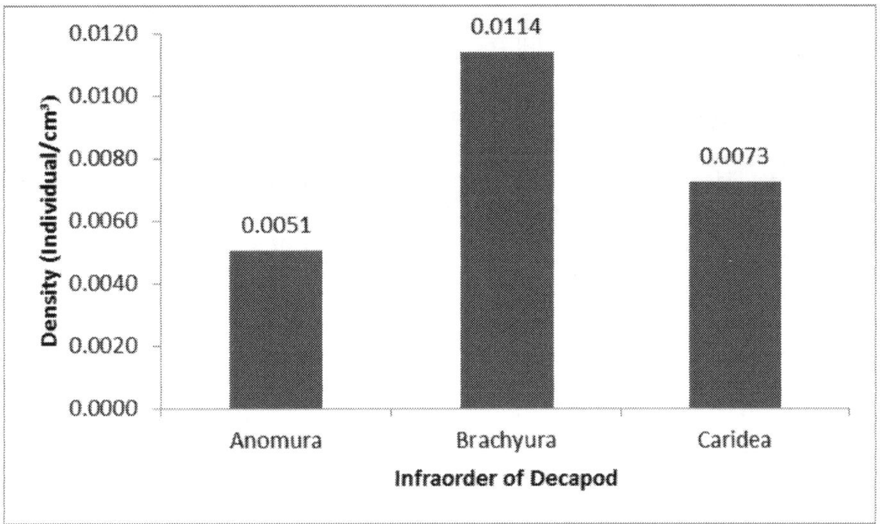

Figure 3. Density of Decapod Infraorder (individual/cm^3) which found within ten dead corals, *Pocillopora* sp., collected at Bunaken Island.

There were 3 infraorders of decapods found on this location. The highest density of infraorder was Brachyura with 0.0114 individual/cm^3. This infraorder was the most common since it was found in all ten dead corals. Brachyura is known as crab which has characteristic of living as semi-terrestrial and host obligate will settle at one location of dead coral *Pocillopora* sp. [25]. On the other hand, Caridea and Anomura are infraorder that moves to other substrates if their habitat is no longer sufficient for their lives, such as room and food availability. Wibowo *et al.* [26] showed that Caridea and Anomura are facultative organisms, they will move to another location such as moving to a deeper location as it grows.

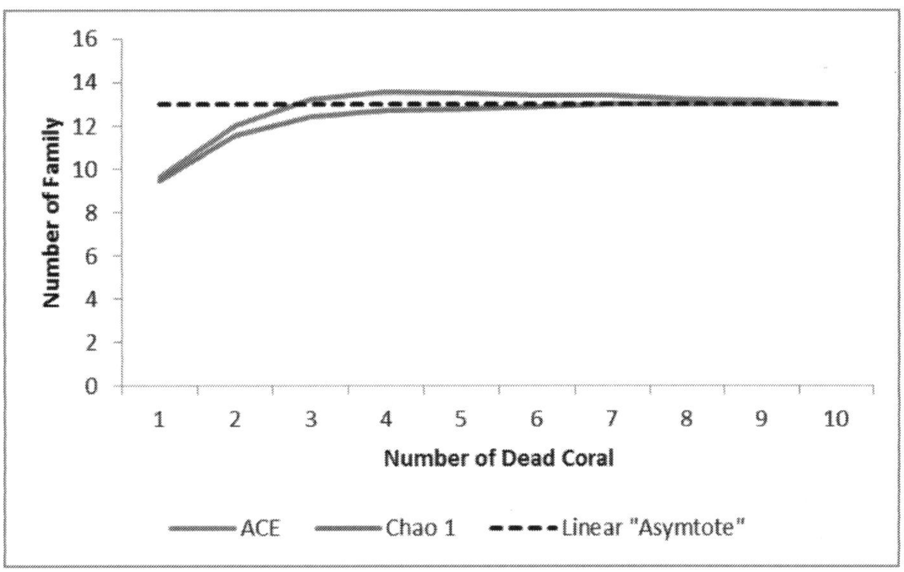

Figure 4. Result of ACE and Chao 1 calculations

Estimation analysis using ACE and Chao 1 showed the maximum number of family found was 13. The rarefaction curve showed an asymptote stage at the 10th dead coral, which indicates that the addition of dead coral samples will not change the number of family found (Figure 4.). Pertiwi *et al.* [15] study shows that at Menjangan Besar, Karimunjawa Island the rarefaction result of ACE and Chao 1 that rarefaction curve an asymptote after sampling by 14 dead corals and only found 11 decapod families. Research conducted by Head *et al.* [27] at Chagos island, Central Indian Ocean stated that decapod abundance variation taken from five dead corals are not effecting to presence or decapod abundance which living on coral reefs at these area. The result of this research may still underestimate the diversity of the entire coral reef within Bunaken Island, since it is the samples were collected only from one location and using only one particular specific organism (decapods). The identification until species level may also change and increase the diversity estimation of an area [28].

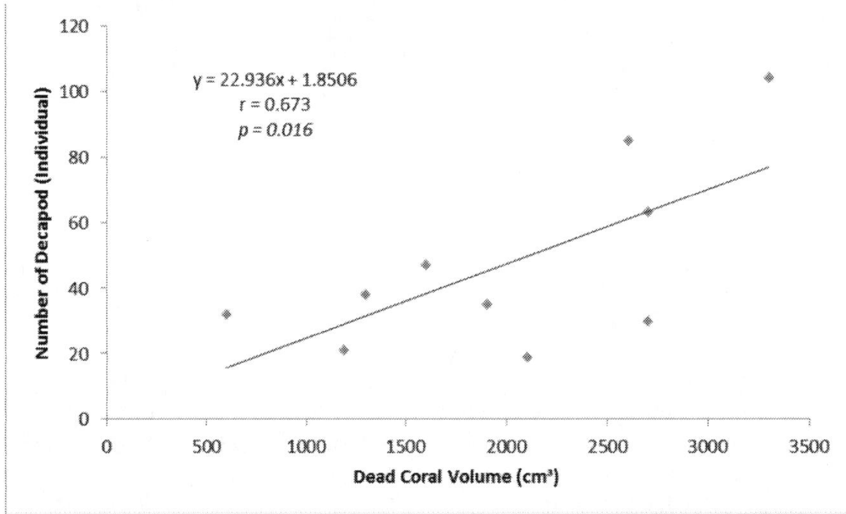

Figure 5. The correlation between the volume of dead corals with number of decapods

There is a significant positive correlation between volume of dead coral with number of decapods found ($p<0.05$). See Figure 4. This correlation clearly shows that the number of decapods is influenced by the size of dead corals. The complexity of coral reefs habitat such as variation sources of food and cavity of dead corals as good habitat are also influencing the number of decapods.

When corals died, the skeleton will be overgrown by sessile organism which is also the source of decapods food, and the skeletons were full of cracked and formed cavity [21]. Research conducted by Abele and Patton [29] showed that a significant correlation between the number of decapods with the sized of coral *Pocillopora damicornis*. Area may influence the species richness in at least three ways by [30] are (1) an increase in area may be correlated with an increase in habitat heterogeneity; (2) an increase in area may permit larger population sized and reduce the probability extinction; (3) an increase in area may simply increase the sample size. Habitats are heterogeneity associated to complexity [28]. Coral reef is complexity habitat with high biodiversity [31].

4. Conclusion

In conclusion, the estimation coral reef association of decapod diversity which found within ten dead corals at Bunaken Island indicated a medium diversity with 13 families. The correlation between number of decapod with the volume of dead coral *Pocillopora* sp. indicated a significant correlation. These results provide an addition to the overall biodiversity coral reef association of decapod at Bunaken Island and basic knowledge with aspect ecology of habitat decapod on coral reef ecosystems.

5. Acknowledgments

We would like to thank I. H. Putra, R. Putri, J. C. Haulussy, B. Tampanguma, M. Annisaqois, M.R. Manzanaris, V. S. Surentu, E. A. Caroles, B. Kohler, IBRC, and BIONESIA for their support and help on this study. This research is founded by National Academy of Sciences (NAS) with program PEER (Partnership for Enhanced Engagement in Research) with No. Grant Award: AID-OAA-A-11-00012.

6. References

[1] Reaka-Kudla M 1997 The Global Biodiversity of Coral Reefs: a Comparison with Rain Forest. In: Reaka-Kudla. M., Wilson DE, Wilson EO (eds) Biodiversity II: Understanding and Protecting our Biological Resources. Joseph Henry Press, Washington, D.C., pp 83-108.
[2] Hutchings PA 1983 Cryptofaunal Communities of Coral Reefs. In: Barnes D.J. (ed.) Perspectives on Coral Reefs. Australian Institute of Marine Science. Townsville. Australia. pp 200-208
[3] Taylor GM 2000 *Proc. Biol. Sci.* **267**: 1475-1480
[4] Gittman RK and Keller DA 2013. *Ecology* **94**: 2709-2718
[5] Plaisance L, Knowlton N, Paulay G and Meyer C 2009 *Coral Reefs*. **28**: 977-986
[6] Ruga L, Langoy M, Papu A, and Beivy K 2014. *J. MIPA UNSRAT*. **2**: 84-86
[7] Rumampuk R 2013 *Lex et Societatis* **1**(5): 54-63
[8] Luasunaung A, Manoppo V and Schaduw JN 2015 *J. SPATIAL-Wahana Komunikasi dan Informasi Geografi* **14**(2): 1-7.
[9] Ambariyanto 2010. *Kebijakan Pengelolaan Organisme Laut di Lindungi: Kasus Kerang Raksasa. Pidato Pengukuhan.* UNDIP
[10] Sutarno and Setyawan DA 2015 *Pros Sem Nas Masy Biodiv Indo* **1**(1): 1-13
[11] Costanza R, Kemp WM and Boynton WR 1993 *Ambio* **2**: 88-96.
[12] Prasetya JD, Ambariyanto A, Supriharyono and Purwanti F 2017 *Adv. Sci. Lett.* **23**: 3277–3282
[13] Barrett N, Seiler J, Anderson T, Williams S, Nichol S and Hill SN 2010 May. Autonomous Underwater Vehicle (AUV) for mapping marine biodiversity in coastal and shelf waters: Implications for marine management. In *OCEANS 2010 IEEE-Sydney* (pp. 1-6).
[14] Ambariyanto 2017 *IOP Conference Series: Earth and Environmental Science* 55(1): 012002
[15] Pertiwi NP, Kurniasih EM, Riantiantoto SA, Dharmawan IG, Mustari TR, Basuki F and Cahyani NKD 2014 *Ilmu Kelautan* **19**(4): 195–201
[16] Poore GCB 2004 Marine Decapod Crustacea : A Guide to Identification. CSIRO. Australia. 617 p
[17] Magurran AE 2004 Measuring Biological Diversity. Blackwell. Oxford, 256 p.
[18] Colwell RK 2013 EstimateS: Statistical estimation of species richness and shared species from samples. Version 9. Persistent URL <purl.oclc.org/estimates>
[19] Patton WR 1964 Decapod Crustacea Commensal with Queensland Braching Corals. Department of Zoology. Ohio Wesleyan University. U.S.A. 271-295
[20] Enochs IC and Hockensmith G 2008 Int. Coral Reef Symp. Ft. Lauderdale, Florida, 7-11 July 2008. **2**: 1375-1379
[21] Krebs CJ 1985 The Experimental Analysis of Distribution and Abundance Third edition. Harper and Row Publishers. New York, 800 p.
[22] Stella JS, Jones GP & Pratchett MS 2010 *Coral Reefs* **29**: 957-973
[23] Hutagalung HP 1988 *Oseana* **XIII**(4): 153 – 164
[24] Sutomo S 2007 *Ilmu Kelautan*, 12(4): 219-226.
[25] Atkinson RJA and Taylor AC 1988. Physiological ecology of burrowing decapods. *Symp. zoll. Soc. London*, 59: 201-226.
[26] Wibowo P, Hartoko A and Ghofar A 2007 *J. Pasir Laut* **2**(2): 18 - 29
[27] Head CE, Bonsall MB, Koldewey H, Pratchett MS, Speight M & Rogers AD 2015 *Coral Reefs* **34**(3): 905-915.
[28] Plaisance L, Caley MJ, Brainard RE, Knowlton N 2011 Plos One. **6** (10) Doi: 10.1371.journal.pone.0025026
[29] Abele LG and Patton WR 1976 *J. Biogeography* **3**: 35-37.
[30] Connor EF and McCoy ED 1979 *Am. Nat.* **113**: 791-833
[31] Rudi E 2005 *Ilmu Kelautan* **10**(1) 50-60

Identification of Antipathogenic Bacterial Coral Symbionts Against *Porites Ulcerative White Spots* Disease

Nor Sa'adah[1*], Agus Sabdono[2] dan Diah Permata Wijayanti[2]

[1]Oceanography Department, Faculty of Engineering and Marine Science, Hang Tuah University, Surabaya, Indonesia
[2]Marine Science Department, Faculty of Fisheries and Marine Science, Diponegoro University, Semarang, Indonesia

Email: nor.saadah@hangtuah.ac.id

Abstract. Coral reef ecosystems are ecosystems that are vulnerable and susceptible to damage due to the exploitation of ocean resources. One of the factors that cause coral damage is the disease that attacks the coral. Porites Ulcerative White Spots (PUWS) is a coral disease found in Indonesia and attacks the coral genera *Porites* allegedly caused by pathogenic microbial attacks. The purpose of this study was to identify the symbiotic bacteria on healthy coral that have antipatogenic potency against PUWS. The method used in this research was descriptive explorative. Sampling was done in Kemujan Island, Karimunjawa. Bacteria were isolated from healthy coral and coral affected by PUWS disease. Streak method was used to purify coral bacteria, while overlay and agar diffusion were used to test antipathogenic activity. Bacterial identification was carried out based on polyphasic approach. The results of this study showed that coral bacterial symbionts have antipathogenic activity against PUWS disease. The selected bacteria NM 1.2, NM 1.3 and KPSH 5. NM1.2 were closely related to *Pseudoalteromonas piscicida*, *Pseudoalteromonas flavipulchra* and *Bacillus flexus,* respectively.

Keywords: Antipathogens, Symbiotic Bacteria, Coral, Coral Disease, PUWS

1. Introduction

Coral reef ecosystems are ecosystem consisting of animals, plants, fish, shellfish and other biota contained in the tropics that require the intensity of sunlight to live [1]. Coral reef ecosystems are ecosystems that are vulnerable and susceptible to damage due to the exploitation of ocean resources such as exploitation of coral reefs building materials, souvenirs, medicinal materials, cosmetic materials and aquarium ornament. Also, household waste, industrial waste, agricultural waste, transportation waste and the used of explosives in fishing. Coral disease infection is one of the major factors that worsen the condition of global coral reefs [2].

Karimunjawa islands have coral reefs dominated by coral of *Porites* sp. Coral *Porites* sp. has a widespread throughout Indonesia [3]. This is due to *Porites* sp. is a coral capable of living in various environmental conditions such as in areas that have high and low sedimentation and areas with high and low salinity [4]. While *Porites* sp. able to live in various environmental conditions, the genus Porites can also get sick if the environmental around the coral has been infected with the disease.

PUWS is a coral disease found in Indonesia and attacks the coral genera Porites allegedly caused by pathogenic microbial attacks. The coral affected by PUWS, the surface of the coral frame will be damaged by the occurrence of white spot regularly [5,6]. In Karimunjawa Islands there are not many studies that study coral diseases especially PUWS. Based on these descriptions it is necessary to have further studies on the presence of PUWS disease in *Porites* sp. in Karimunjawa Island, considering that *Porites* sp is the dominant coral in that area by identifying the symbion bacteria on healthy corals and coral that have antipathogenic potential to corals that are attacked by PUWS.

2. Material and methods

2.1. Sampling and isolation of bacteria symbionts with coral

The coral was collected by scuba diving at depths of 1m to 10m from Kemujan Island, Karimunjawa (S 05° 48' 779" ; E 110° 29' 955"). Several coral colonies were taken using stools and hammers with methods of cutting coral tissue on healthy corals and corals infected disease. Then the sample was labeled into a sterile plastic bag and stored in a cool box [7]. The environmental parameters including salinity and depth at the site around the reef were measured. The tissue samples were scraped off by using a sterile knife. The resultant tissues were then tenfold serially, diluted, spread on the half strong ZoBell 2216E marine agar medium and incubated at room temperature for 48h. On the basis of morphological features, colonies were randomly picked and purified by making streak plates [8].

2.2. Screening of the isolates for anti-pathogenic activity

The bacterial isolates symbionts with healthy coral were tested for anti-pathogenic effect by the agar overlay method (were not presented in this paper) against selected pathogens bacteria. Each of disease bacterial isolated was used for anti-pathogenic effect study. Healthy coral bacteria were taken one ose and planted on ZoBell 2216E marine medium agar on the plates and formed into small spheres. The plates wrapped with plastic wrap and were incubated at room temperature for 48 h. The bacteria causing PUWS disease were grown on liquid ZoBell 2216E for 48 h while shook used a shaker. About 2 mL of the test cultures were suspended in 200 mL of ZoBell 2216E soft medium agar and were poured immediately over the colonies of the healthy coral bacteria on the plates. Then plates were incubated at room temperature for 48 h, antibacterial activity was defined by the formation of inhibition zones around the bacterial colony [8].

All bacterial strains selected from overlay test were rescreened for their activity using agar disk diffusion method. The paper disks diameter 8 mm that previously impregnated with 100 μL of each diseased isolate were placed on the surface of plates. Then, 35 μL of selected isolated were put on each paper disk. The plates were then incubated for 48 h. At the end of incubation period, the zones of inhibition were measured.

2.3. DNA Extraction, PCR amplification and sequencing of 16S rRNA

DNA extraction was conducted by using chelex method (Susilowati *et al.*, 2015). DNA extracts for 16S rRNA gene sequences were amplified by PCR using universal primers 27F (5'-AGAGTTTGATCMTGGCTCAG-3') and 1492 R (5'-TGGTTACCTTGTTACGAC TT-3') (Weisburg *et al.,* 1991; Sabdono *et al.,* 2015). DNA sequencing was conducted at PT Genetika Science, Jakarta. The result of DNA sequences were preliminarily aligned with Clustal W Multiple Alignment and the phylogenetic analyses were performed by using MEGA 5.05 [9].

3. Result and Discussion

Isolates of healthy coral bacteria were collected that provided 53 isolates had antimicrobial activity. After diffusion testing, the most active isolate were NM 1.2, NM 1.3 and KPSH 5 (Figure 1).

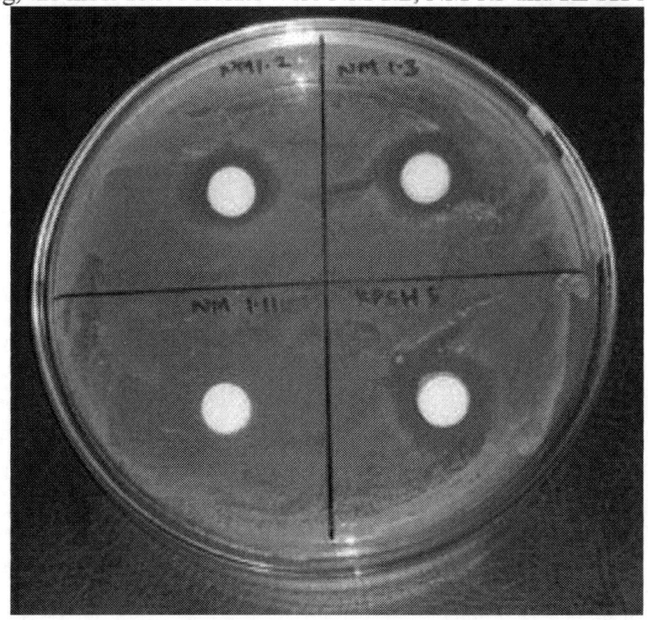

Figure 1. The inhibition zones formed on agar disk diffusion method

The decrease of anti-pathogenic activity using agar disk diffusion method due to penetration of antibacterial coral bacteria into the medium surrounding the disk was too slow to adequately show their potency to inhibit the growth of pathogenic bacteria.

The homology searched results using the BLAST system were presented in Table 1. The precise taxonomic position of coral bacteria and provided more detail information on their phylogenetic position among their closest relatives were showed in Figure 2.

Table 1. Overview of 16S rRNA gene sequences retrieved from anti-pathogenic coral bacteria

Isolate	Length (bp)	Closest Species	Similarity (%)	(BLAST ncbi)	Accession Number (GenBank)
NM1.2	1249	*Pseudoalteromonas piscicida*	92 %	NR113971.1	LC101432
NM1.3	1363	*Pseudoalteromonas flavipulchra*	99 %	NR025126.1	LC101433
KPSH 5	1430	*Bacillus flexus*	99 %	NR024691.1	LC101435

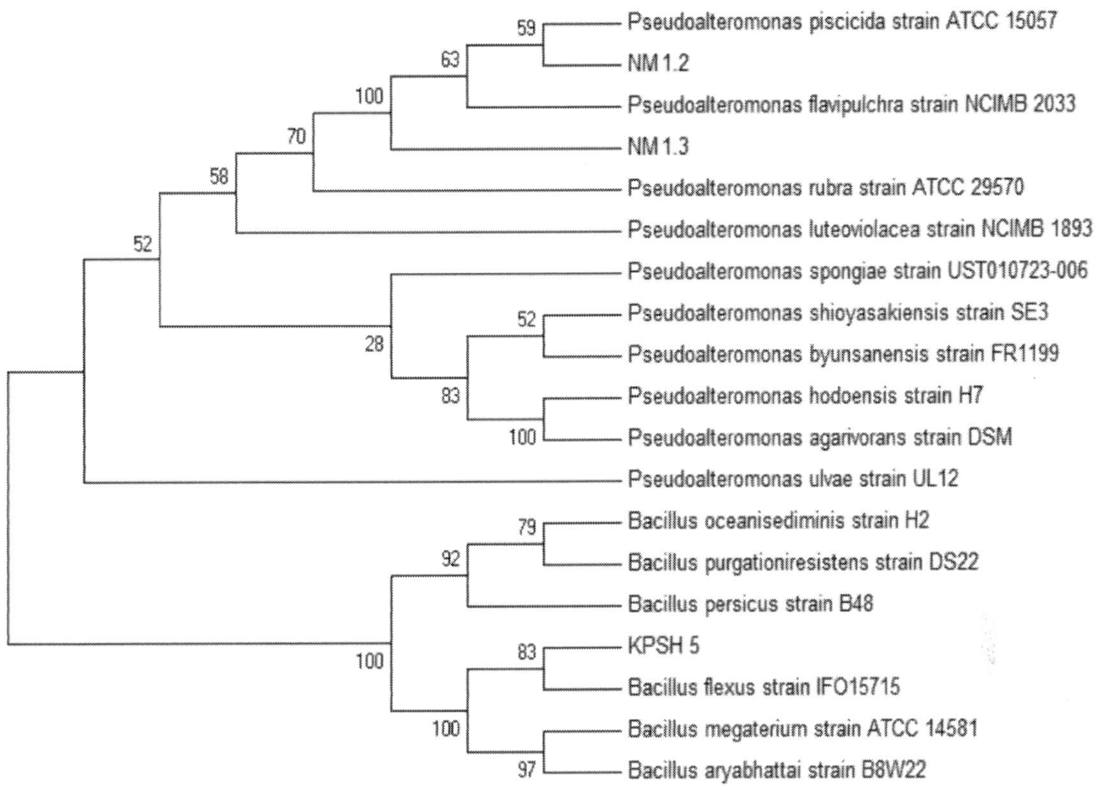

Figure 2. Phylogenetic Tree Isolate NM 1.2, NM 1.3 and KPSH 5 Had Anti-pathogenic Activity

The resulted BLAST homology from NCBI (National Center for Biotechnology Information) showed that the active strain bacteria NM 1.2 92% similarity to *Pseudoalteromonas piscicida* base on 16S rRNA gene sequence. These strain has been submitted to DDBJ (DNA *Data Bank of Japan*) with the accession number LC101432. Isolate NM 1.3, 99% similarity to *Pseudoalteromonas flavipulchra* with the accession number LC101433. Isolates NM 1.2 and NM 1.3 were isolates similarity to *Pseudoalteromonas* sp. Bacteria Pseudoalteromonas sp. was marine bacteria and bacteria that infect most marine organisms. Antibacterial activity of *Pseudoalteromonas luteoviolacea* TAB 4.2 associated with *Acropora* sp. able to inhibit the growth of coral bacteria and pathogens [10].

The resulted BLAST homology showed that the active strain bacteria KPSH, 99% similarity to *Bacillus flexus* and had been submitted to DDBJ (DNA *Data Bank of Japan*) with the accession number LC101435. In marine environments, *Bacillus* sp. can produce metabolites as antimicrobial, anti-fungal or have high toxicity, therefore has a high potential for the search new antimicrobial compounds [11-13].

4. Conclusion

Three bacterias symbionts with corals were able to inhibit *in vitro* bacterial growth of bacteria isolated from disease corals.The molecular identification by partial 16S rRNA gene sequencing revealed that they were closely related to the genera *Pseudoalteromonas* and *Bacillus*. It is important to continue coral disease research such as research on the prevention of coral disease so damage to corals can be reduced.

5. Acknowledgements

Authors are grateful to the staff of Tropical Marine Biotechnology laboratory Faculty of Fisheries and Marine Science of Diponegoro University.

6. References

[1] Siringoringo RM 2007 Oseana **32**(4): 29-37
[2] Weil E, Smith GW and Gil-Agudelo DL 2006 Dis. Aquat. Org. **69**: 1-7
[3] Suharsono 1996 Jenis-Jenis Karang yang Umum di Jumpai di Indonesia P3O-LIPI Jakarta: pp 116
[4] Morton J 1990 The Shore Ecology of the Tropical Pasific Unesco 282
[5] Raymundo LJ, Harvell CD and Reynolds TL 2003 *Dis. Aquat. Org.* **56**: 95–104.
[6] Beeden R, Willis BL, Raymundo LJ, Page CA and Weil E 2008 Underwater Cards for Assessing Coral Health on Indo-Pacific Reefs. Global Environment Fund Coral Reef Targeted Research (GEF-CRTR) Program, Currie Commis-sion, Melbourne.
[7] Sabdono A 2009 Ilmu Kelautan **14**(3): 117-125
[8] Sabdono A, Sawonua PG, Kartika AGD, Amelia JM and Radjasa OK 2015 Elsevier Procedia Chemistry **14**: 15-21
[9] Isnansetyo A and Kamei Y 2003 *Int. J Antimicrob Agents* **21**: 71-74
[10] Radjasa OK, Sabdono A, Junaidi and Zoochi E 2007 J. Pharmaca. **3**(3): 275 – 279
[11] Muscholl-Silberhorn A, Thiel V and Imhoff JF 2007 *Sea Microbial. Ecol.* **55**: 94-106
[12] Sabdono A 2007 J. Coast. Dev. **10**(2): 115-123
[13] Santos OCS, Paula VMLP, Juliana FMS, Guilherme M, Marcia GDM and Marinella SL 2010 Res. Microbiol. **161**: 604-612.

Community Structure of Decapod Inhabit Dead Coral *Pocillopora* sp. in Pemuteran, Bali

N P D Pertiwi[2,4*], M D A Malik[1], N Kholilah[1], E M Kurniasih[1,4], A Sembiring[2,4], A W Anggoro[5], Ambariyanto[1] and C Meyer[3]

[1]Marine Science Department, Faculty of Fisheries and Marine Science, Diponegoro University, Semarang, Indonesia, Jl. Prof. H. Soedarto, S.H., Tembalang Semarang, 50275 Indonesia
[2]Yayasan Biodiversitas Indonesia (BIONESIA), Jl. Sulatri Gg XII No. 4 Denpasar, Bali, Indonesia
[3]National Museum of Natural History, Smithsonian Institution, 10th and Constitution Ave, NW Washington, DC 20560-0163, USA
[4]Indonesia Biodiversity Research Center, Jl. Raya Sesetan Gg. Markisa No. 6 Denpasar, Bali, 80223 Indonesia
[5]Department of Ecology and Evolutionary Biology, University of California Los Angeles, 621 Charles E. Young Drive South, Los Angeles, CA 90095, USA

Email: putudianpertiwi@gmail.com

Abstract. Decapod is one of the cryptic fauna associated with coral reefs, which play an important role and accounts for a major part of the biodiversity in the reef ecosystem. However, this biodiversity is largely overlooked and assessment regarding this biodiversity has not been conducted thoroughly due to lack of resources and the difficulties of sample collection. This research used semi-quantitative sampling methods to assess the community structure of decapod species inhabit dead *Pocillopora sp.* in Pemuteran, Bali. Two dead coral head of *Pocillopora sp.* were used as pilot study and sampled at the 8-12 m depth. All decapod found were collected and identify into family level. The result showed 214 decapods found consisting 12 families, with Xanthidae as the most abundant family. Community Index which consisting diversity, uniformity, and dominancy, indicated medium diversity, stable uniformity, and low dominancy of decapod community. This result also indicated no dominancy between families of decapod. Meanwhile, Species Density Index showed the value of 84.58 ± 2.04 individual per m^3, with mean diversity of 7.05 ± 2.04 individual per m^3 for each family.

Keywords : decapods, dead coral, community, Pemuteran, *Pocillopora sp.*

1. Introduction

Decapoda or decapods are an order of crustacean, which is estimated to contain nearly 15.000 species and around 2.700 genus [1]. Known as one of the cryptic organism, the species of decapod are hardly to identify. Its living characteristic by burrowing on coral and the similar body pattern between taxa was the challenge to identify and measure the diversity of this organism [2-3]. The diversity of decapod is make up of about 66% of all the diversity within coral reef, which make this fauna as one of the best proxy to estimate the biodiversity of the whole organisms on coral reef ecosystems [4]. Decapod also plays an important role in reef ecosystem, including metabolizing and controlling the flow of energy in ecosystems [5-8].

Despite the importance of this organism for the ecosystem, the biodiversity of decapod is largely overlooked and assessment has not been conducted thoroughly due to lack of resources and difficulties of sample collection. In 2014 [9] we manage to conduct the estimation survey on decapod diversity in Karimunjawa using semi-quantitative dead coral head methods and obtained the estimation data of the local biodiversity within Karimunjawa. During this research, we aim to conduct the pilot study to understand the community structure of decapod diversity inhibit dead coral head in Pemuteran, Bali. Pemuteran is known as one of the area in Bali, which located in the center of Coral Triangle [10] - the epicenter of marine biodiversity [11-13]. Escalating threat in marine biodiversity was also risk to affect the diversity in Indonesia's marine biodiversity, due to the global warming and increasing illegal fisheries [14]. Pemuteran is known as one of the famous tourist area in Bali, with its coral reef as one of its main tourist attraction [15]. Although famous for its coral reef area, Pemuteran reef is susceptible with damage due to high intensity, illegal and destructive fishing. Pemuteran also importance for its characteristic of an area that suitable for coral reef, mangrove and seagrass bed ecosystem [16], which is the living habitat for decapod organism and important area to conduct this research.

2. Material and Method

The study was conducted in June 2016 at Pemuteran, Bali (Figure 1). Two dead coral from *Pocillopora* sp. were collected from the depth of 8 – 12 m. Dead coral with a size of a head were bagged and gently broken by its base with a hammer and a chisel to separate it from the coral reef, and quickly placed in a 20-liter bucket underwater.

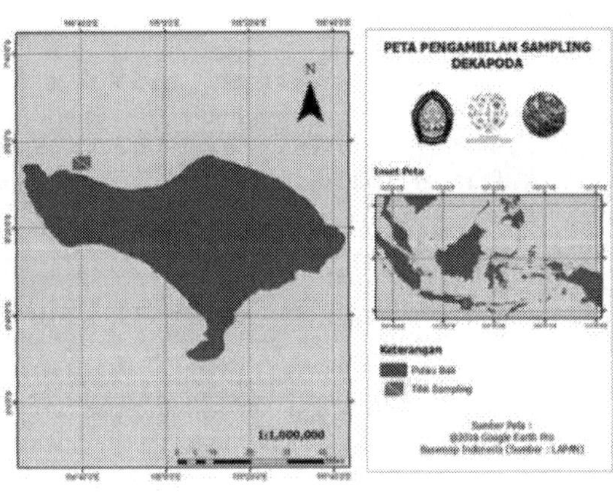

Figure 1. Sampling location.

Length, width and height of the coral head were measured, while the volume was measured using water displacement method. Each branch of coral head was detached carefully and examined closely for motile invertebrates. The remaining rubble were examined a second times for remaining creatures, while the seawater in the bucket was also filtered in order to gather the remaining invertebrates left in the bucket. Only decapod specimen were then collected and identified into family level and recorded along with the abundance of each family. The specimens were then photographed to record its morphological features. This sampling collection and processing is following the method describe in Plaisance *et al.* [3]

Community structure of decapod were analyzed using community index analysis which consisting diversity, uniformity, and dominancy, indicated medium diversity, stable uniformity, and low dominancy of decapod community.

3. Result and Discussion

Result from the two dead coral head of *Pocillopora* sp. was managed to obtain 12 families of decapod from 214 total individual. The highest abundance among all was Xanthidae followed by Alpheidae, while the lowest abundance was Trapeziidae and Diogenidae with only 1 individual found (Figure 2). Xanthidae has been known as a common and dominance family of decapod found living in the coral of Pocilloporidae and Acroporidae, while Trapeziidae is usually found living in the live coral [17] and Diogenidae is associate with the availability of the gastropod shell [18].

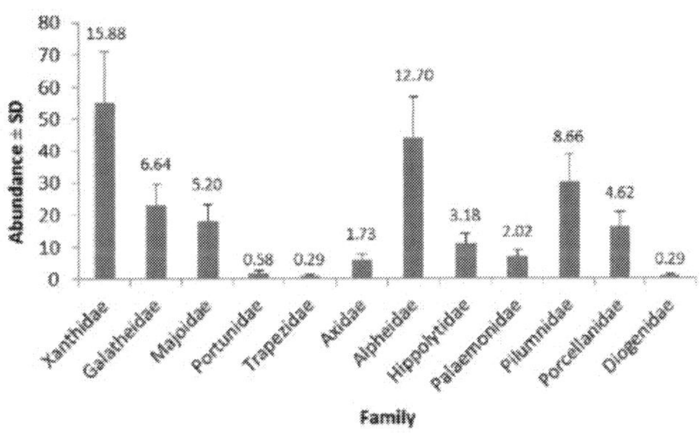

Figure 2. Abundance of each decapod family collected from dead head in Pemuteran, Bali.

Within these results, the community structure of decapod in two dead coral head in Pemuteran, Bali is indicating medium diversity (2.96), stable uniformity (0.83) and low dominancy (0.16). This result also indicated that there is no dominancy between families of decapod. Community structure analyses using Shannon-Wiener diversity indices, index of uniformity and index of dominance indicated as in Figure 3, while the species density was indicated in Table 1.

Diversity index is a mathematical measure of species diversity in a community, which provide the information about the community composition including species richness, rarity and commonness of species in a community [19-20]. Based on Krebs [21], the diversity value of 2,96 was considered as a moderate or medium diversity with the indication of medium environmental pressure. Setyawan *et al.* [22] also stated that the higher the number of diversity indicated a high interaction between species within community, which will also related to the complexity of energy transfer, predation, competition and diversification of the niche.

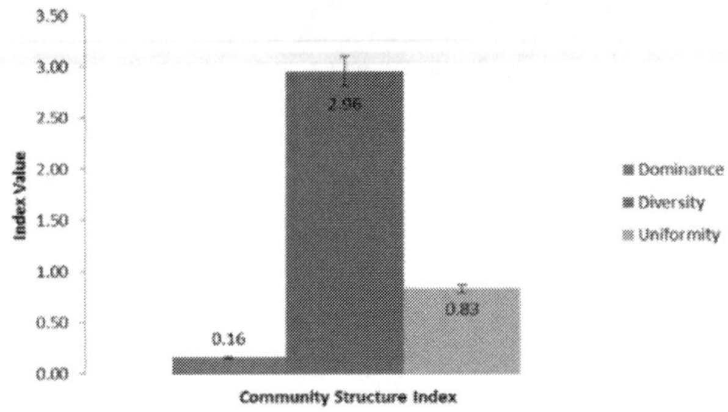

Figure 3. Index of dominance, diversity and uniformity on the decapod structure on dead coral head in Pemuteran, Bali.

Uniformity index is a measure of the similarity or uniform-ness of a community. The uniformity index value of 0.83 in the result is considered as a high uniformity, which indicated a stable community [21]. This uniformity index is inversely proportional with the dominance index. The higher the uniformity, the lower the dominance. This statement was also proven by the result of the dominance index, with the value of 0.16, which considered as a low dominance [21].

Other than those indexes, Species Density Index showed the value of 84.58 ± 2.04 individual per m^3, with mean diversity of 7.05 ± 2.04 individual per m^3 for each family. The highest species density is within Xanthidae and Galatheidae with the value of 21.74 ± 6.28 and 9.09 ± 2.62, respectively (Table 1).

Table 1. Species density of decapod in Pemuteran, Bali.

Family	Value (Individual per m^3)
Xanthidae	21.74 ± 6.28
Galatheidae	9.09 ± 2.62
Majoidae	7.11 ± 2.05
Portunidae	0.79 ± 0.23
Trapezidae	0.40 ± 0.11
Axidae	2.37 ± 0.68
Alpheidae	17.39 ± 5.02
Hippolytidae	4.35 ± 1.26
Palaemonidae	2.77 ± 0.80
Pilumnidae	11.86 ± 3.42
Porcellanidae	6.32 ± 1.83
Diogenidae	0.40 ± 0.11
Mean value of family density	7.05 ± 2.04
Species density of decapod	84.58 ± 2.04

The mean family density value in Pemuteran, Bali is higher than Pari Island, Seribu Islands (0.68 ± 0.22 individual/m^3) [23]. Species density is affected by the availability of crack and crevices in corals, which then lead to the competition for a place for living, especially in the dead coral where the competition is not limited by its interaction with living corals [24]. Other than the competition for space for living, the species density also affected by the competition for food source [23].

4. Conclusion

In conclusion, the community structure of decapod inhabit dead coral head in Pemuteran, Bali showed 12 families of decapod with the indication of medium diversity, stable uniformity, and low dominancy of decapod community. It is also indicated that there are no dominancy happened between families of decapod. This data provide a basic understanding of the community structure of a reef area in Pemuteran, Bali. However, it still needs more additional data in order to give a better representation of a general community structure of a wider area within the coral reefs.

5. Acknowledgement

This study was part of the Biodiversity Inventory project funded by the PEER (Grant Number AID-OAAA-A-11-00012) by NAS and USAID. We would like to thank Dita Cahyani, Astria Yusmalinda, Paul Barber and all the participants of the Biodiversity Inventory Training – Bali for their support and help in collecting data.

6. References

[1] Grave S D *et al* 2009. *Raffles Bulletin of Zoology* **22** 1-109.
[2] Knowlton N 1986. *Journal of Crustacean Biology* **6(3)** 356-363.
[3] Plaisance L *et al* 2009. *Coral Reefs* **28** 977-986.
[4] Stella J S *et al* 2010. *Coral Reefs* **29** 957-973.
[5] Coull B and Bell S 1983. *Behavior and Ecology* 283-319. *In*: Vernber F J and Vernberg W B (eds.). the biology of Crustacea Vol. 7: Behavior and Ecology (New York-Academic Press).
[6] Gittman R K and Keller D A 2013. *Ecology* **94** 2709-2718.
[7] Taylor G M 2000. *Proceeding of Biological Science* **267** 1475-1480.
[8] Bodreau S A and Worm B 2012. *Marine Ecology Progress Series* **465** 195-213.
[9] Pertiwi N P D *et al* 2014. *Ilmu Kelautan* **19(4)** 195-201.
[10] McLeod E *et al* 2010. *Coastal Management* **38(5)** 518-539.
[11] Allen G R and Werner T B 2002. *Environmental Biology Fishes* **65** 209-214.
[12] Veron J E N *et al* 2009. *Galaxea, Journal of Coral Reef Studies*. **11** 91-100.
[13] Tornabene L *et al* 2015. *Molecular Phylogenetic and Evolution*. **82**. 200-210.
[14] Hoegh-Guldberg O *et al* 2009. *The Coral Triangle and Climate Change: Ecosystems, People and Societies at Risk.* WWF Australia, Brisbane.
[15] Putra I N D 2014. *Jurnal Kajian Bali*. **4 (1)** 15-30.
[16] Trialghianty T I and Suadi 2017. *Journal of Coastal Conservation*. **21(6)** 873-882.
[17] Knudsen J W 1967. *Pacific Science* **XXI**.
[18] Scully E P 1979. *Journal of Experimental Marine Biology and Ecology* **31 (2)** 139-152.
[19] Begon M *et al* 1996. *Ecology: Individuals, Populations, and Communities* 3rd edition (Cambridge-Blackwell Science Ltd.)
[20] Rosenzweig M L 1995. *Species Diversity in Space and Time* (New York-Cambridge University Press)

[21] Krebs C J 1985. *Innovative Methods of Marine Ecosystem Restoration* ed T J Goreau and R K Trench (New York-CRC Press).

[22] Setyawan E *et al* 2011. *Laporan Pengamatan Jangka Panjang Terumbu Karang* (Jakarta-Terangi).

[23] Santoso P 2013. *Perbedaan Keanekaragaman dan Kerapatan jenis Dekapoda pada Karang mati antara* Pocillopora verrucosa dan Seriatopora histrix *di Pulau Pari, Kepulauan Seribu, DKI Jakarta*. Skripsi (IPB-Fakultas Perikanan dan Ilmu Kelautan).

[24] Richter C *et al* 2001. *Nature* **413** 726-730.

Amino Acid Profile and Volatile Flavour Compounds of Raw and Steamed Patin Catfish (*Pangasius hypophthalmus*) and Narrow-barred Spanish Mackerel (*Scomberomorus commerson*)

Rusky I Pratama[1], I Rostini[1] and E Rochima[1]

[1] Fisheries and Marine Science Faculty, Padjadjaran University, West Java, Indonesia

E-mail: rusky@unpad.ac.id

Abstract. Fish species and processing methods could affect the volatile flavour composition and amino acid profile of fishery commodity. The objectives of this study were to identify volatile components and amino acid profile of two considered predominant fish species in Indonesia which are freshwater Patin catfish (*Pangasius hypophthalmus*) and marine water fish, Spanish mackerel (*Scomberomorus commerson*). The methods used in this study were to detect volatile compounds using Gas Chromatography/Mass Spectrometry (GC/MS) on fresh and steamed of both species samples (100°C for 30 minutes) and amino acid profile were also analyzed using High Performance Liquid Chromatography (HPLC). The volatile components analysis successfully detects as much as 29 and 59 volatiles compounds in fresh and steamed Patin catfish respectively, while 37 and 102 compounds were detected in fresh and steamed Spanish mackerel samples. Most of detected components derives from hydrocarbons, aldehydes, alcohols and ketone groups which could affected by their chemical composition and resulted from various thermal involved reaction. The amino acids profile identification results showed that glutamic acid was found higher compared to other amino acids standards in both samples. Glutamic acid is non-essential amino acid which is important in umami taste substances.

1. Introduction

With its continuously growing population, Indonesia in the near future will face an increasing foodstuff demand, especially in fishery commodities, considering Indonesia has a vast coastal and ocean area. West Java province has the densest population in Indonesia, thereby developing freshwater and marine water fishery sectors as a source of food will become more important for this province. From 2003 to 2010 fish aquaculture in West Java were increasing from production volume of 230,523 to 622,961 ton [1]. Patin catfish (*Pangasius hypophthalmus*) is one of the strategic targets for aquaculture freshwater fisheries development in Indonesia, aside from dumbo catfish (*Clarias gariepinus*), nile tilapia (*Oreochromis niloticus*), common carp (*Cyprinus carpio*) and gourami (*Osphronemus goramy*) due to its high value as an export commodities and it is a commonly raised fish species in freshwater aquaculture. Patin catfish production has shown a significant growth in number. Its national production was 403,133 tons in 2014 with average increase as much as 39.90% [2]. On the other side, sea catch production in West Java has reached 180,402.14 tons in 2010. Sea catch commodities that were landed in West Java, generally consists of several groups of fish such as pelagic and demersal fish and non-fish (crustacean and molluscs). Economically important fish in pelagic group are dominated by several fish species such as mackerel tuna, chub mackerel, narrow-barred Spanish mackerel, anchovy and big tunas. Spanish mackerel specifically had increased production as much as 17% in 2010, with a production value in West Java alone was 5,071 tons [1].

Fisheries commodities have long been recognized as a valuable nutritional source such as protein and lipids. Each fishery commodity would have differences in their chemical composition and flavour compounds, specifically volatile components, whether if it is still in its fresh or in already processed condition before it is consumed. Information about the chemical composition of a commodity is important to identify the nutritive values, chemical changes and its relation to certain handling, processing or storage method. Such information can also beneficial in order to study and evaluate the potential of a commodity.

Volatile components are groups of compounds that contribute to product flavours specifically aroma. As we all know flavour is an important factor which could affect product acceptance and preference by consumers. Flavour is formed as a result from a combination of experiences and sensation which we perceived on product characteristics [3]. Flavour is generally divided into two categories, one is volatile flavour which contributes to aromas and non-volatile flavour which contributes to taste characteristics.

Many researches have confirmed about the volatile components differences among fisheries products, several of them were [4], [5], [6], [7], [8], [9], [10], [11]. In Indonesia, similar research data concerning flavour composition of fisheries commodities were not easily available and rarely found. Different matters occurred in other countries such as Japan, China, and Scandinavian countries where flavour research, including its identification has been done for more than a decade ago. Many researches object in identifying volatile compounds that were found in Indonesia came from agricultural commodities, but for fishery commodities it is still scarce.

In addition to identifying volatile components on fresh samples, this research also identified volatile components in samples that had already been steamed. Steaming is one of the thermal processing methods which used saturated steam as its heating medium. Steaming could induce physical changes and chemical reactions which in turn could affect product characteristics such as flavour and texture [12]. Steaming process was chosen as processing technique applied due to its minimum impact in sample's nutritional value and it is one of the commonly used household cooking method and traditional processing method in West Java cuisines for making "pepes" or traditional steamed fish with spices. According to [13], steaming process has an advantage compared to other heating methods, because it has smaller risks of losing heat sensitive vitamins and other food compounds. The flavour of processed fish and fishery products differ with the processing technology employed. Such differences exist, although the products prepared from the same species of fish. Therefore the study of flavour components of processed fish and fresh fishery commodities is required as an attempt to specify and identify flavour components of such processed fish [14]

The amino acid profile analysis could provide general information regarding essential and non-essential amino acid composition and also beneficial to indicate which one of the overall amino acids that could affect samples taste characteristics. Heating could cause chemical changes in the amino acid residues, which then later could modify the structural, digestible and functional properties of proteins depending on the applied thermal treatment and processing conditions. According to [15] amino acids and peptides contribute directly to seafood flavour. Non-volatile flavour compounds are usually originated from free amino acids, various peptides, nucleotides such as IMP (disodium 5'-inosine monophosphate), GMP (disodium 5'-guanosine monophosphate), and AMP (disodium 5'-adenosine monophosphate) [10], [16].

The volatile components and amino acids composition results from this research could provide important basic information for more advance flavour research and applications. Thus, the results are also expected to contribute in filling the literature gaps of flavour profile originate from Indonesian fisheries commodities. Therefore an attempt on identification analysis of volatile flavour composition and amino acids of two popular fishery samples is important to carry out. The objective of this research is to identify flavour composition, particularly volatile compounds and amino acid profile of fresh and steamed Patin catfish and Spanish mackerels samples. To our best knowledge, there was still no comprehensive study found in Indonesia that has been conducted in investigating the volatile components of fresh and steamed local Patin catfish and Spanish mackerel.

2. Research Method

Research locations for sampling and analyses were consisting of several areas in West Java. Patin catfish samples were taken from floating net cage complex in Cirata Reservoir, Purwakarta, West Java and Spanish mackerel samples were taken from the fish landing site, Karangsong, Indramayu West Java. Sample preparation was carried out in the Fisheries Product Processing Laboratory, Fisheries and Marine Sciences Faculty, Padjadjaran University. Proximate analyses were carried out in the Inter-University Centre Laboratory, Bogor Agriculture Institute. Amino acid profile analyses were carried out in the Integrated Laboratory, Bogor Agriculture Institute and volatile compound analysis were carried out in Flavour Laboratory, Indonesian Centre for Rice Research, Sukamandi, Subang.

2.1. Samples preparation

After sampling had finished on each location, all samples were then promptly transported in a cool box which contained layers of bulk ice and depart to preparation laboratory. On arrival, samples were then cleaned (washing, descaling, beheading), and weighed before divided into two groups (fresh and steamed). The fish that are intended for fresh groups were filleted before packaging, and the fish that are intended for steamed groups were eviscerated first before steaming and packaging. The steaming process was carried out at (minimum) 100°C for 30 minutes [10], [17]. After all preparation and treatment were completed, a portion of fish meat (white meat portion) samples were selected and weighed adequately for analyses purpose.

All weighed portions of the samples were then packaged into three different packaging layers. The first primary packaging was aluminium foil, the secondary packaging was cling wrap plastics and the last layer tertiary packaging is a zip-lock plastic bags with labels. The purpose of three layered packaging is no other to minimize the changes and degradation to the samples that could be caused by various environmental factors such as air, light and temperatures [18]. Samples that have been finished through packaging steps were then placed in cool box which contain sufficient amount of ice (packed inside plastics) to keep the samples in cool temperature and then promptly transported to each analysis laboratory.

2.2. Proximate analysis

Each sample was then analyzed for their moisture, ash, protein and lipid content. The proximate analysis for all fresh and steamed Patin catfish and Spanish mackerel samples were determined according to [19] standard. Moisture was calculated gravimetrically after complete drying of samples in an oven at 110°C, and total inorganic content (ash %) through combustion of organic matter in a muffle furnace for 24 hours at 450°C. Total protein content was determined by the Kjeldahl method and calculated as % nitrogen x 6.25. Total lipid content was determined by the Soxhlet system by drying samples in an oven (105°C) and refluxed for 8 hours using 150 ml chloroform inside Soxhlet tube and the results expressed in %.

2.3. Amino acids profile

Amino acids profiles of fresh and steamed Patin catfish and Spanish mackerel were carried out based on modification from [20], [21] using High Performance Liquid Chromatography (HPLC) (Shimadzu CBM-20A, Shimadzu Corporation, Japan). HPLC parameters setting used, i.e.: Ultra Techsphere column, 1mL/minute mobile phase flow rate with a fluorescence detector.

2.4. Volatiles compound analysis

Volatiles component from fresh and steamed of both fish were analyzed by procedures according to modification from [22] procedure. The analyses were carried out using waterbath for samples extraction and Gas Chromatography (GC) (Agilent Technologies 7890A GC System) and Mass Spectrometry (MS) apparatus (Agilent Technologies 5975C Inert XL EI CI/MSD) for detecting and identifying the volatile components. Samples extraction method was done by Headspace Solid Phase Micro Extraction (HS/SPME) using DVB/Carboxen/Poly Dimethyl Siloxane fiber. Sample's extraction time used on waterbath was 35°C for fresh samples and 70°C for steamed samples for 45 minutes. Raising temperature increased sensitivity while allowing extraction of more compounds, notably the least volatile of which were not extracted at all at lower temperatures [23]. GC column used was HP-5MS (30 m x 250 μm x 0,25 μm), helium carrier gas, initial temperature was 45°C (hold 2 minutes), temperature's escalation as much as 6°C/minutes, the final device temperature was 250°C (hold 5 minutes) with an overall running time 41,17 minutes.

2.5. Data analysis

The obtained results from proximate analysis samples were calculated and showed as mean value and its deviation standards and then were discussed descriptively. The amino acid profile results were identified using 15 standards of amino acids and quantified in μmol concentration unit based on their peak areas and amino acids standards peak areas.

Samples volatile components mass spectrums that were detected from GC/MS were then compared with the mass spectrum pattern which were available in computer database or NIST (National Institute of Standard and Technology) library 0.5a version. The data then were further analyzed with Automatic Mass Spectral Deconvolution and Identification System (AMDIS) software [24]. The resulting data from volatile compound analysis were discussed descriptively based on identification and the semi quantification intensity of the compounds detected from the analyzed samples.

3. Results and Discussion

3.1. Proximate analysis

Proximate analysis provides information regarding sample's chemical, nutritional compositions and studies their changes. Moisture content, ash, protein and lipid content of fresh and steamed Patin catfish and Spanish mackerel samples were analyzed and the results are shown in Table 1. The proximate composition of both fish were relatively differ. The determination of moisture content is the most frequent general analysis performed on fishery commodities. The amount of water or moisture in food often determines its nutritive value and taste, and its shelf life stability throughout storage [25]. Moisture content analysis showed that fresh samples of both fish were found to contain relatively higher moisture content than the steamed one. Fresh and steamed Patin catfish have 79.65% and 72.26% moisture content, respectively, and fresh and steamed Spanish mackerel have 74.63% and 68.72% moisture content respectively. Moisture content of fish would give an effect on its textural characteristics and if it is too high, then the fish would have a soft and mushy texture [26]. These differences on moisture contents could be influenced by the type of commodities that were analyzed and processing method that the samples had been through [17], [18]. According to [13], processing technique such as steaming could lead to moisture loss from samples inter-cellular spaces and this could become the reason for the lower measurement of steamed samples moisture content.

The analysis result also showed slight differences in steamed samples ash content of both fish species if compared to fresh one. Fresh and steamed Patin catfish have 1.30% and 1.87% ash content, respectively, and fresh and steamed Spanish mackerel have 2.23% and 2.11% ash content, respectively. Ash content measurement was mostly dependent on the mineral contents of each sample, feed, growth phase, seasons and habitat environment. Ash content represents the amount of total mineral of samples measured and inorganic substances that present in samples which were the residue from high temperatures burning of organic samples [17], [18]. According to [27], minerals that had been added to snake skin gourami and short-bodied mackerel when it was processed by salting and sun-drying could result in a much higher ash content ranged from 4.4 to 8.7 g/100 g. This could happen depending of the type of products, amount of minerals added and the salting method used.

Table 1. Proximate analysis of Patin catfish and Spanish mackerel (%)

Parameters	Patin catfish		Spanish Mackerel	
	Fresh	Steamed	Fresh	Steamed
Moisture	79.65±0,26	72.26±0.08	74.63±0.15	68.72±0.14
Ash	1.30±0.02	1.87±0.07	2.23±0.04	2.11±0.01
Lipids	1.01±0.02	1.75±0.04	0.17±0.04	0.24±0.06
Protein	14.47±0.16	20.22±0.05	20.79±0.19	27.98±0.15

*data expressed in mean and its standard deviation, $n=3$

From analysis result, we could see that steamed samples of both fish species have a slightly higher amount of total lipids compared to the fresh one. Fresh and steamed Patin catfish had 1.01% and 1.75% total lipid content, respectively, and fresh and steamed Spanish mackerel had 0.17% and 0.24% lipid content respectively. Similar findings on Patin catfish lipid content from [26] research which showed the lipid content range of 0.89-1.23%. These results were categorized as quite low and could be caused by filleting, which did not include the Patin catfish belly parts. The belly parts of Patin are known to contain high lipid content. These results were also consistent with [27] studies on the chemical composition of several Thai freshwater and marine water fish which showed that most of the marine fish measured had a lower lipid content compared to freshwater fish. The lipid content is found to be influenced by season and geographic location, with lower lipid content in fish from tropical waters. The low lipid content in marine fish is also attributed to environmental factors such as food supply and dietary sources. Water and fat losses during heat treatment could also contribute to the differences of lipid content measurement between fresh and steamed samples. Lipid content in fish flesh directly affect odor and flavour intensity [27], [28], [29].

Protein content analysis showed that steamed samples of both fish species had a higher amount of total protein compared to the fresh samples. Both samples were found to be rich sources of protein. Fresh and steamed Patin catfish had 14.47% and 20.22% protein content, respectively, while fresh and steamed Spanish mackerel had 20.79% and 27.98% protein content, respectively. Protein content in each sample would be affected by fish habitat, seasons, storage time and condition and also processing methods. Moisture content in the samples would also have a major effect on protein content measured in samples. Lower moisture content in the samples will result in higher protein content measured compared to fresh samples [28], [30]. Similar results were experienced on [27], protein content of raw short-bodied mackerel contains 21.1 g/100 g protein, while the steamed and fried one contained higher protein levels of 27.5 g/100 g.

3.2. *Amino acids profile*

The importance of amino acids in fish has been well established from the perspective of nutrition and fish flavours [31]. Amino acid composition will determine the quality of a protein which is among the most important macronutrients in human diet. The amino acid profile analysis could provide us valuable information regarding essential and non-essential amino acid composition which contained in analyzed samples. In addition to that, we could also obtain information regarding amino acids contribution in general to sample's taste attributes. This research used 15 amino acids standards to quantify individual amino acids present in samples. These standards were aspartic acid, glutamic acid, serine, histidine, glycine, threonine, arginine, alanine, tyrosine, methionine, valine, phenylalanine, leucine, lysine and isoleucine. The fresh and steamed Patin catfish and Spanish mackerel were found to be relatively rich in amino acids. Their amino acid compositions are shown in Table 2 and overall the amino acid composition between fresh and steamed samples display slight variation in their values.

Table 2. Amino acids profile of Patin catfish and Spanish mackerel (%)

Amino acids	Patin catfish		Spanish Mackerel	
	Fresh	Steamed	Fresh	Steamed
Aspartic acid	1.98	3.41	2.23	2.88
Glutamic acid	3.28	5.50	3.61	4.63
Serine	0.79	1.31	0.88	1.15
Histidine*	0.45	0.86	1.12	1.26
Glycine	0.81	1.48	1.09	1.48
Threonine*	0.82	1.42	1.05	1.35
Arginine	1.32	2.27	1.46	1.95
Alanine	1,15	1.85	1.37	1.81
Tyrosine	0.69	1.14	0.87	1.04
Methionine*	0.61	0.84	0.74	0.82
Valine*	1.06	1.93	1.30	1.69
Phenylalanine*	0.86	1.46	1.02	1.33
Isoleucine*	1.06	1.84	1.21	1.58
Leucine*	1.66	2.77	1.86	2.42
Lysine*	1.72	3.20	2.37	3.01

*) essential amino acids

Essential amino acids that Patin catfish had from highest to lowest value were lysine, leucine, isoleucine, valine, phenylalanine, threonine, methionine and histidine. There are other types of essentials amino acid, which did not include in the analysis due to its standard was not available such as tryptophan. As the highest amount of essential amino acids contained in the samples, lysine plays an important role in the human body because it is needed as basic composition of blood antibody, strengthen the circulation and maintaining normal cell growth. Together with proline and vitamin C, lysine will decrease excessive blood triglycerides. Methionine is important for lipid metabolism, maintaining liver health, prevent lipid accumulation in the liver and the main artery, preventing allergy and osteoporosis [26]. Leucine is an important molecule that could stimulate the synthesis of muscle proteins and also has a therapeutic role in stress like trauma, burns, etc. Histidine is capable to perform multiple roles in human beings such as protein-protein interaction, precursor of histamine, an important neurotransmitter and also needed for the growth and repair tissues [32]. Deficiency in certain amino acids may hinder healing recovery process [33].

It can be seen from Table 2 that the steamed samples of both species were observed to possess a relatively higher level of amino acids than the fresh samples of both fish. Similar results on amino acid content were reported by [15] with drying treatment of fresh and dried squid samples. The time length of heat treatment and heating method also contributed to the change in amino acid contents of fish samples [33]. Proteolytic reaction that occurred during the heating process could cause free amino acids forming to increase [10], [34].

The highest amount of amino acids detected in fresh and steamed Patin catfish was glutamic acid (3.28%; 5.50%, respectively), whereas for fresh and steamed Spanish mackerels was also glutamic acid (3.61%; 4.63% respectively). The less glutamic acid contained in fish meat it would result in less savory taste of the fish meat [26]. Others amino acids that had a considerable amount in fresh and steamed Patin catfish samples respectively were aspartic acids (1.98%; 3.41%), lysine (1.72%; 3.20%), leucine (1.66%; 2.77%), arginine (1.32%; 2.27%), valine (1.06%; 1,93%), alanine (1.15%;1,85%) and isoleucine (1.06%; 1,84%), whereas for fresh and steamed Spanish mackerel samples respectively were lysine (2.37%; 3.01%), aspartic acid (2.23%; 2.88%), leucine (1.86%; 2.42%) and arginine (1.46%; 1,95%) Leucine, valine, isoleucine and lysine are categorized as essential amino acids on the account of that human body could not produce essential amino acids by itself and have to be obtained from various external food sources.

Each of the amino acids is well known to contribute to the basic taste of a product. Proline mainly contributes to the bitter taste of peptides and the presence of glycine, alanine, valine, leucine, tyrosine and phenylalanine in peptides also impart bitterness. Glutamate imparts umami taste if the concentration in the food product is above the taste threshold and it is likely that umami taste is a signal for protein nutrition [35], [36], [37], [38]. Arginine at sub threshold concentrations significantly enhanced salty taste [37] and give an effect to umami taste in sea urchin [36] at a large amounts in crab and scallops it enrich the sweet taste with complexity and fullness and provide a seafood flavour [35]. Glycine and alanine are taste active components and well known to impart sweetness characteristic in various seafood. Acidic L-amino acids such as glutamate and aspartate impart to umami taste and most of the D-amino acids are dominantly sweet [35], [36]. Aromatic amino acids, basic amino acids and branched amino acids are bitter amino acids. Valine, leucine and histidine are known to impart bitter taste but are not as bitter as phenylalanine [38]. Histidine was known to impart sourness and umami in katsuobushi [40]. According to [36], the amino acids that are present in a product played an important role in the taste of most seafood.

3.3. Volatile flavour components

Volatile compound analysis showed that steamed Patin catfish and Spanish mackerel samples has higher quantities of volatile compounds compared to the compounds that were identified from fresh samples and wide variety of compounds were also observed from both fresh and steamed Patin catfish samples. Volatile compound analysis results from fresh and steamed Patin catfish samples successfully detected and identified as much as 29 and 59 volatile compounds, respectively, which then were categorized into several major groups such as hydrocarbon, aldehydes, alcohols, ketones and others.

Aliphatic, cyclic and aromatic hydrocarbons in fresh and steamed Patin catfish samples (10 and 24 compounds, respectively) were the highest in quantity with heptadecane as the most abundant compounds present in fresh and steamed samples (29.923%; 35.661% respectively). In addition to hydrocarbons, respectively in the fresh and steamed Patin catfish samples, GC/MS were also detecting aldehydes group (7 and 18 compounds) with hexanal (10.130%; 4.963%) which had the highest proportions on both samples, alcohols (7 and 12 compounds) with 1-octanol (11.848%) and 1-octen-3-ol (4.604%) had the highest proportions, ketones (3 and 2 compounds) with 6-methyl-3-heptanone (6.086%) and 2,3-octanedione (1.114%) had the highest proportions in fresh and steamed samples respectively.

In the steamed samples, one ester compound was also detected, Sulfurous acid, dodecyl pentyl ester. Presumably, ester group compound that was found in fish samples, is derived from acids and alcohols esterification which previously formed from lipid metabolism. Esters could derive from lipids thermal degradation products [4], [22]. The volatile component analysis was also detected several nitrogenous group compounds which usually not detected in fresh and steamed fishery samples and need more rigorous identification and 2-pentyfuran which is also previously detected in steamed silver carp and smoked black bream [10], [22]. Furan compounds are heterocyclic and usually derive from glucose dehydration (cellulose thermal degradation), but several of them could also derive from Maillard reaction [4], [41]. The volatile compounds analysis results are shown in Table 3 and Table 4 with their proportions (based on area percentage) sorted from highest to lowest abundance.

Table 3. Volatile compounds in fresh Patin catfish samples.

RT	Groups	Area	Proportion (%)
	Hydrocarbons (aliphatic, cyclic, aromatics)		
21.871	Heptadecane	5251755	29.923
18.0407	Pentadecane	2433484	13.865
5.07	Nonane, 3-methyl-	1618664	9.223

11.056	Naphthalene	371604	2.117
5.595	3-Octene, 2,6-dimethyl-	355593	2.026
20.0012	Hexadecane	346667	1.975
6.9301	Cyclohexene, 1-methyl-4-(1-methylethenyl)-, (S)-	186190	1.061
15.9745	Undecane	162764	0.927
5.2836	Naphthalene, decahydro-	54789	0.312
7.8156	1,3,6-Heptatriene, 5-methyl-	47749	0.272
	Aldehydes		
2.5368	Hexanal	1777914	10.130
6.3149	2,6-Nonadienal, (E,Z)-	360119	2.052
9.2135	Nonanal	331775	1.890
5.423	2-Nonenal, (E)-	144560	0.824
5.3456	2,4-Hexadienal, (E,E)-	118308	0.674
6.3681	Dodecanal	114574	0.653
4.3787	Heptanal	5951	0.034
	Alcohols		
5.8663	1-Octanol	2079349	11.848
4.7512	1-Nonanol	34623	0.197
7.5068	1-Hexanol, 2-ethyl-	16602	0.095
5.3828	1-Heptanol	14889	0.085
4.724	1-Hexanol	9910	0.056
11.6639	2-Hexen-1-ol, (E)-	5996	0.034
19.291	1-Penten-3-ol	1810	0.010
	Ketones		
6.3362	3-Heptanone, 6-methyl-	1068060	6.086
6.3315	2,3-Octanedione	588120	3.351
13.7059	2-Heptanone	15129	0.086
	Others		
23.665	3-Methyl-5-hydroxy-isoxazole	32310	0.184
13.6888	Methylamine, N,N-dimethyl-	1399	0.008

Table 4. Volatile compounds in steamed Patin catfish samples.

RT	Groups	Area	Proportion (%)
	Hydrocarbons (aliphatic, cyclic, aromatics)		
21.841	Heptadecane	85988447	35.661
7.8159	1,3,6-Heptatriene, 5-methyl-	37649504	15.614
18.0174	Pentadecane	37455076	15.533
5.3188	Naphthalene, decahydro-	8242257	3.418
21.426	8-Heptadecene	3944619	1.636
9.822	2,4,6-Octatriene, 2,6-dimethyl-	3535688	1.466
19.9655	Hexadecane	2838388	1.177
10.099	2,4,6-Octatriene, 2,6-dimethyl-, (E,Z)-	2728161	1.131
6.917	2-Methyl-2-bornene	2680450	1.112
11.032	Naphthalene	1546522	0.641
15.9424	Tetradecane	1109115	0.460
11.4618	Dodecane	1043180	0.433
13.7605	Tridecane	930130	0.386
21.947	Hexadecane	813052	0.337
11.982	1H-Indene, 1-ethylideneoctahydro-7a-methyl-, (1E,3aα,7aβ)-	771064	0.320

23.5996 Nonadecane	493220	0,205
7.2834 Cyclohexene, 1-methyl-4-(1-methylethenyl)-, (S)-	489752	0,203
18.488 trans-Calamenene	474284	0,197
5.89 Mesitylene	330758	0,137
21.7028 1-Nonadecene	323625	0,134
8.051 3-Carene	245853	0,102
12.177 Bicyclo[5.1.0]octane, 8-(1-methylethylidene)-	172844	0,072
2.1304 Toluene	154794	0,064
31.3683 Undecane	10486	0,004
Aldehydes		
2.5643 Hexanal	11966421	4.963
9.1949 Nonanal	9702159	4.024
4.3725 Heptanal	2573292	1.067
8.4169 2-Undecenal	1534136	0.636
6.7108 Octanal	1491872	0.619
11.6324 Decanal	383382	0.159
23.9124 Hexadecanal	340726	0.141
1.4622 Pentanal	299246	0.124
22.124 Dodecanal	279610	0.116
4.3371 4-Heptenal	235256	0.098
13.1395 2,4-Heptadienal, (E,E)-	178670	0.074
20.2311 Dodecanal	139235	0.058
9.431 Benzaldehyde, 4-ethyl-	82835	0.034
17.6024 2-Nonenal, (E)-	68753	0.029
1.0914 Butanal	63014	0.026
6.1482 2,6-Nonadienal, (E,Z)-	32278	0.013
11.3632 2-Pentenal, (E)-	24368	0.010
14.7346 2-Heptenal, (E)-	6441	0.003
Alcohols		
6.2627 1-Octen-3-ol	11100825	4.604
6.091 1-Heptanol	809845	0.336
6.3094 2-Octen-1-ol	649970	0.270
6.065 1-Hexanol	273597	0.113
2.2007 1-Pentanol	130539	0.054
8.547 2,6-Nonadien-1-ol	128579	0.053
2.2237 2-Penten-1-ol, (E)-	66845	0.028
7.454 1-Penten-3-ol	58411	0.024
2.2538 2-Penten-1-ol, (Z)-	29513	0.012
13.4453 2-Hexen-1-ol, (E)-	19178	0.008
19.8309 1-Nonanol	12390	0.005
25.2047 1-Octanol	10931	0.005
Ketones		
6.3141 2,3-Octanedione	2686472	1.114
13.6596 2-Heptanone	10661	0.004
Esters		
17,236 Sulfurous acid, dodecyl pentyl ester	255235	0,106
Others		
13.836 Indole	1505303	0.624
14.1732 Furan, 2-pentyl-	6295	0.003

Volatile compound analysis results from fresh and steamed Spanish mackerel samples successfully detected as much as 37 and 102 volatile compounds, respectively, which then were categorized into several major groups such as hydrocarbon, aldehydes, alcohols, ketones and others. Aliphatic, cyclic and aromatic hydrocarbons in fresh and steamed Spanish mackerel samples (12 and 42 compounds) were the highest in quantity with pentadecane as the most abundant compounds present in fresh and steamed samples (38.388%; 18.035% respectively). In addition to hydrocarbons, respectively in the fresh and steamed Spanish mackerel samples, GC/MS were also detecting aldehydes group (10 and 28 compounds) with hexanal (13.215%) and octanal (8.851%) which had the highest proportions, alcohols (9 and 16 compounds) with (Z)-2-Penten-1-ol (5.134%) and 1-octen-3-ol (4.315%) had the highest proportions, ketones (5 and 8 compounds) with 2,3-octanedione (6.103%) and 2-decanone (3.684%) had the highest proportions in fresh and steamed samples respectively. In the steamed Spanish mackerel samples, two ester compounds was also detected with (Z)-3-Hexenoic acid, methyl ester (0,149%) had the highest proportion. The volatile component analysis was also detected 6 compounds from various nitrogenous and furan groups compounds. The volatile compounds of fresh and steamed samples analyses results are shown in Table 5 and Table 6 with their proportions sorted from highest to lowest abundance.

Table 5. Volatile compounds in fresh Spanish mackerel samples.

RT	Groups	Area	Proportion (%)
	Hydrocarbons (aliphatic, cyclic, aromatics)		
18.0164	Pentadecane	12194171	38.388
21.936	Hexadecane, 2,6,11,15-tetramethyl-	2601379	8.189
21.823	Heptadecane	1337526	4.211
11.038	Naphthalene	1099043	3.460
15.9526	Tetradecane	430556	1.355
19.9739	Hexadecane	351285	1.106
7.2736	Limonene	299447	0.943
13.7684	Tridecane	257358	0.810
11.4749	Undecane	110789	0.349
17.237	Heptane, 2,6-dimethyl-	83989	0.264
2.1406	Toluene	24093	0.076
7.3125	Cyclohexene, 1-methyl-4-(1-methylethenyl)-, (S)-	11770	0.037
	Aldehydes		
2.5391	Hexanal	4197730	13.215
4.3573	Heptanal	654198	2.059
9.1975	Nonanal	344298	1.084
6.0988	2-Octenal, (E)-	317901	1.001
13.5813	2-Nonenal, (E)-	135385	0.426
13.5818	Butanal	41607	0.131
11.6385	Decanal	19337	0.061
13.9591	Dodecanal	10359	0.033
14.7554	Octanal	2963	0.009
7.1832	Benzaldehyde, 4-ethyl-	1834	0.006
	Alcohols		
6.2511	2-Penten-1-ol, (Z)-	1630891	5.134
6.2848	2-Octen-1-ol	1012469	3.187
6.2529	1-Octen-3-ol	844424	2.658
6.0846	1-Heptanol	672161	2.116
1.3862	1-Penten-3-ol	186560	0.587
2.1843	1-Pentanol	96996	0.305
2.4966	2-Hexen-1-ol, (E)-	77693	0.245
3.8367	1-Hexanol	49894	0.157

7.4772	1-Hexanol, 2-ethyl-	22449	0.071
	Ketones		
6.3249	2,3-Octanedione	1938801	6.103
13.6716	2-Decanone	421001	1.325
1.4393	2,3-Pentanedione	212674	0.670
9.0534	3-Heptanone, 6-methyl-	30247	0.095
8.9235	2-Heptanone	10263	0.032
	Others		
6.4394	Furan, 2-pentyl-	32385	0.102

Table 6. Volatile compounds in steamed Spanish mackerel samples.

RT	Groups	Area	Proportion (%)
	Hydrocarbons (aliphatic, cyclic, aromatics)		
18.082	Pentadecane	156052623	18.035
21.971	Pentadecane, 2,6,10,14-tetramethyl-	66552586	7.691
21.853	Heptadecane	34976984	4.042
12.006	Cyclohexene, 3-ethenyl-	12372341	1.430
13.7761	Tridecane	10019968	1.158
11.056	Azulene	8313380	0.961
19.9758	Hexadecane	8129937	0.940
15.9515	Tetradecane	7443583	0.860
11.4768	Dodecane	7109119	0.822
8.5505	1,3,6-Heptatriene, 5-methyl-	6705974	0.775
13.163	1,4-Octadiene	5434787	0.628
5.418	Cyclobutane, 1,2-diethenyl-	4767538	0.551
13.393	1,3-Cyclooctadiene, (Z,Z)-	4205079	0.486
9.0718	Undecane	3731451	0.431
8.653	3-Octyne	2869724	0.332
17.237	Heptadecane, 2,6,10,14-tetramethyl-	2453175	0.284
17.609	2-Tridecene, (E)-	2418128	0.279
12.791	1,2-Heptadiene	2287224	0.264
9.615	Bicyclo[4.1.1]oct-2-ene	2207940	0.255
12.662	Oxirane, 3-butenyl-	2113021	0.244
9.863	Cyclopentane, 1-ethyl-3-methyl-, cis-	1885844	0.218
17.868	1-Pentadecene	1879303	0.217
15.755	Z-1,9-Hexadecadiene	1765776	0.204
7.231	β-Cymene	1708841	0.197
9.45	1,3-Cyclooctadiene	1690275	0.195
12.154	Cyclopentane, 2-ethyl-1,1-dimethyl-	1608641	0.186
7.555	2,6-Octadiene, (E,E)-	1373570	0.159
10.837	Dispiro[2.1.2.1]octane	1301766	0.150
7.312	Limonene	1188173	0.137
15.235	Bicyclo[10.1.0]tridec-1-ene	958734	0.111
8.754	Cyclooctene, 3-ethenyl-	890796	0.103
21.44	3-Heptadecene, (Z)-	675850	0.078
25.2988	Nonadecane	487865	0.056
19.834	Z-8-Hexadecene	474294	0.055
7.372	4-Decyne	348494	0.040
23.49	1-Nonadecene	282882	0.033
13.558	6-Tridecene, (Z)-	275222	0.032
14.551	Benzene, 1-methoxy-4-pentyl-	152838	0.018
19.5383	Naphthalene, decahydro-	99718	0.012
17.75	Cyclopentadecane	90797	0.010

2.1726	Toluene	25141	0.003
17.461	Octahydro-1-oxo-cyclopropa[c]indene	8438	0.001
	Aldehydes		
6.7872	Octanal	76583436	8.851
9.2394	Nonanal	75023710	8.670
2.6301	Hexanal	48871819	5.648
4.433	Heptanal	32256893	3.728
11.6468	Decanal	11324770	1.309
13.966	Undecanal	8290141	0.958
10.6137	Benzaldehyde, 4-ethyl-	6819200	0.788
20.245	Tetradecanal	6748297	0.780
16.1723	Dodecanal	6451015	0.746
16.7473	2-Undecenal	6083707	0.703
11.369	cis-4-Decenal	5562383	0.643
23.9233	Hexadecanal	5415633	0.626
18.2703	Tridecanal	5042557	0.583
1.5929	2,4-Hexadienal, (E,E)-	4235587	0.489
22.142	13-Methyltetradecanal	4121475	0.476
6.749	Benzaldehyde	2630618	0.304
14.745	7-Octenal, 3,7-dimethyl-	2043324	0.236
8.0924	2-Octenal, (E)-	1616175	0.187
21.717	E-14-Hexadecenal	1401294	0.162
10.5641	2-Nonenal, (E)-	1040876	0.120
1.2641	Pentanal	892401	0.103
10.3965	2,6-Nonadienal, (E,Z)-	660800	0.076
10.147	Benzaldehyde, 2,4-dimethyl-	336498	0.039
4.5646	Methional	247130	0.029
14.1905	2,4-Heptadienal, (E,E)-	234360	0.027
1.93	2-Pentenal, (E)-	37559	0.004
1.0876	Butanal	33496	0.004
24.6937	2-Heptenal, (E)-	3009	0.0003
	Alcohols		
6.3622	1-Octen-3-ol	37340341	4.315
8.4543	2-Octen-1-ol	14463632	1.672
6.1343	1-Heptanol	9216561	1.065
8.5287	1-Octanol	8040781	0.929
12.549	Z-10-Pentadecen-1-ol	4565570	0.528
13.104	2-Nonyn-1-ol	3057375	0.353
10.9319	1-Nonanol	2702628	0.312
7.809	2,4-Decadien-1-ol	1028914	0.119
15.389	5-Octen-2-yn-4-ol	941351	0.109
13.459	1-Pentanol, 5-(methylenecyclopropyl)-	807643	0.093
25.649	E-2-Octadecadecen-1-ol	614574	0.071
2.283	1-Pentanol	363468	0.042
3.8887	1-Hexanol	288938	0.033
12.927	10-Undecyn-1-ol	236514	0.027
2.319	2-Penten-1-ol, (Z)-	84243	0.010
20.8861	1-Penten-3-ol	7940	0.001
	Ketones		
13.6935	2-Decanone	31873408	3.684
6.3757	3-Heptanone, 6-methyl-	20194967	2.334
8.9342	2-Heptanone	5054230	0.584
16.091	2-Heptanone, 6-methyl-	3028435	0.350

5.637	7-Octen-2-one	1963771	0.227
4.226	2-Hexanone, 5-methyl-	1432620	0.166
18.423	4,5-Heptadien-2-one, 3,3,6-trimethyl-	1371320	0.158
8.995	3,5-Octadien-2-one	371926	0.043
	Esters		
11,192	3-Hexenoic acid, methyl ester, (Z)-	1290392	0.149
17,343	2-Thiophenecarboxylic acid, 2-methylphenyl ester	443560	0.051
	Others		
6,4737	Furan, 2-pentyl-	17717986	2.048
1,6006	Furan, 2-ethyl-	7366048	0.851
13,293	1,3-Benzodioxol-5-amine	3709783	0.429
13,889	Indole	1125118	0.130
11,752	1,3-Benzodioxol-5-amine	1014137	0.117
23,6524	Methylamine, N,N-dimethyl-	32026	0.004

In general, fresh fish are characterized by sweet, mild, green, plant-like, metallic and fishy aromas and volatile compounds contributing to these aromas are generated mainly by oxidative enzymatic reactions and auto oxidation of lipids [6]. According to [26], aroma description range of fresh Patin catfish are fresh, neutral and has species specific aroma as for the steamed one, its aroma description range are a bit fishy, fresh, neutral and species specific aroma. Taste descriptions of steamed Patin catfish are a bit umami, slightly sweet and juicy.

Various volatile compounds that have been detected on both fish species were derived from sample's components, mainly from proteins and lipid content, so that the wide variety of quantities and volatile compounds are related to variation of chemical compounds contained in the samples. Most of those volatile compounds which contribute to commodities aroma were derived from the results of enzymatic reactions, microorganism activities, lipid auto oxidation, resulting substance from various thermal involved reactions and environmental impacts [7]. More varieties of volatile compounds than fresh samples have been detected and identified in processed fishery samples and their types and composition would depend on the sample type, chemical composition and processing methods. Several papers have described these in their research, among them were [14], [15], [17], [18], [22], [27].

In general, it can be assumed that processed fish possessed a higher number of flavour components compared to those of raw fish. One of the reasons is that the further increasing of such components was a result of biochemical pathways of protein and lipid of fish. The other reason may be the concentration of such flavour components was found to be much higher in volatile component analysis due to the reduced moisture content during processing which resulted in a higher concentration of flavour components in the final product. According to [14], flavour components detected in dried horse mackerel were formed probably as a result of oxidation of lipid as well as enzymic hydrolysis of the original components of fish such as lipid and protein. Thermal condition may accelerate retro-aldol degradation of unsaturated aldehydes which lead to altered flavour in these products. Other elements that could affect the numbers and types of compounds discovered in this volatile compound analysis were extraction method, type of samples and GC/MS column and its running parameters [17].

As mentioned above, most of the detected compound group originates from and could be categorized into several groups which were hydrocarbons, aldehydes, alcohols, ketones and their derivatives. Volatile compounds that came from hydrocarbon groups could derive from decarboxylation reaction and the splitting process of fatty acid's carbon chains, a secondary reaction from carotenoid (if present) and unsaturated fatty acids thermal oxidations [4], [5], [10]. Aldehydes group compounds detected could derive from fatty acid carbon's double bonds oxidation whether they were saturated or unsaturated [5], [8], [10], [22], [42], [43]. Alcohols, aldehydes and ketones groups volatiles compounds detected, were also could formed as a result of lipid and fatty acid oxidations and amino acid degradation that occurred during processing [44], [45], [46]. Almost all reaction that could generate or produce volatile compounds would involve saturating and unsaturated fatty acids, which in general were abundantly contained in most of fishery commodities. The sensory differences could be expected among species depending on the lipid content since most of the volatile compounds in fish are derived from the oxidative breakdown of unsaturated fatty acids [29].

Some of the hydrocarbons, aldehydes, alcohols and ketone groups which detected in samples such as limonene, naphthalene, hexadecane, nonadecane, tridecane, dodecane, 1-nonadecene, toluene, nonanal, hexadecanal, pentanal, heptanal, 1-octen-3-ol, 1-nonanol, 1-hexanol, 2-ethyl-1-hexanol, 2-heptanone, 2-decanone were also known being detected in raw, cooked and recooked silver carp [10], wild and cultured sea bream [7] and raw black bream [22]. Ketone compound 2,3-pentanedione were known detected in fresh sardine and has a caramel like flavour characteristic. Aldehydes such as hexanal was detected in all the samples and known to have green-like flavour characteristic, methional that was detected in steamed samples has potato-like flavour characteristics, pentanal has green flavour. (E, E) -2,4-heptadienal detected has cinnamon-like/oxidized oil-like) and (E) -2-pentenal has grass like flavour, they were discovered in both steamed fish samples. (E)-2-octenal has cinnamon like flavour, octanal that was detected in fresh patin and fresh and steamed Spanish mackerel has citrus like flavour characteristics and (E,Z)-2,6)-nonadienal has cucumber like/insect like flavour characteristics. Alcohol compound 1-penten-3-ol detected in all samples is responsible for flavour of fresh marine products and generated from polyunsaturated fatty acid. Toluene is considered as undesirable components because the role in the human body is not known and usually it results from thermal degradation of wood materials [9]. Naphthalene usually originates from plant degradation product by microorganism, phenylalanine pyrolysis [4] or environmental pollutant [5]. Short chain volatile compounds such as (Z)-4-heptenal and E-2-nonenal that have been found in samples were also detected in grey mullet fish and have been related to rancid and bitter taste in fish [29]. Benzaldehyde which detected in steamed Spanish mackerel is known for its almond/fruity/creamy/nutty aroma. Ketones group detected in samples are known to contribute to the sweet aroma of many crustaceans [6].

4. Conclusion

The moisture, ash, protein and lipid content analyses results from fresh and steamed samples of Patin catfish and Spanish mackerel were relatively differ. These measurement results are basically depending on the commodity type, initial chemical composition of the samples, feed, environmental factors and processing technique. Moisture loss during heating process will affect other contents measured. Lower moisture content in samples would result in higher numbers of other contents measurement. Amino acid analysis showed that both species steamed fish samples possess a relatively higher level of amino acids if compared to the fresh one. This increasement could be affected by proteolytic reaction that occurred during steaming and time length of the heat treatment. The highest amount of amino acids detected in fresh and steamed samples of both species is glutamic acid which imparts product's umami taste. Volatile compound analysis showed that steamed Patin catfish and Spanish mackerel samples has higher quantities of volatile compounds compared to the compounds that were identified from fresh samples and most of the detected compound group could be categorized into hydrocarbons, aldehydes, alcohols, ketones and others (esters, nitrogenous compounds, furan). As much as 29 and 59 volatile compounds were detected from fresh and steamed Patin catfish samples, respectively, and 37 and 102 volatile compounds were detected from fresh and steamed Spanish mackerel respectively. It can be assumed from these results that volatile compounds found in marine water fish are higher in quantity compared to freshwater fish. The most abundant compound groups in both samples are aliphatic, cyclic and aromatic hydrocarbons which derive from decarboxylation reaction and the splitting process of fatty acid's carbon chains, a secondary reaction from thermal oxidations of unsaturated fatty acids.

References

[1] Statistical Data and Information Centre 2013 *West Java Province Marine and Fisheries Profile to Support Marine and Fishery Industrialization* (Jakarta: Ministry of Marine and Fisheries) pp 145-221

[2] Statistical Data and Information Centre 2014 *Marine and Fisheries in Figures 2014* (Jakarta: Ministry of Marine and Fisheries) pp 31-44

[3] Burdock A G 2002 *Fenaroli's Handbook of Flavour Ingredients, 4th Edition* (Boca Raton: CRC Press)

[4] Chung H Y, Yung I K S, Ma W C J and Kim J 2002. Analysis of volatile components in frozen and dried scallops (*Patinopecten yesseonensis*) by gas chromatography/mass spectrometry. *Food Research International* **35** 43-53.

[5] Linder M and Ackman R G 2002 Volatile compounds recovered by solid-phase micro extraction from fresh adductor muscle and total lipids of sea scallop (*Placopecten magellanicus*) from Georges Bank (Nova Scotia) *J Food Sci* **67** 2032-37

[6] Morita K, Kubota K and Aishima T 2003 Comparison of aroma characteristics of 16 fish species by sensory evaluation and gas chromatographic analysis. *Journal of the Science of Food and Agriculture* **83** 289-297.

[7] Alasalvar C, Taylor K D A and Shahidi F 2005 Comparison of volatiles of cultured and wild sea bream (*Sparus aurata*) during storage in ice by dynamic headspace analysis/gas chromatography-mass spectrometry *J. Agric. Food Chem* **53** 2616-22

[8] Guillen M D, Errecalde M C, Salmeron J and Casas C 2006 Headspace volatile components of smoked swordfish (*Xiphias gladius*) and cod (*Gadus morhua*) detected by means of solid phase micro extraction and gas chromatography–mass spectrometry *Food Chem* **94** 151-6.

[9] Ganeko N, Shoda M M, Hirohara I, Bhadra A, Ishida T, Matsuda H, Takamura H and Matoba T 2007 Analysis of volatile flavour compounds of sardine (*Sardinops melanostica*) by solid phase micro extraction *Journal of Food Science* **73** 83-8

[10] Liu J K, Zhao S M and Xiong S B 2009 Influence of recooking on volatile and non-volatile compounds found in silver carp *Hypophthalmichthys molitrix Fish Sci* **75**:1067-75.

[11] Miyasaki T, Hamaguchi M and Yokoyama S 2011 Change of volatile compounds in fresh fish meat during ice storage *Journal of Food Science* **76** 1319-25

[12] Lewis M J 2006 Thermal Processing *Food Processing Handbook* ed J G Brennan (Weinheim: Wiley-VCH GmbH & Co. KgaA) chapter 2 pp 33-70

[13] Fellows P 2000 *Food Processing Technology: Principles and Practice* (Cambridge: Woodhead Publ. Ltd.)

[14] Mansur M A, Hossain M I, Takamura H and Matoba T 2002 Flavour components of some processed fish and fishery products of Japan. *Bangladesh J. Fish. Res.* **6** 89-97

[15] Deng Y, Luo Y, Wang Y, and Zhao Y 2014 Effect of different drying methods on the myosin structure, amino acid composition, protein digestibility and volatile profile of squid fillets. *Food Chemistry* **171** 168-176

[16] Chen D W and Zhang M 2006 Non-volatile taste active compounds in the meat of Chinese mitten crab (*Eriocheir sinensis*) *Food Chemistry* **104** 1200-05

[17] Pratama R I, Rostini I and Awaluddin M Y 2013 Flavour composition of fresh and steamed common carp (*Cyprinus carpio*) *Jurnal Akuatika* **4** 55-67

[18] Pratama R I 2011 *Flavour Characteristics of Several Indonesian Smoked Fish* Thesis (Bogor: School of Postgraduate Bogor Agricultural Institute)

[19] Association of Official Analytical Chemist (AOAC) 2005 *Official Methods of Analysis of AOAC International 18th Edition* (Gaithersburg: AOAC International)

[20] Ishida Y, Fujiwara M, Kinoshita T, and Nimura N 1987. *Method of amino acid analysis*. United States Patent 4670403 (http://www.freepatentsonline.com//4670403, accessed: March 8, 2011, 22:15).

[21] Toppe J, Albrektsen S, Hoppe B and Aksnes A 2007 Chemical composition, mineral content

and amino acid and lipid profiles in bones from various fish species *Comparative Biochemistry and Physiology* B **146** 395-401
[22] Guillen M and Errecalde M 2002 Volatile components of raw and smoked black bream (*Brama raii*) and rainbow trout (*Onchorhynchus mykiss*) studied by means of solid phase micro extraction and gas chromatography/mass spectrometry *J Sci Food Agric* **82** 945-52
[23] Duflos G, Moine F, Coin V M and Malle P 2005 Determination of volatile compounds in whiting (*Merlangius merlagus*) using headspace-solid-phase microextraction-gas chromatography-mass spectrometry *Journal of Chromatographic Science* **43** 304-12
[24] Mallard G W and Reed J 1997 *Automatic Mass Spectral Deconvolution and Identification System (AMDIS) User Guide* (Gaithersburg: U.S. Department of Commerce)
[25] Anggo A D, Ma'ruf W F, Swastawati F and Rianingsih, L Changes of amino and fatty acids in anchovy (*Stolephorus* sp) fermented fish paste with different fermentation periods *Procedia Environmental Sciences* **23** 58-63
[26] Suryaningrum D T, Muljanah I, and Tahapari E 2010 Sensory profile and nutrition value of several types of patin catfish and nasutus hybrid *Jurnal Pascapanen dan Bioteknologi Kelautan dan Perikanan* **5** 153-64
[27] Puwastien P, Judprasong K, Kettwan E, Vasanachitt K, Nakngamanong Y and Bhattacharjee L 1999 Proximate composition of raw and cooked Thai freshwater and marine fish *Journal of Food Composition and Analysis* **12** 9-16
[28] Gokoglu N, Yerlikaya P and Cengiz E 2004 Effects of cooking methods on the proximate composition and mineral contents of rainbow trout (*Oncorhynchus mykiss*) *Food Chemistry* **84** 19-22
[29] Lazo O, Guerrero L, Alexi N, Grigorakis K, Claret A, Perez Z A, and Bou R 2017 Sensory characterization, physico-chemical properties and somatic yields of five emerging fish species. http://dx.doi.org/10.1016/j.foodres.2017.07.023
[30] Sebranek J 2009 Basic curing ingredients (*Ingredients in Meat Product Properties, Functionality and Applications*) ed R Tarte (New York: Springer Science) pp 1-24
[31] Antoine F R, Wei C I, Littell R C, Quinn B P, Hogle A D, and Marshall M R 2001 Free amino acids in dark and white muscle fish as determined by O-phthaldialdehyde Precolumn derivatization *Journal of Food Science* **66** 72-7
[32] Vijayan D K, Jayarani R, Singh D K, Chatterjee N S, Mathew S, Mohanty B P, Sankar T V, and Anandan R 2016 Comparative studies on nutrient profiling of two deep sea fish (*Noepinnula orientalis*) and (*Chlorophthalmus corniger*) and brackish water fish (*Scatophagus argus*). *The Journal of Basic & Applied Zoology* **77** 41-8
[33] Oluwaniyi O O, Dosumu O O, and Awolola G V 2010 Effect of local processing methods (boiling, frying and roasting) on the amino acid composition of four marine fishes commonly consumed in Nigeria *Food Chemistry* **123** 1000-06
[34] Toth L and Potthast K 1984 Chemical aspects of the smoking of meat and meat products (*Advances in Food Research*) ed C O Chichester (New York: Academic Press Inc.)
[35] Wongso S, and Yamanaka H 1998 Extractive components of the adductor muscle of Japanese baking scallop and changes during refrigerated storage *Journal of Food Science* **63** 1-5
[36] Kawai M, Uneyama H and Miyano H 2009 Taste-active components in foods, with concentration on umami compounds *Journal of Health Science* **55** 667-73
[37] Zhao C J, Scheber A, and Ganzle M G 2016 Formation of taste-active amino acids, amino acid derivatives and peptides in food fermentations *Food Research International* **89** 39-47
[38] Kato H, M R, Rhue and Nishimura T 1989 Role of free amino acids and peptides in food taste ed R Teranishi Flavour chemistry: trends and developments Wongso S, Yamanaka H 1998 Extractive components of the adductor muscle of Japanese baking scallop and changes during refrigerated storage *J Food Sci* **63** 772-6
[39] Suzuki H, Kajimoto Y, and Kumagai H. 2002. Improvement of the bitter taste of amino acids through the transpeptidation reaction of bacterial γ-Glutamyltranspeptidase *J. Agric. Food.*

Chem **50** 313-8

[40] Kubota S, Itoh K, Niizeki N, Song X A, Okimoto K, Ando M, Murata M and Sakaguchi M 2002 Organic taste-active components in the hot water extract of yellowtail muscle *Food Sci. Technol. Res.* **8** 45-9

[41] Maga J A 1987 The flavour chemistry of wood smoke *Food Review International* **3** 139-83

[42] Sakakibara H, Yanai T, Yajima I and Hayashi K 1988 Changes in volatile flavour compounds of powdered dried bonito (katsuobushi) during storage *Agric Biol Chem* **52** 2731-39

[43] Cha Y J, Baek H H, Hsieh C Y 1992 Volatile components in flavour concentrates from crayfish processing waste *J Sci Food Agric* **58** 239-48.

[44] Yajima I, Nakamura M and Sakakibara H 1983 Volatile flavour components of dried bonito (katsuobushi) II. from neutral fraction *Agric Biol Chem* **47** 1755-60.

[45] Sakakibara H, Ide J, Yanai T, Yajima I and Hayashi K 1990 Volatile flavour compounds of some kinds of dried and smoked fish *Agric Biol Chem* **54** 9-16.

[46] Ho C T and Chen Q 1994 Lipids in food flavours: an overview (*Lipids in Food Flavours*) eds C T Ho and T.G. Hartman (Washington DC: American Chemical Society)

Anti-Pathogenic Activity of Coral Bacteria Againts White Plaque Disease of Coral Dipsastraea from Tengah Island, Karimunjawa

Sakti Imam Muchlissin[1]*, Agus Sabdono[2] Diah Permata W[2]
1) Tropical Marine Biotechnology Laboratory, Diponegoro University, Semarang, Indonesia
2) Faculty of Fisheries and Marine Science, Diponegoro University, Semarang, Indonesia
*email: muchlissin@outlook.com

Abstract. Coral disease is main factor of degrading coral reefs, such as White Plaque (WP) disease that cause loss of epidermal tissue of corals. The purposes of this research were to identify the bacteria associated with White Plaque Disease of coral *Dipsastraea* and to investigate coral bacteria that have antipathogenic potency against White Plaque Disease by Coral *Dipsastraea*. Sampling was carried out by purposive method in Tengah Island, Karimunjawa on March 2015. Streak method was used to isolate and purify coral bacteria, while overlay and agar diffusion method were used to test antibacterial activity. Identification of selected bacteria was conducted by biochemical and molecular methods. Polyphasic identification of bacteria associated with diseased coral White Plague of *Dipsastraea*. It is found that TFWP1, TFWP2, TFWP3 and TFWP4 were closely related to *Bacillus antracis*, *Virgibacillus olivae*, *Virgibacillus salarius* and *Bacillus mojavensis*, respectively. While antipathogen activity bacterial isolates, NM1.3, NM1.8 and NM2.3 were closely related to *Pseudoalteromonas flavipulchra*, *Pseudoalteromonas piscicida,* and *Vibrio azureus*, respectively. Phylogenetic data on microbial community composition in coral will help with the knowledge in the biological control of coral diseases.

Keywords: Coral disease, White Plaque, *Dipsastraea*, Anti-pathogen, and Molecular

1. Introduction

Coral reef ecosystems in Indonesia have highest biodiversity in the world. It is estimated that more than 3,000 species of marine life can be found in the coral reefs ecosystem [1], [2]. Eutrophication, sedimentation and over-exploitation of marine resources are the factors that cause the destruction of coral reef ecosystems. Coral disease is one of the main factors that caused degradation of coral reefs [3], [4], [5].

Coral disease was discovered in the Caribbean in the 1970s, but more recent studies conducted in the 1990s. Black Band Disease (BBD) is the first incident occurred. The discovery of the coral disease until today continues to increase rapidly and it has been reported that more than 29 species of coral syndrome have infected [6], [7], [8], [9]. Based on published literature, the coral diseases in Indonesia were found as the Black Band Disease in Karimun[10]; the coral diseases in the Wakatobi Marine National Park, Southeast Sulawesi[11], [2]; the coral disease surveillance in Spermonde Islands, South

Sulawesi[12], [13];the spread of coral disease in the Lembata, NTT[14]; the distribution and spread of the Black Band Disease (BBD) on coral *Montipora* sp. in the Seribu Islands[15]; White Band Disease on coral *Acropora humilis* and *Acropora turtosa* in Karimunjawa[16]; Pink Blotch Disease in Karimunjawa[17]; the association of bacteria in coral *Pachyseris* sp. infected BBD in Barang Lompo Island, Makassar[18]; about the coral disease in Panjang Island, Jepara[19].

Coral disease problems in Indonesia are very complex as well as the iceberg that is only slightly exposed, but each much a mystery that must be explored and researched. Further research was needed on coral diseases, especially in Balai Taman Nasional Karimunjawa, that coral diseases were found in the other side of Karimunjawa islands. The purpose of this research was to identify the bacteria associated with White Plaque Disease by coral *Dipsastraea*, determined the ability of coral symbiont bacteria activity that could potentially be antipathogen against White Plaque Disease and identifying coral symbiont bacteria biochemical and molecular potentially be antipathogen against White Plaque Disease by coral *Dipsastraea*.

2. Material and methods

2.1. Isolation Bacteria on Healthy and Disease Corals

Seven genera of corals (*Acropora, Isopora, Porites, Montipora, Stylophora, Dipsastraea, and Pocillopora*) were gathered by Scuba Gear at depths of 2 m to 5 m from Tengah Island, Karimunjawa, in March 2015 (S 05° 48,654'; E 110° 30,483'). Identification coral disease and healthy used Underwater Cards for Ascending Coral Health on Indo-Pacific Book [20]. Each specimen was placed separately into Zip-lock Plastic and placed in cool box to evade contact with air. The samples were kept each specimens on coral into Zip-lock plastic and added sea water until processing within a several hours after in the field . Surface tissue of corals were removed from skeleton with Sterilized scrapper. . Tubes used to the homogenized tissues were serially diluted, spread on half-strength ZoBell 2216E marine agar medium and incubated at Incubator (37° C) for 48 hours. On the basis of morphological features, colonies were randomly picked base on cultural characteristic (colors, configurations, margins, elevations) and purified by another plates[21].

2.2 First Screening using the overlay method

The bacterial isolates associated with healthy corals were tested for anti-pathogenic influence of the agar overlay method against selected bacterial pathogens. Each of diseased bacterial isolate was used for anti-pathogenic effect study. The 48 h, healthy corals isolates were spotted on the half-strength ZoBell 2216E marine medium agar and incubated at incubator for 2 d. All diseased isolates were cultured in ZoBell 2216E marine medium liquid and the 48 h cultures were used for the experiments. About 25 μL of the test cultures were suspended in 25 mL of ZoBell 2216E soft-medium agar and were poured immediately over the colonies of the healthy coral bacteria on the plates. The plates were incubated at incubator for 2 d. Antibacterial activity was defined by the formation of inhibition zones around the bacterial colony [22].

2.3 Rescreening of the isolates for antagonistic activity using disc-diffusion method

All bacterial strains selected from overlay test were re-screened for their activity using the agar disc diffusion method. Filter paper discs, 8 mm in diameter (Advantec, Tokyo Roshi, Ltd, Japan) were placed on the surface of plates that previously inoculated with 50 μL of each diseased isolate. Then, 25 μL of selected isolated isolates were put on each filter discs. The plates were then incubated at 37 °C for approximately 72 h. In the end of incubation period, the zones of incubation were measured. Zone measurements were recorded as the distance from the edge of the zone to the edge of the disc [23].

2.4 Microscopic and biochemical characterizations

Selected bacterial strains that consistently active on agar diffused method were grown in ZoBell 2216E agar and underwent further microscopic and biochemical evaluations. Photomicrograph was used to determine the morphology of the isolates. While standard gram staining, motility and biochemical characterizations based on Bergey's Manual of Determinative Bacteriology were used to determine their biochemical properties [24]. Even both microscopic and biochemical characterization were recorded, but were not presented in this paper.

2.5 DNA extraction, PCR amplification and sequencing of 16S rRNA gene fragment
DNA extraction was conducted by Chelex methods [25]. Selected colonies were inoculated in 50-100 μL ddH_2O and 1 mL of 0,5 saponin in PBS 1x (overnight in freezer). The mixture was centrifuged (12000 RPM, 10 min). Supernatant was discarded, then 100 μL ddH_2O and 50 μL of 20% Chelex 100 were added to a final solution and solution and the solution was boiled for 10 min and vortex once after 5 min. The mixture was centrifuged (12000 RPM, 10 min) and stored at -20°C. DNA concentration were quantified and qualified by using NanoDrop 2000 spectrophotometer. DNA extracts for 16S rRNA gene sequences were amplified by PCR using universal primers 27F (5'AGAGTTTGATCMTGGCTCAG-3')[26] and 1492 R (5'TACGGTTAACCTTGTTACGACTT-3') [27]. The PCR mixture consisted KAPA mix Extract (25 μL), Primer 27F (2 μL), Primer 1492R (2 μL), DNA template (2,5 μL), and ddH_2O (18,5 μL). The PCR reaction was performed on Kyrex Thermal Cycler using cycling conditions consisting of an initial denaturation at 95°C for 3 min followed by 30 cycles of denaturation at 95 °C for 1 min, annealing at 55°C for 1 min, and extension at 72°C for 1 min. The Final extension was performed at 72°C for 7 min [28]. The PCR product was analyzed by Agarose 1% gel electrophoresis and the result showed by using UVI-Doc (UVITEC Cambridge). DNA sequencing was conducted at 1st Base, Singapura. The sequences of amplified 16 rRNA genes were deposited in the GenBank database the National Center for Biotechnology Information (NCBI). The sequence was inserted to BLAST search program to identify the sequences of any closely related organisms [29].

2.6 Phylogenic Analysis
The result of DNA sequences was preliminarily aligned with ClustalW Multiple Alignment and the phylogenetic analyses were performed by MEGA6. The phylogenetic trees were determined using the neighbor-joining method with Kimura's two-parameter. The resultant tree topology was evaluated by bootstrap analyses of the neighbor-joining method based on 1.000 resampling[30].

3. Results and discussion

3.1 Isolation and screening of anti-pathogenic coral bacteria
The total associated bacteria on each of the species from Corals was 183 strains marine bacteria associated and four bacteria from white plaque disease in *Dipsastraea* coral. Forty-four (24,04 %) had antimicrobial activity at least one to any pathogenic bacterial tested. These bacterial strains selected for overlay test were rescreened for their activity by using the agar disc diffusion method. The result that showed that 3 isolates were still consistently inhibiting the growth of pathogenic tested (Table 1).

Table 1. Mean and standard deviation (±Sd) of anti-pathogenic assay by using diffusion paper

No.	Isolate Code	Inhibition zone (mm)			
		TDWP 1	TDWP 2	TDWP 3	TDWP 4
1.	NM 1.3	8,18 ± 0,06	21,48 ± 0,17	-	6,02 ± 0,18
2.	NM 1.8	7,40 ± 0,38	22,12 ± 0,51	-	6,37 ± 0,14
3.	NM 2.3	-	-	5,10 ± 0,25	-

IOP Conf. Series: Earth and Environmental Science **116** (2018) 012057 doi:10.1088/1755-1315/116/1/012057

3.2 Molecular characterization and Phylogenetic analysis

In this regard the 16S rRNA sequencing assist resolve the exact taxonomic position of coral bacteria and provided more detailed information on their phylogenetic among their closest relatives (Table 2 and Figure 1).

Table 2. Overview of 16 rRNA gene sequences retrieved from anti-pathogenic coral bacteria

No.	Isolate	Closest species	Similarity (%)	Accession Number
1.	TDWP 1	*Bacillus anthracis*	99%	NR 041248
2.	TDWP 2	*Virgibacillus olivae*	95%	NR 043572
3.	TDWP 3	*Virgibacillus salaries*	95%	NR 041270
4.	TDWP 4	*Bacillus mojavensis*	96%	NR 118290
5.	NM 1.3	*Pseudoalteromonas flavipulchra*	95%	NR 025126
6.	NM 1.8	*Pseudoalteromonas piscicida*	96%	NR 119147
7.	NM 2.3	*Vibrio azureus*	97%	NR 041683

The result from BLAST homology of bacterial strain TDWP 1 and TDWP 4 showed that this bacterium is affiliated to the genus *Bacillus*, and bacterial strain TDWP 2 and TDWP 3 showed that this bacterium is affiliated to the genus *Virgibacillus*. Isolates TDWP 1, TDWP 2, TDWP 3, TDWP 4 come from white plaque disease on coral *Dipsastraea*.

In this study, members of the genera *Bacillus, Virgibacillus, Pseudoalteromonas* were dominant, followed by *Vibrio*. Previus studies indicated the presence of some of these genera, including *Pseudoalteromonas* sp. and *Pseudomonas* sp. Genera *Pseudoalteromonas*, as currently defined [31], comprises heterotropic, gram-negative, aerobic, rod-shaped, polarly flagellated bacteria that are common inhabitants of marine environments (seawater, *algae*, marine invertebrates) [32,33,34].

Several species of the genera *Vibrio* are ubiquitous in aquatic environments, especially the ocean, and are often isolated from various organism ranging from plankton to fish [35,36]

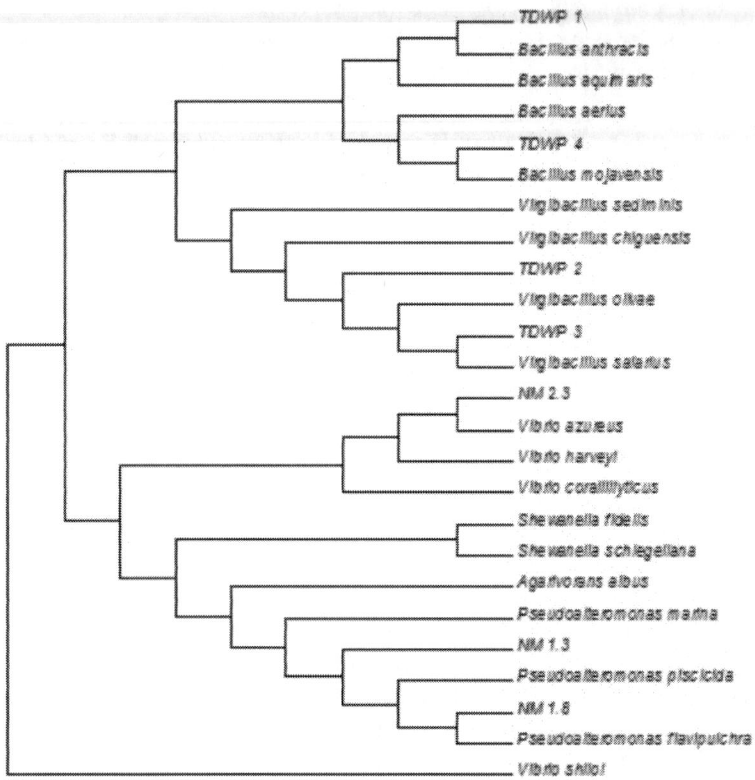

Fig 1. Neighbor-joining phylogenetic tree based on comparative 16S rRNA gene sequence analysis of seven isolates. Selected sequences *Vibrio shiloi* was used as an outer group

4. Conclusion

Three bacterial associated with corals were able to inhabit in vitro bacterial growth of bacteria isolated from white plaque disease on coral *Dipsastraea*. The molecular identification by partial 16S rRNA gene sequencing revealed that they were closely related to genera *Bacillus, Virgibacillus, Pseudoalteromonas,* and *Vibrio*.

Acknowledgements

Author acknowledges fieldwork provides by Handung Nuryadi, Tyas Fajar A., Alinda Cahyani, Nor Sa'adah, Rosa A. Author are grateful for the funding support by Agus Sabdono from PNBP Diponegoro University under Graduated Team Research Grant scheme of Diponegoro University (Hibah Penelitian Tim Pascasarjana, DIPA Undip No :212-02/UN7.5.1/PG/2015).

References

[1] Allen, G R 2007 Conservation Hotspots of Biodiversity and Endemism for Indo-Pacific Coral Reef Fishes *Aquatic Conservation Marine Freshwater Ecosystem*, **18**: 541-556
[2] Haapkyla, J, R K F Unsworth, A S Seymour, J M Thomas, M Flavell, B L Willis, D J Smith 2009 Spatio-Temporal Coral Disease Dynamics in The Wakatobi Marine National Park, South-East, Sulawesi, Indonesia *Disease Aquatic Organisme*, **87**: 105-115
[3] Hughes, T P 1994 Catastrophes, Phase Shifts and Large Scale Degradation of a Caribbean Coral *Reef Science* **265 (5178)**: 1547-1551

[4] Kim, K, C D Harvell 2004 The Rise and Fall of a Six Year Coral Fungal Epizootic *American Naturalist*, **164 (5)**: 52-63

[5] Wilkinson, C R 2004 *Status of Coral Reef of the World*. (Australian Institute of Marine Science: Townsville, Australia)

[6] Edmunds, P J 1991 Extent and Effect of Black Band Disease on a Caribbean Reef *Coral Reefs*, **10**: 161-165

[7] Kuta, K G, L L Richardson 1996 Abundance and Distribution of Black Band Disease on Coral Reefs in The Northern Florida Keys *Coral Reefs*, **15**: 219-223

[8] Green, E P, AW Bruckner 2000 The Significance of Coral Disease Epizootiology for Coral Reef Conservation *Biology Conservation*, **96**: 347-361

[9] Weil, E 2004 Coral Reefs Disease in The Wider Caribbean. *In. Rosenberg E., Loya Y (Eds). Coral Health and Disease*. Berlin: Springer-Verlag, 25-68

[10] Sabdono, A, O K Radjasa 2006 Anti-Bacterial Property of a Coral-Associated Bacterium *Bacillus sp*. Against Coral Pathogenic Black Band Disease (BBD) *Journal of Coastal Development*, **9 (3)**: 175-182

[11] Haapkyla, J, A S Seymour, J Trebilco, D Smith 2007 Coral Disease Prevalence and Coral Health in The Wakatobi Marine Park, South-east, Sulawesi, Indonesia. *Journal Marine Biology Assessment*, **87**: 403-414.

[12] Muller, E M, L J Raymundo, B L Willi, J Haapkylä ,S Yusuf , Joanne R Wilson and Drew C Harvell 2013 Coral Health and Disease in the Spermonde Archipelago and Wakatobi, Sulawesi *Journal of Indonesia Coral Reefs*, **1(3)**: 147-159

[13] Harvell, C D 2013 *Diving into The Coral Triangle* The New York Times Sunday, December 29, 2013

[14] Abrar, M., Bachtiar I., Budiyanto A. 2012. community structure and coral disease in the Lembata, Nusa Tenggara Timur *Indonesia Journal of Marine Science*, **17(2)**:109-118

[15] Johan, Ofri, Bengen, D G, Zamani, N P, Suharsono 2012 Distribution and Abundance of Black Band Disease on Coral *Montipora* sp. in Seribu Islands, Jakarta. *Journal of Indonesia Coral Reefs* **1(3)**:160-170

[16] Hakim, H I, O K Radjasa, M M Khoeri, G A Pratama, D Nasima, Ambariyanto, Sarjito, H Sudoyo 2012 Causative Agents of White Band Disease from Culturable Bacterial Comunity Associated with Healthy and Diseased Corals *Acropora humilis* and *Acropora turtosa* from Karimunjawa Island, Indonesia *Ekologia*, **2 (2)**: 52-59

[17] Soenardjo, Nirwani 2013 Characteristics of bacteria associated with Pink Blotch Disease in Sambangan Island,Karimunjawa *Buletin Oseanografi Marina*, **2** : 58-65.

[18] Jaelani, Irwan 2014 Associated bacteria from coral Pachyseris sp. their infected BBD in Barrang Lompo (Universitas Hasanuddin, Makassar), p 80

[19] Sabdono, A, P H Sawonua, A G D Kartika, J M Amelia, O K Radjasa 2015 Coral Diseases in Panjang Island, Java Sea: Diversity of Anti-Pathogenic Bacterial Coral Symbionts *Procedia Chemistry*, **14**: 15-21

[20] Beeden R, B L Willis, L J Raymundo, C A, Page, E Weil 2008 Underwater Cards for Assessing Coral Health on Indo-Pacific Reefs. Global Environment Fund Coral Reef Targeted Research (GEF-CRTR) Program (Currie Commision, Melbourne, 26 p)

[21] Madigan, M T, J M Martinko, J Parker, and T D Brock *Biology of Microorganisms* (New Jersey, USA: Pretice-Hall;2000)

[22] Isnansetyo, A and Y Kamei 2003 Lysis of methicillin resistant *Staphylococcus aureus* by 2,4-diacetylphloroglucinol produced by Pseudomonas sp. AMSN isolated from marine algae *Int. J Antimocrob Agents*: **21**, 71-74

[23] Radjasa, O K, T Martens, H P Grossart, T Brikhoff, A Sabdono, M Simon 2007 Antagonistic activity of a marine bacterium *Pseudoalteromonas leteoviolacea* TAB4.2 associated with coral *Acropora* sp. *J. Biol.* **7 (2)** 239-246

[24] Holt J G *Bergey's manual of determinative bacteriology* (Baltimore: William & Wilkins) 1994

[25] Walsh, P S, Metzger, D A, Higuchi, R 1991 Chelex 100 as a medium for simple extraction of DNA for PCR-based typing from forensic material *Biotechniques* **10**:506-513

[26] Weisburg W G, Barns S M, Pelletier D A, Lane D J 1991 16S ribosomal DNA amplification for phylogenetic study J. Bacteriol. **173**:697-703

[27] Reysenbach A L, Giver L J, Wickham G S, Pace N R 1992 Differential amplification of rRNA genes by polymerase chain reaction *Appl. Environ. Microbiol.* **58**:3417-3418

[28] Lee, Y K, Hyun Jung Jung and Hong Kum Lee 2006 Marine Bacteria Associated with The Korean Brown Alga, *Unidaria pinnatifida The Journal of Microbiology*, **44 (6)**: 694-698

[29] Sabdono, A 2007 Biodegradation of Chlorpyrifor by a marine bacterium *Bacillus firmus* strain BY6 associated with branching coral *Acropora* sp. J. Coast. Dev.**10(2)**

[30] Susilowati, R, Sabdono, A, Widowati, I 2015 Isolation and characterization of bacteria associated with Brown Algae *Sargassum* spp. from Panjang Island and their antibacterial activities *Procedia Environmental Science* **23** 240-246

[31] Gauthier, G., Gauthier, M. & Christen, R. 1995. Phylogenic analysis of genera Alteromonas, Shewanella, and Moritella using genes coding for small-subunit rRNA sequences and division of the genus Alteromonas into two genera, Alteromonas (emended) and Pseudoalteromonas gen. nov., and proposal of twelve new species combinations. Int J Syst Bacteriol 45, 755-761.

[32] Baumann, P., Gauthier, M. J. & Baumann, L. 1984. Genus Alteromonas Baumann, Baumann, Mandel and Allen 1972. In Bergey's Manual of Systematic Bacteriology, pp. 343-352. Edited by N. R. Krieg & J. G. Holt. Baltomore: Williams & Wilkins.

[33] Gauthier, M. J & Breittmayer, V.A. 1992. The genera Alteromonas and Marinomonas. In The Prokaryotes, pp. 3046-3070. Edited by A. Balows, H. G. Truper, M. Dworkin, W. Harder & K.. H. Schleifer. New York: *Spinger*.

[34] Ivanova,N B, Dimos J T, Schaniel C, Hackney J A, Moore, K A, Lemischka I R 2002 A Stem Cell Molecular *Report* Sciencemag.

[35] Thompson, F. L, Lida, T & Swings, J 2004 Biodiversity of vibrios. M*icrobiol Mol Biol Rev* 68, 403-431.

[36] Yoshizawa S Wada M Kita-Tsukamoto K Ikemoto E Yokota A and Kogure K 2009 *Vibrio azureus* sp. Nov., a luminous marine bacterium isolated from seawater *International Journal of Systematic and Evolutionary Microbiology* 59 1645-1649

Seagrass Parameter Affect the Fish Assemblages in Karimunjawa Archipelago

Endang Sri Susilo[1], Denny Nugroho Sugianto[2,3], Munasik[1], Nirwani[1], Chrisna Adhi Suryono[1]

[1]Department of Marine Science, Faculty of Fisheries and Marine Science, Diponegoro University, Indonesia
[2]Department of Oceanography, Faculty of Fisheries and Marine Science, Diponegoro University, Indonesia
[3]Center for Coastal Disaster Mitigation and Rehabilitation Studies, Diponegoro University

E-mail: endang_susilo@undip.ac.id

Abstract. Seagrass beds promote high species diversity, abundance and biomass, and become important habitats for some economically important fishes. Plants of seagrasses result in structurally highly complex habitats and offering feeding grounds, shelter from predation as well as nursery areas for diverse fish assemblages. However, research on fish communities in Southeast Asian seagrass bed is rarely conducted. In the present study fish assemblages in seagrass beds with different parameters (cover, diversity and similarity indices, domination) was investigated in the Karimunjawa Islands, Indonesia. The purpose of this study were to assess whether fish assemblages differ concerning on the abundance and the species number. This study was conducted on the seagrass bed on Karimunjawa Islands in Java Sea, particularly in the water of Menjangan Besar and Menjangan Kecil Island. Line-quadrant transect was used to assess seagrass data, while the occurrence and individual number of fish harboured in the selected seagrass bed was counted by using underwater visual census in the stationary point count transects. Seagrass cover in Menjangan Kecil Island (41%) with various canopy included both upper and lower canopy was considerable higher than those in Menjangan Besar Island (5%). Fish diversity, species composition and abundance are considerably different between the two study sites. This study revealed that seagrass density or cover and canopy structure affected the fish abundance and species number harboured.

1. Introduction

Seagrass beds promote high species diversity, abundance and biomass, and become important habitats for some economically important fishes [1], [2]. Plants of seagrasses result in structurally highly complex habitats and offering feeding grounds, shelter from predation as well as nursery areas for diverse fish assemblages [3].

Located in Southeast Asian seas, Indonesia is the heart of the coral triangle regions, contain the highest numbers of fish species in the world [4]. In Indonesia, seagrass beds cover shallow water part of coastlines and offshore islands, and can be vegetated by up to ten seagrass species [5] and 80 or more fish species [6], [5] within a single meadow. As their importance for fish species, they become of interest for people.

Most studies of fish assemblages in seagrasses have been performed in the Caribbean. Moreover, diversity of the fauna in Seagrass beds in Southeast Asia is distinctly higher [2], and may therefore generate a differ mechanisms in influencing fish assemblages. However, research on fish communities in Southeast Asian seagrass bed is rarely conducted [7], [8], [9], [10].

In the present study fish assemblages in seagrass beds with different parameters (cover, diversity and similarity indices, domination) was investigated in the Karimunjawa Islands, Indonesia. The purpose of this study were to assess whether fish assemblages differ concerning on the abundance and the species number.

2. Methods

2.1. Study Area

The present study was conducted on the seagrass bed on Karimunjawa archipelago in Java Sea (Fig. 1). The archipelago is about 110 thousand hectare of wide, and consists of numerous islands and outspread reefs. It has a tropical climate, with an annual mean air and water temperature of 32 °C and 28 °C, respectively [11]. May to September is considered as dry, November to March as wet season. The tidal regime is predominantly semi-diurnal with a maximum tidal range at spring tide of 180 cm and an average tidal range of 80 cm (according to the tidal predictions for Semarang harbour). Seagrass bed close to Menjangan Kecil Island and Menjangan Besar Island were chosen as study area. Both islands are patch reefs crowned by corals and shallow subtidal seagrass beds. Both islands are not populated. Two different seagrass meadows of different canopy structure and species composition were chosen.

2.2. Seagrass parameters

Line-quadrant transect according to McKenzie [5] was used to assess seagrass data. A quadrat of 100 cm x 100 cm was randomly thrown along the transect line of 40m which stretched upright to the coast line the selected patches for the visual census. This transect line was then repeated 3 time with 25 m distance inter-transect in parallel with the coat line. The frame was then subdivided into 16 smaller sub-quadrats, each with a side length of 25 cm. The number of shoots of each seagrass species was counted.

2.3. Counting of fish harboured

The occurrence and individual number of fish harboured in the selected seagrass bed was counted by using underwater visual census in the stationary point count transects [3]. In each site, three replicates of 100 m, 60 m and 30 long marked transect lines, were deployed upright to the coast line. A single rope with a length of 5 m was used as a visual guidance for quadrat size. After placing the line, the observer waited 3 min before counting to minimise fish disturbance.

2.4. Data Analysis

Fish were recorded as juveniles when they were smaller than one third of the maximum species' length [12] [5]. Data of both the seagrass and the fish were analysed descriptively and presented in graphs.

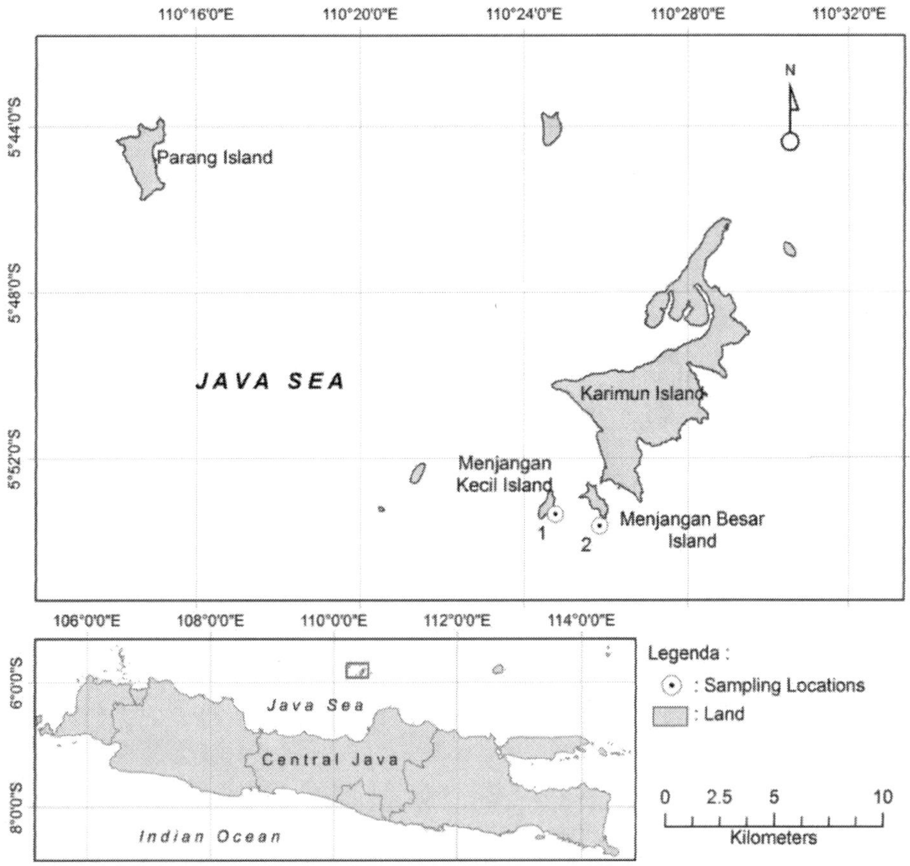

Figure 1. Overview of study area of Menjangan Kecil and Menjangan Besar Parang Islands in the Karimunjawa Archipelago

3. Result

3.1. Study Area

Seagrass cover in Menjangan Kecil Island (41%) was considerable higher than those in Menjangan Besar Island (5%). They were composed of Enhalus acoroides, Thalassia hemprichii, Cymodocea rotundata, C. serrulata, Halodule uninervis, Halophila ovalis and H. minor in different composition between the two sites (Fig.2.). Menjangan Besar Island was comprised only two species of seagrass, while seven seagrass species without any species domination (C=0.21) were found in Menjangan Kecil Island. Diversity of seagrass species as showed from the indices in here was moderate with high similarity (Fig.3.) It result in various canopy included both upper and lower canopy. The upper canopy was mostly generated from E. acoroides shoots. T. hemphrichii comprised the second upper canopy. The rest other species together formed a lower canopy. H. ovalis found in higher stand with bigger leave among the genus. The leave shape was ovate, obovate or oblong. The two other species, H. uninervis and H. minor scarcely occurred.

Different types of canopy were performed in the Menjangan Besar Island. As there were only two species, they formed only high canopy in the seagrass path. E. acoroides and T. hemprichii stand in the relatively same composition and resulted in the low diversity of seagrass meadow.

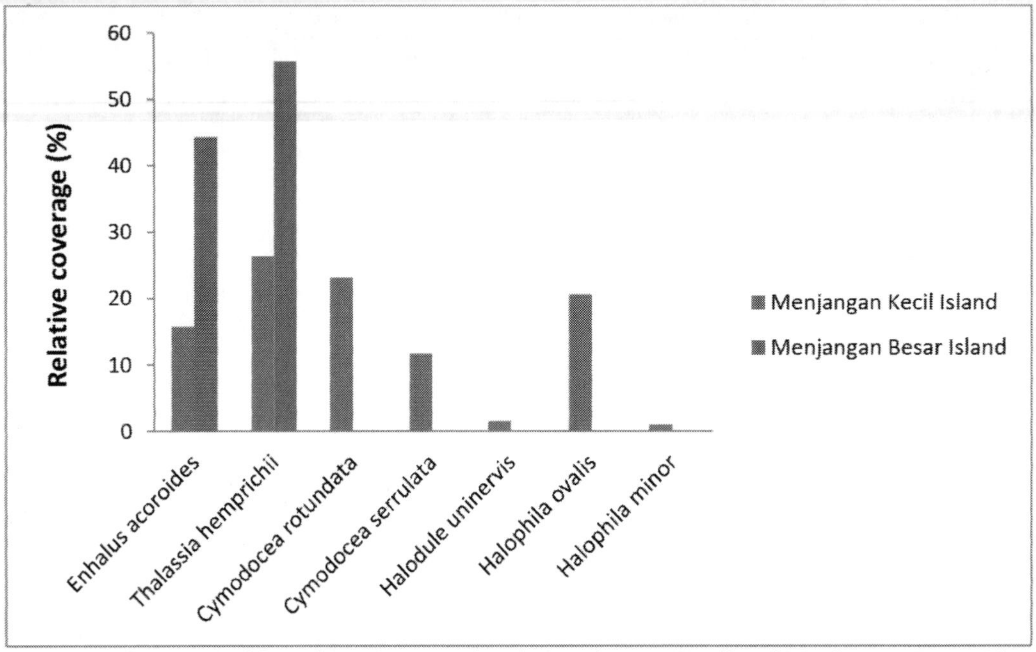

Figure 2. Species composition of seagrass meadow in Menjangan Kecil and Menjangan Besar Islands.

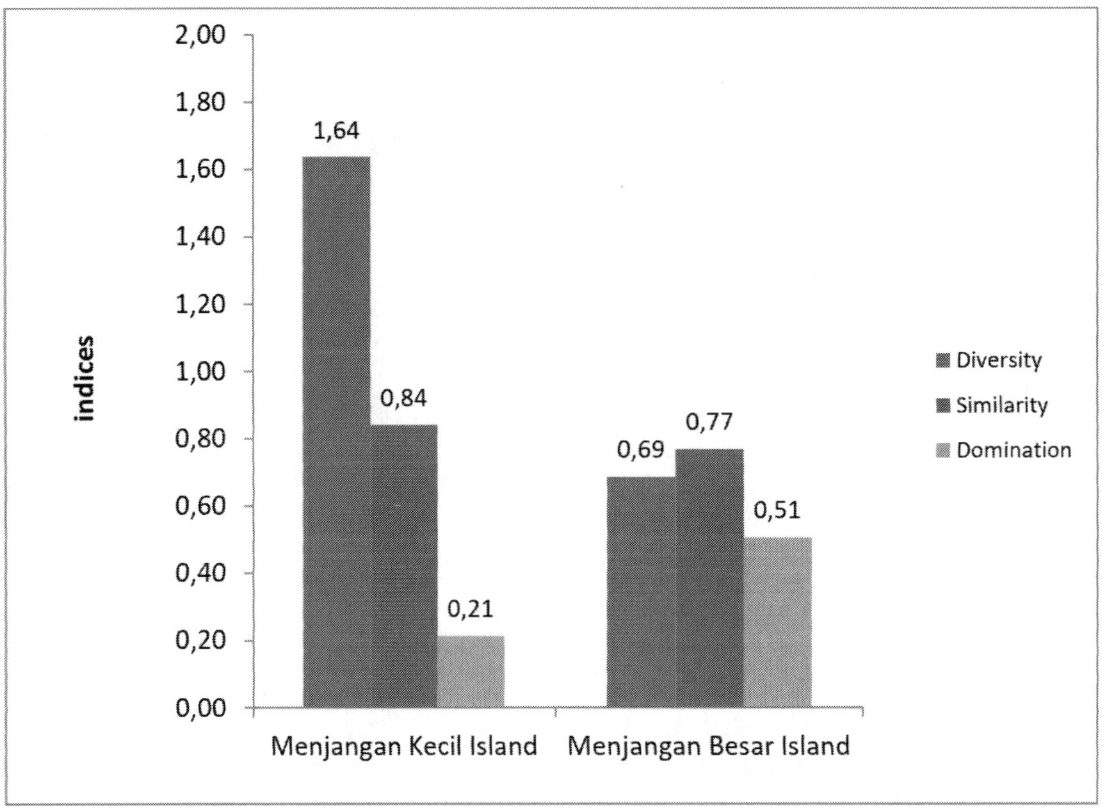

Figure 3. Diversity indices, similarity indices and domination of seagrass meadow in Menjangan Kecil and Menjangan Besar Islands.

3.2. Fish harboured in the seagrass meadow.

It seemed that there was a considerable different in the family, fish and species number between those in Menjangan Kecil Island and Menjangan Besar Island (Table 1.). There were eight species of mangrove/seagrass-associated reef fish found in Menjangan Kecil island. Among the reef fish, *Abudefduf sexfasciatus* and *Sphyraena flavicauda* were the most abundance. In contrast, only 2 species of mangrove/seagrass-associated reef fish belong to Pomacentridae family were in Menjangan Besar Island. Seagrass species from family Engraulidae looked dominated among the fish in both waters.

Table 1. Harboured fish in the seagrass beds

Family/Species	Menjangan Kecil Island	Menjangan Besar Island	Note
Apogonidae			
Apogon margaritophorus	10		Rf
Cheilodipterus isostigmus	5		Rf
Blenniidae			
Petroscirtes variabilis	14	3	Sg
Caesionidae			
Caesio teres	8		Rf
Chaetodontidae			
Chaetodon octofasciatus	5		Rf
Engraulidae			
Encrasicholina heteroloba	384	51	Sg
Stolephorus waitei	136	8	Sg
Monacantidae			
Acreichthys tomentosus	4		Sg
Nemipteridae			
Pentapodus trivittatus	7		Rf
Pomacentridae			
Abudefduf sexfasciatus	23	5	Rf
Dischistodus prosopotaenia	6	3	Rf
Sphyraenidae			
Sphyraena flavicauda	23		Rf
Syngnathidae			
Syngnathoides biaculeatus	5		Sg
Number of individu	630	70	
Number of Species	13	5	
Number of reef fish species	8	2	

4. Discussion

In the present study fish assemblages in seagrass beds with significantly different seagrass parameters were investigated. Fish diversity, species composition and abundance are considerably different between the two study sites, i.e between Menjangan Kecil Island and Menjangan Besar Island, even both are subtidal seagrass bed and close to reefs. This study revealed that seagrass density or cover and canopy structure affected the fish abundance and species number harboured. Moreover, higher diversity and less domination seemed promoted higher fish number as well as species number. Pogoreutz et al. [4] found that fish assemblages in seagrass beds at different water depths and with differing canopy structure defined by significantly different shoot density, biomass and leaf area index (LAI), differed significantly in terms of species number, total abundance, and abundance of most common species. They revealed an effect of water depth and seagrass shoot density on total fish abundance and LAI, seagrass biomass and water depth affected the abundance of most common fish species.

It is generally approved that structurally rich seagrass beds, i.e. beds with high shoot density and/or biomass, provide good conditions, such as advance habitat complexity and food availability, and decreased predation pressure [13], [10]. High fish abundances in relation to high seagrass shoot densities were deal with previous studies [10, [14]. Hovel et al. [13] found that significant relationships between seagrass biomass and animal densities were only detected at high seagrass biomasses, which is in contradiction with the finding of Pogoreutz et al. [4] that showed overall faunal abundance was found to be higher in beds with higher shoot densities and biomasses compared to beds with lower shoot densities, although certain taxa apparently preferred the less structured habitat [10]. The most abundance fish families in the present study were Engraulidae for seagrass species and Pomacentridae and Sphyraenidae for mangrove/seagrass-associated reef fish. Similarly, Pomacentridae comprised the most abundant families in seagrass bed of the Spermonde Archipelago, Indonesia [4]. Conversely, the most abundant family in a seagrass bed at Wakatobi found by Unsworth et al. [2] was Apogonidae. These different results might be due to the sampling methods used or site-specific.

5. Conclusion

Fish diversity, species composition and abundance are considerably different between the two seagrass beds with different species parameters. Higher seagrass density or cover and canopy structure combination with lower domination resulted in the higher fish abundance and species number harboured. Different result in the most abundant family compared to those at Wakatobi might be due to the sampling methods used or site-specific.

References

[1] Parrish J D 1989 Fish communities of interacting shallow-water habitats in tropical oceanic regions vol 58 pp 143–160
[2] Unsworth R K F et al 2007 Diel trophic structuring of seagrass bed fish assemblages in the Wakatobi Marine National Park, Indonesia **72**
[3] Dorenbosch M et al 2005 Seagrass beds and mangroves as potential nurseries for the threatened Indo-Pacific humphead wrasse , Cheilinus undulatus and Caribbean rainbow parrotfish *Scarus guacamaia* **9**
[4] Pogoreutz C et al 2012 Estuarine , Coastal and Shelf Science The in fl uence of canopy structure and tidal level on fi sh assemblages in tropical Southeast Asian seagrass meadows *Estuar. Coast. Shelf Sci.* **107** pp 58–68
[5] McKenzie L et al 2007 Seagrass ecosystems of Papua. In: Marshall A J and Beehler B M (Eds.) *The Ecology of Papua* Part 2 Periplus editions (Hk) Ltd pp 800-823
[6] Hutomo M and Martosewojo S 1977 The fishes of seagrass community on the west side of Burung Island (Pari Islands, Seribu Islands) and their variations in abundances *Marine Research in Indonesia* **17** 147-172.
[7] Nienhuis P H et al 1989 Community structure and biomass distribution of seagrasses and macrofauna in the Flores Sea *Netherlands Journal of Sea Research* **23** 197-214
[8] Erftemeijer P L A and Allen G R 1993 Fish fauna of seagrass beds in south Sulawesi, Indonesia *Records of the West Australian Museum* **16** 269-277

[9] Vonk JA et al 2008 Redefining the trophic importance of seagrasses for fauna in tropical Indo-Pacific meadows *Estuarine, Coastal and Shelf Science* **79** 653-660

[10] Vonk J A et al 2010 Abundance, edge effect, and seasonality of fauna in mixed-species seagrass meadows in southwest Sulawesi, Indonesia *Marine Biology Research* **6** 282-291

[11] Erftemeijer P L A and Herman P M J 1994 Seasonal changes in environmental variables, biomass, production and nutrient contents in two contrasting tropical intertidal seagrass beds in South Sulawesi, Indonesia. *Oecologia* **99** 45-59

[12] Nagelkerken I and Van der Velde G 2002 Do non-estuarine mangroves harbour higher densities of juvenile fish than adjacent shallow-water and coral reef habitats in Curacao (Netherlands Antilles)? *Mar. Ecol. Prog. Ser.* **245** pp 191–204

[13]Gullström, M et al 2002 Seagrass ecosystems in the western Indian Ocean *Ambio* **31** 588-596

[14] Hovel K A 2002. Effects of seagrass landscape structure, structural complexity and hydrodynamic regime on macrofaunal densities in North Carolina seagrass beds *Mar. Ecol. Prog. Ser.* **243** 11-24

Early weaning food for infants (0-6 months old) in madurese people based on transcultural nursing theory

Eka Mishbahatul M. Has[1*], M. Syaltut[1], Tiyas Kusumaningrum[1], Ferry Efendi[1]

[1]*Faculty of Nursing, Universitas Airlangga, Jl. Mulyorejo (Kampus C) UNAIR, Indonesia*
* *eka.m.has@fkp.unair.ac.id*

Abstract. The World Health Organization recommend weaning food is given to infants aged 6-24 months. In Madura culture still exist the practice of giving weaning food before infants reach six months old. The purpose of this study was to analyze the factors of giving weaning food for infants (0-6 months old) based on transcultural nursing theory. Design used was descriptive with cross-sectional approach. Population was Madurese mothers who have infants (0-6 months old) who has been given early weaning food. A total of 61 respondents were chosen by cluster sampling method. Variables in this study were education, economic, political and legal, cultural values and lifestyles, kinship and social, religious and philosophical, and technology. Data were collected by using structured interview and described by using frequency and percentage distribution. The results had showed educational mostly were primary and middle education (92%). Economic mostly come from low economic status (70%). Political and legal mostly positive (54%). Cultural values and lifestyles mostly negative (62%). Kinship and social mostly negative (64%). Religious and philosophical mostly positive (64%). Technology factor dominantly low (56%). Based on transcultural nursing theory it is shown a diversity in positive and negative values. Further research was suggested to reduce the practice of giving weaning food behavior of Madurese mothers which suits with local culture.

Keywords: *early weaning food, transcultural nursing theory, infant (0-6 months old)*

1. Introduction

WHO recommend to exclusively breastfed infants for the first six months of life to achieve optimal growth in development and health. Thereafter, to meet their evolving nutritional requirements, infants should receive nutritionally adequate and safe complementary foods while breastfeeding continues for up to two years of age or beyond [1]. But, in Madura culture still exist the practice of giving weaning food before infants reach six months old [2].

The target of exclusive breastfeeding program in Indonesia based on Ministry of Health's Strategic Plans on 2016 was 42%. At the end of 2016, it was nationally reached 54% for exclusive breastfeeding during 0-5 months old. But, the target was still unreal for exclusively breastfeeding infants until six months old (only 29,5%). While East Java coverage for breastfeeding infants during 0-5 months old was 48,1% (below national target) and 31,3% for six months old (over national target) [3]. Based on health survey in 2015, exclusive breastfeeding coverage for infants 0-6 months old at Kabupaten

Pamekasan was only 59,8%. It was below East Java coverage (69,1%). While data from Community Health Center (Puskesmas) at Proppo, Pamekasan, had shown that at the same year, the exclusive breastfeeding coverage for infants 0-6 months old was only 31,4%. The low coverage of exclusive breastfeeding in this area was caused by Madurese tradition called *lotek* (Madurese language, means giving soft foods) which given before infants reach six months old. They believe this tradition can make baby's stomach full, then the baby can sleep tight [4].

Transcultural nursing theory emphasizing on the nurse's role in giving intervention based on client's culture. Bridging gap between client's folk and professional behavior in maintaining health [5]. Based on this theory, there are seven factors which influence individuals, families, groups, and communities' health behavior, includes: 1) technological; 2) religious and philosophical; 3) kinship and social; 4) cultural values and lifeways; 5) political and legal; 6) economic; and 7) educational [6]. The research result showed that kinship, cultural values, lifeways, and philosophical beliefs influence Puerto Rican infant feeding practices [7]. Research on Javanese mothers had shown that all factors in transcultural nursing influence their infant feeding practices, except education [8]. There were overwhelming research which examine factors affecting infant feeding practices, but rare using transcultural nursing theory as a conceptual framework.

Various efforts have been made by the government to improve exclusive breastfeeding [8]. However, Madurese tradition in early weaning food still hasn't solved yet. Therefore in this paper, we analyze the factors of giving weaning food for infants (0-6 months old) based on transcultural nursing theory on Madurese mothers at Puskesmas Proppo, Pamekasan. So nurses can develop interventions or health promotion strategies which culturally sensitive, in order to promote exclusive breastfeeding and prevent early weaning food before an infant reach 6 months old at Madurese society.

2. Method

Design used was descriptive with cross-sectional approach. Population was Madurese mothers who have infants (0-6 months old) who has been given early weaning food at Puskesmas Proppo Pamekasan. A total of 61 respondents were chosen by cluster sampling method from 16 different village. Variables in this study were education, economic, political and legal, cultural values and lifeways, kinship and social, religious and philosophical, and technology. Data were collected by using structured interview and described by using frequency and percentage distribution.

Table 1. Demographic characteristics of mother and child (n=61)

Variables	Category	n	%
Mother's age	>20 years old	4	7
	20-35 years old	52	85
	>35 years old	5	8
Working status	Yes	31	59
	No	30	41
Number of children	1	23	38
	2	20	33
	3	15	25
	>3	3	5
Number of family's member	3	2	3
	4	9	15
	5	16	26
	>5	34	56
Type of family	Nuclear family	19	31
	Extended family	42	69

3. Result and Discussion

Table 1 presents the characteristics of respondents, mothers who have infants (0-6 months old) who has been given early weaning food. It shows that, mostly mothers' age was classified as young adulthood (20-35 years old), as many as 85% mothers. More than a half was a working mother (51%). Many of them only have one infant (0-6 months old), as many as 38%. Most mothers who have been given early weaning food to their infant were living in an extended family (69%), with more than five family members in one house (56%).

Table 2 Frequency and percentage distribution of respondents based on variables (n=61)

Variables	Category	n	%
Level of education	Primer	30	49
	Middle	26	43
	High	5	8
Level of economy	Low	43	70
	High	18	30
Political and legal	Negative	28	46
	Positive	33	54
Cultural values and lifeways	Negative	38	62
	Positive	23	38
Kinship and social	Negative	39	64
	Positive	22	36
Religiousity and philosophies	Negative	27	44
	Positive	34	56
Technology	Low	34	56
	Fair	15	25
	Good	12	20

Table 2 present that mostly mothers who has been giving weaning food to their infants before 6 months old, had attended primary and middle schools (92%). Mother's level of education can influence their knowledge. Knowledges were important domain to form one's action [9]. As higher as their education, an individual would think by using their rational which supported with scientific evidence. They also can easily adopt a behavior to maintain their health [6]. More information their get, healthier action expected. Based on data, there were mothers with high educational level who has been given early weaning food to their infant before 6 months old. Although the frequency was lower as their level of education was increasing. The transcultural nursing theory had explained that health behavior was formed not only by educational factors, but also various factors which contribute together [5]. Mothers' level of education doesn't specify their behavior in giving early weaning food to their infant (0-6 months old). If there were other factors that more suppress mothers, so their health behaviors were decreasing [10].

Majority mothers came from low economics' level (70%). Their family incomes were lower than standart family monthly incomes at Proppo District. Family incomes mostly obtained from father, because almost all respondents were housewives and only few become entrepreneur (selling goods at their home). Transcultural nursing theory had explained that individuals will use their own resources to maintain health, such as money [6]. In case of infant feeding practice, family financial condition will influence mother's ability to provide good nutrition for themselves [11], to maintain the quality of their breastmilk. Based on interview while fulfill the questionnaire, mothers had said their breastmilk seems not enough to feed their infants. So they add soft food to make baby's stomach full, then the baby can sleep tight [4]. There were 30% mothers from good economics' level who give early weaning food to their infant. It can happen because there were several factors which influence one's health behavior, such as education, occupation, and cultural background [12]. Mothers with high economic level will tend to choose formula milk, instant porridge or biscuits with good quality and nutritious content [13], while families with less economic status will tend to choose rice or bananas as soft food because it was economically cheaper and locally appropriate.

Indonesian government has published regulation Kepmenkes RI No.450, 2014 about breastfeeding that promoted exclusive breastfeeding. Most samples have positive view about the regulation but still giving early weaning food for the infants. According to transcultural nursing theory, rule, and regulation can affect healthy behavior [5]. Mother's knowledge about breastfeeding regulation need to be inquisitive with their outlook and implementation of the exclusive breastfeeding. Mothers with good knowledge about breastfeeding and the regulation still failed to practice exclusive breastfeeding [15]. The same result with this research about breastfeeding can be a sign that a positive view about regulation was not directly affected mother decision about exclusive breastfeeding implementation.

The distribution data of cultural value and life style show the sixty-two percent of samples have negative cultural value and life style in early weaning food for infant. Negative cultural value and life style of early weaning food such as giving coconut water or they called it *ro-moro* to newborn baby in order to make the infant stronger and healthier. Other negative behavior done by most samples was giving smoothed banana or they call it *lotek* to stop the baby crying and sleep faster. Other than that, they also provide infant's formula before the infant reached six months old. People share the culture they learnt and use it as part of their think, act, and decision making [6]. Former research revealed the correlation between cultural value and providing weaning food on infants less than six months old [13].

The result of this research showed 38% respondent have positive cultural value and lifeways about early weaning food for infants. Positive cultural value and lifeways among respondents were belief about breast milk is good for baby's growth and development, breastfeeding was natural way, and performed abstinence was dangerous for baby. However, those respondents also gave early weaning food for their infants. There were several reasons why they still committed early weaning. The first reason was working mother who spent most of the time at their office than at home and their office have less facility to support breastfeed mother. Working mothers usually has less time to interact and breastfeeding their baby, thus made them prefer infant's formula than breast milk. The other reason was some of the respondents didn't finish primary school and had deficient knowledge about breastfeeding.

Negative cultural value promote mother in early weaning for infants. It is usually the parents who teach common cultural values and lifeways they believed to their children. Cultural negotiation might become the right intervention and nursing implementation and help mothers adapt to certain cultures that are more beneficial to their health. Nurses can help clients to choose and define other cultures that are more supportive to reduce the behavior of early breastfeeding and increase exclusive breastfeeding.

Respondent's distribution on kinship and social factor showed that more than half of respondents have negative value toward social and family about early weaning for infants. Community and other family members should function as a support system to improve the health and adapt to changes in the health status of a family members [5]. Community and family support provided could be implied as providing time, attention, and all support required by mother and baby [13].

Negative support from social and family about early weaning might cause by low awareness of women achievement as mother. Negative family concerns caused mothers have negative traits in determining the form and mode of treatment of infants, including in the provision of early weaning. Mother usually obey all family information about baby treatment. There was a correlation between family promotion about early weaning and early weaning behavior on mother with baby under six months old [16]. Negative social and family supports in this study included family who rarely listens to maternal complaints during breastfeeding, always encourage early weaning, and rarely accompanied mothers in consultation to health workers about infant care. Most of the time social and family supports still related with the habits, customs, and beliefs of the family. This resulted in mother's behavior on early weaning.

More than half of respondents come from nuclear family and extended family have negative support from their social and family members. Children from small families receive more attention than children from large families [12]. However, the results of this research have contradictive outcome comparing to Friedman opinion.

From the data distribution we also found respondents who have positive social and family support. Forms of positive support provided were enough food for mother; taking care of the infant while the mother is working; and taking the infant to the health care center for monitoring infant's health. Still those supports did not preclude mother in doing early weaning in infants. The age and number of children may also cause the mother to have less experience with breastfeeding.

Positive support from social and family members is an important part on building up mother role, especially in the Madurese where the family culture is quite strong. Madurese families and communities plays important role in child care especially in the care of babies. Information on exclusive breastfeeding and breastfeeding should not only be given to mothers but also husbands and the other family member. Communities and families need to receive education and training about exclusive breastfeeding in order to provide support for mothers and prevent early weaning for infants less than six months old.

Respondent's distribution based on religiosity and philosophy showed that most of respondents have positive religiosity a philosophy about early weaning, approximately 56% of respondents. According to the theory of transcultural nursing, religiosity provides a powerful motivation for placing truth above all, even above one's own life and lead one's to have a humble and open nature [5]. There are three dimensions in Madurese religiosity. First, the Dimension of experience where man feeling calm with performing religious actions. Secondly, the dimension of religious knowledge, where belief and faith were illustrated by traditions. The last was the dimension of the consequences after performing religious action, between bad and good impacts [17]. In this study the positive religiosity a philosophy of respondents in early weaning such as their perspective about medical treatment, in this case was the early weaning; how the respondent wish for the health of their baby, and providing nutritious food as a form of gratitude and the peace they felt for having the baby. A person's religiosity can affect their behavior including breastfeeding and food weaning.

In this study most of the respondents have positive religiosity & philosophy value, but they still performed early weaning. This was because in this study, religiosity influence more on mother's tranquility when taking care of child, not what being given by mothers to their baby. Maternal behavior in early breastfeeding is not only influenced by maternal religiosity, but there are 6 other factors that influence the mother's behavior based on the culture they choose [6]. There always needs to be modification on other factors influencing mother's behavior about early weaning.

Technology factor dominantly low (56%) on mothers who has been given early weaning food to their infant (0-6 months old). Technology was one of factors which influence an individual's behavior based on their culture [6]. Technology refers to all forms of things or system used to create, store, change, and use information and all its forms. The use of health technology was influenced by health worker's attitudes and the community's need and interest [14]. The development of modern technology can be useful to disseminate information about exclusive breastfeeding, but it also can give a negative impact if the technology was used to advertise formula milk and instant porridge. There was a correlation between technology and mothers' behavior in giving early weaning food on infant (0-6 months old). Less utilization of technology will increase mother's behavior in early weaning food practice [13]. Less technology utilization in Madurese mothers based on research's result was the lack of using printed and electronic media to access information about health, especially about exclusive breastfeeding; lack of using the health service center; and lack of electronic devices to support exclusive breastfeeding, such as breast pump for a working mom, and freezer to store breast milk.

4. Conclusion and Suggestion

Based on transcultural nursing theory, factors which influence the practice of giving weaning food before infants reach six months old in Madurese people were low maternal education, low family economic level, negative cultural values and lifeways, negative kinship and social, and low technology to access correct information about exclusive breastfeeding and the availability of supporting facilities to breastfeed exclusively. Otherwise, positive political and legal factors and positive religiosity and philosophical factors can be the strength of Madurese people to prevent the early weaning food practice.

Madurese mothers and families with infants (0-6 months old) should be more active in accessing information about exclusive breastfeeding. Community health nurses were expected to develop health promotion strategies based on research findings to reduce the practice of giving weaning food behavior of Madurese mothers which suits with local culture.

References

1. WHO. (2002, April 16). *Infant and young child nutrition: global strategy on infant and young child feeding*. The World Health Organization, Secretariat. Geneva: WHO. Retrieved August 21, 2016, from The World Health Organization: http://apps.who.int/gb/archive/pdf_files/WHA55/ea5515.pdf?ua=1
2. Hidayat, A., Dede, N., & Pipit, F. (2013). *Pengembangan model keperawatan berbasis budaya (etnonursing) pada keluarga etnis madura dengan masalah balita gizi kurang di Kabupaten Sumenep*. Konferensi Nasional PPNI Jawa Tengah. Semarang: PPNI Jawa Tengah.
3. Kementrian Kesehatan RI (2017). *Data dan Informasi Profil Kesehatan Indonesia tahun 2016. Kementrian Kesehatan Republik Indonesia, Pusdatin Kemenkes RI*. Jakarta: Pusdatin Kemenkes RI.
4. Launer, L. J., & Habicht, J.-P. (1989). Concepts about infant health, growth, and weaning: A comparison between nutritional scientists and madurese mothers. *Social Science & Medicine*, 29(1), 13-22.
5. Leininger, M., & McFarland, M. (2006). *Cultural care diversity and universality: a worldwide nursing theory*. Sudbury, MA, USA: Jones and Bartlett Publishers.
6. Leininger, M. (2002). Culture care theory: a major contribution to advance transcultural nursing knowledges and practices. *Journal of Transcultural Nursing*, 13(3), 189-192.
7. Higgins, B. (2000, January 1). Puerto rican cultural beliefs: influence on infant feeding practices in western new york. *Journal of Transcultural Nursing*, 11(1), 19-30.
8. Yohmi, E., Marzuki, N., Nainggolan, E., Partiwi, I., Sjarif, B., & Oswari, H. (2015, November 6). Prevalence of exclusive breastfeeding in Indonesia: a qualitative and quantitative study. *Paediatrica Indonesiana (The Indonesian Journal of Paediatric and Perinatal Medicine)*, 55(6), 302-308.
9. Kharismawati, E. (2014). *Analisis faktor pemberian ASI eksklusif pada bayi berdasarkan teori perilaku WHO di Wilayah Puskesmas Mulyorejo Surabaya*. Universitas Airlangga, Faculty of Nursing. Surabaya: Unpublished.
10. Nurhayati, M. (2015). *Analisis faktor pemberian ASI berdasarkan pendekatan sunrise model di wilayah puskesmas kenjeran surabaya*. Universitas Airlangga, Faculty of Nursing. Surabaya: Unpublished.
11. Agustin, E., Ulfiana, E., & Has, E. (2015). *Mother's behavior in the fulfillment of nutrition for underweight toddler based on health promotion model*. The 6th International nursing conference: emphasize the art of nursing in research, education, into clinical and community practice (pp. 55-58). Surabaya: Faculty of Nursing, Universitas Airlangga.

12. Friedman, M., Bowden, V., & Jones, E. (2003). *Family nursing: research, theory, and practice (5th edition)*. N.J.: Prentice Hall.
13. Jannah, S. (2016). *Analisis faktor yang berhubungan dengan tindakan ibu dalam pemberian MPASI pada bayi usia 0-6 bulan berdasarkan teori transkultural nursing*. Universitas Airlangga, Faculty of Nursing. Surabaya: Unpublished.
14. Giger, J. (2012). *Transcultural nursing: assessment and intervention, 6th edition*. NY: Mosby Inc.
15. Saputri, K. (2013). *Alasan ibu memberikan makanan pendamping ASI (MP-ASI) dini dengan pendekatan teori health belief di wilayah kerja puskesmas kecamatan pesanggrahan jakarta selatan*. UIN Syarif Hidayatullah. Jakarta: Unpublished.
16. Tiasna, A. (2015). *Hubungan dukungan keluarga dalam pemberian MP-ASI dengan pemberian makanan pendamping ASI dini pada bayi usia 0-6 bulan di wilayah kerja puskesmas sewon 1 bantul tahun 2015*. STIKES Aisyiyah Yogyakarta. Yogyakarta: Unpublished.
17. Kastono, D. (2015). *Aspek religiusitas masyarakat madura dalam cerpen karapan laut karya mahwi air tawar (kajian sosiologi sastra)*. Universitas Negeri Yogyakarta, Fakultas Bahasa dan Seni. Yogyakarta: Unpublished.

Genetic Diversity and Geographical Gene Flow Patterns of Spawning Broadcast Coral *Lobophyllia corymbosa* in The Sulawesi Waters as A Coral Triangle Area

Widyastuti Umar[1]*, Jamaluddin Jompa[2] and Asmi Citra Malina A.R. Tassakka[3]

[1]*Postgraduate Student of Marine Science and Fisheries, Hasanuddin University*
[2]*Marine Science and Fisheries Departement, Hasanuddin University*
[3]*Fisheries DepartementMarine Science and Fisheries Departement, Hasanuddin University*

*widyastuti160@gmail.com

Abstract. The existence of hard coral is one of the keys to maintain the suistainability of ecosystem in the waters. Currently, the hard coral keeps declining over time due to various disturbances. In addition, hard coral where fish and other organisms live in will directly affect the ecosystem sustainability if the damage still happens. Molecular approach, such as knowing the genetic variation information of coral population can be an informative study to estimate the condition of waters, so that, conservation efforts can be easily done. In this study, we use *Lobophyllia corymbosa* as a spawning broadcast coral to be the sample which is collected from Sinjai and Luwuk Banggai. The study areas are selected based on geographical patterns which are in the South and East of Sulawesi waters. Since they have a considerable distance, it is likely for them to produce high genetic variations. Genome DNA uses mitochondrial genome that is extracted from coral tissue. The result shows that the genetic diversity is high. From the two major groups provided, there have been 8 haplotypes for all locations. In addition, the *L.corymbosa* distribution between Sinjai and Luwuk banggai has a high genetic connectivity with 0.6 fixation index .

Keywords : Genetic Populations, Gene Flow, Hard coral, Conservation Management, *Lobophylliacorymbosa*.

1. Introduction

Genetic variation is a condition when there is genetic diversity in a species, either between the separated population geographically or individual in a population. The genetic variation found in an individual is closely related to the kinship that occurs in the same individual but geographically separated. In this case, biological connectivity that occurs in an organism can be used as a basis in maintaining population and recovering from damages [1]. Biological conservation by looking at the genetic relationship of an organism is based on several main theories of population spawning, genetic origin of organisms, biogeography, and distribution patterns [25]. Besides that, [16] given the same states that the pattern of relationship of an organism is influenced by several factors, such as reproduction, larval distribution, geographical patterns, and physical oceanography.

IOP Conf. Series: Earth and Environmental Science **116** (2018) 012060 doi:10.1088/1755-1315/116/1/012060

Coral is one of the aquatic organisms with a very important role. Nowadays, the existence of coral is increasingly threatened, the damage has become worse which are caused by anthropogenic and natural factors. Could not be denied that the condition occurs in the organism are sixty percent caused by the activities carried out by humans in the water either directly or indirectly [12]. In addition, for coral conditions in the World's Coral Triangle Area, experiencing a higher threat compared to globally, the threat of anthropogenic activity as much as 85% is experienced in the region as the world's highest provider of marine biodiversity [5]. One of the hard corals needed to be preserved today is *Lobophyllia corymbosa*, since it is a kind of ornamental corals that have been widely exported and traded. Based on the data published by the International Union for Conservation of Nature (IUCN) in 2014 [15] which states that *L.corymbosa* species into the vulnerable organism. Seeing the pattern of coral distribution in a waters will provide information about the ability of the organism in adapting to an environment, at least genetic variation is directly related to the ability of an organism to live in a waters [9, 11] Reproduction of *L.corymbosa* that occurs sexually with external fertilization [6] may support the possibility of genetic connectivity in any population of *L.corymbosa* coral or other population with the same reproductive system. It is based on a longer-term developmental phase in the water columns compared to internal fertilized reproducing, so that, the duration during the settlement process of coral larvae will be longer and the spread triggered by current movement will become very easy to happen. The distance of the island and the area contained in Indonesia in general can be a transit area of a larva in its migration until it is finally settle to the suitable substrate. The current oceanographic factor plays a major role in the migration of larvae. Coral life cycles with limited mobility in the larval phase make the movement of these organisms strongly influenced by oceanographic conditions occurred in waters. Wherein its planktonic phase, the larvae migrate follows the movement of the current waters from the hatching site to the area to be alive and growing, and this displacement can take place within a few meters to hundreds of kilometers depending on the current velocity at a site.

The former research [2] has found *Acropora* sp., a species collected from Caribbean waters, geographically distributed as far as 500km and it is seen through a genetic approach. Coral population in Sulawesi waters has a high diversity of organisms, especially in the Sinjai and Luwuk Banggai areas as a Coral Triangle Area. Geographically, the distance from these two areas is ± 700km. This research will look at *L.corymbosa* distribution and genetic variation based on geographical location, so it can be predicted for other hard coral species that have the same reproductive system.

2. Research Methods

This research conducted in Sulawesi waters is done by purposive random sampling. Sample of hard coral *Lobophyllia corymbosa* species were taken using hammer and chisel along 3-7 cm as many as 25-30 coral fragments. We collected the coral sample in a depth of 3-7 meters in Sinjai and Luwuk Banggai as a geographically separated area (Fig.1). The coral fragments that have been taken will be preserved into 96% alcohol and fed into plastic containers for subsequent transport to the laboratory. Alcohol replacement is performed when the sample has been in the laboratory using 100% absolute ethanol which is then stored at room temperature [14].

In this study, there are several stages of the procedure to be performed, including as follows ;

2.1. Molecular analyses (mtDNA)

In this study the sequenced genome is mitochondrial DNA or mtDNA to the coral, at this stage performed several stages in the molecular analysis procedure that begins with DNA genome extraction, PCR (Polymerase Chain Reaction), electrophoresis and sequencing.

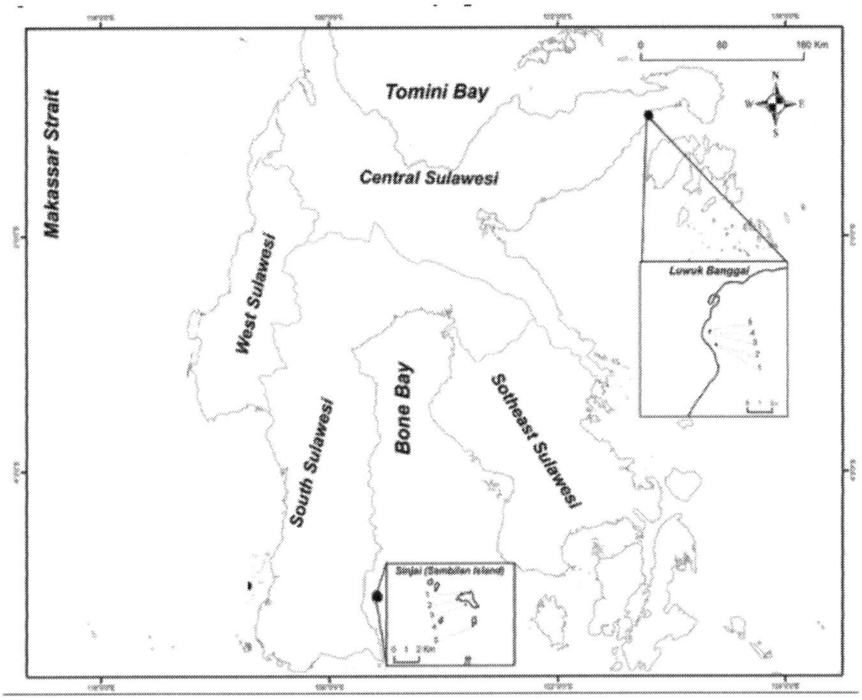

Figure 1. Map of research sites in Sulawesi waters (There are from Sinjai and Luwuk Banggai).

2.1.1. Exraction

DNA genome extraction was performed using an extraction kit (Qiagen Kit), DNA extracted the following the previous reseaarch [14]. The extraction process carried out in principle is the same as network retrieval in the sample. At this stage, coral tissue retrieval by crushing fragments of coral samples that have been collected and weighing 0.10 grams, Then, put in a tube of 1.5 ml. After that, the tissue was given an ATL Buffer solution of 180 μl and added with 20 μl Proiteinase K. The second solution was homogenized using a vortex for 15 seconds, then incubated in a heat block at 56 ° C for 24 hours (over night) to obtain a good lysis result.

The result of lysis has been obtained, then re-vortex before the next reagent. After that, the ALL Buffer is 200 μl, then added with 200 μl ethanol 96%. All reagents that have been mixed always end with stirring using vortex. Furthermore, the supernatant fluid on the tube was removed in a 2 ml Dneasy mini spin column which was then centrifuged for 1 minute at a speed of about 8000 rpm.

At this stage, there is a filtered network on the mini-spin filter paper. The escaped liquid is subsequently discarded along with the tube. Replacement of tube is performed for subsequent reagents. Mini spin column that has been placed on the new tube then given a solution of Buffer AW 1 as much as 500 μl and re-centrifuged for 1 minute at a speed of 8000 rpm. Next, add the Buffer AW 2 solution and centrifuged for 3 minutes at a speed of 14 rpm. For good results, repeat the centrifuge process for 1 minute at 8000 rpm. Then, remove the remaining liquid filter along with it's tube. After that, the spin column was transferred into a 1.5 ml tube which was then added with 100 μl ddH2O for DNA elution. The next liquid is incubated at room temperature for 1 minute and then centrifuged at 8000 rpm in 1 minute. This separation phase of DNA (elution) is repeated once so the volume of DNA template obtained is 200 μl. The extraction results are then preserved in a frezzer at -4°C.

2.1.2 Polymerase Chain Reaction (PCR)

PCR is a technique in the stages of molecular work that serves to replicate DNA enzymatically without the aid of the organism. In this process, the reagents used are MyTaq RedMix (Bioline). Where the reagents contained therein are mixed in one container such as Enzyme, MgCl2,

PCR Buffer and others. At this stage, primer selection is carried out, referring to the ascension numbers contained on the NCBI Web (National Center for Biotechnology Information), specific mitochondrial COI primers can be designed and in BLAST to be suitable for the target sample. The DNA base order obtained for the primers in this sample is mtCOIF CAGGCGCTATGTTAGGAGATG as a forward and mtCOIR CCCGCTAATACAGGCAAAGATA as a reverse. Primer selection is intended to find a primer that can be amplified, because the success at this stage is based on the primary suitability used.

The first thing to do in the PCR process is to prepare 1.5 ml tube as a master mix to mix ddH2O, primer forward and reverse and reagent in one container then mix. Next, the reagent on the master mixed tube was inserted into a 24 μl PCR strip and added 2 μl of a DNA template. Furthermore, the mixing results are spin down in order that there is no bubble and then inserted into the PCR machine with the machine settings referring to the basic protocol to amplify the mitochondrial genome based on research [10]. Then, were performed 38 cycles, each cycle, pre-denatured with 95°C for 120 s, denaturation at 94 °C for 45 seconds, annealing or pasting at 53°C for 30 seconds and extending at 72°C for 90 seconds and Final stage at 72 °C for 5 minutes.

2.1.3. Electrophoresis

Success or failure of PCR process can be known by visualizing PCR products through electrophoresis process. The first stage of electrophoresis is making of 1% agarose gel by 0.75 gram using SB Buffer solvent as much as 75 ml. The solution is entered into a beaker glass which is then cooked using a microwave for 3 minutes or until the agarose powder has dissolved completely. Thereafter, Ethidium Bromide (4 μl) dye was added and then poured into a combed mold and allowed to stand for 15-20 minutes. Subsequently, 4 μl of the product The PCR product was taken and mixed with the Loading dye (1 μl), then inserted in the well in the agarose mold submerged in the SB Buffer solution. Furthermore, a marker added to the well that does not contain PCR products, as a marker in view of the location of the DNA band at the time on visualization steps. Electrophoresis was performed at a voltage of 100 V, 400 Amp for 30 min. The band of electrophoresis can be seen by using ultraviolet light on UV transluminator. Positive PCR results contain the desired DNA, then sent to the Sekuensing Center, Berkeley. DNA sequencing is a technique for determining the sequence of nucleotide bases in DNA molecules, it aims to determine the genetic identity. The principle of this sequence is to determine the order of base ACGT used as a template for later amplified using enzymes [27].

2.2. Data analyses

To achieve the research purpose, there are some data analysis process to be done, referring from the data analysis that has been used in the previous researches.

2.2.1. Gene flow (Genetic Connectivity) and Haplotype Network

Sequencing results have been obtained, then the results are sorted and edited by using MEGA 5.05 software (Molecular Evolutionary Genetic Analysis) [24]. The edited data is then analyzed by the AMOVA (Analysis Molecular of Variance) program found in Arlequin [7]. Analyze the data to obtain the Fixation Index value and the genetic population structure of the coral sample.

In addition, to see the haplotype network in both populations, we used Network 5.0.0.1 software, so that, the haplotype composition will view in each population is presented in the form of connected diagrams.

2.2.2. Genetic Diversity

The diversity of haplotypes and the diversity of nucleotides [18] in the *L.corymbosa* sample, was measured using DNAsp software 5.0 [22], the values obtained in the analysis are then incorporated into the categories stated by Nei [21] in Table 1.

Table 1. Category index of Genetic diversity [21].

Anlaysis	Categories		
	Low	Medium	High
Genetic Diversity	0.1-0.4	0.5-0.7	0.8-1.0

3. Result and Discussion

3.1. Gene flow and Genetic Patterns

The sequencing product had more than 500 bp and after removing and editing the base nucleotides considered to be ambiguous long bases to 450 bp, the mitochondrial genome derived from 60 total samples taken from Sinjai and Luwuk Banggai Waters (each 30 samples per population). Sixty samples, only 30 were successfully amplified and produced an informative sequencing product. Speciments from Luwuk Banggai waters have 10 detected polymorphic sites, which produce 4 haplotypes. These polymorphisms included 9 transitions and 1 transversion. While in Sinjai waters, detected polymorphic sites are 9 with 4 haplotypes. Polymorphisms consist of 8 transitions and 1 transversion. The results of all samples obtained have a similarity about 98% -100% with *L.corymbosa* species on Genebank with an accession number AB117241 [10].

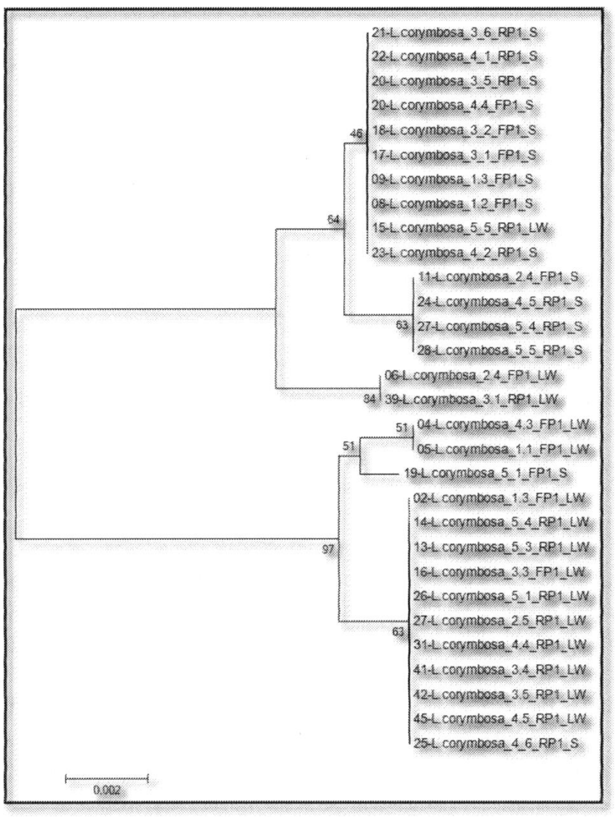

Figure 2. Neighbour-joining tree of mtDNA. Bootstrap values for Sinjai (S) and Luwuk Banggai (LW) are shown above branches.

When it is viewed from the tree phylogeny shown in Figure 1, formed 2 large clusters are divided into several groups. In the first cluster, the cluster branch shows the existence of gene contributions from the Luwuk Banggai population incorporated in a large section of the Sinjai population group, and vice versa. Results demonstrated by phylogeny trees, supporting the value of genetic distance analysis within and between populations of *L.corymbosa* (Table 3) at Luwuk Banggai and Sinjai. The genetic distance is the value that shows the genetic relationship found in a population. In this study, the genetic distance values of both populations fall into the low category based on the value determined by [20] which states that, the lower the genetic distance value in a population indicates that there is a very close genetic relationship in the population. The genetic distance in the population from 2 sampling sites is 0.003 and for the inter population is 0.007. The low of value, indicating that the current flow in both of locations still affect each other.

Sinjai waters entering the Bone Bay area are the waters that are still directly influenced by the current factor comes from Makassar Strait and Flores Sea. The absence of spawning recording data on spawning period of *L.corymbosa* specifically makes predictions of larval distribution of the reef a little difficult. However, if the referring is based on the constant reproduction of reefs that occur in tropical waters, especially on hard corals that reproduce sexually externally (broadcast spawning coral) so that spawning cycle occurs on an annual basis (Single Gametogenic Cycle) and generally reproduce in August – December [13]. The period of reproduction time is directly related to West Monsoon Wind period occurring at the time of spawning, especially in the waters of Sulawesi. In October - April, the wind blows from the Continent of Asia to the Continent of Australia, thus, the movement of the wind that affects the surface currents that occur in the Sulawesi waters blow from North to South Sulawesi waters and receive the flow of Flores Sea flows so it turns to the left (East) Sulawesi Waters. This current flow makes the possibility for the Sinjai and Luwuk Banggai Waters to share genes, because according to expert [4] that current circulation patterns can provide genetic information from an organism that has a dispersion that depends on the movement of the current.

Table 2. Hierarchical analysis of molecular variance (AMOVA) that packaged in Arlequin ver 3.5.2 of mtDNA.

Source of Variation	d.f	Sum of Squares	Variance components	Percentage of Variation
Among Population	1	29.267	1.862 Va	58.09
Within Population	28	37.600	1.343 Vb	41.91
Total	29	66.867	3.204	
Fixation Index	**Fst**	**0.581**		

The value of genetic distance obtained and included in the Nei's categories [20], the category has an index that is directly proportional to the Fixation Index category (Fst) that obtained in the entire population. The value of Fst or the value of population structure is useful to provide a genetic interpretation that shows the genetic flow in a population. Coral population of *L.corymbosa* in Sinjai and Luwuk Banggai showed Fst value of 0.581 or 0.6 (Table 2). [8] further explains that genetic flows that are in the range of 0.6-1.00 fall into the high category. Thus, it can be said that the greater the value of the genetic flow, the greater of genetic flow entering a population. The genetic flow rates obtained at both sites provide information that there is a close familial relationship that may represent genetic connectivity at each location.

Based on the patterns of genetic connectivity obtained from both populations, it can be clearly assumed that there is a high biological linkage to the population. However, if viewed directly at the *L.corymbosa* coral sampling site, these coral waters are classified at an alarming rate, where anthropogenic activities that tend to be damaging are still frequently performed which can threaten sustainability or even break the genetic network of the coral species as well other. According to [19] spatial and temporal connectivity can be used as an implication of coral reef ecosystem management, particularly in designing marine conservation areas. However, if these threats persist, there is no possibility of causing genetic flow termination in both populations.

Table 3. Estimates of Average Evolutionary Divergence over Sequence Pairs Within and Among Groups. (The number of base substitutions per site from averaging over all sequence pairs within each group are shown. Analyses were conducted using the Maximum Composite Likelihood model [23]. The analysis involved 30 nucleotide sequences. All positions containing gaps and missing data were eliminated. There were a total of 450 positions in the final dataset.

Genetic Distance *D*	Location	Luwuk Banggai	Sinjai
Within population	Luwuk Banggai	0.003	-
	Sinjai	-	0.003
Among Population	Luwuk Banggai	-	-
	Sinjai	0.007	-

3.2. Genetic Diversity

The results of analysis performed with DNAsp software, obtained 8 haplotipe (*h*) for the entire population (Table 4). The population of Sinjai waters has 4 haplotypes, as well as Luwuk Banggai's performances of 30 total samples amplified. In addition, for genetic diversity in each population showed a level of haplotypes diversity (*Hd*) in the medium category ranged from 0.552-0.600 based on Nei classification in 1987 [21]. As for the nucleotide diversity (π) ranged from 0.005-0.007.

Table 4. Genetic diversity pututative population of *Lobophyllia corymbosa*.

Site	Specimen	Number Of Haplotype *h*	Haplotype diversity *Hd*	Nucelotida diversity π	Mean pairwise difference	Number of sites with subtitutions *S*
Luwuk Banggai	30	4	0,552	0,007	3,030±1,670	10
Sinjai	30	4	0,600	0,005	2,343±1,354	9

The high genetic diversity/haplotype values in a population reflect the size of the population in an area, while the decrease in population size will reduce genetic diversity or haplotype diversity. The genetic diversity found in Luwuk Banggai and Sinjai is still categorized in the average population size. However, the low genetic diversity category ranges from 0.1-0.4 [21] where, both populations tend to be threatened for impairment, if destructive anthropogenic activity is still common in these waters. One study conducted [26] on the soft coral species *Scleronephthya gracillium* shows the environmental stress that occurs in the coral resulting in a change in gene expression in response to environmental changes and damage. If the change is not lethal and is passed on to the next generation it will increase the variation within the population, however, if the change is lethal it will decrease the population size. Genetic diversity provides the ability for an organism to adapt to environmental, climate change and disease [17]. Therefore, the genetic diversity in a population is a great capital for growth, development and regeneration.

Connectivity is closely related to genetic diversity. A well-preserved genetic connectivity will affect the high genetic diversity formed in a population. According to statement from the expert [3] the genetic diversity of a population will increase if there is genetic input from other populations or commonly referred to as genetic migration and that is the case in the waters of Luwuk Banggai and Sinjai (Figure 3). This is because a large migration will make in cross-breeding and mixing of genes between different populations, so that different gene variations can be obtained and may provide information on migrating pathways of coral larvae during the spawning season. Therefore, it is hoped that there will be special attention to efforts to improve the water quality in Indonesia, one of which is Sulawesi as the center of the world's coral triangle. The preservation of waters, it will also maintain genetic susceptibility to an organism and will also preserve the biodiversity of Indonesia, because genetic is the most basic and also smallest level of biodiversity for a conservation.

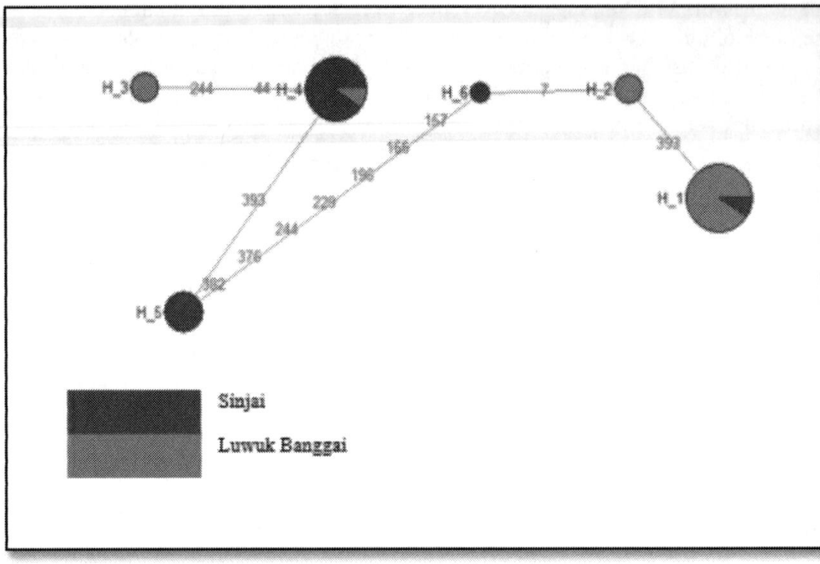

Figure 3. Haplotype network of native (n1 – n6), mtDNA haplotypes found in Sinjai and Luwuk Banggai.

4. Conclusion

The genetic diversity in Luwuk Banggai and Sinjai waters is in the high categories, but, tends to decrease in population size and based on genetic distance values and genetic flows of these two populations have high relation or connectivity and genetic mixing between them. So, the spacing between them is not an obstacle to the occurrence of genetic migration supported by the circulation of currents to connecting them.

5. Acknowledgments

The authors would like to express gratitude for the PMDSU Scholarship given by DIKTI. Samsul Basri Karepesina, which provides assistance to collect the samples in the field. Also to the IBRC Laboratory for providing facilities needed in molecular sample work.

References

[1] Almany G., R, S., R., Connolly, D., D., Heath, J., D., Hogan, G., P., Jones, L., J., McCook, M., Mills, R., L., Pressey and D., H., Williamson. 2009. Connectivity, Biodiversity Conservation and The Design of Marine Reserve Networks For Coral Reefs. *Coral Reefs*, **28**: 339–351.

[2] Baums, I., B., Miller, M.,W. and Hellberg, M., E. 2005. Regionally Isolated Populations of an Imperiled Caribbean Coral *Acropora palmata*. *Mol Ecol.***14**:1377–1390.

[3] Beardmore, J., A. 1983. *Extinction, Survival, and Genetic Variation*. On: *Genetics and Conservation*. Schonewald-cox, C., M., S., Chambers, M., Macbryde, B. and Thomas, W., L. (Editor). The Benjamin/Cummings Publishing Company, Inc. London

[4] Botsford, L.,W., White, J.,W., Coffroth, M.,A., Paris, C.,B., Planes, S., Shearer, T.,L., Thorrold, S.,R. and Jones, G.,P. 2009. Connectivity and Resilience of Coral Reef Metapopulations in Marine Protected Areas: Matching Empirical Efforts to Predictive Needs. *Coral Reefs*. **28**:327–337.

[5] Burke, L., Elizabeth, S., dan Mark, S. 2002. Terumbu Karang yang Terancam di Asia Tenggara. World Resources Institute. USA.

[6] Coexploration. 2017. Life Cycle Story. [online].http://www.coexploration.org.

[7] Excoffier, L. and Lischer, H. 2009. Arlequin ver 3.5 user manual ; An Integrated Software Package For Population Genetics Data Analysis. Swiss Instituteof Bioinformatics.

[8] Exoffier L, Smouse, P., E.and Quattro, J., M. 1992. Analysis of Moleculer Variance Inferred From Metric Distance Among DNA Haplotipes; Application to Human Mitochondrial DNA Restriction Data. *Genetics*. (**131**): 479-491.

[9] Frankham, R. 2005. Stress and Adaptation in Conservation Genetics. *Journal of Evolutionary Biology*. **18**. Issue 4, pp 750-755.

[10] Fukami H., Budd A., Paulay G., sole-Cava A., Chen C.A., Iwao K., Kwlton N. 2004. Conventional Taxonomy Obscurres Deep Divergence Between Pacific and Atlantic Corals. *Nature* **427**:832:834.

[11] Gates, R.,D. and Edmunds, P., J. 1999. The Physiological Mechanisms of Acclimatization in Tropical Reef Corals. *Am Zool* **39**:30– 43.

[12] Goatley, C. H. R., R. M. Bonaldo, R. J. Fox, and D. R. Bellwood. 2016. Sediments and Herbivory as Sensitive Indicators of Coral Reef Degradation. *Ecology and Society* **21**(1):29.

[13] Harrison, P.L. and Wallace, C.C. (1990). Reproduction, Dispersal and Recruitment of Scleractinian Corals. In :Dubinzky, Z. (ed.) *Coral Reefs*. pp. 133-207.

[14] Huang, D., W., Y., Licuanana, A., H., Baird and H. Fukami. 2011. Cleaning Up The 'Bigmessidae': Molecular Phylogeny of Scleractinian Corals From Faviidae, *Merulinidae*, *Pectiniidae* and *Trachyphylliidae*. *BMC Evolutionary Biology*. **471**-2148/11/3.

[15] IUCN. 2014. *Lobophyllia corymbosa*. The IUCN Red List.

[16] Jackson J., B., C. 1986. Modes of Dispersal of Clonal Benthic Invertebrates: Consequences For Species Distribution Sand Genetic Structure Of Local.

[17] Lande, R. 1988. Genetics and Demography in Biological Conservation. *Science*. **241**(4872): 1455-1460.

[18] Lynch, M. and Crease, T., J. 1990. The analysis of Population Survey Data on DNA Sequence Variation. *Moleculer Biology Evolution*, **7**: 337–394.

[19] Munday, P., L., Leis, J., M., Lough, J., M., Paris, C., B., Kingsford, M., J., Berumen, M., L. and Lambrechts, J. 2009. Climate Change and Coral Reef Connectivity. *Coral Reefs*. **28**: 379–395.

[20] Nei, M. 1972. Genetic Distance Between Population. American *Nature,* **106**: 283-292.

[21] Nei, M. 1987. *Moleculer Evolutionary Genetics*. Columbia University. Press. New York. 512 hal.

[22] Rozas, J., Sanchez-DeI, B., J., C, Messeguer, Rozas, X., R. 2003. DnaSP, DNA Polymorphism Analyses by The Coalescent and Other Methods. *Bioinformatics,* **19**: 2496–2497.

[23] Tamura, K., Nei M., and Kumar S. 2004. Prospects for Inferring Very Large Phylogenies by Using The Neighbor-Joining Method. *Proceedings of The National Academy of Sciences* (USA) **101**:11030-11035.

[24] Tamura K., Peterson D., Peterson N., Stecher G., Nei M., and Kumar S. 2011. MEGA5: Molecular Evolutionary Genetics Analysis using Maximum Likelihood, Evolutionary Distance, and Maximum Parsimony Methods. *Molecular Biology and Evolution* (In Press).

[25] Whittaker R. J., M. B., Araújo, P. Jepson, R. Ladle, J. E., Watson and K. J., Willis. 2005. 'Conservation Biogeography: Assessment and Prospect', *Diversity and Distributions*, **11**, pp 3–23.

[26] Woo, S., Sungjin, H., Joon, I., S., Chaolun, A.,C, Dongsung, K. and Seungshic, Y. 2012. Transcriptomic Signature in Soft Coral Exposed to Abiotic Stresses. *Proceedings of The 12th International Coral Reef Symposium* : 1-2.

[27] Zein M. S. A. and Prawiradilaga D. M. 2013. DNA Barcode Fauna Indonesia. Kencana Prenada Media Group. Jakarta. 241 Hal .

Vertical Distribution of Temperature in Transitional Season II and West Monsoon in Western Pacific

Hikari A H Pranoto[1*], Kunarso[1], Endro Soeyanto[2]

[1] Department of Oceanography, Faculty of Fisheries and Marine Science Diponegoro University, Jl. Prof. H. Soedarto, SH, Tembalang, Semarang 50275, Indonesia
[2] Laboratory of Marine Survey, The Agency for Assessment and Application of Technology, Jl. MH. Thamrin 8, Central of Jakarta, Jakarta 10340, Indonesia

*Corresponding author Email: hikariazizah@gmail.com

Abstract. Western Pacific is the water mass intersection from both the Northern Pacific and Southern Pacific ocean. The Western Pacific ocean is warm pool area which formed by several warm surface currents. As a warm pool area and also the water mass intersection, western Pacific ocean becomes an interesting study area. The object of this study is to describe the temperature vertical distribution by mooring buoy and temporally in transitional season II (September – November 2014) and west monsoon (December 2014 – February 2015) in Western Pacific. Vertical temperature and wind speed data that was used in this study was recorded by INA-TRITON mooring instrument and obtained from Laboratory of Marine Survey, BPPT. Supporting data of this study was wind vector data from ECMWF to observe the relation between temperature distribution and monsoon. The quantitative approach was used in this study by processing temperature and wind data from INA-TRITON and interpreted graphically. In the area of study, it was found that in transitional season II the range of sea surface temperature to 500-meter depth was about 8.29 – 29.90 °C while in west monsoon was 8.12 – 29.45 °C. According to the research result, the sea SST of western Pacific ocean was related to monsoonal change with SST and wind speed correlation coefficient was 0.78. While the deep layer temperature was affected by water mass flow which passes through the western Pacific Ocean.

Keywords: *Temperature Vertical Distribution, Western Pacific Ocean, INA-TRITON*

1. Introduction

(*Triangle Trans Ocean Buoy Network*) is one of mooring instrument from JAMSTEC which has been deploying some buoys since 1998. This mooring system buoy sends real-time data to the satellite then received and stored in central data. This mechanism allows the observer to monitor time series data without physical presence in the area. One of the TRITON observation area is Western Pacific Ocean. In September 2014, JAMSTEC (Japan) together with BPPT (Indonesia) deploy INA-TRITON mooring buoy which was using m-TRITON as the buoy. INA-TRITON was made to continue the observation from the previous buoy in Western Pacific, specifically to monitor meteorology condition in Western Pacific. Meteorologists had been monitoring sea surface, especially Sea Surface Temperature (SST) as the main factor that affects low-frequency atmosphere change. The relation between sea and atmosphere is not only about climate system but also the prediction, thus from interdecadal climate variability system showed that sea and atmosphere are related [1]. Western Pacific Ocean as the intersection between North Pacific and South Pacific caused this water is passed by several water masses from both oceans. The mixing from both water masses always affect the salinity variability and especially in thermocline layer. The study area is also known as Indo-Pacific warm pool area which being interesting case to observe the vertical distribution of temperature, the warpool is formed by watermass mixing in the area. Oceanographically, water mass in Western Pacific affect Indonesia Water circulation, such as New Guinea Coastal Current, New Guinea Coastal Undercurrent, Mindanao Current and North Equator Counter Current [2]. Western Pacific also affected by seasonal change in Indonesia water which differs known as West Monsoon (December – March), Transitional Season I (April – May), East Monsoon (June – September) and Transitional Season II (October – November), the seasonal change itself related to the direction change of Monsoon [3]. In Western Tropical Pacific, local wind play important role in driving thermocline anomalies. The topic is accounted for by wind-induced Ekman Pumping anomalies [4]. INA-TRITON time series data was used to observe the needed information of research study. This research study had been aimed to vertical temperature distribution related to Transitional monsoon II and west monsoon condition.

2. Materials and Methods

2.1. Area of Study

The research study was determined by The Agency for Assessment and Application of Technology (BPPT) with purposive sampling method. The consideration of the study area was the Western Pacific has various water masses and also there was previous mooring buoy placed on research study location which is replaced with INA-TRITON. The buoy deployment was located on 1°48'15.87" N and 138°30'26.13" E. the study area was chosen assuming it represents the water variability of Western Pacific [5] and be the sequence of the previous study with TRITON buoy in location. Research Study Location had been described on Figure 1.

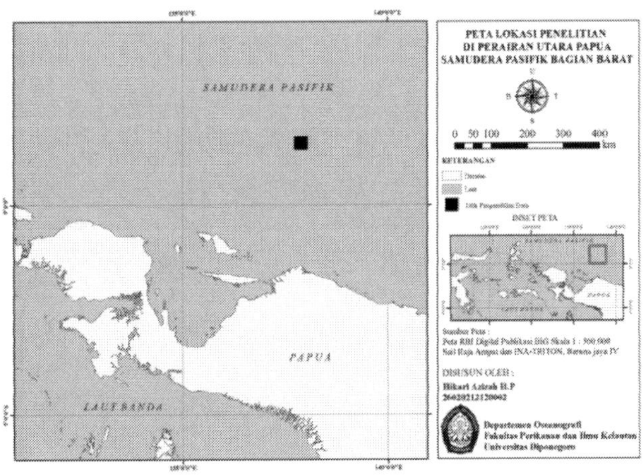

Figure 1. Buoy Deployment and Area of Study

2.2. Period of research

INA-TRITON was deployed in September 2014 in the study area recorded from early September 2014. The data that used in this study was in transitional season II (September – November 2014) and in west monsoon (December 2014 – February 2015).

2.3. Collecting Primary Data

The primary data was conducted by INA-TRITON mooring which recorded every 10 minutes and averaged by the hour. The first primary data was vertical temperature data which recorded by SBE-37-IM from Sea Bird Electronics as CTD (Conductivity, Temperature and

Depth) sensor. The data recorded on 1.5, 25, 50, 74, 100, 125, 150, 175, 200, 250, 300 and 500 meter depth. The second primary data that used was wind surface which recorded by 7050-a from WHOI (with R.M. Young Type 05103 addition). The data was recorded and obtained from Laboratory of Marine Survey, BPPT.

2.4. Collecting Supporting Data

The supporting data was wind surface data that obtained from European Centre for Medium-Range Weather Forecasts (ECMWF). The data was a model from the agency that downloaded in website http://ecmwf.int/ with spatial resolution 13.8 km with level IIIB and resolution $1/12^0x1/12^0$ which consist of component zonal wind (u) and meridional (v).

2.5. Analysis Method

Data from INA-TRITON was recorded every hour as mentioned before, and then it was filtered to sort data from invalid data and null data. Then, for vertical temperature data, every depth was averaged from hourly to daily by Microsoft Excel software, data was averaged again to monsoonal which was in transitional season II and west monsoon. After filtered and averaged, data proceeded by Surfer 10 to interpret vertical temperature distribution temporally in Western Pacific. While wind surface data was averaged per day by Microsoft Excel and interpret by the graphic to describe the windspeed, this data used to observe the correlation between windspeed and sea surface temperature as the monsoonal changes condition.

Wind surface data from ECMWF was obtained in NetCDF (Network Common Data Form) and extracted by ODV 3.4.2 software. The result of the extract was the component of zonal wind (east to west (u)) and meridional (north to south (v)) in altitude reference 10 meters. The data proceeded in Microsoft Excel to obtain wind speed and direction. Then, the data proceed by ArcGIS 10.0 to interpret wind distribution pattern in spatial.

The relation Between SST and wind data from INA-TRITON obtained by find the correlation between both data. The method which used to determine correlation coefficient was Pearson Correlation equation [6]. To interpret between two variables there are criteria according to correlation coefficient value.

Table 1. Coefficient of correlation value from Pearson Correlation

Correlation Coefficient	Relation Interpretation
0 – 0,2	Very Low
0,2 – 0,4	Low
0,4 – 0.7	High
0,7 – 1,0	Very High

3. Result and Discussion

As the result of temperature distribution, according to INA-TRITON data, it was found that temperature distribution pattern in Western Pacific was having colder from September 2014 to February 2015. Vertical temperature profile described the changes of temperature to the depth from surface column up to 500-meter depth. The decreasing of temperature value in the area of study can be seen as the temperature contour was crawling up since September 2014 to February 2015 in Figure 2.

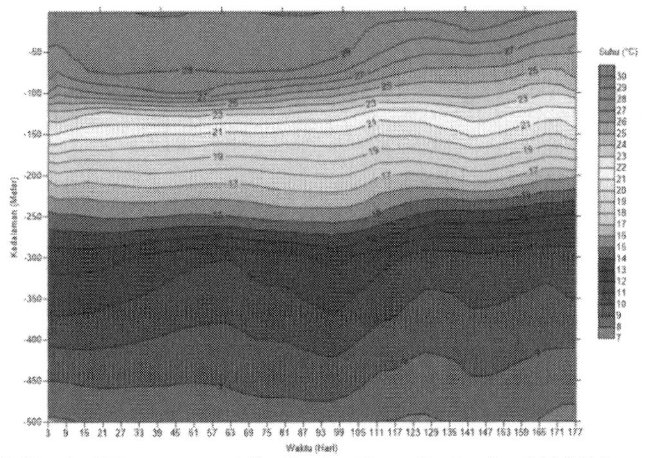

Figure 2. Vertical Temperature Distribution From September 2014-February 2015

According to Figure 2 is represented the averaged temperature was decreasing, in case the temperature average of the surface column was 29.9 °C in transitional season II and become 29.45 °C in west monsoon. The warm temperature is the character of Pacific warmpool, which is warmer than 28 - 29 °C [7] The lowering temperature that exists in sea surface might have a correlation to the monsoonal wind in area of study. The windspeed was stronger in west monsoon than in transitional season II which presented in Figure 3. The increasing of windspeed also might increase the mixing process in the water column. The mixing process will bring the colder water mass from deeper water column to the surface and the sea surface temperature is colder as the result [8]. The decreasing of sea surface temperature in west monsoon was assumed related to upwelling process, that from isotherm line as the indicator which shallowing to the surface in west monsoon (Figure 2). Upwelling process trigger colder water mass from the deeper column to surface water column, this event caused the temperature in the surface water column colder than before due to the entrance of deeper water mass[9].

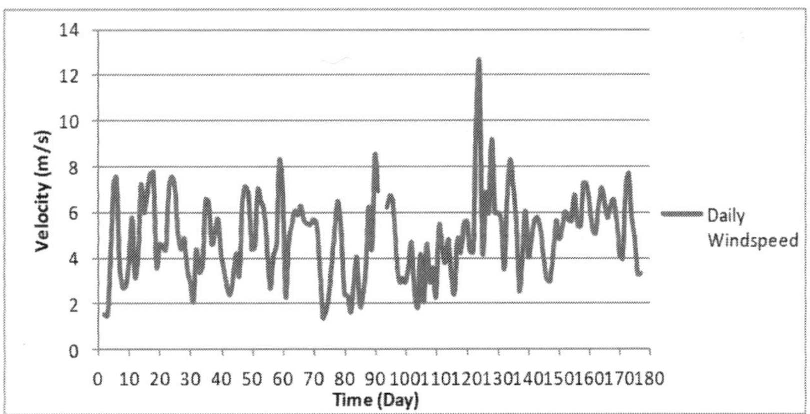

Figure 3. Daily averaged windspeed from September 2014 – February 2015

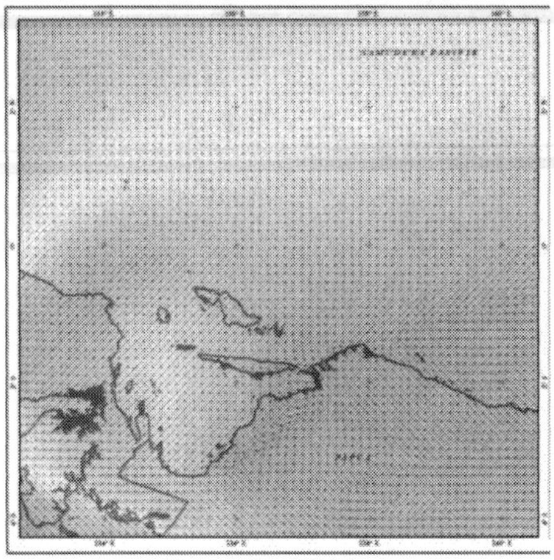

September 2014

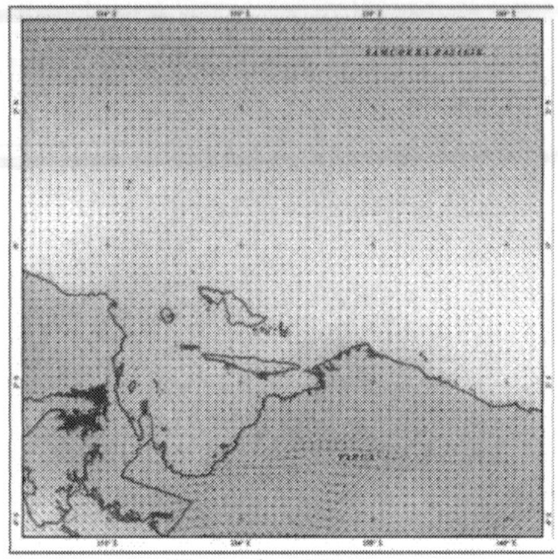

December 2014

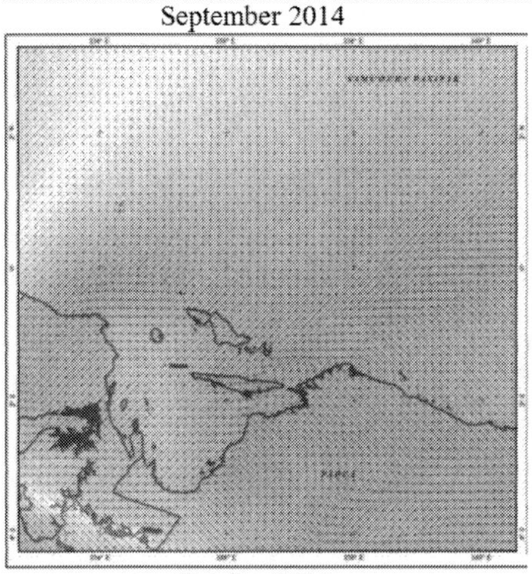

October 2014

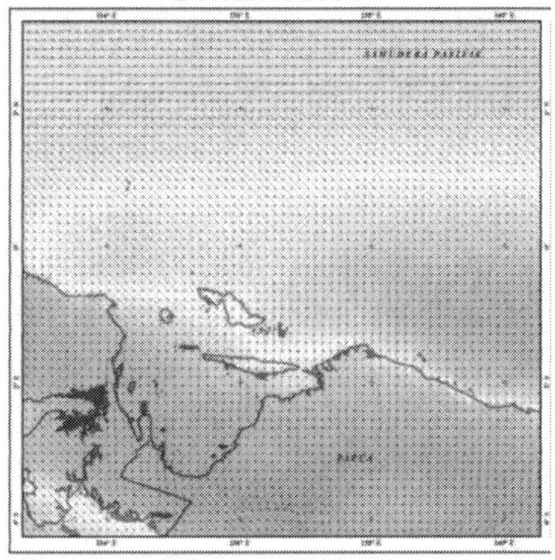

January 2015

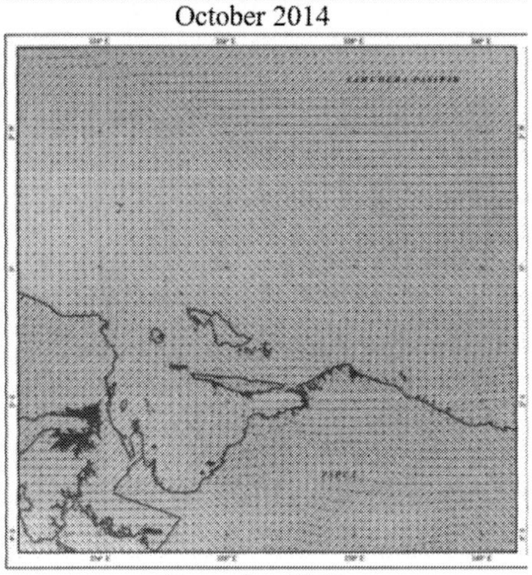

November 2014

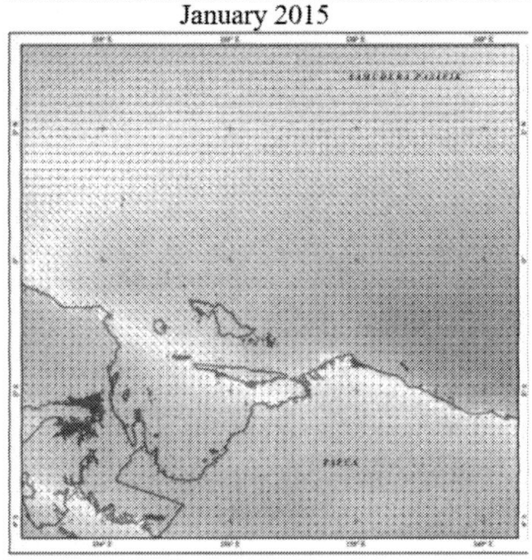

February 2015

Figure 4. Windspeed and direction on September 2014 – February 2015

Upwelling process in Western Pacific was assumed due to current divergent process. As depicted in Figure 4, wind as an indicator had divergent. Wind that was burst from west to east through Western Pacific was having divergent process where some of wind burst turned to the northeast and some of them to the southeast. Wind which burst to the northeast was trigger Ekman current to South. Thus current divergent were formed to east and south, this process allows empty column on water surface which would be filled by its deeper water mass, this event is known as upwelling [10]. This event also mentioned by [5] as the existence of New Guinea Eddy (NGE) that advected by New Guinea Coastal Undercurrent, North Pacific and North Equator Counter Current. Upwelling itself in general terms is vertical displacement of deep waters toward the surface. In reality, upwelling is never takes place deeper than few hundred meters [11].

Table 2. Averaged temperature for depth in transitional season II and west monsoon

Time	Depth (meter)										
	-1.5	-25	-50	-75	-100	-125	-150	-200	-250	-300	-500
Transitional Season II	29.90	29.54	29.26	29.02	27.25	22.96	21.16	17.53	15.06	11.80	8.29
West Monsoon	29.45	28.82	27.64	26.33	24.38	21.89	20.54	17.21	13.95	10.92	8.12

Besides sea surface temperature which had increased contour, the temperature on deeper water column was also increased. This kind of situation present in Table 2, which showed temperature for every sensor depth and averaged in transitional season II and west monsoon. From the table above obtained that The temperature on every recorded depth in west monsoon was colder than transitional season II. This event can be seen in Figure 2, it also showed phenomenon on isotherm 11°C which has wavy pattern that head to the upper water column.the wavy isotherm pattern was showed on ±300-400 meter depth. This wavy pattern was assumed that it indicated the internal wave existence in depth and study area. The internal wave will interfere the boundary layer (the bottom thermocline boundary layer). In the tropical Pacific, the main currents, westward flowing over the upper ocean are forced by the easterly trade winds. In the first 10 meters water depth immediately away from the equator, deflection by the rotation of the earth occurs to the right in the northern hemisphere and the left in the southern hemisphere. This coriolis causes ekman transport [12].The internal wave could be exist due to density difference between thermocline layer and deep water layer. Nontji [13], that internal wave could be identified with the existence of wavy pattern on isotherm profile in deep water. Internal wave that occurs was assumed caused by water mass mixing right below the 12 °C isotherm. Thurman [7], mentioned that surface column of Western Pacific was affected by Subtropical Lower Water which occur on 120 – 200 meter depth, in deeper column was affected by Northern Intermediate Water on 250 – 400 meter depth and also Northern Intermediate Water on 600 – 800 meter depth. Nuriyanto[3], the water masses that exist in area of study was Southern Pacific Tropical Water, Northern Pacific Tropical Water, Northern Pacific Intermediate Water and Antartic Intermediate Water. The influence of water mass mixing to isotherm pattern, that water mass mixing possibly affect the isotherm pattern into wavy pattern [14]. The water mass mixing identified by temperature and/or salinity graphically change from pointed line into slope line [15].

4. Conclusion

Study of vertical temperature distribution in Western Pacific water found that the temperature was related to monsoon change which in this research was during transitional season II and west monsoon. Averaged temperature in every depth was having difference while in transitional season II was higher than west monsoon. The decreased of SST has been founded that in transitional season II was 29.9 °C on the other hand sea surface temperature in west monsoon was 29.45 °C. Mixing on surface water column due to wind divergention which triggered divergent ekman current occurred on the sea surface. On the deep water, the wind burst is not affected the water dynamic, but due to the wavy pattern toward the upper column below 12 °C isotherm, it assumed that internal wave exists caused by water mass mixing on the study area.

5. Acknowledgments

This acknowledges present for Diponegoro University which has supported knowledge for student and this study was conducted by the Japan Agency for Marine-Earth Science and Technology (JAMSTEC) and The Agency for Assessment and Application of Technology (BPPT) project of INA-TRITON. For specifically Laboratory of Marine Survey, BPPT, all crew of RV. Baruna Jaya IV and all participants in the cruise thank you for the research opportunity and data processing. We also thank all collaborated colleagues in data processing activity and in the revision of the manuscript.

References

[1] Habibie, Muhammad Najib 2014 *Jurnal Meteorogikal dan Geofisika* – BMKG
[2] Kashino, Y, M Aoyama, T Kawano, N Hendiarti, Syaefudin, Y Anantasena, K Muneyama, H Watanabe 1996 *Journal of Geophysical Research* **101 (C5)** :12.391 – 12.400
[3] Nuriyanto, Ahmad 2004 *Karakteristik Massa Air di Perairan Utara Papua Pada Bulan Juli 2002* (Bogor : Fakultas Perikanan dan Ilmu Kelautan – IPB)
[4] Iysne, Joann and Clara Deser 2002 *National Research Boulder*, Colorado
[5] Kashino, Y, I Ueki, Y Kuroda, and A Purwandani 2007 *Journal of Oceanography* **63** : 545 – 559
[6] Sudjana 1992 *Metode Statistika Edisi kelima* (Tarsito : Bandung)
[7] Bosc C, Thierry D and Christophe M 2009 *Journal of Geophysical Research* **114**
[8] Wyrtki K 1961 *Physical Oceanography of the Southeast Asian Waters NAGA Report Vol.2 Scripps Institution of Oceanography* (California: University of California) 195 p.
[9] Kunarso, S Hadi, NS Ningsih dan MS Baskoro 2012 *Ilmu Kelautan* **17** (2) : 87-96
[10] Thurman, E.M., D.A. Goolsby, M.T. Meyer, and D.W. Kolpin 1991 *Environment Science Technology* **25** :1794 – 1796
[11] Myberg, Kai and Oleg Andrejev 2003 *Boreal Environment Research* **8** : 97 – 112
[12] H. Dunxin, L Wu, W Cai, AS Gupta, A Ganachaud, B Qiu, AL Gordon, C Lin, Z Chen, S Hu, G Wang, Q Wang, J Sprintall, T Qu, Y Kashino, F Wang, W S Kessler 2015 *Nature* **522** : 299 – 307
[13] Nontji, A 1991 *Laut Nusantara* (Jakarta : Djambatan)
[14] Sidabutar, HC, A Rifai dan E Indrayanti 2014 *Jurnal Oseanografi* **3**(2) : 135 – 141
[15] Hadi, Safwan 2010 *Pengantar Oseanografi Fisis*. Institut Teknologi Bandung : Bandung

The Quality of Edible Film Made from Nile Tilapia (*Oreochromis niloticus*) Skin Gelatin with Addition of Different Type Seaweed Hydrocolloid

H Deanti[1], J M Hulu[1], A. Setyaji[1], R N Eliyanti[1], K Aliya[1], and E N Dewi[2]

[1]Student at Fish Product Technology Department, Faculty Fisheries and Marine Sciences UNDIP, Jl. Prof. Soedarto, SH, Tembalang, Semarang, Jawa Tengah – 50275,
[2]Fish Product Technology Department, Faculty Fisheries and Marine Sciences UNDIP, Jl. Prof. Soedarto, SH, Tembalang, Semarang, Jawa Tengah – 50275,

*Corresponding author Email : hilariadeanti2@gmail.com

Abstract : The functional properties of fish skin's gelatin lower than mammals, hence the gelatin proteins needed a polysaccharides hydrocolloids to form a continuous and more cohesive network of edible film. Polysaccharides hydrocolloid (carrageenan, agar and alginate) containing phenol compounds was oxidized to be converted into quinone and it was expected to act as a cross linking agent. The purpose of this study was to determine the characteristics (thickness, tensile strength, elongation, solubility and water vapour transmition rate) of edible film made from nile tilapia skin gelatin by adding different type polyssacharide hydrocolloid. Edible film was made by mixture of gelatin 5 g and addition of carrageenan (C1), agar (C2), alginate (C3) concentration ; 0,5% (v/w), all the materials were poured into 100 ml destilled water that containing 10% glycerol (w/w). The solution was then heated on a hot plate stirer at 40°C for 30 min and dehydrated in a oven at 50°C. All data were analysed using ANAVA. Based on the result it can be seen that the addition of oxidized polisacarides hydrocolloid have a significant effect on tensile strength (TS), water vapor transmision rate, solubility and elongation at break properties, but did not in thickness. Edible film gelatin with the addition of alginate has better characteristics viewed by tensile strength (23.05 Kgf/cm^2), water vapor transmission rate (0.61 gram/m^2/hour) and thickness (0.16 mm) than carrageenan and agar.

Keywords: Edible Film, Nile Tilapia, Gelatin, Hydrocolloid, Quality

1. Introduction

Packaging with edible coating or film is one of the new relatively preservation techniques of a food. Research undergone on coating of food products with edible coating has been extensively done and proved to be able to lengthen the saving life and improve the quality of food products. Edible film is an organic packaging material made from hydrocolloid and fatty compounds, or their combinations. The compounds of hydrocolloid that can be used are protein and carbohydrate, while fat that enables to be used is candle or wax, glycerol and fatty acids. Starch as a hydrocolloid compound, is a naturally formed polymer in varieties botanical/vegetable sources such as wheat, corn, potatoes, and tapioca. Starch as a renewable natural source is broadly available and easy to gain.

Generally, Edible film is an easily renewable substance. A constantly used biopolymer material is consisting of fat, polysaccharide and protein. Gelatin is a derivative of the amino acid glycine, proline and hydroxy proline. Gelatin can be made from the extraction of fish skin extraction. Edible film resulted from fish skin will have lower functional character if compared with mamals cartilage (cow and pig), so that the chemical and physical treatments are often applied to modify tissue through crosslinking of the polymer chains [1]. Chemicals commonly used as cross linking agents are generally toxic. So it is required the existance of cheap, natural, safe, renewable, and being able to increase the barrier of physical fish gelatin, especially in edible film..

Additional ingredients used in this study were carrageenan, agar, and alginate and glycerol. Carrageenan agar and alginate have the potential as an additional cheap elastic material that can be consumed, and renewable. The addition of the refined glycerol to make the film better becomes flexible and elastic. additionally, the addition of hydrocolloids is also rich because it is sert, safe, biodegradable and its presence is abundant in Indonesia.

The addition of seaweed as cross-linking on edible film will affect in the original properties of the protein. According to Athukorala et al. [2], seaweed extract contains of phenol compounds that can be converted to quinone when oxidized and potentially become cross-clearing agent on the formation of edible film. The existence of these crosslinks will make the pores of the film from the smaller and smaller protein, so that the water absorption of water vapor will be lower.

2. MATERIALS AND METHODS OF RESEARCH

2.1. Material

The sample used in this research was the skin of Nile tilapia fish (*Oreochromis niloticus*) gained from industrial waste. Fish skin obtained was under handling conditions using ice-cooling. The additional substances used in this study were acetic acid (CH_3COOH), glycerol, agar carrageenan and alginate.

2.2. Method of Research

The Nila's fish skin was gained from the processing of waste of fish fillets of Nile tilapia PT Aquafarm Semarang. The early proccess was separating fish skin from the meat and sticking dirts. Then, the seperated fish skin was being cleaned and soaked in hot water for 1-2 minutes. After that, the immersing process of cleaned fish skin was done in CH_3COOH 3% for 12 hours. The fish skin was re-cleaned to netralize the pH. Thereafter, it was done the extraction proccess with aquades at 60-70 celcius degree for 2 hours. The extraction was filtered and dried in the oven at 60 celsius degree. After the dried extraction turned to gelatin sheets. They were being milled until they turned to gelatin powder..The making of gelatin refers to Martianingsih and Atmaja [3] with the modification of the use of acetic acid as a solution of immersion and fish skin that is used as a sample.

2.2.1. The Procedure Making of Edible Film

Edible film was made refers to Junianto *et al.* [4] with the modification of the use of sample using 3 different type of hydrolysed polysaccharides. Glycerol as plasticizer was added of 10% from the total components. The edible film making process of gelatin skin of tilapia fish, 5% (5gram of the weight of the solution) was added to 100mL aquades and heated at 60-70 celsius degree. Then it was added 0,5% (0,5 gram of solution weight) of carrageenan, alginate, agar, and glycerol and heated for 30 minutes. The solution is then printed out in glass plate with the size of 20x20 m and ovened for 24 hours in temperature of 50 celsius degree .Control was made from gelatin skin of tilapia fish without cross linking using

2.2.2. *Testing Parameters*
Edible film testing includes thickness, tensile strength, percent elongation, water vapor transmission rate (WVTR) and solubility.

2.2.3. *The Design of Experiments and Data Analysis*
The design of Experiments used in this research was complete random designs where every treatments was replicatated 3 times. Normality and homogenity testing was done earlierly before ANOVA analysed in order to recognize the data characteristic. The analytical method used was ANOVA variance with Tukey's advanced test to determine if there were any diffrerences between samples.

3. RESULT AND DISCUSSION

3.1. *The Characteristics of Edible Film*
The result of thickness test and the rate of water vapor transmission can be seen on Table 1.

Table 1. The Thickness (mm) and The Rate of Water Vapor Transmission ($g/m^2/hour$) Test

Sample	Thickness (mm)	The rate of water vapor transmission ($g/m^2/hour$)
Control	0,12 0,0045^a	1,002 ± 0,031^a
Agar	0,13 0,0035^a	0,748 ± 0,020^b
Carrageenan	0,14 ± 0,037^a	0,714 ± 0,100^b
Alginat	0,16±0,0055^a	0,610 ± 0,074^b

Note : Value was average of 3 times replications
The same lowercase letters notation showed there was no significant difference between sample($P<0,05$)

3.1.1. *The Thickness*
The results of this study indicated that the different types of using hydrocolloid polysaccharides did not give significant effect to the thickness. The average value of the thickness of the ranged from 0,12-0,16mm. According to Warkoyo [5], this coud be occuring because the addition of active material can cause decrease of molekul density, free space formed on bigger film matrixes. As the result, the formed film is getting thicker. Japannese Industrial Standart (1975) in Santosa et al. [6], recommend the thickness of maximum edible film is 0,25 mm. So, it can be concluded that edible film of research result meet the standart.

3.1.2. *The Water Vapor Transmission Rate*
The results of this study indicated that the different types of the used hydrocolloid polysaccharides have not an effect on the rate of a produced moisture transmission between samples. The further test results show that the valid difference to the value of water vapor transmission rate. The average value of water vapor transmission rate of research results to be ranged from 0,61-1,002 $g/m^2/hour$. The thickness of the lowest edible film is on the control treatment that is with the value by 1,002 mm and the highest is in edible alginate film of 0,61 mm.

Gunawan [7] stated a good edible film has low water vapor transmission rate. It means low water vapor transmission rate can inhibit the loss of water from the product, so the freshness of product was maintained . *Japannese Industrial Standart* (1975) in Santosa et al. [6], stated that the maximum water vapor transmission rate is 10 $g/m^2/$ day, so that it can be concluded that the water vapor transmission rate does not meet the standart.

Water vapor transmission rate affects on the ability of the edible film to held down water vapor. Edible film had low water vapor transmission rate that is suitable for wraps products with high humidity. Narendra *et al.* [8] added the value of high water vapor transmission rate will affect on storing endurance of a product that will be packed. The higher the value of water vapor transmission rate of the used edible film is,the more durable is the packed product.

The rate of water vapor transmission of edible film agar, carrageenan were lower than the control. Agar, carrageenan, and alginate had hydrofylic characteristic. Fatma et al. [9] Agar is a hydrofylic polymer of sulfate Polysaccharide which can be extracted from red seaweed (*Rhodophyceae*). According to Anward et al. [10] films made of alginate has weak mechanical strength, because alginates has only a negatively charged hydroxyl group so the bonds between molecules are weak. Meanwhile, films of alginate had high hydrofylic characteristic. The high hydrofylic characteristic of alginate causes film deriving from alginate enable to absorb water. This statement is made strengthened by Bourtoom [11] carrageenan is used as an edible film raw materials which is included in hydrocolic and generally a poor material for water vapor resistance. This was caused characteristic of carrageenan are hydrofylic which cause the edible film produced could be easily to absorb water vapor.

In addition Pranoto et al. [12] states from his research that on the effect of addition of seaweed extraction to edible film black tilapia fish skin. Decreased WVP (Water Vapor Permeability) in edible films is caused by the free volume of matrix polymers as a result of increased cross-linking through hydrogen bonding and hydrophobic interactions due to the addition of herbal extracts to gelatin films. Gelatin combined with seaweed extract significantly decreases WVP values, compared with control films (without the addition of seaweed).

Water Vapor Transmission Rate The rate of water vapor transmission affects on the ablity of edible film to hold down water vapor. Edible film which has a small water vapor transmission rate is suitable for packaging products that have high humidity. Edible film will inhibit the amount of water vapor released out of the environment so that the product does not dry quickly. Narendra et al. [8] added a high water vapor transmission rate will affect on storing endurance of the product to be packaged. The higher the velocity value of the edible film water vapor is used, the less durable the packaged product is.

3.1.3. The Tensile Strength and Percent of Edible Film Extension

The result of thickness test and the rate of water vapor transmission can be seen on Table 2.

Tabel 2. The Result of The Tensile Strength and Percent of Extension

Sample	The Tensile strenght (Kgf/cm^2)	Percent of Extension (%)
Control	12,51 ± 1,634^a	20,545 ±1,395^a
Agar	13,89 ± 0,152^b	49,905 ± 1,335^b
Carrageenan	18,83 ± 1,067^c	119,725 ± 1,71^c
Alginate	23,05 ± 0,485^d	81,571 ± 0,621^d

Note : Value is average of 3 times reiteration
The value of notation in the same lowercase letters shows invalid difference (P<0,05)

3.1.3.1. The Tensile Strength

The results of this study indicated that the different kinds of the used hydrocolloid polysaccharides have an effect on the resulted tensile strength. The further test results showend that the valid difference to the value of tensile strength. The average value of tensile strength of research results to be ranged from 12,51 to 23,05. The lowest tensile strength of the lowest edible film is on the control treatment with the value by 12,51 Kgf/cm^2 and the highest is in alginate of edible film by 23,05 Kgf/cm^2.

Katili et al. [13] state that edible film with high tensile strength is used for products that require high protection, while edible film with low tensile strength can be used for food products. The value of the edible tensile strength of the film has a standart according to by Krochta and Johnson [14] of 10 kgf/cm^2. So it can be concluded that the tensile strength of edible film research results was meet the standard..

The tensile strength decreased because of reduction of intermolecular interactions of the protein chain so that the matrix of the formed film will be less. The Reduction of intermolecular interactions of protein chains occurs to be caused by the addition of glycerol. Platicizer molecules will disrupt the cohesiveness of starch, the decrease of intermolecular interactions and the increase of polymer mobility.

3.1.3.2. The Percent of Extension

The results of this study indicated that the different kinds of the used hydrocolloid polysaccharides affect on the produced percent elongation. The further test results show that the valid difference to the value of tensile strength. The average percentage of elongation of the results of study outcome to be ranged from 20,545 to 195,725%. The lowest percentage of edible film extension is on the control treatment with the value by 20,545% and the highest on the edible film of carrageenan was 195,725%.

Murdinah *et al.* [15], stated that edible film is categorized less well if the value of percentage of extension is less than 10%. Based on the standart, It can be claimed that percentage of edible film extension is quite good .

The compenent of edible film composition affected the percentage of elongation. The protein contained of gelatin, carrageenan and glycerol as platicizer gave significant effect to the characteristic of edible film. The higher the percent value of elongation of edible film could be observed the better quality. However, it is not the treatment with the addition of carrageenan concentration on the control (without gelatin addition). This was due to the concentration of carrageenan added to the making of edible film will affect on the percent elongation of edible film. The more components of edible film filler will be difficult in the printing process in the glass plate, this is because the edible film will quickly block and when edible surface will be easily damaged.

The addition of an increasingly active ingredient may result in stretching the intermolecular spaces of the film matrix network and decreasing the amount of internal hydrogen bonds, thereby reducing the film's fragility and increasing the elongation percentage.

3.1.4. Solubility

The results of solubility test can be seen on Table 3.

Table 3. The Results of Solubility

Sample	Solubility
Control	99,3 ± 0,608^c
Agar	97,98 ± 0,275^c
Carragenan	92,36 ± 0,865^a
Alginate	95,44 ± 1,025^b

The results indicated that the different kinds of the application of hydrocolloid polysaccharides affected on the solubility. Furthermore the test results showed that there were any significan different to the solubility value. The average value of solubility of the research results to be ranged from 92,36 to 99,3%. The lowest edible film solubility was on the carrageenan addition treatment by 92,36% and the highest is in the control of film edible by 99,3%.

The Edible film made of carrageenan and gelatin mixture was lower than other treatments. This is because carrageenan is a polysaccharide made from hydrophobic seaweed. The addition of carrageenan to the edible film affected the solubility nature of the edible film. Santosa et al. [6], the most common type of carrageenan in food is kappa carrageenan. The hydroxyl and sulphate groups on carrageenan are hydrophilic while the 3,6-anhydro-D-Galactose group is more hydrophobic.

The edible film packaging material that depends on the source of the used raw materials. The lower the solubility value of edible film is, the better it is used as packaging material and biodegradable. Perez-Gazo and Krochta [16] suggest that water solubility is an important feature of edible film. Water film insolubility is needed to improve the product integrity and the resistance to water.

Edible film with a high degree of solubility will be easily degraded in the soil. Zulferiyenni et al. [17] suggests that the higher the solubility , the easier the components degradeted. It can be stated there are hydrophilic compounds hence it can be easier degraded in water and soil, so the ability of the biodegradable film is very well used as packaging material.

The value of edible film solubility is very important to know because it will affect the packed product and packaging function. This was supported by Diova [18], if the application of a film is required as a feasible packing material, high solubility is desired otherwise if the application of edible film is in foods with high water content then it is used insoluble film.

4. Conclusion

Edible gelatin film with the addition of alginate and glycerol has better characteristic when tested from tensile strength (23.05 Kgf /cm^2), vapor transmission rate (0.61 gram/m^2/day) and thickness (0.16 mm) than carrageenan and agar addition. However, the solubility (92.36%) and elongation percentage (119.725%) was the best edible gelatin film with the addition of carrageenan and glycerol.

References

[1]Leuberger,BH 1991 *Food Hydrocolloid* **5** : 353-361
[2] Athukorala Y, Lee K W, Song CB, Ahn CB, Shin TS, Cha YJ, Shahidi F and Jeon 2003 Potential antioxidant activity of marine red alga *Grateulopia filicina* extract. Food Lipids, 10 : 251-265
[3]Martianingsih N dan Atmaja L 2010 Analisis Sifat Kimia, Fisik, Dan Termal Gelatin Dari Ekstraksi Kulit Ikan Pari (*Himantura gerrardi*) Melalui Variasi Jenis Larutan Asam. Prosiding Skripsi.
[4]Junianto N, Kurniawati OS, Djunaidi, and Alexander M A Khan 2012 *African Journal of Food Science*, **6**(5):142-146.
[5]Warkoyo R, B Marseno, DW dan Joko NWK 2014 Jurnal Agritech. **34** (1) : 1-9
[6]Santosa, B., Herpandi., P. A. Pitayati., dan R.Pambayun. 2013. Pemanfaatan Karagenan dan Gum Arabic sebagai Edible Film Berbasis Hidrokoloid. AGRITECH, 33(2): 140 – 146.
[7]Gunawan V. 2009. Formulasi Dan Aplikasi Edible Coating Berbasis Pati Sagu dengan Penambahan Vitamin C pada Paprika (*Capsicum annuum* varietas Athena). Skripsi. Institut Pertanian Bogor. Bogor.
[8]Narendra AP, Alamsyah, MA dan Sudarno 2015 *Jurnal Ilmiah Perikanan dan Kelautan* **7** (2) : 122-125
[9]Fatma, Malaka R, dan Muhammad T 2015 Pengaruh Variasi Persentase Gliserol sebagai Plasticizer terhadap Sifat Mekanik Edible Film Dari Kombinasi Whey Dangke dan Agar. Universitas Hasanudin Makasar
[10]Anward G, Yusuf H, dan Nur R 2013 *Jurnal Teknologi Kimia dan Industri* **2** (3) : 51-56
[11]Bourtoom, T 2009 *Songklanakarin J. Sci. Tech* **30**: 149-165
[12]Pranoto Y, C M Lee, and HJ Park 2006 *LWT Food Science and Technology*. **40**: 766-774
[13]Katili S, BT Harsuno, dan S Irawan 2013 *Jurnal Teknologi*, **6**(1): 29-38
[14]Krochta, JM, and C De Mulder-Johnston 1997 *Food Technology*, **51** (2): 61-74
[15]Murdinah, Darmawan, M dan Dina F 2007 *Jurnal Pascapanen dan Bioteknologi Kelautan dan Perikanan*. **2** (1) : 125-132
[16]Perez-Gazo MB, dan Krochta JM 1999 *Journal of Science* **64** (4):695-698
[17]Zulferiyenny, Marniza, dan Sari EN *Jurnal Teknologi dan Industri Hasil Pertanian* **19** (3): 51-56

[18] Diova D, Darmanto, YS, dan Laras R 2013 *J. Pengolahan dan Bioteknologi Hasil Perikanan*, 2 (3): 149-165

Effect of ENSO on the variability of SST and Chlorophyll-a in Java Sea

Anindya Wirasatriya[1,2*], Indra B Prasetyawan[1,2], Chandra D Triyono[1], Muslim[1], Lilik Maslukah[1,2]

[1]Department of Oceanography, Faculty of Fisheries and Marine Science, Diponegoro University, Semarang, Indonesia
[2]Center for Coastal Disaster Mitigation and Rehabilitation Studies, Diponegoro University, Semarang Indonesia

*Corresponding author E-mail: aninosi@yahoo.co.id

Abstract. Sea surface temperature (SST) and chlorophyll-a (Chl-a) are two parameters often used for identifying the marine productivity. Located at the maritime continent, the variability of SST and Chl-a in the Indonesian seas is influenced by El Niño Southern Oscillation (ENSO). The previous studies showed that the effect of El Niño tend to decrease SST and increase Chl-a in the areas within the Indonesian seas. Using long time observation of satellite data (2003-2016), it was found different result in Java Sea. Since Java Sea has strong seasonal variability influenced by monsoon wind, the effect of ENSO depend on the season. During southeast monsoon season, El Niño (La Niña) tend to increase (decrease) the speed of southeasterly wind cause the decrease or increase of SST. On the contrary, during northwest monsoon season, El Niño (La Niña) tend to decrease (increase) the speed of northwesterly wind cause the increase (decrease) of SST. The dependence of Chl-a on wind speed is only observed in the off shore which exhibit the strong seasonal variation. However, the effect of ENSO on the variability of Chl-a is not robust since the effected amplitude is less than the RMSE of Chl-a data.
Keyword: SST, Chlorophyll-a, Java Sea, ENSO, Monsoon

1. Introduction

Sea surface temperature (SST) and chlorophyll-a (Chl-a) are parameters often used for indicating marine productivity which lead to fisheries productivity. Many studies found that high Chl-a concentration and low SST are correlated with high fish catch [1-3]. The increase of Chl-a concentration will increase the population of zooplankton which attract fish schooling. Thus, SST and Chl-a parameters can be good predictors for determining fishing ground and fishing season [4,5].

Lying on the confluence of the Eurasian Plate, the Indo-Australian Plate, and the Pacific Plate Indonesian Archipelago which is also known as the "Maritime Continent" experience strong seasonal variation caused by monsoon wind [6]. Two monsoon seasons alternates every year, i.e. southeast (SE) monsoon which is associated with easterlies from Australia that carry warm and dry air over the region and northwest (NW) monsoon which is associated with westerlies from the Eurasian continent that carry warm and moist air to the Indonesian region. Thus SE (NW) monsoon bring dry (rainy) season in Indonesia [7]. This monsoon wind then forms the seasonal patterns of SST and Chl-a in Indonesian Seas in general.

The mechanisms on how the monsoon wind influences SST and Chl-a is related to mixing process which determined by the speed of monsoon wind. The faster wind speed, the stronger mixing process occurs. Stronger mixing process can carry colder water (more nutrients) from the deeper water which make the SST (Chl-a) decrease (increase) [8]. Moreover, faster wind speed also cause more heat release from ocean to the atmosphere which cooling SSTs. Thus, the peaks of SST cooling and Chl-a blooming within Indonesian Seas occur during NW monsoon i.e., December, January and February and SE monsoon i.e, Jun, July and August. Besides mixing, surface wind also can generate coastal upwelling which can bring deeper, cooler and richer nutrients water to the surface [9-12].

El Niño Southern Oscillation (ENSO) is one of interannual climate variability which indicated by the anomaly of SST in the central equatorial Pacific. Positive (negative) anomaly indicates the El Niño (La Niña) event. This variability has cycle of about 2-7 years. Synoptic scale interaction between ocean and atmosphere associated with this event determines the ENSO effect on the global climate including Indonesia [13-15].

The impact of ENSO on marine and fisheries productivity within Indonesian Seas has been discussed in many studies. Their analysis showed that the effect of El Niño tend to increase (decrease) Chl-a concentration (SST) in the areas within the Indonesian Seas [10,11,12,16]. Conversely, La Niña tend to decrease (increase) Chl-a (SST). The most prominent impact of ENSO on the variability of Chl-a concentration and SST was found along the southern coast of Java and Sumatra Island. In a normal year, higher Chl-a concentration which is associated with monsoon-generated upwelling during the southeast monsoon cycle spread along the southern coast of Java Island and extend only to the southern tip of Sumatra Island. Anomalous easterly winds during the peak of the 1997/1998 El Niño induced strong upwelling which brought cooler water, and lifted the thermocline much shallower than the climatological mean along the coasts of Java and Sumatra [11]. As a result, higher Chl-a concentrations were observed further northwestward along the Sumatra coast. Moreover, the positive anomaly of Chl-a concentration from its climatological mean during this event reached more than 5 mg/m^3 along southern coast of Java and Sumatra Island.

Located at the center of Maritime Continent, Java Sea has different characteristics with the seas of the southern coast of Java Island. With relatively shallow bathymetry i.e., maximum about 130 m, Java Sea is surrounded by 3 big Islands i.e., Java, Sumatra and Kalimantan in the southern, western and northern part, respectively. In the eastern part, Java Sea is bordered by the deep Banda Sea. Moreover, Java Sea is known as one of the path of monsoon wind blowing from Southern Asia to Australia and vice versa [7]. This condition causes the variabilities of SST and Chl-a in Java Sea are strongly influenced by monsoon [3]. Although many previous studies pointed out the increasing (reducing) Chl-a concentration (SST) occurred during El Niño in many Indonesian Seas, the present study demonstrates the different result. The effect of ENSO on the variabilities of SST and Chl-a in Java Sea is depended on the season.

2. Materials and Method

For Chl-a and SST data, we used MODIS Aqua Lv3 with spatial resolution of 0.04°×0.04° and observation period from 2003 to 2016 [17]. Specifically, MODIS SST 11μm which cover the observation both for day and night were used instead of MODIS SST 4μm which only covers night time SSTs. Both MODIS SST and Chl-a are developed and distributed by the Ocean Biology Processing Group. The algorithm for generating MODIS SST (Chl-a) is described in [18-20]. The quality of these dataset has been examined against in-situ measurements to ensure the best accuracy [21-25].

To investigate the formation mechanism of Chl-a and SST pattern in Java Sea, we used surface wind data obtained from 6 hourly ERA interim data with grid interval of 0.125° [26]. We composed SST, Chl-a and surface wind parameters into monthly and monthly climatology following [16]. The monthly El Niño and La Niña data were also composited, determined by using Oceanic Niño Index (ONI) provided by http://www.cpc.ncep.noaa.gov/. ONI index is the anomalies of SST in the Niño 3.4 region (5°N-5°S, 120°-170°W), based on the basis period of 1971-2000. The threshold for determining El Niño and La Niña period is $\pm$0.5°C.

3. Results and Discussion

3.1. Seasonal variations of SST and Chl-a in Java Sea

The monthly climatology of SST in Java Sea is presented in Figure 1. We show the period of January, April, August and November to represent the season of NW monsoon, transition I, SE monsoon and transition II respectively. In January and August, SST tend to be low ranges about 28°C to 29°C. Lower SST is detected in the north-western and eastern part of Java Island in January and August, respectively. In April and November SST increase for more than 29°C. The peak of SST cooling (warming) occurs in NW and SE monsoon (Transition I and II) season. Thus, semi-annual variation characterize the SST variation in Java Sea. This result is consistent with the SST distribution in Java Sea observed by the previous studies [3,6].

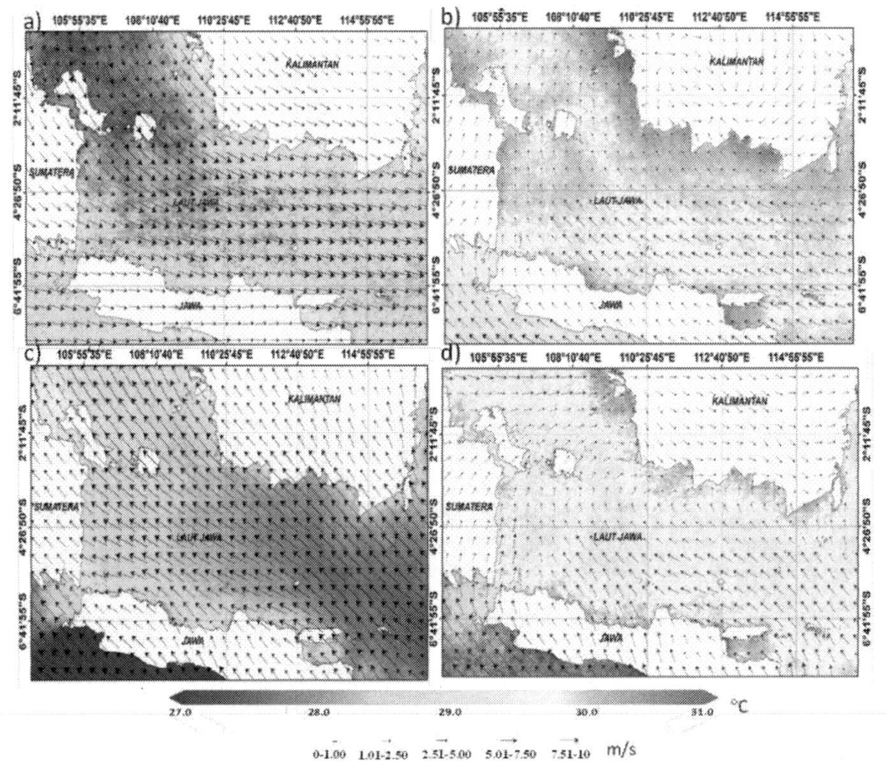

Fig.1. Climatology of SST in Java Sea in (a) January, (b) April, (c) August, and (d) November overlaid with surface wind vector

The mechanisms of the seasonal variation of SST in Java Sea is related to the variation of wind speed. The cooler (warmer) SST in January and August (April and November) correspond to the stronger (weaker) wind speed occurs during those period. In January (August) wind blows north-westerly (south-easterly) cause the lower SST in the north-western (east) part of Java Sea. Thus, the peaks of SST cooling and warming in Java Sea are related to the variation of surface wind.

The spatial distribution of Chl-a in Java Sea shows the different result (Figure 2). Although the peaks of high (low) Chl-a concentration also occur in NW and SE monsoon (Transition I and II) season, the distribution is obviously different between the coastal areas and open seas. Very high Chl-a concentration i.e., more than 1 mg/m^2 is observed only in the coastal areas. Because of this distribution, we sampled 3 areas in Java Sea which represent the coastal area of Java Sumatra and Kalimantan Island and 1 area in open seas for further analysis. Since maximum Chl-a concentration was found in January, Chl-a distribution in January was chosen as a basis for area selection. The edge of area sample in the coastal areas were determined by the Chl-a concentration of 0.6 mg/m^3 and the coastal lines (Figure 3).

IOP Conf. Series: Earth and Environmental Science **116** (2018) 012063 doi:10.1088/1755-1315/116/1/012063

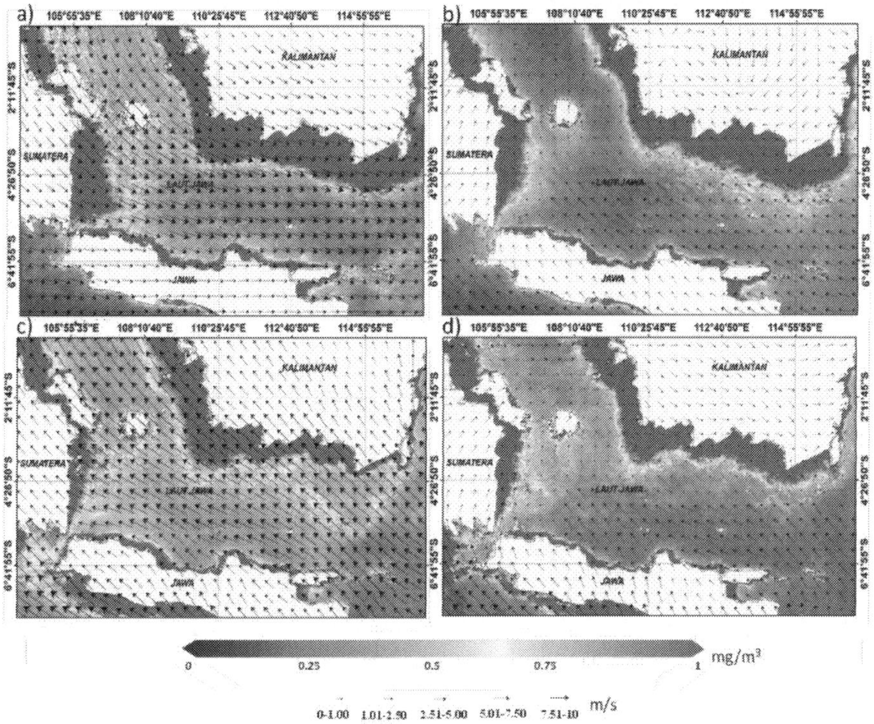

Figure 2. Climatology of Chl-a in Java Sea in (a) January, (b) April, (c) August, and (d) November overlaid with surface wind vector

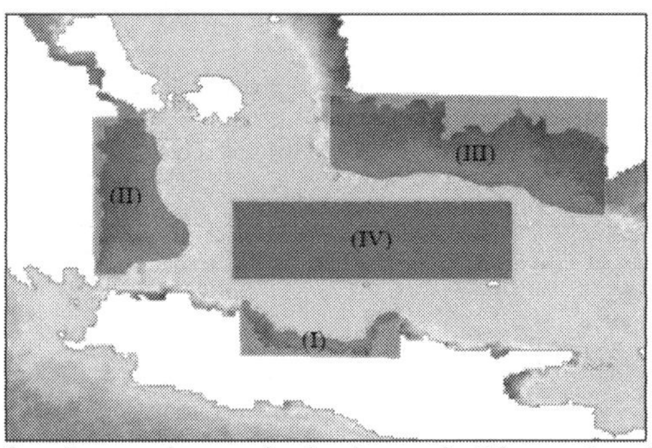

Figure 3. Area selection for time series analysis shown in Figure 4 and 6. See the basis for area selection in text.

The time series analysis of SST and wind speed for all areas is presented in Figure 4a. Semi annual variation is robustly observed for both parameters and all areas. The peak of strong (weak) wind occurred in January and August (April and November) which represent NW and SE (Transition I and II) season, respectively. The strongest wind speed was found in open seas area (area 4) due to the absence of land obstacle which may reduce the wind speed. Conversely, the lowest amplitude of wind speed variation was found in the coastal seas of northern Java (area 1). This may be caused by the position of area 1 which a little covered by the land. Moreover, the peak of strong (weak) wind is negatively correlated with SST variation. The correlations between wind speed and SST for area 1, 2, 3 and 4 are -0.82, -0.72, -0.75, and -0.88, respectively. Stronger wind speed induced stronger mechanical mixing and heat release which reduced SST. However, it is also found the time lag about 1 month between the strongest (weakest) wind speed and lowest (highest) SST in the coastal seas i.e., area 1, 2, and 3. This indicates that other factors may also influence the variability of SST in coastal seas.

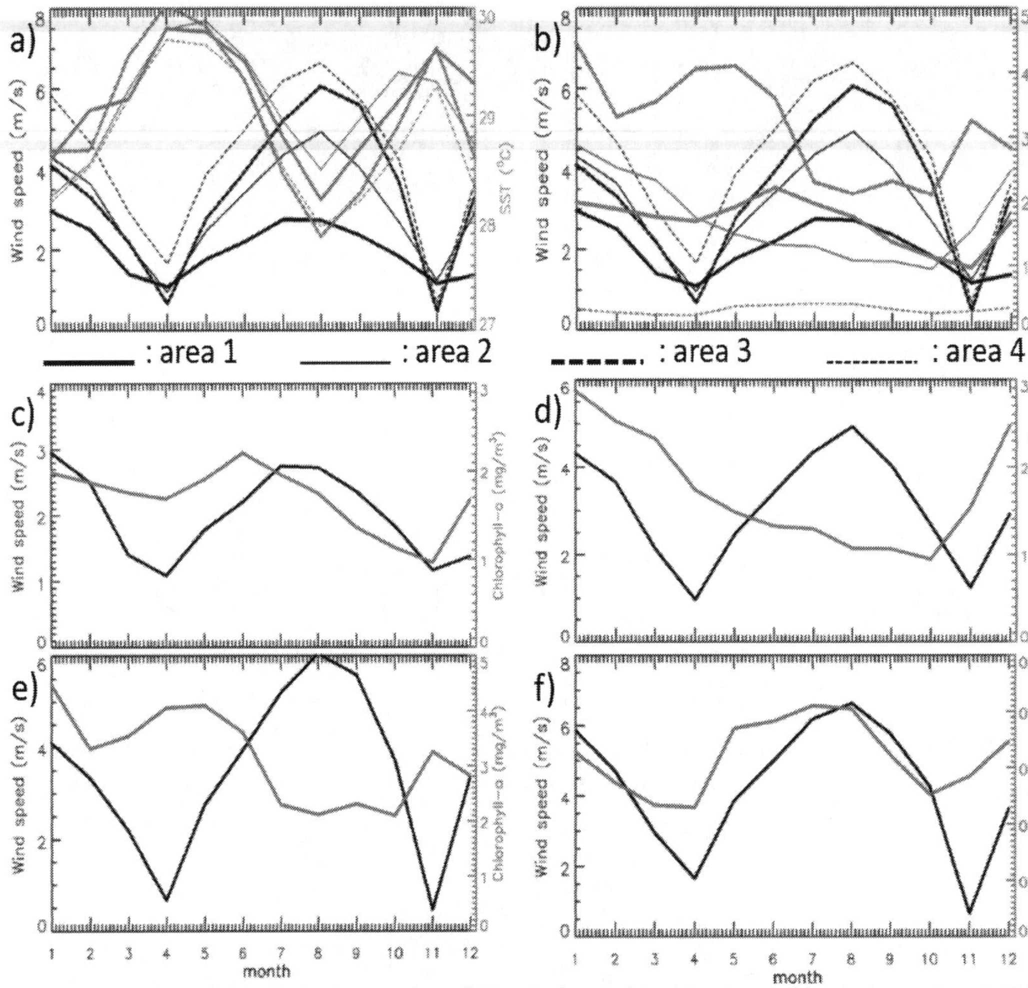

Figure 4. Monthly climatology time series of (a) wind speed and SST and (b) wind speed and Chl-a in area 1, 2, 3 and 4. The adjusted scale axis of wind speed and Chl-a in (c) area 1, (d) area 2, (e) area 3, and (f) area 4.

Chl-a concentration shows the different variation with SST (Figure 4b). The highest (lowest) chlorophyll concentration was found in area 3 (4) ranges from 2.11 mg/m^3 to 4.47 mg/m^3 (from 0.12 mg/m^3 to 0.31 mg/m^3). Area 1 and 2 are in between. In the adjusted scale in Fig. 4c-f, we can see that only Chl-a concentration in the open seas (area 4) has the high positive correlation with the wind speed. The correlations between Chl-a and wind speed for area 1, 2, 3, and 4 are 0.48, -0.002, -0.35, and 0.86 respectively. Although the amplitude of Chl-a variation in area 4 is the smallest one, its highest correlation means that Chl-a variation in the open seas of Java Sea highly depends on the wind speed. Moreover, the small correlations in area 1, 2, and 3 indicate that there are other possible factors influence their Chl-a variability. Since area 1, 2 and 3 are located in the coastal area, the nutrient input transported by the river charge may give significant contribution for Chl-a variabilities in those 3 areas. As discussed by [1,2], the variability of Chl-a in the Lampung Bay is influenced by the nutrient flux transported by the adjacent rivers. Figure 4c-e shows that the high Chl-a concentration mainly occurred during NW monsoon and Transition I season. The high precipitation during these seasons may bring more nutrient rich fresh water to the coastal seas and may increase turbidity in the coastal seas. The fresh water input may also influence the lag correlation between SST and wind speed in this 3 areas. The turbid coastal water may also interfere the Chl-a concentration measured by MODIS sensor due to the contamination of suspended sediment [27]. This possibilities are left for future studies.

IOP Conf. Series: Earth and Environmental Science **116** (2018) 012063 doi:10.1088/1755-1315/116/1/012063

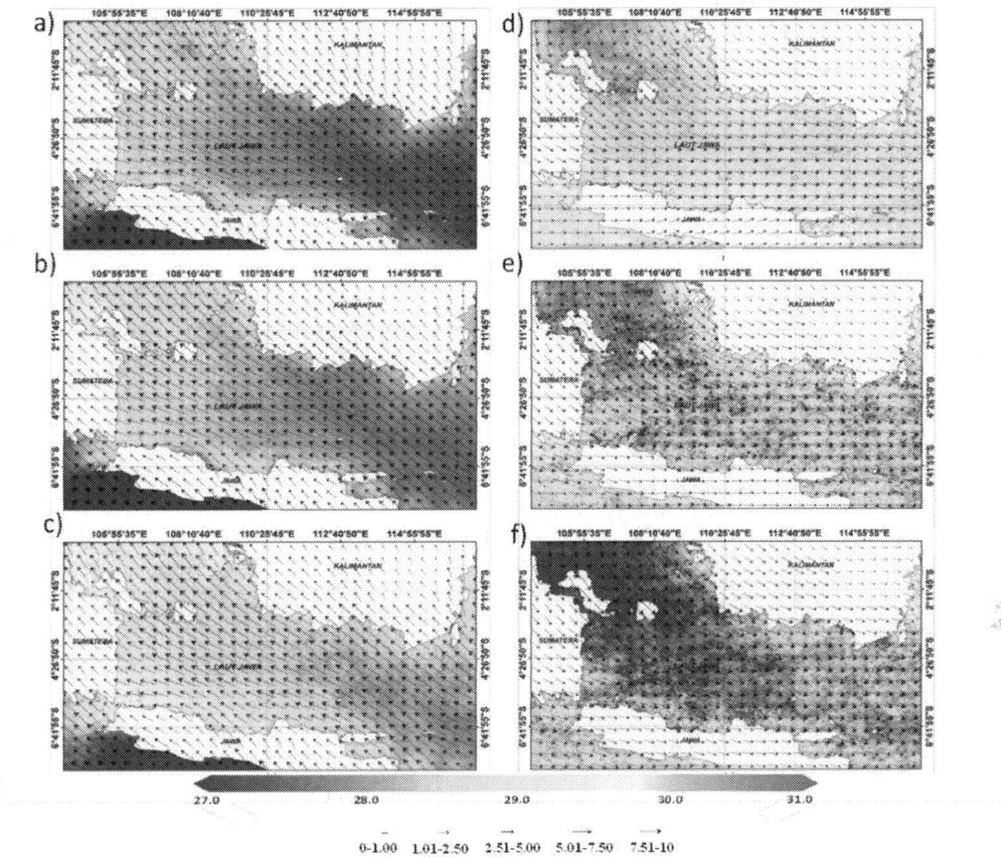

Fig. 5. Composite of SST map overlaid with surface wind vector in (a) August during El Nino, (b) August Climatology, (c) August during La Nina, (d) January during El Nino, (e) January Climatology, and (f) January during La Nina

3.2. Effect of ENSO to the SST and Chl-a variabilities in Java Sea

To investigate the effect of ENSO to the variability of SST in Java Sea, we composed the monthly El Niño and La Niña and compared with monthly climatology data as shown in Figure 5. We chose the period of August and January to be presented in Figure 5 since August and January are the peak of SE and NW monsoon, respectively. In August, SST distribution in Java Sea tent to be cooler (warmer) than its climatology during El Niño (La Niña) (Figure 5a). This is related to the speed of south-easterly wind which is stronger (weaker) during El Niño (La Niña). This result is consistent with the previous studies that found the SST cooling during El Niño in other Indonesian Seas [10-12,16].

The different result is depicted in Figure 5b. In January, SST tent to be warmer (cooler) than its climatology during El Niño (La Niña). This is caused by the weakening of north-westerly wind occurred during El Niño. In the opposite, the north-westerly wind is strengthened during La Niña which makes the decrease of the SSTs. Thus, the present study firstly shows that the effect of ENSO to the variability of SST in Java Sea depended on the season.

For further analysis, we show the time series of SST and wind speed in 4 areas which are described in the previous section. For all areas, the SST difference between the El Niño and La Niña attained more than 0.5°C (Figure 6a-d). It also shows that effect of ENSO to the variabilities of SSTs in all areas depended on the season. In August, El Niño tent to increase wind speed and decrease SST, while in January La Niña took control to increase wind speed and decrease SST. Thus, this result support the analysis in the previous paragraph.

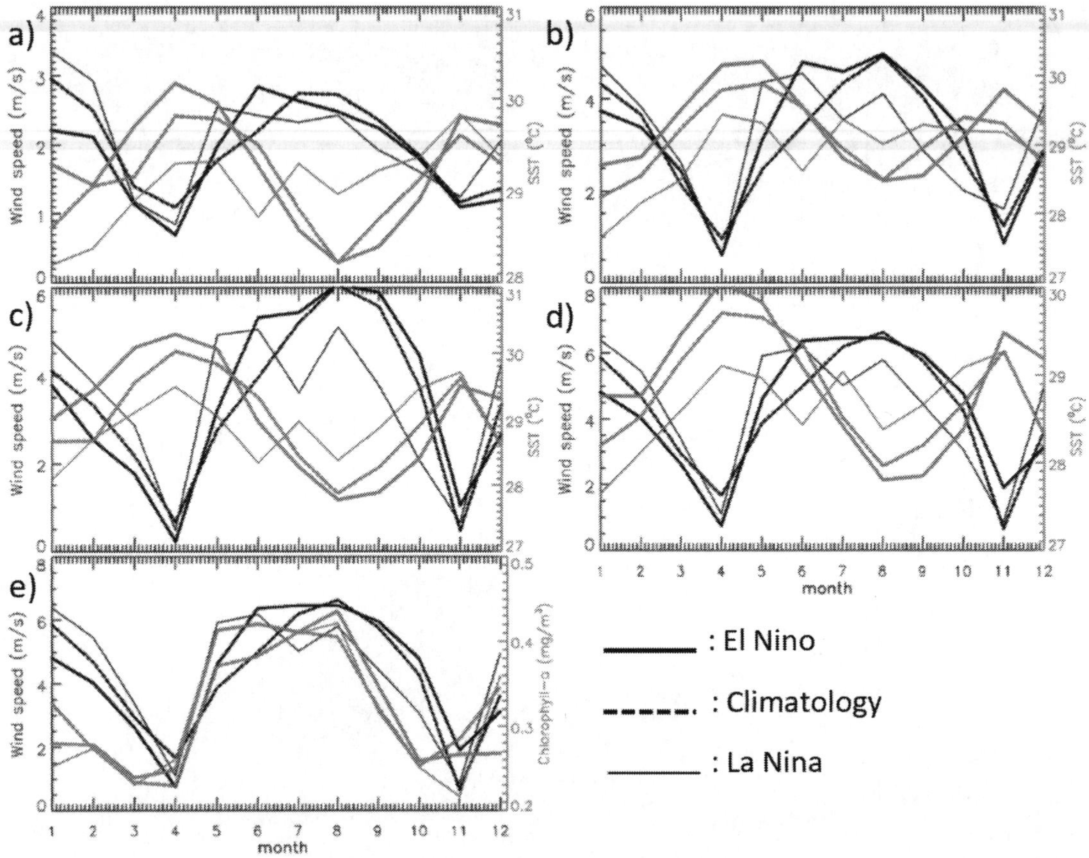

Figure. 6. Time series of (a) wind speed and SST in area 1, (b) wind speed and SST in area 2, (c) wind speed and SST in area 3, (d) wind speed and SST in area 4, and (e) wind speed and Chl-a in area 4 under climatology and ENSO condition

For Chl-a variation, we only examined in the area 4 (Figure 6e) since in the other 3 areas, the variability of Chl-a does not depend only on the wind speed. The complex factors influenced the Chl-a variability in the coastal areas increase the difficulties in investigating the ENSO effect on the Chl-a variation in the coastal areas. This problems is potential for future work. In the open seas (area 4), the effect of ENSO to the variability of Chl-a is not robust. The difference Chl-a concentration between El Niño and La Niña period is less than 0.1 mg/m^3. This difference is much less than the accuracy of MODIS Chl-a used in the present study i.e., 0.3 mg/m^3 [19]. Thus, this analysis should be improved by using the more accurate Chl-a data.

4. Conclusion

The effect of ENSO on the variability of Chl-a concentration and SST in Java Sea has been investigated by using long term observation of MODIS data (2003-2016). The conclusions are as follows:

1. Java Sea shows robust seasonal variation on SSTs. The peaks of high (low) SSTs occur in January and August (April and November) which are influence by the speed and monsoon wind.
2. Seasonal variation of Chl-a in Java Sea only robustly shown in the open seas areas which is strongly influenced by the monsoon wind. In the coastal seas, the variability of Chl-a may be also influenced by the nutrients input from the land carried by the river's charge.
3. The effect of ENSO on the variability of SST in Java Sea depends on the season. During SE monsoon (Jun, July, and August), El Nino (La Nina) tends to decrease (increase) SSTs in Java

Sea. The stronger (weaker) easterly wind during El Nino (La Nina) may enhance (reduce) the mixing process and heat release to the atmosphere which decrease (increase) SSTs. In contrast, During NW monsoon (December, January and February), El Nino (La Nina) tends to decrease (increase) the westerly wind blowing in Java Sea which cause the increase (decrease) SSTs.

4. The effect of ENSO on the variability of Chl-a in the open seas area of Java Sea is not robust since the different of Chl-a concentration between El Nino and La Nina condition 0.1 mg/m^3. This value is much less than the accuracy of MODIS Chl-a used in this study. The more accurate Chl-a data should be used to investigate the effect of ENSO on the variability of Chl-a in Java Sea.

References

[1] Sachoemar SI, Yanagi T, Hendiarti N, Sadly M, and Meliani F 2010 *Coast. Mar. Sci.* **34**(1) 82
[2] Sachoemar SI, Yanagi T, and Aliah RS 2012 *Coast. Mar. Sci.* **37**(2) 75
[3] Hendiarti N et al. 2005 *Oceanography* **18**(4) 112
[4] Dutta S, Chanda A, Akhand A, and Hazra S 2016 *Turkish J. Fish. and Aqua. Sci.* **16** 767 DOI: 10.4194/1303-2712-v16_4_03
[5] Ardianto R, Setiawan A, Hidayat JJ, and Zaky AR 2017 *IOP Conf. Series: Earth and Environmental Science* **54** 1 doi:10.1088/1755-1315/54/1/012081
[6] Susanto RD, Moore II T, and Marra J 2006 *Geochem. Geophys. Geosyst.* **7**(5) 1 doi:10.1029/2005GC001009.
[7] Setiawan RY, and Habibi A 2010 *Indonesian J. Mar. Sci.* **15**(1) 42
[8] Setiawan RY, and Habibi A 2011, *IEEE J. Sel. Topics Appl. Earth Observ. Remote Sens. (JSTARS)*, **4**(4) 944
[9] Setiawan RY, and Kawamura H 2011, *IEEE J. Sel. Topics Appl. Earth Observ. Remote Sens. (JSTARS)*, **4**(1) 241
[10] Susanto RD, Gordon AL, and Zheng Q 2001 *Geophys. Res. Let.* **28**(1) 1.559
[11] Susanto RD, and Marra J 2005 *Oceanography* 18(4) 124
[12] Iskandar I, Rao SA, and Tozuka T 2009 *Int. J. of Rem. Sens.* **30**(3) 663 doi: 10.1080/01431160802372309.
[13] Cane MA 2004 *Earth and Planet. Sci. Let.* **230**(3) 227
[14] McPhaden MJ, Zebiak SE and Glantz MH 2006 Journal Science. **314** 1740
[15] Wang C and Fiedler P 2006 Prog. in Ocean. **69** 239
[16] Wirsatriya A, Setiawan RY and Subardjo P 2017 *IEEE J. Sel. Topics Appl. Earth Observ. Remote Sens. (JSTARS)*, **PP(99)** 1 doi: 10.1109/JSTARS.2017.2745207
[17] Esaias WE *et al.* 1998 *IEEE Transactions on Geoscience and Remote Sensing* 36(4) 1250 doi: 10.1109/36.701076.
[18] Brown OB, and Minnett PJ 2009 http://modis.gsfc.nasa. gov/data/atbd/atbd_mod25.pdf
[19] O'Reilly J, Maritorena S, Mitchell BG, Siegel D, Carder KL, Garver S, Kahru M, and McClain C 1998 *J. Geophys. Res.* **103** 24937
[20] Hu C, Lee Z, and Franz B 2012 *J. Geophys. Res.* **117(C1)** doi: 10.1029/2011jc007395
[21] Moore TS, Campbell JW, and Dowell MD 2009 *Remote. Sens. Environ.* **113** 2424
[22] Zhang CY, Hu CM, Shang SL, Muller-Karger FE, Li Y, Dai MH, Huang BQ, Ning XR, and Hong HS, 2006 *Remote. Sens. Environ.* **102** 250.
[23] Ghanea M, Moradi M, Kabiri K, and Mehdinia A 2015, *Adv. Space Res.* **57**(1) http://dx.doi.org/10.1016/j.asr.2015.10.040
[24] Lee MA, Tzeng MT, Hosoda K, Sakaida F, Kawamura H, Shieh WJ, Yang Y, Chang Y 2010 *Terr. Atmos. Ocean Sci.* **21**(4) 727
[25] Qin H, Chen G, Wang W, Wang D and Zeng L 2014 *Int. J. Remote Sens.*, **35**(11–12) 4315
[26] Dee DP *et al.* 2011 *Q.J.R. Meteorol. Soc.* **137** 553 doi: 10.1002/qj.828
[27] Siswanto E *et al.* 2011 *J. Oceanogr.* **67** 627 https://doi.org/10.1007/s10872-011-0062-z

Integrated Community Based Coastal Management: Lesson From The Field

Sudharto P. Hadi [1*]
[1]Study Program of Environmental Studies and Faculty of Social and Political Science, Diponegoro University Jl. Prof. Soedharto, SH, Tembalang, Semarang
*sudhartophadi@yahoo.co.id and sudhartophadi@undip.ac.id

Abstract. Coastal abrasion has been occurred throughout coastline of Java reaching 745 km at length, account for 44% of total Java's coastline. This phenomena is caused by reclamation, cutting of mangrove, land-use change and other human activities specifically at coastal area. Coastal abrasion stimulates flood or tidal flood, when sea level rise, the sea water flows to the land undated fish pond, settlement and other infrastructures standing at coastal area. Tidal flood destroys settlement lead to significant decrease of property value: land and house. Coastal abrasion caused lose people's job and income. One measure taken by local community is mangrove cultivation intended to prevent sea level rise flowing to the inland. However many efforts taken by community frequently fail because of un-integrated approach. This paper reviews a mangrove plantations in Mangunharjo, district of Tugu, Semarang, Central Java by utilizing an innovative approach integrating environmental, economic and social aspect. These mangrove cultivations environmentally useful to prevent coastal abrasion, economically creating income for local people and socially supported by local community. These three approaches ensure sustainability of mangrove's culture.

Keywords: Coastal abrasion-ecological, economical, social approaches

1. Introduction

Coastal area is a strategically located for various activities such as port, industry, settlement, recreation, fisheries, agriculture etc. These activities are overlapping each other and uncontrollable causing environmental and social impacts. One significant environmental impacts is coastal abrasion predicted caused by reclamation by a company. Reclamation causes the change of current pattern destroying coastal line, in turn, degrading fish ponds. When sea level rises, it undated fish pond, settlement and other infrastructures, disrupting of day to day activities, threatening health and economic of local people. This phenomena has been occurring throughout of Java coast where coastal abrasion reach 745 km at length or 44% out of coast line (Kompas Daily Newspaper, 25 January 2017). Based on data published by Ministry of Marine and Fisheries showing that coastal abrasion reach 1.950 hectare per year (Kompas Daily Newspaper, July 26, 2017). The loss of coastal land during the last fifteen years reached 29.261 hectare. In Mangunharjo a village located at district of Tugu, Semarang, Central Java has been experienced coastal abrasion and tidal flood since 1999 degraded coast line of 3,5 km at length and eroded the main land up to 1,7 km to the settlement of people. This paper reviews the measure and approaches taken by local people in dealing with coastal abrasion and how are they sustained. At a global level, sea levels rise threaten coastal area possibly could be up to 98 cm at the end of this century.

Williams et al (2017: 6) noted the importance of coastal area indicating that within 25 km of the coastline live 1.4 billion people, account for 20% of the world population. Coasts contribute revenue to the GDP of a country indicated by Greek tourism providing 24% of its GDP (Bank of Greece (2015)) and Greece's tourism confederation (SEPE) expected 25 million tourists to visit in 2016. It is also shown by Columbia in which coastal tourism is the third highest source of foreign exchange after oil and coal.

2. Literature Review

Maliao et al (2009: 818) studied coastal resource management (CRM) in the Philippines and concluded that CRM is generally implemented under the co-management framework. They suggest that the key aspect of co-management is the empowerment of the community to participate, control and influence institutional decisions affecting their lives. This CRM strategy is referred to as community based coastal resource management (CBCRM).

Esteban et al (2017: 78) on their studies on awareness of costal floods in Jakarta suggest that to enhance resilience, it is required not only to construct dykes to protect against a tsunami but also to plant mangrove forests. Zagonari (2008: 803) concluded that for achieving coastal quality, an integrated approach is always better than a non-integrated approach. Coastal management must consider all stakeholders involved by focusing at local rather than at national scales, on users rather than on uses.

David Romero Manriqute de Lara, Sera Corral. 2017 (160-161) noted that integrated approach practiced was able to identify the root of the problems and impacts of unsustainability issues and explore the measures from their own view. By facilitating a dialog, local people and public authorities could collaborate a teamwork to reach solutions. These ensure the sustainability of the project by having supports from all stakeholders.

Aheto et al (2017: 53) outlined that due to limited resources and their capacity, local communities cannot manage coastal ecosystems by themselves. Aheto et al viewed that co-management is an approach in which the government shares certain authority, responsibilities and functions of coastal resources management with resources users as partners. Co-management should include rule-making and rule enforcement through local government units by virtue of proximity to the resource management areas

Abreu et al (2017: 17) noted the importance of the interaction between traditional and scientific knowledge for having solutions to environmental and social problems. View of community member having experience of the problems would be importance to have a good solution.

3. Material and Methods

The type of research is descriptive analysis outlining the efforts and approaches taken by local people in dealing with coastal abrasion and how are they sustained. The case study taken is the village of Mangunharjo, sub-district of Tugu, Semarang City. The data gathered through informal interview, physical observation and document analysis. The data then analyzed quantitatively and qualitatively.

4. Results and Discussion

Water The village of Mangunharjo has been experiencing coastal abrasion since 1999 at 3,5 km at length and 1,7 km flows to the settlement destroying 110 hectare of fish pond. Based on research done by Anggoro (2011: 2), the driving factor of coastal abrasion was largely a reclamation done by a wood processing company. The company agreed to negotiate with local people and achieved an agreement to give a part of compensation for impacted local people. Government of Central Java provided seed money for rehabilitating coastal abrasion by building coastal belt. The project was done through self help (mutual assistance) among local people. They designed a coastal belt, utilized materials such as sand, concrete buis, bamboo, stone, etc locally and mobilize the workers from their villages. The seed funding from the Provincial Government could build a coastal belt for 1,2 km. The government provided further funding for the later years. In addition, local people initiated to explore funding from other sources such as the Ministry of Marine and Fisheries,. Up until now the remaining abrasion area is about 600 meter out of 3,5 km. To strengthen the coastal belt, local people initiated to plant a mangrove a long the line of coastal belt. Mangrove grows well and help make fish pond recovered, in turn, it can be planted again with shrimp and milk fish. People now earn from harvesting shrimp and crab at the average of Rp 125.000,- per day. They also earn money by providing boat service for people doing fishing around the fish pond. The self help of local people in building a coastal belt and planting mangrove attracted attention from various pupils and students from various places of Indonesia and abroad. They visit Mangunharjo and help planting mangrove. Students from various countries such as China, Japan,

Thailand, Vietnam, French, Germany, Belgium, South Korea visited Mangunharjo and learned how people doing mutual assistance to deal with their coastal problems. Various private companies also take part in planting and maintaining mangrove.

5. Conclusion

The mutual assistance of people in Mangunharjo dealing with coastal abrasion is the key success of sustainability. Local people involved at the earliest stage of planning in developing coastal belt and planting mangrove. The government provided a seed funding, the academician provide advice and facilitation. Private company also involved in mangrove culture. Local people have a sense of ownership, if the problems with a coastal belt and mangrove occur, local people will do mutual assistance to fix it and replant it. The sense of ownership is also supported by the fact that their fish pond is now can be planted and give income again. Local people also benefit from having many visitors from domestic and abroad. It is recommended for the government to help local people to finish building coastal belt dealing the remaining costal abrasion. The well grown mangrove in Mangunharjo which is now called mangrove forest is one of the few coastal green area in Semarang City. It is required to determine the areas as conservation area through spatial planning revision otherwise the area will be utilized by private sector as industries.

References

[1] Allan Williams , Nelson Guillermo Rangel-Buitrago, Enzo Pranzini , Giorgio Anfuso 2017 The management of Coastal Erosion. *Ocean & Coastal Management* xxx 1-17

[2] Anggoro S 2011 *General View of Reclamation. Unpublished Paper for Reclamation Plan of Jakarta Bay, Indonesia.*

[3] David Romero Manrique de Lara, Sera Corral 2017 Local Community-Based Approach for Sustainable Management of Artisanal Fisheries on Small Islands. *Ocean & Coastal Management* **142** 15

[4] Denis Worlanyo Aheto , Stephen Kankam , Isaac Okyere , Emmanuel Mensah , Adams Osman, Fredrick Ekow Jonah , Justice Camillus Mensah 2017 Community-Based Mangrove Forest Management: Implications for Local Livelihoods and Coastal Resource Conservation along the Volta Estuary Catchment area of Ghana. *Ocean & Coastal Management* **127** 43-54

[5] Juliana Silva Abreu [a], Camila Domit [b], Camilah Antunes Zappes 2017 Is There Dialogue Between Researchers and Traditional Community Members? The importance of Integration between Traditional Knowledge and Scientific Knowledge to Coastal Management. *Ocean & Coastal Management* **141** 10-19

[6] Kompas Daily Newspaper, 25 January 2017 and July 26, 2017

[7] Miguel Esteban, Hiroshi Takag, Takahito Mikami, Aretha Aprilia, Daisuke Fujii, Shota Kurobe, Nuki Agya Utama 2017 Awareness of Coastal Floods in Impoverished Subsiding Coastal Communities in Jakarta: Tsunamis, Typhoon Storm Surges and Dyke-induced Tsunamis. *International Journal of Disaster Risk Reduction* **23** 70–79

[8] Ronald J. Maliao, Robert S. Pomeroy, Ralph G. Turingan 2009 Performance of community-based coastal resource management (CBCRM) programs in the Philippines: A meta-analysis. Marine Policy **33** 818–825

[9] Zagonari, Fabio 2008 Integrated Coastal Management: Top down vs. Community-Based Approaches. *Journal of Environmental Management* **88** 796–804

IOP Conf. Series: Earth and Environmental Science **116** (2018) 012065 doi:10.1088/1755-1315/116/1/012065

Molecular Identification and Genetic Diversity of *Acropora hyacinthus* from Boo and Deer Island, Raja Ampat, West Papua

DP Wijayanti[1*], E Indrayanti[1], H Nuryadi[2,3], RA Dewi[1], A Sabdono[1]
[1]Marine Science Dept, Diponegoro University, Jl. Prof. Soedarto SH, Tembalang, Semarang, 50725
[2]Tropical Marine Biotechnology, Central Laboratory of Research and Services, Diponegoro University, Jl. Prof. H. Soedarto, SH, UNDIP, Tembalang, Semarang, Kotak Pos 1269, Indonesia
[3] Department of Biology, Chemistry and Marine Science, University of the Ryukyus, 1-Senbaru, Nishihara-cho, Okinawa 903-0213

*Corresponding author E-mail:diah_permata@mail.com

AbstractIndonesia lies at the centre of biodiversity for corals. However, the reefs suffered from extensive human exploitation. Marine Protected Areas is thought to be best solution to protect coral reefs ecosystem. Understanding genetic diversity is crucial for effective management of the MPAs, however genetic diversity is rarely been corporate in designing an MPA. Moreover, many MPAs are uneffectively manage due to poor designated and demarcated.Raja Ampat which is located in western tip of West Papua, was designated as a park to mitigatethreatsand protect the valuable marine resources.Scleractinian corals in the genus *Acropora* are among the most dominant distributed in Raja Ampat waters, including the species of *Acroporahyacinthus*.The research aimed to analyze genetic diversity and to describe the kinship relationship of *Acroporahyacinthus* between 2 populations: Boo Island and Deer Island, Raja Ampat. Genetic marker Cytochrome Oxidase I (CO I) of the mitochondrial genome DNA (mtDNA) was used to analyze genetic diversity. Reconstruction of phylogenetic tree and genetic diversity were made by usingsoftware MEGA 5.05 (Moleculer Evolutionary Genetics Analysis). The results of this research indicatecorals *A. hyacinthus* from Boo Island and Deer Island Raja Ampat are in the low category of genetic diversity and overall had a close genetic relationship of kinship. This is likely due to the small size of the population and few numbers of samples that may not represent the population.

1. Introduction

Indonesia lies at the centre of biodiversity for corals [1]. Though the reefs support life of more than 67.500 villages, however it suffered from extensive human exploitation [2]. The occurrence of bleaching events [3] coupled with emergence of coral diseases[4] even in remote area [5] made reef's condition rapidly declining. Therefore, there is an urgent need for reefs protection.

Marine Protected Area (MPA) is thought to be one of the best strategies for protecting biological diversity, increase fisheries yields, and protect particularly vulnerable life stages of marine species [6].

For ensure protection, it is important to understand the genetic connectivity among reefs and identify designated protected areas that are self-sustaining and possibly export young propagules to the adjacent reefs[7].

Raja Ampat is located in the heart of the 'Coral Triangle', an area encompassing Malaysia, Philippines, Solomon, Timor Leste, Papua New Guinea and Indonesia, which has the highest coral diversity in the world. The archipelago is comprised of several large mountainous islands including Waigeo, Batanta, Salawati and Misool, and hundreds of small islands such as Kofiau, Gam, Fam and Ayau. The land and surrounding sea occupy approximately 43.000 km2. While many of these are uninhabited, most of the population (~43,435) lives on 34 islands in the archipelago [8]. Almost all of the populations depend on their marine resources to support their livelihood [9]. Raja Ampat was designated as Marine Proctected Areas (MPA) in order to mitigate threats and protect valuable marine resources. The MPA was established under Goverment Decree No. 27/2008 which protect more than million hectares of coastal and marine in Raja Ampat areas.

Reefs were generally in very good condition compared to most areas of Indonesia with high live coral diversity and minimal stress due to natural phenomenon such as cyclones, predation (i.e. crown-of-thorns starfish), and freshwater runoff [8]. However, 17% of reefs were considered to be in poor condition, mostly because of high levels of silting and the increase explosive use for fishing. Marine resources in Raja Ampat are also under threat from increasing frequency of disturbances associated with climate change (i.e. coral bleaching).

Documenting the diversity of marine life is challenging because many species are cryptic, small, and rare, and belong to poorly known groups10]. A genetic identification system would be particularly useful for such organisms, especially whose identification may be uncertain due to phenotypic plasticity and lack of identification tools at early stages of life [11]. DNA barcoding has become an attractive method to identify species over a wide taxonomic range[12]. Cytochrome c oxidase subunit 1 (COI) has been proposed as the main barcoding gene for metazoans [11]. Though was reported as rapid, inexpensive and need little expertise, however, the method was thought controversial since it simplified various taxonomic characters into one single character of identification only [13].Yet the method was reported can readily discriminate among closely related species across most animal phyla. The method was successfully documented enormous diversity of various cryptic animals that only 10.9% of the sequences were matched to reference barcodes in public databases [10].However, it was suggested that DNA barcode was not approriate for most scleractinian corals because this region evolves very slowly in these organisms and consequently both inter- and intra-specific variation are extremely low[14].In contrast, DNA barcode was successfully revealed the taxonomy of *Stylophora pistillata* by examining COI divergence and phylogeny of the coral. By analyzing Cytochrome Oxidase I sequences, from 241 samples across 4 different geographic ranges, showed this taxon in fact comprises four deeply divergent clades [15].

Acropora, differ from all other extant coral genera has unique form of polyp which formed by one or more central or axial corallites. The axial corallites bud the radial corallites as the coral grow [16].Speciesof *Acropora* are also highly polymorphic andhave always carried problems of identification [17]. The genera also plays a dominant role in the species composition and abundance of many Indonesian reefs [18]. *Scleractinian* corals in the genus *Acropora* are among the most dominant distributedgenera in Raja Ampat waters.This publication aims to determine the connectivity patterns and genetic diversity of *Acropora hyacinthus* of Raja Ampat in order to understand the genetic diversity of the coral to help conservation program of the MPA.

2. Materials and Methods

Sample collections

Acropora hyacinthus was collected from core zone area of Kofiau Marine Park Area. Since sampling was conducted at MPAs area, prior the sampling, a letter of sampling permission was obtained from the Raja Ampat Bureau of National Marine Park (Simaksi, BTN). Samples were taken from the core zone of Kofiau District. Two islands were chosen as sampling location namely Deer Island (-1.155234, 129.846921) and Boo Island (-1.175641, 129.337620). *A. hyacinthus* at the sampling location showed various colony shape, though most of the colony has tabulate form (Fig.1).Twenty fragments were obtained from each island from each sampling location at the two islands.

Fragments are cut from the middle branches of the coral colony using cutting pliers. The length of fragments approximately 1-3 cm. Fragment were then put inside a labelled ziplock plastic. After collection, fragments were removed and put into 15 ml falcon tube and preserved in 95% ethanol. Fragments are then transported to Integrated Lab (CORESDU) Diponegoro University. To avoid to sample clone, distance of one colony to another colony at least 1 m. Both species are capable to reproduce through fragmentation since the colony shape is branching. Samples are preserved until used.

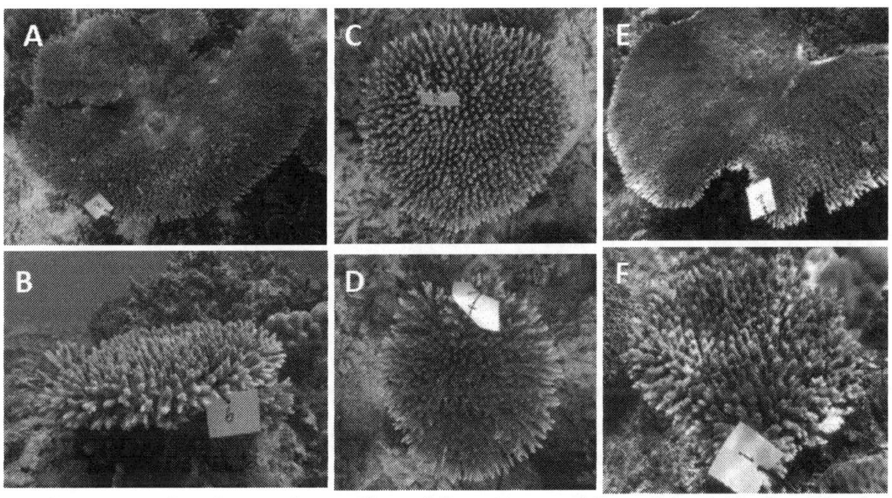

Figure 1. Various type of *A. hyacinthus* collected from Boo and Deer Island, Kofiau District, Raja Ampat. A, B, C, dan D, *A. hyacinthus* collected from Boo Island; E dan D, *A. hyacinthus* from Deer Island

DNA Extraction

A quick and simple DNA extraction method with the resin, Chelex 100 (BioRad) is done. A 0.2mm section of tissues was scrapped using sterile spatula and place into 1.5 ml tubes filled with 300uL 10% Chelex. Vortexed samples in Chelex slurry for 10-15 seconds. Spinned samples briefly (10-15 sec) at high speed in a microcentrifuge. Incubated samples for 20 minutes at 95°C. Vortexed samples again for 10-15 seconds. Spinned tubes again at high speed in microcentrifuge to ensure that all contents are in the bottom of the microcentrifuge tube. Supernatant was directly used for PCR amplification. DNA concentration was then quantified using Nanodrop with spectral length on 260/280 µm and 260/230 µm.

PCR Amplification

Universal primer of DNA barcode were used for this research, there were LCOI1490: 5′-CAAATCATAAAGATATTGG-3′ as forward primer and reverse primer : HCOI2198: 5′-TAAACTTCAGGGTGACCAAAAAATCA-3′ [19]. PCR was performed using the following thermal cycle: 1 cycle of 95 °C for 2 min; 40 cycles of 95 °C for 30 s, 47 °C for 90 s, and 72 °C for 90 s. The amplification reaction used 2.5 µL template DNA, 12.5 µL Promega kit, HCOI primer 2 µL, LCOI primer 2 µL and 16 µL destilled water, in total 25 µL volume. The PCR products will be electrophoresed in a 1% agarose (FMC Bioproduct, Rockland, ME, USA) gel in 1x TAE buffer to assess the yield. The PCR products were then genotype using ABI PRISM 3730xl Genetic Analyzer used standard protocol developed by the manufacturer.

Sequencing and Alignment

Sequencing was conducted at Genetika Science Indonesia. Prominent bands of PCR samples were sent to the company. According to the company protocol, sequencing were conducted using Big Dye Terminator v.3.1 Cycle Sequencing Kit follows the PCR cycle sequen. The formula of sequen procedure were as follows : 8 µl big dye, DNA template, 3.2 pmol primer, add millipore water into the final volume 20 µl. While sequencing cycles are as follows: initial denaturation at 95 °C for 5 minutes, then following by 50 cycles of (denaturation at 95 °C for 30 second; annealing at 48 °C for10 second; and extention at 60 °C for 40 minutes). The results were then kept in 4 °C until purification.

Purification was done by purify the amplified DNA using 5 µl EDTA 100 mMinto PCR products. And then was added60 µl of 100% ethanol. Mixed samples were then be homogenized and were incubated at room temperature for 15 minutes. Samples were then centrifuged at 2000 – 3000 rpmfor 30 minutes. Then added60 µl of 70% ethanol into the samples tube. Samples were then be centrifuged again at 4 °C with 1650 rpm speedfor 15 minutes. Purified DNA then were sequenced using ABI PRISM 3730xl Genetic Analyzer following the protocol available from the Applied Biosystems, USA. The most prominent band of purified PCR products are used as sequencing samples.

The sequenced results were then alignmented and were analyzed using BLAST*(Basic LocalAlignmentSearchTool)* based on genetic data base of NCBI (NationalInstituteforHealth, USA (www.ncbi.nlm.nih.gov). Phylogenetic tree was constructed using MEGA5. *Neighbor Joining Tree* was used to construct a tree to analyse genetic distance. An outgroup was used when constructed the three. Genetic diversity was observed using DNAsp software [20] and Arlequin: http://cmpg.unibe.ch/software/arlequin35/Arlequin35.html.

Results and Discussion

In total, there were 25 sequenced resulted from Deer and Boo Island samples. The resulting Oxidase I (CO I) of the mitochondrial genome DNA (mtDNA) sequences corresponding to the genotype were analyzed for homologies with sequences in the data base using BLAST searching. BLAST analysis showed that all 25 sequences were closely related to *Acropora hyacinthus* under 99% similarity (Table 1). The accession numbers of COI mtDNA of the sources that used as comparison was presented as well in Table 1. All nucleotide sequences of the COI were than deposited to Genebank. All nucleotide sequences have been deposited in the GeneBank database under accession number LC187992 - LC188016 (Table 1).

\

Table 1. BLAST analysis of Deer and Boo Island samples

No.	Sample code	Length of sequence	Closest relative in GenBank (deposited Accession Number)	Homology	Accession Number
1.	PD Ah 7	707 bp	*Acropora hyacinthus* mitochondrion, complete genome LC188011	99%	KF448531.1

2.	PD Ah 6	714 bp	*Acropora hyacinthus* mitochondrion, complete genome LC188010	99%	KF448531.1
3.	PD Ah 4	707 bp	*Acropora hyacinthus* mitochondrion, complete genome LC188009	99%	KF448531.1
4.	PD Ah 3	716 bp	*Acropora hyacinthus* mitochondrion, complete genome LC188008	99%	KF448531.1
5.	PD Ah 20	712 bp	*Acropora hyacinthus* mitochondrion, complete genome LC188016	99%	KF448531.1
6.	PD Ah 16	705 bp	*Acropora hyacinthus* mitochondrion, complete genome LC188015	99%	KF448531.1
7.	PD Ah 15	707 bp	*Acropora hyacinthus* mitochondrion, complete genome LC188014	99%	KF448531.1
8.	PD Ah 13	720 bp	*Acropora hyacinthus* mitochondrion, complete genome LC188013	99%	KF448531.1
9.	PD Ah 12	668 bp	*Acroporahyacinthus* mitochondrion, complete genome LC188012	99%	KF448531.1
10.	PD Ah 1	705 bp	*Acropora hyacinthus* mitochondrion, complete genome LC188007	99%	KF448531.1
11.	PB Ah 9	690 bp	*Acropora hyacinthus* mitochondrion, complete genome LC187998	99%	KF448531.1
12.	PB Ah 6	728 bp	*Acropora hyacinthus* mitochondrion, complete genome LC187997	99%	KF448531.1
13.	PB Ah 5	707 bp	*Acropora hyacinthus* mitochondrion, complete genome LC187996	99%	KF448531.1
14.	PB Ah 4	694 bp	*Acropora hyacinthus* mitochondrion, complete genome LC187995	99%	KF448531.1
15.	PB Ah 3	704 bp	*Acropora hyacinthus* mitochondrion, complete genome LC187994	99%	KF448531.1
16.	PB Ah 20	706 bp	*Acroporahyacinthus* mitochondrion, complete genome LC188006	99%	KF448531.1
17.	PB Ah 2	699 bp	*Acropora hyacinthus* mitochondrion, complete genome LC187993	99%	KF448531.1

18.	PB Ah 19	707 bp	*Acropora hyacinthus* mitochondrion, complete genome LC188005	99%	KF448531.1
19.	PB Ah 18	701 bp	*Acropora hyacinthus* mitochondrion, complete genome LC188004	99%	KF448531.1
20.	PB Ah 16	706 bp	*Acropora hyacinthus* mitochondrion, complete genome LC188003	99%	KF448531.1
21.	PB Ah 15	706 bp	*Acropora hyacinthus* mitochondrion, complete genome LC188002	99%	KF448531.1
22.	PB Ah 14	711 bp	*Acropora hyacinthus* mitochondrion, complete genome LC188001	99%	KF448531.1
23.	PB Ah 12	710 bp	*Acropora hyacinthus* mitochondrion, complete genome LC188000	99%	KF448531.1
24.	PB Ah 10	707 bp	*Acroporahyacinthus* mitochondrion, complete genome LC187999	99%	KF448531.1
25.	PB Ah 1	706 bp	*Acropora hyacinthus* mitochondrion, complete genome LC187992	99%	KF448531.1

Phylogenetic Tree

Reconstruction of phylogenetic tree was done using PAUP* 4.0 (*Phylogenetic Analysis Using Parsimony*) [21]with *parsimony*method and 1000 times bootstrap. Kinship relationship was observed using *TreeView* program.

Phylogenetic construction based on COI sequences found 6 distinct cluster of*A. hyacinthus*. *Outgroup* using DNA data base of *Montipora* sp. Phygenetic tree of *A. hyacinthus* was shown at Figure 2.

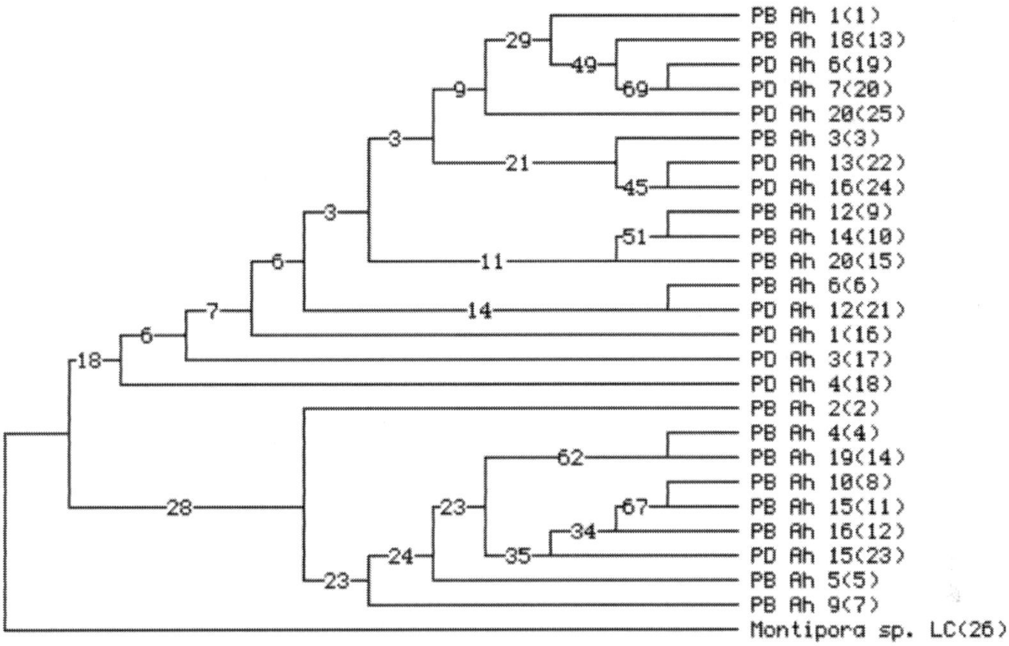

Figure2. Phylogenetic tree construction of *Acroporahyacinthus*collected from Boo and Deer Island, Kofiau District, Raja Ampat. There 6 different clades observed from the tree. *Montipora* sp was use as outgroup; B, represents Boo Isl.; D, represents Deer Isl.

Clade 1 was comprised of mix samples collected from Boo and Deer Island (PB Ah 1, PB Ah 4, PB Ah 5, PB Ah 6, PB Ah 9, PB Ah 10, PB Ah 12, PB Ah 14, PB Ah 15, PB Ah 16, PB Ah 18, PB Ah 19, PB Ah 20, PD Ah 1, PD Ah 4, PD Ah 7, PD Ah 13, PD Ah 15, PD Ah 16, and PD Ah 20) ; clade 2, 3, 4 5, and 6 comprised only one samples (Table 2).

When comparison was made with *A. hyacinthus* population from Karimunjawa Archipelago, similar result was obtained. *A. hyacinthus* of Karimunjawa Archipelago was grouped into 5 differents clades [22]. Samples of *A. hyacinthus* from Kofiau District was collected from Deer Island and Boo Island that was separated by more than 100 km, while samples from Karimunjawa Archipelago was collected from two different cluster islands namely Seruni, Sambangan and Genting cluster and Cilik and Menjangan Kecil cluster islands that separate by most far 60 km only. However, the Menjangan Kecil Island and Cilik Island was set behind the main Karimunjawa Island that face of open sea while Seruni, Sambangan and Genting was set inside lagoonal area of Karimunjawa main island.

Analysis of Genetic Diversity

Genetic diversity of *A. hyacinthus* was observed using DNAsp [20]. Based on statistic analysis of genetic diversity of 25 samples showed 6 haplotypes. Distribution of haplotypes was shown at Table 3. Estimation of haplotype diversity value of *A. hyacinthus* was 0,3667. When haplotype diversity was observed from each island it was showed much greater value. Haplotype diversity of *A. hyacinthus* collected from Boo Island was 0, 9238 (n=15) while haplotype diversity of *A. hyacinthus* collected from Deer Island was 0,9333 (n=10).

Tabel2. Haplotype diversity distribution of *Acropora hyacinthus* from Boo and Deer Island, Kofiau District, Raja Ampat.

IOP Conf. Series: Earth and Environmental Science **116** (2018) 012065 doi:10.1088/1755-1315/116/1/012065

No	Haplotype	Number of samples	Samples code
1.	Haplotype 1	20	PB Ah 1, PB Ah 4, PB Ah 5, PB Ah 6, PB Ah 9, PB Ah 10, PB Ah 12, PB Ah 14, PB Ah 15, PB Ah 16, PB Ah 18, PB Ah 19, PB Ah 20, PD Ah 1, PD Ah 4, PD Ah 7, PD Ah 13, PD Ah 15, PD Ah 16, PD Ah 20
2.	Haplotype 2	1	PB Ah 2
3.	Haplotype 3	1	PB Ah 3
4.	Haplotype 4	1	PD Ah 3
5.	Haplotype 5	1	PD Ah 6
6.	Haplotype 6	1	PD Ah 12
	Total	25	

Using multi-locus nucleotide sequence data, [23] recentlydescribed two, previously unsuspected, cryptic species complexes within the *Acropora* syngameon (i.e., a groups of species connected through genetic exchange). Among the Scleractinia,the genus *Acropora* contains the largest number of species, occurs in all tropical oceans, and dominates most reef habitats [24]. The genus is known for high-levels of shared polymorphism between species, which is due, in part at least, to widespread genetic exchange through introgression ([25]; [26]; [27]; [28]). Distributions of species of the coral genus *Acropora* in the Indonesian archipelago show a duality reminiscent of the Wallace's Line patterns seen in terrestrial animals and plants, rather thanthe concentric pattern predicted by the "centre of origin" model [29]. The duality is due to an overlap of Indian Ocean species distributions diminishing eastwards and Pacific Ocean species distributions diminishing westwards within the archipelago [29]. Additionally, a large number of species with broad Indo-Pacific distribution, as well as some regional endemics, occur within the archipelago. For instance, an endemic Indonesian species *A.* (*Acropora*) *togianensis*, Wallace, 1997, based on its reproductive activity, morphological and genetic analyses was reported belongs to *Isopora* [16]. Thus although Indonesia has a high speciesnumber overall, this is due to the presence of a composite fauna with strong regional differences.

The highest level of endemicity was seen to occur in the Togian Islands, in Central Sulawesi,a location with a particularly unusual *Acropora* fauna [29]. The endemic species of the Togian Islands were hypothesized to indicate a possible late Tethyan relictual fauna [30], a conclusion which is re-assessed here, and the earliest fossil record of the genus is dated at late Paleocene [16]. Because of this, *Acropora* offers an ideal opportunity to follow global colonization through time in detail.If Indonesia were to be regarded as a single entity, (rather than as six separate areas) its diversity would be highest at 91. This record could possibly be challenged by further collecting in the south-western Philippines and other parts of the South China Sea. No single area of Indonesia, however, shows such a high species count. The greatest number of species found in an area of Indonesia is 77, in the Bay of Tomini, Central Sulawesi. This is also the highest species count worldwide. At the sametime, the bay is the smallest area included in the study, being the area between the northern andcentral eastern arms of Sulawesi.

The results of this research indicate that corals *A.hyacinthus* from Boo Island and Deer Island, Raja Ampat are in the low category of genetic diversity and overall had a close genetic relationship of kinship. This is likely due to the small size of the population and few numbers of samples that may not represent the population. Application of more suitable marker for detecting genetic diversity will help to obtain more robust results.

Acknowledgment: The authors thanks to Diponegoro University for provided research grant to DPW through RPI project under decree No. SP DIPA-042.01.2.400898/2016, 7 December 2015 fiscal year 2016. Many thanks go to TNC, Indonesia for logistic and accomodation during sample collection. Special thanks goes to Mr. Purwanto who helped with samples collection. Anonymous reviewers greatly appreciated for their contribution to improve the manuscript.

REFERENCES

[1] Veron JEN, DeVantier LM, Turak E, et al. 2009. Delineatingthe coral triangle.Galaxea.**11**(2), pp. 91–100

[2] Burke L, Reytar K, Spalding M, Perry A. 2012. Reefs at risk revisited. World Resources Institute, The Nature Conservancy, World Fish Center, International Coral Reef Action Network, UNEP world conservation monitoring centre and global coral reef monitoring network, Washington DC

[3] Hughes TP, Baird AH, Bellwood DR, Card M, Connolly SR, Folke C, Grosberg R, Hoegh-Guldberg O, Jackson JBC, Kleypass J, Lough JM, Marshall P, Nystrom M, Palumbi SR, Pandolfi JM, Rosen B, Roughgarden J. 2004. Science 301: 929-933

[4] Wijayanti DP, Hidaka M, Layla F, Munasik, Sabdono A. 2016. J. of Fisheries and Aq Sci11: 232-237 DOI: 10.3923/jfas.2016

[5] Haapkylä J, Seymour AS, Trebilco J, Smith D. 2007. J. Mar BiolAssoc UK. **87**(2) 403-414

[6] Palumbi SR 2003 Ecol Application 13: S146-S158

[7] Baums I B. 2008 Mol Ecol. **17**: 2796-2811

[8] McKenna S, Allen G, Suryadi S2002 A marine rapid assessment of the Raja Ampat islands,Papua province, Indonesia. Conservation International (ISBN: 1881173607)

[9] Mangubhai S, Erdmann MV, Wilson JR, Huffard CL, Ballamu F, Hidayat NI, Hitipeuw C, Lazuardi ME, Muhajir, Pada D, Purba G, Rotinsulu C, Rumetna L, Sumolang K, Wen2012 Marine Pollution Bulletin(http://dx.doi.org/10.1016/j.marpolbul.2012.07.024)

[10] Leray M, Knowlton N 2015DNAbarcoding and metabarcoding of standardized samples reveal patterns of marine benthic diversity. Proc of Nat Acad of Sci of the USA112 : 2076–2081, doi: 10.1073/pnas.1424997112

[11] Hebert PDN, Cywinska A, Ball SL, DeWaard JR. 2003Biologicalidentifications through DNA Barcodes. Proc of the RoyalSoc of London Series B 270 313–321

[12] Valentini, A, F Pompanon, P Taberlet 2008 Trendsin Ecology and Evolution 24, 110–117

[13] Dasmahapatra KK, Mallet J 2006 DNA barcodes: recent successes and future prospects.Heredity 97: 254-255

[14] Shearer TL, MA Coffroth. 2008. MolEcol Res 8, 247–255

[15] Keshavmurthy S, SY Yang , A Alamaru, YY Chuang, M David Obura, S Fontana, SD Palmas, F Stefani, F Benzoni AM Donald, Annika M. E. Noreen, C Chen, CC. Wallace, R M Pillay, V

Denis, AY Amri, J D. Reimer, T Mezaki, C Sheppard , Y Loya , A Abelson , MS Mohammed , AC. Baker , PG Mostafavi , Budiyanto A. Suharsono, C A Chen. 2013. DNA barcoding reveals the coral''laboratory-rat'', *Stylophorapistillata*encompasses multiple identities Sci. Rep 3 : 1520 | DOI: 10.1038/srep01520

[16] Wallace CC, Chen CA, Fukami H, Muir PR 2007 Recognition of separate genera within Acropora based on new morphological, reproductive and genetic evidence from*Acroporatogianensis*, and elevation of the subgenus IsoporaStuder, 1878 to genus (Scleractinia: Astrocoeniidae; Acroporidae). Coral Reefs 26:231–239

[17] Wallace CC, Dai CF 1997 Scleractinia of Taiwan (IV): Review of the Coral Genus Acropora from Taiwan. Zoological Studies **36**(4): 288-324

[18] Wallace CC, Richards Zoe 2001 Pesisir dan Lautan 4, 40-58

[19] Folmer O, Black M, Hoeh W, Lutz R, Vrijenhoek R1994. MolecularMarine Biology and Biotechnology, **3**, 294–299

[20] Rozas J 2009 DNA Sequence Polymorphism Analysis Using DnaSP. Bioinformatics for DNA Sequence Analysis pp 337-35

[21] Swofford DL2002 PAUP 4.10b: phylogenetic analysis using parsimony (and other methods). Sinauer Associates, Sunderland

[22] Wijayanti DP, Indrayanti E, Nuryadi H, Rintiantono S, Sabdono A 2017 DNA Barcode of *Acropora hyacinthus* of Karimunjawa Archipelago. IOP Proceeding

[23] Ladner JL, SR Palumbi 2012MolEcol21(9) 2224-2238 DOI: 10.1111/j.1365-294X.2012.05528.x

[24] Wallace CC, Willis B 1994 Ann Rev of Eco and Syst25: 237-262 https://doi.org/10.1146/annurev.es.25.110194.001321

[25] van Oppen M, B McDonald, B Willis, D Miller 2001 Molecular Biology and Evolution 18, 1315-1329

[26] van Oppen, M, B Willis, T van Rheede, and D Miller 2002 Molecular Ecology 11:1363-1376.

[27] Vollmer SV, SR Palumbi 2002 Hybridization and the Evolution of Reef Coral Diversity. Science 296: 2023-2025

[28]Palumbi SR, S Vollmer, S Romano, T Oliver and J Ladner 2011 Evol. Ecol.26 : 317–335 doi:10.1007/s10682-011-9517-3

[29] Wallace CC 1997 The Indo-Pacific centre of coral diversity re-examined at species levelProceedings of the 8th International Coral Reef Symp

[30] Wallace, C C 1999 Staghorn corals of the world : a revision of the coral genusAcropora (Scleractinia; Astrocoeniina; Acroporidae) worldwide, with emphasison morphology, phylogeny and biogeography. CSIRO, Collingwood, Vic

Seasonal and Non-Seasonal Generalized Pareto Distribution to Estimate Extreme Significant Wave Height in The Banda Sea

Nursamsiah[1,4], Denny Nugroho Sugianto[2,3], Jusup Suprijanto[1], Munasik[1], and Bambang Yulianto[1]

[1] Faculty of Fisheries and Marine Science, Diponegoro University, Jl. Prof. H. Soedarto, SH, UNDIP, Tembalang, Semarang 50275, Indonesia
[2]Center for Coastal Disaster Mitigation and Rehabilitation Studies, Diponegoro University, Jl. Prof. H. Soedarto, SH, UNDIP, Tembalang, Semarang 50275, Indonesia
[3] Departement of Oceanography, Fisheries and Marine Science Faculty, Diponegoro University
Jl. Prof. H. Soedarto, SH, UNDIP, Tembalang, Semarang 50275, Indonesia
[4]Indonesian Agency for Meteorology, Climatology and Geophysics Jl. Angkasa I No.2, Kemayoran, Jakarta10610, Indonesia
nursamsiah533@gmail.com

Abstract. The information of extreme wave height return level was required for maritime planning and management. The recommendation methods in analyzing extreme wave were better distributed by Generalized Pareto Distribution (GPD). Seasonal variation was often considered in the extreme wave model. This research aims to identify the best model of GPD by considering a seasonal variation of the extreme wave. By using percentile 95 % as the threshold of extreme significant wave height, the seasonal GPD and non-seasonal GPD fitted. The Kolmogorov-Smirnov test was applied to identify the goodness of fit of the GPD model. The return value from seasonal and non-seasonal GPD was compared with the definition of return value as criteria. The Kolmogorov-Smirnov test result shows that GPD fits data very well both seasonal and non-seasonal model. The seasonal return value gives better information about the wave height characteristics.

1. Introduction

Information on extreme waves is necessary for marine planning, managing and evaluating. The Occurrence of high waves may disturb the transportation activities [1] and destruct the conservation area of mangroves and coral reefs [2]. Maritime structures in the coastal and offshore region must be designed for an extreme condition such as high waves [3]. The structures designed to stand in extreme wave condition according to expected structure lifespan. The information of maximum waves that may occur in certain periods (return periods) such as 5, 20 or maybe 100 years usually needed for marine planning.

In analyzing extreme wave event, Peaks Over Threshold (POT) method was recommended in the IAHR Working Group on Extreme Wave Analysis [4]. POT uses data that exceeds the threshold as an extreme value. POT was naturally described using Generalized Pareto Distribution (GPD [5], then this method was used widely to predict extreme waves [6],[7],[8].

Wave seasonality or variations that repeat periodically (weekly, monthly, yearly, etc.) were often examined carefully in modeling extreme waves [9] but other researchers only consider the location of the site [6][10]. The seasonal and nonseasonal extreme wave model uses another model (GEV) gave different result according to [11]. The primary objective of this research is to identify the difference between seasonal and nonseasonal GPD model in estimating extreme wave return value for maritime planning and management.

In the Banda Sea, there were inter-island transportation and fishing activity. Good planning and management required to support this activity. Information on extreme waves needs to be estimated to determine the risk of disasters that could interfere with such activities. The wave characteristics in the Banda Sea has associated with the monsoon and vary each month with the highest peak occurring in the Australia monsoon period or in June-July-August (JJA)[12].

2. Research Methods

2.1 Data

This research used significant wave height (Hs) data in the south of Buru Island (4° 30' 0" S, 126° 7' 30" E), provided by BIG (Indonesian Geospatial Information Agency) based on Consortium of Oceanic and Atmospheric Prediction (COAP). The data were obtained from **http://tides.big.go.id/** (access on January 14, 2017). To fit the GPD we used daily maximum Hs from 1991 – 2010, while data from 2011 - 2014 was used as validation data of extreme return period results. The prediction was performed for 2, 5, 10, 25, 50 and 100 years.

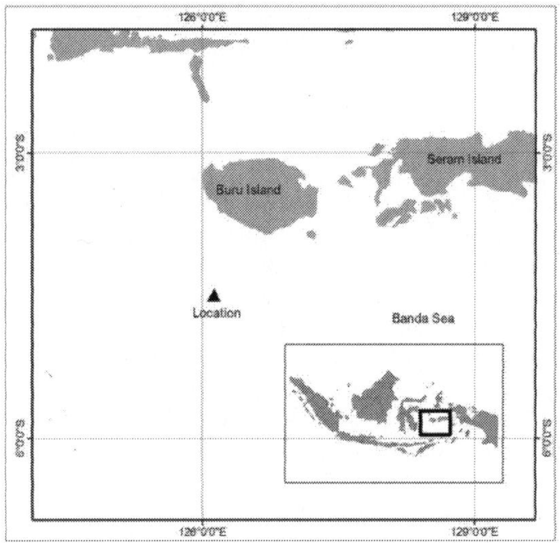

Figure 1. Study location

2.2 Peaks Over threshold

The POT method employed a series of Hs above a defined significant wave height level or threshold level. If threshold value (u) is too low then the exceeded data will produce a biased estimator. On the other hand, if the selected value u was too high then there will not be enough data to fit the model, resulting in large variations. One method of determining the frequently used threshold value was the percentile method. This percentile method was easier and practical, but the resulting threshold determination is accurate [13]. The 95% percentile will be used in this study as an extreme Hs limit. The 95% percentile was also used by [14] and generates the GPD model corresponding to the Hs data used.

2.3 Generalized Pareto Distribution

In general, the parameters of GPD are known as σ scale parameter (σ > 0), k shape parameter ($k \in \mathbb{R}$) dan μ location parameter(μ ∈ ℝ). For threshold u, convergent on GPDs that have a cumulative distribution function (CDF) as follows:

$$F(x) = \begin{cases} 1 - \left(1 - k\dfrac{x-\mu}{\sigma}\right)^{1/k} & k \neq 0 \\ 1 - exp\left(-\dfrac{x-\mu}{\sigma}\right) & k = 0 \end{cases} \quad (1)$$

And probability function (pdf) of GPD :

$$F(x) = \begin{cases} 1 - \left(1 - k\dfrac{x-\mu}{\sigma}\right)^{1/k} & k \neq 0 \\ 1 - exp\left(-\dfrac{x-\mu}{\sigma}\right) & k = 0 \end{cases} \quad (2)$$

The parameter estimation was using the L-moment method in Easyfit software.

Non-seasonal GPD uses all data that exceed the threshold value. The seasonal GPD was created by splitting the model by its seasonal variation in this study we use monthly variation, reference to the study about Indonesian wave characteristics by [12]. The seasonal GPD has twelve model represent monthly variation in the study site. The determination of threshold values and parameter estimation using the same method with non-seasonal GPD model.

2.4 Goodness of fit test

2.4.1. Quantile-Quantile Plot (QQ plot). By plotting empirical and fitted distributions for Hs above the thresholds u against each other, we can see the consistency of the estimated parameter value of selected u. If the QQ plot follows a linear line then the distribution is fitted.

2.4.2. Kolmogorov-Smirnov test. The hypothesis used in Kolmogorov-Smirnov test are :

H_0 : Fn(x) = F0(x) (data follow Generalized Pareto Distribution)
H_1 : Fn(x) ≠ F0(x) (data doesn't follow Generalized Pareto Distribution)

H_0 rejected if $D > D_{1-\alpha/2}$, using a 95% confidence level (α = 5%) with a value of D derived from :

$$D = \underset{x}{Sup}\left|F_n(x) - F_0(x)\right| \quad (3)$$

2.5 Return Value

The return value (x_m) of the extreme wave height exceeding the threshold (u) at least once in the m observation is as follows:

$$x_m = u + \frac{\sigma}{k}\left[(m\zeta_u)^k - 1\right] \quad (4)$$

ζ_u Is the probability of events that exceed the threshold u.

Return value in N-years dan n_y Is the amount of data in a year then$m = N \times n_y$. The equation of N-years return value as follows :

$$x_N = u + \frac{\sigma}{k}\left[\left(Nn_y\zeta_u\right)^k - 1\right] \quad (5)$$

2.6 Model Validation

Validation was performed by comparing the return value with maximum data during the period of 1, 2, 3 and 4 years based on Hs data for the period 2011-2014. The criteria were made based on the definition of return value that is the maximum Hs that occurs at least once during the return period. The return value greater than the maximum Hs was identified as incorrect and for smaller repeat periods of maximum Hs per period was identify as correct. The difference between the maximum Hs and the return period is also considered in model validation.

3. Result and Discussion

The threshold value specified using 95% percentile. Data greater than the threshold value are considered to be extreme values and are used to estimate GPD parameters. The number of extreme data used to estimate nonseasonal GPD parameter was 365 data with threshold value 1.76 meters. The threshold in seasonal GPD varies with the lowest threshold in November (0.74 m) and the highest in July (2.15 m). At the peak of monsoon Australia (JJA) the threshold value is higher than in other months, it means high waves occur in those months. The result of parameter estimation for nonseasonal and seasonal GPD using the L-moment method in Easyfit software was shown in Table 1.

Table 1. Non-seasonal and seasonal GPD parameter

Parameter	Non seasonal	Monthly variation (Seasonal)											
		Jan	Feb	Mar	Apr	May	Jun	Jul	Aug	Sep	Oct	Nov	Dec
Shape (k)	-0,14	0,03	-0,14	-0,28	-0,23	-0,47	-0,15	0,01	-0,14	-0,39	-0,51	0,10	-0,03
Scale (σ)	0,28	0,23	0,27	0,48	0,15	0,32	0,25	0,21	0,21	0,26	0,22	0,15	0,40
Location(μ)	1,77	1,27	1,21	1,14	1,21	1,66	1,40	2,15	2,01	1,49	1,06	0,72	1,02
Threshold(u)	1,76	1,29	1,25	1,09	1,21	1,69	2,12	2,15	2,01	1,52	1,09	0,74	1,05
Number of extreme Hs	365	31	28	31	30	31	30	31	31	30	31	30	31

The deviations in non-seasonal GPD (Figure 2) look significant at a high value of Hs but significant deviations also occur in low-value of Hs in seasonal GPD. QQ plot for seasonal GPD (Figure 3) seen to be around diagonal lines, but it can be seen that considerable deviations on seasonal models in certain months. These results can be taken into consideration in the selection of the best models. Generally, the QQ plots for non-seasonal and seasonal GPD are spreading close to diagonal lines.
Kolmogorov-Smirnov goodness of fit test shows that the value of $D_{1-\alpha/2}$ greater than D value (Table 2). It means that extreme Significant wave height follows GPD with confidence level 95%. According to QQ plot and Kolmogorov-Smirnov test, the non-seasonal and seasonal GPD fit the extreme significant wave height in the location.

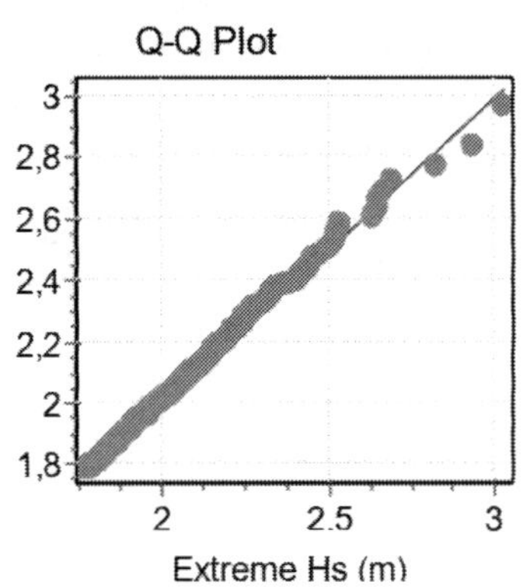

Figure 2. QQ plot for non-seasonal GPD

IOP Conf. Series: Earth and Environmental Science **116** (2018) 012066 doi:10.1088/1755-1315/116/1/012066

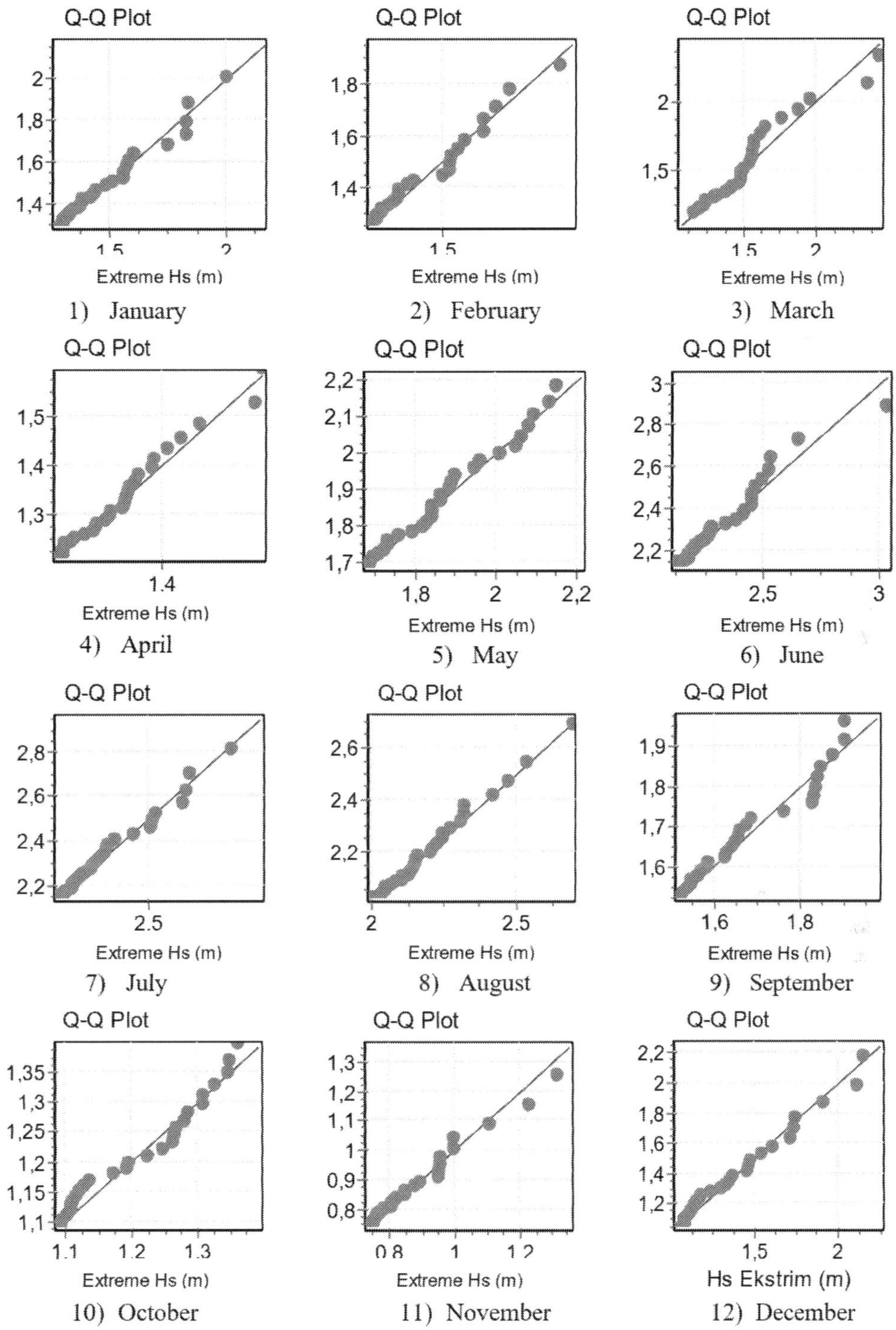

Figure 3. QQ plot for seasonal GPD

Table 2. Kolmogorov-Smirnov goodness of fit test

Parameter	Non seasonal	Monthly variation (Seasonal)											
		Jan	Feb	Mar	Apr	May	Jun	Jul	Aug	Sep	Oct	Nov	Dec
D	0,03	0,10	0,13	0,12	0,09	0,09	0,09	0,07	0,09	0,13	0,14	0,11	0,12
$D_{1-\alpha/2}$	0,07	0,24	0,24	0,25	0,24	0,24	0,24	0,24	0,24	0,24	0,24	0,24	0,24

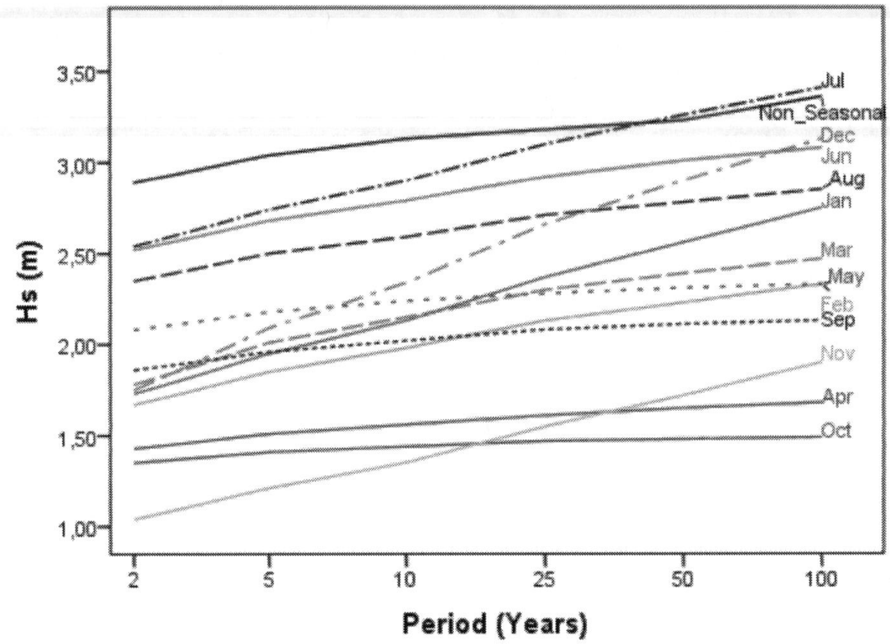

Figure 4. Non seasonal and monthly(seasonal) Return value.

Tabel 3. Model Validation by comparing Hs Return value form nonseasonal and seasonal GPD with maximum Hs. The bold number identify as incorrect.

Period (Years)	Hs Return Value		Maximum Hs (2011-2014)
	Non Seasonal	Seasonal	
1	**2,55**	2,38	2,38
2	2,66	2,54	2,72
3	2,72	2,63	2,72
4	2,76	2,69	2,93

Seasonal return value gives varying value each month while nonseasonal return value only has one value. These monthly variation of seasonal return value gives better information about the characteristics of the study location. The model has similar result with other model [11] in variation but has different in another characteristics. The maximum seasonal return value resulting a lower value in the initial period and increases steadily exceed the non-seasonal return values. The difference between nonseasonal and maximum seasonal return value was very small (less than 1 m). Return values which have positive k (Jan, Jun, and Nov) gave more increases return value in each period than negative k return value.

On comparing with maximum Hs in study location, non-seasonal and maximum seasonal return value in the early period was slightly different (< 0,2 m). The nonseasonal return value has 1 incorrect value for 1 year period, but in 2 – 4 years period Nonseasonal return value closer to maximum Hs than seasonal Hs.

4. Conclusion

The return value from seasonal GPD and nonseasonal GPD was compared to examine the difference. Monthly variation was selected as seasonal variation in study location. From the goodness of fit test, the GPD was fitted to the extreme significant wave height data both seasonal and nonseasonal. The return values from seasonal GPD can be obtained for any month while nonseasonal return value has

one value for each period. Seasonal return value could give better information for each month or season and anticipate the extreme wave height for maritime planning and management.
Model validation show all correct value for the seasonal return value and incorrect return value for the nonseasonal model in 1 years period but closer value for next period. A longer period of return level or other sample site needed to verify consistency of model validation.

References

[1] Sugianto DN, Zainuri M, Darari A, Darsono S, and Yuwono N. 2017. *Int. J. Civ. Eng. Technol.*, 8 (5) :. 604–619.
[2] World Bank. 2016. "Managing Coasts with Natural Solutions," p. 167.
[3] Jonathan P and Ewans K, 2012. *Ocean Eng.*, 62: 91–109.
[4] Mathiesen M, Goda Y, Hawkes PJ, Mansard E, Martín MJ, Peltier E, Thompson EF, and Vledder GV. 1994. *J. Hydraul. Res.*, 32 (6) : 803–814.
[5] Coles S. 2001. *Springer*.
[6] Mazas F, and Hamm L. 2011. *Coast. Eng.*, 58 (5) : 385–394.
[7] You ZJ. 2012. *Coast. Eng.*, 61 (1) : 49–52.
[8] Far SS, Khairi A, and Wahab A. 2016. *Ocean Sci. Discuss.*, pp. 1–25.
[9] Randell D, Feld G, Ewans K, and Jonathan P. 2015. *Environmetrics*, 26: 442–450.
[10] Solari S, and Losada MÁ. 2012. *Coast. Eng.*, 68: 67–77.
[11] Mendez FJ, Menendez M, Luceno A, Medina R, and Graham NE. 2008. *Ocean Eng.*, 35: 131–138.
[12] Kurniawan R, Habibie MN, and Suratno. 2012. *J. Meteorol. Dan Geofis.*, 12 (3): 221–232.
[13] Estiningrum T, Rusgiyono A, and Wilandari Y. 2015. *J. Gaussian*, 4: 141–150.
[14] Agarwal A, Venugopal V, and Harrison GP. 2013. *Renew. Sustain. Energy Rev.*, 27: 244–257.

IOP Conf. Series: Earth and Environmental Science **116** (2018) 012067 doi:10.1088/1755-1315/116/1/012067

Stolephorus sp Behavior in Different LED (Light Emitting Diode) Color and Light Intensities

Fitri Aristi D P[1*], Ramadanita I A[1], Hapsari T D[1], Susanto A[2]

[1]Faculty of Fisheries and Marine Science, Diponegoro University, Semarang, Indonesia
Email: aristidian@fisika.undip.ac.id
[2]Faculty of Agriculture, Sultan Ageng Tirtayasa University, Banten, Indonesia

Abstract. This research aims to observe anchovy (*Stolephorus* sp) behavior under different LED light intensities that affect eye physiology (cell cone structure). The materials used were *Stolephorus* sp taken from the waters off Jepara and 13 and 10 watt light emitting diode (LED). The research method was an experiment conducted from March through August 2015 in the waters off Jepara. Data analysis of visual histology and fish respond was carried out at the fishing gear material laboratory, anatomy and cultivate. Cone cell structure (mosaic cone) of *Stolephorus* sp forms a connected regular square pattern with every single cone surrounded by four double cones, which indicate that anchovies are sensitive to light. The 13 watt LED (628 lux) has faster response than the 10 watt LED (531 lux) as it has wider and higher emitting intensity, which also attracts fish to gather quicker.

Keywords:

1. Introduction

Anchovies (*Stolephorus* sp) are one of the economically important fish in Indonesia. There are 9 types of Anchovy spread throughout the Indonesian waters. Anchovies have positive phototaxis characteristic. This means going toward a source of light. This trait is used ineffectively capturing them. Wiyono (2006)[1] Notanubun and Patty (2010)[2] Fitri *et al* (2017)[3] Fitri *et al* (2017)[4] stated that utilization of light in fishing operation has well developed since the invention of electric light to capture phototaxis positive fish. Most fishermen believe that greater light intensity increases their catch. Therefore, they operate using all kinds of high intensity lights. However, this assumption is not entirely correct as different fish has different light intensity preferences such as the case for in puffer fish that can be caught using either kerosene or mercury pressure lantern Rasyidah (2009)[5].

According to Sulaiman *et al* (2015)[6], LED flood light is the best lighting to replace lamp which has a high cost of maintenance and operation. LED lights run deeper through water and attract fish located far away from the light source. LEDs are also more efficient than the other types of lights. Moreover, they are commercially available in the market. Hence, this research used the 13 watt and 10 watt LED lights.

Behavioral research on the retina of pelagic fish has not been widely conducted. Data on this are very important to know the proper exposure to light during fishing operation. This correlates to the time when fishermen should lift the net as they see schools of fish under the light. This leads to effectiveness in capturing process as prolonged exposure to light may cause burnout on the fish.

The above mentioned facts lead to this research on light intensity to attract anchovies at the laboratory scale. It is aimed at figuring out the proper light duration by analyzing cell cone structure and behavior of anchovies under different LED light intensities. According to Notanubun and Patty (2010)[2], there needs to be a study on the light intensity that can attract fish of all species, as this will lead to more efficient fishing budget.

2. Method

Living anchovies (*Stolephorus* sp) were collected from Jepara waters. Thirteen and 10 watt LEDs were paired at a reflector to focus the lights. The treatment tanks were 4 units made of cement with L x W x T : 319 x 168 x 115 cm, and each tank is given number 1 for nursery, number 2 and 3 for treatment tanks and number 4 for post- treatment tank. Every treatment tank was also given a mark for the 1st zone with light intensity of 737-2217 lux, the 2nd zone with 487-779 lux intensity and the 3rd zone with 216-292 lux intensity. Tank depth was also given three scales of layers; 1-26 cm, 27-52.7 cm and 53.7-77.5 cm, and the water in the tank and the LED lights were separated 34 cm apart.

Measurement of light intensity was carried out using an underwater lux meter with data logger. An illustration of the experiment tank with observation zones in both vertical and horizontal directions is shown in Figure 1.

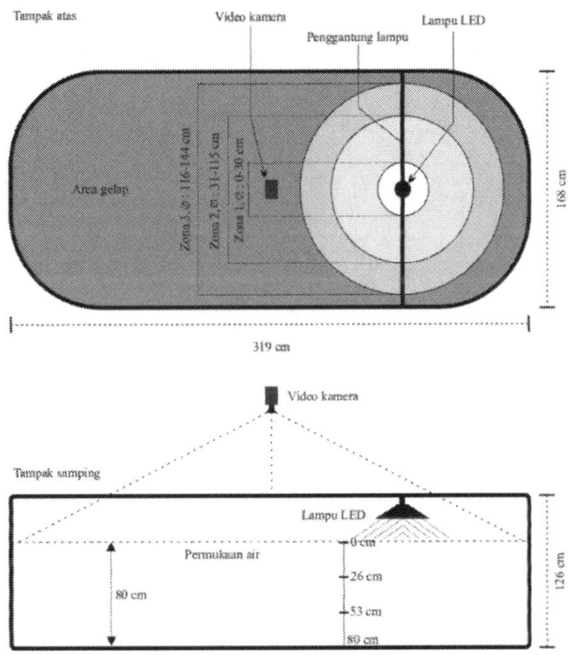

Figure 1. Design of treatment tank with different observation zones.

This study was conducted in 2 stages. The first stage was data collection by observation. Data were taken using digital cameras that observe and record directly. Primary data counted the total fish entering in 1st, 2nd, and 3rd zone per 5 minutes for 1 hour. The treatment started from the moment the lights were switched on and the depths of fish position were used as data. Each treatment tank consisted of approximately 100 anchovies. Prior to the treatment, the fish were adapted for 1 week with 1 feeding/day and the feed was natural feed (zooplankton). Treatments were carried out at night in dark condition and only the LEDs were on. Treatments were done in 6 repetitions for each lighting power. The second treatment was dissection of anchovies' eyes with histological process and each sample was instantly killed to keep the eye receptor unaffected.

Data were arrangements of cone cell structures which describe the structure of cone cell (mosaic cone) as visual cells in the retina. These observations were conducted to know the form of mosaic cones which are arranged in square patterns. Fish that have cone cell in square structures use their visual sense intensively. In most fish species, a double cone is similar to a twin cone and a single cone only consists of one cone cell (Loew and Lyhtgoe, 1978)[7].

3. Results and Discussion

3.1. *Light Intensity*

Results of light intensities from the 13 watt and 10 watt LEDs are shown in Figure 2.

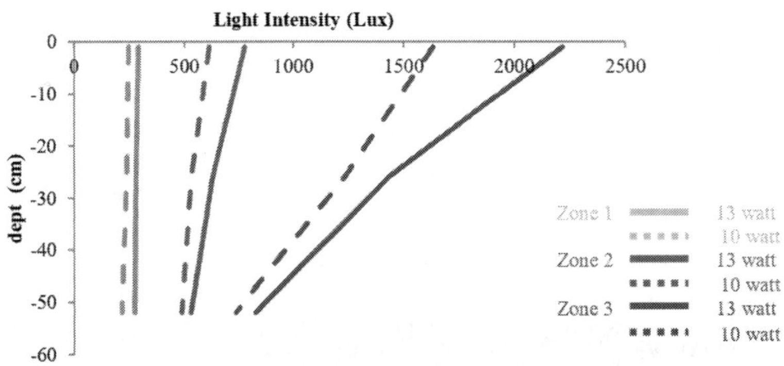

Figure 2. Light intensities of the 13 watt and 10 watt LEDs.

Figure 1 shows that the 13 watt LED has the highest light intensity in zone 3 at 1-26 cm depth with 2217 lux and the lowest light intensity in zone 1 at 52.7-77.5 cm depth with 275 lux. Meanwhile, the 10 watt LED has the highest light intensity in zone 2 at 1-26 cm depth with 1633 lux and the lowest intensity in zone 1 at 1-26 cm depth with 216 lux.

These results are used as a reference to know the ideal light intensity at which Anchovy can adapt perfectly. Light intensity for the 13 watt LED in each zone and depth is higher than for the 10 watt LED. Light intensity tends to decrease with increasing depth, and it tends to increase with closer light source. Yami (1987)[8] explains that the light intensity (lux) of light source will decrease with increasing distance and will decrease when the light goes into the water due to refraction.

3.2. *Fish response time against different light intensities*

Time response against light intensities for the 13 watt and 10 watt LEDs are shown in Figure 3.

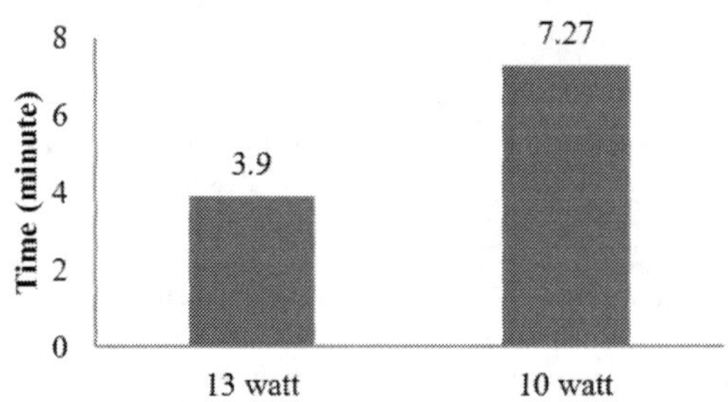

Figure 3. Time Fish response against the light.

Result in Figure 3 show that the fastest time for the 13watt LED is 3.9 minutes, compared to 7.27 minutes for the 10 watt LED. These are the times the fish move from dark toward the lights. The 13 watt LED has faster response than its 10 watt counterpart as the average fish response time is 3.13 minutes, whereas for the 10 watt LED, it is 28 minutes. This is due to the fact that the 13 watt LED has wider and higher emitting intensity. According to Sulaiman *et al* (2015)[6], Fish start to enter the light area under lift net frame after 5-10 minutes. Therefore, the first type of fish entering is small fish, followed by anchovies and the others. Ayodhya and Diniah (1989)[9] argued that the number of fish gathered under the light depends on the power of the light. The greater the power of the light, the more fish it attracts.

3.3. *The number of fish responding to different light intensities*

Table 1. The number of respond fish to the 13 watt LED light intensity.

Minute	Number of Fish		
	829-2217 Lux	527-779 Lux	275-292 Lux
5			4
10			7
15		4	16
20		14	12
25		26	4
30		30	3
35	2	32	1
40	2	48	27
45	4	34	23
50		12	5
55		10	18
60		14	17
Average	2.67	22.40	11.42
Standard Deviation	1.15	13.72	8.65

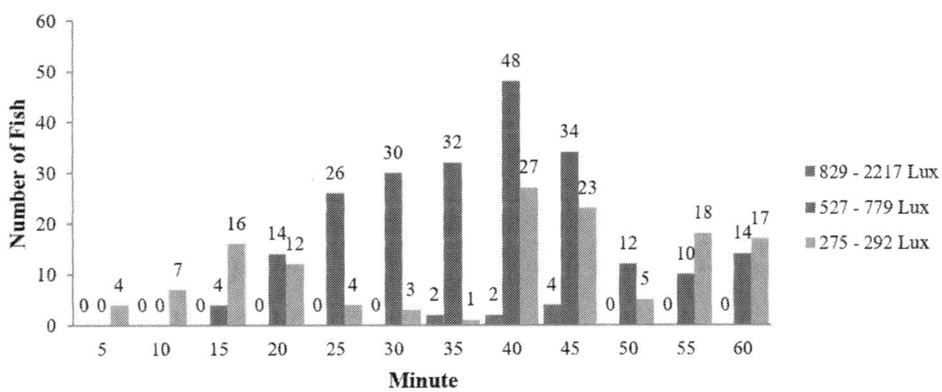

Figure 4. The number of respond fish to the 13 watt LED light intensity.

IOP Conf. Series: Earth and Environmental Science **116** (2018) 012067 doi:10.1088/1755-1315/116/1/012067

Table 2. The number of respond fish to the 10 watt LED light intensity.

Minute	Number of Fish		
	737-1633 Lux	487-613 Lux	216-247 Lux
5			
10			2
15		3	5
20		13	10
25		15	5
30		8	4
35		13	8
40	2	14	2
45		15	9
50	1	12	13
55	2	21	19
60		12	17
Average	1.67	12.60	8.55
Standard Deviation	0.58	4.70	5.79

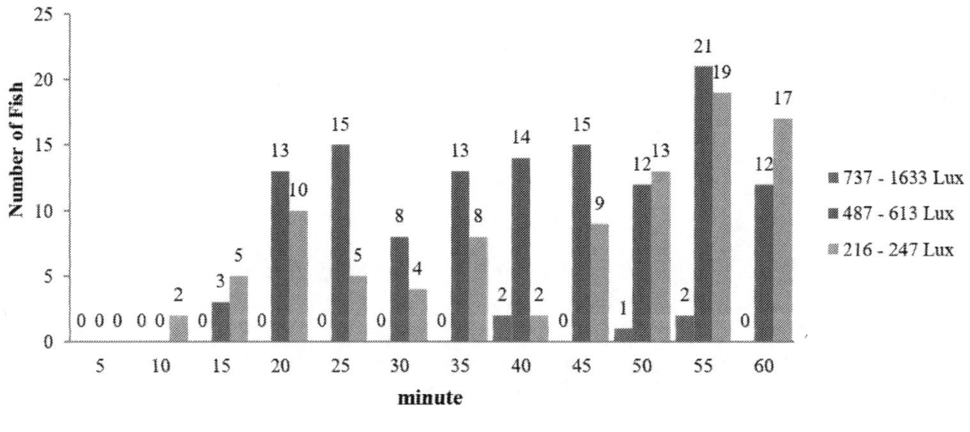

Figure 5. The number of respond fish to the 10 watt LED light intensity.

Results in Figure 4 show that the number of fish in each zone observed every 5 minutes in 1 hour. First, in the 4th minute there are 4 fish gathering in the zone of 275-292 lux intensity. At 15 minutes, there are also 4 fish gathering in the zone of 527-779 lux intensity. The highest number of fish is in the 40th minute in the zone of 829-2217 lux intensity. There are 14 individuals. In the zone of 527-779 lux intensity there are with 48 individuals, and in the zone of 275-292 lux intensity there are 17 individuals. This indicates that the fish began to respond interested in the intensity of the 13 watt lights at 40 minutes after the lights are turned on, and located in zone 2. Zone 2 is the area that has the most preferred lamp intensity of the fish, while zone 1 because the area is right under the light source do not react fast to fish.

It is also observable in Figure 5 that in the 10th minute, the fish start move to the 216-779 lux intensity zone from dark zone, with 2 individual going first. In the 15th minute, the fish move to the 487-613 lux intensity zone, with 13 individuals. The lowest number of fish is in the 55th minute in the 737-1633 lux zone, with only 2 individuals and the highest is in the 487-613 lux zone, with 21 individuals, and the number is even smaller after the 55th minute. Average adaptation time of fish with 10 watt lamp intensity is different with 13 watt lamp intensity, which takes longer adaptation time compared to 13 watt lamp. Zone 2 remains the most dominant area of adaptation compared to zone 1 and zone 3. This is because zone 1 which is right below the light source and zone 3 which is 144 cm area is a zone that does not attract fish to approach and do long adaptation.

3.4. *The number of fish both 13 watt and 10 watt LEDs*

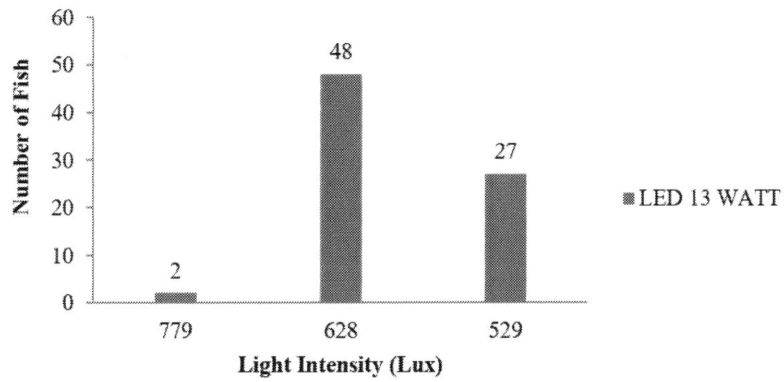

Figure 6. The number of fish (40 minute)

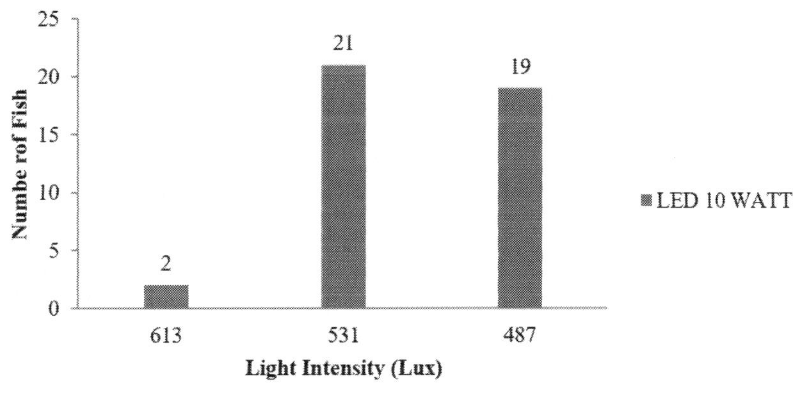

Figure 7. The total number of fish (55 minute)

The highest number of fish in the 13 watt treatment is in the 40th minute at 628 lux intensity with 48 individuals. Light intensity of 529 lux has 27 individuals, while the lowest number of fish is in the 779 lux intensity. Meanwhile, the 10 watt treatment has the highest number of fish in the 55th minutes at 531 lux intensity with 21 fish, whereas at 487 lux intensity there are 19 fish and at 613 lux intensity there are only 2 individuals of anchovy. From Figure 6 and Figure 7, it can be concluded that the light intensity of 628 lux at 13 watt lamp and 531 lux light intensity on 10 watt lamp attract fish more than other intensity because it can make fish to adapt longer on that intensity. In addition, the 13-watt lamp is able to attract more fish to gather than a 10-watt lamp.

3.5. *Anchovy behavior*

This laboratory scale research observes the anchovy's behavior against LED light intensity from 13 watt and 10 watt bulbs. These LED bulbs are available in the market a 10 watt LED has similar light intensity with a 24 watt TL (neon light), while a 13 watt LED equals the light intensity of a 42 watt TL light. Prior to the treatment, the fish were adapted to dark condition. The dark area serves as the starting point before lights were turned on, right after which the fish start to move toward them.

Observations were made every 5 minutes after the light source was turned on, and was done for 1 hour. Every treatment has 6 repetitions. Anchovies tend to swim in the bottom layer irregularly. According to Hermawan (2005)[10], the behavioral response of ''Red snapper'' to different colors and powers of lights can be known if they are treated in dark condition long enough before a treatment is administered.

Observations of the behavior of anchovy under both 13 watt and 10 watt illuminations show that longer lighting period leads to more fish gathering in the area. For the 13 watt LED, the number of fish increases until the 40th minute with 48 individuals in the light intensity of 628 lux. In the light intensity of 529 lux there are with 27 fish, and in the 779 lux intensity, there are 2 fish. After 40 minutes, the number of fish decreases until 60 minutes. This is because longer exposure period increases water temperature that the area is no longer comfortable for fish to gather at. For the 10 watt LED, the highest number of fish is in the 55th minute with 19 individuals at 487 lux light intensity. At 531 lux intensity there are 21 fish, and at 613 lux intensity there are 2 fish.

Anchovy's movements in both 13 watt and 10 watt LED illumination are twisting and surrounding the light area. According to Sudirman *et al* (2004)[12], anchovy tends to choose higher exposure, swim at the surface of water, and quickly enter area so that need no long period of adaptation to light is required. Instead, the longer period of exposure saturates Anchovy and this keeps them away from the light source.

Based on visual observations, Anchovy prefers the area of 27-52.7 cm depth with light intensity between 281-628 lux. This is consistent with an earlier research by Sulaiman *et al* (2015)[6], which explained that anchovy swims in areas of 1-15 m depth with 16-650 lux illumination intensity. According to Godo *et al* (2004)[13], anchovy swims in the area of 1-15 m because this area has positive photo taxis characteristic. This study argues that a 13 watt LED attracts Anchovies faster than its 10 watt counterpart. Therefore, it can be applied in the field to capture fish with an optimum light intensity of 281-628 lux, and an exposure time of 40 minutes.

Anchovy can adapt optimally in the adaptation process but this should be observed further in a laboratory scale. Sudirman (2003)[11] explains that from the aspect of light adaptation process, anchovies' retina tend to like high intensity lighting. This is based on observations of their cone cells, which shows that their pigments quickly adapt to light intensity. Normally, the number of hauling processes in a trip is 3 times. For the case of high number of anchovy present, then the number of hauling process can be add by 4 – 5 times a night. In other words, attracting this type of fish only requires 2 hours of light immersion into the water.

The 13 watt and 10 watt LEDs do not have differences in attracting fish. The only factor affecting attraction to fish is illumination as more fish are observed to be attracted to the lighting zone of 281-628 lux intensity. The number of fish attracted seems to be influenced by the range of lighting by the diameter of the LEDs. The light used in this study has different diameters for each lighting zone. The 13 watt LED has 116 cm diameter and is wider than the 10 watt LED, which is put at 27-51.7 cm depth with 108 cm of diameter. The range of lightings influenced the number of fish gathering under the light. According to Ayodhya and Diniah (1989)[9], the number of fish which gather under the light depends on lighting power. The stronger the light, the faster fish approach the light source.

3.6. *Cone cell structure*

Figure 8 shows the cone cell structure of anchovy's retina (*Stolephorus* sp). It can be seen that the composition of eye cell model is of single or double cone and is mosaic in shape. Fish with mosaic cone cell response faster to light. According to Razak *et al* (2005)[14], fish that have mosaic cone cell in line or square shape are very intensive in the use of their visual sense. Other than that, a mosaic eye cell structure (single or double cone cell) indicates that the fish is sensitive to light, both against its intensity and spectrum. Therefore, fishing operation using lift net at night should consider the effect of lighting, because light is key for the success of catching photo taxis fish such as anchovy.

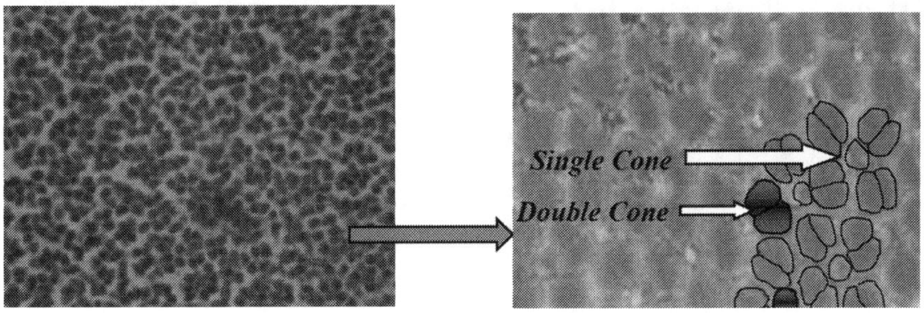

Figure 8. Structure of the cone cell.

4. Conclusion

There are difference is the behaviors of *Stolephurus* sp against different light intensities from LEDs. A 13 watt LED gets faster response from this type of fish (3.9 minutes), whereas a 10 watt LED gets a slower response (7.27 minutes). More fish gather under the illumination of a 13 watt LED with 628 lux intensity, compared to those gathering under the illumination of a 10 watt LED with 531 lux intensity. These differences in behaviors are also affected by the eye cell structure of *Stolephurus* sp with its characteristic mosaic pattern, which indicates that this type of fish is sensitive to different light intensities.

Acknowledgement

The authors wish to thank The Laboratory of Coastal Area Development (LPWP) Jepara, The Laboratory of Aquaculture of The Faculty of Fisheries and Marine Science, Diponegoro University Semarang, and Aziz Afriandi for his assistance.

References

[1] Wiyono E 2006. *J. of Tropical Fisheries* Science **18** 2
[2] Notanubun, J and Patty 2010 Thesis Sam Ratulangi University Manado
[3] Fitri A D P, Boesono H, Sabdono A, Nadiaadlina N 2017 **55** 1 012008
[4] Fitri A D P, Boesono H, Sabdono A, Supadminingsih F N, Nadiaadlina N 2017 *AACL Bioflux* 10 2 pp 191199
[5] Rosyidah 2009 *Indonesian Journal Marine Science* **2** 1 1907
[6] Sulaiman M, Mulyono S B, Am A T Sugeng H W, and Roza Y 2015 *J of Tropical Marine Science and Technolog* 7 1
[7] Loew and Lyhtgoe 1978 The ecology of vision Clarendon Press Oxford
[8] Yami B M 1987 Fishing With Light. Agriculture Organization of The United Nation by Fishing News Books Ltd Farnham Surrey England pp 121
[9] Ayodhya and Diniah R 1989 Animal Food (Post-Mortem Physiology and Technology) Graha Ilmu Yogyakarta pp 103
[10] Hermawan R 2005 Thesis Faculty of Fisheries and Marine Science Diponegoro University
[11] Sudirman 2003 *J. of Aquatic Sciences and Fisheries Indonesia* **3** 28
[12] Sudirman, M S Baskoro, A Purbayanto, D R Monintja, M Jufri and T Arimoto 2004 *J. of Aquatic Sciences and Fisheries Indonesia* **10** 2 85
[13] Godo O R V, Hjellvik S A, Iversen A S, Lotte E Tenningen and T Torkelsen 2004 *J. ICESJMS* **61** 7 1093
[14] Razak A, K Anwar and M S Baskoro 2005 Physiology Fish Eye Faculty of Fisheries and Marine Science Bogor Agriculture Institute Publisher. Bogor pp 111

Inclusive blue swimming crab fishery management initiative in Betahwalang Demak, Indonesia

A Ghofar[1], S Redjeki[2], H Madduppa[3,4], M Abbey[5], N Tasunar[6]

[1] Department of Aquatic Resources Management, Faculty of Fisheries and Marine Science, University of Diponegoro, Semarang, Indonesia;
[2] Department of Marine Science, Faculty of Fisheries and Marine Science, University of Diponegoro, Semarang, Indonesia;
[3] Department of Marine Science and Technology, Faculty of Fisheries and Marine Sciences, Bogor Agricultural University, Bogor 16680, Indonesia
[4] Indonesian Association of Blue Swimming Crab Processors, Gresik, Indonesia
[5] National Oceanographic and Atmospheric Administration, NOAA Fisheries,1315 East-West Highway, Silverspring, United States MD 20910
[6] District Marine and Fisheries Service of Demak, Indonesia

E-mail: aghofar099@gmail.com

Abstract. There has been a growing interest in the sustainability of the blue swimming crab (*Portunus pelagicus,* BSC) fisheries in Indonesia. The fishery is operated on a small-scale basis and yet it significantly contributes to the Indonesia's fisheries as the third biggest export commodities following tuna and shrimp. The project inclusively (i) brings together coastal and fishing communities, university, the private sector, government at various levels and international agencies, (ii) bottom up approach is integrated with top-down (government policy) approach and (iii) integration of conservation into fisheries management.
This approach resulted in better understanding and participation among the coastal fishing communities on sustainable fisheries and the necessity to perform fisheries management. This led to the establishment of BSC fishery management body (legally support by Village Regulation – *No.06/2013* on BSC fishery management in 2013, followed by a District Regulation *No.523/0166/2014* on BSC fishery management in 2014. More recently, the Governor of Central Java issued a Governor Regulation No. 33/2017 on Crab and Lobster fisheries management and a Governor Decree No. 523/93/2017 on the establishment of the BSC fisheries management committee in Central Java. Further impacts have been raised awareness in sustainable BSC fishery management in surrounding districts in other provinces, namely East Java and Southeast Sulawesi.
There remains, further needs to strengthen fishery governance by means of integrating national and local government effort in sustaining the fisheries, including the Issuance and effective implementation of the provincial decree on BSC fishery management for Central Java, that will enable the use of province's resource to implement fisheries management and strengthen law enforcement. To help improve the stock, a plan for stock enhancement should also be developed with proper monitoring program and community commitment to avoid "put and take" practices.

Keywords: *inclusivity, conservation-fisheries integration, blue swimming crab, village regulation, district regulation*

1. Introduction

The blue swimming crab (BSC) occurs in nearly half of the Indonesian marine environment. The biggest landings are made from the north coast of Java and Madura, followed by Lampung and Southeast Sulawesi. The BSC fishery in Indonesia is a small-scale fishery employing gillnets and collapsible trap from less than 10GT fishing boats. The BSC has been the Indonesia's third biggest, most valuable export commodity following tuna and shrimp. US is the largest market to Indonesian BSC, contributing to 40% of the whole US import from all countries in 2013 (NOAA, 2014). It fetches high price, rising from IDR2,000 in 1990 to IDR20,000/kg in 2012 and nearly IDR100,000/kg

in 2014, for whole crab at fishers' level. As a result, BSC resource is currently under heavy fishing pressure. Since 2015 global BSC market was oversupplied, and pushed the price down to an average of IDR35,000. Pre-assessment by the Marine Resources Assessment Group (MRAG) in 2009 showed that there are serious issues in three areas: stock status, ecosystem impacts and fisheries governance in this fishery. Furthermore, BSC is subject to exploitation by environmentally unfriendly fishing methods such as mini-trawl '*arad*' which are small mesh-size bottom seines). Observation shows the occurrence of small sized, immature BSC (<8 cm of carapace width) in fishermen's catches. Catching BSC at the size below first maturity may lead to not only overfishing, but also, more seriously, to l depletion of the BSC stock. This practice should therefore be stopped, and the fishery should be managed sustainably. Some effort to trace origin of meat process in the mini plant and processor level in order to comply government regulation on minimum landing size and banned trawl (Madduppa *et al*, 2016).
Since then, there has been a collaborative effort to overcome these issues. As a result, fisheries improvement project (FIP) has been imposed following the assessment. The Marine Stewardship Council (MSC) describes a FIP as: "a deliberate, managed and sustained effort to improve the environmental performance of a fishery towards sustainability. A FIP is often a partnership effort that can involve number of players and interests" (MSC, 2012). SFP further defines FIP for BSC as a series of improvement as shown in Table 1.

Table 1. The improvement tracker for Fisheries Improvement Project (SFP, 2012b).

Stages	**Indicators**
1 Launch FIP	Sustainability evaluation, best practices guidance or Fisheries Improvement recommendations publicly available
2 Form FIP	FIP members are organized and are evaluating the fishery
3 Encourage improvement	Work plan with annual improvement milestones is publicly available and FIP members are engaging regulators
4 Improve policies/ practices	Fishery is achieving annual improvement milestones and fishery policies or fishery practices changed or fishery management system became more precautionary or managers are following scientific advice more closely
5 Improvements in the water	Fish stock biomass increased or fishing mortality has decreased or
6 MSC certification (optional)*	

*This refers to NFI Crab Council policy focusing on effort to sustainability and leaving the certification as a business decision

An estimated 65,000 fishermen and 13,000 pickers are directly employed in over 400 miniplants and cooking stations throughout the country (APRI, 2017; SFP, 2012a). In addition, several thousands of other people are involved in the fishery, including middlemen, operators of mini-plants (where initial processing takes place) and final exporters (processors) who export the meat products (APRI, 2017; SFP, 2012a).

In an effort to address this issue, in early 2013 an agreement was achieved between the Faculty of Fisheries and Marine Science of the Diponegoro Univesrsity (UNDIP), US National and Oceanic and Atmospheric Administration (NOAA) Asia Pacific Capacity Building and the NFI Crab Council and Indonesian Blue Swimming Crab Association (APRI) to take the initiative for a collaborative fishery

IOP Conf. Series: Earth and Environmental Science **116** (2018) 012068 doi:10.1088/1755-1315/116/1/012068

management pilot program that involves BSC fisher groups as well as other key stakeholders at the local level. Betahwalang was identified as an important village in the north coast of Java, where the fishermen fish exclusively for BSC.

2. Approach used

The project applies an inclusivity approach throughout its processes in addressing the above issues, thus enhancing people ownership and responsibility. This means that integration should be applied, that is: (i) conservation is integrated into fisheries management; (ii) bottom up approach is integrated with top-down (government policy) approach. It also implies that in an effort to solve the existing issues, a root cause(s) is/are identified first, before other factor(s). The project brings together coastal and fishing communities, university, the private sector, government at various levels and international agencies. A baseline study on marine affairs and fisheries of Demak was started at UNDIP in the early 1980s. Trust building is the key to the success in community works, and develop partnerships and networks. Therefore it is also applied throughout the process. This approach requires that it start from the people, involving community-based process in the first instance, then introducing sustainable laws and policies, so that bottom-up and top-down approaches are synergized. It also integrates BSC conservation into fishery management by means of establishing BSC conservation zone within the fisheries management area.

In practice, it starts from people, involving community-based process, and introduce sustainable laws and policies.

3. Applying inclusivity: some results

The BSC fishery in Betahwalang employs 229 crab traps, 222 gillnets and 82 mini-trawls. There are a total of 743 crab fishers consisting of 447 boat owners and 296 labor fishers. In the fishing community, there are formal village leader as well asinformal religious leaders. Initial assessment of the community status and perception of these 743 fishers related to BSC fishery sustainability was carried out in 2013 and revealed the following features (Table 2).

Table 2. Assessment of the community perception (N= 743 fishers) related to BSC fishery sustainability conducted in 2013

Status/Proportion	Yes	No	Do not know	Reason/ Challenge
Importance of the BSC fishery for their livelihood?	100%			
Needs for fishery sustainability?	100%			
Fishery needs to be managed?	10%	70%	20%	Belief that crab is given and will be there all the time
Their dependency to crab collectors?	90%		10%	Availability for support
Role of district fishery service	40%	50%	10%	Distribution issue
Whom do you listen for advice: formal leader?	40%	60%		Only for formal purposes

Whom do you listen for advice: informal leader?	90%	10%	Availability for support
Have children at village school?	90%	10%	

In July 2013, when the project started, the understanding and awareness on long term BSC sustainability was lacking, and the fishermen were merely taking short-term economic profit from the BSC resource, without local rules that govern the fisheries. Developing a knowledge base and changing fishermen's attitude were therefore the initial key challenges faced by the project.
The BSC FIP participants think the BSC population is currently overfished. They consider this is a problem and therefore they wanted to work towards sustainability. They chose the FIP concept as a means to do this, because it provided guidelines for achieving a sustainable state.

Awareness raising started with enhancing the understanding and awareness of the community toward a sustainable fishery. This action involved the whole community rom children, with designed games, a BSC drawing competition and intervention through school, to fishermen and their families through social and religious events. This approach resulted in increased understanding of the coastal fishing communities on sustainable fisheries and the necessity to perform fisheries management. This led to an agreement to establish a BSC fishery management body (FMB) by the fishing community, locally named "Sustainable BSC Fisheries Management Institue (LP2RL)" in December 2013, which was then legally supported by the Village Regulation *No.06/2013* on BSC fishery management. Eight months later in 2014, the district of Demak issued a District Regulation *No.523/0166/2014* on BSC fishery management. The FMB includes a Scientific Committee (ScC, whose main task is to assess BSC resource and fishery as the basis for fisheries sustainability) and a Stakeholders Committee (StC, whose main task to provide general policy and guidance).

The Diponegoro University (UNDIP) supports such assessment and community guidance. It also provides scholarships in Fisheries Studies to young students of Betahwalang Demak origin, that requires they should return back to their village/district after the completion of their studies, to take the lead in sustainable BSC fishery management. Following people's agreement on sustaining BSC fishery, village government was convinced that a village regulation is required to sustain the fishery. Further impacts has been raised awareness in sustainable BSC fishery management in surrounding districts in three provinces, namely Central Java, East Java and Southeast Sulawesi.

Establishing a BSC conservation area enhances fish and shellfish abundance in the area, in addition to providing natural barrier for the entrance of destructive fishing from outside. These results had been influencing other villages and encourages the whole district to follow the process already undertaken, and has led to the issuance of District (Bupati) Decree of Demak on sustainable BSC fishery management in 22 August 2014, which also included the establishment of a District BSC fishery management team (BSCWG) Two district meetings followed this decree in order to implement it. The first meeting was held on September 19, 2014 in the District Secretary Building, with main objective to prepare the implementation of the Decree, and to coordinate among the BSCWG members. The second meeting was held on September 25, 2014 in Demak District Hall, during which BSC fisheries management in Demak District was formally informed to public. The latter meeting was attended by BSC fising communities and collectors from two districts (i.e. Demak and Rembang), Demaks District officers, APRI, BBPPI, JICA, US-NOAA, UNDIP and press.

To sum up, this project resulted in the following outputs:
a. Establishment of BSC FMB (legally support by Village Regulation – PERDES or *Peraturan Desa No.06/2013* on BSC fishery management, which is the first FMB established

in Indonesian fisheries), in which BSC conservation is a vital component.

b. As part of public communication, and in order to ensure that the PERDES is understood by village people, key messages of the PERDES was made available in all mosques in Betahwalang. This was the initiatives of the Village BSC fishery management body (LP2RL);
c. Increasing awareness and commitment on fisheries sustainability amongst the people from the village to district and an increasing willingness among the fishermen on fishing gear exchange to sustainable fishing gear – "*bubu*" (trap), to be administered through LP2RL;

Three short videos were produced by the involvement of Betahwalang community on BSC conservation education (https://www.youtube.com/watch?v=SQPU6BU9tio) and on BSC fishery of Betahwalang (https://www.youtube.com/watch?v=QjnMlR7kFhs) and subsequent visit of US Ambassador to Betahwalang (https://youtu.be/ADFmUi-z20M).

The BSC project has also been connected to several international events, such as the IOPAC meeting in 2013 where the project finding was presented and received positive attention, and the Marine Stewardship Council (MSC) based in London. Following a presentation on Indonesian BSC fishery in its annual meeting in Madrid in 2013, MSC approved a data-poor-stock-assessment pilot study for BSC fisheries in three Southeast Asian countries including Indonesia. In the same year, MSC also provided a scholarship to a Wageningen University of the Netherland student, Ms Floor Bokkes, to undertake a socio-economic study on BSC fishery in Betahwalang, under joint-supervision of Dr. Simon R. Bush, Mr. Paul Zwieten (Wageningen) and Dr. Abdul Ghofar (UNDIP).
The inclusivity approach had strengthen community ownership; at the end of 2014 the village of Betahwalang built a gate at the village entrance, named BSC Fishery Management Village ("*Betahwalang Desa Pengelolaan Rajungan*"). Furthermore, a workshop is prepared for sharing the awareness and commitment with all 13 districts in the north coast of Central Java.

4. Impacts

The approach this project developed are now followed nation-wide. In September 2014 UNDIP, in collaboration with US-NOAA held a training workshop on sustainable BSC fishery management. Participants from other north Java coastal districts, notably Semarang, Rembang, Pati, Jepara and Lamongan districts, were also involved in this meeting. There is now a general increase of commitment to sustain the BSC fishery in the three districts. Furthermore, participants are at present also preparing to deliver messages from the workshop to their surrounding communities.

As described above, the fisheries management initiative in Betahwalang village has a multiplying impacts to surrounding villages of Demak, and further move to other districts, including:

a. District Bupati Decree on sustainable BSC fishery management (see *Surat Keputusan Bupati Demak No.523/0166/2014* on BSC fishery management), which wasformally announced to public by the District Secretary during public meeting on September 25, 2015.
b. Following engagement of four Central Java districts (Demak, Jepara, Pati, Rembang), a series of discussions has been held bewteen UNDIP and the Provincial Marine and Fisheries Office since 2015, on the need of BSC fishery management throughout the province. In August 2017 the Governor Regulation No. 33 of 2017 on the management of crab and lobster was issued in August 2017 followed by a Governor Decree No. 523/93/2017 on the establishment of BSC fisheries management committee in November 2017.
c. The Diponegoro University (UNDIP), in collaboration with Japan International Cooperation Agency (JICA) also organized overseas trainings for potential candidate from Demak district that can in the long term help sustain and manage their BSC fishery on their return. From September

to December 2014, Mr. Nanang Tasunar of the District Fisheries Office in Demak is participating in JICA training on sustainable fishing techniques in Yokohama, Japan. It is aimed that in the long term Mr. Tasunar and his staff will take the responsibility of sustainably managing the BSC fishery throughout the Demak District. In September 2014 UNDIP, in collaboration with US-NOAA held a training workshop on sustainable BSC fishery management. Participants from other districts, notably Rembang and Semarang, were also involved in this meeting. There is now a general increase of commitment to sustain the BSC fishery in the three districts. Furthermore, participants are at present also preparing to deliver messages from the workshop to their surrounding communities. The village of Betahwalang has built a gate at the village entrance, named BSC fishery management village ("Betahwalang Desa Pengelolaan Rajungan"). The district provided a budget for sustainable BSC fisheries management, starting from 2015.

d. Furthermore, in November 12, 2014, UNDIP received a verbal request from the provinces Southeast Sulawesi, so that the province may learn from fisheries management model of Betahwalang Demak for their further adaptive implementation. In 2017 the Governor Decree of Southeast Sulawesi on sustainable BSC fisheries management was also issued.

These outcomes and impacts are summarized in Figure 1.

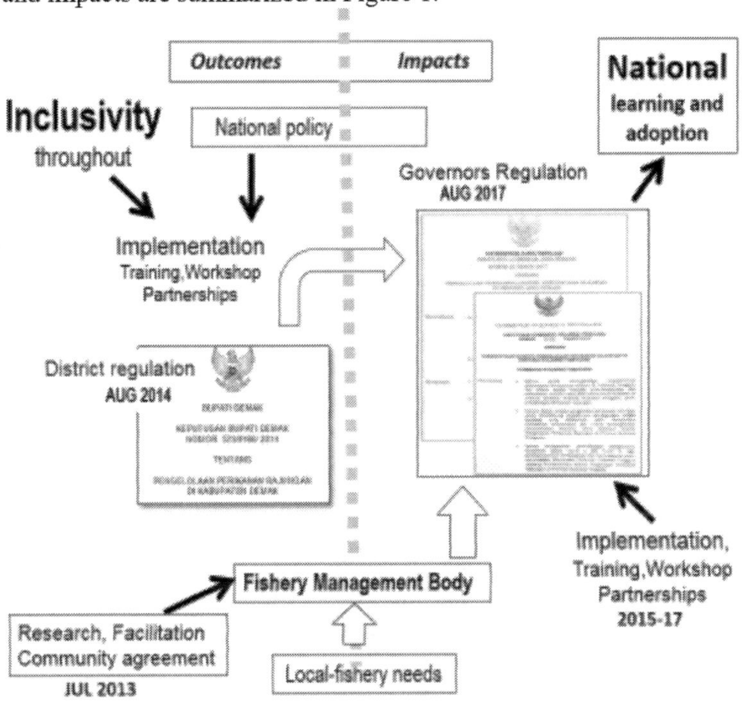

Figure 1. A summary chart showing the inclusivity approach, activities, outcomes and impacts of the BSC fishery development in Betahwalang Demak from 2013-2017

There are challenges toward sustainable BSC fishery management in the study area that may apply for the whole country, including:

a. Effective law enforcement. Ministerial Decrees No.01/2015 on banning the use of trawl and seine, and No. 02/2015 and No. 56/2016 on minimum size of captured BSC, mangrove crab and lobster were issued in 2015 but have not been effective, due to lack of law enforcement. In late 2014, interviews were carried out to fishers in Betahwalang (Bokkes, 2014), revealing that the regulations, has not been fully implemented until January 2015, which makes it still a quasian open- access system and at risk of overexploitation, following Hardin's argumentation about the tragedy of the commons (Hardin, 1968; Feeny *et al.*, 1990). On the one hand the nature of the resource is such that it is difficult to exclude people from using it and therefore it can be commonly accessed. On the other hand, one person or actor group using the resource will always have a negative effect on another actor's ability to exploit the resource (Feeny et al., 1990). The BSC FIP participants think the BSC population is being overfished at the moment. They consider this a problem and therefore they wanted to work towards sustainability. They chose the FIP concept as a means to do this, because it provided guidelines for achieving a sustainable state.

b. More recently in March 2016, several interviews with BSC fishers in the north coast of Java clearly indicate this. More depressingly is the fact, that fishers who has been engaged in sustainable practice by obeying the rules until the end of 2015, accidentally follow those who break the rules in favour of more immediate income. The respond such as "we do obey as rules say, but why are others allowed to break them?" is quite common.
c. More coordinated partnership. There has been an increasing interests in BSC fisheries as indicated by international as well as in-country's partnership. Lessons learnt from Betahwalang strongly suggests that such involvement can cause overexposure to outsiders, which may further lead to unnecessary confusion, apatism and dependence to outside support (Ghofar et al, 2008).
d. As in many part of the developing world, stock enhancement is not usually followed by sufficient monitoring program. To help improve the stock, a plan for BSC stock enhancement is also developed, from already stalled BSC conservation area, with proper monitoring program and community engagement to avoid "put and take" practices. With that, distribution and dispersal of young BSC may be traced and analysed for fishery management purpose.
e. Synchronizing the Fisheries Law: In Article 1, fisheries is defined as a comprehensive business sector including fish (including BSC) resources, aquaculture, post harvest part, trade, markets and social factors. However, fisheries management as defined in Article 7 tends to limit the fisheries to merely fish resources. Fisheries management, and FIP, should therefore be put in a broader fisheries context that connects to business sector reality.

[1] Conclusion and way forward

Implementing inclusivity in BSC fishery management in Betahwalang provides positive impacts in enhancing participation and commitment among the fishing community and fisheries stakeholders, both locally in the study area and extended areas including other villages, districts and provinces. Apart from the local and central government of Indonesia, there are international agencies, which have been involved later, including US National Oceanic and Atmospheric Administration (NOAA), National Fisheries Institute's (NFI) Crab Council, Sustainable Fisheries Partnership (SFP), Environmental Defense Fund (EDF) and Japan International Cooperation Agency (JICA).
The issuance of Ministerial Decree on BSC minimum size limit of 10 cm carapace length and non-berried females has not been effectively enforced locally, due mainly to the absence of provincial regulation that provide the basis for support for such enforcement and resource for implementing the fisheries management.

Implementation of inclusive fisheries management is relatively recent to give impacts to the improvement of the BSC stocks. The following approach is suggested for further improvement:

- Reconcile legal definition of fisheries. Indonesia's Fisheries Law defines fisheries as "all activities relating to the cultivation and utilization of fish resources and the environment thereof, starting from pre-production, production, cultivation, up to marketing, implemented in a system of fishery business" (Article 1:1), whereas fisheries management as focusing more on biological fish resources (Article 1:7).

- Further develop in-country partnership to make sure that this project will be financially sustainable in the long run. This includes further engaging government at all levels as above mentioned. The District of Demak had provided a budget for the implementation of sustainable BSC fishery management, followed by an interest to follow the BSC sustainability program by some Districts of Rembang, Jepara, Pati of Central Java Province and Lamongan of the East Java Province. The involvement of the government is vital, as they are the authority which by law is responsible to sustain the fisheries.
- Further develop an integrated partnership with international agencies and NGOs which are concerned about marine and fisheries sustainability. Discussion on an international training-workshop is ongoing between JICA, US-NOAA, SFP and UNDIP to further advance sustainable BSC fishery management which also involved BSC fishery stakeholders working in Demak. Particular attention is given on avoiding overlaps in partners programs, which in turn can damage community's participation already built. Fatique and confusion among the community members usually are the intermediate indicators to such avoidable consequence.
- Existing partnership should also support the plan for BSC stock enhancement following the already stalled BSC conservation area, with proper monitoring program and community engagement to avoid "put and take" practices. With that, distribution and dispersal of young BSC may be traced and analysed for fishery management purpose.

Acknowledgement

The authors acknowledge the collaboration and support from the fishing community in Betahwalang, the district government of Demak, the NFI Crab Council, Indonesian Blue Swimming Crab Association (Asosiasi Pengelolaan Rajungan Indonesia-APRI), NOAA, SFP and JICA.

References

[1] APRI 2017. Asosiasi Pengelolaan Rajungan Indonesia Fisheries Improvement Project (FIP). website: www.apri.or.id

[2] Bokkes F 2013. MSc Thesis on BSC fishery of Bethahwalang (Wageningen University, The Netherlands), visit http://edepot.wur.nl/292820

[3] Feeny D, Berkes F, McCay B J, Acheson J M 1990. The Tragedy of the Commons: Twenty-Two Years Later. *Human Ecology* 18, (1): 1-19.

[4] Ghofar, A, Schorr D K and Halim A 2008. Selected Indonesian Fisheries Subsidies: Quantitative and Qualitative Assessment of policy coherence and effectiveness. The Nature Conservancy. 89p.

[5] Hardin G 1968. The Tragedy of the Commons. *Science* 162: 1243-1248.

[6] MSC 2011. Kroger, Costco, Supervalu and Walmart include MSC in seafood sustainability commitments (http://www.msc.org/newsroom/news/kroger-costco-supervalu-and-walmart-include-msc-in-seafood-sustainability-commitments. Last update: May 12, 2011. Date of access: 19 September 2013).

[7] MSC 2012. Next Steps / Project Plan – Fisheries in Transition and MSC's Strategic Plan. April 1, 2012.

[8] Madduppa H, Zairion, Nuraini S, Nugroho K, Nugraha B 2016. Setting Up Traceability Tools for the Indonesian Blue Swimming Crab Fishery: A Case Study in Southeast Sulawesi. Fisheries and Aquaculture in the Modern World, 09/2016; , ISBN: 978-953-51-2686-7, DOI:10.5772/64252

[9] NOAA 2014. Ecosystem approach to managing blue swimming crab in Indonesia: developing a fisheries management framework. 26p.

[10] Pauly D, Watson R, Alder J 2005. Global trends in world fisheries: impacts on marine ecosystems and food security. *Philosophical Transactions of the Royal Society B* 360: 5-12.

[11] SFP 2011. Sustainable Fisheries Partnership Discussion Paper SE Asia Blue Swimming Crab Minimum Size Implementation plan. Date of access: 30 October 2012.
[12] SFP 2012a. [Online on] http://www.sustainablefish.org/fisheries-improvement/crabs/indonesian-blue-swimming-crab. Last update: December 2012. Date of access: 16 January 2013.
[13] SFP 2012b. Fishery Improvement Projects. http://www.sustainablefish.org/fisheries-improvement. Last update: January 2012.
[14] SFP 2013. About SFP. http://www.sustainablefish.org/about-us. Last update: 2013.

The Growth and Mortality Rate of Mullet (*Mugil dussumieri*) on Seagrass Beds of The Teluk Awur Bay, Jepara

L K Pinandita, I Riniatsih, and I Irwani

Departement of Marine Science, Faculty of Fisheries and Marine Science, Diponegoro University. Jl. Prof. H. Soedarto, SH, Tembalang, Semarang 50275, Indonesia.

Email: laras.kinan@yahoo.com

Abstract. Seagrass beds that have relatively high primary productivity are used as habitat for many marine species. Fish use seagrass as feeding, nursery, and spawning grounds. This research aimed to determinate the growth and mortality rates of mullet (*Mugil dussumieri*) on seagrass bed ecosystems of Teluk Awur Bay water, Jepara, Central Java. The descriptive method was applied in this research with the purposive method for sampling. Microsoft Excel software and FISAT II of FAO were used for data analyses, and the samples of 347 mullet (*M. dussumieri*) were taken from October until December 2016. The results of this research showed that length of fish ranges 8 – 28.9 cm with weight range 5 – 248 grams. The growth coefficient value (K) was 0.33 with asymptotic length ($L\infty$) 30.24 cm, and the value of t_0 was -0.305, which will be reaching for 11 years. The rate of total mortality (Z) was 0.854 per year, the value of natural mortality (M) was 0.706 per year and the value of fishing mortality (F) was 0.148 per year. Exploitation ratio (E) was 0.173 per year, it indicated that only 17.3% of mullet's (*M. dussumieri*) deaths in Teluk Awur Bay waters caused bycatch. It can be estimated that the death of mullet in Teluk Awur Bay waters affected more by the condition of the waters, in this case, the decreasing density of seagrass in research location is expected to affect the growth of mullet.

Keywords: Seagrass, Mullet (*Mugil dussumieri*), Growth, Mortality, Exploitation

1. Introduction

Seagrass beds is one of the most productive biomass and there is some function that seagrass can provide for biotas who lived in it, such as provide foods in coastal food webs, an organic carbon export to adjacent ecosystems, and habitat for invertebrates and vertebrates. Especially for fishes, including those in commercial fisheries, seagrass beds play several roles like nursery for juveniles, shelter, or feeding grounds [1].

The importance of seagrass bed as a habitat and food source for marine animals is expected to vary with the species composition of seagrass. Some researchers found the differences in fish assemblages in seagrass bed represented by different species of seagrass. The richness of seagrass species may influence faunal assemblages because more diverse seagrass provide greater structural complexity and therefore more niches for the associated plants and animals. The physical nature of the seagrass canopy is thought to play a major role, potentially influencing available shelter, food, and protection from predators. This fact is raising a concern about the role of seagrass diversity on their ecological function in marine ecosystems, in particular, because there is a tendency of declining and/or extinction of certain species of seagrass due to climate change and other factors [2].

In fishing and aquaculture, mullets are important economic resources that support many small communities. Mullets live as semi-catadromous, which the juveniles being recruited to lagoons and estuaries following a period of offshore spawning [3]. Previous research said that Mugilids are generally considered as herbivorous, omnivorous, plankton feeders, or even microcrustacean predators. The tropic behavior of mullets has been expressed by different authors using extensive terminology which categorized feeding patterns of these species. It includes algae feeders, micro and meiobenthos feeders, interface-feeders, deposit feeders, benthic microphagous omnivores and limnobenthofagous [4].

During the last decades, the problems of seagrass degradation have received increased attention worldwide. Seagrass loss, however, mainly occurs due to human impacts and the most eutrophication, this also may affect the density and composition of associated fish species [5].

The waters of Teluk Awur Bay, Jepara has a relatively fertile seagrass communities. The condition of coastal morphology will affect the density and the type of seagrass contained therein. However, the existence of human activities and fishing around the waters of Teluk Awur Bay allegedly may affect the density of the seagrass and the growth of fish in these waters. According to the Ministry of Marine Affairs and Fisheries in 2011 informed that there has been an increase in the number of catching mullets every year ranged from 2.4% from 36.077 tons in 2000 to 44.905 tons in 2010 [6].

Based on that conditions it is necessary to do research that aims to determine the rate of growth and mortality of mullet fish (*Mugil dussumieri*) which lives on seagrass in the waters of Teluk Awur Bay, Jepara. Existing data are expected to be useful and serve as basic data for the management of capture fisheries and seagrass ecosystems in Teluk Awur Bay waters, Jepara.

2. Materials and Methods

2.1. Time and Place

This research used mullet (*M. dussumieri*) as main material and descriptive method. This research was conducted in October - December 2016. Purposive sampling method was used to determine the sampling location, in this case, consideration of research location based on seagrass density conditions. Sampling was carried out in the waters of Teluk Awur Bay, Jepara for 10 hours from 06.00 a.m. – 16.00 p.m., followed the activities of fishermen.

Primary data was collected for three months with a two-week sampling time interval. Mullet fishing grounds located at coordinate 6°37’13.3” S dan 110°38’12.6” E. Fishing location is directly near to Marine Station, Teluk Awur Village, Jepara, Java Sea.

2.2. Sampling Method

Shore operated stationary lift nets that are settled in the waters was used to collect the fish data in this research. Lift nets frame mounted at a height of 8 meters above sea level, it had a net with overall width size is 6x6 meters and the mesh size is 1 cm. All the caught fish were taken to the mainland to be separated between mullets and other types of fishes. All of the fresh mullets were put into a cool box, and then it brought to the Diponegoro University Biology’s laboratory for the fish identification and the measurement of weight and length. The fish identification based on Saanin’s book (1968). Observation of water quality parameters measured in-situ together with mullets and seagrass sampling. The observed parameters are including temperature, dissolved oxygen (DO), salinity, pH, depth, and brightness.

2.3. Growth Parameter Analysis

The measurement of the total length of fish used a ruler with 0.1 cm accuracy, while the measurement of fish weight used a digital scales with 0.01 grams accuracy.

Data processing of distribution frequency in this research used Microsoft Excel software. The long fish-based growth model formula was exposed by Von Bertalanffy, and known as Von Bertalanffy's Model is [7]:

$$Lt = L\infty\ (1 - \exp(-K(t - t_0))$$

Where L∞ is the mean length the fish would reach if they were to grow to a very old age (asymptotic length) (cm), K is a growth coefficient (cm), t_0 is an imaginary age of fish when they had zero length, and Lt is the length at the age t.

The value for L∞ and K were obtained from the calculation results by ELEFAN 1 method contained in the program FISAT II, meanwhile to get the value of t_0 can use von bertalanfy's empirical formula [8]:

$$\text{Log}\,(-t_0) = 0.3922 - 0.2752(\log L\infty) - 1.0380(\log K)$$

2.4. Fish Mortality Analysis

The rate of total mortality (Z) based on Beverton and Holt [6] defined as:

$$Z = \frac{K\,(L\infty - \bar{L})}{\bar{L} - L'}$$

Where L∞ and K are parameters of von Bertalanffy growth equations, $\bar{L}$ is mean length of fish in the catch (cm), and L' is the smallest length of fish in the catch (cm).

The calculation of the natural mortality value (M) is known through the empirical formula by Pauly and the calculation of fish habits that clustered in the value of natural mortality multiplied by the number 0.8. So that species are clustered have estimated value 20% lower [8]:

$$\ln M = -0.0152 - 0.279 \ln L\infty + 0.6543 \ln K + 0.463 \ln T$$

$$M = 0.8 * \exp(\ln M)$$

Where L∞ and K are parameters of von Bertalanffy growth equations and T is mean temperature (°C) of the water in wich fish lives.

Total mortality value (Z) is the sum of the natural mortality value (M) and fishing mortality value (F), so that F is obtained from the following calculation results:

$$F = Z - M$$

If the fishing mortality (F) and natural mortality (M) has been obtained, so exploitation ratio (E) of fish may be computed from [8]:

$$E = \frac{F}{F + M}$$

3. Result and Discussion

3.1. Description of Condition of Research Sites

The caught fish on seagrass beds in Teluk Awur Bay mostly was the fish that are not the permanent inhabitant, even there were also fish that come from the coral reef. Types of fish that are often caught by fishermen in the form of mullets (*Mugil* sp.), 'lencam' fish (*Lethrinus obsoletus*), 'baronang' fish (*Siganus oramin*), and *Lutjanus lutjanus*. However, the most commonly found fish with high economic value is mullets (*Mugil* sp.). Beside *Mugil dussumieri* as mullet species which was analyzed in this research, fishing gear also caught other types of mullets, such as *M. subviridis, M. tade, M. troscheli, M. seheli*, dan *M. vaigiensis*.

Based on the previous study, it was known that *M. dussumieri* was a dominant mullet species in some waters of Indonesia and the larvae were found in coastal water near river estuaries. It was suitable with the Teluk Awur Bay conditions, where there were seagrass and mangrove ecosystems in the coastal waters which can be used as a place of life for mullet (*M. dussumieri*).

3.2. Seagrass Condition

Based on this research for seagrass density in the sites (**Table 1**), there were three species of seagrass that had been found, i.e. *Enhalus acoroides, Thalassia hemprichii*, and *Cymodocea serrulata*. The most common type of seagrass was *T. hemprichii* with the density value was 1576 stands/m^2 and the least found was *C. serrulata* it only 22 stands/m^2, while for *E. acoroides* the density value was 171 stands/m^2. If it compared with the results of previous research in 2015, in same sites and same seagrass ecosystem areas, it was recognized that the most dominant seagrass species was *Enhalus* sp. with the density value was 1943 stands/m^2 and the least found was *Thalassia* sp. 321 stands/m^2, and the density value for *Cymodocea* sp. was 1908 stands/m^2, and the presence of seagrass species *Halophila* sp. with the density value was 220 stands/m^2 [9].

Table 1. Types and Density of Seagrass in Research Sites

Spesies	Density (stands/m^2)	Relative Density (RD) (%)
Enhalus acoroides	171	9.67
Thalassia hemprichii	1576	89.09
Cymodocea serrulata	22	1.24

Thus, it can be seen that seagrass density in the waters of Teluk Awur Bay was decreased for these 2 years. The continued consequences of the destruction of the seagrass ecosystem can affect the distribution and abundance of fish and another biota that associated with the seagrass.

3.3. Conditions of Water

Fish data was collected from morning to evening, the conditions of water on this research sites can be seen from the observation of environmental parameters. The environmental parameters data collection was performed simultaneously with the fish data, waters parameters data were presented in **Table 2** below.

Table 2. Data of Environmental Parameter Range in Research Sites.

Parameters	Range
Temperature	29-30 °C
Salinity	30-31 ppt
DO	4.1-4.3 mg/L
pH	7-8
Current	0.16-0.19 m/s
Depth	50-75 cm
Brightness	38-60 cm

3.4. Frequency Distribution

The samples that observed for three months was mullets type *M. dussumieri* amounted to 347 fish, it was done every 2 weeks interval. The number of mullet fish caught in small amounts also affected by an uneven distribution of frequencies, so that not all classes of hoses can be represented by the caught mullet.

Body length class interval of mullet *M. dussumieri* was 1.3 cm, starts from the smallest class hose was 8 – 9.3 cm until the longest class hose was 27.6 – 28.9 cm. The most common frequencies of mullet found in class hose 10.8 – 12.1 cm, with a total of 61 fish that dominantly have the length of the body.

The frequency distribution of fish body size was also done in measuring body weight of mullet fish, body weight mullet type *M. dussumieri* caught in this research had interval 16.2 grams, started from the smallest class hose was 5 – 12.2 grams until the longest class hose is 233.2 – 249.4 gram. A total of 40 mullet fish have the dominant body weight in an nterval of 5 – 12.2 grams. Based on this frequency distribution data it was known that the average fish size was relatively small. These fish was estimated to be young, so the seagrass ecosystem suspected as the enlargement and feeding ground, while spawning was done at sea. However, the data distribution based on weight of the fish was not as determinant as of the rate of growth of mullet, because according to von Bertalanffy's model body length is a function of age [7].

The growth of mullet can be seen from the frequency distribution chart which was then arranged in order so that it can be seen the cohort. Fish body length considered as fish age, so that the cohort chart used to find out a growth of mullet was fish length-based cohort [8].

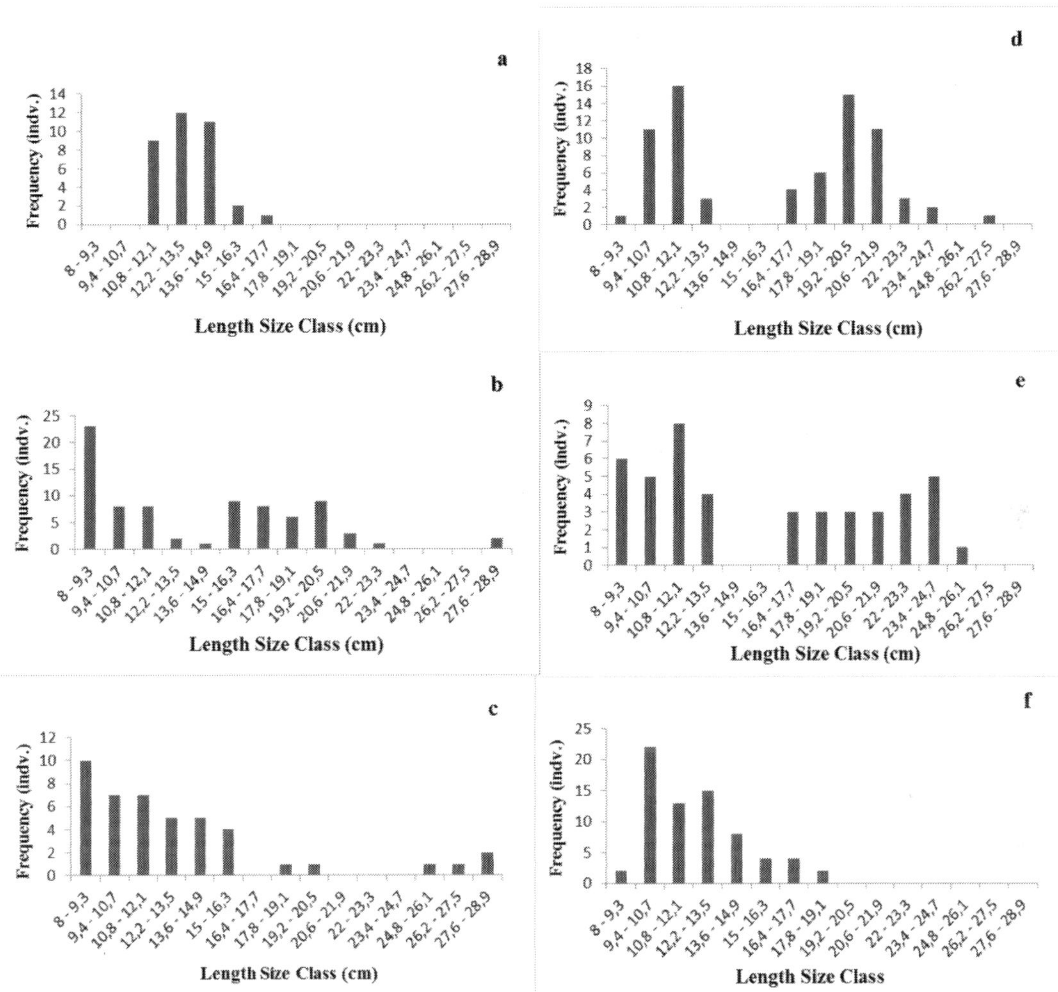

Figure 1. Length Cohort Chart of Mullets (*M. dussumieri*); a) 16 October 2016; b) 30 October 2016; c) 13 November 2016; d) 27 November 2016; e) 11 December 2016; f) 25 December 2016

3.5. Growth Parameter Estimation

Growth parameter was analyzed by FISAT II software to determine the value of K and L∞. The value of growth parameter on mullets represented by the following **Table 3**.

Table 3. The Value of Growth Parameter on Mullet (*M. dussumieri*)

Parameters	Value
K	0.33 per year
L∞	30.24 cm
t_0	-0.305 per year

According to growth parameter shown on the table, von Bertalanffy equation became $Lt = 30.24 (1\text{–}exp(-0.33(t+0.305)))$ with graphic growth curve was represented by **Figure 2**.

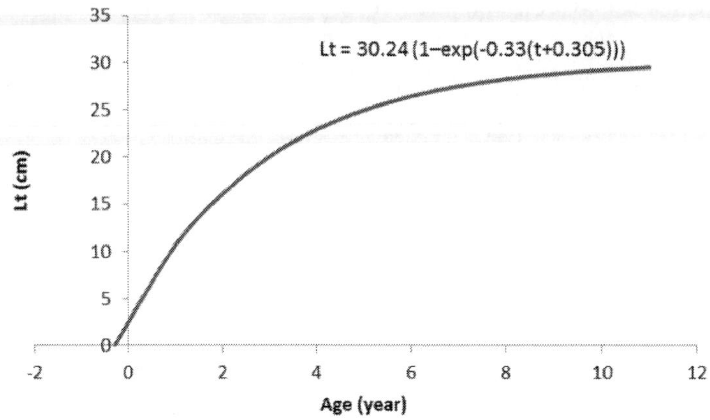

Figure 2. Growth Curve of Mullet (*M. dussumieri*).

The fish growth parameter (K) connected to asymptotic length (L∞), that represents the maximum length until the fish full-grown, and t_0 is the estimation period when its 0 cm length. The minimum K value indicates that Mullet fish had slow growth rate and long life period (around 11 years) to reach its asymptotic length.

If it compared with the previous research, the asymptotic length of mullet in the waters of the Banger River Estuary was 29.1 cm [10], while the mullet in Ujung Pangkah Water had K value 0.82 per year and L∞ was 39.39 cm, as the result mullets could reach its asymptotic length by 5.5 years [11].

The fundamental models used are based on four parameters: Growth, recruitment, natural and fishing mortality. Age and growth are particularly important for describing the status of a fish population and for predicting the potential yield of the fishery. It also facilitates the assessment of production, stock size, recruitment to adult stock and mortalities [12]. The growth is the result of all the physiological mechanisms by which any living organism increases its substance which is continuous when it lasts. Growth is a process of major fish biology and is a key process in structured model focusing on length [13]

The difference between growth parameter values caused several factors, factors that act to increase the biomass of a fish stock include growth, reproduction, and recruitment. Meanwhile, many factors in the marine environment act to reduce the chances of survival of individuals in a population. These include adverse conditions, lack of food, competition and, perhaps most important of all in marine species, predation [14]. The difference between L∞ and K value caused by the maximum length that obtained on the samples, the fishing ground, the samples amount, stocks and different recruitments.

3.6. Mortality Value Estimation

According to this research it can be known that average length of the samples ($\bar{L}$) was 14.2 cm and length of the smallest fish in the sample (L') is 8 cm, so it could be determined the total mortality rate (Z), and from this calculation it can discover Natural Mortality value (M), Fishing Mortality (F) and its Exploitation Rate (E).

The fish's growth is a positive aspect of stock's dynamic, and the negative aspects are disease and mortality. The key parameter to describe the fish mortality is mortality rate. Z is the coefficient that used to describe the total amount of death, and the coefficient of death caused by capture is F, and the death caused by nature is M [7]. The total mortality rate from mullet in Teluk Awur Bay water was 0.854 per year, while the natural mortality (M) was 0.706 per year and the fishing mortality (F) was

0.148 per year. Empirical equation represents that the small fish had high natural mortality and when the temperature around the area is rising, it will affect the natural mortality [8].

Based on this data it can be known that the natural mortality was higher than the fishing mortality.High mortality rates imposed on mature individuals, through factors such as exploitation, can alter these characteristics and ultimately change the response of fish populations to their physical environment [15]. The higher natural mortality is attributed to factors like diseases, old age, predation, spawning stress, and starvation. Most of these causes are connected with the ecosystem where it lives. The same species in different places may have different rate of natural mortality depending on the density of detritus and prey whose wealth is swayed by fishing activities [16].

The exploitation rate could explain the state of fish stock in the water. If the exploitation rate exceeds the optimum number which is 0.5, it means that area is overfishing. The optimum exploitation happens when the fishing mortality (F) is equal to its natural mortality (M) [7]. The exploitation rate is an index, which estimates the level of utilization of a fishery. The value of exploitation rate is based on the fact that sustainable yield is optimized when the fishing mortality coefficient is equal to natural mortality [12]. The mullet exploitation rate in Teluk Awur Bay was around 0.173 per year, it still in safe-exploitation rate, which means only around 17.3% of the mortality in Teluk Awur Bay caused by human catch activity.

4. Conclusion

As the results, mullet (*M. dussumieri*) growth rate in Teluk Awur Bay, Jepara were, for growth coefficient was 0.33 per year with asymptotic length ($L\infty$) 30.24 cm, and the imaginary fish period when its 0 cm length (t_0) was -0.305. The mullet mortality rate was more caused by the natural mortality (M) with 0.706 per year from the total mortality (Z) which was 0.85 per year and fishing mortality (F) only 0.148 with exploitation rate 0.173 which means that only 17.3% mullet mortality in Teluk Awur Bay, Jepara caused by human catch activity.

Acknowledgments
We wish to thank all those who have helped from the start of the research to the completion of this article.

References

[1] Dorothe'e K, Yolande B N, Max L, David M, and Claude B 2010 *J. Coastal Research* **26** (6) 1133-1141
[2] Rohani A R, Muhammad N N, Husain L, Dmitry L L 2013 *J. Ecology* **3** (5) 331-342
[3] Rafael J A and Francisco G A 2010 *J. Zoologia* **27** (3) 331-340
[4] Kaniz F, Wan M W O, and Mansor M 2013 *J. Life Sciences and Technologies* 1 **(1)** 47-50
[5] S F Mehanna and M M Hegazi 2013 *International Conference on " Oceanography & Sustainable Marine Production: A Challenge of Managing Marine Resources under Climate Change, ICOSMaP"* 530-539 Kuantan- Malaysia
[6]. Director General of Capture Fisheries 2011 *Indonesia Fishing Capture Statistics in 2010*, Ministry of Maritime and Fisheries Affairs of The Republic of Indonesia 11 **(1)** 1-190
[7] P Sparre and S C Venema 1999 *Introduction to Tropical Fish Stock Assessment - Part 1: Manual* (FAO: Roma) pp 66-157
[8] D Pauly 1984 *Some Simple Methods for The Assesment of Tropical Fish Stocks* (FAO: Roma) pp 20-31
[9] D K Prasetya, Ruswahyuni, Widyorini N 2015 *Diponegoro Journal Of Maquares* **4** (4) 155-163
[10] A Okfan, M R Muskananfola, Djuwito 2015 *Diponegoro Journal Of Maquares* **4** (3) 156-163
[11] Sulistiono 2001 *Indonesian Journal of Ichthyology* **1** (2) 39-47

[12] Ezekiel E N and Abowei J F N 2014 *J. Annals of Biological Sciences* **2** (2) 18-26
[13] Sarr S M, Leye B, and Kabre J A T 2015 *J. Agricultural Policy and Research* **3** (2) 67-76
[14] Michael K 2007 *Fisheries Biology, Assessment and Management* Ed. 2 (Australia: Blackwell Publishing Ltd)
[15] Kuo-Wei L, Chang I Z, Hee J K, Long-Jing W, and Li-Jhih L 2017 *J. Marine and Coastal Fisheries* **9** (1) 271-280
[16] Murugan S, Ashiq Ur Rahman M, Ajmal K S, and Lyla P S 2014 *J. Thalassas*, **30** (2) 47-56

Conditions of Decapods Infraorders in Dead Coral *Pocillopora* sp. at Pemuteran, Bali: Study Case 2011 and 2016

Nenik Kholilah[1*], Muhammad Danie Al Malik[1], Eka Maya Kurniasih[1,3], Andrianus Sembiring[2,3], Ambariyanto Ambariyanto[1], Christopher Mayer[4]

[1]Marine Science Department, Faculty of Fisheries and Marine Science, Diponegoro University, Jl. Prof. H. Soedarto, S.H., Tembalang Semarang, 50275 Indonesia
[2]The Indonesia Biodiversity Research Center, Jl. Raya SesetanGg. Markisa No. 6 Denpasar, Bali, 80223 Indonesia
[3]Yayasan Biodiversitas Indonesia (BIONESIA) Foundation, Jl. SulatriGg XII No. 4 Denpasar, Bali, Indonesia
[4]National Museum of Natural History, Smithsonian Institution, 10th and Constitution Ave, NW Washington, DC 20560-0163

*e-mail: nenikkholilah@student.undip.ac.id

Abstract

Decapods are marine organism which have burrowing-life characteristic and tend to live in the hard coral, such as *Pocillopora* sp. Pemuteran district is located in West Bali with high marine biodiversity. In 2016 almost all of the coral reefs in this area have bleached. This research investigates the condition of decapods before and after coral bleaching in Pemuteran. Dead corals, *Pocillopora* sp., were taken from 8-12 meters depth in 2016. All organisms within those corals were collected and identified until infraorder and family level. Comparison was done with data collected in 2011. This study found 12 families with a total of 5 infraorder which are equal to the previous data. The number of individual has increased from 88 into 214 individual. The mean presence increased from 6.2875 ind/fam to 15.2875 ind/fam. While the density also increased from 23.68 ind/L to 42.09 ind/L. Uniformity and dominance indices for all infraorder is low. These results show that there is an increase of the density of decapods after coral bleaching event, but the diversity of decapods was slightly changed.

KEYWORDS: Decapods, Dead Corals, Bleaching, *Pocillopora* sp., Pemuteran

1. Introduction

Indonesia is part of coral triangle and also known as the center of marine biodiversity [1-4]. This area alone represents 76 % of global coral diversity [5] and very diverse fishes [4][6]. Understanding biodiversity of marine organisms is very important for management and conservation practices to avoid declining natural populations and extinction [7-8]. However, corals and fishes represent only a fraction of biodiversity associated with coral reefs. There are many organisms associated with coral reefs and can be used as indicators of biodiversity, especially those which lives in the coral reefs cracks and gaps

[8] such as decapods. It was found that 86% of organisms taken on the dead coral *Pocillopora* sp. were decapods [9].

This crustacean not only has many members, but also has important ecological and economical values [10]. Ecologically, decapod is an important food source for fish and other organisms, and in the form of zooplankton larvae, they are also key components of the marine food chain [11]. In addition, some decapods have been reported to have important economic value, for example crabs [12-13] and crayfish [14].

Increasing human activity in coastal areas gives pressure on marine ecosystems including decapods [15]. The pressure that comes from nature such as global warming also can affects the coral reef ecosystem. In 2016 coral bleaching occurred in Indonesian waters including in Pemuteran, Bali which resulted in coral reefs mass death. Coral bleaching not only affected to coral, but also to associated organism [16-17] such as decapods. This study aimed to compare the conditions of decapods infraorders living in dead coral *Pocillopora* sp. in 2011 and 2016 after coral bleaching at Pemuteran, Bali which is also known as popular tourist destination.

2. Material and Methods

Samplings were conducted on August 2011 and June 2016 in Pemuteran, Bali. See Figure 1.Two dead corals with similar diameter about ±30 cm were taken at a depth of 8-12 m. The method used for retrieval of dead coral in the sea was based on those used by several studies [9][18-19].

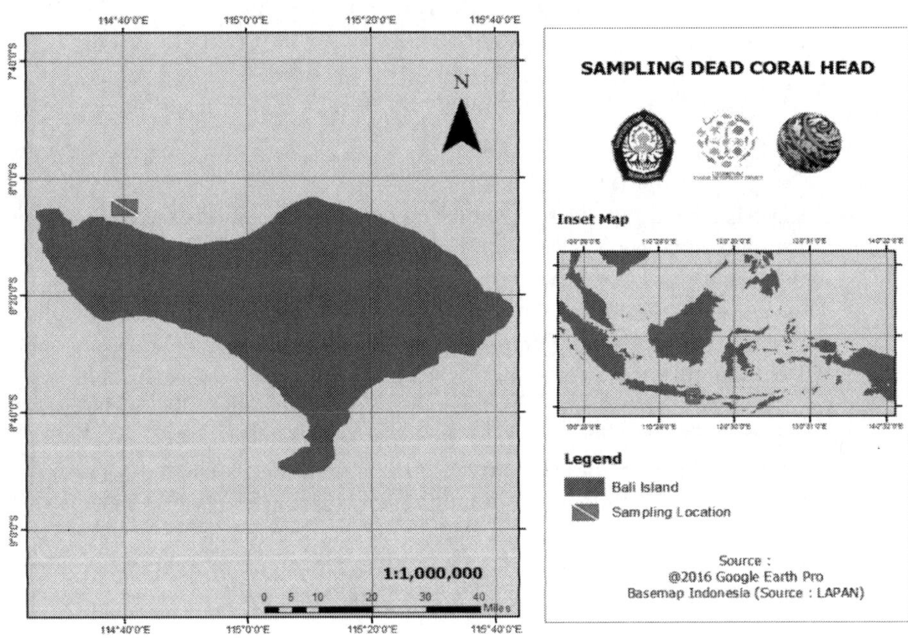

Figure 1. Sampling location at Pemuteran, Bali.

The dead corals were pulled out from the substrate by hammer and chisel. They were put into plastic bags individually and brought to the laboratory for further process. Volume of each dead coral was measured by using water displacement method in a bucket. These dead corals then were broken into small pieces and collected all organisms found. These organisms were sorted base on morphological characters and identified up to family levels including infraorder. Identification of decapods was done based on Crustacean Guide of the World [20]. These organisms then were put in clove oil and were photographed for documentation.

Abundance, relative abundance, diversity index, uniformity index, dominancy index, and density were calculated and categorized based on the formula of Odum [21], Krebs [22], and English *et al.* [23].

3. Result and Discussion

The results show that total number and density (ind/L) of decapods found in dead corals of *Pocillopora* sp. in 2016 was higher than in 2011. See Figure 2. The number of family decapods found in 2011 and 2016 was the same i.e. 12 family, but with different diversity. The family of Dromiidae and Thalassinidae found in 2011 but not discovered in 2016, while Majoidea and Axiidae found in 2016 but did not appear in 2011. See Figure 3. The number of decapods individual of each family generally increased in 2016 compared to 2011. The highest increase in number was Xanthidae.

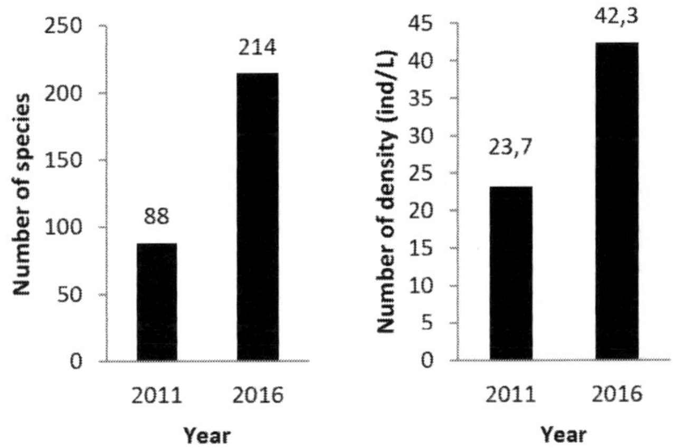

(a) (b)

Figure 2. Total number **(a)** and density **(b)** of decapods found in dead corals, *Pocillopora* sp. collected at Pemuteran, Bali in 2011 and 2016

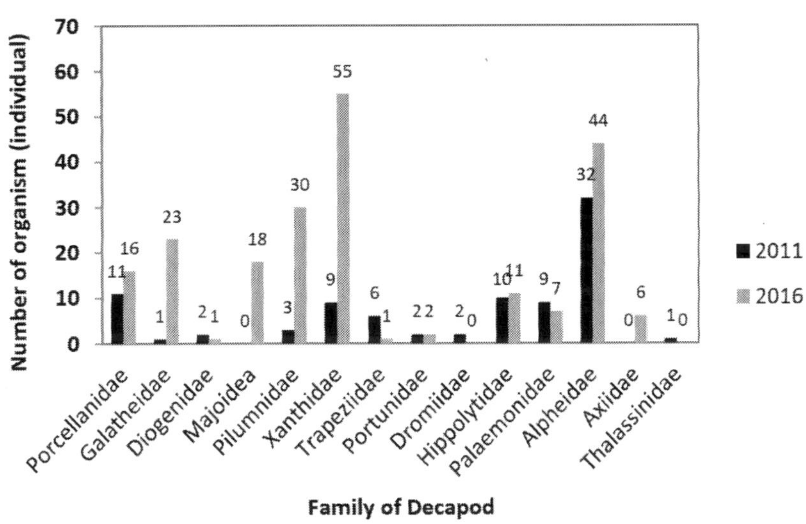

Figure 3. Number of decapods from different family found in dead corals, *Pocillopora* sp. collected at Pemuteran, Bali in 2011 and 2016

There were in total five infraorder found in 2011 and 2016 i.e. Anomura, Brachyura, Macrura, Axiidae, and Gebiidae. However, Axiidae, and Gebiidae were not found in 2011 and 2016, respectively. See Figure 4. The number of individual infraorders generally also increased in 2016 compared to 2011. The highest increment of the number was Brachyura from 22 in 2011 to 106 in 2016.

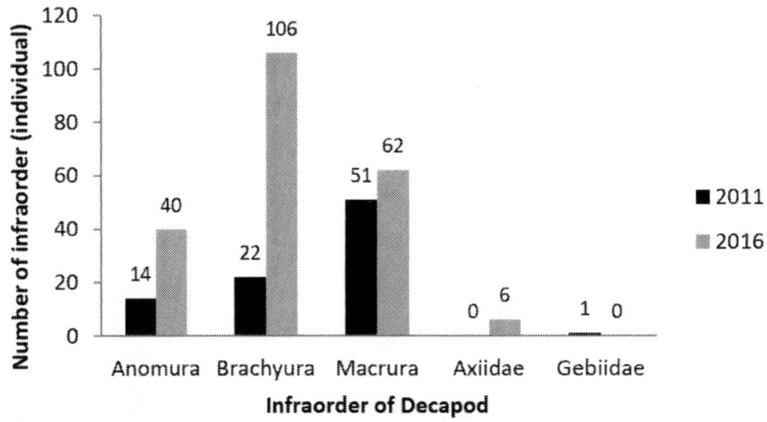

Figure 4. Number of individual decapods from different infraorder found in dead corals, *Pocillopora* sp. collected at Pemuteran, Bali in 2011 and 2016

The relative abundance of the decapods infraorders Anomura, Bracyura, and Axiidae increased from 2011 to 2016, however in Macrura and Gebiidae infraorders experienced a decrease (Figure 5). Axiidae and Gebiidae lower than other family because they generally life with burrowed into the sand or mud [24] and they were rarely found in reefs or cavities of sessile animal such as sponges and coral [25]. While Diversity index (H') of decapods on each infraorder has different values. Brachyura in 2016 and Macrura in 2011 are included in medium category (1 <H'<3) and the others are low category (H' <1). See Table 1.

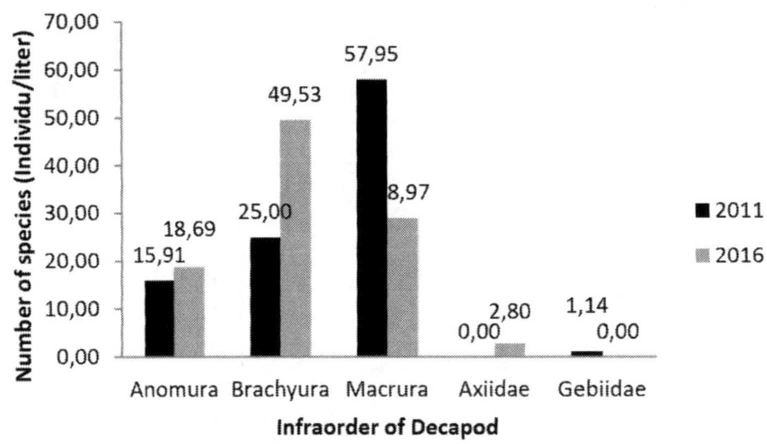

Figure 5. Relative abundance of decapods from different infraorder found in dead corals, *Pocillopora* sp. collected at Pemuteran, Bali in 2011 and 2016

Table 1. Diversity index (H) of decapods infraorders found in dead corals, *Pocillopora* sp. collected at Pemuteran, Bali in 2011 and 2016

No	Infraordo	Diversity (H')		Uniformity (E)		Dominance (C)	
		2011	2016	2011	2016	2011	2016
1.	Anomura	0.57	0.66	0.22	0.18	0.02	0.02
2.	Brachyura	1.01	1.30	0.33	0.28	0.02	0.09
3.	Macrura	1.22	0.85	0.31	0.21	0.16	0.05
4.	Axiidae	0.00	0.14	0.00	0.08	0.00	0.00
5.	Gebiidae	0.07	0.00	0.00	0.00	0.00	0.00
Average		0.58	0.59	0.17	0.15	0.04	0.03

The diversity index value is influenced by the diversity of individuals obtained in each year. The average value of the decapods infraorder diversity index in 2011 was 0.58 and in 2016 is 0.59. The average index of diversity has increased, but the value is included low category. When the value of diversity index H'<1 means that organism community is not stable or the waters quality is heavily polluted. The diversity is not only a synonym of the number of species, but also character of community which is determined by the number of species and evenness of individual abundance of each species [21].

Uniformity index is composed of individuals of each species in a community [21]. The value of uniformity index of all decapods infraorders in 2011 and 2016 includes in the low category. The average value of the uniformity index in 2011 was 0.17 and the 2016 is 0.15. There was a decrease in the value of uniformity index, but both are still included in the low category 0.00 <E <0.50, it indicates if the community under stressed because of the individual richness ownedby each infraorder is much different.

The dominancy index of each decapods infraorder in 2011 and 2016. The average dominance index of each Decapods's infraorders in 2011 was 0.037 and in 2016 was 0.031, the value of both indices showed a low dominance values 0.00 <E <0 05.The density index of decapods found generally increased except for Gebiidae which was experiencing decrease in 2016. The highest increase occured in Brachyura as presented on Figure 6.

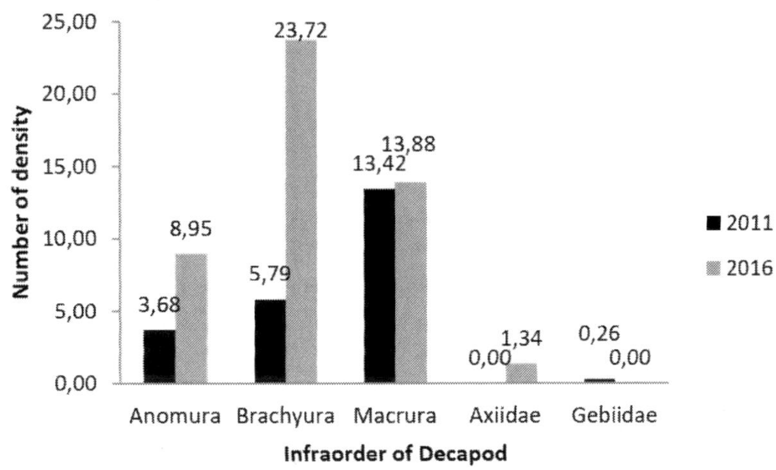

Figure 6.Density of decapods infraorders found in dead corals, *Pocillopora* sp. collected at Pemuteran, Bali in 2011 and 2016

The results of this study indicate an increase in the number and density of decapods present in dead corals in conditions after bleaching in 2016. These findings differ from the general opinion that bleaching will damage the coral ecosystem, including its associated organisms [17]. The increase in the density can be caused by several factors such as the recruitment, mortality, or movement (migration) [26]. All three factors are also influenced by environmental conditions, especially water quality and substrate conditions where decapods live. The results of the study [9] show that the abundance of decapods on living corals is actually lower than that of dead corals, as the availability of holes and gaps in dead corals is even greater. This condition provides the availability of a place for the decapods to live in it. Bleaching causes coral mortality, which actually provides a place of life for the decapods.

While the influence of temperature rise that causes the occurrence of bleaching, resulted in changes in the type of decapods that can live on the dead coral. A slight temperature change can cause a great impact on the vitality, growth and reproduction rate of sea organisms [16]. Infraorder Axiidae appears in 2016 after bleaching, indicating that the conditions are more suitable for life. In contrast, Gebiidae found in 2011, although with a small amount, was not found after bleaching.

4. Conclusion

There is an increase in total number and density of decapods in 2016 after coral bleaching event, possibly is caused by the occurrence of more places for decapods to live. However, there was a slight change on the diversity as well as decrease in evenness and dominance indices with no dominance individual organism.

Acknowledgement

Thanks go to everyone involved in this study. Thanks also to the National Science Foundation (NSF) Partnership for Enhanced Engagement in Research (PEER) Program [Number Grant award: AID-OAA-A-11-00012] for providing fund this study.

Refferences

[1]. Myers N, Mittermeier R.A., Mittermeier CG, da Fonseca GAB, and Kent J 2000 *Nature* **403**: 853-858.
[2]. Roberts CM, McClean CJ, Veron JEN, Hawkins JP, Allen GR, McAllister DE, Mittermeier CG, Schueler FW, Spalding M, Wells F, Vynee C, and Werner TB 2002 *Science* **295**: 1280-1284.
[3]. Allen GR and Adrim M 2003 *Zoological Studies* **42** : 1-72.
[4]. Allen GR 2008 *Mar. Freshwater Ecosystems* **18**: 541-556.
[5]. Veron JEN, Devantier LM, Turak E, Green AL, Kininmonth, S, Stafford-Smith M and Peterson N 2009 *Galaxea.* **11** (2): 91-100.
[6]. Allen GR and Werner TB 2002 *Environ. Biol. Fishes.* **65** (2): 209-214.
[7]. Ambariyanto 2017 Conserving endangered marine organisms: causes, trends and challenges. *IOP Conference Series: Earth and Environmental Science* **55**(1): 012002
[8]. Barber PH, Ablan-Lagman MCA, Ambariyanto A, Berlinck RG, Cahyani D, Crandall ED, Ravago-Gotanco R, Juinio-Meñez MA, Mahardika IGN, Shanker K., Starger CJ, Toha AHA, Anggoro AW and Willette DA 2014 *Bull. Mar. Sci.* **90** (1): 187-210.
[9]. Plaisance L, Knowlton N, Paulay G and Meyer C 2009 *Coral Reefs* **28**: 977-986.
[10]. Martin JW and Davis GE 2006 *Crustaceana* **70**: 1347-1368.
[11]. Pratiwi R and Astuti O 2012 *Ilmu Kelautan* **17** (1): 8-14.
[12]. Rhyne AL, Penha-Lopes G and Lin J 2005 *Aquaculture* **245** (1): 183-191.
[13]. de Magalhães HF, Neto EMC and Schiavetti A 2012 *J. Ethnobiology and Ethnomedicine.* **8** (1): 13.
[14]. Hobbs Jr HH and Whiteman M 1987 *Proc. Biological Soc. Washington.* **100** (2): 403-411.
[15]. McLeod E, Moffitt R, Timmermann A, Salm R, Menviel L, Palmer MJ, Selig ER., Casey KS and Bruno JF 2010 *Coast. Manag.* **38** (5): 518-539.

[16]. Hoegh-Guldberg O 1999 *Mar. Freshw. Res.* **50** (8): 839-866.
[17]. Stella JS, Munday PL and Jones GP 2011 *Trapezia cymodoce. Coral Reefs.* 30(3): 719-727.
[18]. Knowlton N, Brainard RE, Fisher R., Moews M, Plaisance L and Caley MJ 2010 Coral reef biodiversity. In: A.D. McIntyre (Ed.) *Life in the World's Oceans: Diversity, Distribution, and Abundance. Part II: Ocean Present – Geographic Realms, Chapter 4,* (Blackwell Publishing Ltd) pp: 65-78.
[19]. Plaisance L, Caley MJ, Brainard R.E and Knowlton N 2011 *Plos One.* **6** (10): e25026
[20]. Debelius H 2001 *Crustacean Guide of The World.* (Conch Books. Frankfurt) p320
[21]. Odum EP 1971 *Fundamentals of Ecology* (Philadelphia. Saunders) p574
[22]. Krebs, C.J. 1989. Ecological Methodology. Harper and Row Publisher. New York: 654 pp.
[23]. English S, Wilkinson C and Barker V 1994 *Survey Manual for Tropical Marine Resources, Australian Institute of Marine Science* (Townville-Australia)
[24]. De Grave S, Pentcheff ND, Ahyong ST, Chan TY, Crandall KA, Dworschak PC, Felder DL, Feldmann RM, Fransen CHJM, Goulding LYD, Lemaitre R, Low MEY, Martin JW, Ng PKL, Schweitzer CE, Tan SH, Tshudy D and Wetzer R 2009 *Raffles Bull. Zool.* **21**: 1-109.
[25]. Silva AC and Lemos JMM 2012 *Zoologi* **29** (2):144-158.
[26]. Begon. M and Parker GA 1986 *Oikos* **47** (3): 293-302.

The Effect of Different Feed and Stocking Densities on Growth And Survival Rate Of Blue Swimming Crablets (*Portunus pelagicus*)

R W Ariyati[1], S Rejeki[1], R H Bosma[2]

[1] Aquaculture Department, Faculty of Fisheries and Marine Sciences, Diponegoro University, Jl. Prof. H. Soedarto, SH, Tembalang, Semarang 50275
[2] Aquaculture and Fisheries group, Wageningen University, De Elst 1, 6708 WD Wageningen, The Netherland

resti_wisnoe@yahoo.com

Abstract. Blue swimming crab is targeted by commercial fisheries because of the high economic value, good taste, and attractive colors. As a result, the stock is overexploited and fisherman catch market also juveniles. The most sustainable solution would be to stop fishing for commercial trade and to culture this crab from brood to market size. This study aimed to find the best feed and stocking density for the on-growing of crablets. In 20 tanks juvenile crabs with a carapace width ±1 cm were stocked in three densities; 40, 60 and 80 crablets /m^2, and fed ad-libitum twice a day with either trash fish or pellets of shrimp feed, for 8 weeks. The circular (⌀ 1.6 m x 1 m) tanks with 1 m^3 of water were aerated, and temperature, salinity, dissolved oxygen and pH recorded daily. In the end, growth and survival rates were determined. In general, feeding the crablets at a density of 40 m^{-2} gave the highest growth. For crablest fed with pellets, the density of 40 m^{-2} gave significantly better growth than 80 m^{-2}. The crablets at a density 40 m^{-2} having a fish diet with the density 40 m^{-2} grew better than 60 and 80 m^{-2}. There was no significant difference between fed used among different densities. The lower densities resulted in higher survival, either on crablets fed with pellets or fish. But, crablets fed the fish diet and cultured in the lowest density (40 m^{-2}) had the highest survival rate.

Keywords. Feed, stocking density, blue swimming, crablet

1. Introduction

Blue swimming crab (*Portunus pelagicus*) is one of the fishery commodities exported mainly to the United States of America, which reaches 60% of the total catch of crabs. Landings of this crab in 2014 reached 212.571 tonnes [1]. Recently, crab commodities are ranked third or fourth of the total export value of Indonesian fishery products after shrimp, tuna, and seaweed. However, the number of blue swimming crab catch from the wild is declining now and the cultivation of this species is an alternative to be developed.

In term of developing culturing blue swimming crab, several studies have been conducted to develop of the crab aquaculture system, e.g, cannibalism [2], water quality requirement for BSC culture [3-5], diet and feeding behaviour [6-7]. Moreover, the introduction of the artificial hatchery and nursery P. *pelagicus* in Thailand [8] and Indonesia [9] is developing. For instance, the hatchery that developed

by Centre for Brackish Water Development and Cultivation (BBPBAP) Jepara has been successfully produced first crab stages in mass production. In this first crab
Stage, the megalops have metamorphosed to be young crab but is only 1/5 of an inch from point to point [10]. However, the successful cultivation of the next stage, i.e. at the stage of on-growing, has not achieved success yet. Factors of production and cultivation methods and also environmental influences that affect the successful cultivation of crab production for consumption size have not known for certain so far.

Growth in crabs is influenced by many things including given feed, stocking density, seed selection, proper cultivation techniques, and water quality and much more. Stocking density is one aspect of cultivation that needs to be known because it can affect the rate of growth, synthesis, and production levels [11]. Increased growth and survival of crabs can also be done through feed with a certain amount of protein or different types of feed. Feed is the main component needed by crabs to maintain its survival and growth. Nutritional equipment in the feed is absolutely necessary to keep the growth of crabs can take place normally. Nutritional requirements that include protein, fat, carbohydrates, vitamins and minerals for the growth of fish differ by type and size [12].

This study aims to determine the appropriate stocking and feeding density on growth performance and survival rate of blue swimming crab (*P. pelagicus*) which is cultivated in a tarpaulin tanks. This research is expected to contribute in the form of information to the community about the technical aspect of on-growing crab system.

2. Material and methods

2.1. Research location

This research was conducted from December 2016 to March 2017 in Brebes District, Central Java, Indonesia. The experimental station is located in Pandan Sari Village, in the sub-district of Kaliwlingi. For this trial, 20 tanks were used to test the effect of different feeds and stocking density on growth and survival rate of blue swimming crabs.

2.2. Research materials

The materials used in this research are young crabs (first crab stages) with size ± 1 cm length. The seeds were obtained from the Center for Brackish Water Development and Cultivation (BBPBAP) Jepara. The crabs are cultured with different stocking density and types of feed. Pellet which was used is pellet for culturing shrimp while the fish meat is trash product from fish processing which was called with "surimi"

2.3. Research design

The method used in this research is the experimental method using the Complete Design (RAL). The cultivated construction ponds were randomized and each treatment was repeated 3 times. The treatments provided are:

A: 40 stocking density and pellet feed
B: 60 stocking density and pellet feed
C: 80stocking density and pellet feed
D: 40stocking density and fish feed
E: 60stocking density and fish feed
F: 80stocking density and fish feed

The crab seeds that have been selected based on the size then weighed and then stocked based on the treatment used, i.e. with 40 ind, 60 ind, and 80 ind/m^2 and two types of feed, i.e. shrimp pellets and fish meat (surimi). Furthermore, after being stocked the crabs were kept in a circular tarpaulin tanks with a diameter of ± 2.4 m and height of tank is ± 1 m meanwhile the water depth is 0.5 m

Crabs were fed with 5 % fix feeding of the total weight twice a day. Below is the proximate analysis comparing between of shrimp pellet and fish meat.

Table 1. Nutritional composition of pellet and surimi

PARAMETER	*UNIT*	*Pellet (dry weight)*	*Surimi (wet weight)*
Water	%	8.7	76.0
Ash	%	9.5	1.3
Fat	%	3.9	3.5
Protein	%	35.7	18.4
Fiber	%	4.9	0.1
Carbohidrat	%	37.3	0.7

Source. Center for Industrial Pollution Prevention Technology Laboratory result, 2017

2.4. Growth measurement

All crabs were weighed and measured (length) individually every two weeks. Weighing was done with the A&D® HL-100 electronic weighing scale with a precision of 0.01 g. The Total Growth (G) was calculated with the formula:

$$SGR = Wt - Wo \quad (1)$$

Where BW_t is the final body weight (g); BW_o is the initial body weight (g).

The specific growth rate (SGR) was calculated with the formulae:

$$SGR = \frac{Ln\, Wt - Ln\, Wo}{Ln\, Wo\, x\, t}\, x\, 100\% \quad (2)$$

Where SGR is specific growth rate (% BW day-1); BW_t is the final body weight (g); BW_o is the initial body weight (g); and T is the duration of the experiment (days) [13].

The survival rate will be measured by counting the removed (residues of) dead crabs and counting during weighing. The survival rate (SR) will be calculated with the formulae:

$$SR = \frac{Nt}{No} \times 100\% \quad (3)$$

Where SR is the survival rate (%); N_t is the number of crabs collected at sampling time t; and N_o is the number of crabs initially stocked [13].

2.5. Water quality parameters

The following water quality parameters were recorded:

- Temperature (by using an electronic thermometer with a read-out precision of 0.1 °C) (twice daily 6 and 18 am)

- pH (by using HANNA® HI98129 pH meter with a precision of 0.01) (daily at 6 am)
- salinity (by using ATAGO® PAL-06S refractometer giving a precision of 1 ppt) (daily at 6 am)
- Dissolved oxygen (DO) (by using a YSI® Pro20 dissolved oxygen meter; read-out in 0.1 mg L-1) (twice daily 6 and 18 am)

2.6. Data analysis
The data analyzes included total weight, total length, specific growth rate (SGR weight and length), survival (SR). The variables were obtained then analyzed using ANOVA. While the data about water quality was described descriptively.

3. Result

Growth rates of crabs (Table 1) were rather slow and final body weight ranged from 5.63 to 6.98 g among the different treatments. The highest final body weight was found in the tank which has density 40 ind and fed with fish meat (surimi), followed by a pellet in the same number of stocking density, while the lowest body weight was found in the tank which has 80 ind of stocking density which fed with a pellet. In terms of SGR, blue swimming crab cultured in 40 ind of stocking density also showed the highest result (2.6 %), while blue swimming crab cultured in 80 ind of stocking density showed the lowest (2.3 %), both of them fed by surimi.

Table 2. Growth of blue swimming crabs

	Total growth		SGR		SR
TREATHMENTS	*Weight (g)*	*Length (g)*	Weight (%)	Length (%)	(%)
A (pellet)					
40	6.94 ± 0.3^{ab}	3.58 ± 0.2^{ab}	4.083 ± 0.1^{a}	2.547 ± 0.1^{a}	37.67 ± 26.6^{a}
60	6.33 ± 0.2^{a}	3.32 ± 0.1^{a}	4.079 ± 0.1^{a}	2.507 ± 0.04^{a}	21.67 ± 19.4^{a}
80	6.22 ± 0.6^{ab}	3.22 ± 0.5^{ab}	4.077 ± 0.2^{a}	2.446 ± 0.2^{a}	17.0 ± 7.9^{a}
B (surimi)					
40	6.98 ± 0.9^{ab}	3.64 ± 0.4^{ab}	4.103 ± 0.2^{a}	2.591 ± 0.1^{a}	31.33 ± 7.1^{a}
60	6.66 ± 0.4^{a}	3.18 ± 0.1^{a}	3.964 ± 0.1^{a}	2.416 ± 0.04^{a}	28.0 ± 25^{a}
80	5.63 ± 0.8^{ab}	3.01 ± 0.4^{ab}	3.796 ± 0.2^{a}	2.347 ± 0.2^{a}	24.0 ± 25.5^{a}

Different superscripts indicate significant differences ($P < 0.05$).

In brief, the total growth of crabs was influenced by stocking density but not influenced by the type of feed. Further, the total weighs and length of crabs have a significant difference between stocking density treatments ($P < 0.05$) and crabs who treat in lowest stocking density have the fastest growth.
The water quality during the treatment was recorded in the following table.

Table 3. Water Quality

Parameters	Average
Salinity (ppt)	15 - 25
Temperature (°C)	20 – 30.3
Dissolved Oxigen (mgL^{-1})	3.3 – 4.8
pH	7.49 – 8.6
Amonia mg L^{-1}	< 0.02

The amount of each water quality parameter above was appropriate to support blue swimming crab growth.

4. Discussion

4.1. Growth

The result of statistical analysis for the total weight and the total length of crabs gave significant effect ($P < 0.5$) between treatments but gave no significant effect for SGR. The highest total weight of crab is in the 40 stocking density and fed surimi, which was 6.98 ± 0.9 gram and the total weight of the lowest crabs is on the density of stocking of 80 ind and surimi was 5.63 ± 0.8 grams. The highest total length of crabs obtained in the treatment with stocking 40 ind and fed surimi (3.64 ± 0.37 cm) and the lowest total length obtained in treatment with stocking 80 and fed surimi (3.01 ± 0.4 cm).

The best stocking density for the cultivation of crabs found in this study is 40 ind /m^2. This is due to the cannibalism and crab required a wide range of motion. High stocking density will result in limited space and more competition in getting the feed and also the cannibalism of the crab increased resulted in mortality. According to Suharyanto and Tahe [11], stocking density affects competition against space, food needs, and environmental conditions which in turn can affect growth and synthesis that characterize production. The high stocking density will increase the risk of death and decrease the individual growth rate.

Growth is one of the important factors in the development of aquatic organism which can be seen based on the growth of increasing weight and length. Growth is the result of strong regulation between the factors contained in the environment, such as the availability of feed, temperature, photoperiod that will impact settings or metabolism in the fish body. Specific weight growth rate obtained during the treatment can be seen that the highest weight SGR value is obtained in 40 ind stocking treatment and fed surimi with the value of 4.1 ± 0.2 % / day, while the lowest value is obtained in 80 ind stocking treatment in the same feed that is equal to 3,8 ± 0,2 % / day. In addition, for the specific growth rate of length, it can be seen that the highest value of SGR is in the 40 stocking density and surimi fed is 2.59 ± 0.1 % / day, and for the lowest SGR value is obtained in the density of stocking 80 in the same feed that is equal to 2,37 ± 0,2 % / day. Differences in the growth rate of weights and width of crabs caused by the favorite of crabs in response to feeding. It is seen that crabs that consume surimi respond quicker than crabs which eat shrimp pellets. in nature, crabs are used to prey on live feed so the need to adapt when fed pellets. It affected the crabs who fed pellet have lower growth rate compare to crabs who eat surimi.

The existence of daily growth rate in the end observation indicates the existence of blue swimming crab. The growth means that the consumption of crab exceeds the energy needed for basic needs, such as maintenance and other body activities. Differences in the rate of growth are also thought to be caused by differences in feed digestion in the digestive tract associated with the amount of feed needed and the time to digest. According Suharyanto [14] states, that the frequency of feeding on fish will increase the rate of food flow in the digestive tract. In addition to the digestibility of food that causes growth differences are also caused by the quality of the food itself.

4.2. Survival rate

The survival rate is the parameter of the success of a cultivation activity. The survival rate is a comparative value between the numbers of early organisms when the dispersion is expressed in percent where the greater the percentage value indicates more organisms that live during maintenance [15].

The survival rate of crabs is still low, ranging from 17% - 37%. It can be influenced by many things, including the quality of seeds, the stocking density, water quality maintenance media, surrounding environment and disease. Currently, the very low crab survival rate is one of the obstacles to the cultivation of small crabs. The main constraint in the cultivation of crabs is the low life graduation rate is around 4% -29% [16].

4.3. Water quality

Water quality management of maintenance media is very important to note because it affects the growth during cultivation period of crabs. Water quality has an influence on the physiological balance of the fish body. The dissolved oxygen level (DO) on during the experiment was 3.3 to 4.8 mg / L due to adequate aeration. This value was still in reasonable range because according to Susanto et al. (2006) the decent range is 4.5-6 mg / L. The pH value during the study took place as much as 7.49 – 8.6. According to Susanto et al. [17], the pH value of a decent range of 8-8.5. Incorrect acidity (pH) values can lead to stress crabs, decreased metabolism, slow growth, and death.

The temperature on maintenance media is 27,2-31,9°C, the value is still in the unsuitable range for the crab because the temperature is 28-34°C [18]. An increase in temperature can lead to an increase in the rate of metabolism and respiration of aquatic organisms. The coastal area in Brebes district are found at temperatures ranging from 27 ° C to 30 ° C, salinity 30 ppt to 33 ppt and water depths ranging from 3 m to 12 m with substrate conditions dominated by sand fractions, mud fractions and clay fractions [19].

Salinity on maintenance media ranges from 15 - 25 ppt. This value is considered feasible because of the optimal value for the cultivation of crabs according to Erlinda et al.[20], salinity for crabs ranges from 30-40 ppt. This low salinity is one of the causes of the low lifespan of crabs. According to Chande and Magya [7], that crabs can survive up to 9 ppt salinity. But if the salinity is always low will cause mortality because it does not reach the optimal value so that the crab metabolism will also be disrupted. The low salinity is caused by continuous rain so that the water source from the pond becomes brackish and the salinity continues to decline.

Ammonia levels in the media during this study were 0.1 – 0.2 mg / L. The ammonia levels are still feasible for culturing crabs. According to Serang [21], in general, the concentration of ammonia in water should not be more than 1 mg / L. Ammonia concentrations of 0.4 - 2 mg / L in a short time can cause death in fish. The degree of ammonia toxicity is affected by the pH and temperature of the aquatic environment, where the ammonia concentration increases with increasing pH and temperature.

5. Conclusion

The difference of stocking density gave significant effect for total growth, both of weight and length but the type of feed given did not have a significant effect (P> 0,5) on total weight growth, total length, SGR weight, and survival of crab seeds. Both of shrimp pellet and the surimi can be used for survival rate and growth of blue swimming crab.

Acknowledgment

The authors would like to give great appreciation is addressed to NWO-TRO who funded PASMI project, the Center for Brackish Water Development and Cultivation (BBPBAP) Jepara which provides the blue swimming crab seeds and all colleagues who have assisted in the implementation of this study.

References

[1] FAO 2017 Species Fact Sheets *Portunus pelagicus* (Linnaeus, 1758).
[2] Castine S, Southgate P C and Zeng C 2008 *Aquaculture* **281**(1) : 95-99.
[3] Romano N and Zeng C 2006 *Aquaculture* **260** (1) : 151-162.

[4] Romano N and Zeng C 2007Aquaculture **266** (1) : 246-254.
[5] Romano N and Zeng C 2010 Comparative Biochemistry and Physiology Part C: Toxicology & Pharmacology, 151(2), 222-228.
[6] Sumpton W and Williams L 2002 *Queensland's Fisheries Resources: Current Condition And Recent Trends 1988-200, Blue Swimming Crab* (Department of Primary Industries, Brisbane) pp 119-123.
[7] Chande I and Mgaya Y D 2003 *J. Mar. Sci.* **2** (1): 75-84.
[8] Oniam V, Chuchit L and Arkronrat W 2012 *Songklanakarin Journal of Science & Technology* **34** (4) : 31-43
[9] Ruliati L, Anindiastuti and Kaemudin 2010 Produksi Baby Crab (Portunus pelagicus) dengan Sistem Modular Pertemuan Indonesian Aquaculture 2010 Bandar Lampung 6 Oktober 2010
[10] Romano N and Zeng C. (2008). Blue swimming crabs — emerging species in Asia, Global Aquaculture Advocate, pp. 34–36.
[11] Suharyanto dan Suwardi T. 2007. Jurnal Riset Akuakultur. **2** (1) : 19 – 25.
[12] Atifah Y 2016 *Jurnal Eksakta* **1** : 42-49
[13] Busacker P G, Adelman R I and Goollish M E 1990 Growth 363-387. Methods for Fish Biology. American Fisheries Series, 13.
[14] Suharyanto 2012 *Biosfera* **29** (2) : 93 – 101.
[15] Effendi. 2002. Biologi Perikanan. Cetakan Kedua. (Yogyakarta: Yayasan Pustaka Nusantara) 163 hlm.
[16] Supriyatna A 1999 Pemeliharan Larva Rajungan (*Portunus Pelagicus*) dengan Waktu Pemberian Pakan Artemia yang Berbeda. Prossiding Seminar Nasional Puslitbangkan Bekerjasama Dengan JICA ATA. 173178.
[17] Susanto, B., I. Setyadi, Haryanti, dan A. Hanafi. 2004. Pedoman Teknis Teknologi Pembenihan Rajungan (*Portunus Pelagicus*). Pusat Riset Perikanan Budidaya. Jakarta Selatan
[18] Zaidin M Z, I J Effendy and K Sabilu 2013 *Jurnal Mina laut Indonesia* **1** (1) : 112 – 12.
[19] Hamid, A. 2015. Habitat, Biologi Reproduksi dan Dinamika Populasi Rajungan (*Portunus Pelagicus* Linnaeus 1758) Sebagai Dasar Pengelolaan Di Teluk Lasongko, Sulawesi Tenggara. [Skripsi]. Institut Pertanian Bogor. Bogor.
[20] Erlinda S, L Sara, N Irawati 2015 *Jurnal Manajemen Sumber Daya Perairan* **1** (1): 29-38
[21] Serang, A. M. 2006. Pengaruh Kadar Protein dan Rasio Energi Protein Pakan Berbeda terhadap Kinerja Pertumbuhan Benih Rajungan (*Portunus pelagicus*). [Tesis]. Institut Pertanian Bogor.

The Estuaries Contribution for Supplying Nutrients (N and P) in Jepara Using Numerical Modelling Approach

Lilik Maslukah[1], Sri Yulina Wulandari[1], and Indra Budi Prasetyawan[1]

[1]Department of Oceanography, Faculty of Fisheries and Marine Science, Diponegoro University, Jl. Prof. H. Soedarto, SH, Tembalang, Semarang 50275, Indonesia

E-mail: lilik_masluka@yahoo.com

Abstract. Coastal water is dynamic area since it is influenced by both ocean and land. It has high primary productivity that determined fishing ground area. Increased supply of nutrients in coastal water is significantly influenced by seasons and the presence of the river estuaries carrying water masses from the mainland. This study focused on the rivers (Serang, Wiso, Grenjengan Mlonggo and Pasokan rivers) contributed nutrients supply spatially and temporally to Jepara water using numerical modeling. The results showed nutrients content of N (Nitrate) and P (Phosphate) from those rivers were 39.19 tons N/month and 2.26 tons P/month in June, 19.94 tons N/month and 1.96 tons P/month in August. From simulation modeling nutrient of N and P showed that the distribution pattern of N and P was larger during the neap tide than the spring tide. Furthermore, compared with the other rivers, Serang river was the highest nutrient supplier to Jepara water.
Keyword: Contribution, N and P nutrients, estuaries, numerical modeling, Jepara waters

1. Introduction

The Indonesian waters as part of the tropical sea are characterized by sufficient sun exposure but have low nutrient concentrations [1]. It leads to an extremely low primary productivity. However, in some coastal waters have high primary productivity, uniqueness, ecological and economic value [2]. The high primary productivity in waters makes potentially to be fishing ground areas.

Fishing ground is a region or an area in which a population of an organism is abundant. The abundance of these organisms is strongly influenced by physical and chemical parameters of the seas such as temperature, salinity, pH, brightness, current, depth, seabed topography, dissolved oxygen, and nutrient. The increase of the nutrients supply in the fishing ground is positively correlated with the primary productivity as the basis of the food chain [3].

The dynamic of nutrient supply in coastal waters is greatly affected by the presence of the river estuaries that carry water mass from the mainland. The river estuary that plays a significant role as a major exporter of nutrients into coastal waters also considered as fertilizing agents in coastal areas [4]. According to Libes [5], the land is the largest contributor to the various chemical compounds in the marine environment. The distribution of the nutrients from the land that are discharged to the river is strongly influenced by hydro oceanography processes including currents and tides. Then, It is distributed through a dilution process and accumulated through deposition process in the sediment. The precipitated nutrients may release to the water column through the resuspension process by current and waves [6].

The increased upstream area utilization, particularly for agriculture, transportation, and settlement has led to the enrichment of organic matter and nutrients. Nitrogen (N) and Phosphorus (P) nutrients are key parameters of quality of the waters. The concentration of these nutrients will vary depending on the terrestrial environment, the season and the condition of geology [7]. N and P are directly or indirectly affect the growth of phytoplankton [8] and oxygen concentration.

The Jepara water gets fresh water supplies from several rivers that have the different characteristic of the region and the level of human activities. The rivers that flow into the Jepara Water

is divided in two zones: (1) under the administration of Jratun Seluna river basin (located in the west) and (2) Wiso Gelis for rivers that are in the east [9]. The existence of rivers that run into the Jepara water provides an overview of the influence of land regarding supply of nutrients and so influence the existence of fishing grounds. On the other hand, the water quality of each river is also influenced by the upstream. The characteristics and the level of nutrients will vary from one location to the others. It is also affected by the upstream and the season. Due to the intensive rainfall, the flow rate of the river during the west season is higher compared with the dry season [10][2]. During the west season, the water bodies receive more nutrients leading to the increment of primary productivity in a wider area.
Nutrients are inorganic materials which are the result of organic materials decomposition by decomposers. The existence of nutrients in the waters is very important to enhance the primary productivity. Some elements of the main nutrients have an important role in primary productivity are Nitrogen (N) and Phosphorus (P). Some research regarding the measurement of N and P had been carried out in the coastal waters and the sea, however, in conjunction with the amount of supply and fluctuation of each river had never been carried out yet. Therefore, this study was aimed to determine the contribution from four rivers (Serang, Wiso, Grenjengan Mlonggo and Pasokan) toward the supply of nutrients (especially N and P) into the around water.

2. Materials and Methods

The experiment was conducted in June and August 2016 in the waters of Jepara. The data collected were seawater samples, bathymetry, tides and ocean currents. To determine the location of the research, purposive sampling method was used. At this stage, the water sampling station was determined by considering the presence of watershed management. There are four locations namely Serang river, the Wiso river, Grenjengan Mlonggo river, and the Pasokan river. The field data of ocean current measurement at one station will be used for current data verification of model at the same station. The ocean current measuring station is determined by considering that the station may represent the state of overall waters.Water samples were taken using Nansen bottles from the surface. After arriving at the laboratory, filtering is done with Millipore filter paper (size 0.45 m) followed the analysis of phosphate and nitrate [2].

2.1. Laboratory analysis of the phosphor: Spectrophotometric phosphomolybdate blue method was used in determining phosphate. The principle works are ammonium molybdate, and potassium antimonyl tartrate react with orthophosphate (in acid condition) to form orthophosphomolibdate acid. Phosphomolybdate acid is then reduced by ascorbic acid up to blue color appeared. This color is proportional to the concentration of phosphorus. Absorbance reading at a wavelength of 880 nm [11][12].

2.2. Laboratory analysis of the nitrate: Analysis of Nitrate in the water samples were carried out in the laboratory using a method of [13] with UV-Vis spectrophotometer with a wavelength of 543 nm.

Preservation and Storing Water Samples: Nansen bottle was used to take water samples and then the samples were inserted into the sample container. The sample container such as a bottle PET (Poly Ethylene Terephthalate) needed to be washed with distilled water before the storing stage. Moreover, water samples should be refrigerated at 4°C in a coolbox to avoid samples damaged due to temperature change, and they could last for 48 hours. Water samples obtained from the field and then taken to the laboratory for analysis of phosphate and nitrate content [15].

2.3. The Measurement of River Flow Rate: The calculation of flow rate was performed by measuring the flow velocity and cross-sectional area [14]. Calculation of river flow rate was done at The river using formula as follows:

$$Q\text{d} = F\text{d} \times V\text{d}$$

$$F\text{d} = \text{b} \times \frac{c + 2d + e}{4}$$

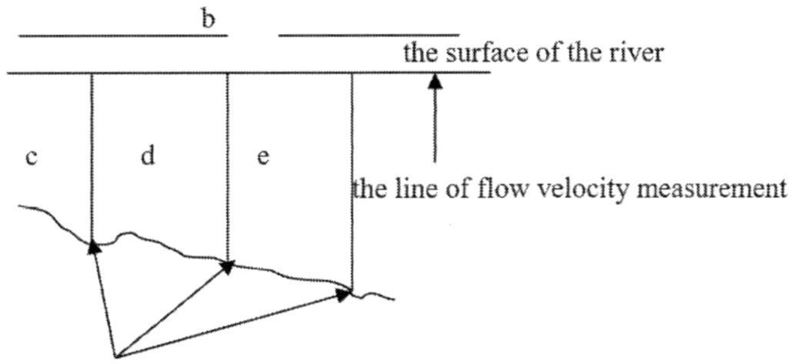

Fig 1. Lines of depth and flow velocity measurements by Waller and Yitayew [14]

Qd is stream flow, Fd is the cross-sectional area between line measurement of water depth c and e, Vd is average flow velocity in the flow line d, b is the width of the river and c, d, e are the depth of waters. The lines of depth measurements were carried out according to a method that has been standardized by Waller and Yitayew [14]. A cross-section of the river was divided into four cross-sections at three different depths, as shown in Fig 1. Measurements carried out at a depth of the second flow (d). River discharge measurements in this study were calculated from cross sections of the river.

2.4. Calculation of Flux and Nutrient Load of N and P

The Flux (F) rate of N and P are calculated by multiplying the value of the concentration with the current velocity at that section.

The nutrient load input (L) of N and P from the river is calculated using the following formula [13]

- $L = Q \times C$
- $Q = A \times V$

The flux rate was calculated using the formula:

$$F = v \times C$$

In which, L is nutrient load input in monthly basis (ton/month), F is flux ($m^2.s^{-1}$), Q is flow rate ($m^3.s^{-1}$), A is cross-section of wet river stream (m^2), V is velocity ($m.s^{-1}$), 2592000 is conversion factor (month to second), 1000000 is conversion factor (ton to gram), C is the concentration of each type of nutrients which are N dan P in monthly basis ($mg.L^{-1}$)

2.5. Hydrodynamic Modelling in the Jepara Waters

Hydrodynamic modeling in this study was executed using a software module of Hydrodynamics DHI MIKE 21 Flow Model Flexible Mesh. DHI MIKE is a mathematical program that can provide a solution for a hydraulic simulation in a wide range of waters environments including lakes, estuaries, bays, coastal, and marine areas. In the simulation model, bathymetric and tides data were needed as model inputs. Bathymetric data were extracted by digitizing the depth point from Bathymetry map published by Dishidros Navy. Tidal forecasting was used to define the input of boundary condition. Hydrodynamic simulations using a hydrodynamics module simulates the water level variations and flows. Two-dimensional unsteady flow is applied to identify the numerical solution by using equation conservation of mass and momentum [16].

2.6. The Nutrient Distribution Model

The simulation of nutrient distribution in Jepara waters with inputs of nutrients from four different estuaries was done using numerical modeling approach of equation advection-diffusion material. According to the results of numerical modeling using this module, the distribution of nutrients N and P were influenced by currents. The distribution of nutrient N and P in waters of Jepara in June and August 2016 will be modeled in this research. The nutrients of N and P in this study is assumed has a conservative character. Therefore the distribution pattern is influenced by the current. The equation used in numerical modeling of the distribution of nutrients N and P are as follows [17]:

$$\frac{\partial C}{\partial t}+\frac{\partial uC}{\partial x}+\frac{\partial vC}{\partial y}+\frac{\partial wC}{\partial z}=\left[\frac{\partial}{\partial x}\left(D_h\frac{\partial}{\partial x}\right)+\frac{\partial}{\partial y}\left(D_h\frac{\partial}{\partial y}\right)\right]C-k_pC+C_sS$$

In which u, v is the velocity of current at the direction of-x (east-west) and direction of-y (north-south), C is nutrient concentration, D_h is coefficient of horizontal diffusion, k_p is the rate of linear shading, C_s is concentration of the nutrients at the source point

3. Result and Discussion

The concentration of nitrogen in some waters in Jepara was identified based on the concentration of nitrate (NO_3^-) and nitrite (NO_2^-). Meanwhile, the phosphorus concentrations were represented by the amount of PO_4. The element nitrogen (N) and phosphorus (P), which is considered as nutrients, are required by the flora (marine plants) as the growth booster. The importance of the availability of phosphate and nitrate for the growth of phytoplankton can be seen from the general equation of chemical photosynthesis occurring in the sea which was described by [18] as follows:

$$106\ CO_2 + 16\ HNO_3 + H_3PO_4 + 122\ H_2O + \text{trace elements \& vitamin} \rightarrow C_{106}H_{263}O_{110}N_{16}P + 138\ O_2$$

According to the result of the measurement, the concentration of N was ranging from 0,336-3,444 $mg.L^{-1}$ and P is ranging from 0,0201-0,0528 $mg.L^{-1}$. The more detail result can be seen in Table 1.

Table 1. The Concentration of N and P in Different Flow rate

No.	Location		Flow rate ($m^3.s^{-1}$)	Concentration of N ($mg.L^{-1}$)		Consentration of P ($mg.L^{-1}$)	
				June	Agust	June	Agust
1	Grenjengan, Mlonggo	1	13.14	0.609	0.371	0.0344	0.0227
		2	11.07	0.77	0.378	0.0528	0.0238
		3	-	0.69	0.385	0.0390	0.0227
		Mean		0.6895	0.378	0.0420	0.0231
2	Wiso	1	5.64	3.444	0.42	0.0505	0.0465
		2	3.7	1.673	0.434	0.0528	0.0475
		3	-	1.659	0.434	0.0517	0.0455
		Mean		2.2587	0.4293	0.0517	0.0465
3	Serang	1	72.75	0.336	0.364	0.0344	0.0486
		2	49.2	0.35	0.378	0.0321	0.0465
		3	-	0.392	0.378	0.0528	0.0455
		Mean		0.3593	0.3733	0.0398	0.0468
4	Pasokan	**1**	-	-	0.469	-	0.0217
		2	14.48	-	0.462	-	0.0207
		3	-	-	0.441	-	0.0217
		Mean			0.4573		0.0214

Based on Table 1, overall, the concentration of N and P fluctuate in every area of the river mouth. Wiso river which is located in the urban center of Jepara has a fairly high concentrations of N and P compared with other rivers. In June, the river Wiso had a concentration of N five times higher than in August. This condition was caused by the rain that was occurred during sampling. The existence of resident area along the upstream to the downstream of the river caused domestic sewage was discharged into coastal waters consistently. The water traffic of fishing boat around the river mouth of Wiso caused resuspension process in this river resulting in the increase of nutrient from the sediment into the water column rose significantly [19].

Ignatiades et al [20] Classified waters by the tropical level based on the concentration of N and P. Based on this classification, the seas around Jepara were included in the eutrophic category. The concentration of N nutrient in the Jepara waters (1.89 mg.L^{-1}) was nearly two hundred and seventy times the concentration of N nutrient in eutrophic waters (0.007 mg.L^{-1}) and the mean concentration of phosphate (0.036 mg.L^{-1}) was almost three times the mean average concentrations of phosphate in eutrophic waters (0.011 mg.L^{-1}). Furthermore, when compared with the seawater quality standard of Ministry of Environment in 2004 (that was 0,015 mg.L^{-1} for phosphate and 0.008 mg.L^{-1} for nitrate), so the average concentration of phosphate (0.036 mg.L^{-1}) and nitrate (1.86 mg.L^{-1}) in the Jepara water in June and August 2016 was higher. This means that the Jepara water was rich in nutrients and considered as fertile waters. The conditions and processes that occurred in water related to the ratio between N and P can be seen in Table 2.

Table 2. The Ratio Value of Nutrient N & P

Location	ratio	
	June	August
Grenjengan	16.41667	16.36364
Wiso	43.68859	9.232258
Serang	9.027638	7.976496
Pasokan	**	21.36916

The ratio value of N and P in Wiso river was the highest in June (Table 2). A high ratio illustrated that N nutrient was abundant compared with P nutrient. However, this ratio was lightly decreased in August. The high ratio in June was related to the rainy seasons that had brought the run-off from the mainland.

To understand the contribution of N and P from several rivers into the Jepara water, the flux values and nutrient loads of N and P came from the river to the sea were presented in Fig. 2 and 3.

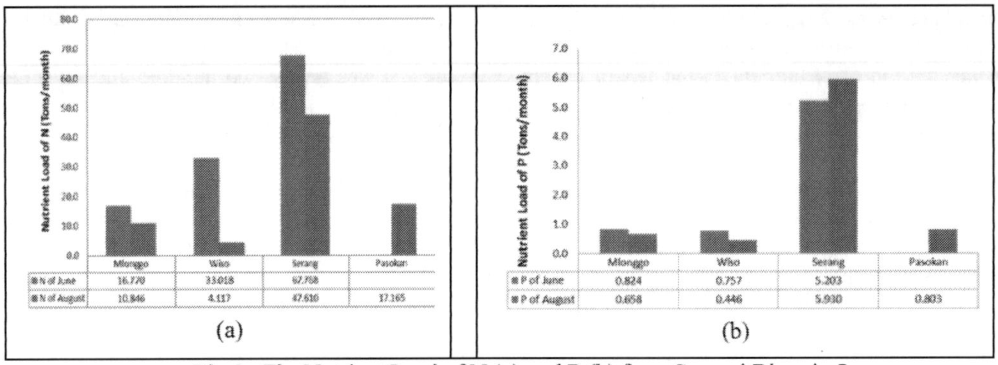

Fig 2. The Nutrient Load of N (a) and P (b) from Several River in Jepara

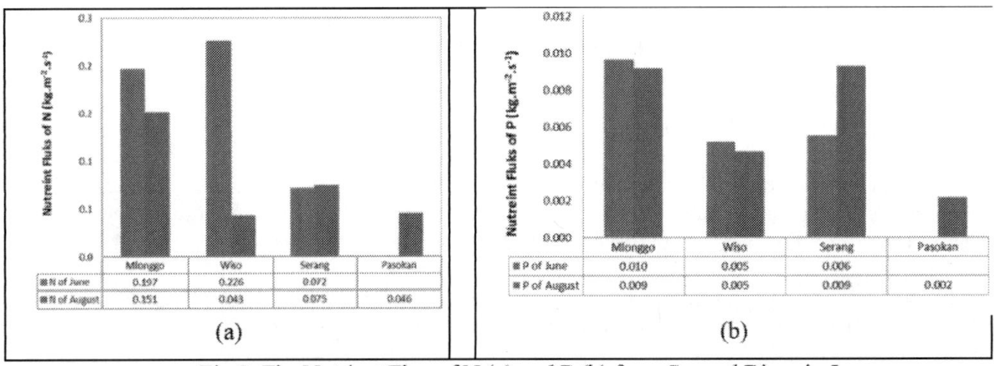

Fig 3. The Nutrient Flux of N (a) and P (b) from Several River in Jepara

Based on fig 2 and 3, loads of nutrient input of N ranged 4.177 to 67.76 ton/month with a mean value of 29.56 ton/month, and P ranged from 0.446 to 5.90 ton/month to an average value of 2.11 ton/month. The N flux value was between 0.043 to 0.226 $kg.m^{-2}.s^{-1}$ and P between 0.002 to 0.010 $kg.m^{-2}.s^{-1}$. Compared with the results of the research that have been conducted by Zhang *et al.* [21], the N value load in Coastal water of Jepara is a relatively smaller than total nitrogen and phosphorus loadings into the Gulf of Finland from the Neva River and St.Peterburg. Annual nutrient flux of total nitrogen and phosphorus from the Nova River 219 and 7.64 $kg.km^{-2}.yr^{-1}$. Mayor rivers discharging into the Baltic sea, ranging 594-798 $kg.km^{-2}.yr^{-1}$ (TN) and 15-56 $kg.km^{-2}.yr^{-1}$ (TP) [22].

Table 3. showed the results of calculation of nutrient loads carried by [23] in some of the waters of the Jakarta Bay.

Table 3. The total Nitrogen load in Several River, Bay Jakarta [23]

The Name of River	Total Nitrogen Loads (tons/yr)
The Citarum river	18.1
The Ciliwung river	7.1
The Sunter river	3.7
The Cisadane river	2.5
The Cengkareng river of Jang	1.9

Numerical model of Nutrient N & P

The results of numerical model simulations show the movement of currents in vector form and distribution of the concentration of nitrate and phosphate concentrations in June and August 2016. The simulation models adapted to the tidal conditions in the waters of Jepara and divided into two tidal

IOP Conf. Series: Earth and Environmental Science **116** (2018) 012072 doi:10.1088/1755-1315/116/1/012072

conditions, i.e. neap tides and the spring tides in which are divided into two periods of high tide and low tide.

Based on Fig 4 and 5, the river of Serang supplies a load of nutrients N and P to Jepara water in the highest rate compared to other rivers, causing Jepara waters near the Serang became more fertile. These waters include the waters around Kedung District, Demak Regency. Another fishing ground is situated in the district Keling and Donorejo, Jepara Regency. In these areas, perch, pomfret, tuna, mackerel, mackerel, Banyar fish, crab, and squid were easy to find. The high production of fish in these waters was assumed related to the nutrient input of N and P from Pasokan river, which at second ranks.

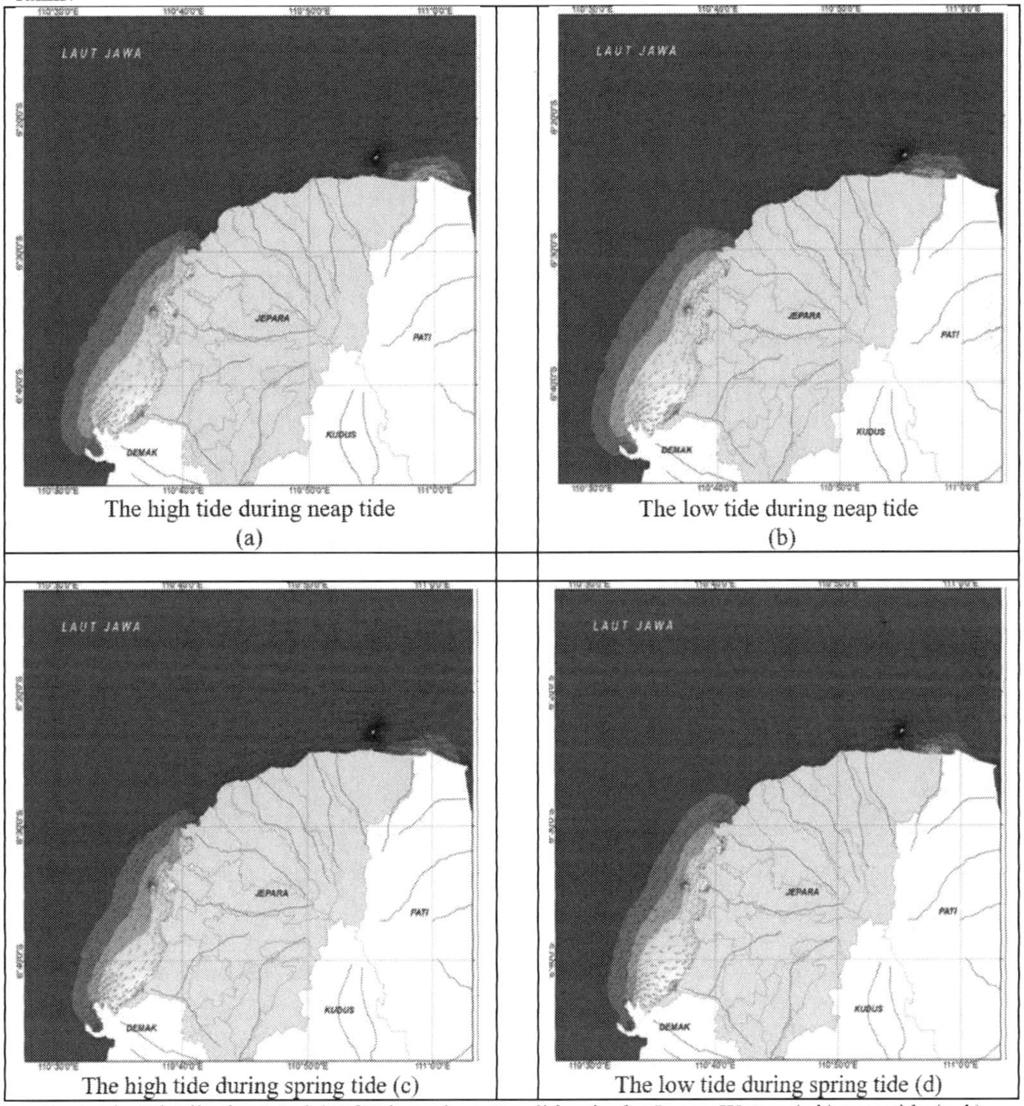

The high tide during neap tide (a)

The low tide during neap tide (b)

The high tide during spring tide (c)

The low tide during spring tide (d)

Fig 4. The Distribution Model Of N in various Condition in the Jepara Waters (a,b) neap tide (c,d) spring tide

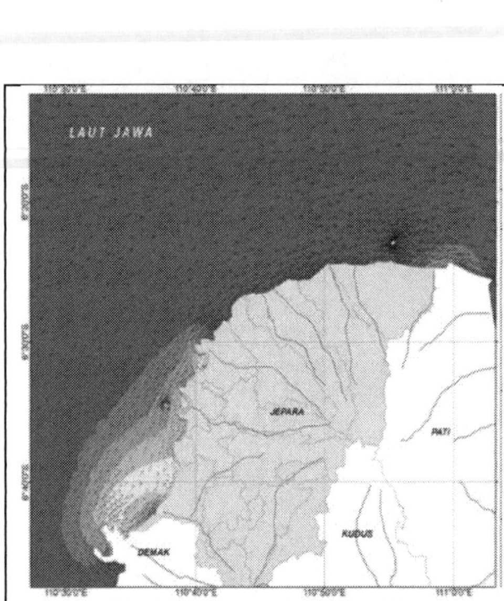

The high tide during neap tide
(a)

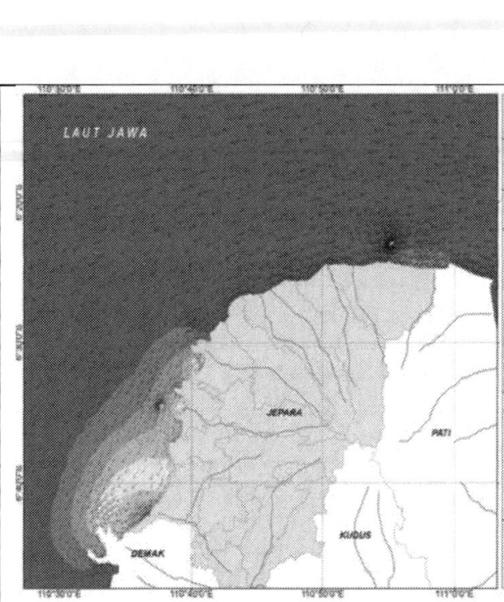

The low tide during neap tide
(b)

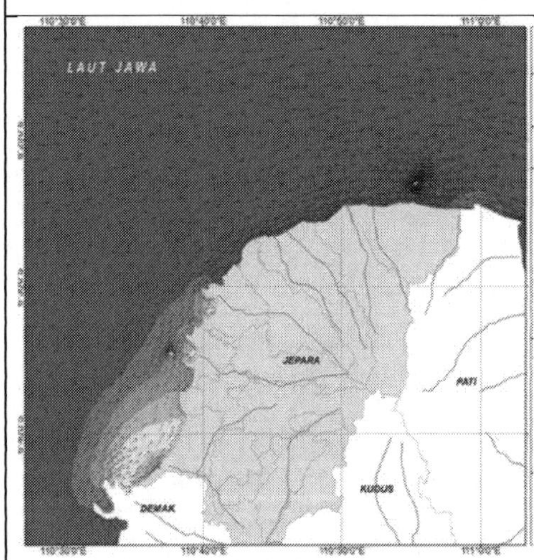

The high tide during spring tide
(c)

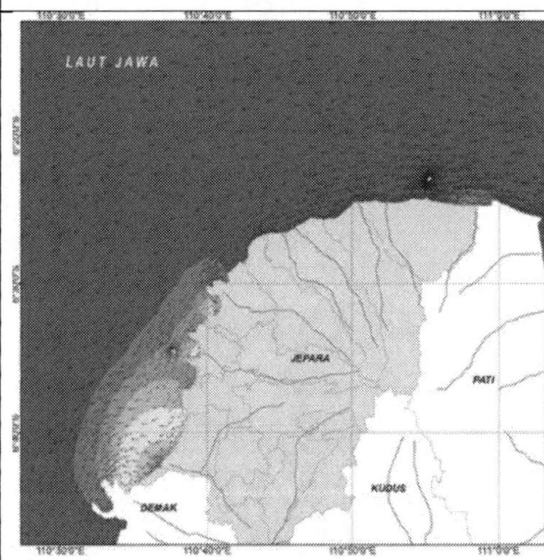

The low tide during spring tide
(d)

Fig 5. The Distribution Model Of P in various Condition in the Jepara Waters (a,b) neap tide (c,d) spring tide

Fig. 4 and 5 also illustrated the distribution pattern of N and P nutrient from modeling simulation during a neap tide and the spring tide. The simulation results showed that during in the neap tide, the distribution area of nutrients N and P is wider than in the spring tide. This was due to the velocity of tidal currents that was higher during the spring tide than the neap tide that caused the water intake from the river on the full tide conditions more restrained in coastal waters [24]. According to research that was conducted by Dick et al [25] in the Wadden Sea, the process of hydrodynamics waters would affect the distribution of nutrient concentrations as well as the water quality.

4. Conclusion

The results showed that, contributing nutrients N (Nitrate) and P (Phosphate) from four rivers in Jepara were 19.94-39.19 ton N/month with a mean value of 29.56 tons N/month and 1.96-2.26 tons P/month with a mean value of 2.11 tons P/month respectively. It was also identified, that N load entering the seas around Jepara in June was higher than in August. The results of simulation modeling nutrient N and P showed that during the neap tide, the distribution pattern of N and P is wider than the spring tide. Furthermore, Serang River was the highest supplier of nutrients compared with Wiso, Grenjengan Mlonggo, and Pasokan river. It could be concluded that a number of nutrients supply was high related to the width of the river as well as its flow rate.

References

1] Moore C M et al 2013. www.nature.com/naturegeoscience. acces on January[22th] 2017
[2] Li R H et al 2014 *Biogeosciences*. **11** 481–506. doi:10.5194/bg-11-481-2014.
[3] Bouwman A F, Bierkens M F P, Griffioen J, Hefting M M, Middelburg J J, Middelkoop H and Slomp C 2013 *Biogeosciences*.**10** 1–22 doi:10.5194/bg-10-1-2013
[4] Wang X, Olsen L M, Reitan K I, Olsen Y 2012 *Aquacult. Environ.Interact*. **2** 267-283
[5] Libes S M 1992 An Introduction to Marine Biogeochemistry (New York: Wiley) p 734
[6] Zhang L et al 2014 *J. Earth Sci.* ***25*** 197–206 doi: 10.1007/s12583-014-0413-y.
[7] Jia J J, Gao J F, Liu Y F, Gao S and Yang Y 2012 *J. Asian Earth Sci.* **52** 158–168.
[8] Thongdonphum B, Meksumpun S, Meksumpun C, Thawonsode N, Sawasdee B 2014_*Int. J. of Environ and Rural Develop*., 5-2. January[5th] 2017.
[9] http://bpsda-seluna.jatengprov.go.id/database/das.php
[10] Struyf E, Van Damme S, Meire P 2004 *Est., Coast. and Shelf Sci.* **60** 649-661
[11] Murphy J and Riley J P 1962 *Analitica Chimica Acta* **27** 31-36.
[12] Liu S M, Zhang J and Li D J 2004 *Est. Coast. and Shelf Sci.* **59** 209-218.
[13] Parsons T R, Maita Y and Lalli C M 1984. *A manual of Chemical and Biological Methods for Seawater Analysis* (USA: Pergamon Press) p 173
[14] Waller P, Yitayew M 2016 Irrigation and Drainage First ed (USA: Springer) p 742
[15] Gordon N D, Mahon M T A and Finlason B L. 1992. *Stream Hidrology and Introduction for Ecologists* (England: Chichester)
[16] Plawsky J L 2014 *Transport Phenomena Fundamentals* 3rd ed (USA: CRC Press) p 838
[17] DHI Water and Environment, 2012. *MIKE 21 Flow Model FM, Hydrodinamic and transport Module Scientific Documentation* (Denmark) p14
[18] Paytan A and Laughlin K M 2007 *Chem. Rev.* ***107*** 563–576 doi: 10.1021/cr0503613
[19] Dzialowski A R, Shih-Hsien W, Niang-Choo L, Beury J H and Huggins D G 2008 *Lake Res. Manag.* **24** 313-320 doi: 10.1080/0743814080935 4841.
[20] Ignatiades L, Karydis M and Vounatsou P 1992. *Mar. Poll. Bull.* **24** 238-243.
[21] Zhang C Y et al 2006 *Remote. Sens. Environ* **102** 250.
[22] Stålnacke P, Grimvall A, Sundbland K, Tonderski A 1999 *Environ. Monit. Assess.* **58** 173-200
[23] Wulp S A, Damar A, Ladwig N, Hesse K J 2016 *Mar. Poll. Bull.* **110** 675–685.
[24] Suntoyo H et al 2015_*Procedia Earth and Planetary Sci.* **14** 144 – 151.
[25] Dick S et al 1999 *German J.l Of Hydrography* **51** 182-219.

Seasonal Variations of Oceanographic Variables and Eastern Little Tuna (*Euthynnus affinis*) Catches in the North Indramayu Waters Java Sea

Mega Syamsuddin*, Sunarto, and Lintang Yuliadi

Department of Marine Sciences, Faculty of Fishery and Marine Science
Padjadjaran University, Jl. Raya Bandung–Sumedang KM. 21 Jatinangor 45363
Indonesia

*Email : vegha16@gmail.com

Abstract. The remotely derived oceanographic variables included sea surface temperature (SST), chlorophyll-*a* (Chl-*a*) and Eastern Little Tuna (*Euthynnus affinis*) catches are used as a combined dataset to understand the seasonal variation of oceanographic variables and Eastern Little Tuna catches in the north Indramayu waters, Java Sea. The fish catches and remotely sensed data were analysed for the 5 years datasets from 2010-2014. This study has shown the effect of monsoon inducing oceanographic condition in the study area. Seasonal change features were dominant for all the selected oceanographic parameters of SST and Chl-*a*, and also Eastern Little Tuna catches, respectively. The Eastern Little Tuna catch rates have the peak season from September to December (700 to 1000) ton that corresponded with the value of SST ranging from 29 °C to 30 °C following the decreasing of Chl-*a* concentrations in September to November (0.4 to 0.5) mg m^{-3}. The monsoonal system plays a great role in determining the variability of oceanographic conditions and catch in the north Indramayu waters, Java Sea. The catches seemed higher during the northwest monsoon than in the southeast monsoon for all year observations except in 2010. The wavelet spectrum analysis results confirmed that Eastern Little Tuna catches had seasonal and inter-annual variations during 2012-2014. The SST had seasonal variations during 2010-2014. The Chl-*a* also showed seasonal variations during 2010-2011 and inter-annual variations during 2011-2014. Our results would benefit the fishermen and policy makers to have better management for sustainable catch in the study area.

Key words: Eastern Little Tuna, Oceanographic variables, Seasonal variations

1. Introduction

Oceanographic conditions in the Java Sea is a consequence of the unique position lying between the Indonesian islands of Kalimantan to the north, Karimata strait to its northwest, links it to the South China Sea. Due to its unique position, there is important pathway that allows water entering the Java sea from the South China Sea through the Karimata Strait [1]. The oceanographic condition in the study area is affected by the monsoon wind reversal drives the surface layer of the Java Sea southeastward, bringing relatively cool, fresh South China Sea water into the region [2]. The northwest monsoon peaks in December to February and the southeast monsoon reaches its peak on June to August [3]. The region is rich with sources of interannual, seasonal, intraseasonal ocean variations that have direct impacts on the oceanographic conditions and therefore change the pelagic fish distribution.

The Eastern Little Tuna (*Euthynnus affinis*) represents a dominant catch in the Java Sea. The Eastern Little Tuna is an epipelagic, neritic species inhabiting water temperatures ranging from 18 °C to 29 °C and tend to form multispecies schools by size. The Eastern Little Tuna is largely confined to continental shelves and islands of the western Pacific and the Indian Ocean [3]. It is a highly migratory species and frequently forms large schools which are often mixed with other scombrid species. Although sexually mature fish may be encountered throughout the year, there are seasonal spawning peaks varying according to regions: i.e. March to May in Philippine waters; from the middle of the North West monsoon period to the beginning of the South East monsoon (January to July) off East Africa; and probably from August to October off Indonesia [4].

The behaviour and survival of Eastern Little Tuna have been related to a range of environmental and ecological factors, necessitating the need to combine several oceanographic factors to investigate the effects of oceanographic factors on the Eastern Little Tuna catch in the Java sea. As proxies of potential pelagic fish fishing grounds, the sea-surface temperature (SST) have been used to investigate productive frontal zones [5]. Thermal or color gradients in satellite images that arise from the circulation of water masses often indicate areas of high productivity [6]. Chlorophyll-*a* (Chl-*a*) data can also be used as a valuable proxy for water mass boundaries that can influence pelagic fish distribution in a region.

This study focus on how the ocean variations in seasonal time scale influenced the Eastern Little Tuna catch and distribution in the north Indramayu, Java sea. The satellite data included SST and Chl-*a* and Eastern Little Tuna catches are used as a combined dataset to understand the seasonal variation of oceanographic variables and Eastern Little Tuna catches in the north Indramayu waters, Java Sea. Therefore, understanding the effect oceanographic conditions and Eastern Little Tuna catches is an essential step towards the sustainable management of Eastern Little Tuna resources in the Java Sea.

2. Study area

The study area was located in the north Indramayu waters, Java Sea in the Indonesian Seas region focusing on geographical coordinates of 3 °S to 7 °S and 108 °E to 110 °E (Figure 1). Java Sea is relatively shallow water with average depth about 40 m. It will result in good mixing of water mass creating homogenous layer from the surface to the bottom [3].

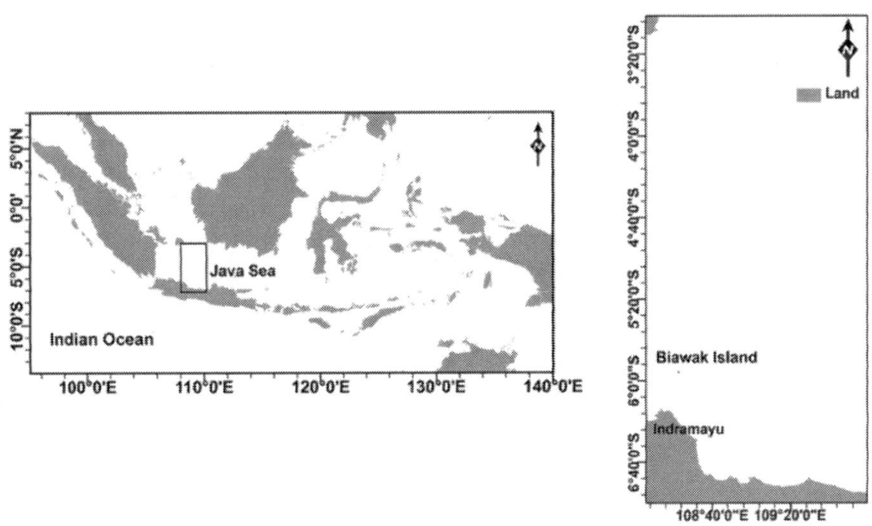

Figure 1. Map of Indonesian Seas with the inset box representing the study area

IOP Conf. Series: Earth and Environmental Science **116** (2018) 012073 doi:10.1088/1755-1315/116/1/012073

3. Data and methods

3.1 Data

A time series of Eastern Little Tuna catch data were obtained from fishing logbooks provided by the Fishing Port Indramayu, West Java and the Ministry of Marine Affairs and Fisheries Indonesia. Data included fishing position (latitude and longitude), operational dayas, fish weight (in kilograms) and number of vessel. The remotely sensed data consist of sea-surface temperature (SST) and Chlorophyll-*a* (Chl-*a*). The SST and Chl-*a* were derived from satellite imagery of Aqua Modis and downloaded from http://oceancolor.gsfc.nasa.gov. The SST and Chl-*a* data had spatial and temporal resolutions of 4 km and monthly, respectively. For data analysis, all SST, and Chl-*a* were composed into monthly data. In this study, the fish catch and satellite remotely sensed data were analyzed for the 5 years datasets from January 2010-December 2014.

3.2 Methods

3.2.1 Wavelet Spectrum Analysis

In this study, the wavelet spectrum analysis was applied on the monthly Eastern Little Tuna catch data, SST and Chl-*a* data. The wavelet transform analysis (WT) was performed in order to obtain a more detailed structure of the variability in time and frequency domain. Wavelet analysis is used to perform a time-frequency analysis of the dominant signal and decompose the signal in different time scales and identify how these signals vary in time. The cross wavelet spectrum (XWT) was applied to find region in time frequency space where the time series show high common power .

The spectra are significant at the 95% confidence level assuming a background red noise. High energy variances are represented in red and low values are in blue as shown in the colorbars that represent different wavelet amplitude. The black line (solid contour) shows the 95% confidence level and a thin black line shows the cone of influence (COI) and are the region of the wavelet spectrum in which edge effects become important [7].

4. Result and Discussion

4.1 Mean monthly climatology of Eastern Little Tuna cathes and oceanographic variables

In more detailed understanding of catches variations and their association with oceanographic variables, we examined mean monthly climatology of SST and Chl-*a* in relation to Eastern Little Tuna catch. This study has shown the effect of monsoon inducing oceanographic condition in the west Java Sea. Seasonal change features were dominant for all the selected oceanographic parameters of SST and Chl-*a*, and also Eastern Little Tuna catches, respectively. Seasonal change features were dominant for all the selected oceanographic. The time series plots on Figure 2 showed that the Eastern Little Tuna catch rates have the peak season in March and October (900 to 1000) ton that corresponded with the value of SST ranging from 29 °C to 30 °C following the decreasing Chl-*a* concentrations in September to November (0.4 to 0.5 mg m^{-3}). The declined catches of Eastern Little Tuna in September were observed for SST values 31°C and Chl-*a* ranging from 0.6-0.65 mg m^{-3}.

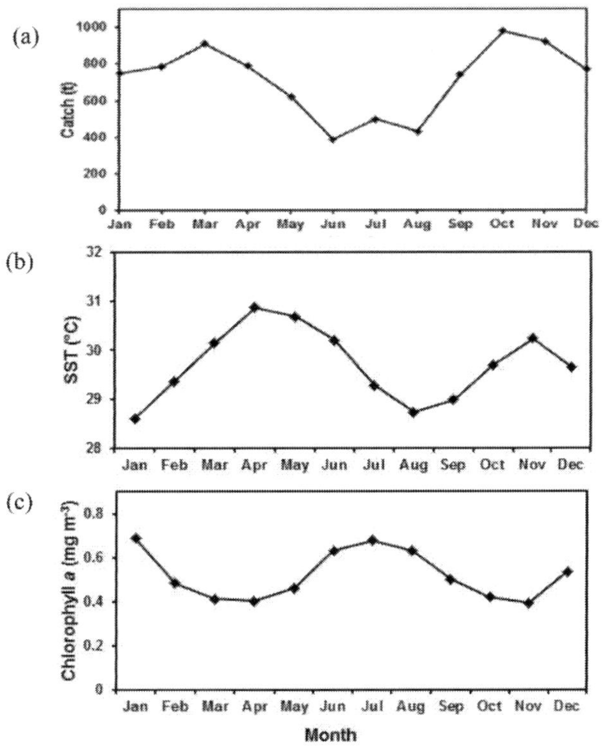

Figure 2. Mean monthly of (a) catch rate of Eastern Little Tuna, (b) SST, (c) Chl-*a* concentrations during 2010-2014. The x-axis represents the month, and y-axis shows the variable value.

4.2 Seasonal variations of Eastern Little Tuna Catches and distributions

The seasonal catch variations in Eastern Little Tuna is presented in Figure 3. In general, the total catch in northwest monsoon is higher (131000 kg) than southeast monsoon. Eastern Little tuna seemed higher during the northwest monsoon (September-February) than in the southeast monsoon (March-August) for all year observations except in 2010 due to La Niña event. SST during La Niña in November 2010 (northwest monsoon) showed warmer SST (30 °C to 33 °C) concentrating along the coast and offshore of the Java Sea, and resulted in lower catch during northwest monsoon This inferred that different ocean climate events might cause different oceanographic conditions favorable to Eastern Little Tuna catches in the Java Sea. The La Niña event was less favourable for Eastern Little Tuna catches [8].

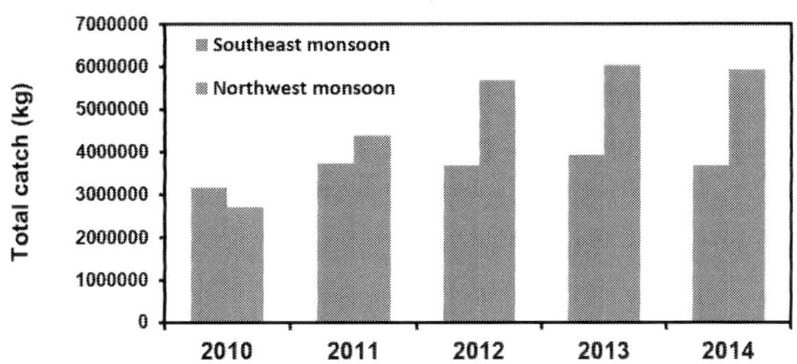

Figure 3. The seasonal variations of Eastern Little Tuna catches during 2010-2014

IOP Conf. Series: Earth and Environmental Science **116** (2018) 012073 doi:10.1088/1755-1315/116/1/012073

Based on the seasonal pattern, we differentiated between southeast and northwest monsoon over the data period 2010-2014 that displayed together in Figure 4. Eastern Little Tuna fishing locations were consistently observed between 3°15'00"S - 6°40'00"S and 108°15'00"E - 110°15'00"E, with minor variability in latitude and longitude between season and years. During northwest monsoon, Eastern Little Tuna tuna fishing location having more conducted location as sparse in 3°15'00"S - 6°40'00"S and 108°15'00"E - 110°13'00"E, than during southeast monsoon between 3°40'00"S - 6°40'00"S and 108°0'00"E - 110°15'00"E.

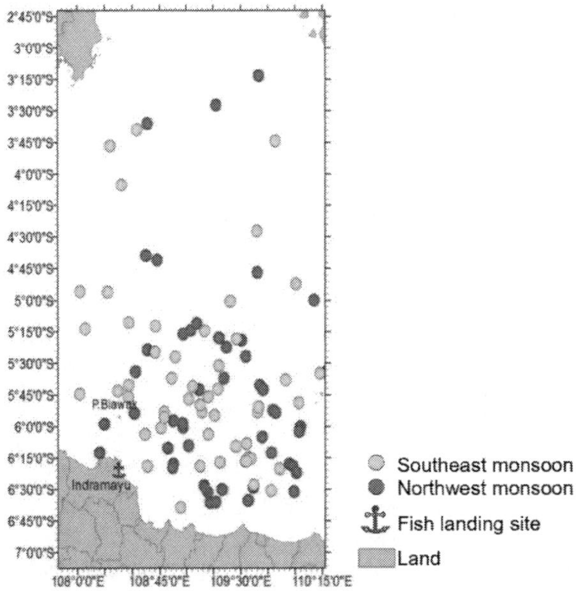

Figure 4. Seasonal variations in fishing position of Eastern Little Tuna

4.3 Wavelet spectrum analysis (CWT)

In order to clarify the physical characteristics of the seasonal change, we performed the wavelet spectrum analysis to confirm dominant signals related to amplitude (variance) and phases of Eastern Little Tuna catch, SST and Chl-*a* during 2010-2014 (Figure 5). The wavelet analysis revealed the seasonal (6 months) signal for all parameters. Eastern Little tuna catch showed seasonal signal during 2013-2014 and Chl-*a* during 2010-2011. The SST had also seasonal signal along the period of observation (2010-2014). Eastern Little Tuna showed strong interannual signal during 2013-2014. Also, Chl-*a* showed strong interannual signal occurrence for the years 2011-2014.

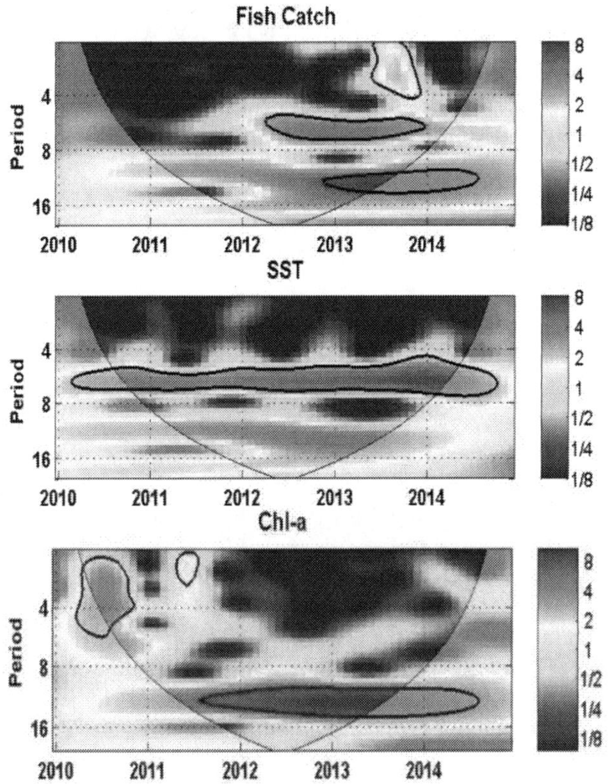

Figure 5. Wavelet Spectrum Analysis during 2010-2014

The Cross Wavelet Transform (XWT) finds regions in time frequency space where the time series show high common power. The XWT showed correlations of Eastern Little Tuna catch with SST and Chl-*a* from 2010 until 2014. The XWT signal revealed the interannual signal with the period of 16 months for all parameters (Figure 6). The XWTs between Catch and Chl-*a* indicated correlations within interannual signal of period 16 months during middle of the year of 2011-2014. The XWT between Catch and SST were also showed that there was a strong correlation within seasonal signal of period 4-8 months and interannual signal of 16 months during the year of 2012-2014. It was interesting to note here that all parameters revealed seasonal signal, and interannual signal could only been seen on Eastern Little Tuna catches and Chl-*a*. The interannual signal related to the climate regime such El Niño Southern Oscillation (ENSO) [9].

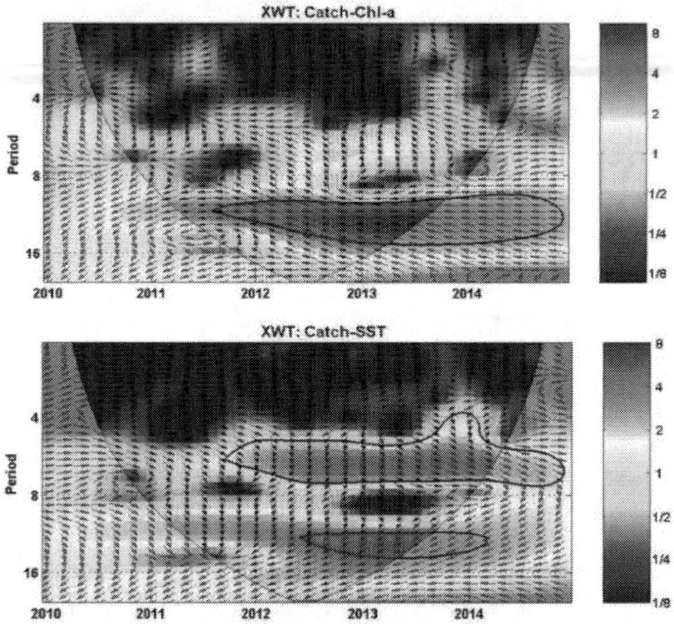

Figure 6. The cross wavelet spectrum (XWT) for Eastern Little Tuna catch in relation to SST and Chl-*a* during 2010-2014. The y-axis is represented a period (monthly) and x-axis showed the time (year). Colorbars represented different wavelet amplitude, the black line (solid contour) showed the 95% confidence level and a thin black line showed the cone of influence.

The monsoonal system plays a great role in determining the variability of oceanographic conditions and catch in the north Indramayu waters, Java Sea. Our results clearly indicated that the northwest monsoon had the most favorable oceanographic conditions to increase the Eastern Little Tuna catches in the study area. Our findings also could be used to point out the fishing ground and its changing from season to season, thus this result will be useful to know when is the best fishing season and the information will be useful for the policy maker to have sustainable fishing management of Eastern Little Tuna in the study area.

5. Conclusion

This study has shown the effects of monsoonal systems induced oceanographic conditions on Eastern Little Tuna catch and distribution in the Java Sea. The continuous wavelet analysis confirmed that all parameters (Eastern Little Tuna, SST and Chl-*a*) had seasonal variations. The cross wavelet spectrum analysis showed very clear relationship between Eastern Little Tuna catches and all oceanographic parameters to seasonal change and the interannual time scale. We recommend to have further investigations through the use of long-term historical time series to predict fishing ground locations and an emerging need to improve our climate understanding and forecast skill to conserve small pelagic catches.

Acknowledgements

The authors wish to thank the Ministry of Research, Technology and Higher Education of the Republic of Indonesia for the research grant. We thank Padjadjaran University, Marine Affairs and Fisheries Indramayu, the Goddard Space Flight Center/NASA, Physical Oceanography Distributed Active Archive Center, NOAA Climate Prediction Center, and ICTCRED committee.

References

[1] Qu T, Du Y, Strachan J, Meyers G, and Slingo J 2005 *Oceanography.* **18** 50–123.

[2] Nagara G, Sasongko N, Olakunle O 2007 *Introduction to Java Sea* (Norway: University of Stavanger) p 14

[3] Pepperell J 2010 *Fishes of The Open Ocean* (Chicago: The University of Chicago Press) pp 75-80

[4] FAO 2017 *Species Fact Sheet Euthynnus affinis Food and Agriculture Organization of the United Nations* (Rome: FAO Fisheries and Aquaculture Department)

[5] Zainuddin M, Saitoh S, and Saitoh K 2004 *Geophys. Res. Lett.* **31** L20311

[6] Saitoh S *et al* 2009 Remote sensing applications to fish harvesting in Remote sensing in fisheries and aquaculture. *Reports of the International Ocean-Colour Coordinating Group (IOCCG)* (Canada: IOCCG) pp 57–76

[7] Torrence C, and Compo G P 1998 *Bull. of the Am. Met. Soc.* **79** : 61-78

[8] Syamsuddin M, Sunarto, Yuliadi L 2016 *Jur. Teknol.* **78** : 139-143.

[9] Syamsuddin M, Saitoh S, Hirawake T, Syamsudin F, and Zainuddin M 2016 Interannual variation of bigeye tuna (*Thunnus obesus*) hotspots in the eastern Indian Ocean off Java *Int. J. of Rem. Sen.* p 14

The Influenced of *Lactobacillus plantarum* Starter Addition and The Length Time of Fermentation Process on The Activity of Seaweed Antioxidant *Ulva lactuca* from Krakal Beach, Yogyakarta

N D Ambarsari[1], I R P A Rushanti[1], A Setyaji[1], T R Ningsih[1], N Nurhana, I Subekhi[1], and E N Dewi[2]

[1]Student at Fisheries Processinng Department, Faculty Fisheries and Marine Sciences UNDIP, Jl. Prof. Soedarto, SH, Tembalang, Semarang, Jawa Tengah – 50275,
[2]Lecture at Fisheries Processinng Department, Faculty Fisheries and Marine Sciences UNDIP, Jl. Prof. Soedarto, SH, Tembalang, Semarang, Jawa Tengah – 50275,

*Corresponding author: +6285 729926313, Email : nataliadesiambarsari@gmail.com

Abstract: Seaweed contains phenol compound functioning as antioxidant. *Lactobacillus plantarum* starter addition in a fermentation process was expected will increase the activity of antioxidant. The purpose of this research was to determine the influence of *L. plantarum* addition and the length of fermentation on the activity of antioxidant in *U. lactuca.* The experiment was conducted with factorial design. The first treatment consisted 2 different factors namely without *L. plantarum* addition and *L. plantarum* addition. While the second treatment were the different length fermentation time: 0, 12, 24, and 36 hours. Each treatment were done in thriplicate. The data was analyzed using ANOVA and BNJ test was applied if there any differences betweenthe treatments. The results showed that the fresh *U. lactuca* with *L. plantarum* addition for 36 hours fermentation had TPC BAL 9,83 CFU/ml, pH 4,26, phenol 231 ppm and antioxidant activity IC_{50}1375,12 ppm. Dried *U. lactuca* with *L. plantarum* addition that was fermentized for 36 hours had TPC BAL 9,10 CFU/ml, pH 4,75, phenol 166,24 ppm and antioxidant activity IC_{50}4070,32 ppm. The fresh *U. lactuca* with *L. plantarum* addition for 36 hours fermentation was the best treatment since the antioxidant activity is IC_{50}1375,12 ppm. Although the antioxidant activity was categorized as weak but it was still showed an increase compared to the result of antioxidant activity with maceration method using n-hexana dissolver which was 11213,076 ppm, ethyl acetate 9770,285 ppm, and ethanol extact 4921,79 ppm.

Keywords: *Ulva lactuca*, Phenol, Antioxidant, *Lactobacillus plantarum,* Fermentation Length Time

1. Introduction

The usage of antioxidant compound recently is developing quickly, either for food or medical treatment. The usage as medicine and food are developing as equal as the knowledge development of free radical activity toward some degenerative diseases such as heart and cancer. The antioxidant sources can be classified into two, which are synthetic antioxidant (antioxidant which is obtained from chemical reaction synthesis result) and natural antioxidant (antioxidant which is the result of natural sources extraction). There are many synthetic commercial antioxidants such as butylated hydroxyl toluene (BHT), butylated hydroxy anisole (BHA), tert-butylhydroquinone (TBHQ) and propyl gallate (PG) which are used to reduce the harmful effects of free radicals. However, these synthetic antioxidants may have other harmful effects [1].

Fermentation with lactat acid bacterium addition that was frequently used as a starter in probiotic drink production such as Lactobacillus plantarum which is able to increase the amount of antioxidant compound content, one of them is phenol compound. Phenol is the result of sugar destruction in seaweed *U. lactuca* becomes poliphenol. According to Primurdia and Kusnadi [2] the more fermentation time, the more lactat acid bacterium will reproduce, so that the bacterium ability to rend glucose becomes primary metabolit (lactat acid) and secondary metabolit (poliphenol) will be more.

Seaweed is one of the fishery results that contains phenol compound functioning as antioxidant, rich of Iodium fiber and other important minerals. Antioxidants are substances that prevent the process of oxidation by scavenging the free radicals in body cells, and may reduce potential mutations and thereby help prevent cancer and heart diseases [1]. *U. lactuca* is one of green algae that is mostly found in Waters Krakal Beach, Yogyakarta. The seaweed is named sea lettuce and it is categorized in father seaweed which is eatable seaweed, having antioxidant content, antibacterium, antifungal, and antitumor. Seaweed produces various types of antioxidant that counteract environmental stresses. Hence it is considered as a potential source of novel antioxidants [3].

The antioxidant activity increase in seaweed *U. lactuca* can be done with fermentation and starter bacterium addition such as L. plantarum that is able to increase functional value in seaweed *U. lactuca*. According to Aminin et al. [4] in his research about fermentation of seaweed *Eucheuma cottoni* by adding bacterium starter *L. plantarum, Streptococcus thermophilus* can increase antioxidant content in seaweed. One of organic acids that is produced by BAL during fermentation process is lactat acid. Bisson [5] mentioned that lactat acid bacterium has the ability to produce phenol compound. One of organic acids that is produced by BAL during fermentation process is lactat acid. According to Kruszewska et al. [6] the antioxidant activity increase is also caused by the existence of secondary metabolit from bacterium metabolism.

2. Materials dan methods

2.1. Samples and Starter Preparation

The samples of sea lettuce *U. lactuca* was taken from waters Krakal Beach, Gunung Kidul, Yogyakarta. The samples were then washed by flowing water to remove the muck away. The samples was then placed in an open place and winded for more or less one week to get the dried samples. The samples were chopped in a small size more or less 1x1 centimeter. Starter bacterium used in this research was *L. plantarum* FNCC-0027 that is obtained from PAU UGM in the form of standing algae in a reaction tube. The recultured was done twice by taking one ose *L. plantarum* that was grew in 10-mL sterile MRS Broth media. It was then incubated in a temperature of 37°C for 24 hours inside the incubator.

2.2. The Calculation of Bacterium Cell Number

The calculation of bacterium cell number was done by using the method according to Gupta et al. [7]. Inoculum was prepared with 1 ml *L. plantarum* bacterium added 25 ml MRS Brorth and incubated for 24 hours. The media containing MRS, algae, CaCO3, Na-azida that were incubated for 24 hours with a temperature of 37°C. Colony was counted by the amount of BAL total to know the number of bacterium colony *L. plantarum* in 1000 μl.

2.3. Fermentation Process U. lactuca with Bacterium L. plantarum

The fresh and dried seaweed *U. lactuca* are chopped in a small size at more or less 1 centimeter. Jar bottle filled with 24 ml aquades is heated by autoclave for 15 minutes with a temperature of 95°C. Five grams of fresh and dried seaweed *U. lactuca* are put into the different jar bottle. Then it was added 5% of starter *L. plantarum* that has been incubated for 24 hours in every glass jar containing aquades and seaweed *U. lactuca*. Fermentation process was done for 36 hours then it is filtered by filter paper and is done the test of the total amount of BAL, pH, total content of phenol, and antioxidant IC50 at 0, 12, 24, and 36 hours.

2.4. Data Analysis

The data of antioxidant IC50, the total amount of BAL, pH test, and total content of phenol test were statistically analyzed using one way ANOVA. Normality and homogenity test are done first before ANOVA analysis to know the data characteristics [8]. If there any differences between treatments , the test was continued with BNJ in order to find out the variations.

2.5. Test Parameter

2.5.1. Total Amount of BAL Test

Total BAL content test was done by using the mothod according to Fardiaz [9]. The production of lactat acid by bacterium was done by using pour plate method using algae MRS media (MRS broth + agar 1,5% + CaCO3 0,80% + Na-azida 0,01%). 5 ml of samples was given the solution of 45 ml sterile NaCl 0,85% and homogenized in stomacher for one minute. 1 ml of samples that has been homogenized was put into the reaction tube containing 9 ml sterile solution (0,85%). The next step was, from each three series of the last thinning degree, we take 1 ml and put it into the sterile petry dish (every thinning uses twice replications). Then we pour 10-12 ml of algae MRS with a temperature of 45°C. After that, the homogenant and algae media are mixed by rubbed the dish on the table forming the number of eight (8). When media get solid, the dish was turn back and incubated it in a temperature of 37°C for 48-72 hours in a reverse position. The number of colony formed is calculated by colony counter.

2.5.2. pH Test

pH (acidity degree) test was using the pH meter that was based on pH measurement method of SNI 06-6989-11-2004.

2.5.3. Phenol Content Test

Phenol content was tested by using the method based on Orak [10]. Five grams of samples was put into the erlenmayer 100 ml. By using aquades, the samples was thinned in the measure jar until the volume reached 100 ml. The solution was filtered or centrifuged until get clear filtrate. 1 ml of solution or clear filtrate was put into reaction tube, then 0,5 ml of Follin Denis (1:1),poured and continued 1 ml of Na_2CO3 solution , left it for 10 minutes. Another 10 ml aquades was poured , then the solution was vortexed until it becomes homogeneus. The samples absorbation was read by using spectrofotometer which the length of the wave 730 nm.

2.5.4. Antioxidant IC50 Test

a. *Reagen Preparation*

The solution of DPPH (2,2-difenil-1-pikrilhidrazil) was measured by Molyneux [11]. 1,97 mg of DPPH was dissolved by methanol in the measure jar until it becomes 100 ml so the solution which the number of concentration was 50 μm.

b. *Samples Preparation*
The samples was calculated in an amount of 5 grams and poured 24 ml of methanol that will be left for 3 days, that process was used as the stock solution.
c. *The Antioxidant Activity Test*
The antioxidant activity was analyzed by method based on Hanani [12]. Ten mg of extract was dissolved with 10 ml of methanol until the volume was reached 10 ml and the concentration number of 1 mg/ml. The methanol was added so samples concentration in a number of (50, 100, 150, dan 200 μg/ml). The antioxidant activity determination from each concentration was taken by using the drop pipette with a number 0,2 ml of samples solution with micro pipette and put it into the vial, then pour 3,8 ml of DPPH solution 50 μM. The solution was homogenized and left for 30 minutes in a dark place, the absorbtion was measured by using spektrofotometer UV – Vis with the wave length is 515 nm. As the comparison, ascorbic acid (concentrations of 2,3,4,5,6 μg/ml) was used with the same treatment for the curve standard samples. The determination of the DPPH maximum wave length [13]. The solution of DPPH 50 μM was taken by using a pipette in an amount of 3,8 ml and we pour 0,2 ml of methanol. After leaving it for 30 minutes in a dark place, the solution absorbtion was measured by spectrofotometer UV – Vis with the wave length was 400-800 nm.

3. Result and discussion

3.1. *Total Amount of Lactat Acid Bacteria Test*
The total amount of lactic acid bacteria was presented in Table 1.

Table 1. The Result of Total Amount Lactic Acid Bacteria in Fermented *U. lactuca* with *L. plantarum* (CFU/mL) Starter

Treatment	Length Time of Fermentation (hours)			
	0	12	24	36
A (fresh)	8,55±0,05^{ab}	8,71±0,07^{b}	9,50± 0,06^{d}	9,83± 0,10^{e}
B (dried)	8,31±0,05^{a}	8,54±0,11^{ab}	8,71± 0,13^{b}	9,10± 0,11^{c}

Note : The data was average from three time replications
The data followed by different lowercase letters indicates highly significant different (P<0.05)

ANOVA test showed an interaction between the addition of *L plantarum* starter with the length time of fermentation had significantly effect the total amount of lactic acid bacteria (p<0,05)

The length time of fermentation significantly affected the growth of lactic acid bacteria. Table 1 showed that the amount of BAL value from fresh and dried seaweed *U. lactuca* seaweed sampless at the beginning of fermentation has a value that was not significantly different but at the end of fermentation process has a significantly different value. When fermentation was started , the result of BAL amount on fresh sampless is 8,55 CFU/mL and in the dried sampless is 8,31 CFU/mL. At the end of fermentation (36 hours), the result of BAL amount on fresh sampless was 9,85 CFU/mL and dried sampless was 9,10 CFU/mL.

Table 1 showed that the fresh and dried sampless in fermentation have a value that was not significantly different. This may happen because there is no activity that BAL does in the samples yet, so that when the test of BAL total was done, the ones counted is the amount of BAL that has just been put in the samples. In the end of fermentation, the fresh and dried sampless have a value significant difeerent. This may happen since BAL has done metabolism in the samples so that the amount of BAL increased. The difference of BAL amount in the end of fermentation in the fresh and dried sampless may happen because the difference of growth media. The fresh samples has more BAL value compared with the dried one so that BAL was easier to grow in the fresh sampless. The fresh sampless has more nutrition content for the growth media of BAL rather than in dried seaweed samples. According to Mallesha et al. [14], the growth of lactat acid bacteria increased by increasing the incubation time, the optimalization of temperature, humidity, light, pH, and nutrition. This resulted was match with the research of Luhur [15], about fermentation of seaweed *Glacilaria* sp. With starter *L. plantarum*, the amount of lactic acid bacteria in the end of fermentation increased because BAL has already activated metabolism and done the growth phase.

The samples of fresh seaweed *U. lactuca* has more amount of nutrition compared with the dried one. The nutrition was used for the growth media of lactat acid bacteria because in the fresh samples, of *U. lactuca*, there was no process yet that caused nutrition contained in *U. lactuca* becomes destroyed and decreased. In other hand different with dried samples of *U. lactuca* that has been done the heating process in *U. lactuca*, so that the approximation of nutrition contained in *U. lactuca* is decreased compared with the fresh *U. lactuca*. This is strengthened by the result of the research in table 1, that the more fermentation time, the more amount of lactat acid bacteria grow. In the dried samples, lactat acid bacteria is also increasing but not as much as in the fresh samples.

3.2. pH Test

The pH test was performed to determine the pH value of *U. lactuca* during fermentation seaweed process. The pH value was presented in Table 2.

Table 2. The Result of pH Test on Fresh *U. lactuca* and Dried *U. lactuca* Fermented with *L. plantarum* Starter

Treatment	Fermentation Length (hours)			
	0	12	24	36
A (fresh)	5,08 ±0,14^d	4,71 ± 0,15^b	4,49 ±0,07^a	4,26 ±0,05^a
B (dried)	5,24 ±0,05^e	5,04 ±0,10^d	4,87 ±0,08^c	4,75 ±0,05^c

Note : The data was average from three times replications
The data followed by different lowercase letters indicates highly significant different ($P<0,05$)

ANOVA test showed the was an interaction between samples (fresh and dried *U. lactuca*) on addition of *L. plantarum* with the length time of fermentation and it given significantly effect on the pH ($p<0,05$).

The result of pH value in the beginning time of fermentation, dried samples of *U. lactuca* fermentized with the addition of *L. plantarum* bacteria has pH value that was significant different with dried samples of *U. lactuca* which is 5,08 and 5,24. This is because in the beginning of fermentation, the microorganisms inside either the seaweed or in starter haven't done the metabolism process yet to produce acid. In the end of fermentation process, the pH value in the fresh samples was significant different with the dried samples.

The data presented in Table 2 showed that the pH value at the beginning fermentation process in the fresh samples was lower than the dried one. This can be explained since in fresh form there was much nutrition that can be used by acid bacteria to proceed the metabolism process to produce acid compound, while in the dried samples, lactic acid bacteria have been decreased because of the nutrition decreased that can be used by the bacteria to grow because of the drieding process. The pH both fresh and dried samples, were decreased during the fermentation process from the beginning until it was finished . The increased of fermentation length time in the fresh and dried samples with the addition of starter *L. plantarum* bacteria give significant different for the pH.

The result of ANOVA test showed there was an interaction between the samples type used and fermentation length time so that those are significant influenced toward the pH value decreased

($p<0,05$). Kumalaningsih et al. [16] stated that pH decrease happens because of microorganism activity producing lactat acid from metabolism result. The acumulated lactat acid causes pH value becomes decreased during fermentation time. *L. plantarum* can decrease the pH of fermentation environment by changing the sugar becomes lactic acid (primary metabolite). This is strengthened by Nahariah et al. [17] that the pH decrease was caused by the fermentation activity that changes carbohydrate or sugar in food sources becomes acid and water and other final products. *L. plantarum* can digest complex compound becomes simple compound with the final result is lactic acid. This result just the same with Pratama et al. [18], on his research about casava fermentation with *L. plantarum* addition toward the total of lactic acid and pH. The longer fermentation time, the lesser pH value, because the bacteria metabolism so that they produce lactic acid.

3.3. The Amount of Phenol

The test of amount of phenol done to find out how much the amount of phenol from the fermentation of *U. lactuca* seaweed. The result of phenol content was presented in Table 3.

Table 3. The Results of Total Number of Phenols from *U. lactuca* Fermented with *L. plantarum* (ppm) Starter

Treatment	Fermentation length (hours)			
	0	12	24	36
A (fresh)	132,31±17,10^{c}	146,15±15,63^{cd}	170.8 ±7,15^{e}	231±14^{f}
B (dried)	52,9±5,06 ^{a}	97,59 ± 1,11^{b}	140,1±7,15^{cd}	166,24±7,35^{de}

Note : The data was average from three times replications.
The data followed by different lowercase letters indicated a highly significant different between samples ($P<0,05$)

Length time of fermentation was really influenced toward the amount of phenol. Table 3 showed that the amount of phenol in the fresh and dried samples of seaweed *U. lactuca* in the beginning and the end of fermentation has the value that significantly different. This can be seen from from phenol value in the fermentation result of fresh and dried *U. lactuca*, that there was a significant difference of phenol amount. The longer fermentation time, the more amount of *U. lactuca* phenol. This can be seen that in the hour of 0, the amount of phenol in fresh *U. lactuca* is 132,31 ppm, then it increased in the end of fermentation, which was 231 ppm. The dried *U. lactuca* that in the hour of 0, the amount of phenol is 52,9 ppm and it increased until the end time of fermentation. The increased of phenol amount was caused by the bacteria activity that digest the sugar contained in the samples that produces primary metabolit (lactic acid) and secondary metabolit (poliphenol), so that the content value is increasing. Fermentation length gives a significant influenced toward the increase of phenol value ($p<0,05$). According to Primurdia and Kusnadi [2], the longer fermentation time, the more lactic acid bacteria that was reproduced, so that the bacteria ability to digest glucose becomes primary metabolit (lactic acid) and secondary metabolite (polyphenol) becomes higher. This result agreed with Hitayezu et al. [19], where polyphenolic and phenolic compounds are bioactive non-nutrient secondary metabolites present in fruits, vegetables, and cereals. These compounds protected the plant against pathogen attack or ultraviolet radiation. In food, these metabolites contributes not only to antioxidant activity but also to color, astringency, or bitterness.

3.4. Antioxidant IC50 Test

The IC50 antioxidant test was performed to find out how *U.lactuca* sea lettuce antioxidant activity to preventing free radicals at the beginning of fermentation and in the end of fermentation process. The results of IC50 antioxidant activity were presented in Table 4.

IOP Conf. Series: Earth and Environmental Science **116** (2018) 012074 doi:10.1088/1755-1315/116/1/012074

Table 4. The Results of Antioxidant Amounts of IC50 *U. lactuca* Sea Lettuce Fermented with *L. plantarum* (ppm) Starter

Treatment	Fermentation Length (hours)			
	0	12	24	36
A (fresh)	1917,18±1,23[d]	1739,22±0,93[c]	1532,33±0,97[b]	1375,11±0,9 [a]
B (dried)	16512,12±0,83 [h]	15960,17±0,97[g]	5571,21±0,83[f]	4070,31± 0,59[e]

Note : The data was average from three times replications.
The data followed by different lowercase letters indicated there was a highly significant different (P<0,05).

The length time of fermentation is significantly influenced toward the amount of antioxidant IC50. Table 4 showed that the amount of antioxidant in the fresh and dried seaweed *U. lactuca* in the beginning and the end of fermentation has the value that is significant different. This can be seen from the phenol value in fermentation result of the dried and wet *U. lactuca*, that there was a different amount of antioxidant IC50 that is significant. The longer fermentation time, the more antioxidant IC50 activity of *U. lactuca* that is increased. This can be seen that in the hour of 0, the wet *U. lactuca* has the value of 1917,18 ppm and it is increasing in the end of fermentation, which is 1375,11 ppm. It is same as the dried *U. lactuca*, that the amount of antioksidan IC50 in the hour of 0 is 16512,12 ppm and it is increasing until the end of fermentation time. The increase of antioksidan IC50 amount is caused by bacteria activity that digests the sugar contained in the samples that produces primary metabolit (lactat acid) and secondary metabolit (poliphenol). Phenol compound is the former compund of antioxidant. The more phenol amount, the more antioxidan IC50 activity. Although antioxidant IC50 activity increased during the fermentation process, but the antioxidant IC50 activity in seaweed *U. lactuca* was categorized very weak. This can be seen from the number of IC50, while the smaller number of IC50, the antioxidant activity is stronger. We can predict that the weak antioxidant content is caused by the fermentation time that is not enough, or the the concentration of starter bacteria has to be increased. According to Arditiana et al. [20] Inhibition Concentration (IC50) can be defined as samples compound concentration that will cause a reduction toward the DPPH activity 50%. The smaller number of IC50 means the antioxidant activity is higher. A compound can be considered as a very-strong antioxidant if the value of IC50 is less than 50 ppm. The antioxidant activity was strong if the IC50 value was between 50-100 ppm, it was average level if the IC50 value was between 100-150 ppm, and it was weak activity if the IC50 value was between 150-200 ppm.

Antioxidant activity of *U. lactuca* with *L. plantarum* addition, although it is still categorized weak, but it is increasing during the fermentation process and the activity was higher if it is compared with the result of antioxidant activity test with maceration method and using n-hexana solvent, which is 11213,076 ppm, ethyl acetat 9770,285 ppm and ethanol extract 4921,79 ppm [21]. It can be said that starter *L. plantarum* bacterium addition and the length time fermentation were significant influenced toward the actioxidant *U. lactuca* activity. This was agreed with Luhur [15] about fermentation of *Glacilaria* sp. with starter *L. plantarum* additon, he stated that the longer fermentation time, the bigger antioxidant (DPPH IC50) activity in defending the free radical.

4. Conclusion

Fermentation with starter *L. plantarum* increased the antioxidant amount of *U. lactuca*. The best value/number of fermentation was available in the hour of 36 of the fresh U. lactuca with the result of antioxidant IC50 test, which is 1375,117 ppm, and although the antioxidant activity is still categorized weak, but it was increased compared with the result of antioxiant activity test with maceration method using n-hexana solution, which was 11213,076 ppm, etil asetat 9770,285 ppm, and ethanol extract 4921,79 ppm.

References

[1] Maryam K, Yousefzadi M, Ali Ahmadi A, Feghhi MA, and Keshavarz M 2013 *Journal of the Persian Gulf* **4** : 45-50.
[2] Primurdia EG, J Kusnadi 2014 *Jurnal Pangan dan Agroindustri* **3**
[3] Sharmila S and LJ Rebecca 2014 *Int. J. Pharm Bio.* **5** :830-834.
[4] Aminin A L N, B D Wandasari, R A Pratiwi, S R Utomo, P R Sarjono and N S Mulyani 2014 *Antioxidant and Alpha-Glucosidase Inhibitor Capacity of Fermented Eucheuma cottoni from Lombok, NTB.* Compendia of Abstracts Int. Symp. on Food and Argo-biodiversity.
[5] Bisson LF, AL Waterhouse, SE Ebler, MA Walker and JT Lapsey 2001 *Nature* 696-699
[6] Kruszewska D, Lan J, Lorca G, Yanagisawa N, Marklinder I and Ljungh A 2002 *Microecol Ther* **29** : 37–49
[7] Gupta S, R Gaurav and Abu-Ghannam N 2010 *Int. J. Food Sci. Technol.* **45** : 482–489.
[8] Kusriningrum RS 2008 *Perancangan Percobaan.* (Surabaya: Airlangga University Press)
[9] Fardiaz S 1993 *Analisa Mikrobiologi Pangan.* (Jakarta : PT. Raja Grafindo Persada)
[10] Orak HH 2006 *Electronic Journal of Polish Agricultural University Food Science and Technology* **9** (1): #18
[11] Molyneux P 2004 *J. Sci. Technol.* **26** : 211-219.
[12] Hanani E, A Mun'im, R Sekarini 2005 *Majalah Ilmu Kefarmasian.* **II** (3) : 127 - 133
[13] Okawa M, J Kinjo, T Nohara and M Ono 2001 *Biol. Pharm. Bull.* **24** :1202-1205
[14] Mallesha, Shylaja R. and Selvakumar DJH 2010 *Rec. Res. Sci. Technol.* **2** : 42-46.
[15] Luhur N 2015 *Pengaruh Penambahan Starter Lactobacillus plantarum dan Lama proses Fermentasi Terhadap Aktivitas Antioksidan Rumput Laut Glacilaria sp.* (Teknologi Hasil Perikanan. UNDIP: Semarang)
[16] Kumalaningsih, S., Wignyanto., V.R. Permatasari., and A. Triyono. 2014. *Pengaruh Jenis Mikroorganisme dan pH Terhadap Kualitas Minuman Probiotik Dari Ampas Tahu.* (Program Pasca Sarjana. Universitas Brawijaya, Malang: Surabaya)
[17] Nahariah, A M Legowo, E Abustam, A Hintono, Y B Pramono and F N Yuliati 2013 *Jurnal Ilmu dan Teknologi Peternaka* **3** :33-39.
[18] Pratama A Y, R N Febriani, and S Gunawan 2013 *Jurnal POMITS* **2**
[19] Hitayezu R, Baakdah M M, Kinnin J, Henderson K, and Tsopmo A 2015 *Journal of Cereal Science* **63** :35-40.
[20] Arditiana A, N Rochmawati, P Widinugroho and R D Puspitasari *2015 Jurnal Pangan dan Agroindustri* **3**
[21] Febriansah E M, E R E Sakti and R A Kodir 2015 *Proc. SPeSIA Unisba.*

Contribution of Golden Apple Snail Flour to Enhance Omega-3 and Omega-6 Fatty Acids Contents in Weaning Food

D D Marsyha[1*], H S Wijayanti[1,2], Nuryanto[1,2], G Anjani[1,2]**

[1]Department of Nutrition Science, Faculty of Medicine Diponegoro University, Jl. Prof. H. Soedarto, SH, Tembalang, Semarang 50275, Indonesia
[2]Centre of Nutrition Research (CENURE) Diponegoro University, Jl. Prof. H. Soedarto, SH, Tembalang, Semarang 50275, Indonesia
* Presenting author
** Corresponding author; E-mail: gemaanjani@gmail.com ; Tel: 0812-8537-6785

Abstract. The case of undernourished children in Grobogan District (15.3%) is caused by children nutrients intake less than the Recommendation Dietary Allowance (RDA). To enhance children nutrients intake, be required formulation of weaning food using high-nutrient local food such as golden apple snail (*Pomacea canaliculata*). Golden apple snail flour contains high contents of zinc, iron, omega-3 and omega-6 fatty acids. This study aims to analyze the effect of golden apple snail flour substitution on nutrients content and organoleptic properties of weaning food (baby porridge). This is an experimental research by substitution of golden apple snail flour in the making of weaning food with four treatments of substitution (0%, 5%, 10%, 15%). Substitution of golden apple snails flour could affect the nutrient content levels of fat, zinc, iron (p=0.0001), carbohydrate (p=0.011), water (p=0.003), ash (p=0.001), omega-3 and omega-6 fatty acids. Whereas, it could not affect the content of energy (p=0.678), protein (p=0.129) and fiber (p=0.482). Furthermore, the substitution could affect the organoleptic properties include color, texture and taste (p=0.0001) while not for the aroma (p=0.798). Based on nutrient content analysis, substitution of golden apple snail flour could enhance the zinc, iron, omega-3 and omega-6 fatty acids contents of weaning food.

Keywords: Weaning food, golden apple snail flour, zinc, iron, omega-3, omega-6.

1. Introduction

Malnutrition, especially undernourished, is one of the biggest nutritional problems which is not resolved in some developing countries such as Indonesia[1]. Based on Riskesdas (Riset Kesehatan Dasar) 2013[2], the prevalence of under nutrition in Indonesia continues to increase in every year. The prevalence of undernourished toddler (based on the value of Z-Score weight-for age ≤ -2SD) is 19.6% in 2013which increased compared in 2010 amounted to 17.9%.

Grobogan District is one of the Districts in Central Java which had undernourished prevalence in medium category by 15.3%[3]. Undernourished classification based on WHO in high category if the prevalence more than 20%[4]. The highest incidence of undernourished is in children under five years[2]. Based on some previous researches, it can be concluded that undernourished in Central Java, especially in Grobogan, caused by children nutrients intake less than the Recommendation Dietary Allowance (RDA).

The government has conducted several programs of interventions against malnutrition problems. Nutritional improvement movement in Indonesia focused on the group of 1000 first day of life (1000 HPK) which is Scaling Up Nutrition (SUN Movement) at the global level. Undernourished intervention carried out since born until the age of two years[5]. Beside the improvement of nutritional status pregnant women, exclusive breastfeeding and weaning food after birth are also very important.

The making of weaning food is prominent local food with ingredients that are inexpensive and easily obtained in the local area[6]. Then weaning food should be high energy and nutrients both macronutrients and micronutrients. One of the foodstuffs in Grobogan with high nutrients but its utility is still lacking is the golden apple snails (*Pomacea canaliculata*). Golden apple snail protein and minerals content such as zinc and iron are quite high so it can improve the children nutritional status and prevent nutrients deficiency in the metabolism process[7]. In addition, golden apple snail also contains omega-3 and omega-6 which is good for the immune system and brain development in children growth[8].

Prevention of undernourished children should be done early with the consumption of weaning food for > 6 months children are vital. One type of weaning food that is appropriate for children aged 6-12 months is baby porridge. Therefore, this research aim to analyze the effect of golden apple snails flour substitution on nutrients content include protein, fat and fatty acids (omega 3 and omega 6), carbohydrates, fiber, water, ash, zinc, iron and also the organoleptic properties.

2. Methods

This is an experimental research with complete randomized design by substitution of golden apple snail with skimmed milk in the making of weaning food (baby porridge). This research divided into two stages, the initial stage of processing golden apple snail flour to be analyzed and as the basic for formulation determination. The second stage is making the weaning food by substitution of golden apple snail based on the formulation determination in the early stage then done to processing of nutrient analysis and organoleptic properties test by panelist acceptance test of the weaning food. This research was carried out in the Integrated Laboratory of Diponegeoro University, Food Technology Laboratory of Soegijapranata Catholic University, Integrated Testing and Research Laboratory of Gadjah Mada University, and Central Laboratory of Pollution Prevention Industry (BBTPPI).

The first step is to process the golden apple snail meat into flour by purging the meat and boiled for approximately 30 minutes, then cut the flesh to thin shapes and dried using a dry oven for 24 hours with a temperature of $\pm$ 50^{0}C, the dried flesh will be milled using a grinder and filtered using 1000 μm sieve then conducted nutrient content analysis of energy, carbohydrates, fat, protein, fiber, iron, zinc, omega-3 fatty acids and omega-6 fatty acids. In addition to heavy metal analysis that is lead (Pb).

The formulation were determined by the result of golden apple snail flour nutrient analysis, then calculates with the other nutrient contents of the other ingredients (rice flour, skimmed milk, powdered sugar, vegetable oil) which is obtained from the list of foodstuffs compositions book (DKBM) and adjusted to the terms of government regulations (based on Indonesian Ministry of Health Decision in 2007 about instant powder of weaning food for children aged 6-12 months). There were four treatments of golden apple snail flour substitution by 0%, 5%, 10%, and 15%.

The second stage was making the weaning food with golden apple snail flour substitution by 5%, 10%, and 15% and mixed with the other ingredients. The rice flour will be gelatinized first, by boiled glutinous rice flour using water with a ratio of 1:4 then pat dried in dry oven and milled using grinder. After that process done, continued to mixing all the ingredients of weaning food (baby porridge) based on formulation determination using blender and filtered on the 1000 μm sieve or 18 mesh sieve. That powdered baby porridge will be cooked using water with a ratio of 1:8 for approximately 10-15 minutes.

Furthermore, the powdered baby porridge will be analyzed the nutrient contents with minimal two times repetitions. Analysis of the energy content method using Bomb Calorimeter, carbohydrate content analysis using calculations Carbohydrate-by-difference method, protein content analysis using Bradford method, fat content analysis using Soxhlet method, fiber and ash content analysis using Gravimetri method, water content analysis using Oven method, iron and zinc content analysis using AAS (*Atomic Absorption Spectrophotometry*) method, omega-3 fatty acids and omega-6 fatty acids using GCMS (*Gas Chromatography Mass Spectrometry*). Assessment of 25 panelist acceptance test using organoleptic test on color, aroma, texture, and flavor with four scales of assessments namely 1 (very dislike), 2 (dislike), 3 (like), and 4 (really like). 25 rather-well trained young adult panelists are nutrition science's students of Diponegoro University in Semarang.

The data that has been collected will be analyzed using statistical software. The effects of variation of golden apple snail substitution on nutrient content was tested using one way ANOVA test with 95% confidence degree, and followed by Tukey Posthoc Test to know the real difference between the treatments group, while the acceptance level test using Friedman test and followed by Wilcoxon Test.

3. Results

3.1 Formulation Determination

Nutrients content analysis of golden apple snail flour is required to determine the weaning food formulation then those analysis results could be calculated with the other ingredients of weaning food making process based on the specification requirement of Indonesian Ministry of Health Decision (Keputusan Kementerian Kesehatan RI) in 2007 about instant powder of weaning food for children aged 6-12 months. This formulation determination is aim to determine the best formulation and eligible weaning food (based on the organoleptic test) which can fulfill the children nutrition requirements at age 6-12 months to prevent malnutrition specifically undernourished. In this research, the nutrients content that could be seeded is the fatty acids (omega-3 and omega-6) and also mineral content (zinc and iron). The result of nutrients content analysis of golden apple snail flour is presented in Table 1.

Table 1. Nutrient content of golden apple snail flour (100 grams)

Test parameters	Content	Unit
Energy	363.98	Kkal
Protein	12.73	Gram
Fat	13.67	gram
Carbohydrate	49.68	gram
- Fiber	0.87	gram
Ash	18.78	gram
Water	4.26	gram
Zinc	76.77	mg
Iron	66	mg
Omega 3	1.50	gram
Omega 6	3.07	gram
Lead	<0.00004*	ppm

*Terms of lead level ≤ 1.14 ppm

The determination of formulation aim to increase the nutritional value of weaning food, especially in fatty acids content (omega-3 and omega-6 fatty acids) and mineral content (zinc and iron). Based on calculations of the nutrients content analysis of golden apple snail flour and the nutrient content of other ingredients which is obtained from the list of foodstuffs compositions book (DKBM), the formulation determined is substitution of golden apple snail as much as 5%, 10% and 15%. Experimental formulation in this research is presented in Table 2.

Table 2. Experimental formulation (%)

Ingredients	KM 0 (Control)	KM I	KM II	KM III
Golden apple snail flour	-	5	10	15
Rice flour	35	35	35	35
Skimmed milk	50	45	40	35
Powdered sugar	5	5	5	5
Vegetable oil	10	10	10	10
Total	100	100	100	100

Description: KM is an acronym of Keong Mas

3.2 Nutrient Contents

The results of statistical tests indicate that substitution of golden apple snail flour could affect the fat content and mineral content (zinc and iron) (p=0.0001), carbohydrate content (p=0.011), water content (p=0.001), while there is no effect on the energy content (p=0.682), protein content (p=0.129), and fiber content (p=0.482). Based on Indonesian Ministry of Health Decision about instant powder of weaning food for children aged 6-12 months, the nearest experimental formulation is formulation KM I (5% of golden apple snail flour substitution), all the nutrient content (energy, fat, carbohydrate, fiber, ash, and iron) have been qualified except protein, zinc and water content. Protein and zinc content is below the predetermined standards whereas, water content is over the predetermined standards. The nutrients content analysis of weaning food (four treatments) is presented in Table 3 and Figure 1.

Table 3. Average Result of Energy and Nutrients Content Weaning Food with Substitution of Golden Apple Snail Flour

Average Results	Formulation				*p value*
	KM 0	KM I	KM II	KM III	
Energy	423.47±59.11	418.05±15.20	421.04±131.57	413.13±90.52	0.682*
Protein	10.77±0.17	13.84±2.30	11.43±1.41	11.49±1.12	0.129*
Fat	11.04±0.78^a	11.88±0.84^{ab}	13.16±0.74^b	20.31±0.98^c	0.0001*
Carbohydrate	69.22±1.76^a	62.61±2.78^{ab}	63.69±2.09^a	55.66±0.80^b	0.011*
Crude Fiber	1.08±0.69	1.55±0.91	1.21±0.20	1.81±0.59	0.482*
Water	5.01±0.64^a	5.48±0.25^a	6.53±0.10^b	6.22±0.35^b	0.003**
Ash	2.84±0.01^a	3.25±0.16^b	3.58±0.08^a	4.16±0.04^c	0.001*
Zinc	1.64±0.00^a	1.88±0.00^b	1.98±0.00^c	2.39±0.10^d	0.0001*
Iron	1.04±0.00^a	6.85±0.91^b	12.18±0.10^c	17.76±0.02^d	0.0001*

Note: numbers are followed by different superscript letters (a, b, c, d) show real differences
*Analysis Test using *One-way ANOVA* ; ** Analysis Test using Kruskal-Wallis

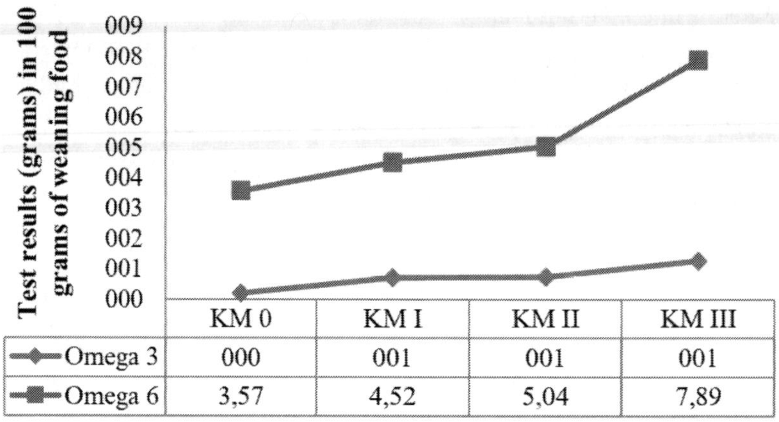

Figure 1. Graphic of Increased omega-3 and omega-6 fatty acids content of weaning food

Since there is no repetitions data for omega-3 and omega-6 fatty acids, the data is presented in descriptively by graphic. Based on the results of omega-3 and omega-6 fatty acids content analysis, there is an increase of omega 3 and omega 6content of weaning food with substitution of golden apple snail flour by 5%, 10% and 15%. Substitution of golden apple snail flour with skimmed milk can increase the content of omega-3 fatty acids by ± 0.5 gram in each formulation, while the omega 6 contents increases by ± 0.5-2 grams.

3.3 Acceptance Level

Based on organoleptic properties test indicate that substitution of golden apple snail flour could affect the organoleptic properties include color, texture and taste (p=0.0001) while not for the aroma (p=0.798). Rather-well trained panelists who are young adults tend to choose control formulation (KM 0 with 0% golden apple snail flour substitution) while weaning food with substitution of golden apple snail flour which is acceptable is formulation KM I (5% of golden apple snail flour substitution). The average rate of acceptance level results of weaning food (four treatments) is presented in Table 4.

Table 4. The average rate of acceptance level results of weaning food (four treatments)

Formulations	Color		Aroma		Texture		Flavor	
	Average	Info	Average	Info	Average	Info	Average	Info
KM 0	$3.24{\pm}0.66^{a}$	Like	2.72±0.67	Like	$2.96{\pm}0.73^{ac}$	Like	$3.08{\pm}0.57^{abc}$	Like
KM I	$2.08{\pm}0.75^{b}$	Dislike	2.56±0.58	Like	$2.16{\pm}0.68^{ab}$	Dislike	$2.20{\pm}0.64^{ab}$	Dislike
KM II	$2.12{\pm}0.66^{c}$	Dislike	2.52±0.71	Like	$2.44{\pm}0.65^{b}$	Dislike	$2.04{\pm}067^{b}$	Dislike
KM III	$1.60{\pm}0.65^{d}$	Very Dislike	2.56±0.76	Like	$2.16{\pm}0.68^{c}$	Dislike	$1.96{\pm}0.73^{c}$	Very Dislike
p value	p = 0.0001*		p = 0.798*		p = 0.0001*		p = 0.0001*	

Note: KM is an acronym of Keong Mas, numbers are followed by different superscript letters (a, b, c, d) show real differences

4. Discussions

4.1 Formulation Determination
Based on the nutrient content analysis of golden apple snail flour, omega-3 and omega-6 fatty acids also zinc and iron content could be seeded. In addition the heavy metals content such as lead has qualified based on government regulation which is safe to be an ingredient of making weaning food. Formulation determination is done based on the estimation of nutrient content calculations from the main ingredients of weaning food (rice flour, skimmed milk, powdered sugar, vegetable oil) and substitution ingredient (golden apple snail flour) that has been qualified. However, there are some nutrients which are not appropriate with the initial formulation determination. It occurs due to processing such as heating or interaction between the presence nutrients which can cause the existence of increase or decrease in the content of certain nutrients.

4.2 Nutrient Contents

4.2.1. Energy
Energy content of the 4th formula of weaning food ranges between 413.13-423.47 kcal/100 grams are qualified the specification requirement (400-440 kcal/100 grams)[9]. The energy content can fulfill 11% from the Recommendation Dietary Allowance (RDA) of children aged 7-11 months in one serving of 20 grams and in one day child can consume weaning food in four times so it could be fulfill 44% of RDA in one day. Children energy needs which amounted to 725 kcal/day[10] also fulfill by breastfeeding.

4.2.2. Protein
In this research, the protein content is not stable caused by the use of blender is not constant during the mixing process. Previous research shows that the milling process with blender could lower quality of protein[11]. Dissolve protein content of weaning food with the golden apple snail flour substitution ranged from 11.43-13.84 gram/100 grams that reach approximately 80% of the specification requirement of Indonesian Ministry of Health Decision is 15-22 grams/100 grams. The protein content can fulfill 12-15% RDA of children aged 7-11 months in one serving of weaning food (20 grams)[10].

Protein content decreased along with the increasing of golden apple snail flour substitution is caused by protein content of golden apple snail flour (12.73 grams/100 grams) which is lower than the protein content of skimmed milk (22.5 grams/100 grams), it does not comply with previous research (62%/100 grams).[12] In addition, the heating at temperatures $\geq 60^{o}C$ proved to be able to reduce the protein content in food by as much as $\pm$ 1 gram/100 grams[13]. Golden apple snail is known to contain essential amino acids such as histidine, leusin, isoleucine, threonine, methionine, lysine, phenylalanine, and valin, which were involved in the process of children linear growth[14][15].

4.2.3. Fat
The fat content of golden apple snail flour amounted to 13.67 grams/100 grams which higher compared to the fat content of skimmed milk that only amounted to 1 gram/100 grams. Golden apple snail flour substitution could increase the fat content of weaning food. However, formulation KM III (20.31 grams/100 grams) exceeded the specification requirement of 10-15 grams/100 grams, whereas other formulations are qualified[9].

In the making of weaning food, there is animal fat content from golden apple snail apple flour and vegetable fat content from vegetable oil. Vegetable fat such as phospholipids contain lecithin (phosphatidylcholines) that can increase the absorption of nutrients into the cell through its function as cell membrane[15] protector while the animal fat accounted to the most of the essential fatty acids such as omega-3 and omega-6 fatty acids.

4.2.4. Omega-3 and omega-6 fatty acids

Omega-3 fatty acids content of golden apple snail flour amounted to 1.50 grams/100 grams and omega-6 fatty acids amounted to 3.07 grams/100 grams or 10.96% omega-3 and 22.45% omega-6 of the fat content that high compared to with other foodstuffs such as poultry, meat and some types of freshwater fish[7][16]. The omega-3 content of weaning food can fulfill 38-51% RDA while the omega-6 can fulfill 20-35% RDA of children aged 7-11 months in one serving (20 grams)[10]. Omega-3 fatty acids have an importance role in the brain intelligence and bone growth[17][18] while the omega-6 fatty acids can prevent the energy loss during metabolic process as to prevent energy deficiency[19] and the role of both fatty acids in increasing the concentrations of Growth Hormone (GH)[19][20].

4.2.5. Carbohydrate

Carbohydrate content of weaning food range between 55-69 grams/100 grams, it contributes about 1.08-1.81 grams/100 grams of crude fiber. There is no specific specification of carbohydrate content in weaning food, but there is only specification requirement for the fiber content that should not exceed 5 gram/100 grams. Carbohydrate content of four treatments of weaning food can fulfill 13-16% RDA of children aged 7-11 months[10]. Carbohydrate content of weaning food is decreased along with the increasing of golden apple snail flour substitution caused by carbohydrate content of skimmed milk is higher than golden apple snail flour[21].

4.2.6. Fiber

Fiber content on animal food tends to be lower than on vegetable food. Fiber content of golden apple snail flour amounted to 0.87 grams/100 grams is equivalent to freshwater fish which is only about 0-1%[22]. The varieties of fiber golden apple snails among others are cellulose, hemicellulose and lignin. The fiber content of golden apple snail is influenced by their living environment. Rice field environment contain organic material such as lignin and cellulose[23]. The fiber content in 100 grams weaning food has qualified the specification requirement because it's not exceeding 5 grams/100 grams[9].

Children should be introduced by weaning food at aged > 6 months. At that age, children have digestive systems adaptation especially in the absorption phase in the intestine. The fiber cannot fully ferment. Therefore, there are two types of fiber digestion results in small intestine. Firstly, the fermented part which has the prebiotic function as oligosaccharides in breast milk, then the non-fermented part which has a role in the feces softening[21]. Fiber content of weaning food should be qualified the specification requirement of Indonesian Ministry of Health Decision because it can cause digestive system disorder and mal-absorption of certain nutrients in the intestine[24].

4.2.7. Water

Golden apple snail flour has high water content that is ± 4% or 4 grams/100 gram. Golden apple snail flour substitution can increase the water content of weaning food. The content of weaning food water ranges 5.48-6.53 grams/100 grams which tend to exceed the specification requirement limit of Kepmenkes (4 grams/100 grams). Besides the substitution of golden apple snail flour that contain high water content, the high water content of weaning food is caused by inadequate drying process which uses dry oven or hot air drying (HAD). The evaporation process of water in the dry oven is very minimal and slow which caused the moisture of the material tends to still high[13][25]. The high water content is related to the food shelf life because of the increase of microbial activity in that food. It can be overcome by storing the weaning food in the closed and airtight container, also save it in relatively low temperature place[26].

4.2.8. Ash

High content of mineral in golden apple snail flour is related to its ash content which is also quite high[27]. The ash content of golden apple snail flour is 18.78 grams/100 grams, which contribute about 3.25-4.16 grams/100 grams of weaning food. There is no specification requirement of Kepmenkes for ash content of weaning food, but according to the Indonesian National Standards (SNI-01-7111.1-2005) about instant powder weaning food, ash content in baby food aged 6-12 months does not exceed 3.5 grams/100 grams, so that the formulation KM III (15% of golden apple snail substitution) is not eligible[28].

4.2.9. Zinc

Zinc content of 100 grams golden apple snail flour is very high at 76.7 mg/100 grams which is equivalent to the oysters zinc content[29]. Golden apple snail flour substitution proved to increase the zinc content of weaning food, but inconsistency with the initial formulation estimation which is only the last formulation (KM III with 15% substitution of golden apple snail flour) in accordance with the specification requirement (2.5-4 mg/100 grams). However, the zinc content can fulfill 12-15% RDA of children aged 7-11 months in one serving (20 grams)[10].

The lack of zinc content could be caused by the mixing process with the other ingredients, especially with vegetable oil, which is high in phytoestrogen content like ligand, resulting in a binding process between zinc and ligand causing the zinc content of weaning food decreasing[30]. Children aged > 6 months require zinc intake with high bioavailability[31]. Animal proteins in golden apple snail tend to contain less phytic content, thus increasing the absorption of zinc in cell membrane[32][33].

4.2.10. Iron

The amount of iron content in 100 grams weaning food is 66 mg which is known to be higher than the previous research (48 mg/100 grams)[7]. The iron content of weaning food with golden apple snail flour substitution ranged 6.85-17.76 mg/100 grams that can fulfill 19-5-% RDA of children aged 7-11 months in one serving (20 gram)[10]. Iron content in the last two formulation (KM II and KM III which has 10% and 15% golden apple snail substitution) higher than the specification requirement (8 grams/100 gram)[9]. The results are also higher than the initial formulation estimation which ranged 3.98-10.52 mg/100 grams, it may due to the processing that tends to use iron-cookware[34].

The iron content of golden apple snail flour included in heme irons that can easily absorption sue to the low content of phytic in animal food[35]. Excessive intake of iron in children can lead to poisoning and indigestion such as nausea, vomiting and chronic diarrhea while for long-term can cause liver damage[34].

4.3 Acceptance Level

4.3.1. Color

The statistical results show that there are color differences from the four treatments of weaning food with golden apple snail flour substitution. The panelist tend to prefer formulation KM 0 (control) and very dislike the color of formulation KM III (15 % substitution of golden apple snail flour) because of the darker color (tend to approach the color of gray sand). This is because the color of golden apple snail meat is dark (blackish brown)[36] and the drying process causes the color of the golden apple snail flour to be darken[37].

4.3.2. Aroma

There is no effect of golden snail flour substitution with the aroma of weaning food. Almost all panelists like the four treatments of weaning food. The flavor of golden apple snail flour that was originally fishy can be removed after cooking process for ± 15-20 minutes. The typical scent of golden apple snail caused by compounds such as volatile fatty acids and amino acids, the aroma is the result of enzymatic reactions and the content of the components derived from the living environment[37]. The antioxidant content of the golden apple snail is known to prevent the reduction of aroma quality[38].

4.3.3. Texture

The golden apple snail flour is significantly affect the texture of weaning food. Panelists tend to like the texture of formulation KM 0 (control), the higher percentage of golden apple snail flour substitution the more thicken the texture. The coagulated texture is produced form actomiosin content of the golden apple snail protein[37]. It is also due to the high carbohydrate content in golden apple snail that decreases the protein thermodynamic affinity in the aqueous solution and increases the gelling capacity of the weaning food[39]. The cooking process of the weaning food with water causes the crystallization or starch to produce a coagulant texture. Therefore, this baby porridge should be consumed quickly in a warm state so that the texture is not increasingly agglomerate[25].

4.3.4. Flavor

In flavor assessment, the panelist also selected formulation KM 0 (control) as the most preferred flavor and formulation KM III (15% golden apple snail substitution) as the least favored weaning food flavor. There are significant differences in ach formulation. The combination of the savory flavor and the sweet taste of the powdered sugar is still no acceptable to young adult panelists. The process of cooking the weaning food causes the taste becomes more tasty, because the denaturation of protein into amino acids. The glutamate amino acids contained in the golden apple snail cause tasty taste in the weaning food[37]. The content of the glutamate amino acid in the golden apple snail is 10%/100 grams[14].

4.4 Formulation Selection

Weaning food with golden apple snail flour substitution has approached the specification requirement of Indonesian Ministry of Health Decision is formulation KM I with 5% of golden apple snail flour substitution, while other formulation (KM II and KM III) are not suitable because of the water, iron and fat content are exceeding form the specification requirement. For the acceptance level test of weaning food with substitution of golden apple snail flour, the formulation that could be acceptable besides the control formulation (KM 0) is the formulation KM I. So the selected formulation is KM I by 5% golden apple snail flour substitution. The formulation KM I is also selected based in the presentation graphic of the nutrient content data and the acceptance level test of weaning food presented in Figures 2 and 3.

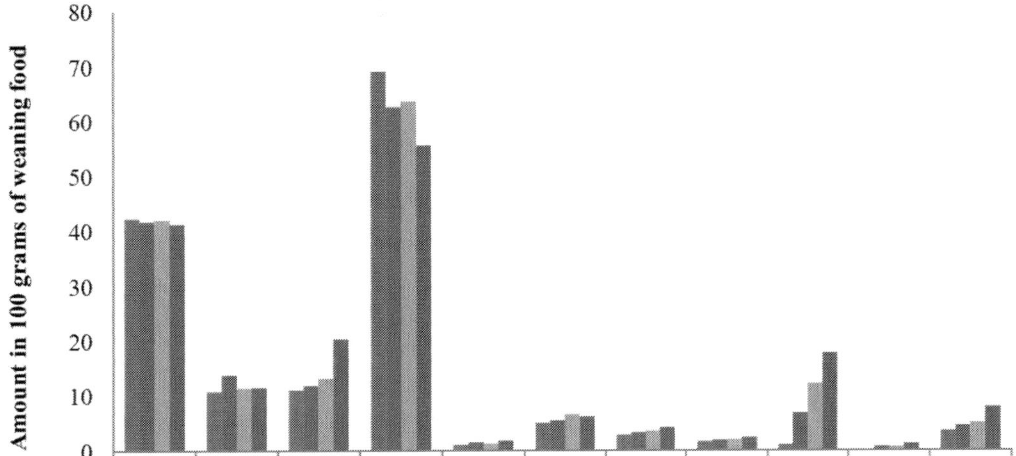

	Energy	Protein (g)	Fat (g)	CH (g)	Fiber (gr)	Water (g)	Ash (g)	Zn (mg)	Fe (mg)	O-3 (g)	O-6 (g)
KM 0 (0%)	42,34	10,77	11,04	69,22	1,08	5,01	2,84	1,64	1,04	0,2	3,57
KM I (5%)	41,8	13,84	11,88	62,61	1,55	5,48	3,25	1,88	6,8	0,71	4,52
KM II (10%)	42,1	11,42	13,16	63,69	1,21	6,53	3,58	1,98	12,18	0,75	5,04
KM III (15%)	41,3	11,49	20,31	55,66	1,81	6,22	4,16	2,39	17,76	1,28	7,89

Figure 2. Nutrients Content Analysis Test Results of Weaning Food

	Color	Aroma	Texture	Flavor
KM 0	3,24	2,72	2,96	3,08
KM I	2,08	2,56	2,16	2,2
KM II	2,12	2,52	2,44	2,16
KM III	1,6	2,56	2,16	1,96

Figure 3. Acceptance Level Test Results of Weaning Food

4.5 Contribution of Nutritional Adequacy

Contribution in one serving (20 gram) of weaning food with golden apple snail flour substitution has fulfill RDA of children aged 7-11 months. In addition, it is in accordance with commercial instant baby porridge product with chicken and fish flavor for children aged 6-12 months while higher on the omega-3 and omega-6 fatty acids. So that, the formulation KM I can compete with the other commercial products. The presentation of contribution data on nutritional adequacy is presented in Table 5.

Table 5. Presentation of Formulation KM I Data Adjusted with Indonesian Ministry of Health Decision and % Fulfillment of Children RDA

	Formulation KM I (5%)	Specification Requirement of Indonesian Ministry of Health Decision	% Fulfillment of RDA Children in One Serving (20 gram)
Energy	418.05 kcal	400-440 kcal	11.5%
Protein	13.84 gram	15-22 gram	15.3%
Fat	11.88 gram	10-15 gram	6.6%
Omega 3	0.71 gram	-	28.4%
Omega 6	4.52 gram	Min. 1.4 gram	20.5%
Carbohydrate	62.61 gram	-	15.2%
Fiber	1.56 gram	5 gram	3.1%
Water	5.48 gram	4 gram	-
Ash	3.25 gram	3.5 gram (SNI terms)	-
Zinc	1.88 gram	2.5-4 gram	12.5%
Iron	6.85 gram	5-8 gram	19.5%

5. Conclusion

1. Golden apple snail flour substitution with four treatments level (0%, 5%, 10%, 15%) could increase water, ash, fat, zinc, iron, omega-3 fatty acids and omega-6 fatty acids contents while decreases carbohydrate content. There is no relation between golden apple snail flour substitution and the content of energy, protein, fiber of weaning food.
2. There are differences in panelist acceptance level test on the color, texture and flavor of the weaning food with organoleptic properties test, while there is no effect on the assessment of aroma of the weaning food.
3. Selected formulation (KM I with 5% of golden apple snail flour substitution) can comply 80-100% the specification requirement of Indonesian Ministry of Health Decision and panelists of acceptance level test are reasonably accepted

Acknowledgement
The authors would like to thank the funding of this research by the PNBP Grant from the Faculty of Medicine, Diponegoro University, Semarang. The author would like to thank for the guidance and advice given by Mrs. Gemala Anjani, SP, MSi, PhD and Mr. Nuryanto, S.Gz, M.Gizi as mentors, and to Mrs. Ninik Rustanti, S.TP, M.Si as the examiner. The authors also thank the family and all those who have supported this research.

References

[1] UNICEF 2013 *Improving Child Nutrition: The achievable imperative for global progress* (New York: United Nations Publication) p 2-4
[2] Kementerian Kesehatan RI 2013 *Riset Kesehatan Dasar 2013* (Jakarta: Badan Penelitian dan Pengembangan Kesehatan)
[3] Kementerian Kesehatan RI 2015 *Buku Saku Pemantauan Status Gizi dan Indikator Kinerja Gizi Tahun 2015* (Jakarta: Direktorat Gizi Masyarakat Kemenkes RI)
[4] WHO 2010 *Nutrition Landscape Information System (NLIS) Country Profile Indicators: Interpretation Guide* (Switzerland: Department of Nutrition for Health and Development WHO)
[5] Kementerian Kesehatan RI 2012 *Pedoman Perencanaan Program: Gerakan Nasional Percepatan Perbaikan Gizi dalam Rangka Seribu Hari Pertama Kehidupan/Gerakan 1000 HPK* (Jakarta: Bidang Kesejahteraan Rakyat Kemenkes RI)
[6] Depkes RI 2006 *Pedoman Umum Pemberian Makanan Pendamping Air Susu Ibu (MP-ASI) Lokal Tahun 2006* (Jakarta: Ditjen Bina Kesehatan Masyarakat Depkes RI)

[7] Nurhasan M, Maehre H K, Malde M K, Stormo S K, Halwart M, James D, and Elvevoll E O 2010 *Journal of Food Composition and Analysis* **23** 205-213
[8] Ferreira A C and Capitulo A R 2016 *Studies on Neotropical Fauna and Environment* **52** 95-102
[9] Kementerian Kesehatan RI 2007 *Keputusan Menteri Kesehatan RI Nomor: 224/Menkes/SK/II/2007: Spesifikasi Teknis Makanan Pendamping Air Susu Ibu (MP-ASI) Bubuk Instan untuk Bayi 6-12 Bulan* (Jakarta: Kemenkes RI)
[10] Badan Pengawasan Obat dan Makanan 2014 *Mengenal Angka Kecukupan Gizi (AKG) Bagi Bangsa Indonesia* (Jakarta: InfoPOM)
[11] Augustin MA, Riley M, Stockmann R, Bennett L, Kahl A, Lockett T, Osmond M, Sanguansri P, Stonehouse W, Zajac I, and Cobiac L 2016 *Food Science and Technology* **56** 115-125.
[12] Hamid S A, Halim N R A and Sarbon N M 2015 *International Food Research Journal* **22** 1615-1623
[13] Chen X, Li X, Mao X, Huang H, Miao J, and Gao W 2016 *Food and Bioproducts Processing* **98** 266-274.
[14] Dao L, Xidong M, Hongmei S, Dongen G, Yexin Y, Xuejie W, Jianren L, Yinchang H, and Jiaen Z 2012 *Chinese Journal of Ecology* **31** 2004-2010.
[15] Semba R D, Shardell M, Sakr F A, Moaddel R, Trehan I, Maleta K M, Ordiz M I, Kraemer K, Khadeer M A, Ferrucci L, and Manary M K 2016 *EBioMedicine* **6** 246–252
[16] Oliviera I S d, Lourenco L d F H, Sousa C L, Joele M R S P, Ribeiro S d C d A 2015 *Food Control.* **50** 38-44
[17] Mulder K A, King D J and Innis S M 2014 *PLos One* **9** 1-9.
[18] Koren N, Simsa-Maizel S, Shahar R, Schwartz B, and Monsonego-Oman E 2014 *The Journal of Nutritional Biochemistry* **25** 623-633.
[19] Jumbe T, Comstock S S, Hahn S L, Harris W S, Kinabo J, and Fenton J I 2016 *PLoS One* **11** 1-19
[20] Young LR, Kurzer MS, Thomas W, Redmon J B, and Raatz S K 2013 *Nutr Res* **33** 565-571.
[21] Boehm G 2013 *Danone Research for Specialised Nutrition*
[22] PERSAGI 2005 *Daftar Komposisi Bahan Makanan (DKBM)* (Jakarta: Perpustakaan Nasional Republik Indonesia)
[23] Isnaningsih N R and Marwoto R M 2011 *Biodiversity Journal Portal* **10** 441-447
[24] Kranz S, Brauchla M, Slavin J L, and Miller K B 2012 *Advances in Nutrition* **3** 47-53
[25] Wang R, Chen C, Guo S 2017 *Journal of Food Engineering* **207** 81-89
[26] Muchtadi T R and Sugiyono 2013 *Prinsip Proses dan Teknologi Pangan* (Bandung: Alfabeta) p
[27] Sundari D, Almasyuhri, and Lamid A. 2015 *Media Libangkes* **25** 236-242.
[28] Badan Standarisasi Nasional. (BSN) 2005 *Makanan Pendamping Air Susu Ibu Bagian 1 : Bubuk Instan. (SNI 01-7111.4-2005)* (Jakarta)
[29] Gallagher M L 2008 *The Nutrients and Their Metabolism In Mahan LK, Escott-Stump S. Krause's Food, Nutrition, and Diet Therapy. 5th edition* (Philadelphia: Saunders)
[30] Sandstead H H and Freeland-Graves J H 2014 *Journal of Trace Elements in Medicine and Biology* **28** 414-417
[31] Umeta M, West C E, Verhoef H, Haidar J, and Hautvast J G A J 2003 *The Journal of Nutrition* **3** 1064-1068
[32] Pambudi N D, Salamah E, and Puwaningsih S 2011 *e-Journal Institut Pertanian Bogor* **1**
[33] Gregory P J, Wahbi A, Adu-Gyamfi A, Heiling M, Gruber R, Joy E J M, and Broadley M R 2017 *Global Food Security*
[34] Whitney E and Rolfes S R 2011 *Understanding Nutrition* (United States of America: Wardsworth Cengage Learning)
[35] Magallanes-López AM., Hernandez-Espinosa N, Velu G, Posadas-Romano G, Guadalupe Ordoñez-Villegas VM, Crossa J, Crossa J, ammar K, and Guzman C 2017 *Food Chemistry.* 2017 **237** 499-505
[36] Kaensombath L and Ogle B 2005 *Research Center of the National Agriculture and Foresty Institute of Laos*

[37] Mualim A, Lestari S, and Hanggita S 2013 *Fishtech* **2** 74-81.
[38] Giraud-Billoud M, Vega I A, Tosi M E R, Abud M A, Calderon M L, and Castro-Vasquez A 2013 *The Journal of Experimental Biology* **216** 614-622
[39] Chiu Y W, Wu J P, Hsieh T C, Liang S H, Chen C M, Huang D J. 2014 *Journal of Environtmental Biology* **35** 667-673

Phycocyanin stability in microcapsules processed by spray drying method using different inlet temperature

L Purnamayati[1], EN Dewi[1]and R A Kurniasih[1]

[1]Department of Fish Product Technology, Faculty of Fisheries and Marine Science, DiponegoroUniversity, Jl. Prof. H. Soedarto, SH, Tembalang, Semarang50275, Indonesia
Email: lukita_anandito@yahoo.com

Abstract: Phycocyanin is natural blue colorant which easily damages by heat. The inlet temperature of spray dryer is an important parameter representing the feature of the microcapsules.The aim of this study was to investigate the phycocyanin stability of microcapsules made from *Spirulina* sp with maltodextrin and κ-Carrageenan as the coating material, processed by spray drying method in different inlet temperature. Microcapsules were processed in three various inlet temperaturei.e. 90°C, 110°C, and 130°C, respectively. The results indicated that phycocyanin microcapsule with 90°C of inlet temperature produced the highest moisture content, phycocyanin concentration and encapsulation efficiency of 3,5%, 1,729% and 29,623%, respectively. On the other hand, the highest encapsulation yield was produced by 130°C of theinlet temperature of 29,48% and not significantly different with 110°C. The results of Scanning Electron Microscopy (SEM) showed that phycocyanin microcapsules with 110°C of inlet temperature produced the most rounded shape. To sum up, 110°C was the best inlet temperature to phycocyanin microencapsulation by the spray dryer.

Keywords: Inlet temperature, Spray Drying, Phycocyanin, Microcapsules

1. Introduction

Food products usually use color to improve their appearance, promotion, and selling. Several food products are known to contain synthetic food coloring and textile dyes which hazardous for health. Brilliant Blue FCF is the most common synthetic colorant used in food, where its use is mixed with tartrazine [1]. One of the efforts to reduce the using of synthetic coloring is by using safer natural coloring.

Spirulina contains phycocyanin blue pigment [2] which known as an antioxidant, able to reduce oxidative stress [3], free radical [4], and trap nitrate oxide [5]. However, phycocyanin will damage during storage as it sensitive to high humidity, light, temperature, and pH [6,7]. Therefore, phycocyanin is processed into microcapsule by microencapsulation technology.

Microencapsulation is a process that turns liquid into the powder with size around 1-1000 μm [8]. Spray drying is one of the methods used in the microencapsulation process because its equipment is easy to find and the cost is quite affordable. Spray drying is used in several microencapsulation processes [9, 10,11]. Spray drying turns the liquid form into solid (powder) through droplet dispersion from the product in the chamber which is in contact with hot air, where the coating material is added as encapsulating agents [12]. Maltodextrin is a coating material that commonly used because of its ability to protect bio-active compounds [13]. Carrageenan able to form viscous gel and stable [14].

The spray drying condition especially the inlet temperature able to affect the physical characteristic of produced microcapsules [15]. Otherwise, it is also expected to affect the bio-active compounds in microcapsules.The aim of this study was to investigate the effect of inlet temperature to the phycocyanin microcapsule stability which coated with maltodextrin and carrageenan.

2. Methods

2.1. Material

Phycocyanin extracted from *Spirulina* sp. powder (PT. Neoalga, Sukoharjo, Indonesia) [16] with modification. Coating materials maltodextrin DE 10 (CV. Multi Kimia Raya, Semarang, Indonesia) and k-Carrageenan (PT. Selalu Lancar Maju Karya, Jakarta, Indonesia).

2.2. Phycocyanin Microencapsulation

Phycocyanin microencapsulation [10] with modification was performed using homogenization of phycocyanin extract with the ratio 9% maltodextrin:1% k-carrageenan (w/v) toward the phycocyanin extract using homogenizer (Ultraturrax T50 Basic Ika Werke, Germany) at 4500 rpm for 2 minutes. The sample then put into spray dryer (Plant Lab, England) with different inlet temperature: 90°C, 110°C, and 130°C. Microcapsule produced then stored in dark glass bottle and coated with aluminum foil.

2.3. Yield

The microcapsule yield was calculated based on obtained phycocyanin microcapsule divided by total dissolved solid from phycocyanin microparticle solution multiplied by 100% [17,18].

2.4. Moisture Content (%)

Moisture content was measured using the gravimetric method by heating the sample at 105C for 24 hours. The moisture content percentage was obtained after reaching the constant weight [19].

2.5. Phycocyanin Concentration

Around 40 mg of phycocyanin microcapsule was diluted in 10 ml buffer phosphate pH 7, then homogenized and spectrophotometrically measured at 620 nm.

$$\%\ \text{Phycocyanin} = \frac{\text{A620 x (ml solvent)}}{\text{3,39 x (mg sample) x sample's dry weight}} \text{ x } 100\%$$

3,39 = C-phycocyanin coefficient at 620nm wavelength [20, 10]

2.6. Encapsulation Efficiency/ EF (%)

Encapsulation efficiency calculated based on percent ratio of phycocyanin level before coated and phycocyanin level after microencapsulation process [6].

2.7. Morphology

Phycocyanin microcapsule was observed its morphology by using Scanning Electron Microscopy [21].

2.8. Statistical Analysis

This study using Completely Randomized Design with one factor which is inlet temperature. The data was obtained from assay triplication and analyzed using SPSS 17. The advanced test was performed using Tukey analysis.

3. Results and Discussions

Table 1. Phycocyanin concentration and phycocyanin microcapsule moisture content

No	Phycocyanin Microcapsule with inlet temperature	Yield (%)	Moisture Content (%)	Phycocyanin Concentration (%)	Encapsulation Efficiency/ EF (%)
1	90°C	18,240±1,115^{a}	3,500±0,361^{b}	1,729±0,041^{c}	29,623±0,700^{c}
2	110°C	27,700±0,548^{b}	3,200±0,200^{ab}	1,458±0,050^{b}	24,963±0,873^{b}
3	130°C	29,480±1,506^{b}	2,633±0,153^{a}	1,078±0,047^{a}	18,457±0,798^{a}

Note: The data was average of triplication ± Standard deviation
Different superscript on the same column indicates significant different on level α 0,05

3.1. Yield

Phycocyanin microcapsule with maltodextrin-carrageenan as coating materials processed by spray drying method with different inlet temperature produced microcapsule yield. At 90°C of inlet temperature, the lowest phycocyanin microcapsule was produced for about 18,240% and its result was significantly different with the yield processed at 110°C and 130°C. Moreover, microcapsule produced at 110°C and 130°C was not significantly different 27,700% and 29,480%, respectively. This result was similar with Sarabandi et al. [22] that the higher inlet temperature used, the higher microcapsule yield produced.

Carrageenan as coating materials was easy to form gel [23]. At 90°C of Inlet temperature, phycocyanin microcapsule emulsion can't dry perfectly and partly stuck at the spray dryer chamber wall which results in the lower microcapsule yield. Suzihaque et al. [24] stated that the loss of spray drying result due to it stuck at the chamber wall.

3.2. Moisture Content

Moisture content is an important factor on a food product. Moisture content also an important factor to keep the quality of microcapsule. The higher moisture content on microcapsule depends on drying condition especially temperature [19]. The higher inlet temperature will decrease the moisture content of microcapsule [22]. Table 1 showed that the higher inlet temperature used on spray drying process produced a low moisture content of phycocyanin microcapsule. At 90°C of inlet temperature, the moisture content of phycocyanin microcapsule was 3,500% and it was significantly different with phycocyanin microcapsule produced at 130°C. This result was lower compared to Dewi et al. [10] where phycocyanin microcapsule with 80°C of inlet temperature produced 8,36% of moisture content. Inlet temperature produced hot air which able to evaporate the moisture level on the droplet and turn it into powder [25].

3.3. Phycocyanin Concentration

Table 1 showed the phycocyanin concentration on microcapsule where the higher inlet temperature used, the lower phycocyanin microcapsule concentration produced. The decreasing of phycocyanin content due to phycocyanin was susceptible to temperature, where the increasing of temperature able to decrease the phycocyanin content [7]. Phycocyanin microcapsule concentration with 90°C of inlet temperature was 1,729% and this result was significantly different with phycocyanin concentration on phycocyanin microcapsule with higher inlet temperature. This result was lower compared to Dewi et al. [10] that produced 2,83% of phycocyanin concentration at 80°C of inlet temperature.

3.4. Encapsulation Efficiency/ EF (%)

Total phycocyanin which able to encapsulated was showed with encapsulation efficiency [18]. Table 1 showed that phycocyanin microcapsule with higher inlet temperature produces low encapsulation efficiency. Encapsulation efficiency related to phycocyanin concentration. Phycocyanin microcapsule produced at 90°C of inlet temperature had the highest encapsulation efficiency about 29,623% and it was significantly different with phycocyanin microcapsule with higher inlet temperature. This result was in line with phycocyanin concentration where microcapsule with 90°C of inlet temperature also produced the highest phycocyanin concentration. This showed that low inlet temperature able to maintain the phycocyanin content.

Encapsulation efficiency produced on this research was higher than Dewi et al. [10] please use onlu number as suggested for IOP format that produced 12,89% of encapsulation efficiency. However, this was lower and inversely proportional with Venil et al. [26] please did the same as suggested which perform pigment microencapsulation with carrageenan as the coating materials with 140°C of inlet temperature and produced 2,054% of encapsulation efficiency and keep increasing along with the increasing of inlet temperature until 180°C.

3.5. Morphology

Figure 1. Scanning Electron Microscopy of phycocyanin microcapsule (a) 90°C inlet temperature (b) 110°C inlet temperature (c) 130°C inlet temperature should be mention in main text

Phycocyanin microcapsule was measured using Scanning Electron Microscopy (SEM). Phycocyanin microcapsule SEM images showed in Figure 1. On all SEM images, showed that most of the phycocyanin microcapsule had an irregular shape of microcapsule and cracked. Phycocyanin microcapsule with 110°C and 130°C of inlet_temperature, produced more rounded shape than microcapsule with 90°C of inlet temperature. The crack appearance on microcapsule was affected by the liquid evaporation speed on the spray dryer. The similar result also showed by microcapsule produced at 180°C of inlet temperature with different coating materials [26].

4. Conclusion

Phycocyanin microencapsulation with different inlet temperature showed that the higher inlet temperature used,will increase the microcapsule yield, where at 130°C of inlet temperature produced the highest yield about 2,948% and was not significantly different with 110°C of inlet temperature but significantly different with 90°C of inlet temperature. This result was inversely proportional to moisture content, phycocyanin concentration, encapsulation efficiency of 2,633%, 1,078%, and 18,457%, respectively. As can be seen from its morphology, phycocyanin_microcapsule at 110°C of inlet temperature produced the most rounded shape. In the end, 110°C was the best inlet temperature to produce phycocyanin microencapsulation by the spray dryer.

5. Acknowledgement

This research funded by Diponegoro University through research scheme of Riset Pengembangan dan Penerapan (RPP) year 2017.

References

[1] Saleem N, Umar Z N, and Khan S I 2013 Survey on the use of synthetic food colors in food samples procured from different educational institutes of Karachi City. *The Journalof Tropical Life Science*. **3**(1) 1-7.

[2] Walter A, de Carvalho J C, Soccol V T, de Faria A B B, Ghiggi V and Soccol C R 2011 Study of phycocyanin production from *Spirulina platensis*under different light spectra. *Brazilian Archives of Biology and Technology*. **54**(4) 675-682.

[3] Bertolin T E, Farias D, Guarienti C, Petry F T S, Colla L M and Costa J A V 2011 Antioxidant effect of phycocyanin on oxidative stress induced with monosodium glutamate in rats. *Brazilian Archives of Biology and Technology*.**54**(4) 733-738.

[4] Seo Y C, Choi W S, Park J H, Park J O, Jung K H and Lee H Y 2013 Stable isolation of phycocyanin from *Spirulina platensis*associated with high-pressure extraction process. *Int. J. Mol. Sci.* **14** 1778-1787.

[5] Kamble S P, Gaikar R B, Padalia R B and Shinde K D 2013 Extraction and purification of C-phycocyanin from dry spirulina powder and evaluating it's antioxidant, anticoagulation and prevention of DNA damage activity. *Journal of Applied Pharmaceutical Science*.**3**(08) : 149-153.

[6] Yan M, Liu B, Jiao X and Qin S 2014 Preparation of phycocyanin microcapsules and itsproperties. *Food and Bioproducts Processing*.**92** 89-97.

[7] Chaiklahan R, Chirasuwan N and Bunnag B 2012 Stability of phycocyanin extracted from *Spirulina* sp.: influence of temperature, pH, and preservatives. *Process Biochemistry*.**47** 659–664.

[8] Chanana A, Kataria M K, Sharma Mand Bilandi A 2013 Microencapsulation: advancements in applications. *International Research Journal of Pharmacy*.**4**(2) 1-5.

[9] Hernandez-Carranza P, Lopez-Malo A and Jimenez-Munguia M T 2014 Microencapsulation quality and efficiency of *Lactobacillus casei* by spray drying using maltodextrin and vegetable extracts. *Journal of Food Research*. **3**(1) 61-69.

[10] Dewi E N, Purnamayati L and Kurniasih R A 2016 Antioxidant activities of phycocyanin microcapsules using maltodextrin and carrageenan as coating materials. *Journal of Technology.* **78**(4-2) 45-50.

[11] Lewandowski A, Czyzewski M and Zbicinski I 2012 Morphology and microencapsulation efficiency of foamed spray-dried sunflower oil. *Chemical and Process Engineering.* **33**(1) 95-102.

[12] Costa S S, Machado B A S, Martin A R, Bagnara F, Ragadalli S A and Alves A R C 2015 Drying by spray drying in the food industry: Micro-encapsulation, Process Parameters, and Main Carriers Used. *African Journal of Food Science.* **9**(9) 462-470.

[13] Tan S P, Kha T C, Parks S, Stathopoulos C and Roach P D 2015 Optimizing the ncapsulation of an aqueous bitter melon extract by spray-drying. *Foods.* **3** 400-419.

[14] Rojas-Nery E, Garcia-Martinez I and Totosaus A 2015 Effect of emulsified soy oil with different carrageenans in rennet-coagulated milk gels. *International Food Research Journal.* **22**(2) 606-612.

[15] A-Sun K, Thumthanaruk B, Lekhavat S and Jumnongpon R 2016 Effect of spray drying condition on physical characteristics of coconut sugar powder. *International Food Research Journal.* **23**(3) 1315-1319.

[16] Chaiklahan R, Chirasuwan N, Loha V, Tia S and Bunnag B 2011 Separation and purification of phycocyanin from *Spirulina* sp. using a membrane process. *Bioresource Technology.* **102** 7159–7164.

[17] Nunes I L and Mercadante A Z 2007 Encapsulation of lycopene using spray drying and molecular inclusion processes. *Brazillian Archives of Biology and Technology an International Journal.* **50**(5) 893-900.

[18] Dewi E N, Purnamayati L and Kurniasih R A 2017 Physical characteristics of phycocyanin from spirulina microcapsules using different coating materials with freeze drying method. *IOP Conf. Series: Earth and Environmental Science.* **55** 1-7.

[19] Mehrad B, Shabanpour B, Jafari S M and Pourashouri P 2015 Characterization of dried fish oil from menhaden encapsulated by spray drying. *Aquaculture, Aquarium, Conservation & Legislation International Journal of The Bioflux Society.* **8**(1) 57-69.

[20] Boussiba S and Richmond A E 1979 Isolation and characterization of phycocyanins from the blue-green alga *Spirulina platensis*. *Archives of Microbiology.* **120** 155–159.

[21] Devi N and Kakati D K 2013 Smart porous microparticles based on gelatin/sodium alginate polyelectrolyte complex. *Journal of Food Engineering.* **117** 193-204.

[22] Sarabandi K, Peighambardoust S H and Shirmohammadi M 2014 Physical properties of spray dried grape syrup as affected by drying temperature and drying aids. *International Journal of Agriculture and Crop Sciences.* **7**(12) 928-934.

[23] Dewi E N, Darmanto Y S and Ambariyanto 2012 Characterization and quality of semi-refined carrageenan (SCR) products from different coastal waters based on fourier transform infrared technique. *Journal of Coastal Development.* **16**(1) 25-31.

[24] Suzihaque M U H, Hashib S A and Ibrahim U K 2015 Effect of inlet temperature on pineapple powder and banana milk powder. *Procedia Social and Behavioral Sciences.* **195** 2829-2838.

[25] Phisut N 2012 Spray drying technique of fruit juice powder: some factors influencing the properties of product. *International Food Research Journal.* **19**(4) 1297-1306.

[26] Venil C K, Khasim A R, Aruldass C A and Ahmad W A 2016 Microencapsulation of flexirubin-type pigment by spray drying: characterization and antioxidant activity. *International Biodeterioration & Biodegradation.* **113** 350-356.

Freshwater Clams (*Pilsbryoconcha Exilis*) as an Potential Local Mineral Sources in Weaning Food to Overcome Stunting in Grobogan, Central Java, Indonesia

S R Putri[1,*], Gemala Anjani[1,2], Hartanti Sandi Wijayanti[1,2], Nuryanto[1,2,]**

[1]Department of Nutrition Science, Faculty of Medical, Diponegoro University, Jl Prof. Soedarto,SH, Tembalang, Semarang 50275, Indonesia
[2]Center of Nutrition Reaserch (CENURE), Diponegoro University, Jl Prof. Soedarto,SH, Tembalang, Semarang 50275, Indonesia
*Presenting author
**Corresponding author; E-mail: nyt_gizi@yahoo.com; Tel: 0813-9024-0288

Abstract. Based on a preliminary survey in Grobogan Central Java, one of the local food that has a prospective nutrient content developed is freshwater clams. Results of PSG report in 2015, especially in Grobogan district with stunting problem of 31.5% consisting of 7.2% very short and 24.1% short. From the stunting event factor, the role of weaning food in the form of baby porridge is the most important component in the fulfillment of infant nutrition. The research is divided into two stages of the research is the initial stage to make the formulation process by way of freshwater clams, nutrient analysis, and formula calculation. The second stage is the process of formulation weaning food by way of formula making, nutrient analysis, and acceptance test of weaning food on the baby slurry is substituted by the freshwater clams. The influence statistically different substitution of freshwater clams on protein content (p=0.014), fat (p=0,041), carbohydrate (p=0,039), water (p=0,0001), ash (p=0.0001), iron (p=0.0001), and zinc (p=0.0001). There is no influence substitution of freshwater clams on energy content (p=0,129) and fiber (p=0,225). The percentage of favorite level on the acceptance test of color (p=0,005) and taste (p=0,006) was statistically different, but there was no difference flavour (p=0,913) and texture (p=0,198). freshwater clams flour is rich in iron and zinc content, so it can serve as local food, especially in Grobogan Central Java as a stunting handling. Obtained from the results of several formulations substituted of freshwater clams in formulation 2 is a formula that can meet the standards of KEPMENKES except on the content of protein and moisture content. Formula 2 acceptable to the panelists, except for the assessment of taste.

Keywords: freshwater clams, Weaning food, Value of nutrients, Stunting

A. Introduction

Nutrition problems in Indonesia are increasingly complex, where the problem of malnutrition is a problem that must be handled. The National Medium Term Development Plan for 2010-2014 states that improving the community's nutritional status is one of the priorities by reducing the prevalence of short toddlers (stunting) to 32%. Based on the results of Riskesdas Central Java, especially Grobogan District, is one of the problems with high stunting occupy the second position in the province. The result of the report from Riskesdas of Central Java Province in 2013 has stunting rate of 36.7% which consists of

16.8% very short and 19.8% short [1]. According to the Determination of Nutritional Status (PSG) 2015 Grobogan District [2], with stunting problems of 31, 5% consisting of a very short 7.2% and a short 24.1 %. Public health problems are categorized as severe if the short prevalence is 30-39% and is seriously categorized when the prevalence is ≥40% [3].

Of the factors that may affect the incidence of stunting, the role of the family in the care of the child greatly determines the nutritional status and development of children. Maternal care patterns closely related to the pattern of consumption of child feeding, the quality and quantity of breastfeeding (breast milk) and weaning food (food companion breast milk). The quality and quantity of good weaning food is the most important component, with nutrient content that plays a role in child growth [4]. Weaning food is a food given to under-fives under two years, especially 6 to 24 months. Provision of weaning food over 6 months due to the fulfillment of breast milk alone can not meet the needs so the need for introduction to toddlers to family food. Breastfeeding foods can be baby porridge, which contains various nutrients such as carbohydrates, proteins, fats, vitamins, and other minerals [5].

According to the Decree the Minister of Health the Republic of Indonesia, the manufacture of baby milk powder in powder form must meet the nutritional needs, with energy content per 100 grams of 400 to 440 kcal; Protein content of 15 to 22 grams; Fat 10 to 15 grams; Fiber with a maximum content of 5 grams; Iron content of 5 to 8 mg; Zinc of 2.5 to 4 mg; and water content up to 4 grams [6]. The preparation of weaning food generally comes from a mixture of rice flour, skim milk, refined sugar and vegetable oil [7]. To increase the nutrient content, these substances can be substituted with locally sourced foods of protein and substances iron. One of the local foods sourced from protein and iron that can be utilized as weaning food based on preliminary survey in Grobogan area that is shellfish (*Pilsbryoconcha exilis*) [8]. Local food is shellfish is an appropriate alternative way as one way of fulfilling the nutrition of toddlers In the area.

freshwater clams belong to Pelecypoda animals that can live in rivers, freshwater, lakes, and ponds [9]. This material has a protein content of 7.37 grams and iron of 31.02 mg [8]. Protein in freshwater clams is one of the nutrients Macro consisting of more complete amino acids than vegetable proteins [10]. Freshwater clams are used by the community as food and as a source of animal protein, usually available in the freshest form ready to be cooked and processed [8].

Based on the above background, the research was conducted to find out the nutrient content of freshwater clams flour and powder formulation of weaning food, as well as acceptance test of infant porridge in weaning food substituted freshwater clams (*Pilsbryoconcha exilis*) with various formulations as the manifestation of breastfeeding food stunting Children Below Two Years.

B. Methods

The research belongs to the field of food production or food technology. The research is divided into two stages, namely the initial stage of determining the formulation by way of freshwater clams, nutritional analysis, and formula planning. The second stage is done formulation process by making the formulation of weaning food substituted flour freshwater clams, nutrient analysis, and panelist acceptance test. Implementation of the research was carried out for seven months, which was conducted in Integrated Laboratory of Diponegoro University, Faculty of Agricultural Technology Study Program of Food Technology, Soegijapranata Catholic University, and Central Laboratory of Pollution Prevention Industry (BBTPPI). Each treatment performed at least two repetitions on the Bomb Calorimeter (energy), gravimetric (ash and fiber), AAS/Atomic Absorption Spectrometry (iron and zinc), ICP/Inductively Copled Plasma (lead), and carbohydrate by difference (carbohydrates). Minimum of three repetitions on Bradford (protein), oven (air), and soxhlet (fat).

The initial phase of the research was done by determining the formulation process with the process of processing the freshwater clams into flour from Kedung Ombo Reservoir, Grobogan Central Java. Initial process is done that is shell washing twice as many stages, drain the shell, boil the shell for 30 minutes, separate the shell with mussel meat, cut the shell meat thinly, dry the mussel meat with dry oven with temperature 50-60°C for 24 hours, Grind the dried mussels using a grinder, and sift the flour using a 1000 micrometer sieve. Flour that has been through the stage until sieving, the process of testing the content of nutrients in the form of energy, protein, fat, carbohydrates, fiber, ash, water, zinc, iron, and lead levels. Planning uses calculations from the analysis of nutrient content of flour shellfish with a list of food composition (DKBM). The initial composition of the baby slurry used was 35% rice flour, 50% skim milk, 10% vegetable oil, and 5% refined sugar [11]. So that four treatments (F1/control, F2, F3 and F4) were obtained in the form of a combination of substitution of freshwater clams flour by 5%, 10% and 15%.

The second stage with the formula of planning results, then carried out the stage of formulation. Prior to the process of mixing the formulation, the flour of rice was processed by gelatinization. Where the gelatinization stage of rice flour is to weigh the rice and air flour with a ratio of 1: 4, then stir the dough until evenly, heat the dough until the form of a gray transparent gel, put the dough on the baking sheet to do drying process using dry oven with 50- 60◦C for 24 hours, the dry dough of the grinding process using a grinder, and the sieving process using a 1000 micrometer. Furthermore, the process of making weaning food powder through a stage such as weighing the materials according to the formulation plan, mixing the material using blander for 5 minutes, drying the formulation using dry oven for 1 hour. Formulations with various concentrations, conducted testing of nutrient analysis in the form of energy content, proteins, fats, carbohydrates, fiber, air, ash, zinc, and iron.

Assessment of acceptance of weaning food on infant porridge with various formulations on color, aroma, texture, and taste using hedonic test with 4 rating scale 1 (very like), 2 (like), 3 (dislike), 4 (very dislike). Assessment of acceptance of baby slurry is ready to be served to 25 rather skilled panelists who are undergraduate students of Nutrition Science Program Diponegoro University of Semarang. The process of serving the infant porridge substituted shellfish by brewing the baby powder with the addition of water (treatment 1: 8) for 8 minutes using a stove until smooth textured porridge in the absence of clots.

The data that has been collected is then analyzed using a computer. The effect of variation of freshwater clams substitution on nutrient content was tested using one way ANOVA with 95% confidence degree followed by Tukey posthoc test to know the real difference between treatment, as well as acceptance test was done statistical analysis using Friedman test.

C. Result

1. Determination of formulation

Based on a preliminary survey in the Grobogan area on the freshwater clams (Pilsbryoconcha exilis) is an appropriate alternative way as one way of fulfilling the nutrition of children under five in the area. This material has high iron and zinc content [12]. The following is the result of laboratory tests on freshwater clams flour presented in Table 1.

Table 1. The content of freshwater clams flour (100 grams)

Nutrition	Content	Unit
Energy	350.4	Kkal
Protein	8.1	Gram
Fat	5.8	Gram
Carbohydrate	60.8	Gram
- Fiber	2.2	Gram
Ash	19.5	Gram
Water	3.7	Gram
Irin	170	Mg
Zinc	48.6	Mg
Lead	0.4	ppm*

*Terms of lead level not more than 1.14 ppm[6]

The formulation aims to increase the nutritional value of weaning food, therefore the freshwater clams flour iron and zinc are substituted with skim milk. Based on calculations from the analysis of freshwater clams flour content in preliminary test and list of food ingredients composition (DKBM), it is determined substitution of freshwater clams flour by 5%, 10%, and 15% which is attached in Table 2.

Table 2. Composition of MP-formula formulation of baby powder

Information	F1 (Control)	F2	F3	F4
freshwater clams flour	0%	5%	10%	15%
Skim milk	50%	45%	40%	35%
Rice flour	35%	35%	35%	35%
Vegetable oil	10%	10%	10%	10%
Refined sugar	5%	5%	5%	5%
Total	100%	100%	100%	100%

Description: F is a formula

IOP Conf. Series: Earth and Environmental Science **116** (2018) 012077 doi:10.1088/1755-1315/116/1/012077

2. Weaning Food with Substitution of Freshwater Clams Flour

a) Nutrient Contents in Weaning Food

Tabel 3. Results of nutrient content between treatments

Treatment	Energy (kkal)	Protein (gram)	Fat (gram)	Carbohydrate (gram)	Fiber (gram)	Ash (gram)	Water (gram)	Iron (mg)	Zinc (mg)
F1 (Control)	424.5±1.9	10.7 ± 0.5^{a}	11.9 ± 0.7^{ab}	67.1 ± 1.1^{a}	2.09±0.6	2.69 ± 0.0^{a}	5.23 ± 0.1^{a}	1.1 ± 0.0^{a}	2.8 ± 0.0^{a}
F2	425.4±4.8	13.1 ± 0.8^{b}	10.6 ± 0.5^{a}	65.2 ± 0.9^{a}	2.70±0.6	3.38 ± 0.0^{b}	5.36 ± 0.1^{a}	7.9 ± 0.0^{b}	2.8 ± 0.0^{a}
F3	410.6±5.1	12.1 ± 0.8^{ab}	11.6 ± 0.3^{abc}	61.8 ± 1.0^{b}	3.65±0.2	3.94 ± 0.0^{c}	6.11 ± 0.2^{b}	14.5 ± 0.1^{c}	3.7 ± 0.0^{b}
F4	410.1±10.4	11.5 ± 0.1^{ab}	12.1 ± 0.5^{ac}	63.7 ± 1.3^{ab}	1.69±1.2	4.53 ± 0.0^{d}	6.25 ± 0.2^{b}	20.5 ± 0.1^{d}	4.7 ± 0.0^{c}
Terms of KEPMENKES	400-440	15-22	10-15	-	Maximum 5	-	Maximum 4	5-8	2.5-4.0
p value	0.129	0.014	0.041	0.039	0.225	0.0001	0.0001	0.0001	0.0001

Information : F is a formula, Figures followed by different superscript letters (a, b, c, d) show real differences
*Testing with one way ANOVA

There was influence of freshwater clams substitution with 0%, 5%, 10%, and 15% on protein content (p=0.014), fat (p=0.041), carbohydrate (p=0.039), water content (p=0.0001), ash (p=0.0001), iron (p=0.0001) and zinc (p=0.0001). Meanwhile, in the energy content (p=0.129) and fiber (p=0.225) there was no influence with the freshwater clams substitution of 0%, 5%, 10%, and 15%.

The requirements of KEPMENKES regarding the specification of weaning food on formula 1 (F1 / Control) that have fulfilled the requirements include energy, fat, fiber, iron and zinc content. Formulation 2 (F2) with 5% shellfish substitution includes energy, fat, fiber, iron and zinc content. The shell substitution is 10% in formulation 3 (F3), including energy, fat, fiber and zinc content. Meanwhile, the formulation of 4 (F4) with the shellfish substitution of 15% includes the energy, fat, and fiber content. In carbohydrate content, there is no special requirement on the specification but the highest content of the freshwater clams substitution in formulation 2 and the lowest content of formulation 4.

b) *Receive power test*

Formulation 2 (F2) with 5% freshwater clams substitution includes energy, fat, fiber, iron and zinc content. The shell substitution is 10% in formulation 3 (F3), including energy, fat, fiber and zinc content. Meanwhile, the formulation of 4 (F4) with the freshwater clams substitution of 15% includes the energy, fat, and fiber content. In carbohydrate content, there is no special requirement on the specification but the highest content of the freshwater clams substitution in formulation 2 and the lowest content of formulation 4.

Table 4. Results of trial analysis of weaning food baby slurry with subchiefs of freshwater clams

Treatment	Color		Taste		Flavour		Texture	
	Average	Info.	Average	Info.	Average	Info.	Average	Info.
F1 (Control)	$2.92{\pm}0.57^{a}$	Like	$2.56{\pm}0.76^{abc}$	Like	2.64±0.63	Like	2.60±0.81	Like
F2	$3.16{\pm}0.74^{b}$	Like	$2.20{\pm}0.70^{a}$	Dislike	2.76±0.72	Like	2.92±0.86	Like
F3	$2.44{\pm}0.71^{b}$	Dislike	$1.80{\pm}0.76^{b}$	Dislike	2.68±0.98	Like	2.68±0.80	Like
F4	$2.44{\pm}0.82^{a}$	Dislike	$2.56{\pm}0.71^{c}$	Like	2.68±0.85	Like	2.48±0.82	Dislike
p value	p=0.005		p=0.006		p=0.913		p=0.198	

Information: F is a formula, Figures followed by different superscript letters (a, b, c, d) show real differences

The color of the infant formula formulation in weaning food made from the freshwater clams flour, obtained in formulation 2 (F2) is favored by the panelists of the other formulations. In the taste of a baby porridge, in formulation 1 (F1 or control) in the absence of addition of freshwater clams flour and formulation 4 (F4), it may be accepted by the panelists included in the likes category. Meanwhile, on the aroma of baby porridge on the weaning food all data formulations received by the panelist with the likes category. Similarly, in the texture, obtained formulations 1 to 3 can be favored by the panelists, while in the formulation of 4 with subchiefs of flour shellfish of 15% is not favored by the panelists. The percentage of substitution preference level on color (p=0.0005) and taste (p=0.0006), statistically obtained there was a significant difference in subchief of flour shell flour. While in the aroma (p=0.913) and texture (p=0.198), there is no difference in the preferred level of substitution of freshwater clams flour.

D. Discussion

1. Determination of formulation

Food diversity program is currently being encouraged, one of the efforts that can be done is to make food that has high nutrient content. One type of product that has a high iron and zinc content of freshwater clams flour. The content of freshwater clams flour per 100 grams is 350.4 kcal on energy, 8.1 gram protein, 5.8 gram fat, 60.8 gram carbohydrate, 2.2 gram fiber, 19.5 gram ash content, 3.7 gram water content, 170 mg iron content, 48.6 mg zinc , and 48.6 ppm lead. Research on lead levels has met the requirements according to KEPMENKES, where the lead content (metal contamination) for weaning food does not exceed 1.14 ppm.

One of the efforts of food diversity in this research is to make a product of weaning food in the form of baby porridge as an alternative food of baby companion, especially in Grobogan area in handling stunting problem, which has high nutritional and economic value. Before making a formula product, it is necessary to have a formulation planning stage. The result of the analysis of the content of the freshwater clams flour that has been obtained is then calculated with a list of food composition (DKBM) to find out the content contained in skimmed milk powder, refined sugar, vegetable oil, and rice flour.

2. Weaning Food with Substitution of Freshwater Clams Flour

a) Nutrient Content

Energy

The energy content of baby powder in weaning food with substitution of freshwater clams ranges between 410,11-424,53 kcal per 100 gram. Based on the infant powder requirements of weaning food according to KEPMENKES in which the energy content of 400-440 kcal in 100 grams, all of the infant powder treatment in the weaning food meets the requirements [6]. Formula 2 is the formula with the highest energy content, while formula 4 is the content The lowest energy of the four formulations.

Energy requirements in infants are 24-30% greater than in infants ages 3-5 months [13]. Infants aged 6-12 months, complementary feeding is a transitional period from milk-based intakes to semi-solid foods. So that breastfeeding food given to the baby must meet the requirements of energy adequacy [14]. In the human body, energy is shattered due to the burning of carbohydrates, proteins and fats. Thus in order to be fulfilled his energy needs needed intake with enough nutrients into the body. If the body is in a state of lack of energy source substances such as carbohydrates and fats, then the body will use proteins to form energy and replace its main function as a builder substance. In toddlers this condition has a disruption to growth [15].

Protein

The results showed that infant powder protein content of breastfeeding food (weaning food) with freshwater clams substitution ranged from 10.71 to 13.18%. Where is the requirement of protein content of weaning food according to KEPMENKES equal to 15-22 gram in 100 gram of baby powder in weaning food [6]. Protein content in baby powder produced not yet able to reach weaning food milk supply, but can pass the protein content of 50- 70% of the standard weaning food. This is because the protein content of the freshwater clams flour in 100 grams only contributed by 8.1 grams. Formula 2 is the most approximate formulation of KEPMENKES, and where formulation 1 is the formula with the lowest protein content.

According to the Nutrition Adequacy Rate (AKG) at the age of 6 months and over the needs of infants 12-18 grams per day, where the formulation 2 on the weaning food that has been made to help the fulfillment of infant nutrition [16]. Age 6 to 12 months, the baby is still breastfed where ASI contents can contribute 1.5 grams of protein in 100 grams so that the child's needs according to AKG from formulation 2 can be fulfilled [17].

Warming can cause both expected and unexpected reactions. This reaction is influenced by the temperature and duration of heating, pH, the presence of oxidizing agents, antioxidants, radicals and other active compounds, especially carbonyl compounds. Most food proteins are denatured when heated to moderate temperatures (60-90◦C) for one hour or less so as to decrease the protein content [18].

Age 6 to 12 months is a critical period, where rapid growth occurs at the age of 6-12 months and at this time the baby begins to rely on supplementary food [19]. Protein, in addition to the energy source also serves as a body building agent and regulatory substances in the body. One of the main functions of protein in the body is the growth and maintenance of tissues. In the body, there is a growth hormone called somatotrophic hormone (SH) or somatotropin [20]. It can be concluded that, lack of protein intake in children is more at risk of stunting [21].

Fat

The specification of fat content in weaning food according to KEPMENKES requires the fat content of 10-15 grams in 100 grams of baby powder in weaning food [6]. The fat content of baby powder in weaning food made from freshwater clams ranges from 11.96-12.12%. The baby powder formulations produced from the four formulas contained the fat content in the vulnerability required by KEPMENKES. The highest fat content significantly in powdered infant infant powder was found in the formula 4 (F4) of 12.12 grams with the composition of freshwater clams flour by 15%. While on the lowest fat content in the formulation 2 (F2) with freshwater clams substitution of 5%.

Fat is an efficient source, the infant's limited stomach capacity energy density in weaning food can be achieved by the addition of fat. Essential fatty acid content is essential for brain growth and childhood mining [21]. The results of the study show that the proportion of children with low fat intake stunting more than the proportion of children with adequate fat intake.[22] The presence of fat may also be helpful in the absorption of fat soluble vitamins (Vitamins A, D and E) so low fat intake can lead to deficiency of fat soluble vitamins [18].

Carbohydrate

The quality of good food intake is an important component in the diet of children because it contains a source of macro nutrients such as carbohydrates, where macro nutrients, especially carbohydrates play a role in the growth of children. The main function of carbohydrates is to provide energy for the body, because carbohydrates are the main source of energy in the body. Carbohydrates for infants are a major role, so carbohydrates should at least meet 52-54% of energy needs [23].

Carbohydrate content is calculated using carbohydrate by difference method, where the calculation is strongly influenced by other nutrient content such as water content, ash, fat, protein and fiber. Results of carbohydrates with various formulations of 61.80-67.14 grams per 100 grams of baby powder on the weaning food. There is no requirement regarding the range of carbohydrate content in infant powder according to KEPMENKES. Formulation 2 (F2) has higher carbohydrate content than other substituted freshwater clams formulations, obtained 65.25% carbohydrate content with freshwater clams substitution of 5%. The lowest carbohydrate content is found in formulation 3 (F3).

Fiber

Weaning food fiber content in infant powder with various formulations obtained a coarse fiber content of 1.69-3.65 grams per 100 grams of packaging. According to KEPMENKES, maximum fiber content of 5 grams per 100 gram of packaging. Obtained from all four formulations have fulfilled the requirements of weaning food on baby powder. Formulation 3 (F3) is the highest formula, whereas in formula 4 is a formula with substitution of freshwater clams on the lowest fiber content.

Fiber is mostly contained in vegetables, fruits, cereals and seeds. The content of crude fiber in baby food according to KEPMENKES does not exceed 5 grams per 100 grams of baby powder packaging. When a food product contains high crude fiber, the food product is relatively harmful because the crude fiber is potentially disruptive in the absorption of nutrients such as proteins, fats, vitamins and minerals the body needs. High fiber content can cause the baby's stomach to feel full faster, because the fiber has a high water absorption [24].

Ash

The ash content in the foodstuff has a related with the mineral content which is an inorganic substance. The burning process causes the organic material to burn out, while the inorganic material does not burn out. The result of this combustion is referred to as ash [25]. The ash content of baby food powder of breast milk, with substitution of flour shell of river equal to 2,69-4,53%. There is no ash content requirement according to KEPMENKES, but there are conditions under SNI 01-7111.42005 where ash content does not exceed from 3.5 gram per 100 gram per pack. Obtained, the formulation of 2 (F2) of 3.38 gram has fulfilled the requirement of weaning food according to SNI 01-7111.42005.26 Infant powder with 15% shellfish flour formulations on Formulation 4 (F4) contains the highest ash content of 4.53%. While on the formula 1 (Control) is the lowest ash content.

Ash content has something to do with the minerals of a material. Ash is an organic waste residual organic waste material of an organic material [27]. The higher temperature of the vacuum drying will increase the ash content due to the corresponding temperature increase in a drying process does not result in the destruction of nutrients of food, especially minerals, only reduces the water content of food alone [28].

Water

Water content in food is very influential on the quality of a food. Where water content is very important in determining the durability of food, because it will affect the physical properties, chemistry, microbiological changes, and changes of enzymatic food [29]. Water content in infant milk powder feeding food range between 5.23-6.25%. Standard of weaning food according to KEPMENKES in 100 gram that is maximal 4 gram.

Of the four formulations of water content exceeding KEPMENKES standards because the raw materials used contain 3.7 grams of water, so it can be expected to have a shorter shelf life. This can be handled by making air-tight packaging such as aluminum foil to decrease mold growth and pass through heating to reduce water content [29]. Where the highest water content is found in formula 4 whereas formula 1 (Control) has the lowest moisture content.

Iron

The results of iron content obtained ranged from 1.13-20.53 mg per 100 grams of baby powder on the weaning food substituted freshwater clams flour. Terms of weaning food in the form of baby powder, iron content ranges from 5-8 grams per 100 grams. Of the four formulas, there are two formulations that qualify for the manufacture of weaning food in infant powder that is formula 1 and formula 2 (F1/control and F2). According to the Nutrition Adequacy Ratio (AKG), children aged 7 months and over require an iron intake of 7 mg per day where in formlation 2 with substitution of freshwater clams flour by 5% can meet the needs of children in the day with breastfeeding. The higher the formulation of freshwater clams flour makes the iron content increased, this is because the iron content contained in freshwater clams flour for 170 mg in 100 grams. Obtained in Formulation 4 (F4) has the highest iron content, whereas in formula 1 (Control) has the lowest iron content.

Iron deficiency is also associated with decreased immune function as measured by changes in some components of the immune system that occur during iron deficiency. The consequences of immune function changes are susceptible to infectious diseases. Iron deficiency anemia is most often found specifically in children under five, the state of anemia will slowly inhibit the growth and development of intelligence, children will be more susceptible to disease due to decreased endurance. Iron deficiency anemia in children can delay physical and mental development and decrease resistance to disease [30].

Zinc

Specification of infant powder making on weaning food, zinc content according to KEPMENKES of 2.5-4,0 mg in 100 gram of packing. Zinc infant powder levels with substitution of the resulting freshwater clams range from 2.8-4.76 mg per 100 grams. Therefore, from the four formulations that fulfill the requirements of KEPMENKES are formulations 1 to 3. The content of 15% freshwater clams flour substitution in Formulation 4 (F4) contains the highest iron content, whereas in formula 1 (Control) and 2 have low iron content.

According to the nutritional adequacy rate (AKG) of zinc content in infants aged 7 months and over by 3 mg per day, where formulations with 5% and 10% of freshwater clams substitutes (F2 and F3) can meet the standard needs of infants in a day [15]. Zinc Important in terms of structure and function of biomembrane, zinc becomes an important component of several enzymes that regulate cell growth, protein and DNA synthesis, energy metabolism, regulation of gene transcription, hormone levels and growth factor metabolism. Not only that, zinc plays a role in human immune function. The content of zinc is an essential nutrient and has received quite serious attention lately. Zinc deficiency in childhood can cause stunting [30].

b) Selection Formulation According to KEPMENKES

Based on the figure below, it shows that formula 2 with subchannel freshwater clams has nutrient content according to KEPMENKES standard except on protein and water content. Of the four formulations, there is a formulation with the highest nutritional content of formula 4. However, in the formula 4 protein content, moisture content, iron, and zinc can not meet the requirements according to KEPMENKES. Ash content in the formulation 4 according to SNI in 2005, ash content exceeds the requirements limit.

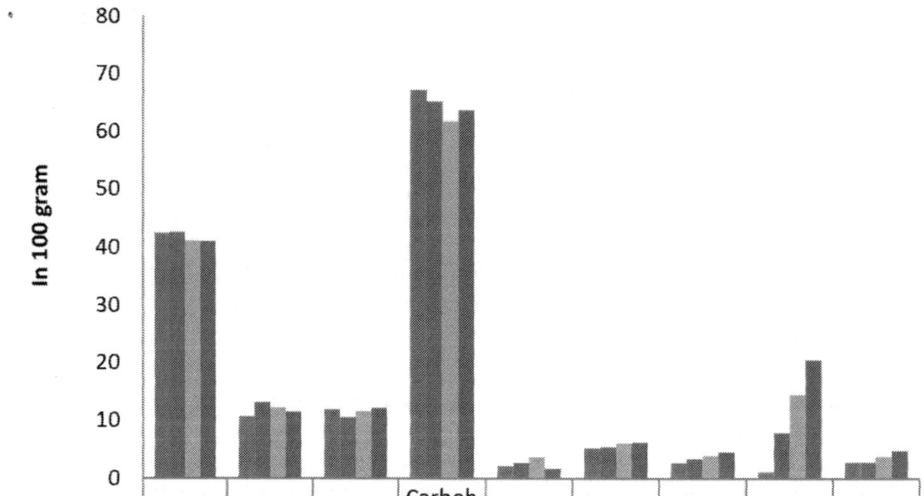

	Energy	Protein	Fat	Carbohydrate	Fiber	Water	Ash	Iron	Zinc
F1 (Control)	42,45	10,71	11,96	67,14	2,09	5,23	2,69	1,13	2,8
F2	42,55	13,18	10,6	65,25	2,7	5,36	3,38	7,91	2,8
F3	41,06	12,16	11,65	61,8	3,65	6,11	3,94	14,55	3,79
F4	41,01	11,5	12,12	63,72	1,69	6,25	4,53	20,53	4,76

Figure 1. Results Formulation of Nutritional Ingredients Weaning Food on Baby Powder

c) *Receive power test*

1) *Color*

Receive power test, color is the first time assessment of the eyes to the product. Color holds the most important role, if the color of a food is unattractive even though its complete nutritional content will reduce the level of panelist acceptance of the product. So color is the quickest and easiest response to give a good impression of a food product [31].

Breastfeeding foods in baby porridge generated from various variations of freshwater clams flour are found in Formulation 2 (F2) with 5% freshwater clams flour composition preferably panelists. The addition of flour shellfish streams in the formulations 3 and 4 (F3 and F4), will cause the color of the baby porridge tend to be darker. This is because the colors contained in the freshwater clams flour tend to be brownish. Not only that, when the process of cooking protein that joins the sugar in a hot atmosphere will cause the color to darken this is due to the reaction maillard [31].

2) *Flavour*

In flavour is one of the organoleptic aspects that greatly affect the acceptability or the level of the panelist's preference for a product. That is, a product although physically has an attractive color, good texture and aroma, but if it feels bad then cause the product is rejected or not favored by panelis [32].

Based on the assessment of the flavur aspect, substitution of the freshwater clams with a concentration of 15% is highly favored by the panelists. The panelist acceptance decreases in substitution of freshwater clams flour with a concentration of 5-10% because the addition does not significantly affect taste by a rather well-trained panelist (adult panelist). Thus, the higher concentration of freshwater clams flour added to weaning food formula, will affect the panelist's acceptance of the freshwater clams flour.

3) *Aroma*

According to the food industry, the aroma is important because one of the decisive factors in the quality of a food product. The presence of aroma substances that dissolve in water and fat that is volatile (yawn) [33]. Panzer preferences level to the scent, all formulations in the baby slurry can be accepted by panelists with the likes category because all dominant formulations are not fishy from freshwater clams flour. The aroma produced in each formulation, resulting in a fragrant aroma. This is influenced by the basic material used in the formulation of skim milk. The addition of skim milk contains lactose and lactic acid, where lactic acid is one of the odor-forming components. Although skim milk is milk that has reduced fat content but the fat content is still there to play a role in the odor formation of baby slurry products in complementary foods ASI [34].

4) *Texture*

Suitable breastfeeding supplements for ages 6 months and above are semi-solid textured foods [35]. According to KEPMENKES standards set on infant powder standards, when mixed with water will produce fine pulp in the absence of clots with viscosity that allows spoon-feeding. The resulting form on the 100% powder passes the filtering test using a 1000 micrometer filter.

Brewing process, baby powder added with water (ratio 1: 8). The selection of baby powder comparison with water was done experiment first. Where in the ratio of 1: 6 used other commercial products, weaning food with subchief of freshwater clams has a texture on the flour with a low level of maturity because the product of weaning food in this study has not been at the stage of instant baby porridge. While at a ratio of 1:10, the texture of weaning food formula product becomes more fluid. So we get 1: 8 ratio when mixed with water obtained texture of fine porridge and does not leave a clot with viscosity that allows the giving by using a spoon to the baby.

The acceptance rate of baby porridge texture on complementary feeding of ASI acceptable to panelists is found in formulations 1 to 3 under the likes category. While the formulation of 4 levels of panelist preferences decreased to dislike, this is due to the texture produced in the formulation 4 tend to be thick compared to other formulations that tend to be smooth.

5) *Selection of Formula According to Panelists*

Based on the figure below, it can be concluded that formulation 2 (F2) has the most acceptable power test result by a rather well trained panelist that is adult panelists except on the assessment of taste. The dominant assessment of panelist flavor tends to choose formula 1 (Control) and 4.

	Color	Taste	Flavour	Texture
F1 (Control)	2,92	2,56	2,64	2,6
F2	3,16	2,2	2,76	2,92
F3	2,44	1,8	2,68	2,68
F4	2,44	2,56	2,68	2,48

Figure 2. Acceptance Results of Weaning Food Panel on Baby Porridge

3. Contribution to Nutritional Adequacy

Contribution in a serving quantity (20 gram), formulation 2 (F2) can meet 11.74% energy adequacy, 14.67% protein adequacy, 5.89% fat adequacy, 15.91% carbohydrate adequacy, 5.4% fiber adequacy, 18.67% zinc adequacy, and 22.28% Iron adequacy. According to WHO, adequate breastfeeding at the age of 6 months and above can contribute 413 kcal of energy so that with the provision of weaning food in formulation 2 can meet the needs of children in day.[36] The results of nutritional contributions by AKG and KEPMENKES in Formulation 2 (F2) Briefly presented there are Table 5.

Table 5. Nutritional Contribution of Formulation 2 per Serving Meal of Baby Porridge (20 gram)

Formulation 2 (F2)				
Content	Results of nutrients (100 gram)	Terms of KEPMENKES	Results of nutrients (20 gram)	% AKG (20 gram)
Energy	425.46 kkal	400-450 kkal	85.09 kkal	11.74%
Protein	13.18 gram	15-22 gram	2.64 gram	14.67%
Fat	10.60 gram	10-15 gram	2.12 gram	5.89%
Carbohydrate	65.25 gram	-	13.05 gram	15.91%
- Fiber	2.70 gram	Maximum 5 gram	0.54 gram	5.4%
Water	5.36 gram	Maximum 4 gram	1.07 gram	-
Ash	3.38 gram	-	0.68 gram	-
Zinc	2.80 mg	2.5-4.0 mg	0.56 mg	18.67%
Iron	7.91 mg	5-8 mg	1.58 mg	22.28%

E. Conclusions

1. The influence of substitution of freshwater clams with treatment level 0%, 5%, 10%, and 15% on protein content (p=0.014), fat (p=0.041), carbohydrate (p=0.039), water content (p=0.0001), ash (p=0.0001), iron (p=0.0001) and zinc (p=0.0001). Meanwhile, in the energy content (p=0.129) and fiber (p=0.225) there was no influence with the freshwater clams substitution of 0%, 5%, 10%, and 15%.
2. The result of nutrient content of the four formulas, substitution of 5% freshwater clams flour in formula 2 (F2) can fulfill the requirement of baby powder nutrient specification in weaning food according to KEPMENKES except protein and water content.
3. Obtained from various subchiefs of freshwater clams flour in weaning food baby slurry, level of panelist acceptance of color; aroma; and texture of formula 2 (F2) with 5% freshwater clams substitution preferred by panelists. However, on the taste assessment, the freshwater clams substitution of formula 2 (F2) is less favored by a slightly trained panelist, since the addition does not significantly affect the taste.

Acknowledgments

The authors would like to thank the funding of this research by the PNBP Grant from the Faculty of Medicine, Diponegoro University, Semarang.

References

[1] Ministry of Health RI 2013 Basic Health Research Results (RISKESDAS) Central Java Province Ministry of Health Republic of Indonesia Jakarta
[2] Ministry of Health of the Republic of Indonesia 2015 Pocket Book Monitoring of Nutrition Status and Indicators of Nutrition Performance 2015 Ministry of Health Republic of Indonesia Jakarta
[3] Rahayu LS 2011 Relationship hight Parent With Change Status Stunting from Age 6-12 Months to 3-4 Years Postgraduate Program Gajah Mada University Yogyakarta
[4] Istiftriani N 2011 Relationship of Breast Feeding and Other Factors with Baduta Nutrition Status in Depok Sub-District Pancoran Mas Sub-District Depok Year 2011 Faculty of Public Health University of Indonesia Depok
[5] Husna EA, Affandi DR, Kawiji, Anandito RBK 2012 Journal of food Teknosains **1** Avaliable online at www.ilmupangan.fp.uns.ac.id
[6] Ministry of Health of the Republic of Indonesia 2017 Decree of the Minister of Health of the Republic of Indonesia Number: 224 / Menkes / SK / II / 2007 Concerning Technical Specification of Breast Milk Companion Food (MP-ASI) Jakarta
[7] Larasati D, Wahjuningsih SB, Pratiwi E 2008 Journal of Food Technology and Agricultural Products 5 112-118 Semarang State University of Semarang
[8] Anisa, Adi AC 2013 Effect of Added Meat Meat Shrimp (Pilsbryoconcha Exilis) and Carrot (Daucus Carota L) to the Acceptance and Nutritional Feeding of Mocaf (Modified Cassava Flour) Nutrient Fluid Indonesian Nutrition Media 9 84-88
[9] Suhardi 1983 Avertebrata evolotion UI Press Jakarta
[10] Muchtadi D 1989 Evaluation of Food Nutrition Value Inter-University Center for Food and Nutrition, Bogor Agricultural University Bogor
[11] Andarwulan N and Fatmawati S 2004 Proceedings of the National Seminar and Congress of Indonesian Food Technology Experts Association
[12] Prihartini W 1999 Diversity of Types and Ecobiology of Freshwater Shellfish Family Unionidae (Mollusca: Bivalvia) Multiple Situ in Bogor District and Municipality Bogor Postgraduate Program Bogor Agricultural University
[13] World Health Organization 2000 Complementary Feeding: Family Foods for Breastfed Children Department of Nutrition and Development WHO Geneva
[14] Baisden B, Bunyapen C, Bhatia J 2008 Feeding The Premature Infant In: Berdanier CD, Dwyer

J, Feldman EB, editors Handbook of Nutrition and Food 2nd ed New York: CRC press **79** 236, 259, 281
[15] Almatsier S 2006 Basic Principles of Nutritional Sciences Gramedia Pustaka Utama Jakarta
[16] Ministry of Health of the Republic of Indonesia 2014 nutritional adequacy rate Jakarta
[17] LIPI 2004 List of Foodstuff Composition LIPI Jakarta:
[18] Sundari Dian, Almasyhuri, Lamid Astuti 2015 Effect of Cooking Process on Food Nutrient Composition of Biomedical Protein Source and Basic Technology of Health Ministry of Health of the Republic of Indonesia
[19] Parízková J 2010 Nutrition, Physical Activity, and Health in Early Life 2nd edition CRC Press USA
[20] Guyton, C 2007 Arthur Physiology of Medicine EGC Jakarta
[21] Ernawati F, Rosmalina Y, Permanasari Y 2013 nfluence of maternal protein intake and baby's length born to stunting event in 12 month old child in Bogor District 36 1-11
[22] Torun B, PSW Davies, MBE Living Stones, M Paolisso, R Sackett, GB Spur and MPE de Gusman 2006 European Journal of Clinical Nutrition (EJCN) **50**
[23] Xiaoli W, Beng H, Sufang G, Shusheng L, Wenyuan Z, Yan W 2009 The Nutrition Society **12** 1991—1998
[24] Bahmat Dian, Bahar Herwanti, Jus'at Idrus 2010 Jurnal Fakultas Kesehatan Masyarakat Universitas Esa Unggul
[25] Sulaeman A 2003 Development of Toddler Food Product Formula From Base Material of Flour Cassava Flour and Banana Flour IPB Bogor
[26] National Standards Agency 2005 Maternal Breeding Food (MP-ASI) Part 1: Instant Powder SNI 01-7111.42005 Jakarta
[27] Sudarmadji S, B Haryono, Suhadi 1989 Food and Agricultural Food Analysis Publisher Liberty Yogyakarta In collaboration with Inter-University Center for Food and Nutrition University of Gadjah Mada Yogyakarta
[28] Harris, R. S dan E. Karmas 1989 Evaluation of Nutrition In Food Processing ITB Bandung
[29] Winarno FG 2002 Kimia Pangan dan Gizi PT Gramedia Pustaka Utama Jakarta
[30] Anugraheni HS, Kartasurya MI 2012 Journal of Nutrition College **1** 30-37
[31] PJ Fellows 2000 Food Processing Technology Principle and Practice Wood Publishing in Food Science and Technology Cambridge England
[32] Nadimin, Ayu SD, Rauf S 2012 Consumer Acceptance Against Dodol Multigizi Media Food Nutrition Vol.XIII Edition 1
[33] Marliyati and Anna S 2002 Household Food Processing Department of Education And Culture Directorate General of Higher Education Inter-University Center for Food and Nutrition Institution Institut Pertanian Bogor Bogor
[34] Hadiwiyoto 1994 Testing Quality of Milk and Processe PT.Liberty Jogyakarta
[35] Baxter SD 2005 Introducing Solid Foods to Infants In Bhatia J, Perinatal Nutrition Optimizing Infant Health and Development Marcel Dekker New York
[36] Dewey KG and Brown KH 2003 SAGE Journals **24**

Content Heavy Metal Pb, Cd In *Perna viridis* And Sediments *In Semarang Bay*

D Suprapto[1], S Suryanti[1], N Latifah[1]

[1]Department of Aquatic Resources, Faculty of Fisheries and Marine Sciences, Diponegoro University Jl. Prof Soedarto, SH, Tembalang, Semarang, Indonesia 50275

Email: djs_suprapto@gmail.com, suryantidr@gmail.com, nurullatifah271@gmail.com

Abstract. Waste disposal from human activities, generally contain heavy metals such as Pb and Cd which derived from industrial activities. The aims of the study were to know the concentration of Pb and Cd heavy metals contained in *Perna viridis* tissue, sediment and water at Semarang Bay. This study was conducted in May 2017 at Semarang Bay. - Samples were collected using purposive sampling method. The heavy metal content in the water and clam was observed using- APHA method and was analyzed using AAS (Atomic Absorption Spectrophotometer). The results showed that concentration of heavy metal of Pb in the water was 0.00-50.5mg/L and the Cd content was of 26.9-51.7 mg/L, whereas the concentration of Pb in the sediment is 445.5-2.053.0mg/L and Cd 963.3-2,150.0 mg/L. Pb content in soft tissue of *Perna viridis* - is 67.1-1.933.9 mg/L and the concentration of Cd was 203.5-5.787.3 mg/L. The analysis of Pb and Cd in seawater, sediment and soft tissue of *Perna viridis* according to Enviroment Ministerial decree (KepMenLH) number 51 of 2004 and applied by NOAA 1999 does not exceed the quality standard, that meant that the *Perna viridis* has been contaminated by metal Pb it is controversial with the above sentence and Cd. It concluded that the metal content of Pb and Cd in *Perna viridis* tissue exceeds the quality standard, so it is not suitable to be consumed, especially in high quantity

Keywords: *Perna viridis*, Pollution, Heavy metal, Sediment.

1. Introduction

Population growth and industrial development along the coast have greatly affected the pollution loads to the coastal zone as well as marine environment [1]. In the case of Indonesia, the high land-based activities are major source of the pollutions of the coastal waters. Based on the studies conducted in coastal areas, heavy metal contaminations are among the persistent issues in the country [2]. These heavy metals have brought the negative impacts to the health of marine living resources and consequently the people who consume seafood products which collected from those areas [3]

Industrial heavy metal pollution as well as from domestic sources can affect coastal waters, while the heavy metals were positively correlated to the organic matter within the local waters [4]. Heavy metals from waste of fuels burning have influenced to the fish production of the fisherman and it seems that, this is due to the Pb contamination [5]. It is likely that the activities in the port basin indicated as the source of heavy metal pollution in the area [6].

Metals are natural constituents of the sediments in the coastal zones and they may be as an oligo-elements, such as zinc, which is the natural element essential for living organisms if present in small amounts, or do not have any known biological roles, as in the case for the lead. However, both can become toxic for living organisms at high concentrations [7]. Much of those human activities located in the fluvial watersheds and in the margins of estuaries being important areas for the concentration of contaminants, due to the coastal industrial activities and human settlement.

Lead, or Plumbum (Pb), has atomic number 82, atomic weight 207.2 which describe as a small size molecule, solid and gray [8]. Lead is widely used in industries such as the production of storage batteries for cars, ammunition, cable coatings, pipes, dyes and blends in the manufacture of ceramic coatings and fuels [9]. Lead could penetrate to the human body through inhalation, digestion, skin pores and eyes [8].

Semarang estuary water is narrow and muddy in the bottom, dirty and has high suspended load [10]. These area mostly used as aquaculture activities and disposal of various waste from land. If this conditions occurs in the long term, it will lead to heavy metal pollution. Based on the monitoring of some researchers in 1984/1985 and 1986/1987, it were found that coastal waters of Semarang were polluted by heavy metals such as Pb and Cd [2]. It could be found in the waters of Babon and Seringan estuaries [11].

Pollution of heavy metals in this water can cause heavy metal accumulation in sediments as well as in aquatic biota. Sediments in the waters of Tanjung Emas harbor, Dermaga (harbor) Nusantara, and Pulau Panjang contains Cd, Cr, Cu, Pb and Zn while in the bay of Semarang containing Hg metal that possibly could accumulates in the body of *Perna viridis* [12]. This study aimed to determine the heavy metal concentration in the body/the tissue of *Perna viridis,* in the sediment and in the column of the water.

2. Materials and methods

The samples of water, sediment, and *Perna viridis* tissue were collected at the location of research the waters of Tambak Lorok, Semarang and was conducted in May 2017. There were two replications, where the first was taken at low tide time and the second one was taken at high tide moment. Samples were obtained in two replication, at low tide and high tide.

Water samples were collected using Nansen Bottle Sampler from the bottom of the water column, while the sediment samples were taken using a grab sampler. *Perna viridis* soft tissue samples were classified into - three different shell size i.e. small size (< 2.5 cm), medium size (2.5-3 cm), and large size (3-5 cm).

The level of heavy metal Pb and Cd in those samples were determined by ICP (Inductively Coupled Plasma) [13-14]. Environmental condition variables measured were: water temperature, water salinity, pH, dissolved oxygen, and water depth.

3. Result and discussion

Based on the measurements of heavy metals on water samples during - high -and low tide -, showed that the heavy metal concentration of Pb at 4 stations with 3 replications in waters lorok semarang exceeds the quality standard, as well as -the heavy metal concentration of Cd in all stations (Table 1).

Table 1. Result of heavy Metal concentration of Pb and Cd Measurement in Waters

Station	Pb (mg/L)		Cd (mg/L)	
	Low tide	High Tide	Low Tide	High Tide
I	50.5	34.6	42.8	26.9
II	1.6	-23.5	44.7	28.8
III	-34.1	17,7	36.1	29.8
IV	-31.8	-53.1	51.7	30.6
KepMen LH no. 51	≥0.008	≥0.008	≥0.001	≥0.001

Heavy metals in seawater can be divided into two major components, i.e., metals in soluble form and metals that bound to suspended particles or sediments (phytoplankton, zoo- plankton, debris, clay and silts) [15]. Toxic trace metals in coastal waters, especially Hg, Pb and Cd have been measured in some components of coastal waters but knowledge on other trace metals such as As, Cr and Sn is still very limited [16]. Understanding of toxic metals in water could be mainly represented by measuring metals insolution form which was commonly found to be very low in concentration [17].

The overall result shown that Cd metal contamination was greater than Pb. This is thought to be derived from port and the industrial activities around the port that dumps Cd waste into the waters.

Please give some references to ensure that this is a real situation in the field. This high contamination level can be dangerous for marine life. Since the waters is widely used for fish cultivation, it caused less productive ponds, and consequently the growth rate of fish is not optimal [18]. Heavy metals naturally have low concentrations in the waters. High and low concentrations of heavy metals are caused by the amount of metal waste that enters into the waters as well as the pH, since it will affect the solubility of the heavy metals in the waters [19].

Lead solubility is low in waters so that the level of lead concentration in the water is relatively low [17]. Both the content and toxicity of lead are influenced by hardness, pH, alkalinity, and oxygen concentration. Lead toxicity is lowered down by the increase of hardness as well as dissolved oxygen [20]. It is categorized into IV-A group in periodic table of chemical elements, atom number of 82 with atom weight of 207.2, concentrated in deposit, such as metal ores. In battery industry, lead is used as a grid, compound of bismuth metal (Pb-Bi) with ratio 93:7. Pb compound and Cr, Mo and Cl are widely used as pigment chrome, lead silicate compound (Pb-silicate) used as one of ceramic polished materials, and used as refractory material [21].

Based on the Quick Reference Tables Screening or Squirts [22], should be numbered sediments in coastal waters and the West Flood Canal estuary to identify potential impacts of waste impact on coastal resources and habitats. Pb and Cd metal content in sediment which exceeds the limit will affect the biota. This influence according to [23], in addition can be deadly, also sublethal that inhibits growth, development, and reproduction, causing morphological changes, and change organism behavior.

Table 2. Heavy Metal Pb and Cd Measurement Results in Sediment Samples

Station	Pb (mg/L)		Cd (mg/L)	
	Low tide	High Tide	Low Tide	High Tide
I	651.6	783.2	2150	1614.4
II	1675.7	445.5	1434.1	1254.2
III	633.1	2053	1733.6	1348.8
IV	321.9	683.9	963.3	1546
NOAA 1999	30.24	30.24	0.676	0.676

From the measurement of heavy metal content of Pb and Cd on sediment samples can be seen that both exceed the existing quality standard. In addition, the content of Pb and Cd at the study sites is very high (Table 2). This is generally due to the development of industry in a region. Sediments can be used as an indicator due its role as a sink of various contaminants. Measurement of heavy metals in sediments provided better indicator than those in solute form (in water), and the concentration of heavy metals in sediments was relatively stable. In contrast to the toxic trace metals in solute form, the concentrations of metal in sediments showed an increasing trend during the period of 1985-2000 and concentrations of trace metal such as Pb and Cd were relatively higher compared to background levels, especially in coastal areas bordering the provincial cities in northern coast of Java Island, namely Jakarta, Semarang, and Surabaya [4]. Increasing industrial areas,harbour developments and population growth contribute to the increasing trend of metal contaminants during early 1980s to 1990s [17].

Highly toxic trace metals in sediments indicated that metals in water column might mostly be found to particulate matter and settled not far from the shore. Once contaminants settled in sediments, they are potentially absorbed by benthic organisms that can transfer the metals from sediments to higher trophic levels of food chain [24]. In the long run, this will become potential problems to the quality of pelagic system. Recent studies showed that metal concen- trations in sediments do not fully provide information on metal bioavailability. Metal bioavailability to benthic organ- nisms depends on various factors, such as geochemical his- tory of sediments[25], animal behaviour[5] and exposure duration. Early works on metal speciation in Semarang coastal waters were done by [26] in Semarang harbor.

Based on the measurements of heavy metals that have been done on *Perna viridis* soft tissue, it is found that the heavy metal content of Pb and Cd in all stations exceeds the quality standard (Table 3).

Table 3. Heavy Metal Pb and Cd Measurement Results in *Perna viridis* Soft Tissue Samples

Station	Pb						Cd					
	Low Tide			High Tide			Low Tide			High Tide		
	S	M	L	S	M	L	S	M	L	S	M	L
I	128.6	125.2	188.4	296.6	210.5	318.7	237.7	512.4	498.1	1972.3	1515.8	644.1
II	1933.9	730.7	216	67.1	90.4	239.6	1248.7	5787.3	1233.4	355.7	314.2	422
III	138.1	208.9	131.9	542.8	270.3	124.9	307.7	431.6	219.9	210.6	659.9	244
IV	642.3	286.5	333.1	122.7	101.5	149.3	203.5	3918	5475.6	229.1	242.8	255
SK DirJen POM	2	2	2	2	2	2	0.1	0.1	0.1	0.1	0.1	0.1

- S : Smal (< 2,5 cm)
- M: Medium (2,5-3 cm)
- B : Big (3-5 cm)

Table 4. Environmental Variabels Measurement Results

Station	Temperature (°C)		pH		Salinity (ppt)		Depth (cm)		DO (ppm)	
	Low Tide	High Tide	Low Tide	High Tide	Low Tide	High Tide	Low Tide	High Tide	Low Tide	High Tide
I	27.8	27.7	7.92	8.12	30	31	350	500	3.29	6.22
II	27.5	27.8	8.05	8.01	30	30	550	700	6.82	7.94
III	27.4	28	8.09	7.96	30	32	550	600	6.3	8.2
IV	27.1	27.9	8.02	8.11	30	35	350	500	7.2	6.2

According to [27], the concentration of Cd accumulated in a shell tissue can reach 100,000 times greater than the concentration in the waters of its habitat. In a normal seawater, the concentration of Cd could range between 0.05 ppb. to 0.1 ppb [28].

Pb can enter into the body of marine life through the food chain, gills and diffusion of the skin surface. The accumulation of Pb metal in shells can occur through the absorption of water, particles and plankton by filtering (filter feeder). Accumulation of Pb metal in the tissue of shell fish is possible because this type of organism can not excrete it well. Consequently it will accumulates continuously in the tissues, corresponding to the increase in metal Pb concentration in the water. In the case of mollusk (the soft animal),the Pb will be accumulated in the tissue of the animals.

According to [29] the accumulation of heavy metals in a shells is influenced by internal (body size) and external factors Spawning). Previous studies, many authors reported a negative relationship between the size of the shell with the accumulation of heavy metal concentration. For example, high levels of Cd, Pb and Zn are found in smaller individuals, and this may be due to the average metabolic activity in different organism ages. The increase in metabolic rate, is influenced by uptake of heavy metal wich depend on the body size [30].

The size of animals will influence the content of heavy metals in the soft tissues of mussels, using measurement of concentration factor. This measures the ability of aquatic organisms in taking direct pollutants from the waters and shows the tendency of organisms to accumulate heavy metals [31]. Smaller animal shown a greater accumulation capability than the larger one [32]. The high content of heavy metals in the body of aquatic organisms, depends on the physiological, inc;luding the absorption and the excretion process of the organism itself.

The temperature and salinity range in all stations in the research location is still within the tolerable for marine life(it is still in the range of quality standards of Ministerial decree 2004) ie. 28-30 °C and and of 0,5-30 ppt. respectively. The heavy metal content of Pb and Cd was higher at low tide than in pairs. This was because the pH value at low tide is higher than at high tide.The degree of toxicity of heavy metals in waters can also be affected by pH., it will increase as the pH increases. It is due to that the metal is difficult to dissolve in the high pH and in turn will be deposited in the bottom of the water [33].

4. Conclusion

There was interaction between type of fish and body part on moisture, protein, ash and carbohydrate content ($P < 0.05$), but no interaction on fat content and energy ($P > 0.05$). The body of cultured eel had higher protein than cultured one. The wild eel had higher fat content and energy than cultured one, while the fat content and energy in body and tail were higher than in head.

5. Acknowledgment

The authors would like to express our sincerely thanks to all of those who have provided support, so that the research as well as this article publication can be completed.

6. References

[1] Sánchez-Q, David T S, Antonio 2015 *Environment International* **83** 158-170
[2] Arifin, Z, Puspitasari, Rahma M, Nobuyuki 2011 *Coastal Marine Science* **35** 227-233
[3] Rodríguez-Hernández Á, Camacho M, Henríquez-Hernández LA 2016 *Science Total Environment* **575** 919 - 931
[4] Arifin Z, Puspitasari R, Miyazaki N 2012 *Coastal Marine Science* **35** 227 - 233.
[5] Olafisoye OB, Adefioye T, Osibote OA 2013 *Polish Journal Environmental Studies* **22** 1431 - 1439.
[6] Mali M, Dell'Anna MM, Mastrorilli P, Damiani L, Piccinni AF 2017 *Marine Pollution Bulletin* **114** 768-777
[7] Duruibe J O, Ogwuegbu M O C and Egwurugwu J N 2007 *International Journal of Physical Sciences* **2** 112-118
[8] Assi MA, Hezmee MNM, Haron AW, Sabri MYM and Rajion MA 2016 *Journal Veterinary World* **9** 660 - 671
[9] Sudarmaji, Mukono J and Coire I. P 2006 *Journal of Environmental Health* **2** 129 - 142
[10] Tjahjono A, Bambang A N and S. Anggoro 2017 *Ecological Engineering Journal* **18** 7 - 11
[11] Wulandari S Y, Yulianto B, Santosa G W and Suwartimah K 2009 *Marine Science* **14** 70 - 175.
[12] Santoso B, Sabikis and Prayitno, A 2009 *Determination of Mercury (Hg) On Perna viridis In Semarang Bay Using Atomic Absorption Spectrophotometry* (MIPS).
[13] Wijaya A R O, Aaron K T, Kentaro C, Michael D S, Sittipun S, Ryuichi O, Shigeru 2013 *Journal of Geochemical Exploration* ***126-127*** *45-54*
[14] Moor C L, Theopisti D, Volker J 2001 *Microchimica Acta* ***136*** *123-128*
[15] Robin R S , Pradipta R M, Vishnu V K, Ganguly D, Abhilash K R, Balasubramanian T 2012 *American Journal of Chemistry* **2** 191-208
[16] Vaezi A. R, Karbassi A. R, Habibzadeh S, Kokabi, Heidari M, Valikhani S A R 2016 *NISCAIR Online Periodicals Repository* **45** 1017-1023
[17] Arifin, Z 2004 *in* Rachmawati, R., Aldrian, E., Hendiarti, N. and Tejakusuma *Proceeding of The International Workshop on Coastal Resources Explorationand Conservation* **1** 16-21
[18] Purnomo T and Muchyiddin 2007 *Journal of Neptunus* **14** 68 - 77
[19] Puspita F, Melanisa R and Santoso B 2012 e-Print *School of Electronics and Computer Science* 1-10
[20] Effendi H 2006 *Telaah Kualitas Air Bagi Pengelolaan Sumber Daya Dan Lingkungan Perairan*

Kanisius Yogyakarta
[21] Rhidowati S 2013 *Mengenal pencemaran logam Graha Ilmu Yogyakarta*
[22] NOOA 2004 *Sediment Quick Reference Tables (SQUIRT)*
[23] Budiarto A, Kusreni and S Musinah 2010 *Proceedings of the National Seminar on Science and Technology*
[24] Rees J G, Setiapermana D, Sharp V A, Weeks J M and William T M 1999 *Oceanol Acta Journal* **22** 627 - 640
[25] Wang W X, Yan Q L, Fan W and Xu W Y 2002 *Marine Ecology Progress Series Journal* **240** 27 - 38
[26] Takarina N D, Browne D R and Risk M J 2004 *Marine Pollution Bulletin* **49** 854 - 874
[27] Goksu MZL, Mustafa A, Fatma C and Oziem F 2005 *Artikel ilmiah* 89-93.
[28] Sanusi H S, Hutagalung H and Razak H 1984 *The relationship between age, mercury (Hg) and cadmium (Cd) levels accumulated by Green shrub Mytilus viridis L, which is cultivated in the waters of Jakarta Bay* (Bogor Agricultural Institute.Bogor)
[29] Kamaruzzaman B Y 2011 *International Journal of Chemistry* 5: 54-60
[30] Syarifudin A R, Maddusa S S, Akili R H 2017 KESMAS **6** 1-10
[31] Putri L S E P, Prasetyo A D and Arifin Z 2012 *Journal of Environmental Research and Develompment* **3** 389 - 396
[32] Otchere FA 2003 *African Journal of Biotechnology* **2** 280 - 287
[33] Rochyatun E, Lestari and Rozak A 2005 *Oceanography and Limnology in Indonesia* **38** 23 - 46

Safely Intake Number of *Macridiscus* sp. (Kerang Ceplos) from Tambak Lorok Waters, Semarang, Central Java, Indonesia

Eduard Meirenno Tielman[1], **Jusup Suprijanto**[1], **Ita Widowati**[1]

[1]*Department of Marine Sciences, Faculty of Fisheries and Marine Sciences, Diponegoro Univesity Tembalang, Semarang 50275 Telp/Fax. 024-7474698*

Telp/Email: +62 822 4275 0098 / tielmannino@gmail.com

Abstract. The dynamics pollution that supposed to be derived from industrial activities around Tambak Lorok waters will affect the quality of waters, and also biota such as *Macridiscus* sp. mussels (Kerang Ceplos) that live and accumulate pollutants such as heavy metals (Pb, Cu, Al, Mn and Fe). However, *Macridiscus* sp. mussels which have been contaminated by heavy metals is usually sold for consumption by the people and if they consume it in excess, it will be toxic in the people's body. So that, this study was to analyze Safely Intake Number of *Macridiscus* sp. from Tambak Lorok waters. This study used AAS (*Atomic Absorption Spectrophotometry*) method to analyze the accumulation number of the pollutant (Pb, Cu, Al, Mn and Fe). Safely Intake Number calculation is used MWI (*Maximal Weekly Intake*) and MTI (*Maximal Tolerable Intake*) calculation method. The results of AAS showed that the highest numbers of pollutant was Al (reached 534,51 mg/kg in the body of *Macridiscus* sp. that taken in February, 2016) and has exceeded the safely intake number (MWI Al = 1 mg/kg, based on WHO/FAO) so that it's MTI values was low (0,08 kg/week/person). It means that *Macridiscus* sp. was not safe to be consumed excessively at that time.

Keywords: Ceplos, *Macridiscus* sp., MWI-MTI, Pollution, Tambak Lorok

1. Introduction

Heavy metal distribution in aquatic environments are usually correlated to population density [1] and urbanization [2] via the large uncontrolled input from industrial activates [3]. Heavy metal input to the aquatic environment is also from anthropogenic activities [4]. This problem also happened in marine environments and can affect the life of benthic organisms.

Tambak Lorok is actually one of fisherman village which is located on the shoreline of The Central Java Sea (North Shoreline of Java Sea), Indonesia. One of major function of Tambak Lorok is for the specific trade of fisheries products such as fish, shrimps, and mussels that taken from the Tambak Lorok waters. However, as the time goes by, Tambak Lorok waters are threatened by pollutants such as heavy metals (Pb, Cu, Al, Mn and Fe) from industrial and anthropogenic activities around those waters. Heavy

IOP Conf. Series: Earth and Environmental Science **116** (2018) 012079 doi:10.1088/1755-1315/116/1/012079

metal is able to accumulate in marine sediments and may affect human health through the food chain via benthic organisms [2].

Mussels are one of benthic organisms that live in marine environments. There are some variant of mussels. One of them is *Macridiscus* sp. [5] / Kerang Ceplos (local name) that is sold in Tambak Lorok Market for consumption. This organism is one of favorites fisheries product that traded in Tambak Lorok for consumption because they are a good source of protein [6], and also good for bone health [7]. It contains high vitamins (vitamin B6, niacin, thiamine and riboflavin) and essential minerals (such as Ca, Fe) [8].

Meanwhile, mussels are also a potential bioaccumulator of heavy metals such as Pb, Cu, Al, Mn and Fe. Green Mussels (*Perna viridis*) can accumulate heavy metals and suitable as a biomonitor of heavy metals pollution [9]. It means that *Macridiscus* sp. is also a bioaccumulator of heavy metals. It can be dangerous for people health if they don't know safely intake of mussels such as *Macridiscus* sp. which has been contaminated by heavy metals from Tambak Lorok Waters. The heavy metals can be toxic to the human body if they consume in excess and leading to various diseases [10], [11].

A research to analyze heavy metals accumulation and Safely Intake Number of *Macridiscus* sp. from Tambak Lorok waters is important due to get information about how much mussels which have contaminated by heavy metals that can be eaten by people. So that people will not be poisoned by heavy metals.

2. Methods

Sampling of *Macridiscus* sp. were conducted in December 2015 and February 2016. The location of sampling was at Tambak Lorok, Semarang, Central Java. The determination of Sampling Points used Purposive Sampling method [12], due to getting information about the existence/abundance of *Macridiscus* sp in Tambak Lorok waters. There were 3 (three) points we had chosen for sampling and you can see the picture below.

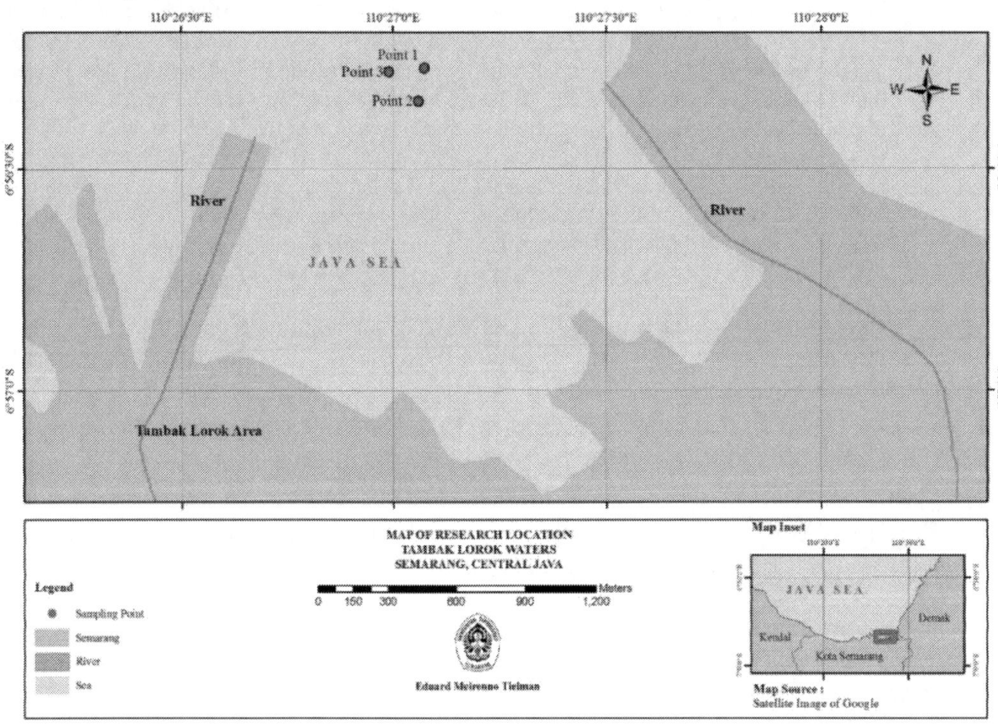

Figure 1. Sampling Point of *Macridiscus* sp.

Other paragraphs are indented (BodytextIndented style). The analysis method of heavy metals in the sample was used AAS (*Atomic Absorption Spectrophotometry*) (APHA, 1995) method at Balai Besar Teknologi Pencegahan dan Pencemaran Industri (BBTPPI), Central Java. Before using AAS, we have to prepare the sample. Put as many as 250 grams sample of *Macridiscus* sp. tissue in the evaporator cup and heated it in an oven at temperature 105°C for 12 hours. Then, grind it until becoming homogeny. After that, use the furnace to heat it with temperature 550°C ± 8 hours. Then, chilled it. Next, take 4 grams of the sample from the sample we got for destruction in beaker glass. After that, filter it with Whatman paper no. 40. Then, pour it into measurement glass as many as 50 ml and let it for a night before analyzing with AAS.

After we got the accumulation number of heavy metals by AAS, we analyzed the Safely Intake Number by using MWI (*Maximal Weekly Intake*) and MTI (*Maximal Tolerable Intake*) [14] calculation method. The formula of MWI is :

$$\boldsymbol{MWI = Body\ Weight^{\ a)}\ x\ PTWI^{\ b)}} \qquad (1)$$

a. Body weight assumption [45 kg (female) and 60 kg (male)]
b. PTWI (*Provisional Tolerable Weekly Intake*) by Joint FAO [15]: 0.025 mg/Kg

And the formula of MTI :

$$\mathbf{MTI} = \frac{\mathbf{MWI}}{\mathbf{Ct}} \qquad (2)$$

- MWI : *Maximum Weekly Intake*
- Ct : heavy metal accumulation in the body of *Macridiscus* sp.

3. Result and Discussion

3.1. Result

3.1.1. Heavy Metals Accumulation in the body of Macridiscus sp. The picture below is the results of the analysis of heavy metals (Pb, Cu, Al, Mn and Fe) that has been carried out :

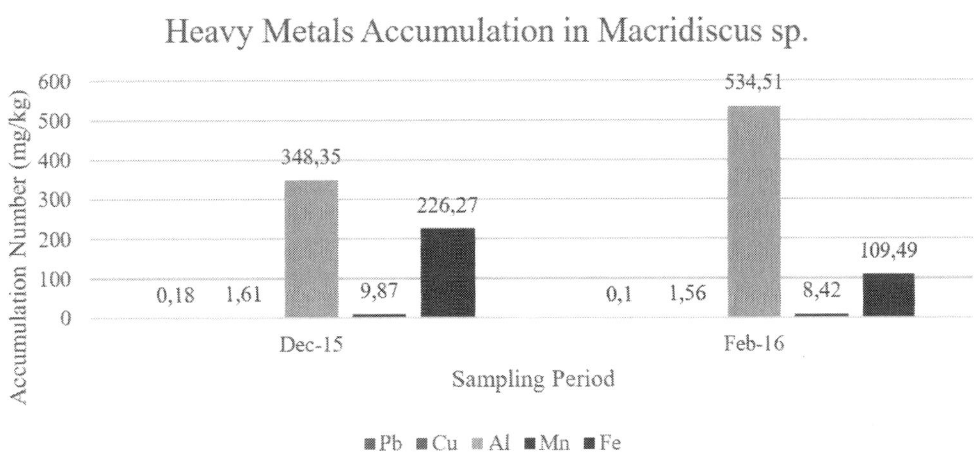

Figures 2. Heavy Metals Accumulation

3.1.2. Safely Intake Number of Macridiscus sp. (MWI and MTI). The table below shows us about the number of MWI and MTI of Macridiscus sp. that can be safe for consumption by the people.

Table 1. Safely Intake Number of *Macridiscus* sp.

No	Heavy Metals	Heavy Metals Accumulation in Biota (mg/Kg)		PTWI (mg/bw based on WHO / FAO)	MWI (mg/week/body)		MTI (Kg/week/body)			
							45 Kg		60 Kg	
		Dec-15	Feb-16		45kg	60 kg	Dec-15	Feb-16	Dec-15	Feb-16
1	**Pb**	0.18	0.1	0.025	1.125	1.5	6.25	11.25	8.33	15
2	**Al**	348.35	534.51	1	45	60	0.13	0.08	0.17	0.11
3	**Fe**	226.27	109.49	5.6 (PTMDI 0.8 mg/bw)	252	336	1.11	2.30	1.48	3.07
4	**Cu**	1.61	1.56	3.5 (PTMDI 0.5 mg/bw)	157.5	210	97.83	100.96	130.43	134.62
5	**Mn**	9.87	8.42	0.98	44.1	58.8	4.47	5.24	5.96	6.98

3.2. Discussion

3.2.1. Heavy Metals in Macridiscus sp. Based on the result above, accumulation average of heavy metals in the body of Macridiscus sp. is decreased from December 2015 to February 2016. It probably caused by precipitation (seasonal factor), temperature and salinity dynamics [16] that happening in December 2015 until February 2016 which affected the accumulation of heavy metals by *Macridiscus* sp.

3.2.2. Safely Intake Number. Every heavy metal has their own standard for being consumed. It determined by MWI and MTI values. MWI values are to determine how much heavy metals contamination in the body of biota that being consumed by the people in a week per body weight (45 kg for female and 60 kg for male). MTI values determine how much mussels you can consume in a week per body weight.

- Lead (Pb)

Exposure to Pb can occur from many sources but usually arises from industrial use that is around Tambak Lorok Waters like Tanjung Mas Harbor [15]. It can occur from fishing vessels which are accidentally shedding their vessels fuel into the waters, too.

Based on the result of the analysis, accumulation of Pb in the body of *Macridiscus* sp. is not exceeding the MWI value yet. MTI values mussels which contaminated by Pb is high. It means the mussels are still safe for Pb consumption.

- Aluminum (Al)

Al is a major component of the earth's crust. It is released to the environment both by natural processes and from anthropogenic sources. We will find Al use in a wide variety of applications including structural materials in construction, automobiles and aircraft, packaging materials, various containers and kitchen utensils and pharmaceuticals [15] .

The result showed us that accumulation of Al in the body of *Macridiscus* sp. is exceeding its MWI value. Therefore, we can only eat *Macridiscus* sp. in a slight amount. FAO/WHO said that Al compound can affect the reproductive system and it will be dangerous if we consume it in excess.

- Iron (Fe)

Fe is one of the essential heavy metals that needed by the human body to produce the enzyme cytochrome oxidation and respiratory pigments/hemoglobin [15]. Fe can occur in waters by the corrosion from ships and piling port that readily rusts.

The result of this study showed that accumulation of Fe in the body of *Macridiscus* sp. is exceeding MWI value. so it's not safe if we consume it in excess. The excess of Fe consumption cause symptoms such as vomiting, damage the intestine, cardiomyopathies and much more.

- Copper (Cu)

Copper is both an essential nutrient and a drinking-water contaminant. It has many commercial uses. It is used to make pipes, valves, and fittings and is present in alloys and coatings [15].

The result of this study showed that Cu accumulation in the body *Macridiscus* sp. is not exceeding the MWI value. However, we must be careful. High levels consumption of mussels that contaminated by Cu can cause symptoms of acute toxicity, including nausea, abdominal discomfort (diarrhea), emesis, hemoglobinuria and/or haematuria, jaundice, oliguria/anuria, hypotension, coma, and death.

- Mangan (Mn)

Mn is one of the important components of the human body in a slight amoun t[15]. Mn can occur from an active substance in battery which had been used and discharged into the river and coastal. It may also caused by people that live at Tambak Lorok Village who discharged it to Tambak Lorok waters. In this study, accumulation of Mn in the body of *Macridiscus* sp. is not exceeding MWI value.

4. Conclusion

This study has provided us that *Macridiscus* sp. that taken from Tambak Lorok water is not safe to be consumed by people at that time. It has contaminated by heavy metals such as Pb, Al, Fe, Cu, and Mn. It can impact to body health if we consume it in excess. Besides that, This study may also provide an evaluation of relevant government policy related to waste and fishing village management environmentally.

Acknowledgment

The author would like to thank Dr. Ir. Jusup Suprijanto, DEA and Dr. Ir. Ita Widowati, DEA as a supervisor for this research. This research is part of project research with the title "Valorization of Potential Sea Product toward Sustainable Resource Management in 2015" by Dr. Ir. Jusup Suprijanto, DEA and Dr. Ir. Ita Widowati, DEA through *Skema Hibah Kompetensi* which has obtained financial support by Kemenristek Dikti in 2016. The author also says thanks to Balai Besar Teknologi Pencegahan Pencemaran Industri (BBTPPI) Central Java as part of Industrial and Trading Ministry of Indonesia (Kemenperindag Indonesia).

References

[1] Conrad CF, D Fugate, J Daus, C J Chisholm-Brause and S A Kuehl 2007 *Mar. Pollut. Bull.* **54** (4) 385–395
[2] Shang Z, J Ren, L Tao and X Wang 2015 *Environ. Monit. Assess.* **187** (3) 79
[3] Lourino-Cabana B, L Lesven, A Charriau, G Billon, B Ouddane and A Boughriet 2011 *J. Hazard. Mater.* **186** (2) : 2129–2137
[4] Rath P, U C Panda, D Bhatta, and K C Sahu 2009 *J. Hazard. Mater.* **163** (2) : 632–644
[5] Kong L, A Matsukuma, I Hayashi, Y Takada and Q Li 2012 *J. Molluscan Stud.* **78** (1) : 1–11
[6] Choo S E and C S Ng 1990 *Singapore J. Prim. Ind.* **18** : 48–53
[7] Zalloua PA, Hsu YH, Terwedow H, Zang T, Wu D, Tang G, Li Z, Hong X, Azar ST, Wang B, Bouxsein ML, Brain J, Cummings SR, Rosen CJ, Xu X 2007 *Maturitas* **56** : 1–11
[8] Yoshino K, Mieko Inagawa, Kazumasa Yamagishi, Miyuki Oshima, Takeshi Tanigawa, Kimiko

Yokota, Shinichi Sato, Mitsumasa Umesawa, Takashi Shimamoto, Ma Enbo and Hiroyasu Iso 2005 *J. Epidemiol.* **15** (2) : 29–37

[9] Kamaruzzaman BY, MS Mohd Zahir, B Akbar John, KCA Jalal, S Shahbudin, SM Al-Barwani and JS Goddard 2011 *International Journal of Biological Chemistry* **5**: 54-60

[10] Hashmi DR, S Ismail and G H Shaikh 2007 *Pakistan J. Bot.* **39** (3) : 747–751

[11] Anhwange BA, J A Kagbu, E B Agbaji and C E Gimba 2009 *Electron. J. Environ. Agric. Food Chem.* **8** (11) : 1150-1155

[12] Bernard H R 2002 *Research methods in anthropology: Qualitative and quantitative methods . Walnut Creek* (CA: AltaMira Press)

[13] A. P. H. A. (APHA) 1995 *Standard Methods for the Examination of Water and Wastewater*

[14] Turkmen H, A Turkmen and Y TEPE 2008 *J. Chil. Chem. Soc.* **5** (1) : 1424–1428

[15] F. A. O. Joint 2014 *WHO food standards programme codex committee on contaminants in foods*," in *Fifth Session [displayed 10 February 2014]. Available at ftp://ftp. fao. org/codex/meetings/CCCF/cccf5/cf05_INF. pdf*, 2011.

[16] Hamed M A and A M Emara 2006 *J. Mar. Syst.* **60** (3–4) 220–234

Water Quality Improvement of Media Culture for Tilapia *(Oreochromis niloticus)* with Cleaner Production Method

Haeruddin*, Supriharyono, S Febrianto

Department of Aquatic Resource, Faculty of Fishery and Marine Science, Diponegoro University, Jl Prof Soedarto SH Tembalang, Semarang 50275, Indonesia.

Email: haeruddindaengmile@yahoo.co.id

Abstract. The tilapia (*Oreochromis niloticus*), is known as a high adaptability and brackish water tolerance fish. This fish is also has a meat with high protein content, that ranges about 65 -75%. Generally the tilapia is cultured using a conventional system with high density. It is caused degradation of water quality of media culture, and finally increase mortality rate of fish cultured. The application of tilapia cultivation with cleaner production method by giving enzyme into the feed to upgrade the efficiency of feed utilization, presumed that could improve the water quality of cultivation media. It is due to the lower of feed and feces residues. Therefore the concentration of toxic compounds, such as ammonia, nitrite and sulfide, will be lower.

The experiments were conducted for 35 days with a completely factorial randomized design. The first factor was the dosage of enzyme in the feed, consisting of 4 dosages, and the second factor was the duration of the test fish maintenance (5 weeks). Water quality variables examined included ammonia, nitrite and sulfide. The results showed that enzyme dosage had no significantly impact on ammonia, nitrite and sulfide concentrations in the test media culture. However, the feeding with enzyme in low dosage, resulted less concentration of ammonia, nitrite and sulfide than it was without enzyme). The duration of fish cultured has significantly effect on the concentration of ammonia, nitrite and sulfide in the test media. While it is no significantly correlation between dosage and duration of maintenance.
Keywords: Oreochromis niloticus, cleaner production, ammonia, nitrite and sulfide

1. Introduction

Tilapia (*Oreochromis niloticus*) is a potentially of cultivated fish. Since this fish is very tolerant in any environmental condition. Tilapia is still *survives*, even if it is maintained at high density [1]. In addition this fish is has a high protein content, that is about 65 -75% meat [2]. The genus *Oreochromis* has a high adaptability and tolerance to water quality over a wide range, including in brackish waters [3].

Tilapia production increased year to year in Indonesia, since the year 2010 to 2014, according to Statistics of Production Directorate, General Directorate of Cultivation. Increasing of tilapia production among others due to fishery intensification, with a high density of dissemination per unit area of land. Such fish intensification, however, may resulted a big problem with water pollution in the cultivation media, due to liquid waste from the cultivation activities.

The Increase of awareness or partisanship of the world community towards the effort of environmental conservation, emphasizes the importance of applying cultivation patterns that can maintain environmental preservation, especially related to waste management [4]. The Increase of freshwater fish and marine fish demands, and the limitation availability of water and land for fish cultivation, causing the need for modernization and revision of cultivation technology [5].

There are several ways that can be done in order to control water pollution. One of them is a control by using cleaner production technology known as cleaner production, that is continuous application and integrated prevention strategy which are applied to various processes, products and services to get economic value, social, health, security and environment benefit [6].

The application of water pollution control by using cleaner production method so far is only limited in manufacturing industry, while the application over aquaculture industry is still very rare [5].

Actually, this method is also possible to be applied in the aquaculture industry. It is mainly in order to reduce the amount /volume and concentration of waste by using raw material and machine production which contained in fish and its environment.

The raw materials in the cultivation industry consist of seed and feed, while the machine production is media cultivation for fish. That is consists of water and other requirement, i.e. freshwater pond, brackish water pond or other ponds. In order to control pollution in media cultivation, among other it is used recirculation and bio filter systems.

The higher the price of raw materials to make feed, encourage the use of feed in low-protein rate (about 15%) which is given enzyme supplements to improve the digestibility of feed, so that more feed can be converted into fish meat biomass [7, 8]. As a result of the application of low-protein feed technology with enzyme supplementation, it is estimated that the volume of feces produced less, so as the concentration of toxic compounds derived from the process of feces decomposition and the rest of high protein feed will also be less.

This study is conducted to review the use of enzyme that has been used to improve feed efficiency, to suppress the production of liquid waste in the aquaculture system, as the implementation of cleaner production principles.

2. Materials and methods

The material used in this study was tilapia which was obtained from Siwarak Fish Hatchery, Semarang Regency. Tilapia used had a length of 2 ± 0.5 cm. During the experiment the tilapia was fed artificial feed / mill feed with a dose of 5% of body weight, with the frequency of giving 2 times a day namely in the morning and evening. Before given to tilapia, formerly the feed was mixed with enzyme with dosage 0.0125%, 0.025%, 0.050% and 0.075%. During the experiment fish was kept in an aquarium with capacity 20 liters, and was filled with 15 liters of water. Maximum maintenance density was 1 g biomass fish/ liter of media.

Water quality measurement was performed on physical parameters, included temperature, chemical parameters including: pH, dissolved oxygen, ammonia (NH_3-N), nitrite (NO_2-N) and Hydrogen Sulfide (H_2S) concentrations. Measurement of water quality was conducted twice a day, in the morning and afternoon, except for ammonia, nitrite and hydrogen sulfide, the measurement was taken once a week.

Table 1. Equipment and methods used in the measurement of water quality variables

No	Variable	Measuring instrument	Method
Physics			
1	Temperature	Digital Thermometer	Conductivity meter
Chemistry			
3	pH	pH meter digital	Conductivity meter
4	Dissolved oxygen	DO meter digital	
5	Ammonia (NH_3N)	AAS	Spectrophotometry
6	Nitrite (NO_2-N)	AAS	Spectrophotometry
7	Sulfide (H_2S)	AAS	Spectrophotometry

The experimental design used was a completely randomized design with factorial pattern. Factor I was the influence of enzyme dose in feed (consisting of 4 treatments level and 1 control, each treatment was repeated 3 times) against the quality of fish cultivation media. Factor II was the influence of fish culture duration towards the quality of fish cultivation media. The fishes test were treated for 35 days and they were weighted every 7 days.

The data were analyzed quantitatively by using the analysis of variance and correlation regression analysis. The analysis of variance was conducted to assure the effect of treatment to liquid waste production. While the regression analysis was conducted in order to find out the pattern and relationship closeness between treatment with ammonia, nitrogen and sulfide concentration variables in the water of fish culture media.

IOP Conf. Series: Earth and Environmental Science **116** (2018) 012080 doi:10.1088/1755-1315/116/1/012080

3. Results and discussion

3.1. Results

Analysis water sample of media of fish test resulted data of ammonia concentration (NH_3-N) as presented in Table 2.

Table 2. Average and standard deviation of ammonia concentration (NH3-N, mg / l) of fish test media culture during the experiment

Treatment	Average and SD	The duration of fish maintenace (Day) 0	7	14	21	28	35
A	Average	0.2225	5.6627	2.715	2.4332	0.2239	0.2195
	SD	0.1642	0.3138	2.1662	1.9231	0.1478	0.0922
B	Average	0.2909	5.2973	4.6869	3.6208	0.0549	0.1704
	SD	0.0967	1.1535	3.0353	3.4540	0.0217	0.0261
C	Average	0.1932	5.276	8.2153	9.7967	0.103	0.1893
	SD	0.0452	0.8309	1.3927	5.3897	0.0815	0.0360
D	Average	0.1541	5.4795	5.5372	7.6083	0.112	0.1682
	SD	0.1103	0.7528	4.0558	7.7192	0.0731	0.0336
K	Average	0.3315	5.966	5.8095	4.2125	0.1321	0.2398
	SD	0.1106	0.6967	3.9225	4.3897	0.0617	0.0766

The result of analysis of variance shows that the different dosage of enzyme in feed has no significant effect on ammonia concentration produced by fish test, but maintenance duration has significantly effect on the ammonia concentration produced. However, the combination of different dosage of enzyme in feed together with duration of maintenance of fish test, both of them do not have any significantly effect on ammonia concentration produced by the fish test.

Although the dosage was not significantly effect on ammonia concentration, but the Figure 1 shows that the highest concentration of ammonia found in the control treatment and the lowest in treatment A. The ammonia concentration during the study tends to increase from the beginning to maximum on day 12, then drops to the end of the study. As shown in Figure 2, the relationship between age /duration of fish culture with ammonia concentration in the test media stated in the equation $y = 0.001x3 - 0.0698x2 + 1.2078x + 0.1617$; $R^2 = 0.4323$.

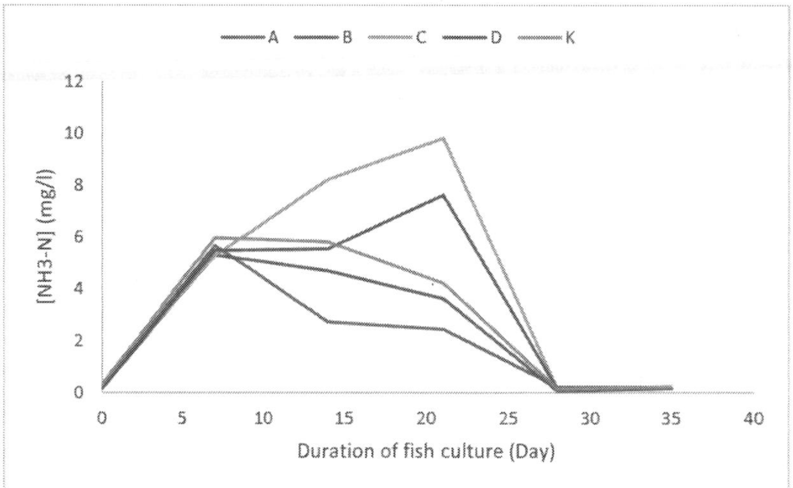

Figure 1. The curve of average Ammonia concentration in the test media during the experiment.

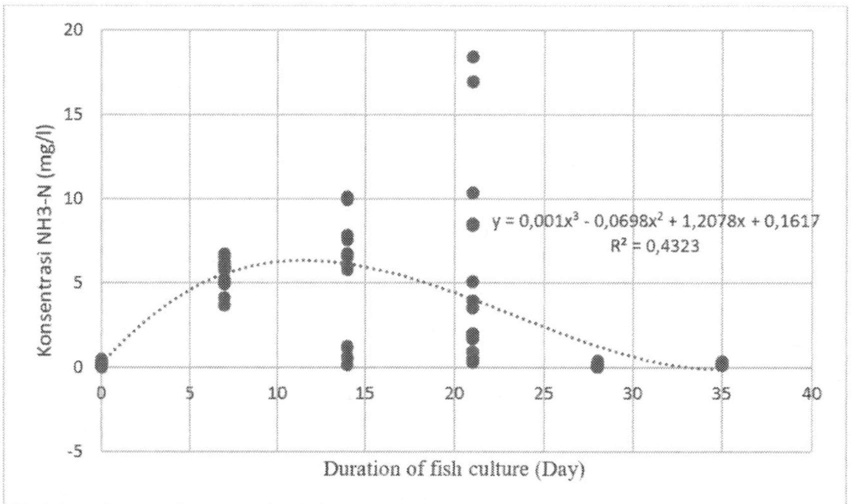

Figure 2. Curve of relationship between age/duration of fish culture with ammonia concentration in test media.

The analysis of water samples conducted during the study resulted nitrite concentration (NO2-N) as presented in Table 3. The result of analysis of variance shows that the enzyme giving has no significantly effect on nitrite concentration in the test media, but the duration of culture has significant effect. As well, the combination of both the enzyme treatment and duration of culture do not have significant effect on the nitrite concentration in the test media.

Although the enzyme concentration was not significantly effect on the nitrite concentration in the test media, but the average data indicates that the highest nitrite concentration is in control (without enzyme) and the lowest is in treatment B. The nitrite concentration tends to increase by the increasing of fish maintenance duration, the nitrite concentration during the experiment ranges 0.03 up to 13.24 mg / l. The relationship between the nitrite concentration and the duration of fish maintenance stated by exponential equation y = 0.0447e0.1036x, $R^2 = 0.368$.

Table 3. Average and standard deviation of nitrite concentration (NO2-N, mg / l) of fish maintenance media during the experiment

Treatment	Average and SD	The duration of fish maintenance (Day)					
		0	7	14	21	28	35
A	Average	0.3093	0.2611	0.7378	0.4085	9.6642	9.234
	SD	0.3791	0.2709	0.0881	0.2805	13.0338	6.5711
B	Average	0.0958	0.2199	0.1495	0.6643	4.6713	0.9945
	SD	0.1028	0.2338	0.1716	0.1788	5.3524	0.3327
C	Rataan	0.0379	0.0247	0.025	0.54	6.636	0.592
	Average	0.0066	0.0142	0.0043	0.1879	5.8034	0.1254
D	Average	1.2018	0.1643	0.1113	0.6718	2.8178	2.6288
	SD	1.6710	0.1823	0.0829	0.0635	2.5005	2.1132
K	Average	0.0577	0.1564	0.3181	0.1742	13.2472	6.4487
	SD	0.0374	0.1617	0.0982	0.0901	8.7028	8.8187

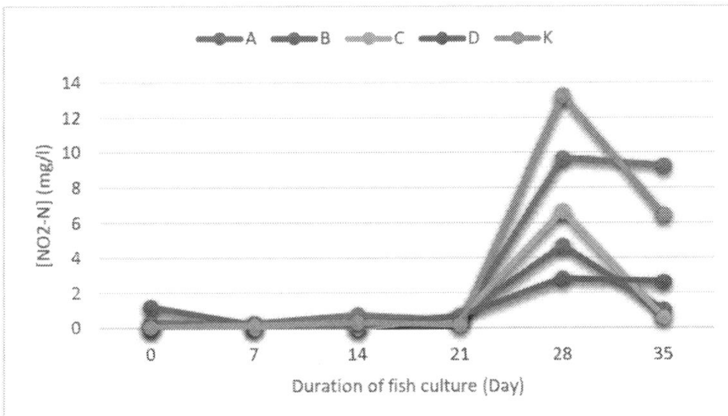

Figure 3. The average curve of nitrite concentration in the test media during the experiment

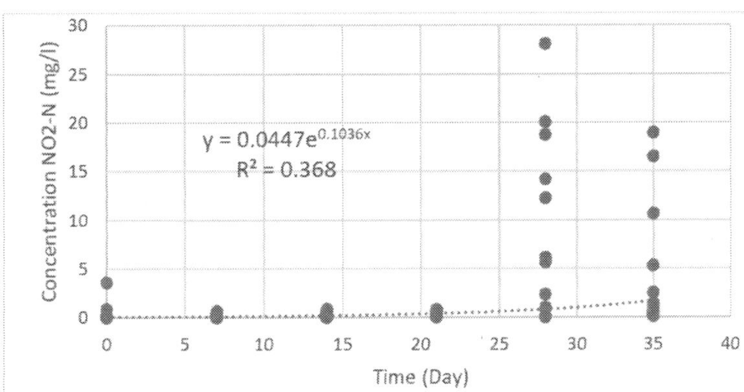

Figure 4. Curve of relationship between age / duration of fish culture with nitrite concentration in test medium.

The analysis of water samples conducted during the study resulted sulfide concentration (H2S, mg/l) as presented in Table 4.

Table 4. Average and standard deviation of sulfide concentration (H2S, mg/l) on culture media of fish test during the experiment.

Treatment	Average and SD	Duration of fish maintenance (Day)					
		0	7	14	21	28	35
A	Average	0.002	0.013	0.057	0.054	0.034	0.018
	SD	0	0.013	0.001	0.003	0.016	0.003
B	Average	0.002	0.033	0.069	0.051	0.032	0.023
	SD	0.0005	0.003	0.008	0.009	0.028	0.007
C	Average	0.002	0.041	0.073	0.058	0.036	0.017
	SD	0.0005	0.001	0.005	0.006	0.029	0.003
D	Average	0.002	0.034	0.048	0.048	0.039	0.019
	SD	0.001	0.011	0.034	0.004	0.031	0.009
K	Average	0.002	0.055	0.055	0.052	0.022	0.022
	SD	0.0005	0.007	0.008	0.005	0.014	0.008

The result analysis of variance shows that enzyme has no significant effect on the sulfide concentration in the test media, but the duration of maintenance has significant effect. As well, the combination of both treatment enzyme given and duration of maintenance do not significantly effect on the sulfide concentration in the test media. Nevertheless, the average of each treatment shows that the lowest H_2S concentration in treatment A and the highest in treatment C (Figure 5). Moreover, the concentration of Hydrogen Sulfide in the fish culture media tends to increase at the beginning of the experiment and reaches the highest concentration on the 14th day of fish culture, then drops until the end of the experiment. The H_2S concentration during the experiment ranges from .0.002 to 0.055. mg / l. The relationship between the nitrite concentration and the duration of fish maintenance stated by the equation y = -3E-08x5 + 3E-06x4 - 9E-05x3 + 0.0011x2 + 6E-05x + 0.005, R^2 = 0.5832.

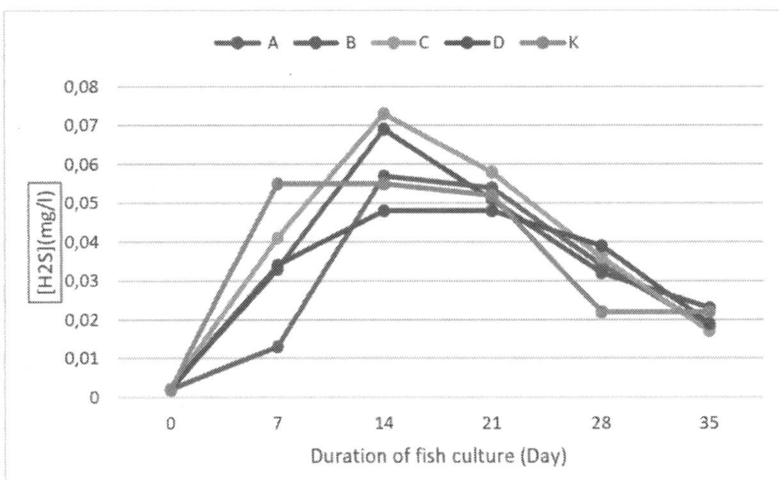

Figure 5. Curve of H_2S average concentration in the test media during the experiment

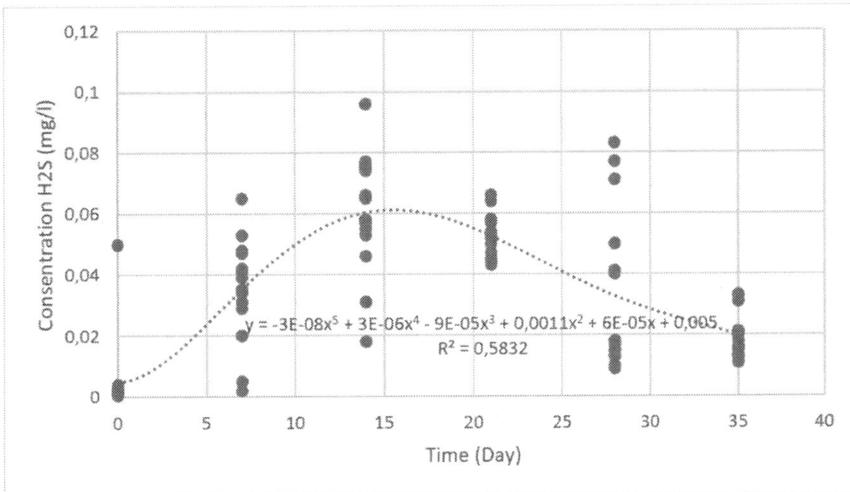

Figure 6. Relationship curve between age/duration of fish culture with sulfide concentration in media test.

3.2. Discussion

This study indicates that enzyme which given into the feed can improve the water quality of culture media of fish test. Although it is significantly effect, but the enzyme may appeared to improve the water quality of test media. The ammonia, nitrite and sulfide concentrations of test media for test fish fed with enzymes are generally better than test media for test fish fed without enzymes, especially test media for test fish fed with low enzyme content (0.0125% - 0.025%). The improvement of water quality in the test media which was given was low-dose enzyme may occurred as a result of the increasing efficiency on feed utilization, therefore the solid waste in the form of feed and feces residual became lower.

Several types of enzymes have been utilized in the activity of fish cultivation, such as complex enzymes which consist of fungal xylanase, β-glucanase, pentosonase, β-amylase, fungal β-glucanase, hemicellulase, pectinase, cellulase, and cellubiase [8]. Fitase enzyme for improving phosphor digestibility on Tilapia [9], complex enzyme EZplus containing protease, lipase, amylase, pepsin, trypsin and chemotrypsin on tilapia [10]. The complex enzyme EZplus has also been tested on milkfish by [7].

Those Various enzymes are proven to improve the digestibility of feed, and reduce the feces volume produced by fish. [10]. , resulted that tilapia (*Oreochromis niloticus*) which is fed with lower protein content (26%) produces a faster growth than fish fed with higher protein content (32%) with the addition of enzyme to fish feed with the highest enzyme level (5%). The enzyme used is EZplus containing protease, lipase, amylase, pepsin, trypsin and chemotrypsin. Study on the addition of Farmazyme enzyme (containing enzyme fungal xylanase, β-glecanase, pentosonase, β-amylase, fungal β-glecanase, hemicellulose, pectinase, cellulose and cellubiase) performed by [8] on African catfish, *Clarias gariepenus*, shows better SGR (Specific Growth Rate), FCR (Food Conversion Ratio) and PER (Protein Efficiency Ratio). In this study, enzymes used containing protease, lipase, amylase, pepsin, trypsin and chemotrypsin.

Soni *et al* [7] combines the use of feed with low protein level and the complex enzyme EZplus. The use of feed with low protein level aimed to press the production cost of milkfish cultivation, while the application of biotechnology product is undertaken to improve the efficiency of milkfish digestibility, so that more feed converted into meat than wasted with fish feces and reduce the contamination / waste generated from feces done by cultivated fish [7].

The residue of inedible feed and fish feces containing the protein will be overhauled by the bacterial decomposer contained in the water into toxic organic compounds namely ammonium (unionized ammonia NH_4^+) and ammonia (NH_3-N). Unionized ammonia is harmful to fish, but ammonia is highly toxic. *Nitsomonas sp* bacteria will convert ammonia to nitrite (NO_2-N), and Nitrobacter converts nitrite to nitrate in the nitrogen cycle. This process occurs for 2 weeks until 2 months, so it can be understood from the various curves presented that the concentrations of ammonia, nitrite and sulfide in the test media tends to increase from the beginning of the trial to the second week and decrease after more than 2 weeks up to the end of the experiment.

Hydrogen Sulfide is known as a colorless toxic gas, flammable and smells like rotten egg is also found in the water. In the water this gas is found as a result of disintegration of organic matter, when bacteria analyzes the organic matter in an anaerobic atmosphere. The sulfides contained in the test media are presumed to be the result of feces decomposition and the remains of feed which are not consumed by the fish.

Concentrations of ammonia, nitrite and H_2S in the culture media of fish test tended to increase with the increasing of fish test age. This occurred as a result of the increase of volume, feces weight and the feed residuals produced, due to the increase of feed amount given along with the increase of the test fish weight. During the experiment, the test fish were given feed in form of pellet with a portion of 5% from body weight. The research results performed by the previous writer showed that the feces weight and the residual of inedible feed went up by the increase of the fish weight.

4. Conclusion

The results of this study indicated that the enzyme usage although it is not significantly effect on the quality of media culture of tilapia maintained for 35 days, but there is a tendency that the usage of low-dose enzymes (0.0125% to 0.025%) tended to have ammonia, nitrite and sulfide that were lower than the maintenance media of fish maintained with non-enzyme feed. The more increasing the fish age cultivated, the more increasing and different the ammonia, nitrites and sulfides concentrations in maintenance media in each week.

The results of this study open the hope of the possibility of water quality improvement on aquaculture by the application of cleaner production cultivation method using enzymes mixed into the feed. The method of cultivation with cleaner production methods should continue to develop, while improving the various deficiencies occurred.

Aknowledgement

The authors are very grateful to all parties who have assisted to prepare, implement of this study.

References

[1] Melo CV Neto R V R Costa A C de Freitas R T F Freato T A Souza U N 2013 *Acta Sci Anim Sci* **35** (4)
[2] Kurniasih S 2013 *Aquaculture Research Journal* **Vol 8** (1)
[3] Mashaii N Rajabipour F Mohammadi M Sarsang Hi Bitaraf A Hossein H Zadeh 2016 *Journal of Applied Aquaculture* Volume **28** Issue 1
[4] Jensen K O Denver S Zanoli R 2011 *J Life Sci* 58: 79 – 84.
[5] Unger J M Schumann A Brinker 2015 *Aquacult Environ Interact* vol **7**: 223-238
[6] UNEP 2011 Towards a life cycle sustainability assessment: making informed choice on products United Nation Environment Programme
[7] Sony AFM E Sutikno I.M Suitha 2014 Milkfish (*Chanos chanos* Forskal) Culture in fish pond environment friendly
[8] Yildirim Y B and Turam F 2010 *Journal of Animal and Veterinary Advances* **9**: 327 – 331
[9] Amin M Yusadi D Mokoginta I 2011 *Fisheries Science & Technology Journal* **Vol 6** (2): 52 – 60.
[10] Arafat M Y N Abdulgani R D Devianto 2015 *ITS Journal of Sciences and Arts* **Vol 4** (1): 2337 – 3520

Epidemiology of Child Tuberculosis (A Cross-Sectional Study at Pulmonary Health Center Semarang City, Indonesia)

L D Saraswati[1*], P Ginandjar[1], B Widjanarko[1], R A Puspitasari[2]

[1]Faculty of Public Health, Diponegoro University, Jl. Prof. H. Soedharto,SH, Tembalang, Semarang 50275, Indonesia
[2]District Health Office Semarang, Jl. Pandanaran No.79, Mugasari, Semarang 50249, Indonesia

e-mail address: lintang.saraswati@live.undip.ac.id

Abstract. Mycobacterium tuberculosis is an acid-resistant bacterium that caused tuberculosis. The children might suffer tuberculosis by direct transmission of adult smear positive. The analytic observational study with a cross sectional was conducted. Samples were pediatric patients from January 2015 until December 2016. The 344 subject as total sampling were included as samples. Data were obtained through interviews and direct observations then analyzed with Chi-Square method. The results of these study were the majority of subjects were 3 years old (51.7%), the majority were female (55.2%), already immunized BCG (82.6%), and had moderate nutritional status (40.7%). More than half subjects had no contact with adult smear positive (64.5%). Most of subject had house population density $\geq 8m^2$ per person (99.4%), the level of humidity 40-70% (99.4%), the presence of ventilation 10% from floor area (75.6%), and all the composition of the floor made from tiles. In contrary, the lighting levels were not qualified <60 lux (99.5%). Most of the subject had family members who smoked inside (61.6%). The education level of the parents majority completed high school (41.9%), relatively had a good knowledge about tuberculosis (98%), and almost respondents were housewives (89.5%) with family income above the minimum wage Semarang (53.4%). The results of Chi-Square test showed that there was a relationship between the gender of the child (OR = 0.445; 95%CI = 0.241-0.821) and the presence of smokers (OR = 2.007; 95%CI = 1.074-3.751). It was suggested to educate the parents about smoker as the risk factor of child tuberculosis.
Keywords: tuberculosis, children, smoker

1. Introduction

Tuberculosis is an infectious disease of the respiratory system caused by *Mycobacterium tuberculosis* to the part of the lung parenchym. Tuberculosis is transmitted through the air (droplet nuclei) [1]. A person diagnosed as a suspect tuberculosis if found on his/her main clinical symptoms such as cough with phlegm for more than three weeks, coughing up blood, shortness of breath and chest pain [2] Tuberculosis (TB) has existed for millennia and remains a major global health problem. In 2015, 6.1 million new TB cases were notified to national authorities and reported to WHO. Notified TB cases increased from 2013–2015, mostly due to a 34% increase in notifications in India. However, globally there was a 4.3 million gap between incident and notified cases, with India, Indonesia and Nigeria accounting for almost half of this gap [3].

Infection typically occurs when a child inhales *Mycobacteria tuberculosis* exhaled by an individual with pulmonary or laryngeal TB. Primary infection due to skin inoculation, ingestion into the gastrointestinal tract or congenital infection due to transplacental transmission occurs very rarely. After inhalation, a primary reaction often occurs consisting ofa small parenchymal lesion in the lung with regional lymphaden- opathy, often perihilar or mediastinal. Bacilli may then spread throughout the lungs and pleura, which can result in intrathoracic TB [4].

Tuberculosis is arguably one of the major deadly diseases affecting children in developing countries, causing the death of approximately >80,000 children annually, and with over 500,000 new cases each year, reflecting ongoing transmission within communities. At least 1 million children become ill with TB each year. Children represent about 10-11% of all TB cases. Only 5 countries do not yet report to WHO the notification of cases among children [5,6]. There was an increase in the case notification rate of TB among children between 2011 and 2014 the Lagos State TB and Leprosy Control Programme (LSTBLCP) shown that the incidence of pediatric tuberculosis is higher than the number of notifications, particularly in young children [5,6]. *Global Tuberculosis Report* 2013 published by WHO states that among 530,000 cases of tuberculosis in the world, 74,000 cases are children tuberculosis cases [7] In 2015, 210,000 children died of TB, including 40,000 TB deaths among children who were HIV positive. Researchers estimate that 67 million children are infected with TB latent TB) and are therefore at risk of developing disease in the future. Researchers estimate that 25,000 children develop multidrug-resistant TB every year [3]

According to Tuberculosis Recent Update of Indonesia January-December 2012 issue states that the proportion of children tuberculosis in Indonesia is 8.2%. While for the Central Java Province in 2012, the proportion of children tuberculosis is amounted to 11% and for Semarang in 2011 is amounted to 13% of all tuberculosis cases. The Center for Pulmonary Health Semarang is one of health services focuses on pulmonary health. BKPM Semarang also focuses on tuberculosis treatment. According to BKPM Profile in 2012 shows that there is an increasing number of children tuberculosis case in 2011-2013. This number indicates there is an increasing of cases found. The number of cases in 2011 amounted to 3.53%, then by 7% in 2012, and by 11.57% in 2013.

Childhood tuberculosis has been neglected by health programmers and academics, because the cases are few, non-infectious and the perception that effective TB control in adults will prevent TB in children. Childhood tuberculosis is underdiagnosed in many countries with high TB burden [8–10].

Factors that contribute to low case detection among children include inability of children to produce sputum. Even when sputum is produced by a child, cheap diagnostic tests, such as smear microscopy, can only accurately detect 30% of the cases. Other factors include low capability among healthcare staff to diagnose childhood TB, reliance on smear microscopy as the primary method of TB detection, ineffective family-centered contact tracing, paucity of printed workers aid/guidelines, dearth of adequate training on childhood TB screening and diagnosis, and lack of data-capturing tools [11].

Children are more susceptible to developing active TB compared with adults, and acquire the infection mostly from adults within their household, including their parents [12,13]. Lack of relevant data on age-related trends of TB in children. The understanding of the global epidemiology of childhood TB could have been better, and planning of effective control measures easier if relevant data were available [5].

2. Methods

2.1. Study site

The research was conducted at The Center for Pulmonary Health Semarang, Central Java Province, Indonesia.

2.2. Selection subject

For this study, cases of TB are defined in accordance with WHO recommendation. Initial screening criteria used to identify suspected TB cases included low grade fever and weight loss. Additional criteria were cough for more than 3 weeks for pulmonary TB, and localizing signs/symptoms including palpable lymph nodes, headache/vertigo, and backache. Children suspected of having pulmonary TB are encouraged to provide morning sputum for microscopy. Three morning sputum samples are examined for each child. Chest Xray and tuberculin skin test (TST) is carried out by the clinic staff (treating physician) in those unable to produce sputum and those found to be sputum smear negative on microscopy. TST is performed using five tuberculin units (TU) of tuberculin purified protein derivative (PPD)-S with .5 mm induration as cut off.

2.3. Study design

The analytic observational study with a cross sectional was conducted. Samples were pediatric patients from January 2016 until August 2017. The 344 subject as total sampling were included as samples. Data were obtained through interviews and direct observations then analyzed with Chi-Square method.

2.4. Ethic statement

Ethical issues (Including plagiarism, Informed Consent, misconduct, data fabrication and/or falsification, double publication and/or submission, redundancy, etc.) have been completely observed by the authors. Ethical approval was obtained from the Committee of Public Health Research Ethics, Diponegoro University (126/EC/FKM/2015). Informed consent was acquired from parent or guardian.

3. Result and discussion

Table 1 shows the distribution incidence of Child Tuberculosis. From this table it can be seen that half of the subject suffered tuberculosis.

Table 1. Distribution Incidence of Child Tuberculosis

Incidence of Child Tuberculosis	Total (f)	Percent (%)
Positive of Child TB	172	50
Negative of Child TB	172	50

After a long period of neglected by health program planner in childhood tuberculosis due to lack of childhood TB cases, difficulty of diagnosis, non-infectious and the belief that effective TB control in adults will prevent TB in children [8,9,14], in this study, we found half of subject suffered from childhood tuberculosis (Table 1). TB in children is increasingly becoming an important cause of global child morbidity and mortality in developing countries. At least 1 million children fall ill with tuberculosis (TB) each year. Children represent about 11% of all TB cases [3,6]. All reflecting ongoing transmission within communities [15]. The estimated proportion of total childhood tuberculosis for each country ranged from 4% to 21% [6]. The burden of childhood TB infection and TB disease is an important indicator of the overall trend of disease and ongoing transmission within the community [15,16]. Estimating the TB burden in the children population, in any level, is important to encourage authorities to facilitate the next process such as new drugs and market appropriate pediatrics formulations [17].

Most of the children who are subject of this study are mostly 0-4 years old at (65.7%) and (55.2%) of them are female. A total of (82.6%) have been immunized with BCG classified as moderate nutritional status as much as (40.7%). It was similar with a result that stated uniform distribution of ages in the household model of exposure led to disease estimates that were 11% higher than the distribution based on adult age and sex notifications [6]. An assumption that the efficacy of BCG vaccination did not wane towards the equator resulted in disease estimates that were 27% lower for all methods of estimating infection [6].

Most children in this study also showed that (64.5%) had no contact history with positive BTA patients and do not have a history of contact with neighbors (79.7%). In contrary with other results that stated children are more susceptible to developing active TB compared with adults, and acquire the infection mostly from adults within their household, including their parents [12,18].

A study showed that 1.8% of children in household contacts of adult TB patients also had TB, higher proportion of TB in female children which showed vulnerability of the female child and considerable role of mother in disease transmission [19,20]. From our result we know that gender of the child were associated with the incidence of childhood tuberculosis (OR=0.445; 95%CI= 0.241-0.821), this in line

IOP Conf. Series: Earth and Environmental Science **116** (2018) 012081 doi:10.1088/1755-1315/116/1/012081

with the study conducted by Codlin et al in Pakistan, they found that TB is highly prevalent, translating, into thousands of excess female cases per year [21]. Higher frequency of disease in the females may reflect poor nutritional status of the girl child in making them more vulnerable to the disease [22].In crowded houses, a greater degree of shared airspace increases exposure to *M. tuberculosis,* which can even be increased by limited air movement in closed spaces—hence a greater risk of infection and tuberculosis disease [12,13]. There are many risk factors related to TB like living in the same house with close family members or friends who had active TB is significantly [8.69 confidence interval (CI): 3.00-25.18] associated with TB [23]. Large family size and small house predisposes 9 times a risk to develop LTBI (CI: 3.00-25.28) [24].

In terms of environmental factors (Table 2), (99.4%) of subject have qualified house environment regarding the population density, but the lighting levels are not qualified by (96.5%). Humidity level of house environment is already qualified by (99.4%) and already has qualified ventilation by (75.6%). The whole floor is already plastered or tiled. TB is also more prone to develop where ventilation and sunlight are inadequate because *Mycobacterium* would not easily move away or killed [25,26]. Meanwhile, there are (61.6%) smokers who live with subject under the same roof. Most smokers smoke in the living room (32%). The education level of the parents by (41.9%) completed high school with relatively good knowledge about tuberculosis by (57%) As many as (89.5%) respondents are not working or can be said only as a housewife with family income per month by (63.4%) are above minimum wage. TB adult and childhood TB is prevalent in poor and marginalized population which associated with overcrowding and poor nutrition [12,27].

In this study some variables associated with the incidence of child tuberculocis are child age (*p value* 0,023), sex (*p value* 0,000), nutritional status (*p value* 0,033), contact history with positive BTA patients (*p value* 0,018), and prescence of smokers (*p value* 0,033). Age and sex variations in the prevalence of tuberculosis infection and disease have been reported worldwide, related to biologic mechanisms, socioeconomic and cultural factors may play a role in determining age and sex differences in rates of infection, progression to disease, and treatment outcome [28–30].

While immunization status, history of neighbor contact, population density, lighting level, humidity level, prescence of ventilation, education level, knowledge level, and income level not significantly related with the incidence of child tuberculocis. For more details can be seen in Table 3. Vaccination provides a highly effective strategy to control vaccine-preventable diseases; however, the fact that only a few people develop tuberculosis after *Mycobacterium tuberculosis* infection differentiates tuberculosis from classic vaccine-preventable diseases. The BCG vaccine offers some protection against disseminated forms of tuberculosis in young children (and against leprosy), but does not consistently protect against adult-type tuberculosis [31]. Disseminated BCG disease need careful consideration during genetically modified BCG vaccine development. Additionally, prominent T-cell epitopes are genetically highly conserved, similar to essential housekeeping genes, suggesting that *M tuberculosis* subverts some T-cell-mediated immune response to benefit its own survival and spread. An accurate correlate or biomarker of protection has not yet been identified [32]. The important roles played by innate immune responses and localized (non-circulating) T-cell populations have only recently been described, and these might lead to new discoveries. The diverse range of pathological abnormalities related to the ontogeny of the immune response in children presents opportunities to characterise important immunological mechanisms. Immunohistological studies could also advance understanding of protective immune responses in children with latent *M tuberculosis* infection and provide answers to many unresolved beliefs about latent tuberculosis infection [33,34].

Improvements in the mechanistic understanding of tuberculosis disease and protection are greatly needed, as is a reduction in major policy–practice gaps, since most children in tuberculosis endemic areas are unable to access effective tuberculosis care [35–38].

Table 2. Distribution of House Environment of the Respondents.

Factors Related	Total (f)	Percent (%)
Population Density		
1. Unqualified	2	0.6
2. Qualified	342	99.4
Lighting Level		
1. Unqualified	332	96.5
2. Qualified	12	3.5
Humidity Level		
1. Unqualified	2	0.6
2. Qualified	342	99.4
Presence of Ventilation		
1. Unqualified	84	24.4
2. Qualified	260	75.6
Composition of the Floor		
1. Half or whole is soil	0	0
2. Whole plastered/ tiled	344	100
Presence of Smokers		
1. Present	212	61.6
2. None	132	38.4
Smoking area		
1. No one smokes	164	47.7
2. Terrace	18	5.2
3. Living room	110	32.0
4. Family room	52	15.1

Our study found that present of smoker in home were related to the incidence of tuberculosis (OR 0.498;95%CI=0.320-0.776). The present of smokers in home is an indisputable fact as risk factor of tuberculosis in many studies. Study also proved that family history of smoking related to incidence of TB (OR=2.8, 95% CI:2.3-18.2). Exposure of second-hand smoke increase for both the risk of TB infection and development of active TB disease, among children or adults. Smoking inside home will make their family also exposed. Smoking is also contributing factor for latent infections [odds ratio (OR) 3.20 (1.30-8.20)] [39]. A meta-analysis produced evidence that smoking is a risk factor for TB infection and TB disease [40,41]. TB infection and TB disease are biologically increased by smoking. Smoking causes histological changes in the lower respiratory tract, including peribronchial inflammation, fibrosis, vascular intimal thickening, and destruction of alveoli. This leads to alterations in the epithelial function, such as reduced ciliary activity, decreased clearance of inhaled substances, and abnormal vascular and epithelial permeability, alterations in macrophage number and response [42,43], and decrease in CD4 and CD8 cells that produce interferon gamma and TNF alpha [43].

Table 3. Factors related to the incidence of childhood tuberculosis

Factor Related	Diagnosis of Child Tuberculocis				*p value*	POR	CI (95%)
	f	%	f	%			
Child Age					0.023[a]	1.674	1.093-2.564
a. <3 years old	94	56.6	72	43.4			
b. ≥3 years old	78	43.8	100	56.2			
Sex					0.000[a]	0.445	0.288-0.686
a. Male	60	39.0	94	61.0			
b. Female	112	58.9	78	41,1			
Immunization status					0,118 [a]	1,632	0.926-2.876
a. No BCG	24	40.0	36	60.0			
b. BCG	148	52.1	136	47.9			
Nutritional status					0,033 [a]	0.602	0.386-0.939
a. Malnutrition	52	41.9	72	58.1			
b. Moderate-good nutrition	120	54.5	100	45.5			
Contact History					0,018 [a]	1.757	1.123-2.748
a. Present	72	59.0	50	41.0			
b. None	100	45.0	122	55.0			
History of neighbor contact					0.082 [a]	1.662	0.974-2.833
a. Present	42	60.0	28	40.0			
b. None	130	47.4	144	52.6			
Population Density					0.499 [b]	0.497	0,447-0.553
a. Unqualified	0	0.0	2	0.0			
b. Qualified	172	50.3	170	100.0			
Lighting Level					0.378 [a]	0.488	0.144-1.652
a. Unqualified	168	50.6	164	49.4			
b. Qualified	4	33.3	8	66.7			
Humidity Level					0.499 [b]	0.497	0.447-0.553
a. Unqualified	2	100.0	0	0.0			
b. Qualified	170	49.7	172	50.3			
Presence of Ventilation					0.380 [a]	0.777	0.474-1.273
a. Unqualified	46	54.8	38	45.2			
b. Qualified	126	48.5	134	51.5			
Presence of Smokers					0.003 [a]	0.498	0.320-0.776
a. Present	92	43,4	120	56,6			
b. None	80	60,6	52	39,4			
Smoking area					0.013 [a]	0.569	0.371-0.873
a. Terrace/ Living room/ Family room	78	43.3	102	56.7			
b. No one smokes	94	57.3	70	42.7			
Education Level					0.174 [a]	1.445	0.887-2.354
a. Completed kindergarten,elementary school, junior high school	50	56.8	38	43.2			
b. Completed high school, diploma/ bachelor/ master	122	47.7	134	52.3			
Knowledge Level							
a. Poor knowledge	78	52.7	70	47.3	0.446 [a]	1.209	0.789-1.854
b. Good knowledge	94	48.0	102	52.0			
Income Level					0.057 [a]	1.573	1.011-2.448
a. Below minimum wage	72	57.1	54	42.9			
b. Above minimum wage	100	45.9	118	54.1			

[a]*Continuity correction*; [b] *fisher's exact test*

Conclusion

The results of Chi-Square test showed that there was a relationship between the gender of the child (OR = 0.445; 95%CI = 0.241-0.821) and the presence of smokers (OR = 2.007; 95%CI = 1.074-3.751). It was suggested to educate the parents about smoker as the risk factor of child tuberculosis.

Acknowledgments

All authors acknowledge Diponegoro University for funding this research (number: SP DIPA /UN7.5.1/PG/2017). The authors also thank the study participants and The Center for Pulmonary Health Semarang for their cooperation in facilitating the study. The assistance of Putri Septyarini and Hafidzoh Najwati in enrollment of study participants, and in data collection are also highly appreciated.

References

[1] Palomino J C, Leão S C and Ritacco V 2007 *Tuberculosis* ed Juan Carlos Palomino, Sylvia Cardoso Leão and Viviana Ritacco (Brazil: Bernd Sebastian Kamps and Patricia Bourcillier)

[2] Basterrechea M, Sancho R, Idígoras P and Temprano Y M 2009 [Characterization of tuberculosis cases in the foreign- and native-born population in Guipúzcoa (Spain) from 2003-2007]. *Gac. Sanit.* **23 Suppl 1** 74–9

[3] World Health Organization 2017 *Global tuberculosis report 2016* (Geneva, Switzerland: World Health Organization)

[4] Morris S K, Demers A-M, Lam R, Pell L G, Giroux R J and Kitai I 2015 Epidemiology and clinical management of tuberculosis in children in Canada. *Paediatr. Child Health* **20** 83–8

[5] Daniel O J, Adejumo O A, Abdur-Razzaq H A and Ebunoluwa J O 2015 Trend of childhood TB case notification in Lagos, Nigeria, 2011–2014 *Int. J. Mycobacteriology* **4** 239–44

[6] Dodd P J, Gardiner E, Coghlan R and Seddon J A 2014 Burden of childhood tuberculosis in 22 high-burden countries: a mathematical modelling study *Lancet Glob. Heal.* **2** e453–9

[7] World Health Organization 2013 *Global Tuberculosis Report 2013* (Geneva, Switzerland)

[8] Lamb G S and Starke J R 2017 Tuberculosis in Infants and Children *Microbiol. Spectr.* **5**

[9] Singh J P N, Mohammed R, Thayyullathil T, Al-Khal A and Al-Suwaidi Z 2016 Childhood tuberculosis in Qatar. *Int. J. mycobacteriology* **5 Suppl 1** S162

[10] World Health Organization 2015 *Childhood tuberculosis* (Geneva, Switzerland: World Health Organization)

[11] Oshi D C, Chukwu J N, Nwafor C C, Meka A O, Madichie N O, Ogbudebe C L, Onyeonoro U U, Ikebudu J N, Ekeke N, Anyim M C, Ukwaja K N and Aguwa E N 2016 Does intensified case finding increase tuberculosis case notification among children in resource-poor settings? A report from Nigeria *Int. J. Mycobacteriology* **5** 44–50

[12] Karim M R, Rahman M A, Mamun S A, Alam M A and Akhter S 2012 Risk factors of childhood tuberculosis: a case control study from rural Bangladesh *WHO South-East Asia J. Public Heal.* **1** 76–84

[13] Zafar M 2014 Prevelence of latent tuberculosis and associated risk factors in children under 5 years of age in Karachi, Pakistan *J. Assoc. Chest Physicians* **2** 16

[14] Marais B J and Pai M 2007 Recent advances in the diagnosis of childhood tuberculosis. *Arch. Dis. Child.* **92** 446–52

[15] Brent A J 2012 Childhood TB Surveillance: Bridging the Knowledge Gap to Inform Policy *J. Trop. Med.* **2012** 1–6

[16] Lestari T, Probandari A, Hurtig A-K, Utarini A, Starke J, Marais B, Pai M, Sakundarno M, Nurjazuli N, Jati S, Sariningdyah R, Purwadi S, Alisjahbana B, Werf M van der, Bachtiar A, Miko T, Machmud R, Besral B, Yudarini P, Mehta F, Chadha V, Basri C, Loprang F, Jitendra R, Nelson L, Wells C, Kesehatan R D, Irawati S, Arias N, Prihatini S, Rintiswati N, Voskens J, Kimmerling M, Teo S, Alfaham M, Watson J, Berman S, Kibel M, Fourie P, Strebel P, Rie A van, Beyers N, Gie R, Kunneke M, Zietsman L, Donald P, Marais B, Gie R, Schaaf H, Hesseling A, Enarson D, Beyers N, Lobato M, Cummings K, Will D, Royce S, Nelson L, Schneider E, Wells C, Moore M,

Harries A, Hargreaves N, Graham S, Mwansambo C, Kazembe P, Broadhead R, Maher D, Salaniponi F, Marais B, Gie R, Schaaf H, Hesseling A, Obihara C, Starke J, Enarson D, Donald P, Beyers N, Smieja M, Marchetti C, Cook D, Smaill F, Aissa K, Madhi F, Ronsin N, Delarocque F, Lecuyer A, Decludt B, Remus N, Abel L, Poirier C, Delacourt C, Sharma S, Sarin R, Khalid U, Singla N, Sharma P, Behera D, Lolekha R, et al 2011 High caseload of childhood tuberculosis in hospitals on Java Island, Indonesia: a cross sectional study *BMC Public Health* **11** 784

[17] Coghlan R, Gardiner E, Amanullah F, Ihekweazu C, Triasih R, Grzemska M and Sismanidis C 2015 Understanding Market Size and Reporting Gaps for Paediatric TB in Indonesia, Nigeria and Pakistan: Supporting Improved Treatment of Childhood TB in the Advent of New Medicines ed S Esposito *PLoS One* **10** e0138323

[18] Frydenberg A R and Graham S M 2009 Toxicity of first-line drugs for treatment of tuberculosis in children: review *Trop. Med. Int. Heal.* **14** 1329–37

[19] Batra S, Ayaz A, Murtaza A, Ahmad S, Hasan R and Pfau R 2012 Childhood tuberculosis in household contacts of newly diagnosed TB patients. *PLoS One* **7** e40880

[20] Ruwende J E T, Sanchez-Padilla E, Maguire H, Carless J, Mandal S and Shingadia D 2011 Recent trends in tuberculosis in children in London *J. Public Health (Bangkok).* **33** 175–81

[21] Codlin A J, Khowaja S, Chen Z, Rahbar M H, Qadeer E, Ara I, McCormick J B, Fisher-Hoch S P and Khan A J 2011 Short report: Gender differences in tuberculosis notification in Pakistan *Am. J. Trop. Med. Hyg.* **85** 514–7

[22] Schaaf H S, Cilliers K, Willemse M, Labadarios D, Kidd M and Donald P R 2012 Nutritional status and its response to treatment of children, with and without HIV infection, hospitalized for the management of tuberculosis. *Paediatr. Int. Child Health* **32** 74–81

[23] Escott S and Newell J 2007 Don't forget the bigger picture ed A Green *J. Health Organ. Manag.* **21** 506–18

[24] Rafiza S, Rampal K G and Tahir A 2011 Prevalence and risk factors of latent tuberculosis infection among health care workers in Malaysia *BMC Infect. Dis.* **11** 19

[25] Kibret K T, Yalew A W, Belaineh B G and Asres M M 2013 Determinant factors associated with occurrence of tuberculosis among adult people living with HIV after antiretroviral treatment initiation in Addis Ababa, Ethiopia: a case control study. *PLoS One* **8** e64488

[26] Bouscaillou J, Evanno J, Prouté M, Inwolé A, Kabran M, N'Guessan T, Djé-Bi S, Sidibé S, Thiam-Niangoin M, Badou R N, Blanchetière P and Luhmann N 2016 Prevalence and factors associated with HIV and tuberculosis in People who use drugs in Abidjan, Ivory Coast *Int. J. Drug Policy*

[27] Boccia D, Hargreaves J, De Stavola B L, Fielding K, Schaap A, Godfrey-Faussett P and Ayles H 2011 The association between household socioeconomic position and prevalent tuberculosis in Zambia: a case-control study. *PLoS One* **6** e20824

[28] Güler M, Ünsal E, Dursun B, AydIn Ö and Capan N 2006 Factors influencing sputum smear and culture conversion time among patients with new case pulmonary tuberculosis *Int. J. Clin. Pract.* **61** 231–5

[29] Mukherjee A, Sarkar A, Saha I and Chowdhury R 2012 Gender differences in notification rates, clinical forms and treatment outcome of tuberculosis patients under the RNTCP *Lung India* **29** 120

[30] PrayGod G, Range N, Faurholt-Jepsen D, Jeremiah K, Faurholt-Jepsen M, Aabye M G, Jensen L, Jensen A V, Grewal H M S, Magnussen P, Changalucha J, Andersen A B and Friis H 2011 Weight, body composition and handgrip strength among pulmonary tuberculosis patients: a matched cross-sectional study in Mwanza, Tanzania. *Trans. R. Soc. Trop. Med. Hyg.* **105** 140–7

[31] McShane H, Jacobs W R, Fine P E, Reed S G, McMurray D N, Behr M, Williams A and Orme I M 2012 BCG: Myths, realities, and the need for alternative vaccine strategies *Tuberculosis* **92** 283–8

[32] Comas I, Chakravartti J, Small P M, Galagan J, Niemann S, Kremer K, Ernst J D and Gagneux S 2010 Human T cell epitopes of Mycobacterium tuberculosis are evolutionarily hyperconserved *Nat. Genet.* **42** 498–503

[33] Mudenda V, Lucas S, Shibemba A, O'Grady J, Bates M, Kapata N, Schwank S, Mwaba P, Atun R, Hoelscher M, Maeurer M and Zumla A 2012 Tuberculosis and Tuberculosis/HIV/AIDS-Associated Mortality in Africa: The Urgent Need to Expand and Invest in Routine and Research Autopsies *J. Infect. Dis.* **205** S340–6

[34] Zumla A, Atun R, Maeurer M, Mwaba P, Ma Z, O'Grady J, Bates M, Dheda K, Hoelscher M and Grange J 2011 Viewpoint: Scientific dogmas, paradoxes and mysteries of latent Mycobacterium tuberculosis infection *Trop. Med. Int. Heal.* **16** 79–83

[35] Wallis R S, Kim P, Cole S, Hanna D, Andrade B B, Maeurer M, Schito M and Zumla A 2013 Tuberculosis biomarkers discovery: developments, needs, and challenges *Lancet Infect. Dis.* **13** 362–72

[36] Bivas-Benita M, Gillard G O, Bar L, White K A, Webby R J, Hovav A and Letvin N L 2012 Airway CD8+ T cells induced by pulmonary DNA immunization mediate protective anti-viral immunity *Mucosal Immunol.* **6**

[37] Ottenhoff T H M 2012 The knowns and unknowns of the immunopathogenesis of tuberculosis [State of the art] *Int. J. Tuberc. Lung Dis.* **16** 1424–32

[38] Donald P R, Marais B J and Barry C E 2010 Age and the epidemiology and pathogenesis of tuberculosis *Lancet* **375** 1852–4

[39] Den Boon S, Van Lill S W P, Borgdorff M W, Verver S, Bateman E D, Lombard C J, Enarson D A and Beyers N 2005 Association between smoking and tuberculosis infection: a population survey in a high tuberculosis incidence area *Thorax* **60** 555–7

[40] Bates M N, RD R, KR S, DC C, S B and R P 2007 Risk of Tuberculosis From Exposure to Tobacco Smoke *Arch. Intern. Med.* **167** 335

[41] Patra J, Bhatia M, Suraweera W, Morris S K, Patra C, Gupta P C and Jha P 2015 Exposure to Second-Hand Smoke and the Risk of Tuberculosis in Children and Adults: A Systematic Review and Meta-Analysis of 18 Observational Studies ed M Pai *PLOS Med.* **12** e1001835

[42] Hodge S, Hodge G, Ahern J, Jersmann H, Holmes M and Reynolds P N 2007 Smoking Alters Alveolar Macrophage Recognition and Phagocytic Ability *Am. J. Respir. Cell Mol. Biol.* **37** 748–55

[43] Shang S, Ordway D, Henao-Tamayo M, Bai X, Oberley-Deegan R, Shanley C, Orme I M, Case S, Minor M, Ackart D, Hascall-Dove L, Ovrutsky A R, Kandasamy P, Voelker D R, Lambert C, Freed B M, Iseman M D, Basaraba R J and Chan E D 2011 Cigarette Smoke Increases Susceptibility to Tuberculosis—Evidence From In Vivo and In Vitro Models *J. Infect. Dis.* **203** 1240–8

Vaccines Cold Chain Monitoring: A Cross Sectional Study at Three District In Indonesia

L D Saraswati [1]*, P Ginandjar [1], Budiyono[1], Martini[1], A Udiyono[1], and Kairul[2]

[1]Faculty of Public Health, Diponegoro University, Jl. Prof. H. Soedharto,SH, Tembalang, Semarang 50275, Indonesia
[2]District Health Office of Sorolangun, Jambi Province, Indonesia

Email: lintang.saraswati@live.undip.ac.id

Abstract. Vaccine cold chain is a procedure that is used to keep vaccines at a certain temperature. The aim was to describe the vaccine cold chain management of basic immunization program in health centers district. The study design descriptive observational. The samples was Health Centers (HCs); 12 HCs in Sarolangun Jambi Province, 16 HCs in Brebes Central Java Province, and 24 HCs in Temanggung Central Java Provice. Basic immunization vaccines were BCG, DPT-HB-HIB, Polio, and Measles. The results showed proportion of officers graduated from college in Sorolangun, Brebes, and Temanggung were 66.7%, 81.3%, and 52.0% respectively. Proportion of HC that did not have thermometer and fridge freeze was mostly found in Temanggung (52%) and in Sorolangun (91.7%). The heat-sensitive vaccines arranged near the evaporator mostly found in Temanggung (88%), while freeze-sensitive vaccines prepared away of the evaporator mostly in Brebes (100%). Freezer temperature recording chart is not available mostly found in Sorolangun and Brebes (50%), In Sorolangun 41.7% of the officers monitoring 2 times a day and mostly (91.7%) the refrigerator thermostat tape was not isolated. The officers did not perform daily maintenance (50%), weekly (66.7%), and montly (33.3%) mostly found in Sorolangun. From this study we can conclude there is no vaccine immunization program management in Sarolangun, Brebes, and Temanggung that managed according to Ministry of Health Regulations number 42/2013 on the Implementation of immunization. Improvement oversight, control over management of vaccine and management personal, also managing the temperature of the vaccine were recommended.
Keywords : immunization, cold chain Vaccine, Vaccine, Vaccine Potency

1. Introduction

Immunization is widely acknowledged as one of the most successful and cost-effective public health interventions in history, saving two to three million lives every year [1,2]. Immunization is given by using the vaccine as the main components can improve immunity against disease contagious, for that matter its availability must be guaranteed up to targets and still effective. To achieve immunization targets, there are two important thing must be considered. First is the delivery of potent vaccines to children through properly maintained cold chain systems and other is immunization coverage.

In Indonesia, national immunization programs continue to face delivery challenges in terms of sustainably closing the immunization coverage gap. Today, vaccine storage and handling process are important for immunization programs. Vaccine cold chain and logistics systems are central to addressing some of these challenges [3–5]. Every single process of Vaccine cold chain has potential errors such as maintaining vaccine temperature during storage and handling process. Vaccines gradually lose their potency, which is permanent and irreversible, when stored out of the recommended range of temperature [6,7] This notice advises vaccine providers of the importance of proper cold chain management practices.

Cold chain is a procedure to maintain vaccine in certain temperature, in order to ensure the vaccine has potent state, from the manufacturer to the person being immunized [8–11]. Infrastructures of cold chain consist of cool room, freezer, refrigerator, cool box, cool pack, vaccine carrier, and generator [12,13]. The cold chain is important for national immunization programs in tropical climates countries. Ideally, high coverage of vaccination resulted in high immunity. So, primary health care must have adequate knowledge to manage the cold chain because vaccine as main component to yield immunogenicity toward program of diseases that can be prevented by immunization need specific cold chain [14–16]. Good practices to maintain proper vaccine storage and handling can ensure that the full benefit of immunization is realized. This study described vaccines cold chain in all primary health care in Sarolangun, Temanggung and Brebes.

2. Method

2.1. Study site

The study was carried out in all primary health cares in three district health offices, namely: Sorolangun, Brebes and Temanggung. There were a total of 128 primary health centre in those districts; 12 in Sorolangun, 16 in Brebes, and 24 in Temanggung.

2.2. Study Design

This was an observational descriptive study, to describe management of cold chain three district health cares. Variables consisted of education level of vaccine manager, training, budget of cold chain, temperature in refrigerator, availability of cool pack in refrigerator, distance between vaccines, distance between heat sensitive vaccine with evaporator, distance between freeze sensitive vaccine with evaporator, availability of thermometer to monitor temperature in refrigerator, availability of freeze tag to monitor expose of cold temperature to vaccine, availability of recording graph, thermostat of refrigerator, number of cool pack in the refrigerator, number of vaccine carrier, maintenance of refrigerator, and test of vaccine potency using shake test.

2.3. Ethic Statement

Ethical issues (Including plagiarism, Informed Consent, misconduct, data fabrication and/or falsification, double publication and/or submission, redundancy, etc.) have been completely observed by the authors. Ethical approval was obtained from the Committee of Public Health Research Ethics, Diponegoro University (112/EC/FKM/2016). Informed consent was acquired from cold chain officers.

3. Results

There was no differences in Sorolangun, Temanggung and Brebes for percentage of budgeting tools of cold chain (p value 0.0), percentage of source of funding (p value 0.0), percentage of number activities related with cold chain (p value 0.002), percentage of fee management for 1 months (p value 0.0), percentage received funding for cold chain (p value 0.005), p percentage of fee management for 1 year (p value 0.01), temperature monitor (p value 0.0), number of cool pack (p value 0.007), the existence of addition such as box (p value 0.013), the use of soap for laundry of vaccine carrier (p value 0.001), and maintenance for one week (p value 0.008).

There were no differences between three district for percentage of budget cold chain (p value 0.333), temperature of refrigerator (p value 0.342), calibration (p value 0.69), frequency of calibration (p value 0.569), prove of calibration (p value 0.61), position of cool pack (p value 0.926), number of cool pack (p value 0.302), another materials related with cold chain (p value 0.59), space between boxes of vaccines (p value 0.064), vaccine sensitivity (p value 0.302), expired dates (p value 0.302), VVM vaccines (p value 0.302), FIFO (p value 0.302), EEFO (p value 0.302), HS vaccines (p value 0.302), vaccine freeze sensitive (p value 0.302), number of reeze tag (p value 0.07), monitoring of freeze tag (p value 0.332), flag documentation (p value 0.323), temperature graph (p value 0.053), frequencies of recording (p value 0.464), routine (p value 0.489), Thermostat isolator (p value 0.066), the use of cool pack (p value 0.302), the use of ice cube (p value 0.145), freezing 12 hours (p value 0.138), thermometer (p value 0.394), the laundry of vaccine carrier (p value 0.138), maintenance per day (p value 0,55), maintenance per months (p value 0.138), and the result of shake test (p value 0.302).

Table 1. Description of each Indicator of Cold Chain in Sorolangun, Brebes and Temanggung Sub District

Indicator	Sarolangun	Brebes	Temanggung
All vaccines keeped in temperature + 2°C s/d 8°C	The temperature was above 8°C, it is 21,2°C and there are 3 primary health care that the temperature was unknown.	The temperature was between 2°C - 8°C. So it is suitable with the regulation of Ministry of Health number 42/2013	There was no thermometer inside.
The bottom of refrigerator was put cool pack as cooler resistant	All primary health care have coolpack more than 8	All primary health care have coolpack more than 8	All primary health care have coolpack more than 8
There was cool pack inside of refrigerator	There was cool pack inside refrigerator (100%)	There was cool pack inside refrigerator (100%)	There are no cool pack inside refrigerator
There must be a range for vaccine box inside of refrigerator (1-2 cm)	There was no space inside the refrigerator between 1 box to another	The vaccine was arrange properly inside refrigerator. There was space between 1 box to another	There was no space inside the refrigerator between 1 box to another
Vaccine position didn't arrange by the kind of vaccine (heat sensitive or freeze sensitive)	The vaccine was not arrange based on the sensitivity of the vaccine	All vaccine was arrange properly based on its sensitivity	The vaccine was not arrange based on the sensitivity of the vaccine
There was thermometer inside of refrigerator	there was no thermometer inside	there was no thermometer inside	there was no thermometer inside
There was graphic charts of temperature	50% have graphic charts	50 % have graphic charts	80% have graphic chart
Thermostat taped isolated in order to the temperature of refrigerator		All primary health care in Brebes have isolated thermostat	There are fifteen primary health care that don't have isolated thermostat
There was vaccine carrier			Nine primary health care in Temanggung didn't have vaccine carrier more than 3
There was maintenance of the refrigerator	There were a lot of frost inside the refrigerator. It means the maintenance of the refrigerator didn't do regularly (daily, weekly, or monthly)	There were no frost inside the refrigerator. It means the maintenance of the refrigerator do regularly (daily, weekly, or monthly)	There were a frost inside the refrigerator. It means the maintenance of the refrigerator didn't do regularly (daily, weekly, or monthly)

We also found three vaccines of *Tetanus Toxoid* was suspected freezing from three refrigerator from different health cares, then conducted shake test and founded that vaccines had been freezing before so they might be impotent vaccines. Damage from accidental freezing can result in potency loss

IOP Conf. Series: Earth and Environmental Science **116** (2018) 012082 doi:10.1088/1755-1315/116/1/012082

for freeze-sensitive vaccines such as diphtheria, tetanus, pertussis, liquid *Haemophilus influenzae* type b (Hib), hepatitis B, and inactivated polio virus [14–16].

Table 2. Cold Chain Requirements in Sorolangun, Brebes and Temanggung Sub District.

No	Cold Chain Requirements	Percentage (%) Sorolangun	Brebes	Temanggung
1	Minimum Educational Background is senior high school	33,3% cold chain officer, educational background is senior high school, and 66,7% university	18.8% cold chain office is senior high school, 81.3% bachelor degree	48% is senior high school
2	Had been trained about cold chain vaccine	100% officer had been trained minimum one time	62.5% had been trained minimum one time	28% had been trained minimum one time
3	Funding of cold chain	100%	43.8%	84%
4	All vaccines keeped in temperature + 2°C s/d 8°C	66,7%	100%	4%
5	The bottom of refrigerator was put cool pack as cooler resistant	100%	100%	88%
6	There was cool pack inside of refrigerator	100%	100%	16%
7	There must be a range for vaccine box inside of refrigerator (1-2 cm)	58,3%	100%	72%
8	Vaccine position heat sensitve put nearby evaporator	33,3%	100%	88%
9	Vaccine position freeze sensitive put away form evaporator	75%	100%	88%
10	There was thermometer inside of refrigerator	75%	50%	48%
11	There were graphic charts of temperature	50%	50%	80%
12	Thermostat taped isolated in order to the temperature of refrigerator	8,3%	100%	40%
13	There was vaccine carrier	100% had vaccine carrier > 3 pieces	100% had vaccine carrier > 3 pieces	64% had vaccine carrier > 3 pieces
14	There was maintenance of of refrigerator (Daily)	50%	100%	72%
	There was maintenance of of refrigerator (Weekly)	33,3%	100%	88%
	There was maintenance of of refrigerator (Monthly)	66,7%	100%	80%

Figure 1(a) There was no thermometer inside

Figure 1(b) Temperature was found 21˚C, contained vaccine.

Figure 2(a) A broken thermometer inside

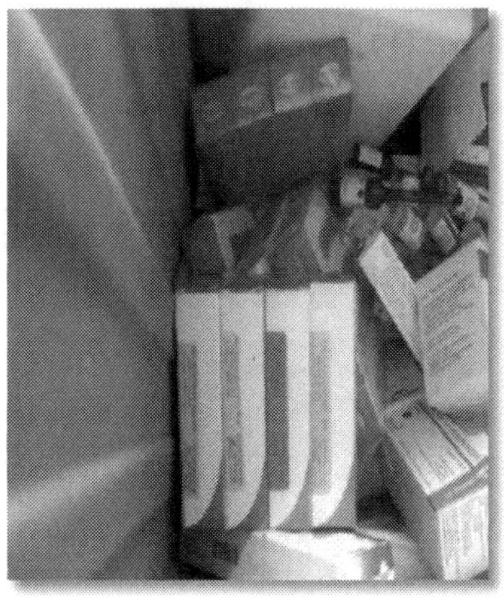

Figure 2(b) Arrangement of vaccine: no space and scrambled

4. Discussion

Vaccines (HB Uniject, BCG, Polio, DPTHB, Measles, DPT-HB-Hib, TT, DT and TD vaccines) taken from the Provincial Health Office then stored in the vaccine storage room, stored in refrigerators and freezers while solvents and dropper (pipette) is stored on a separate rack. In primary health care, the vaccine has been taken from the District Health Office, before it is given to the patient, first stored in the freezer as well as the vault as the storage of solvents and dropper (pipette).

Monitoring of storage temperature vaccine is very important in setting quickly whether the vaccine is still feasible used or not, always pay attention to vaccine vial monitor (VVM) that exist on each one of vaccine to find out if vaccine is still feasible to use. Vaccine potency depends on correct storage temperatures [14,17,18]. This research indicates that the availability of cold chain is not satisfied and did not meet criteria from Ministry of Health of Indonesia or WHO. For a vaccination program to be effective and provide maximum protection, it is essential for vaccines to be stored in optimum conditions. If vaccines become too hot or too cold, their effectiveness may be compromised. Vaccines naturally biodegrade over time and this process may be accelerated if they are stored outside the recommended temperature range. Vaccines must never be frozen because in addition to vaccine deterioration, this increases reactogenicity by irreversible denaturing of the proteins in the vaccine. This causes the emulsions in the vaccine to become unstable and produces hairline cracks in the ampoule, vial or prefilled syringe, potentially contaminating the contents. Any glass spicules produced may also cause serious local adverse reactions. These factors may result in the failure of the vaccine to stimulate the intended immune response [9,14,15,19,20]. The series of storage locations and transport vehicles needed to get vaccines from manufacturers to vaccine recipients), and can traverse great distances and different climates. Many locations do not have reliable refrigerators or freezers (i.e., cold storage) or reliable means of monitoring temperatures in these devices [10,21–23].

The result of this study similar with studies in India[16,23,24]. Furthermore, it also similar with the result about the capacity of officers in charge to properly monitor the availability of vaccine storage equipment in all health facilities at the North West Region of Cameroon that still limited [25].

Air circulation in the room around the refrigerator should also be good, can be done by spacing between the refrigerators with the back wall and between refrigerators about 10-15 cm and the refrigerator should not be exposed to direct sunlight. Primary health cares some were found not to kept cool packs at the bottom of the storage so that if there is a power outage the temperature in the storage cabinet can immediately change whereas according to the guidelines on the bottom of the refrigerator should be placed cool pack in order to withstand cold and keep the temperature stability in the event of a power outage. The vaccine box has a spacing of at least 12 cm, this is to keep the temperature circulation inside the storage cabinet. According to the guidelines, a heat sensitive vaccine is placed near or attached to the refrigerator wall while the freeze sensitive vaccine is placed not to stick to the refrigerator wall, which if not done can affect the quality of the vaccine and may even cause vaccine damage [10,14].

Optimal immunization programs depend on a mix of stable vaccines used optimally in accordance with their stability, fitted with appropriate temperature monitoring devices on each vial, and by assessing the potential threat of temperature excursions using a risk management approach [10]. One approach of improving the integrity of the vaccine delivery system is to remove the obstacles in maintaining a refrigerated supply chain, which can be approached by a combination of three methods: developing ways to prevent vaccine freezing, for example, by using cold water packs instead of ice for transport; using more effective time–temperature monitors that can go on every vial of vaccine from the time it leaves the manufacturer until it reaches the recipient; and by better control and regulation of the supply chain [10].

The door rubber were not sealed, which should be according to the guidelines of the refrigerator/freezer gate should be tightly closed to prevent exhausted cold air which can affect the temperature of the storage cabinet. Some are known because they do not have internal or external temperature gauges that are damaged so that the temperature record is not done and does not have freeze tags. If it does not have a temperature gauge it can cause the vaccine temperature to be unknown so that if the vaccine is given the vaccine it is not able to stimulate the body's immune optimally, abnormal vaccine temperature is one of the main problems in vaccine storage. According to the Management Guidelines Cold Chain refrigerator/freezer should always be available inside and outside the thermometer in order to control the temperature to record the temperature of the vaccine 2 times a day and at holidays [21, 26].

Recording conditions of cold chain (logbook) were not made by some primary health cares, this is due to negligence of the officers. This logbook report must be made to determine the condition of the cold chain [22].

Vaccine stocks at primary health cares are often damaged due to the fact that no vaccines are taken and the travel time is far away, frequent congestion, no temperature monitoring device in cold box, limited cool pack in cold box so it cannot maintain vaccine temperature during the trip. According to the guidelines, packing of freeze sensitive vaccines should be equipped with a freezing indicator in order to know whether during the course of the vaccine has clotted or not so that testing can be done immediately [13, 26].

Treatment of cool packs to be used does not conform to the Cold Chain Management Guidelines ie before use should be put into the freezer for 24 hours at 2°C-8°C. This can damage the quality of the vaccine because the cool pack that is inserted into the freezer for 24 hours at 2°C-8°C can maintain the temperature of the vaccine during the trip to the primary health care's [10].

Cold Boxes should be cleaned before and after use and contain 4 pieces of cool pack [10,27] but seen in the field of each cold box there is only 1 cool pack and cold box was never cleaned before or after use but directly stored it on the storage. This is because the limited places in the cold box and cold box are always in a clean state so no need to be cleaned again.

5. Conclusion

There is no cold chain of vaccine in all primary health centers in 3 Districts (Sorolangun, Temanggung, and Brebes) meet the requirement of health ministry (*Permenkes* 42/2013 about immunization). Recommended for DHO to allocate budget for replace the broken tools of cold chain such as freeze tag, thermometer, vaccine carrier, rubber in refrigerator door or else, also budgeting for calibration such as thermometer and electricity of refrigerator.

Acknowledgments

The authors thank Diponegoro University and UNICEF Indonesia for funding this study (number: SP DIPA -1051-35/UN7.5.1/PG/2016). The authors also thank the study participants, District Health Office of Sorolangun, Temanggung, and Brebes with all their Primary Health Centre for their cooperation in facilitating the study. The assisstance of Puji Kurniasih and Tyas Wigati in enrollment of study participants, and in data collection are also highly appreciated.

References

[1] Chan M 2014 The contribution of immunization: saving millions of lives, and more. *Public Health Rep.* **129 Suppl 3** 7–8

[2] Toumi M and Ricciardi W 2015 The Economic Value of Vaccination: Why Prevention is Wealth. *J. Mark. access Heal. policy* **3**

[3] Zaffran M, Vandelaer J, Kristensen D, Melgaard B, Yadav P, Antwi-Agyei K O and Lasher H 2013 The imperative for stronger vaccine supply and logistics systems *Vaccine* **31** B73–80

[4] Humphreys G 2011 Vaccination: rattling the supply chain *Bull. World Health Organ.* **89** 324–5

[5] Kaufmann J R, Miller R and Cheyne J 2011 Vaccine Supply Chains Need To Be Better Funded And Strengthened, Or Lives Will Be At Risk *Health Aff.* **30** 1113–21

[6] World Health Organization 1998 *Global programme for vaccines and immunization: expanded programme on immunization: safe vaccinehandling: cold chain and immunizations* (Geneva, Switzerland)

[7] World Health Organization 2004 *Immunization in practice: module 3: the cold chain* (Geneva, Switzerland)

[8] Rogers B, Dennison K, Adepoju N, Dowd S and Uedoi K 2010 Vaccine Cold Chain *AAOHN J.* **58** 391–400

[9] Rogers B, Dennison K, Adepoju N, Dowd S and Uedoi K 2010 Vaccine Cold Chain: Part 1. Proper Handling and Storage of Vaccine *AAOHN J.* **58** 345–6

[10] Kartoglu U and Milstien J 2014 Tools and approaches to ensure quality of vaccines throughout the cold chain. *Expert Rev. Vaccines* **13** 843–54

[11] Nelson C M, Wibisono H, Purwanto H, Mansyur I, Moniaga V and Widjaya A 2004 Hepatitis B vaccine freezing in the Indonesian cold chain: evidence and solutions. *Bull. World Health Organ.* **82** 99–105

[12] Kartoglu U and Milstien J 2014 Tools and approaches to ensure quality of vaccines throughout the cold chain *Expert Rev. Vaccines* **13** 843–54

[13] Karp C L, Lans D, Esparza J, Edson E B, Owen K E, Wilson C B, Heaton P M, Levine O S

and Rao R 2015 Evaluating the value proposition for improving vaccine thermostability to increase vaccine impact in low and middle-income countries *Vaccine* **33** 3471–9

[14] Kumru O S, Joshi S B, Smith D E, Middaugh C R, Prusik T and Volkin D B 2014 Vaccine instability in the cold chain: Mechanisms, analysis and formulation strategies *Biologicals* **42** 237–59

[15] Yakum M N, Ateudjieu J, Walter E A and Watcho P 2015 Vaccine storage and cold chain monitoring in the North West region of Cameroon: a cross sectional study *BMC Res. Notes* **8** 145

[16] Naik A K, Rupani M P and Bansal R K 2013 Evaluation of vaccine cold chain in urban health centers of municipal corporation of surat city, Western India. *Int. J. Prev. Med.* **4** 1395–401

[17] Plotkin S, Orenstein W and Offit P 2013 *Vaccines* (Saunders)

[18] WHO, Dept. of Vaccines and UNICEF 2002 Ensuring the quality of vaccines at country level : guidelines for health staff *IRIS* 40

[19] Hefti K and David G 2013 Safe handling of vaccines: the rewards of rigorous routines. *S. D. Med.* **Spec no** 119–22

[20] McColloster P J 2011 US vaccine refrigeration guidelines: loose links in the cold chain. *Hum. Vaccin.* **7** 574–5

[21] Kartoğlu Ü, Nelaj E and Maire D 2010 Improving temperature monitoring in the vaccine cold chain at the periphery: An intervention study using a 30-day electronic refrigerator temperature logger (Fridge-tag®) *Vaccine* **28** 4065–72

[22] Sinha A K, Verma A R, Chandrakar A, Khes S P, Panda P S and Dixit S 2017 Evaluation of cold chain and logistics management practice in Durg district of Chhattisgarh: pointer from Central India *Int. J. Community Med. Public Heal.* **4** 390

[23] Krishnappa L, Anniappan A B, Voderhobli N H, Kumar Krishna S, Yathiraj S and Sreekantaiah P 2014 Evaluation Of Cold Chain Practices In Urban Health Centers Of A Metro City In India *Natl. J. Community Med.* **5**

[24] Singh A, Grover S, Sinha S, Das M, Somvanshi P and Grover A 2017 Mechanistic Principles Behind Molecular Mechanism of Rifampicin Resistance in Mutant RNA Polymerase Beta Subunit of Mycobacterium tuberculosis *J. Cell. Biochem.*

[25] Yakum M N, Ateudjieu J, Walter E A and Watcho P 2015 Vaccine storage and cold chain monitoring in the North West region of Cameroon: a cross sectional study *BMC Res. Notes* **8** 145

[26] Rogie B, Berhane Y and Bisrat F 2013 Assessment of cold chain status for immunization in central Ethiopia. *Ethiop. Med. J.* **51 Suppl 1** 21–9

[27] Kristensen D D, Lorenson T, Bartholomew K and Villadiego S 2016 Can thermostable vaccines help address cold-chain challenges? Results from stakeholder interviews in six low- and middle-income countries *Vaccine* **34** 899–904

The Evidence of Imposex in Turbo sp. from Ujungpiring Waters of Jepara

RAT Nuraini*, R Hartati, H Endrawati, Widianingsih, MJ Rachma, RT Mahendrajaya

Marine Science Departement, Faculty of Fisheries and Marine Science, Diponegoro UniversityJl. Prof. Soedharto, SH, Tembalang, Semarang 50275

Email: riaazizahtn@yahoo.co.id

Abstract. Imposex is an endocrine disruption syndrome, in which females of marine gastropods develop sexual characteristics of males (penis and/or vas deferens). This syndrome is caused by tributyltin (TBT) or triphenyltin (TPT), toxic organotin compounds found in naval paints used as antifouling system in boats and artificial structures. The main objective of this study was to assess the incidence and severity of imposex in *Turbosp*.A hundred individus of *Turbo sp* was collected from Ujung Piring waters of Jepara, and observed for their sex and imposex. The results showed that the ratio male and female was 1.13 : 1, and the frequency of imposex females was 64.58%. The relative penis size were longer in imposex female than in male. The possibility of pollution was clearly shown from high evidence of imposex female.
Keywords:gastropod, imposex, *Turbo sp*, Jepara

1. Introduction

A shipping activity in coastal areas is an example of the causes of marine pollution. Generally, The shipping activity happens on the ships which use antifouling paint containing of tributyltin on the body structure and ship propellers. The effects of the use of tributyltinare very hazardous to the incident conditions including macrozoobenthic organisms existence in it.

According to [1] the concentration of tributyltin 1 µg./l itself can lead to the imposex. For example, in 2003 in Jakarta Bay It was found that the concentration of tributyltin in sea air was reaching 15 µg./l and the bladder in sediment attained 649 µg./g. So the imposex level occured in Jakarta Bay is very high.

The regulation in the use of tributyltin was applied in many countries in the late 1980s, which was aimed to reduce the vessels further so, the result of the regulation could bring down the concentration of tributyltin in the estuary water column normally, which was dominated by the small vessels. However, it does not mean concentration reducing of tributyltin was over. The contamination happened previously remained enough high level at some events and the chronic effects of sensitive marine organisms. This is due to the attraction between tributyltin and sediment so it can settle and locate in the sediment. The deposition of tributyltin in the sediments may affect on benthic organisms that can accumulate tributyltin from sediments in the gastropods so anomalies occur in the anatomy of the gastropods [2].

One of the biota affected by tributyltin is gastropods. Tributyltin can lead to the occurrence of imposex on the female and male antelope, that result in the destruction of the reproductive system so it can be able to affect their population. This study is aimed to observe the occurrence of the nozzle on the body of the gastropod Turbo sp. Which is located in Ujung Piring waters which has a shipping activity in Mlonggo Sub-district, Jepara Regency.

2. Material and method

The research material used in this research is gastropod organism typed of Turbo sp. It was obtained from Ujung Piring water, Mlonggo sub-district, Jepara regency. The number of gastropod samples used is 100 pieces, with heights of ranging approximiately from 1.2 cm to 2.5 cm. Samples were acquired at each point and then inserted into a plastic zip lock and stored in an ice cool box to keep the sample fresh until it is brought to the laboratory for the further analysis.

Sample of gastropods that have been compiled and cleaned in the laboratory, then placed on the tray for identification. Sampling can be done by viewing at the morphology of the gastropod shell, such as the shape of the shell, the color of the shell, and the shape of the operculum, which is then matched with the gatropod guiding book [3]. The entangled samples in the height of the shell, the width of the shell, and the width of the operculum use a sliding span having a precision of 0.1mm. Once measured, the sample is weighed out by its total weight and the sample of shell is cracked by using a hammer cautiously. The body gastropods from the shell isweighed to get the weight data.

The imposex observation on the gastropod samples were carried out by using petri dishes and the help of loops. The first thing to observe is to know the sex of the gastropod by looking at the color of the gonads. Female sex can be determined based on the presence of a cream or white capsule gland (cg), albland gland (ag) that is dark brown or black, sperm ingesting gland (sig) white and yellow ovary. While the male sex can be ascertained based on the existence of prostate gland (pg) is yellow or brownish orange and testis are yellow or orange. In the male gastropods can be found seminalis vesicle (sv) forming channels resembling a white tube winding [4].

Then the next step is to see the position of the gastropod's tentacle, because the location of the gastropod reproduction is at the lower right part equivalent to the location of the tentacles and close to the gonads [5]. Individual which has the female's imposex, can be inferred to have pseudopenis and or vas deferens at the vaginal opening [1]. While individuals which have the male imposex can be inferred through their small prostate gland and the addition of a large penis to the seminal vesicle [6]. While the ranging height of pseudopenis is the total height of pseudopenis (mm) divided by the number of measurements [1].

The determination of the imposex can be determined by calculating the Relative Penis Size Index (RPSI) according to the manner researched by [7] and [4]. Relative Penis Size Index (RPSI) is used for each population defined as the most of the average female penis formulated as a percentage of most average male penis [8]. RPSI calculation formula as follows:

$$RPSI = \{(\text{Average height of female's pseudopenis})^3/(\text{Average height male's penis})^3\} \times 100\%$$

According to [5], the imposex can be seen through the value of the frequency of the imposex (Fi) which can be obtained from the following formula :

$$Fi = (\text{No. of Imposex female/No. Total female}) \times 100\%$$

3. Result and discussion

The result of observation on 100 indvs. *Turbo* sp. from Ujung Piring Waters, Sekuro Village,JeparaRegency showed that there were three gastropods condition, i.e. normal male (53 indvs.),normal females (16 indvs.) and imposexfemales (31 indvs.)

The height and width of samples were ranged of 14.35 – 24.7 mm (height) and 13 – 21.6 mm (width), while the total and body weight were 2.12– 8,7 gr.and 0,85–2,78 gr. The lenght of male's penis and female's pseudopenis were 1,1–8.7 mm.and 1.5–7.2 mm.(Table 1).

Table 1.The morphometry of *Turbo* sp. From Ujungpiring waters.

No.	Status	Total	Shell height (mm.)	Shell Width (mm.)	Total Weight (gr.)	Body Weight (gr.)	Penis or Pseudopenis Height (mm)
1.	Male	53	14.35-24.6	12.85-21.6	1.74-8.34	0.98-2.73	1.1-8.7
2.	Female	16	14-24.7	13-21.2	2,2-8,7	0.85-2.78	-
3.	Imposex female	31	14.65-22.65	13-20.3	2,43-6.5	1,3-2.25	1.5-7.2

There was significant different on ratio of shell height of imposex female and normal female (Table 2). The evidence of imposex female *Turbo* sp. was 64.58%. Relative Penis Size Index (RPSI)of *Turbo* sp. From Ujungpiring Waters was 187.86%. The average penis height on normal male and pseudopenisof imposex female was shown in Table 3.

Table 2. Average shell height imposex female and normal female and ratio of imposex female and normal female *Turbo* sp.

Status of female	Total	Average shell height $\pm$ SD	Ratio of imposex female and normal female
Imposex Female	31	18.33 $\pm$ 2.42	1,82
Normal Female	16	10.09 $\pm$ 2.69	

Table 3. Average penis height on normal male and pseudopenisof imposex female.

JenisBetina	Average penis or pseudopenis height $\pm$ SD (mm)	RPSI(%)
Male	3.46 $\pm$ 1.5	187.96%
Imposex female	4.26 $\pm$ 1.3	

Evidence to date suggests that imposex is generally caused by tributyltin (TBT) [8] [9] [10], a manufactured chemical massively used in self-polishing antifouling paints, which until recently was the most effective solution to prevent fouling on boat hulls [11]. Numerous investigations have linked imposex in Neogastropods with TBT contamination from antifouling paint in the marine environment [12] [13] [14] [15].

As the evidence of 64.58% imposex among female *Turbo* sp it is suggested that TBT from antifouling paints on small ships seems the most likely source of imposex in the populations of Ujungpiring Waters, Jepara. The present results corroborate a previous study and confirm the high contamination detected in Demak watres [16]. This phenomenon is known to be induced by TBT at very low ambient concentrations, just a few ng/l are sufficient and imposex development seems dose-dependent [17]. Furthermore, high correlations between likely sources of TBT and imposex induction have been confirmed in several studies by chemical analyses of water, sediments and gastropod tissues. For instance, significant correlations between TBT concentration in seawater and imposex indices have been reported for *Nucella lapillus* [18] [19],*Thaïs clavigera* [20], *Hexaplex trunculus* [21], *S. haemastoma* [22], and *Stramonita haemastoma* [23].This sexual aberration is mainly attributable to TBT in seawater [22] [24]. Imposex levels and organotin body burden can be assumed to be highest in populations nearest to major shipping activities and gradually lower with increasing distance from these sources [25] [26].

4. Conclusion

In summary, imposex was recorded in *Turbo sp.* from three sites of the Ujungpiring waters of Jepara Regency. The evidence of imposex female *Turbo* sp. was 64.58%. Relative Penis Size Index (RPSI)of

Turbo sp. From Ujungpiring Waters was 187.86%. The smallest shell height of female imposex was 14,65 mm suggested the young age of gastropod affected by TBT.

References

[1] Widianwari, P. 2010. *Ilmu Kelautan* **02**: 399 – 414.
[2] Sudaryanto, A. 2001. *JurnalTeknologiLingkungan.* **02 (03)**: 241 – 246.
[3] Oliver, A.P.H. 2004. Guide to Seashells of The World. London : Octopus Publishing Group Ltd.
[4] Tan, K. S.1997. Mar. Pollut. 1997. Bul., **34:** 577-581
[5] Kusumastuti, R., Widianingsih, dan R.A.T. Nuraini. .2013. *Journal of Marine Research.* **2(3):**114-122.
[6] Yusuff, M. F., Zulkifli, S. Z., Ismail, A. 2011. Journal of Tropical Marine Ecosystem **2**: 1-9.
[7] Bryan, G. W., Gibbs P. E., Hummerstone, L. G. R. 1986. *J. Mar. Biol. Assoc. U.K.,.* **66(3)**: 611-640.
[8] Gibbs P. E., P.L. Pascoe, and G.R. Burt. 1991. *J. Of The Mar. Biolo. Ass. Of the U.K.* **67**:507-523.
[9] Spooner N., Gibbs P.E., Bryan G.W. & Goad L.J. 1991. *Mar.Environ. Res.*, **32:** 37-49.
[10] Sayer C.D., Hoare D.J., Simpson G.L., Henderson A.C.G., Eleanor R. & Liptrot E.R. 2006.. *Environ. Sci. Technol.*, **40(17)**: 5269-5275.
[11] Terlizzi A., Scuderi D., Faimali M., Minganti V. & Geraci S.1997. *Biol. Mar. Medit.*, **4(1)**: 496-499.
[12] Bryan G.W., Gibbs P.E., Burt G.R. & Hummerstone L.G. 1987.. *J. Mar. Biol. Assoc. U.K.*, **67**: 525-544.
[13] Gibbs P.E, Pascoe P.L & Burt G.R. 1988.. *J. Mar. Biol. Assoc. U.K.*, **68**: 715-731.
[14] Oehlmann J., Stroben E. & Fioroni P. 1996a. *Malacol. Rev. Suppl.*, **6**,: 149-155.
[15] Oehlmann J., Fioroni P., Stroben E. & Market B. 1996b. *Sci. Total Environ.*, **188**: 205-223.
[16] Basuki, L.R., Widianingsih, R.A.T.Nuraini. 2016. J. Mar. Res. **23(4):**1-8
[17] Bryan G.W., Gibbs P.E., Hummerstone L.G. & Burt G.R. 1986. *J. Mar. Biol. Assoc. U.K.*, **66**: 611-630.
[18] Stroben E., Oehlmann U.S., Fioroni P. and Oehlmann J. 1995. *Haliotis*, **24**: 1-12.
[19] Birchenough, A.C., Barnes N., Evans S.M., Hinz H., Kronke I.C.& Moss C. 2002. *Mar. Pollut. Bull.*, **44**: 534-543.
[20] Horiguchi T., Shiraishi H., Shimizu M. & Morita M. 1994. *J. Mar. Biol. Assoc. U.K.*, **74**: 651-669.
[21] Axiak V., Micallef D., Muscat J., Vella A. & Mintoff B. 2003. *Environ. Int.*, **28**: 743-749.
[22] Fernandez A.M., Limaverde A.M., Castro I.B., Almeida A.C.M. & Wagener A.L.R. 2002., *Cadernos de Saúde Pública*, Rio de Janeiro.**18(2):** 463-476.
[23] El Mortaji, H., N.Elkhiati, A. Benhra, B. El Haimeur, M. Bouhallaoui, S. Benbrahim, M. Kabine and M. Ramdani. 2011. *Bulletin de l'Institut Scientifique,* Rabat, section Sciences de la Vie **33 (1)**: 13-18.
[24] Trigui El Menif N., Lahbib Y., Ramdani M., Boumaiza M. & Le Pennec M. 2007. *Vie Milieu*, **57(1-2):** 37-42.
[25] Leung K.M.Y., Kwong R.P.Y., Ng W.C., Horiguchi T., Qiu J.W., Yang R., Sang M., Jiang G., Zheng G.J. & Lam P.K.S. 2006. *Chemosphere*, **65**: 922-938.
[26] Lemghich I. & Benajiba M.H. 2007.. *Ecol. Indic.,* **7:** 209-214.

IOP Conf. Series: Earth and Environmental Science **116** (2018) 012084 doi:10.1088/1755-1315/116/1/012084

Antibacterial Activity Symbiotic Fungi of Marine Sponge *Axinella* sp., *Aspergillus Sydowii* on Four Growth Medium

S Widyaningsih[1], A Trianto[2,4], OK Radjasa[2], K Wittriansyah[3]

[1]Oceanography Department, Faculty of Engineering and Marine Science, Hang Tuah University, Surabaya, Indonesia
[2]Marine Science Department, Faculty of Fisheries and Marine Science, Diponegoro University, Semarang, Indonesia
[3]Fishery Mechanical Engineering, Cilacap, Indonesia
[4]Natural Product Laboratory, Centre of Research and Services (Laboratorium Terpadu) Diponegoro University, Indonesia

Email: sekar.widyaningsih@hangtuah.ac.id

Abstract. Many infectious diseases caused by *Escherichia coli* and *Staphylococcus aureus* which turned into a resistant pathogen. A symbiotic fungi of marine sponge *Axinella* sp., *Aspergillus sydowii* from the waters of Riung, East Nusa Tenggara, Indonesia showed antibacterial activity, cultured on the four media, MEB (ST), Noni Juice Media (MG), avocado leaves media (AL), and Soursop leaves media (SR). The symbiotic fungi was cultured for 14 days on each media. The largest weight of symbiotic fungi biomass on ST media 138,95gr and at least 99,12gr of AL media. Purification of bioactive compound is carried out using separatory funnel, and column chromatography. The highest rendemen of extracts on SR media was 3,67%, while the lowest in ST media was 1,22%. The bioactive test used diffusion agar method. Fungi extracts from four mediums have bioactivity against, *E. coli* and *S. aureus*. The biggest inhibition zone obtained from the extract of MG KN-15-3-1-3, with inhibition zone 10.71mm and 10.98mm against *E. coli* and *S. aureus*.
Keywords: antibacterial, sponge, symbiotic fungi, *Aspergillus sydowii*

1. Introduction

Axinella sp. is one of marine sponge that have many biological activities [1], such as antifungi, anti-microbial, aniviral, insecticidal activity, cytotoxicity, and antioxidant [2-7]. Axinella is also source of bioactive compounds [1]. Some compounds from marine sponge *Axinella* sp. are Axinastatin, Cyclonellin, and Daminin [8-10].

Only few of them passed to the preclinical and clinical stages [11]. Supply is a major problem for drug development. Therefore, another alternative is required, by utilizing microorganisms such as fungal symbionts [12-13]. Kelecom (2001) [14] mentioned that symbiotic microorganisms like fungi symbiont produce secondary metabolites that are similar to those produced by its host. Fungi symbiont can be isolated and cultured [15].

Growth media is one of the major factors in isolating and growing microorganisms. This is because it can affect the content of compounds in these microorganisms [16-17]. Atalla et al. (2008) [18], stated that the formation of a secondary metabolite in micro-organisms will be affected by conditions and growth media.

This paper describes the effect of growth media on the biomass and antibacterial activity of symbiotic fungi of marine sponge *Axinella* sp., *Aspergillus sydowii* from the waters of Riung, East Nusa Tenggara, Indonesia.

2. Materials and methods

2.1. Sample collection

Symbiotic Fungi of Marine Sponge *Axinella* sp., *Aspergillus sydowii* was sample collection of Natural Product Laboratory, Faculty od Fisheris and Marine Science, Diponegoro University, Semarang.

2.2. Antibacterial screening

The MDR bacteria, *Escherichia coli* and *Staphylococcus aureus,* was obtained from Microbiology Laboratory, Kariadi Hospital, Semarang. Each MDR bacteria was cultured in the logarithmic phase, be mixed in Zobell soft agar medium (1% v/v), and poured on the agar surface that previously inoculated with fungi symbiont. The inhition zone shown after 24 h incubation and the isolates active as antibacterial [19]. (the data were not shown in this paper).

2.3. Isolation and Cultured of Fungi

Isolation and cultured of fungi symbiont *Aspergillus sydowii* conducted with modification methods of Trianto *et.al.* [20]. There are four mediums that used for mass cultured, Malt Extract Broth/MEB (ST), Noni Juice Media (MG), Avocado leaves media (AL), and Soursop leaves media (SR). Symbiotic fungi *Aspergillus sydowii* was cultured on growth media for 7 days or until the maximum growth at ambient temperature.

2.4. Extraction

The media was filtered to obtain the micellium that was extracted with methanol at room temperature. The solvent was filtered using filter paper and then dried with rotary evaporator [21].

2.5. Purification

2.5.1. Separatory funnel

The solvents used in the separatory funnel method are water and ethyl acetate with a ratio of 1: 3, and dried with rotary evaporator [22].

2.5.2. Thin layer chromatography (TLC)

Thin layer chromatography were performed on silica-gel 60 TLC plate and developed using different solvent systems to obtain best separation. There were two methods to visualize the compound. Using UV-light and *vanilin sulfuric acid* that sprayed at plat TLC and heated on hot plate until the color changes on the TLC plate [23].

2.5.3. Open Column chromatography (OCC)

Purification by column chromatography conducted with modification methods of Jafarzade *et al.* [24]. The solvents used were n-hexane (1), chloroform (2), chloroform: ethyl acetate (1: 1) (3), ethyl acetate (4), and ethyl acetate: methanol (2:1) (5).

2.6. Bioassay

Bioassay based on the method proposed by Safaeian *et al.* [25] with slight modification. The extracts were tested against *S.aureus* and *E.coli* using difussiun agar method at concentration 100 and 50 µg/disk with two times repeated. The pathogenic bacteria were inoculated by spread method on MEA media, after 30 minutes incubation, the paper disks contained the extract were placed on the agar surfaced.

3. Result and discussion

Fungi of marine sponge, *Aspergillus sydowii,* code KN-15-3 is isolate collection of Natural Product Laboratory, Faculty of Fisheris and Marine Science, Diponegoro University, Semarang. The isolate was be morphology and molecular identify. Further studies were done after antagonistic test (The process was not performed in this paper).

The isolate fungi was isolated dan cultured on four liquid mediums, Malt Extract Broth/MEB (ST), Noni Juice Media (MG), Avocado Leaves Media (AL), and Soursop Leaves Media (SR) for 14 days (on optimum growth). Then be extracted with methanol.

The result of isolation and mass cultured, known that the largest wet weight of fungi symbionts were 138.95gr on ST medium, and the least were 99.12gr on AL medium with filtrate volume 750mL. The largest rendement extract from SR medium (Table 1).

Table 1. Crude extracts of symbiontic fungi on four growth media

Isolate	Media	Weight wet of fungi (gr)	Filtrate volume (mL)	Crude extract (gr)	Rendement (%)
KN 15-3	ST	138.95	750	1.707	1.23
	AL	99.12		3.122	3.15
	SR	103.07		3.783	3.67
	MG	126.59		4.244	3.35

The crude extract were separated again with separatory funnel. From the separation, fraction of ethyl acetate AL medium had the highest amount, and the lowest was ST medium (Table 2). The result of separation was checked with TLC plate (Figure 1), and bioassay test result on Figure 2.

Table 2. Separatory funnel (fraction extract)

Isolate	Media	Crude extract (gr)	Fractions	Weight of fraction extract (gr)
KN-15-3	ST	1.7068	Ethyl acetate	0.5384
			Water	1.1684
	AL	3.1221	Ethyl acetate	0.9569
			Water	2.1652
	SR	3.7834	Ethyl acetate	1.9192
			Water	1.8642
	MG	4.2442	Ethyl acetate	0.8461
			Water	3.3981

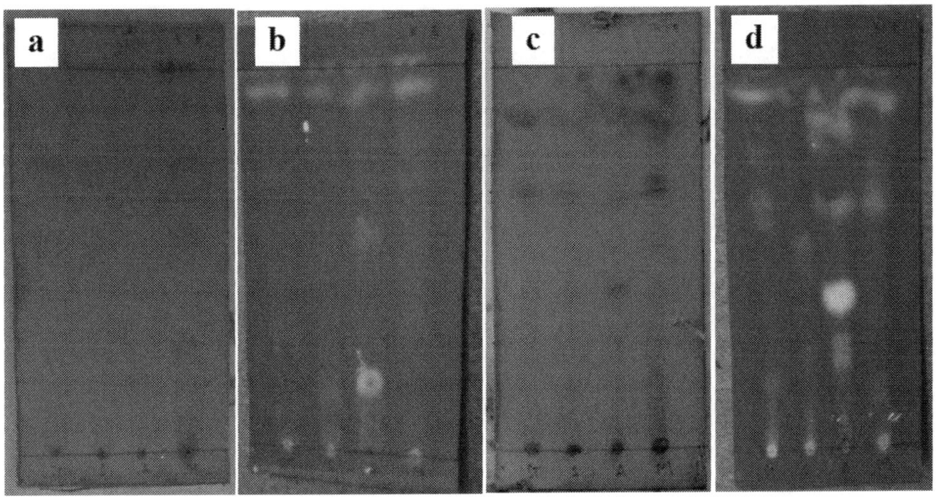

Figure 1. Visualization of TLC, vanilin sulfuric and UV rays (left: ST, SR, AL,and MG)
(a, b) Kloroform : Etil Asetat (4 : 1); (c, d) N-Heksan : Etil Asetat (1 : 1).

The visualization of TLC plate showed the spot change color and and the spots formed on each media were almost identical. The Spot colors on the TLC plate are dominated by blue, green, and red (Figure 1). The blue green spot on the TLC plate indicated fraction containing steroid and red spot were indicating terpenoid compounds [26].

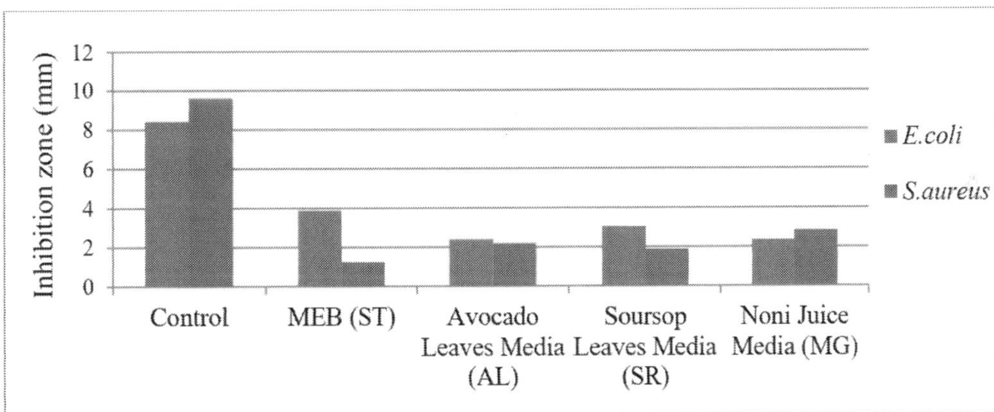

Figure 2. Diagram Inhibition zone of *separatory funnel* ethyl acetat fractions (KN-15-3-1).

The biggest *E.coli* inhibition zone was ST-KN-15-3-1 of 3.88mm, and the biggest *S.aureus* inhibition zone was MG-KN-15-3-1 of 2.83mm (Figure 2). All of the ethyl acetate fractions were separated again using open column chromatography (Table 3). The largest weigth was MG-KN-15-3-1-3 of 383.8mg and the smallest was ST-KN-15-3-1-4 of 34mg.

The different between wet weight and rendement extract on four growth were influenced by the nutrient contents in each growth medium (Table 1-3). The nutrient contents on each growth media affected the growth quantity of microorganisms [27].

Table 3. Open Column Chromatography (fraction extract)

Isolate	Media	Ethyl acetate fraction (separatory funnel) (gr)	Fraction (OCC)	Weight of the OCC fraction extract (mg)
KN-15-3-1	ST	0.5384	1	120.9
			2	75.7
			3	183.8
			4	34
			5	116.1
	AL	0.9569	1	157.4
			2	103.9
			3	329.8
			4	35.6
			5	64.3
	SR	1.9192	1	210
			2	245.9
			3	168.9
			4	262.2
			5	148.4
	MG	0.8461	1	120.1
			2	178.4
			3	383.8
			4	47.1
			5	105.2

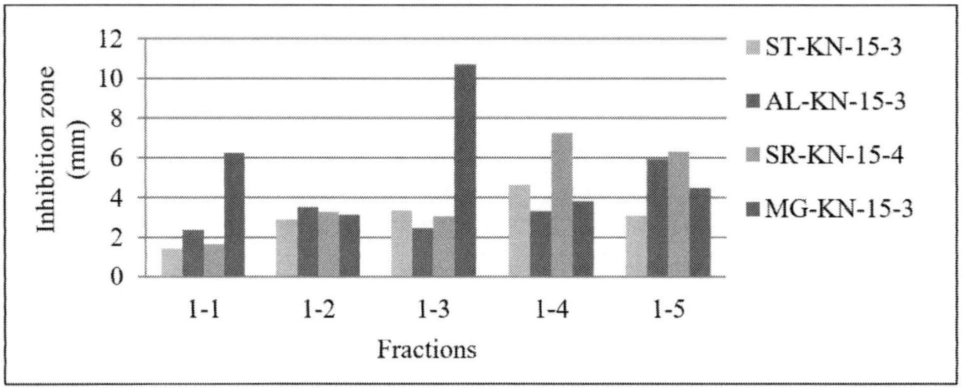

Figure 3. Diagram inhibition zone of Open Column Chromatography against *E.coli*.

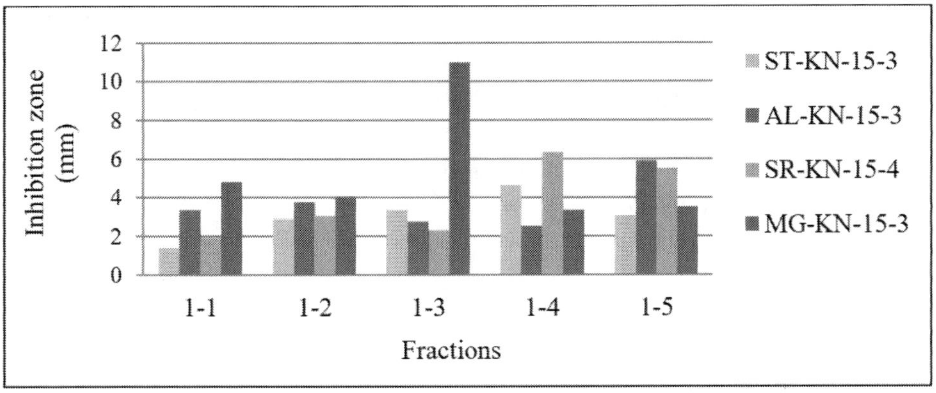

Figure 4. Diagram inhibition zone of Open Column Chromatography against *S.aureus*.

The biggest *E.coli* and *S.aureus* inhibition zone were MG-KN-15-3-1-1 of 10,71mm and 10,98mm (Figure 3-4). Growth media also affected the quality and ability to inhibit pathogenic bacteria, *E.coli* and *S.aureus*, extracts of symbiotic fungi of marine sponge *Axinella* sp., *Aspergillus sydowii* (Figure 2-4). The content of nutrients in growth media affected the content of bioactive compounds from microorganisms [28-29].

4. Conclusion

Symbiotic fungi of marine sponge *Axinella* sp., *Aspergillus sydowii* could be grown on four media, Malt Extract Broth/MEB (ST), Noni Juice Media (MG), Avocado leaves media (AL), and Soursop leaves media (SR). Growth media affected the quantity and quality of the symbiotic fungi extract, *Aspergillus sydowii* and its ability to inhibit pathogenic bacteria, *E.coli* and *S.aureus*.

References

[1] Yalçın F N 2007 Biological activities of the marine sponge *Axinella Hacettepe* University, *J Faculty of Pharmacy*, 27(1), 47-60.

[2] Tsukamoto S, Tane K, Ohta T, Matsunaga S, Fusetani N, and Van Soest R W M 2001 Four new bioactive pyrrole-derived alkaloids from the marine sponge *Axinella brevistyla*. *J. Nat. Prod.*, 64, 1576-1578.

[3] Wratten S J, and Meinwald J 1981 Antimicrobial metabolites of the marine sponge *Axinella polycopella*. *Cell. Mol. Life Sci.*, 37, 13-14. https://link.springer.com/article/10.1007/BF01965540

[4] Hamill P, Hudson D, Kao R Y, Chow P, Raj M, Xu H, Richer M J, and Jean F 2006 Development of a red shifted fluorescence based assay for SARS coronavirus 3CL protease: identification of a novel class of anti-SARS agents from the tropical marine sponge *Axinella corrugata*. *Biol. Chem.*, 387, 1063 – 1074.

[5] Supriyano A, Schwarz B, Wray V, Wittle L, Muller W E, Van Soest R, Sumaryono W, and Proksch P 1995 Bioactive Alkaloids From the Tropical Marine Sponge *Axinella carteri*. *Z. Naturforsch* 50, 669 – 674.

[6] Funel C, Berrue F, Roussakis C, Rodriguez R F, and Amade P 2004 New cytotoxic steroids from the Indian Ocean sponge *Axinella cf. bidderi*. *J. Nat. Prod.*, 67, 491 – 494. https://www.ncbi.nlm.nih.gov/labs/articles/15043442/

[7] Aktas N, Genc Y, Gozcelioglu B, Konuklugil B, and and Harput U S 2013 Radical scavenging effect of different marine sponges from Mediterranean Coasts. *Rec. Nat. Prod.* 7(2), 96-104.

[8] Pettit G R, Gao F, Cerny R L, Doubek D L., Tackett L P, Schmidt J M, and Chapuis J C 1994 Antineoplastic agents. 278. Isolation and structure of axinastatins 2 and 3 from a Western Caroline Island marine sponge. *J. Med. Chem.*, 37(8), 1165–1168.

[9] Milanowski D J, Rashid M A, Gustafson K R, O'Keefe B R, Nawrocki J P, Pannell L K, and Boyd M R. 2004 Cyclonellin, a new cyclic octapeptide from the marine sponge *Axinella carteri*. J. Nat. Prod. 67(3), 441 – 444.

[10] Aiello A, D'Esposito M, Fattorusso E, Menna M, Müller W E G, Perovic-Ottstadt S, Tsuruta H, Gulder T A M, and Bringmann G 2005 Daminin, a Bioactive Pyrrole Alkaloid From the Mediterranean Sponge *Axinella damicornis*. *Tetrahedron*, 61, 7266 – 7270.

[11] Radjasa, O K, Sabdono A, Junaidi and Zocchi E 2007. Richness of Secondary Metabolite-Producing-Marine Bacteria Associated with Sponge Haliclona sp. *Int. J. Pharmacol.*, 3(3): 275-279.

[12] Radjasa O K, Martens T, Grossart H P, Brinkhoff T, Sabdono A, and Simon M. 2007. Antagonistic Activity of a Marine Bacterium *Pesueodaltromonas luteoviolacea* TAB 4.2 Associated with Coral *Acropora* sp. *J. Bio. Sci.*, 7(2): 239-246.

[13] Avila E, Carballo J L, and Cruz-Barraza J A 2007 Symbiotic relationships between sponges and other organisms from the Sea of Cortes (Mexican Pacific coast): same problems, same solutions. *Porifera Res. Biodiv., Innov. Sustain,*. 147-156

[14] Kelecom A 2001. Secondary Metabolites from Marine Microorganism. *Ann. Braz. Acad. Sci.*, 74 (1, 151-170.

[15] Nofiani R, Kurniawan R A, Mabella T, Ardiningsi P, Zahara T A and Sunarto 2012 Study of Media Effect toward Antimicrobial, Antiplasmodial, Herbicidal Activities and Toxicity of Extract Produced by Acropora dibranchiata-Associated *Moraxella* sp. RA15. *J. App. Pharmaceutical Sci.*, 2(9), 010-014.

[16] Rajeswari P, Jose P A, Amiya R, Robinson S, and Jebakumar D 2015 Characterization of saltern based *Streptomyces* sp . and statistical media optimization for its improved antibacterial activity. *Frontiers in Microbiol*, 5(1), 1–11.

[17] Wang Y, Fang X, An F, Wang G, and Zhang X 2011 Improvement of antibi- otic activity of *Xenorhabdus bovienii* by medium optimization using response surface methodology. *Microb.Cell Fact.*, 10, 98.

[18] Atalla M M, Zeinab H K, Eman R H, Amani A Y, and A. Abd. El Aty Abeer 2008 Production of some biologically active secondary metabolites from marine-derived fungus *Varicosporina ramulosa*. *Malay. J. Microb.*, 4(1): 14-24.

[19] Murti P D B and Radjasa O K 2012 Antibacterial activity of bacterial symbiont of soft coral lobophytum sp. Against mdr bacteria *Escherichia coli* and *Staphyllococcus aureus*. *J of Coastal Develp.* 15 (3):279-302

[20] Trianto A, Widyaningsih S, Radjasa O K, and Pribadi R 2016 Symbiotic Fungus of Marine Sponge *Axinella* sp. Producing Antibacterial Agent. *IOP Conf. Ser.: Earth Environ. Sci.*, 55, 1-7.

[21] Almeida AP, Dethoup T, Singburaudom N, Lima R, Vasconcelos MH, Pinto M and Kijjoa A 2010 The in vitro anticancer activity of the crude extract of the sponge-associated fungus *Eurotium cristatum* and its secondary metabolites. *J. Nat. Pharm.* 1, 25–29.

[22] Berlinck R G S, Lira S P, and de Vita-Marques A M 2008 A multi-screening approach for marine-derived fungal metabolites and the isolation of cyclodepsipeptides from *Beauveria feline*. *Quim. Nova*. 31(5) : 1099-1103.

[23] Touchstone J C and Murrell F D 1983 Practice of Thin Layer Chromatography 2nd Ed. John wiley & sons,inc. 404pp.

[24] Jafarzade M, Yahya NA, Shayesteh F, Usup G, and Ahmad A 2013 Influence of culture conditions and medium composition on the production of antibacterial compounds by marine *Serratia* sp. WPRA3. *J. of Microbiology*. 51(3):1-7.

[25] Safaeian S, Hosseini H, Abbas P A A, and Farmohamadi S 2009 Antimicrobial activity of marine sponge extracts of offshore zone from Nay Band Bay, Iran. *J. Med. Mycology*, 19(1), 11–16.

[26] Santiago M and Strobel S 2013 Thin Layer Chromatography. *Methods in Enzymol.*, 533: 303–24.

[27] Pitt J I and Hocking A D 2009. Fungi and Food Spoilage (Google eBook), 536pp. http://books.google.com/books?id=-B1s6GhOlzkC&pgis=1

[28] Kumar S N, Siji J V, Ramya R, Nambisan B, and Mohandas C 2012 Improvement of antimicrobial activity of compounds produced by *Bacillus* sp. associated with a *Rhabditid* sp. (entomopathogenic nematode) by changing carbon and nitrogen sources in fermentation media. *J. of Microbiology, Biotech, Food Sci.* 1(6) : 1424 – 1438.

[29] Bill G F, Platas G, Fillola A, Jimenez M R, Collado J, Vicente F, Martin J, Gonzalez A, Bur-Zimmermann J, Tormo J R, and Pelaez 2007 Enhancement of antibiotic and secondary metabolite detection from filamentous fungi by growth on nutritional arrays *J of Applied Microbiology*, 104 : 1644 – 1658.

The Biophysical Characteristics Of Hatching Habitat Of Lekang Turtle (*Lepidhochelys olivacea*) Eggs In Turtle Conservation And Education Center, Bali

Suryono[1*], R Ario[1], E Wibowo[1], G Handoyo[2]

[1]Department of Marine Science, Faculty of Fisheries and Marine Science, Diponegoro University Jl. Prof. Soedharto, SH, Tembalang, Semarang
[2]Department of Oceannography, Faculty of Fisheries and Marine Science, Diponegoro University Jl. Prof. Soedharto, SH, Tembalang, Semarang

Email: suryono1960@gmail.com

Abstract. *Lekang* turtle (*Lepidhochelys olivacea*) is one of the fauna that is protected as an endangered population. This marine reptile was able to migrate in great distance along the Indian Ocean, the Pacific Ocean, and South East Asia. Its existence has long been threatened, either by nature or human activities that endangered the population directly or indirectly. The decreasing number of sea turtle population that nest in Bali area is one indication of the reducing number of *Lekang* turtle in Indonesia. If left unchecked, it will result in the loss of *Lekang* turtle. This study aims to determine the successful percentage of conservation techniques and *Lekang* turtle hatching eggs (olive ridley sea turtle) in TCEC, Bali. The method used in this research is the method of observation or direct observation done in the field. Data collection is done by direct observation in the field. The results showed that the turtle breeding site is located in an area that is less strategic because too far from the sea, so that the temperature and humidity cannot be stable. Water content is most an important factor in the growth of embryo and egg hatching. This will lead to the decrease of hatching percentage of turtle eggs.
Keywords : Hatching, Lekang turtle *(Lepidhochelis olivacea)*, TCEC Bali

1. Introduction

It is recorded that six out of seven species of sea turtle in the world live in the waters of Indonesia. They are Green Turtle (*Cheloniamydas*), Hawksbill (*Retmochely simbricata*), Grey Turtle / *Lekang* turtle (*Olive ridley* sea turtle), Flat Turtle (*Natator depressus*), Star Turtle (*Dermochelys coriacea*), and the loggerhead Turtle (*Caretta caretta*).This amount is still being debated because Nuitja (1992) [1] said that there are only five species as *Caretta caretta* does not exist in Indonesia. However, some researchers revealed that *Caretta caretta* has home ranges that include Indonesia [2].

Various policies related to the management of turtle have been done so far, either by Ministry of Forestry, Ministry of Environment, and the Ministry of Marine Affairs and Fisheries. Even the government is continuously developing appropriate policies in turtle conservation management efforts by doing regional cooperation with some institutions such as the IOSEA-CMP, SSME, and BSSE. The emergence of Law No. 31 of 2004 on fisheries and Government Regulation No. 60 of 2007 on Conservation of Fish Resources brings new nuance in the management of turtle conservation [3].

Nevertheless, the granting of protection status alone is clearly insufficient to restore or at least maintain the population of turtle in Indonesia. A comprehensive, systematic, and measurable conservation management should be implemented immediately, including providing technical knowledge on the management of sea turtle conservation for the parties concerned, especially the executors in the fields. But so far, there are not many books containing information on the management of turtle conservation that is easily understood by all people. Therefore, in line with the management of turtle conservation efforts in Indonesia, a technical manual for the perpetrators of turtle conservation management is indispensable

2. Material and methods

The research was conducted from August 4th until 22nd, 2014 in TCEC (Turtle Conservation and Education Center) conservation area, Bali. The method used in this activity is the method of observation or direct observation made in the field to determine the existing activities in TCEC (Turtle Conservation and Education Center) conservation area. The equipment used in this study include thermometer, gloves, buckets, cameras, hoes, and stationery.

Temperature measurement in semi-natural nest was done twice a day in a vulnerable specified time during egg incubation period, which is 40 days. Temperature measurement in semi-natural nest was conducted at 08.00 pm. To measure the temperature, it was used a thermometer with an accuracy of 0.5°C. It was inserted into the semi-natural spring nest to a depth of approximately 20cm and then the temperature was recorded. Then the egg nest was dug up to take the rotten eggs using gloved hands. The rotten eggs were separated and then buried. The discussion or interview was conducted with relevant parties regarding *Lekang* turtle conservation and development.

The calculation of *Lekang* Turtle hatching success percentage was calculated based on the ratio between the number of hatched eggs and the number of incubated eggs in the nest (Dobbs, *et. al.*, 1999) [4], it is as follows:

$$HSs = \frac{JS}{JS + TM} \times 100\,\%$$

Description:
HSS = hatching success rate (%)
JS = the number of hatched eggs
TM= the number of eggs that failed to hatch

3. Results and discussion

Water As one effort to conserve turtle, it was built TCEC (Turtle Conservation and Education Center), a turtle breeding conservation in Serangan, Bali. This is used as a conservation, semi-natural nest, and hatching maintenance until the turtle babies are released back to the sea. This turtle breeding location is less strategic because it is too far from the sea, so that the temperature and humidity are not stable. Water content is an important factor in the growth of embryo and egg hatching [5]. This will lead to the decrease of hatching percentage of turtle eggs.

The data of nest temperature taken during the study, the number of eggs, the success of hatched eggs, and the eggs that failed to hatch can be seen in Table 1.

Table 1. The Hatching Results of *Lekang* Turtle (*Lepidhochelys olivacea*) Eggs.

DAYS TO-	DATE	TEMPERATURE (°C)	NUMBER OF EGGS (grain)	HATCH (grain)	FAIL HATCH (grain)
1	06/8/2014	29	-	-	-
2	07/8/2014	30	-	-	-
3	08/8/2014	31	-	-	-
4	09/8/2014	30	-	-	-
5	10/8/2014	31	100	59	41
6	11/8/2014	32	100	25	75
7	12/8/2014	31	-	-	-
8	13/8/2014	33	146	49	97
9	14/8/2014	30	97	72	25
10	15/8/2014	30	-	-	-
11	16/8/2014	29	-	-	-
12	17/8/2014	30	-	-	-
13	18/8/2014	31	-	-	-
14	19/8/2014	30	-	-	-
15	20/8/2014	29	-	-	-
16	21/8/2014	30	122	96	26

The following table is the percentage result Lekang turtle hatching eggs (olive ridley sea turtle) that hatch early.

Table 2 . The Percentage Result of *Lekang* Turtle (*Lepidhochelys olivacea*) Eggs.

Early eggs	NestTemperature (°C)	Eggs hatch	Hatching percentage %
100	31	59	59
100	32	25	25
146	33	49	33
97	30	72	74
122	30	96	78

Nest condition greatly affects the success of turtle eggs to hatch. There are two main factors that directly affect the success of turtle eggs to hatch during the incubation period, they temperature and water content in the nest. The result of nest temperature measurement ranges in 30° - 33°C with the average temperature is in Table 2. According to Marquez (1990) [6], if the temperature during the incubation period is much lower or higher than the optimum temperature of 28° - 32°C, the eggs will hatch less than 50%. The tolerable temperature for the hatching success of turtle eggs is 25° - 34°C. The low and high nest temperature will determine the sex of the turtle babies [7], [8].

The results of temperature measurement do not show daily fluctuation. This proves that the metabolism process in each nest relatively produces the same heat. In other words, there is not any nest experiencing lower or higher metabolism rate. There is higher temperature range by the time the eggs hatch. This shows that by the time the eggs hatch the metabolism process increases. Nuitja (1992) [1] said that 5 days before the eggs hatch, the nest temperature will increase a few degrees due to the higher metabolism process of the eggs.

The factors that affect the hatching success are the nest temperature, nest depth, the number of eggs implanted, the predators, the egg itself, and also the quality of the sand. The depth of semi-natural spring hatchery in TCEC is about 30 – 40 cm. The depth of the nest is measured without any certain measurement, it is made based on the approximate measurement of the turtle egg growers. This indicates a less optimal hatching in the semi-natural hatchery. Substrate sand that is too dry causes the withdrawal of fluid out from the eggs resulting embryonic death [7], [8]. Excessive moisture content may also cause high humidity. High humidity in the nest increases the growth of fungi and bacteria that can cover the pores of the eggshell. This coverage can disrupt the following process of egg respiration and cause the restriction of the embryo growth even lead to death [9].

If the data of hatching eggs in TCEC is compared to the data in Sukamade Beach, there is a significant difference. The result of hatching eggs in Sukamade Beach is much higher than the result of hatching eggs in TCEC. This may happen because there several factors, the different temperature of the nest, the depth of the egg nest, the quality of the sand in the nest, the humidity of the nest, the number of eggs, the diameter of the nest.

On August 10th, the total amount of eggs got was 100 eggs or 59%, while on August 11th the total amount of eggs got was 100 eggs with the hatching percentage is 25%. This was because the eggs in the nest were already rotten caused by predators. On August 13th, the total amount of eggs got was 146 eggs with the percentage of 33%. The small percentage was caused by the number of hatched eggs was divided by the number of initial and on August 13th there were no eggs that failed to hatch, and the remain of hatching continued on the next day on August 14th. On August 14th the total amount of eggs got was 97 eggs with hatching percentage of 74% and on August 21st the total amount of eggs got was 122 with hatching percentage of 78%.

Semi-natural hatchery conditions must be made as similar as the original nest so that the percentage of the turtle egg hatching success increases. To make the semi-natural hatchery conditions similar to the original ones is very difficult since the natural condition changes every time. The water content in natural nest is more than the water content in semi-natural hatchery since the location of natural nests is closer to the coast line than the location of semi-natural hatchery. Besides that, the semi-natural hatchery is closed (roofed) that will block the water when it rains and will shade it from the sunlight. The difference of the nest condition affects the turtle hatching success.

Semi-natural egg hatching is done inside a building measuring 3 x 3 m with a substrate of beach sand. This is done to reduce the risk of failure of turtle eggs to hatch in the natural nest, the risk of predators, the risk from people who still want to hunt turtle eggs, and also the risk of the existing coastal vegetation. Turtle eggs are taken from the natural nest and moved to the semi-natural nest. The eggs are planted and will hatch in the average time of 1.5 – 2 months. The planted eggs are marked with plates / boards containing information on the types of turtle, the number of eggs, and the egg planting time. Management activities in the semi-natural egg hatching include recording the number of eggs that successfully hatch, the number of dead turtle babies, and also keeping the hatching house clean. The eggs that are not ready to hatch are replanted in the nest until they ready to hatch completely. Eggs will be collected that are stated to fail to hatch are collected and then buried. This is done in order not to cause odor and do not call other animals to come to prey on the eggs that still exist in the semi-natural nest.

Besides that, another outside factor that cause the failure of hatching is the character defect. This is caused by the position of eggs that are stacked and pressed together in the nest so that the eggs below get the pressure most and get less oxygen. This can cause the differentiation process of embryonic turtle in the eggs disrupted resulting in the formation of organs is not maximum. Then it can be stated that the egg abnormalities cause the body hatching growth defects and unhealthy, and can lead to premature death. There is also a predator in the sand that is used as the medium for the hatching. This is one obstacle to the successful hatching of turtle eggs. When the nest is dismantled, there is a kind of animal that inhibit the growth of turtle eggs, they are red ants. Than ants eat turtle eggs in the nest so that the eggs cannot develop and hatch.

Dismantling the nest is done to clean the nest for reusing and also to assess how many eggs that cannot hatch into hatching. Based on the data obtained, the failure of egg hatching in the semi-natural hatchery is due to the internal factor of the turtle egg itself. Turtle eggs that cannot hatch do not have embryos.

Besides that, the water content in the sand is also an important factor in the growth of turtle eggs and egg hatching. The more the embryos grow in the eggs, the more they absorb water from the eggs. This makes the diameter of eggs expanded and enlarged several millimeters. On the contrary, dry sand will absorb water from the eggs it has a higher salt concentration than the eggs. This water absorption by sand can cause the embryo in the egg does not develop and die [5].

According to Darmawan (1995)[10], turtles make holes for nesting that are not close to the beach so that the sea water cannot reach the nest and the temperature and humidity will be maintained. The eggs exposed to the sea water will usually be damaged and fail to hatch. Therefore, the distance of semi-natural hatchery in Sukamade Beach is not too close to the beach to avoid excessive intrusion of sea water that can destabilize either the temperature or the humidity of the nest that will affect the success of the turtle hatching.

4. Conclusion

Considering the condition of the surrounding area, *Lekang* turtle (*Lepidochelys olivacea*) conservation technique in TCEC Bali is good enough. The success percentage of *Lekang* turtle egg (olive ridley sea turtle) hatching observed is quite high.

References

[1] Allan Nuitja, INS, 1992. Biology and Ecology Conservation Sea Turtle. Bogor Agricultural University (IPB): Bogor.
[2] Limpus, C. and P.C. Reed. 1992. The Green Turtle, *Chelonia mydas*, in Queensland, "A Preliminary Description of the Population Structure in a Coral Reef". Feeding Ground, Reprinted from the Biology of Australian Brogs and Reptiles pp 45-52.
[3] Government Regulation No. 60 of 2007
[4] Dobbs, K.A., J.D. Miller, C.J. Limpus, and A.M.Jr. Landry. 1999 Hawksbill Turtle, *Eretmochelys imbricata*, Nesting at Milman Island, Northem Great Barrier Reef, Australia. Chelonian Conservation and Boilogy. Vol. III. Pp 344-361.
[5] Alam Lestari Foundation. 2000. Getting to know turtles. Alam Lestari Foundation and Keidanren Nature Conservation Fund (KNCF) of Japan. 81 pgs
[6] Marquez, M.R. 1990. Sea Turtle of The World. An Annotated and Illustrated Catalogue of Sea Turtle Species Known to Date. FAO Fisheries Synopsis 125, Vol.11. Rome. 43 p.
[7] Mustika, I. 1987. Differentiation of Embryonic gonads Males and Females Enduring turtle, olive ridley sea turtle were incubated at High and Low. Department of Biology ITB. Bandung (S2 unpublished thesis). 65Pgs
[8] Anonymous, 2000. Know turtle. Alam Lestari Foundation, Jakarta (Translated by Akil Yusuf). 82 Pages.
[9] Solomon, S.E. and Baird, T. 1980. Effect of Fungal The Penetration of the Eggshell of the Green Turtle. Electron Microscopy volume II.
[10] Darmawan, R. 1995. Influence of Difference Time Decision eggs from nest to nest Natural Natural Semi Against Enduring Turtles Hatching success (*Olive ridley* sea turtle) in the National Park Alas Purwo-South Banyuwangi. Papers; Research and Management PENYU in Indonesia. Directorate General of Forest Protection and Nature Conservation, Bogor. pgs: 243-258.

Mapping of Nitrate, Phospat And Zooxanthelae With Abundance Of Sea Urchins on Massive Coral Reef in Karimunjawa Island

S Suryanti, C Ain, N Latifah

Department of Aquatic Resources, Faculty of Fisheries and Marine Sciences, Diponegoro University Jl. Prof Soedarto, SH, Tembalang, Semarang, Indonesia 50275

Email: suryanti@gmail.com

Abstract. Coral reefs have high organic productivity because coral reefs can withstand nutrients and accommodate all external inputs. Many factors affect the life of corals, which is nitrate, phosphate and zooxanthellae. The purpose of this study are to know mapping of the content and the relationship between of nitrate, phosphate, zooxanthellae and abundance of sea urchins on massive coral reefs in Karimunjawa Islands. This research was conducted in May - June 2017 in three stations are Karimunjawa, Menjangan Kecil and Cemara Kecil Island. The method used in this research is survey method with quantitative approach. Results of mapping of nitrate contents on massive corals on all three islands showed the highest nitrate content on Cemara Kecil Island and lowest on Karimunjawa island, with a range of values 5.078-212.853 mg/kg. In mapping the distribution of phosphate content in the three islands showed the highest phosphate content in Menjangan Kecil island and the lowest on Karimunjawa island, with a range of values from 6.78–19.35 mg/kg. Zooxanthelae map shows that the highest and lowest distribution of zooxanthela content on Karimunjawa island, with a range of values 2.84–8.88 cell/cm^2. The sea urchins found in Karimunjawa Islands during the study were *Diadema setosum* and *Echinothrix calamaris* with a range of values 5–147. Based on multiple regression analysis showed that the relationship between nitrate, phosphate and zooxanthela with abundance of sea urchins showed a strong correlation result with correlation value (r) is 0.64. These results can be an indicator of coastal environmental health, especially coral reef ecosystems.
Keywords: Nitrate; Phospate; Zooxanthellae; Coral; Sea Urchins

Introduction

Coral reefs are the most diverse of all marine ecosystems. They teem with life, with perhaps one quarter of all ocean species depending on reefs for food and shelter [1]. This is a remarkable statistic when you consider that reefs cover just a tiny fraction (less than one percent) of the earth's surface and less than two percent of the ocean bottom. Because they are so diverse, coral reefs are often called the rainforests of the sea. Coral reefs are also very important to people. The value of coral reefs has been estimated at 30 billion U.S. dollars and perhaps as much as 172 billion U.S. dollars each year, providing food, protection of shorelines, jobs based on tourism, and even medicines [2].

Coral reefs and other coastal marine ecosystems in the tropics, provide rich reservoirs of natural resources. However, terrestrial and climate impacts, and scientific and governance limitations, have resulted in severe degradation of many nearshore tropical ecosystems, and continued threats to many more. In fact, coral reefs might become the first marine ecosystem to be driven to extinction by anthropogenic activity, perhaps within the next century [3]. Unfortunately, people also pose the greatest threat to coral reefs. Overfishing and destructive fishing, pollution, warming, changing ocean chemistry, and invasive species are all taking a huge toll. In some places, reefs have been entirely destroyed, and in many places reefs today are a pale shadow of what they once were.

Indonesia's coral reef resources are among the richest and most diverse in the world. Eastern Indonesia lies at the centre of diversity for corals [4], mollusca, reef fishes and other reef organisms, along with the Philippines and the north coast of Papua New Guinea. This wealth in biodiversity emphasizes Indonesia's importance in global efforts to conserve marine resources and preserve biodiversity [5].

Coral reefs are the only totally biogenic ecosystem, although deep-sea vents and rainforests approach this condition. With the exception of basaltic intrusions in locations such as Hawaii, terrigenous sediments in some reefs near continental coastlines, and exposed underlying bedrock in subtropical locations where coral growth is scant, everything on a coral reef is produced by living organisms. Many Pacific reefs perch on top of thousands of meters of biogenic limestone deposited on a slowly subsiding base over thousands of years. Coral reef formation is tightly constrained by the narrow tolerances of corals and other calcifiers such as calcareous reefbuilding algae [6]. While there are differences among species, corals generally require a firm substratum, depths less than 50m (unless the water is exceptionally clear), oceanic salinity and pH, temperatures between about 18°C and 35°C, and minimal pollution or sediment load [7].

Like ecosystem services, the nature and magnitude of threats to coral reefs, mangroves and seagrasses vary considerably among regions. However, there are attributes common to all coral reefs and associated habitats making them especially vulnerable to certain threats and impacts. Many of these attributes also allow them to provide such diverse ecosystem services [8]. The longterm evolution of reef systems that has generated such high ecological complexity and diversity also makes them vulnerable to rapid and irregular changes relative to those experienced over their evolution. Their shallow and nearshore location that allows easy accessibility also places these systems in direct contact with marine, coastal and land-based anthropogenic activity relative to ecosystems in deeper water and further offshore [9]. We address a broad range of threats and impacts, but again without attempting a comprehensive review. For a thorough review of the long-term decline of a reef system and assessment of the primary drivers [10].

Herbivory is an important process shaping natural communities and transferring energy through food webs in autotrophic ecosystems [11]. However, there are significant differences between terrestrial and aquatic systems in the mechanism of energy transfer from primary producers to higher trophic levels [12]. In contrast to terrestrial systems, where large vertebrates still perform an important role as vegetal consumers [13], in marine ecosystems they no longer represent strong trophic links in communities [14]. This is because overfishing in coastal systems has decimated herbivorous megafauna, such as sirenians and turtles [15]. Consequently, fishes and sea urchins are now the most important herbivores in reef ecosystems [16]. However, their relative influence on primary producers differs along a latitudinal gradient [17]. On tropical reefs, fish and sea urchins are both important herbivores, while sea urchins predominate in temperate reef systems [18]. Abundance of sea urchins is found in living corals according to their habitat, so there is a significant difference between the number of sea urchins on coral reefs and seagrass beds [19]. The functional importance of sea urchins in temperate seas is demonstrated by their formation of vast barren areas when at high densities, decreasing habitat complexity and affecting benthic cover and local community dynamics [20].

Located in the middle of the Java Sea, Karimunjawa National Park is one of only seven national marine parks in all of Indonesia. A chain of 27 islands dot the waters, most of them uninhabited. Karimunjawa Islands as one of the districts in the Jepara Regency, lies in the Java Sea about 115 km north of Semarang. The area can be reached by air as well as by sea from Semarang and by sea from Jepara. The islands is composed of 27 islands and cover an area of 7,120 ha with an interior sea of 107.225 ha. The five largest islands are Karimunjawa, Kemujan, Parang, Genting and Nyamuk support permanent settlement [21]. Since 2005 Karimunjawa has been confirmed as Marine National Park which a total area of 111.625 hectares which have 51 genera with more than 90 species of coral biota and 242 species of ornamental fish. Two protected biota species, black coral (*Antiphates* sp) and organ pipe coral (*Tubipora musica*) and divided into 7 zones. Aims for this research were to document coral diseases in Karimunjawa National Park (KJNP) that never recorded in order to develop a species list of corals that are affected by specific disease. Also, to document disease prevalence was compared among coral communities at different sites in KJNP [22].

Coral reef ecosystems have a very important role in the Karimunjawa Islands. In addition to being used as a marine tourism object, coral reef is one of the coastal ecosystems. The existence of coral reefs in their habitats is influenced by several things such as Nitrate, Phosphate, and one of its natural predators, the sea urchins. The purpose of this study are to know mapping of the content and the relationship between of nitrate, phosphate, zooxanthellae and abundance of sea urchins on massive coral reefs in Karimunjawa Islands.

Materials and methods

The study was conducted in May-July 2017. The study was divided into 3 observation stations consisting of Karimunjawa Island, Menjangan Kecil Island, and Small Pine. Each station is divided into 6 substations that surround the station.

Coral samples in this study were taken using Line Transect sampling. Line intercept transect (LIT) is used to determine the percentage cover of benthic communities. It can be used on its own or in combination with other methods, such as quadrats. The LIT is the standard method recommended by the GCRMN to determine percentage cover and colony size for management level monitoring. Information obtained: Percentage cover of benthic communities e.g. hard coral, soft coral, sponges, algae, rock, dead coral. Medium to detailed information can be collected from growth forms (shape) to family, genus or species level depending on objectives or expertise available. (Hill and Wilkinson, 2004). The content of nitrate, phosphate, and zooxanthellae measured was taken from a sample of masiv coral reef, while the abundance of pig hair was obtained from the number of pigs found at the study site.

The measurement results are then analyzed using Multiple Regression Analysis to know the correlation of all factors. Mapping of nitrate, phospat, zooxanthelae used to software ArcGIS with IDW method.

Result and Discussion

Results of mapping of nitrate contents on massive corals on all three islands showed the highest nitrate content on Cemara Kecil Island and lowest on Karimunjawa island, with a range of values 5.078-212.853 mg/kg. In mapping the distribution of phosphate content in the three islands showed the highest phosphate content in Menjangan Kecil island and the lowest on Karimunjawa island, with a range of values from 6.78–19.35 mg/kg. Zooxanthela map shows that the highest and lowest distribution of zooxanthela content on Karimunjawa island, with a range of values 2.84–8.88 cell/cm^2. The sea urchins found in Karimunjawa Islands during the study were *Diadema setosum* and *Echinothrix calamaris* with a range of values 5–147 .

Based on the results of Nitrate Measurements in Coral Massive at the Study Sites, it can be seen that the highest nitrate concentrations are found in Cemara Kecil Island while the lowest is on Karimunjawa Island. This is allegedly because zooxanthellae conditions on Small Pine Island is better due to the depth of the waters more shallow than in Karimunjawa Island so that sunlight is more optimum. With a better zooxanthellae condition the intake nitrate also works better.

Nitrogen is thereby conserved within the symbiosis and, once assimiliated by the symbiont, may be returned to the host in the form of translocated amino acids [22]. In addition to internal recycling of ammonium, the zooxanthellae have also been shown to be capable of removing net quantities of ammonium and nitarte from seawater, both, in the intact symbiosis [23] and in isolation [24] The capacity for simultaneous net uptake of both nitrate and ammonium is suprising. Nitrate uptake by microalgae is generally supressed in the presence of sgnificant amounts of ammonium [25] and zooxanthellae might be expect to be exposed to elevated tissue ammnoium concentrations due to host excretion. Its possible that uptake by the tartion in the tissue or, as have suggested, reef corals may have evolved low rates catabolism, with little ammonium production [26].

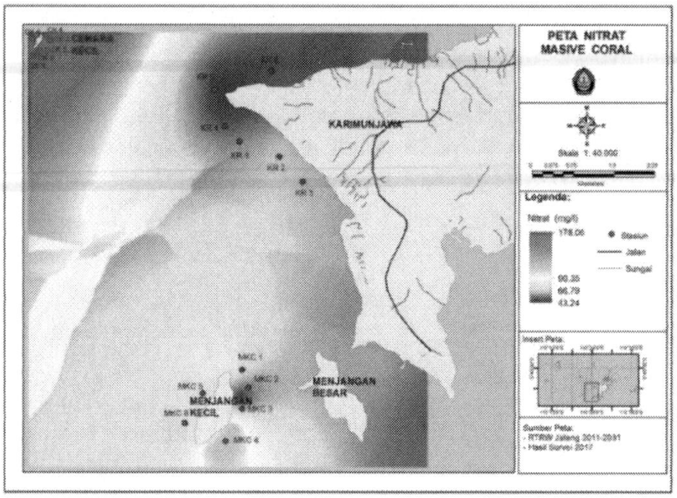

Figure 1. Mapping of Nitrat Distribution

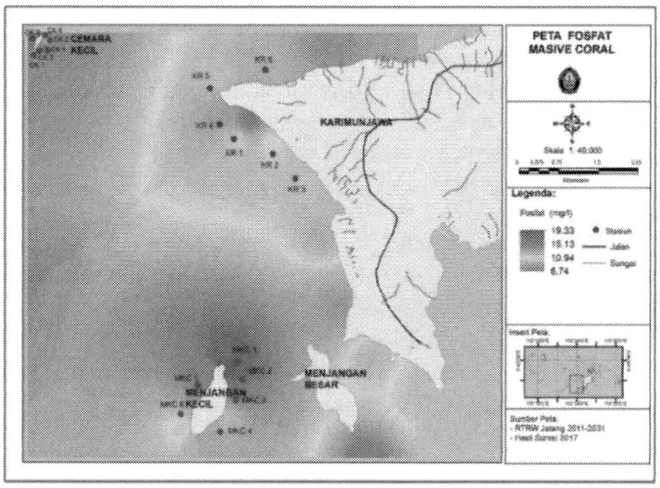

Figure 2. Mapping of Phosphat Distribution

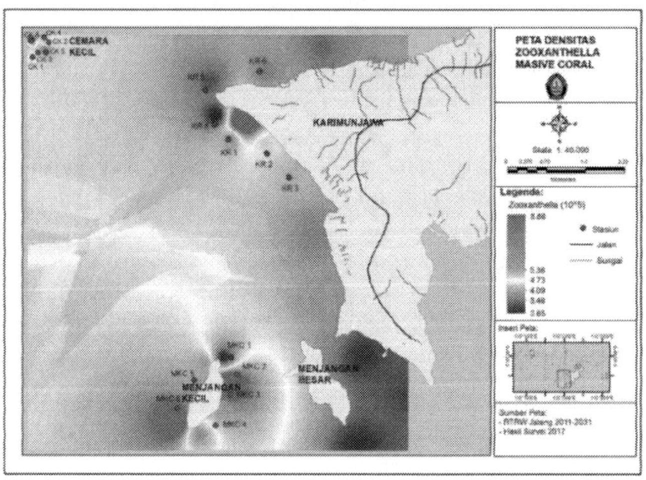

Figure 3. Mapping of Zooxanthelae Distribution

IOP Conf. Series: Earth and Environmental Science **116** (2018) 012086 doi:10.1088/1755-1315/116/1/012086

Based on the results of phosphate measurements that have been done, itcan be seen that Menjangan Kecil Island has the average value of the highest phosphate content than other islands. This is presumably due to zooxanthellae on Menjangan Kecil Island is quite higher than other islands.

Zooxanthellae require nutrients (P and N), dissolved inorganic carbon (DIC), and solar radiation for photosynthesis. Hermatypic corals require, minimally, DIC, ATP, and calcium for calcification [27]. Average phosphorus concentrations on natural reefs are generally low (0.01 mg L^{-1}, [28]), but there appears to be a trend towards increasing values in recent years. For example, concentration up to 0.266 mg L^{-1} have been documented on some reefs in the Southern Gulf of Mexico [29]. Elevated phosphate levels are known to affect several aspects of coral health, such as growth rate [30], skeletal density [31], re- production [32], mortality [33], disease susceptibility [34], and zooxanthellae density [31]. Phosphate can stimulate phytoplankton population growth, potentially reducing the availability of light to corals [33], or stimulating growth of benthic filamentous algae that can overgrow corals [35].

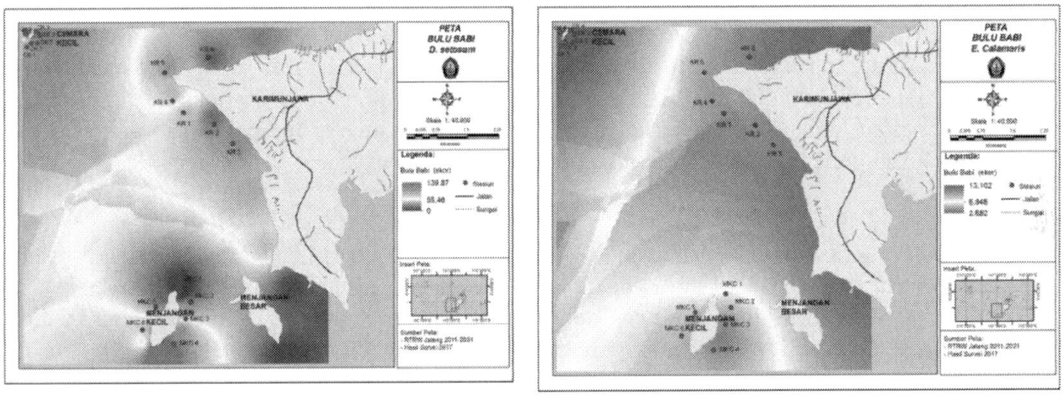

Figure 4. Mapping of Sea Urchin Distribution

Based on the measurements of the number of sea urchins found at the study site, two species of sea urchin are *Diadema setosum* and *Echinotrix calamaris*. The most abundant number of sea urchins is located in Pulau Cemara Kecil.

The presence of *D. setosum* on the coral reef ecosystem has a significant effect on the ecological balance [36]. Suggest that sea urchins are commonly found in coral reef ecosystems, especially species of D. setosum, because the abundance of species popolates is important for coral reefs as a counterweight [37].

The presence of species populations Diadema setosum is important for coral reefs as a counterweight. Diadema population equilibrium will keep the equilibrium of algae and coral populations. Diadema setosum species research is most dominant because the species is one type of sea urchin (Echinoidea) that lives in coral reef and seagrass ecosystems. According to [38], coral reef is a complex ecosystem and has a high aesthetic value, and inhabited by various types fauna, including echinoderms of one of the sea urchins which is a dominant coral reef dweller. Furthermore [39] stated that *D. setosum* has a place to live in the coral reef ecosystem, where this species can occupy the average sand, algae growth areas, coral fragments and dead corals. Pig feathers (Echinoidea) that live in sand-aligned zones, algae growth areas, and coral reefs usually live in clusters in large clusters whereas in coral reefs (Echinoidea) live in small groups or alive life in dead and fractured coral holes coral.

Based on multiple regression analysis showed that the relationship between nitrate, phosphate and zooxanthela with abundance of sea urchins showed a strong correlation result with correlation value (r) is 0.64. These results can be an indicator of coastal environmental health, especially coral reef ecosystems.

Table 1. Result of Multiple Regression Analysis

Regression Statistics	
Multiple R	0,643810756
R Square	0,414492289
Adjusted R Square	0,289026351
Standard Error	30,53351541
Observations	18

Basically the phosphate is occupied in the sediments or is present in coral polyps. On the reef itself, phosphate is required by zooxanthellae as a nutrient and affect the growth of corals. The effects of phosphate on coral condition and/or growth have been studied both in the lab and the field. The results of past studies, however, were not consistent, par-ticularly with respect to coral growth. Growth is often considered an indicator of coral health [40-41] and is com- monly quantified via measurement of linear extension [42] or changes in buoyant weight [43]. Although the buoyant weight technique is attractive because it allows for non-destructive measurement of complete colony growth, it was not used in our study, because it may produce highly variable data [44]. Phosphates have an indirect effect with the abundance of sea urchins. this is due to the phosphate affecting the abundance of zooxanthellae which is the source of food of the sea urchins.

Conclusion

There was interaction between type of fish and body part on moisture, protein, ash and carbohydrate content (P <0,05), but no interaction on fat content and energy (P> 0,05). The body of cultured eel had higher protein than cultured one. The wild eel had higher fat content and energy than cultured one, while the fat content and energy in body and tail were higher than in head.

Acknowledgments

We thank to Head of Research and Community Service Centre of Diponegoro University (Undip) for the research fund from PNBP DIPA Undip in the 2017 fiscal year of, Research Development & Application (Riset Pengembangan dan Penerapan) Scheme, to head of BTNKJ for the permit and information

References

[1] Knowlton N, Jackson J 2013 Corals and Coral Reefs *Encyclopedia of Biodiversity* 330–346

[2] Van Zanten B T, Van Beukering P J H, Wagtendonk A J 2014 Coastal Protection by Coral Reefs A Framework for Spatial Assessment and Economic Valuation *Ocean Coastal Manage* **96** 94–103

[3] Graham N A J, Cinner J E, Norström A V, Nyström M 2014 Coral Reefs as Novel Ecosystems EMBRACING New Futures *Current Opinion Environmental Sustainable* **7** 9–14

[4] Glynn P W, Manzello D P 2015 Bioerosion and Coral Reef Growth a Dynamic Balance In *Coral Reefs in the Anthropocene* 67–97

[5] Lira-Noriega A, Soberón J 2015 The relationship Among Biodiversity, Governance, Wealth, and Scientific Capacity at a Country Level Disaggregation and Prioritization *AMBIO* **44** 391–400

[6] Adam T C, Burkepile D E, Ruttenberg B I, Paddack M J 2015 Herbivory and the resilience of Caribbean Coral Reefs Knowledge Gaps and Implications for Management *Marine Ecology Progress Series* **520** 1–20

[7] Cunning R, Yost D M, Guarinello M L, Putnam H M, Gates R D 2015 Variability of Symbiodinium Communities in Waters, Sediments, and Corals of Thermally distinct Reef Pools in American Samoa *PLoS One* **10**

[8] Fenner D 2012 Challenges for Managing Fisheries on Diverse Coral Reefs *Diversity* **4** 105–160

[9] Elko N, Feddersen F, Foster D, Hapke C, Mcninch J, Mulligan R 2015 The Future of Nearshore Processes Research. *Shore and Beach* **83** 13–38
[10] De'ath G, Fabricius K E, Sweatman H, and Puotinen M 2012 The 27-Year Decline of Coral Cover on the Great Barrier Reef and its Causes *PNAS* **109** 17995–17999
[11] Choat J, Clements K 1998 Vertebrate Herbivores in Marine and Terrestrial Environments a Nutritional Ecology Perspective *Annual Rev Ecol Syst* **29** 375–403
[12] Burkepile D E 2013 Comparing aquatic and terrestrial grazing ecosystems is the Grass Really Greener *Oikos* **122** 306–312
[13] Asner G P, Levick S R 2012 Landscape-scale Effects of Herbivores on Treefall in African Savannas *Ecology Letters* **15** 2008–2011
[14] Pandolfi J M, Bradbury R H, Sala E, Hughes T P, Bjorndal K A, Cooke R G, McArdle D, McClenachan L, Newman M J H, Paredes G, Warner R R, Jackson J B C 2003 Global Trajectories of the Longterm Decline of Coral Reef Ecosystems Science **301** 955–958
[15] Jackson J 1997 Reefs since Columbus Coral Reefs **16** 23–32
[16] Hughes T P, Bellwood D R, Folke C S, McCook L J, Pandolfi J M 2007 No-Take Areas, Herbivory and coral Reef Resilience *Trends Ecol Evol* **22** 1–3
[17] Gaines, S. and Lubchenco, J 1982 A Unified Approach to Marine Plantherbivore Interactions II Biogeography *Annual* Review of *Ecology* and Systematics **13** 111–138
[18] Tuya F, Boyra A, Sanchez-Jerez P, Barbera C, Haroun R J 2004 Relationships Between Rocky-Reef Fish Assemblages, The Sea Urchin *Diadema antillarum* and Macroalgae Throughout the Canarian Archipelago *Marine Ecology Progress Series* **278** 157–169
[19] Suryanti dan Ruswahyuni 2014 Differences of Abundance Sea Urchin on Coral and Seagrass Ecosystems in Pancuran Belakang, Karimunjawa Jepara *Journal of Fisheries Science and Technology* **10** 62-67
[20] Wright J T, Dworjanyn S A, Rogers C N, Steinberg P D, Williamson J E, Poore A G B 2005 Density-dependent Sea Urchin Grazing Differential Removal of Species, Changes in Community Composition and Alternative Community States *Marine Ecology Progress Series* **298** 143–156
[21] Puwanti F 2001 Tourism And Conservation Management Options For Karimunjawa Marine National Park (Case Study and Reviews) *Journal of Coastal Development* **4** 99–106
[22] Luthfi O M T Nakamura and Tsuchiya M 2012 Coral Disease Assessment in Karimunjawa National Park, Central Java Indonesia. Faculty of Science University of The Ryukyus Nishihara Okinawa Japan
[23] Muscatine L and D'Elia C F 1978 The Uptake, Retention, and Release of Ammonium by Reef Corals *Limnology Oceanography* **23** 725–734
[24] D'Elia C F 1983 Dissolved Nutrients in Coral reef Waters. In Barnes, D J Perspective on Coral Reefs Pub 200 Aust Inst Mar Sci Townsville 56–68
[25] Syrett P J 1981 Nitrogen Metabolism of Microalgae *Can Bull Fish Aquat Sci* **210** 182–210
[26] Szmant-Froelich A and Pilson M E Q 1984 Effects of Feeding Frequency and Symbiosis with Zooxanthellae on Nitrogen Metabolism and Respiration of The Coral *Astranginadanae Mar Bio* **81** 153-162
[27] Dunn J G, Sammarco P W and Junior G L 2012 Effects of Phosphate on Growth and Skeletal Density in The Scleractinian *Acrophora nuricata* A Controlled Experimental Approach *Journal of Experimental Marine Biology and Ecology* **411** 34–44
[28] Kleypas J A, McManus J W, Meñez, L A B 1999. Environmental Limits To Coral Reef Development Where do we Draw The Line *Am Zool* **39** 146–159
[29] Cruz-Piñón G, Carricart-Ganivet G P, Espinoza-Avalos J 2003 Monthly skeletal extension Rates of the Hermatypic Corals Montastraea annularis and Montastraea faveolata Biological and Environmental Controls *Marine Biology* **143** 491–500
[30] Renegar D A, Riegl BM, 2005 Effect of Nutrient Enrichment and Elevated CO_2 Partial Pressure on Growth Rate of Atlantic Scleractinian Coral *Acropora cervicornis Marine Ecology Progress Series* **293** 69–7

[31] Koop K, Booth D, Broadbent A, Brodie J, Bucher D, Capone D, Coll J, Dennison W, Erdmann M, Harrison P, Hoegh-Guldberg O, Hutchings P, Jones G B, Larkum A W D, O'Neal J, Steven A, Tentori E, Ward S, Williamson J, Yellowlees D, 2001 ENCORE The Effect Of Nutrient Enrichment On Coral Reefs Synthesis of Results and Conclusions *Marine Pollutan Bulletin* **42** 91–120

[32] Harrison, PL, Ward S 2001 Elevated Levels of Nitrogen and Phosphorus Reduce Fertilization Success of Gametes From Scleractinian Reef Corals *Marine Biology* **139** 1057–1068

[33] Walker D I, Ormond R F G 1982 Coral Death From Sewage and Phosphate Pollution at Aqaba, *Red Sea Marine Pollutan Bulletin* **13** 21–25

[34] Bruno J F, Petes LE, Harvell CD, Hettinger A 2003 Nutrient enrichment can increase the Severity of Coral Disease *Ecology Letters* **6** 1056–1061

[35] Done T J 1992 Phase Shifts In Coral Reef Communities and Their Ecological Significance *Hydrobiol* **247** 121–132

[36] Thamrin S, Y J Siregar S H 2011 Analysis of *Diadema setosum* (Echinoidea) Density on Different Coral Reef Conditions in Mapur Village Riau Island *Journal Environmental Science Riau* **5** 45–53

[37] Suryanti and C A'in 2013 Differences of Sea Urchin Abundance in Different Substrates in Legon Legon Boyo Karimunjawa Jepara *Prosiding SEMNAS III Fishery and Marine Products FPIK UNDIP Semarang* **4** 165–172

[38] Clark A M 1976 Echinoderm of Coral Reefs, In O A Jones and R Endean (eds) Geology and Ecology of Coral Reefs 3 Academy Press New York 95 –123

[39] Sugiarto dan Supardi 1995 Reports of Sea Urchin Famili Diadema *Oseana* **20** 34–41

[40] Neudecker S 1983 Growth and Survival of Scleractinian Corals Exposed to Thermal effluents in Guam Proc 4th *International Coral Reef Symposium Manila Philippines* **1**173–180

[41] Rosen B R 1981 The Tropical High Diversity Enigma The Coral's Eye View Chance Change, and Challenge The Evolving Biosphere Brit Mus (Nat Hist) Cambridge University Press Cambridge London

[42] Jinendradasa S S, Ekaratne S U K 2002 Linear Extension of *Acropora muricata* at Selected Reef Locations in Sri Lanka Proc 9th *International Coral Reef Symposium Bali Indonesia* **1** 417–420

[43] Davies P S 1989 Short-Term Growth Measurements of Corals Using An Accurate Buoyant Weighing Technique *Marine Biology* **101** 389–395

[44] Bucher D J, Harrison P L 2002 Growth Response of the Coral Reef *Acropora longicyathus* to Elevated Inorganic Nutrients Do Responses to Nutrients Vary Among Coral Taxa Proc 9th *International Coral Reef Symposium Bali Indonesia* **1** 443–448

The Assessment of Biological and Pollution Index of Estuaries Around Port of Tanjung Emas Semarang

A Tjahjono[1*], O Wahyuni[2] and S Purwantini[3]

[1]Engineering Department, Semarang Merchant Marine Polytechnic, Singosari 2 a Semarang, Central Java, Indonesia
[2&3]Port & Shipping Department, Semarang Merchant Marine Polytechnic, Singosari 2 a Semarang, Central Java, Indonesia
*+62 0813 2610 5514, a_agus_tjahjono70@yahoo.co.id

Abstract. Estuary is a place of accumulation of the population's actitivites produced by domestic, industry or agriculture. This research was conducted to three of estuary of the rivers around the waters of Port of Tanjung Emas Semarang (PTES). They were estuaries of Baru river, Banjir Kanal Timur (BKT) and Siangker in west monsoon from October to December 2015. The purpose of this research was to analyze pollution index, the abundance of microorganisms either phytoplankton or zooplankton, the content of heavy metal in sediment and sea water, biological index that included diversity (H), uniformity (e), dominance (D), Saprobik Index (SI), and the Total of Saprobik Index (TSI) in the waters either HTL (High Tide Level) or LTL (Low Tide Level). The concentration of heavy metal in both sea water and sediments were analyzed by using Atomic Absorption Spectrophotometer (AAS). The result obtained from 12 parameters which were tested showed that the three waters can be categorized at heavily polluted condition at each value from 12.52 to 24.98. The concentration of heavy metal at sea water during HTW and LTW ranging from Cd is around 0.033 and 0.048 mg/kg, Cu 0.047 and 0.07 mg/kg, Pb 0.48 and 0.71 mg/kg, and Zn 0.043 and 0.057 mg/kg. The saprobity value index based on the existence of phytoplankton or zooplankton was ranging of Oligosaprobik at low pollution or has not been polluted yet.

Keywords : Port of Tanjung Emas Semarang (PTES), pollution index, heavy metal, biological index, west monsoon.

1. Introduction

Coastal areas and river estuaries are susceptible area of pollution from upstream. Pollution in the river estuary area is caused by frequent activity by both domestic and shipping industries in the region [1].

Kali Baru river estuaries located in the Port of Tanjung Emas Semarang (PTES), beside the domestic pollution from upstream, port activities such as shipyards, power plants, ship activities in the region have affected the estuary. Banjir Kanal Timur estuary is an area that is also close to PTES waters and a fishing area where the water quality is influenced by fishing activities and discharges from domestic households in the area. Siangker Estuary is a river estuary located in the west side of PTES close to the airport that will affected to the river estuary.

The purpose of this research is to analyze the value of Kali Baru river estuary pollution, BKT (Banjir Kanal Timur) and Siangker in terms of diversity (H '), uniformity (e), dominance (D), Saprobik Index (SI) and Total Saprobik Index (TSI) in both phytoplankton and zooplankton at HTL (High Tide Level) and LTL (Low Tide Level). Analysing the heavy metal concentrations Cu, Cd, Pb and Zn in both seawater and sediment. The last calculation of the pollution index on the territorial waters

2. Material and Methods

The research was conducted at the coastal estuaries around the Port of Tanjung Emas Port Semarang (PTES) in 3 areas with 3 research station, 1st station (estuary Kali Baru), 2nd station (estuary Banjir Kanal Timur) and 3rd station (estuary Siangker). Research conducted from October to December 2015.

Phytoplankton identification includes phytoplankton endemic in port waters, environmental parameters other than depth, current velocity, direction of current at each station also physical parameters (brightness, BOD_5, DO, temperature), chemical parameters (pH, salinity), while phytoplankton was analyzed using individual abundance (N), index of species diversity (H'), fairness index (E), dominance index (D), Saprobic Index (SI) and Total Saprobic Index (TSI) [2,3,7].
Abundance of individuals (N)

$$N = \frac{1}{A} x \frac{B}{C} x \frac{D}{E} x F \quad (1)$$

Where
N = abundance (individual/liter),
A = volume of filtered water (liter),
B = volume of water in sample (125 ml),
C = volume of preparation at identification (1 ml),.
D = breadth of glass cover (mm^2),
E = field view area (mm^2),
F = average number of observed individuals
The diversity index and equitability index by Shannon-Wier (1949),

$$H' = - \Sigma (pi \ln pi) \quad (2)$$

Where
H '= index of species diversity,
Ni = species abundance at rank -1,
N = total abundance
The evenness index (e)

$$E = \frac{H^1}{H^1 maks} \quad (3)$$

Where
e = (Index of evenness or stability),
H_{max} = log 2 S = 3.3219 log10S,
S = the number of taxa in a community.
To know the dominance of certain types in the waters can be used Simpson dominance index,

$$D = \Sigma S_{i=1} \left| \frac{n_i}{N} \right|^2 \quad (4)$$

where
D = Simpson dominance index,

N_i = individual type of rank -i,
N = total number of individuals,
S = number of genera.
Index of dominance between 0-1; D = 0, means that no species dominate other species or community structures in a stable state;
D = 1, there is a species that dominates other species or unstable community structures, because of the ecological pressures.
Saprobik Index (SI) Persoone & de Pauw is used to determine the level of contamination of organic matter in water [2][7].

$$SI = \frac{C+3D+B-3}{A+B+C+D}. \quad (5)$$

where,
SI = saprobic coefficient (-3 to 3),
A = Ciliata organism group
C = Chloroccales and Diatomae organism group
D = Peridinae, Chrysophyceae and Conjugaeceae organism group
A, B, C and D = the amount of different organism of each group
Index Saphrobic Trophic is

$$TSI = \frac{1(nC)+3(nD)+(nB)-3(nA)}{1(nA)+3(nB)+1(nC)+1(nD)} \text{ x } \frac{nA+nB+nC+nD+nE}{nA+nB+nC+nD} \quad (6)$$

where
N = number of individual organisms in each group of saprobity,
NA = the number of individual composers of the polysaprobik group,
NB = the number of individual composers of the α-Mesosaprobic group,
NC = the number of individual composers of the β-Mesosaprobic group,
ND = the number of individual composers of the Oligosaprobik group,
NE = the number of individual composers other than A, B, C and D
Index Pollution method is use to determine the level of relative pollution toward the parameter of water quality needed with formula (Nemerow and Sumitomo, 1970)

$$IP_j = \sqrt{\frac{\left(C_i/L_{ij}\right)_M^2 + \left(C_i/L_{ij}\right)_R^2}{2}} \quad (7)$$

In which:
IP_j = index of pollution for j
C_i = parameter of consentration of water quality i
L_{ij} = parameter of quality which submitted in quality j water
M = maximum
R = average

There are 4 IP (Index Pollution) class with score $0 \leq IP \leq 1.0$ fullfilled quality standard, $1.0 \leq IP \leq 5.0$ slightly polluted, $5.0 \leq IP \leq 10$ moderately polluted and $IP \geq 10$ status of the water heavily polluted. The observed parameter consist of 12 parameters, there are turbidity, pH, sulfida, phenol, DO, BOD_5, PO_4 (phosphat), nitrat , metal Cd, Cu, Pb and Zn.

3. Results and Discussion

The mean average concentration of heavy metal under sea water at three estuaries when LTL is bigger than when its HTL, but for Cu concentration when LTL smaller than when its HTL. Heavy metal concentration, both when its LTL or HTL each are Cd around 0.033 and 0.048 mg/kg, Cu around 0.046 and 0.070, Pb around 0.48 and 0.71 mg/kg and Zn around 0.043 and 0.057 mg/kg (Figure 1).

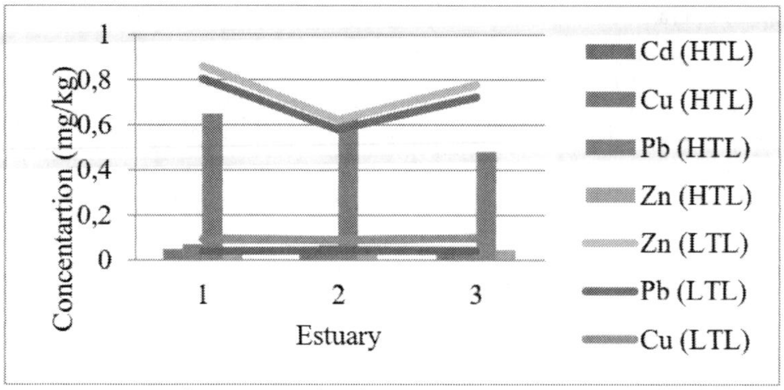

Figure 1. Heavy metal concentration in sea water at three estuaries around PTES

Zn consentration at three estuaries each Zn around 23.253 and 53.957 mg/kg, Pb around 7.583 and 18.0 mg/kg, Cu around 6.947 and 22.03 mg/kg meanwhile Cd around 0.247 and 0.351 mg/kg. Consentration Zn is on highest rank then followed by Cu, Pb and last, Cd. All of metal element consentration are on highest rank at Kali Baru estuary (Figure 2).

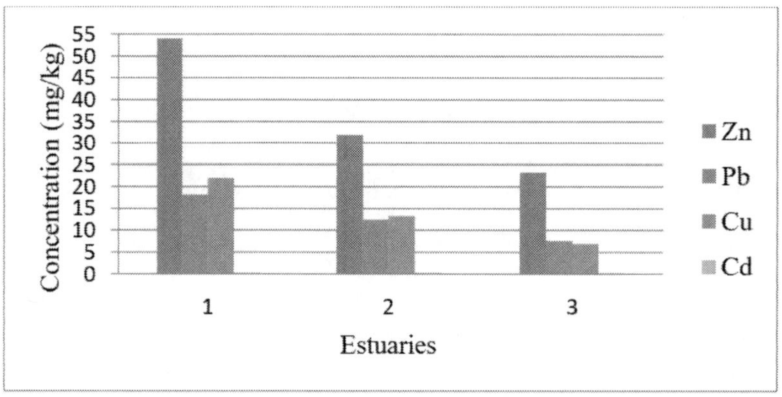

Figure 2. Heavy metal consentration in sediment at three estuaries around PTES

H' index (diversity) at three estuaries when its HTL around 0.65 and 1,65 meanwhile when LTL around 0.975 and 1.88 which shows when its LTL, the estuary of Kali Baru is heavily polluted, BKT moderate and Siangker heavily polluted. When the condition HTL and each is heavily polluted, moderate and moderate. The uniformity index (e) at LTL ranges from 0.37 and 0.66 while in HTL 0.425 and 0.71. The higher the uniformity index value indicates that all species are more abundant while the lower shows the smaller [5]. The dominance index (D) at LTL ranges from 0.32 and 0.7 whereas when the tide ranges from 0.21 and 0.565. Based on the index, the three river estuaries showed a heavily polluted condition when LTL, while when HTL of Kali Baru river, BKT and Siangker each showed heavily, moderate and heavily polluted (Figure 3).

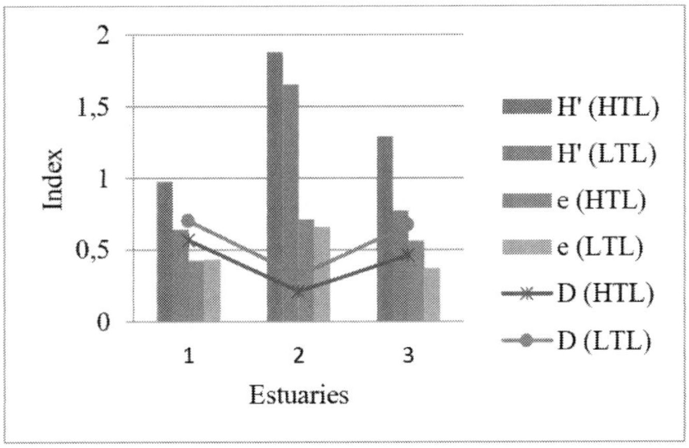

Figure 3. Index value H', e and D phytoplankton at October 2015 at three estuaries around PTES

Saprobity Index (SI) value is when it's HTL or LTL close to similar of the three estuary. Meanwhile TSI value at three estuary get several difference in which when it's HTL the value will be bigger compared when it's LTL (Table 1.). At three estuaries, the value of saprobity counted as oligosaprobic waters, in which slightly polluted or haven't polluted [3].

Table 1. SI and TSI phytoplankton at three estuary at October 2015 around PTES

Stations/Tides		Baru	BKT	Siangker
SI	LTL	1.925	1.33	1.8
SI	HTL	1.925	1.28	1.8
TSI	LTL	3.255	4.08	3.22
TSI	HTL	3.545	6.12	3.98

There are four classes of phytoplankton found in October 2015 named *Bacillariophyceae, Dinophyceae, Chrysophyceae* and *Cyanophyceae* where the abundance at HTL is greater than at LTL. The abundance of *Skeletonema* when HTL or LTL dominate between the other genus (Table 2)

Table 2. The abundance of phytoplankton in HTL or LTL on three estuaries on October 2015

No.	Genus	LTL			HTL		
		Estuaries					
		Baru	BKT	Siangker	Baru	BKT	Siangker
A.	Bacillariophyceae						
1	*Skeletonema sp.*	13822	122	20255	32197	34	13248
2	*Sydnedra sp.*	127.5	8	255	2357	20	637
3	*Rhizosolenia sp.*	95.5	8	637	732.5	4	1083
4	*Asteroinella sp.*	605	-	-	1306	1	-
5	*Chaetoceros sp.*	286.5	4	573	1242	1	955

6	*Thalassionema sp.*	159	14	-	10350	6	
7	*Thalassiothrix sp.*	478	30	2357	1337.5	52	1720
8	*Navicula sp.*	32	-	-	32	-	-
9	*Cyclotella sp.*	63.5	3	-	700.5	1	-
10	*Stephanopyxis sp.*	63.5	-	-	63.5	-	-
11	*Guinardia sp.*	-	27	-	0	18	191
12	*Coscinoduscus sp.*	-	-	382	318.5	3	955
13	*Pleurosygma sp*	-	-	-	-	3	-
14	*Bacteriastrum sp.*	32	-	-	-	-	-
15	*Pleurosygma sp.*	-	6	127	-	-	-
16	*Hemiaulus sp.*	-	-	-	-	-	-
17	*Eucampia sp.*	-	4	-	-	-	-
18	*Biddulphia sp.*	-	4	-	-	2	-
19	*Nitzschia sp.*	6.5	-	-	-	-	-
20	*Asterionella gracialis*	-	-	-	-	-	-
21	*Amphora sp.*	-	-	-	-	-	-
22	*Thallasiosira sp.*	-	-	-	-	-	-
23	*Gyrosigma sp.*	-	-	-			
B.	*Dinophyceae*						
24	*Ceratium sp.*	-	-	382	-	-	446
25	*Protoperidinium sp.*	-	-	-	-	-	-
26	*Noctiluca sp.*	-	-	-	-	1	382
27	*Ornithocercus sp*	-	-	-			
C.	*Chrysophyceae*						
28	*Peridinium sp.*	-	-	-	-	-	191
29	*Pyrocystis sp.*	-	-	-	-	-	-
30	*Dictyocha sp.*	-	-	-	-	-	-
D.	*Cyanophycea*						
31	*Anabaenopsis sp.*	95.5	1	-	63.5	2	-
Total		15866.5	231	24968	50700	148	19808

The zooplankton biodiversity index in October at HTL ranged from 1.92 and 2.26 while at LTL ranged from 1.13 and 2.22 indicating that Kali Baru estuary and BKT were slightly polluted while the Siangker estuary was slightly polluted. The uniformity index (e) shows that the species condition is abundant except for the LTL condition at the Kali Baru estuary. The dominant index on the three river estuaries indicates the absence of a species dominance in these waters (Figure 4).

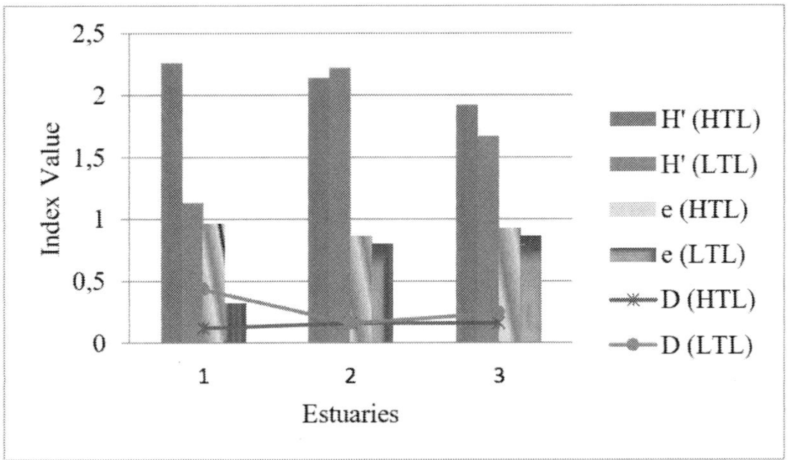

Figure 4. The index values of H ', e and D zooplankton in October 2015 in three estuaries around PTES

The *Acartian* genus dominates on three estuaries compared to *Microstella sp.* and *Trocophora sp.* Where the abundance of zooplankton is more at the Kali Baru estuary than in other waters. Crustacean class is more dominant than other classes, where in that class found 21 species of genus while in class Ciliata only found 13 genus while class Cirripida least found (Table 3).

Table 3. The abundance of zooplankton at HTL and LTL on three estuaries in October 2015 around PTES waters

No.	Genus	HTL			LTL		
		Station					
		Baru	BKT	Siangker	Baru	BKT	Siangker
	Crustacea						
1	*Nauplius*	2.12	-	-	-	-	-
2	*Corycaeus*	0.64	-	-	-	-	-
3	*Acartia*	3.48	7.64	13	28.5	6	6
4	*Calanus*	0.64	2.55	25	12.5	2	19
5	*Oithona*	0.85	6.37	38	19	5	51
6	*Euterpina sp.*	-	-	6	16	-	-
7	*Eurytemora sp.*	-	5.1	25	22.5	4	19
8	*Undinopsis sp.*	-	-	-	-	-	-
9	*Oncaea sp.*	-	1.27	-	-	-	-
10	*Eucalanus sp.*	-	-	-	-	-	-
11	*Microsetella sp.*	-	-		-	1	-
12	*Parvocalanus sp.*	-	-	13	-	-	-
13	*Evadne sp.*	-	-	-	-	-	-
14	*Euphausia sp.*	-	-	-	-	-	-
15	*Tigriopus*	-	1.27	-	-	4	-
16	*Paracyclopina sp.*	-	-	-	-	-	-

17	*Canthocalanus sp*	-	-	-	-	-	-
18	*Acrocalanus sp.*	-	-	-	-	-	-
19	*Podon sp.*	-	-	-	-	-	-
20	*Sergia sp.*	-	1.27	-	-	1	-
21	*Sapphirina sp.*	-	-	-	-	-	-
Ciliata							
22	*Eutintinnus*	6.115	-	-	-	-	-
23	*Salpingella*	1.27	-	-	-	-	-
24	*Codonellopsis*	5.48	-	-	-	1	-
25	*Epicycloclysis*	3.32	-	-	-	-	-
26	*Achanstostomella*	5.32	1.27	-	-	8	-
27	*Leprotintinnus sp.*	-	2.55	6	3	2	19
28	*Scolecithrix sp.*	-	-	-	-	-	-
29	*Tintinnopsis sp.*	11	17.84	25	32	30	-
30	*Parafavella sp.*	-	-	-	-	3	6
31	*Favella sp.*	-	5.1	-	-	7	
32	*Xystonellopsis sp*	6.5	-	-	-	-	-
33	*Rhabdonella*	-	-	-	-	-	-
34	*Dictyoscysta sp*	-	-	-	-	-	-
35	*Trocophora sp.*	-	-	-	-	1	-
	Scyphomedusae						
36	*Trichocerca sp.*	-	-	-			
37	*Branchionus sp.*	-	5.1	-	-	13	
	Tentaculata						
38	*Notholca sp.*	-	-	-			
39	*Salpinengacantha*	5.48	-	-	-	-	-
40	*Rhodonella sp.*	5	-	-	-	-	-
41	*Trocophora sp.*	-	-	-	-	-	-
42	*Balanus*	-	-	-	3	3	6
43	*N Eurytermora*	-	-	-	-	-	-
44	*N Euphasia*	-	-	-	-	-	-
45	*N Oithona*	-	-	-	-	-	-
46	*N Acartia*	-	-	-	-	-	-
47	*Branchiiolaria sp.*	8	-	-	-	-	-
	Cirripeda						
48	*Balanus*	-	-	-	-	-	-
Total		65.215	57.33	151	136.5	91	126

Diversity value in Kali Baru estuary at HTL or LTL show that biota community is not stable or heavily polluted while at another water area is`slightly polluted to moderate. These things seen at the value of homogeneity at Kali Baru estuary which is equally relative of homogeneity while for another estuary teritory low species homogeneity is occured. Ecological pressure is occured at Kali Baru estuary rather than other water in BKT and Siangker because domination value that close to one. This thing is because of the pressure form domestic waste and shipyard around the estuary (Figure 5).

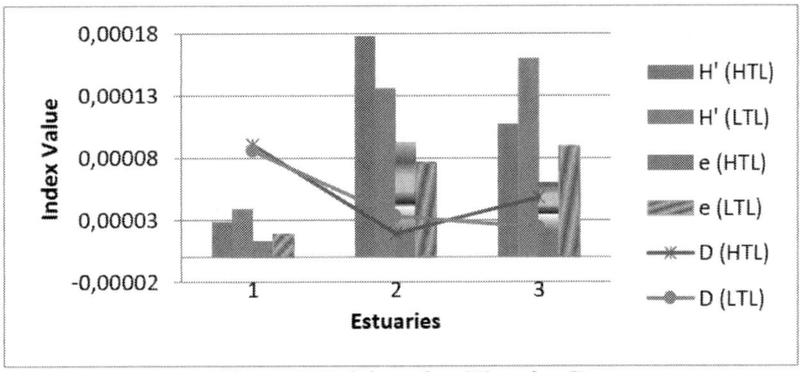

Figure 5. Indeks value H', e dan D
phytoplankton in November 2015 in three estuaries around PTES

Saprobity water condition is in oligosaprobik condition or slightly polluted or haven't polluted at all except at HTL at BKT is in condition β-mesosaprobik which is polluted condition at low to medium level (Table 4).

Table 4. SI dan TSI phytoplankton in three estuaries
in November 2015 in around PTES

Station/Tide		Baru	BKT	Siangker
SI	HTL	1.59	0.6	1.5
SI	LTL	1.665	1.5	1.67
TSI	HTL	3.01	1.38	2.48
TSI	LTL	2.955	2.57	2.95

In November there is amount of class that fewer than in the earlier month at west season, there are only three classes which are *Baciilariophyceae*, *Dinophyceae*, *Chrysophyceae* dan *Cyanophycea*. Skeletonema genus is still dominating followed by *Synedra sp.* and the last is *Thallassiothrix sp.* and *Pyrocystis sp.* (Table 5).

Table 5. The abundance phytoplankton at HTL and LTL
at three estuaries in November 2015

No.	Genus	LTL			HTL		
		Station					
		Baru	BKT	Siangker	Baru	BKT	Siangker
A.	Bacillariophyceae						
1	*Skeletonema sp.*	16798.3	3375.8	3694.27	11648.62	1528.66	7133.76
2	*Sydnedra sp.*	4.775	-	-	11648.62	1528.66	7133.76
6	*Thalassiothrix sp.*	31.845	-	-	31.845	-	-

7	*Coscinoduscus sp.*	3.185	-	1464.97	79.62	-	-
8	*Pleurosygma sp*	-	445.86	1082.8	38.215	382.17	-
9	*Biddulphia sp.*	-	-	-	-	445.86	-
B.	*Dinophyceae*						
10	*Ceratium sp.*	767.515	318.47	828.03	98.725	828.03	1401.27
11	*Noctiluca sp.*	-	-	505.55	0	382.17	191.08
C.	*Chrysophyceae*						
12	*Peridinium sp.*	159.235	382.17	-	95.54	-	191.08
13	*Pyrocystis sp.*	-	-	-	31.845	-	-
D.	*Cyanophycea*						
14	*Anabaenopsis sp.*	170.38	-	-	41.4	-	-
15	*Oscallatoria sp.*	63.695	-	-	-	-	-
Total		18674.09	6305.74	9677.53	24106.15	8025.49	17770.7

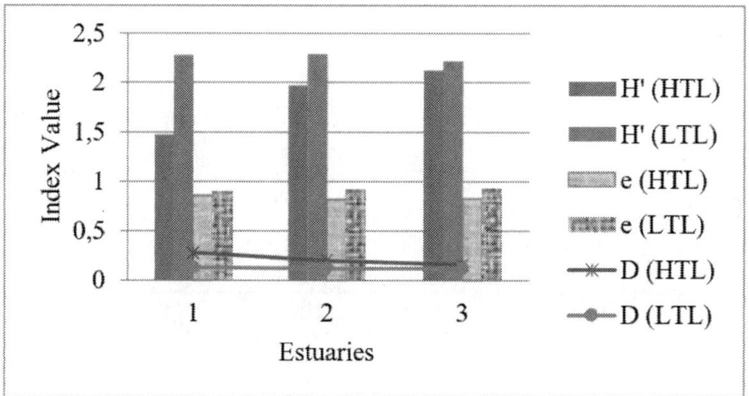

Figure 6. Index value H’, e and D
zooplankton in November 2015 in three estuaries around PTES

Table 6. Zooplankton abundance when HTL and LTL in all three estuaries in November 2015

No.	Genus	HTL			LTL		
		Station					
		Baru	BKT	Siangker	Baru	BKT	Siangker
A.	Crustacea						
1.	*Acartia*	25.4775	15.92	35.03	19.11	15.92	9.55
2.	*Calanus*	3.1845	6.37	22.29	7.96	9.55	19.11
3.	*Oithona*	19.1085	41.4	73.23	25.48	12.74	22.29
4.	*Euterpina sp.*	0	3.18	0	1.59	3.18	0
5.	*Eurytemora sp.*	19.1085	9.55	22.29	12.735	12.74	9.55
6.	*Eucalanus sp.*	0	0	3.18	0	9.55	9.55
7.	*Macrostella sp.*	0	0	3.18	0	3.18	0
8.	*Parvocalanus sp.*	6.3695	15.92	38.22	3.185	3.18	12.74

9.	*Tigriopus*	0	3.18	0	3.185	0	0
10.	*Undinopsis sp.*	0	0	0	1.59	0	0
11.	*Ctenocalanus sp.*	0	0	0	0	6.37	6.37
12.	*Metridia sp.*	0	0	0	1.59	0	0
13.	*Brachionus sp.*	0	3.18	6.37	0	0	0
14.	*Cytemnestra sp*	0	0	0	3.185	0	0
15.	*Globigerina sp.*	0	0	0	0	0	3.18
B.	*Ciliata*						
16.	*Codonellopsis*	0	0	0	3.185	0	0
17.	*Achanstostomella*	0	0	0	3.185	0	0
18.	*Leprotintinnus sp.*	6.3695	3.18	15.92	31.845	0	0
19.	*Tintinnopsis sp.*	12.7395	0	0	14.33	0	0
20.	*Parafavella sp.*	0	3.18	9.55	0	6.37	22.29
21.	*Favella sp.*	0	6.37	6.37	6.365	0	0
C.	*Tentaculata*						
22.	*Notholca sp.*	0	0	0	0	22.29	0
D.	*Rotatoria*						
23.	*Balanus*	0	0	0	6.37	3.18	9.55
E.	*Mollusca*						
24.	*Trocophora sp.*	0	0	0	3.185	0	0
F.	*Cirripeda*						
25.	*Balanus*	3.1845	0	6.37	0	0	0
Total		95.542	111.43	242	148.075	108.25	124.18

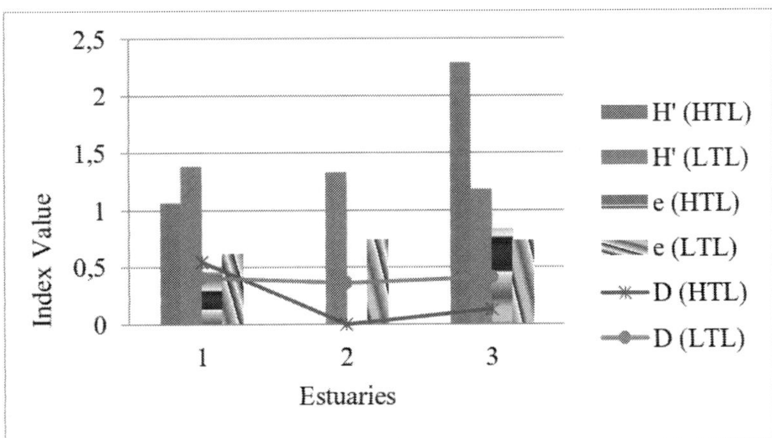

Figure 7. Index value H', e and D phytoplankton in December 2015 in three estuaries around PTES

Table 7. SI and TSI phytoplankton in three estuary in December 2015 around PTES

Station/Tide		Baru	BKT	Siangker
SI	HTL	1	0	1
SI	LTL	1	1.67	1
TSI	HTL	2.745	0	1.13
TSI	LTL	2.62	2.7	2.32

Table 8. Phytoplankton abundance when HTL and LTL in all three estuaries in December 2015

No.	Genus	HTL			LTL		
		Station					
		Baru	BKT	Siangker	Baru	BKT	Siangker
A.	Bacillariophyceae						
1	*Skeletonema sp.*	16410.83	-	2886.4	5391.72	3949.04	9235.67
2	*Rhizosolenia sp.*	63.695	-	828.03	63.695	764.33	1592.36
3	*Asteroinella sp.*	3.185	-	4012.74	-	-	-
4	*Chaetoceros sp.*	713.38	-	382.17	264.33	-	-
5	*Thalassionema sp.*	-	-	573.25	-	-	-
6	*Thalassiothrix sp.*	35.03	-	-	127.39	-	-
7	*Cyclotella sp.*	334.395	-	318.47	1436.305	-	1783.44
8	*Stephanopyxis sp.*	-	-	-	-	254.78	-
9	*Guinardia sp.*	-	-	1146.5	130.575	-	-
10	*Coscinoduscus sp.*	73.25	-	127.39	130.575	-	-
11	*Pleurosygma sp*	82.805	-	764.33	63.695	318.47	-
12	*Hemialus sp.*	-	-	191.08	-	-	-
13	*Biddulphia sp.*	168.79	-	1910.83	512.74	-	445.86
14	*Helosira sp.*	-	-	191.08	-	-	-
B.	*Dinophyceae*						
15	*Ceratium sp.*	254.775	-	700.64	226.115	1401.27	2292.99
C.	*Chrysophyceae*						
16	*Peridinium sp.*	63.695	-	191.08	63.695	-	-
D.	*Chyanophycea*						
17	*Anabaenopsis sp.*	63.695	-	127.39	-	-	-
18	*Oscillatoria sp.*	191.08	-	1719.75	194.27	-	-
Total		18458.605	0	16071.13	8605.105	6687.89	15350.32

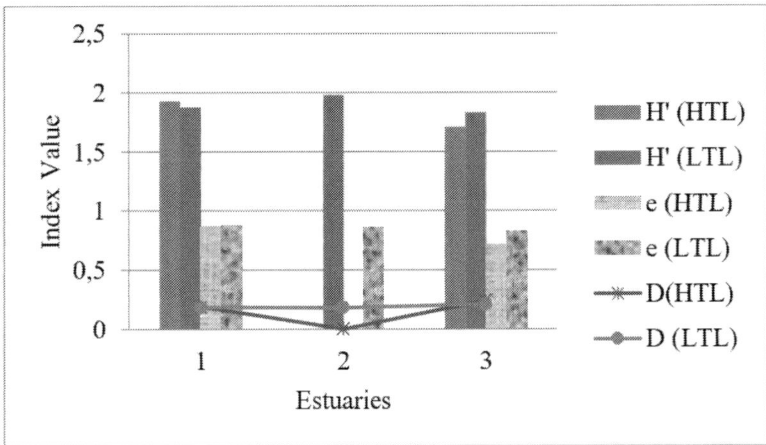

Figure 8. index value H', e and D
zooplankton in December 2015 in the three estuaries around of PTES

Table 9. Zooplankton abundance HTL or LTL
on the three river estuary in December 2015

No.	Genus	HTL			LTL		
		Station					
		Baru	BKT	Siangker	Baru	BKT	Siangker
A.	***Crustacea***						
1	*Acartia*	11.4655	0	42.04	8.9175	10.191	3.82
2	*Calanus*	5.096	0	11.46	0.637	1.274	1.27
3	*Oithona*	14.0125	0	91.72	14.014	6.369	1.27
4	*Euterpina sp.*	3.185	0	0	6.37	0	1.27
5	*Eurytemora sp.*	4.459	0	3.82	3.1845	2.548	0
6	*Oncaea sp.*	0	0	6.37	0	0	0
7	*Parvocalanus sp.*	5.0955	0	0	4.459	0	0
8	*Tigriopus*	1.274	0	0	3.185	1.274	0
9	*Acrocalanus sp.*	3.185	0	0	0	0	0
10	*Sergia sp.*	0.637	0	0	0	0	0
11	*Lucicutia sp.*	1.274	0	2.55	3.185	0	0
12	*Eutintinus*	0	0	2.55	0	0	0
13	*Canthocalanus sp*	0	0	0	0.637	0	0
B.	***Ciliata***						
14	*Leprotintinnus sp.*	10.192	0	15.29	21.021	2.548	8.92
15	*Tintinnopsis sp.*	14.0135	0	57.32	16.2405	2.548	7.64
16	*Parafavella sp.*	0	0	0	0	1.274	0
17	*Favella sp.*	0	0	1.27	4.775	0	1.27
C.	***Mollusca***						

18	*Trocophora sp.*	0	0	5.1	0	0	0
D.	***Sarcodina***						
19	*Globigerina sp.*	0	0	0	0	0	1.27
E.	***Maxillopoda***						
20	*Balanus*	3.822	0	0	0	1.274	1.27
F.	***Monogononta***						
21	*Brachionus sp.*	0	0	0	0	1.274	0
Total		77.711	0	239.49	86.6255	30.574	28

Table 10. Physical-chemical waters parameters

Parameter	Unit	Baru		BKT		Siangker		Quality [6]
		HTL	LTL	HTL	LTL	HTL	LTL	
Temperatur	^0C	27.82	30.18	18.43	32.9	30.3	32.33	
Depth	meter	6.257	6.33	1.26	1.25	1.62	1.12	
Brightness	meter	1.08	1.27		0.71	0.63	0.51	
TDS	mg/l	303.7	508.5	426	507	447.67	505.67	
Salinity	‰	33.35	33.2	22.3	26.2	27.67	28.43	
pH		7.887	7.812	5.327	7.587	7.94	7.737	7 - 8.5
DO	mg/l	5.573	4.157	3.33	6.91	5.75	5.4	< 5
Nitrate	mg/l	0.19	0.14	0.0365	1.233	0.53	0.237	0.008
Phosphat	mg/l	0.955	0.455	0.84	1.13	0.483	0.72	0.015
Turbidity	NTU	33.125	36.87	27.25	26.845	16.59	56.85	< 5
Sulfide	mg/l	0.065	0.088	0.075	0.07	0.063	0.06	0.01
Phenols	mg/l	0.018	0.022	0.033	0.022	0.034	0.024	0.002
BOD_5	mg/l	3.265	1.617	2.175	1.98	1.557	1.05	20

llution contamination index on the three river estuaries shows highly polluted conditions
T and Siangker estuaries are higher at HTLwhile at LTL at the estuary of the BKT river
hest value (Table 11).

Table 11. IP values on the three estuaries around PTES waters from October to December 2015

No.	Parameters	HTL			LTL				IP HTL			IP LTL		
		C_{imax}	C_{imax}	C_{imax}	C_{imax}	C_{imax}	C_{imax}	L_{ij}	Baru	BKT	Siangker	Baru	BKT	Siangker
1	Turbidity	62.5	32.1	19.5	113	39.8	98.5	5						
2	pH	7.96	8.07	7.95	7.93	7.9	7.88	8						
3	Sulfide	0.07	0.08	0.08	0.23	0.11	0.08	0.01						
4	Phenol	0.027	0.04	0.037	0.03	0.032	0.04	0.002	12.544	14.998	19.495	12.4	24.908	12.521
5	DO	5.7	5.2	5.4	5.2	8.94	5.9	5						
6	Cd	0.05	0.049	0.045	0.04	0.058	0.052	0.01						
7	Cu	0.072	0.067	0.061	0.06	0.066	0.077	0.05						

8	Pb	0.69	0.67	0.61	1.01	0.58	0.71	0.05
9	Zn	0.05	0.07	0.07	0.11	0.05	0.07	0.1
10	Nitrate	0.23	0.73	1.49	0.31	1.32	0.31	0.008
11	BOD_5	3.3	3.38	2.98	3.35	3.19	1.71	20
12	PO_4	1.23	1	0.87	0.84	1.52	0.89	0.015

4. Conclusion

Pollution index from all three estuaries shows the most heavily polluted condition which are in BKT and Siangker the value is higher when the HTL, and when the LTL BKT estuary has the highest value. The result obtained from 12 parameters which were tested showed that the three waters can be categorized in heavily polluted condition at each value from 12.52 to 24.98. the concentration of heavy metal at sea water during HTL and LTL ranged from Cd is around 0.033 and 0.048 mg/kg, Pb 0.48 and 0.71 mg/kg, and Zn 0.043 and 0.057 mg/kg. the saprobity value index based on the existence of phytoplankton or zooplankton was ranged of Oligosaprobik at slightly polluted or haven't been polluted.

Acknowledgement

Thank you for all the vicinity research team from Fisheries and Oceanology Faculty, Diponegoro University that being held in October until December 2015. Also thank you to Vian that never tired to do identification and calculation of plankton in laboratory and also to all staff in Health Laboratory of Semarang Region

References

[1] Ojekunle O Z, Ojekunle O V, Adeyemi A A, Taiwo A G, Sangowusi O R, Taiwo A M and Adekitan A A 2016 Evaluation of surface water quality indices and ecological risk assessment for heavy metals in scrap yard neighbourhood *Springer Plus* **5**: 560, p. 1-16.

[2] Effendi H 2003 *Water quality study for resource management and aquatic environments* (Penerbit Kanisius : Yogyakarta) [In Indonesian].

[3] Anggoro S 1983 *Water subsidence problems for increased fish production in ponds* Faculty of Animal Husbandry, Diponegoro University of Semarang.

[4] Nemerow N L and Sumitomo H 1970 Benefits of water quality enhancement report no. 16110 DAJ, prepared for the U.S. Environmental Protection Agency.

[5] Fachrul M F 2007 *Methods of bioecological sampling* (Bumi Aksara: Jakarta) [In Indonesian].

[6] Republik Indonesia 2014 *The degree of the Minister of the Environment number 51/2004 about sea water quality standards* [In Indonesian].

[7] Wibisono M S 2005 *Introduction to marine science* (Gramedia Widiasarana Indonesia/Grasindo: Jakarta) [In Indonesian]

[8] Anggoro S 1988 T*ropical-saprobic analysis (Trosap) to assess the feasibility of marine aquaculture location, in: sea water cultivation workshop of Central Java* Coastal Development Laboratory. Prof. Dr. Gatot Rahardjo Joenoes. University of Diponegoro, Semarang, pp. 66-90 [In Indonesian].

[9] Sachlan M 1982 *Planktonology correspondence course centre* Directorate General of Fisheries, Ministry of Agriculture Jakarta, 141 p.

Prospective Source of Antimicrobial Compounds From Pigment Produced by Bacteria associated with Brown Alga (Phaeophyceae) Isolated from Karimunjawa island, Indonesia

A T Lunggani[1,2,*], Y S Darmanto[3], O K Radjasa[3,4], A Sabdono[3]

[1]Coastal Resources Management, Faculty of Fisheries and Marine Sciences, Diponegoro University, Jl. Prof H. Soedharto, SH, Tembalang, Semarang 50275, Indonesia
[2]Faculty of Science and Mathemathics Diponegoro University, Jl. Prof H. Soedharto, SH, Tembalang, Semarang 50275, Indonesia
[3]Faculty of Fisheries and Marine Sciences, Jl. Prof H. Soedharto, SH, Tembalang, Semarang 50275, Indonesia
[4]Ministry of Research, Technology and Higher Education of the Republic of Indonesia

Email: aryna_tri@yahoo.com

Abstract. Brown algae or Phaeophyceae characterized by their natural pigments that differ from other important algal classes. Several publications proves that brown algae - associated bacteria have great potential in developing marine pharmaceutical industry since they are capable to synthesized numerous bioactive metabolite compounds. However the potency of marine pigmented microbes associated with brown alga to produce natural pigments and antimicrobials has been less studied. Marine pigmented bacteria associated with brown algae collected from Karimunjawa Island were successfully isolated and screened for antimicrobial activity. The aim of this research was evaluated of the antimicrobial activity of pigments extracted from culturable marine pigmented bacteria on some pathogenic bacteria and yeast. The results showed that all isolates had antimicrobial activity and could be prospectively developed as antimicrobial agent producing pigments. The 6 marine pigmented bacteria was identified to genus level as Pseudoalteromonas, Sphingomonas, Serratia, Paracoccus, Vibrio.
Keywords: Marine pigmented bacteria, brown algae, antimicrobial activity, Karimunjawa

1.Introduction

Pigments are widely used in food [1], pharmaceutical [2], textile [3], cosmetics [4]. The need of pigment from natural sources has increased because of awareness of consumers about hazardous of synthetic colors [5]. Therefore, it is imperative to find an alternative pigment which can be biodegraded and easily available at a minimal production cost [4-5]. Carotenoids, one of the main pigments, can be synthesized from plant and microbial sources. Algae are marine photosynthetic organisms, whose pigment profiles are used as a basis for taxonomic classification into green, red and brown algae [6]. Brown algae pigments have been reported that have great pharmaceutical prospects [7-10], neutraceuticals [11] and cosmeceuticals [12-13]. Brown algae and the pigmented bacteria are independently studied. Marine and terrestrial microbes differ from each other due to the influence of their respective environmental conditions. Marine bacteria, is a very interesting organism for researchers because it potentially produces compounds with unique biological properties [14].

The uniqueness of the antimicrobial component produced by marine microbes is influenced by the host species as well as the environmental conditions in which the organism grows. Based on the phenomenon, microbial symbionts, especially the symbionts of brown algae that grow in Karimunjawa Island, is expected to answer the problem of limitations of brown algae use as a source of pigment. Therefore, marine pigmented microbes associated with brown algae having the ability to produce unique antimicrobial components can be an alternative source for exploring marine natural products with promising pigment as a bioactive compounds.

2. Methodology

2.1. Algae sampling

Algae samples (Genera of Sargassum and Padina) were collected from Karimunjawa Island, Jepara Indonesia at 0,5 – 2 m depth into the ocean.. The collected samples were rinsed with sterile seawater and put into sterile ziplock plastic bag labelled contains sterile seawater and were stored in cool box.

2.2. Isolation and purification of marine microbes

Isolation of microbes was carried out by spread and pour plate method using Zobell Marine Agar 2216 media (Hi-Media), Nutrient Agar (Merck), R2A Agar (Oxoid). Approximately 5 g (wet weight) of sample was suspended in 200 mL of sterile seawater in erlenmeyer flask. This was kept in a rotary shaker at 120 rpm for 15 min. [3] with slight modification. Then the water samples were serially diluted and plated in duplicates. Plates were incubated at room temperature for 48 - 168 h. Distinct pigmented colonies were selected and purified. The pure isolate was used for next steps.

2.3. *Screening antimicrobial activity*

2.3.1. Pigment extraction

The pigment was extracted according to the method of [15-16], with some modifications. The bacterial cells were first grown for 24 - 48 h followed by centrifugation at 7500 rpm for 10 min. The cells in the mixed solution were broken up by adding 5 sterile glass beads (0.4 mm diameter). The suspension was then separated by centrifugation at 5,000 g for 20 minutes at 4^0C in the ratio of 1: 5 (supernatant) or until the pellet was colorless, i.e., complete pigment extraction has been achieved. The bacterial cell pellet was then discarded, while the supernatant is then dried with N2 gas. The pigment concentration process was carried out until around 1% (v/v) of the initial solvent volume was left in the evaporation flask. The concentrated pigment was then transferred onto glass Petri dishes prior to drying for 3 days at 25^0 C. Crude extracts are stored at -18° C until used.

2.3.2 Antimicrobial activity of the extracted pigment

Antimicrobial activities of Extract pigment were studied against *Staphylococcus aureus, Bacillus subtilis, Escherichia coli,* and *Candida albicans* obtain from Biotechnology Lab of Biologi Dept Diponegoro University. . Microorganisms were grown in Zobell Marine Broth (for bacteria) and PDB (for yeast) at 28°C for overnight. The overnight culture measured equivalent to 0.5 mc Farland standard. The dilution was used as the inoculums for antimicrobial activity assay. 0,1 ml pathogenic bacteria were inoculated by spread method on MA media for bacteria and PDA for yeast. The paper disks contained the carotenoid extract were placed on the agar media surface. Inhibition zones were recorded after overnight incubation at 28 °C [17-18].

2.4. Identification of the isolate

2.4.1. PCR amplification 16S rDNA gene sequence

DNA extraction for bacteria was obtained by chelex method with slight modification [19] Supernatant which contains DNA used as a template for PCR analysis. Primers used for PCR amplification were universal primer 27F(5'-AGAGTTTGATCMTGGCTCAG-3) and 1492 R (5'-GGTTACCTTGTTACGACTT-3') [20]. The PCR mixture consisted of *forward* primer (27 F) 1 μl,

Reverse primer (1492 R) 1 µl , DNA template 2 µl, 10 x Buffer 5 µl, 10 mM dNTP mix 1 µl, 25 mM $MgCl_2$ 3 µl , ddH2O 34 µl. The total volume measured 50 µl. The PCR condition was pre denaturation at 95^0 C for 3 min , denaturation at 94^0 C for 1 min, anneling at 54^0 C for 1 minute, extention at 72^0 C for 30 second and post cycling at 72^0 C for 7 minute. All condition were repeated 30 times. The PCR products were loaded in agarose gel (1%) for electrophoresis then visualized by UVI Doc HD5

2.4.2. DNA sequencing
Sequencing of extracted DNA was sent a sequencing service company. The allignment of sequences result were done using MEGA 6, then allignment result was inserted to the BLAST program to create homology of any closely related bacteria in the gene bank database [21]

3. Results and discussion

3.1. Isolation and purification of brown algae pigmented symbiont
Brown algae are considered to be nutrient-rich organisms that trigger high competition among microbial communities [22-24]. Recent studies suggest that microbial communities associated with algae are significantly different from those in seawaters [25-26]. Isolation of pigmented bacteria was grown using 4 types of general media, Zobell Marine Agar, Nutrient Agar, R2A and Potato Dextrose Agar at the first step. Associated microbes were extracted from *Sargassum sp* and *Padina* sp thallus and cultivated on all four media, but on subsequent observations Zobell Marine Agar media provide suport for pigmentation and growth for bacteria.

A total of 36 types of pigmented microbes asociated with brown algae based on morphological colony differences were found in the present study. All isolate were selected based on distinct pigmented colonies, micmorphological characteristics, their ability to grow and the stability of the pigments produced on the media used. Selection results obtained 6 types of bacterial isolates that will be used in the next stage of research.There were 6 brown algae -associated bacteria as shown at Fig 1.

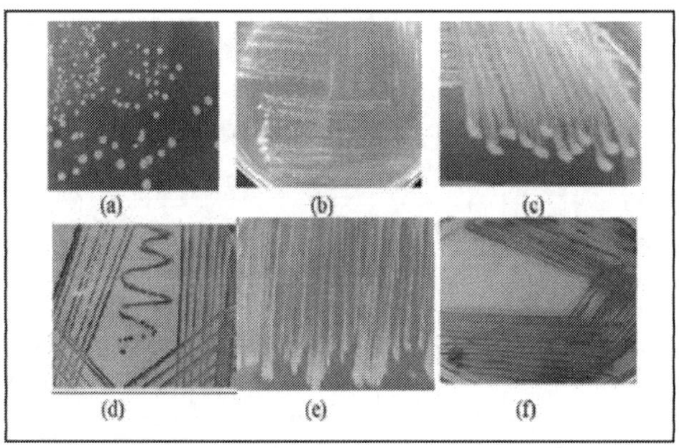

Figure 1. *Characteristics and growth of bacteria morphotypes on media Zobell Marine Agar*
a – f: Genera isolate bacteria brown algae symbionts.

3.2. Screening of bacteria producing antimicrobial compounds
Brown Algae which are phototrophs, is one of the important members of the marine ecosystem not only individually as a food source, but also at the level of the ecosystem as a habitat-forming and primary producers [26]. Studies on brown algae suggest that brown algae produce bioactive components that exhibit anti-obesity, anti-inflammatory, anti hypertensive, anticholesterol [8] activities to prevent osteoporosis and rheumatoid arthritis [27] and anticoagulant [28]. Symbiotic microbes are known to be

capable of producing the same compounds as their hosts, although they may not be identical to the original components produced by their host. Recent publications report that most of the active compounds isolated from marine organisms, produced by symbiotic microbes [29].

Genera of Sargassum and Padina are known to have antimicrobial activity and are prospective agents for the development of antimicrobial components [30-31]. The screening procedure to investigate antimicrobial activity in this study used microorganisms test of both Gram positive and Gram negative as a target strains, as well as pathogenic yeast. Those microorganism test are *Staphylococcus aureus, Bacillus subtilis, Escherichia coli., and Candida albicans*. Based on medical point of view, these pathogenic test strain get important attention because they are well represented the causative agents for human infection [32].

Based on the results obtained, pigment extracts from bacterial isolates that had previously been observed with a uv vis spectrophotometer were belonged in the carotenoid group. According to the result, most of the carotenoid extract of isolates were able to produce the antimicrobial activity on the all of the microbial test strain shown by the inhibition zone on agar plate. Isolates are more able to inhibit the growth of gram-positive bacteria than gram-negative bacteria. Data presentation refers to research conducted [32] (Table 1).

Table 1. Screening of antimicrobial activity of carotenoid extract brown algae associated bacteria against pathogenic microbes.

NO	Inhibition of test strains	Activity pattern	Number of isolates	Percentage
1	Only *E. coli*	a	3	50
2	Only *S. aureus*	b	4	83,3
3	Only *B. subtilis*	c	6	0,67
4	Only *C. albicans*	d	4	0,67
5	*S. aureus* and *B. subtilis*	e	4	0,67
6	*S. aureus* and *C. albicans*	f	3	50
7	*B. subtilis* and *C. albicans*	g	3	50
8	*E. coli* and *S. aureus*	h	3	50
9	*E. coli* and *B. subtilis*	I	3	50
10	*E. coli* and *C. albicans*	j	3	50

The result of this research is consistent with previous research done by [32] that revealed the antibiotic activity of bacteria associated with brown algae *Laminaria saccharina* and found that more than 50 % associated bacterial strains inhibited the growth of either gram-negative or gram-positive bacteria or yeast.

Several studies on bacterial symbionts have shown that they have potential to function as natural pigment sources and bioactive compounds. It has been reported that several new pigmented strains of bacteria are explored from various habitats, and have beneficial secondary metabolites, such as: *Exiguobacterium* sp. The yellow pigment-producing bacteria isolated from Marina Beach [33], *Janthinobacterium lividum XT1,* the pigment-producing violet producer of glacier in China [34], a psychophysiologic bacteria RT102 suspected to be a *Janthinobacterium* bacteria producing violacein have high anti-bacterial activity and anti protozoa and antifungal [35,36], *Pseudoalteromonas tunicata* marine bacteria producing dark green pigments, potentially as antifouling [37]. Marine bacteria *Pseudoalteromonas piscicida* H1.7 isolated from Kakaban Land-Locked Marine Lake showed antimicrobial activity towards *Staphyllococcus aureus* and produced a yellow pigment [38].

Based on the facts mentioned above, reinforced by the results of antimicrobial activity screening has been done. In principle we know that symbiotic microbes are capable of producing prospective secondary metabolite components, but they may be produced by specific conditions as well because the marine environment is a unique environment such as certain salt concentrations, hydrostatic pressures, competition between microbes and nutrients [39] These results are likely to be different if performed on a laboratory scale, whose is well controlled and generally tries to use microbial cultivation on liquid medium as a main methods and also depend on secondary metabolite product. [40-41]. Although the microbial cultivation method based on fermentation in the laboratory is a promising effort in the discovery of new compounds [41].

3.3. Analysis of PCR products

PCR products (2µl) were analyzed by agarose gel electrophoresis Electrophoresis was performed at 100V for 30 min. The gels were stained for 45 min with gel red nucleic acid gel stain (Biotium) and photographed under UV illumination.
All isolates were successfully amplified by PCR (Figure 2).

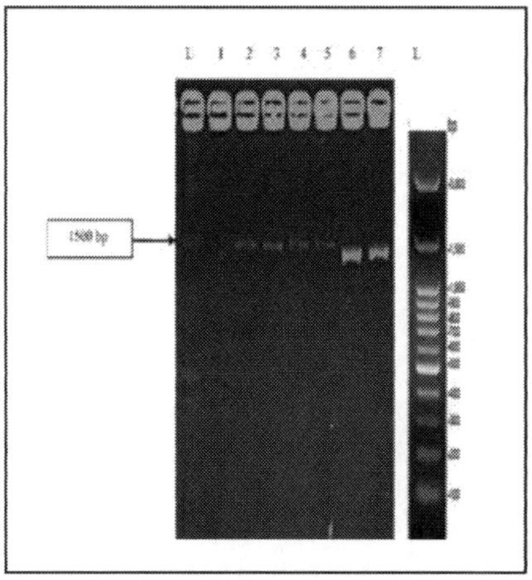

Figure 2. Visualization of PCR product
where: L = Ladder; 1 = negative control 1-7 : isolates

The closest phylogenetic relatives of each isolate were identified by comparison of the 16S rRNA gene sequence to the National Center for Biotechnology Information (NCBI) GenBank database using the Basic Local Alignment Search Tool (BLAST) analysis tools. Blast homology search result of the potential pigmented microbes was shown in Table 2.

Table 2. Homology of *brown algae pigmented symbiont*

NO	CODE	Closest Relative	E Value	Homology
1	KR11	*Pseudoalteromonas flavipulchra strain ES8*	0,0	99 %
2	KR124	*Sphingomonas desiccabilis Strain CP1D*	0,0	99 %
3	KR16	*Serratia rubidaea strain N22*	0,0	99 %
4	KR19	*Serratia rubidaea strain NBRC103169*	0,0	97 %
5	KR120	*Paracoccus marcusii*	0,0	97 %
6	KR121	*Vibrio fortis strain 9G1*	0,0	97 %

In the recent decades, considerable intensive research on bacterial communities associated with algae has been conducted. It is directed to study bacterial communities associated with algae, an understanding of the succession of community structures, in terms of the interactions between microbes and algae. and no less important is the exploration of natural bacterial products associated with algae [22][43]. However, based on a comprehensive study with sequences of 16S rRNA gene and denaturing gradient gel electrophoresis (DGGE) fingerprinting, it is known that microbial communities associated with algae differ from planktonic bacterial communities [21][43-44].

PCR amplification of 16S rDNA from six bacterial isolates were showed positive results with the presence of appropriate base length of approximately 1300 - 1500 bp (Fig. 1). Brown algae marine pigmented microbial associant that have been isolated are included in the phyla Proteobacteria (Alpha and Gamma). This result is consistent with the study that members of Alphaproteobacteria and Gammaproteobacteria have a more global distribution and are found to be associated with eukaryotic in marine and coastal areas. Meanwhile, other taxa that are also often found are Bacteroidetes, Actinobacteria, Planctomycetes and Chloroflexi [22][46-47]

Genus Pseudoalteromonas that reported capable of producing yellow and orange pigment. epiphytic *Pseudoalteromonas tunicata* and *Roseobacter gallaeciensis* also have been found in association with green alga *U. australis* and capable producing a range of extracellular inhibitory compounds against marine fungi, bacteria, invertebrate larvae, and algal spores [48-49]. Bacteria of the genus Pseudoalteromonas play an important role in marine environments because of its ability to produce highly variable metabolites including bioactive components and enzymes. These bacteria are spreading cosmopolitan because they can adapt and survive in a poorly nutrient marine environment and extreme pressures [48][50-51]. The study conducted [52] stated that genus Pseudoalteromonas is of great interest to the scientific community because of (i) its prolific metabolite-producing capacity and (ii) its usual association with macroorganisms, leading to a suspected and sometimes documented ecological significance. [53] highlighted a Pseudoalteromonas isolate as producer of a novel antimicrobial protein, which inhibited human pathogenic strains causing dermatologic diseases.

Studies that have been done also found two types of serratia, namely *S. marcecens* and *S. rubideae*. Characteristics of Serratia that successfully isolated is a convex-shaped colonies and produce red pigment. This bacteria is facultative and able to grow well in aerobic and anaerobic conditions in artificial media. The red pigment produced by *S. marcescens* is a secondary metabolite known as prodigiosin from the tripyrrole family which generally contains 4-methoxy, ring 2-2 bipyrolle ring system [54]. Prodigiosin is a promising drug due to its reported, antibacteria , antifungal, immunosuppressive and anti-proliferative activities [55-56]. in this study, serratia able to inhibit bacteria either from gram positive or gram negative bacteria.

Other pigmented bacteria found in this study were from the genera Sphingomonas, Vibrio and Paracoccus. Genus sphingomonas have been isolated from a variety of habitats, such as rhizospheres, soil, aquatic habitats and clinical material, but little is known about Sphingomonas strains dwelling in marine environments, particularly associated with animals. The publication of the Sphingomonas isolation from the marine that has been successfully done are *Sphingomonas molluscorum* sp. nov., a novel marine isolate associated with marine mollusc *Anadara broughtoni* [57], the crab *Paralithodes camtschatica*, a marine crustacean [58] and *Sphingomonas jejuensis* sp. nov., isolated from marine sponge *Hymeniacidon flavia* [59]. Several representatives of genus Sphingomonas have been reported to produce anticyanobacterial compounds such asargimicins and antimicrobial compounds [56].

The genera of pigmented bacteria also found in this study is that of the genus Paracoccus. The genus Paracoccus consists of Gram-negative, oxidase and catalase-positive, capable of producing astaxanthin and great antimicrobial activity. At present, the genus Paracoccus includes 17 recognized species [31][60].

4. Conclusion

Despite the great challenge and difficulty in isolating and harvesting marine pigmented bacteria, especially those bacteria are associated with marine organisms, but the progress of exploration in this field is promising. In particular, based on studies antibacterial activity of pigmented bacteria associated with brown algae and types of isolates that were successfully obtained, the development of bioactive components of these bacteria can be developed more widely and continue to provide promising avenues for both fundamental sciences and applied biomedical research.

5. Acknowledgement
This research is supported by the grant from Ministry of Research, Technology and Higher Education, through Doctor Dissertation Research Scheme No. 345-13/UN7.5.1/PG/2017. We wish to thank Siu Semar Saratu Langden and Dyah Wulandari for her assist thru the research.

6. References

[1] Dufossé L 2006 *Food Technol Biotech* **44** pp.313-323
[2] Venil C K, Zakaria Z A and Ahmad, W A 2013 *Process Biochem* **48** pp.1065-1079.
[3] Tuli H S, Chaudhary P, Beniwal V and Sharma A.K 2015. *J. Food Sci. Technol* **52** pp.4669-4678
[4] Suganya K, Preethi P.S, Suganya M and Nanthini A UR 2016. *Management* **4**
[5] Wrolstad RE, Culver CA 2012. *Annu. Rev. Food Sci. Technol* **3** 59-77.
[6] Yolmeh M, Habibi Najafi MB, Farhoosh R 2014. *Food Chem* **155** 319–24
[7] Cho Y J, Park J P, Hwang H J, Kim S. W, Choi, J. W and Yun J W 2002 *Lett Appl Microbiol* **3** 195–202
[8] Takaichi S. 2011. *Mar Drugs* **9** 1101-1118
[9] Pangestuti R, and Kim S K 2011. *J Func Foods* **3** 255-266
[10] Sugawara T, Matsubara K, Akagi, R, Mori M and Hirata T 2006 *J Agric Food Chem* **54** 9805-9810
[11] Holdt S L, and Kraan S 2011 *J. Appl. Phycol* **23** 543-597.
[12] Wijesinghe W A J P and Jeon Y J 2011. *Phytochemistry Reviews* **10** 431-443.
[13] Fitton J H, Irhimeh M and Falk N *Cosmetics and toiletries* **122** 55
[14] Soliev A B, Hosokawa K and Enomoto K 2011. Bioactive pigments from marine bacteria: applications and physiological roles. *Evidence-Based Complementary and Alternative Medicine* Article ID 670349, 17 pages
[15] Setiyono E, Delianis P, Heriyanto, Monika N U, Prihastyanti, Yuzo Shioi and Tatas H P B 2016. Prosiding Seminar Nasional Tahunan Xiii Hasil Penelitian Perikanan Dan Kelautan. Yogyakarta.
[16] Ahmad WA, Ahmad W Y W, Zakaria Z A and Yusof N Z 2012. Isolation of Pigment-Producing Bacteria and Characterization of the Extracted Pigments. In *Application of Bacterial Pigments as Colorant* (pp. 25-44). Springer Berlin Heidelberg
[17] Yolmeh M, Hamedi H and Khomeiri M 2016 *Zahedan J Resc Med Sci* **18**
[18] Radjasa O K, Salasia, S I O, Sabdono A, Wiese J, Imhoff J F, Lämmler C and Risk M J 2007. *J of Pharmacol* **3** pp.170-174.
[19] de Lamballerie X, Zandotti C, Vignoli C, Bollet C and de Micco P 1992 *Res. Microbiol.* **143** 785–90
[20] Weisburg WG, Barns SM, Pelletier DA and Lane DJ 1991 **173** 697–703.
[21] Atschul S F, Madden TL, Schaffer, Zhang J, Zhang Z, Miller W and Lipman D J 1997*Nucleid Acid Res* **25** 3389-3402
[22] Singh R P and Reddy C R K 2014 *FEMS Microbiol Ecology* **88** pp.213-230.
[23] Amstrong E, Yan L, Boyd K G, Wright P C, and Burgess J G 2001 *Hydrobiologia* **461** 37 – 40
[24] Penesyan A, Kjelleberg S, and Egan S 2010 *Marine Drugs* ***8*** 438-459.
[25] Lachnit T, Blümel M, Imhoff J F, and Wahl M 2009 *Aquatic Biology* **5** 181-186
[26] Egan S, Harder T, Burke , Steinberg P, Kjelleberg S and Thomas T 2013 *FEMS Microbiol Rev* **37**
[27] Das S K, Ren R, Hashimoto T and Kanazawa K 2010 *J Agric Food Chem* **58** 6090-6095.

[28] Athukorala Y, Lee K W, Kim S K and Jeon Y J 2007 *Bioresc Technol* **98** 1711-1716.
[29] Blockley A, Elliott D R, Roberts A andMichael S 2017 *Diversity* **9** 49
[30] Vijayabaskar P, Vaseela N and Thirumaran G 2012. *Chinese J Natl Med* **10** 421-428.
[31] Ismail A Ktari L, Ahmed M, Bolhuis H, Boudabbous A, Stal LJ, Cretoiu MS and El Bour M 2016. *Front. Microbiol* **7** 1072.
[32] Wiese J, Thiel V, Nagel K, Staufenberger T and Imhoff J F 2009. *Marine Biotechnol* **11** pp.287-300.
[33] Balraj J, Pannerselvam K and Jayaraman A 2014 *Inter J Scientific Technol Res* **3** 375 – 384
[34] Lu Y Wang, Xue L, Zhang Y, Xing C, Lou and Su Z 2009 *Biochem Engineering J* **43** 135-141
[35] Nakamura Y, Asada and Sawada 2003 *J. Biotech and Bioprocess Eng* **8** 37-40
[36] Pantanella F, Berlutti, F, Passariello C, Sarli S, Morea C and Schippa S 2007 *J Applied Microbiol* **102** 992-999.
[37] Franks A, Haywood P, Holmström, Egan C, Kjelleberg and N. Kumar 2005 *Molecules* **10** 1286–1291
[38] Radjasa O K, Limantara L, and Sabdono A 2008. *J. Coast. Dev* **12** 100-104
[39] Brinkmann C. M, Marker A and Kurtböke D İ *Diversity* **9** 40
[40] Goddard J M, Hotchkiss J H 2007 *Prog. Polym. Sci* **32** 698–725.
[41] Bowman J P 2007 *Mar Drugs* **5** 220–241
[42] Jimenez JT, Šturdíková M, Šturdík E 2009 *Acta Chim Slov* **2** 63–74
[43] Kouzuma A and Watanabe K 2015 *Curr Opinion in Biotech* **33** 125-129
[44] Goecke Saavedra F 2013. Associations between microbes and macroalgae: Host, epiphyte and environmental effects. Dissertation.
[45] Grossart H P, Simon M 2007 *Aquatic Microbial Ecology* **47** 163
[46] Giovannoni S J , Stingl U 2005 Nature 437:343–348.
[47] Burke C, Steinberg P, Rusch D, Kjelleberg S, Thomas T 2011 *P Natl Acad Sci USA* **108** 14288–14293.
[48] Holmström C and Kjelleberg S 1999 *FEMS Microbiol Eco* **30** 285-293
[49] Rao D, Webb JS, Holmstr€om C, Case R, Low A, Steinberg P and Kjelleberg S 2007 *Appl Environ Microbiol* **73** 7844–7852.
[50] Ivanova E P, Bakunina I. Y, Sawabe, T, Hayashi K, Alexeeva Y V, Zhukova N Vand Mikhailov, V V 2002 *Microbial Ecology* **43** 242-249.
[51] Nam E and Ahn J 2011 *Brazilian J Microbiol* **42** 927-936.
[52] Offret C, Desriac F, Le Chevalier, P, Mounier J, Jégou C and Fleury Y 2016 *Marine drugs* **14** 129
[53] Longeon A, Peduzzi J, Barthelemy M, Corre S, Nicolas JL, Guyot M 2004 *Mar Biotechnol* **6** 633–641
[54] Boger DL and Patel M 1988 *J Org Chem* **53** 1405–1415
[55] Casullo de Araújo, K. Fukushima G M Campos Takaki *Molecules* **15**
[56] Anita K, Mahnaz MA, Fatemeh AF 2006 *Online J. Biol. Sci* **6** 1-13.
[57] Romanenko L A, Uchino M, Frolova G M, Tanaka N, Kalinovskaya N I, Latyshev N and Mikhailov V V 2007 *Inter J Syst Evol Microbiol* **57** pp.358-363.
[58] Romanenko L.A, Tanaka N, Frolova G M and Mikhailov V V 2009 *Inter J Syst Evol Microbiol* **59** 1179-1182.
[59] Park S, Yokota A, Itoh T and Park J.S 2011. *The J of Microbiol* **49** 238-242.
[60] Lee J H, KimY S, Choi T J, Lee W J and Kim, YT 2004. *Intern J System Evolutionary Microbiol* **54** 1699-1702.

The Effect of Fermentation Time with Probiotic Bacteria on Organic Fertilizer as *Daphnia magna* Cultured Medium towards Nutrient Quality, Biomass Production and Growth Performance Enhancement

Vivi Endar Herawati*[1], Ristiawan Agung Nugroho[1], Pinandoyo[1], YS Darmanto[2]; Johannes Hutabarat[1]

[1]Department of Aquaculture, Faculty of Fisheries and Marine Science, Diponegoro University, Jl Prof Soedarto,SH, Tembalang, Semarang 50275, Indonesia
[2]Department of Food Science, Faculty of Fisheries and Marine Science, Diponegoro University, Jl Prof Soedarto,SH, Tembalang, Semarang 50275, Indonesia

E-mail: anshinvie@yahoo.com

Abstract. The nutrient quality and growth performance of *D. magna* are highly depend on the organic fertilizer which is used in its culture medium. The objective of this study was to identify the best fermentation time by using probiotic bacteria on organic fertilizer as mass culture medium to improve its nutrient quality, biomass production, and growth performance. This study was conducted using completely randomized experimental design with five treatments and three repetitions. Organic fertilizers used cultured medium with chicken manure, rejected bread and tofu waste fermented by probiotic bacteria then cultured for 0, 7, 14, 21 and 28 days. The results showed that medium which used 25% chicken manure, 25% tofu waste and 50% rejected bread cultured for 28 days created the highest biomass production, population density and nutrient content of *D. magna* those are 233,980 ind/L for population density; 134.60 grams for biomass production, 0.574% specific growth rate; 68.06% protein content and 6.91% fat. The highest fatty acid profile is 4.83% linoleic and 3.54% linolenic acid. The highest essential amino acid is 53.94 ppm lysine. In general, the content of ammonia, DO, temperature, and pH during the study were in the good range of *D. magna* life. The conclusion of this research is medium which used 25% chicken manure, 25% tofu waste and 50% rejected bread cultured for 28 days created the highest biomass production, population and nutrient content of *D. magna*.
Keywords: *D. magna*, fermentation time, nutrient value, probiotic bacteria, various wastes

1. Introduction

Daphnia sp. is a zooplankton which is the best natural food for freshwater fish larvae and ornamental fish culture because the nutrient content and size of *Daphnia* sp. suitable with mouth opening and nutrient needs of nile tilapia larvae. The nutrient content is highly depends on the culture medium where it grows and breeds [1, 2]. Nutrient content of *D. magna* highly depends on its culture medium as the growth place of phytoplankton as *D. Magna*'s feed [3, 4]. Recently, not only chicken manure is used as common culture medium for *D. magna* [5], but also the used of goat, and cow manure as culture medium [3]. Another medium used is a combination of chicken dung, bran, and copra waste [6] and different animal wastes those are chicken, goat and quail manure [4]. The organic matters in bran are highly nutritious for the growth of *D. magna*.

The research using various animal waste with fermentation time for 14 days has been done in 2016, the study result stated that chicken manure is the best culture medium for *Daphnia magna*'s nutrient quality and growth performance. The use of organic fertilizers in culture media including the wastes/faeces of chicken, and quail mixed with the rejected bread and tofu waste based on different fermentation time with the probiotic bacteria has been conducted as the use of organic fertilizer could impact the growth performance and nutrient content of *D. magna*. The highest nutrients - particularly for the content of N, P and Ca in organic fertilizer are the food sources of *D. magna*. Chicken waste contained N (4.75%); P (3.57%) and Ca (4.80%) and quail waste contained N (4.06%); P (2.96%) and Ca (2.57%) [2]. Furthermore, the analysis on the dried materials of tofu waste has been done by previous study and the results showed that it contained crude protein (27.09%), crude fiber (22.85%), fat (7.37%), ash (35.02%), and extract material without nitrogen/BETN (6.87%) [4][7]. Rejected bread contained crude protein (12.63%), crude fiber (0.13%); crude fat (4.63%); ash (4.19%) and the extract material without nitrogen (58.42%) [2][8].

Fermentation of fertilizer has been proven to be effective to increase nutrient of culture medium. The aims of the fermentation are to produce a product (food materials) that contains the nutrients, to have a longer storage time, and to have a better organoleptic characteristics and nutriental components [2]. Probiotic bacteria are really supportive for the health of organisms. It also serves to decompose and ferment organic matters [9]. Decomposition is a biological process that makes bacteria produces growth substances, hormones, vitamins, and other enzymes [5][10].

The objective of this study was to identify the best fermentation time by using probiotic bacteria on organic fertilizer as mass culture medium to improve nutrient quality, biomass production, and growth performance of *Daphnia magna*. The fertilizer itself is fermented with probiotic bacteria (*Lactobacillus casei* and *Sacchoramyces cerevisiae*).

2. Materials and Methods

2.1. Fermentation Stage

Fermentation stage is the preparation of molasses ratio, water and pro-biotic. The ratio used was 1 : 1, i.e. 1 mL of molasses, 1 mL of probiotic bacteria and 100 mL of solvent. The organic matter used was a combination of 25% chicken manure + 50% rejected bread + 25% tofu waste then it was dried. The treatments used in this research are as follow: A. 0 day fermentation; B. 7 days fermentation; C. 14 days fermentation; D. 21 days fermentation; E. 28 days fermentation. Pro-biotic bacteria (*L. casei* and *S. cerevisiae*) that were already activated for 3 hours were given to fertilizer with a combination weight of 200g/L [4][9][11]. They were left for fungi to grow and acidic smell to develop. Once the fertilizer was ready, the samples were put into the culture medium and aerated for 14 days. And when it was ready, 100 ind/L of *D. magna* was inoculated [3].

2.2. Water Quality

Water quality during the study was maintained at 28-30^{0}C temperature, 0.3 ppm DO and 8.1-8.2 pH, which is ideal. This is in line with the previous study that the proper temperature for *D. magna* culture is 25-30°C, DO at 0.3-0.6 ppm, and pH at 6.5-9 [4][12]. Excellent water quality helps the growth of phytoplankton and algae for *D. magna* to stimulate its growth.

2.3. D. magna Culture

The 100 ind/L *D. magna* was spread for each pool containing 200 g/L fermented organic fertilizer. Observation for the abundance of *D. magna* was conducted every two days to monitor the population of *D. magna*. The water (20-25%) of this culture was replaced, and its pH level was monitored every morning at around 7 a.m. to maintain the quality. pH level was maintained at its optimum range with the addition of 1 L of dolomite /1.000 L of water.

IOP Conf. Series: Earth and Environmental Science **116** (2018) 012089 doi:10.1088/1755-1315/116/1/012089

2.4. Statistical Analysis

This study had a completely random design with five treatments and three repetitions. The biomass weight was analyzed using variant analysis to determine differences among treatments. Parameters analyzed were growth, biomass production, and nutrient content of *D. magna*.

Proximate analysis. The proximate chemical composition of the samples was determined using a standard procedure [4][13]. Crude protein content was calculated by multiplying total nitrogen factor. Carbohydrate content was estimated by the difference.

Essential amino acid profile. The amino acid composition of the samples was determined using HPLC (Shimadzu LC-6A) [4][13].

Fatty acid profile. The fatty acid composition of the samples was determined using a gas chromatograph (Shimadzu) [4][13].

3. Results and Discussion

This study is the development of previous study about *D. magna* mass culture using chicken manure, rice bran and coconut oil cake fermented by probiotic bacteria in 2015 and 2016 with the use of various manure in *D. magna* mass culture media. The results of the study found that organic matters in the form of fermented various chicken manure mixed with rejected bread and tofu waste had improved nutrient quality. To determine the quality of *D. magna* cultured medium, nutrient analysis through N, P, and K and medium nutrient through proximate analysis were done. Cultured medium N, P and K nutrient content before fermented at different times is presented in Table 1.

Table 1. N, P, and K nutrient content of organic fertilizer before fermented at different times

Parameters	Results (%)					Methods
	A	B	C	D	E	
Nitrogen	1.05	2.31	2.75	2.86	3.07	Kjeldhal
Potassium (K)	0.18	0.22	0.09	0.22	0.44	AOAC 983.02.2000
Phosphor	0.54	0.54	0.58	0.60	0.62	AOAC 983.02.2000

Study result found that cultured medium before fermented which has highest Nitrogen content is cultured medium with 28 days fermentation (E) about 3.07 % and the lowest Nitrogen content is at cultured medium without fermentation (A) about 1.05%. Cultured medium which has the highest Potassium (K) and Phosphor contents is cultured medium with 2 days fermentation (E) those are 0.44% and 0.62%, respectively. N, P, and K nutrient content of organic fertilizer after fermented at different times is presented in Table 2.

Table 2. N, P, and K nutrient content of organic fertilizer after fermented at different times

Parameters	Results (%)					Methods
	A	B	C	D	E	
Nitrogen	1.58	2.96	3.06	3.79	4.86	Kjeldhal
Potassium (K)	0.10	1.43	1.09	1.61	2.65	AOAC 983.02.2000
Phosphor	0.47	1.50	1.49	2.12	2.97	AOAC 983.02.2000

The results showed that enhancement of cultured medium nutrient quality after fermentation at different times occurred. The highest enhancement occurred at cultured medium fermented in 28 days (E) about 4.86%. It increased about 1.79% than before and cultured medium nutrient quality average enhancement for Nitrogen was about 0.53% - 1.79%. The highest Potassium (K) and Phosphor content occurred at cultured medium fermented in 28 days (E) those are 2.65% and 2.97%. Organic fertilizer nutrient content (Proximate) before and after fermented at different times is presented in Table 3.

Table 3. Organic fertilizer Nutrient content (Proximate) before and after fermented at different times

Materials	Content in 100% Dried Matter					Methods
	Ash (%)	Crude Fat (%)	Crude Fiber (%)	Crude Protein (%)	Carbohydrate (%)	
A (Without Fermentation)	30.57	11.59	17.32	17.67	22.85	
B (7 days Fermentation)	29.62	11.61	18.84	19.04	20.89	
C (14 days Fermentation)	19.26	13.75	19.28	28.79	18.92	Kjeldhal
D (21 days Fermentation)	17.99	14.78	16.80	29.69	20.74	
E (28 days Fermentation)	17.58	15.98	13.20	33.69	19.55	

The highest proximate analysis result was at cultured medium fermented for 28 days (E) for protein and fat are 33.69% and 15.98% respectively; while the lowest result was cultured medium without fermentation (A) those are 22.85% and 11.59%. The improvement of nutrient quality on fermented medium is an anaerobic dissimilation process of organic compounds by the activity of microorganisms or extracts from the cells of microorganisms.

Differences between nutrient and nutrient content contained in the culture medium *D. magna* caused by *Lactobacillus* sp. bacteria as fermenter during fermentation process. Lactobacillus sp. increase the protein content of the ingredients proved by the increase of nutrient and nutrient content in the culture medium. Time difference in the fermentation process has an effect on the amount of bacteria that develops. This study results are accordance with previous study stated that fermentation process can increase energy, protein and crude fiber content [14]. Microbes used in the fermentation process can produce enzymes that will degrade complex compounds to be simpler and synthesize proteins. The statement is strengthened by previous study stated that Lactobacillus sp. as fermenter has unique characteristic because it can neutralize acidic or alkaline organic matters [15]. Proteolytic microbes are capable of producing protease enzymes that will breakdown proteins. Protein breakdown is converted into polypeptide, then becomes a simple peptide, then this peptide will be breakdown into amino acids. These amino acids will be utilized by microbes to multiply themselves. The number of microbial colonies that are the source of single cell proteins increases during the fermentation process.

Improvement of nutrient medium, especially nitrate, serves to determine the amount of phytoplankton present in the culture medium as a source of *D. magna* feed other than bacteria and detritus. Based on the results of the study, the abundance of plankton that grows and dominates the culture medium comes from the phylum Chlorophyta, Euglenozoa, Nematoda, Ciliophora, and Rotifera. *D. magna* is a non-selective filter feeder that feeds on unicellular algae and a variety of organic detritus including protists and bacteria, even in adult sizes able to eat small crustaceans and rotifers, thus the more phytoplankton exist the faster the growth of *D. magna*. This study results is strengthened by previous studies stated that the more abundance of phytoplankton and organic matter in medium, growth rate of *Daphnia* sp. will occur faster [3]. This study results strengthened by previous study stated that phytoplankton population enhancement and growth in water related with nutrient availability especially nitrate and sunlight, nitrate will increase water fertility characterized by high number of phytoplankton exists [16-17].

Population density graph of *D. magna* cultured using organic fertilizer based on fermentation time difference is presented in Figure 1.

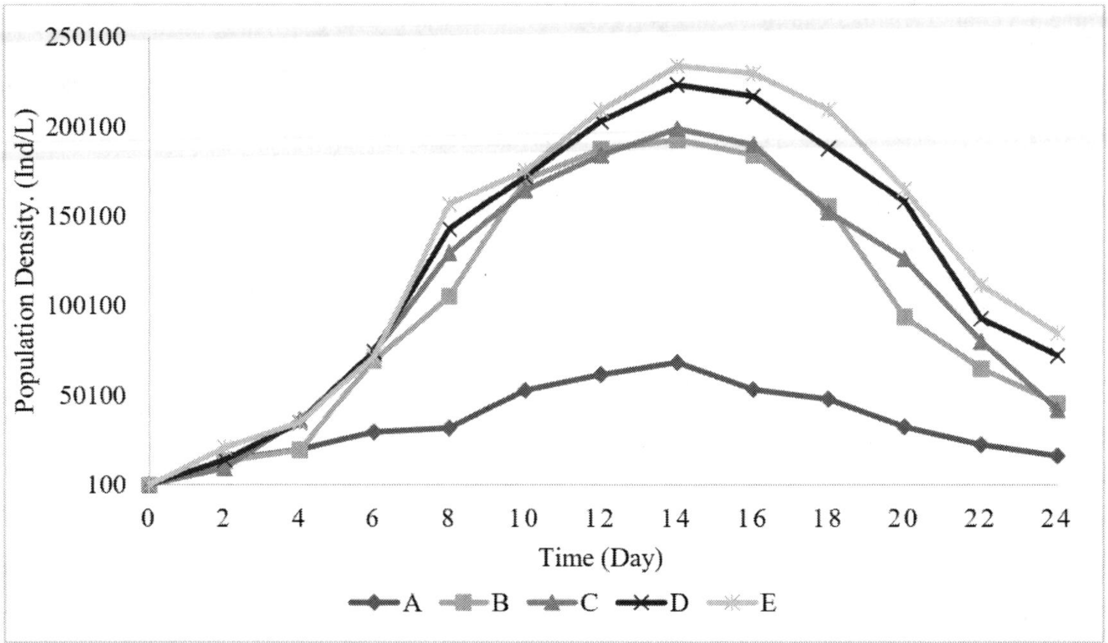

Figure 1. Population density graph of *D. magna* cultured using organic fertilizer based on fermentation time difference

D. magna growth phase cultured using organic fertilizer based on fermentation time difference is presented in Table 4.

Table 4. *D. magna* growth phase cultured using organic fertilizer based on fermentation time difference

Phase	Treatments (ind/L)				
	A	B	C	D	E
Lag	63,490.00[a]	102,280.00[a]	119,640.00[a]	123,330.00[a]	127,500.00[a]
Exponential	61,620.67[a]	187,590.33[b]	184,390.33[b]	203,200.33[b]	209,370.33[b]
Stationary	68,570.33[a]	192,920.00[b]	199,120.33[b]	223,500.33[b]	233,980.33[b]
Death	23,813.33[a]	68,123.33[ab]	82,776.67[ab]	107,820.00[b]	220,426.67[b]

Notes: [a]different between treatments; [b]not significantly different between treatments; [ab]significantly different between treatments

D.magna population growth phase during cultivation consists of adaptation phase (*lag phase*), exponential phase, stationary phase, and death phase. Lag phase occurred on day 4 with highest population density of 127,500.00 ind/L; exponential phase occurred on day 10 with highest population density of 209,370.33 ind/L; stationary phased occurred on day 14 with highest population density of 233,980.33 ind/L; death phase occurred on day 16 with highest population density of 220,426.67ind/L. *D. magna* growth rate cultured using organic fertilizer based on fermentation time difference is presented in Figure 2.

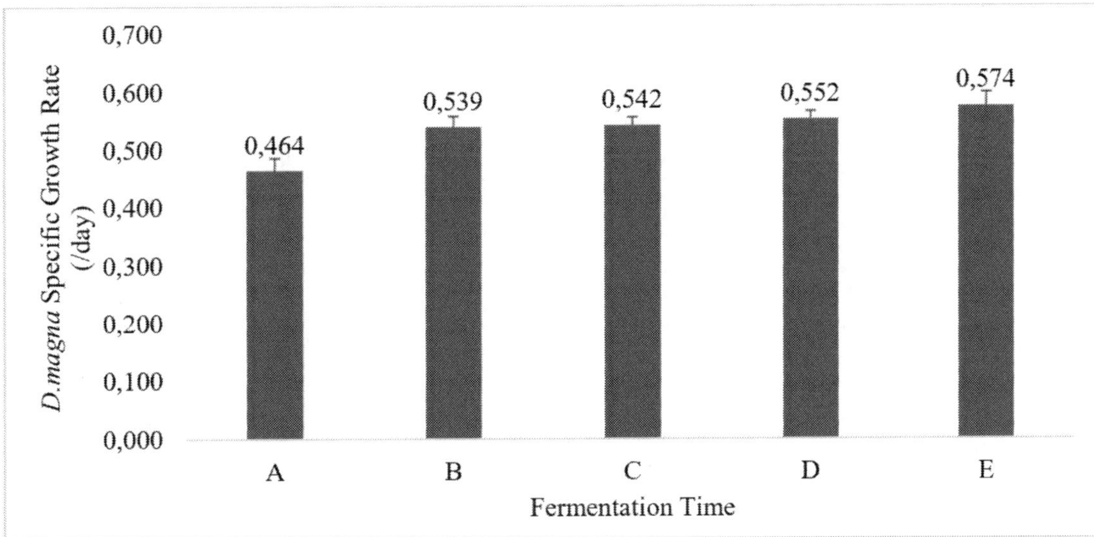

Figure 2. *D. magna* growth rate cultured using organic fertilizer based on fermentation time difference

Based on the results, the highest *D. magna* growth rate occurred in *D. magna* which cultured using organic fertilizer fermented for 28 days (D) is 0.574 ± 0.005/day. The lowest *D. magna* growth rate occurred in *D. magna* without fermentation (A) is 0.464 ± 0.002/day.

Growth of mass-cultured *D. magna* use 25% Chicken manure + 50% rejected bread + 25% tofu waste with different fermentation length gave no significant effect between treatment at lag phase ($P > 0.01$), lag phase happened on day 4 and the highest by 25% Chicken manure + 50% rejected bread + 25% tofu waste with 28 days fermentation (E), that is 127,500.00 ± 0.01 ind/L. This is due to *D. magna* begins to adapt to the new environment at lag phase, if cultured medium concentration is the same with natural medium, it will make *D. magna* grows faster. However, if there are differences between culture medium concentration, *D. magna* needs longer time to grow. The difference in concentration of culture medium and liquid cells in plankton will have an effect on restitution of enzyme and the concentrate substrate to a further extent for growth and presence of nutrients in cells through the diffusion process as a result of the difference in concentration between the culture medium and the liquid body [2][18].

Exponential phase occurs on day 10, with the highest population is 209,370.33 ind/L stationary phase occurred on day 14, with the highest population is 233,980.33 ind/L; The death phase occurred on day 16, with the highest population is 220,426.67 ind/L. Exponential phase is the phase where the nutrient content in *D. magna* is highest while growth is not maximized and *D. magna* began to increase. The exponential phase of the study took place on the day 10, the results were in line with previous study stated that *Daphnia* sp. is in lag or exponential phase on day 9 and 10 [19]. The exponential and stationary phase in this study gives a very real effect between treatments ($P < 0.01$). The length of the stationary phase is correlated with the duration of adapting the *D. magna* phase with the new culture medium. This is because the length of the stationary phase affects the absorption of nutrients in the culture medium by *D. magna*. The results were in line with previous study which showed about exponential phase stop because of nutrient lack in cell density enhancement [2][20].

D. magna biomass weight cultured using organic fertilizer based on fermentation time difference is presented in Figure 3.

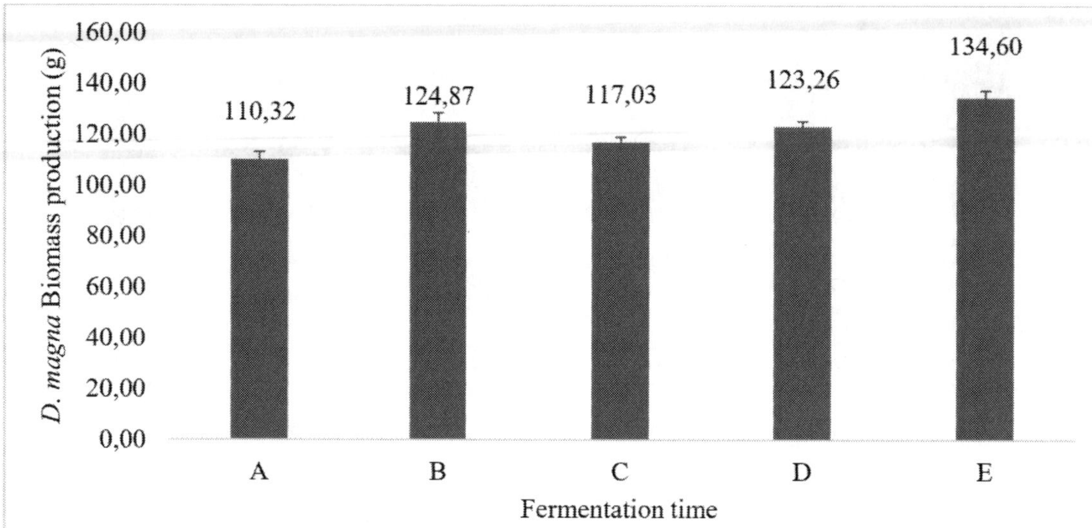

Figure 3. *D. magna* biomass weight mass cultured using organic fertilizer based on fermentation time difference

The highest *D. magna* biomass weight was *D. magna* cultured using organic fertilizer which fermented for 28 days (D) that is 134.60 ± 0.005 g and the lowest *D. magna* biomass weight was *D. magna* cultured using organic fertilizer which fermented for 0 day (A) that is 110.32 ± 0.001g. Quality of nutrient culture medium give effect to the supply of plankton and bacteria to increase population growth and biomass of *D. magna*. High organic matters can have an effect on the density and biomass of *D. magna* [3]. Proximate analysis result of *D. magna* cultured using organic fertilizer based on fermentation time difference is presented in Table 5.

Table 5. Proximate analysis of *D. magna* cultured using organic fertilizer based on fermentation time difference

Treatments	Carbohydrate (%)	Ash (%)	Crude fat (%)	Crude fiber (%)	Crude protein (%)
A (Without Fermentation)	21.87	5.73	5.59	4.03	62.78
B (7 days Fermentation)	18.00	6.36	6.76	4.31	64.58
C (14 days Fermentation)	15.09	5.34	6.58	5.50	67.49
D (21 days Fermentation)	16.30	4.94	6.61	4.50	67.65
E (28 days Fermentation)	13.76	6.81	6.91	4.46	68.06

The highest *D. magna* protein and fat nutrient quality analysis result was *D. magna* cultured using organic fertilizer fermented for 28 days (D) those are 68.06% and 6.91%; the lowest protein and fat nutrient content was *D. magna* cultured using organic fertilizer fermented for 0 day (A) those are 62.78% and 5.59%. *D. magna* total fatty acid profile analysis result cultured using organic fertilizer based on fermentation time difference is presented in Table 6.

Table 6. *D. magna* total fatty acid profile analysis result cultured using organic fertilizer based on fermentation time difference

Fatty acids profile (%)	Seed	A	B	C	D	E
Miristic	0.28 ± 0.04	0.49 ± 0.04	0.48 ± 0.09	0.41 ± 0.02	0.50 ± 0.05	0.52 ± 0.05
Pentadecanoic	0.10 ± 0.08	0.18 ± 0.06	0.15 ± 0.08	0.17 ± 0.304	0.08 ± 0.02	0.09 ± 0.06
Palmitic	2.01 ± 0.06	2.29 ± 0.08	2.59 ± 0.04	1.97 ± 0.08	2.12 ± 0.01	3.14 ± 0.09
Stearic	0.41 ± 0.07	1.65 ± 0.02	2.91 ± 0.09	0.52 ± 0.03	2.08 ± 0.05	2.71 ± 0.07
Oleic/ω9	1.62 ± 0.05	0.95 ± 0.03	2.61 ± 0.01	0.89 ± 0.08	2.55 ± 0.03	3.07 ± 0.02
Linoleic/ω6	1.54 ± 0.02	2.46 ± 0.07	3.37 ± 0.02	2.49 ± 0.07	3.75 ± 0.02	4.83 ± 0.09
Linolenic/ω3	0.19 ± 0.06	2.38 ± 0.09	3.32 ± 0.01	2.39 ± 0.03	2.56 ± 0.07	3.54 ± 0.05
Arachidic	0.02 ± 0.07	2.83 ± 0.02	3.05 ± 0.03	1.02 ± 0.04	1.25 ± 0.05	2.30 ± 0.08
Arachidonic	0.07 ± 0.01	0.15 ± 0.05	0.13 ± 0.08	0.15 ± 0.02	0.06 ± 0.08	0.07 ± 0.02
Eiksapentaenoic	0.27 ± 0.05	2.53 ± 0.09	3.52 ± 0.06	0.50 ± 0.04	0.05 ± 0.02	0.06 ±0.07
Omega 3	0.03 ± 0.02	4.01 ± 0.04	5.91 ± 0.04	3.99 ± 0.08	2.02 ± 0.04	4.06 ± 0.08
Omega 6	0.63 ± 0.08	5.64 ± 0.07	7.53 ± 0.08	5.68 ± 0.02	5.03 ± 0.05	6.05 ± 0.02
Omega 9	0.62 ± 0.05	0.95 ± 0.01	0,61 ± 0,01	0.56 ± 0.08	0.90 ± 0.03	1.48 ± 0.02
Unsatturated fatty Acid	0.88 ± 0.09	4.46 ± 0.06	3.56 ± 0.07	4.24 ± 0.02	2.68 ± 0.04	3.65 ± 0.08
Satturated fatty acid	0.97 ± 0,01	1.09 ± 0.09	2.0 ± 0.08	1.49 ± 0.07	0.97 ± 0.02	1.24 ± 0.02
Mono unsaturated fatty acid (MUFA)	1.63 ± 0.05	2.642 ± 0.02	3.97 ± 0.06	2.40 ± 0.04	3.57 ± 0.05	3.93 ± 0.03
Polyunsaturated fatty acid (PUFA)	1.24 ± 0.08	2.62 ± 0.08	3.58 ± 0.04	2.83 ± 0.02	3.17 ±0.01	4.31 ± 0.09
AA	0.08 ± 0.09	0.15 ± 0.04	0.13 ± 0.07	0.15 ± 0.09	2.18 ± 0.03	2.71 ± 0.03
DHA	0.06 ± 0.02	0.08 ± 0.04	1.07 ± 0.03	0.07 ± 0.01	0.39 ± 0.08	0.83 ± 0.05
EPA	0.27 ± 0.03	0.53 ± 0.02	1.52 ± 0.06	1.50 ± 0.07	1.09 ± 0.02	2.08 ± 0.08

The highest *D. magna* linoleic acid and linolenic acid profile analysis result was *D. magna* cultured at medium fermented for 28 days (E) those are 4.83% and 3.54%; the lowest linoleic acid and linolenic acid profile analysis result was *D. magna* cultured using organic fertilizer fermented for 0 day (A) those are 1.54% and 0.19%. The highest *D. magna* amino acid profile cultured using organic fertilizer fermented for 28 days (E) those are non-essential amino acid aspartic acid 10.70 ppm and essential amino acid Lysine 53.94 ppm. The lowest amino acid for this treatment (E) was leucine 12.40 ppm. The lowest fatty acid profile was non-essential amino acid aspartic acid 8.25 ppm, essential amino acid Lysine 30.14 ppm, and amino acid leucine 5.82 ppm at *D. magna* cultured using organic fertilizer fermented for 0 day (A). *D. magna* amino acid profile mass cultured using organic fertilizer based on fermentation time difference is presented in Table 7.

Table 7. Amino acid profile of *D. magna* cultured using organic fertilizer based on fermentation time difference

Amino Acid Profile	Seed	A (ppm)	B (ppm)	C (ppm)	D (ppm)	E (ppm)
Histidine	30.85 ± 0.07	38.14 ± 0.03	44.52 ± 0.03	39.85 ± 0.05	48.92 ± 0.08	50.94 ± 0.01
Serine	10.76 ± 0.05	15.40 ± 0.07	16.63 ± 0.07	16.76 ± 0.02	15.61 ± 0.03	18.62 ± 0.01
Arginine	50.36 ± 0.03	57.39 ± 0.02	64.35 ± 0.05	58.36 ± 0.07	76.61 ± 0.04	79.37 ± 0.07
Glycine	15.33 ± 0.09	18.48 ± 0.05	19.78 ± 0.09	17.33 ± 0.02	19.36 ± 0.04	20.19 ± 0.01
Aspartic Acid	7.65 ± 0.01	8.25 ± 0.09	9.70 ± 0.07	8.65 ± 0.03	9.78 ± 0.03	10.70 ± 0.01
Glutamic Acid	16.85 ± 0.05	19.76 ± 0.05	28.89 ± 0.04	22.85 ± 0.07	20.51 ± 0.04	27.28 ± 0.01
Threonine	15.47 ± 0.09	17.37 ± 0.09	23.56 ± 0.03	20.47 ± 0.07	19.02 ± 0.09	23.37 ± 0.01
Alanine	30.51 ± 0.03	33.21 ± 0.02	35.98 ± 0.05	30.51 ± 0.01	40.65 ± 0.05	42.51 ± 0.09
Proline	15.08 ± 0.04	18.98 ± 0.07	19.44 ± 0.04	20.08 ± 0.09	20.25 ± 0.05	23.00 ± 0.06
Cystine	23.72 ± 0.02	25.74 ± 0.04	28.13 ± 0.02	25.72 ± 0.03	30.24 ± 0.05	32.87 ± 0.04
Lysine	25.89 ± 0.05	30.14 ± 0.05	45.22 ± 0.05	30.89 ± 0.06	48.92 ± 0.08	53.94 ± 0.01
Tyrosine	27.59 ± 0.06	29.86 ± 0.06	30.99 ± 0.03	33.59 ± 0.06	39.57 ± 0.04	44.16 ± 0.01
Metheonine	15.87 ± 0.02	19.95 ± 0.09	21.20 ± 0.07	20.87 ± 0.02	19.97 ± 0.03	22.79 ± 0.04
Valine	31.47 ± 0.05	33.67 ± 0.04	35.46 ± 0.07	32.47 ± 0.05	32.44 ± 0.05	36.88 ± 0.05
Isoleucine	12.41 ± 0.07	15.62 ± 0.03	18.97 ± 0.05	16.41 ± 0.07	15.49 ± 0.07	19.98 ± 0.10
Leucine	7.85 ± 0.05	5.82 ± 0.01	7.40 ± 0.05	6.85 ± 0.05	10.10 ± 0.08	12.40 ± 0.04
Phenylalanine	14.76 ± 0.02	15.40 ± 0.05	17.63 ± 0.07	15.76 ± 0.02	15.61 ± 0.03	17.62 ± 0.01
Tryptophan	42.36 ± 0.07	45.39 ± 0.03	47.35 ± 0.01	44.36 ± 0.05	46.61 ± 0.03	52.37 ± 0.04

The difference between highest result and lowest result of non-essential amino acid aspartic acid is 2.45 ppm, essential amino acid lysine is 23.8 ppm, and essential amino acid leucine is 6.58 ppm.

Nutrient content based on proximate analysis as shown in Table 5, found that the highest protein and fat of *D. magna* are on mass-cultured *D. magna* using organic fertilizer which is fermented for 28 days (D) that is 68.06% and 6.91%; the lowest protein and fat are on mass-cultured *D. magna* using organic fertilizer which is fermented for 0 day (A) that is 62.78% and 5.59%. the result is lower than previous study which reached 73.90% ± 0.04 protein [2], higher than other previous studies which reached 66.12% [21] and 4% from wet weight [12]. The highest fat content about 8.84% ± 0.01i higher than previous study which reached 7.89% ± 0.02 [2]. A high protein content and low fat from the study due to the high nutrient in the culture medium *D. magna* where the higher nitrate and phosphate levels. The higher the N and P content, the higher the protein in the culture medium [22]. The protein content is always the opposite of fat, because the fat in the body works twice as much as the protein [23].

Factors that affecting biomass and nutrient content *D. magna* are quality of nutrient medium, availability of phytoplankton, bacteria, and detritus as food, and the environment [2][3]. Organic matter contained in fermented medium can increase the number of bacteria and organic particles of decomposition results, thus increasing the availability of nutrients in the medium, this affects growth of population and biomass of *D. magna*. Fermentation aims to multiply the number of microorganisms as well as intensify the metabolism in food, resulting in new food products using microorganisms [1].

Total fatty acid profile found the highest linoleic and linolenic fatty acid profile of *D. magna* are by mass-cultured using organic fertilizer fermented for 28 days (E) that is 4.83% and 3.54% (Table 6); the lowest are by mass-cultured using organic fertilizer fermented for 0 day (A) that is 1.54% and 0.19%.

This results were higher than previous study which reached 0.2% [2]. Linoleic fatty acids serve as a base substrate in the formation of PUFA long chains. Linoleic fatty acids act as a base substrate forming long chains of Omega 6 and Omega 3 [2][24-25].

The study result the highest of amino acid profile of *D. magna* are by mass-cultured using organic fertilizer and fermented for 28 days (E), the result are non-essential amino acids aspartic acid are 10.70 ppm, lysine essential amino acids are 53.94 ppm, and the amino acid leucine that is 12.40 ppm; the lowest amino acid profile of *D. magna* are by mass-cultured using organic fertilizer and fermented for 0 day (A), the result are non-essential amino acids aspartic acid are 8.25 ppm , lysine essential amino acids are 30.14 ppm, and and the amino acid leucine that is 5.82 ppm. The function of the lysine amino acid is as frame of vitamin b1 and anti-virus, helps in the absorption of calcium, stimulates appetite, and helps in the production of carnitine to convert fatty acids into energy [2][22][26].

4. Conclusion

Mass-cultured *D. magna* using fermentation medium of 25% chicken manure + 50 % rejected bread + 25% tofu waste combination was dried for 28 days fermentation (E) provide the highest improvement to growth and biomass production. The highest nutrient quality based on proximate, fatty acid profile, and amino acid profile analysis of *D. magna* was also showed in *D. magna* cultured by the same treatment (E).

5. Acknowledgements

The authors acknowledge APPIHIS (Semarang Ornamental Fish Cultivators and Merchants Association) which provided pond facilities for *D. magna* mass culture for the study.

6. References

[1] Nwachi OF 2013 *J Fish Aquat Sci* **8** 30–32
[2] Herawati VE, Hutabarat J, Pinandoyo and Radjasa OK 2015 *HAYATI J Biosci* **22** 169–173
[3] Damle DK and Chari MS 2011 P *J Fish Aquat Sci* **6** 57–61
[4] Herawati VE, Nugroho RA, Pinandoyo and Hutabarat J 2017 *IOP Conf Ser Earth Environ Sci* **55** 12004
[5] Zahidah, Gunawan W and Subhan U 2012 *J Aquat Sci* **3** 84–94
[6] Herawati VE and Agus M 2014 *J Sci Tech 26 1e11* **26** 1–11
[7] Liswahyuningsih E, Khotimah AU and Febriana DT 2011 *J Ind* **2** 57–66
[8] Purbowati E, Sutrisno CI, Baliarti E, Budhi SPS and Lestariana W 2007 *Pros Semin Nas Teknol Peternak dan Vet* 394–401
[9] Yuniwati M, Iskarima F, Padulemba A 2012 *J Tech* **5** 172–181
[10] Asadi Rad M, Zakeri M, Yavari V and Mousavi SM 2012 *J Pers Gulf* **3** 15–24
[11] Abu-Elala N, Marzouk M and Moustafa M 2013 *Int J Vet Sci Med* **1** 21–29
[12] Schlotz N, Sørensen J and Martin-Creuzburg D 2012 *Comp Biochem Physiol A Mol Integr Physiol* **162** 449–454
[13] Association of Official Analytical Chemists 2005 *Official Methods of Analysis, 18th ed* (Washington DC: AOAC International) p 25
[14] Hersoelistyorini W, Sumanto D and Najih L 2010 *J Pangan dan Gizi* **1** 24–34
[15] Siswati N dyah, Theodorus H and S PWE 2009 *J Buana Sains* **9** 63–68
[16] Sumich JL 1992 *Marine Life, 5th ed.* WMC Brown Publishers, printed in USA
[17] Witty LM 2004 *Practical guide to identifying freshwater crustacean zooplankton, 2nd ed* (Sudbury: Cooperative Freshwater Ecology Unit) p 66
[18] Pangkey H 2009 *J Perikan dan Kelaut Trop* **5** 33–36
[19] Sitohang R V, Herawati T and Lili W 2012 *J Perikan Kelaut* **3** 65–72
[20] Fogg GE 1965 *Algal Cultures and Phytoplankton Ecology, 2nd ed.* (Madison: University of Wisconsin Press) p 126
[21] Mokoginta I, Jusadi D and Pelawi T 2003 *J Aquat Ind* **2** 7–11

[22] Valverde JC, Martínez-Llorens S, Vidal AT, Jover M, Rodríguez C, Estefanell J, Gairín JI, Domingues PM and Rodríguez CJ, García BG 2013 *Aquac Int* **21** 413–433
[23] Lim C, Yildirim-Aksoy M and Klesius P 2011 *J Aquac* **73** 188–193
[24] Pratiwi AR, Syah D, Hardjito L, Panggabean LMG and Suhartono *HAYATI J Biosci* **16** 151–156
[25] Zengin H, Vural N and Çelik VK 2013 *J Fish Aquat Sci* **13** 397–405
[26] Ovie SO and Eze SS 2013 *J Fish Aquat Sci* **8** 94–100

Behind the Slow Road to Progress: Addressing Myriad Causes of the Persistence of Relatively High Maternal Mortality in Brebes Regency after the Post EMAS Program

Sri Kusumo Habsari[1*], Sofiah Sofiah[1], Sumardiyono Sumardiyono[1]

[1]Sebelas Maret University, Indonesia
*Email: skhabsari@staff.uns.ac.id

Abstract. The purpose of this article is to discuss the restricting factors which hinder the Brebes regency's goal of reducing maternal and new born mortality, especially in the aspects of communication strategy which has been applied by the local district government. The location of the research was Bulakamba sub-district which has applied the system of "*desa siaga madya*" (mid-size alert village) but unfortunately has the highest maternal mortality in Brebes regency. Through analyzing data which have been collected by making observation, doing interviews, conducting focus group discussion and studying documents using an interactive data analysis technique, the results show that there are some complex obstacles which hinder the success of the program. Although the local government has attempted to produce health regulations as an intervention, to improve the quality of the health services and to develop special communication strategy, the rate of maternal mortality is still relatively high in this sub-district. However, the cultural change as the impact of modernization and cultural mobility, especially in the coastal area of the regency could not be blamed as one of the myriad causes of the persistence. It still needs a special address from the government to intervene, especially to prepare the society to face the modern life with all of its complexities.

1. Introduction

Early research on the post implementation of EMAS (Enhancing Maternal and New Born Survival) program in Brebes has shown that this program has subsequently not been able to bring down the rate of maternal and new born mortality as the target planned. In 2015, the highest rate of maternal mortality in Central Java was in Brebes with 47 cases although compared to the 2014 rate, it was a 45% reduction rate. In 2016, Brebes maternal and new-born mortality was still in a crucially high state, with five cases each month[1]. The number of the cases has caused dissatisfaction among health practitioners in the regency and led to curiosity towards the hindering factors of the still relatively high maternal mortality. One case worth questioning has been that the highest maternal mortality rate predicate was actually given to a sub-district which was considered to be highly knowledgeable about maternal health.

Ending preventable maternal mortality is an internationally accepted goal with the target of global average of less than 70 per 100,000 live births by 2030. USAID has developed the Vision for Action for Ending Preventable Maternal mortality applying integrated, comprehensive, and holistic approach with putting on equity and respect for women and context specifics supported with "strong guiding principles and good governance. [2][3] The target plan was to reduce 75% of Maternal Mortality Ratio (MMR) by 2015. However, until September 2015 of the time of the end of the era of Millennium Development Goals (MDGs), the global decline of MMR was only 44%. [4] Every country has different pathways to reduce maternal mortality. Challenges include poverty, geographic features limiting access to health services, cultural norms and/or weak health systems.

In Indonesia, the hindering factors are relatively complex. It is often not a medical condition, such as suffering hemorrhaging, eclampsia, sepsis or unsafe abortion identified globally as the main causes.[5] The Women Research Institute has identified that a lack of access to adequate reproductive health facilities is often the main cause of the persistence of Maternal Mortality Rate in Indonesia. Poverty is the main reason why pregnant women do not try to get help although they are categorized as being in high risk. Most of them deliver their baby at home assisted by untrained traditional birth assistant (*dukun bayi*). Findings show that childbirths attended by untrained traditional birth assistance also encounter other problems such as the lack of clean water, electricity for lighting, sterile space and equipment, and necessary medicines which could be risky for the health of the mother. However, many pregnant women refuse to get service from doctors or midwives although they acknowledge that they have better medical expertise because the cost of traditional midwives is more affordable and their services are from the pre-natal and delivery process until the post-delivery care to the infant and the family. Another hindering factor is the geographical condition, such as the considerable distance between the community health center to the regional public hospitals compounded by poor road infrastructure and a lack of public transportation. The other cause is the myths which are relatively still strong in the society such as traditional midwives having supernatural powers who could give them a comfortable feeling and is still reinforced with the strong belief that delivering a baby is natural for women. Cultural norms also prescribes women's lack of access to their body: women could not make their own decision about where they will deliver the baby and who will assist them in the process of delivery.[6]

In 1996 Indonesia developed a program to promote safe motherhood called *Gerakan Sayang Ibu* (Love Mom Movement). Some strategies developed are creating a program of the Making Pregnancy Safer Movement and making an effective partnership through a cooperative cross-sector program. In 2006 the government made a policy through Ministry of Health to develop a program of *Desa Siaga* (Village Alert System) to encourage the local government to provide early assistance to keep out the three delays. However, WRI found out that some districts did not have enough political will to allocate funds for health programs which in consequence the GSI program did not show an expected success to reach the target of MMR reduction.[5] In 2012 Indonesia has developed EMAS program with the support of USAID. EMAS is a five-year program with the goal to contribute to an overall 25% reduction in maternal and neonatal mortality. It works through the collaboration of five institutions, Jhpiego as the lead partner, Lembaga Kesehatan Budi Kemuliaan Muhammadiyah, Save the Children and RTI International. EMAS Program interventions involve emergency maternal and new-born care at hospitals, district referral systems, maternal and new-born care at health centres (*puskesmas*) and local government and civil society accountability in which all interventions interact synergistically between community, civil society, health facilities, and district health system. 300 districts and cities in six provinces in Indonesia acquired the EMAS program and Brebes regency is one of them.[7] Although EMAS has reported that the program has successfully reduced the MMR in Indonesia, some regencies still have high MMR including Brebes.

In 1987 in concern with the high rate of maternal mortality the program of Safe Motherhood Initiatives was brought about to call for local, national and international level to design actions with mutual agreement aiming at improving women health status and women's access to health services during pregnancy and the quality of medical care for treating complications during pregnancy and delivery. There were many interventions proposed and they were the subject of the debates among health practitioners and policy makers. However, they had one agreement that the interval between the assault of the obstetric complications and the provision of the timely obstetric care was the potential causes of maternal deaths. [8] Meanwhile, an in depth analysis of multifactorial causes is pointed to demographic, behavioural, nutritional, and health services related factors.[9] The post-2015 Sustainable Development agendas have broaden to include health, gender equality, and sexual and reproductive health and rights to cover the limitations of MDGs which has too focus only on agenda of reducing maternal mortality rate and do not consider the interdependence between health and human right. The unachieved target of MDGs in related to MMR reduction is due to the action of the government in providing grossly under-funding health rather than making long standing commitment. The paradigm supported by World Bank is Universal Health Coverage (UHC) as the health goal which means that the poorest and the most marginalised people should have access to healthcare. However, the implementation of UHC also raises an issue of financing.[10]

Maternal mortality has become a focus of attention in research and policy, especially since it is the goal of number 5 in MDGs. Various researches have been done in looking at the causes of the relatively persistence of the MMR especially in many developing countries and the impact of interventions.[11] Various policies and strategies have also been developed and implemented to reduce the MMR, such as Every Newborn: cash transfer and voucher scheme in South Asia or An Action Plan to End Preventable Death (ENAP) and Strategies toward ending Preventable Maternal Mortality (EPMM).[3][8][12] Some have reported their success through "the combination of a right-to-health approach with learning and capacity building, community networking, popular mobilization and legal action" and intervention standards.[11][13] However, most researches still reported that the enormous number of maternal death in developing countries stayed the same as 100 years ago and direct obstetrics are the main causes of maternal mortality.[3][14]

This paper provides an overview of the implementation of EMAS and the implication of the program in Brebes regency with interests in the sub-district located in the coastal area of Brebes which is reported having the highest number of maternal mortality cases in 2016. It reviews especially the implementation challenges and limitations based on currently available data. This study in some way is an attempt to identify other unexpected factors behind the slow progress of the target planned. Points to develop in this paper are addressed to urban issues, as most previous studies have more focused on rural setting categorized as the poor without considering the deprivation or urban areas as an impact of globalization.[15]

2. Material and Methods

This article uses a qualitative approach which considers people's attitudes, norms, and values as data of this research. The location of the research was Brebes regency, especially the Bulakamba sub-district. The data were obtained from FGD with pregnant women and community leaders which were conducted separately, participants' observation, documentation survey and in depth interviews with governmental staffs of community health centers of the subdistrict and the regency of Brebes. Verbal and non- verbal behaviors were also used to understand cultural norms, patterns, values, etc. In depth interview, in the form of personal communication was also employed to some participants when non-verbal behaviors indicated conflicting information or needed further clarification. The data collection employed triangulation to avoid the limitations of a single method, to help create the reliability and validity of the data. All data were then managed carefully.

3. Results and Discussion

3.1. The Coastal Culture of Brebes Regency

Brebes regency is located in the north-western part of Central Java province. It comprises 17 sub-districts which spread from the mountainous area of the mount Slamet to the coastal area of the north-western part of Central Java. Majority of Brebes people speak in Javanese language with a special accent which shows the intersection of Javanese and Sundanese accent and culture. With the population of about 1.7 million people (2016), Brebes' economy is supported by several occupations, with the main focus on agriculture and fisheries. This fact is supported by the high rainfall, up to 18,94 mm per month and a tropical climate. Brebes regency has about 37,7% wetland area, consisting of rice fields and swamps, supporting up to 70% of the population.

Concentrated fisheries and agriculture areas are mainly in Bulakamba sub-district. Bulakamba, one of the most densely populated sub-district in Brebes regency, is located west of the capital of Brebes regency, and is 3 km above the sea level. It borders the Java Sea in the north, Larangan and Ketanggungan sub-district in the south, Tanjung and Kersana sub-district in the west, and Wanasari sub-district in the east. Bulakamba sub-district uses 7411 Ha of land for agricultural needs (wet lands), and the remaining 2882 Ha are non-wet lands. Aside from the vast wet lands, three rivers flow through Bulakamba sub-district, which are Pakijangan River, Bangsri River, and Kluwut River. These rivers are what supports Bulakamba as an agricultural sub-district. However, some villages are located on the coastal area with fisheries as the main income of the people. Some people work as a fisherman and others which are mostly female work on the traditional fish processing. Information from reliable person informally acquired mentions that some young people work as sailors however this occupation is not specifically reported in *Brebes dalam Angka* (Brebes in Number).

As reported in 2015 Brebes in Number, there are 5 government and private hospitals. To provide the health service of the community as well as to ease the community to access the service, there are 38 main community health centers and 60 supporting community health centers available in every sub-district. Some sub-districts, such as ones located in remote areas or covering wider area with more dense population, it is even more than one community health centers, such as in Bulakamba sub-district. In relation to provide services of women reproductive health, the community health center in Bulakamba has provided some basic emergence of obstetric and neonatal services.

3.2. Behind the Slow Road to Progress

Maladies that cause maternal mortality such as haemorrhage, sepsis, hypertensive disorders, illegal abortions, and obstructed labour can be treated successfully as long as there is no delay of seeking care, reaching care and receiving care to public health specialists. However, other result from in depth analysis to the multifactorial causes of MMR points to other factors, such as socio-demographic, cultural, and communal factors, beside health service related factors. As various studies have reported, the persistence of the relatively high maternal mortality rate (wherever) is caused by various factors such as poverty, limited access to educational and economic opportunities, gender-based stereotypes and discrimination and harmful traditional practices such as early marriage.

Finding on the research on the persistence of Brebes MMR has shown that Brebes has designed various intervention with EMAS support to reduce the number of maternal mortality. In line with Safe Motherhood Initiatives, Brebes has also developed interventions and strategic drivers to reduce maternal mortality and shorten the three delays. The setting of Brebes which spreads from the mountainous to coastal area with the centre of the local government and the regional public hospital in between is potential for causing delay to reach and receive care because of geographical constraints. It is still possibly compounded with socioeconomic conditions and the financial environment that obstruct access to money and information as well as cost of services. Although it is sometimes not considered by the community of the remote areas or the low income people because of their low esteem, the scope and organization of medical services and the quality of care are still questionable, especially in the eyes of the health practitioners of the regency. 2015 research on the implementation of EMAS in Brebes, for example has identified that the referral exchange system is still problematic despite the policy produced to improve communication among health centres and regional public hospital in order to stabilize, refer, and prepare for emergency referrals.

In conversations with some pregnant women and their mothers, separately conducted, they acknowledge the professional ability of the midwives on their knowledge, skills, and competence in assisting delivery. They also acknowledge that good facilities of health centre is important for pregnancy checks and delivery process. However, when questions on who has checked their pregnancy and who will help their delivery, reluctantly they answer that they still go to the traditional birth assistance (*dukun bayi*). Reasons are mentioned that *dukun bayi* does not only provide pregnancy check and delivery process, but also in looking after other pregnancy disturbing conditions, such as giving massage and making herbal potion for creating fresh feeling. In some way, the function of *dukun bayi* is to replace the position of a mother in looking after the pregnant women, who are still relatively young (in their 20s). *Dukun bayi* could satisfy their need for attention and care that unfortunately could not be provided by midwives. *Dukun bayi* could also serve any time as needed. Aside pregnancy related care, *dukun bayi* also looks after the baby after the delivery. Because of the complete service of *dukun bayi* and also the flexibility of the payment of the service, some pregnant women or their mother prefers to get service from *dukun bayi.* Such conditions produce difficulties in monitoring the condition of the pregnancy. Brebes regency has produced regulation to stop this practice by giving penalty to the *dukun bayi* with some of money established as a fine for helping the process of delivery, however, it does not work. Women who get services from *dukun bayi* report that they deliver the baby on the way to the health centre. It can be said that they cover the practice of *dukun bayi* because they consider the importance of the service. As the target of reducing the number of maternal death is more important rather than the implementation of the regulation, the community of health centre of the sub-district Bulakamba facilitates monthly regular meeting between *dukun bayi* and health practitioners to disseminate various informations on pregnancy health especially to detect the high risk pregnancy that needs serious attention and care from health centre. Health communicators of the sub-districts also work hand in hand with *ustadz,* Islamic preachers to insert some informations of the importance of maternal health on their preaching.

The contrastive geographical setting produces also unique cultural differences between the two areas. Some sub-districts are located on the mountainous area which are geographically remote and the culture of the sub-districts is very communal. Meanwhile, some are located on the coastal area which is close to the highway connecting Central Java to West Java and the characteristics of the people is more closely to urban in attitude and values. The complex characteristics of the community encourage the regency of Brebes to develop programs with some considerations of the local wisdom of the sub-districts. Thus, the socio-cultural milieu, such as values, beliefs, and attitudes of the community can be considered in designing and producing interventions.

Finding also shows that Brebes has been successful in enabling and mobilizing individuals and communities though the establishment of *Forum Masyarakat Madani*, a civic forum which becomes the medium between the EMAS program and the community. This forum works with community organizations, sometimes door to door, to educate the community, all women and families, to have adequate knowledge about maternal health. This forum also has a role to communicate the need of quality health care which includes access to services, goods, and information. As a medium between the government and community, this forum could develop a dialogue between the government and community and it can result in changes of procedures and policy from the feedbacks given by the community. In some way, this forum has successfully reduced the MMR to zero cases such as in the sub-district of Sirampog, despite the geographical barriers that are the winding and steep roads and considerably far between the location of the community health centres of the sub-district to the regional public hospitals. The problems of the traditional practice of early marriage, and also the community myth that early pregnancy should be kept out of public have also been tackled through the strategic communication developed by the civic forum. Their intervention is to escort the pregnant women from the early pregnancy to the delivery process through educating the whole member of the family on the importance of the quality of services and health care which are only provided by the community health centre. Even, the develop "*kelas ayah*" group of fathers to educate them about all issues related to stage of pregnancy and also informations on new born which should be recognized by father.

This research also identifies the different attitude of the pregnant women related to finding information on pregnancy health. The characteristics of the community on the coastal area have someway been infused with some aspects of modernization and it influences the perspectives of the pregnant women on the importance of maternal health and health facilities. Some of them say that they go to specialist for their pregnancy check and plan to have the delivery on private hospital. Some of them go to community health centre for regular check as suggested by the person they trust. Some of them use internet to get various information on reproductive health while others still rely on their relative's suggestion. It can be said the pregnant women in the coastal area of Brebes are diverse on their individuality and perspective on pregnancy. This phenomenon shows the urban advantage of the women in accessing the reproductive health information.[15] Looking at the literacy of the pregnant women on reproductive health information and the characteristics of the individuals in finding informations, the local government of Brebes does not provide EMAS support to reduce the number of maternal mortality. Thus, the community health centers on Bulakamba do not create *Forum Masyarakat Madani* as a mediator between the government and the community. It is ironical that the highest rate of maternal death is reported from Bulakamba.

Finding also shows that sub-districts in Bulakamba have complex problems. Despite high risk pregnancy because of too old and too young for pregnancy, some of them have low income and no healthcare. Although pregnancy health check could be performed in community health centre for free, they could not get further obstetric treatment from regional public hospital until they can show their healthcare card. Unfortunately, healthcare in Indonesia is not free. Individual should pay certain amount of money every month to get health cover which unfortunately is still not affordable for some low income people. Cases on neonatal death are also high in Bulakamba. Data show that low income makes some women reluctantly check their pregnancy health and pay attention to the development of the fetus. That is why, anaemia and poor nutrition are the main causes of maternal and neonatal death. Still in line with previous research conducted in Africa, Asia and Latin America in 2010 assessing the "urban advantage" in maternal health care, this research also identifies the unequal access of the urban poor to the health services and barriers to access in different contexts which suggest to do more interventions from both the central and local government.[15]

Low income could encourage individuals to be migrant workers. In Bulakamba, informal information accepted that some individuals work as sailor on foreign ships who are in some way experience cultural mobility. By working as a sailor on foreign ships, somehow they could improve their financial condition. However, they also bring back some unexpected perspectives and attitudes which are still morally unacceptable for the community. The biggest issues raised by the staff of the community health centre in Bulakamba as the unexpected impact brought to home by migrant workers are sexual permissiveness and homosexuality which are still morally and religiously unacceptable in Indonesia. Issue needs to tackle seriously by the local government is the number of people with HIV that shows an increase in recent years. Also, some women unfortunately become pregnant because of sexual permissiveness practice, which is shameful not only for the women but also for the community. Meanwhile, cultural norm, elsewhere in Indonesia is still relatively strong which produces such a question as related to the existence of the husband when a pregnant woman checks her maternal health in public health facilities. In a culture which is to require the participation of the husband during the pregnancy checks, a single woman attending community health centre for a reproductive check could raise many questions that can be disturbing. It is not surprising if a woman who finds herself pregnant without a husband prefer to hide her pregnancy rather than accepting disturbing questions and comments from the people around. Also, there is an issue of who will take responsibility for the baby. It is plausible that these women want to terminate their pregnancy with any meant. Unfortunately, abortion is illegal by law, religion and culture. Five cases of maternal mortality are caused by the unintended pregnancy and the reluctance of the pregnant women to have their pregnant health checks. Rumours spread around that these women purposively abandon their pregnancy by not having pregnancy health checks which lead to the unhandled obstetric complication. Some of these pregnant women unfortunately are economically weak and uninformed on the knowledge of woman reproduction. In such cases, the position of the staffs of the community health centre and the head of the community is dilemmatic. Giving information on access to modern contraceptive, for example is still considered as controversial issue because some people still understand it as supporting to sexual permissiveness. It creates barriers for both the staffs of community health centre and the head of the community as the local leader to manage on how to educate the community on sexual health and access to modern contraceptive. Meanwhile, modernization has also eroded the community's emotional connection to their local leader which further reduces the degree of interaction between the leaders and the community members. As people become more individualized, the power of the local leader to interfere when unacceptable behaviours are noted gradually diminishes. Despite various interventions addressing to both clinical and social problems, cultural issues need to have more attentions, especially to factors accompanying modernization.

4. Conclusion

In an attempt to reduce the number of maternal mortality cases, the regency of Brebes has conducted various intervention efforts with the support of EMAS program. Looking at the intervention programs developed by the local government of Brebes hand in hand with the community from many sub-districts, it shows that the programs are the implementation of Safe Motherhood Initiatives. However, the development of the program is still narrowed on reducing the MMR only although all of the implementations have shown the collaboration between donors, health system, health policy, and civil society. MDGs has been over and post-2015 is the era of SDG which promotes universal access to quality health services. Although on the level of sub-districts, the local government hand in hand with the community have escorted pregnant women from early pregnancy to safe delivery, when it comes to obstetric complication which needs referral to regional public hospital, there is still an issue who will finance the medical treatment. The problem of the regency of Brebes and also elsewhere in Indonesia is poverty. Unfortunately, many low income families still do not have health insurance which leads to the lack access to health care. Other unexpected issue which makes the slow progress of reducing the number of maternal death is the case of unintended pregnancy resulted from the life style change of the community. What happen to the coastal area of Brebes is a common issue in Indonesia as a transition of the society from communal into modern as an impact of modernization. Special strategy needs to design to meet the complexity of the society while still putting on the first on achieving the target of reducing MMR to 75%. Thus, further research needs to conduct to look at the cultural issues which become the hindering factors of the slow progress to reduce the maternal mortality rate, such as through ethnographic studies to certain communities.

References

[1] Dhani 2016 *Angka Kematian Ibu dan Bayi di Brebes Masuk Zona Merah* [cited 2017, 28 July]; Available from: http://brebesnews.co/2016/05/angka-kematian-ibu-dan-bayi-di-brebes-masuk-zona-merah/
[2] USAID 2014 *Ending Preventable Maternal Mortality: USAID Maternal Health Vision for Action* (Washington: U.S. Agency for International Development) p 5
[3] Chou D 2015 Ending preventable maternal and newborn mortality and stillbirths. *BMJ* **351** 4255
[4] Moran A C, Jolivet R R, Chou D, Dalglish S L, Hill K, Ramsey K, Rawlins B and Say L 2016 A

common monitoring framework for ending preventable maternal mortality, 2015–2030: phase I of a multi-step proces *BMC Pregnancy and Childbirth* **16** (250) 1035

[5] Bucagu M, Kagupare J M, Basinga P, Ngabo F, Timmons B K and Lee, A C 2012 Impact of health systems strengthening on coverage of maternal health services in Rwanda, 2000-2010: a systemic review *Reproductive Health Matters* **20**(39) 50-61

[6] Noerdin E. 2011 *In Search of a Spearhead to Reduce Maternal Mortality in Indonesia* (Jakarta: Woman Research Institute) pp 4-6

[7] USAID 2016 *Quarterly Report for Expanding Maternal and Neonatal Survival: Year 6, Quarter 1: October-December 2016* p 3

[8] Thaddeus S and Maine D 1994 Too far to walk: maternal mortality in context *Soc. Sci. Med.* **38** 1091

[9] Tessema G A, Laurence C O, Melaku Y A, Misganaw A, Woldie S A, Hiruye A, Amare A T, Lakew Y, Zeleke B M and Deribew A 2017 Trends and causes of maternal mortality in Ethiopia during 1990–2013: findings from the Global Burden of Diseases study 2013 *BMC Public Health* **17** 160

[10] Berer M 2013 A new development paradigm post-2015, a comprehensive goal for health that includes sexual and reproductive health and rights, and another for gender equality *Reproductive Health Matters* **21**(42) 4-12

[11] Pollard S L, Mathai M and Walker N 2013 Estimating the impact of interventions on cause-specific maternal mortality: a Delphi approach *BMC Public Health* **13**(3) 512

[12] Jehan K, Siney K, Smith H and de Costa, A 2012 Improving access to maternity services: an overview of cash transfer and voucher schemes in South Asia *Reproductive Health Matters***20**(39) 142-54

[13] Ray S, Madzimbamuto F and Fonn S 2012 Activism: working to reduce maternal mortality through civil society and health professional alliances in sub-Saharan Africa *Reproductive Health Matters* **20** (39) 40-49

[14] Pacagnella R C, Cecatti J G, Osis M J and Souza J P 2012 The role of delays in severe maternal morbidity and mortality: expanding the conceptual framework *Reproductive Health Matters* **20** (39) 155-63

[15] Matthews Z, Channon A, Neal S, Osrin D, Madise N and Stones W 2010 Examining the 'urban advantage" in maternal health care in developing countries *PLoS Medicine* **7**(9) 1-7

Proximate content of wild and cultured eel (*Anguilla bicolor*) in different part of body

I Wijayanti[1]* and E S Susilo [2]

[1]Department of Fisheries Product Technology, Faculty of Fisheries and Marine Sciences, Diponegoro University Jl. Prof Soedarto, SH, Tembalang, Semarang, Indonesia 50275
[2]Department of Marine Sciences, Faculty of Fisheries and Marine Sciences, Diponegoro University Jl. Prof Soedarto, SH, Tembalang, Semarang, Indonesia 50275

Email: imasetianto@gmail.com

Abstract. Proximate content in fish varies depends on intrinsic and extrinsic factors. Intrinsic factors include species, sexual maturity, size and body parts. Extrinsic factors include habitat, season and type of food (diet). This study aimed to know the effect of fish body parts (intrinsic factor) on proximate levels in wild and cultured eel (extrinsic). The experimental design used factorial completely randomized design with two factors 2x3. The first factor is the habitat of eel (wild and cultured) and the second factor is the part of the body (head, body and tail) with five replications. The result of statistical analysis showed that there was interaction between fish habitat and body part on moisture, protein, ash and carbohydrate content ($P < 0.05$), but no interaction on fat content and energy ($P > 0.05$). The highest water content (67.02%) was found in head of wild and the lowest one (59.44%) in the tail of wild eel; The highest protein content (18.09%) was found in the body of cultured eel and the lowest one (15.72%) was in the body of wild eel; The highest ash content (3.73%) was the head of wild eel and the lowest (1.32%) was in the body of cultured eel; The highest carbohydrate (3.73%) was found in the head of cultured eel and the lowest one (0.16%) was in the body of cultured. The wild eel had higher fat content and energy than cultured one, while the fat content and energy in body and tail were higher than in head.
Keywords: *Anguilla sp,* proximate, part of body

1. Introduction

Fish is one of the nutrient-rich foods. High protein content and low fat are preferred by all levels of society. Fish protein contains the complete essential amino acids needed by the body. Fat in fish contains EPA and DHA essential fatty acids for both children and adults. Mineral content in fish is also quite complete, consuming enough fish can meet the body's need for essential minerals.

The content of nutrients or proximate in fish is very diverse. Internal influencing factors include species [1-2], size/age [3-4], sex [3], sexual maturity [5] and body parts [6]. External influencing factors include habitat [7-9], seasonal variations [10-13] and feed [14-16].

Eel is one of Indonesia's native fish that has high protein and fatty nutrients. Proximate composition of this type of fish allegedly also varied. Eel is an export commodity that is in great demand by Japan, Korean and China. The eel farming is continuously developed to meet increasing demand. In addition to cultivation eel fish are also caught in their habitat in the river. Different feeding significantly affected the proximate composition of eel [17]. However comparison of proximate composition of wild and cultured eel has not been extensively investigated.

Eel are exported in fresh or processed form. In processed form, eel is filleted without head. The eel's head may contain nutrients that could be used as valuable products. The composition of the different parts of the eel body (head and tail) should be known to provide information and utilize eels according to their nutritional content. The purpose of this research is to know the difference of proximate content of wild and cultured eel in different body parts.

2. Materials and methods

2.1. Research materials

The materials used in this study were some fresh wild and cultured eel of consumption size or weight of about 300 grams. Samples were collected from Cilacap in the southern area of Central Java. Then they were transported to the laboratory for further analysis. Each parameter was tested three samples. The chemical materials used for proximate analysis were H_2SO_4, NaOH, HCl, $KMnO_4$, Boric Acid, and n-hexane (Merck, Gemany)

The equipments used in this study consisted of those for sample handling of eel from the site to the laboratory (Styrofoam boxes) and those for the proximate analysis (oven, Kjeltec system, furnace, soxlet apparatus, desiccator, filter paper, reaction test tubes, spectrophotometers and GC (FOCUS Gas Chromatograph with autosampler A 2000, equipped with Polaris Q MS detector (Thermo Scientific,USA).

2.2. Sample preparation

Samples were brought from Cilacap waters to the lab in living condition. In laboratory both samples of wild and cultured eel were immediately killed and cut into 3 parts, namely the head, flesh of the body and flesh of the tail. Further samples were analyzed for proximate content (moisture, protein, lipid, carbohydrates and energy content).

2.3. Design

This study was aimed to describe the proximate content of both wild and cultured Eel in three different part of body (tail, body and head). The experimental design of this study was completely randomized design (CRD) Factorial with two factor (2x3). The first factor were type of fish comprises two levels (wild and cultured) and second factor were different part body (tail, body and head) and repeated 3 times.

2.4. Variable observed

The research variable observed in this study was nutritional characteristic of eel. The analysis of nutritional characteristics carried out within wild and cultured eel were consisted of proximate analysis (moisture, protein, fat, ash, and carbohydrates.

2.5. Analysis procedures

2.5.1. Moisture content

Determination of water content was based on samples weight before and after drying [18]. An empty cup was dried in an oven for 1 hour at 105 ° C temperature, and then it was put in a desiccator for 15 minutes and afterward it was weighed. One gram sample was inserted into the cup and then it was dried in an oven at 105 ° C temperature until its weight was constant (Drying prosess was approximately done for 6 hours) . Afterward, the cup was inserted into the desiccator for 30 minutes. Later, it was weighed again. The water content was determined by the formula of:

$$\text{Water Content} = \frac{\text{Final Samples Weight (B)}}{\text{Early Samples Weighr (A)}} x100\% \quad (1)$$

2.5.2. Protein content

Analysis of protein content was conducted according to Kjeldahl method[18]. The principles of the method are that how to do the oxidation of carbonaceous materials and the conversion of nitrogen to ammonia by sulfuric acid. Then, ammonia reacts with the excess of acid to form ammonium sulfate. Later, formed ammonium sulfate is elaborated and the solution is made to bealkaline with NaOH. Evaporated ammonia is then going to be tied with boric acid. The quantity of nitrogen contained in the solution is determined by titration using standard solution of acid.

Five grams of dried samples was placed in a 100 ml Kjeldahl flask, followed by addition of 0.25 grams of selenium and 3 ml of concentrated H_2SO_4 in it. Furthermore, the destruction was done (heating through boiling process) for 1 hour until the solution was clear. Then, 50 ml of distilled water and 20 ml of 40% NaOH were added and then they were distilled. Distillation result was escrowed in Erlenmeyer flask containing a mixture of 10 ml of 2%H_3BO_3 and 2 drips of pink Brom Cresol Green-Methyl indicator. When the destillate reached a volume of 10 ml and become bluish-green colour, distillation process was stopped. Then, the distillate was titrated using 0,1N HCl until the colour become pink. The same treatment was also done against the blank. The protein content was calculated by the formula of:

$$\text{Nitrogen Content (\%)} = \frac{(B-A) \times C \times 14{,}007 \times 100\%}{D} \quad (2)$$

$$\text{Protein Content (\%)} = \text{Nitrogen Content (\%)} \times \text{FK} \quad (3)$$

Where :
A = HCl volume for blank titration
B = HCl volume for samples titration (ml)
C = normality level of HCl used (N)
D = samples weight (mg)
FK = conversion factor (6.25 for fishery products)

2.5.3. Fat content

Fat content analysis was according to AOAC [18]. Two grams of eel meat (W1) were spread out over the cotton which was reposed on filter paper and then it was rolled up to be a thimble. Wrapped samples were inserted into a fat flask that hadbeen weighed before (W2) and it was connected to Soxhlet tube. Later, fatsheath was inserted into the tube Soxhlet extractor chamber and doused with fat solvent (n-hexane).

Then, reflux process was done for 6 hours. Fat solvent in the fat flask was distilled until all it was evaporated. During distillation process, the solvent will be accommodated in an extractor chamber, and then it was discarded so it reenter into the fat flask. Afterward, fat flask was dried in an oven at temperature of 105 ° C. Subsquently, fat flask was put in a desiccator until reached constant weight (W3). The fat content was determined by the formula of:

$$\text{Fat Content (\%)} = \frac{W3 - W2}{W1} x100\% \quad (4)$$

Where :
W1 = Samples Weight (gram)
W2 = Fat Flask Weight without fat (gram)
W3= Fat Flask Weight with fat (gram)

2.5.4. Ash content
Ash content was analyzed according to AOAC [18]. The cup was cleaned and then dried in an oven for 30 minutes at the temperature of 105 °C, following by storing indesiccators and weighing. Five grams of samplewasthen weighed and put in the cup. The sample was then burned in the electric stove. When there was no longer smoke come out from the stove, the sample was put into the incinerating furnace with a temperature of 600 ° C. After 7hours the cup was inserted in a desiccators and then was weighed. The ash content was determined by the formula of :

$$\text{Ash Content (\%)} = \frac{\text{Ash Weight}}{\text{Dried Samples Weight}} x100\% \quad (5)$$

3. Result and discussion

3.1. Moisture content
Moisture content data is presented in Table 1. Moisture content of wild and cultured eel in different parts of the body vary from 59.44% - 67.02%. Statistical data analysis shows that there was interaction between fish type and body part (P <0.05) on moisture content of eel. The highest moisture content is in the head of wild eel, but it was not significantly different with the head of cultured one. The lowest moisture content was in the tail of the wild eel.

Table 1. Moisture Content (%) of Wild and Cultured Eel in Different Part of Body

	Head	Body	Tail
Wild	67.02 ± 0.86^{a}	60.5 ± 0.36^{cd}	59.44 ± 0.82^{d}
Cultured	66.55 ± 1.16^{ab}	63.57 ± 0.59^{bc}	63.71 ± 1.51^{b}

Values with different superscript are significantly different (P<0.05)

Moisture content on the tail wild eel showed the lowest value. Other study reported that the moisture content of the tail (caudal) in the fish Hilsa (Tanualosa ilisha) was lowest compared to the dorsal and ventral[6], which is line with our finding.

The results of this study indicate the water content in the body and tail of cultured fish was higher than wild eel. The results of this study were similar to the results of Onyia et al [8] which showed that moisture content of cultured African catfish (Heterobranchus bidorsalis) was higher than wild one.

3.2. Protein content

Protein content of wild and cultured eel in different parts of the body are presented in Table 2. The results of data analysis indicate there are interaction between different types of fish and body parts.

Table 2. Protein Content (%) of Wild and Cultured Eel in Different Part of Body

	Head	Body	Tail
Wild	16.79 ± 0.48^{bc}	15.72 ± 0.28^{d}	15.75 ± 0.19^{d}
Cultured	16.07 ± 0.49^{cd}	18.09 ± 0.81^{a}	17.15 ± 1.54^{ab}

Values with different superscript are significantly different (P<0.05)

The highest protein content was found in body of cultured eel but not significantly different with protein content in the tail. The lowest protein content was in the head of cultured eel but not significantly different with head of wild eel. In general, protein content of cultured eel were higher than natural eel. Mahboob et al [7] reported that the protein content of wild Labeo rohita in various sizes were higher than in the cultured one.

Protein content of cultured fish were higher than wild one which is likely caused by higher protein intake of cultured eel than wild one. High protein feed will be converted to protein in fish meat. Seo et al reported that eel that fed with the highest protein content showed the highest protein levels in flesh[17].

Protein content in wild eel fish indicate that the head has the highest protein content, while the body and tail were not significantly different. Fish such as other animals exhibit bio-accumulation of specific nutrients in various organs of the body[19]. From this study, it has been observed that wild eel protein was more concentrated in the head than any other part of the body in this study. Kefas et al reported protein levels in the head of the fish *Synodontis clarias* and *Oreochromis niloticus* that live in reservoirs were higher than in body parts[19].

3.3. *Lipid content*

Lipid content of eel is presented in Table 3. Lipid content of wild and cultured eels in different sections showed a high value (above 5%). Lipid content in this study ranged from 10.36% to 20.36%. Data analysis showed no interaction between fish type and fish body part, but each factor had significant effect on lipid content. lipid content in wild eel was lower than in cultured one. Among the part of body showed the lowest lipid content was in the head while the body and tail were not significantly different.

Table 3. Lipid Content (%) of Wild and Cultured Eel in Different Part of Body

	Head	Body	Tail
Wild	12.22 ± 0.66^{Ab}	20.36 ± 1.75^{Aa}	19.29 ± 0.45^{Aa}
Cultured	10.36 ± 1.66^{Bb}	16.84 ± 1.66^{Ba}	17.15 ± 0.77^{Ba}

Superscript with different capital letter show significantly different between wild and cultured (P<0.05); superscript with small letter show significantly different among part of body (P<0.05)

Lipid content of wild and cultured eels were significantly different due to the scarcity of food in nature which results in reduced fish growth due to limited food supply to natural fish[7]. Onyia et al reported that lipid content of wild catfish was higher than cultured one[8], that similar with our finding.

The lipid content of the heads was lower than that in the body and tail. Kefas et al reported similar results that the lipid content in the head of Synodontis clarias and Oreochromis niloticus was lower than that of the body[19]. Lipid content content of wild and cultured eel in all parts of bady showed a high value (> 10%) indicating that this fish is an excellent source of fish oil. Fish fats, especially the content of unsaturated fatty acids are needed by humans[20].

The ash content of eel is presented in Table 4. Statistical analysis shows there were an interaction between fish type and fish part (P <0.05). The highest ash content was found in the head of the wild eel, while the lowest one was in the body of cultured eel but was not significantly different with the tail of the cultivated fish and the body part of wild one (P> 0.05).

Table 4. Ash Content (%) of Wild and Cultured Eel in Different Part of Body

	Head	Body	Tail
Wild	3.73 ± 0.28^{a}	1.44 ± 0.30^{d}	2.38 ± 0.17^{c}
Cultured	3.29 ± 0.11^{b}	1.32 ± 0.09^{d}	1.59 ± 0.21^{d}

Values with different superscript are significantly different (P<0.05)

The ash content in the head of wild eel showed the highest content, it is likely that the head contains many minerals in the form of calcium. The results of research by Kefas et al show the same trend that the ash content of fish head of Synodontis clarias and Oreochromis niloticus was higher than in body or flesh[19].

3.4. Carbohydrate content

Carbohydrate content of both wild and cultured are eel relatively small (0.23% -3.73%). Carbohydrate levels vary greatly (Table 5). The highest carbohydrate is achieved by cultured eel fish on the head, while the lowest one in the body of cultured eel.

Table 5. Carbohydrate Content (%) of Wild and Cultured Eel in Different Part of Body

	Head	Body	Tail
Wild	0.23 ± 0.29^{c}	1.89 ± 0.26^{c}	3.09 ± 0.17^{ab}
Cultured	3.73 ± 0.83^{a}	0.16 ± 0.23^{c}	0.88 ± 0.14^{bc}

Values with different superscript are significantly different (P<0.05)

3.5. Energy content

Energy content of wild and cultured eel in different parts of the body varies between 176.7 -258.98 Kcal/100g (Table 6). Data analysis showed no interaction between fish type and fish body part, but each factor had significant effect on the amount of energy. Energy in wild eel was higher than in cultured eel.

Table 6. Energy Content (Kcal/100g) of Wild and Cultured Eel in Different Part of Body

	Head	Body	Tail
Wild	183.3 ± 4.8^{Ab}	258.98 ± 10.59^{Aa}	253.84 ± 2.02^{Aa}
Cultured	176.73 ± 6.29^{Bb}	230.39 ± 11.56^{Ba}	226.15 ± 8.97^{Ba}

Superscript with different capital letter show significantly different between wild and cultured (P<0.05); superscript with small letter show significantly different among part of body (P<0.05)

The energy contained in wild eel fish was higher than cultured one. It is closely related to water content in wild eel fish which was lower than cultured fish. In the eel's body also showed higher energy than the head. It is also closely related to moisture content of the body which were lower than in the head. Ball et al suggests that increased energy density in fish may be due to reduced water supply [21].

4. Conclusion

There was interaction between type of fish and body part on moisture, protein, ash and carbohydrate content (P <0,05), but no interaction on fat content and energy (P> 0,05). The body of cultured eel had higher protein than wild one. The wild eel had higher fat content and energy than cultured one, while the fat content and energy in body and tail were higher than in head. Our research found that different body part of wild and cultured eel had different proximate content.

5. Acknowledgment

Authors acknowledge the Diponegoro University for funding this research through Beginner Lecturer Research (Riset Dosen Pemula) Scheme for fiscal year of 2016

6. References

[1] Fawole O O, Ogundiran M A, Ayandiran T A and Olagunju O F 2007 *Internet J. of Food Safety*, **9**: 52-55
[2] Marichamy G, M A Badhul Haq, R Vignesh, V Sedhuraman and AR Nazar 2012 *International Journal Of Pharma And Bio Sciences* **3** (2): 54-64
[3] Alemu L A, Melese A Y and Gulelat D H 2013 *American Journal of Research Communication* **1**(11): 405-410
[4] Dewan BK, Mia MS, Yeasmin F, Sarker SC, Siddky MNSM and Kamal M 2015 *International J. of Natural and Social Sciences*, **2**(5): 52-55
[5] Kailasam S, K Immaculate Jeyasantha, Hermina Giftson and Jamila Patterson 2012 *Sky J. of Food Science* **4**(7), pp. 108 – 115
[6] Shamim AH MD, Ahmed K MD and Abdullah A BTM 2011 *Dhaka Univ. J. Biol. Sci.*, **20**(2): 109-115
[7] Mahboob S, Fauzia Liaquat, Sadia Liaquat, Muhammad Hassan and Muhammad Rafique 2004 *Pakistan J. Zool.*, **36**(1), pp. 39-43
[8] Onyia L U, Michael KS, Manu, J.M. and Sabo, M. 2011 *Nigerian J. of Fisheries and Aquaculture* **1**(1): 7 – 12
[9] Jalili S, Navid Zoriasatein, Foad Pour 2014 *International J. of Ecosystem* **4**(3): 119-123
[10] Zafar, A. and M. Ashraf 2011 *Int. J. Agric. Biol.* **13**: 701–706
[11] Boran G and Karaçam H 2011 *Turkish J. of Fisheries and Aquatic Science* **11**: 1-5
[12] SusmitaDey and Goswami U.C. 2016 *IOSR J. of Dental and Medical Sciences (IOSR-JDMS)* **15**(2): 75-80
[13] Ashwini. L, S. Benakappa, H.N. Anjanayappa, Akshay. L. 2016 *International J. of Engineering Science and Computing* **6**(6): 7400-7403
[14] Yildiz M, Erdal Sener and Metin Timur 2006 *Turkish J. of Fisheries and Aquatic Sciences* **6**: 99-104
[15] Bilguven M anf Ibrahim AK 2013 *Proc. Annual Int. Interdisciplinary Conf. AIIC* (Azores, Portugal) p 807
[16] Iqbal K J, M Ashraf, A Javid, N A Qureshi, N Khan, H Azmat, M Altaf, H Majeed, M S Haider, W Shehzad I and M Y Zahoor 2015 *The J. of Animal & Plant Sciences,* **25** (3 Supp. 2): 514-518
[17] Seo Jae-Sung, Jae-Hee Choi, Ji-Hun Seo, Tae-Ho Ahn, Won-Seog Chong, Seung-Hoon Kim, Hye-Sun Cho and Jun-Cheul Ahn. 2013 *Fish Aquatic Science* **16**(2), 85-92
[18] [AOAC] Association of Official Analytical Chemist 2005 *Official Methods of Analysis of The Association of Official Analytical Chemist* (Arlington, Virginia USA : AOAC Inc)
[19] Kefas, M., Michael, K.G Abubakar, K.A, Edward, A. and Wahide, J.A. 2014 *Global J. of Biology, Agricultural and Health Science,* **3**(3):116-121
[20] Oyebanmiji, O.F; Fagbohun,T.R and Olubanjo,O.O 2008 *Turkish J. of fisheries and Aquatic sciences* **8**: 7-13
[21] Ball JR, Daniel Esler, Joel A Schmutz 2007 *Polar Biol.* **30**:699–708

Determination of Soft Lithology Causes The Land Subsidence in Coastal Semarang City by Resistivity Methods

Sugeng Widada[1], Sidhi Saputra[1], Hariadi[1]
[1]Department of Oceanography, Faculty of Fisheries and Marine Sciences, Diponegoro University, Jl. Prof. H. Soedarto, SH, Tembalang, Semarang 50275 , Indonesia

E-mail: swidada1@gmail.com

Abstract. Semarang City is located in the northern coastal plain of Java which is geologically composed of alluvial deposits. The process of the sediment diagenesis has caused a land subsidence. On the other hand, the development of the industrial, service, education and housing sectors has increased the number of building significantly. The number of building makes the pressure of land surface increased, and finally, this also increased the rate of land subsidence. The drilling data indicates that not all layers of lithology are soft layers supporting the land subsidence. However, vertical distribution of the soft layer is still unclear. This study used Resistivity method to map out the soft zone layers of lithology. Schlumberger electrode configuration with sounding system method was selected to find a good vertical resolution and maximum depth. The results showed that the lithology layer with resistivity less than 3 ohm is a layer of clay and sandy clay that has the low bearing capacity so easily compressed by pressure load. A high land subsidence is happening in the thick soft layer. The thickness of that layer is smaller toward the direction of avoiding the beach. The improvement of the bearing capacity of this layer is expected to be a solution to the problem of land subsidence.
Keywords: Land subsidence, Semarang, Resistivity

1. Introduction

Semarang City which is located in Central Java Province is one of the cities with many problems regarding with the process of land subsidence. As the effect of the land subsidence, some areas have been in below the sea level so it will be flooded with coastal flooding (rob) [1]. It results in disturbing the activity of the people and several infrastructures facing a fast defect. The handling of this situation has been managed to be conducted by arranging the function of the area by paying attention to all behaviors of land subsidence, but it is not completely succeed yet.

The land subsidence rate that happens in Semarang is varied between one place and another, and the biggest one reached 16 cm/year [1]. The measurement of land subsidence in Semarang City based on *Levelling, Interferometric Synthetic Aperture Radar (InSAR)* methods, *Microgravity* and survey using *Global Positioning System (GPS)* method, showed that the rate of land subsidence in Semarang City is more than 19 cm/year from 1999 to 2011 [2]. That rate of land subsidence was affected by the process of alluvial consolidation and added by the exploitation of overflowing ground water and the building construction in some points [3],[4]. Although there area fault geological structure in Semarang City [5] [6]. Tectonic activity is not related to the land subsidence so that the bedrock does not decrease [7]. The effect of land subsidence is predicted to cause the flooded area increases from 2,162.5 Ha (5.6%) to 3,896.3 Ha (10.1%) in the next ten years [8]. Even though the previous research had stated some factors of land subsidence, it was not yet explained the type and thickness of lithology layer which is contributing in the land subsidence.

This research aimed to map the spread of lithology with soft and asunder characteristics, so it is consolidated which resulting in the land subsidence by identifying the soft lithology layer and mapping the spread spatially according to the data of resistivity measurement.

2. Methods

Resistivity method is one of the geophysical methods that uses resistivity as characteristic of rock to observe subsurface of earth condition [9]. Measurement is conducted on the land surface by injecting electricity current (in milliamperes) along two current electrodes (C1 and C2) then gaining potential difference (in millivolts) from two potential electrodes (P1 and P2). There is a positive connection between the rare electrode with the current penetration into the earth. The widest electrodes space could get a deep current penetration. Hence the physical characteristics of deeper rock layer can be analyzed [10]. The location points of resistivity measurement can be seen in Figure 1.

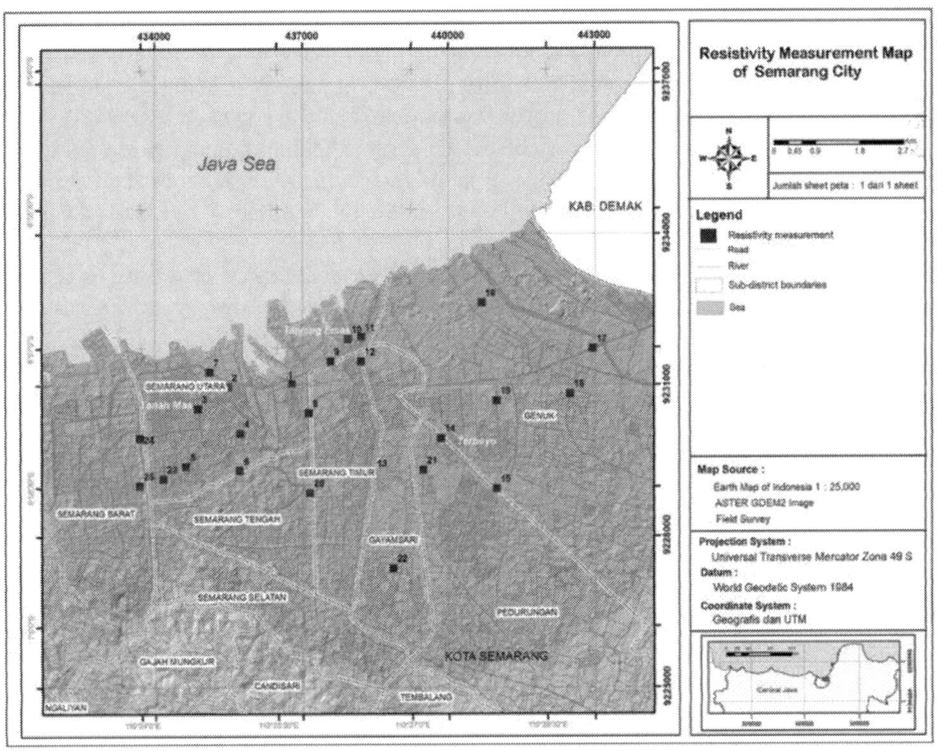

Figure 1. Resistivity measurement map

In the field implementation using Schlumberger configuration [9]. The distance of the electrode is arranged so R1 = R4 = (b – ½ a) and R2 = R3 = (b + ½ a), where a is the distance between the two potential electrodes and b is the distance from the center point to the current electrode (Figure 2). In this state, the range of potential electrode P1 – P2 is started from 1/3 of a range of the current electrode C1 – C2. Furthermore, the measurement was done by only moving the current electrode to a distance where the measurement result of the potential difference P1 – P2 is already small, then P1 – P1 is extended step by step.

IOP Conf. Series: Earth and Environmental Science 116 (2018) 012092 doi:10.1088/1755-1315/116/1/012092

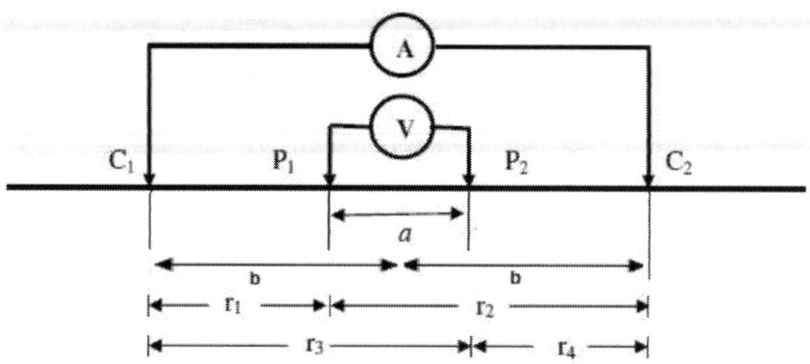

Figure 2. The scheme of the electrode arrangement of Shclumberger configuration [11].

The field data that were obtained from the resistivity measurement were the distance of current electrode, a distance of potential electrode, current (i), and the potential difference (ΔV). The resistivity of the medium is calculated from the quotient of the $\Delta V/I$ using a modified Ohm's law. Based on the data, the apparent resistivity was measured for every current injection, so the value of apparent resistivity was obtained for every electrode spread. The graphic of the spacing electrode and the apparent resistivity then were used as the basic in determining the thickness and the true resistivity by using the Ip2Win software. The results found were the thickness and the resistivity of every lithology layer [9].

The resistivity of every lithology layer that was obtained in the data processing has then interpreted the type of lithology by referring to the resistivity of lithology type from the references and correlated to the type of lithology based on the geotechnical drilling data. In the geotechnical drilling data, besides the description of the type of lithology, also listed the physical characteristics of lithology related to its ability to accept the burden so that it could be interpreted the behaviors of lithology related to the land subsidence. From the interpretation result in every point of resistivity measurement, a contour map was made to depict the position of soft lithology layer causing the land subsidence in the research area.

3. Result and Discussion

The result of resistivity measurement showed that there were several types and thickness of lithology in Semarang City. According to the correlation of the drilling data (Figure 3), it was found that the soft lithology had a resistivity < 3 ohm, which was layers of clay and sandy clay. That lithology was soft to medium consistency as a part of coastal alluvial deposit in forms of clay, silt, sand, and granule [12]. The layer of clay will be prone to have smaller resistivity than the sand, especially if it contains water [13]. The thickness of that layer in each resistivity measurement point pairing with the data of land subsidence measurement [4] was presented in Table 1.

The map of land subsidence overlayed by the map of soft layer thickness with the resistivity value less than 3 ohms is presented in Figure 4. Based on the figure, it can be seen that the rate of land subsidence in the location around Tanah Mas, Tanjung Emas Port, and Terboyo are higher than another areas and generally the farther from the coast, the rate of land subsidence is getting the smaller. The same pattern can also be found on the map of soft layer thickness, which is the highest thickness in those three locationss and decreasing to the south avoiding the coast.

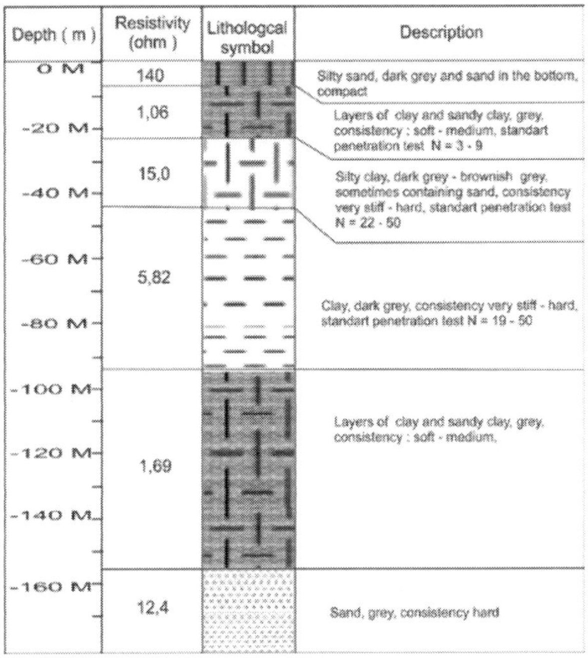

Figure 3. The correlation of the resistivity data and the drilling data in point 09

Table 1. The thickness of soft lithology layer based on the resistivity data and the rate of land subsidence in Semarang City

Point No.	East	North	Lithology thickness with $\rho < 3$ Ohm	Rate of land subsidence (cm/year)
1	433697	9229943	67.15	5.77
2	435485	9230954	71.98	3.50
3	434868	9230527	109.00	5.28
4	435731	9230038	54.75	8.51
5	434629	9229383	70.90	10.69
6	435719	9229303	21.08	0.47
7	435105	9231261	74.19	9.90
8	437136	9230440	68.53	3.50
9	437583	9231474	82.30	10.45
10	437942	9231918	49.10	3.50
11	438217	9231957	41.98	9.80
12	438215	9231470	47.62	10.45
13	438464	9229248	80.81	7.00
14	439857	9229941	39.50	7.00
15	440994	9228939	26.80	5.09
16	442496	9230815	61.80	5.09
17	442960	9231722	34.70	8.81
18	440693	9232628	79.80	8.58
19	440992	9230686	50.40	5.76
20	437161	9228864	7.87	4.02
21	439486	9229312	20.80	7.00
22	438870	9227356	21.69	5.09
23	434170	9229137	52.10	10.69
24	433697	9229943	24.70	9.90
25	433697	9229000	Not recorded	0.50

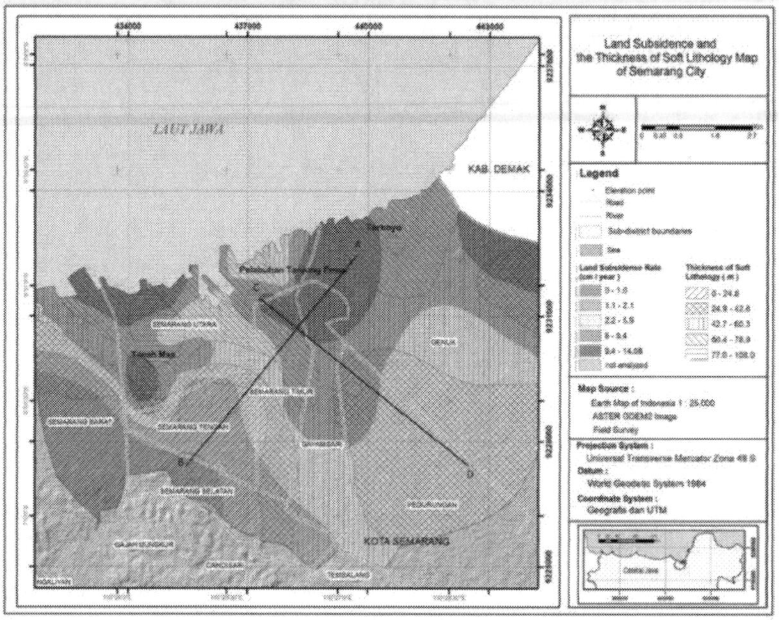

Figure 4. Map of the land subsidence and the thickness of soft lithology in Semarang City

That figure was also showed that the land subsidence was about 1-14.08 cm/year. The location around Tanjung Emas Port and Tanah Mas in North Semarang Sub-district had the higher land subsidence than the other areas at 9.4 - 10.08 cm/year. In that location, it was found that the lithology layer thickness with the resistivity < 3 ohm was also thicker than another area at 60-108 m in Tanah Mas and 42 – 108 m in around Tanjung Mas Port. In the point 23 of resistivity measurement located in Krobokan with the land subsidence of 10.69 cm/year also had a lithology layer thickness with the resistivity of < 3 ohms more than 60 m. The location with a slightly higher rate of land subsidence was also found in around Terboyo, at 8.58 cm/year, and it had a lithology with resistivity < 3 ohms for 79.8 m. The spread of soft layer can be found in all research areas with the thickness varied to 108 m.

The slice by the AB and CD points as shown in Figure 5 showed that the higher rate of land subsidence, the thicker the layer of soft lithology.

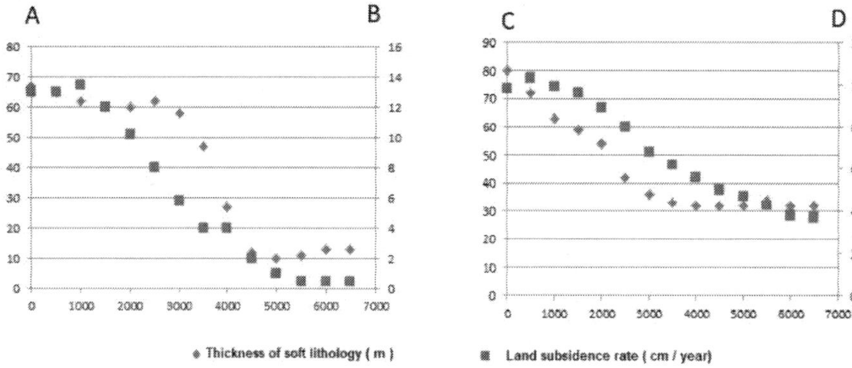

Figure 5. Graphic of land subsidence and the thickness of soft lithology layer

A high land subsidence is happening in the thick, soft layer. It happened because that soft layer is compressed when given the load from the above layer or several buildings and infrastructures above the ground [2], [4] and [15]. The consolidation process in the alluvial deposit is natural and surely happen (13). The core drilling data illustrates that on layers of clay and sandy clay there is the presence of sand entering the clay, so the compression became higher. It made the rate of land subsidence higher in the area of clay and sandy clay thick.

4. Conclusion

The rate of land subsidence in Semarang City is getting higher in the location with the thickness of soft lithology layer detected as the lithology with resistivity < 3 ohms. The thickness of that layer is smaller toward the direction of avoiding the coast. The thickness of the soft layer in Tanah Mas, Tanjung Emas Port and the area around Terboyo were about 42 – 108 m, and the rate of land subsidence was about 9.4 - 10.08, higher than other areas. The consolation process and inclusion of sand into soft clay layers beneath became the cause of the land subsidence.

5. Acknowledgement

We thankfully appreciate for any supports from Department of Oceanography Diponegoro University's members who helped us in data measurement, processing and discussion. Research is supported by PNBP Faculty of Fisheries and Marine Science, Diponegoro University, fiscal year 2017.

References

[1] Marfai M A and King L 2007 Environ **Geol 53** 651–659
[2] Gumilar I, Abidin H Z, Sidiq T P, Andreas H, Maiyudi R, Gamal M, and Fukuda Y 2013 *Indonesian Journal of Geospatial* Vol. **2** Number. **2** 26 – 41
[3] Abidin H Z, Andreas H, Gumilar I, Sidiq T., Gamal M, Murdohardono D, Supriyadi, and Fukuda Y 2010 *Proceeding FIG Congress* 1 – 15
[4] Abidin H Z, Andreas H, Gumilar I, Sidiq T P, and Fukuda Y 2012 **Geomatics, Natural Hazards and Risk 3** Vol **4**, Number **3** 226 - 240
[5] Yulianto T, Gernowo R, and Widada S 2017 *Advanced Science Letters* **Vol. 23**, 6635–6637
[6] Nurwidyanto M I, Yulianto T, Widada S 2017 *Advanced Science Letters* **Vol. 23**, 6623-6626
[7] Wardhana D D, Harjono H, and Sudaryanto 2014 *Riset Geologi dan Pertambangan* Vol. **24** No. **1** 53 – 64.
[8] Suhelmi 2012 *Jurnal Ilmiah Geomatika* Vol. **18** No. **1** 9 – 16
[9] Widada S, Satriadi A, Rochaddi B 2017 *Jurnal Kelautan Tropis* Vol. **20** No. **1** 35 - 41
[10] Setyawan A, Najib, Aribowo Y, Trihadini A, Hastuti D, Ramdhani F, Waskito F, Febrika G, and Virgiawan G 2017 IOP Conf. Series: Earth and Environmental Science **55** 012053.
[11] Dobrin M B, 1976 *Introduction to Geophysical Prospecting*. Mc Graw-Hill, New York
[12] Thanden R E, Sumadirdja, Richards P W, Sutisna K, and Amin T C 1996 *Geological Map Of The Magelang And Semarang Sheets Jawa*. Second Edition (Bandung: Geological Research and Development Centre)
[13] Yulianto T, Gernowo R, and Krisna W 2016 *International Journal of Applied Environmental Sciences* Vol. **11**, Number **6** 1363-1373
[14] Reddish D J and Whittaker B N 1989 *Subsidence* Vol. 57 1st Ed. Elsevier Science.
[15] Lubis M A, Sato T, Tomiyama N, Isezaki N, and Yamanokuchi T 2011 *Journal of Asian Earth Science* **40** 1079 - 1088

The Preliminary Study of Organochlorine Pesticide Residues on Sediments of Bivalvia Fishing Ground at Eastern Part of Coastal Semarang

Chrisna Adhi Suryono[1]*, Subagyo[1], Wilis Ari Setyati[1], Endang Sri Susilo[1], Baskoro Rochaddi[2] and Robertus Triaji Mahendrajaya[1]
[1]Department of Marine Science, Faculty of Fisheries and Marine Science, Diponegoro University, Jl. Prof. H. Soedarto,SH, Tembalang, Semarang 50275, Indonesia
[2]Department of Oceanography, Faculty of Fisheries and Marine Science, Diponegoro University, Jl. Prof. H. Soedarto,SH, Tembalang, Semarang 50275, Indonesia
Email : chrisna_as@yahoo.com

Abstract. This paper presents the occurrence level of organochlorine contamination in marine sediments of Semarang coastal areas as a fishing ground of Bivalvia. Five compounds (Heptachlor, Aldrin, Endosulfan, Endrin and pp-DDT) of contaminant have been determined in the sediments surface of Semarang coastal waters. The samples were then analyzed by using gas chromatography and followed by using the method of Standard Method Examination. The result showed that the average concentration of organochlorine pesticides Heptachlor, Aldrin, Endosulfan, Endrin and pp-DDT were 25.5, bellow detected, 7.1, 37.2, 28.6 μ g/L, respectively. The high concentration of pesticide showed on Endrin (65,84 ppb), pp-DDT (29,53 μ g/L), and heptachlor (25,45 μ g/L). The low concentrations were detected on aldrin (bellow detected) and endosulfan (0,92 μ g/L). The concentration of organochlorine pesticides in these areas might contribute on four rivers which deposited the sediment in these coastal waters

1. Introduction

Persistent organochlorine compounds including organochlorine pesticides (OCPs) and polychlorinated biphenyls (PCBs) were widely used in industry and agriculture. There are two kinds for OCPs, one is made of benzene including hexachlorocyclohexane (HCH) and dichlorodiphenyltrichloroethane (DDT), and another one is cyclodiene including heptachlor, aldrin, dieldrin, and endrin. PCBs were produced as dielectric and hydraulic fluids on an industrial scale since 1929, with particularly large volumes made in the 1950s–1970s [1]. Persistent organochlorine compounds are thought as an important component of xenobiotic chemicals. Their residues were causing the hazard to the environment and had the potential to elicit endocrine disruption in biota [2];[3]. Although organochlorine pesticides (OCPs) such as DDT, HCH, and Aldrin were banned in developed countries over 25 years ago, their continued use has been reported in several Asian countries [4]. The use of such pesticides in tropical countries is often justified for the eradication of disease vectors such as the malaria-carrying mosquito. OCPs are still used in large quantities especially in developing countries for the control of agricultural pests [5]. The bans of PCBs and OCPs were imposed in many countries after their toxicity, and environmental persistence was realized [6]. Due to high toxicity, long environmental half-lives and long scale transportation ability, they are of significant concern and have been studied extensively in the last decades [7]. Organochlorine compounds were detected in many environmental media such as water [8]; [3], sediment [3]; [7], air [9], biota [10] and in bivalve [11]. Organochlorine pesticides are one of the important components of chemical pollutants present in the marine environment. They are toxic to all living organisms [11]. Organochlorine pesticide enters into the marine environment through different sources, more particularly from estuaries [12]. Estuarine and marine sediments are the ultimate global sinks for most of the pollutants. Contamination of soil and sediments of coastal areas with persistent organochlorine compounds may be related to point sources, e.g., industrial discharges and domestic sewage, but more frequently, as in the case of pesticide pollution, it can also be attributed to precipitation, agricultural runoff, etc. [3]. Therefore, the study on organochlorine pesticides in sediment which is the bivalve habitat in Semarang coastal areas is very important to be conducted to discover how far the contamination level of those materials in the sediment. Because in the sediment, the bivalve gets the food by filtering the sediment.

2. Materials And Methods

2.1. Samples collection

Sampling points were located on coastal waters areas at the east part of Semarang. Figure 1 shows the sampling sites, the samples of sediments were collected in polyethylene plastic bags bottles from different stations (4 stations). The samples were collected in 3 kg plastic bags. Then, plastic bags were labeled properly and sealed tightly. All the samples were brought to the laboratory for the organochlorine pesticide to be analyzed. The parameters analyzed were psycho-hydrochemical properties of marine waters and organochlorine contents of sediments. The salinity, total dissolved solids, pH, dissolved oxygen, and temperature were measured at the time of samples collection using water quality checker (Horiba Co. Ltd., Japan)

2.2. Sample analysis

The analytical procedure applied was the method of Standard Method [13]. The analytical procedure of Organochlorine Pesticide (OCPs) in sediment samples was modified according to the method described previously [14]; [15]. 15 g of freeze-dried and homogenized sediment sample was placed into a Soxhlet apparatus and extracted with 250 ml of hexane-dichloromethane in n-hexane and capillary columns. One-liter samples were extracted with a solvent mixture and then concentrated in a Kuderna-Danish apparatus. The extracts were cleaned up with Florisil column. The final pesticide extract was obtained in 5 mL of hexane. The samples were then analyzed by using gas chromatography Model Hitachi 163 FPD (Flame Photometer Detector), and nitrogen High Pure (HP) was used as the carrier gas. A 2 in a glass column (3 mm ID) packed with 3% Silicon OV1 on 80-100 mesh Supelcon was used. Gas flow at 30 mL min', column temperature at 160-230°C, detector temperature at 290°C and the injector temperature at 290°C were maintained.

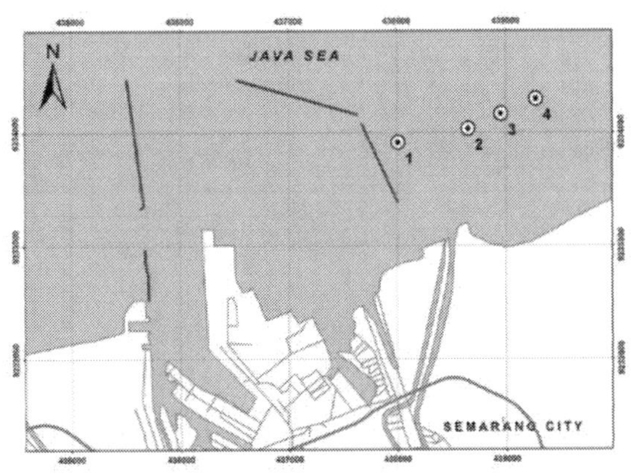

Figure 1. The sampling areas at the eastern part of Semarang Coastal waters

3. Results And Discussion

The eastern part of Semarang coastal waters is a fishing ground area for several biotas such as fish, shrimp, crab, and bivalve. The high activities of fishermen in this area are because in this area the waters is very fertile, some rivers are flowing through this area such as East Banjir Kanal River, Banger, Babon and Sri Mulyo which all of them are flowing the nutrient and pollutant whether in solid, suspension and solution [16]. The high production of fisheries in that area is supported by a good quality of water so it can support the life of sea organisms [16]. It is also proved by the research result on recent water quality that can be seen in table 1. From the table, it can be seen that the water quality can support the life of sea organisms to live and grow. The range of salinity between (31 – 33 ppt), total dissolved solid between (30.5 – 31.5 mg/L), pH between (7.3 – 7.5), dissolved oxygen (5.6 – 6.4 mg/L) and temperature (29.5 – 29.9 °C).

Table 1. Physico-chemical properties of water samples of marine waters in the sampling areas (Eastern part of Semarang coastal waters)

Water Quality	ST 1	ST 2	ST 3	ST 4
Salinity (ppt)	33	33	31.5	31
Total Dissolved Solid (mg/ L)	31.5	30.5	31.5	31
pH	7.3	7.3	7.5	7.3
DO (mg/L)	5.9	5.9	6.4	5.62
Temperature (°C)	29.8	29.5	29.8	29.9

The inclusion of nutrient in the waters is also followed by the inclusion of pollutant such as heavy metal, pesticide, bacteria, etc. Several previous researches conducted in the Semarang waters showed that Organochlorine Pesticides such as Heptachlor, Aldrin, Endosulfan, Endrin, and pp-DDT were found in the coastal waters in the West part of Semarang [17]. The research of organochlorine pesticides accumulation in the sea sediment had also been conducted in many countries like India [3], China [15], Singapore [7], Korea [18]. From all of those countries, organochlorine pesticides were found in sediment.

The recent research result by taking the sediment samples in the bivalve fishing ground areas in the east part of Semarang showed that the sediment in that area was found the concentrations of heptachlor, endosulfan, endrin, and pp-DDT exclude aldrin. The range of pesticides residues concentration from the highest to the lowest as followed endrin (8.55 – 65.84 μ g/L) detected in two samples, then pp-DDT (27.66 – 50 μ g/L) detected in two samples, heptachlor (25 μ g/L) only detected in one sample, Endosulfan (0.92 -13.18 μ g/L) detected in three samples while Aldrin on the four samples showed the number of below detection. The result can be seen in Table 2 and Figure 2.

Table 2. Range, frequency of occurrence and mean ± standard deviation of organochlorine pesticide residue levels (μg/L) from bivalvia fishing ground at eastern part of coastal Semarang

Sediment	Heptachlor	Aldrin	Endosulfan	Endrin	pp-DDT
ST 1	*bd*	*bd*	1.56	*bd*	*bd*
ST 2	25.45	*bd*	13.18	8.55	bd
ST 3	*bd*	*bd*	0.92	65.84	27.66
ST 4	*bd*	*bd*	*bd*	*bd*	29.53
% detection	25	0	75	50	50
Range	25	0	0.92 -13.18	8.55 - 65.84	27.66 - 50
Mean + SD	25.45	0	5.22 ± 6.9	37.2 ± 40.51	28.59 ± 1.32

Bd = below detection

Evens until now in Indonesia there are no concentration criteria of organochlorine pesticide in the sea sediment for the feasibility of living organisms but OCPs is a persistent material and can be accumulated in the tissues of marine organisms. Moreover, the bivalve from cockles type is normally found in muddy areas near the river mouth [19]. They are sedentary filter feeder, taking in food particulates through their gills when their shells are open [20]. Besides, the sediment in the research location comes from several river currents which flowing through agricultural, residential and industrial areas estimated to still using OCPs. It is estimated later that the higher the OCPs concentration in sediment will also be followed by the increase of the material in bivalve tissue.

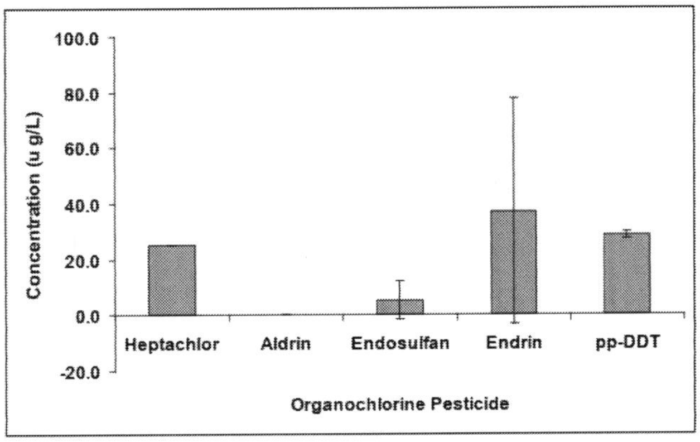

Figure 2. Mean ± SD of occurrence of organochlorine pesticide residue levels (μ g/L) from bivalvia fishing ground at eastern part of coastal Semarang

4. Conclusion

Some kinds of organochlorine pesticide have been contaminating sediment in the Semarang coastal area which is a fishing ground of bivalve. The OCPs including Heptachlor, Endosulfan, Endrin, and pp-DDT. That was the high concentration of pesticide showed on Endrin (65,84 ppb), pp-DDT (29,53 μ g/L), and heptachlor (25,45 μ g/L). Futher more the low concentrations were detected on aldrin (bellow detected) and endosulfan (0,92 μ g/L).

5. Acknowledgment

This article is part of the research which supported by the grant from PNBP 2017 Faculty of Fisheries and Marine Sciences Diponegoro University.

6. References

[1] Wang H, He M, Lin C, Quan X, Guo W, Yang Z. *Environ Monit Assess.* 2007; **133** (1):231–42.
[2] Basheer C, Obbard JP, Lee HK. *J. Chromatogr A.* 2005; **1068** (2):221–8.
[3] Rajendran RB, Imagawa T, Tao H, Ramesh R. *Environ. Int.* 2005; **31** (4):503–12.
[4] Kannan K, Tanabe S, Giesy JP, Tatsukawa R. *Springer*; 1997. p. 1–55.
[5] Pandit GG, Rao AMM, Jha SK, Krishnamoorthy TM, Kale SP, Raghu K. *Chemosphere.* 2001; **44** (2):301–5.
[6] Feng K, Yu BY, Ge DM, Wong MH, Wang XC, Cao ZH. *Chemosphere.* 2003; **50** (6):683–7.
[7] Wurl O, Obbard JP. *Chemosphere.* 2005;58(7):925–33.
[8] Booij K, Hoedemaker JR, Bakker JF. *Environ. Sci. Technol.* 2003; **37** (18):4213–20.
[9] Jaward FM, Di Guardo A, Nizzetto L, Cassani C, Raffaele F, Ferretti R. *Environ. Sci. Technol.* 2005; **39** (10):3455–63.
[10] Carlson DL, Hites RA. *Environ Sci Technol.* 2005; **39** (19):7389–95.
[11] Sundar G, Selvarani J, Gopalakrishnan S, Ramachandran S. *Environ. Moni.t Assess.* 2010; **160** (1):593–604.
[12] David B V. 1992. pp. 225–50.
[13] Association APH. 1992.15.
[14] Doong R-A, Peng C-K, Sun Y-C, Liao P-L. Taiwan. *Mar Pollut Bull.* 2002; **45** (1):246–53.
[15] Yang R, Lv A, Shi J, Jiang G. *Chemosphere.* 2005; **61** (3):347–54.
[16] Suryono CA, Rochaddi B. *Jurnal Kelautan Tropis* 2017; **20** (1).
[17] Suryono CA, Rochaddi B, Irwani I. *Bul. Ose. Mar.* 2017; **5** (2):101–6.

[18] Hong SH, Yim UH, Shim WJ, Oh JR, Lee IS. *Mar Pollut Bull.* 2003; **46** (2):244–53.
[19] Pathansali D. *L Malay. Agric. J.* 1963; **44**:18–25.
[20] McCoy EW, Tuaycharoen S, Vakily JM, Boonchuwong P. ICLARM. 1988.

Hierarchical Synthesis of Coastal Ecosystem Health Indicators at Karimunjawa National Marine Park

Johan Danu Prasetya [1,2], Ambariyanto [2,3], Supriharyono [2], Frida Purwanti [2]

[1]Environmental Engineering, the Universitas Pembangunan Nasional Veteran Yogyakarta. 55283, Indonesia.
[2]Faculty of Fisheries and Marine Sciences, Diponegoro University, Semarang 50275, Indonesia
[3]Coastal Disaster Rehabilitation and Mitigation Center, Diponegoro University, Semarang 50275, Indonesia.

Email: johan.danu@upnyk.ac.id.

Abstract. The coastal ecosystem of Karimunjawa National Marine Park (KNMP) is facing various pressures, including from human activity. Monitoring the health condition of coastal ecosystems periodically is needed as an evaluation of the ecosystem condition. Systematic and consistent indicators are needed in monitoring of coastal ecosystem health. This paper presents hierarchical synthesis of coastal ecosystem health indicators using Analytic Hierarchy Process (AHP) method. Hierarchical synthesis is obtained from process of weighting by paired comparison based on expert judgments. The variables of coastal ecosystem health indicators in this synthesis consist of 3 level of variable, i.e. main variable, sub-variable and operational variable. As a result of assessment, coastal ecosystem health indicators consist of 3 main variables, i.e. State of Ecosystem, Pressure and Management. Main variables State of Ecosystem and Management obtain the same value i.e. 0.400, while Pressure value was 0.200. Each main variable consist of several sub-variable, i.e. coral reef, reef fish, mangrove and seagrass for State of Ecosystem; fisheries and marine tourism activity for Pressure; planning and regulation, institutional and also infrastructure and financing for Management. The highest value of sub-variable of main variable State of Ecosystem, Pressure and Management were coral reef (0.186); marine tourism pressure (0.133) and institutional (0.171), respectively. The highest value of operational variable of main variable State of Ecosystem, Pressure and Management were percent of coral cover (0.058), marine tourism pressure (0.133) and presence of zonation plan, regulation also socialization of monitoring program (0.53), respectively. Potential pressure from marine tourism activity is the variable that most affect the health of the ecosystem. The results of this research suggest that there is a need to develop stronger conservation strategies to facing with pressures from marine tourism activities.

Keywords: Ecosystem, Conservation, Karimunjawa, Indicator, AHP

1. Introduction

Indonesia is the world's largest archipelagic state. All provinces and more than 80% of the districts and municipalities have coastal areas [1]. Indonesia is also known as the center of marine mega-biodiversity [2], [3]. In Indonesia, there are about 590 coral species consist of 82 genera [4]. Indonesia has coral reef area with an estimated area of about 51,000 km^2, which is the largest coral reef area in Southeast Asia. Some 15 plant families, with 18 genera and 41 species, and 116 associated species are found in the mangrove ecosystems in Indonesia. Indonesian mangrove forest covers an area of 35.337 km^2, which is 76% of the total mangroves in the Southeast Asia. There are also 13 species of seagrass, covering an area 30.000 km^2 [1].

Increasing threats to the Indonesian coastal ecosystem was widely reported and contributes to the degradation of ecosystem and resources [5]. Threats are mainly occur on mangroves, seagrass and also coral reefs ecosystem. Mangrove facing various threats, large areas of mangrove have been converted for aquaculture [6]; timber extraction and the expansion of urban areas [7] and also oil palm plantations [8]. Seagrass in Indonesia also have been degraded, about 30–40% of the seagrass beds have been lost in the last 50 years, with as much as 60% being destroyed around Java [9]. Among Coral Triangle Countries, Indonesian coral reef was the most at risk. For about 50% of these reefs, the level of threat is high or very high. Only about 12% is low risk [1].

Healthy coastal ecosystem provides maximum benefits for biodiversity. Maintaining ecosystem in a healthy condition is crucial to maintain biodiversity [10]. Marine Protected Area (MPA) is one of the main strategy to maintain ecosystem in a healthy condition and conserve biodiversity. MPA play fundamental roles in the conservation of biodiversity [11]. Effectiveness conservation management based on MPA strategy need to be evaluated continuously, based on ecosystem condition. Ecosystem monitoring is one of the methods to providing information about effectiveness conservation management. Monitoring is a fundamental part of resource management [12]. Monitoring the condition of the ecosystems in MPA periodically to ensure ecosystem in a healthy condition [10]. A set of systematic and consistent indicators in monitoring program, need to be investigated to indicate the real condition of ecosystem. Present study investigates a set of indicators namely Coastal Ecosystem Health Indicators (CEHI) and perform indicators weighting using hierarchical synthesis. Selected indicator included consistent weighting use to evaluate coastal ecosystem health of Karimunjawa National Marine Park (KNMP).

2. Method

Coastal Ecosystem Health Indicators (CEHI) were identified and selected based on field observation and reviewing reference. The weights of importance of the variables on the health of coastal ecosystems were determined through the Analytic Hierarchy Process (AHP) developed by Thomas L. Saaty. The AHP is a basic approach to decision making [13]. The indicators list that consist of Main variables, Sub-variables and Operational variables arranged into a decision tree using the AHP. Four steps are conducted in using the AHP i.e. setting up the decision hierarchy; collecting input data by pairwise comparisons of decision elements; using the eigenvalue method to estimate the relative weights of decision elements and aggregating the relative weights of decision elements [14]. Expert judgments were organized to conduct pairwise comparisons of strength each variables compared other variables. The expert judgments were divided into representative groups namely, coastal manager, local government, academics and non-government organization.

Three main variables were identified namely, State of Ecosystem, Pressure and Management. Each Main variables consist of several Sub-variables, each Sub-variables divided into some Operational variables. The strength of preference is evaluated through a 9-point intensity scale (see table 1). The application of AHP considers consistency in judgment through consistency ratio (CR). Consistency indicates that: if $a > b$; $b > c$ then $a > c$ [15]. The value of CR ranges from 0 to 1, a CR of 0.10 or less is considered acceptable [13].

Table 1. Intensity scale of importance [13]

Intensity of Importance	Definition	Explanation
1	Equal Importance	Two Activities contribute equally to the objective
3	Weak importance of one over another	Experience and judgment slightly favor one indicator over another
5	Essential or strong importance	Experience and judgment strongly favor one criterion/indicator over another A
7	Demonstrated importance Absolute importance	A indicator is strongly favored and its dominance is demonstrated in practice
9	Absolute importance	The evidence favoring one indicator over another is of the highest possible order of affirmation
2, 4, 6, 8	Intermediate values between the two adjacent judgments	When compromise is needed

3. Results and Discussion

3.1. Main and Sub-Variables of Coastal Health Ecosystem

The variables of coastal ecosystem health indicators in this synthesis consist of 3 level of variable, i.e. main variable, sub-variable and operational variable. As a result of assessment, coastal ecosystem health indicators consist of 3 main variable, i.e. State Of Ecosystem, Pressure and Management. Main variable State of Ecosystem and Management obtain the same value i.e. 0.400, while Pressure value was 0.200 (see **Figure 1**).

Main variable State of Ecosystem consist of 4 sub-variables i.e. Coral reef, Reef fish, Mangrove and Seagrass. Coral reef gain the best value (0,184), followed by Mangrove (0,090), Coral reef fish (0,069) and Seagrass (0,058), respectively (see **Figure 2**). In the other main variable, Pressure, the sub-variable Pressure of tourism activity (0,133) get the superior value because of their importance based on rating from expert judgment, while lower value obtain by Pressure of fishery activity (0,067) (see **Figure 3**). Main variable Management situated two sub-variables i.e. Planning & regulation and also Institutional (0,16) as the highest sub-variable, while the lowest was Infrastructure & Financing Facility (0,08) (see **Figure 4**). The figures of Operational variables and Sub-variables show original value from Expert Choice software, before adjusted by each Sub-variables values.

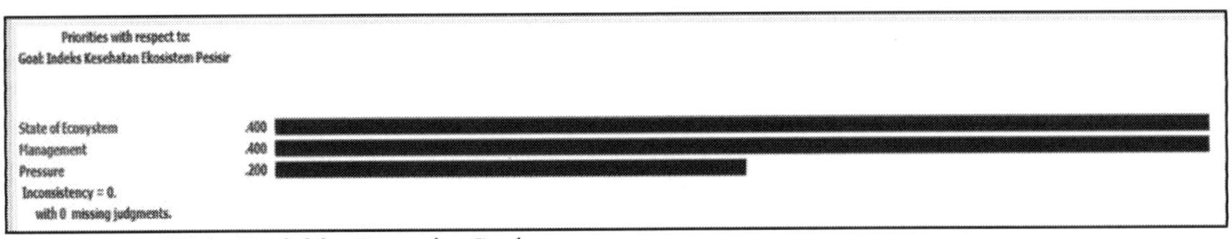

Figure 1. Main Variables Intensity Scale

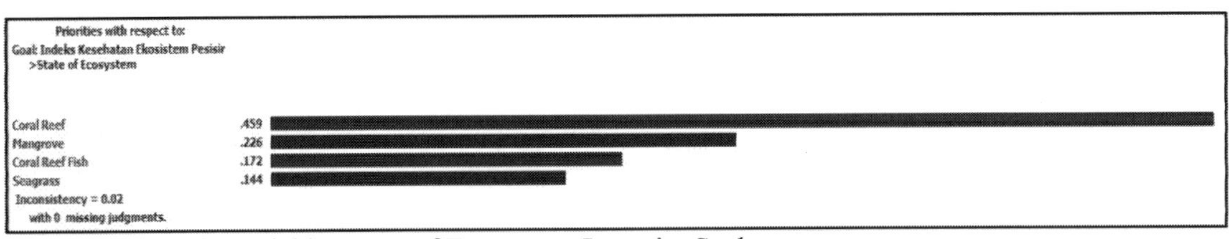

Figure 2. Sub-Variables State of Ecosystem Intensity Scale

IOP Conf. Series: Earth and Environmental Science **116** (2018) 012094 doi:10.1088/1755-1315/116/1/012094

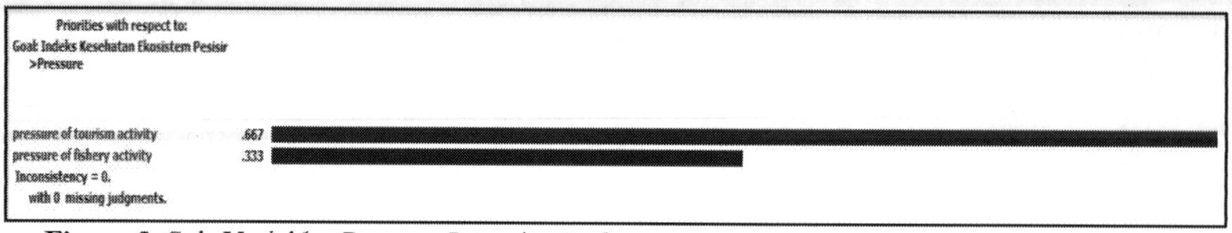

Figure 3. Sub-Variables Pressure Intensity Scale

Figure 4. Sub-Variables Management Intensity Scale

3.2. Operational Variables of Coastal Health Ecosystem

Sub-variables were divided into several Operational variables. Sub-variables Coral reef and Coral reef fish consist of 5 Operational variable while sub-variables Mangrove and Seagrass consist of 6 Operational variable. Operational variables of Coral reef i.e. Coral reef percent cover; Coral reef biodiversity; Coral reef similarity; Coral reef dominance and Coral reef cover. The highest and the lowest importance value were Coral reef percent cover (0,058) and Coral reef dominance (0,017), respectively (see **Figure 5**). Operational variables of Coral reef fish i.e. Species abundance per family, Number of species, Coral reef fish biodiversity, Coral reef fish similarity and Coral reef fish dominance. Coral reef fish biodiversity (0,025) was the highest variables, while Coral reef fish dominance and Coral reef fish similarity, obtain the same value (0,007) (see **Figure 6**). Operational variables mangrove consist of 6 Operational variable, i.e. Mangrove percent cover; Mangrove density; Mangrove biodiversity; Mangrove similarity; Mangrove dominance and Mangrove cover. The highest and the lowest importance value were Mangrove percent cover (0,019) and Mangrove similarity (0,009), respectively (see **Figure 7**). Operational variables of Seagrass i.e. Seagrass percent cover; Seagrass density; Seagrass biodiversity; Seagrass similarity; Seagrass dominance and Seagrass cover. Seagrass density (0,012) and Seagrass similarity (0,005) were the highest and the lowest variables (see **Figure 8**).

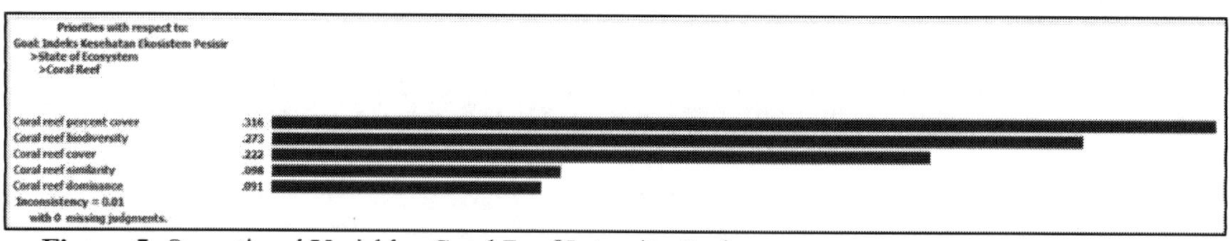

Figure 5. Operational Variables Coral Reef Intensity Scale

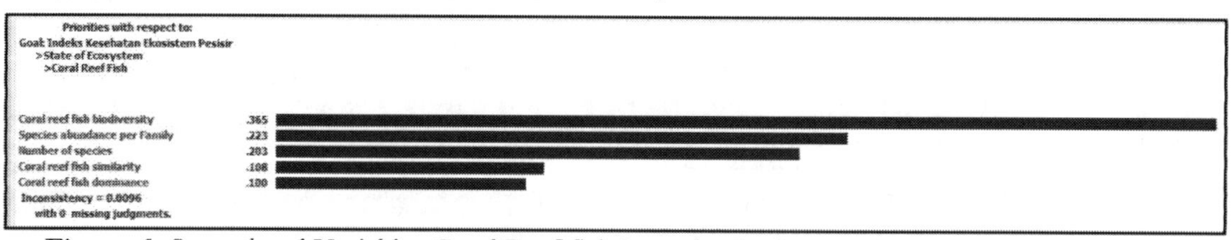

Figure 6. Operational Variables Coral Reef fish Intensity Scale

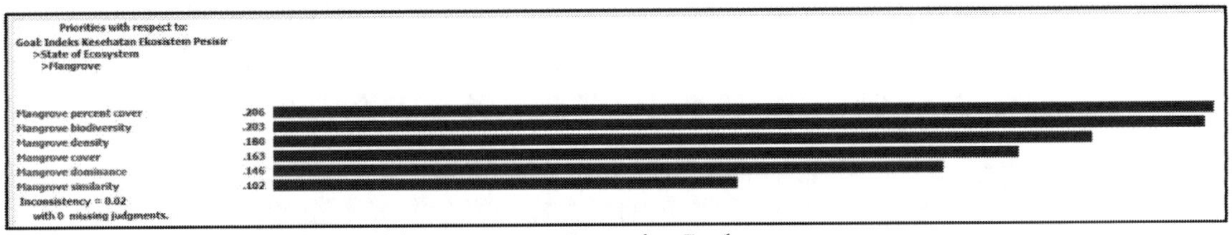

Figure 7. Operational Variables Mangrove Intensity Scale

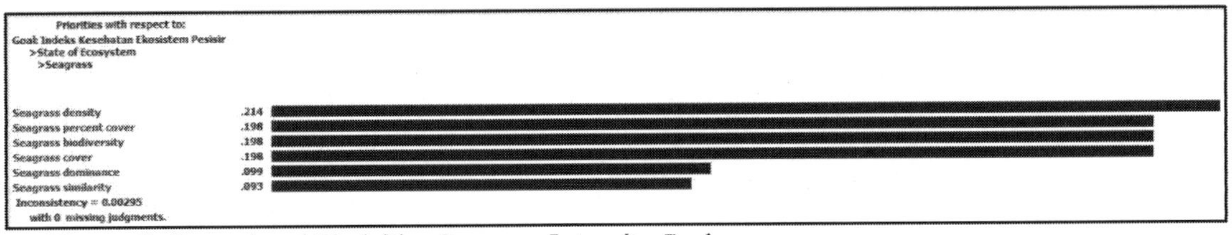

Figure 8. Operational Variables Seagrass Intensity Scale

Sub-variables Planning & regulation consist of 3 Operational variable while sub-variables Institutional and Infrastructure & financing facility consist of 4. Sub-variables Planning & regulation and Infrastructure & financing facility have Operational variables which have same value that is 0,053 and 0,020 respectively. The highest and the lowest Operational variables in each Sub-variable Institutional were Coordination (0,048) and Number of employees (0,034) (see **Figure 9**).

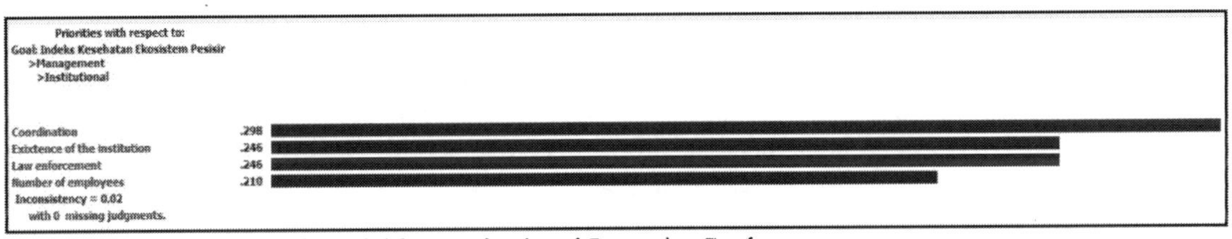

Figure 9. Operational Variables Institutional Intensity Scale

3.3. *Overall Variables Analysis of Coastal Health Ecosystem*

A set of variable that will be used as indicators of coastal ecosystem health have been determined and weighted. The variables consist of 3 level i.e. Main variables, Sub-variables and Operational variables. Afterward, operational variables will be used as indicators of coastal ecosystem health assessment. Based on integrated analysis of AHP to all of variables, 3 variables get the highest value i.e. Pressure of tourism activity (0,064); Coral reef percent cover (0,059) and Coral reef biodiversity (0,051) (see **Figure 10**).

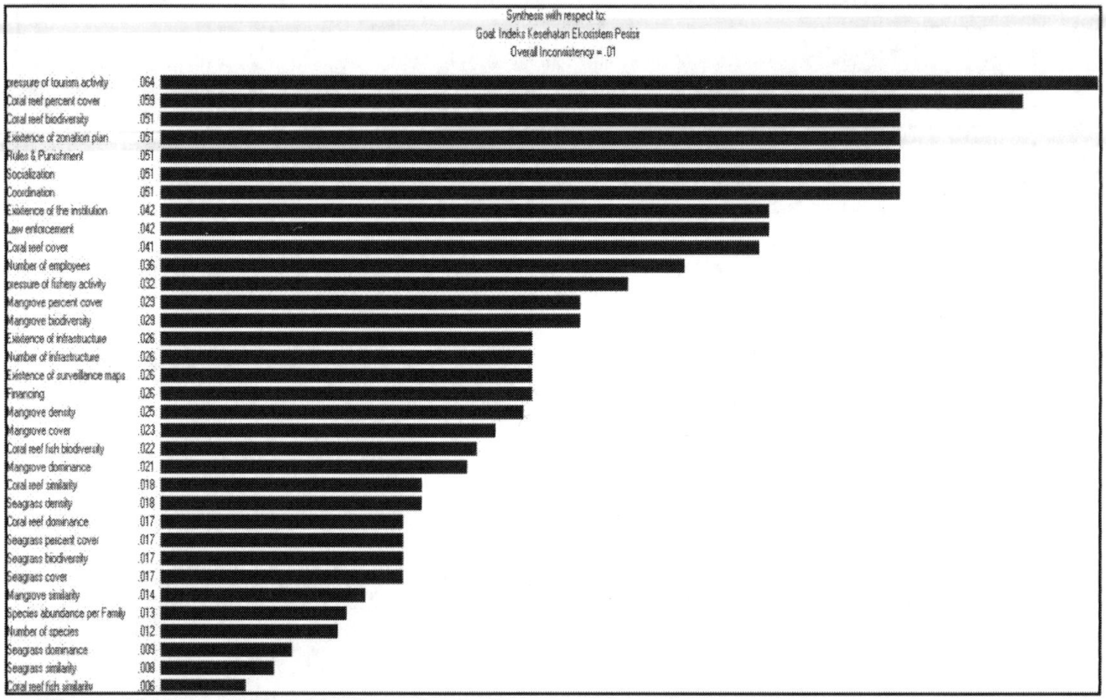

Figure 10. Overall Variables Intensity Scale

Most of the expert respondents agreed that there is an increasing number of potential pressures derived from tourism activities in Karimunjawa National Marine Park (KNMP). It is in line with the fact that there is an increase in the number of visitors. The number of visitors' karimunjawa in 2014 increased by nine times greater when compared in 2004 [16]. Increased intensity of tourist visit has the potential to threaten coral reef ecosystem in TNMP [17], [18].

The threat of degradation of ecosystem quality in KNMP arrives particularly from fishery activities and marine tourism activities that are not environmentally friendly [18]. Destructive fishing activities in the form of not environmentally friendly fishing gear and zoning transgressions have been monitored. Moreover, some transgressions in fishing activities have been punished. Contrary with monitoring fisheries activities which has been done, supervision of destructive marine tourism activity has not been optimally monitored. It is seen from the absence of reports of transgressions in marine tourism activity whereas there has been some ecosystem damage due to marine tourism activity. Broken branching corals and massive corals that die from being trampled by tourists found in locations with high marine tourism activity [18]. The high biodiversity of KNMP makes this area as a potential as a tourist destination. Destructive marine tourism activities will cause ecosystem damage and threaten biodiversity. There is a need to develop stronger conservation strategies to facing with pressures from marine tourism activities.

4. Conclusion

Pressure of tourism activity is the variable that most affect the health of the ecosystem in KNMP, based on expert respondents. Tourism activities that are not environmentally friendly have not been optimally monitored. There is a need to develop stronger conservation strategies to facing with pressures from marine tourism activities.

References

[1] Asian Development Bank 2014 State of The Coral Triangle: Indonesia. Asian Development

Bank, Mandaluyong City, Philippines:1-84.
[2] Ambariyanto 2010 Kebijakan pengelolaan organisme laut dilindungi: kasus kerang raksasa. Undip Press, Semarang.
[3] Barber P H, Ablan-Lagman M C A, Ambariyanto, Berlinck R G S, Cahyani D, Crandall E D, Ravago-Gotanco R, Juinio-Meñez M A, Mahardika I G N, Shanker K, Starger C J, Toha A H A, Anggoro A W and Willette D A 2014 Advancing biodiversity research in developing countries: the need for changing paradigms. *Bull. Mar. Sci.*, 90(1): 187–210.
[4] Veron J E N 1995 Corals in space and time. The biogeography and evolution of the scleractinia. UNSW Press, Sydney.
[5] Ambariyanto 2017 Conserving endangered marine organisms: causes, trends and challenges. *IOP Conf. Ser. Earth Environ. Sci.*, 55: 1-10.
[6] van Oudenhoven A P E, Siahainenia A J, Sualia I, Tonneijck F H, van der Ploeg S, de Groot R S, R. Alkemade and Leemans R 2015 Effects of different management regimes on mangrove ecosystem services in Java, Indonesia. *Ocean Coast. Manag.*, 116: 353–367.
[7] Giesen W, Wulffraat S, Zieren M and Scholten L 2007 Mangrove guidebook for Southeast Asia. Dharmasarn Co., Ltd: 1-781.
[8] Clough B 2013 Continuing the journey amongst mangroves. ISME Mangrove Educaitonal Book Series. International Society for Mangrove Ecosystem (ISME), Okinawa, Japan and International Tropical Timber Organization,Yokohama, Japan. 1: 1-86.
[9] Mimura N, Ed. 2008 Asia-Pasific Coasts and Their Management. Springer, 11: 1-365.
[10] Prasetya J D, Ambariyanto, Supriharyono and Purwanti F 2017 Mangrove health index as part of sustainable management in mangrove ecosystem at Karimunjawa National Marine Park Indonesia. *Adv. Sci. Lett.*, 23(4): 3277-3282.
[11] Juffe-Bignoli D, Burgess N, Bingham H, Belle E M S, De-Lima M G, Deguignet M, Bertzky B, Milam A N, Martinez-Lopez J, Lewis E, Eassom A, Wicander S, Geldmann J, van Soesbergen A, Arnell A P, O'Connor B, Park S, Shi Y N, Danks F S, MacSharry B, Kingston N 2014 Protected planet report 2014. Cambridge, UK: 1-68.
[12] Flower J, Ortiz J C, Chollett I, Abdullah S, Castro-Sanguino C, Hock K, Lam V and Mumby P J 2017 Interpreting coral reef monitoring data: A guide for improved management decisions. *Ecol. Indic.*, 72: 848-869.
[13] Saaty T L and Vargas L G 2012 Models, methods, concepts & applications of the Analytic Hierarchy Process. Springer Science+Business Media New York, 175: 1-341.
[14] Zahedi F 1986 The Analytic Hierarchy Process: A survey of the method and its applications, *Interfaces (Providence)*. 16(4): 96–108.
[15] Soma K 2003 How to involve stakeholders in fisheries management a country case study in Trinidad and Tobago. *Mar. Policy*, 27: 47–58.
[16] DKP. Prov. Jateng 2014 Identifikasi potensi dan pemetaan pulau-pulau kecil di pulau Menjangan Kecil Karimunjawa. Dinas Kelautan dan Perikanan Provinsi Jawa Tengah.
[17] Kartawijaya T, Prasetia R, Ripanto and Jamaludin 2011 Pengembangan ekowisata berbasis masyarakat di Taman Nasional Karimunjawa. Wildlife Conservation Society-Indonesia Program.
[18] Pardede S, Tarigan S A R, Setiawan F, Muttaqin E, Muttaqin A and Muhidin 2016 Laporan Teknis Monitoring Ekosistem Terumbu Karang Taman Nasional Karimunjawa 2016. Wildlife Conservation Society-Indonesia Program.

Reef Development on Artificial Patch Reefs in Shallow Water of Panjang Island, Central Java

Munasik[1], Sugiyanto[2], Denny N Sugianto[3], Agus Sabdono[1]

[1]Department of Marine Sciences, Diponegoro University, Jl. Prof. Soedharto, SH, Tembalang, Semarang
[2]Department of Mechanical Engineering, Diponegoro University, Jl. Prof. Soedharto, SH, Tembalang, Semarang
[3]Department of Oceanography, Diponegoro University, Jl. Prof. Soedharto, SH, Tembalang, Semarang
* munasik_m@undip.ac.id

Abstract. Reef restoration methods are generally developed by propagation of coral fragments, coral recruits and provide substrate for coral attachment using artificial reefs (ARs). ARs have been widely applied as a tool for reef restoration in degraded natural reefs. Successful of coral restoration is determined by reef development such as increasing coral biomass, natural of coral recruits and fauna associated. Artificial Patch Reefs (APRs) is designed by combined of artificial reefs and coral transplantation and constructed by modular circular structures in shape, were deployed from small boats by scuba divers, and are suitable near natural reefs for shallow water with low visibility of Panjang Island, Central Java. Branching corals of *Acropora aspera*, *Montipora digitata* and *Porites cylindrica* fragments were transplanted on to each module of two units of artificial patch reefs in different periods. Coral fragments of Acropora evolved high survival and high growth, Porites fragments have moderate survival and low growth, while fragment of Montipora show in low survival and moderate growth. Within 19 to 22 months of APRs deployment, scleractinian corals were recruited on the surface of artificial patch reef substrates. The most recruits abundant was Montastrea, followed by Poritids, Pocilloporids, and Acroporids. We conclude that artificial patch reefs with developed by coral fragments and natural coral recruitment is one of an alternative rehabilitation method in shallow reef with low visibility.

Keywords: Keywords: artificial patch reefs, reef development, Panjang Island, Central Java

1. Introduction

The coral reefs of tropical waters ecosystems are very important in order to support in food security. However, recently status of coral reefs is threatened due to the practice of fisheries that are not environmentally friendly and the impact of the development of the coastal area [1]. Artificial reef is one of the tools to rehabilitate or restoration coral reefs [2-3]. A variety of artificial reefs have been deployed for a variety of purposes, as the home of fish, marine tourism and rehabilitation of coral reefs. A lot of material that was applied in the manufacture of artificial reef construction either natural or artificial materials. Natural materials made of stone, bamboo and wood, while artificial materials, namely tires, concrete blocks, a former vehicle, iron, etc. [4]. In Indonesia, the artificial reef that was applied for the rehabilitation of coral reefs are generally using concrete blocks with various shapes: pyramid, pipe, rod and loop and cubic forms [5]. All it aims to replace or mimic natural reefs [6]. Unfortunately, most are not generally successful. Failure caused by unstable structures and unsuccessful methods in increasing biomass of coral on the substrate of artificial reef.

Some have applied to do a transplant coral reef rehabilitation [2-3]. Coral transplantation is usually performed using nursery shelves but not until coral fragments transferred to rehabilitation areas because many deaths. The combination of artificial reef with coral transplantation be alternatives in supporting the successful rehabilitation of coral reefs in the coral biomass increase. The success of the artificial reef is determined also by the proximity of natural reef, it can support recruitment of hard corals [5] and increased biomass of coral, the occurrence of recruitment in a natural increase of biota associated with artificial reef [7]. This paper will discuss the development of reefs on the artificial reef which is combined with the coral transplantation technique later called artificial patch reefs (APRs) have been applied in the reef flat of Panjang Island, Central Java which evolved shallow and murky [8-9].

2. Material and Methods

The APR is a structure of artificial reef in circular shaped which constructed modules (modular) and made by concrete blocks and consists a multilevel up to 5 levels (Figure 1A). The artificial reef structures placed on the East side of Panjang Island reefs at a depth of 3-5 meters (6° 34' 34.8" LS and 110° 37' 52.9" BT; Figure 1B). APRs deployment were built by inserting a concrete block module (weight 30-40 kg) one by one using SCUBA Dive equipment, each structure composed of 180 modules. Two unit of the APRs (APR_1 and APR_2) were installed adjacent to natural reefs on June 8th, 2015 and September 22, 2015. Each module attributed holes in order to coral transplantation using underwater cement. APR_1 which deployed on 8 June were transplanted by using branching coral Acropora while APR_2 which deployed on September 22, and then were transplanted using branching coral Montipora and after death in the West Monsoon then transplanted with Branching coral Porites in mid-June 2016. Observation on growth of Acropora coral fragments on the APR_1 performed on the 100 and 145, Montipora coral fragments on growth of the APR_2 on a day-to-55 days whereas growth of Porites coral fragments was conducted in the first week of transplanting until 8th weeks (June-August 2016). Observations of the recruitment of corals and reef fish were performed in early June 2017, after the APR_1 was 2 years old and the APR_2 was 18 months.

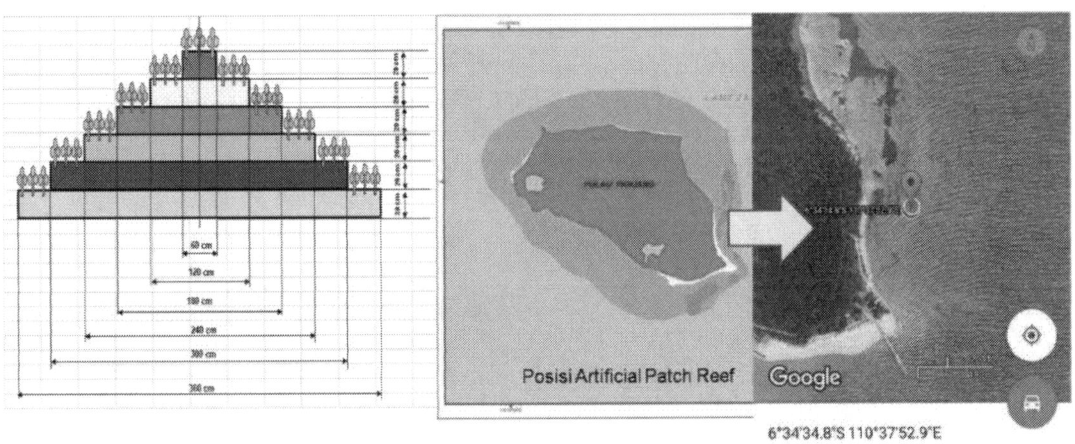

A B

Figure 1. A. Structure of artificial reef APR; B. Deployment site of APR at Pulau Panjang, Central Java

3. Results and Discussion

3.1. The growth of coral fragments

Artificial reef APR_1 deployed on June 2015 which is composed by branching *Acropora aspera* whereas APR_2 deployed on September 2015 consists of fragments of branching coral *Montipora digitata* (Figure 2). The selection of this type of reef with consideration of time period reproduction coral Acropora and election time also with consideration of water quality. June is the start of the season, and September is the end of the season and entered the transitional season. Combined with design/construction APR that supports regeneration of coral branching resulting in the survival of moderate coral transplant ranges from 25% to 57% in the observation day-155. The results also showed that the coral colonies have the highest survival is at the top of the circle to the middle of the otherwise the lowest survival is in the circle of the base.

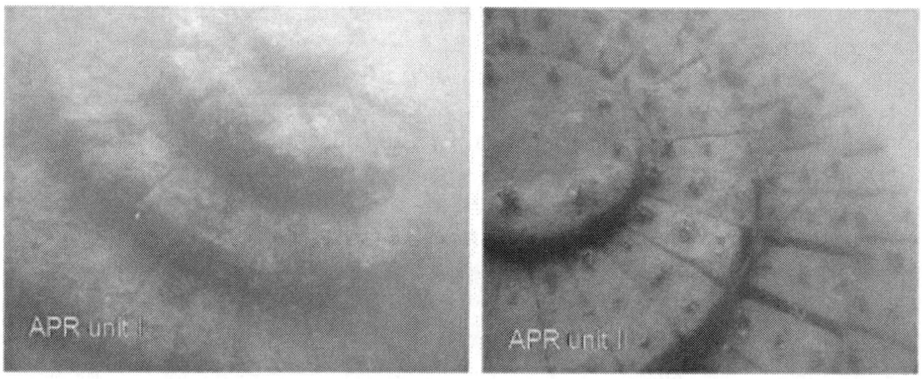

Figure 2. Reef condition of Artificial Patch Reef I (APR_1) deployed on 8 Juni 2015 and APR_2 deployed on 21 September 2015.

Condition of APR_1 after 100-days show reef development indicated by increasing biomass of coral, a colony of coral branches grows rapidly, covered CCA (Crustose Coralline Algae), macro algae as Padina. Thus, APR have been promoting a new habitat in initial period. Fragments of coral branching Acropora can grow significantly on the substrate of APR using cement. It is indicated that structure of APR is suitable for coral propagation, particularly branching coral Acropora on (Figure 3). Observation on the growth of coral colonies showed 145 Acropora rapidly increased up to 15 cm in diameter, the growth of coral fragments up to more than 1000%.

Figure 3. Growth of biomass of coral propagation Acropora on APR_1 which deployed on 8 June 2015 (A. initial propagation; B. 100 days; C. 145 days).

Propagation of Acropora coral fragments on artificial reef APR_1 can increase biomass of coral *Acropora aspera*, one of the extinct coral population in of Panjang Island. As many as 50% of coral fragments Acropora have survive until 155 days (Figure 4). When coral fragments that are planted on June 8th, 2015 as much as 360 then until 2015, November 14, observations have been obtained 180 colony of coral branching *Acropora aspera*. Fragments of coral can grow rapidly into large coral colonies with a size reached Ø 16 cm on the 155 or 5 months (Figure 5). However not all colonies grow rapidly, some coral colonies obtained measuring Ø 6 cm. These results indicated that coral *Acropora aspera* can evolved asexual reproduction in order to increase population by coral fragmentation on artificial reef APR [10].

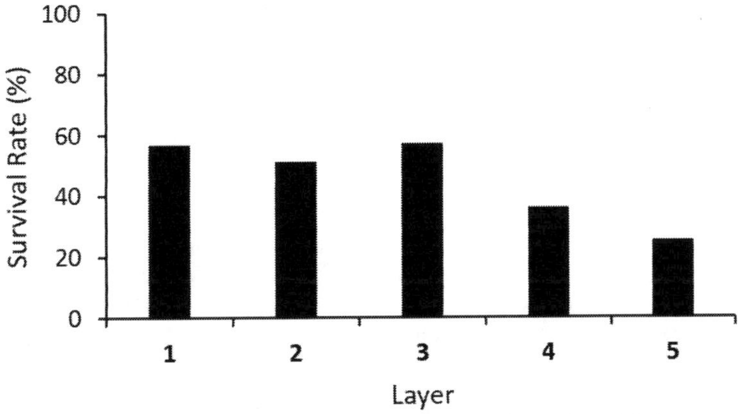

Figure 4. Survival rate of coral propagation Acropora on APR_1 after 155 days.

Figure 5. Growth of biomass of coral propagation Acropora on APR_1 after 45 days.

The deployment of artificial reefs APR_2 on September 21, 2015 has also provided a new habitat (Figure 6). Substrate concrete block has been covered with CCA (Crustose Coralline Algae), however these macro algae are not found in APR_1. Initially, reef development of APR_2 on 11 October 2015 showed that fragments *Montipora digitate* grow rapidly. Growth of Montipora fragments reach in various size in this periods. However, all of the coral fragments died during in west monsoon (December-February 2016).

IOP Conf. Series: Earth and Environmental Science **116** (2018) 012095 doi:10.1088/1755-1315/116/1/012095

Figure 6. Growth of biomass of coral propagation Montipora on APR _2 deployed on 21 September 2015

After mortality of Montipora fragments, coral branching of Porites fragments replaced onto artificial reef APR_2. The coral fragments can survive longer on APR_2, coral fragments can last up to 1 year of observation later. Early growth of fragments of branching coral *Porites cylindrica* of large range 47-350% during 6 weeks however slowly growth occurred in the coral fragments after a year of transplanted (Figure 7).

Figure 7. Growth of coral propagation Porites on APR_2 transplanted mid of June 2015.

3.2. Recruitment of hard corals

Reef development of APR is also indicated by the occurrence of coral recruitment showed settlement of hard corals. The viable of corals juvenile were found in the second year of deployment. Genera of Scleractinian coral were found settled to the surface of substrate of the APR is Montastrea, Porites, Acropora, and Pocillopora Goniastrea (Figure 8). In order to highlighted in composition of the genus coral recruits, Montastrea was highest followed by Porites (Figure 9). Constituent types of observations on natural reefs around the APR is Faviid and Poritiid [11]. This suggests that coral recruitment in accordance with the composition of the genus corals there are nearby [6].

Figure 8. Generic composition of juvenile corals recruit on APR after 18-24 months (A. Goniastrea; B. Porites; C. Pocillopora; D. Montastrea)

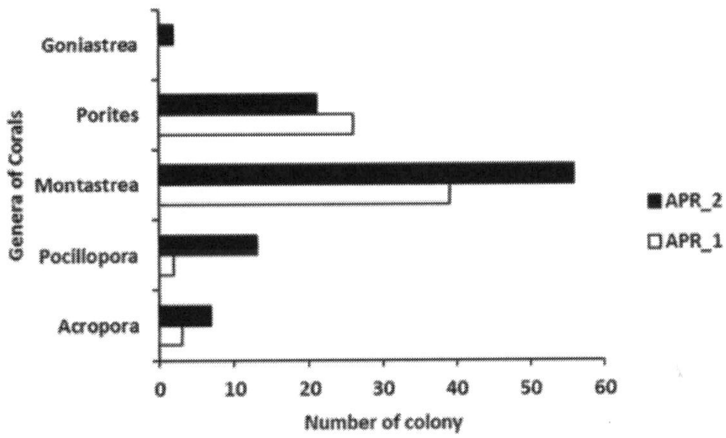

Figure 9. Abundance of juvenile corals recruit on APR after 18-24 months

Coral recruitment occur two of APRs, number of corals recruited in APR_2 was higher than in APR_1. Coral recruits were generally found on level 2 and 3 of APRs. Montipora are found on horizontal surface of the substrate, Pocillopora settled on vertical surface of substrate. Branching coral Porites can settle on both horizontal and vertical surface of substrate (Figure 10). Generally, coral settlement of Pocilloporid in the column water [12], while other coral taxa can settle in horizontal plate in the bottom of the sea [13] (Munasik, 2008b). Larva of Scleractinian corals have settlement preference in various morphotypes [14], substrate structure and substrate composition [15-17]. Local current pattern of Panjang Island waters may also effect on distribution of adult corals and juvenile corals [18]. The combination of tidal, monsoon system, bathymetry depth resulting may increase coral recruitment in APR of Panjang Island.

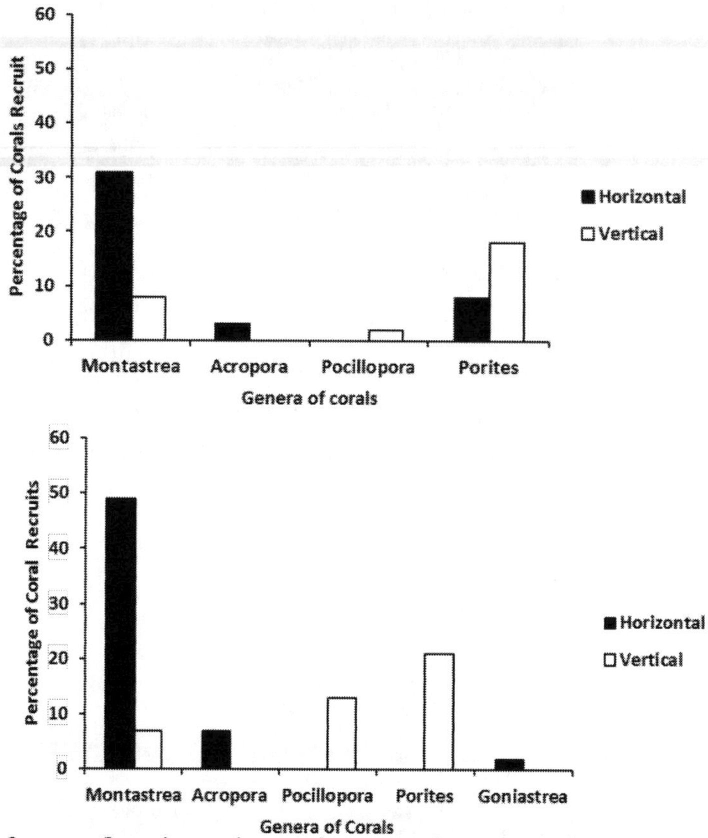

Figure 10. Settlement preference of coral generic on substrate surface of APRs after 18-24 months (A. APR_1 dan B. APR_2).

3.3. Recruitment of Reef Fish

Fish health indicators of coral reefs of the Caetodontidae have also been found around the colonies of coral transplantation results. Juvenile of butterflyfish *Chaetodon octofasciatus* has been found associated with colonies of coral Acropora branch at the age of more than 100 days APR. In addition, the artificial reef APR has also been presenting schooling group Caesionidae who often travels in a natural reef environment. These results suggest that artificial reef presents a new habitat has been the APR in shallow waters and along with increasing time will be able to increase the biodiversity gradually (Figure 10). The second artificial reef has been presenting new habitat on the 20th day of monitoring is indicated by the presence of reef fish Pomacentridae (Figure 11).

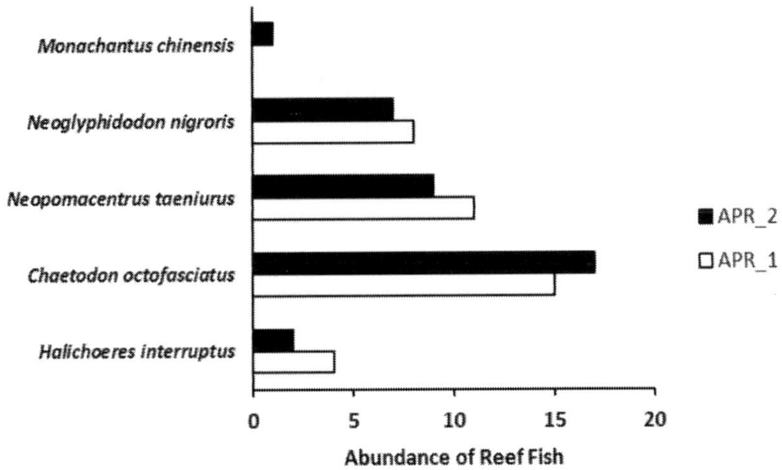

Figure 11. Abundance of reef fish which associated with APRs at Panjang Island, Central Java

4. Conclusion

APRs have provided new habitats in shallow waters indicated by increased live coral coverage from both propagation of coral fragments and larval trapped. Reef development generally occurs in the middle of APRs, indicated by the high survival of coral fragments and high coral settlement. Branching coral Acropora is one of best species in order to increase coral biomass in ARs located in the shallow water. Composition of coral species which settled in APRs seems to be similar to nearby natural reefs..

References:

[1] Wilkinson CR 2000 Executive summary In C R Wilkinson (Ed.), Status of coral reefs of the world: 2000 (pp. 7–19). Australia: Australian Institute of Marine Science.

[2] Rinkevich B 2005 *Environ. Sci. Technol.* **39** 4333-4342

[3] Abelson Avigdor 2006 *Bulletin of Marine Science* **78** (1) 151-159

[4] Baine M 2001 *Ocean & Coastal Management* **44** 241–259

[5] Munasik 2008a *Kondisi terumbu buatan berbahan beton pada beberapa perairan di Indonesia.* In: Simposium Munas Terumbu Karang II, 20 Nopember 2008, Hotel Bidakara Jakarta

[6] Perkol-Finkel, S N Shashar, Y Benayahu 2006 *Marine Environmental Research*, **61** (2) 121-135

[7] Meesters, Erik HWG, Sarah R Smith, Leontine E Becking 2015 A review of coral reef restoration. Report to IMARES Wageningen, UR

[8] Edinger EN, Jompa J, Limmon GV, Widjatmoko W, Risk MJ 1998 *Marine Pollution Bulletin* **37** 617-630

[9] Sabdono, A. et al. 2014. An Early Evaluation of Coral Disease Prevalence on Panjang Island, Java Sea, Indonesia. *Int. J Zool. Res.* **10**: 20-29.

[10] Munasik, W Widjatmoko 2004 *ILMU KELAUTAN: Indonesian Journal of Marine Sciences* **9**(4) 211-216

[11] Munasik et al. Sebaran Spasial Karang Keras (Scleractinia) di Pulau Panjang, Jawa Tengah. *Buletin Oseanografi Marina*. April: Vol. I:16-24. Tahun 2012

[12] Munasik, Suharsono, J Situmorang, Kamiso H N 2014 *ILMU KELAUTAN: Indonesian Journal of Marine Sciences* **19**(3) 171-180

[13] Munasik. 2008b. Reproduksi Karang *Pocillopora darnicornis* (Linnaeus) di Pulau Panjang, Jawa Tengah. *Disertasi*. Fakulatas Biologi Universitas Gadjah Mada.

[14] Munasik, Diah Permata Wijayanti, Retno Hartati, Rudi Pribadi, Yoko Nozawa 2016 *ILMU KELAUTAN: Indonesian Journal of Marine Sciences* **21**(4) 185-190

[15] Nozawa Y 2008 *J. Exp. Mar. Bio. Ecol.* **36** 127-130

[16] Nozawa, Y. 2010. Survivorship of fast-growing coral spats depend less and refuge structure: the case of *Acropora solitaryensis*. *Galaxea* 12(1) 31-36.
[17] Yadav S, Rathod P, Alcoverro, T & Arthur, R 2016 *Coral Reefs* **35** 211-222
[18] Munasik, Denny N Sugianto, Widodo S Pranowo, Suharsono, Jesmant Situmorang, Kamiso HN. 2006 *ILMU KELAUTAN: Indonesian Journal of Marine Sciences* **11**(1) 11-18

IOP Conf. Series: Earth and Environmental Science **116** (2018) 012096 doi:10.1088/1755-1315/116/1/012096

Saprobic analysis to Marina coastal, Semarang city

D M Nuriasih[1], **S Anggoro**[2] **and Haeruddin**[2]
[1] Post Graduate Program of Coastal Management, Diponegoro University, JL Prof Sudarto, Semarang-Indonesia
[2] Lecturer of Faculty of Fisheries and Marine Science, Diponegoro University, JL Prof Sudarto, Semarang-Indonesia
Indonesia
Email: dianminggus@gmail.com

Abstract. Semarang city is one of coastal city in Indonesia, that antropogenic activities have impact to coastal of Semarang, including Marina beach. Therefore, it is important to study the quality of seawater related with antropogenic activities. The research purpose was to analyze the saprobic level of Marina beach as an indicator of marine pollution. This case study research used survey method. Purposive method was used for sampling the seawater at five stations at the beach. This research can be concluded that TSI (Tropic Saprobic Index) higher than standard that indicated the Marina Beach Seawaters polution at level of β – Mesosaprobic.

1. Introduction

Semarang city is one of coastal city in Indonesia, where is located in north coastal of Java, with coastal area of Semarang city is 37.376 ha [1]. Semarang city is the capital city of Central Java Province, and become the centre of trade and industry in Central Java Province. Coastal areas are central to human activities, including centre of trade, industry, aquaculture, capture fisheries, tourism and houses. So, coastal areas are susceptible to pollution, including Pantai Marina Semarang as one of the coastal tourism sites in Semarang City.

The quality of the waters at Marina Semarang Beach is influenced by anthropogenic activity in the surrounding environment. Some of these anthropogenic activities are related to settlements, industries, and fisheries, both along the West Flood Canal River, as well as the waters east and west of Marina Beach. In this study, pollution estimation due to tourism activities in the coast of Marina Semarang beach is done by approaching saprobic level. The purpose of this research was to analyze the saprobic level of Marina beach as an indicator of marine pollution.

2. Research Methods

This research was conducted in March-April 2017. There were 5 observed stations, i.e.:

a. Station 1 (located at 6°56'44.29.5"S and 110°23'35.0"E) is at the estuary of 'Banjir Kanal Barat' river. The activities of households and industries that are located around 'Banjir Kanal Barat' river give an impact to river and coastal pollution.
b. Station 2 (located at 6°56'44.0"S and 110°23'33.7"E) is on the open coastal at Marina Beach where there are several the tourism activity, including fishing, jogging and cullinary.

c. Station 3 (located at 6°57'00.1"S and 110°23'23.6"E) is located on artificial lagoon. This location is relatively protected seawaters and there are several tourism activities, including cullinary, public toilets, and boat base.
d. Station 4 (located at 6°56'57.1"S and 110°23'11.6"E) is near the new reclamation site of Marina Beach.
e. Station 5 (located at 6°56'44.1"S and 110°23'03.0"E) is in open seawaters at the new reclamation site of Marina Beach.

Determination of seawater pollution level used saprobic index (SI) and tropical saprobic index (TSI), using the following formula [2,3]:

$$SI = \frac{1(nC) + 3(nD) + 1(nB) - 3(nA)}{1(nA) + 1(nB) + 1(nC) + 1(nD)} \quad (1)$$

$$TSI = \frac{1(nC) + 3(nD) + 1(nB) - 3(nA)}{1(nA) + 1(nB) + 1(nC) + 1(nD)} \times \frac{nA + nB + nC + nD + nE}{nA + nB + nC + nD} \quad (2)$$

That 'n' is the number of individual organisms in each group of saprobic, 'nA' is the number of individual organisms of the polysabrobic group, 'nB' is the number of individual organisms of the alpha-mesosaprobic group, 'nC' is the number of individual organisms of the beta-mesosaprobic group, 'nD' is the number of individual organisms of the oligosaprobic group and 'nE' is the number of individual out of A, B, C, and D groups.

3. Results and Discussions

Marina Beach is an artificial lagoon and one of coastal tourism location in Semarang City. According to Miththapala [4], the lagoon is a shallow coast, that separated from the sea by barriers. According to Satriadi [5], Marina Beach has a depth of 0 to 9 m with a slope of 0.178 percent to 0.200 percent. The distribution of sediments of Marina Beach consists of 3 types, i.e. sand 28.125% of where is located on the beach, especially the western part, silty sand of 56.25 percent where on the deeper sea and sandy silt where on more deeper sea with a depth of 5 to 9 m.

3.1. Anthropogenic activities

There are several anthropogenic activities that give pollution in Marina Beach seawaters. The several of the anthropogenic activities is include:

a. Housing at located in the river zone of 'Banjir Kanal Barat' have the potential to become a source of household, agricultural and industrial waste pollution.
b. Tourism activities at Marina Beach. There are several activities that can be a source of pollutants, including the food and beverages waste, and activity of boat tours. Marina Beach is also used as a place to throw the ashes and flowers for Chinese ethnic.
c. Port of 'Tanjung Mas'. 'Tanjung Mas' Port is a large-scale port with a vessels visit of 2 413 ship per year, with goods of 3 023 416 tons and passengers of 7 379 978 persons [6]. Passenger activity, and loading and unloading of goods contributes to pollutants in the seawaters
d. Aquaculture. There are aquaculture activities at around of the west of Marina beach, including at 'Mangkang Kulon' village. The brackish-water ponds in 'Mangkang Kulon' village is dominated by milkfish culture and a small portion of shrimp culture. There are several pollutant source in aquaculture, including fertilizer, artificial feed and fish feces.
e. Capture fisheries. There are several fishing bases and fishing grounds in Semarang coastal. Capture fisheries in Semarang City is dominated by small scale fisheries that use boat under 30 GT. The fishing bases are found in District of North-Semarang and District of Genuk (east of Marina Beach) and District of Tugu (west of Marina Beach). There are also lift net on the eastern waters of Marina Beach. Capture fisheries activities can contribute to pollution, including oil and fuel spills, feed and waste of fish caught in the seawaters.

f. Industries. Industrial activities located in the port area, and industrial areas, both District of Tugu and Genuk can contribute to pollution in the seawaters of Marina Beach through the ocean currents. There are also several industries in the river of 'Banjir Kanal Barat'. According to Maslukah [7], the 'Banjir Kanal Barat' river is a potentially polluted by several industries, including industries of textile, metal and machinery, chemical, and ceramics.
g. Office, education, trade and services. The activities of office, education, trade and services can also provide pollutants in the seawaters of Marina Beach. It is several types of pollutants from offices, education, trade and services, inclusing organic waste, and packaging materials (plastic).

3.2. Homogenity dan diversity of plankton

Plankton could become as indicator of water pollution. If the value of diversity is less than 2.3026, so it is mean that the seawater have a small diversity and low stability of community [2]. This research proved that the level of plankton diversity have the value less than 2.3026, that is in the range of 0.902-2.110. That is means if the diversity of plankton in Marina Beach seawaters is relatively small and the community stability is low. Human activities are the main factor (including plastic debris) that reduce the biodiversity, including the biodiversity in the marine environment [8, 9, 10]

If the value of diversity of 2.0 or less indicates that a seawater has been polluted [3]. The smaller the value of diversity so the level of pollution tends to increase. The findings of this research indicate that seawaters of Marina Beach have been polluted, given the average value of plankton diversity in each station less than 2.0.

Several types of plankton that found in the seawaters of Marina Beach are including *Skeletonema* sp., *Rhizosolenia* sp., *Synedra* sp., *Chaetoceros* sp., *Planktoniella* sp., *Pleurosygma* sp., *Dictyocha* sp., *Nitzschia* sp., *Bacteriastrum* sp., *Guinardia* sp., *Coscinodiscus* sp., *Ceratium* sp., *Thalassiothrix* sp., *Peridinium* sp., *Cyclotella* sp., *Trichodesmium* sp., *Eucampia sp. Biddulphia sp.* and *Dinophysis* sp. There are several factors that influence the biomass of plankton, including nitrogen, phospor, silica, and eutrofication. The booming of toxic dinoflagellate is one of indicator if the aquatic environment is the bad condition, including *Nitzschia* sp and *Dinophysis* sp [11].

Table 1. Homogenity and diversity of plankton

	Stations				
	1	**2**	**3**	**4**	**5**
March 2017					
Sampling 1					
Species (S)	12	12	12	10	15
H Max	2.485	2.485	2.485	2.303	2.708
Diversity (H')	1.481	1.150	1.085	1.350	1.125
Homogenity (E)	0.596	0.463	0.436	0.586	0.415
Sampling 2					
Species (S)	9	10	10	12	15
H Max	2.197	2.303	2.303	2.485	2.708
Diversity (H')	1.440	1.623	2.007	1.089	1.128
Homogenity (E)	0.656	0.705	0.808	0.438	0.417
April 2017					
Sampling 1					
Species (S)	14	13	14	13	12
H Max	2.639	2.565	2.639	2.708	2.485
Diversity (H')	1.944	1.429	1.430	0.902	1.717
Homogenity (E)	0.737	0.557	0.542	0.333	0.691
Sampling 2					

	Stations				
	1	**2**	**3**	**4**	**5**
Species (S)	11	12	13	14	10
H Max	2.398	2.485	2.565	2.639	2.303
Diversity (H')	1.779	1.875	1.625	0.956	1.652
Homogenity (E)	0.742	0.747	0.634	0.362	0.718

3.3. Saprobic Index and Tropic Saprobic Index

The Results of research on saprobic index and tropical saprobic index indicate that seawaters of Marina Beach have been polluted. If the value of TSI in the range of 0.5 to 1.5, then a seawater has a level of β - Mesosaprobic, that means there are low to moderate levels of pollution [2, 3]. While if the value of TSI ≥ 1.5, then seawaters are at the level of Oligosaprobic, that is classified as low-contamination to uncontaminated. The research results showed that the research sites were dominated by β - Mesosaprobic, and a small part has categories of Oligosaprobic.

Table 2. Saprobic index and tropic saprobic index of plankton

SI and TSI	Stations				
	1	**2**	**3**	**4**	**5**
March 2017-Sampling 1					
SI	1.26	1.15	1.13	1.20	1.18
TSI	1.47	1.21	1.30	1.48	1.29
March 2017-Sampling 2					
SI	1.31	1.21	1.45	1.16	1.15
TSI	1.48	1.50	2.50	1.33	1.18
April 2017-Sampling 1					
SI	1.22	1.04	1.04	1.03	1.07
TSI	1.62	1.20	1.21	1.10	1.44
April 2017-Sampling 2					
SI	1.20	1.24	1.02	1.02	1.12
TSI	1.60	1.55	1.23	1.13	1.33

If the value of SI is in the range of 1.0 - 1.5, then a seawaters could be categorized of β-Mesosaprobic to Oligosaprobic, that is means if the seawaters are polluted by organic and inorganic materials [2]. The research results showed that all research stations had a range of 1.0-1.5 (β - Mesosaprobic to Oligosaprobic). The findings of this research indicate that the seawaters of Marina Beach have been polluted. Therefore, it is important to pollution control to reduce the level of pollution.

3.4. Pollution control

Several studies showed that the waters of the rivers and coastal of Semarang city have been polluted. According to Suhartono [12], The BOD level of Banjir Kanal Barat-Semarang river reached of 24 mg per L, while in its estuary reached of 45 mg per L. Whereas the BOD limit for class IV waters is 12 mg per L based on Government Regulation No. 82 of 2001 about 'Water Quality Management and Water Pollution Control'. Class IV is a water that designation could be used for irrigation, and farming. BOD could be used to measure natural biodegradation process. Hight content of organic waste in the water, so BOD is more high. The COD in the 'Banjir Kanal Barat' River-Semarang reached of 36 mg per L, while in the estuary reached of 236.5 mg per L. While the COD threshold for class IV waters is 100 mg per L. According to Yusuf [13] at study on the estuary of 'Babon' River-Semarang also indicate that this river has been contaminated, and there were several parameters have exceeded the quality standards, including: total suspended solids (TSS), dissolved oxygen, BOD_5, COD, ammonia, detergents, Cr^{6+} and Cd. The research of Pratama et al [14] showed that the metal content of Pb in 'Tapak' River-Semarang exceeded the quality standard (more than 0.03 ppm).

The above description shows that the waters in Semarang city have experienced pollution. Each of these polluted rivers brings pollutants to the coastal and is aggravated by anthropogenic activity on the coast. This research also proved that the Semarang coastal, including Marina Beach, has been polluted, so it is important to push the pollution control.

The marine pollution is one of major challenge to marine tourism. In case of Geoje Island, marine pollution (debris) made revenue lost in tourism about USD 29 millions and USD 37 millions in 2010 and 2011, and also decreased the tourist from 890 435 person in 2010 to 330 207 person in 2011 [15]. The water pollution is also can reduce the human health relate with the growth of pathogenic-resistent bacteria [16].

There are several efforts that can be done to control pollution in the coastal of Semarang city. This several efforts are including environmental waste management, WWTP (wastewater treatment plant), campaigns of clean river and coastal, periodic water quality monitoring, installation of water quality monitoring boards, extension and enforcement of regulations. WWTP can become a key driver of water pollution control. WWTP is not only an obligation for the company or factory, but it is necessary for households and brackish-water ponds with different specifications. For households, It can push to have a communal WWTP. While WWTP in brackish-water ponds can utilize biofilter by using shellfish, seaweed, fish and mangrove, that before the water output will be discharged to the river or coastal need to be checked first the quality.

According to Prameswari and Purnomo [17], communal WWTP can use Anaerobic Baffled Reactor (ABR) technology, that investment costs of IDR. 518 888 444 and it can be used to serve 672 HHs. The result of the calculation for maintenance and operation, the cost of maintenance and operation of communal WWTP was IDR 5 493 per month per house hold. The management of Marina Beach should also set and operate the WWTP and waste management.

4. Conclusion

The research results proved that the seawaters of Marina Beach have been polluted, both base on SI and TSI values. Based on TSI, seawaters of Marina Beach have category of β-Mesosaprobic (low to moderate contamination). Meanwhile, based on SI, the seawaters of the Marina Beach is classified as β – Mesosaprobic or Oligosaprobic, that it is mean the seawaters are low-polluted by organic and inorganic materials. Therefore, it is proven that the seawaters of Marina Beach has been polluted with tendency on β-Mesosaprobic

References

[1] Central Java Government *2016 Report on Coastal Resource Assessment in The Revision of Regional Regulation of Coastal and Islands Zonation Plan of Central Java Province.* Central Java Government and PT Indotama Mahesa Karya (in Indonesian).

[2] Zahidin M 2008 Water quality assessment in estuary of Pekalongan River base on macro-benthos diversity index and plankton saprobic index. Thesis. Master of Coastal Resources Management Program. Diponegoro University, 86 p. (in Indonesian).

[3] Suryanti 2008 Study of saprobic level in estuary of Morodemak River at the time of tide and low tide. *Jurnal Saintek Perikanan* **4** (1): 76-83. (in Indonesian).

[4] Miththapala S 2013 *Lagoons and Estuaries*. Coastal Ecosystems Series (Volume 4). Sri Lanka: International Union for Conservation of Nature, 73 p.

[5] Satriadi A 2012 Study of batimetry and type of seabed in the Seawater of Marina, Semarang, Central Java. *Buletin Oseanografi*, **1**: 53-62. (in Indonesian).

[6] BPS Kota Semarang 2015 *Semarang City in Figure 2014.* Bappeda Kota Semarang and Badan Pusat Statistik (BPS) Kota Semarang, 680 p. (in Indonesian).

IOP Conf. Series: Earth and Environmental Science **116** (2018) 012096 doi:10.1088/1755-1315/116/1/012096

[7] Maslukah L 2007 Heavy metals concentrations dissolved of Pb, Cd, Cu, and Zn, in seston, and in sediment of 'Banjir Kanal Barat' Estuary, Semarang. *Akuatik-Jurnal Sumberdaya Perairan* **2**(1): 1-4 (in Indonesian).

[8] Derraik JGB 2002 The pollution of the marine environment by plastic debris: a review. *Marine Pollution Bulletin* **44** (2002): 842–852.

[9] Andrady AL 2011 Microplastics in the marine environment *Marine Pollution Bulletin* **62** (2011): 1596–1605.

[10] Cole M, Lindeque P, Halsband C, and Galloway TS 2011 Microplastics as contaminants in the marine environment: a review. *Marine Pollution Bulletin* **62**: 2588–2597

[11] Manna S, Chaudhuri K, Bhattacharyya S and Bhattacharyya M 2010 Dynamics of Sundarban estuarine ecosystem: eutrophication induced threat to mangroves. *Saline Systems* **6** (8): 1-16.

[12] Suhartono E 2009 Identification the waters quality of coastal that affected by domestic waste on the east monsoon by method of pollution index: case study in Jakarta, Semarang and Jepara. *Wahana Teknik Sipil* 14(1): 51-62 (in Indonesian).

[13] Yusuf M 2011 Assessment of pollution impacts on water environment quality and community structure of macro-zoobenthos organisms in estuary of 'Babon' River, Semarang. *Buletin Oseanografi Marina* **1**: 27 – 35 (in Indonesian).

[14] Pratama A G, Pribadi R, and Maslukah L 2012 The content of heavy metal Pb and Fe in water, sediment, and *Perna viridis* in 'Tapak' River at Tugurejo Village, District of Tugu, Semarang City. *Journal of Marine Research* **1**(1): 118-122 (in Indonesian).

[15] Jang YC, Hong S, Lee J, Lee MJ, and Shim WJ 2014 Estimation of lost tourism revenue in Geoje Island from the 2011 marine debris pollution event in South Korea. *Marine Pollution Bulletin* **(2014)**: 1-6

[16] Ahmad I and Dar MA 2013 Perspectives in water pollution. *InTech, Janeza Trdine* **9**, 51000 Rijeka, Croatia, 114 p.

[17] Prameswari R P A and Purnomo A 2014 Planning of communal wastewater service in Krasak Village, District of Jatibarang, Indramayu City. *Jurnal Teknik Pomits* **3** (2): 81-84 (in Indonesian).

Effect of Tourist Characteristic, Marine Tourism Demand, and Number of Visits to the Value Perceptions and Willingness to Pay to Environmental Marine Tourism in Ambon City

Renoldy L Papilaya[1]

[1] Study Program of Social Economics of Fisheries and Marine Science, Pattimura University
E-mail: recodyo.p18@gmail.com

Abstract. Development of tourism in this era must balance between supply and demand aspects. The tendency of policy makers pay more attention to aspects of supply than demand aspect will lead to the development of tourist products and services do not get good results. This research examined the relationship between marine tourism demand, characteristics and number of visits to the level of perception and willingness to pay (WTP) for tourists on a marine tourism destination in Ambon city. Respondents come from overseas tourists, domestic and local, amounting to 140 people. The analysis was performed descriptively and further using SEM analysis Amos 19.00 with path analysis. The analysis shows the close relationship between marine tourism demand variables, characteristics, and the number of tourist visits to the perception of value and WTP for tourists. Variable marine tourism demand when connected directly to the tourist perception and WTP tends to correlate negatively than when using a variable as a variable number of tourist visits as a intervening variables. Instead tourist characteristic variables are positively correlated directly or indirectly with the perception of value and WTP for tourists. It is hoped that with this study will motivate tourism policy makers and local communities to be concerned and studied aspects of perception, WTP, marine tourism demand, the number of tourist visits and the characteristics that turned out to be related to each other.

Keywords: Tourist characteristics, Demand, Perception, WTP, Marine Tourism

1. Introduction

Tourism is one sector whose development began in demand in recent decades because it has the prospect of an increase in national income and welfare of the surrounding community. One of the preferred interests and tourism today is frequented marine tourism. As the capital of the province of Maluku, Ambon city has marine tourism potential is quite large with panoramic coastal and ocean surface or bottom of the sea. But in reality, this potential can not be maximized by the government not only to increase revenue but primarily on the welfare of the surrounding community effort. Currently in Ambon city there are thirty-six (36) potentials for marine tourism destination and a newly developed eight (8) destination. Destination management is done by the government, the private sector and also by the local community (private and group) [1].

Discussing about marine tourism in Ambon city, the tendency of governments and other stakeholders more frequently discussed than the supply side as the potential and the quality of coastal/marine owned, accommodation, transport and others are also in the fact its development although still not done. Aspects of demand of marine tourism rarely mentioned let alone to be developed, whereas the demand side it will largely determine the success of the development of a product, either goods or services products, including marine tourism services. As with the request (demand) for goods and services in general, demand for tourism products was influenced by various factors, and categorized into potential demand and actual demand [2]. During the run tourism activities, tourists will be spending, so it leads directly to the market demand for goods and services (Tourism Final Demand). Furthermore, the final demand tourists indirectly raise demand for capital goods and raw materials (Investment Derived Demand) to produce meet tourist demand for goods and services. Attempt to meet tourist demand required investment in transport and communications, hospitality and other accommodations, craft industry and consumer product industries, service industries, restaurants, and others [3].

Along with the change of the paradigm of world tourism since the 1980s , the bulk of the tour is to be the pattern traveled individuals or small groups, more flexible on the way traveled and tourists can interact with nature and the higher culture [4-6]. In Indonesia, the shift is seen from the number of tourists interested in ecotourism began to utilize sea, beaches, tropical forests, rivers, lakes, and other forms of the landscape other [7]. Nature tourism is a type of prospective tourist, because in addition to being a source of income for the region in the form of local revenue is also on the state in the form of foreign exchange. Source of income is derived from the money spent by tourists (expenditure) and measured from the time of stay (length of stay), as well as nature tourism can guarantee the preservation of nature and create prosperity for the people [8-9]. Weiler and Hall [10] states that one of the motivations special interest travelers are seeking quality. Quality is motivation in seeking the search for forms of objects and attractions that can provide significant value for the benefit of tourists (rewarding), enrichment or personal development value (Enriching), the value of the challenge or adventure (adventuresome), as well as the value of knowledge or insight new (learning). Damanik [11] also stated that one of the salient trends in the development of the global tourist market is shifting activities and tourist destinations due to changes in global tourist psychographics. This is manifested in the form of changes in tastes, interests, expectations and consumption patterns [12-13]; views or tourist image of a tourist destination [14-16] and on a wider scale changes due to special tourist satisfaction problems [17-18]. Tourism widely regarded as having multidimensional activities of a series of development process. Development of the tourism sector concerning socio-cultural, economic and political

The aims of this research were to find a model relationship of the characteristics of tourists, marine tourism demand, the number of tourists visiting to the tourist perception and willingness to pay for environmental marine tourism in the city of Ambon. It is expected that the study will be useful in the development of marine tourism in the city of Ambon

2. Methodology

2.1 Research Methods

The method used in this study is a survey method, is a survey investigation performed to obtain the facts and symptoms that exist, looking for good information about the real basis of social institutions, economics or politics of a group/region. The rationale for this method because the location is fairly extensive research that requires a special approach anyway. Type of survey method is to implement direct interviews with the respondents and the use of questionnaires as the main data collection instrument.

2.2 Study Site

Location of the study are set intentionally (purposive) considering that the site is the location of marine tourism is included in Ambon city tourism development since 2010, can be broken down as follows: Air Manis and Amahusu beach (Outer Ambon Bay) ; Waimahu, Namalatu, Santai and Pintu Kota beach (South coast of Ambon city) ; Hutumury and Hukurila beach (East coast of Ambon city) can be seen in Figure 1.

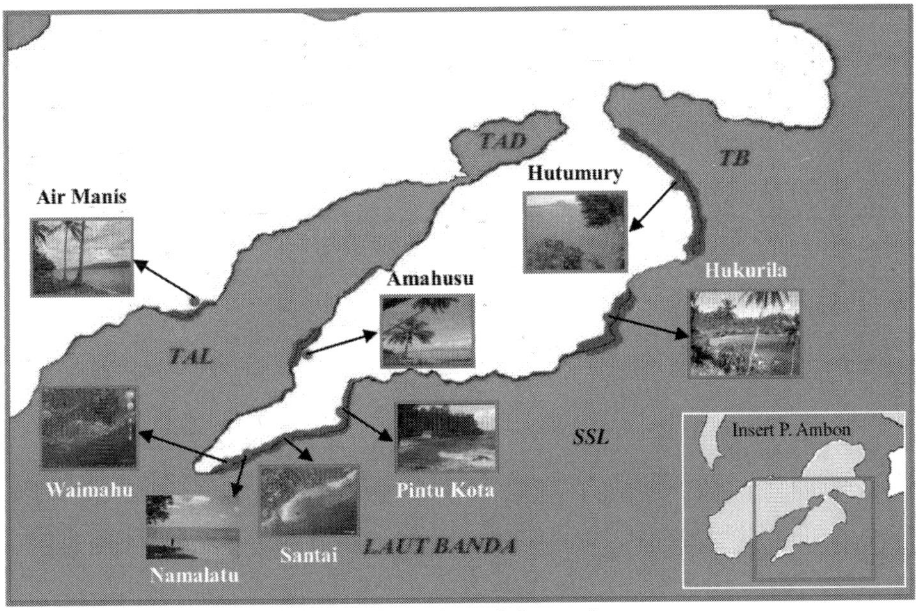

Figure 1. Location of the study

2.3 Sampling Techniques and Data Collection

Population is a generalization region consisting of the object / subject that has certain qualities and characteristics are determined by the investigator to be studied and then drawn conclusions. The sample is part of the number and characteristics possessed by the population. As per the respondents that the purpose of the study will be examined consist of three (3) categories as shown in Table 1.

Table 1. Respondents Catagorical of Toursts in Ambon City

No.	Tourist Origin	Number (person)	Percentage (%)
1	Foreign	26	18.57
2	Domestic	41	29,29
3	Localy	73	52.14
	Respondent Total	140	100

Sourcess :Primary data, Processed, 2012

Sampling size has an important role in the interpretation of the data, since this study uses data analysis with Structural Equation Modeling (SEM), the sample size provides a basis for estimating the sampling error (Byrne, 2010). Hair et al (2010) provide a guide number of samples required to estimate the SEM based on a number of variables, indicators and number of communities. The number of samples should by 5 (five) times the number of indicators that are observed and technical indicators used Maximum Likelihood Estimation (MLE) requires samples ranged 100-200 samples. In a study of tourists there are 17 indicators are researched to fit his criteria then 17 x 5 = 85 samples, so that tourist respondents numbering 140 people have more than the minimum criteria required. Observation of the tourist respondents rely entirely done with the questionnaire as the main data collection that has previously been tested for validity and reliability in a preliminary study. The questionnaire includes questions about the characteristics of various tourists, tourists demand function, the perception of tourists and the economic value of environmental WTP marine tourism (Existence value and Bequest value).

2.4 Processing and Data Analysis

To review and find models the relationship between the characteristics of tourists, tourists demand function and the number of tourist arrivals to the value / level of perception and WTP for tourists (Existence value and Bequest value) of marine tourism destinations in the city of Ambon. Furthermore, to more clearly seen in the form of path diagrams of SEM Amos 19:00 as in Figure 2.

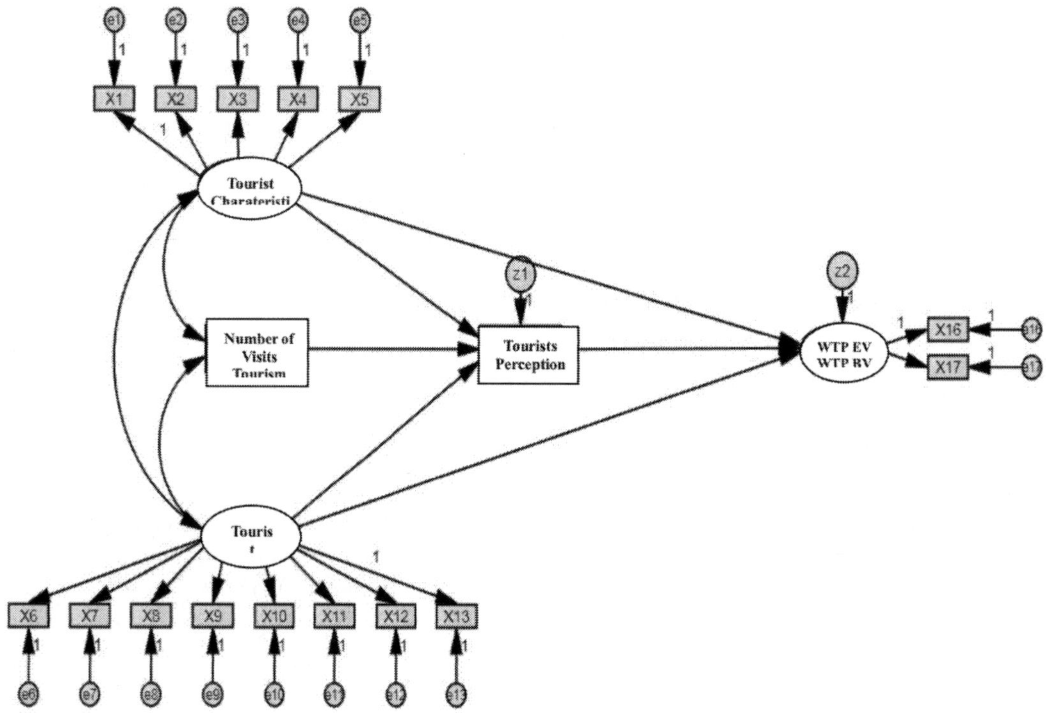

Figure 2. path diagrams of SEM Amos

Where : X_1 = Age rating, in units of years
X_2 = Tourists Education, in units of years
X_3 = Tourists Income per month, in units IDR/Month
X_4 = Gender/Sex (1 = Female ; 2 = Male)
X_5 = Job type rating (1 = Student ; 2 = Household mothers ; 3 = Civil Official ; 4 = Private employees/self-employed and 5 = Lecturer/Researcher
X_6 = Distance shelter travelers with tourist sites (Km)
X_7 = Time tourists traveled in location (hours)
X_8 = Number of Tourists Group i to location j (person)
X_9 = Categories of Stay (1= family ;2=homestay ;3=Lodge Bed/Inn; 4= Hotel)
X_{10} = Main Goals tourists traveled to the location (1 = Panoramic ; 2 =Recreation/ Sports Beaches ; 3 = Water Sports ; 4 = Culinary ; 5 = Snorkling/Diving)
X_{11} = Leisure time in units of days/month
X_{12} = Travel Cost to locations such as marine tourism (IDR/trip)
X_{13} = Travel Cost Subtitution: Travel Expenses of other region (Liang Beach,

Central Maluku District) in units IDR/trip
X_{14} = Number of Tourist Arrivals during the year 2012 at the tourist sites Ambon (frequency/year)
Z_1 = Average value tourists perception with Likert Scale (1 = Not Strongly support ; 2 = Not support ; 3 = Self-support ; 4 = Support ; and 5 = Strongly supports).
Z_2 = Willingness to pay (WTP) consisting of :
X_{16} = Average WTP for Existence value in units of IDR/month .
X_{17} = Average WTP to value the heritance(Bequest value) in units of IDR/month.

Structural equation model (SEM) was a second-generation multivariate analysis technique that combined factor analysis and path analysis that allows researchers to simultaneously test and estimated the relationship between exogenous and endogenous multiple variables with many indicators [19-23]

3. Results and Discussion

3.1. Respondents Characteristic

In accordance with the purpose of the first study examines the characteristics of travelers on marine tourism destination in the city of Ambon then there were 140 tourists who responded to the study with characteristics as summarized in Table 2.

Table 2. Tourists characteristics data recapitulation respondents in Ambon, 2015

No	Categories	Information					
1	**Tourists origin**	Foreigns	Domestics	Local	-	-	-
	person	26	41	73	-	-	-
2	**Age Rating(years)**	15 - 24	25-34	35-44	45-54	55-64	> 64
	person	28	36	40	22	10	4
3	**Education (years)**	Junior HS	Senior HS	Diploma	Scholar	Double-degree	-
	person	8	59	21	40	12	-
4	**Occupation**	Student	Housewife	Entreprnuer	Civil Servant	Employee free	Lecturer/ Researcher
	person	28	7	32	39	25	9
5	**Income (IDR)**	< 2499999,-	< 4999999,-	< 7999999,-	< 9999999,-	> 9999999,-	-
	person	19	43	21	16	20	-
6	**Gender**	Male	Female	-	-	-	-
	person	76	64	-	-	-	-
7	**Familly status**	Fathers	Mothers	Child	-	-	-
	person	48	38	54	-	-	-

Sources : Primary Data, Processed in 2015

To work seen varies with the highest respondent worked as a teacher and civil servant is 39 (27.86 %) while the lowest is the retired and housewives (7 people or 0.05%). Fixed income earned from work respondents also varied with the largest revenue in the range of IDR.2.499.999,- 4,999,999,- was 43 respondents (30.71%). Gender and family status of the respondent is a factor that also plays an important role in one's tour. The biggest unknown is traveling men by 76 people while women numbered 64 people, while the status of the child respondents in the largest family traveled with 54 people, 48 people attended the respondent 's father, while the housewife by 38 people.

3.2 Description of Travel Demand, Total Visits, Perception and WTP Tourists

Tourists in tourist travel has its own preferences satisfy her liking. Tourists desire will be realized when managing marine tourism destination is able to accommodate the wishes and expectations. To determine travel demand, the number of tourist arrivals, tourist perception and WTP to the marine environment of tourist destinations in the city of Ambon, can be seen in Table 3. It is seen a variety of data about tourists who will show the actual demand patterns tourists. The results showed that tourists prefer destinations that can be accessed by both and has maritime tourism facility. Location was the most visited coastal Latuhalat village contained three marine tourism destinations (Santai, Namalatu and Waimahu). Distance of residence will affect tourists with access to a designated tourist destination.

Table 3. Data Recapitulation Tourists Demand, Number of Visits, Perception and Environment WTP value Tourists in Ambon City, 2015

No	Categories	Information				
1	Travelled location	Air Manis	Amahusu	Pintu Kota	Latuhalat	Hutumury
	person	14	19	14	61	18
2	Dwelling distance (Km)	< 500	500 - 1500	1500 - 2500	2500 - 3500	> 3500
	person	71	31	19	03	16
3	Travelling long time (Minute)	≤ 120	121 - 180	181 - 240	241 - 300	> 300
	person	14	24	36	49	17
4	Total group	≤ 2	3 - 4	5 - 6	7 - 8	> 8
	person	25	54	36	8	7
5	Stay Category	Family	Homestay	Lodge/Inn	Hotel	-
	person	51	37	29	23	-
6	Main goals trips	Panoramic view	Beach sports	Water sports	Culinary	Snorkling/ Diving
	person	20	33	19	38	30
7	Leisure times	≤ 5 days	6 - 8 days	9 - 11 days	12 - 14 days	> 14 days
	person	40	30	37	22	11
8	Travel cost Ambon(IDR)	≤ 250.000,-	< 750000	< 1250000	<1750000	>1750000
	Orang	54	51	10	05	20
9	Travel cost substituted (IDR)	≤ 250.000,-	< 750000	< 1250000	<1750000	>1750000
	person	64	45	7	12	12
10	Number of visits	1	2-4	5 -7	8-10	>10
	person	30	61	20	15	14
11	Tourists Perception	140 - 250 (Not Strongly S.)	252 - 364 (Not Suport)	364 - 476 (Self Suport)	476 - 588 (Suport)	588 - 700 (Strongly S.)
	person	-	-	26	86	28
12	WTP of the Existance value (IDR)	≤ 25.000,-	25000 - 100000	100000- 175000	175000 - 250000	> 250000
	person	60	54	19	05	02
13	WTP of the Bequest value (IDR)	≤ 25.000,-	25000 - 100000	100000- 175000	175000 - 250000	>250000
	person	55	60	12	11	02

Sources : Primary data , processed in 2015

Variable number of tourist arrivals was used as an intervening variable to bridge the demand characteristic variables tourists and tourist travelers with the assumed level of perception will affect travelers WTP value marine tourism on the environment. There are 30 respondents who first visited tourist destinations in the city of Ambon marine while most tourists have visited the city of Ambon 2-4 times the 61 people. The number of visits over 10 times and only had 14 people who are local tourists .

There were 15 questions to be indicators of the traveler 's perception. This study uses Measurement Scale (Likert Scale) with 5 (five) criteria for assessment of the lowest score of 1 (Strongly Supports) to the highest score of 5 (Very Supports). Respondents rating the sample in this study was (n) = 140 was the lowest score of 1 x 140 = 140 and the highest total score was 5 x 140 = 700. From the analysis of the data turned out to be the largest number of tourist perceptions within the range of values that "It supports" the 86 respondents, while for Scoring value " Not Very Supports and Not Supports " was not answered by the respondents .

WTP values of the respondents about the Existance value environment based on the standard tourist revenue each respondent and submitted the game offers (Bidding Games) by first giving statements about the existence of marine tourism for the community and environment. After that respondents give the appropriate level of income. From the values given by the value of the largest number of turns on < IDR.25.000,-/month by 60 respondents, while only 2 respondents were WTP the value of the environment where marine tourism over IDR 250.000,-. The average value where the respondent is IDR.57.814,29. As was the case with Existence values, the WTP value inheritance (Bequest Value) also have a value that is not too different. They differ only in the number of respondents who were WTP the value of the inheritance. In the range of < IDR.25.000,- from 60 respondents on the existence value to 55 people for the value of the inheritance. Different changes also in the range of values IDR.175.000 - 250,000,- which amounts to 5 to 12 respondents in the WTP values of respondents inheritance. (bequest value). The average Bequest value was IDR.62.210,71.

3.3 SEM for Research Marine Tourism

Study aspects of travel demand in the form of a causal relationship of the indicators of the exogenous variables travelers personal aspects, demand and the number of tourist arrivals, tourist perceptions about marine tourism destinations in the city of Ambon, and the WTP tourists for marine tourism environment changes due to the utilization of the limited the existence value and bequest value. For answering the goal of this study also used path analysis with SEM with Amos SPSS 19.00, because in addition to examine the relationship of exogenous and endogenous variables will also see a causal relationship between variables. The results obtained path analysis can be seen in Figure 3.

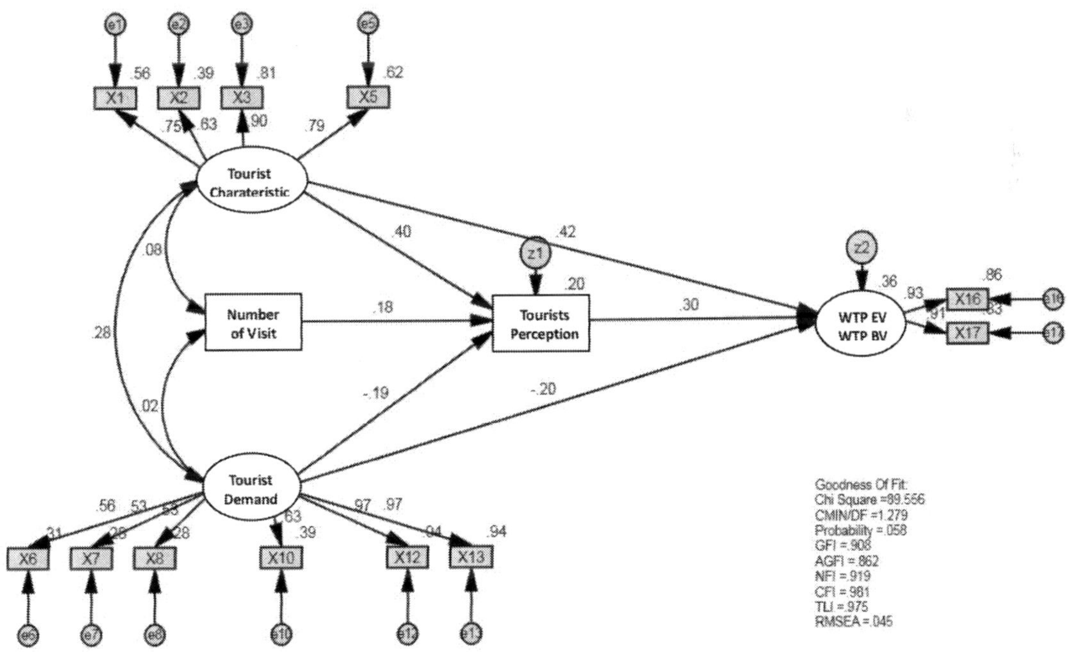

Figure 3. Casual relationship of Tourists with SEM Amos 19.00

Based on Figure 3, shown the existed of a relationship between variables that influence - the personal aspects of the exogenous variables (X_1 - X_5) and tourist demand (X_6 - X_{13}) with the perception of the endogenous variables (Z_1) and WTP tourists (Z_2) the existence of WTP values (X_{16}) and bequest value of tourism environment (X_{17}). It was assumed that the size of the level of perception and tourist WTP values are very dependent on the number of tourist visits conducted on marine tourism destination in the city of Ambon, because it was so in this study, there was one variable that mediated endogenous variables intervening intended. The results that intervening variables have a positive influence on the level of perception which in turn influence the WTP variable tourists. There were 3 indicator for weakening the overall corrected model was built because it has a low loading factors (< 0.5), which is an indicator of gender (X_4) ; category of stay (X_9) and the leisure time (X_{11}). Effect of latent constructs aspects of tourist demand in the picture above gives a negative effect on the perception and willingness to pay for tourists, this suggests that the higher demand of tourists with a variety of existing indicators that the lower the WTP and the perception of tourists on the environment of marine tourism in the city of Ambon. Instead latent constructs personal aspects tourists with indicators (age, income, occupation and education) have a positive influence on tourist perceptions and WTP.

Correlational relationships also occur in tourists latent constructs personal aspects and aspects of tourist demand with a variable number of tourist visits, this suggests that these three variables have a relationship that mutually affect one another especially between aspects of the personal aspects tourists demand. In the picture above also looks confirmatory factor analysis (CFA) with the results of goodness of fit which indicates that the overall rating is a model of research on "fit model" is acceptable.The results of the CFA model testing tourists construct the boundary value and criteria can be seen in Table 4.

Table 4. Variable Testing Results CFA Tourists

No	Goodness Of Fit Index	Cut off value	Result	Criteria
1	$_x2$ - Chi Square	< 91.236	88.56	Good Fit
2	Significance Probability	≥ 0,05	0,06	Good Fit
3	DF	≥ 0	70	Over Identified
4	GFI	≥ 0,90	0,91	Good Fit
5	AGFI	≥ 0,90	0,90	Good Fit
6	CFI	≥ 0,90	0,98	Good Fit
7	TLI	≥ 0,90	0,98	Good Fit
8	CMIN/DF	≤ 2,0	1,28	Good Fit
9	RMSEA	≤ 0,08	0,05	Good Fit

Sources : Primary data, processed in 2015

Furthermore, to determine the influence of indicators endogenous constructs in research on these tourists can be seen in Table 6. of the regression weights. Based on Table 5, seen a significant influence on the overall indicator constructs endogenous, it is characterized by the value of $P < 0.05$ or there are asterisks (***). Observed endogenous construct indicators that give the highest positive value of Critical Ratio is Travel Cost marine tourism to the city of Ambon with the value 28.968 (X_{12}) followed by indicators of WTP (Bequest value) marine tourism environment (X_{17}) = 11.117 and tourist revenues (X_3) at 10.275. In this study also looks a casual relationship between endogenous constructs (latent variables) that negatively impact tourists that demand aspects (indicators X_6 - X_{13}) with a value of -2.27 for indicators of tourist perception and value of -2.412 relation to the latent constructs WTP to value the existence and bequest value. It can be interpreted, in the event of tourists such as changes in demand for one unit, then the perception of tourists and willingness to pay on the value of the marine environment of tourist destinations to be decreased by 2,272 tourist perception and assessment of IDR.2,412 for WTP value.

Table 5. Variable Regression Weights perception, personal aspects, aspects of demand, the number of tourist visits and WTP at Marine Travel Destinations.

Variable		Variable	Estimate	S.E.	C.R.	P	Label
Tourists Perception	<---	Number of Tourists Visits	.024	.010	2.366	.018	par_11
Tourists Perception	<---	Tourists Demand	-.075	.033	-2.272	.023	par_12
Tourists Perception	<---	Tourists Characteristic	.020	.004	4.434	***	par_13
WTP EV_dan BV	<---	Tourists Perception	.661	.181	3.658	***	par_10
WTP EV_dan BV	<---	Tourists Characteristic	.045	.011	4.142	***	par_14
WTP EV_dan BV	<---	Tourists Demand	-.173	.072	-2.412	.016	par_15
X1 Tourists Age	<---	Tourists Characteristic	1.000				
X2 Tourists Education	<---	Tourists Characteristic	.204	.028	7.198	***	par_1
X3 Tourists Income	<---	Tourists Characteristic	.099	.010	10.275	***	par_2
X5 Tourists Job	<---	Tourists Characteristic	.138	.015	9.027	***	par_3
X13 Travel Cost subtituted	<---	Tourists Demand	1.000				
X12 Travel Cost Ambon	<---	Tourists Demand	.980	.034	28.968	***	par_4
X10 Main Goal Traveled	<---	Tourists Demand	.566	.062	9.144	***	par_5
X8 Total Group	<---	Tourists Demand	1.523	.214	7.130	***	par_6
X7 Long Time Traveling	<---	Tourists Demand	.138	.019	7.184	***	par_7
X6 Dwelling Distance	<---	Tourists Demand	1.168	.152	7.691	***	par_8
X16 Bequest Value	<---	WTP EV_dan BV	1.000				
X17 Existance Value	<---	WTP EV_dan BV	.962	.087	11.117	***	par_9

Sources : Primary data, processed in 2015

4. Conclusions

There 4 of 5 (five) characteristics of respondents rating, which has significant influence were age, education, income and employment, while gender of respondents had no influence in the perception and WTP for tourists. There is a positive correlation between the characteristics of tourists with the number of tourists and tourism demand. Tourist Demand constructs have a significant influence on the perception and WTP. Construct categories of stay and leisure time, does not have a significant influence. Casual relationship model latent constructs respondent characteristics, travel demand and the number of tourist visits as intervening variable the level of perception and WTP or tourists showed a very close relationship positively (construct characteristics of respondents and the number of tourist visits) or negatively (construct tourist demand).

Research constructs such as tourists characteristics and travel demand should be a custodian of the data base for policy decisions on programs and activities of marine tourism in the city of Ambon which is often overlooked when compared to marine tourism supply function is always performed. Knowledge of consumer opinion of the tourists perceptions about tourist destinations in the city of Ambon and the WTP value (existence and Bequest value) of marine tourism environment shows a strong intention of tourists to the situation and condition of marine tourism in the city of Ambon. This needs to be a constructive input for the development of marine tourism.

Acknowledgements

Thank you especially to my promoter, Professor Lachmuddin Sya'rani MSc. who has guided during the doctoral education process, also to Professor Azis Nur Bambang and Professor Supriharyono as Co Promoters. To Ambon Tourism Office and Mollucas Provincial Tourism Office which has supported this research.

References

[1] Badan Pusat Statistik Kota Ambon, 2015. Kota Ambon dalam Angka, 2015. Sektor Pariwisata Kota Ambon.
[2] Indonesia Tourism Care 2012. Mass Tourism and Implications in Indonesia Tourism Destination (accesed date at 12 December 2016 dari htttp://www. Indonesiatourism care. com)
[3] Spillane J C 2004. Indonesian Tourism : Economic Strategy and Culture Engineering. Publisher : BP. Kanisius - Yogyakarta.
[4] Kontogeorgopoulos N 2004 *Geographic J.* **61** (1) : 1–11.
[5] Cole S 2006 *J. Sustain. Tourism.* **14** (6) 629 - 644
[6] Koens J F, C Dieperink, and M Miriam 2009. *J. Environ. Dev. Sustain.* **11** (3) 1225 – 1237.
[7] Fandeli Ch 1999. Ekowisata dalam Paradigma Baru Pariwisata. Makalah Ilmiah pada semiloka "Sustainable Tourism Development"di Universitas Jendral Sudirman. Purwokerto, tanggal 22 - 25 Februari 1999. Purwokerto.
[8] Okazaki E 2009 *J. Sustain. Tourism.* **16** (5) 511 – 529.
[9] Loraine N and B Thapa 2010 *J. Environ. Dev. Sustain.* **12** (4) 839–857
[10] Weiler B and C M Hall 2006. (Eds) Special Interest Tourism. Publisher. Belhaven - London.
[11] Damanik J 2013. Manajemen Destinasi Pariwisata. (Sebuah Pengantar Ringkas). Edisi Revisi. KEPEL Press Yogyakarta.
[12] Scherrer P, A Alonso, and L Sheridan 2009 *Int. J. Tourism Res.* **11** (2) 451–463
[13] Balesteros C 2011 *J. Sustain. Tourism.* **19** (4) 1018 – 1029.
[14] Simpson P M and J A Siguaw 2008 *J. Travel Res.* **47** (2) 167 - 182.
[15] Tsung H L 2009 *J. Sustain.Tourism.* **17** (6) 727 – 745.
[16] Stuart L and D Hawkins 2010 *J. Business Ethics.* **89** (2) 569 – 585.
[17] Dolezal C 2011 *ASEAS - Austrian J. South-East Asian Studies.* **4** (1) 129-138.
[18] Prayag G 2012 *Int. J. Tourism Res.* **14** (2) 1 - 15.
[19] Gefen D, D Straub, and M C Boudreau 2006 *J. Communication of the Association for Information System.* **4** (7) 1 - 77.
[20] Chin W W, R A Peterson, and S P Brown 2008 *J. Marketing Theory and Practice.* **16** (4) 287 - 298.
[21] Bollen K.A and J B Kirby 2009 *J. Socio Methodolog.* **39** (1) : 327 - 355.
[22] Hwang H, N K Malhotra, Kim Y, and M A Tomiuk 2010 *J. Marketing Res.* **67** (1) 699 - 712.
[23] Gefen D, E E Rigdon, and D Straub 2011 *J. MIS Quarterly* **35** (2) 3 - 14.

Identification Sponges-Associated Fungi From Karimunjawa National Park

Agus Trianto.[1,2*], Agus Sabdono[1], Baskoro Rochaddi[1], Desy Wulan Triningsih[2], Dewi Seswita Zilda[3]

[1] Department Marine Science, Faculty of Fisheries and Marine Science, Diponegoro University, Semarang, Indonesia. Jl. Prof. Soedarto, SH, Tembalang, Semarang, Central Java, Indonesia.
[2] Natural Product Laboratory, Centre for Research and Services, Diponegoro University. Jl. Prof. Soedarto, SH, Tembalang, Semarang, Central Java, Indonesia.
[3]Research Center for Marine and Fisheries Product Processing and Biotechnology, Jl.KS Tubun Petamburan Pusat, Jakarta.
*agustrianto@undip.ac.id

Abstract. Marine sponges are rich sources of bioactive substances with various pharmacological activities. Previous studies have shown that most bioactive compounds were originally produced by associated-microorganisms. Fungi associated with the marine sponges collected off Karimunjawa National Park were isolated and identified by morphological characteristics and molecular level analyses based on internal transcribed spacer (ITS) regions. A total of 2 isolates which were characterized, the fungi *Penicillium spinulosum* and *Trichoderma virens* have been revealed.

Key words. Associated fungi, *Penicillium spinulosum, Trichoderma virens,* sponge

1. Introduction

Our investigation on marine sponges have shown that the sponges are potential sources of anticancer compounds [1]. Many research have also proven that the sponges have pharmacological potency as antibacterial, antifungal, antimalarial, and antioxidant [2-6]. However, among the bioactive compounds, only few of them passed to the preclinical and clinical test stage, as they are source limited [7]. Extracting a large amount of a particular compound from nature will inevitably disturb the local ecology. Chemical synthetics are the preferred method for producing the bioactive compounds; however, in many cases, this method is not feasible due to the complexity and chirality of the target compounds [8].

Sponges are well known to contain a large number of microorganisms within the mesophyll that some of them permanently associated with the host sponges [9]. Some microsymbiont provide the bioactive compounds [10]. Several studies indicated that some metabolites were produced by associated microorganisms [11].

In recent years, some fungi have been shown to be sources of bioactive compounds and enzymes that important for industrial or ecological properties. Some bioactive compounds produced by the genus *Fusarium* have various biological activities, i.e. antifungal, antibacterial, and antimalarial [12-14].

In this paper, we describe the isolation and molecular identification of the sponge-associated fungi from marine sponges collected from Karimunjawa National Park, Indonesia.

2. Materials and Methods

2.1. Sponges collection
The marine sponges were collected from the Karimunjawa National Park, Karimunjawa, Indonesia by SCUBA diving at 3-15 m depth. The specimens were kept cool until the inoculation process [15].

2.2. Isolation of the associate fungus
Isolation of the fungus from the sponges was conducted using the method proposed by Safaeian *et al.* with modification of the sponge size and media. The sponges were washed with sterilized seawater before inoculation process to remove any associated microorganisms from their surface [16]. Each specimen was then cut into small piece (approx. 1 cm x 1 cm x 0.3 cm). The fungus was isolated by putting each piece of the sponge directly on the surface of malt extract agar (MEA) medium; one agar plate contains one sample to avoid contamination and confusion regarding the source of fungi. After incubation for three days, the fungal colonies were separated based on their morphological characteristics, and each colony was inoculated on a new agar plate containing MEA. This process was repeated until a pure sample of each isolate was obtained [17].

2.3. Sponge identification
The sponge host of the potential isolates was identified based on morphological and spicules characteristics according to a method proposed by Hooper [18].The features were examined for morphological characteristics including colony shape, color, texture, consistency, and surface type. For the spicules preparations, a small fragment of the sponge was placed in a small Erlenmeyer flask, then a small quantity of commercial bleach (sodium hypochlorite) was added. After the organic components dissolved leaving only the mineral skeleton, the bleach was diluted with ethanol and removed carefully. The washing process was repeated until the clean spicules were obtained. The clean spicules suspension was transferred onto a glass slide and observed and photographed at 100x and 400x magnification.

2.4. Characterization of LB-18-1 and LB-19-4 isolates

2.4.1. Phenotypic characterization. The fungal isolates were grown on MEA medium (5 days) before analysis. The photograph the morphological characteristics under a microscope. Isolate identification was conducted using a fungal taxonomical book and taxonomical guidance [19].

2.4.2. Molecular identification of the fungal using 18S rRNA gene analysis. The DNA isolate was extracted with Chelex following the instruction provided by the company (Bio-Rad, US) [20]. In brief, the mycelia were added with 50 µL-100 µL aquabides (ddh2O) and 1 mL of diluted saponin (Bio-Rad, US) in PBS (Phosphate Buffer Saline). The mixture was incubated for 24 hours in 4 °C prior to centrifugation at 13,000 rpm for 1 minute for supernatant separation. A 250 µl of 10% chelex (Bio-Rad, US) was added to the DNA extract and homogenized under vortex for 20 seconds prior to centrifugation at 13,000 rpm for 2 minutes. Then, the mixture was heated at 95 °C for 45 minutes and homogenized using Vortex homogenizer. The mixture was centrifuged again at 13,000 rpm for 2 minutes. After removing the supernatant, the DNA was extracted with isopropanol (Merck, Germany) for further analyses with electrophoresis on agarose gel (Bio-Rad, US). Electrophoresis was run at 100 volts using TAE (Tris-Acetate-EDTA) (Sigma-Aldrich, US) as a running buffer. Gels were observed under UV light after staining with ethidium bromide (Bio-Rad, US). Amplification PCR was conducted using universal primer for 18s rRNA (Bio-Rad, US). Sequencing of the DNA was conducted in Genetika Science Jakarta, Indonesia.

2.4.3. Analysis of sequence alignment of 18S rRNA gene and phylogenetic analysis. Phylogenetic analysis was conducted using BLASTn software provided by Gen Bank (NCBI), followed by alignment process using Clustal X software. The phylogenetic tree was developed using a MEGA version 5 software.

3. Results and Discussion

A total of 2 fungal symbionts from seven sponges collected off Karimunjawa National Park were isolated and identified. On the previous study, it was reported that Karimunjawa National Park provided antifungal sources from the sponge-associated fungi [21]. The fungal identification was carried out based on the morphological and molecular level.

The fungus LB-18-1 have white colony on MEA plate and the fungus LB-19-4 showed green on morphology. Based on the microscopic feature, it belongs to the genus *Trichoderma*. The occurrence of anamorphic or telomorphic phase of the fungal species [22], require the molecular level to identify the fungi. Genotypic characterization was done employing fungal 18S rRNA.

The genomic DNA from LB-18-1 isolate was amplified by PCR. This amplified gen product was used for further identification. The consensus sequence of 678bp of 18S rRNA (Figure 1) gen was generated from forward and reverse sequence data and searched through the BLAST homology data and the result showed 99% similarity with *Penicillium spinulosum* (GenBank accession no. JQ717353.1). Hence, the fungus LB-18-1 could be identified as *P. spinulosum*. The PCR product of LB-19-4 was sequenced and showed the length 623bp. After the BLAST homology searched, this fungus had 99% similarity with *Trichoderma virens* with accession no. KT803076.1.

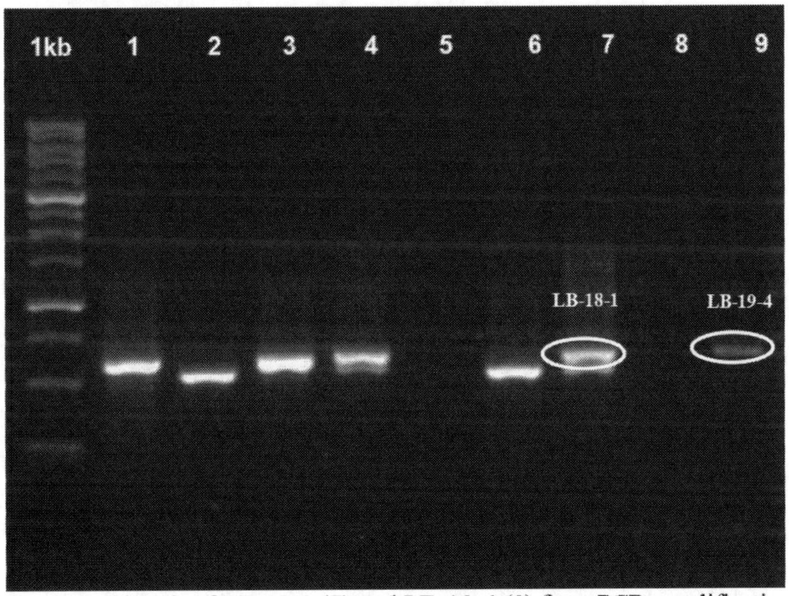

Figure 1. The fungal bands of LB-18-1 (7) and LB-19-4 (9) from PCR amplification products

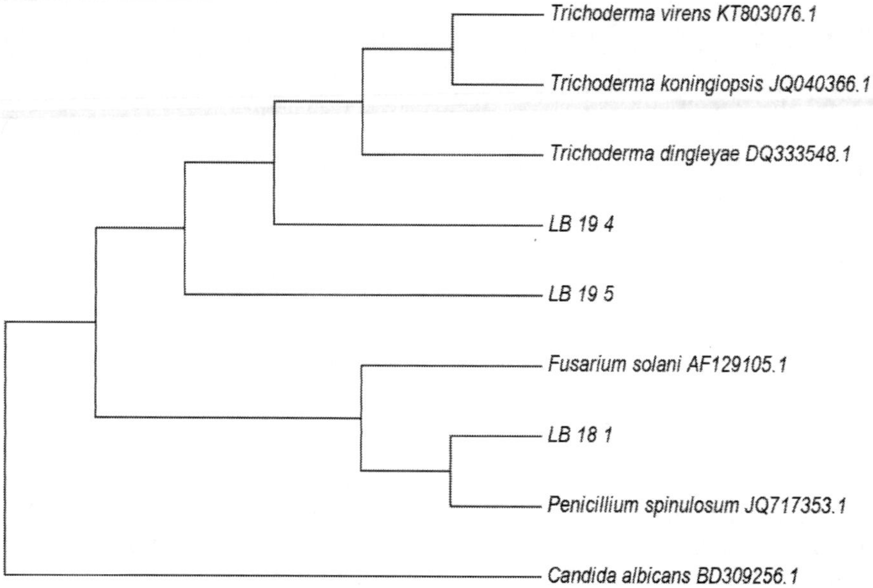

Figure 2. The phylogenetic tree of LB-18-1 and LB-19-4

Penicillium belongs to the sponge-generalist genera as it can be found in more than one sponge and dominantly isolated from sponges [23]. *Penicillium* is particularly known for their ability to produce important bioactive compounds. Alkaloids from *P. aurantiogriseum* have been reported to have cytotoxic activity against several cell lines [24].

The genus of *Trichoderma* has frequently been reported as the marine-derived fungus. This genus also widely discussed as terrestrial fungi mostly isolated from soil [25]. The serious observation and identification of marine-derived fungi have been done in a significant progress [26]. The marine isolates have been studied to have taxonomically close relation to species that are from terrestrial environments [27]. On the other hand, it produces different and unique secondary metabolites compare to the terrestrial one [28]. In the marine area, *Trichoderma* species were found ranging from the tidal zone and wetland [29]. The members of *Trichoderma* produce numerous metabolites and have been evaluated for its bioactivity on a various study as biocatalisator [30], antioxidant [31], biological control agent [32], anticancer [33], and antibiotic [34]. The recent report provides the information that the fungi from Karimunjawa National Park could be utilized as the source of the bioactive substance.

4. Conclusion

A total of 2 isolates which were characterized, the fungi *Penicillium spinulosum* and *Trichoderma virens* have been identified from sponges collected off Karimunjawa National Park. The further study about its bioactivities needs to be done to reveal the potential of Karimunjawa National Park biodiversity.

5. Acknowledgement

The authors deliver gratitude to Directorate of Research and Community Development, Directorate General of Straightening for Research and Development, Ministry of Research, Technology and Higher Education of the Republic of Indonesia for the funding through the Postgraduate Team Scheme (Hibah Tim Pasca Sarjana.

References

[1] Trianto A, Ridhlo A, Triningsih DW, and Tanaka 2017 *J IOP Conf. Series: Earth and Environmental Science* 55, 012069
[2] Qaralleh H, Idid S, Saad S, Susanti D, Taher M, and Khleifat K 2010 *J. Mycol. Med.* **4** 315–20.
[3] Sik W, Ki H, Young K, Am S, Soo Y, and Hee I 2006 *FEBS Lett.* **5** 1490–6
[4] Fattorusso C, Persico M, Basilico M, Taramelli D, Fattorusso E, Scala F, and Taglialatela-Scafati O 2011 *Bioorganic Med. Chem.* **1** 312–20
[5] Trianto A, Hermawan I, de Voogd N J, Tanaka J 2011 *Chem. Pharm. Bull.* **10** 1311–13
[6] Hadas E, Shpigel M, and Ilan M 2005 *Aquaculture* **1-4** 159–69
[7] Dunetz J R, Julian L D, Newcom J S, Roush W R, and Florida S 2008 *J. Am. Chem. Soc* **48** 16407–16
[8] Li Z Y, He L H, Wu J, and Qun J 2006 *J. Exp. Mar. Biol. Ecol.* **1** 75–85
[9] Visamsettia A, Ramachandrana S S, Kandasamy D 2016 *Microbiol. Res.* 185: 55–60
[10] Dunlap W C, Battershill C N, Liptrot C H, Cobb R E, Bourne D G, Jaspars M, Long P F and Newman D J 2007 *Methods* **4** 358–76
[11] Altomare C, Perrone G, Zonno M C, Evidente A, Pengue R, Fanti F and Polonelli L 2000 *J. Nat. Prod.* **8** 1131-5
[12] Shu R G, Wang F W, Yang Y M, Liu Y X and Tan R X 2004 *Lipids* **39** 667–3
[13] von Bargen K W, Niehaus E M, Bergande, K, Brun R, Tudzynski B and Humpf, H-U 2013 . *Nat. Prod.* **76** 2136–40
[14] Li Q and Wang G 2009 *Microbiol. Res.* **2** 233–41
[15] Safaeian S, Hosseini H, Asadolah A P and Farmohamadi S 2009 *J. Mycol. Med.* **1** 11–16
[16] Terkina I A, Parfenova V V and Ahn T S 2006 *Appl. Biochem. Micro.* **42** 173–6
[17] Hooper J N A. 2000 *Sponge Guide : Guide to Sponge Collection and Identification.* Queensland Museum, South Brisbane. (online) http://www.spongeguide.org/ speciesinfo.php? species
[18] Seifert K. F 1996 *Fusarium* Interactive Key. Her Majesty The Queen in Right of Canada *Agr. & Agri - Food Canada*. 65 pages.
[19] Walsh P S, Metzger D A and Higuchi R 1991 *BioTechniques* **10** 506-13.
[20] Trianto A, Sabdono A, Rochaddi B and Triningsih D W 2017 *Asian J. Microbiol. Biotechnol. Environ. Sci.* **3** (*preprint* AJ– F- 105)
[21] Taylor J W, Jacobson D J, Kroken S, Kasuga T, Geiser D M, Hibbett D S, et al. 2000 Fungal Genet. Biol. **31** 21–32.
[22] Li Q and Wang G. 2009.164: 233—241.
[23] Xin Z H, Fang Y, Du L, Zhu T, Duan T, Chen J, Gu Q and Zhu W 2007 *J. Nat. Prod.* **5** 853–5.
[24] Ahmed Abdel-Lateff. 2008 *Z. Naturforsch C* **9-10** 631-5
[25] Jones E B G, Sakayaroj J, Suetrong S, Somrithipol S, and Pang K L 2009 *Fungal Divers* **35** 1–187
[26] Ho¨ller H W, Wright A D, Mathee G F and Schulz B J 2000 *Mycol. Res.* **104** 1354–65
[27] Pruksakorn P, et al. 2010 *Med. Chem. Lett.* **20** 3658–63
[28] Saravanakumar K, Yu C, Dou K, Wang M, Li Y, Chen J 2016 *PLOS ONE* **12** e0168020
[29] Martins M P, Mouad A M, Boschini L, Regali Seleghim M H, Sette L D, Meleiro Porto A L. 2011 *Mar. Biotechnol. (NY)* **2** 314-20.
[30] Fang, F, Zhao J, Ding L, Huang C, Naman C B, He S, Wu B, Zhu P 2017 *Mar. Drugs* **8** 260
[31] Gal-Hemed I, Atanasova L, Komon-Zelazowska M, Druzhinina I S, Viterbo A and Yarden O 2011 *Appl. Environ. Microbiol.* **15** 5100–9
[32] Garo E, Starks C M, Paul R. Jensen, Fenical W, Lobkovsky E, and Clardy J 2003 *J. Nat. Prod.* 3 423-6
[33] Liaw C C 2015 *Planta Med.* **81** - PM_148

Preliminary Study Contamination of Organochlorine Pesticide (Heptachlor) and Heavy Metal (Arsenic) in Shallow Groundwater Aquifer of Semarang Coastal Areas

Baskoro Rochaddi[1], Chrisna Adhi Suryono[2], Warsito Atmodjo[1] and Alfi Satriadi[1]
[1]Oceanography Department, Diponegoro University, Semarang, Indonesia
[2]Marine Science Department, Diponegoro University, Semarang, Indonesia
Email : rochaddi@ymail.com

Abstract. The present study was conducted to assess the level of pesticide and heavy metal contamination in shallow aquifer of Semarang coastal areas. Results indicated that Heptachlor and Arsenic were detected in the water samples in the range 0.023-0.055 μg L^{-1} and 0,03-1,63 μg L^{-1} , respectively. Compared to the standard limits of the organochlorine contents in the water sample by World Health Organization (WHO) limits and Indonesian Drinking and Domestic Water Quality Standard for Ground Water (IWQS), groundwater of Semarang Coastal Areas was contaminated with pesticide and heavy metal. This study has proven the presence of organochlorine and heavy metal contamination of some shallow aquifer supplies in the coastal areas of Semarang.

Keyword: Pesticide, Heavy Metal, Shallow Aquifer

1. Introduction

Semarang coastal land and water has been contaminated by heavy metals and pesticides. The anthropogenic heavy metals have found in coastal water and land in western part of Semarang [1]. Furthermore, found several heavy metals in marine sediments such as Cr, Pb, and Cu [2]. Some of the organochlorine pesticides on marine water (Heptachlor, Aldrin, Endosulfan, Endrin, and pp-DDT) has been investigated, and it showed that all of the marine water samples contaminated with organochlorine pesticides[3].The previous research proved that the coastal are of Semarang has been contaminated by heavy metals and organochlorine both on the marine water and sediment. The heavy metals in water and sediments there is predicted come from outside as well as inside of Semarang coastal water. Semarang coastal area has been surrounded by many rivers such as Semarang river, West Banjir Kanal River, East Banjir Kanal River, Babon and Banger river. While the heavy metals which come from inside coastal water are assumed come from metal corrosion of ships and buildings.

Pesticides are usually used in agriculture since their benefits to mankind, and the environment outweigh the risks associated with their application[4]. However, rice cultivation, especially in low-lying paddy fields, is inherently risky to the environment since paddy water normally flows out into the surrounding water[5]. There have been several cases found where rice pesticides moved out of their intended area and contaminated the surrounding environments like rivers and lakes[6-8]. Consequently, large-scale application of these toxic materials in agriculture areas can contribute to the presence of those compounds in surface and groundwater, lakes, estuary and ultimately in the coastal areas. Most of these compounds are recalcitrant to biodegradation and their entry into the aquifer might be, poses many challenges to the existing of drinking well water from pollution. The possibility of pesticides polluting the coastal or marine environment is considered unlikely since paddy effluents normally travel long distances before they reach coastal areas. By then, most of the residues are degraded by physicochemical and biological factors. However, this is only true for rice paddy fields located upstream or more inland. Nowadays, nevertheless, humans are forced to utilize even the coastal lowlands for rice cultivation [9] due to the limited area of agricultural lands. Consequently, this increases the possibility of pesticides ending up in coastal environments

Arsenic (As) is one of the most important toxicants that is widely distributed in nature and occurs in the form of inorganic and organic compounds. Exposure to inorganic compounds may occur in a variety of ways through certain industrial effluents, chemical alloys, pesticides, wood preservative agents, combustion of fossil fuels, catalysts, glass, fire retardant, occupational hazards in mining that are anthropogenic and natural sources [10]. Arsenic is known to cause arsenicosis due to its manifestation in drinking water, the most common species being arsenate As (V) and arsenite As (III) [11]. Arsenic contamination of groundwater has led to a massive epidemic of arsenic poisoning in Bangladesh [12-13] and neighboring countries [14-15]. It is estimated that approximately 57 million people in Bangladesh [16] and 11 million people in Vietnam [14] are drinking groundwater with arsenic concentrations elevated above the World Health Organization's standard of 10 parts per billion [17]. This tragedy was occurring as the high arsenic levels present in drinking water. Bangladesh maintains an allowable limit of 50 ppb, and the limit in Canada is 25 ppb[18]. The arsenic found in Bengali and Canada drinking water is present naturally in the sediment but is released only by humans, primarily from the digging of shallow wells. Similarly unacceptably high arsenic levels are found in shallow-well domestic drinking water in the American Midwest [19].

Recently there has been increasing anxieties concerning arsenic-related problems. Occurrence of arsenic contamination has been reported worldwide, for example, Mongolia [20], USA[19][21-22], Bangladesh [12-13], Taiwan [23], Spain [24], *et cetera*. In Indonesia, no significant As contamination has been reported yet. However, local contaminations in soils, sea, and groundwaters with high As and Hg levels have frequently been revealed particularly around Buyat Bay mining areas [25-26].

Clean water is probably the biggest problem in many houses of coastal urban settlement. Based on field investigation, many homes in this region are not connected to the city water supply, and wells are often unpredictable. Most houses do not have overhead tanks for catchment or storage but, instead, have large ground storage tanks from which the water is pumped on demand by a small electric pump. The most prevalent of these is the shallow aquifer (dug well) which has become the major source of water in the coastal inhabitants. However, groundwaters and surface waters are never pure that probably contain a variety of organic compounds [27-28]. Organic chemicals such as polychlorinated biphenyls, PAHs and organochlorine pesticides have been identified in the surface water [29-31] and groundwater [32-34]. Contaminants from residential areas, industries and agricultural practices through migration and infiltration enter to ground waters [36-37]. Therefore, a study of the distribution and concentration of organochlorine in the shallow aquifer become the most relevant. This research aims to examine the organochlorine pesticide (Heptachlor) and Heavy metal (Arsenic) content of the shallow aquifer in Semarang coastal areas and assess how far conform to WHO/ Indonesia recommendation for fresh water/ drinking water

2. Material And Methods

Water sample collection: Sampling points were located on coastal areas flowing from the west part to the east part of Semarang. Figure 1 shows the sampling sites; the water samples were collected in polyethylene plastic bottles from 30 different dug wells across those cities. The water samples were collected in 1000 mL plastic bottles. Then, bottles were properly labeled and tightly sealed. All the water samples were brought to the laboratory for performing organochlorine pesticide (Heptachlor) and Heavy Metal (Arsenic) analyses. The parameters analyzed were psycho-hydrochemical properties and organochlorine pesticide (Heptachlor) and Heavy Metal (Arsenic) contents of shallow groundwater. The pH, salinity, and temperature were measured at the time of samples collection using pH meter, hand refractor salinometer and a standard centigrade thermometer (Horiba Co. Ltd., Japan. The analytical Arsenic procedure applied was the method of the Standard Method for the Examination of Water and Waste Water [37] using 15% methylene chloride in n-hexane and capillary columns. One-liter samples were extracted with a solvent mixture and then concentrated in a Kuderna-Danish apparatus. The extracts were cleaned up with Florisil column. The final pesticide extract was obtained in 5 mL of hexane. The samples were then analyzed by gas chromatograph Model Hitachi 163 FPD

IOP Conf. Series: Earth and Environmental Science **116** (2018) 012099 doi:10.1088/1755-1315/116/1/012099

(Flame Photometer Detector), and nitrogen High Pure (HP) was used as the carrier gas. A 2 $^{m\ a}$ glass column (3 m$_{i}$n ID) packed with 3% Silicon OV1 on 80-100 mesh Supelcon was used. Gas flow at 30 mL min', column temperature at 160-230°C, detector temperature at 290°C and the injector temperature at 290°C were maintained. Arsenic metal analysis. The water samples were acidified with an aliquot of HNO_3. About 100 mL each of the well-mixed acidified water was digested with five mL HCl. The digested samples were then analyzed for As and Fe using atomic absorption spectrometer Perkin Elmer Model 5000 [38]. Statistical analysis was conducted among the standard parameters using Linear correlation method. This was performed to know whether any relationship among the parameters existed or not

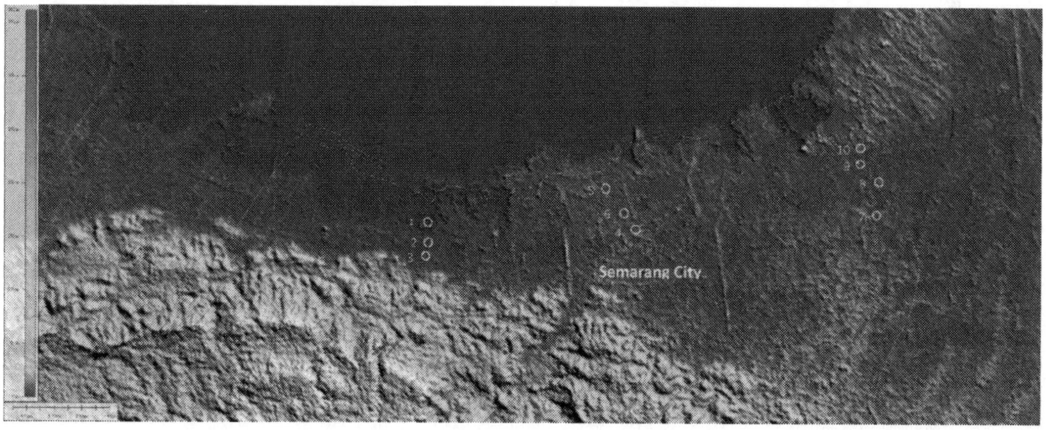

Figure 1. The map of sampling site location

3. Results And Discussion

Semarang coastal areas are divided into three areas (West, central and East). This city was chosen for study because of their representative for the main industrial and human center of northern coastal urban. Hence, those cities have a great probability of the existence of declining water quality resources. Pollution of groundwater has been reported for a number of urban throughout the world on a wide range of pollutants, such as organochlorine, heavy metals, nitrate, fecal, bacteria, virus, domestic waste, pesticide [39-47]. This is mainly due to a higher population density and more intensified agricultural and industrial activities [48-50]. In the present study, the results of some physicochemical parameters and selected priority Organochlorine Pesticide (Heptachlor) and Heavy Metal (Arsenic) in Shallow Groundwater Aquifer of Semarang Coastal Areas are preserved in Table1 and 2, respectively.

Table 1. Physico-chemical properties of water samples of shallow groundwater in the Semarang coastal areas

Sampling Site No	Physico-chemical Parameters					
	Colour	Odor	Teste	pH	Salinity	Temp
1	Brownish	Offensive	Tasteless	7	2	29
2	Bluish	Offensive	Tasteless	7	0	29
3	Colourless	Offensive	Tasteless	7	2	29
4	Colourless	Offensive	Tasteless	7	2	29
5	Greysh	Offensive	Tasteless	7	0	29
6	Colourless	Offensive	Tasteless	7	2	29
7	Colourless	Offensive	Tasteless	7	0	29
8	Colourless	Offensive	Tasteless	7	5	29

9	Colourless	Offensive	Salty	7	7	29
10	Colourless	Offensive	Salty	7	9	29

This study indicates that low variation existed among some physicochemical parameter (temperature, salinity, pH). Colour, taste, and odors were detected in water samples. Some of the water samples, particularly some water samples, did not comply with the standard limits for drinking. The water samples have color and offensive odor. The water temperature values obtained in this study varied slightly ranged from 28.5-29.5°C with a mean of 28.93°C. This is the most common water temperature in the tropical zone. This temperature range of water samples is supposed to be affected by the intensity of the sunlight as the temperature rose from 28.5-29.5°C on relatively hot day afternoon. Some workers reported that the temperature of water from tropical climate was little affected by seasonal variations of temperature [51-52]. Salinity range was varied from 0 to 9.0%o with a mean of 2.1±2.51%0. The highest desirable level for pH (7.0) is within the range of 6.5-8.0 values for drinking purposes [53].

Table 2. Range, frequency of occurrence and mean ± standard deviation of organochlorine pesticide (Heptachlor) and Heavy Metal (Arsenic) residue (μg L^{-1}) from shallow groundwater Semarang coastal areas

Number	Contaminant		Methods	Remark
	Heptachlor (ppb) μg/l	Arsen (ppb)		
1	0.038	0.54		
2	*bd*	0.13	GLC-ECD	1.Heptachlor
3	0.058	*bd*		α OD = 7.04
4	0.041	1.63	AAS Flame	
5	*bd*	0.5		2. As
6	0.025	6.21		α OD = 0.002
7	*bd*	3.42		ppb
8	*bd*	0.03		
9	0.023	*bd*		
10	0.027	*bd*		
Range	0 - 0.058	0 - 6.21		
Means + SD	0.0228 ± 0.02	1.2463 ± 2.05		
WHO/Ina	0.03/0.03	0.5		

Bd: bellow detected

Water samples from coastal areas of Semarang showed the presence of selected organochlorine pesticide (Heptachlor) (Table 2). Ten samples (60%) contained residues of heptachlor, ranging from 0 to 0.055 μg L^{-1}, and 70 % of samples were Arsenic contaminated whereas ranging from 0 to 6.21μg L^{-}. Comparison of the organochlorine contents in the water sample with World Health Organization (WHO) limits and Indonesian Drinking and Domestic Water Quality Standard for Groundwater showed that some of the household wells were contaminated with organochlorine.

The mean levels of heptachlor were exceeded the maximum permissible levels for drinking water. Variation in residual levels in these samples could be the due location of the sampling. The present study sufficiently shows the presence of organochlorine residues in most of the samples. Comparison of data on organochlorine residues of Semarang areas water samples showed that the highest levels and frequency of occurrence on heptachlor. Although we have not analyzed surface water samples for organochlorine residues (Heptachlor) in the present investigation, comparing the results reported here with other regional studies on the river and sea water contamination [54], different contamination levels with heptachlor. Heptachlor was not detected. It is not known whether this is due to irresponsible insecticide handling practice or groundwater transport processes.

The appearance of arsenic in groundwater and drinking water is a worldwide concern since it has been found that arsenic is one of the most toxic and carcinogenic chemical elements. In the present study As metal concentration was ranging from 0 - 6.21μgL^{-1} respectively in three different areas of Semarang coastal land. The maximum As concentration was observed in Cental part of Semarang (6.21 μgL^{-1}) and (3. 42 μgL^{-1}) in East part of Semarang. This condition could be explained that first, arsenic can enter the water supply from natural deposits in the earth or industrial and agricultural pollution. It is a natural element used for a variety of purposes within industry and agriculture. It is also a by-product of copper smelting, mining, and coal burning. Industries in Semarang were suspected release much more arsenic into the environment every year. Once released, arsenic remains in the environment for a long time. However, in the present results of this study when it was compared to the previous results in Buyat Bay for Drinking Water[53], it was lower. Because it may be stated that the environmental parameters like salinity, temperature, dissolved oxygen, and pH also have some effect on the accumulation of trace metals.

4. Conclusion

Characteristics of selected physicochemical parameters and organochlorine pesticide (Heptachlor) and Heavy Metal (Arsenic) residues in shallow groundwater on Semarang coastal areas has been determined in this study. Heptachlor was above the maximum permissible value recommended by WHO and Indonesian Water Quality Standard. Low concentrations of arsenic in water which are continuing to be used need to be monitored over time by authorities concerned to ensure that concentrations of arsenic in water of these shallow aquifer remain at an acceptable level over time. Some of these aquifer needs further purification to ensure their sustainability for consumption of coastal urban inhabitants. It is necessary to study further in these areas during the wet and dry period due to fully evaluate the organochlorine and heavy metal impact of pollution on shallow aquifer in seasonal variation

5. Acknowledgment

This article is part of the research which supported by the grant from non APBN 2017 Faculty of Fisheries and Marine Sciences Diponegoro University

6. Reference

[1] Suryono CA 2016 Polusi Logam Berat Antropogenik (As, Hg, Cr, Pb, Cu dan Fe) pada Pesisir Kecamatan Tugu Kota Semarang Jawa Tengah. *J. Kelaut. Trop.* **19** (1):37–42.
[2] Suryono CA 2016 Akumulasi Logam Berat Cr, Pb dan Cu dalam Sedimen dan Hubungannya dengan Organisme Dasar di Perairan Tugu Semarang. *J. Kelaut. Trop.* **2** (19):143–9.
[3] Suryono CA, Rochaddi B, Sabdono A, Susanti BT 2007 Physico-chemical characteristics and heavy metal contents in shallow groundwater of Semarang coastal region. *Ilmu Kelautan : Ind. J. Mar. Sci.* **12** (4):227–32.
[4] Cooper J, Dobson H. 2007 The benefits of pesticides to mankind and the environment. Crop Prot. **26** (9):1337–48.
[5] Miao Z, Padovani L, Riparbelli C, Ritter AM, Trevisan M, Capri E 2003 Prediction of the environmental concentration of pesticide in paddy field and surrounding surface water bodies. *Paddy Water Environ.* **1** (3):121–32.
[6] Numabe A, Nagahora S 2006 Estimation of pesticide runoff from paddy fields to rural rivers. *Water Sci Technol.* **53** (2):139–46.
[7] Kawakami T, Ishizaka M, Ishii Y, Eun H, Miyazaki J, Tamura K 2005 Concentration and distribution of several pesticides applied to paddy fields in water and sediment, from Sugao Marsh, Japan. *Bull Environ Contam Toxicol.* **74** (5):954–61.
[8] Nakano Y, Miyazaki A, Yoshida T, Ono K, Inoue T 2004 A study on pesticide runoff from paddy fields to a river in rural region—1: field survey of pesticide runoff in the Kozakura River,

IOP Conf. Series: Earth and Environmental Science **116** (2018) 012099 doi:10.1088/1755-1315/116/1/012099

Japan. *Water Res.* **38** (13):3017–22.
[9] Kyuma K. 2003 Soil resources and land use in tropical Asia. *Pedosphere.* **13** (1):49–57.
[10] Smedley PL, Kinniburgh DG 2002 A review of the source, behaviour and distribution of arsenic in natural waters. *Appl geochemistry* **17** (5):517–68.
[11] World Health Organization 1993 Guidelines for Drinking Water Quality, Vol 1. Recommendations. World Health Organization.
[12] Anawar HM, Akai J, Komaki K, Terao H, Yoshioka T, Ishizuka T 2003 Geochemical occurrence of arsenic in groundwater of Bangladesh: sources and mobilization processes. *J. Geochemical Explor.* **77** (2):109–31.
[13] Chowdhury AMR 2004 Arsenic crisis in Bangladesh. *Sci Am.* **291** (2):86–91.
[14] Berg M, Tran HC, Nguyen TC, Pham HV, Schertenleib R, Giger W 2001 Arsenic contamination of groundwater and drinking water in Vietnam: a human health threat. *Environ Sci Technol.* **35** (13):2621–6.
[15] Nickson R, McArthur J, Burgess W, Ahmed KM, Ravenscroft P, Rahmanñ M 1998 Arsenic poisoning of Bangladesh groundwater. *Nature.* **395** (6700):338.
[16] Yu WH, Harvey CM, Harvey CF 2003 Arsenic in groundwater in Bangladesh: a geostatistical and epidemiological framework for evaluating health effects and potential remedies. *Water Resour Res.* **39** (6).
[17] World Health Organization 2011 Guidelines for drinking-water quality. Edition Four. **38** WHO chronicle 104-108 p.
[18] Nordstrom DK 2002 Worldwide occurrences of arsenic in ground water. *Science* **296** (5576):2143–5.
[19] Kim M-J, Nriagu J, Haack S 2002 Arsenic species and chemistry in groundwater of southeast Michigan. *Environ Pollut.* **120** (2):379–90.
[20] Gong Z, Lu X, Watt C, Wen B, He B, Mumford J et al 2006 Speciation analysis of arsenic in groundwater from Inner Mongolia with an emphasis on acid-leachable particulate arsenic. *Anal Chim Acta.* **555** (1):181–7.
[21] Welch AH, Westjohn DB, Helsel DR, Wanty RB 2000 Arsenic in ground water of the United States: occurrence and geochemistry. *Ground Water* **38** (4):589–604.
[22] Hudak PF 2008 Distribution of arsenic concentrations in groundwater of the Seymour Aquifer, Texas, USA. *Int. J. Environ. Health Res.* **18** (1):79–82.
[23] Chen S-L, Dzeng SR, Yang M-H, Chiu K-H, Shieh G-M, Wai CM 1994 Arsenic species in groundwaters of the blackfoot disease area, Taiwan. *Environ. Sci. Technol.* 28 (5):877–81.
[24] Garcia-Sanchez A, Moyano A, Mayorga P 2005 High arsenic contents in groundwater of central Spain. *Environ. Geol.* **47** (6):847–54.
[25] Khalik A, Wulandari F. Buyat Bay tests show high mercury levels. Jakarta Post July. 2004
[26] KLH. Environmental Quality Assessment of Buyat Bay and Totok Bay. 2004.
[27] Fitch A, Du J 1995 Solute transport in clay media: Effect of humic acid. *Environ. Sci. Technol.* **30** (1):12–5.
[28] Nelson SD, Letey J, Farmer WJ, Williams CF, Ben-Hur M 1998 Facilitated transport of napropamide by dissolved organic matter in sewage sludge-amended soil. *J. Environ. Qual.* **27** (5):1194–200.
[29] Jaynes DB, Hatfield JL, Meek DW 1999 Water quality in Walnut Creek watershed: Herbicides and nitrate in surface waters. *J. Environ. Qual.* **28** (1):45–59.
[30] Louchart X, Voltz M, Andrieux P, Moussa R 2001 Herbicide transport to surface waters at field and watershed scales in a Mediterranean vineyard area. *J. Environ. Qual.* **30** (3):982–91.
[31] Rupp DE, Peachey RE, Warren KL, Selker JS 2006 Diuron in surface runoff and tile drainage from two grass-seed fields. *J. Environ. Qual.* **35** (1):303–11.
[32] Muszkat L, Lahav D, Ronen D, Magaritz M 1993 Penetration of pesticides and industrial organics deep into soil and into groundwater. *Arch. Insec.t Biochem. Physiol.* **22** (3-4):487–99.
[33] Barbash JE, Thelin GP, Kolpin DW, Gilliom RJ 2001 Major herbicides in ground water. *J. Environ. Qual.* **30** (3):831–45.
[34] Lapworth DJ, Gooddy DC, Stuart ME, Chilton PJ, Cachandt G, Knapp M 2006 Pesticides in

groundwater: some observations on temporal and spatial trends. *Water. Environ. J.* **20** (2):55–64.

[35] Ilani T, Schulz E, Chefetz B 2005 Interactions of organic compounds with wastewater dissolved organic matter. *J. Environ. Qual.* **34** (2):552–62.

[36] Yusuf KA 2007 Evaluation of groundwater quality characteristics in Lagos-City. *J. Appl. Sci.* **7** (13):1780–4.

[37] Federation WE 2005 Association APH. Standard methods for the examination of water and wastewater. Am Public Heal Assoc Washington, DC, USA.

[38] Grasshoff K, Kremling K, Ehrhardt M 2009 Methods of seawater analysis. (John Wiley & Sons)

[39] Mille CJ, Wilson LG, Amy GL, Brothers K 1993 Fate of organochlorine compounds during aquifer storage and recovery: the Las Vegas experience. *Groundwater* **31** (3):410–6.

[40] Somasundaram M V, Ravindran G, Tellam JH 1993 Ground-Water Pollution of the Madras Urban Aquifer, India. *Groundwater* **31** (1):4–11.

[41] Zhang WL, Tian ZX, Zhang N, Li XQ 1996 Nitrate pollution of groundwater in northern China. *Agric. Ecosyst. Environ.* **59** (3):223–31.

[42] Kaçaroğlu F, Günay G 1997 Groundwater nitrate pollution in an alluvium aquifer, Eskişehir urban area and its vicinity, Turkey. *Environ. Geol.* **31** (3–4):178–84.

[43] Mazari-Kriart M, Torres-Beristain B, Velazquez E, Calva JJ, Pillai SD 1999 Bacterial and viral indicators of fecal pollution in Mexico City's southern aquifer. *J. Environ. Sci. Heal. Part A.* **34** (9):1715–35.

[44] Reddy MR, Dunn SJ 1984 Effect of domestic effluents on groundwater quality: A case study. *Sci. Total Environ.* **40** (1):115–24.

[45] Howard G, Pedley S, Barrett M, Nalubega M, Johal K 2003 Risk factors contributing to microbiological contamination of shallow groundwater in Kampala, Uganda. *Water Res.* **37** (14):3421–9.

[46] Downs TJ, Cifuentes-Garcia E, Suffet IM 1999 Risk screening for exposure to groundwater pollution in a wastewater irrigation district of the Mexico City region. *Environ. Health Perspect.* **107** (7):553.

[47] Sabdono A, Rochaddi B, Chrisna AS, Susanti BT 2008 Persistent organochlorine residues in household wells of Java coastal urban areas, Indonesia. *J. App. Sciences.* **8** (12):2318-2323

[48] Baudino OM, Suero EA, Augusto M, Gimenez ME, Flores N 2003 Monitoring organochlorine pesticides in surface and ground water in San Juan Argentina. *J. Chil. Chem. Soc.* **48** (2):7–12.

[49] Fatoki OS, Awofolu OR 2004 Levels of organochlorine pesticide residues in marine-, surface-, ground-and drinking waters from the Eastern Cape Province of South Africa. *J. Environ. Sci. Heal. Part B.* **39** (1):101–14.

[50] Mukherjee I, Gopal M. Organochlorine 2002 insecticide residues in drinking and ground water in and around Delhi. *Environ. Monit. Assess.* **76** (2):185–93.

[51] Efe SI, Ogban FE, Horsfall MJ, Akporhonor EE 2005 Seasonal variations of physico-chemical characteristics in water resources quality in western Niger Delta region, Nigeria. J Appl Sci Environ Manag. **9** (1):191–5.

[52] Ravindra, K., A. Meenakshi, M. Rani and A. Kaushik 2003 Seasonal variations in physico-chemical characteristics of River Yamuna in Haryana and its ecological best-designated use., *J. Environ. Monit.* **1 5**: 419-426

[53] OWorld Health Organization 2004 Guidelines for drinking-water quality. Vol. **1.** World Health Organization;

[54] Ratnaningsih D, Helmy M, Bagus BE, Nety W, Heni P A 2002 Survey on water pollution by Endocrine Disrupter compounds; Monitoring of Organochlorine Pesticides, Phenols and Phthalates In: The Coastal Hydrosphere of Indonesia. United Nations University International Symposium on Endocrine Disrupting Chemicals (EDCs), April 17-18, University of Malaya, Kuala Lumpur, Malaysia. pp: 313-316.

Proliferative Activity of Mammary Carcinoma Cells by AgNOR Count in C3H mice Receiving Ethanol Extract of Sponge *Haliclona sp*

Lanceria Sijabat[1], Neni Susilaningsih[1], Agus Trianto[2,4], Retno Murwani[3,4*]

[1]Department of Histology, Faculty of Medicine Diponegoro University
[2]Laboratory of Marine Biotechnology, Faculty of Fisheries and Marine Science, Diponegoro University
[3]Faculty of Animal and Agriculture Science, Laboratory of Physiology dan Biochemistry, Diponegoro University
[4]Natural Product Laboratory, Centre of Research and Services, Diponegoro University

* rmurwani@gmail.com

Abstract. Quantification of argyrophilic nucleolar organizer region (AgNORs) was considered as one of markers of proliferative activity of cancer cells. Sponge *Haliclona sp* extract contains anticancer bioactive compounds and our previous study showed that the extract was able to improve histological grade of induced mammary adenocarcinoma in mice. The following research was conducted to study the extract administration on the proliferative activity of the carcinoma cells represented by AgNOR count in mice. This experimental study applied post test only control group design. Twenty C3H mice were divided into four groups namely C (control), H_1, H_2 and H_3. Each group was given 0, 0.15, 1.5, and 15 mg *Haliclona sp* extract respectively. After three weeks of extract administration, mice were inoculated with breast cancer cells from donor mice. The extract administration were continued for another three weeks. AgNOR count was performed on tumor sections and expressed as mean of AgNOR (mAgNOR) and percentage of AgNOR (pAgNOR). Means of mAgNOR in C, H_1, H_2 and H_3were 4.070, 3.195, 3.450, and 3.190 respectively. Means of pAgNOR in C, H_1, H_2 and H_3 were 34,40, 25,40, 38,40 and 19,80 respectively. The lowest means of mAgNOR and pAgNOR which is an indication of lower proliferative activity of the cancer cells was found in H3. However no significant difference can be found among treatment groups ($p>0.05$). Using AgNOR count, the ethanol extract of *Haliclona sp* could not show significant reduction in proliferation of mammary carcinoma cells of C3H mice. This finding support the view that AgNOR alone could not be used to determine pathology of cancer cells.

Keyword : breast tumor, mammae, sea, marine, soft coral

1. Introduction

Breast cancer remains as a disease with high prevalence worldwide.[1] Chemotherapy is an important element of breast cancer treatment either as an adjuvant, neoadjuvant, or in paliatif treatment.[2,3] In other hand there is breast cancer which is resistant to chemotherapy and attempts to discover new cancer drug with low toxicity is continously being sought. One of novel source for antitumor substance is derived from marine biota.[4,5] Sponge and soft coral contains bioactive compounds which function as a defense system aginst natural enemies. The compounds belong to secondary metabolites and they have anticancer acitivity. The compounds were then being synthesized chemically[4-6].

Previous study has shown that sponge Haliclona sp from Labuan Bajo waters in Flores Indonesia can inhibit L1210 cells (leukemia derived cell line).[7] On the basis of its IC_{50} value, toxicity studies of graded dosage of Haliclona sp extracts had been carried out in vivo mice.[8] The extract of *Haliclona sp* was considered to be safe to be given by oral route. Therefore, subsequent study was conducted to test further the anti tumor activity of the extract in vivo laboratory animals. There has been no research regarding the anticancer activity of Haliclona sp extract against *adenocarcinoma mammae in vivo*.[9]

In the following study a graded level *Haliclona sp* extract was tested against *adenocarcinoma mammae* in C3H mice. The study used argyrophilic nucleolar organizer region (AgNOR) staining to measure the proliferative activity of cancer cells. AgNORs expression reflect the speed of cell cycle and the the doubling time of tumor cells. Proliferative activity determined the rate of tumor mass growth thereforAgNOR is considered as one of markers of proliferative activity of cancer cells.[10].

2. Material and Methods

Feed pellet for CH3 mice used commercial feed (CP781) with 31 – 33 % protein, 3 - 5 % fat, 4 – 6 % crude fibre, 10 - 13 % ash, and 11 – 13 % moisture. Feed was mixed with Halicona sp (Hs) extracts by weighing 150 grams for each treatment. Extracts were weighed according to three treatments namely 0.3, 3, and 30 mg extracts per mouse per day. The extracts were dissolved in 200 mL ethanol 96% until homogenous. The feed pellet was mixed with the dissolved extract and stood until it was absorbed completely. Control feed pellet was mixed with 200 mL ethanol only with no extracts. The feed pellet was dried by vacum and heated at 40°C to complete the drying and stored in the freezer until use.

2.1. Donor cancer cells

Cancer cells were obtained from donor mice. The cancer lumps were surgically removed from the donor mouse and transferred to sterile physiological salt containing petri dish. The tissue was disintegrated using special scissor to release the cancer cells. These cells were ready to be transplanted into healthy C3H3 mice.

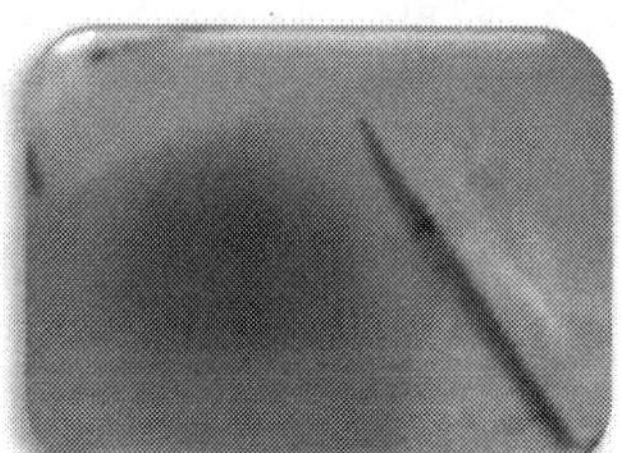

Figure 1. Donor cancer cells in physiological salt

2.2. Haliclona sp (Hs) Extracts administration to adenocarcinoma bearing C3H3 Mice

Mice were adapted for 7 days to the lab environment where they were kept and feed pellet and drinking water were provided in free access. Fresh feed pellet, drinking water, and cleaning were done daily. The experiment was designed as "*The Post Test Only Control Group Design*" [3] with one control and three treatments [4]. Twenty of 8-10 weeks of age mice with an initial weight of 20-25 gram were randomly allocated into: control group (C), receving 0.15 mg Hs extracts (H-1), and receving 1.5 mg Hs extracts (H-2), 15 mg Hs extracts (H-3) per mouse per day respectively for three weeks. Cancer cells were introduced to all groups from a donor cancerous mouse. The donor cancer cells of 0.2 mL were injected to each mouse via left or right axila and allowed to grow. The extract administration were continued for another three weeks after which termination is made on all mice. Terminations were done by ether inhalation and followed by collar bone dislocation. Finally the entire adenocarcinoma mammae tissue of each mouse was removed and fixed in a 10% buffered formalin. Histology preparation of adenocarcinoma mammae was prepared by AgNOR staining.

The proliferative activity of adenocarcinoma mammae was scored using AgNOR count according to ploton method. According to Derenzini et al in Miranti, the method was used to calculate the total interphasic AgNOR per cell in 100 cells.[11] The preparations are then examined under light microscope with 100x magnification.[12] AgNOR appears as a black or dark brown dot with yellowish nucleus in the background. AgNOR count was obtained by summing the points in the nucleus one by one when the points can be separated. If the points are united then they are counted as a single point.[13] Lymphocytes are mostly present only as one point, determined by internal control reactions.[14] AgNOR points from stromal cells[12], inflammatory cells[11], and overlapping cells or cells in necrotic areas should be removed.[14] The viewing field for the observation of AgNOR point was taken from the edge and the area where the maximum number of AgNOR points indicated the most anaplastic area. Then the mAgNOR (mean of AgNOR) and pAgNOR (percentage of AgNOR) are summed from the AgNOR point count. mAgNOR is determined from the average number of AgNOR points per cell or per nucleus, whereas pAgNOR is the number of cells containing five or more AgNOR points in its nucleus in 100 cells.[13] The microscopic reading of the preparations was performed by two observers to avoid subjective judgment.

Distribution and homogeneity of data were tested using Saphiro Wilk and Levene respectively.When mAgNOR and pAgNOR values are normally distributed and the variance is normal, they were analyzed by one way ANOVA. If both conditions are not met the data were analysed by nonparametric test using Kruskal Wallis. The data were analyzed using SPSS 15.0 for Windows program. A p value less than 0.05 indicates a significant difference.

3. Results and Dissusion

Normality test by Saphiro Wilk produced abnormal distribution results for mAgNOR and normal distribution for pAgNOR so that for mAgNOR data are described using median as the maximum concentration and minimum size as the size of spread. As for pAgNOR data is described using the mean as a centralized measure and standard deviation as the size of the spread. An overview of the number of mAgNORs and pAgNORs are shown in table 1.

Table 1. mAgNOR and p AgNOR values

Kelompok	N	mAgNOR		
		Median	Minimum	Maksimum
H1	5	3,1950	2,91	4,47
H2	5	3,4500	3,26	6,18
H3	5	3,1900	2,71	3,40
K	5	4,0700	3,45	4,35
			pAgNOR	
			Average	SD
H1	5		25,40	13,87
H2	5		38,40	23,68
H3	5		19,80	4,43
K	5		34,40	7,34

Levene test showed that mAgNOR variance is homogeneous with p value 0.063 (> 0.005) and homogeneous pAgNOR with p value 0.000 (<0.05). It means that in pAgNOR there are at least two groups that have significantly different data variance.

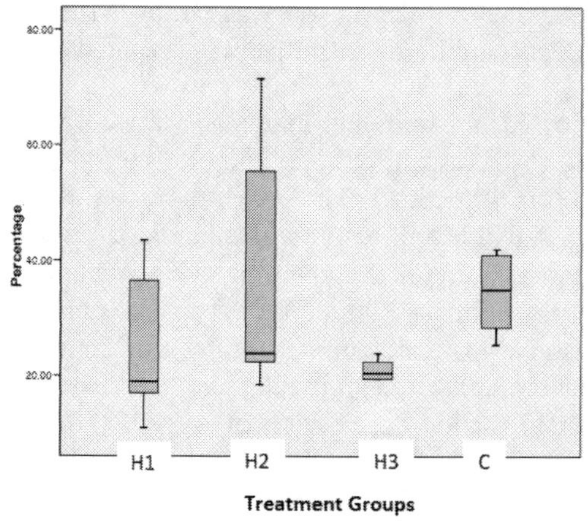

Figure 2. Box Plot Distribution of pAgNOR

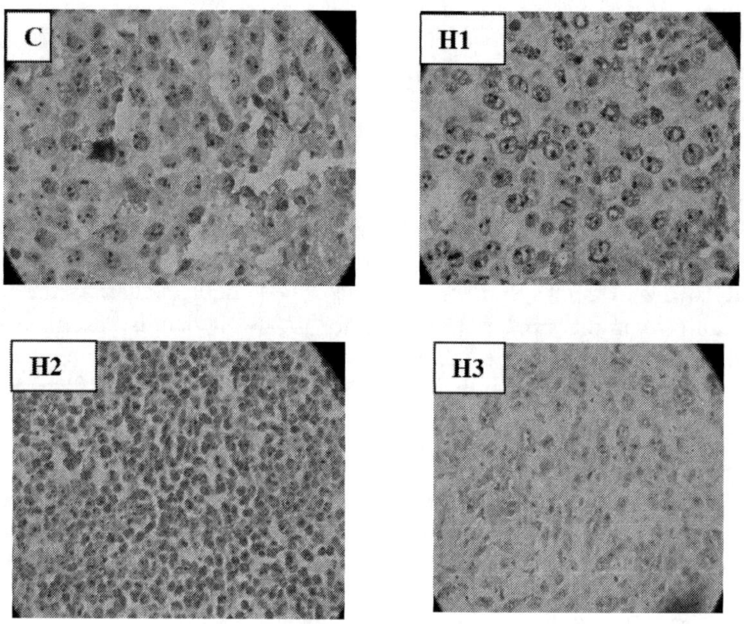

Figure 3. AgNOR staining with 100x magnification, C=control, H1, H2, and H3

Nucleolar Organizer Region (NOR) is a special part of DNA called rDNA. NOR contains the ribosome gene. Using NOR RNA polymerase enzymes encode ribosomal RNA transcription. rRNA is part of the ribosome responsible for protein synthesis. It is thought that there is a relationship between NOR and cell proliferation because protein synthesis is an important stage in cell proliferation [10].

NOR active transcription is associated with high non-histone argirophilic acid protein known as B23, C 23 (nucleolin) and RNA polymerase 1. Nukleolin and RNA polymerse 1 can be painted with silver colloidal technique so that NORs can be seen using a light microscope. NORs whose proteins are painted with silver are called argyrophilic nucleolar organizer regions (AgNOR). [15] The amount of AgNOR is directly related to the rate of ribosomes biosynthesis where cell proliferation is closely related to the duration of the cell cycle. [16,17] There is increasing evidence supporting the mechanisms that control cell proliferation by controlling the progress of the cell cycle also regulate the speed of the biosynthesis of the ribosome. Increased protein synthesis through increased regulation of ribosomal production rates is necessary in cell proliferation to ensure that cell children have sufficient numbers of cell elements to survive and function normally. [18] So the AgNOR value can measure the rate of cell proliferation. [17,18]

There are two common parameters in the AgNOR count, i.e. the average point of AgNOR per cell or per nucleus (mAgNOR) and percentage of AgNOR (pAgNOR). [13] mAgNOR is related to cell ploidy reflecting which DNA content depends on the phase in the cell cycle. The cell cycle in cancer cells is disorganized as well as normal cells so that cell cycles become out of sync within the cell (aneuploid cells) and increase cell size in phase S (DNA synthesis). In general, measuring and analyzing cell proliferation activity is based on cell cycle. In other words, the ploidy that reflects the cell's DNA content is one of the parameters for determining cell proliferation activity. [19] pAgNOR as well as mAgNOR also represents cell proliferation activity as it relates to the percentage of cells undergoing phase S in the cell cycle. [13]. The results of this study indicate that there is no significant difference in the value of mAgNOR and pAgNOR control group with the treatment group. This indicates that the Haliclona sponge extract has not been shown to decrease the amount of mAgNOR and pAgNOR of adenocarcinoma mammae cells in C3H mice.

Sponge *Haliclona sp* produces secondary metabolites namely *alkaloids, steroids, terpenoids, and phenols.* [9] The chemical compounds released are toxic to their environment [5], but these secondary metabolites have the potential to be anti-viral, anti-bacterial, anti-malarial, anti-inflammatory, anti-oxidant and anticancer. [4,5]. The two types of alkaloids isolated from the *Haliclona sp* species are Haliclonacylamine A and Haliclonacylamine B. [20] Other sources call it *halicynone A* and *halicynone B*. [21] *Haliclonacylamine A* ($C_{32}H_{56}N_2$) is made up of 8 methines (4 of which are alkenes) and 24 metylene. *Haliclonacylamine B* is an isomer of *Haliclonacylamine A*. [20]. In 3 different fractions, Haliclona sponge sp showed anticancer activity against leukemia cancer cell (L1210 *cell line*) with IC_{50} 3,25, 2,37, and 2,90 μg / ml. [7] The results of this study showed that Haliclona sponge did not significantly inhibit the activity of adenocarcinoma mammary proliferation in C3H mice, although the number of mAgNOR and pAgNOR counts of the H3 group was lower than the treatment group but not significantly different. This can be caused by several things, namely the sponge extract of *Haliclona sp* which is not a specific compound (crude extract) so that the anticancer compound Haliclona sp is still mixed with other compounds in *Haliclona sp* which can affect the anticancer compound. Another thing that influences the results of this study is the reading of preparations by calculating the point AgNOR is still too subjective and requires the accuracy of preparatory readers, this has been tried to be overcome by using two readers or observers.

4. Conclusion

Using m and pAgNOR count, there was no significant difference in the proliferation activity of adenocarcinoma mammae cells of C3H mice given Haliclona sponge extracts compared to controls. This finding support the view that AgNOR alone could not be used to determine pathology of cancer cells.

5. Acknowledgement

Thankyou to Dr. Ika Pawira Miranti, M.Kes, Sp.PA. for her advice in AgNOR reading method. This research was funded by The Ministry of Reseach and Technology through the RUT grant with contract number: 66/RT/PPK/II/2008

References

[1] Garcia M, Jemal A, Ward EM, Center MM, Hao Y, Siegel RL, *et al.* Global cancer fact and figures 2007 [Online]. 2007 [cited 2008 Jul 26]; Available from: URL:http://www.cancer.org/downloads/STTGlobal_Cancer_Facts_and_Figures_2007_rev.pdf
[2] Makhoul I. Breast Cancer [Online]. 2006 Sept 20 [cited 2008 Jul 29]; Available from: URL: http://www.emedicine.com/med/topic2808.htm
[3] American Cancer Society. Detailed guide: breast cancer chemotherapy [Online]. 2007 Sept 9 [cited 2008 Aug 5]; Available from: URL: http://www.cancer.org/docroot/CRI/content/CRI_2_4_4X_Chemotherapy_5.aspsitearea
[4] Schawartsmann G, Rocha AB, Berlink RGS, Jimeno J 2001 *The Lancet Oncology* **2**(4):221-5
[5] Arif JM, Al-Hazzam AA, Kunhi M, Al-Khodairy F 2004 *Journal of Biomedicine and Biotechnology* **2** p 93-8.
[6] Murniasih T 2005 *Oseana* **30** p 19-27
[7] Ambariyanto, Murwani R, and Trianto A 2003 *J Biotech* p 667-80
[8] Trianto A, Ambariyanto, and Murwani R 2004 *Jurnal Ilmu Kelautan* **9** p 120-4
[9] Trianto A, Murwani R, and Susilaningsih N. Studi aplikasi ekstrak sponge *Haliclona sp* dan gorgonian *Isis hippuris* sebagai obat anti kanker alami. Laporan Akhir Tahun I Program Insentif Riset Terapan. Semarang: Lembaga Penelitian Universitas Diponegoro; 2007
[10] Dirven CMF 1998 The pilocytic astrocytoma, immunohistochemical and genetic studies in relation to tumor behavior [tesis]. Amsterdam: *Thesis Publishers*
[11] Miranti IP 2007 Korelasi hitung AgNOR dan status HER-2 pada karsinoma duktus invasif payudara (NOS) [thesis]. Semarang: Universitas Diponegoro
[12] Menolulu SHC, Kaya H, Kotuloulu E, Erdem GI, Demuray E, and Ekucuoulu GIN 2001 *Turk J Med Sci* **31** p 523-8.
[13] Khanna AK, Yadav SK, Dixit VK, and Kumar M. 2005*J Pancreas* **6** p 575-80.
[14] Pich A, Margaria E, and Chiusa L. 1994 *Am J Pathol* **145** p 481-9.
[15] Flynn JM, Shelton RM, and Landini G 2002 *Biomaterials* **23** p 19–26.
[16] Abboud P, Lorenzato M, Joly D, Quereux C, Birembaut P, and Ploton D 2008 *Am J Obstet Gynecol* **199** p 146.e1-146.e7.
[17] Derenzini M, Ceccarelli C, Santini D, Taffurelli M, and Trere´ D. 2004*J Clin Pathol* **57** p 755–61.
[18] Montanaro L, Trere´ D, and Derenzini M. Nucleolus, ribosomes, and cancer. 2008 *Am J Pathol* **173** (2) p 301-10.
[19] Tjahjono 2002 *Media Medika Indonesiana* **37**(1) p 1-8.
[20] Charan RD, Garson MJ, Brereton IM, Willis AC, and Hooper JNA 1996 *Tetahedron* **52**(27) p 9117-20
[21] Zhou GX and Molinski TF 2003 *Marine Drugs* **1** p 46-53

Ethanol Extract of *Haliclona* sp. Improved Histological Grade of Mammary Adenocarcinoma in C3H Mice

R Murwani[1,4,*], A Trianto[2,4], E Wijayanti[2], A Ridlo[2], N Susilaningsih[3]

[1]Faculty of Animal and Agriculture Science, Laboratory of Physiology dan Biochemistry, Diponegoro University, Jl. Prof. Soedharto, SH, Tembalang, Semarang
[2]Department of Marine Science, Faculty of Fisheries and Marine Science, Diponegoro University Jl. Prof. Soedharto, SH, Tembalang, Semarang
[3]Department of Histology, Faculty of Medicine Diponegoro University, Jl. Prof. Soedharto, SH, Tembalang, Semarang,
[4]Natural Product Laboratory, Centre for Research and Services, Diponegoro University, Jl. Prof. Soedharto, SH, Tembalang, Semarang,

Email: rmurwani@gmail.com

Abstract. The sponge *Haliclona* sp. contains secondary metabolites belong to alkaloids which are cytotoxic to the human tumor. The following research was conducted to study the effect of a graded level of *Haliclona* sp. extract on mammary adenocarcinoma in C3H mice. *Haliclona* sp. was obtained from Bandengan water in Jepara, and the crude extract was prepared by maceration with ethanol. Fifteen C3H mice with an initial weight of 20-25 gram were assigned into control, H-1, and H-2 groups. Control, H1, and H2 groups each received the ethanol extract of 0, 0.15, and 1.5 mg per mouse per day respectively for two weeks. Cancer cells were introduced to all groups from a cancerous donor mouse. The donor cancer cells were injected into each mouse via left or right axilla and allowed to grow. The cancer mass was removed and processed for histological examination, and the cancer growth was determined according to Elston and Ellis criteria. The result showed that histological grade of cancer mass from control group was in grade 2 or differentiated moderately. The histological grade of cancer mass from H-1 and H-2 groups were in grade 1 or similar to a normal cell. Analyses of the data by Kruskal-Wallis showed a significant difference ($p<0,05$) between control and treated groups. No significant difference was found between H-1and H-2 groups. The results suggest the potential of active substances in the ethanol extract of *Haliclona* sp as an anti-cancer drug.
Keywords: tumor, mammae, bioactive, sea, marine, soft coral

1. Introduction

Breast cancer is the second cancer-related death in women. Breast cancer metastasis is the main cause of death in breast cancer patients post the eradication of the primary tumor that cancer can still occur in other areas of the body [1]. The sponge *Haliclona* sp. is abundant in nature however it has not been utilized for the source of the anticancer drug. The sponge contains secondary metabolites belong to alkaloids which are cytotoxic to human breast cancer cell lines in vitro. We had investigated the effect of subchronic administration (3 months) of the *Haliclona* sp. extracts on liver histopathology of Swiss mice and showed no significant difference to control groups which indicated its safe use via the oral route. *Haliclona* sp. extract has LD_{50} value >2000mg/kg and it is categorized as low toxicity [2].

However, to obtain an effective dosage of Haliclona sp extract against cancer cells efficacy data in the animal test is needed.

2. Material and methods

2.1. Extract of Haliclona sp.

Sponge *Haliclona* sp. was sampled purposively form Bandengan water, Jepara Regency, Central Java Province, Indonesia. An amount of 1,350 kg was cut into smaller pieces, and soaked in ethanol and stood at room temperature for 24 hours. The solution was filtered using a paper filter, and the filtrate was dried by rotary evaporator under vacuum. The extraction process was repeated three times. The dry extracts were stored in the refrigerator until the further experiment [3].

2.2. Preparation of Hs extracts containing feed

Feed pellet for C3H mice used commercial feed (CP781) with 31 – 33 % protein, 3 - 5 % fat, 4 – 6 % crude fiber, 10 - 13 % ash, and 11 – 13 % moisture. The feed was mixed with the extract with three concentration i.e. 0.00, 0.03, and 0.3 mg/g. The extracts were dissolved in ethanol 96% until homogenous. Then, the pellet was mixed with the dissolved extract and stood until it was absorbed completely. The control pellet was mixed with 200 mL ethanol only. Then, the pellet was dried under vacum and heated at 40 °C to complete the drying process and stored in the freezer until use.

2.3. Donor cancer cells

Cancer cells were obtained from donor mice. The cancer lumps were surgically removed from the donor mouse and transferred to sterile physiological salt containing petri dish. The tissue was disintegrated using a special scissor to release the cancer cells. These cells were diluted in physiological salt solution, then ready to be transplanted into healthy C3H mice [4].

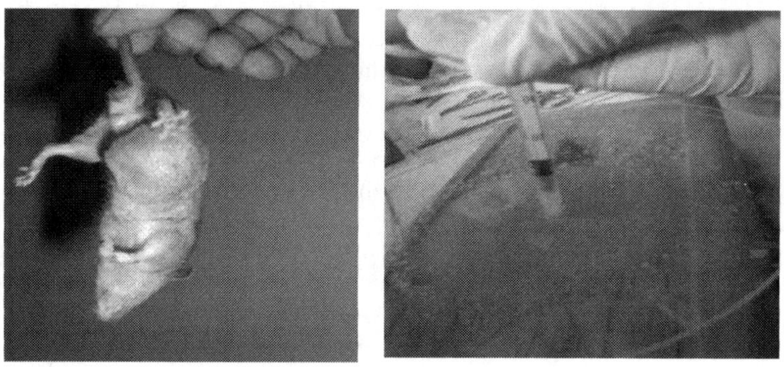

a. b.

Figure 1. Donor mice with cancer lumps (a), and the cancer cells in physiological salt (b).

2.4. Haliclona sp. extracts administration to adenocarcinoma bearing C3H mice

Mice were adapted for seven days to the lab environment. The feed pellet and drinking water were provided in *ad libitum* that provided daily. The experiment was designed as *"The Post Test Only Control Group Design"* with one control and three treatment [5]. Fifteen mice with an initial weight of 20-25 gram were randomly allocated into: control group (C), receiving 0.15 mg extracts (H-1), and receiving 1.5 mg extracts (H-2) per mouse per day for two weeks. The cancer cells were introduced to all groups from a cancerous donor mouse. The 0.2 mL donor cancer cells were injected into each mouse via left or right axilla and allowed to grow up to one month.

IOP Conf. Series: Earth and Environmental Science **116** (2018) 012101 doi:10.1088/1755-1315/116/1/012101

2.5. Histological score of tumor mass

The mice were terminated at the end of the experiment to surgically removed the tumor lump, and they were processed for histological preparation and *Hematoxylin-Eosin* staining. *Hematoxilyn* a nucleus blue, while *eosin* stains cytoplasm red [6].

Figure 2. Paraffin blocks of tumor tissue from experimental mice.

Histological examination of tumor mass was done by observation of cancer cells grade microscopically to determine structural abnormalities of the tissue. Observation under the microscope was done using 400x magnification from 5 viewing fields, i.e. one in the middle and four in each corner. Histological grades of *Adenocarcinoma Mammae* in C3H mice are classified into the table 1.

Table 1. Classification of histological grades of *Adenocarcinoma Mammae* in C3H mice.

Features	Score
Tubule and gland formation:	
Majority of tumor (>75%)	1
Moderate degree (10-75%)	2
Little or none (<10%)	3
Nuclear pleomorphism:	
Small, regular uniform cells	1
Moderate increase in size and variability	2
Marked variation	3
Mitotic counts:	
Dependent on microscope field area	1-3

Breast cancer cells are scored using Elston and Ellis (known as Nottingham Grading System) criteria [7]. The cancer was classified into three grades. The grade 1 (well differentiated) – the cancer cells look most like normal cells and are usually slow-growing with a score of 3-5. The grade 2 (moderately differentiated) – the cancer cells look less like normal cells are growing faster with a score of 6-7. And, the grade 3 (poorly differentiated) – the cancer cells look most changed and are usually fast-growing with a score of 8-9. The histological characteristic base on Nottingham Grading System is determined from the tubular structure, the tubular structure, and mitotic count. The tubular structure observed the % tumor tissue which has a normal duct or tubular structure. The nuclear pleomorphism is base on an evaluation of the size and shape of the nucleus in the tumor cells. While, the mitotic counts is base on the presence of the number dividing cells, which represent the growth rate of the cancer cells.. These characteristics were each scored 1 to 3 (1 for the best and 3 for the worst), then each score is added to reach a score of 3 to 9. With this method, breast cancer is scored from 3 to 9 dan classified into one of the three histological grades.

2.6 Data analyses
All data were analyzed qualitatively and quantitatively using tables, figures, or photos of histological examinations and compared to control. Data analyses were assisted using SPSS 16, and normality of the data was tested using Kolmogorov-Smirnov. When data were normally distributed, they were analyzed by ANOVA and tested further with Post Hoc. When data were not distributed normally, they were analyzed by nonparametric Kruskal-Wallis and Man Whitney tests. When $p<0{,}05$ the difference was significant when $p>0{,}05$ there was no significant difference [8].

3. Results and discussion
3.1. Cancer induction in C3H mice
The transplanted tumor in control mice which receive no sponge extracts appeared first (palpable) compared to mice receiving the sponge extracts. The tumor in control mice and the sponge extracts treated mice appeared eight days and 14 days respectively after transplantation. One month after tumor transplantation all mice had tumor lumps and they were terminated, and the tumor was surgically removed.

The sponge extract inhibited the tumor growth for six days in the experiment. The sponge is known contain the bioactive compounds belong to the renieramycins and mimosamycins that active againts the cancer cell [9,10].

3.2. Histological grade of tumor
During the administration of the extract, all mice were normal in their behavior, activity, and movement. The result in Figure 3 and Table 2 showed that histological grade of cancer mass from the control group was in grade 2 or differentiated moderately, from H-1 and H-2 groups were in grade 1 or similar to a normal cell. Analyses of the data by Kruskal-Wallis showed a significant difference ($p<0{,}05$) between control and treated groups. No significant difference was found between H-1and H-2 groups.

Table 2. Histological Grade of *Adenocarsinoma mammae* of C3H Mice.

	Treatment Groups		
Replicates	C	H1	H2
1	7,2	5,4	5,4
2	7,8	5	5,2
3	7,4	4,6	5,2
4	7	5,4	5
5	7,2	5,2	5,2
Average	7,32 [b]	5,12 [a]	5,2 [a]
SD	0,30	0,33	0,14
Significant $p<0{,}05$	(grade 2)	(grade 1)	(grade 1)

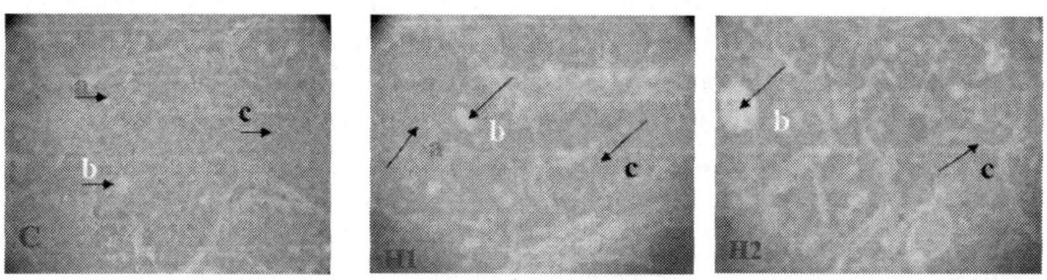

Figure 3. Observation of cancer tissue from C3H mice (400x magnification.
a = mitosis, cell division varies; b = tubulus, ductus;
c = pleomorfic nuclei, size and shape of nuclei

The results which showed that Hs extract could delay the appearance by six days suggested that the extracts can reduce the speed of tumor formation and hence has anticancer activity in vivo C3H mice. The sponge contains the bioactive compounds i.e. renieramycin, mitomycin and their derivates which have bioactivity as antileishmania, anticancer[11,12].

4. Conclusion

Ethanol extracts of *Haliclona* sp. improved the histological grade of mammary adenocarcinoma in C3H mice which suggest the potential of active substances in the ethanol extract of *Haliclona* sp. as an anti-cancer drug.

References

[1] Song HK et al. 2017. Crotonis Fructus Extract Inhibits 12-O-Tetradecanoylphorbol-13-Acetate-Induced Expression of Matrix Metalloproteinase-9 via the Activator Protein-1 Pathway in MCF-7 Cells *J. Breast Cancer.* **20** 234-239.

[2] Trianto, A, N. Susilaningsih, R. Murwani, Widagdo. 2008. Subchronic administration of ethanolic extract of sponge Haliclona sp. in combination with Gorgonian Isis hippuris does not affect liver histopathology of Swiss mice. Konperensi Bioteknologi Indonesia. Bogor 5-7 Agustus.

[3] Trianto A, de Voogd NJ, Tanaka J, 2014. Two new compounds from an Indonesian sponge Dysidea sp. *J. Asian Nat. Prod. Res.* **16** 163-168.

[4] Kurniawan, K., Purnawati, RD. 2005.. Karya Tulis Ilmiah. FK-Universitas Diponegoro. Semarang.

[5] Rodolfo JRA, Neil S Lipman, Elyn R R, and Felix R W. 2010. Treatment and Eradication of Murine Fur Mites: I. Toxicologic Evaluation of Ivermectin-Compounded Feed *J Am Assoc Lab Anim Sci* **49** 564–570.

[6] Fischer AH, Jacobson KA, Rose J, Zeller R. 2008. *CSH Protoc.* 2008 May 1; pdb. prot 4986.

[7] John Hopkins Medicine. 2015. Breast Cancer and Breast Pathology. http://pathology.jhu.edu/breast/grade.php (accessed October 2017).

[8] Becker E-M, Herrfurth C, Irmisch S, Köllner TG, Feussner I, Karlovsky P, Splivallo R, 2014. Infection of corn ears by *Fusarium* spp. induces the emission of volatile sesquiterpenes *J. Agric. Food Chem.* **62** 5226–5236

[9] Parameswaran PS, CG Naik, SY Kamat. 1998. Renieramycins H and I, two novel alkaloids from the sponge Haliclona Cribricutis Dendy. *Indian J. Chem.* **37** 1258-1263

[10] Rashid MA, Gustafson KR, Boyd MR. 2001. A new isoquinoline alkaloid from the marine sponge Haliclona species. *J Nat Prod.* **64** 1249-50.

[11] Singh N, Mishra BB, Bajpai, Singh RK, Tiwari VK. 2014. Natural product based leads to fight against leishmaniasis. *Bioorg. Med. Chem.* **22** 18–45

[12] Oku N, Matsunaga Soest R, Fusetani N. 2003. Renieramycin J, a highly cytotoxic tetrahydroisoquinoline alkaloid, from a marine sponge Neopetrosia sp. *J Nat Prod.* **66** 1136-9.

Reduction of fecal parasites by *Arecha catechu L.* seed and *Anredera cordifolia* (Ten) *Steenis* leaves powder in laying hens

Endang Kusumanti[1], Retno Murwani[1,2*]

[1] Laboratory of Physiology and Biochemistry-Faculty of Animal and Agriculture Science, Diponegoro University, Jl. Prof. Soedharto, SH, Tembalang, Semarang
[2] Natural Product Laboratory, Centre of Research and Services, Diponegoro University. Jl. Prof. Soedarto, SH, Tembalang, Semarang, Central Java, Indonesia.
*rmurwani@gmail.com

Abstract. *Arecha catechu L* (bettle nut) seed and *Anredera cordifolia (Ten) Steenis* ("Binahong") have been shown to have anti helmintic and wound healing activities respectively. Their combine use as phytogenic additives in layers have been shown to reduce serum transaminase. Further study was conducted to evaluate the effect of *A. catechu* seed and *A. cordifolia* leaves powder supplementation on fecal parasites number and the performance of laying hens obtained from outdoor small scale layer farmers. Forty eight of 42 weeks old laying hens were allocated randomly into 4 treatment groups i.e. no supplementation (T_0), supplemented with 0.025% *A. catechu* seed and *A. cordifolia* leaves powder ($T_{0.025\%}$), 0.05% *A. catechu* seed and *A. cordifolia* leaves powder ($T_{0.05\%}$), 0.1% *A. catechu* seed and *A. cordifolia* leaves powder ($T_{0.1\%}$). Each treatment consisted of six replicates with two hens per replicate. Supplementation were carried out by administering alternately *A. catechu* seed powder for 3 days followed by *A. cordifolia* leaves powder for another 3 days. The alternate supplementation for each groups was conducted for 18 days. Feed consumption, egg production, egg weight, hen day production (HDP) were recorded daily. Parasites counts were sampled and enumerated at the beginning and at the end of supplementation. The result showed that alternate supplementation of 0.0.025% *A. catechu* seed and *A. cordifolia* leaves powder up to 18 days to 42 weeks old laying hens reduced fecal parasites without affecting performance.

Key words: heat stress, microbial load, A. galli, coccidia, Eimeria sp., antibiotic free

1. Introduction

A large number of small scale layer farmers in tropical country like Central Java Indonesia used commercial laying hens to produce commercial eggs. The hens start to produce eggs at about 16 to 20 weeks of age depending on the strain. Egg production peaks of about 24 to 30 weeks after which production gradually declines. When production declines and no longer profitable, the hens will be sold as meat chicken. Small scale layer farmers often extend production time which consequently extend environmental and disease challenges such as worm *Ascardia galli* and coccidia. For feed, farmers commonly use commercial complete feed as it is readily available and it has included antibiotics and coccidiostat to prevent endo-parasites infestation [1]. The present climate change such as daily sudden change between rainy and hot-dry weather has added environmental stress and disease challenge which may no longer be able to be prevented by the existing in feed medication. In order to maintain the efficacy the existing one and prevent increasing the dosage of in feed medication, phytogenic additives can be used to help to maintain poultry productivity and health [2, 3, 4, 5]. *Areca catechu* has been shown to have antimicrobial, antihelmintic, and antioxidant activity and *Anredera cordifolia* leaves are known to assist wound healing [6, 7, 8, 9]. Their activities therefore could be beneficial to reduce endoparasites and ameliorate parasite associated internal wound. Earlier study using their combine porwder in laying hens reduced serum transaminase which indicated reduced liver cell damage [10]. Therefore we studied further supplementation of *A. catechu* seed and *A. cordifolia* leaves powder in laying hens to reduce endoparasite number and maintain performance.

2. Material and Methods

2.1. Collection of plant material and preparation of powder

Fresh dry *A.catechu* seed were collected from local market and the pericarp was discarded. Fresh *A. cordifolia* leaves were obtained from local farmers and air dried. Each of the dried plant materias were ground to a coarse powder and filtered through 25-50 mesh screen to obtain a homogenous powder. The dried plant powder was stored in refrigerated container for further use.

2.2. *Supplementation of A. catechu seed and A. cordifolia leaves powder in laying hens*

Laying hens used for this study was obtained from a small scale layer husbandry farmers. The hens were raised in an open battery housing and they were selected randomly on the basis of body weight and age. Forty two weeks old of 48 Isa Brown laying hens with an average body weight of 1.82 ± 0.17 kg were allocated randomly into 4 treatment groups i.e. T_0 : control with no supplementation; $T_{0.025\%}$: supplemented with 0.025% *A. catechu* seed and *A. cordifolia* leaves powder; $T_{0.05\%}$: supplemented with 0.05% *A. catechu* seed and *A. cordifolia* leaves powder; $T_{0.1\%}$: supplemented with 0.1% *A. catechu* seed and *A. cordifolia* leaves powder. Each group consisted of 6 replicates with two hens per replicate. Each hen was given 120 g commercial layer diet (17% protein from Charoen Popkpand) per day and free access to drinking water. Supplementation in each group was carried out by mixing it in the diet. *A. catechu* seed powder was supplemented for 3 days followed by *A. cordifolia* leaves powder for 3 days, and the alternate administration was conducted for 18 days. Egg production were recorded daily from all birds in the groups.

2.3. *Collection of fecal samples, parasites identification and enumeration*

On day 7 fecal samples were collected from fresh drops from one bird from each replicates. The samples were immediately preserved in 10% formalin and taken to Laboratory of Animal Health Service, Office of Animal Husbandry and Health, Semarang-Central Java Province for parasite identification and enumeration. The parasite determination was done by on duty laboratory staff expert. In brief, three gram of feces were dissolved in aquadest, stirred to homogenize, filtered to separate dirt and debri, and the filtrate was collected in plastic tubes. The filtrate in plastic tubes was stood at room temperature for about 15 minutes to let the sample settling in the bottom. The upper part of aquadest was then partly discarded carefully. The remaining sample in the tube was stirred to homogenize, and finally transferred into petri dish to be examined under microcope [11]. The microscope was equipped with camera and connected to computer screen and moved horizontally and vertically to scan the whole sample for parasites identification and enumeration.

2.4. Experimental design and statistical analysis:

A Completely Randomized Design with 4 treatments and six replicates was employed. Each replicate consisted of 2 hens. Parasites data from each supplemented group were analyzed by simple grouping into parasites types and number and analyzed descriptively. Statistical analyses were assisted using Excel. Feed consumption, egg number and weight, and hen day production (HDP) were analyzed by

ANOVA and when means were significantly different ($p<0.01$) Duncan's multiple range test was carried out.

3. Results and Discussion

The results of fecal parasites types and number before and after 18 days supplementation of *A. catechu* seed and *A. cordifolia* leaves powder were shown in Table 1. It showed that before supplementation the types of internal parasites found were A. galli, T. americana sp. , S. trachealis, and Eimeria sp, and Ascaridia sp. The number of each types of fecal parasites varied in each group. For A. galli, the highest number was found in $T_{0.05\%}$ (46) followed by T_0 (31), $T_{0.1\%}$ 17), and the lowest in $T_{0.025\%}$ (5). For T. americana sp or Ascaridia sp they were found only in $T_{0.025\%}$ (2). For S. trachealis, they were found only in T_0 (5) and $T_{0.025\%}$ (3). For Eimeria sp, they were found only in $T_{0.025\%}$ (8) and $T_{0.05\%}$ (1). Before supplementation $T_{0.025\%}$ was the only group infested with all types of parasites, albeit in relatively low number.

Table 1. Total number and types of fecal parasites after 18 days supplementation of *A. catechu* seed and *A. cordifolia* leaves powder alternately in laying hens.

Endoparasites	T_0	$T_{0.025\%}$	$T_{0.05\%}$	$T_{0.1\%}$
Before supplementation :				
Ascaridia galli	31	5	46	17
Tetrameres americana sp	0	2	0	0
Syngamus Trachealis	1	3	0	0
Eimeria sp	0	8	1	0
Ascaridia sp	0	5	0	0
	2 types	5 types	2 types	1 type
After supplementation :				
Ascaridia galli	2	5	9	2
Tetrameres americana sp.	0	0	0	0
Syngamus Trachealis	0	0	0	0
Eimeria sp	0	0	0	0
Ascaridia sp	0	0	0	0
	1 type (A galli)	1 type (A galli)	1 type (A galli)	1 type (A galli)
Reduction of endoparasites (%) :				
A. Galli	93,5	No change (5)	80,4	88,2
Tetrameres americana sp.	No change (0)	100	No change (0)	No change (0)
Syngamus Trachealis	100,0	100	No change (0)	No change (0)
Eimeria sp	No change (0)	100	100	No change (0)
Ascaridia sp	No change (0)	100	No change (0)	No change (0)

After 18 days alternate supplementation of *A. catechu* seed and *A. cordifolia* leaves powder, only *A. galli* were persistently found in all groups in a relatively low number i.e 2 (T_0), 5 ($T_{0.025\%}$), 9 ($T_{0.05\%}$), and 2 ($T_{0.1\%}$) or equivalent to reduction of 92.3%, 0% (no change), 80.4%, and 88.2% respectively.The increasing dosage of supplementation appeared to be unrelated to the percentage of reduction. For example the initial A. galli in T_0 was 31 with no supplementation, and it was reduced to 2 after 18 days. On the other hand higher dosage of supplementation in $T_{0.1\%}$ and $T_{0.05\%}$ only reduced to 88.2% and 89.4% respectively, which were lower than control group. Such result could be due to the severity and the health status of individual hens being studied which can not be controlled. The hens for this study were obtained from local farmers and the hens were raised in an open (outdoor) battery housing. They were selected randomly on the basis of body weight and age. Thus the severity and parasites infestation could not be controlled. The persistent finding of fecal *A. galli* at the end of supplementation albeit in low number may reflect that this infestation is not easily eliminated in laying hens which are producing reaching 45 weeks. This was supported by the fact that $T_{0.025\%}$ which had the lowest number of A.galli (5) before treatment remained the same after 18 days treatment. *Ascaridia galli* infestation occur worldwide and are very common in chicken and much more abundant in traditional farming (open cage house) than in industrial production facilities (close house). Birds become infected after eating infective eggs from contaminated feed, water, equipment, or through infected earthworms [12]. *Ascaridia galli* larvae enter and stays in gastrointestinal tissue and harms the lining of gastrointestine. Their number in intestinal lumen could be fewer and therefore the excreted number in feces may not reflect the actual total number.

From all groups, $T_{0.025\%}$ was the only group with four other types of parasites i.e. *Tetrameres americana sp, Syngamus trachealis, Eimeria sp, and Ascaridia sp*. They were initially present as many as 8 before supplementation and they were all eliminated after 18 days supplementation. *Eimeria sp* in $T_{0.05\%}$ (higher dosage) which was initialy present as many as 2 (lower than $T_{0.025\%}$) was also eliminated. The data suggested that lower dosage was sufficient to eliminate Eimeria sp. It indicated that these phytogenic additives have anti coccidia activity. Further study is needed to provide more evidence.

Production performance of the hens were presented in Figure 1 and statistical analyses results showed that there was no effect of 18 days supplementation on feed consumption, egg weight and number, and HDP.

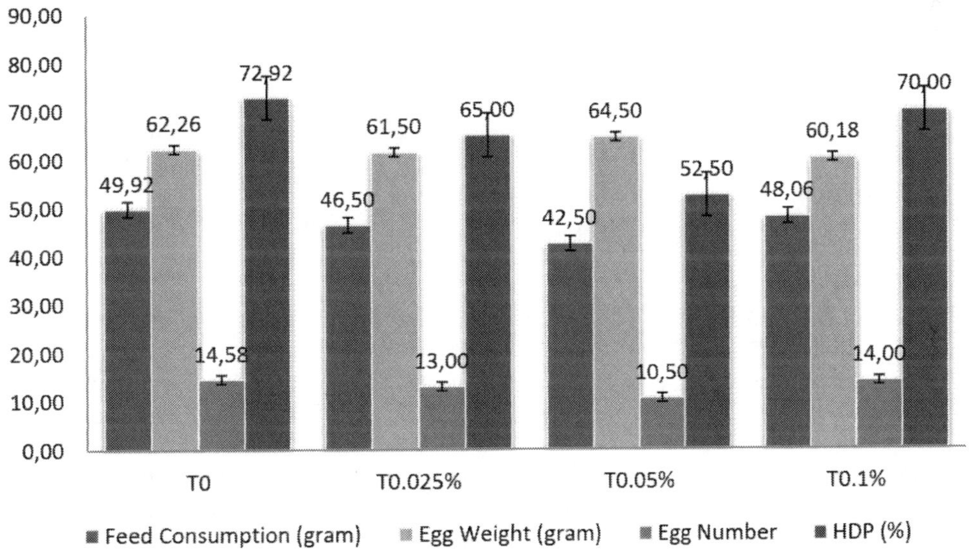

Figure 1. Production performance of 42 weeks laying hens after 18 days alternate supplementation of *Arecha catechu L.* seed *and Anredera cordifolia (Ten) Steenis* leaves powder

The supplementation appears to be safe to be adminstered to older laying hens without affecting production performance. Further study to dissect the mechanism of *A. catechu* seed and *A. cordifolia* leaves powder supplementation on parasites reduction using histopathology is needed. In situ outdoor layer farming study will be benefecial for local farmers.

4. Conclusion

Supplementation of *A. catechu* seed and *A. cordifolia* leaves powder up at 0.025% reduced fecal parasites types and number without affecting egg production performance.

References

[1] Murwani R and Bayuardhi B 2007 Broilers serum cholesterol and glutamic oxaloacetic transaminase and their relation to antibiotic in feed and medication programs in four broiler producers in Semarang region-Central Java, Indonesia. International Journal of Poultry Science, Vol. 6, No.4, pp. 266-270. Retrieved September 9th, 2016, from http://scialert.net/abstract/?doi=ijps.2007.266.270

[2] Murwani R and Murtini S 2009 Effect of chlortetracycline additive in broilers fed local diets on antibody titers to NDV vaccine. International Journal of Poultry Science, Vol. 8, No.8, pp. 755-759. Retrieved September 9th, 2016, from http://scialert.net/abstract/?doi=ijps.2009.755.759

[3] Murtini, S., Murwani, R., Satrija, F. and Handaryani, E.. 2010. Anti Marek's Disease Virus Activity of *Scurrula oortiana* (Tea Mistletoe) Stem Extract in Embryonic Chicken Eggs. International Journal of Poultry Science, 9 (9): 879-885.

[4] Murwani R, Indriani A, Wihardani K, Wahyuningrum M A, Tawakal N R, Mulyono and Kusumanti E 2011 Blood biochemical indices and productivity of broilers on diet supplemented with mannan oligosacharide, baker yeast, or combined baker yeast and Noni leaves extracts. International Journal of Poultry Science, Vol.10, No.12, pp. 990-997. Retrieved September 9th, 2016, from http://scialert.net/abstract/?doi=ijps.2011.991.998

[5] Tanod, W.N.L., Murwani, R., Susanti, S., Kusumanti, E. 2015 Addition of cashew (*Annacardium occidentale*) apple powder into diet can increase body weight and intestinal relative weight in broiler. Pakistan Journal of Nutrition 14(9), pp. 629-631

[6] Lim S K, Hong S P, Jeong S W, Kim B, Bak H, Ryoo H C, Lee S H and Ahn S K 2007 Simultaneous effect of ursolic acid and oleanolic acid on epidermal permeability barrier function and epidermal keratinocyte differentiation via peroxisome proliferator-activated receptor-α. The Journal of Dermatology, Vol. 34, pp. 625-634.

[7] Tangalin, M.G.G. 2011. Anthelmintic Effect of Processed Mature Betel Nut as Dewormer to Native Chicken and Small ruminant (Sheep and Goat). Asian Journal of Health.Vol.1.No.1., pp. 230-243.

[8] Prasetyo A T and Herihadi E 2013 The application of moist exposed burn ointment (MEBO) and binahong leaves in treating partial thickness burn : A case report. Jurnal Plastik Rekonstruksi (The Indonesian Journal of Plastic Reconstructive and Aesthetic Surgery), Vol. 3, pp.142-146. Retrieved September 9th, 2016, from http://www.jprjournal.com/index.php/jpr/article/view/178/139

[9] Baby A A and Raphael K R 2014 Potential antimicrobial, anthelmintic, and antioxidant properties of *Areca catechu L.* Root. International Journal of Pharmacy and Pharmaceutical Sciences, Vol. 6, No.6, pp. 486-489. Retrieved September 29th, 2016, from www.ijppsjournal. com/Vol6Issue6/9731.pdf

[10] Marlani, H.P., Kusumanti, E., Murwani, R. 2017. Arechaatechu L. Seed and anredera cordifolia (ten) steenis leaf powder supplementation reduced serum transaminase in laying hens. Livestock Research for Rural Development, 29(5), from http://www.lrrd.org/lrrd29/5/rmur 29094.html

[11] Parfitt, J. W. and Banks A. W. 1970. A Method for Counting Fasciolu Eggs in Cattle Faeces in The Field. Vet. Rec. 87 : 180-182.

[12] Désirée S. Jansson , Ann Nyman , Ivar Vågsholm , Dan Christensson , Magnus Göransson ,Oddvar Fossum & Johan Höglund 2010. Ascarid infections in laying hens kept in different housing systems. Avian Pathology Volume 39, - Issue 6

The application of Environmental Friendly Technique For Seagrass Transplantation

Ita Riniatsih*, Retno Hartati, Hadi Endrawati, Robertus Mahendrajaya, Sri Redjeki, Widianingsih Widianingsih,

Marine Science Departement, Faculty of Fisheries and Marine Science, Diponegoro University Jl. Prof. Soedharto, SH, Tembalang, Semarang. Phone & Fax: +62-24-7474698

Email: iriniatsih@yahoo.com

Abstract. Many attempts have been made to solve tremendous seagrass losses the marine environment in many area in the world. Artificial transplanting of shoots and spreading of seeds from intact meadows to non-vegetated coastal sediment are the most applied techniques. The study was aimed to apply environmentally seagrass tranplantation techniques in less vegetated area. Frame and small tube made from bamboo were used to do seagrass tranplantation in Jepara Waters. Vegetative shoots (springs) of *Enhalus acoroides dan Cymodocea serrulata* were collected from a healthy donor bed located nearby the transplantation site, and planted into unvegetated areas. Test transplant survival was assessed every month for three months. The result revealed that the survival rates of transplants were varied with the area. It might have due to difference in sediment characteristics of transplanting sites and transplant technique are discussed.
Keywords: seagrass, transplantation, bamboo, tranplant technique

1. Introduction

Seagrasses are plants with roots, stems and leaves adapted to living in the marine environment and capable of producing flowers, fruits and seeds. These plants are more evolved and complex than seaweed, which have a more simple structure, although the two species are often confused. Seagrass beds occur extensively in shallow waters and can reach up to depths of 40m or more, if the environmental conditions permit photosynthesis. Beside play a role in carbon cycle in the atmosphere [1][2], seagrass ecosystems have important roles as a source of primary productivity and a foraging and nursery ground for some marine biota [3]. Twelve out of 69 world seagrass species were found in Indonesia [4]. In Jepara waters, especially in Bandengan, Teluk Awur [5] and Karimunjawa [6] waters there were 5, 7 and 6 species of seagrass respectively. They live in variety habitats such as sand, sandy mud, mud, and rubble substrate.

Seagrass ecosystem conditions in some parts of Indonesia are found to be under threats from human activities such as tourism, ports, aquaculture, and sand mining. It was estimated that 58% seagrass ecosystems in the world has decreased the extents number [7]. According to [8], seagrass bed areas in Indonesia are declined by about 30-40% , and the biggest damage of seagrass beds is found in Java Island [4]. The coverage of seagrass in Teluk Awur and Bandengan waters, Jepara, also tends to decrease from 79% in 2000 [9] to 53% in 2012 [10]. This reduction occurs continuously every year as the effect of anthropogenic pressures.

Because seagrass habitats are so valuable, the protection of seagrass ecosystems needs to be done immediately as well as seagrass restoration. Seagrass habitat restoration can be accomplished by improving the habitat quality and transplanting seeds or adult shoots. Habitat improvement may be a very slow process and consequently, requires long periods of time whereas seagrass transplantation might be a rapid way to restore seagrass habitats. Transplanting can establish seagrass habitat before natural processes might permit recolonization [11]. Artificial transplanting of shoots and spreading of seeds from intact meadows to non-vegetated coastal sediment are the most applied techniques. Planted seagrass beds can function and grow exactly as natural beds. However, very few seagrass transplantation studies have been made In Indonesia to restore the damaged. Among of them were done by [12] at reef flat of Pari Island, [13]at Banten Bay, [11] at Westcoast of South Sulawesi, and [14] at Prawean waters, Jepara but are mostly done for *Enhalus acoroides* none for *C. serrulata*.

Seagrass transplantion have been conducted either transplanting adult plants, which is traditionally the most widely used method, probably because habitats are immediately created or planting seeds because of its reduced costs as collection, maintenance, transportation and planting processes are easier and more cost-effective. The transplantation is conducted by several method such as plug methods, in which seagrass and attached sediment are harvested using core tubes of various sizes; staple method, in which plants are dug up using shovels; peat pot method, in which sediment blocks are removed as when using the plug methods; and TERF or frame systems method, which is a modification of the staple method. All methods are involved metal material such as core tubes (plug methods), staple and metal wire (staple method), metal frames (frame systems) that have to be removed from the site, otherwise they pollute the environment [15][16]. Therefore an attempt to look for more environmentally material is urgently needed. In this study the bamboo was selected as material to transplant seagrass because they are easilly available in Indonesia, able to degrade in certain of time therefore no waste produced and relatively cost effective. The objectives of present work are to compare the performance (growth and survival rate) of *Enhalus acoroides* and *Cymodocea serrulata* transplantation done by different transplantion method (tube and frame method) using bamboo material in different locations.

2. Material and method

Two seagrass species *Enhalus acoroides* and *Cymodocea serrulata* were replanted in two different location, i.e. Bandengan and Teluk Awur waters, Jepara. Bottom sediment of both locations are characterized as sanddy to sandy mud (Bandengan) and sandy mud (Teluk Awur) sediment with coral fragments. The locations are semi-closed and protected from waves, the depth of water were ranged of 0.5-1 m and 2-2,5 m during lowest and highest tides respectively.

Seagrass (*E. acoroides and C. serrulata*) was randomly retrieved from well developed seagrass beds in the same area where they will be replanted. The collector followed the blades of the shoot to the substrate, uprooted approximately 15-20 cm of the rhizome by digging under the rhizome by hand, and snapped the rhizome to remove the plant. This technique allowed us to minimize disruption of root-rhizome layer. The plants then were temporarily stored in large coolers with a small amount of seawater to prevent exposure.

Transplantation of *E. acoroides* and *C. serrulata* were performed in June-September 2017 using two methods, i.e. modified plug and frame methods of [15][16]. In both, environment-friendly transplantation material i.e. bamboo as tubes and frames were used. Transplantation at each site was done in triplicate to create three patch sizes. A plot size of 3 m x 3 m was designed with spacing of 0,5 m intervals between planting units within a row. For convenience seagrasses were transplanted at both stations during low tide. For plug method, since the rhizome size of each species of seagrass could be different, the donor were planted in 10 and 5 cm length bamboo cut for *E. acoroides* and *C. serrulata* respectively. After the donor were planted into bamboo tubes, then the bamboo tube were inserted in the bottom substrate and filled with sediment/sand taken from nearby. For frame method transplantation, a 3 meter x 3 meter bamboo frame is lowered into the water until it rests on the bottom. The donor shoots are attached with cotton rope ties, aligned in parallel with distance of 50 cm in the frame. The anchor were tied up in the corner of the frame so it is steady in the position at the bottom of the waters and not be carried away by the current.

Growth and survival rates of transplanted seagrasses were assessed monthly. Growth was measured using leaf marking technique where twenty five plants at each transplanting plot were marked on leaf sheath. After a period of 2-3 d, leaves were remarked again. Growth was then measured as the difference between the first mark and the initial baseline after the second mark has been done. The total number of remaining plants at each plot was counted to examine the survival rate. Sediment samples were collected from transplanted sites using hand corer to determine grain size distribution, total organic matter, phosphat and nitrate analyses. Oceanographic condition of planting sites were also monitored every month during the experimental period by measuring seawater temperature, salinity, current speed, and light penetration.

3. Result and discussion

In the seagrass transplantation, to select transplantation habitats carefully, and to optimize the transplantation techniques, it is necessary to consider on selection of an appropriate donor population, spreading of risks, and ecosystem engineering effects [17]. Therefore, two transplantation locations with almost same the similar physical properties between donor sites and transplanted sites were chosen. Sediment characteristic which support seagrass growing during the initial transplanting period is regarding to be one of the key factors contributed to seagrass transplantation success [18] although it is apparent that sediment physical properties cannot always predict the suitability of the areas for seagrass transplantation [17]. Grain size analysis of sediment in the transplantation and control location revealed that the sediment was mainly composed of sand (69,04-85,41%) to form sand to sandy muddy substrat (Table 1). The concentration on nitrate and phosphat we in in the range of 0,0143-0,5750 and 0,0131-0,0262 mg.gr^{-1} respectively. Sediments are important parameters that have a major influence on the growth and spread of seagrass and therefore affect the success of seagrass transplantation [19]. Moreover, sediment characteristics also an important and critical factor in choosing a suitable transplanting method for a specific area.

Table 1. The sediment characteristic and nutrient of transplantation locations.

Locations	Fraksi Substrat (%)			Clasification	Nitrate (mg.gr^{-1})	Phosphate (mg.gr^{-1})
	Gravel	Sand	Silt			
Control Teluk Awur	3,54	76,32	20,14	Sandy muddy	0,2393	0,0262
Control Bandengan	8,96	74,31	16,73	Sandy muddy	0,5750	0,0171
Teluk Awur1	2,46	81,38	16,16	Sandy muddy	0,1790	0,0220
Teluk Awur2	11,13	69,04	19,83	Sandy muddy	0,1964	0,0171
Teluk Awur3	12,9	68,65	18,45	Sandy muddy	0,1857	0,0128
Bandengan 1	5,81	85,41	8,78	Sand	0,3857	0,0210
Bandengan 2	21,11	71,66	7,23	Sandy muddy	0,0143	0,0131
Bandengan 3	16,35	76,76	6,89	Sandy muddy	0,2500	0,0184

The water quality measured during present work were presented in Table 2. All measured parameters fall within the range of seagrass. Seagrass growth was influenced not only external factors such as temperature, salinity, turbidity, bottom substrate and nutrient resources[20] but also internal factors including physiology and metabolism processes of seagrass. Tropical seagrasses grew optimum at temperature of 28-30 ^{0}C and salinity of 20-35 $^0/_{00}$. Organic matter content was almost the same for both locations (3,3-3,6%). [20] Benham et al (2016) stated that high sediment thickness and soft-sediment contain organic material needed for seagrass growth. Light intensity reached the bottom (0,75-1,70 meter) supporting seagrass growth. Seagrasses require a minimum of 20% of surface light to survive [21].

Tabel 2. Physical and chemical parameter of Teluk Awur and Bandengan waters.

Parameters	Teluk Awur Waters	Bandengan Waters
Temperature (°C)	28-29,5	28,5-30
Salinity ($^0/_{00}$)	32-33	32-33
Current speed (m.s^{-1})	1,2	1,1
Organic matter (%)	3,3	3,6
Depth (m)	0,75-1,50	0,85-1,70
Light penetration (m)	0,75-1,50	0,85-1,70

The average growth rate of tranplanted *E. acoroides* dan *C. serrulata* in different location are represented in Figure 1, and showed that the highest growth rate was *E. acoroides* (0,313 %.day^{-1}) transplanted with bamboo tube methods at Bandengan waters, and *C. serrulata* transplanted with same method shows the highest growth rate (0,041 %.day^{-1}). Compare to control location, they have lower growth rate. The growth rate of *E. acoroides* and *C. serrulata* of control location in Teluk Awur waters were 0,365 and 0,048 mm.day^{-1}. In control location Bandengan waters, growth rate of *E. acoroides* and *C. serrulata* were 0,428 and 0,047 mm.day^{-1}. This slower growth rate of transplant seagrass compared to plants at the donor sites [22] suggesting that *seagrass* transplants needed time to acclimatize to new environment.

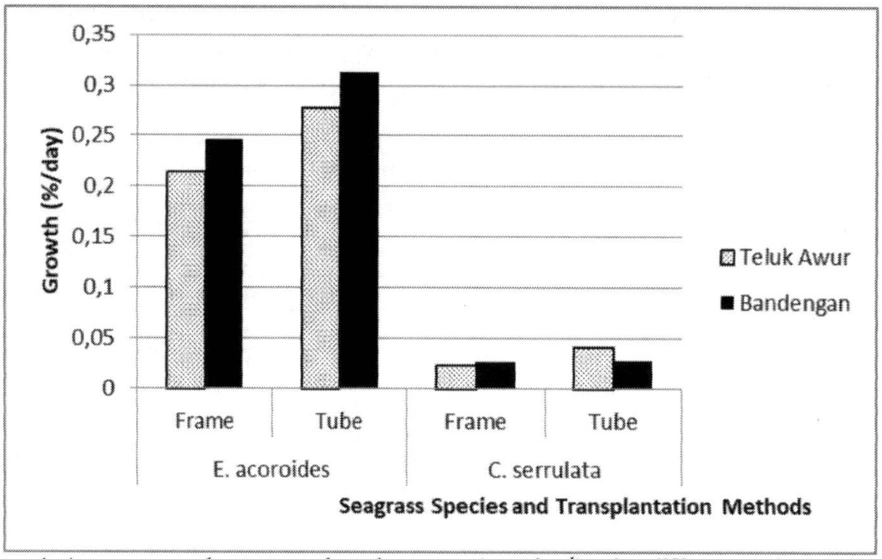

Figure 1. Average growth rate transplanted seagrass (mm.day^{-1}) using different method and locations (FE = Frame *E. acoroides*, TE = Tube *E. acoroides*, FC = Frame *C. serrulata*, TC = Tube *C. serrulata*)

Figure 2 illustrates the transplantation survival rate of *E. acoroides* and *C. serrulata* at Teluk Awur and Bandengan waters. It showed that transplantation using bamboo tube were more survived than on the frame method. The highest survival rate of both species was showed in Teluk Awur waters using bamboo tube (45% and 24,6% respectively) and while the lowest survival rate of both species was transplanted at Bandengan waters using bamboo frame (13,4 and 0% respectively).

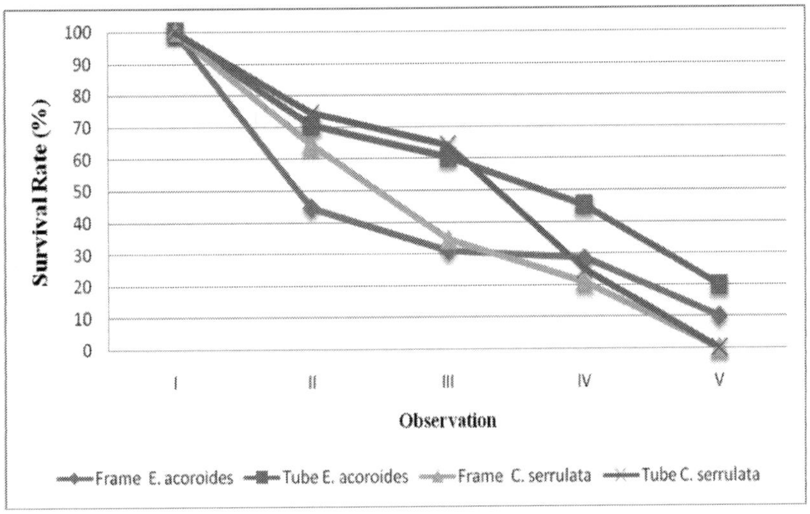

Figure 2. Survival rate (%) transplanted seagrass in the end third month using different method at Teluk Awur

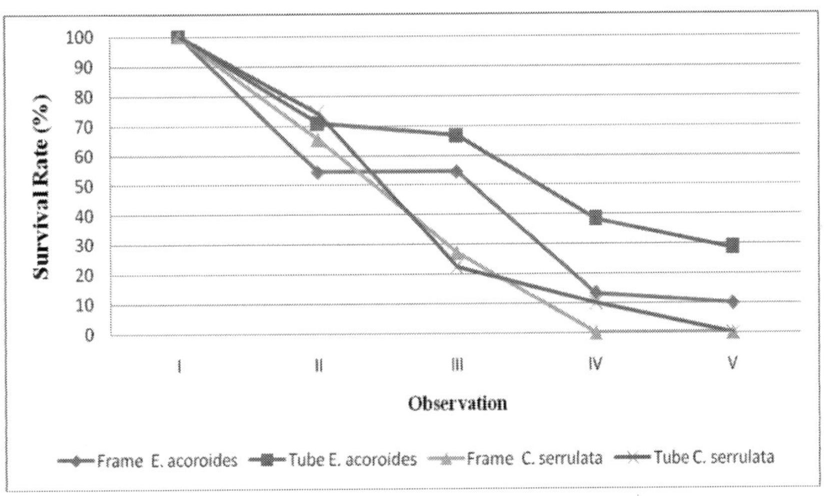

Figure 3. Survival rate (%) transplanted seagrass in the end third month using different method at Bandengan

The survival rates of seagrass transplanted in bamboo tube were relatively high during the first month of transplantation, i.e. more than 70% in both transplantation location. First one-month period is a phase of adaptation for the transplants to their new substrate and environments. High survival rate after one month transplantation indicated that no critical physical and biological disturbances to the transplants at transplanting sites [11]. But in the third month the survival rate were decreased considerably especially whom transplanted with frame method. Figure 2 also showed that transplantation using bamboo tube were better than frame method. The tube gave better protection for rhizome from waves and currents. Bamboo frame methods weaker than bamboo tube, since many seagrasses were seen lifted from the sea floor. This condition lead to decrease the growth rate, and some seagrass could not grow well and some died. Data shows that the highest growth rate for seagrass transplanted at Bandengan was 0.313 mm.day^{-1} for *E acoroides*. It was expected that physical and chemical oceanography characteristic of Bandengan suited for the growth of *E acoroides*. The differences in seagrass covers that was influenced by their growth rate were caused by several factors including seagrass topography, physical and chemical water condition, coastal community activities around the seagrass, and distribution of seagrass species [23]. It can be seen that coastal community in Bandengan Waters was more activities (fishing and tourism) than Teluk Awur Waters. According to [24], if the location of seagrass ecosystems was very close to the mainland it had the potential to receive the negative effects of anthropogenic activities from that mainland. In addition to the base of substrate and water conditions, number of residents in a region, which was correlated with anthropogenic stress, was another factor affecting seagrass cover and distribution. [25][26].

Assuming appropriate light levels, sediments are important parameters that have a major influence on the growth and spread of seagrass and therefore affect the success of seagrass transplantation [19]. Sediment can affect the stand material of transplanted seagrass material through a process of erosion/deposition of sediment. Therefore sediment stabilization play important role to keep sediment and transplants in place and being not swept away by strong currents and waves [11]. The efficiency of anchoring device used in the present study especially for frame method is also an important factor determining the success of seagrass transplantation. The efficiency of anchoring device may depend on the sediment type at the transplanting sites. Coastal tidal flat of Teluk Awur and Bandengan Waters composed of sand and silt therefore anchoring device can not be pressed deeper into the sediment so that the bamboo frame can not hold seagrass transplants firmly against strong currents and waves. The efficiency of anchoring device used in the present study is also an important factor determining the success of seagrass transplantation.

In general, the growth and survival rate of *E. acoroides* is higher than *C. serrulata*. *E. acoroides* was characterized by long shoot, root and rhizome with slow rhizome branching frequency. Growth of *E.acoroides* varied with season and location where leaf growth rate can be varied from less than 1 cm d-1 up to more than 2 cm d-1 [27][28]. Comparing to other transplantation of *E. acoroides* by [29] and [11], transplantation in this present work had lower the survival rate. C. serrulata was rather robust seagrasses with ribbon-like curved leaves with serrated leaf tip, smooth herbaceous rhizomes and well developed leaf sheaths [30]. The shoot of *C. serrulata* has open leaf sheath scars. Transplantation of *Cymodocea rotundata* has been conducted by [31], with plug and sprig method revealed that the survival and growth rate were higher than *C. serrulata* in present work. There lower survival rate might be due to the different method and physical characteristic of locations.

4. Conclusion

It could be concluded that transplantation of seagrasses *Enhalus acoroides* and *Cymodocea serrulata* using tube method showed better growth and survival rate than bamboo frame methods. The transplantation locations had significant impact to transplantation success rate.

Acknowledgement

The authors thank to Dean Faculty Fisheries and Marine Science who provided the research funding through PNBP program.

References

[1] Duarte CM, Middelburg JJ, Caraco NF 2005 Biogeoscience **2(1)** 1-8
[2] Kennedy H, Beggins J, Duarte CM, Fourqurean JW, Holmer M, Marbà N, Middelburg JJ 2010 Global Biogeochem Cycles **24** GB4026, doi: 10.1029/ 2010GB003848.
[3] Erftemeijer PLA, Osinga R, Mars AE 1993 Aquat Bot **46** 67-90
[4] Kawaroe M, A H Nugraha, Juraij, I A Tasabaramo 2016 *BIODIVERSITAS* **17(2)** 585-591
[5] Hartati R, Trianto A, Widianingsih. 2017a. IOP Conf. Series: Earth and Environmental Science **55** 012041 doi: 10.1088/1755-1315/55/1/012041

[6] Hartati R, Widianingsih, Trianto A, Zainuri M, Ambariyanto 2017b. BIODIVERSITAS **18(3)** 947-953
[7] Waycott M, McMahon K, Mellors J, Calladine A, Kleine D 2004. A Guide to Tropical Seagrasses of the Indo-West Pacific. James Cook University, Townsville.
[8] Vo ST, Pernetta JC, Paterson CJ 2013 Ocean Coast Manag **85** 153-163
[9] Riniatsih I, Hermawan I, Sedjati S. 2000. ILMU KELAUTAN (Indonesian Journal of Marine Science). **5** (19)
[10] Riniatsih I, Widianingsih W, Redjeki S, Endrawati H. 2013 ILMU KELAUTAN (Indonesian Journal of Marine Science) **18(2)** 84-90
[11] Lanuru M. 2011. Bottom Sediment Characteristics Affecting the Success of Seagrass (*Enhalus acoroides*) Transplantation in the Wescoast of South Sulawesi (Indonesia). 2011 3rd International Conference on Chemical, Biological and Environtmental Engineering IPCBEE vol. 20 (2011) IACSIT Press, Singapore.
[12] Azkab, M.H. 1988. Seagrass (*Thalassia hemprichii* Ehrenb Aschers) Transplantation at Reef flat at Pari Island Seribu Islands Jakarta Bay: Biology, Culture, Oceanography, Geology and Water Cindition) (Transplantasi Lamun *Thalassia hemprichii* (Ehrenb.) Aschers di Rataan Terumbu Pulau Pari, Kepulauan Seribu: Teluk Jakarta : Biologi, Budidaya, Oseanografi, Geologi dan Kondisi Perairan). P3O-LIPI, Jakarta. 1988
[13] Kawaroe M, Erlangga DK, Rahadian NP 2010 *Jurnal Mar. Res. Indonesia* **35(2)**
[14] Febriyantoro Febriyantoro, Riniatsih I, Endrawati H. 2013. Buletin Oseanografi Marina **2(1)** 23-26
[15] Calumpong HP, Fonseca MS 2001 Seagrass transplantation and other seagrass restoration methods. in : Global Seagrass Research Methods. Eds : Short FT and Coles RG. Elsevier Science. chapter **22**, pp: 425-442
[16] Ganassin C, PJ Gibbs 2008. A review of seagrass planting as a means of habitat compensation following loss of seagrass meadow . NSW Department of Primary Industries - Fisheries Final Report Series No. 96 . 41P.
[17] van Katwijk MM, Bos AR, de Jonge VN, Hanssen LSAM, Hermus DCR, de Jong DJ 2009 *Marine Pollution Bulletin* **58** 179-188.
[18] Vichkovitten T, Intarachart A, Khaodon K, Rermdumri S 2016 *GMSARN International Journal* **10** 113 – 120
[19] Newell RIE, Koch EW 2004 *Estuaries* **27(5)** 793-806.
[20] Benham, CF, SG Beavis, RA Hendry, EL Jackson. 2016. Marine Pollution Bulletin **109** 461-470.
[21] Dennison WC 1987 *Aquatic Botany.*, 27: 15–26.
[22] Lepoint G, Vangeluwe D, Eisinger M, Paster M, van Treeck P, Bouquegneau J.-M, Gobert S. 2004 *Marine Pollution Bulletin* **48** 465-470.
[23] UNEP 2008. National Reports on Seagrass in the South China Sea.UNEP, Bangkok.
[24] Ambo-Rappe R. 2014 Mar Pol Bull **86**:298-303.
[25] Campbell SJ, McKenzie LJ. 2004. Estuar Coast Shelf Sci. **60** 477-490
[26] Waycott M, Longstaff BJ, Mellors J. 2005. Mar Pol Bull **51** 343-350.
[27] Vichkovitten T. 1998. *Kasetsart Journal (Natural Science)* **32** 109-115.
[28] Kenyon RA, Conacher CA, Poiner IR 1997 *Marine and Freshwater Research* **48** 335-342.
[29] Kawaroe M, Jaya I, Indarto HS. 2008. Technology Application of Seagrass Trabsplantation *Enhalus acoroides* at Seribu Islands Space Region of Jakarta. (Rekayasa Teknologi Transplantasi Lamun pada Jenis *Enhalus acoroides* dan *Thallassia hemprichii* di Kepulauan Seribu, DKI Jakarta). LPPM. IPB Bogor. 26 hal.
[30] Larkum AWD, Orth RJ, Duarte CM. 2006. Seagrasses Biology, Ecology and Conservation. Springer. Netherland.
[31] Riniatsih I, Endrawati H. 2013. Buletin Oseanografi Marina **2** 34 - 40

The Ethanolic Extracts The Gorgonian *Isis hippuris* Inhibited the Induced Mammary Carcinoma Growth In C3H Mice

Agus Trianto[1,4], Yogi Andriyas[1], Ali Ridlo[1], Sri Sedjati[1], Neni Susilaningsih[2], Retno Murwani[3,4*]

[1]Department of Marine Science, Faculty of Fisheries and Marine Science, Diponegoro University, Jl. Prof. Soedharto, SH, Tembalang, Semarang
[2]Department of Histology, Faculty of Medicine Diponegoro University
[3]Faculty of Animal and AgricultureScience, Laboratory of Physiology dan Biochemistry, Diponegoro University, Jl. Prof. Soedharto, SH, Tembalang, Semarang
[4]Natural Product Laboratory, Centre for Research and Services, Diponegoro University, Jl. Prof. Soedharto, SH, Tembalang, Semarang

* rmurwani@gmail.com

Abstract. The gorgonian *Isis hippuris* contains secondary metabolites gorgosterol and hippuristanol which are capable of inhibiting cancer cells. However, *in vivo* test of the gorgonian *Isis hippuris* extract as the anticancer drug has not been conducted. The research to study of the effect of ethanolic extract of the gorgonian on the induced tumor growth in C3H mice. The *I. hippuris* was obtained from Karimunjawa water in Jepara. The extract was prepared by maceration using ethanol. A total 20, 8-10 moths old of C3H mice with an initial weight of 20-25 gram were assigned into control, Ih-1, Ih-2, and Ih-3 groups. Control, Ih-1, Ih-2, and Ih-3 groups each received 0, 0.15, 1.5, and 15 mg extract per mouse per day respectively for two weeks. Cancer cells were introduced to all groups from a donor cancer mouse by injection via left or right axilla and allowed to grow. The cancer mass was removed and processed for histological examination, and cancer growth was determined according to Elston and Ellis criteria. The result showed that histological grade of cancer mass from the control group was in grade 2 or differentiated moderately. The histological grade of cancer mass from Ih-1, Ih-2, and Ih-3 groups were in grade 1 (low grade) or similar to a normal cell. Statistical analysis by Kruskal-Wallis test showed a significant difference ($p<0,05$) between control and treated mice. Mann-Whitney test found no significant differences among Ih-1, Ih-2, and Ih-3 treated mice. The results indicated the potential of active substances in the ethanol extract of *I. hippuris* as an anti-cancer drug.

Keywords: tumor, mammae, bioactive, marine biota, soft coral

1. Introduction

Breast cancer is a major health problem and was among the top three most common malignancies globally in 2012 [1]. Surgery, chemotherapy, or radiation are standard therapies with varying success depending on the stage of the breast cancer. The cost of breast cancer therapy is high with side effects which often put the patients to give up treatment.The side effect of chemotherapy is due to less specificity of the drug which kills not only cancer cells but also normal cells. There is also breast cancer cells which develop resistance to the various drug and named as *Multidrug Resistance* (MDR) breast cancer cells [2]. The presence of MDR breast cancer cells is linked to breast cancer recurrence and prevent recovery. Breast cancer is a global burden, and various lines of approach have been made to reduce it. New drugs are constantly being sought and searched to improve specificity, efficacy, and reduce toxicity with the goal to improve prognosis and complete recovery [3].

There is a great diversity of marine organisms from which a large number of pharmacologically active compounds have been isolated. Some of the compounds have anticancer activity. Ascididemin, lamellarin-D, and kahalalide F are the examples of bioactive compounds from a marine source that inhibit the human MCF7 breast cancer cells[4].

Gorgonian *Isis hippuris* is one of the marine organisms that has potent as a source of anticancer compounds. In Indonesia, *I. hippuris* can be found in Jepara, Makasar and Flores waters, or in the islands of Karimunjawa and Krakatau. However, their potency has not been fully explored. Previous in vitro study on hippuristanol and gorgosterol have shown that they can inhibit cancer cells [5] [6] however in vivo study is still rare. Therefore the following study was carried out to examine the effect of *Isis hippuris*extracts on *adenocarcinoma mammae in vivo* C3H mice.

2. Material and Methods

2.1. Extract of I. hippuris

I. hippuris was collected purposively from Karimunjawa water, Jepara. An amount of 2,5 kg was cut into small pieces and extracted with ethanol at room temperature for 24 hours with three repetitions. The extract solution was filtered on the paper filter and concentrated with a rotary evaporator under vacuum. The extracts were transferred into vials and dried further using a vacuum pump. The extract was kept in the refrigerator until further experiment [7].

Figure 1. Isis Hippuris from Karimunjawa

2.2. Preparation of I. hippuris extracts containing Feed

The feed for C3H mice used commercial pellet feed (CP781) with 31 – 33 % protein, 3 - 5 % fat, 4 – 6 % crude fiber, 10 - 13 % ash, and 11 – 13 % moisture. The pellet was mixed with the extract with three concentrations, i.e., 0.00, 0.03, and 0.3 mg/g. The extracts were dissolved in ethanol 96% until homogenous, then, the pellet was mixed with the dissolved extract until absorbed completely. The control pellet was mixed with 200 mL ethanol only. Then, the pellet was dried under vacuum and heated at 40 ^{0}C to obtain a dry pellet.

2.3. Donor cancer cells

Cancer cells were obtained from donor mice. The cancer lumps were surgically removed from the donor mouse and transferred to sterile physiological salt containing petri dish. The tissue was disintegrated using a special scissor to release the cancer cells. These cells were ready to be transplanted into healthy C3H3 mice. Both cancer donor and experimental mice were obtained from Universitas Indonesia, Jakarta.

2.4. Isis hippuris (Ih) extracts administration to adenocarcinoma bearing C3H3 Mice

Mice were adapted for seven days to the lab environment, feed pellet and drinking water. The pellet and drinking water were provided everyday *ad libitum* [8]. The experiment was designed as "*The Post Test Only Control Group Design*" with one control and three treatment [9]. Fifteen mice with an initial weight of 20-25 gram were randomly allocated into control group (C), receiving 0.15 mg Ih extracts (Ih1), receiving 1.5 mg Ih extracts (Ih2), and 15 mg Ih extracts (Ih3) per mouse per day respectively for two weeks. Cancer cells were introduced to all groups from a cancerous donor mouse. The donor cancer cells of 0.2 mL were injected to each mouse via axilla and allowed to grow.

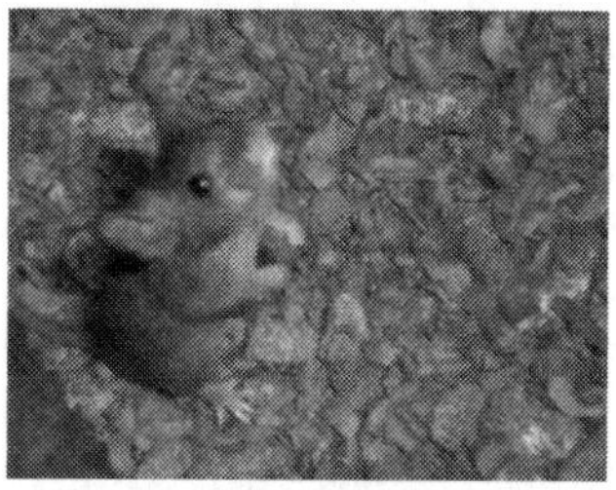
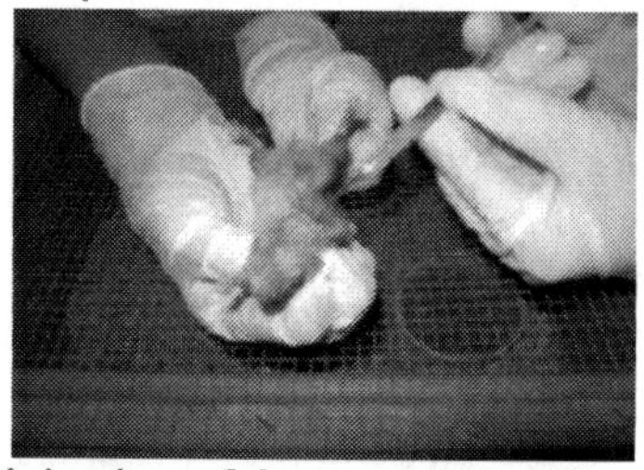

Figure 2. C3H mouse (left) and injection of donor cancer cells

2.5. Histological score of tumor mass

Mice were terminated at the end of the experiment to surgically removed the tumor lump, and they were processed for histological preparation and *Hematoxylin-Eosin* staining. *Hematoxylin* stains nucleus blue, while *eosin* stains cytoplasm pink [10]. Histological grade of tumor mass was determined by microscopic observation of cancer cells consisting of structural abnormalities of the tissue [9]. Microscopic observation was done using 400x magnification from 5 viewing fields, i.e., one in the middle and four in each corner. Histological grades of *Adenocarcinoma Mammae*of C3H mice are classified into the following scoring :

Features	*Score*
Tubule and gland formation:	
Majority of tumour (>75%)	*1*
Moderate degree (10-75%)	*2*
Little or none (<10%)	*3*
Nuclear pleomorphism:	
Small, regular uniform cells	*1*
Moderate increase in size and variability	*2*
Marked variation	*3*
Mitotic counts:	
Dependent on microscope field area	*1-3*

The Elston and Ellis (known as Nottingham Grading System) criteria were utilized for the breast cancer cells scoring that classified the cells into three classes/grade. The grade 1 (well differentiated): the cancer cells look most like normal cells and are usually slow-growing with score of 3-5. The grade 2 (moderately differentiated): the cancer cells look less like normal cells are growing faster with a score of 6-7. The grade 3 (poorly differentiated): the cancer cells look most changed and are usually fast-growing with a score of 8-9. The histological characteristic in Nottingham Grading Systems was determined base on the tubulus structure (% tumor tissue which has normal duct or tubular structure, nuclear pleomorphism (an evaluation of the size and shape of the nucleus in the tumor cells), mitotic counts (the number of divided cells, which indicated the growth rate of the tumor cells and divide). These characteristics were each scored 1 to 3 (1 for the best and 3 for the worst), then each score is added to reach a score of 3 to 9. With this method, breast cancer is scored from 3 to 9 dan classified into one of the three histological grades [11].

2.6 Data Analyses
All data were analyzed qualitatively and quantitatively using tables, figures, or photos of histological examinations and compared to control. Data analyses were assisted using SPSS 16, and normality of the data was tested using Kolmogorov-Smirnov. When data were normally distributed, they were analyzed by ANOVA and tested further with Post Hoc. When data were not distributed normally, they were analyzed by nonparametric Kruskal-Wallis and Man Whitney tests. When $p<0.01$the difference was very significant, when $p<0.05$ there was a significant difference when $p>0.05$ there was no significant difference [12].

3. Results and Discussion

3.1. Cancer induction in C3H Mice
Cancer transplantation by injecting cancer cells donor via axilla was successful and shown by the presence of lump around axilla. There is a lymphatic system in axilla which transports the cancer cells into breast gland. The first lump was observed 19 days after transplantation. All transplanted mice were 100% successfully formed tumor lumps. After one-month extract administration, all mice were sacrificed and terminated to remove the tumor.

3.2. ObservationofTumor Histological Grade
Microscopic examination for histological grade according to Elston dan Ellis criteria are shown in Figure 3 and summarized in Table 1.

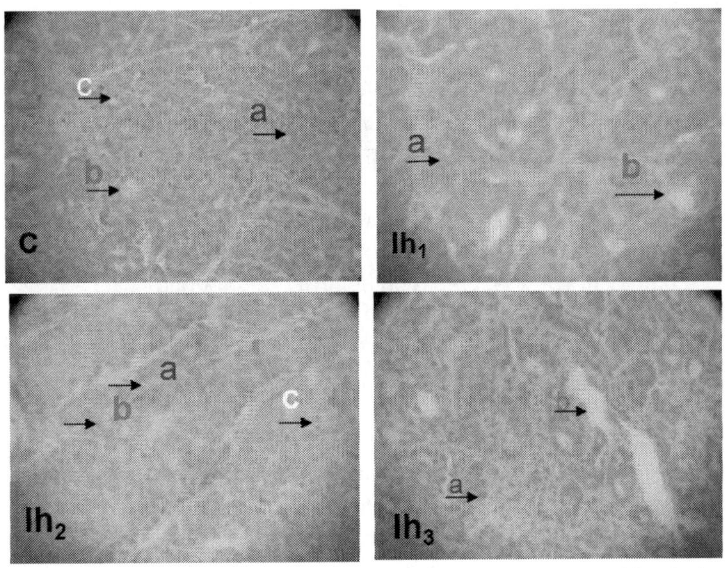

Figure 3. Observation of cancer tissue from C3H mice (400x magnification)
a = pleomorfic nuclei, size and shape of nuclei varies,b = tubulus/ ductus; c = mitosis, cell division.

Figure 3.Ih1 is in grade 1 with pleomorphic nucleus (2), tubular (2), dan mitosis (1) and the total score (5). Ih2 is in grade 2 with pleomorphic nucleus(3), tubular (2), dan mitosis (1), and the total score

(6). Figure3.Ih3 is in grade 1 with pleomorphic nucleus(2), tubular (2), dan mitosis (1), total score (5). C is in grade 3 with pleomorphic nucleus (3), tubular (3), dan mitosis (2), total score (8).Histological grade of each replicate sample (from five viewing fields) was averaged and displayed in Table 1. For statistical analyses, each score of a, b, and c (1 to 3) was used.

Table 1. The average of histological grade scores of*Adenocarsinoma mammae*of C3H mice from 5 viewing fields

Samples	C	Ih_1	Ih_2	Ih_3
1	7,2	5,4	5,4	5,4
2	7,8	4,6	4,4	5,6
3	7,4	6	5,4	5,4
4	7	4,6	5,2	5,4
5	7,2	5,2	4,6	4,6
Average	7,32	5,16	5	5,28
SD	0,30	0,59	0,47	0,39
Significa	b (grade 2)	a (grade 1)	a (grade 1)	a (grade 1)

On average Ih1, Ih2, and Ih3 showed that the cancer tissue cells had similarity to normal cells with good differentiation (grade 1). For control group without administration of *I. hippuris* extract the histological grade was 2 or 3 with a total score between 6 to 8. It indicates that *I. hippuris* extracts at a dosage of 0,15 mg/g, 1,5 mg/g, dan 15 mg/g per mouse can inhibit adenocarcinoma mammae cell mitosis and therefore reduced cancer growth.

Gorgonian *Isis hippuris* contain anticancer compounds especially belong to gorgosterol and hippuristanol groups. Gorgonian *Isis hippuris* from Okinawa water, Japan has been studied to contain gorgosterol and a new polyoxygenated steroid which showed antitumor and anticancer activity. Another compound from *I. hippuris* i.e. hippuristanol also showed potency as anticancer [13][14]. It also selectively inhibits *eukaryotic initiation factor* (eIF4A) which plays a role in the activity of RNA helicase [15]. It is possible that hippuristanol in the extracts may inhibit the activity of *eukaryotic initiation factor* 4A (eIF4A) which is found in RNA helicase. The eIF4A is a *prototypical* member of DEAD-box on RNA helicase. eIF4A functions to open the helical structure of mRNA using ATP [15][16]. Inhibition of translation on RNA prevent the formation protein, and therefore interphase in cell cycle can not proceed and consequently prevents mitosis.

The potency of *I. hippuris* extracts to inhibit and kill cancer cells has lent itself as the candidate of an anticancer drug from the ocean.

4. Conclusion

The ethanolic extract of *I. hippuris* inhibited the cancer cells growth and improved the histological score of mammary adenocarcinoma in C3H mice indicated. Therefore, the gorgonian has potency as an anti-cancer drug.

Acknowledgments

We thank the MDC member for their assistance during collection of the gorgonian *I. hippuris*. This research was funded by The Ministry of Reseach and Technology through the RUT grant with contract number: 66/RT/PPK/II/2008

References

[1] International Agency for Research on Cancer. Estimated number of incident cases, worldwide (top 10 cancer sites) in 2012. [Accessed August 12th, 2017]. http://gco.iarc.fr/today
[2] Veltri L and Abraham J. 2012. Breast Cancer Management. **1**(1), 5–7.
[3] Martin HL, Smith L, Tomlinson DC. 2014. *Breast Cancer.* 2014; **6** 1–13.
[4] Bayet-Robert M, Lim S, Barthomeuf C, Morvan D. 2010. *Biochem. Pharmacol.* **80** (8) p 1170-1179.
[5] Trianto A, Ambariyanto, Murwani R. 2004. *J. Ilmu Kelautan.* **9** (3) p 120-124.
[6] Ambariyanto, Murwani R, Trianto A., 2003. *J Biotech.* 667-680
[7] Trianto A, Tanaka J. and Higa T 2001 *Jurnal Imu Kelautan.* Special edition. pp 215-221
[8] Azizah T, Gandjar IG, Sugiyanto. 2002. *Pharmacon: J. Farmasi Indonesia.* **3**(1) p 5-12
[9] Harris AD et al. 2006. J Am Med Inform Assoc. **13** p 16–23.
[10] Fischer AH, Jacobson KA, Rose J, Zeller R. 2008. *CSH Protoc.* 2008 May 1; pdb. prot 4986.
[11] John Hopkins Medicine. 2015. Breast Cancer and Breast Pathology. http://pathology.jhu.edu/breast/grade.php (accessed October 2017).
[12] Vermote A, G. Brackman, M. D. P. Risseeuw, T. Coenye, and Van Calenbergh A, *Eur. J. Med. Chem.*, **127**, pp. 757–770, 2017.
[13] Higa T, Tanaka J, Tsukitani, Y., Kikuchi, H. Chem. Lett. 1981, 1647-1650.
[14] Uddin MH, Hanif N, Trianto A, Agarie, Higa T, Tanaka J. 2011. *Nat. Prod. Res.* **25** (6) p 585–591
[15] Wagner G, Tanaka J, and Pelletier J. 2006. 2 (4) p 213-220.
[16] Bordeleau, M E, A Mori, M Oberer, L Lindqvist, L S Chard, T Higa, G J Belsham. 2006. *Nat Chem Biol.* **2**(4) p 213-20.

IOP Conf. Series: Earth and Environmental Science **116** (2018) 012105 doi:10.1088/1755-1315/116/1/012105

Effectivity Test Of Crude Protein Spore of *Myxobolus koi* as Materials Development For Sub Unit Vaccine To Prevent the Gold Fish (*Cyprinus carpio,* Linn) Dead by Myxobolusis

Kismiyati[1] and G. Mahasri[2]
[1] Department of Fish Health Management and Aquaculture Faculty of Fisheries and Marine, Universitas Airlangga, Kampus C Mulyorejo, 60115
[2] Department of Fish Health Management and Aquaculture Faculty of Fisheries and Marine, Universitas Airlangga, Kampus C Mulyorejo, 60115

email : kismiyati@fpk.unair.ac.id
ORCID ID : 0000-0003-3627-2895

Absract. Production of fisheries culture according to totality estimated is increasing about 4.9% per year. Gold fish is one of species that have economically value in Indonesia, but there are problem because of myxobolusis. The aims of this research are : Isolation of crude spore protein of *Myxobolus koi* by using SDS-PAGE and to analyze profile of *Myxobolus koi* protein that will be developed for sub unit vacsine that can protect koi fish infected by *Mxobolus sp.* The method of this research used experimental method and belonged to stage : Culture of *Myxobolus* by using spore scattered, characterization of Myxobolus proteine spore by using SDS-PAGE, Antibody polyclonal production by using ELISA methode to find out optical density of this antibody, and vaccination was done by dipping, with dose 600 μl per liter. The result showed that there were 5 bands of crude protein spore of *Myxobolus koi* with molecule weight (MW) 41.1 kDa, 51 kDa, 89.8 kDa, 121.7 kDa and 230,1 kDa. The highest Optical Density (0.699) from polychlonal antibody production by using crude spore protein of *Myxobolus koi* happened on the 42nd day. The result of the chellence test of the protein showed that the highest of leucocyte count happened on the fish immunized by crude spore protein and not infected by *Myxobolus koi*, but can increase the survival rate (SR) from 20% to 86,37%, so the crude spore protein of Myxobolus can be used for developtment of material for sub unit vaccine to prevent the myxobolusis.

Keywords : *Crude* protein spora, Leukocyte, *Myxobolus*, Myxobolusis, Optical Density.

1. Introduction

One of the species of freshwater fish that have a bright prospect is the goldfish (*Cyprinus carpio* Linn). Based on document from the Ministry of Fisheries and marine in 2010, it is stated that Indonesia recently mastered 7.5% freshwater fish trading, still lower if compared to Singapore which has reached 22.5%. This led to the development of the carp increased mainly through farming good intensively or taditional. However, many obstacles can cause a failure in the culture, in which the main constraint of very noteworthy is the emergence of an attack of the disease.

One of the diseases that often become parasiter is a Protozoan disease caused by *Myxobollus koi* called myxobolusis. Myxobolusis is a disease parasiter on fish caused by the sporozoa among other *Myxobolus* sp. Generally disease-causing organisms is known as its spore morphology, number, and location of the polar filament. The disease can cause serious problems on the goldfish of koi and can lead to the death of up to 60-90% with prevalence reaching 100%. Furthermore it is said that in 1974 and 1978 there was a case of *Myxobolus* attact in Indonesia that led to the deaths of up to 100% of the koi fish mainly on stadia seeds.

Fish Stricken by the disease of myxobolusis show difficulty to breathe because there was a nodule or cyst or nodule on the Gill filaments. In Blitar it was reported by farmers that in 2010 there was myxobolusis desease occuring on carp koi that was 3-5 cm with mortality reaching the 90%. The spread of this parasite occurs due to the transfer of parasites from one fish to another, either directly or through the host between certain phases of the life cycle of the parasite [8]. In 2002 mass death of goldfish occurred in Kulon Progo and Sleman, caused by a parasite *Myxobolus* sp. and *Henneguya* so that the loss experienced by the fish farmers was quite large. *Myxobolus* sp. is also found in the Ngrajek area of Magelang Regency in 2006 with prevalence reaching 91%, and then in goldfish ponds koi in Blitar with prevalence reaching 86% in 2010.

Prevention efforts and countermeasures against myxobolusis have been done using a disinfectant as well as other chemicals, but not yet able to meet the target and can even cause resistance and residues in the body of the fish. It is necessary to look for alternative prevention efforts that do not cause a negative impact. One of the efforts that have already begun to be developed at this time is by vaccination that can be done by means of immersion or injection [3]. [2] managed to find a antigenic carbohydrate isolated from glycoprotein.

Characterization of protein can be done by using the method of *Sodium Dodecyl Sulfate-Poly Acrylamide Gel Electrophorosis* (SDS-PAGE) where the results can be known whether protein molecular weight is suitable for use as a vaccine. A protein core and anti-*Zoothamnium arbuscula* with cytoplasmic protein spasmin-1 on spasmonema. Immunoblotting analysis of results showed that the protein has a molecular weight of antigen of 68 kDa, 55 kDa and 71 kDa. This protein is thought to play a role in parasitic infestations on the host where the major protein is as an important ligand bonding in Ichthyiophthirius multifiliis and is the parasite's entry bridge into the cell.

Protein can be a good antigen when it has a molecular weight greater than 1,000 Dalton and the complexity of the structure. The antigen that is also referred to as imunogen is a material that can stimulate the immune response or material that may react with antibodies that are already there without regard to its ability to stimulate the production of antibodies.

The protein of spasmin on *Zoothamnium arbuscula* spasmonema can already be created Polyclonal antibody in hela cells. Immunoblotting analysis of the results shows that the location of the protein of antigens at molecule weight is 68/69 kDa, 55 kDa and 70 kDa. Proteins of the membrane *Tetrahymena* sp. which is the one with the *Ichthyophthyrius* familia *multifiliis* had most likely developed as a sub unit vaccine against the disease white spot on the fish. The antigenic membrane protein in Paramecium by SDS-PAGE and Western Blotting, with 61 kDa molecular weight, 63 kDa and 65 kDa.

Based on the background above, searching sub-unit of myxobolus material needs to be done to be developed as a sub unit vaccine that an provide protection against attacks on the carp myxobolusis, until the death of fish in the pond can be suppressed.

2. Research Methodology

This research was carried out from March to October by 2015, with the location of the research in the laboratory of Dry *(Kering)*, Fisheries and Marine Faculty, Airlangga University and the laboratory of molecular biology, the Faculty of Sciences Brawijaya University of Brawijaya, Malang.

2.1 Materials and Equipments

The main material for the isolation and characterization of a protein used are physiological NaCl, ethanol, solvents percoll gradient, pepsin, HCl, Ethylenediamenetetraacetic (EDTA), KI, Na2HPO4, KH2PO4, trypsin, sodium citrat, NaHPO4, H2), NaHCO3, glucosa, phenols red 0.5%, NaOH, filter 0.22 um, bovineserum albumin, dextrosa, Ethyl Alohol (ETOH), proteinase, ForwardERIB1 primer pairs 5 '-ACCTGGTTGATCCTGCCAG-3 ' (2-20) and Reverse ERIB10 5 '-CCTCCGCAGGTTCACCTACGG-3 ' (2069-2070), 400 UM DNTP, 3 um Mgso4 , yellow and blue dye, agarose, Tribase Acetic and EDTA (TAE) buffer, sybrsafe, 100 DNAladder bp and bp 1, loading dey, tris-HCL, 2-mercaptoethanol, sodium dodecyl sulfate (SDS), bromophenol blue, glycerol, SDS loading buffer, polyacrylamide, stacking gels, ammonium peroxidaisulphate (APS), glycine and Tetramethylenediamine (TEMED).

Materials for the production of Polyclonal Antibodies are rabbit from the Center Pusvetma/Veterinaria Farma, Surabaya, with an average weight of 2.385 kg 5 – 6 months and the male gender, Freund's complete adjuvants (Sigma), and Incomplete Adjuvants Freund's (Sigma).

The main equipment used are Haemocytometer, rubber pollicemen, microscope tube micrometer, autoclaf, centrifuse, swinging rotors, water bath sonicator and 1 set of tools of SDS-PAGE electrophoresis.

2.2 Research Methode

Research methods used in this research is descriptive method, where the blood picture and examination stages of counting of survival rate of carps are performed using 4 pieces of aquariums with a volume of 10 litres, for not exposed by a spore protein spores (given PBS) and not infected (K1), 1 for treatment of exposure to infection with spores and protein *Myxobolus koi* (K2), 1 piece of the Aquarium for treatment not exposed with protein and infected *Myxobolus koi* (K3) and 1 piece of akurium for the treatment of infection not exposed with protein and *Myxobolus koi* (K4). Dose used in this research is 600 micro litres per fish through injection (IM). Each of the aquariums is filled with 10 goldfish.

2.3 Sample preparation of *Myxobolus* sp. Spore.

The main ingredients of this research are nodules containing spores of *Myxobolus* sp. stadia adults are obtained from goldfish stricken by myxobolusis. The infected fish or there are nodules on gills first are cleaned using aquadest so that dirt stuck in the body of the fish is lost. Retrieval of nodules containing spores of *Myxobolus* is done manually by first taking the gills of carp and then using tweezers nodules on the gills slowly so that the nodules are not destroyed and turned so that the spores can be drawn wholly from the stricken organs (gills). Nodules containing spores of *Myxobolus* sp. are collected and washed with aqudest in order to be free from dirt and slime, and then broken down and crushed and then centrifugeded with speed 1,500 rpm. Spores will precipitate out. Deposits of spores are washed again with centrifugedion again with speed 1,500 rpm for 5 minutes. The deposits are taken and spores are counted in numbers and saved and given PBS.

2.4 Isolation of Spore Protein Crude of *Myxobolus* sp.

Spores that have been counted then arew ashed with PBS and centrifuged at speed 5000 rpm for 10 minutes. Leaching process is repeated twice. Pellet is resuspenced with buffer Lysis (0.32 M scrose (sucrosa) 54.8 gram, 1% v/v Triton X-100 0.5 ml, 10 mM Tris HCL pH 7.4 10 ml) so as to obtain a concentration ± 11ml. then on pellet suspension added 0.5 ml of EDTA 1 μ M and PMSF (phenylmethylsulfonyl fluoride) 5 mM as much as 0.5 ml. then sonication is done with ice in a sonicator waterbath (1 minute sonication, ½ minute break) repetitive 10 times. The results of the sonication is vortexed (½ minute vortex 1 minute rest) in ice, carried out repeatedly 15 times. Solution of sonication results is crude protein.

Analysis of protein concentration on research uses the method of Biuret and read with a UV-Visible spectrophotometer. Reaction with biuret method is the formation of the complex which is purple. This complex is formed when one or more peptides reacts with NaOH and CuSO4. Colour arising is due to Cu^{2+} ions complex with four nitrogen atoms that come from four of the ring peptide. Protein levels are calculated by converting on Bovine Serum Albumin curve (BSA) whose concentration is known.

2.5 The production of Polyclonal Antibody against Crude of Spores Protein Of Immunization of Spore Protein in rabbits

Adult healthy male Rabbit is injected intramusculerly with whole protein with a dose of 50-100 μg which was previously added adjuvants complete (Sigma) with the same comparison so that the final volume by as much as 500 μl. Injection is done under the skin at the four locations of the body that have loose skin. Injection was repeated with the same protein by the addition of adjuvants incomplete (Sigma) at 2 weeks after the first injection. Reinjection the next reset used the same protein with adjuvat the same way incomplete onwards until a high antibody titer was obtained. Prior to injecting, the first capture serum negative control in test as ELISA. Taking the serum before further carried out booster is to see any antibody response after injection with the same test.

2.6 Determination of Antibody Titer by ELISA Method

Titer of antibodies derived from rabbit serum that has been on immunization is determined by the method of ELISA. Negative control is taken from rabbits that are not in immunization by gender and age. The Cup of microtiter used is mikrotiter cup sumuran 96. Every sumuran is in the content with 100 μl of the antigen solution with the concentration 10 μg/ml in buffer coating and in incubation at a temperature of 20^0 C for 24 hours. The cup of mikrotiter is washed one time with wash buffer and then in each sumuran is added 100 μl of serum of rabbits immunizated and diluted with PBS. Rabbit serum is diluted in series, namely 2^0, 2^{-1}, 2^{-2}, 2^{-3}, ..., 2^{-10}. The cup of mikrotiter is incubated at temperature 37^0 C for one hour and continued to be washed three times with wash buffer. Every sumuran is in the content with 100 μl of the conjugated solution of IgG goat antimouse peroxidase. Conjugate is diluted with PBS 10,000 X and BSA is added 1%. The cup of mikrotiter is incubated in temperature 37^0 C for one hour and then is washed three times with wash buffer. Each sumuran is added as many as 150 μl ABTS substrate μ 1 (one mg/ml in buffer substrate ABTS and 0.3 μl of hydrogen preokscide). The cup of mikrotiter is incubated in room temperature for 30-45 minutes. antibody titer is read with a spectrophotometer for ELISA (ELISA reader) at a wavelength of 405 nm. Antibody Titer indicates negative if the results of the reading of ELISA between control rabbit and the immunized rabbit gets almost the same optical density (OD) values immunization. Antibody Titer indicates positive when value OD immune rabbit higher or at least twice the value of the OD of control rabbit control.

2.7 Examination of Erithrosit Number

Hemosit is taken on the ventral part of the second abdominal segments using needle 25 G one syringe one ml entered 0.2 ml of solution modified with cold Alsever (AS 19.3 mM; Na citrate 239.8 mM NaCl 182.5 and glucose 6.2 mM EDTA; pH 7.2) as an anticoagulant. Counting the number of haemocyte is done with methods of May Grunwald-Giemza, namely using light microscopy (LM) with an enlarged 1000 times, then calculated with the Coulter counter ZM SUNDAY model (Counter Electronic Ltd), hemosit particle size range 0.4-800 μm, as supporting data can also be observed with the electron microscope (EM) by first centrifugede 700 X gravity at a temperature of 4^0 C for five minutes.

2.8 Examination Of The Differential White Blood Cells

The blood is shed on glass objects and made blood review with giemsa staining, later the type of cell is identified . Differential hemosit aims to find out the amount, type and percentage of cell hemosit. The number of hemosit is calculated up to 100 cells and look for the percentage of each type is found, done from the beginning to the end of the study.

2.9 Determination of Survival Rate (SR)

Determination of level of survival rate was conducted to analyze the ability of crude spore protein protection on goldfish. It is expressed in the form of a percentage of the number of koi fish that lives up to the experimental treatment against 30pasca the total number of fish kept. Fish survival is calculated by using the formula :

$$SR = \frac{Nt}{No} x100\%$$

Description:
SR = survival rate
NT = the number of fish that live at the end of the observation
No = number of fish that live in the beginning of the test

2.10 The parameters of the research and Data analysis

The main parameter of this research is the protein profile of *Myxobolus koi*, carp blood Picture (the number of Leukocytes and Erythrosit) and Survival Rate of the koi fish exposed with protein of *Myxobolus koi*. Data from the study results will be presented in the form of quantitative data and analyzed with Analyzed of Variance (ANOVA) [12].

3. Results And Discussion

The results of the examination and identification of spores of *Myxobolus koi* of koi fish can be seen in Figure 1 and 2. While the results of the determination of the characterization protein conducted using SDS-PAGE method can be seen in Figure 3.

Figure 1. Nodules on the gills of fish infected by *Myxobolus koi* (A)
and Nodules that contains the spores of *Myxobolus koi*

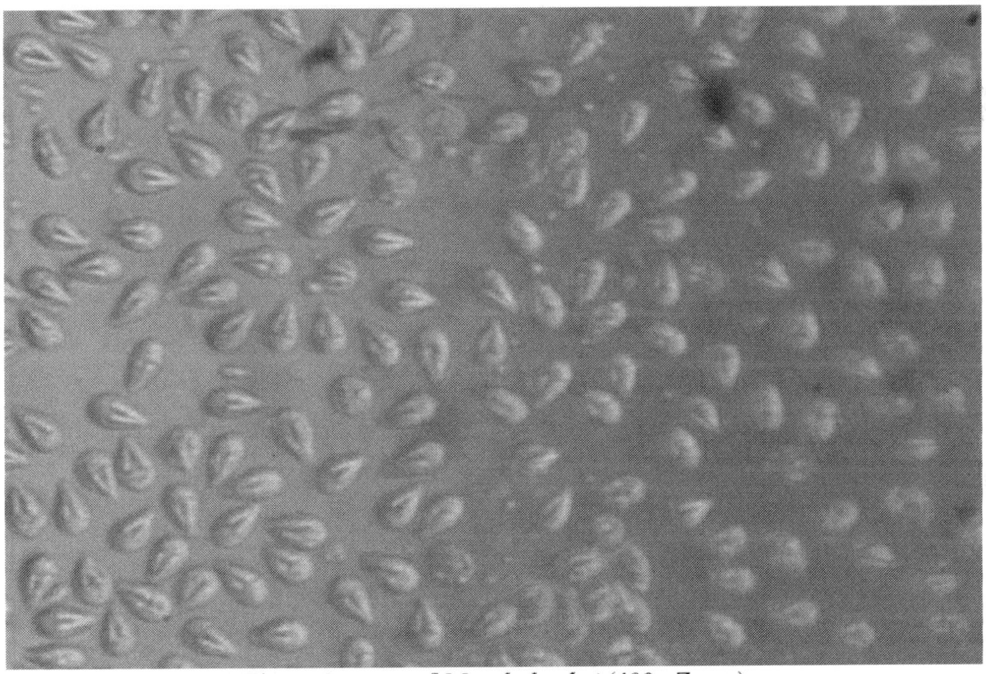

Figure 2. spore of *Myxobolus koi* (400x Zoom)

3.1 Isolation of Crude Protein

Samples of nodules containing spores of *Myxobolus* sp was obtained from Blitar, Tlogo village, Kanigoro district as many as 154. Isolation of proteins was carried out by taking 1 gram of tissue samples of nodules that were then made into crude protein and stored in a refrigerator with a temperature of 4 ° C in a solution of Tris-HCl 20 mM.

Crude protein concentration analysis is performed using the method of solution concentration using Biuret BSA 0 – 10000 ppm. Determination of the wavelength of the BSA was done in concentration of BSA at 4000 ppm so the obtained maximum wavelength was 541 nm.

3.2 Analysis of Crude Protein of Spores By The Method Of SDS-PAGE

Protein analysis is carried out using the method of SDS-PAGE which is a method to solve the protein based on molecular weight. SDS-PAGE uses the stacking gel used 3% and 12% separating gel and was run on 200 Volt voltages and 2 mA. The results of the SDS-PAGE in the form of band of proteins, where the band can be calculated from the weight of the molecule so that the character of the protein from the spores of *Myxobolus* sp. can be known. (Figure 3)

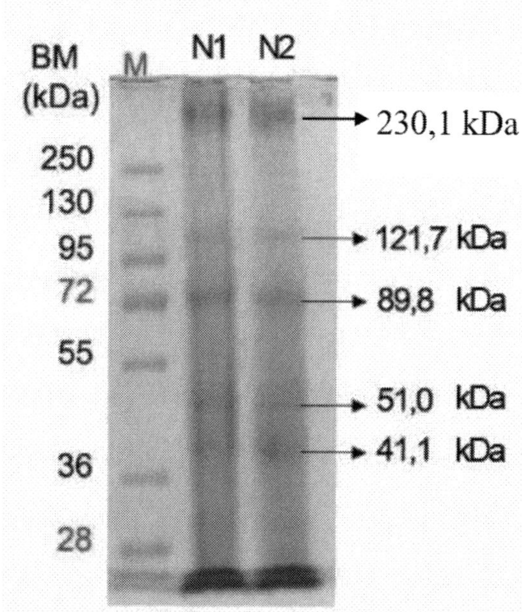

Figure 3. Showed that results of spore protein characterization of of *Myxobolus* sp. Using SDS-PAGE is invalidated 5 bands of protein with molecular weight (BM) 230.1 kDa, 121.7 kDa, 89.8 kDa, 51.0 kDa, 41.1 kDa.

3.3 Production of Polyclonal Antibodies

The result of the production of Polyclonal Antibodies from spore protein *Myxobolus koi* obtained Value OD that on day 1 is 0.233 and has increased up to day to 42. Every time the blood taking is done. boster OD Values on day 56 though done booster was done, it declined to be a 0547.

Table 1. The value of the Optical Density (OD) of Spore Proteins Polyclonal Antibody in rabbits by Indirect ELISA

Day to	Value of OD		
	The Rabbit serum infected by *Myxobolus*	Control Serum (-)	PBS control (-)
1	0.233	0.154	0.103
14	0.489	0.157	0.101
28	0.531	0.158	0.102
42	0.699	0.151	0.103
56	0.547	0.132	0.102

Value OD on day 42 reached 0.699 that shows more than 2 times the value of OD control. This is in accordance with the opinion of the Tizzard (1988) who says that animals given repeated vaccine will stimulate the formation of antibodies.

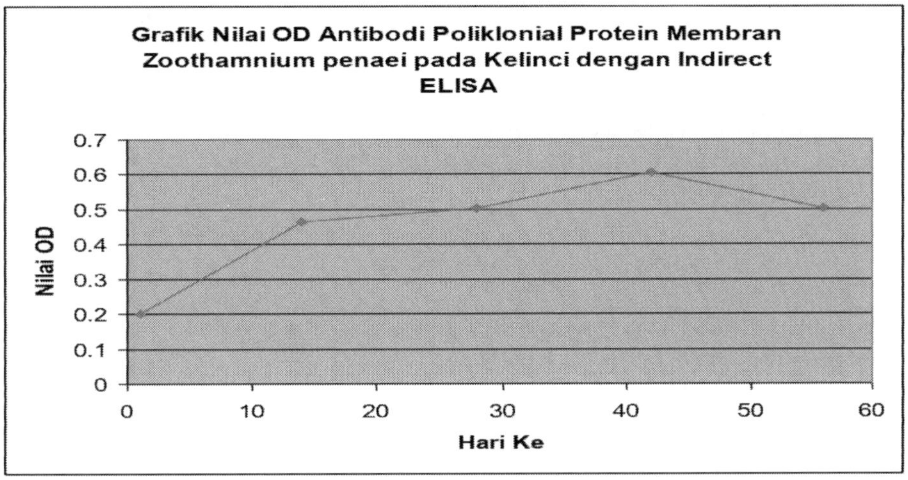

Figure 4. Graph of OD Value of Spore Protein Polyclonal Antibody of *Myxobolus koi* in rabbits by Indirect ELISA

ELISA test results showed that the crude spore proteins of *Myxobolus koi* can induce the formation of antibodies on goldfish. Table 1 shows the highest optical density values occusr on day 42. The increase in the value of the OD on the fish exposed later infected with crude spore protein *M. koi* shows the ability of the protein protection against myxobolusis infection. The mechanisms of the immune response in fish, when antigen (vaccine) enters into the body of the fish then the antigens is presented by MHC, antigen will be captured by receptors on T helper cells (2), and helper T cells (2) going to secrete cytokines, namely IL-2, IL-4, IL-6 aimed at the differentiation and proliferation of B cells, differentiation of B cells and plasma cells will produce memory cells. Next the cell plasma will synthesize specific antibody and bind antigens so as to prevent the movement of antigen and to ease the process of phagocytosis.

3.4 Examination results of Blood picture of Goldfish Exposed with Spore Protein
The results of the determination of the blood picture of Goldfish exposed with the spore proteins after 14 days of observance, with a dose of 600 µl is presented in table 2. Table 2 shows that the number of Erythrosit occurs on a control (K1) and lowest on K3 (infected by *Myxobolus koi*, not exposed by spore protein) namely 30.4% and the highest occurred in control (K1) by 91,13%, while on treatment of K3 (exposed by protein and infected by *Myxobolus koi*) amounted to 30.4%. For the number of leukocytes occurs on K3 that is 69.36% and lowest in 8.87% of K1, whereas in the treatments of K3 66,58%.

Table 2. The results of the determination of the Blood picture of a Goldfish exposed by spores protein after 14 days of maintenance.

Treatments	Image of Carp Blood after 14 days of maintenance (%)	
	Erythrosit	Leukosit
Control, no infection and no exposed Protein (K1)	91.13[a]	8.87[e]
Infected by *Myxobolus* and not exposed by Spores Protein a dose of 600 μ l/fish (K2)	32.42[b]	66.58[f]
Infected by *Myxobolus* and not exposed by Spores Protein with a dose of 600 μ l/fish (K3)	30.64[c]	69.36[g]
Not Infected by *Myxobolus* and not exposed by Spores Protein with a dose of 600 μ l/fish (K3)	69.33[d]	30.67[h]
Normal	96.5 – 98,0	14 - 36

Description: different Superscript on the same column indicates the existence of the real difference ($p < 0.05$)

3.5 Results of determination of Differensial Goldfish Leukocytes Exposed by Crude Spore Proteins

The results of the differensial determination of white blood cells in the goldfish exposed with crude spore protein after 14 days of observance, with a dose of 600 μl are presented in table 3. Table 3 shows that the results of the counting of the lymphocytes in the blood of carp at the highest is treatment K3 (fish without being vaccinated but infected with 80 spores/fish *M. koi*), namely in the amount of 86.37% and the lowest percentage is in the treatment of K namely 70.46%. The results of the counting of monocytes in the blood of carp at each treatment obtained the highest percentage in the treatment of K3 namely 23.33% and the lowest is K4 treatments at 9.24%. Percentage of observations of heterofil in blood of koi fish showed the highest percentage in the treatment of K4 namel 13.4% and lowest is in treatment of K2 (fish were vaccinated and infected by 80 spores/fish *M. koi*) of 6.7%. Observation of the number of eosinophils demonstrated the highest percentage on K3 treatment of 6.71% and lowest in the treatment of the K1 (control) of 6.8%. Observation of the number of basophils indicates the highest percentage in the treatment of 5.93% K3 and K1 is at the lowest treatment 2.23%.

Table 3. Results of determination of Differensial Leukocytes Carp exposed by Crude spore Protein after 14 days of maintenance.

Treatment	Differensial of Leukocytes Carp exposed by Crude spore Protein after 14 days of maintenance (%)				
	Limfosit	Basofil	Heterofl	Eosinofil	Monosit
Control, not infected and not exposed by protein (K1)	73.12	2.23	13.4	3.65	10.25
Infected by *Myxobolus* and exposed by spore Protein Spora dose 600 μl/fish (K2)	82.23	3.95	6.7	5.19	1.53
Infected by *Myxobolus* and not exposed by spore protein dose 600 μl/fish (K3)	86.37	5.93	12.6	6.71	23.33
Not infected by *Myxobolus* and exposed by spore Protein Spora dose 600 μl/fish (K4)	70.46	2.53	8	3.27	9.24
NORMAL	**54**	**2**	**14**	**7**	**23**

3.6 Results of determination of survival rate of Carps exposed by spore Protein

Survival rate of carp is expressed in percentage and calculated on day 14 presented in table 4. Table 4 shows that in spore protein protection test of *M. koi*, there is a noticeable difference ($p < 0.05$) between the treatment of survival rate between exposed and not exposed with spores of *Myxobolus koi* keeping after 14 days.

The results of calculation show that the highest at the treatment of survival rate is in K4 (fish are not infected and fish are exposed with spores of *M. koi*) namely 96.00% and treatment (control) of K1 94,00%, while the lowest treatment is K3 (fish are infected with 80 spores/fish of *M. koi* and not exposed) with the value of SR for 20%.

Table 4. The Results Of The Determination Of survival rate of Koi exposed by The Spores Protein after 14 days of maintenance

Tratment Group	Survival rate of koi fish after vaccination
Control, not infected and not exposed by protein (K1)	$94.00^{a} \pm 2.13$
Infected by *Myxobolus* and exposed by spore Protein Spora dose 600 μl/fish (K2)	$83.00^{b} \pm 6.43$
Infected by *Myxobolus* and not exposed by spore protein dose 600 μl/fish (K3)	$20.00^{c} \pm 3.54$
Not infected by *Myxobolus* and exposed by spore Protein Spora dose 600 μl/fish (K4)	$96.00^{a} \pm 2.46$

Description: different Superscript on the same column indicates the existence of the real difference ($p < 0.05$)

3.7 Water quality of maintenance medium

Water quality is the most important factor in fish farming. Water quality is affected by a variety of chemical substances that are dissolved in the water, that are temperature, pH, dissolved oxygen and ammonia.

4. Discussion

The results of the identification of spores shows that *Myxobolus* sp. obtained on goldfish gills in this research is *Myxobolus* sp. nodule samples obtained subsequently are made into protein samples by way of physical breakdown by crushing and disonicating to break down the cell. Sonication can be used to speed up the separation of particles in the sample by breaking Intermolecular interaction and then centrifuged with high speed and enough time to separate between the spores, nodules chains, and spore protein, and nodule protein. Sentrifus is a tool that is used to rotate the sample at high speeds and forced heavy particles accumulated into the base tube of sentrifus.

Crude protein concentration of nodules containing spores of *Myxobolus* sp. which is measurable by using UV-Visible Spectrophotometry is 5116.25 μg/ml. States that the lowest concentration of protein is needed for protein analysis of 1.2 μg/ml, so it can be inferred that the crude proteins of *Myxobolus* sp. protein analysis can be conducted using SDS-PAGE electrophoresis.

Stacking gel SDS-PAGE technique used was 3% and separating gel 12 % and being run on with 200 Volt voltages and 2 mA. This study used separating gel 12% protein because the spores of *Myxobolus* sp. ranged between 7 – 130 kDa. The protein which has a molecular weight of 20 – 200 kDa packed used separating gel 12%.

The results of the SDS-PAGE in this study showed 5 bands of protein with molecular weight (BM) 41.1 kDa, 51 kDa, 89.8 kDa, 121.7 kDa and 230.1 kDa. Protein band results obtained is not yet known which the spore proteins and band proteins are and which the protein from tissue nodules is. The protein form spore of *Myxobolus cerebralis* has BM 7 kDa, 45 kda, 60 kDa and 130 kDa. Therefore, on this research protein bands with BM 41.1 kDa, 51 kDa, 89.8 kDa and 121.7 kDa and 230.1 kDa are crude of proteinof *Myxobolus* sp. which most likely is a highly immunogenic protein

Changes in the total number of leukocytes and erythrosit can be used as an indicator of the existence of the attack of infectious diseases in fish. Leukocytes in blood components is one that serves as a specific body defenses that will neutralize and destroy pathogens through phagocytosis. Lymphocytes is one form of leukocytes. figure 3 shows the percentage of lymphocytes in each treatment. The observations of lymphocytes showed the highest percentage of 86.37% in the treatment of K3 (not being exposed and infected with 80 spores/fish of *M. koi*) when compared with the treatment of K1 (control) of 73.12%. The increased number of lymphocytes in the fish infected with spores of *M. koi* is the body's defense system response of fish over the entry of pathogens. The antibody-producing lymphocytes function as immune to interference from disease.

Lymphocyte cell consists of two populations i.e. B cells and T cells. B cells have the ability to transform into plasma cells i.e. cells that produce antibodies. T cell immunity is as instrumental in cells interaction (cytotoxic T cells) and control the immune response (suppressor T cells). After the binding of an antigen with antigen receptor cell lymphocytes, lymphocyte cells will then divide and differentiate into effector cells and memory cells.

Monocyte percentage of each of the treatments is shown in Table 3. There was a decline in the number of monocytes in ttreatment K4 i.e. 9.24% compared with K1 (control) i.e. 10.25%. The decline was caused by the monocytes which fagositosised spores of *M. koi* that went in the body of the fish. This is in accordance with the opinion that the monocytes with macrophage will fagosit agents causing deseases that enter the body. The monocytes function as phagocytes against foreign objects which act as the agent of the disease.

The amount of heterofil in treatment K2, K3 and K4 has decreased when compared with the treatment of the K1 (control) as shown in Table 1. This is due to the increase in lymphocytes and monocytes, so heterofil is declining. Cell heterofil not do not play a role mainly in responding to an infection caused by a parasite and is more instrumental in infections caused by bacteria.

The results of the calculation against the number of eosinophils demonstrated the highest percentage on K3 treatment of 6.71% compared with treatment K (control) of 3.65%. Spore infection of M. koi increases the number of eosinophils in the blood of fish. The eosinophils is one of the body's defense cells that are dominant in the blood and will increase sharply in number in case of infection disease parasiter.

The results of the calculation showed that the lowest percentage basophils is 2.23% in K1 (control), then treatment K4 also shlows low value namely 2.53%. Tthe percentage of basophils in the blood of fish ranged from 0.17-0.194% and measuring 8-12 μm. The existence of basophils in blood circulation has been observed only in a small number of species of fish. Even more rarely basophils are found on examination of blood compared to eosinophils. The basophil granules contain eosinophils chemotaxis factor and type 1 hypersensitivity mediator when there is stimulation from allergens that cause the occurrence of snapping allergens on basophils, the content of basophils will release.

The results showed that spore proteins of *Myxobolus koi* were able to protect (protective nature) koi fish against myxobolusis. This is evident by the presence of increased survival of carp from 20.00% up to 83.00% to harvest takes place, so that it can be interpreted as a decline of the death of goldfish.

The results of measurements of water quality: temperature, pH and dissolved oxygen on maintenance as indicated in Table 3 show the average temperature of the medium maintenance of each treatment was about 27-28 ° C, pH of water 8 and dissolved oxygen of 5 ppm. It indicates that the condition of water quality of maintenance medium is still in normal conditions for the survival of the fish. The goldfish can live on temperature range 8-30 ° c. The levels of ammonia (NH3) before and after the treatment showed a difference in which ammonia before treatment of was 0 ppm while the after treatment was 0.5 ppm. The content of ammonia in the maintenance media tends to high caused by the presence of remnants of feed and fish metabolism excretion results.

5. Conclusion

5 profiles of Crude spora protein of *Myxobolus koi* are retrieved with molecular weight of 41.1 kDa, 51 kDa, 89.8 kDa, 121.7 kDa and 230.1 kDa. Crude spore Protein of *Myxobolus koi* can be developed as sub unit vaccine ingredients to prevent the death of goldfish, because it can improve the immune response of carp. Crude spore Protein *Myxobolus koi* can give protection to carps, so it can improve the survival rate of 20% up to 85%.

References

[1] Eszterbauer, E., 2004. *Genetic Relationship Among Gill-Infecting Myxobolus Species (Myxosporea) of Cyprinids: Molecular Evidence of Importance of Tissue-Specificity.* Dis.Aquatis Org. 58:35-40.

[2] Feizi, T and R.A. Childs. 1987, *Carbohydrates as Antigenic Determinants of Glycoprotein,* Biochem J, 245(1):1-11.

[3] Fiala, I. 2006. *The Phylogency of Myxosporea (Myxozoa) Based on Small Subunit Ribosomal RNA Gene Analysis.* Int. J. Parasitol 36: 1521-1534.

[4] Hanson, L.A, D. Lin, L.M. Pote and R. Shivaji, 2001. *Small Subunit rRNA Gene Comparisons of Four Actinosporean Species to Establish a Polymerase Chain Reaction Test for The Causative Agent of Proliferative Gill Disease in Channel Catfish.* J. of Aq. An. Health, 13:117-123.

[5] Lowers, J.M. and J.L. Bartholomew, 2006, *Detection of Myxozoan Parasites in Oligochaetes Imported as Food for Ornamental Fish,* Journal of Parasitology, 70:84-91.

[6] Mahasri, G. 2007. Protein Membrane Imunogenik *Zoothamnium penaei* Sebagai Bahan Pengembangan Imunostimulan Pada Udang Windu (*Penaeus monodon* Fab.) Untuk Mencegah Zoothamniosis. Disertasi, Program Pascasarjana,Universitas Airlangga, Surabaya.

[7] Mahasri, G. 2012. Respon Imun Dan Kelulushidupan Udang Vannamei *Lithopenaeus vannamei*) Yang Dimunisasi Dengan Protein/Membran Imunogenik *Zoothamnium penaei. Article of Konggres and International Seminar IFS, Can Tho University, Vietnam.*

[8] Ryce, E.K.N., 2003. *Factors Affecting The Resistance of Juvenile Rainbow Trout To Whirling Disease.* Ph.D. Thesis, Montana State University, Bozeman, MT.

[9] Schlegel, M., J. Lom, A. Stechmann, D. Bernhard, D. Leipe, A.I. Dykova and M.L. Sogin, 1996, *Phylogenetic*